Developments in Aquaculture and Fisheries Science - 31

BIOLOGY OF THE HARD CLAM

DEVELOPMENTS IN AQUACULTURE AND FISHERIES SCIENCE

The following volumes are still available:

9. WATER QUALITY MANAGEMENT FOR POND FISH CULTURE
   By C.E. Boyd 1982 xii+318 pages

17. DISEASE DIAGNOSIS AND CONTROL IN NORTH AMERICAN MARINE AQUACULTURE
    Edited by C.J. Sindermann and D.V. Lightner 1988 xv+412 pages

18. BASIC FISHERY SCIENCE PROGRAMS: A COMPENDIUM OF MICROCOMPUTER PROGRAMS AND MANUAL OPERATIONS
    By S.B. Saila, C.W. Recksiek and M.H. Prager 1988 iv+230 pages

19. CLAM MARICULTURE IN NORTH AMERICA
    Edited by J.J. Manzi and M. Castagna 1989 x+462 pages

22. FRONTIERS OF SHRIMP RESEARCH
    Edited by P.F. DeLoach, W.J. Dougherty and M.A. Davidson 1991 viii+294 pages

23. MARINE SHRIMP CULTURE: PRINCIPLES AND PRACTICES
    By A.W. Fast and L.J. Lester 1992 xvi+862 pages

24. THE MUSSEL MYTILUS: ECOLOGY, PHYSIOLOGY, GENETICS AND CULTURE
    By E. Gosling 1992 xiv+589 pages

25. MODERN METHODS OF AQUACULTURE IN JAPAN (2ND REVISED EDITION)
    Edited by H. Ikenoue and T. Kafuku 1992 xiv+274 pages

26. PROTOZOAN PARASITES OF FISHES
    By J. Lom and I. Dyková 1992 xii+316 pages

27. AQUACULTURE WATER REUSE SYSTEMS: ENGINEERING DESIGN AND MANAGEMENT
    Edited by M.B. Timmons and T. Losordo

28. FRESHWATER FISH CULTURE IN CHINA: PRINCIPLES AND PRACTICE
    Edited by J. Mathias and S. Li 1994 xvi+446 pages

29. PRINCIPLES OF SALMONID CULTURE
    Edited by W. Pennell and B.A. Barton 1996 xxx+1040 pages

30. STRIPED BASS AND OTHER MORONE CULTURE
    Edited by R.M. Harrell 1997 xx+366 pages

31. BIOLOGY OF THE HARD CLAM
    Edited by J.N. Kraeuter and M. Castagna 2001 xix+751 pages

32. EDIBLE SEA URCHINS: BIOLOGY AND ECOLOGY
    Edited by J.M. Lawrence 2001 xv+419 pages

Developments in Aquaculture and Fisheries Science - 31

BIOLOGY OF THE HARD CLAM

Edited by

JOHN N. KRAEUTER
Haskin Shellfish Research Laboratory
Institute of Marine and Coastal Sciences, Rutgers University
6959 Miller Avenue, Port Norris, New Jersey
U.S.A.

MICHAEL CASTAGNA
Virginia Institute of Marine Science, Virginia
U.S.A.

2001

ELSEVIER
Amsterdam - London - New York - Oxford - Paris - Shannon - Tokyo

ELSEVIER SCIENCE B.V.
Sara Burgerhartstraat 25
P.O. Box 211, 1000 AE Amsterdam, The Netherlands

First edition 2001

Library of Congress Cataloging in Publication Data
A catalog record from the Library of Congress has been applied for.

ISBN: 0-444-81908-8
Transferred to digital print 2007
Printed and bound by CPI Antony Rowe, Eastbourne

Preface

The hard clam, *Mercenaria mercenaria*, has been an important recreational and commercial species for as long as shellfish have been harvested from the inshore waters of the Atlantic coast of the United States. In 1998 coastwide landings of 7,193 millon pounds were worth $41,775,000 ($5.80 per pound). Despite this importance, scientific efforts and popular literature on hard clams has been overshadowed by that on the oyster, *Crassostrea virginica.*

Historically, during the oyster season, clams were considered by-catch. In the southern coastal states these clams supplemented the crew's meager earnings. At the end of the oyster harvesting season some watermen or baymen harvested clams full time while others switched to other fisheries or shore employment. Many clams were consumed locally, but considerably more were processed for clam chowder by companies such as Campbells and Bordens. After WWII the discovery and exploitation of the surf clam and ocean quahog resources reduced the demand for chowder size clams. In spite of the loss of this large market, hard clam harvesting continued to be an important shellfishery.

From the 1950s to present, and the coastwide dollar value of hard clam harvests approaches that of surf clams and ocean quahogs. During the last three decades hard clam farming has become a major industry in the Atlantic and Gulf coasts of the United States. We have included a chapter on this aspect of hard clam biology, but clam culture has been covered more thoroughly in the "Developments" series entitled Clam Mariculture in North America. This is one of the few widespread economically successful shallow water marine aquaculture systems that has been developed in the past 40 years. It accounts for an increasing portion of marine aquaculture products produced in the US.

We would be remiss if we did not mention the varied common or commercial names assigned to this species. *Mercenaria mercenaria* has several market names that depend upon the size of the animal: littleneck (neck or nick); top neck; cherrystone and chowder. In addition there are a number of common names for this species including hard clam, hard shell clam, quahog, quahag, and most recently the American Fisheries Society (1988) has suggested that northern quahog be utilized as the common name for this species. While this latter name has wide use in a portion of New England it is not the most common name in use throughout the species range. Hard clam has a wider geographical range of use, is an important commercial name for the species, and is more readily understood by the public. Given the overwhelming use of hard clam for this species we have accepted that as the preferred common name.

Our knowledge of the species *M. mercenaria* is diverse and extensive. (J.L. McHugh et al. and McHugh and Sumner's bibliographies contains 2693 citations all published before 1989). In view of the long history of work on this ecologically, commercially and recreationally important species, it may be surprising that the information has never been gathered in a single volume. Again the dominance of oysters in commercial exploitation, the competition for research funds, and lore has served to relegate hard clams to secondary status. In spite of this, it is still surprising that many of the authors of this review have found major gaps in basic information about the biology of this important species. This book offers a single volume, that we hope will serve as a first synthesis of the accumulated data on the hard clam. To achieve this end we have gathered chapters written by the acknowledged experts in each field. We feel fortunate to have worked with these individuals and appreciate their perseverance in developing this volume from information that was more often than not hidden in obscure publications, gray literature or Masters and Ph.D. documents. We would also like extend our sincere gratitude for all those individuals who spent their time and effort peer reviewing the chapters. Their input has served to substantially improve the overall quality of the book.

Lastly, one of our most esteemed colleagues, J. Lauren McHugh died after submitting his chapter, and we have included it without updates as his last effort on hard clam biology and fisheries. We believe it is only fitting that this volume be dedicated to him.

REFERENCE

American Fisheries Society, 1988. Common and Scientific Names of Aquatic Invertebrates from the United States and Canada: Mollusks. American Fisheries Society Special Publication 16.

John N. Kraeuter
Michael Castagna
July, 2000

List of contributors

WILLIAM S. ARNOLD	*Department of Environmental Protection, Florida Marine Research Institute, 100 Eight Avenue SE, St. Petersburg, FL 33701-5095, USA*
V. MONICA BRICELJ	*Institute for Marine Biosciences, National Research Council, Halifax, NS B3H 3Z1, Canada*
MELBOURNE R. CARRIKER	*College of Marine Studies, University of Delaware, Lewes, DE 19958, USA*
MICHAEL CASTAGNA	*School of Marine Science, Virginia Institute of Marine Science, College of William and Mary, Wachapreague, VA 23480, USA*
KENNETH K. CHEW	*College of Ocean and Fishery Sciences, University of Washington, Seattle, WA, USA*
ALBERT F. EBLE	*College of New Jersey, Department of Biology, Trenton, NJ 08628, USA*
ARNOLD G. EVERSOLE	*Department of Aquaculture, Fisheries and Wildlife, Clemson University, Clemson, SC 29634-0362, USA*
STEPHEN R. FEGLEY	*Corning School of Ocean Studies, Maine Maritime Academy, Castine, ME 04420, USA*
SUSAN E. FORD	*Rutgers University, Institute of Marine and Coastal Sciences, and New Jersey Agricultural Experiment Station, Haskin Shellfish Research Laboratory, 6959 Miller Avenue, Port Norris, NJ 08349, USA*
LOWELL W. FRITZ	*US Department of Commerce, NOAA/NMFS Alaska Fisheries Science Center, 7600 Sand Point Way NE, Seattle, WA 98115-0070, USA*
RAYMOND E. GRIZZLE	*Jackson Estuarine Laboratory and Zoology Department, University of New Hampshire, Durham, NH 03824, USA*
M.E. HARTE	*1180 Cragmont Avenue, Berkeley, CA 94708, USA*
THOMAS J. HILBISH	*Department of Biological Sciences, and Marine Science Program, University of South Carolina, Columbia, SC 29208, USA*

JOHN N. KRAEUTER — *Rutgers University, Institute of Marine and Coastal Sciences, and New Jersey Agricultural Experiment Station, Haskin Shellfish Research Laboratory, 6959 Miller Avenue, Port Norris, NJ 08349, USA*

CLYDE L. MacKENZIE, Jr. — *James J. Howard Marine Sciences Laboratory, Northeast Fisheries Science Center, National Marine Fisheries Service-NOAA, Highlands, NJ 07732, USA*

J.L. McHUGH — *Deceased*

CHARLES H. PETERSON — *University of North Carolina at Chapel Hill, Institute of Marine Sciences, Morehead City, North Carolina 28557, USA*

SANDRA E. SHUMWAY — *Natural Science Division, Southampton College of Long Island University, Southampton, NY 11968, USA*

DAVID L. TAYLOR — *Division of Marine Fisheries, North Carolina Department of Environment, Health and Natural Resources, P.O. Box 709, Morehead City, NC 28557, USA*

Contents

Section 1

Descriptive Biology

Biology of the Hard Clam
J.N. Kraeuter and M. Castagna (Eds.),

Chapter 1

Systematics and Taxonomy

M.E. Harte

1.1 TAXONOMY

While the specific status of the hard clam is widely accepted, its generic and subfamily placement is still open to controversy among some systematists. A number of common names exist arising from its commercial importance and popularity, and several varieties have been described.

1.1.1 Names

Mercenaria mercenaria (Linnæus, 1758)
Subfamily **Chioninae** (Frizzell, 1936)
Family **Veneridae** (Rafinesque, 1815)
Superfamily **Veneroidea** (Rafinesque, 1815)
Order **Veneroida** (Adams and Adams, 1857)
Subclass **Heterodonta** (Neumayr, 1883)
Class **Bivalvia** (Bonnani, 1681)
Phylum **Mollusca**

Mercenaria mercenaria is the type species for the genus *Mercenaria* Schumacher (1817), which Keen (1969) classified within the Chioninae of Veneridae.

The validity of Chioninae as a separate subfamily is arguable. Chioninae was differentiated from the very similar subfamily Venerinae solely on the absence in the left valve of an anterior lateral tooth (Frizzell, 1936), which is weakly to moderately developed within the Venerinae. Fischer-Piette (1975) observed that several venerine species with a weakly developed anterior lateral tooth sometimes contain individuals with no anterior lateral tooth, or structures so obsolete that they do not conform to the definition of an anterior lateral tooth. For example, he noted that some individuals of *Venus foliaceo-lamellosa* Schröter (1788) have no anterior lateral tooth, while others have only a protuberance occurring along the ventral margin of the hinge plate below and slightly separated from the anterior cardinal. Conversely, he observed that some individuals of the chionine species *Venus lamellata* Lamarck (1818) [a synonym of *Bassina disjecta* (Perry, 1811)] have a protuberance along the ventral margin in the area where an anterior tooth would be located. Fischer-Piette and Vukadinovic (1977) observed that some individuals of *M. mercenaria* have such a protuberance. He also observed that the venerine species *Venus punctigera* (Dautzenberg, 1910) was so similar to the chionine species *Chione paphia* (Linnæus, 1767) that one could consider them two varieties of the same species, although this could be a case of evolutionary parallelism (e.g., Harte, 1992b).

Ultimately Fischer-Piette (1975) concluded that rather than being a discrete character, the left anterior lateral tooth existed as a gradation between the two subfamilies.

While the fossil record for Veneridae indicates a general trend towards loss of anterior lateral teeth in more recently evolved forms, the fossil record does not support the evolution of Chioninae from Venerinae. Fossils of Venerinae (*Dosina (Hina)*) first appear in the late Eocene, 43 Mya; chionine fossils (*Tawera (Turia)* s.s.; Placamen) first appear in the mid Eocene, 46 Mya (Beu and Maxwell, 1990).

Wells (1957) argued that rather than being a part of a gradation between the subfamilies, the reduction of the left anterior lateral tooth within Venerinae represented convergent evolution towards the chionine state. Furthermore, examination of hard clams with an anterior protuberance of the ventral hinge margin reveals that the protuberance appears to be created by an underlying bulge beneath the hinge plate formed by the insertion of the pedal muscle just anterior to the bulge. Thus it is not structurally related to an anterior lateral tooth. A slighter protuberance of the ventral hinge plate margin sometimes occurs at the anterior end of the pedal scar insertion. Despite the ambiguity of the anterior lateral tooth as a useful character for discrimination, Fischer-Piette (1975) retained the two subfamilies, reinforcing this separation by observing differences in the posterior cardinal in the left valve between the two subfamilies. Within Venerinae, he observed, it tends generally to be horizontal and long and within Chioninae, more vertical and shorter. Although these differences have yet to be verified statistically, the hard clam remains classified as a chionine clam.

M. mercenaria (Linnæus, 1758) is known by a variety of common names. Some of the most common include the quahog or quahaug, the hard clam, the little neck clam, the topneck clam, the cherrystone clam, the chowder clam and hogs (Tom Kehoe, 1993 and George Noonan, 1993, personal communications). The last five names refer, respectively, to small, medium and large sizes of the clam, which are prepared as food in different ways. In Connecticut, cherrystone and chowder clams are known collectively as bigs (George Noonan, 1993, personal communication); formerly, the hard clam was known as the common round clam (Perkins, 1869) or simply, the round clam (Sumner et al., 1913).

Of the above common names, quahog is probably the oldest one still used, and is derived from its use by native Americans. Corrupted from the Algonquin plural form poquahock, into quahaug or quauhog, the quahog was also known variously by English settlers as the poquau, the hen, hens-poquahock, and Poquahauges (Gould, 1870). In 1634, Connecticut Governor John Winthrop noted that it was called wampampeege (Gould, 1870), which probably referred to its use as a source of native American money, wampum. Apparently the natives broke off only the purple part and made it into black money, suckauhock, which was twice the value of pompom, or white money, made chiefly from a periwinkle (Gould, 1870). In recent decades, the species has been called the Northern Quahog (Abbott, 1974; Turgeon, 1988) to distinguish it from a similar, more southerly species called the Southern Quahog.

1.1.2 Synonymies

In their list of synonymies for *Mercenaria mercenaria* L., Fischer-Piette and Vukadinovic (1977) placed *Mercenaria campechiensis* (Gmelin, 1791), which is now generally considered a distinct species, in synonymy with *M. mercenaria*, while keeping *Mercenaria kennicottii* Dall (1872) separate. The latter, it was later discovered, referred to living *M. mercenaria* that were introduced to the northwest coast of North America. When references to *M. campechiensis* are deleted and those of *M. kennicottii* and the synonymies from Palmer (1927) added, the list is still a formidable one:

Pectunculus albus creberrimis faciis acutis exasperatus. Campeche, Lister, 1685, pl. 283, fig. 121.

Venus mercenaria, Linnæus, 1758, ed. X: 686, no. 99;
Winkley, 1907, XXI: 74;
Wood and Wood, 1927, XLI: 12, 14;
Dexter, 1944, LVIII: 71.

Venus mercenaria L., Linnæus, 1767, ed. XII: 1131, no. 123;
Gmelin, 1791, 3271;
Lamarck, 1818, V: 601 (591);
Wood, 1828a,b, 35, pl. 7, fig. 4;
Deshayes, 1832, Vers, III: 1117;
Deshayes, 1835, ed. 2, VI: 346;
Gould, 1841, 85, fig. 67;
Hanley, 1843, in Hanley, 1842–1856: 115;
de Kay, 1843, Zool. New York, V, Moll.: 217, pl. XXVII, fig. 276;
Mighels, 1843, IV: 320;
Philippi, 1845, 69;
Chenu, 1847, *Venus*, pl. 8, fig. 5;
Stimpson, 1851, 19;
Sowerby, 1853, II: 733, pl. CLXII, fig. 204, 205, 206;
Römer, 1858, 36;
Reeve, 1864, XIV, pl. II, fig. 4a,b;
Römer, 1865, 135;
Pfeiffer, 1869, ed. 2, XI (1): 123, pl. 2, fig. 1, 2;
Gould, 1870, ed. 2, Moll.: 1133, text-fig. 445;
Verrill, 1874, part I: 681, pl. XXVI, fig. 184;
Verkruzen, 1878, 211;
Carpenter, 1888, II: 102;
Dall, 1889, 37: 54, pl. 55, fig. 7, pl. 71, fig. 1, 3;
Ford, 1889, III: 29;
Johnson, 1890, IV: 5;
Winkley, 1891, IV: 113;
Baker, 1891, XLIII: 47;
Carpenter, 1891, IV: 138;
Dall, 1902b, XXVI: 376;
Chadwick, 1906, XIX: 103;
Weeks, 1908, XXII: 98;
Cary and Spaulding, 1908, 15;
Winkley, 1909, XXIII: 87;
Johnson, 1915, 7(3): 70;
Winkley, 1916, XXIX: 110;

Johnson, 1916, XXX: 90;
Jacot, 1919, XXXII: 92;
Jacot, 1920, XXXIII: 112;
Maury, 1921, 8(34): 108;
Jacot, 1924, XXXVIII: 49;
Palmer, 1927, 394, pl. 63, figs. 2, 3, 4, 5, 7;
Clench, 1928, XLI: 120;
Procter, 1929, XLII: 102;
Johnson, 1934, 40: 49;
Richards, 1935, XLIX: 132;
MacGinty, 1936, L: 5;
Smith, 1937, 54, pl. 21, figs. 3, 6;
Lamy and Fischer-Piette, 1938, 401;
Harry, 1942, I: 5;
Dexter, 1942, LVI: 60;
Jacobson, 1943, LVI: 142;
Hackney, 1944, LVIII: 58;
Russell, 1946, LIX: 97;
Speck and Dexter, 1946, LX: 34;
Dodge, 1952, 00, art. 1: 97;
La Roque, 1953, 29, Biol. ser. no. 44: 69;
Morris, 1956, 70, pl. 21, Figs. 2, 3;
Jacobson and Emerson, 1961, 88, fig. 89;
Shikama, 1964, II: 79, pl. 48, fig. 8;
Vilas and Vilas, 1970, 41, pl. XII, fig. 1a,b.

Venus mercenaria Linnaei, Chemnitz, 1788, X: 352, pl. 171, fig. 1659, 1660.
Venus meretrix Bolten, 1798, 126. No. 287.

Mercenaria violacea, Schumacher, 1817, 135, pl. X, fig. 3.
Deshayes, 1853, I: 113.
Adams and Adams, 1857, II: 419.
Holmes, 1860, 33, pl. VI, fig. 11.
Meek, 1864, VII, no. 183: 9.

Mercenaria notata Adams and Adams, 1857, II: 419.

Mercenaria cancellata Gabb, 1860, 4: 376, pl. 67, fig. 25.

Crassivenus mercenaria Perkins, 1869, 13: 147.

Mercenaria kennicottii Dall, 1872, VII: 147, pl. XVI, fig. 1.

Mercenaria mercenaria Tryon, 1874, 158, Figs. 388-390.

Venus mercenaria var. *antiqua* Verrill, 1875, Am. J. Sci., 3rd Ser. X: 371
(*non Venus antiqua* King and Broderip, 1832, *non Venus antiqua* Munster.)

Venus kennicottii Dall, 1902a, XXIV: 560, pl. XL, fig. 7;
Dall, 1902b, XXVI: 396;
Dall, 1916, 33;
Dall, 1921, 12: 42;
Oldroyd, 1924, I: 155, pl. 14, fig. 7;
Burch, 1944, 45: 15;

La Roque, 1953, Bull. 129: 69.
Venus mercenaria var. *alba* Dall, 1902b, 26: 377;
Dall, 1903 in Dall, 1890–1903, Trans. Wag. Inst. III (6): 1314.
non Venus radiata Dillwyn, 1817, 189.
Venus mercenaria var. *notata* Dall, 1902b, 26: 376;
Dall, 1903, in Dall, 1890–1903 III (6): 1312;
Johnson, 1915, VII (13): 70.
Venus mercenaria var. *radiata* Dall, 1902b, 26: 377.
Venus notata, Say, 1822, 2: 271.
Venus mercenaria Lmk., Bory de St. Vincent, 1827, 152.
Venus obliqua, Anton, 1837, 1: 284.
Venus cyprinoides, Anton, 1839, 9.
Venus notata Say, Gould, 1841, 86, fig. 67;
de Kay, 1843, V, Moll.: 218, pl. XXVII, fig. 278;
Mighels, 1843, IV: 320;
Philippi, 1844, I: 128, pl. II, fig. 3;
Reeve, 1864, XIV, pl. 2, fig. 4a;
Römer, 1865, 136;
Binney, 1870, ed. 2, Moll.: 135, text-fig. 446.
Mercenaria violacea Schum., Deshayes, 1853, 113.
Mercenaria notata Say, Deshayes, 1853, 114.
Venus mercenaria alba Dall, Palmer, 1927, 395.
Venus mercenaria subradiata Palmer, 1927, 395.
Venus submercenaria Palmer, 1927, 394.
Venus mercenaria notata Say, Baker, 1950, LXIII: 124.
Mercenaria mercenaria L., Abbott, 1954: 406, pl. 32, fig. h;
Bousfield, 1960, 31, pl. VIII, fig. 81;
Moore, 1961, I (1): 44;
Dexter, 1962, LXXVI: 68;
Porter and Tyler, 1972, 4, fig. 43;
Shoemaker, 1972, LXXXV: 117;
Porter, 1974, 81;
Abbott, 1974, ed. 2: 523, fig. 5861.
Venus (Mercenaria) kennicottii Dall, Grant and Galc, 1931, I: 324.
Burch, 1944, 42: 10.
Venus kennicottii, Keen, 1937, 26.
Venus ziczac Pearse, 1936.

1.1.3 Conchological Description

Mercenaria mercenaria (Linnæus) is a large species, with some specimens attaining a length of 15 cm, and the species has all the typical characteristics of Veneridae: three cardinal teeth in each valve, a lunule and escutcheon, a pallial sinus (Fig. 1.1 [1]), and predominantly

[1] Photographs of all specimens include a centimeter scale, subdivided into millimeters.

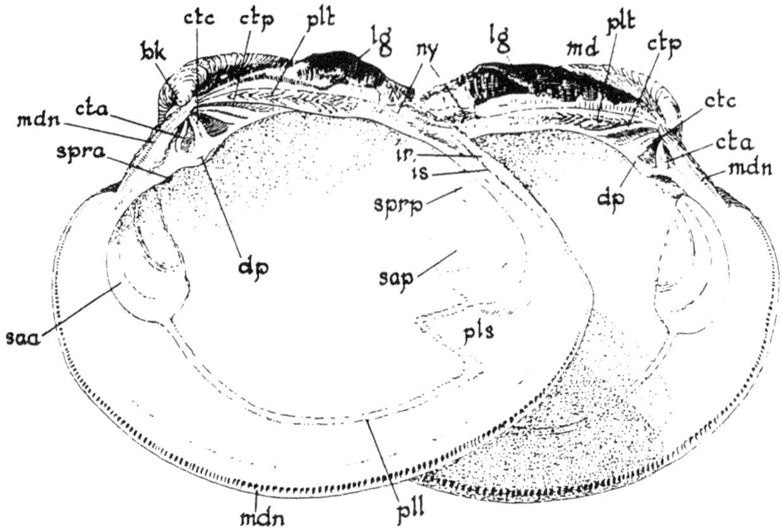

Fig. 1.1. *Mercenaria mercenaria*, interior of right and left valves, after Jones (1979). bk = umbo, cta = anterior cardinal tooth, ctc = central cardinal tooth, ctp = posterior cardinal tooth, dp = dental platform (i.e., the hinge plate), ir = interlocking ridge, is = interlocking slot, lg = ligament, md = dorsal margin, mdn = marginal denticle (i e., crenulation), ny = nymph, pll = pallial line, pls = pallial sinus, plt = the rugose area of the nymph, saa = scar of anterior adductor muscle, sap = scar of posterior adductor muscle, spra = scar of anterior pedal retractor muscle, sprp = scar of posterior pedal retractor muscle.

concentric sculpture (Fig. 1.2) on its valves, which are subovate, with a slightly pointed posterior end and a prosogyrous umbo.

The cardinal teeth are solid and slightly arching, located on a moderately wide hinge plate. The right anterior and median cardinal teeth, as well as the left median and posterior cardinal teeth are bifid and wedge shaped, while the left anterior and right posterior cardinal teeth are thin and lamellar. The cardinal teeth are attached dorsally to the umbone, and not to each other, unlike many pitarine genera. The lunule is large, cordate, impressed, and defined by a sharply incised line. The escutcheon is an elongate, concave area bordered by a dorsal ridge that is often sharply defined in the left valve and usually moderately defined in the right valve. The escutcheonal ridge is accentuated by a slight furrow just ventral to it; the furrow is more pronounced in the right valve. The escutcheon is slightly smaller in the right valve, which overlaps the left valve slightly (Fig. 1.7), and an internal dorsal groove exists along the inner margin of the right valve where the left valve meets it.

The pallial sinus is free of the pallial line and acutely triangular; the apex of the sinus is oriented towards the lower half of the anterior adductor muscle scar. The sinus is moderately short, extending roughly a quarter of the distance between the anterior and posterior adductor muscle scars.

The sculpture of the valves consists of low, widely spaced concentric lamellae on juveniles that are usually worn away with age. Interspersed between the lamellae are fine growth lines. In adults these lamellae are more closely spaced and well defined at the sides, developing towards the middle into thicker, concentric ridges that coalesce to form smooth, sometimes

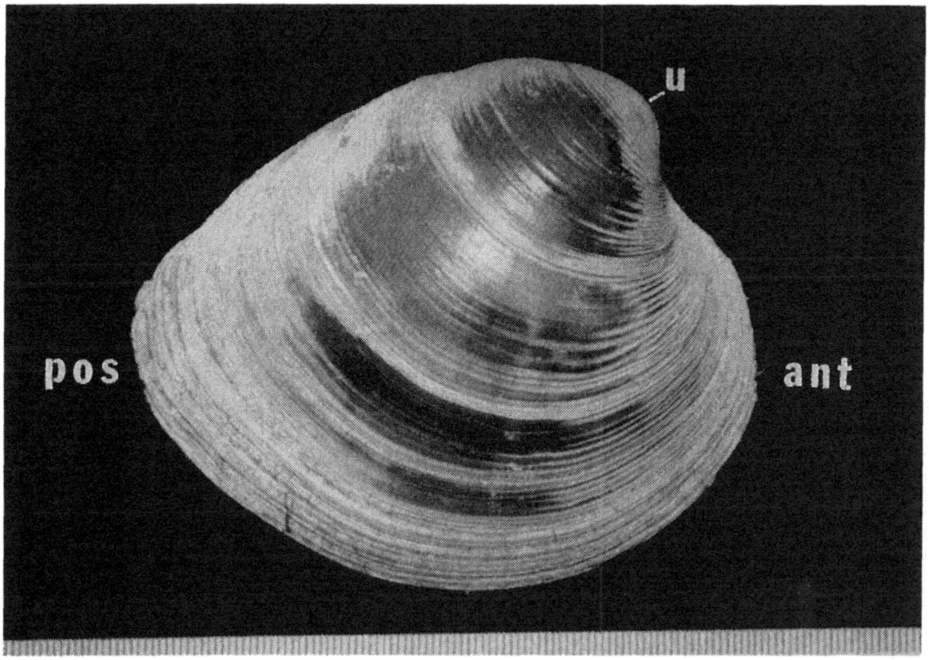

Fig. 1.2. Exterior (right valve) of a typical hard clam (Academy of Natural Sciences of Philadelphia [ANSP] specimen 373468). ant, pos = anterior and posterior ends; u = umbo.

polished medial areas, obscuring both the fine and coarser concentric sculpture. Groups of lamellae are periodically separated by a slightly deeper sulcus that indicates the end of a period of growth. Fine radial striae overlapping the concentric sculpture are often obvious across the polished medial area or sometimes on the umbone.

The chionine characteristics of the hard clam include the absence of any anterior lateral teeth, and crenulate margins. Its sculpture is relatively weak compared to the fairly strong sculpture characteristic of other chionine species. The crenulations of the internal margin begin as fine, slightly oblique grooves at the lunule and continue along the ventral margin as coarser, radial, and slightly irregular grooves that fade posteriorly. These grooves are pronounced at the margin but fade dorsally towards the pallial line, an indication that they are a continuation of a fine radial element existing throughout the shell and visible externally as fine radial sculpture (Dall, 1903).

A prominent rugose nymph, located just posterior to the cardinal teeth on the hinge plate, is the distinguishing characteristic of *Mercenaria* s.s. and the type species of the genus, *M. mercenaria*. The nymph is the area of the hinge plate behind which the ligament is inserted. Even in juveniles the nymph is long, and it continues to lengthen as the animal matures. In adults, the nymph is roughly thrice the length of the posterior cardinal tooth. The rugose area first appears as a series of irregular creases on the posterior flank of the posterior teeth in juveniles. It then develops and expands onto the nymph, often subsequently obscuring the presence of the left posterior cardinal tooth; it is separated by a smooth socket from the posterior cardinal tooth in the right valve. The rugose area in mature specimens covers much of the anterior part of the nymph with irregular, rough chevrons and ridges radiating from the umbone. Dall (1902a) proposed that this area acted as a supplemental cardinal tooth.

The anterior adductor scar is usually slightly smaller than the posterior adductor scar and often slightly narrower. Both scars are roughly semilunular, shaped somewhat like teardrops. The pedal muscle scar is a small, semi-ovate impression inserted on the underside of the hinge plate beneath the right anterior cardinal tooth and the area anterior to the left cardinal teeth, respectively. Just posterior to the scar, the ventral margin of the hinge plate bulges directly beneath the right medial cardinal tooth and its socket.

Small, olive brown threads of a thin periostracum often glisten between the ventral concentric ridges of dried valves, the only vestiges of the periostracum evident in this species. The shells are chalky white and gray, the smooth medial area often stained with rust or black. A few thin brown zigzag lines are sometimes present near the umbones. The interior is predominantly white, although the posterior end, sometimes including the posterior adductor scar and extending around the area ventral to the pallial line, is usually colored a deep bluish purple.

1.1.4 Type Material

The holotype of the hard clam, *Venus mercenaria* Linnæus (1758), resides in the collection of the Linnean Society of London (Fig. 1.3). Dodge (1952) noted that Linnæus (1758) originally described the type locality as Pennsylvania, a state with no coastline, which at that time was united with Delaware, a state with a generous coastline; thus, Dodge (1952) suggests, the type locality should be amended to the coast of Delaware. The type was collected by P. Kalm.

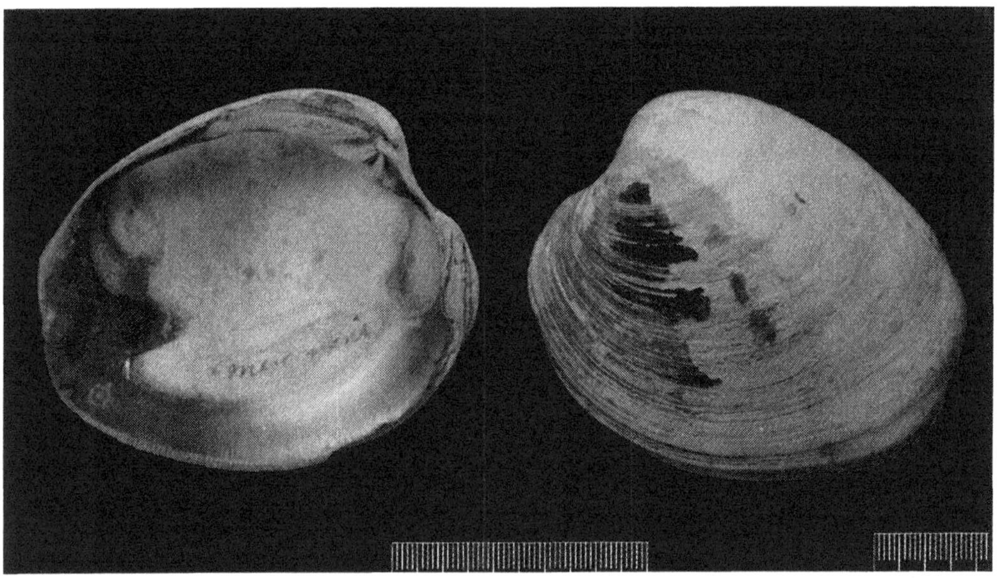

Fig. 1.3. Internal and external views of the holotype (a left valve) of *Mercenaria mercenaria* (Courtesy of the Linnaean Society).

1.1.5 Distribution

Mercenaria mercenaria (Linnæus) first appears in the fossil record during the Upper Miocene, 5–11 Mya from Massachusetts to Florida (Dall in Palmer, 1927). The recent distribution (Fig. 1.4) extends from the Bay of Chaleurs, Gulf of St. Lawrence and Sable Island south to the Florida Keys and, according to Palmer (1927), west to Corpus Christi, Texas, although Andrews (1971) does not list it as present along the Texas coast, and authors might have mistaken *Mercenaria campechiensis* var. *texana* for *Mercenaria mercenaria*, including recent workers that assert its presence in Puerto Rico (Juste and Cortes, 1990). In addition, it was introduced, usually unsuccessfully, to the eastern Atlantic along the coasts of England and France during the mid to late nineteenth century and early twentieth century (Heppell, 1961), although colonies now exist in both countries. The species has been recorded in Dutch waters (Kaas, 1937) and from Belgium (Tebble, 1966). Introductions were also made to the northwest coast of North America in Washington and California, of the United States (Hanna, 1966) (see Chapter 16).

The hard clam occurs in groups ranging from small patches to extensive beds at intertidal and subtidal depths, from sand to muddy sand sediments (see Chapter 8). Sumner et al. (1913) noted that it was abundant just below low tide level, especially in sheltered areas, such as bays and estuaries. They recorded shells from depths of 4–26 m, with living specimens taken in depths as great as 12 m. Besides inhabiting relatively bare substrate, hard clams occur in eel

Fig. 1.4. The geographic distribution of *Mercenaria mercenaria* (solid line; large dots denote successful introductory sites), and congeneric species, *Mercenaria campechiensis* (dashed line) and *Mercenaria kellettii* (dashed and dotted line).

grass beds (Peterson and Beal, 1989), and in the shelly bottoms near oyster beds, as well as between and under oysters (Wells, 1961). In England, Tebble (1966) noted that the hard clam lives in mud with stones and shells, intertidally to depths of a few fathoms. Well known as a marine species, *Mercenaria mercenaria* is tolerant of low salinities (Wells, 1961). Hard clams were found living upstream in a North Carolina estuary, where salinity averaged 19‰ (Wells, 1961), roughly half that of normal marine conditions. Failing to detect a low salinity death point for mature animals, Wells (1961) hypothesized that the animal must rely on its store of glycogen for anaerobic respiration when the valves are closed, the amount of stored glycogen determining how long the clam can survive under saline stressful conditions.

Both its wide tolerance of salinity and temperature probably contribute to the ubiquity of the hard clam within its range. The hard clam survives between 0 and 30°C (McHugh, 1984; Van Winkle et al., 1976). North of Cape Cod, Massachusetts, the hard clam is restricted to bodies of water where oceanographic conditions favor spawning (McHugh, 1984).

1.1.6 Infraspecific Variation

Variations of *Mercenaria mercenaria* occurred with the first appearance of the species in the Miocene. *Mercenaria mercenaria* var. *cancellata* (Dall, 1902b) was described from a rare Miocene specimen and is occasionally found living (Dall, 1902b). In this form, the medial

Fig. 1.5. Right valves of juvenile and young adult *Mercenaria mercenaria* (ANSP 24766). The 'notata' markings are evident on the left specimen.

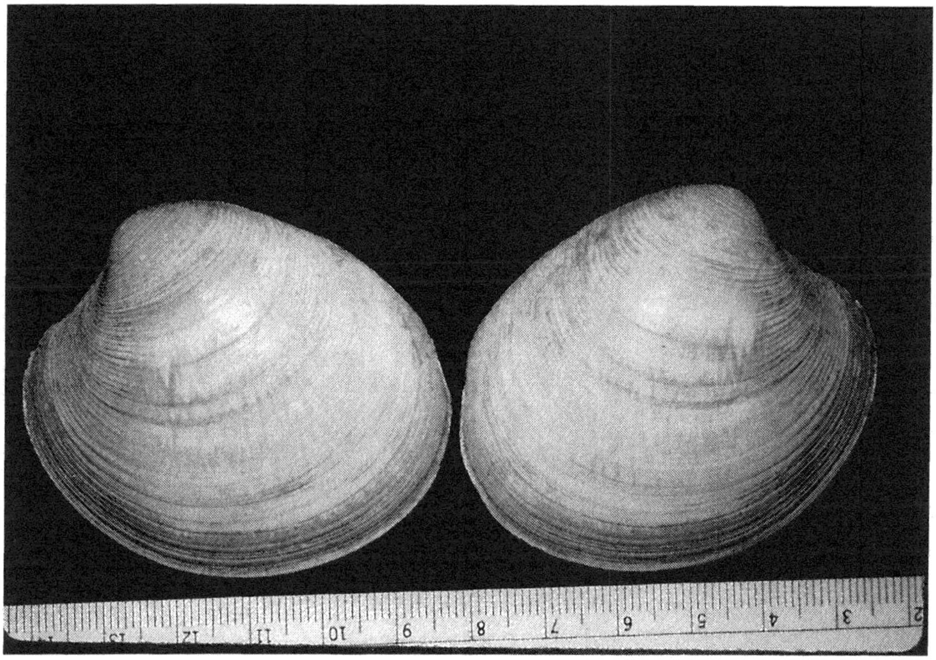

Fig. 1.6. External view of *Mercenaria mercenaria* var. *notata* (ANSP 145694). Left valve (left), right valve (right).

smooth space of the valve is sculpted with flat concentric ribs that are intersected by unusually distinct fine radial sculpture. *Venus submercenaria* Palmer 1927 is a synonym. *Mercenaria mercenaria* var. *radiata* (Dall, 1902b) is tersely described as similar to the above form, except that the smooth medial area lacks concentric ribs. *Venus subradiata* (Palmer, 1927) is a synonym. Besides these polymorphic variations, sculpture varies with age, as noted above. Juvenile hard clams lack any smooth medial area, and have low, thin, moderately well-spaced lamellae interspersed with fine growth lines (Fig. 1.5).

Mercenaria mercenaria var. *alba* (Dall, 1902b) refer to specimens that lack any internal purple coloration but are otherwise normal. This condition occurs in younger specimens, and frequently among older specimens; Gould (1870) noted that in very old specimens the purple color is often obscured by a thick white glaze. He also noted fishermen's observations that specimens lacking purple were found near Chatham, Massachusetts.

Mercenaria mercenaria var. *notata* (Say, 1822) is marked by thin zigzag brown blotches and lines (Figs. 1.6 and 1.7), and lacks the internal purple coloration. It appears to be caused by a heterozygous condition of a binary genetic locus for white (no color) and reddish brown (color) (Chanley, 1961). Aquaculturists (see Chapter 15) have taken advantage recently of this form, selectively breeding and reseeding their beds with the variant clams as a means of determining the rate of success of cultured juveniles in the wild (David Reylea, 1993, personal communication). It occurs throughout the natural range of the species, and is particularly obvious on younger, smaller specimens. Synonyms of this form include *Venus obliqua* Anton (1837), and *Venus cyprinoides* Anton (1839; Dall, 1902b).

Fig. 1.7. A posterior view of the escutcheon of a specimen of *Mercenaria mercenaria* var. *notata* (ANSP 53082) highlighted by markings that often occur on species of *Chione* but occur rarely on the hard clam. Note the slightly overlapping valves.

All of these forms appear to represent variations of the conchological characters that occur throughout the range of the species and do not demarcate any particular geographic subspecies. Beyond the distinctive variation described above, there is some individual variation in shape, with slightly longer, blunter, or more quadrate specimens occurring occasionally (Figs. 1.8 and 1.9). According to Newcombe et al. (1938), the animals tend to grow thicker shells in warmer, more southerly waters.

Although *Mercenaria campechiensis* var. *texana* (Dall, 1902b) has been considered as a southern subspecies of *M. mercenaria* by some authors, both genetic and morphometric data indicate that it belongs to *M. campechiensis* (Dillon and Manzi, 1989a).

Hybrids of *Mercenaria mercenaria* and its closely related congener, *Mercenaria campechiensis* (Gmelin, 1791), display a wide range of conchological variation that involves a mixture of sculpture, interior color and thickness of the valve (Fig. 1.10).

Malformed specimens of *M. mercenaria* are fairly rare. A few show an irregular rippling of the shell as if it were affected from nestling among rocks (Fig. 1.11). One intriguing malformation causes one valve to become flattened, while the other forms an obese cup (Fig. 1.12); it is unclear whether the origins are pathogenic or genetic. A specimen of the closely related southern hard clam (Fig. 1.13) shows a radial rippling reminiscent of certain malformed specimens of the extinct *M. corrugata* (Lamarck, 1818), below.

Fig. 1.8. An ovate specimen (right valve) of *Mercenaria mercenaria* (ANSP 176316).

Fig. 1.9. A slightly elongate specimen (right valve) of *Mercenaria mercenaria* (ANSP 19148).

Fig. 1.10. Hybrids (right valves) of *Mercenaria mercenaria* × *Mercenaria campechiensis* (ANSP 396994).

1.1.7 Conchological Comparisons to Congeners

Mercenaria mercenaria (Linnæus) is most closely related to *Mercenaria campechiensis* (Gmelin, 1791), the southern hard clam, a more southerly but geographically overlapping species that ranges from the Chesapeake Bay south to Cuba and west to the Yucatan (Palmer, 1927). It often hybridizes with *M. mercenaria* (see Chapters 5 and 6); the hybridization decreases northward, a result of increasingly distinct seasonal cues which increasingly isolates them genetically (Dillon, 1992). The cues, both spatial and temporal, ultimately isolate what are otherwise very closely related species that probably evolved from a common ancestor (perhaps *M. prototypa* Maury, 1924, below) during the Miocene. *M. campechiensis* occurs in the shallow estuarine areas of southern Florida (excepting the Florida Keys), Mexico and elsewhere in the Caribbean, as well as offshore further north, uncommonly as far north as Cape May, New Jersey

Fig. 1.11. Crumpled, malformed specimens of *Mercenaria mercenaria*. Left: an anterior view of the deformed lunule on a left valve. Middle: a profile of the opposing right valve (ANSP 175782). Right: a right valve with slight rumpled ventral section (ANSP, no number).

(Merrill and Ropes, 1967). *M. campechiensis* closely resembles *M. mercenaria* in general conchological form and has been considered even recently by some workers as conspecific with *M. mercenaria* (Fischer-Piette and Vukadinovic, 1977). There are, however, distinctive, if sometimes subtle, conchological differences. The adult shell of *M. campechiensis* lacks a smooth medial area and the internal purple color; it is "shorter, rounder, larger, and much thicker than that of *V. mercenaria* ... the escutcheon better defined and wider, the disk wholly covered with fine close lamellation, which is not ... colored red-brown or black; the lower posterior angle of the pallial sinus is generally more acute, the crenulation of the inner margin finer, and the disposition of the cardinal teeth less fan-like than in *V. mercenaria*" (Dall, 1903). Its low, sharp lamellae are thicker (Palmer, 1927; Dillon and Manzi, 1989a), and the finer, interstitial concentric threads are distinct and sharp. Although *M. campechiensis* was thought to have a wider lunule than *M. mercenaria* (Palmer, 1927), this character was never considered to be very consistent. Dillon and Manzi (1989a) suggested that the fault was in comparing lunule width to lunule height; they showed that a more consistent difference was that *M. campechiensis* (and its variety *texana* Dall, 1902b) had a shorter lunule length than *M. mercenaria* when compared to overall shell length, width, and height. The typically coarser concentric lamellae of *M. campechiensis* is probably an adaptation to burrowing in the coarser carbonate sands offshore, in peninsular Florida, and the Caribbean Sea (Dillon and Manzi, 1989a).

Dillon and Manzi (1989b) note that *M. campechiensis* is represented in the bays and inlets of the northern Gulf of Mexico by a subspecies, *M. campechiensis texana* (Dall, 1902b). The sculpture of this variety more closely resembles that of *M. mercenaria*: "the concentric

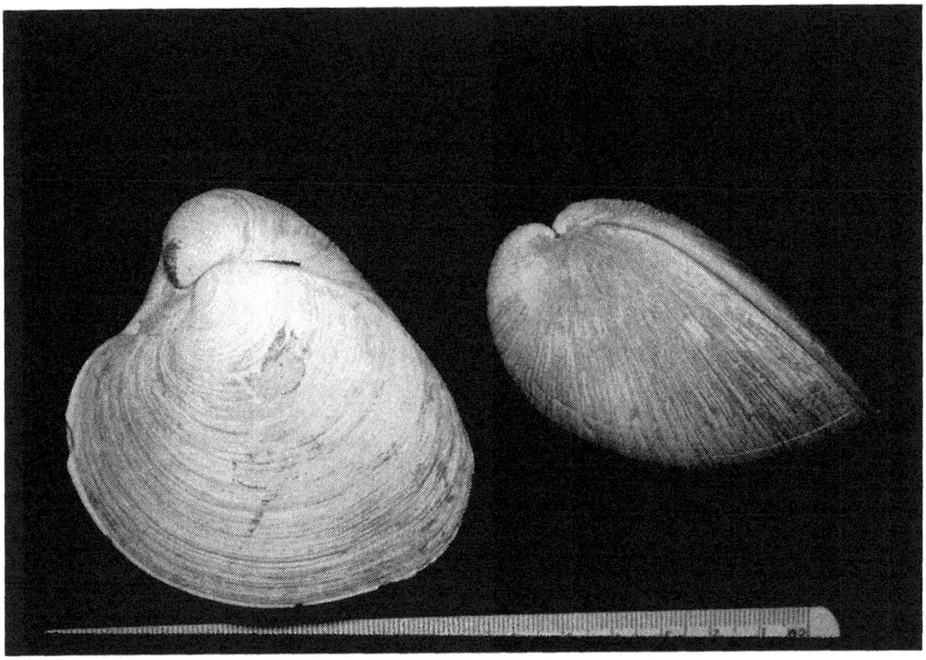

Fig. 1.12. Malformed 'cup' specimens of *Mercenaria mercenaria*; the right valve has formed into a cup, with the left valve flattened like a lid over it. Left: profile (ANSP 342298) of flattened left valve with the umbo of the right valve projecting above it. Right: anterior view (ANSP 119127) of the convex right valve and flattened left valve.

lamellae toward the middle of the disk coalescent, forming broad, more or less inosculating, low, flat-topped ribs with polished tops, sometimes showing the brown lineations of the younger stages" (Dall, 1902b). Some later workers designated this a variety of *Mercenaria mercenaria* (e.g., Menzel, 1970; Abbott, 1974) although data from several genetic loci as well as the morphometric data discussed above indicate that *texana* is a variety of *M. campechiensis* (Dillon and Manzi, 1989), and not a naturally occurring hybrid of *M. mercenaria* and *M. campechiensis* as surmised by Menzel (1970). Dillon and Manzi (1989a) hypothesize that the thin, easily erodable lamellae common to both *texana* and *M. mercenaria* might be "an adaptation for burrowing in the fine, terrigenous silt and mud found in the estuaries of the American Atlantic and northern Gulf coasts." They further note that preliminary hybridization results indicate that sculpture is controlled primarily by heredity. A minimal evolutionary scenario that explains these observations is, they suggest, one in which the thicker, heavier ribs of *M. campechiensis* evolved after *M. mercenaria* split off from *campechiensis*.

Young *M. campechiensis* shells resemble *M. mercenaria* more than the adults, being more elongate and having a smooth area located on the lower middle of the valve. Some specimens of *M. campechiensis* var. *texana* have markings of zigzag lines, as compared to the zigzag blotches and lines present on specimens of *Mercenaria mercenaria* var. *notata*.

The eastern Pacific *Mercenaria kelletii* (Hinds, 1845) is a smaller, more elongate and tropical species; its lamellae flare out laterally, and the smooth area often dominates the rest of the valve (Harte, 1992a: Fig. 1.2). It lacks any internal purple color.

Fig. 1.13. The deformed radial ripples in this right valve of *Mercenaria campechiensis* somewhat resemble the more massive forms of the Miocene species, *Mercenaria corrugata.*

1.2 ADAPTATIONS AND EVOLUTION

The evolutionary success of the hard clam can be traced in part to a variety of conchological adaptations to its environment. Cladistic analyses of conchological data indicate the genus *Mercenaria* to be most closely related to *Securella* Parker (1949; Harte, 1998). Paleontological and conchological data indicate that the ancestors of *Mercenaria* originated in the northwest Pacific and belonged to the taxon *Securella,* which in turn was probably derived from an Eocene–early Oligocene member of the genus *Placamen* Iredale (1925). *Securella* migrated probably along the northern part of the North Pacific Current into the Americas at the beginning of the Oligocene. *Securella* and possibly *Mercenaria* s.s. first appeared in the lower (early) Oligocene in southeastern North America. *Mercenaria* s.s. radiated into several species during the Miocene (Table 1.1) and subsequently declined; three species exist today.

1.2.1 Conchological Adaptations [2]

Much has been observed about bivalve adaptations (e.g., Carter, 1968; Stanley, 1970; Seilacher, 1974; Savazzi, 1985). The burrowing paradigm of Seilacher (1974) required that valve sculpture be perpendicular to the direction of burrowing, asymmetrical in cross section,

[2] Except for the paragraph on pigmentation, this discussion is derived from Harte (1992b).

TABLE 1.1

Fossil species of *Mercenaria*

Species	Stratigraphic range
Mercenaria (Mercenaria)	
campechiensis	
var. *capax* (Conrad, 1843)	Upper Oligocene, North Carolina–lower Miocene, Virginia
var. *mortoni* (Conrad, 1837)	Pleistocene, South Carolina
var. *carolinensis* (Conrad, 1875)	Upper Pliocene–Pleistocene, North Carolina
ducateli (Conrad, 1838)	Lower Miocene, New Jersey–Maryland
halidona (Dall, 1900)	Upper Oligocene–lower Miocene, Florida
langdoni (Dall, 1900)	Lower Miocene, Florida
blakei (Ward, 1992)	Lower mid Miocene, Maryland
nannodes (Gardner, 1926)	Middle Miocene, Florida
prototypa (Maury, 1924)	Lower Miocene, Brazil
?*altilaminata* (Palmer, 1927)	Miocene
cuneata (Conrad, 1867)	Mid Miocene, Maryland–Virginia
tetrica (Conrad, 1867)	Upper Miocene, Maryland–Virginia
druidi (Ward, 1992)	Upper Miocene, Virginia–North Carolina
corrugata (Lamarck, 1818)	Upper Pliocene, Virginia–Florida
inflata (Dall, 1903)	Lower Pliocene, Virginia
kellettii (Hinds, 1845)	Pliocene, Ecuador
campechiensis	
var. *texana* (Dall, 1902b)	Recent
var. *alboradiata* (Sowerby, 1853)	Recent
Mercenaria (Securella)	
mississippiensis (Conrad, 1848)	Lower Oligocene, Mississippi
craspedonia (Dall, 1903)	Lower Oligocene, Mississippi
perbrevisformis (Dockery, 1982)	Lower Oligocene, Mississippi
carmanahensis (Clark, 1925)	Upper Oligocene, Washington and British Columbia–lower Miocene, Japan
cryptolineata (Clark, 1918)	Upper Oligocene, California and Alaska–Lower Miocene, Japan
vancouverensis (Clark and Arnold, 1923)	Upper Oligocene, British Columbia
alaskensis (Clark, 1932)	Miocene, California and Alaska
clallamensis (Reagan, 1909)	Miocene, Washington and Oregon
ensifera (Dall, 1909)	Miocene, Washington, Oregon and Siberia
juanensis (Loel and Corey, 1932)	Miocene, California
montesanoensis (Weaver, 1912)	Miocene, Washington
panzana (Anderson and Martin, 1914)	Miocene, California
margaritana (Anderson and Martin, 1914)	Miocene, California
diabloensis (Clark, 1915)	Miocene, California
pabloensis (Clark, 1915)	Miocene, California
valentinei (Wiedey, 1929)	Miocene, California
postostriata (Kanno, 1958)	Early–mid Miocene, Japan
chitaniana (Yokoyama, 1926)	Miocene, Japan
kurosawai (Kanno, 1958)	Miocene, Japan
moriyensis (Tanaka, 1961)	Miocene, Japan
sigaramiensis (Makiyama, 1927)	Miocene, Japan
yiizukai (Kanehara, 1937)	Miocene, Japan
yokoyamai (Makiyama, 1927)	Miocene, Japan
stimpsoni (Gould, 1861)	Late Miocene or early Pliocene, Japan
bisculpta (Dall, 1909)	Pliocene, Oregon
elsmerensis (English, 1914)	Pliocene, California
securis (Shumard, 1858)	Pliocene, California, Oregon

and reduced medially (perimeter smoothening). Later experimentation and observations have supported this paradigm (Stanley, 1977; Stanley, 1981; Savazzi, 1985). A clam burrows anteriorly, and assumes a life position with the posterior end closest to the sediment surface. From this it is logical to assume that the anterior end will facilitate burrowing and anchorage. The posterior end, especially of shallow burrowers, coming into contact with the substrate only towards the end of burrowing, can contribute little towards burrowing. Being the point closest to the surface and predators, however, the posterior end probably functions more towards predatory defense and reducing surface scour of sediment around the shell, thereby preventing disinterment. These are useful perspectives for analyzing venerid adaptations.

Most chionine clams burrow sluggishly and shallowly, with the posterior tip positioned within 1 cm of the sediment surface (Stanley, 1970). The shells are moderately thick, prosogyrous, and subovate with a slightly angular posterior end; most have strong valve ornamentation. In each species, the unique set of variations among these characters reflects a unique balance and compromise of adaptive strategies.

M. mercenaria is a large, thick shelled, moderately rapid burrower (Stanley, 1970) of subdued, predominantly concentric sculpture. Such sculpture aids burrowing, while size and thickness help keep it anchored in the sediment (Kauffman, 1969). Thickness aids in deterring borers and crushers, while size restricts the spectrum of possible predators.

The clam adjusts burrowing depth (1–2 cm between posterior end and sediment surface) and life position to sediment type, and inhabits an unusually broad range of environmental conditions (Stanley, 1970); this ability to adapt to sediment changes probably accounts at least partly for its wide exploitation of habitats.

The crenulate margins of the hard clam might confer several survival advantages. Hypotheses on defensive functions of crenulate margins include increasing resistance of the shell to compression from shell-crushing predators (Waller, 1969), restricting predatory access of starfish, and creating a tight seal (Carter, 1968) thereby preventing release of diagnostic chemicals into the environment, and increasing survival times in a predator's digestive tract (Vermeij, 1987). Restricting predatory access and creating a tight seal might be effected equally by marginal folding, and function similarly. Jones (1979) observed, for example, that marginal folding effectively keeps the posterior dorsal margin closed while siphons are extended, and suggested that the resulting marginal overlap might deter polydorid polychaete pests. Additionally, both marginal interdigitation and folding can thicken the marginal juncture, discouraging boring predators.

The predominance of concentric sculpture within the hard clam and the rest of the family Veneridae might in part be due to the ontogenetic ease with which strong, well spaced concentric sculpture can be modified into structures that aid anchorage and defense (lamellae) or burrowing (cords or ridges), facilitating evolution into different life strategies. Indeed, this transition can be seen within the hard clam, where juveniles, more vulnerable to disinterment, have widely spaced low, anchoring, concentric lamellae (Pratt and Campbell, 1956) that gradually become closely spaced, more subdued threads, medially worn smooth in adults, traits which aid burrowing more than anchorage.

The internal purple coloration usually present posteriorly in the hard clam might represent yet another anti-predatory adaptation. Such coloration is present in most species of chionine clams. Shell pigments are thought to be the waste products of metabolism, secreted in the shell as a means of disposal (Comfort, 1951) and are sometimes intimately associated with conchiolin layers. In corbulids, however, the pigmented conchiolin layer has been shown to effectively deter borers (Lewy and Samtleben, 1979), and this might also be true for *M. mercenaria*. The toxicity often associated with metabolic waste products might be utilized within the purple pigment of the hard clam as a chemical deterrent for boring fouling organisms, rather than boring predators, as in corbulids. Deposition of the pigment in the area most vulnerable to boring foulers, i.e. the posterior end, supports this hypothesis, while the absence of purple pigment in juveniles might reflect a biochemical inadequacy that changes upon maturation.

1.2.2 Evolutionary Origin

When and where did *Mercenaria* arise?[3] While the fossil record indicates that representatives of Veneridae have existed since the early Cretaceous 190 Mya (Keen, 1969), a major evolutionary change occurred during the transition from the tropically warm Eocene to the relatively cooler Oligocene, roughly 38 Mya. This was also a period of global oceanic circulatory changes: the pan-equatorial east-to-west current began to be broken up by the increasing uplift of Central America between South and North America and the increasing nearness of Africa to Europe (Hansen, 1987, 1992). Towards the end of the Eocene 84% of all bivalve species went extinct apparently due to global climatic cooling (Hansen, 1987, 1992). Subsequently, a major development of the family Veneridae occurred in the Oligocene with the expansion of Venerinae and Chioninae in North America and elsewhere. Until the end of the Eocene, Veneridae had been dominated by clams with well developed anterior lateral teeth, generally smooth inner margins and often fine concentric sculpture, belonging to Pitarinae, Dosiniinae and related subfamilies. The evolution of Chioninae and Venerinae, members of which have small or no anterior lateral teeth, crenulated inner margins and often strong sculpture that incorporates both radial and concentric elements, represented a significant evolutionary development within the family. Roughly 30% of all extant venerid species (over 500 species) are venerine or chionine. It was during the lower Oligocene that the first recognizable members of *Mercenaria* appeared.

Several possible ancestors to *Mercenaria* exist. Two west Atlantic pre-Oligocene genera with rugose nymphs are described in Palmer (1927). Palmer (1927) noted that *Omnivenus* Palmer (1927) of the Eocene combined both pitarine and venerine–chionine characters: the anterior lateral tooth of Pitarinae, and the crenulate margins and rugose nymphs of certain chionines, including *Mercenaria*. Stenzel (1955) speculated that *Mercenaria* evolved from a southern North American Eocene pitarine genus, *Rhabdopitaria* Palmer (1927) (Fig. 1.14). Stenzel et al. (1957) expanded on this, noting that the discovery of better preserved fossil material indicated that the genus contained crenulated margins and rugose nymphs and thus was not significantly different from *Omnivenus*. They suppressed the genus name of *Omnivenus*, the sole member of which occurs rarely in the fossil record, in favor of the more abundant fossil genus *Rhabdopitaria*.

Stenzel (1955) noted further that *Rhabdopitaria* was the only pre-Oligocene venerid genus that shared rugose nymphs, thick shells with crenulate margins, and a peculiar radiating

---

[3] The following discussion derives significantly from Harte, 1998.

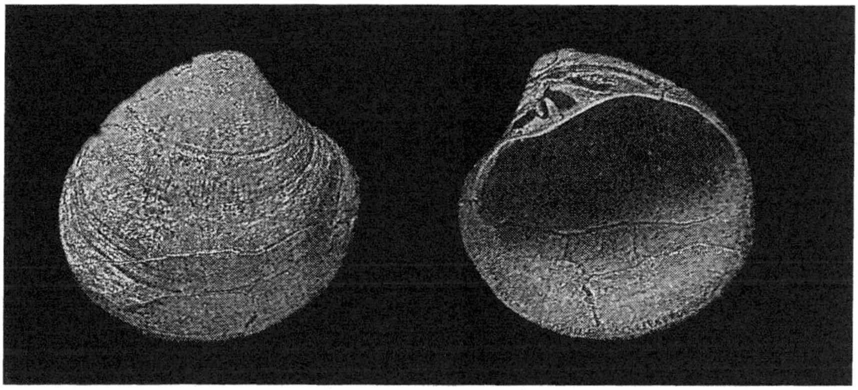

Fig. 1.14. The holotype (a right valve) of *Rhabdopitaria texangulina* Stenzel (1957) (Bureau of Economic Geology Collection, University of Texas at Austin), after Stenzel et al. (1957); length = 26 mm.

shell-wall structure with *Mercenaria*. Radiating shell wall structures are present in many venerid species that have crenulate margins (Eocene *Gemma*, e.g.), however, and are present in several diverse venerid subfamilies (Venerinae, Chioninae, Pitarinae, Sunettinae, Gemminae, and Gafrariinae), as well as in the pre-Oligocene chionine genera *Placamen* and *Tawera* (Keen, 1969; Beu and Maxwell, 1990). Rugose nymphs also occur in diverse subfamilies (Chioninae, Pitarinae, and Cretaceous Dosiniinae). This indicates strongly the propensity of these traits to evolve in parallel, and thus weakens their utility as conservative characters for tracing evolutionary lineages. Assuming erroneously that *Mercenaria* was restricted to the Gulf and Atlantic Coastal Plains, Stenzel (1955) concluded that the ancestor of *Mercenaria* must have originated there. Several significant changes to the characteristically pitarine hinge would have to occur for *Mercenaria* to evolve from *Rhabdopitaria*: loss of the lateral teeth, an anterior shift of the cardinal teeth, a change in the angular position and shape of the right cardinal teeth, and dissolution of the arch connecting the right anterior and posterior cardinal teeth.

What other possible ancestors to *Mercenaria* exist? The second genus discussed by Palmer (1927), *Cyprimeria* Conrad (1864) of the Cretaceous (65–136 Mya), immediately follows the section of *Mercenaria (Venus)* in Palmer (1927), possibly indicating the author's belief that they might be linked evolutionarily. *Cyprimeria* has no lateral teeth (Figs. 1.15 and 1.16). It has a rugose nymph, but lacks crenulate margins, and its general form is lentiform, almost reminiscent of the venerid genus *Sunetta*. It is compressed, fairly thin, with a smooth, almost glossy surface, and a sharply demarcated, deeply sunken escutcheon, such as that present in *Sunetta*. Despite the disparity of form, it could be an ancestor, evolving over tens of millions of years towards *Mercenaria*. No intermediate fossil forms have been discovered.

Species of the Eocene venerid genus *Mercimonia* Dall (1902b) (Tapetinae), recorded from the Paris basin (Cossmann and Pissaro, 1904–1906) and west coast of North America (Turner, 1938; Vokes, 1939), superficially resemble *Mercenaria* in form and were originally classified under *Mercenaria* (Cossmann, 1886). Conceivably, the pan-equatorial east to west current of the Eocene could have facilitated the migration of this taxon from Europe to North America. *Mercimonia* lacks crenulated margins, and has a typical tapetine hinge, with relatively smaller,

Fig. 1.15. The exterior (left valve) of *Cyprimeria depressa* (Conrad) (ANSP 20165).

Fig. 1.16. The interior (right valve) of *Cyprimeria depressa* (Conrad) (ANSP 20165). The white bar indicates the narrow rugose area of the nymph.

thinner cardinal teeth than *Mercenaria*. The hinge of *Mercimonia* is not as deep or massive as *Mercenaria*, and does not extend beyond the termini of the cardinal teeth; the sockets are more excavated. The sculpture is of fine, smooth, indistinct concentric threads; no radial elements appear externally or beneath the surface of the shell.

Cladistic and statistical analyses (Harte, 1998) and to a certain extent paleontological evidence, however, support yet a fourth hypothesis: the ancestor of *Mercenaria* evolved in the northwestern Pacific, perhaps in Japan, migrating across the northern North Pacific Ocean to North America, and south along its west coast, entering the Caribbean during the early (i.e. lower) Oligocene. Fossil faunal assemblages indicate that tropical seas extended to southern Alaska during the Eocene, and that the succeeding Oligocene experienced a general cooling towards a mild, temperate climate (Marincovich, 1991). These relatively mild climatic conditions, combined with the eastern flow of the northern part of the North Pacific Current, provided the migratory corridor for the ancestor of *Mercenaria*, and, as indicated by the fossil record, several other molluscan taxa (Masuda, 1986; Marincovich and Kase, 1986; Marincovich, 1988).

Securella Parker, 1949 (Fig. 1.17) is very similar to *Mercenaria* and has an extensive fossil record dating from the Oligocene on the Pacific coast of North America. Its presence in the Miocene of Japan supports this trans-Pacific migration. In western North America, several species of *Securella* were initially classified as *Chione* or *Venus*. Parker (1949) records three species from the upper Oligocene. Addicott (1976) noted that *Securella* ranged through the Miocene and died out before the Quaternary at the end of the Pliocene.

Fig. 1.17. Left valve of *Securella ensifera* (Dall) (UCMP 36096).

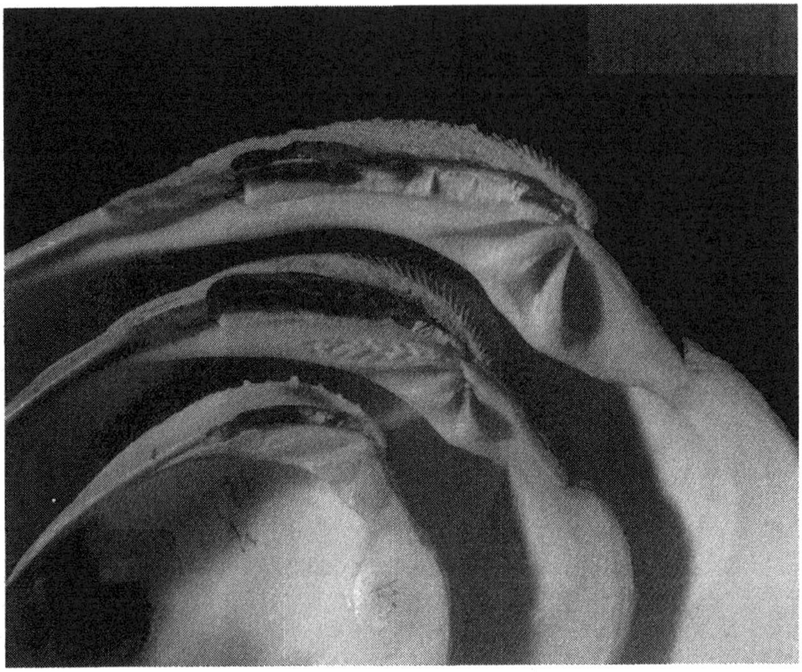

Fig. 1.18. From top to bottom, hinges of the left valves of *Mercenaria (Securella) stimpsoni*, *Mercenaria (Mercenaria) mercenaria*, and *Placamen berrii* (Wood).

That *Securella* might be closely related to *Mercenaria* is not entirely a new idea. Moore (1963) observed that the extant *Mercenaria stimpsoni* (Gould, 1861) of northern Japan "seems to be more closely related to *Securella* than to *Mercenaria*." In such conchological respects as profile, sculpture, hinge dentition, and pallial sinus, *Securella* resembles *Mercenaria* except that it lacks a rugose area on the nymph (Fig. 1.18).

Two upper Oligocene species, *Securella carmanahensis* (Clark, 1925) and *Securella cryptolineata* (Clark, 1918), were reported by Arai and Kanno (1960) to co-occur in the Chichibu Basin, Japan, and in northwestern North America. The formations of the Chichibu Basin have been subsequently redated to the early and medial Miocene (Tsuchi, 1981), however, so the Japanese species are probably distinct, although apparently very similar to the North American species. Parker (1949) indicated that a Miocene species, *Securella ensifera* (Dall, 1909), co-occurred in Siberia and northwestern North America.

From whence might the prototypic Oligocene *Securella* have evolved? The only pre-Oligocene chionine fossils known are a member of Turia Marwick (1927), an extinct subgenus of the extant *Tawera* Marwick (1927), and a member of the extant *Placamen* Iredale (1925); these fossils occur in the western South Pacific in Eocene strata of New Zealand (Keen, 1969; Beu and Maxwell, 1990). Members of *Turia* are very similar to *Tawera* s.s.: they are ovate, with central, moderately prosogyrate umbos and an external sculpture of distinct concentric cords. Extant *Tawera* are restricted to the South Pacific and South Atlantic Oceans. Extant *Placamen* (Fig. 1.19) occur in the tropical and subtropical western parts of the South and North Pacific, including Japan, and throughout the Indian Ocean.

Fig. 1.19. Right valve of *Placamen placidus* (University of California Museum of Paleontology [UCMP]), the type species of the genus.

Placamen is much more similar conchologically to *Securella* than *Turia*. Like *Securella, Placamen* specimens have strongly prosogyrate umbos located slightly anteriorly, a subovate profile with an angular posterior end, and concentric sculpture of distinct cords or lamellae separated by fine interstitial concentric threads. Many species of *Placamen* have nodulose instead of bisected cardinal teeth, which at least indicates the genetic potential within the genus for secondary sculpture to occur on the hinge plate. I have observed some nodulose secondary sculpture occurring on the nymph of at least one specimen of *Placamen berrii* (Wood, 1828a). The overall conchological similarities, the presence of secondary sculpture on the hinge plate and the geographical distribution of *Placamen* render this genus a more plausible ancestor of *Securella* than *Turia*. In such a scenario, a member of *Placamen* in the northwest Pacific became the ancestor of the first *Securella*. One prediction resulting from this hypothesis is that *Placamen* would be more similar to the more immediate descendent *Securella* than to *Mercenaria.*

A comparison of the left hinges of extant species of *Placamen, Mercenaria (Securella)* and *Mercenaria (Mercenaria)* indicates just that – a morphological transition from *Placamen* to *M. (Securella)* to *M. (Mercenaria)* (Fig. 1.18), which is supported by statistical and cladistic analyses (Harte, 1998). *Placamen* and *M. (Securella)* retain similarities in some of their cardinal teeth, whereas *M. (Securella)* and *M. (Mercenaria)* illustrate similar relative nymphal development, profile and sculpture. If the lengths of the cardinal teeth, nymphs and the valves are compared, an interesting pattern emerges (Harte, 1998). The left medial and posterior cardinal teeth significantly shorten from *Placamen* to *Securella*, whereas the relative lengths of the left anterior and medial teeth remain the same. The nymph enlarges on an absolute scale, but not significantly, although the shortening of the adjacent posterior cardinal tooth augments the appearance of this enlargement. In the transition from *Securella* to *Mercenaria* a shortening of the left anterior tooth relative to its adjacent tooth occurs, as well as a significant enlargement of the nymph. Cardinal teeth in the right valve shorten significantly from *Placamen* to *Securella* to *Mercenaria*.

Thus, *M. (Securella)* and *M. (Mercenaria)* have extensive nymphs relative to the posterior cardinal teeth (roughly three or more times the length of the left posterior cardinal tooth) as compared to *Placamen* (less than twice the length of the posterior cardinal tooth), in addition to much greater similarities of sculpture and profile; the bisections of the cardinal teeth are similar. Hence, the transition from *Placamen* to *M. (Mercenaria)* involves a shortening of all hinge teeth, but a relative shortening of the left anterior cardinal teeth compared to the adjacent medial tooth, and an extension of the nymph. The transition in profile involves an expansion of the anterior end, and elongation of the posterior end. The sculpture changes from one of discrete, well spaced thickened ribs to closely spaced threads that can become closely spaced lamellae.

If *Securella* evolved from *Placamen* in the northwest Pacific and then migrated across the Pacific and down the west coast of North America at the beginning of the Oligocene, then the development of an enlarged rugose nymph occurred after the migration of *Securella* into the Caribbean and the evolution of *Mercenaria*. Ontogenetic data support this: the relative nymph length of juvenile *Mercenaria mercenaria* is not significantly different from that of extant *Securella* (Harte, 1998). That the nymph lengthens and becomes more rugose as *Mercenaria mercenaria* matures (Section 1.1.3) is another indication that the direction of transition to be from an initially smaller smooth nymph (*Securella*) to a larger rugose one (*Mercenaria*). *Mercenaria kellettii* (Hinds, 1845), the sole extant east Pacific species, has a less rugose nymph than its extant Atlantic congeners, *M. campechiensis* and *M. mercenaria*.

What is the earliest species of *Mercenaria* known? As defined by Schumacher (1817), the group includes those chionine clams with prominent rugose nymphs. Palmer (1927) classified all fossil and Recent species of *Mercenaria* under *Venus* and observed a small rugose area on the nymphs of some specimens within the Cornell Paleontological collection for the earliest species she described, *Venus mississippiensis* (Conrad, 1848) an uncommonly occurring fossil of the lower Oligocene (Dockery, 1993, personal communication). The poor condition of the holotype, composed of fragments coarsely glued together, precludes a definitive description of the condition of the nymph. No rugose areas are present on the excellently preserved specimens of the species illustrated in Dockery (1982), who classified the species under the chionine genus *Chamelea* Moerch (1853).

Is *Chamelea* or *Securella* a more plausible classification for this species? Extant *Chamelea* consists of two very similar Recent species (Backeljau et al., 1994), one of which is the type species, *Chamelea gallina* (Linnæus, 1758); they are restricted to the east Atlantic and Mediterranean. Keen (1969) noted that the fossil record of the genus extended back to the Oligocene and into eastern North America, although her description of the sculpture, "narrow close concentric lamellae, no radial sculpture", does not match the type species and this might have distorted her definition (and that of other workers) of the extent of the fossil record. The sculpture of both the type and its sister species of *Chamelea* is primarily concentric but differs from fossil and Recent *Mercenaria* and *Securella*, consisting of fine radial striae overlying distinct, sometimes somewhat flattened concentric cords that often anastomize laterally (Fischer-Piette and Vukadinovic, 1977; personal observations), rather than the fine sharp concentric threads that characterize both *Mercenaria* and *Securella* and that are only occasionally overlain or interrupted by fine radials threads. In contrast, the sculpture of *Securella* is identical to that of *Mercenaria* and specifically *Venus mississippiensis*, and its stratigraphic and paleontologically geographic range is close to that of *Mercenaria*. These data indicate that *V. mississippiensis* should be classified under *Securella*.

Statistical data support this, indicating that the mean relative nymph lengths and lengths of the cardinal teeth in the right valve of *Venus mississippiensis* are closer to those of Recent *Securella* and *Placamen* than Recent *Mercenaria* (Harte, 1998). If one accepts *Securella mississippiensis* (Conrad, 1848) as the most plausible classification, the current fossil record demonstrates that these two taxa are contiguous and sequential, with *Securella* preceding *Mercenaria*. The first chionine species with a clearly rugose nymph appear to be *Mercenaria campechiensis* var. *capax* (Conrad, 1837) from the late Oligocene of North Carolina (Ward, 1992) and *Mercenaria halidona* (Dall, 1900) from the late Oligocene of Florida (Ward, 1993, personal communication).

The migration hypothesis of *Securella* is only partly supported by the fossil record. The oldest fossils of *Securella* or *Mercenaria* reported from Japan date from the Miocene. The lack of Oligocene fossil data in Japan might be due to lack of preserving conditions, which includes relatively little land combined with the earthquakes and volcanism associated with the active plate tectonic movements occurring along the Asian rim of the North Pacific Ocean. The earliest record of *Securella* from the west coast of North America dates from the Upper Oligocene, where plate tectonic movements have also resulted in repeated inland flooding and uplift of marine fossil strata onto land. In contrast to both these conditions, however, the source of the earliest known chionine fossils in North America, the Vicksburg Group in Mississippi, is known as containing possibly the best preserved lower Oligocene marine fauna in North America (Dockery, 1982), and occurs in a much less active tectonic area.

Chione craspedonia Dall (1903) co-occurs with and is very similar to *Mercenaria mississippiensis* in the lower Oligocene Vicksburg formation. Although Dall (1903) noted that the profile was quite distinct from the holotype of *M. mississippiensis*, other specimens of *M. mississippiensis* appear so similar to *C. craspedonia* that Dockery (1982) detailed minutely the differences separating the two species. Unlike members of *Chione* s.s., *C. craspedonia* exhibits no radial sculpture. In this and other conchological characters it is more similar to and clearly belongs to *Securella*. A second co-occurring species, *Chione perbrevisformis* Dockery (1982), also lacks radial sculpture and might well be a member of *Securella*.

Harte (1998) proposed that *Securella* Parker (1949) be subsumed as a subgenus under *Mercenaria*. Although the hinges are not illustrated, the photographs of exterior views of various described Japanese species of Miocene and Pliocene *Mercenaria* indicate that they are congeneric with *Securella* (e.g., *Mercenaria sigaramiensis* and *Mercenaria yiizukai* (Kanehara, 1937) in Hayasaka and Uozumi, 1954). In the original description of *Securella*, the ligament, bifurcations of the cardinal teeth and details of the sculpture fall well within the generic description of *Mercenaria* (see below). The deeply sunken ligament is just as deeply sunken within *Mercenaria*. While Parker asserted that radial sculpture only exists beneath the external concentric sculpture in this group of species, I have observed fine radial sculpture overlying some of the non-eroded concentric sculpture of various members of this group,

Fig. 1.20. Left valve of *Securella securis* (Shumard) (UCMP 36099).

e.g., *Securella ensifera* (Dall, 1909) (UCMP hypotype 36096), *Securella cryptoliniata* (Clark, 1918) (UCMP 11178), and the type species of the genus *Securella securis* (Shumard, 1858) (UCMP 36099, Fig. 1.20). Parker (1949) did not mention the pronounced nymph that is twice or more the length of the left posterior cardinal tooth, but it is present in the hinges of the *Securella* illustrated in his work and in all the *Securella* material I have examined.

I also observed a narrow, slightly rugose area on the posterior flank of the left posterior cardinal in a specimen of *Securella securis* (UCMP 36099; Fig. 1.21). I have been unable to discriminate a rugose area from the scraping marks left from specimen preparation on other *Securella* specimens that I have examined, however, and it is absent in most species.

1.2.3 Evolutionary Forebears [4]

From the first recognized species in the early Oligocene *Mercenaria* expanded into several species during the Miocene, declining in diversity through the Pliocene to the present. A review of the fossil history of *Mercenaria* demonstrates conservation of form and sculpture; no radical changes are evident. The nymphs were smooth in early Oligocene species, becoming rugose in late Oligocene species. The rugose area of the nymph, the nymph itself,

[4] Most of the descriptions of fossil species are derived from Ward (1992), Palmer (1927) and Dall (1903) for Atlantic species, Parker (1949) for east Pacific species, and personal observations of holotypic material. Readers are referred to these works for their excellent illustrations. Parenthetical lengths of all species discussed are holotypic measurements, unless designated as representing a medium, large, or adult specimen, or a range of lengths.

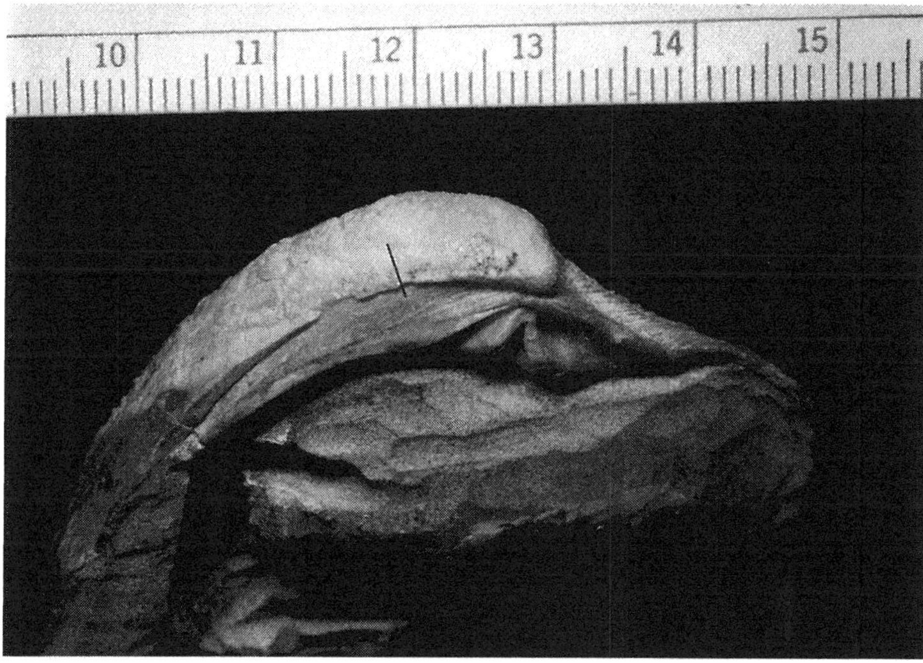

Fig. 1.21. Hinge of a left valve of *Securella securis* (UCMP 36099) with rugose area (note black bar) on the nymph.

and size of the shell increased in the Miocene. The variety in sculpture expanded from prominent concentric ribs to evenly weak concentric threads or lamellae occasionally overlain by weak fine radial threads. Table 1.1 summarizes the recognized species and subspecies in the fossil record.

In the west Atlantic Ocean *Mercenaria (Securella) mississippiensis* (Conrad, 1848) and *Mercenaria (Securella) craspedonia* (Dall, 1903), the earliest species of *Mercenaria* known (see above), occur in the early Oligocene, 33–36 Mya (Figs. 1.22 and 1.23). Both have fairly prominent concentric ribs that are regular and sometimes well spaced, lack radial sculpture and are relatively small (45 mm, adult length). The holotype of *Cytherea mississippiensis* appears to have well spaced concentric ribs. In this, it resembles Recent *Lirophora* Conrad (1862), but differs in having a broader posterior end and a prominent nymph. *M. (M.) mississippiensis* also has a slightly angular posterior end as in *M. (M.) mercenaria*; Dall (1903) noted that the young resemble the young of *M. (M.) campechiensis*.

Mercenaria s.s. blossomed in the Miocene, radiating into ten or more west Atlantic species. They included species that grew large, as well as the first appearance of the relatively weakly sculptured extant Atlantic species, *M. (M.) mercenaria* and *M. (M.) campechiensis*, while other species exhibited clear links with the Oligocene. Ward (1992) noted that one species, *M. (M.) capax* (Conrad, 1843), overlapped the two epochs, occurring from the late Oligocene of North Carolina to the early Miocene of Virginia. *M. (M.) capax* (45 mm, length), treated by Palmer (1927) and Dall (1903) as a subspecies of *M. (M.) campechiensis*, is suborbicular, with a small, pointed pallial sinus, and Ward (1992) argued that it might be identical to

Fig. 1.22. The exterior of the holotype (a right valve) of *Cytherea mississippiensis* Conrad, 1848 (ANSP 30660).

Fig. 1.23. The interior of *Mercenaria mississippiensis* (Conrad) (Holotype, ANSP 30660).

M. (M.) ducateli (Conrad, 1838) [*ducatelli* of Conrad and authors; amended to *ducateli* in the plate caption of Conrad, 1838]. *M. (M.) ducateli* (60 mm, length) occurs in the lower Miocene of New Jersey and Maryland and is sculptured with prominent, moderately to closely spaced, recurved concentric lamellae that are more elevated and lamellate posteriorly; it is orbicular, deep and compressed, with a blunt rather than angular posterior end, and has a narrow rugose area on the nymph.

Another Miocene suborbicular species with prominent concentric sculpture is *M. (M.) nannodes* (Gardner, 1926) (synonym *Venus alumbluffensis* Palmer, 1927). *M. (M.) nannodes* (46 mm, length) occurs in the middle Miocene of Florida (Ward, 1993, personal communication); it is sculptured with prominent, moderately to closely spaced, recurved concentric lamellae that are more elevated and lamellate posteriorly. It is orbicular but with a slightly pointed posterior end and is more obese than *M. (M.) ducateli*, which it most closely resembles. Palmer (1927) described a third orbicular species, '*Venus altilaminata* Conrad', at the end of her section of *Venus* species that comprised *Mercenaria*. The name *V. altilaminata* (77 mm, adult length) is based on specimens labeled Miocene by Conrad but lacking locality data; these are in the Academy of Natural Sciences, Philadelphia. Ward (1993, personal communication) has specimens from the Kirkwood Formation, New Jersey. The specimens most closely resemble *ducateli* but the shape is more quadrate and the umbos are larger. Despite the assertion of Palmer (1927) that the specimens have rugose nymphs, I found it difficult to ascertain whether or not they truly have rugose nymphs due to the condition of the specimens, thus complicating their proper classification.

Mercenaria (M.) halidona (Dall, 1900) from the upper Oligocene–lower Miocene of Florida (Ward, 1993, personal communication) has sculpture similar to that of *M. (M.) mississippiensis* of the lower Oligocene but the lamellae are slightly flattened posteriorly. It is smaller (32–34 cm, length), more compressed, has a broad, subquadrate posterior end, and like *M. mississippiensis* lacks radial sculpture. *Mercenaria (M.) halidona* resembles the young of the early to mid Miocene *Mercenaria (M.) langdoni* (Dall, 1900) of the redated Chipola formation in Florida (see Weisbord, 1971), although young *M. (M.) langdoni* (88 mm, adult length) have more numerous, thinner and less prominent lamellae (Dall, 1903). Adult *M. langdoni* have distant, numerous, thick, elevated recurved concentric ribs, which are flattened posteriorly (Dall, 1900); the interstitial concentric striae are almost lamellose. A radial element is evident in the internal shell structure, but not externally. The prominent concentric lamellae resemble those of several Recent *Lirophora* species.

A prototype of *M. (M.) mercenaria* and *M. (M.) campechiensis*, *Mercenaria (M.) prototypa* (Maury, 1924) appeared in the early Miocene in northeastern Brazil (Maury, 1924). Although the state of the fossils prevented description of the hinge, the sculpture, shape and size indicate a species very similar to *M. (M.) campechiensis* and *M. (M.) mercenaria*. Several varieties of *M. (M.) campechiensis* appeared further north: (1) *tetrica* (Conrad, 1838) (90 mm, length), from Maryland and Virginia lower upper Miocene beds (Ward, 1992), is longer and more compressed than typical, with close, prominent lamellae; (2) *cuneata* Conrad, 1867 (82 mm, length), from lower middle Miocene beds in Maryland and Virginia to upper middle Miocene beds in Maryland (Ward, 1992), is more trigonal than typical, with a more pronounced anterior end and a tendency for the concentric lamellae to flatten out; (3) *mortoni* (Conrad, 1837; syn. *submortoni* Orbigny, 1852) (115 mm, length) from the Pleistocene of South Carolina (Ward, 1993, personal communication), is more elongate and has a larger umbonal area than typical;

the posterior end is broader and more rounded than that of *tetrica*; (4) *carolinensis* (Conrad, 1875) (125 mm, length), ranging from the late Pliocene to Pleistocene in North Carolina (Ward, 1993, personal communication), is an elongated, very inequilateral form with an umbonal area sculptured with coarse, flattened, uneven, imbricated ridges; and (5) *permagna* (Conrad, 1838) is based on very large specimens of what Palmer (1927) asserted were typical *campechiensis* from the lower Pleistocene (Ward, 1993, personal communication) of North and South Carolina.

Campbell (1993) retained this as a subspecies in his description of Pliocene specimens, and suggested that *mortoni* might be a synonym for it. As in *M. mercenaria*, some of these varieties appear to represent polymorphisms (Dall, 1903 labeled them 'mutations') and not distinct subspecific entities.

Ward (1992) treated the *cuneata* and *tetrica* forms as distinct species, however, and placed *Mercenaria plena* Conrad (1869b) (53 mm, length) in synonymy with *cuneata*. Both Dall (1903) and Ward (1992) describe two forms for the same species: a small, rounded form (juvenile?) which, contends Ward (1992), has often been labeled *Mercenaria plena* or *Mercenaria capax*, and a large, massive, more trigonal form (the described *cuneata*). Palmer (1927) described the smaller form, *M. plena*, as moderately thin and sculptured with very close, lamellar, concentric ribbing. My observations of the holotype of *M. plena* indicate that the sculpture is similar to *M. (M.) campechiensis*. Palmer (1927) noted a peculiar dorsal posterior profile on other, more complete specimens of *M. plena*, with a dorsal edge that "extends almost straight from the umbo, then slopes obliquely to half the distance from the beak to the posterior tip and then slopes downward at about an angle of 45°." This change in dorsal profile might indeed indicate a change in form from a juvenile to an adult state that encompasses the two described forms.

Ward (1992) noted that *M. (M.) druidi* Ward, 1992 (synonym: *Venus (Mercenaria) berryi* Gardner, 1943) (63 mm, length) occurs in the upper Miocene in Virginia and North Carolina. Gardner (1943) described the species with a sculpture similar to *M. (M.) ducateli* (above) as well as a somewhat similar profile, but having fused lamellae that can be so pervasive as to sometimes approach the sculpture of *M. (M.) mercenaria*. Ward (1992) noted that *M. (M.) druidi* was smaller and less massive than a thinner, more compressed form of *M. (M.) tetrica* found in the lower Upper Miocene beds of Maryland.

Mercenaria (M.) blakei Ward, 1992 (81 mm, length) from the lower Middle Miocene beds of Maryland, is elongate–ovate, thin, compressed, with very finely impressed, concentric lamellae separated by equally fine concentric striae; fine internal radial sculpture underlies the external concentric sculpture. It has previously been confused with *M. (M.) corrugata* (Lamarck, 1818), a Pliocene species.

The first recognizable members of *Mercenaria (Securella)* in the Pacific Ocean appear during the late Oligocene (24–30 Mya) and include *M. (S.) carmanahensis* (Clark, 1925) of Washington, British Columbia and the lower Miocene of Japan (35 mm, length), *M. (S.) cryptolineata* (Clark, 1918) of California, Alaska and Japan (65 mm, length, UCMP hypotype 36090), and *M. (S.) vancouverensis* (Clark and Arnold, 1923) of British Columbia (40 mm, length). Parker (1949) noted that given the poor condition of the types for *M. (S.) carmanahensis*, further collection of better material might prove it to be a synonym for the Miocene *M. clallamensis* (Reagan, 1909), below. The holotype of *M. (S.) carmanahensis* has a profile typical of *Chione* s.s.: a fairly deep, triangularly subovate shape. *Mercenaria cryptolineata* is triangularly subovate, sculptured with moderately spaced low concentric lamellae with fine interstitial concentric threads. It is similar to *M. clallamensis*, but it has a much heavier hinge and shell. Parker (1949) speculated that it might be a variant of *M. clallamensis*, adapted to a presumably different environment of coarser sediment and rougher water. *Mercenaria vancouverensis* is also similar to *M. cryptolineata* and *M. clallamensis*; the small, eroded types differ only in having a narrower ligamental groove (Parker, 1949). In addition, Parker (1949) placed in synonymy with *M. (S.) cryptolineata* two other described species, *M. (S.) lineolata* (Clark, 1918) and *M. (S.) mediostriata* (Clark, 1918), noting that all three forms were found in association and that differences appeared to be due to degrees of weathering and individual variation.

In the eastern Pacific, ten species of *Mercenaria (Securella)* have been described from the Miocene: *M. (S.) alaskensis* (Clark, 1932) of California and Alaska (35 mm, length), *M. (S.) clallamensis* (Reagan, 1909) of Washington and Oregon (45 mm, length), *M. (S.) juanensis* (Loel and Corey, 1932) of California (69 mm, length), *M. (S.) ensifera* (Dall, 1909) of Washington, Oregon and Siberia (75 mm, maximum recorded length in Parker, 1949), *M. (S.) montesanoensis* (Weaver, 1912) of Washington, (29 mm, a sample length in Parker, 1949), and *M. (S.) panzana* (Anderson and Martin, 1914) of California (80 mm, length), as well as several similar California species previously synonymized under *M. (S.) panzana* by Parker (1949) but later separated as distinct by Adegoke (1969): *M. (S.) margaritana* (Anderson and Martin, 1914), *M. (S.) diabloensis* (Clark, 1915), *M. (S.) pabloensis* (Clark, 1915) and *M. (S.) valentinei* (Wiedey, 1929). As in the Atlantic, these represented the emergence of larger forms of *Mercenaria*.

Mercenaria (S.) alaskensis is triangularly subovate and has moderately spaced concentric ridges, underlain by fine radial sculpture. Parker (1949) noted that it was similar to the Pliocene *Mercenaria (S.) securis* (Shumard, 1858) but had a more pouting lunule, and finer, more numerous radial ribs. No hinges were available for comparison. *Mercenaria (S.) clallamensis* is very similar to *M. (S.) cryptolineata*, but is thinner, and less obese. It is also similar to *M. securis*, but the right anterior cardinal appears partially connected to the hinge margin in *M. (S.) securis*, but well separated and curved in *M. (S.) clallamensis*. *Mercenaria (S.) juanensis* closely resembles *M. (S.) cryptolineata*, but because material available for study by Parker (1949) was eroded and deformed, he did not attempt to subsume it under that species. *Mercenaria (S.) panzana* is similar to the Miocene species *M. (S.) ensifera* but is more elongate; the closely similar species delineated by Adegoke (1969) above represent the more elongate representatives of Miocene *Mercenaria (Securella)*. *Mercenaria (S.) ensifera*, in turn, is very similar to *M. (S.) securis*, but is slightly heavier and shorter, with a more inflated umbo. *M. ensifera* is also similar to *M. (S.) cryptolineata*, but has a narrower finer hinge, and the sculpture is of fine concentric threads with occasional sulci, as in *M. (M.) mercenaria*, but lacking medial smoothness.

Mercenaria (Securella) first appeared in Japan during the Miocene and is represented by seven separate Miocene species: *M. (S.) chitaniana* (Yokoyama, 1926), *M. (S.) kurosawai* (Kanno, 1958), *M. (S.) moriyensis* (Tanaka, 1961), *M. (S.) sigaramiensis* (Makiyama, 1927), *M. (S.) yiizukai* (Kanehara, 1937), *M. (S.) yokoyamai* (Makiyama, 1927), and later, in the late Miocene or early Pliocene, *M. (S.) stimpsoni* (Gould, 1861). Hayasaka and Uozumi (1954) placed *M. yokoyamai* (Makiyama, 1927) in synonymy with *M. chitaniana*, although this was not accepted by some later Japanese workers (Kaseno and Matsura, 1965; Ogasawara, 1973).

Hayasaka and Uozumi (1954) compared the profiles of *M. chitaniana*, *M. yiizukai* and *M. stimpsoni*, and concluded that the relative height of the shell had decreased through geologic time.

In addition to the *Mercenaria (Securella)* described above by Arai and Kanno (1960) as occurring both in Japan and North America, *M. (S.) postostriata* (Kanno, 1958) was described from lower to middle Miocene Japanese deposits and has a pronounced anterior end, somewhat resembling the later Pliocene California species, *M. (S.) elsmerensis* (English, 1914).

In the Pliocene, beginning roughly 5 Mya, *M. (Mercenaria) corrugata* (Lamarck, 1818), *M. (M.) inflata* (Dall, 1903), possibly *M. (M.) blakei* (Ward, 1992; Campbell, 1993), *M. (M.) mercenaria*, *M. (M.) campechiensis* and some of their varieties (e.g., *M. (M.) campechiensis permagna* (Conrad, 1838; see Campbell, 1993) occurred in the west Atlantic. *Mercenaria (M.) blakei* had previously been confused with *M. (M.) corrugata* (Lamarck, 1818) (synonyms: *Cyprine tridacnoides* (Lamarck, 1818); *Venus deformis* (Say, 1824); *Venus percrassa* (Conrad, 1867); *Venus rileyi* (Conrad, 1838); Wilson (1983) discussed the taxonomic resurrection of *M. corrugata*). *Mercenaria corrugata* (93 mm, length), occurring in the upper Pliocene from Virginia to Florida (Ward, 1993, personal communication; Palmer, 1927), is more compressed than *M. mercenaria* with smaller, less prominent umbos. The shell varies in thickness, and is sculptured with heavy, coarse, crowded lines of varying widths, not distinct lamellae. The concentric sculpture is often eroded, exposing fine radiating striae, a common phenomenon in *M. mercenaria*. The synonym *tridacnoides* refers to a thickened, often radially rippled form that occurs throughout the stratigraphic range of *M. corrugata*. Thomas (1993) noted similarly rippled forms of the bivalve *Glycymeris americana* (DeFrance, 1829) occurring in the same stratigraphic range, and postulated that both occurrences represented unusual phenotypic expression triggered by changes in various environmental factors associated with the closing of the Panamanian isthmus. *M. (M.) inflata* was based upon two forms of York River, Virginia specimens that all (1903) had described. He called the smaller, more orbicular form inflata and the larger, more trigonal form *nucea*, and classified both as forms of *M. (M.) plena*, above. Campbell (1993) elevated both forms to a distinct Pliocene species, *Mercenaria inflata* (Dall, 1903), arguing that *M. (M.) plena* was shown by Ward (1992) to be restricted to the middle Miocene.

Recent Atlantic *Mercenaria* include *M. (M.) mercenaria* and *M. (M.) campechiensis*. The extant varieties of *M. (M.) campechiensis* include *mortoni* and *carolinensis* as well as *texana* (Dall, 1902b) and *alboradiata* (Sowerby, 1853), described by Dall (1902b) as a "Shell with broad brownish rays on a paler background." Extant *Mercenaria* are described in Sections 1.1.6 and 1.1.7.

In the west Pacific, *Mercenaria (Securella) stimpsoni* (Gould, 1861) continued through the Pliocene. In the east Pacific, *M. (S.) bisculpta* (Dall, 1909) of Oregon (45 mm, length), *M. (S.) elsmerensis* (English, 1914) of southern California (95 mm, length), and *M. (S.) securis* (Shumard, 1858) of California, Oregon and Washington (76 mm, length) appeared in North America, while another extant species, *Mercenaria (Mercenaria) kellettii* (Hinds, 1845) appeared off Ecuador, South America (Pilsbry and Olsson, 1941) (41 mm, their specimen length). Parker (1949) noted that the type of *M. (S.) bisculpta* is greatly weathered and he had no access to a well-preserved specimen; the type, a right valve, is subtrigonal and deeper than many *Securella* and might well belong to *Protothaca* or *Chione*. *Mercenaria (S.) elsmerensis* has a pronounced anterior end as in *M. (S.) panzana*, above, and is very similar to it, but its umbo is more centrally located and its underlying radial ribs are larger and fewer. *Mercenaria (S.) securis* is very similar to *M. (S.) ensifera*, above, but *M. (S.) securis* is slightly longer and thinner, with less pronounced umbos, and a straighter posterior dorsal margin. Of all these Pacific Pliocene species, only *M. (S.) stimpsoni* and *M. (M.) kellettii* are extant and are described below.

1.3 THE SYSTEMATICS OF *MERCENARIA*

There are three extant species of *Mercenaria* s.s., ranging from the east Atlantic to the east Pacific (Fig. 1.4). Conchological, anatomical, and biomolecular data all indicate *Mercenaria* to be more closely related to *Anomalocardia* and *Lirophora* than to *Chione*. Paleontological and conchological data indicate that *Securella* is closely related to *Mercenaria*. *Anomalocardia* was created simultaneously with *Mercenaria*, which has the longer paleontological record. This supports placing *Anomalocardia*, *Lirophora* and *Securella* as subgenera under the taxon *Mercenaria*.

1.3.1 Phylogeny

Conchological, anatomical, and biochemical comparisons have been made of *Mercenaria mercenaria* and other chionine clams. Harte (1992a) examined conchologically species that represent all generic units allied with *Anomalocardia* and *Chione* and compared them with *Mercenaria*. I observed that a rugose nymph occurred in *Anomalocardia* s.s. (Fig. 1.24), and in the *Chione* subgenera of *Iliochione* Olsson, 1961 and *Lirophora* Conrad (1863). This link between *Mercenaria* and these taxa was supported by a transition in valve shape and sculpture extending between *Mercenaria* and *Anomalocardia* s.s. (Fig. 1.25), with *Mercenaria kelletii* (Hinds, 1845), *Iliochione*, and *Lirophora* (Fig. 1.25) forming good intermediaries. The slightly angular posterior end in *Mercenaria* attenuates through this transition into a rostrate posterior end in *Anomalocardia* s.s. In *Mercenaria mercenaria*, well-spaced intervals of the valve are demarcated by sulci that are slightly deeper than those separating other concentric threads. In *M. kellettii*, these sulci demarcate areas that sometimes appear slightly swollen and terminate in pronounced lateral flanges, and in *Iliochione*, *Lirophora* and *Anomalocardia* s.s. these areas are pronounced swollen lirae that are often flattened posteriorly. Faint radial sculpture occurs distinctly in *Mercenaria* to indistinctly in *Anomalocardia* s.s. In contrast, other subgenera of *Chione*, including *Chione* s.s., lack a rugose nymph, exhibit no transition towards an attenuated posterior end, and have prominent radial sculpture, a trait lacking in the above transition.

These conchological data are supported by immunological data (Harte, 1992b). When the shell proteins of *Mercenaria mercenaria* are compared to those of the type species of the genera *Anomalocardia* and *Chione* s.s., the resulting immunological distances indicate that *Mercenaria* and *Anomalocardia* are more closely related to each other than to *Chione* s.s., although the resulting phylogenetic tree constructed from the data show *Chione* and *Mercenaria* sharing a common, unique node.

The above relationship is more weakly supported by anatomical data (Jones, 1979), gathered from a comparison of *Mercenaria mercenaria*, species of *Chione* s.s., *Lirophora*

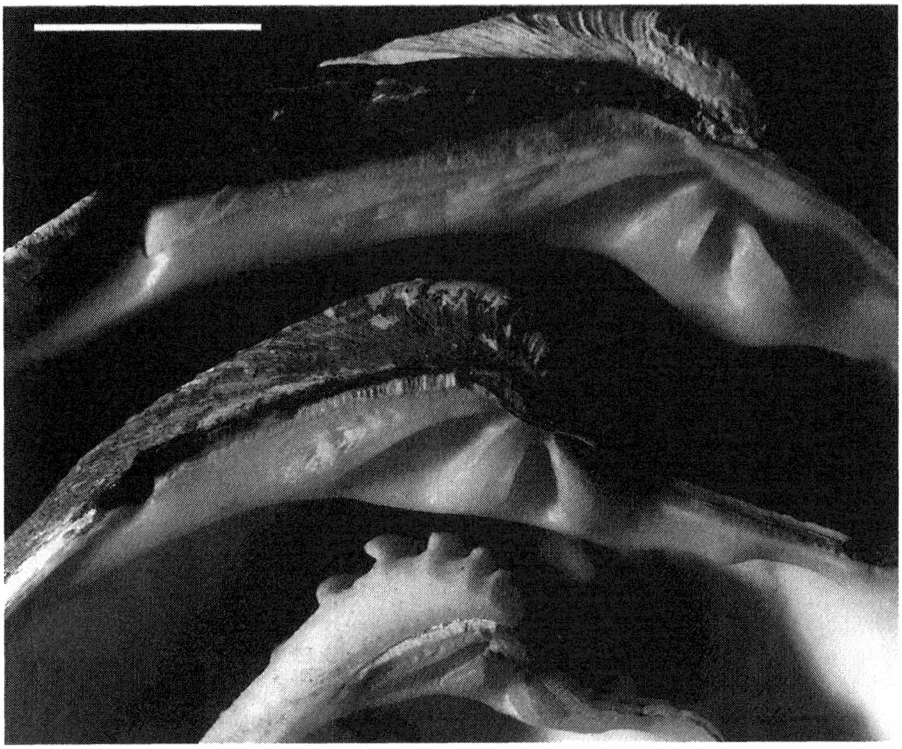

Fig. 1.24. Hinges of the left valves, from top to bottom, of *Mercenaria mercenaria*, *Mercenaria kellettii*, and *Anomalocardia flexuosa* (after Harte, 1992b; white bar = 1 cm).

(represented by *Chione paphia*), and *Austrovenus*. A summary of anatomical differences between species (Jones, 1979: Table 1.1) shows that *Mercenaria* is slightly more similar to *Lirophora* than to *Chione* s.s.

Paleontological data on *Chione*, *Lirophora*, *Securella* and *Mercenaria* indicate that these taxa originated in the lower Oligocene (Dockery, 1982). *Anomalocardia* first appears in the silex beds near Tampa, Florida, which were originally dated as Oligocene (Dall, 1900), but have been redated as lower Miocene (Weisbord, 1973). Assuming common origin, fossil data precludes any more precise dating of when each taxon diverged. Conchological data within a framework of minimal evolution might indicate the most probable scenario, however. Cladistic analyses (Harte, 1998) support a transition from Placamen (Eocene to Recent) to *Securella* to *Mercenaria* s.s. (see above). A phylogeny, based on the conchological, anatomical and biomolecular data presented above and consistent with the paleontological data and cladistic analyses, is proposed for *Mercenaria* and related taxa in Fig. 1.26.

1.3.2 Taxonomic Status

The hard clam was commonly classified within the genus *Venus* until 1936, when Frizzell (1936) proposed the breakup of Venerinae into Venerinae and Chioninae. Concomitantly, he

Fig. 1.25. The exteriors and interiors of *Anomalocardia flexuosa* (UCMP), top, and *Lirophora latilirata* (UCMP), bottom. Left: right valves. Right: left valves.

proposed the separation of *Mercenaria* from *Venus*. *Mercenaria* was accepted as a distinct genus in subsequent works (Keen and Frizzell, 1939; Keen, 1951; Keen, 1969; Abbott, 1974; Fischer-Piette and Vukadinovic, 1977). This was supported through a clearer definition of the genus *Venus* resulting from the designation of its type species, *Venus verrucosa* (Linnæus, 1767), by a ruling of the International Commission on Zoological Nomenclature (1954).

Anatomical, immunological and conchological data indicate that *Mercenaria* is more closely related to *Anomalocardia*, *Lirophora*, and *Iliochione* than to *Chione* s.s. Although *Anomalocardia* was created simultaneously with *Mercenaria*, the fossil record indicates that *Mercenaria* is the oldest taxon. These data support classifying *Anomalocardia* and *Lirophora* as subgenera under *Mercenaria*; conchological data support subsuming *Iliochione* under *Lirophora*.

Schumacher (1817) described the genus *Anomalocardia* simply as "two cardinal teeth in each valve, the right anterior and left posterior teeth are triangular. The nymphs are very small and interiorly crenulated." He actually described the two median cardinals. The ones he failed to observe, the left posterior and the right anterior teeth, are small, thin, and lamellar in *Anomalocardia flexuosa* (Linnæus, 1758) (Fig. 1.25), the type species he designated for the genus, and thus, easy to overlook.

Authors have variously applied the name *Anomalocardia* to any chionine with an elongate, rostrate posterior end (e.g., Kira, 1962; Keen, 1969; Habe, 1977; Woodring, 1982) or to various east Pacific tropical species with a similar sculpture of predominantly concentric folds (e.g., Grant and Gale, 1931; Olsson, 1932). Harte (1992a) proposed subsuming *Mercenaria* and

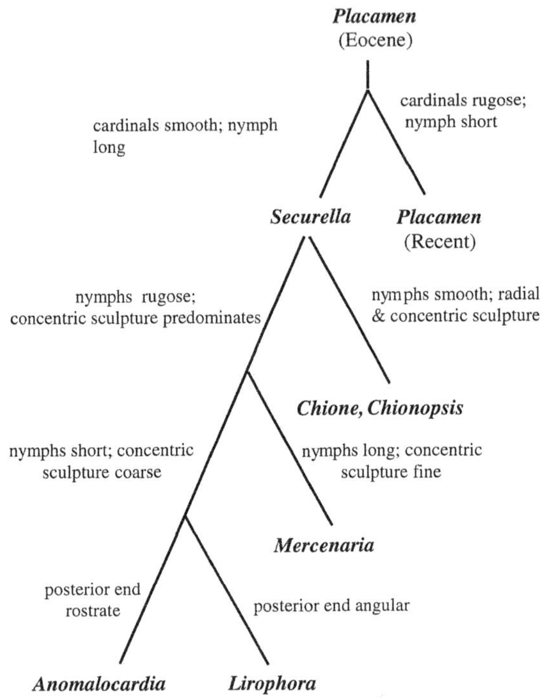

Fig. 1.26. A proposed phylogeny of *Mercenaria* and related taxa. Long nymphs are twice the length of the posterior cardinal tooth; short nymphs are less than that. All taxa are presented sensu stricto.

Anomalocardia under *Chione*, but doing so fails to demarcate sufficiently the similarities of the taxa *Mercenaria*, *Anomalocardia* and *Lirophora* that separate them from the rest of *Chione*.

Conchological data support subsuming *Iliochione* under *Lirophora*. Olsson's (1961) original description of *Iliochione* does not compare it to *Lirophora*, although its sculpture, hinge and pallial sinus all fall within his description of *Lirophora* (rounded concentric folds, hinge as in *Chione* s.s., pallial sinus small). The detailed description of the round, concentric, sometimes irregular or obsolete folds of *Iliochione* and the observations that member species were often assigned to *Anomalocardia* indicate strongly that Olsson (1961) felt this particular sculpture indicated subgeneric status for an east Pacific taxonomic counterpart of *Anomalocardia*. Although Olsson (1961) specifically stated that *Iliochione* was more closely allied to *Chione* than *Anomalocardia*, he failed to present any supporting data.

1.3.3 Taxonomic Definition and Description

(*Mercenaria* (Schumacher, 1817)
Crassivenus (Perkins, 1869) Proc. Boston Soc. N. Hist. 13: 147).

Schumacher (1817) originally described *Mercenaria* (in Latin and French) as: "Triangular shell in the form of a heart, equivalve, more or less rounded. Hinge: in each valve three cardinal teeth, compressed, erect, diverging little; the anterior ones in the left valve and the posterior ones in the right are almost divided in two; the others are practically lamellar and simple. The interior nymphs are pronounced, their surfaces correspondingly flat, obliquely wrinkled or grooved; the wrinkles waved." It is unclear why he renamed *Venus mercenaria* Lin. as *Mercenaria violacea* when he designated it the type species of the genus, although he apparently did this with many taxa (Coan, 1993, personal communication).

Most of this description applies to much of Chioninae: the form and the hinge teeth can match the descriptions of a few American chionine taxa. He does not mention shell sculpture or size. Species of *Anomalocardia* and *Lirophora* (see above) have rugose nymphs. Ultimately, it is his emphasis on a *pronounced*, rugose nymph and a designation of a type species that leaves no room for doubt. The nymph of *Mercenaria* extends to about thrice the length of the posterior cardinals, whereas in other chionine species (with the exception of *Protothaca*, which has smooth nymphs) it ends near the ventral termini of the posterior cardinals.

Utilizing this criterion, it is possible to discern at least two other living species of *Mercenaria* s.s.: *M. campechiensis* (Gmelin, 1791), described above, and *M. kellettii* (Hinds, 1845) of the tropical eastern Pacific (see below). *Mercenaria apodema* (Dall, 1902b) Keen (1971) and Olsson (1961) was based upon a single worn valve that does not exhibit the generic characters of *Mercenaria* (Harte, 1992a). The east Pacific *M. kellettii* and some specimens of the west Atlantic *M. campechiensis* have a smaller, thinner rugose area on the nymph than most *M. mercenaria* (Harte, 1992a).

The most conchologically similar extant ally to *Mercenaria* s.s. is *Mercenaria (Securella) stimpsoni* (Gould, 1861) of Japan. Whereas Keen (1951) classified it as a subgenus of *Chione*, subsequent workers classified it under *Mercenaria* (e.g., Habe, 1977). Moore (1963) noted that *M. (S.) stimpsoni* was more conchologically similar to *Securella* than *Chione*. *Mercenaria stimpsoni* has a low lamellate concentric sculpture similar to that of *M. campechiensis* with fine radial sculpture sometimes apparent between, but its anterior end is more pronounced and the umbos more pointed (Dance, 1974); the valves are more compressed than either *M. campechiensis* or *M. mercenaria*. Thus, in most respects, it fits the generic definition, leading Harte (1992a) to include those chionine species having a pronounced nymph with a rugose area on the nymph or the left posterior cardinal tooth. In fact, *M. (S.) stimpsoni* has no rugose area on the hinge; the previously observed rugosity on the left posterior tooth (Harte, 1992a) resulted from damage to the tooth. Based on conchological similarities, it is now proposed (above) that *Securella* be placed as a subgenus of *Mercenaria*. The distribution of *Mercenaria*, then, is much greater than that of *M. mercenaria*, extending, through introductions, from the east Atlantic to the west Pacific.

From the above group of species, it is possible to construct a more comprehensive generic definition of valve sculpture and other conchological features for *Mercenaria*. The sculpture of *Mercenaria* is primarily concentric, with a radial element that is usually fine, weak and sometimes absent externally, showing up only as weathering strips away the outer shell layers. Fine concentric threads can be polished smooth medially, become lamellate, or coalesce to form prominent strong ribs or recurved thickened lirae. The sculpture of *Mercenaria* s.s. consists primarily of concentric, closely spaced threads, which can be slightly lamellate, even scaly, especially laterally and in juveniles, or coalesce to form prominent, thick lamellae in some fossil species. Widely spaced, slightly deeper sulci can occur among the threads, and the concentric sculpture is sometimes polished smooth medially. Radial structure, when present, is fine and weak. The sculpture of *Securella* is similar. In *Lirophora*, the concentric sculpture is of swollen, thick, often recurved lirae. In *Anomalocardia*, the concentric sculpture ranges from concentric lirae to concentric waves.

The pallial sinus is short in *Mercenaria* and acutely triangular. The ligament is partially sunken between the valves in a moderately but not always sharply defined escutcheon. The lunule is moderately obese, slightly protrusive medially, and distinctly incised. The general profile for Recent *Mercenaria* s.s. is prosogyrous, with a slightly swollen umbo, and subovate with an angular posterior end. In *Mercenaria*, it ranges from suborbicular and subovate (fossil and Recent *Mercenaria* s.s., *Lirophora* and *Securella*) to attenuate (*Anomalocardia*) or ovate (*Securella* and fossil *Mercenaria* s.s.). The nymph ranges from relatively short, not extending far beyond the posterior cardinal tooth (*Anomalocardia* and *Lirophora*) to pronounced, more than twice as long as the left posterior cardinal tooth (*Mercenaria* and *Securella*), and often rugosely sculptured (*Mercenaria* s.s., *Anomalocardia*, and *Lirophora*).

In *Mercenaria* s.s. the anterior and medial cardinal teeth are shorter than the posterior cardinal teeth. The right anterior and left posterior teeth are thin and lamellar, whereas the others are stouter and bisected dorsally. The left posterior tooth is rugosely sculptured along its posterior flank and sometimes is obscured by the rugosity of the nymph. The inner margins extending from the dorsal anterior end of the umbo to the posterior end of the ventral side are finely crenulate.

Anomalocardia and *Lirophora* will not be described here. Descriptions and species lists of these taxa can be found in Palmer (1927) and Fischer-Piette and Vukadinovic (1977).

1.4 CONCLUSIONS

Paleontological, conchological, anatomical and biomolecular data indicate that *Mercenaria* is a major chionine genus, encompassing four subgenera that occur primarily in the northern parts of the Pacific and Atlantic Oceans, with most extant species occurring along the coasts of the American continents. Paleontological and conchological data indicate that its ancestors might have originated in the western Pacific and migrated to the Americas during the Lower Oligocene. *Mercenaria* s.s. radiated into several species in the Miocene, but only three exist today. Of those three, however, *Mercenaria mercenaria* is the most abundant, and has formed a significant fishery for more than three centuries, as evidenced by the many names given to it by present and past cultures.

1.5 ACKNOWLEDGMENTS

I am grateful to Dr. Eugene Coan, Dr. Louie Marincovich, Dr. Barry Roth, Dr. LouElla Saul, Dr. Richard Squires, Dr. Thomas Waller and Dr. Lauck Ward for their constructive comments and criticisms. I thank the Academy of Natural Sciences at Philadelphia, the California Academy of Sciences, the U.S. National Museum, and the University of California Museum of Paleontology for allowing me to examine their specimens, both fossil and Recent, of *Mercenaria*. I am grateful to Dr. Kathleen Way, who provided photographs of the holotype of *Mercenaria mercenaria* L. with the permission of The Linnaean Society. I thank the journal, *Malacologia*, for permission to reprint the illustration of *Mercenaria* in Fig. 1.1. I thank the Bureau of Economic Geology, Balcones Research Center, University of Texas, Austin, for permission to reprint the figures of *Rhabdopitaria texangulina*.

REFERENCES

Abbott, R.T., 1954. American Seashells. van Nostrand, New York, 541 pp., Figs. 100.

Abbott, R.T., 1974. American Seashells, 2nd ed. van Nostrand Reinhold, New York, 663 pp., 24 Pls., 6405 Figs.

Adams, H. and Adams, A., 1857. The Genera of Recent Mollusca, Vol II. John van Voost, London.

Adegoke, O.S., 1969. Stratigraphy and paleontology of the marine Neogene formations of the Coalinga Region, California. University of California Publications. Bull. Dept. Geol. Sci., 80: 1–241.

Addicott, W.O., 1976. Neogene molluscan stages of Oregon and Washington. In: Fritsche, A.E., Ter Best H., Jr., and Wornardt, W.W. (Eds.), The Neogene Symposium. Society of Economic Paleontologists and Mineralogists. Pacific Section. Neogene Symposium, San Francisco, pp. 95–115.

Anderson, F.M. and Martin, B., 1914. Neogene record in the Temblor Basin, California, and Neocene deposits of the San Juan District, San Luis Obispo County. California Academy of Sciences Proceedings, 4th Series, 4, pp. 15–112, Pls. 1–10.

Andrews, J., 1971. Sea Shells of the Texas Coast. Univ. Texas Press, Austin, TX, 285 pp.

Anton, H.E., 1839. Verzeichniss der Conchylien. Anton, Halle, 110 pp.

Anton, H.E., 1837. Diagnosen einiger neuen Conchylien-Arten. Archiv fur Naturgeschichte, 3 (1): 280–284.

Arai, J. and S. Kanno, 1960. The Tertiary System of the Chichibu basin, Saitama Prefecture, Central Japan. Part II. Palaeontology. Japan Society for the Promotion of Science, pp. 123–396, Pls. 31–51.

Backeljau, T., Bouchet, P., Gofas, S. and Debruyn, L., 1994. Genetic variation, systematics and distribution of the venerid clam *Chamelea gallina*. Journal of the Marine Biological Association of the United Kingdom, 74: 211-223.

Baker, B.B., 1950. Some mollusks near St. Petersburg. Nautilus, 58 (4): 123–125.

Baker, F.C., 1891. Notes on a collection of shells from southern Mexico. Proceedings of the Academy of Natural Sciences of Philapelphia, 43: 45–55.

Beu, A.G. and Maxwell, P.A., 1990. Cenozoic mollusca of New Zealand. New Zealand Geological Survey Paleontological Bulletin 58: 518 pp. many illustrations.

Binney, W.G., 1870. Report on the Invertebrata of Massachusetts. Wright and Potter, Boston, 2nd ed., 517 pp., 27 Pls.

Bolten, J.F., 1798. Museum Boltenianum, sive catalogus cimeliorum e tribus regnis naturaequae olim collegerat J.F. Bolten. Pars secunda continens conchylia sive testacea univalvia, bivalvia and multivalvia. John. Christi. Trappii: Hamburgi. viii + 199 pp.

Bonnani, F., 1681. Ricreatione dell'occhio e dela mente nell' osservation delle Chiocciole. Rome, 400 pp.

Bory de St. Vincent, J.B.M., 1827. Tableau encyclopédique et méthodique des trois règnes de la nature. Vers, Coquilles, Mollusques et Polypiers, Vol. I. Mme. V. Agasse, Paris, 180 pp., 95 Pls.

Bousfield, E.L., 1960. Canadian Atlantic Sea Shells. Department of Northern Affairs and National Resources, National Museum of Canada, Ottawa, xiii pp. + 72 pp., 8 Pls.

Burch, J.A., 1944. Additions and Corrections. Minutes of the Conchological Club of and Washington Southern California, No. 45, 20 pp.

Campbell, L.D., 1993. Pliocene molluscs from the Yorktown and Chowan River Formations in Virginia. Virginia Division of Mineral Resources, Publication No. 127, 259 pp., 43 Pls.

Carpenter, H.F., 1888. The shell-bearing mollusca of Rhode Island. The Conchologists' Exchange, 2 (8): 101–102.

Carpenter, H.F., 1891. Edible mollusks of Rhode Island. Nautilus, 4 (12): 137–139.

Carter, R.M., 1968. On the biology and paleontology of some predators of bivalved Mollusca. Palaeogeography, Palaeoclimatography, Palaeoecology, 4: 29–65.

Cary, L.R. and Spaulding, H.M., 1908. Further Contributions to the marine Fauna of the Louisiana coast. Gulf Biologic Station Cameroun, LA, Baton Rouge, LA, 21 pp.

Chadwick, G.H., 1906. Shells of Prince Edward Island. Nautilus, 19 (9): 103–104.

Chanley, P.E., 1961. Inheritance of shell markings and growth in the hard clam, *Venus mercenaria*. Proceedings of the National Shellfisheries Association, 50: 163–169.

Chemnitz, J.H., 1788. Neues Systematisches Conchylien Cabinet, Vol X. Raspe, Nuremberg, 376 pp., 173 Pls.

Chenu, J.C., 1847. Illustrations Conchyliologiques. Vol. II: *Venus*, 8 Pls.

Clark, B.L., 1915. Fauna of the San Pablo Group of middle California. University of California Publications, Bulletin of the Department of Geological Sciences 8 (22): 385–572, Pls. 42–71.

Clark, B.L., 1918. The San Lorenzo Series of middle California. University of California Publications, Bulletin of the Department of Geological Sciences 11 (2): 45–234, Pls. 3–24.

Clark, B.L., 1925. Pelecypoda from the marine Oligocene of western North America. University of California Publications, Bulletin of the Department of Geological Sciences 15 (4): 69–136, 15 Pls.

Clark, B.L., 1932. Fauna of the Poul and Yakataga Formations (upper Oligocene) of southern Alaska. Geological Society of America Bulletin 43 (3): 797–846. 9 Pls.

Clark, B.L. and Arnold, R., 1923. Fauna of the Sooke Formation, Vancouver Island, with description of a new coral by T. Wayland Vaughan. University of California Publications, Bulletin of the Department of Geological Sciences 14 (5): 123–124.

Clench, W.J., 1928. *Venus mercenaria* var. *notata* Say. Nautilus, 41 (4): 120–122.

Coan, E., 1993. Research Associate, Department of Invertebrate Zoology, California Academy of Sciences, Golden Gate Park, San Francisco, CA 94118, USA, personal communication.

Comfort, A., 1951. The pigmentation of molluscan shells. Biological Reviews, 26: 285–301.

Conrad, T.A., 1867. Descriptions of new marine shells, from upper California, collected by Th. Nuttall. Journal of the Academy of Natural Sciences at Philadelphia 7 (2): 227–268, Pl. 19.

Conrad, T.A., 1838. Fossils of the Tertiary Formations of the United States. Philadelphia Part 1: 1–32, 17 Pls.

Conrad, T.A., 1843. Descriptions of nineteen species of Tertiary fossils or Virginia and North Carolina. Proceedings of the Academy of Natural Sciences at Philadelphia 1: 323–329. (Read in 1843, published in 1844.)

Conrad, T.A., 1848. Observations on the Eocene formation, and descriptions of one hundred and five new fossils of that period from the vicinity of Vicksburg, Mississippi; with an appendix. Descriptions of new Eocene fossils [from South Carolina] in the cabinet of Lardner Vanuxem. Journal of the Academy of Natural Sciences at Philadelphia, Series 2, 1: 111–134, Pls. 11–14. (orig. publ. in 1847, Proceedings of the Academy of Natural Sciences at Philadelphia 3: 280–299.)

Conrad, T.A., 1862. Catalogue of the Miocene Shells of the Atlantic slope. Proceedings of the Academy of Natural Sciences at Philadelphia, 14: 559–582.

Conrad, T.A., 1863. Descriptions of new Recent and Miocene shells. Proceedings of the Academy of Natural Sciences at Philadelphia, 14: 583–586.

Conrad, T.A., 1864. Notes on shells, with descriptions of new fossil genera and species. Proceedings of the Academy of Natural Sciences at Philadelphia, 16: 211–214.

Conrad, T.A., 1867. *Mercenaria cuneata* Con. sp. nov. p. 138. In: Cope, E.D. (Ed), An addition to the vertebrate fauna of the Miocene period, with a synopsis of the extinct cetacea of the United States. Proceedings of the Academy of Natural Sciences at Philadelphia 19: 138–156. (Described in Conrad, 1869a.)

Conrad, T.A., 1869a. Descriptions of and references to Miocene shells of the Atlantic Slope and descriptions of two new supposed Cretaceous species. American Journal of Conchology 4 (4): 278–279, Pls. 19–20.

Conrad, T.A. 1869b. Descriptions of new fossil mollusca, principally Cretaceous. American Journal of Conchology 5: 96–103, pl 9.

Conrad, T.A., 1875. Descriptions of new genera and species of fossil shells of North Carolina in the state cabinet at Raleigh. pp. 1–13, Pls. 1–2. Appendix: p. 20. in: Kerr, W.C., 1875. Geology of North Carolina, Vol. 1.

Cossmann, M., 1886. Catalogue des coquilles fossiles de l'Eocene des environs de Paris. Annales de la Societe Royale Malacologique de Belgique 21: 17–186 (4th series, Vol. 1).

Cossmann, M. and Pissaro, G., 1904–1906. Iconographie complete des coquilles fossiles de l'Eocene des environs de Paris, Vol. I. Pelecypodes, Paris.

Dall, W.H., 1872. Descriptions of sixty new forms of mollusks from the west coast of North America and the North Pacific ocean, with notes on others already described. American Journal of Conchology 7: 93–169, Pl. 16.

Dall, W.H., 1889. A preliminary catalogue of the shell-bearing marine mollusks and brachiopods of the southeastern coast of the United States, with illustrations of many of the species. Bulletin of the U.S. National Museum 37: 221 pp., 74 Pls.

Dall, W.H., 1900. Contributions to the Tertiary fauna of Florida. Transactions of the Wagner Free Institute of Science 3 (5): 949–1218.

Dall, W.H., 1902a. Illustrations and descriptions of new, unfigured or imperfectly known shells, chiefly American, in the U.S. National Museum. Proceedings of the U.S. National Museum 24: 499–566, Pls. 27–40.

Dall, W.H., 1902b. Synopsis of the family Veneridae, and of the North American Recent species. Proceedings of the U.S. National Museum 26: 335–412, Pls. 12–16.

Dall, W.H., 1890–1903. Contributions to the Tertiary fauna of Florida. Transactions of the Wagner Free Institute of Science 3 (1–6): 1654 pp., 60 Pls.

Dall, W.H., 1909. The Miocene of Astoria and Coos Bay, Oregon. U.S. Geological Survey Professional Paper 59: 1–279, 23 Pls.

Dall, W.H., 1916. Checklist of the recent Bivalve Mollusks (Pelecypoda) of the northwest coast of America from the Polar Sea to San Diego, California. Bulletin of the Southern California Academy of Science, Los Angeles, 44 pp.

Dall, W.H., 1921. Summary of the marine shellbearing mollusks of the northwest coast of America, from San Diego, California, to the Polar Sea, mostly contained in the collection of the U.S. Nat. Mus., with illustrations of hitherto, unfigured species. Bulletin of the U.S. National Museum 112: 217 pp., 22 Pls.

Dance, S.P. (Ed.) 1974. The Collector's Encyclopedia of Shells. McGraw Hill, New York, 288 pp.

Dautzenberg, P., 1910. Contribution à la faune malacologique d'Afrique occidentale. Actes de la Societé linneenne de Bordeaux 64: 174 pp., 4 Pls.

DeFrance, M.J.L., 1829. Dictionnaire des Sciences Naturelles, Vol. 39. L.G. Levrault, Paris, 225+ pp.

Deshayes, G.P., 1832. Encyclopedie Méthodique. Vers, 3: 1150 pp.

Deshayes, G.P., 1835. Histoire Naturelle des Animaux sans Vertebres, Vol. VI. 2nd ed., Bailliere, Paris, 600 pp.

Deshayes, G.P., 1853. Catalogue of the Conchifera or Bivalve Shells in the collection of the British Museum. Part I. Veneridae, Cyprinidae, Glauconomidae, Petricolidae. London, 292 pp.

Dexter, R.W., 1942. Notes on the marine mollusks of Cape Ann, Massachusetts. Nautilus, 56 (1): 57–61.

Dexter, R.W., 1944. *Venus mercenaria* not at Cape Ann. Nautilus, 58 (2): 71.

Dexter, R.W., 1962. Further studies on the marine mollusks of Cape Ann, Massachusetts. Nautilus, 76 (2): 63–70.

Dillon Jr., R.T., 1992. Minimal hybridization between populations of the hard clam *Mercenaria mercenaria* and *Mercenaria campechiensis*, co-occurring in South Carolina. Bulletin of Marine Science, 50 (3): 411–416.

Dillon Jr., R.T. and Manzi, J.J., 1989a. Genetics and shell morphology of hard clams (Genus *Mercenaria*) from Laguna Madre, Texas. Nautilus, 103 (2): 73–77.

Dillon Jr., R.T. and Manzi, J.J., 1989b. Genetic and shell morphology in a hybrid zone between the hard clams *Mercenaria mercenaria* and *M. campechiensis*. Marine Biology, 100: 217–222.

Dillwyn, L.W., 1817. Descriptive Catalogue of recent Shells, arranged according to the Linnean method; with particular attention to the synonymy. I. J. and A. Arch, London, 580 pp.

Dockery, III, D.T., 1982. Lower Oligocene bivalvia of the Vicksburg Group in Mississippi. Mississippi Department of Natural Resources Bureau of Geology Bulletin 123, 261 pp, 62 Pls.

Dodge, H., 1952. A historical review of the mollusks of Linnæus. Part I. The classes Loricata and Pelcypoda. Bulletin of the American Museum of Natural History 100: 263 pp.

English, W.A., 1914. The Fernando Group near Newhall, California. University of California Publications, Bulletin of the Department of Geological Sciences 8 (8): 203–218, Pls. 23.

Fischer-Piette, E., 1975. Revision de Venerinae (Mollusques Lamellibranches). Memoires du Museum national d'Histoire naturelle. Nouvelle Serie, Serie A, Zoologie, 93: 1–64.

Fischer-Piette, E. and Vukadinovic, D., 1977. Suite de revisions des Veneridae (Mollusques Lamellibranches) Chioninae, Samarangiinae et complement aux *Venus*. Memoires du Museum national d'Histoire naturelle. Nouvelle Serie, Serie A, Zoologie, 106: 1–186.

Ford, J., 1889. List of shells of the New Jersey coast south of Brigantine Island. Nautilus, 3 (3): 27–29.

Frizzell, D.L., 1936. Preliminary reclassification of veneracean pelecypods. Bulletin du Musée Royal d'Histoire naturelle de Belgique, 5: 1–84.

Gabb, W.M., 1860. Descriptions of new species of American Tertiary and Cretaceous fossils. Journal of the Academy of Natural Sciences at Philadelphia, Series 2, 4: 375–406.

Gardner, J.A., 1926. The molluscan fauna of the Alum Bluff Group of Florida, Part IV. Veneracea. U.S. Geological Survey Professional Paper 142-D: 151–184, Pls. 24–28.

Gardner, J.A., 1943 [1944]. Mollusca from the Miocene and Lower Pliocene of Virginia and North Carolina. U.S. Geological Survey Professional Paper 199-A., 178 pp. 23 Pls. (Original date, 1943, amended in a later publication to January, 1944.)

Gmelin, J.F., 1791. Caroli a Linne, Systema Naturae, 8th ed.

Gould, A.A., 1841. Report on the Invertebrata of Massachusetts, Cambridge, 373 pp.

Gould, A.A., 1861. Descriptions of shells collected by the North Pacific Exploring Expedition. Proceedings of the Boston Society of Natural History , 8: 14–40.

Gould, A.A., 1870. Report on the Invertebrata of Massachusetts, 2nd ed. Wright and Potter, State Printers, Boston. 517 pp., 27 Pls.

Grant, U.S. and Gale, H.R., 1931. Catalogue of the marine Pliocene and Pleistocene mollusca of California and adjacent regions. Memoires of the San Diego Society of Natural History 1: 1036 pp., 32 Pls.

Habe, T., 1977. Systematics of Mollusca in Japan: Bivalvia and Scaphopoda. Zukan no Hokuryukan: Tokyo.

Hackney, A.G., 1944. List of mollusca from around Beaufort N. Carolina, with notes on Tethys. Nautilus, 58 (2): 56–64.

Hanley, S., 1842–1856. An illustrated and descriptive catalogue of recent Bivalve Shells. Williams and Norgate, London, 334 pp. Appendix: 335–392, Pls. 9–24.

Hanna, G.D., 1966. Introduced mollusks of western North America. Occasional Papers of the California Academy of Sciences No. 48., 108 pp., 85 Figs., 4 Pls.

Hansen, T., 1987. Extinction of late Eocene to Oligocene molluscs: relationship to shelf area, temperature changes, and impact events. Palaios, 2: 69–75.

Hansen, T., 1992. The patterns and causes of molluscan extinction across the Eocene/Oligocene boundary. pp. 341–348. In: Prothero, D.R. and Berggren, W.A. (Eds.), Eocene–Oligocene Climatic and Biotic Evolution. Princeton University Press, Princeton. xiv + 568 pp.

Harry, H.W., 1942. List of mollusca of Grand Isle, Louisiana, recorded from the Louisiana State University Marine Laboratory 1929–1941. Occasional Papers of the Marine Laboratory of Baton Rouge, 1: 13 pp.

Harte, M.E., 1992a. An eastern Pacific *Mercenaria* and notes on other chionine genera (Bivalvia: Veneridae). Veliger, 35: 137–140.

Harte, M.E., 1992b. A new approach to the study of bivalve evolution. American Malacological Bulletin, 9 (2): 199–206.

Harte, M.E., 1998. The evolution of *Mercenaria* Schumacher, 1817 (Bivalvia: Veneridae). pp. 305–316. In: Johnston, P.A. and J.W. Haggart (Eds.), Bivalves: An Eon of Evolution, Paleobiological Studies Honoring Norman D. Newell. University of Calgary Press, Calgary, xiv + 461 pp.

Hayasaka, I. and Uozumi, S., 1954. Fossil species of genus *Mercenaria* from the Cenozoic deposits of Hokkaido. Transactions and Proceedings of the Palaeontological Society of Japan, New Series No. 15: 165–172, Pls. 22.

Heppell, D., 1961. The naturalization in Europe of the quahog *Mercenaria mercenaria* (l.). Journal of Conchology, 25 (1): 21–34.

Hinds, R.B., 1845. Mollusca. In: R.B. Hinds (Ed.), The Zoology of the Voyage of HMS Sulphur, During the Years 1836–1842. Volume II. Smith, Elder and Co., London, 72 pp. 21 Pls.

Holmes, F.S., 1860. Post-Pleiocene fossils of South Carolina. Russel and Jones, Charleston, SC, xii + 122 pp., 28 Pls.

International Commission on Zoological Nomenclature., 1954. Opinion 195: Designation, under the Plenary Powers, of a types species for the genus *Venus*, Linnæus, 1758 (Class Pelecypoda) in harmony with accustomed usage. Opinions and Declarations 3 (14): 191–198. Also in Bulletin of Zoological Nomenclature 4: 304–305, 1950.

Iredale, T., 1925. Mollusca from the continental shelf of eastern Australia. Records of the Australian Museum, 14 (4): 243–270.

Jacobson, M.K., 1943. Marine mollusca of New York city. Nautilus, 56 (4): 139–144.

Jacobson, M.K. and Emerson, W.K., 1961. Shells of the New York city area. Argonaut Books, Larchmont, NY, 142 pp.

Jacot, A., 1919. Some marine mollusca about New York city. Nautilus, 32 (3): 90–94.

Jacot, A., 1920. On the marine mollusca of Staten Island, N.Y.. Nautilus, 33 (4): 111–115.

Jacot, A., 1924. Marine mollusca of the Bridgeport, Connecticut region. Nautilus, 38 (2): 49–51.

Johnson, C.W., 1915. Fauna of New England. List of the mollusca. Occasional Papers of the Society of Natural History of Boston 7 (13): 231 pp.

Johnson, C.W., 1916. Consistency in population names of shells. Nautilus, 30 (8): 87–91.

Johnson, C.W., 1934. List of marine mollusca of the Atlantic coast from Labrador to Texas. Proceedings of the Boston Society of Natural History 40 (1): 204 pp.

Johnson, C.W., 1890. Annotated list of the shells of St, Augustine, Fla. Nautilus, 4 (1): 4–6.

Jones, C.C., 1979. Anatomy of *Chione cancellata* and some other chionines (Bivalvia: Veneridae). Malacologia, 19 (1): 157–199.

Juste, V. and Cortes, R., 1990. Distribution and biological aspects of the hard clam *Mercenaria mercenaria*

(Linnæus), *Mercenaria mercenaria notata* (Say), and *Mercenaria campechiensis* (Gmelin) in Puerto Rico. Caribbean Journal of Science, 26 (3/4): 136–140.

Kaas, P., 1937. *Venus mercenaria* L., een nieuwe mollusk voor de Nederlandshe Fauna. Basteria, 2 (4): 58–60.

Kanehara, K., 1937. Neogene shells from the Etaibets oil field, Hokkaido. Journal of the Geological Society of Japan 44 (527): 791–796, Pls. 25.

Kanno, S., 1958. New Tertiary molluscs from the Chichibu Basin, Saitama Prefecture, Japan. Science Reports Tokyo Kyoiku Daigabu, Section C, 6 (55): 157–229.

Kaseno, Y. and Matsura, N., 1965. Pliocene shells from the Omma Formation around Kanazawa City, Japan. Science Reports, Kanazawa University 10 (1): 27–62, Pls. 1–20, 3 Figs.

Kauffman, E.G., 1969. Form, function, and evolution. In: R.C. Moore (Ed.), Treatise on Invertebrate Paleontology. Geological Society of America and University of Kansas, Lawrence, KS, pp. N129-N205.

de Kay, J.E., 1843. Zoology of New York or the New York Fauna. Part V, Mollusca. Carroll and Cook, Albany, NY, 271 pp., 40 Pls.

Keen, A.M., 1937. An abridged check list and bibliography of West North American marine Molluska. Stanford University Press, Stanford, 84 pp.

Keen, A.M., 1951. Outline of a proposed reclassification of the pelecypod family Veneridae. Minutes of the Conchological Club of Southern California 113: 2–11.

Keen, A.M., 1969. Veneridae. In: R.C. More, (Ed.), Treatise of Invertebrate Paleontology, Geological Society of America and University of Kansas: Lawrence, KS, pp. N671–N688.

Keen, A.M., 1971. Sea Shells of Tropical West America. Stanford University Press: Stanford, CA, xiv + 1064 pp., 22 Pls., numerous Figs.

Keen, A.M. and Frizzell, D.L., 1939. Illustrated key to West North American pelecypod genera. Stanford University Press, Stanford, CA, 28 pp.

Kehoe, T., 1993. K and B Seafood, P.O. Box 502, Northport, NY 11768, USA, personal communication.

King, P.P. and Broderip, W.J., 1832. Description of the Cirrhipeda, Conchifera and Mollusca, in a collection formed by the officers of H.M.S. *Adventure* and *Beagle* employed between the years 1826 and 1830 in surveying the southern coasts of South America, including the Straits of Magalhaens and the coast of Tierra del Fuego. Zoological Journal, 5: 332–349.

Kira, T., 1962. Shells of the Western Pacific in Color. Hoikusha, Osaka, Japan, 224 pp., 72 Pls.

Lamarck, J.B., 1818. Histoire Naturelle des Animaux sans Vertebres, Volume V. A. Lanoe, Paris, 612 pp.

Lamy, E. and Fischer-Piette, E., 1938. Notes sur les especes lamarckiennes de *Venus s. s.* et de *Chione* (Moll. Lamellibr.). Bulletin du Museum national d'Histoire naturelle, Paris, Series 2, 10 (4): 401–405.

La Roque, A., 1953. Catalogue of the recent mollusca of Canada. National Museum of Canada Bulletin 129: 406 pp.

Lewy, Z. and Samtleben, C., 1979. Functional morphology and palaeontological significance of the conchiolin layers in corbulid pelecypods. Lethaia, 12: 341–351.

Linnæus, C., 1758. Systema Naturae, 10th ed. I. Salvius, 824 pp.

Linnæus, C., 1767. Systema Naturae. 12th ed. I (II). Salvius, 1327 pp.

Lister, M., 1685–1687. Historiae sive Synopsis Methodicae Conchyliorum. London, 1057 Pls.

Loel, W. and Corey, W.H., 1932. The Vaqueros Formation, Lower Miocene of California. I. Paleontology. University of California Publications, Bulletin of the Department of Geological Sciences 22 (3): 31–410, Pls. 4–65, 2 maps.

MacGinty, P.L., 1936. A canoe trip in the ten thousand islands to collect *Liguus*. Nautilus, 50 (1): 1–8.

Makiyama, J., 1927. Preliminary report on the Tertiary Fossils from Kamimanochi-gun Shinano. Chikyu (The Globe), 8 (3): 181–188.

Marincovich Jr., L., 1988. Late middle Eocene mollusks of the Tolstoi Formation, Alaska Peninsula, Alaska, and correlations with faunas from California to the Far-Eastern U.S.S.R. In: Filewicz, M.V. and R.L. Squires (Eds.), Paleogene Stratigraphy, West Coast of North America, Pacific Section, Society of Economic Paleontologists and Mineralogists, West Coast Paleogene Symposium, Vol. 58, pp. 265–281.

Marincovich Jr., L., 1991. Origin of the modern Gulf of Alaska cold-water molluscan fauna. Western Society of Malacologists Annual Report, 23: 8–9.

Marincovich Jr., L. and Kase, T., 1986. An occurrence of *Turritella (Hataiella) sagai* in Alaska: implications for the age of the Bear Lake Formation. Bulletin of the National Science Museum, Tokyo, Series C (Geology and Paleontology), 12 (2): 61–66.

Marwick, J., 1927. The Veneridae of New Zealand. Transactions and Proceedings of the New Zealand Institute 57: 567–635, Pls. 34–54.

Masuda, K., 1986. Notes on origin and migration of the Cenozoic pectinids in the Northern Pacific. pp. 95–110. In: T. Kotaka (Ed.), Japanese Cenozoic Mollusca: their Origin and Migration. Palaeontological Society of Japan, Special Paper No. 29. vii + 255 pp., 21 pp. of Pls.

Maury, C.J., 1921. Recent molluscs of the Gulf of Mexico and Pleistocene and Pliocene species from the gulf states. Bulletins of American Paleontology, 8 (34): 35–147.

Maury, C.J., 1924. Fosseis Terciarios do Brasil. Servico Gelogico e Mineralogico do Brasil, Monographia 4, 665 pp., 24 Pls.

McHugh, J.L., 1984. Fishery management. Lecture notes on coastal and estuarine studies, No. 10. Springer-Verlag, New York, 207 pp.

Meek, F.B., 1864. Check list of the invertebrate fossils of North America. Miocene. Smithsonian Miscellaneous Collection 7 (183): ii + 32 pp.

Menzel, R.W., 1970. The species and distribution of quahog clams *Mercenaria*. Proceedings of the National Shellfisheries Association 60: 8 (abstract).

Merrill, A.S. and Ropes, J.W., 1967. Distribution of southern quahogs off the middle Atlantic coast. Commercial Fisheries Review, 29: 62–64.

Mighels, J.W., 1843. Catalogue of the marine, fluviatile and terrestrial shells of Maine and adjacent ocean. Boston Journal of Natural History, 4: 308–350.

Moerch, O.A.L., 1853. Catalogus Conchyliorum quae reliquit d. Alphonso d'Aguirra & Gadea Comes de Yoldi. Hafniae, II, 74 pp.

Moore, D.R., 1961. The marine and brackish water mollusca of the state of Mississippi. Gulf Coast Research Laboratory 1 (1): 58 pp.

Moore, E.J., 1963. Miocene marine mollusks from the Astoria formation in Oregon. U.S. Geological Survey Professional Paper No. 419, iv + 109 pp. 33 Pls.

Morris, P.A., 1956. A Field Guide to the Shells of our Atlantic and Gulf Coasts. Houghton Mifflin, The Riverside Press, Cambridge:Boston, 236 pp., 45 Pls.

Neumayr, M., 1883. Zur Morphologie des Bivalvenschlosses. Akademie der Wissenschaften Wien. Sitzungsberichte, 88 (1): 385-418.

Newcombe, C.L., Thompson, S.J. and Kessler, H., 1938. Variations in growth indices of *Venus mercenaria* L. from widely separated environments of the Atlantic coast. Canadian Journal of Research D, 16: 1–5.

Noonan, G., 1993. Echo Bay Seafoods, 32 Fulton Fish Market, New York, NY 10038, USA, personal communication.

Ogasawara, K., 1973. Molluscan fossils from the Nishikurosawa Formation, Oga Peninsula, Akita Prefecture, Japan. Science Reports, Tohoku University, 2nd Series, Special Volume, No. 6 (Hatai Memorial Volume): pp. 137–155, Pls. 12–13. 4 Figs.

Oldroyd, T.S., 1924. The Marine Shells of the West coast of North America. Stanford University Publications. University Series. Geological Sciences 1 (1): 247 pp., 57 Pls.

Olsson, A.A., 1932. Contributions to the Tertiary paleontology of Northern Peru; pt. 5. The Peruvian Miocene. Bulletins in American Paleontology 19 (68). 264 pp., 24 Pls.

Olsson, A.A., 1961. Mollusks of the tropical eastern Pacific; Panamic-Pacific Pelecypoda. Paleontological Research Institute: Ithaca, NY, 574 pp., 86 Pls.

Orbigny, A.d'., 1852. Prodrome de Paleontologie Stratigraphique Universelle, Vol 3. Victor Masson, Paris, 190 pp.

Palmer, K.V.W., 1927. The Veneridae of Eastern America; Cenozoic and Recent. Palaeontographica Americana 1 (5): 209–522, 76 Pls.

Parker, P., 1949. Fossil and Recent Species of the pelecypod genera *Chione* and *Securella* from the Pacific Coast. Journal of Paleontology, 23 (6): 577–593.

Pearse, A.S., 1936. Estuarine animals at Beaufort, North Carolina. Journal of the Elisha Mitchell Scientific Society, 52 (2): 174–222.

Perkins, G.H., 1869. The molluscan fauna of New Haven. Proceedings of the Boston Society of Natural History, 13: 139–163.

Perry, G., 1811. Conchology, or the Natural History of Shells. W. Miller, London, 61 Pls.

Peterson, C.H. and Beal, B.F., 1989. Bivalve growth and higher order interactions: importance of density, site and time. Ecology, 70 (5): 1390–1404.

Pfeiffer, L., 1869–1872. Die Familie der *Venus* muscheln, Veneracea. In: Martini, F.H.W. and J.H. Chemnitz. (Eds.) Systematisches Conchylien Cabinet. Vol. XI (1), 2nd ed. Bauer and Raspe, Nuremburg, 302 pp., 42 Pls.
Philippi, R.A., 1844. Abbildungen und Beschreibungen neuer oder wenig gekannter Conchylien. II. Venus: 127–129; 175–178, Pls. I; III.
Philippi, R.A., 1845. Bemerkunger uber die Mollusken-fauna von Massachusetts. Zeitschrift fur Malakozoologie, 2: 68–79.
Pilsbry, H.A. and Olsson, A.A., 1941. A Pliocene fauna from western Ecuador. Proceedings of the Academy of Natural Sciences at Philadelphia, 93: 1–79.
Porter, H.J., 1974. The North Carolina and Estuarine Mollusca: an Atlas of Occurrence. University of North Carolina, Institute of Marine Sciences: Morehead City, 351 pp.
Porter, H.J. and Tyler, J., 1972. Sea Shells common to North Carolina. North Carolina State University: Moorehead City, NC 36 pp., 121 Figs.
Pratt, D.M. and Campbell, D.A., 1956. Environmental factors affecting growth in *Venus mercenaria*. Limnology and Oceanography, 1: 2–17.
Procter, W., 1929. *Venus mercenaria* at Mt Desert, Maine. Nautilus, 42 (3): 103.
Rafinesque, C.S., 1815. Analyse de la nature. Palermo, 225 pp.
Reagan, A.B., 1909. Some notes on the Olympic Peninsula, Washington. Kansas Academy of Science Transactions 22: 131–238, map.
Reeve, L.A., 1864. Conchologia Iconica: or, Illustrations of the Shells of Molluscous animals. Vol 14. Reeve, London.
Reylea, D., 1993, Frank M. Flower and Sons, Oyster Bay, NY, personal communication.
Richards, H.G., 1935. Some shells from the North Carolina. Nautilus, 44 (4): 130–134.
Römer, E., 1858. Kritische untersuchung der Arten des Molluskengeschlechts *Venus* bei Linne und Gmelin. Cassel. xiii + 135 pp.
Römer, E., 1865. Kritische Uebersicht sammtlicher Arten der zur Gattung *Venus* gehorenden Untergattungen Mercenaria und Gemma. Malakozoologische Blatter: 134–141.
Russell, H.D., 1946. Ecologic notes concerning *Elysia chlorotica* Gould and Stiliger fuscata, Gould. Nautilus, 59 (3): 95–97.
Savazzi, E., 1985. Adaptive themes in cardiid bivalves. Neuew Jahrbuch fur Geologie und Palaeontologie Abhandlunger, 170: 291–321.
Say, T., 1822. Account of some of the marine shells of the United States. Journal of the Academy of Natural Sciences at Philadelphia , 2 (2): 257–276.
Say, T., 1824. An account of some of the fossil shells of Maryland. Journal of the Academy of Natural Sciences of Philadelphia 4 (1): 124–155, Pls. 7–13.
Seilacher, A., 1974. Fabricational noise in adaptive morphology. Systematic Zoology, 22: 451–465.
Schröter, J.S., 1788. Namen Register in Systematisches Conchylien Cabinet, I. Nuremberg., 124 pp.
Schumacher, C.F., 1817. Essai d'un nouveau systeme des habitations des Vers Testaces. Schultz, Copenhagen, 287 pp., 22 Pls.
Shikama, T., 1964. Selected Shells of the world. Illustrated in colours, Vol. II. Hokuryu-kan: Tokyo, 212 pp., 70 Pls.
Shoemaker, A.H., 1972. Reef mollusks of South Carolina. Nautilus, 85 (4): 114–120.
Shumard, B.F., 1858. Descriptions of new fossils from the Tertiary formation of Oregon and Washington Territories and the Cretaceous of Vancouver Island. Academy of Sciences at St. Louis Transactions, 1: 120–125.
Smith, M., 1937. East Coast Marine Shells. Edwards Brothers, Ann Arbor, MI, 150 pp., 74 Pls.
Sowerby, G.B., 1853. Thesaurus Conchyliorum or Monographs of Genera of Shells. II. *Venus*: pp. 703–742, Pls., 152–157.
Speck, F.G. and Dexter, R.W., 1946. Molluscan food items of the Houma Indians. Nautilus, 60 (1): 34.
Stanley, S.M., 1970. Relation of shell form to life habits of the Bivalvia (Mollusca). Geological Society of America Memoirs, 125: 1–296.
Stanley, S.M., 1977. Coadaptation in the Trigoniidae, a remarkable family of burrowing bivalves. Palaeontology, 20: 869–899.
Stanley, S.M., 1981. Infaunal survival: alternative function of shell ornamentation in the Bivalvia (Mollusca). Paleobiology, 7 (3): 384–393.
Stenzel, H.B., 1955. Ancestors of the quahog. Journal of Sedimentary Petrology, 25: 145.

Stenzel, H.B., Krause, E.K. and Twining, J.T., 1957. Pelecypoda from the type locality of the Stone City Beds (Middle Eocene) of Texas. The University of Texas Publication No. 5704, 237 pp., 22 Pls.

Stimpson, W.M., 1851. A revision of the synonymy of the Testaceous Mollusks of New England. Phillips, Sampson and Co., Boston, 56 pp., 2 Pls.

Sumner, F.B., Osburn, R.C. and Cole, L.J., 1913. A biological survey of the waters of Woods Hole and vicinity. Pt. II. Sect. III. A catalogue of the marine fauna. U.S. Dept. Commerce and Labor Bureau of Fisheries Bulletin, 31 (1911): 547–794.

Tanaka, K., 1961. Studies on the molluscan fossils from central Shinano, Nagano Prefecture, Japan. Part 6. Molluscan fossils from the Moriya Formation. Bulletin of the Faculty of Education, Shinshu University No. 12: 61–97, Pls. 1–2.

Tebble, N., 1966. British Bivalve Seashells. Alden Press, London. 212 pp.

Thomas, R.D.K., 1993. Paleobiological and biogeographic significance of polymorphism in Pliocene shells of the arcoid bivalve *Glycymeris americana*, from the southeastern U.S.A. Geological Society of America Annual Meeting, Abstracts with Programs 25 (6): A-56.

Tryon, G.W., 1865. Description of a new species of *Mercenaria*. American Journal of Conchology 1: 297, Pl. 26.

Tryon, G.W., 1874. American marine conchology: or, descriptions of the shells of the Atlantic coast of the United States. G.W. Tryon, Philadelphia. (1873 ed: vi + 208 pp. 44 Pls.)

Tsuchi, R. (Ed.), 1981. Neogene of Japan. Its Biostratigraphy and Chronology. IGCP–114 National Working Group of Japan: Shizuoka.

Turgeon, D.D. (Ed.), 1988. Common and scientific names of aquatic invertebrates from the United States and Canada. Mollusks. American Fisheries Society Special Publication No. 16, vii + 277 pp., 12 pp. of Pl.

Turner, F.E., 1938. Stratigraphy and mollusca of the Eocene of western Oregon. Geological Society of America Special Paper 10, 130 pp. 22 Pls.

Van Winkle, W., Feng, S.Y. and Haskin, H.H., 1976. Effect of temperature and salinity on extension of siphons by *Mercenaria mercenaria*. Journal of Fisheries Research Board of Canada, 33 (7): 1540–1546.

Verkruzen, T.A., 1878. Zur Fauna von Neu-Schotland (Nova Scotia) und Neufundland. Jahrbuch der Malakozoologie Gesammten. pp. 208–230.

Vermeij, G.J., 1987. Evolution and Escalation: An Ecological History of Life. Princeton University Press, Princeton, NJ, 527 pp.

Verrill, A.E., 1874. The report of Prof. S.F. Baird, Commissioner of Fish and Fisheries, On the condition of the sea fisheries of the south coast of New England in 1871 and 1872. pp. 295–778. In: Verrill, A.E. and Smith, S.I. (Eds.), Report upon the invertebrate animals of Vineyard Sound and the adjacent waters, with an account of the physical characters of the region. Part I(8). Washington Government Printing Office, vi + 778 pp. 38 Pls.

Verrill, A.E., 1875. Brief contributions to zoology from the museum of Yale College, No. XXXVI. On the post-Pliocene fossils of Sankoty Head, Nantucket Island. American Journal of Science [3rd series, volume 10] 110: 364–375.

Vilas, C.N. and Vilas, N.R., 1970. Florida Marine Shells. Ch. E. Tuttle Co., Rutland, VT, 170 pp., 14 Pls.

Vokes, H.E., 1939. Molluscan faunas of the Domengine and Arroyo Hondo Formations of the California Eocene. Annals of the New York Academy of Sciences 28: 246 pp. 22 Pls.

Waller, T.R., 1969. The evolution of the *Argopecten gibbus* stock (Mollusca: Bivalvia) with emphasis on the Tertiary and Quaternary species of eastern North America. Journal of Paleontology, Supplement to No. 5, 43: 1–125.

Ward, L.W., 1992. Molluscan biostratigraphy of the Miocene, Middle Atlantic coastal plain of North America. Virginia Museum of Natural History Memoir No. 2, 159 pp., 26 Pls.

Ward, L., 1993. Ph.D. Virginia State Museum, Martinsville, VA 24112, USA, personal communication.

Weaver, C.E., 1912. A preliminary report on the Tertiary paleontology of western Washington. Washington Geological Survey, Bulletin 15, 80 pp., 12(+) Pls.

Weeks, W.H., 1908. A collecting-trip at Northport, N.Y.. Nautilus, 22 (9): 98–99.

Weisbord, N., 1971. Corals from the Chipola and Jackson Formations of Florida. Florida Bureau of Geology Bulletin No. 53, 100 pp.

Weisbord, N., 1973. New and little known corals from the Tampa formation of Florida. Florida Bureau of Geology Bulletin No. 56, 146 pp.

Wells, H.W., 1957. Status of the name *Venus*. Ecology, 38 (1): 160–161.

Wells, H.W., 1961. The fauna of oyster beds with special reference to the salinity factor. Ecological Monographs, 31: 239–266.

Wiedey, L.W., 1929. New Miocene mollusks from California. Journal of Paleontology, 3 (3): 280–289.

Wilson, D., 1983. The Lee Creek Enigma, *Mclellania aenigma*, a new taxon in fossil Cirrhipedia. In: Ray, I.C.E. (Ed.), Geology and Paleontology of the Lee Creek Mine, North Carolina, pp. 483–493. Smithsonian Contributions to Paleobiology, No. 53, 529 pp.

Winckley, H., 1891. Edible mollusks of Maine. Nautilus, 4 (10): 112–113.

Winkley, H.W., 1907. Cape Code notes. Nautilus, 21 (7): 74–75.

Winkley, H.W., 1909. Essec County notes. Nautilus, 23 (7): 86–89.

Winkley, H.W., 1916. Collecting at Nantucket and Martha's Vineyard. Nautilus, 29 (10): 109–110.

Wood, A.E. and Wood, H.E., 1927. A quantitative study of the marine mollusks of Cape May Country, New Jersey. Nautilus, 41 (1): 8–16.

Wood, W., 1828a. Index Testaceologicus; or a Catalogue of Shells, British and Foreign, 2nd ed. London, 212 pp., 38 Pls.

Wood, W., 1828b. Supplement to the Index Testaceologicus; or a Catalogue of Shells, British and Foreign. London. 59 pp., 8 Pls.

Woodring, W.P., 1982. Geology and paleontology of the Canal Zone and adjoining parts of Panama, descriptions of Tertiary mollusks (Pelecypods: Propeamussiidae to Cuspidariidae; additions to families covered in P-306-E; additions to gastropods; cephalopods). U.S. Geological Survey Profession Paper 306-F: 541–759, Pls. 83–124.

Yokoyama, M., 1926. Tertiary mollusca from southern Totomi. Journal of the Faculty of Science, Imperial University of Tokyo, section 2, 1 (9): 352–353, Pls. 39, Fig. 13.

Biology of the Hard Clam
J.N. Kraeuter and M. Castagna (Eds.),

Chapter 2

Shell Structure and Age Determination

Lowell W. Fritz

"... patterns very rarely represent continuous growth; they are episodic and record calcification events that are rhythmically synchronized to environmental changes." Pannella (1976)

2.1 INTRODUCTION

Patterns within the shell of the hard clam, *Mercenaria mercenaria*, have been well studied throughout its range (Ansell, 1968), but particularly along the western shores of the Atlantic Ocean. While certainly the species' commercial value contributed to many of these studies, the hard clam has many characteristics that make it ideally suited for shell microstructure analyses. It is equivalve and has a regular, consistent ovoid shape. Shell material is added concentrically along the valve margins and across the interior surface, permitting almost the entire ontogenetic growth history of an individual hard clam to be exposed by a single section of the shell. It has a widespread distribution both geographically and within many estuarine/coastal habitats. Most importantly, however, the hard clam records within its shell information about its growth cycles of various frequencies, from annual/seasonal to tidal/daily. Analyses of these records have been useful in: (1) molluscan biology/ecology and fishery management, for age determination of individuals and populations and determination of habitat-specific growth rates (Pannella and MacClintock, 1968; Rhoads and Pannella, 1970; Thompson, 1975; Fritz and Haven, 1983; Peterson et al., 1983; Grizzle and Lutz, 1988; Jones et al., 1989, 1990; Arnold et al., 1991; Richardson and Walker, 1991); (2) paleoanthropology, for determination of seasonal movements of early man from season of death of shells in middens (Clark, 1979; Quitmyer et al., 1985; Bernstein, 1990; Claasen, 1990); (3) paleontology/paleoecology, for making inferences about paleoenvironments and duration of the day, seasons, and year (Barker, 1964; Pannella et al., 1968; Clark, 1974; Berry and Barker, 1975; Pannella, 1976); and (4) the environmental sciences, for determination of the effects of natural and anthropogenic events on individual and population growth (Cunliffe and Kennish, 1974; Kennish and Olsson, 1975; Kennish, 1977, 1978, 1980).

The shell of the adult hard clam is composed of three aragonitic layers. Proceeding from the shell interior to exterior, these layers are: (1) an inner "homogenous", or complex crossed-lamellar layer; (2) a middle "homogenous", or complex crossed-lamellar layer; and (3) an outer composite prismatic layer (Pannella and MacClintock, 1968; Taylor et al., 1973; Kennish, 1980) (Fig. 2.1). The term "homogenous" had been used to describe shell layers, such as the inner layer of *M. mercenaria*, with structural units too small to distinguish or which do not repeat. As will be discussed in subsequent sections, the microstructure of at least the outer and middle layers varies seasonally and with age, such that a single description may be inappropriate. Within the inner and middle layers are myostraca, thin layers of columnar

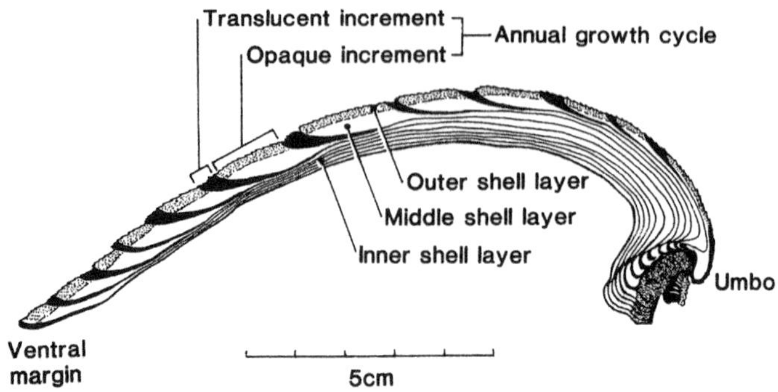

Fig. 2.1. Schematic of a radial section of a *M. mercenaria* shell showing the position of the three primary carbonate layers and the banding pattern typical of clams from the southeastern U.S. Translucent increment is synonymous with dark band, opaque increment with light band. Reprinted from Jones et al. (1990) with authors' permission. Growth is to the left.

prisms to which the mantle (pallial myostracum) or muscles (adductor myostracum) are attached to the shell. The outer shell layer is deposited by portions of the mantle located between the pallial line and the shell margin, while the inner and middle layers are formed by the inner mantle (Crenshaw, 1980).

Bivalve growth is generally described in terms of increases in shell dimensions, particularly those of length (anterior–posterior dimension), height (greatest distance from the umbo to the ventral margin), and width (greatest thickness through both valves; Fig. 2.2). In hard clams, the hinge and ligament occupy the dorsal surface, the valve margins form the ventral surface, and the height axis is not perpendicular to the length axis.

2.2 LARVAL SHELL MORPHOLOGY

The larval and early post-larval shell morphology of several *Mercenaria* and venerid species was recently described by Goodsell and Eversole (1992) and Goodsell et al. (1992). Their scanning electron micrographs, reproduced here in Figs. 2.3–2.5, show the development of the hinge dentition and the evolution of the adult shell shape. The first larval shell, the prodissoconch I (P_I), is formed within the first 24–48 h after the egg is fertilized (Fig. 2.3). At this point, the larva becomes a free-swimming planktotrophic veliger and has exhausted its yolk supply. The P_I of *M. mercenaria* has a mean length of 103 μm (Goodsell et al., 1992) and ranges in length in various *Mercenaria* taxa from 91 to 116 μm (Goodsell and Eversole, 1992). The P_I has a symmetrical D-shape with no umbo, and a straight hinge with no dentition (Fig. 2.4). During the 2-week planktonic phase (when the prodissoconch II (P_{II}) is formed), larval length increases from about 100 μm to 250 μm and the first two cardinal teeth appear in the hinge (Fig. 2.5).

Ockelmann (1965) first suggested that the size of the P_I is related to the size of the egg and hence, the larval developmental strategy employed by a bivalve species. Large eggs (150–200 μm in diameter) have greater amounts of yolk and are most often associated with lecithotrophic larvae, while species with smaller eggs have either planktotrophic larvae (from 40 μm to 85 μm eggs), or those with a combination of the two trophic types (from 90 μm

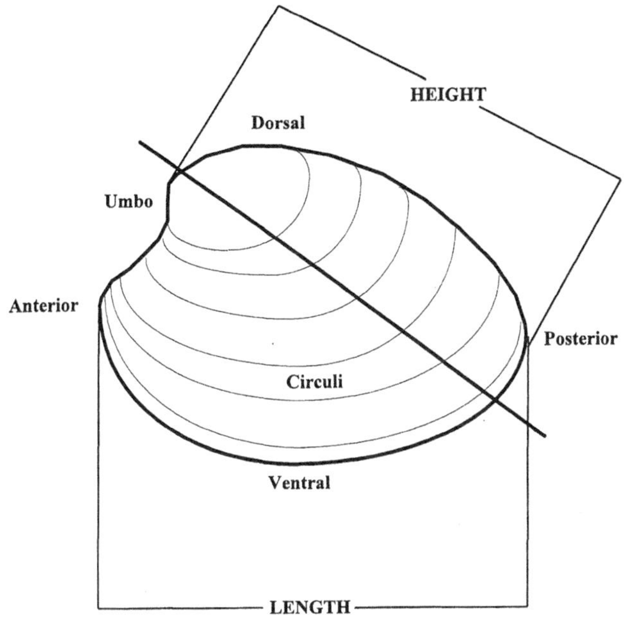

Fig. 2.2. Outline of left valve of *M. mercenaria* showing principal valve (lower case) and measurement (upper case) axes. Growth lines, or circuli, on the shell exterior, and the location of the cut for the radial section in Fig. 2.1 is shown.

to 140 μm eggs). Larval trophic strategy and egg size are inextricably linked with larval dispersal capability and ecology, as noted by Jablonski and Lutz (1980). Lecithotrophic, generally demersal larvae (with P_I > 250 μm in length) generally have poorer dispersal capabilities than smaller planktotrophic larvae with P_I lengths ranging from 100 μm to 150 μm. The hard clam, with its small P_I and predominately planktotrophic larvae, fits this paradigm. Even within a species, however, egg and P_I sizes are related. Goodsell and Eversole (1992) found that differences in egg size (and presumably yolk supply) as small as 10 μm were reflected in significant differences in P_I lengths, which could be directly related not only to female condition prior to spawning but also to larval survival potential.

Settlement of the larvae to the bottom is reflected in the shell at the boundary of the P_{II} and adult shell, or dissoconch. It is after settlement that the umbo becomes prominent (350 μm), the remaining two cardinal teeth are formed in the hinge (at about 430 and 675 μm), circuli or concentric growth lines become prominent on the shell exterior, and the valves elongate antero-posteriorly into the adult shell shape (Figs. 2.3 and 2.5).

2.3 ADULT SHELL MICROSTRUCTURE AND AGE DETERMINATION

Shell microstructures of bivalves have been investigated with a variety of techniques, such as unassisted visual observations, and both light (reflected and transmitted) and scanning electron microscopy. All methods involve sectioning or fracturing the shell along various shell axes to expose each of the shell layers. To reveal the growth history of an individual, a valve is

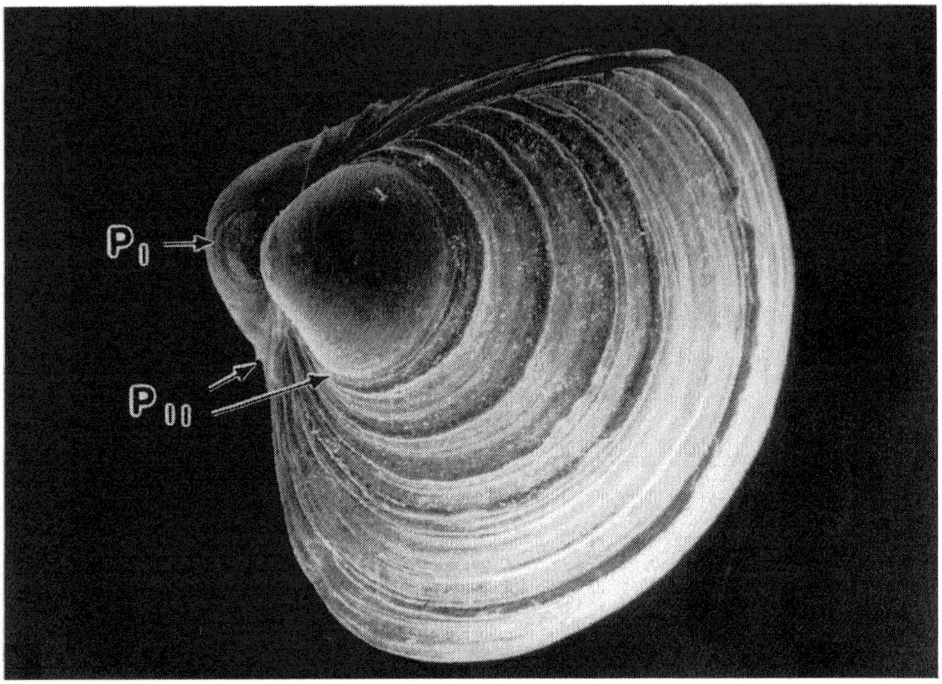

Fig. 2.3. Scanning electron photomicrograph of a *Mercenaria campechiensis* juvenile 753 μm in shell length illustrating prodissoconch I (P_I) and prodissoconch II (P_{II}) boundaries. Reprinted from Goodsell and Eversole (1992) with authors' permission.

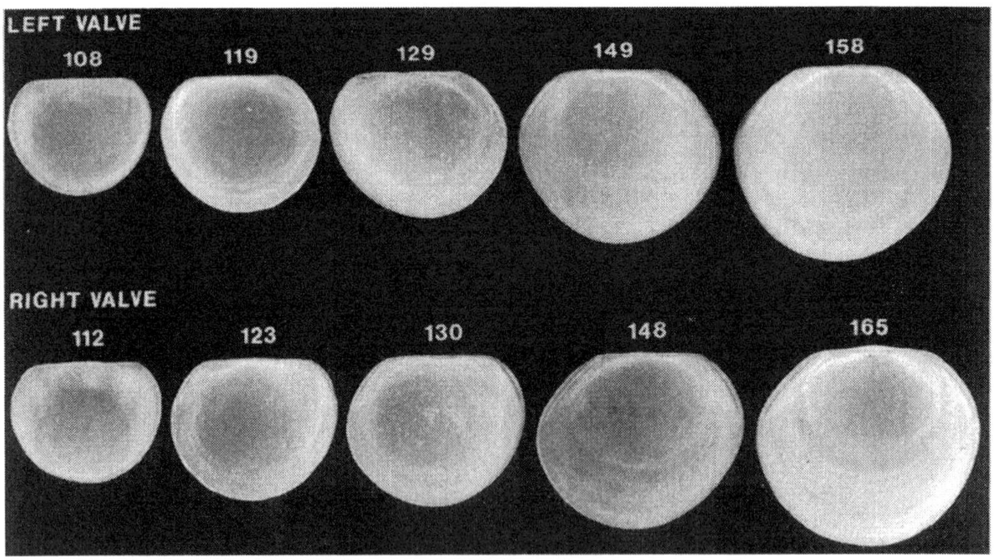

Fig. 2.4. Scanning electron photomicrographs of left and right disarticulated valves of *M. mercenaria* larvae. Numbers indicate shell length in μm. Reprinted from Goodsell et al. (1992) with authors' permission.

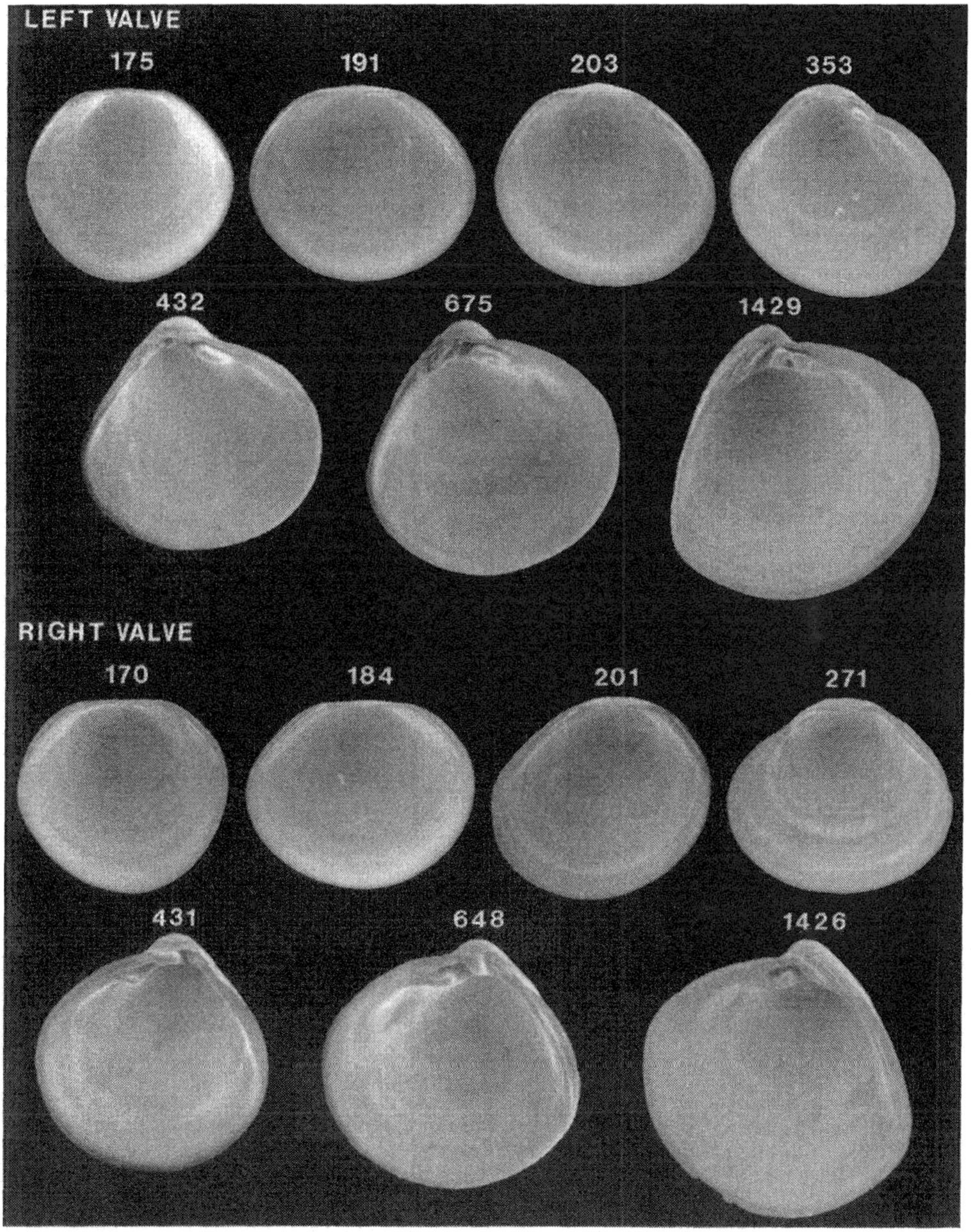

Fig. 2.5. Scanning electron photomicrographs of left and right disarticulated valves of post-larval *M. mercenaria*. Numbers indicate shell length in μm. Reprinted from Goodsell et al. (1992) with authors' permission.

sectioned from the umbo to the ventral margin along the height axis (Fig. 2.2), revealing not only each of the three primary carbonate layers, but also prominent growth lines in the middle and outer layers and within the umbo/hinge area (Fig. 2.1). These growth lines reflect hard clam growth rhythms of various periodicities, each resulting from a different response of the organism to its environment. Light microscopic methods involve transmitting light through shell replicas (e.g., acetate peels) or thin shell sections. The advantage of these techniques is that they enhance the visibility of periodic growth structures in both the middle and outer shell layers of hard clams. Acetate peels are made by polishing the radial surface of a sectioned valve and etching it (with a dilute acid, such as 1% hydrochloric acid, or a chelator, such as 0.1 M ethylene diamine tetra-acetic acid, or EDTA). The etchant removes carbonates from the radial shell section at different rates revealing not only the edges of crystalline units but also periodic growth structures (Kennish et al., 1980). Thin sections are made by finely polishing a thin section of the shell (along the height axis) which has been glued to a glass slide permitting light transmission (Clark, 1980). Polished and etched shell sections are also highly suited for scanning electron microscopic analysis, as are fracture sections. Fracture sections do not reveal the edges of periodic growth structures as well as polished and etched sections since the shell has been broken randomly, exposing the edges of crystalline units, some of which may span more than one growth increment.

2.3.1 Shell Microstructure of New Jersey Hard Clams

The microstructure of the outer and middle shell layers of *M. mercenaria* changes seasonally, with latitude, and can reflect patterns of disturbance. To describe these changes, details of the seasonal pattern of outer and middle layer shell growth will be described for hard clams collected from various populations in New Jersey waters. The patterns observed in shells of clams from these mid-Atlantic coastal waters will be compared with those of clams collected from other locations to the north and south along the U.S. Atlantic coast.

Microstructure and seasonal growth patterns in the outer and middle shell layers of *M. mercenaria* were examined in collections from three water bodies along New Jersey's Atlantic coast: Sandy Hook Bay, Barnegat Bay and Little Egg Harbor. Specimens were collected from Sandy Hook Bay (dredged from 4 m depth at low water) and Barnegat Bay (collected by hand rake from 1 m depth at low water) on six dates from November 1987 to November 1988, while samples from Little Egg Harbor were obtained from a commercial clam dealer in Tuckerton, NJ, on six dates from March 1986 to January 1987. Sandy Hook Bay is contiguous with Raritan Bay in northern New Jersey and has open access and exchange with waters of the New York Bight along the mid-Atlantic coast of North America. By contrast, Barnegat Bay is a long, narrow, shallow semi-enclosed embayment separated, by a string of barrier islands, from the Atlantic Ocean and exchanging water with it through small inlets. Little Egg Harbor, connected to the southern end of Barnegat Bay, is somewhat intermediate between Sandy Hook Bay and Barnegat Bay in its rate of exchange with the Atlantic Ocean.

2.3.1.1 Outer layer microstructure

The outer shell layer of *M. mercenaria* is primarily composed of prismatic aragonitic units oriented perpendicular to the inner shell surface (Figs. 2.6–2.8). Outer shell layer prisms are circular or ovoid cylinders approximately 10–60 μm in diameter or length (dorso-ventral). However, the outer shell layer is not uniformly composed of prismatic microstructure throughout a radial shell section, but may also contain regions of crossed-lamellar and columnar prismatic microstructure. The type of microstructure formed varies seasonally, with individual age, and between sites sampled.

When the outer layer is analyzed in polished and etched section (either as an acetate peel replica or directly using the scanning electron microscope), microgrowth increments become visible. Microgrowth increments record growth rhythms of individual organisms, and may result from alternating periods of shell deposition and dissolution (Crenshaw and Neff, 1969; Lutz and Rhoads, 1977, 1980; Gordon and Carriker, 1978). In *M. mercenaria*, microgrowth increments may be deposited (during periods of relatively uninterrupted growth by young specimens) at the rate of one per solar day (Pannella and MacClintock, 1968; Rhoads and Pannella, 1970; Kennish and Olsson, 1975; Fritz and Haven, 1983) or tidal cycle (Pannella, 1976), allowing detailed analyses and correlation of individual growth histories with environmental records. Microgrowth increment boundaries are thought to contain a greater percentage of organic material (Lutz and Rhoads, 1977) or be composed of a more etchant-resistant microstructure (Dieth, 1985) than the shell between the boundaries. Measurement of microgrowth increment widths and the relative number, placement and width of growth cessation marks (thick microgrowth increment boundaries) have been used to reconstruct growth histories and determine rates and causes of mortality (Pannella and MacClintock, 1968; Rhoads and Pannella, 1970; Kennish and Olsson, 1975; Pannella, 1976; Kennish, 1977, 1980; Fritz and Haven, 1983).

In New Jersey waters in spring, the outer layer was composed of prismatic microstructure in all but the oldest specimens collected from all three sites (Fig. 2.6A,B, Fig. 2.7A and Fig. 2.8A,E,F). Specimens which had not resumed growth in spring by May had either a sublayer of columnar prisms or modified crossed-lamellar units along the inner surface of the outer layer (Fig. 2.6C,D). By May, all specimens collected from Barnegat Bay and Little Egg Harbor, regardless of age, had resumed growth, while only the three youngest specimens (2, 2, and 8 years old) collected from Sandy Hook Bay had resumed growth. This was almost certainly due to the much earlier warming of Barnegat Bay and Little Egg Harbor than Sandy Hook Bay. In radial section, winter growth cessations were recorded as thick microgrowth increment boundaries (Fig. 2.7B), sublayers of columnar prisms (Fig. 2.8B,H), or as modified crossed-lamellar units (Fig. 2.6C,D). There was often a V-shaped notch in the outer layer associated with winter growth cessation marks (Fig. 2.7A and Fig. 2.8A) which, toward the exterior, contained thin organic sublayers (Fig. 2.8C). The organic sublayers are most likely periostracum trapped within the notch upon growth resumption in spring. A thin sublayer of crossed lamelles was deposited prior to prisms in early spring by some specimens (Fig. 2.8C), but not by all (Fig. 2.7B).

Throughout the remainder of spring and early summer, the outer layer of all but the oldest specimens was composed of prismatic microstructure. In summer, prismatic microstructures also dominated in all individuals 8–10 years of age and younger, but with a tendency for the inclusion of secondary elements around each prism (Fig. 2.6E,F). This may be a hybrid of the prismatic and crossed-lamellar microstructures, the latter of which increasingly dominated the outer layer in late summer and with increasing age (Fig. 2.7C and Fig. 2.8A,D,E,G,H). The inclusions of crossed-lamellar microstructure first appeared in the lower (toward the shell interior) portions of the outer layer (Fig. 2.7A,C). When the specimen was young (less than 8–10 years old), crossed-lamellar inclusions did not always extend to the shell exterior: in Fig. 2.7D, the outer region of the outer layer is composed of prisms, while the inner

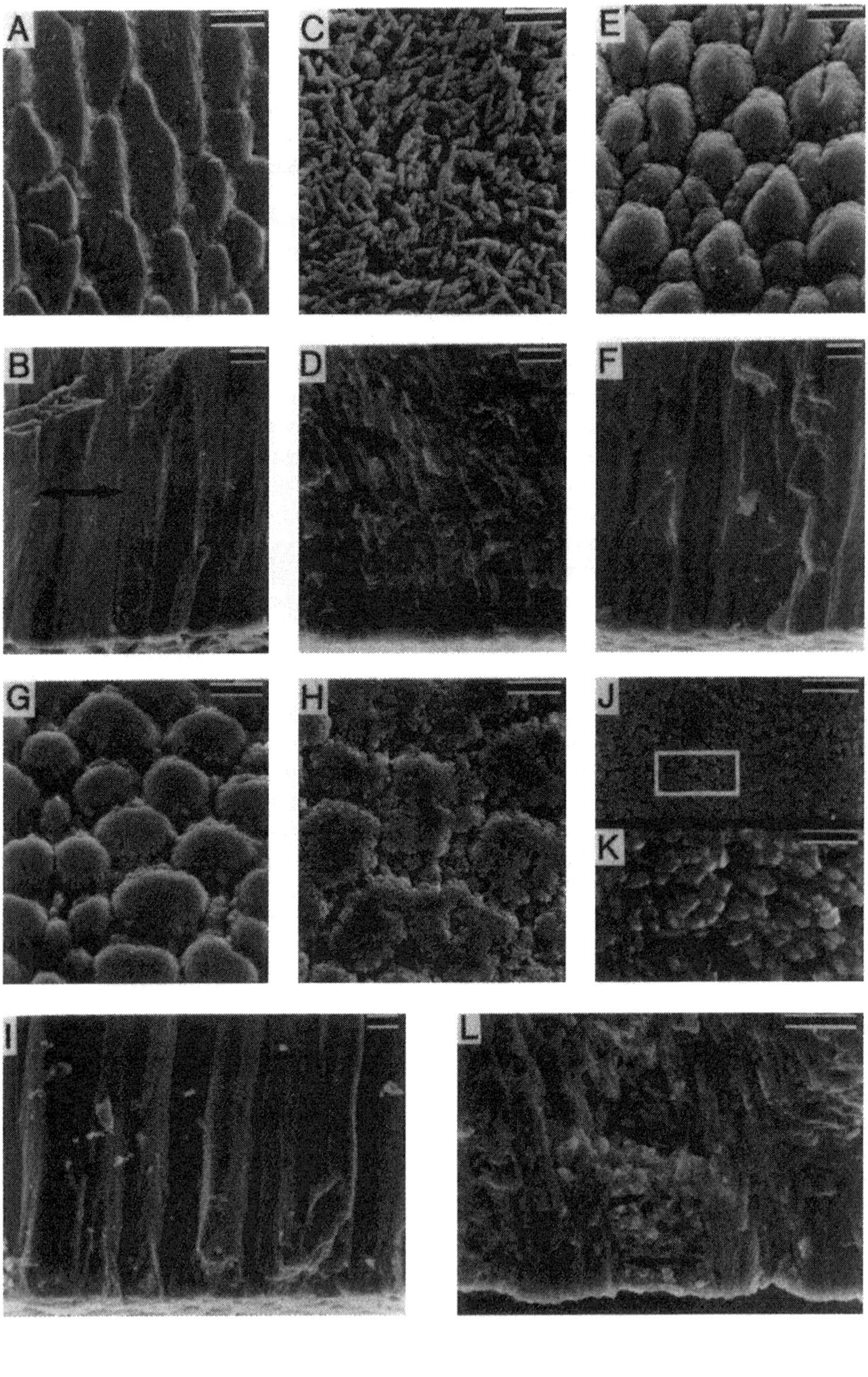
A
B
C
D
E
F
G
H
J
K
I
L

portion is composed of crossed-lamelles of contemporaneous deposition. The association of crossed-lamellar microstructures in the inner portion of the outer layer with closely spaced microgrowth increments in the outer portion strongly suggests that both were formed during a period of reduced shell growth rates in summer. Furthermore, the similarity of the middle and outer layer microstructures in Fig. 2.7C,F suggests that regions of (complex?) crossed-lamellar microstructure within the outer layer may be extensions of the middle layer. Regions of crossed-lamellar microstructure within the outer layer often extended farther from the middle layer in late summer (when water temperatures were warmer) than in early summer (compare cl1 through cl4 in Fig. 2.8A). Specimens collected from Barnegat Bay, where temperatures in summer were the warmest of the three sites, had the greatest tendency to form crossed-lamellar microstructures throughout summer (and fall), and at a younger age, than specimens collected from Sandy Hook Bay or Little Egg Harbor.

Hybrid prismatic and crossed-lamellar microstructures, or those composed solely of the latter, were common in specimens of all ages collected in fall from all three sites. Three manifestations of the hybrid fall microstructure are shown in Fig. 2.6G–L, which ranged from predominantly prismatic with inclusion of 2° elements (Fig. 2.6G), an even mixture of the two (Fig. 2.6H,I), to almost entirely crossed-lamellar (Fig. 2.6J–L). In polished and etched shell sections, it is not possible to distinguish the first two hybrids (Fig. 2.6G–I) from true prismatic microstructure. Thus, the outer layer on the left in Fig. 2.7A, deposited in fall, looks similar to that deposited in spring, on the right. There was also a direct correlation with the deposition of true crossed-lamellar microstructure within the outer layer in fall, and throughout the year, with increasing age (Fig. 2.8A,D,E,G).

2.3.1.2 Middle layer microstructure

The middle shell layer of *M. mercenaria* has been termed "homogenous" by some authors (Pannella and MacClintock, 1968; Rhoads and Pannella, 1970) due to the apparent lack of crystalline element organization within a repeating microstructure. Light and dark bands within the middle layer, as observed on polished shell sections (Pannella and MacClintock, 1968; Rhoads and Pannella, 1970; Peterson et al., 1983; Grizzle and Lutz, 1988), thin sections (Clark, 1979) and acetate peels (Fritz and Haven, 1983; Richardson and Walker, 1991), however, are formed seasonally and have been used to determine age. Clark (1979) termed

Fig. 2.6. Scanning electron photomicrographs of inner (A,C,E,G,H,J,K) and radial fracture surfaces (B,D,F,I,L) of the outer prismatic shell layers of specimens of *M. mercenaria* collected at various times from three locations in New Jersey. The direction of growth is up in the inner surface micrographs, and to the right in fracture surface micrographs. The type of middle layer growth band (light = opaque = fast growth, or dark = translucent = slow or no growth) present along the inner growing margin at the time of collection is listed. (A and B) 5-year-old specimen collected from Barnegat Bay on 13 May 1988; light band. Scale bars in both micrographs 20 μm. (C and D) 31-year-old specimen collected from Sandy Hook Bay on 13 May 1988; dark band. Scale bars in both micrographs 5 μm. (E and F) 6-year-old specimen collected from Little Egg Harbor on 15 July 1986; dark band. Scale bars in both micrographs 10 μm. (G) 3-year-old specimen collected from Little Egg Harbor on 17 November 1986; dark band. Scale bar 10 μm. (H and I) 8-year-old specimen collected from Sandy Hook Bay on 1 November 1988; light band. Scale bars in both micrographs 10 μm. (J,K and L) 8-year-old specimen collected from Barnegat Bay on 26 October 1988; light band. Scale bars in J and L 10 μm; in K 2 μm. K is a detail of J.

middle layer bands observed in thin shell sections opaque and translucent, depending on their optical properties. Fritz and Haven (1983) showed that opaque bands in thin section were equivalent to light bands in polished section and acetate peels, while translucent

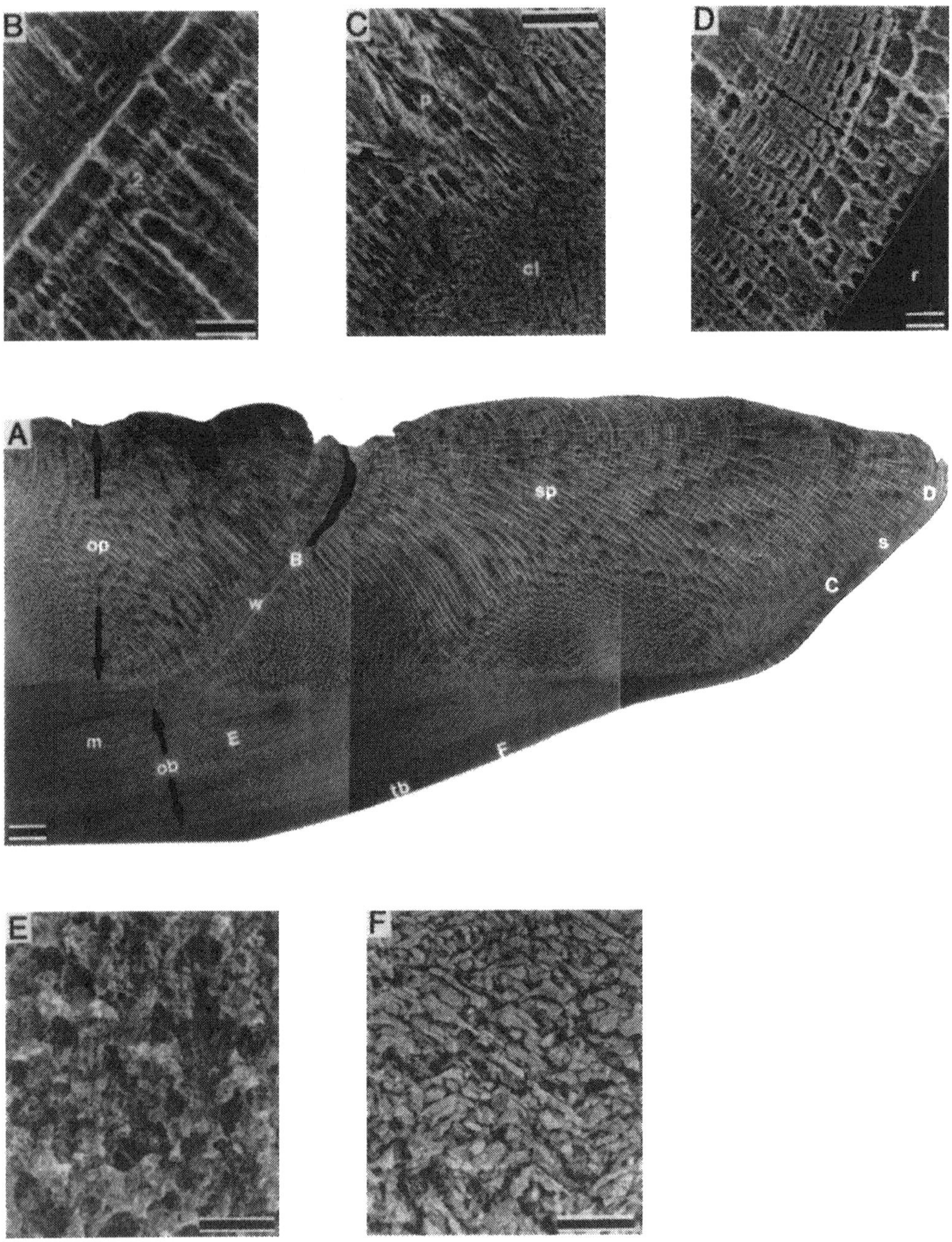

bands were equivalent to dark bands. Consequently, the microstructure of the middle layer is not "homogenous" throughout the year, but alternates between light (opaque) and dark (translucent) forms which have clearly different optical properties (Fig. 2.1).

In samples collected in New Jersey, young individuals (less than 8–10 years old) formed light middle layer bands in spring and fall and dark bands in summer and winter. Older individuals tended to form light bands in spring (if at all) and dark bands from summer through winter. In some years, however, a summer dark band formed by an old specimen was separated from the winter growth cessation in the middle layer (described below) by a light band formed in fall. Light bands were present along the inner surface of the middle layer in May in specimens of all ages collected from Barnegat Bay and Little Egg Harbor. In specimens collected from Sandy Hook Bay, only the youngest individuals were forming a light band in May. By June, all specimens collected from Sandy Hook Bay were forming a light middle layer band. This difference in timing of band formation was most probably due to the slower warming of Sandy Hook Bay compared with Barnegat Bay and Little Egg Harbor. Dark bands first appeared in June in specimens from Barnegat Bay, or one month before they first appeared in Sandy Hook Bay specimens. From July through September, dark band formation predominated at all sites. In fall and early winter, older specimens at all sites tended to have dark bands, while younger individuals had light bands along the inner surface of the middle layer.

Light bands were generally associated with prisms and relatively wide microgrowth increments in the outer layer (Fig. 2.7A,E and Fig. 2.8A,E). In young specimens, dark bands were formed contemporaneously with relatively narrow outer layer microgrowth increments and prisms, with possible inclusions of crossed-lamellar microstructure in lower portions of the outer layer (Fig. 2.7A,C,D,F). Older specimens tended to have regions of crossed-lamellar microstructure in the outer layer, within which microgrowth increments are generally not apparent, associated with dark middle layer bands (Fig. 2.8A,E).

The microstructure of the middle layer is a form of crossed-lamellar (Fig. 2.9). The lack of precise crossed-lamellar organization (such as that of the outer layer of *Rangia cuneata*; Fritz et al., 1990) was evident when the middle layer was analyzed in radial fracture section, as shown by the similarity of the microstructures formed by some individuals in spring (Fig. 2.9B), summer (Fig. 2.9F) and fall (Fig. 2.9H,J,L). The texture of the inner surface of

Fig. 2.7. Scanning electron photomicrographs of polished and etched radial sections of portions of the outer prismatic (op) and middle (m) shell layers of a 6-year-old *M. mercenaria* specimen collected 15 July 1986 from Little Egg Harbor. Specimen was forming a dark band in the middle layer when collected. Growth in each micrograph is to the right. (A) Low-magnification montage of micrographs showing portions of the outer and middle shell layers deposited from fall 1985 (left) to the date of collection (right). B–F mark the locations of micrographs B–F, respectively. w = growth cessation mark resulting from winter 1985–86. sp and s = portions of the outer shell layer deposited in spring and summer 1986, respectively. ob and tb = opaque (= light) and translucent (= dark) bands deposited in spring and summer, respectively, within the middle layer. Scale bar 200 μm. (B) Growth cessation marks (1 and 2) in the outer prismatic layer resulting from winter 1985–86. Scale bar 20 μm. (C) Junction of prismatic (p) and crossed-lamellar (cl) microstructures in the outer layer. Scale bar 20 μm. (D) Closely spaced increments labelled with double-headed arrow are contemporaneous with the crossed lamellar microstructure labelled in C. Scale bar 20 μm. (E) Poor organization of crystalline elements in middle layer light band. Scale bar 5 μm. (F) Ordered arrangement of crystalline elements in middle layer dark band. Scale bar 5 μm.

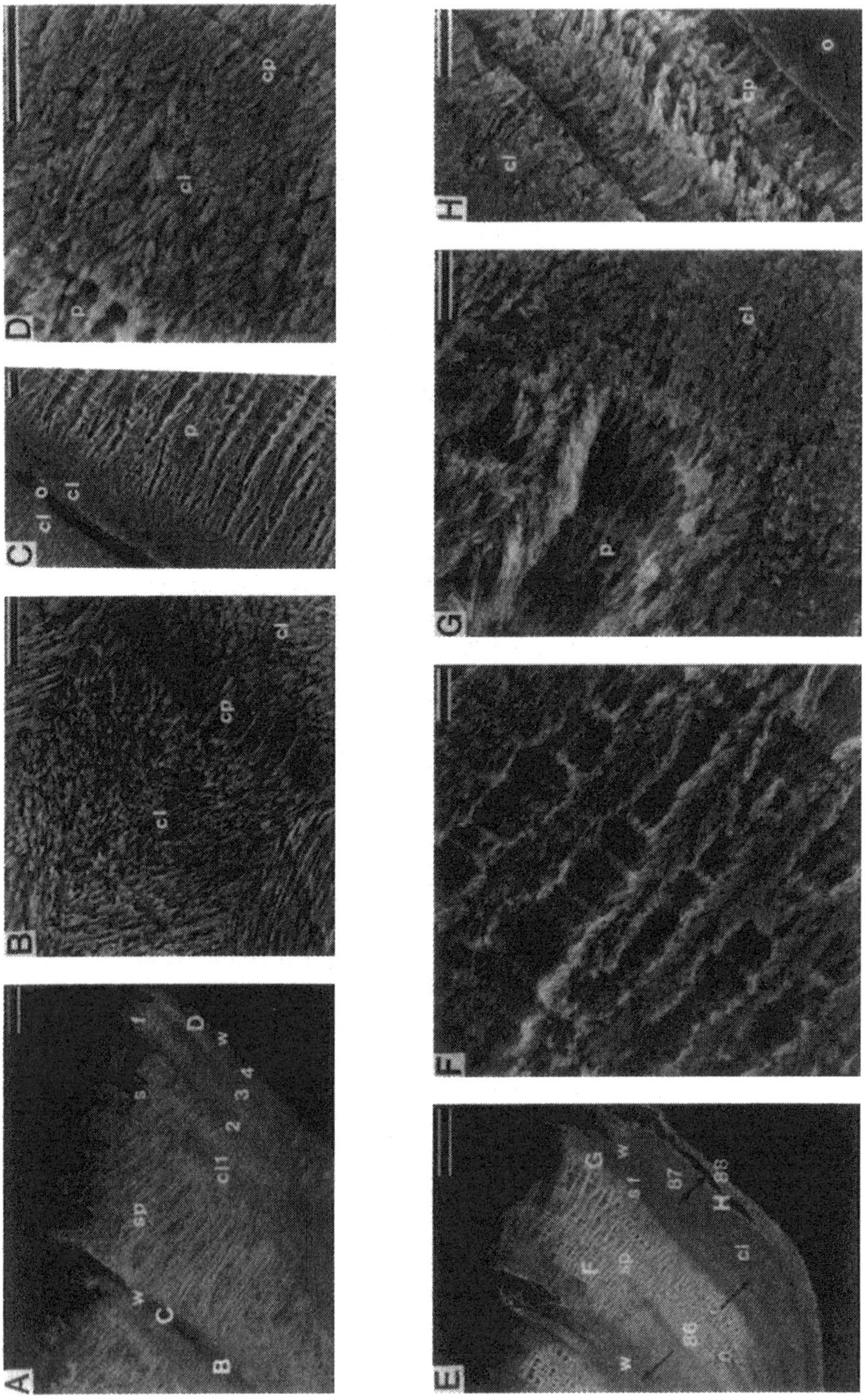
A
B
C
D
E
F
G
H
cl
cp
p
w
o

the middle layer most resembled crossed-lamellar microstructure and showed more seasonal variability than that observed on the fracture surface. In May (Fig. 2.9A), the inner surface of the middle layer was composed of an irregular arrangement of laths in those individuals which had resumed growth. These included all individuals collected from Barnegat Bay and Little Egg Harbor along with three young specimens collected from Sandy Hook Bay. Those specimens, which had not resumed growth by May, had either a sublayer of columnar prisms (Fig. 2.9C,D) or a zone of etched and redeposited crossed-lamelles along the inner surface of the middle layer. The formation of columnar prismatic sublayers during winter occurred with greater frequency with increasing age at all three sites. Winter growth cessations in the middle layers of younger specimens usually consisted of a thin dark band associated with a growth cessation mark in the outer layer (Fig. 2.7A). In summer and fall, the texture of the inner surface of the middle layer varied from blocky (Fig. 2.9E) or smooth (possibly etched and redeposited; Fig. 2.9G), to the irregular lath arrangement seen in spring (Fig. 2.9I,K). Based solely on analyses of fracture sections, the microstructural basis for the difference in optical properties of light and dark bands was not clear. However, in polished and etched radial section, it can be seen that light bands (Fig. 2.7E) were composed of relatively unorganized crystalline elements while dark bands (Fig. 2.7F) were composed of a regular, but complex arrangement of laths. This may in part explain the difference in optical properties of the two middle layer band types: light (opaque) bands, because of their irregular arrangement of laths (Fig. 2.7E and Fig. 2.9A,B,I,K,L) reflect light in polished and thin sections, while the more highly organized elements within dark (translucent) bands (Fig. 2.7F) transmit light.

2.3.1.3 Age determination

Individual age is determined by counting the number of annual increments in the outer and/or middle shell layers in radial section from the umbo to the ventral margin. As described

---

Fig. 2.8. Scanning electron photomicrographs of polished and etched radial sections of portions of the outer shell layer of a 14-year-old specimen of *M. mercenaria* collected 13 May 1988 from Sandy Hook Bay (A–D) and of a 22-year-old specimen of *M. mercenaria* collected 13 May 1988 from Barnegat Bay (E–H). Growth in each micrograph is to the right. (A) Low-magnification micrograph near ventral margin (right) showing growth from the winter 1986–87 growth cessation (w on left) through spring (sp), summer (s), fall (f) and winter 1987–88 (w on right). Four inclusions of crossed-lamellar microstructure (cl1, 2, 3 and 4) in the outer layer deposited in summer 1987 are noted. B–D mark locations of each micrograph, B–D, respectively. Scale bar 200 μm. (B) Crossed-lamellar (cl) inclusion in the outer layer formed in fall 1986 (upper left) and early spring 1987 (lower left). Winter growth cessation mark is represented by a sublayer of columnar prisms (cp). Scale bar 10 μm. (C) Crossed-lamellar inclusions in the outer layer formed in fall 1986 and early spring 1987 similar to B. Note the gap in mineralized structures (with the thin organic (o) sublayers) between the crossed-lamellar inclusions. The crossed-lamellar microstructure formed in early spring grades into prismatic microstructure (p). Scale bar 20 μm. (D) From left to right, prismatic, crossed-lamellar, and columnar prismatic microstructures formed in fall and winter 1987–88. Scale bar 10 μm. (E) Low-magnification micrograph near ventral margin (right) showing growth from the winter 1985–86 growth cessation (w on left) through spring (sp), summer (s), fall (f) and winter 1986–87 (w on right), to spring 1988. Crossed-lamellar microstructure (cl) was formed in the outer layer from summer 1986 through spring 1988. F–H mark locations of each micrograph, F–H, respectively. Scale bar 200 μm. (F) Prismatic microstructure formed in spring 1986. Scale bar 10 μm. (G) Prismatic and crossed-lamellar microstructure formed in summer 1986. Scale bar 5 μm. (H) Crossed-lamellar and columnar prismatic microstructures, and an inclusion of organic periostracum (o) formed in fall 1986 and winter 1986–87. Scale bar 5 μm.

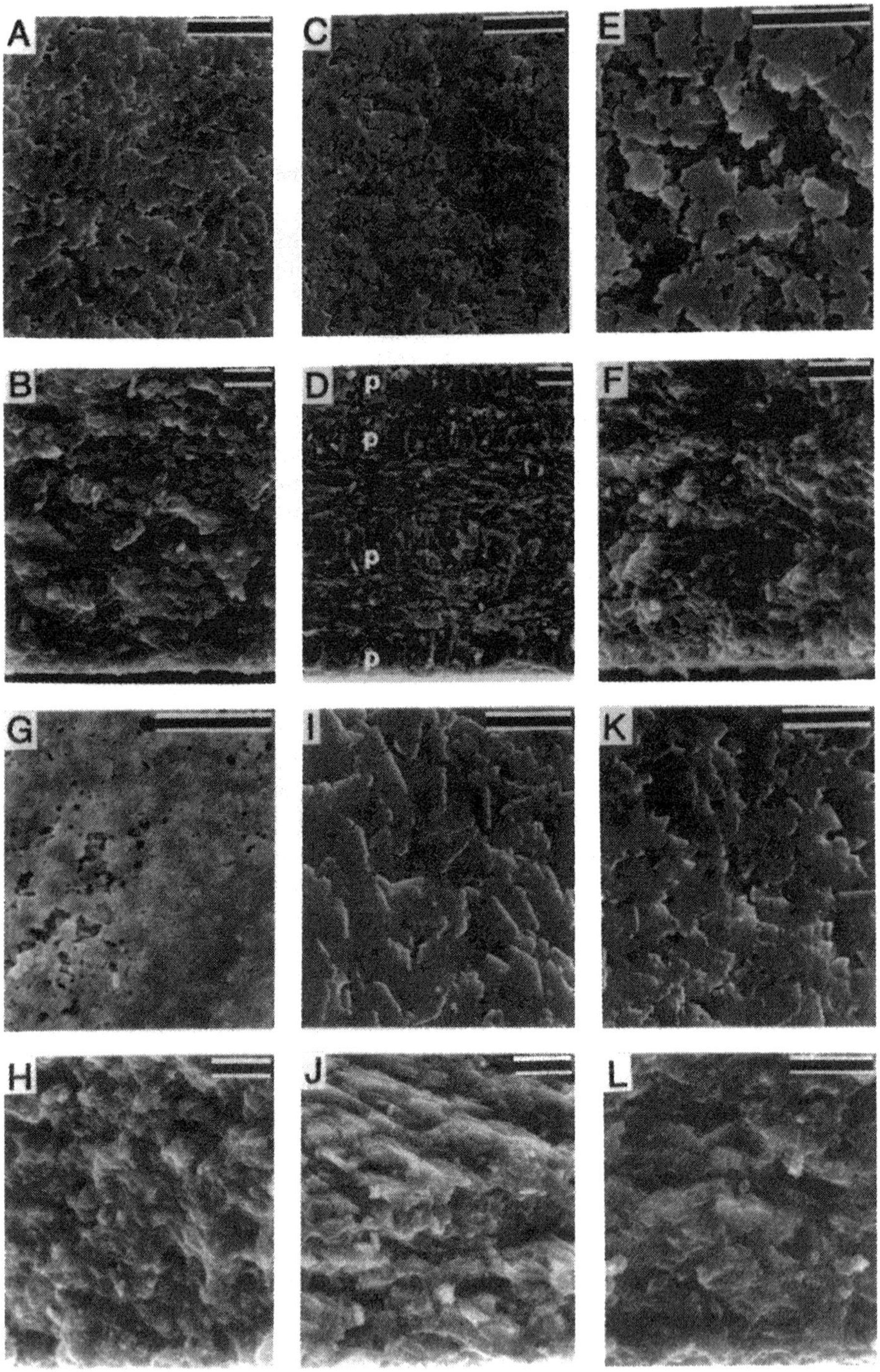
A
B
C
D
p
p
p
p
E
F
G
H
I
J
K
L

above, annual increments in the two shell layers contain winter growth cessation marks in the outer (and middle) shell layers, and middle layer dark (summer/fall) and light (spring and sometimes fall) bands. In the first 1–3 annual increments in specimens from Barnegat Bay, the first 1–5 from Little Egg Harbor, and the first 1–8 in specimens from Sandy Hook Bay, winter growth cessations within the outer and middle layers and in the umbo were the most clearly defined annually produced microstructural feature (Fig. 2.10A,B). Winter growth cessation marks in the first several annual increments usually consisted of a series of thick microgrowth increment boundaries associated with a thin dark band in the middle layer (Fig. 2.7A and Fig. 2.10). Summer was rarely represented within the middle layer as a single, consolidated band in the first several annual increments, but as a series of thin dark bands separated by light bands (Fig. 2.10B).

In "middle" age (annual increments 4–8 in specimens from Barnegat Bay, 6–12 from Little Egg Harbor, and 9–15 from Sandy Hook Bay), the summer dark band was only rarely broken by thin light bands. Fall light bands were less common with increasing age at each site. Thus, the annual sequence of middle shell layer bands in "middle" age was generally a spring light band and a summer/fall dark band, with the winter growth cessation mark in the middle and outer layers consisting of a sublayer of columnar prisms. In "old" age, annual increments were often only thin dark bands separated by sublayers of columnar prisms in both the middle and outer shell layers (Fig. 2.9D). This sequence of middle layer band formation and the changes associated with age are similar to those described by Grizzle and Lutz (1988) for young (1–5 years old) specimens of *M. mercenaria* in New Jersey and by Fritz (1982) and Fritz and Haven (1983) for populations in Virginia (Fig. 2.11).

The last 19 years of growth in the outer and middle shell layers by a 33-year-old specimen from Sandy Hook Bay are shown in Fig. 2.12. During these years, total annual increment widths (as measured in the outer layer along the height axis) varied only from 0.1 to 1.3 mm, but there was considerable variation in the structure of each annual increment, especially in the outer layer. Most outer-layer annual increments had more than one growth cessation mark. From 1972 to 1974, for instance, there were at least two growth cessation marks each year while in 1975, there were three. The dorsal-most marks each year consisted of inclusions of crossed-lamellar microstructure within the outer layer, and were formed in summer. The ventral-most mark was the winter growth cessation mark consisting of a well-defined thin line in the middle and outer layers (which is how columnar prismatic sublayers are represented

---

Fig. 2.9. Scanning electron photomicrographs of inner (A,C,E,G,I,K) and radial fracture surfaces (B,D,F,H,J,L) of the middle shell layer of specimens of *M. mercenaria* collected at various times from three locations in New Jersey. The direction of growth is to the right in each micrograph. The type of middle layer growth band (light = opaque = fast growth, or dark = translucent = slow or no growth) present along the inner growing margin at the time of collection is listed. Scale bars in each micrograph except D, 5 μm; in D, 10 μm. (A and B) 9-year-old specimen collected from Barnegat Bay on 13 May 1988; light band. (C and D) 31-year-old specimen collected from Sandy Hook Bay on 13 May 1988. Columnar prisms (p) were present along the inner surface of the middle layer when specimen was collected. Three other annually formed columnar prismatic sublayers are marked. (E and F) 6-year-old specimen collected from Little Egg Harbor on 15 July 1986; dark band. (G and H) 3-year-old specimen collected from Little Egg Harbor on 17 November 1986; dark band. (I) 8-year-old specimen collected from Barnegat Bay on 26 October 1988; light band. (J) 10-year-old specimen collected from Barnegat Bay on 26 October 1988; dark band. (K and L) 8-year-old specimen collected from Sandy Hook Bay on 1 November 1988; light band.

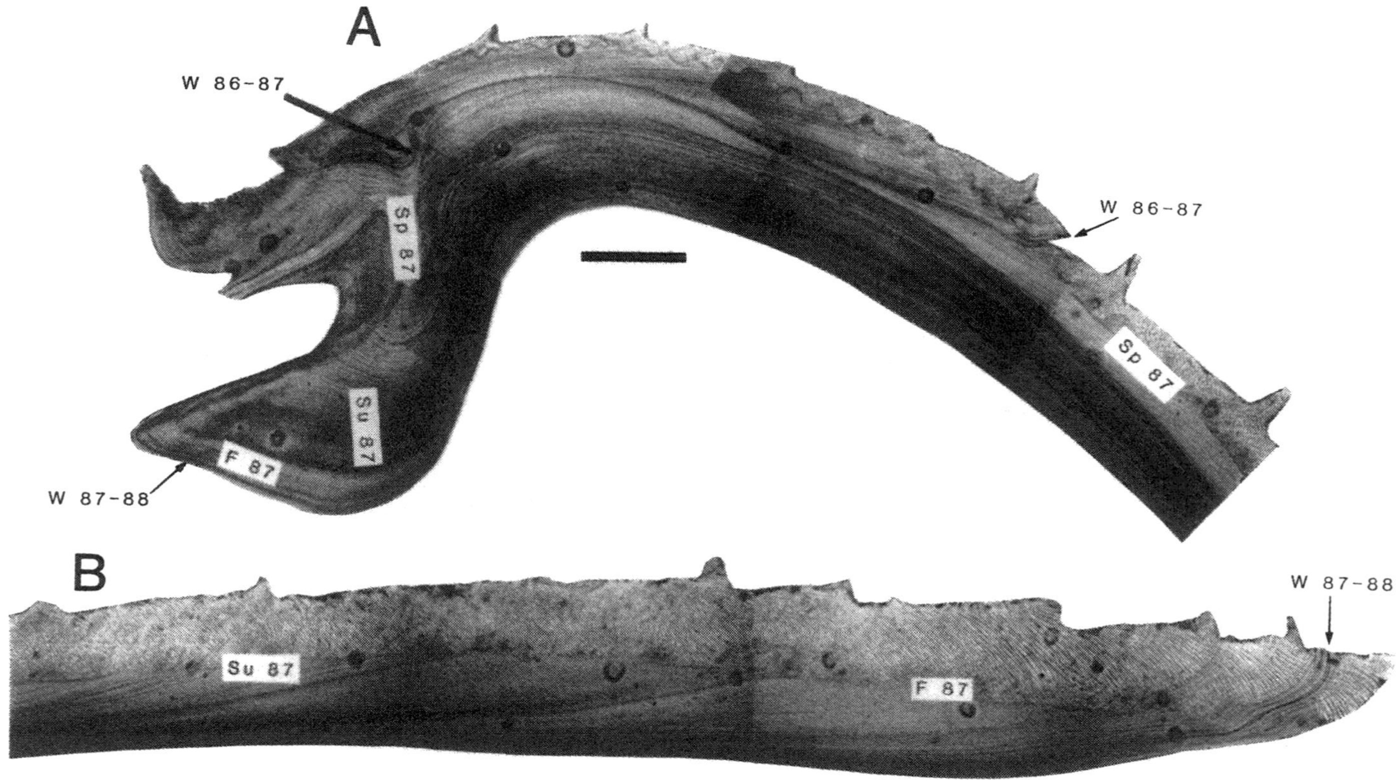

Fig. 2.10. Montage of light photomicrographs of an acetate peel of a polished and etched radial shell section of a 2-year-old specimen of *M. mercenaria* collected from Sandy Hook Bay on 13 May 1988. Regions of the inner, middle and outer shell layers are labelled by their season (W = winter; Sp = spring; Su = summer; F = fall) and year of formation in both A (dorsal portion of section near umbo) and B (ventral portion of section). Summer 1987 is represented within the middle layer as a series of dark sub-bands. Growth is to the right; scale bar 0.5 mm.

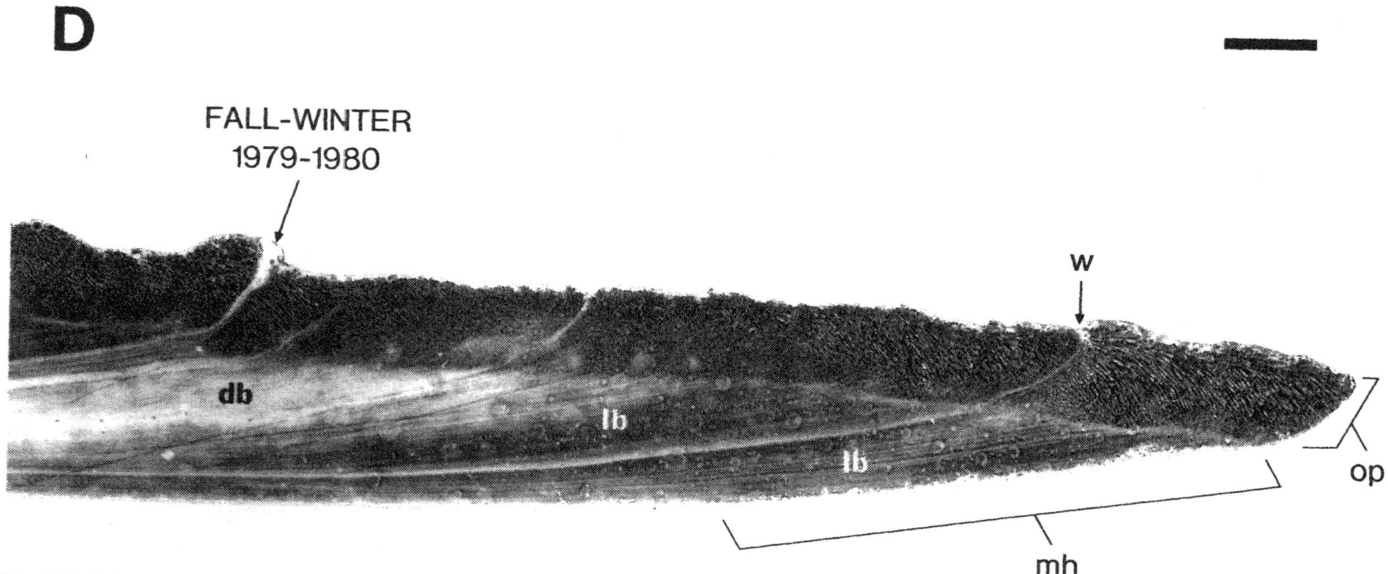

Fig. 2.11. Enlargement of an acetate peel of a polished and etched radial shell section of a 2-year-old specimen of *M. mercenaria* transplanted in October 1979 and collected on 31 May 1981. Specimen did not resume growth until spring 1980, leaving a thick growth cessation mark (fall–winter 1979–1980). In summer 1980, a dark band (db) was formed in the middle "homogenous" shell layer (mh). The light band (lb) formed in fall 1980 and spring 1981 was bisected by a winter growth cessation mark (w), which extended through the outer prismatic shell layer (op) to the shell exterior. Growth is to the right; scale bar 1 mm. Reprinted from Fritz and Haven (1983).

Fig. 2.12. Montage of light photomicrographs of the ventral portion of an acetate peel (polished and etched shell section) of a 33-year-old specimen of *M. mercenaria* collected from Sandy Hook Bay on 30 March 1988. Each annual increment from 1969 to 1987 is labelled near the junction of the outer prismatic (op) and middle (m) shell layers. Winter growth cessation marks (*w*) are labelled at the ventral edge of three annual increments (1975, 1983, and 1985) in which they were separated from the summer dark band (narrow outer layer microgrowth increments) by a fall light band (wide microgrowth increments). Other annual increments with two or more growth cessations each year include 1971–1974, 1976, and 1979–1980. Growth is to the right; scale bar 0.5 mm.

in acetate peels). Extremely small annual increments, such as those deposited in 1978 and from 1982 to 1984 by this specimen, were distinguished by thin columnar prismatic sublayers visible across the entire section of the middle layer and associated with winter growth cessation marks in the outer layer.

2.3.2 Effects of Latitude/Temperature and Age on Seasonal Shell Microstructure

Shell growth patterns of hard clams from the mid-Atlantic U.S. coast, as described above, are a mixture, or hybrid, of those found to the south and north. In warmer climates to the south (coastal embayments of the Florida, Georgia, and North Carolina coasts), a light band, representing relatively fast shell growth, was formed in the middle shell layer in winter. During the remainder of the year, but particularly during late summer and fall, growth was slower, resulting in the formation of a dark band in the middle layer (Clark, 1979; Peterson et al., 1983, 1985; Jones et al., 1990; Arnold et al., 1991). By contrast, in northern, cooler regions (along the coast from Connecticut to Massachusetts and in England), the opposite pattern of shell growth has been observed. Slow growth in the cold winter months results in a dark middle layer band and a growth cessation mark in the outer layer, while warmer water temperatures in the remainder of the year are represented by a light middle layer band (Pannella and MacClintock, 1968; Rhoads and Pannella, 1970; Jones et al., 1989; Bernstein, 1990; Richardson and Walker, 1991). In the northern mid-Atlantic (e.g., New Jersey as described above), hard clam shell growth patterns are similar to the northern pattern during the first several years of life, but become increasingly "southern" with age. Similarly, in the southern mid-Atlantic (Virginia), summer dark bands predominate at all ages, but distinct winter dark bands can be formed by a small percentage (15%) of younger clams (Fritz, 1982; Fritz and Haven, 1983).

The generalized latitudinal model of middle layer banding patterns described above is a direct result of the effects of temperature on hard clam shell growth. Ansell (1968), in a review of hard clam growth and activity throughout its geographic distribution, concluded that the optimum temperature range for shell growth was 15–25°C. Growth slows at temperatures below 15°C and above 25°C, and ceases at 9°C and 31°C. Ansell (1968) found no evidence that this relationship between temperature and shell growth rate changed within the hard clam's distribution.

In studies of hard clam shell growth patterns in Georgia (Jones et al., 1990) and Florida (Arnold et al., 1991), water temperatures were 25°C or above from May through October, resulting in slowed shell growth and the formation of a dark middle layer band. During the remainder of the year, water temperatures remained above 10°C, which were associated with faster shell growth rates and light middle layer bands.

The annual range in water temperatures in the mid-Atlantic states is greater than to the north or south, resulting in a hybrid annual shell growth cycle. In Virginia hard clams, dark bands were also associated with water temperatures above 25°C (June through September), just like populations to the south. However, winter water temperatures in Virginia can remain below 10°C for extended periods, resulting in little or no shell growth and the formation of distinct winter growth cessation marks. As described by Fritz (1982) and Fritz and Haven (1983), distinct winter marks are formed only when there is a period of rapid shell growth in the fall. If the individual clam does not respond to the decreasing water temperatures after summer by forming a light middle layer band, then the winter growth cessation will merge with the summer dark band and a single annulus, or dark band, will be formed. This happens increasingly with age, resulting in an annual pattern of spring light band and summer–fall–winter dark band formation.

In northern areas (Narragansett Bay, RI, and England), water temperatures in summer are cooler than those to the south, remaining within the optimum range for a longer period; this results in the formation of a light middle layer band most of the year (Pannella and MacClintock, 1968; Rhoads and Pannella, 1970; Jones et al., 1989; Richardson and Walker, 1991). However, water temperatures in winter are below the optimum for longer periods than to the south, which is reflected in winter growth cessation marks in the middle and outer shell layers, preceded and followed by periods of slowed growth (narrow outer layer microgrowth increments).

Local environmental conditions can modify the general latitudinal pattern described above. Shallow, protected embayments subject to solar heating may warm up earlier in the spring, resulting in earlier resumption of growth (and the formation of a light band in the middle layer) than for clams living on exposed coasts in the northern and mid-Atlantic states. However, temperatures may also rise above the optimum for shell growth in summer earlier in shallow embayments, slowing growth and causing middle layer dark band formation sooner than in other nearby areas. Bernstein (1990) described just an occurrence in comparing seasonal band formation by clams in Greenwich Cove, RI, with those collected from the main stem of nearby Narragansett Bay. Similarly, seasonal shell growth patterns of clams from Barnegat Bay, NJ, were affected by the local bathymetry and hydrography when compared to nearby Sandy Hook Bay, which is subject to greater cooling influence of waters from the Atlantic Ocean.

Fritz (1982) showed that the percent agreement between the number of microgrowth increments formed in the outer prismatic layer between two growth cessation marks of known formation time, and the number of days separating the events decreased with increasing age. Assuming that microgrowth increments are formed with regular periodicity at a given site (each solar day (Pannella and MacClintock, 1968; Rhoads and Pannella, 1970; Fritz, 1982; Fritz and Haven, 1983; Bernstein, 1990) or tidal cycle (Pannella, 1976)), then it can be inferred that old clams deposit shell on fewer days than younger clams: growth slows with age. There are marked seasonal differences in growth rate with age that are reflected in growth patterns within the shell. As described in detail for clams from New England (Pannella and MacClintock, 1968; Rhoads and Pannella, 1970; Bernstein, 1990), the mid-Atlantic (Fritz, 1982; Fritz and Haven, 1983; Grizzle and Lutz, 1988), and the southeastern U.S. (Clark, 1979; Peterson et al., 1983, 1985; Jones et al., 1990; Arnold et al., 1991), fast shell growth (light middle layer band) becomes increasingly limited to spring in the north and winter in the south with age. Peterson and Fegley (1986) suggested that slower winter shell growth rates of adults than juvenile clams in North Carolina might reflect the partitioning of energy by adults into gametogenesis in preparation for spring spawning. Jones et al. (1989) suggest a similar reason for differing seasonal shell growth rates and middle layer band formation with age for clams in Narragansett Bay, RI.

2.3.3 Growth Cessation Marks in Outer Layer Microstructure

Kennish and Olsson (1975) and Kennish (1980) discuss in detail a wide variety growth cessation marks that are distinguishable within the outer shell layer of *M. mercenaria*, including those caused by cold, heat, abrasion, and spawning. Each may have its own characteristic pattern of microgrowth increment thickness before and after the growth cessation, as well as the presence or absence of cross-lamellar microstructures within the outer layer prisms. One of the most common growth cessation marks is that caused by cold water temperatures in winter in northern climates (described in detail by Kennish (1980) and Richardson and Walker (1991)). The winter growth cessation mark is characterized by a slowing of growth prior to the mark, which is revealed by the decreasing width of microgrowth increments. The growth cessation mark itself is a thick microgrowth increment boundary, reflecting a long period of valve closure, possible anaerobiosis, and carbonate dissolution along the shell interior (Lutz and Rhoads, 1977). The mark may also include a columnar prismatic sublayer, as shown in Fig. 2.8B,H. Distal from the mark along the radial shell section are microgrowth increments of gradually increasing width, reflecting resumption of growth in spring. The mark may also be associated with a V-shaped notch in the shell exterior and inclusions of organic periostracum.

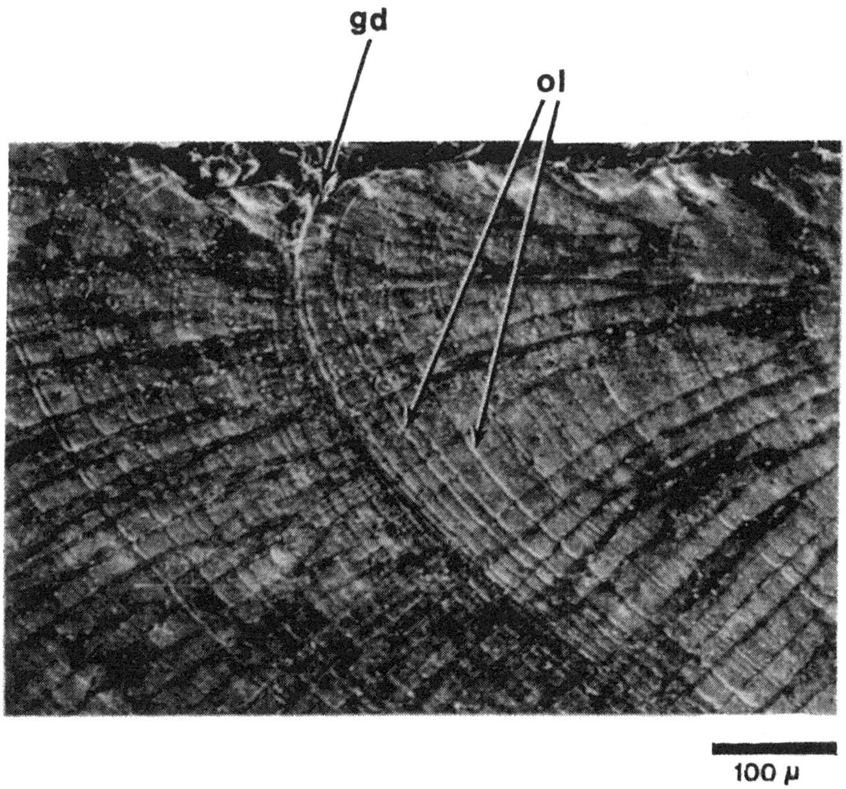

Fig. 2.13. Scanning electron photomicrograph of the outer prismatic layer of a specimen of *M. mercenaria* showing microgrowth increments separated by regions of more etchant resistant carbonate (ol) and a growth disturbance or cessation mark (gd) induced by cold-shock (described in text). Growth is to the left and the shell exterior is at the top. Reprinted from Fritz (1982).

Inducing the formation of growth cessation marks can be of great use in situ growth studies by marking dates within the outer layer microstructure (Fritz, 1982; Fritz and Haven, 1983; Peterson et al., 1985). A growth cessation mark can be made by simply removing the clam from the water and keeping it cool (4°C) and moist for 24 h prior to replanting. If this is attempted when the clams are active and growing (spring), then the growth cessation mark is usually preceded by wide microgrowth increments, and followed by increments of gradually increasing width (Fig. 2.13).

2.4 CONCLUSIONS

There is an abundance of information on the growth history of individual animals stored within the shells of bivalve molluscs. Analyses of many of these records can provide long time series' of individual growth, including periods of time that precede the inception of particular studies. This recording of growth, once deciphered, interpreted, and placed in the context of other time series (e.g., climatic, oceanographic), can provide valuable data in a wide variety of studies, from marine biological (e.g., ecology, paleoecology, population dynamics, environmental sciences) to anthropological. The hard clam, *M. mercenaria*, by virtue of its widespread distribution, its cultural importance and use for thousands of years, and its shell macro- and microstructure, is well suited for such studies.

2.5 ACKNOWLEDGMENTS

I am indebted to Lisa Wargo for preparation of the plates used in Figs. 2.6–2.10 and 2.12, to the authors who kindly permitted their work to be reproduced here, to John Grazul of Rutgers University for his assistance with the SEM analyses, to Bruce Ruppel and Tom Belton of New Jersey Department of Environmental Protection for their financial support of research on seasonal shell microstructure of bivalves, and to Michael Kennish, John Kraeuter (both of Rutgers University) and Michael Castagna (Virginia Institute of Marine Science) for their reviews of the manuscript. This work would also not have been possible without the inspiration and guidance of Dexter Haven (Virginia Institute of Marine Science), Richard Lutz (Rutgers University), and Alyce Fritz (my wife).

REFERENCES

Ansell, A.D., 1968. The rate of growth of the hard clam (*Mercenaria mercenaria* (L.)) throughout the geographical range. J. Cons. Perm. Int. Explor. Mer, 31: 364–409.

Arnold, W.S., Marelli, D.C., Bert, T.M., Jones, D.S. and Quitmyer, I.R., 1991. Habitat-specific growth of hard clams *Mercenaria mercenaria* (L.) from the Indian River, Florida. J. Exp. Mar. Biol. Ecol., 147: 245–265.

Barker, R.M., 1964. Microtextural variations in pelecypod shells. Malacologia, 2: 69–86.

Bernstein, D.J., 1990. Prehistoric seasonality studies in coastal southern New England. Am. Anthropol., 92: 96–115.

Berry, W.B.N. and Barker, R.M., 1975. Growth increments in fossil and modern bivalves. In: G.D. Rosenberg and S.K. Runcorn (Eds.), Growth Rhythms and History of the Earth's Rotation. Wiley, London, pp. 9–27.

Claasen, C., 1990. Investigations of monthly growth in shellfish for application to archaeology. Final Rep. NSF Grant BNS-8507714, Appalachian State University, Boone, NC.

Clark II, G.R., 1974. Growth lines in invertebrate skeletons. Annu. Rev. Earth Planet. Sci., 2: 77–99.

Clark II, G.R., 1979. Seasonal growth variations in the shells of recent and prehistoric specimens of *Mercenaria mercenaria* from St. Catherines Island, Georgia. Anthropol. Pap. Am. Mus. Nat. Hist., 56: 161–179.

Clark, G.R., II, 1980. Study of molluscan shell structure and growth lines using thin sections. In: D.C. Rhoads and R.A. Lutz (Eds.), Skeletal Growth of Aquatic Organisms: Biological Records of Environmental Change. Plenum Press, New York, pp. 603–606.

Crenshaw, M.A., 1980. Mechanisms of shell formation and dissolution. In: D.C. Rhoads and R.A. Lutz (Eds.), Skeletal Growth of Aquatic Organisms: Biological Records of Environmental Change. Plenum Press, New York, pp. 115–132.

Crenshaw, M.A. and Neff, J.M., 1969. Decalcification at the mantle–shell interface in molluscs. Am. Zool., 9: 881–885.

Cunliffe, J.E. and Kennish, M.J., 1974. Shell growth patterns in the hard-shelled clam. Underwater Nat., 8: 20–24.

Dieth, M.R., 1985. The composition of tidally deposited growth lines in the shell of the edible cockle *Cerastoderma edule*. J. Mar. Biol. Assoc. UK, 65: 573–581.

Fritz, L.W., 1982. Annulus formation and microstructure of hard clam (*Mercenaria mercenaria*) shells. M.A. Thesis, College of William and Mary, Williamsburg, VA, 161 pp.

Fritz, L.W. and Haven, D.S., 1983. Hard clam, *Mercenaria mercenaria*: shell growth patterns in Chesapeake Bay. Fish. Bull., 81 (4): 697–708.

Fritz, L.W., Ragone, L.M. and Lutz, R.A., 1990. Microstructure of the outer shell layer of *Rangia cuneata* (Sowerby, 1831) from the Delaware River: Applications in studies of population dynamics. J. Shellfish Res., 9 (1): 205–214.

Goodsell, J.G. and Eversole, A.G., 1992. Prodissoconch I and II length in *Mercenaria* taxa. Nautilus, 106 (3): 119–122.

Goodsell, J.G., Fuller, S.C., Eversole, A.G., Castagna, M. and Lutz, R.A., 1992. Larval and early postlarval shell morphology of several venerid clams. J. Mar. Biol. Assoc. UK, 72: 231–255.

Gordon, J. and Carriker, M.R., 1978. Growth lines in a bivalve mollusc: subdaily patterns and dissolution of the shell. Science, 202: 519–521.

Grizzle, R.E. and Lutz, R.A., 1988. Descriptions of macroscopic banding patterns in sectioned polished shells of *Mercenaria mercenaria* from southern New Jersey. J. Shellfish Res., 7 (3): 367–370.

Jablonski, D. and Lutz, R.A., 1980. Molluscan larval shell morphometry: ecological and paleontological applications. In: D.C. Rhoads and R.A. Lutz (Eds.), Skeletal Growth of Aquatic Organisms: Biological Records of Environmental Change. Plenum Press, New York, pp. 323–377.

Jones, D.S., Arthur, M.A. and Allard, D.J., 1989. Sclerochronological records of temperature and growth from shells of *Mercenaria mercenaria* from Narragansett Bay, Rhode Island. Mar. Biol., 102: 225–234.

Jones, D.S., Quitmyer, I.R., Arnold, W.S. and Marelli, D.C., 1990. Annual shell banding, age, and growth rate of hard clams (*Mercenaria* spp.) from Florida. J. Shellfish Res., 9 (1): 215–226.

Kennish, M.J., 1977. Growth increment analysis of *Mercenaria mercenaria* from artificially heated coastal marine waters: a practical monitoring method. In: Proc. XII Int. Soc. Chronobiol. Conf., Washington, DC, Casa Editrice Il Ponte, Milano, pp. 663–669.

Kennish, M.J., 1978. Effects of thermal discharges on mortality of *Mercenaria mercenaria* in Barnegat Bay, New Jersey. Environ. Geol., 2 (4): 223–254.

Kennish, M.J., 1980. Shell microgrowth analysis: *Mercenaria mercenaria* as a type example for research in population dynamics. In: D.C. Rhoads and R.A. Lutz (Eds.), Skeletal Growth of Aquatic Organisms: Biological Records of Environmental Change. Plenum Press, New York, pp. 255–294.

Kennish, M.J. and Olsson, R.K., 1975. Effects of thermal discharges on the microstructural growth of *Mercenaria mercenaria*. Environ. Geol., 1: 41–64.

Kennish, M.J., Lutz, R.A. and Rhoads, D.C., 1980. Preparation of acetate peels and fractured sections for observation of growth patterns within the bivalve shell. In: D.C. Rhoads and R.A. Lutz (Eds.), Skeletal Growth of Aquatic Organisms: Biological Records of Environmental Change. Plenum Press, New York, pp. 597–601.

Lutz, R.A. and Rhoads, D.C., 1977. Anaerobiosis and a theory of growth line formation. Science, 198: 1222–1227.

Lutz, R.A. and Rhoads, D.C., 1980. Growth patterns within the molluscan shell: an overview. In: D.C. Rhoads and R.A. Lutz (Eds.), Skeletal Growth of Aquatic Organisms: Biological Records of Environmental Change. Plenum Press, New York, pp. 203–254.

Ockelmann, W.K., 1965. Developmental types in marine bivalves and their distribution along the Atlantic coast of Europe. In: L.R. Cox and P.F. Peake (Eds.), Proc. First European Malacol. Congress, 1962, London. Conchological Soc. Great Britain and Ireland and Malacol. Soc. London, London, pp. 25–35.

Pannella, G., 1976. Tidal growth patterns in recent and fossil mollusc bivalve shells. Naturwissenschaften, 63: 539–543.

Pannella, G. and MacClintock, C., 1968. Biological and environmental rhythms reflected in molluscan shell growth. J. Paleontol., 42 (5; Suppl.): 64–80.

Pannella, G., MacClintock, C. and Thompson, M.N., 1968. Paleontological evidence of variations in length of the synodic month since Late Cambrian. Science, 162: 792–796.

Peterson, C.H. and Fegley, S.R., 1986. Seasonal allocation of resources to growth of shell, soma, and gonads in *Mercenaria mercenaria*. Biol. Bull. (Woods Hole), 171: 597–610.

Peterson, C.H., Duncan, P.B., Summerson, H.C. and Safrit Jr., G.W., 1983. A mark-recapture test of annual periodicity of internal growth band deposition in shells of hard clams, *Mercenaria mercenaria*, from a population along the southeastern United States. Fish. Bull., 81 (4): 765–779.

Peterson, C.H., Duncan, P.B., Summerson, H.C. and Beal, B.F., 1985. Annual band deposition within shells of the hard clam, *Mercenaria mercenaria*: consistency across habitat near Cape Lookout, North Carolina. Fish. Bull., 83 (4): 671–677.

Quitmyer, I.R., Hale, H.S. and Jones, D.S., 1985. Paleoseasonality determination based on incremental shell growth in the hard clam, *Mercenaria mercenaria*, and its implications for the analysis of three southeast Georgia coastal shell middens. Southeast. Archaeol., 4: 27–40.

Rhoads, D.C. and Pannella, G., 1970. The use of molluscan shell growth patterns in ecology and paleoecology. Lethaia, 3: 143–161.

Richardson, C.A. and Walker, P., 1991. The age structure of a population of hard-shell clam, *Mercenaria mercenaria* from Southampton Water, England, derived from acetate peel replicas of shell sections. ICES J. Mar. Sci., 48: 229–236.

Taylor, J.D., Kennedy, W.J. and Hall, A., 1973. The shell structure and mineralogy of the Bivalvia. II. Lucinacea–Clavagellacea: Conclusions. Bull. Br. Mus. (Nat. Hist.) Zool. Suppl., 22 (9): 253–294.

Thompson, I., 1975. Biological clocks and shell growth in bivalves. In: G.D. Rosenberg and S.K. Runcorn (Eds.), Growth Rhythms and History of the Earth's Rotation. Wiley, London, pp. 149–163.

Biology of the Hard Clam
J.N. Kraeuter and M. Castagna (Eds.),

Chapter 3

Embryogenesis and Organogenesis of Veligers and Early Juveniles

Melbourne R. Carriker

3.1 INTRODUCTION

The veliger, arising from the ubiquitous trochophore larva, probably emerged as a typical feature of the biphasic molluscan life cycle within the Monoplacophora, subsequently becoming the dominant larval phase of the Phylum Mollusca (Rieger, 1994). Today, a veliger stage characterizes all descendant classes, but the Cephalopoda (Stasek, 1972; Crisp, 1974), and enhances wide genetic exchange. While the veliger reflects an adaptational response to the immediate needs of a plankter in a common aquatic environment (Russell-Hunter, 1979), its ontogenetic successor, the pediveliger, evolved specializations for leaving the water column and transforming into the benthic juvenile. The high degree of specialization of some of the veliger–pediveliger organs can thus be viewed as aiding in broad dispersal, searching for, and settling on suitable substrata (Carriker, 1990). The striking similarity that characterizes the general anatomic level of organization of pediveligers of the many different species of marine bivalves is therefore not unexpected (Bayne, 1971). Although resemblances in morphology and behavior of pediveligers of different taxa at the time of metamorphosis are to be expected, it is changes following metamorphosis that establish recognizable specific adult differences (Yonge, 1959; Bayne, 1971).

As the paucity of literature on the early stages of venerid bivalves suggests, emphasis by investigators has been placed on the more applied aspects of their biology (Menzel, 1989; Rice, 1992). Hence, extensive, comparative, morphological, functional researches that integrate systematic, developmental, ecologic, behavioral, and physiological aspects of their biology are generally wanting.

Despite (or perhaps because of) the commercial importance of *Mercenaria mercenaria* (Linnaeus, 1758) (hard clam, quahog, quahaug, Zinn, 1973; formerly known as *Venus mercenaria*, see Wells, 1957), details of the biology of its early life history are poorly known. Main publications on the early stages of *M. mercenaria* include those of Belding (1912), Loosanoff (1937), Loosanoff and Davis (1950), Turner and George (1955), Carriker (1961), Keck et al. (1974), Gallagher (1988), Menzel (1989); on related venerid clams, Quayle (1952) (*Venerupis pullastra* (Montagu)), Ansell (1962) (*Venus striatula* (Da Costa)), D'Asaro (1967) (*Chione cancella* Linne) and Sastry (1979) (bivalves, less ostreids).

In the present chapter, I emphasize the embryogenesis and organogenesis, and in Chapter 7 the functional morphology and general behavior of veliger, pediveliger and byssal plantigrade stages of *M. mercenaria*. Where informational voids exist for this species, available knowledge, mostly on other species in the family Veneridae, will be interleafed.

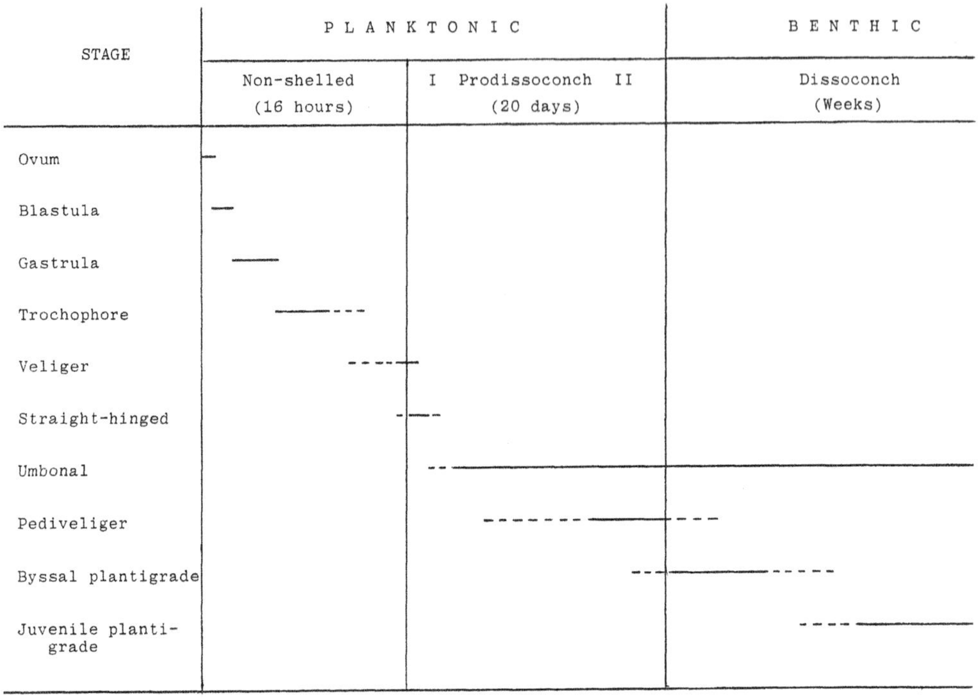

Fig. 3.1. Approximate duration of early stages of development of *Mercenaria mercenaria*. Solid lines, usual duration, which (dotted lines) may be shorter or longer.

3.2 DEFINITIONS

During larval development and metamorphosis, the young of *M. mercenaria* pass through a series of well-defined recognizable stages (Fig. 3.1, Table 3.1). Stages are defined principally on the basis of the valves, locomotor organs, byssus, siphons, and the spatial position of the bivalve in its native habitat (see also Levin and Bridges, 1995). Valve dimensions accompanying the appearance, and in some cases subsequent disappearance, of larval organs are only generally diagnostic of larval and post-larval stages because depending on ecological and genetic factors, shell dimensions of successive stages overlap broadly in different individuals. Hence, length limits recorded in Table 3.1 are offered only as rough indicators of the stages, and those given for early post-settled stages are incomplete. The position and behavior of young *M. mercenaria* in their native habitat change with each successive major anatomical development; it follows that accurate identification of stages is important in their mariculture and in studies of their functional morphology, physiology, behavior, and autecology.

The following terms are defined as background for descriptions in the sections that follow (Carriker, 1996):

Anterior and *posterior*: directions parallel to the hinge. As prodissoconch II shell develops, valves become slightly more pointed anteriorly and more rounded posteriorly. This trend is

TABLE 3.1

Summary definitions of early stages of *Mercenaria mercenaria* based on laboratory-reared individuals (see Fig. 3.1)

A. Planktonic stages

1. *Non-shelled.* Development from fertilized ovum through blastula, gastrula, and trochophore to nonshelled veliger stage, occurs in about 16 h. Early stages float passively in water masses; trochophores and naked veligers swim actively by use of a strongly ciliated velum.
2. *Shelled (prodissoconchs).*
   a. *Straight-hinged veligers.* Possess smooth valves; velum grows in size, larvae are strong swimmers; length range approximately 90–140 μm, age range 1–3 days. Prodissoconch I valves are thin, uniform, translucent, secreted by surface of the mantle, and appear during first 24 h of life.
   b. *Umbonal veligers.* Smooth valves continue to grow symmetrically, but now a gently sloping umbo projects above the middle of the hinge line; length range approximately 140–220 μm, age range 3–20 days. Prodissoconch II valves are secreted onto Prodissoconch I valves by the edge of the mantle, and are still smooth but with faint commarginal growth striae.

B. Pediveligers (swimming–crawling stage)

At an age varying from 6 to 20 days and a length ranging from about 170 to 240 μm, veligers develop a foot, and thereafter alternate swimming in the water and crawling on the bottom; this stage exists for a variable period, and terminates at a shell length of about 200–230 μm when the velum is lost.

C. Plantigrades (crawlers, dissoconchs, benthic stage)

After the velum is lost and beginning at a length of about 200–300 μm, metamorphosing larvae are limited to crawling over the bottom on the foot:

1. *Byssal plantigrades.* Soon after — and sometimes before — loosing the velum, plantigrades attach to substrata by a byssus (the settling or spatting stage); thereafter for a number of weeks (to a shell length of about 9 mm), they alternate byssal attachment and active crawling, remaining on, or superficially in fine sediment; deposited shell now takes the form of commarginally ridged dissoconch valves, small primary ridges being followed by more conspicuous secondary ridges.
2. *Juvenile plantigrades.* Young individuals approximately 9 mm and longer, the byssus gland no longer functional, and the byssus no longer formed; plantigrades maintain their position beneath the surface of sediment by the foot and valves alone; definitive siphons are fully formed.

amplified in plantigrade valves. In the plantigrade, mouth and anterior adductor muscle are located anteriorly, and anus and posterior adductor muscle, posteriorly.

Beak: the earliest part of each dissoconch valve.

Commarginal: sculptural or structural features of the shell, or of internal organs, that parallel shell margins or previous traces of the shell margin.

Commissure: the line along which edges of the valves are in contact.

Dissoconch: the part of the valves that begins at the metamorphic line after metamorphosis, ends the prodissoconch II stage, and continues for the duration of adult sessile existence.

Dorsad: toward the larval and adult hinge.

Height: maximal dorsoventral dimension perpendicular to the hinge.

Juvenile: essentially the *byssal plantigrade*, in which the byssal gland is active and the plantigrade can attach to substrata by the *byssus*; from metamorphosis to a shell length of about 9 mm.

Length: maximal anterior–posterior dimension parallel to the hinge line.

Pediveliger: the swimming–crawling larval stage (Carriker, 1961) that develops toward the end of the prodissoconch II stage and serves as the transitional form between planktonic and benthic existence.

Prodissoconch I: extends from the first appearance of the larval shell material to the first meeting of valve edges, forming a straight-hinged, D-shaped shell.

Prodissoconch I/II boundary: a narrow, faintly commarginally striated transitional band sandwiched between prodissoconch I and II regions of the valves.

Prodissoconch II: extends ontogenetically from the prodissoconch I/II boundary to *metamorphosis*; the latter is identified by the *metamorphic line* in each valve. An abrupt change occurs beyond this line in shell microstructure and mineralogy, marking the appearance of the *plantigrade* stage with a functional *foot* for crawling and a *byssus* for attachment to substrata. Prodissoconch II valves include new shell deposited on the inside of prodissoconch I valves as well as that added beyond the margins of prodissoconch I valves. The exterior shell surface of prodissoconch II bears conspicuous commarginal growth striae that clearly distinguish it from the faint microsculpture of the shell surface of prodissoconch I. The prodissoconch (I plus II) is the entire larval shell formed before metamorphosis. The term *veliger* refers to the larva within the prodissoconch valves, which bears the swimming–feeding organ, the *velum*.

Radial: elements of each valve or of soft tissues that radiate from the umbo.

Umbo: (pleural, *umbones* or *umbos*), the rounded elevated oldest part of each valve located to each side of the hinge and atop the beak.

Ventrad: away from the hinge.

Width: maximal dimension between exterior surfaces of right and left valves. The terms 'depth' and 'convexity' are sometimes used synonymously with 'width'.

3.3 EMBRYOGENESIS

When first discharged through the exhalant siphon by the female, ova of *M. mercenaria* vary in diameter from about 60 to 85 μm, are spherical in shape, granular, and slightly grayish in color, packed with yolk granules, and covered by a distinct primary envelope (the vitelline membrane, Wourms, 1987). Outside this envelope there is present a conspicuous, hyaline, gelatinous envelope (fertilization envelope? Wourms, 1987) about 25 μm in thickness (Loosanoff and Davis, 1950). Sperm–egg binding and blocks of polyspermy are discussed by Longo (1983) and Rosati and Focarelli (1996) in other bivalve species, and the gametogenic cycle in *M. mercenaria* is briefly described by Eversole (1997). The gelatinous envelope, a distinguishing feature of *M. mercenaria*, soon swells in seawater, and in some 4 h can attain a thickness of about 95 μm, giving a large ovum an overall diameter of approximately 275 μm. The gelatinous covering protects and aids in the flotation of the ovum (see also de Severeyn et al., 1994). A few ova may occur free of the envelope, possibly unfertilized, and occasionally two ova may be present within one envelope, separated from each other and of unequal size. The majority of ova, however, remain within a single envelope until they are freed as ciliated trochophores (Belding, 1912; Loosanoff and Davis, 1950). Veligers and juveniles from large ova survive better than those from small ones (Kraeuter et al., 1982).

Development of embryos is indirect and by unequal spiral cell division. Duration of cell divisions from time of fertilization at room temperature (about 22°C) is generally: to the two-cell stage, 45–50 min; 4 cells, 90–110 min; 8 cells, 148 min; 16 cells, 185 min; 32 cells, 200 min. Cellular division continues and in 6 h the ciliated rotating blastula, a compact mass of small cells surrounding a group of larger nutritive cells is formed. In 9–10 h, the embryo, now a young gastrula, acquires minute cilia over its exterior surface. By the late gastrula stage, oval in shape, the embryo whirls rapidly on its longer axis within the gelatinous envelope (Belding, 1912; Loosanoff and Davis, 1950).

In 12–14 h, at room temperature, the embryo, now an early pear-shaped trochophore, and more elongated in form than the gastrula, escapes and swims away from the gelatinous envelope in a spiral path. How escape occurs, has not been determined. The trochophore possesses a dense circlet of long cilia, the prototroch, over its anterior region, a velar tuft of longer prominent apical flagellae, and begins to form a primitive mouth and shell gland. At this stage, the trochophore is approximately 90 μm long and 65 μm in diameter. The apical flagellar tuft is almost as long as the trochophore itself. The trochophore swims rapidly through seawater, holding the apical flagellae straight in front of it. When held captive under a microscope cover glass, the larva periodically whips the tuft basally. In about 16 h, the trochophore, not yet feeding, its nourishment provided by nutritive cells, becomes more or less spherical in shape. A broad zone of long velar cilia, 7 μm in length, forms around the base of the apical flagellar tuft. By this stage, the larva is a strong swimmer, and continuously rotates on its longitudinal axis as it swims spirally.

In about 24–36 h after the ovum has been fertilized, a thin transparent shell develops gradually over the larva, extending from the dorsum ventrally. The shell completely envelops the soft organs and results in the straight-hinged stage with a highly developed velum (Fig. 3.2a) (see Carriker, 1996 for discussion of valve formation in *Crassostrea virginica*). A straight-hinged veliger of average dimensions when first fully clothed by prodissoconch I, is 98 μm long, 78 μm high, 48 μm wide, and has a hinge line 65 μm long. Although ratios of these dimensions remain relatively constant, the size of straight-hinged larvae varies considerably, lengths as small as 86 μm being recorded by Loosanoff et al. (1951) and Carriker (1961). Goodsell and Eversole (1992) found that prodissoconch I lengths of *M. mercenaria* veligers developing from ova 80 μm in diameter were significantly greater (109.7 ± 0.7 μm) than those (101.9 ± 0.7 μm) from ova 70 μm in diameter, lengths varying consistently with size of egg.

In about 40 h, many veligers reach an approximate size of 110 μm long and 90 μm high; in 4 days, approximately 120 × 98 μm, still a straight hinge (Fig. 3.2b); in 6 days, an early umbo approximately 150 × 140 μm (Fig. 3.3); in 8 days, approximately 195 × 175 μm; in 10 days, many reach the late umbo stage, approximately 215 × 190 μm, and after 12 days, some of the ready-to-set pediveligers may be as large as 225 × 210 μm. In laboratory cultures, many pediveligers may settle in about 12 days at 24°C, the smallest larva being 170 μm long, and the largest 240 μm long. The size of pediveligers at the time of settlement can vary widely with conditions of temperature, crowding, food, and other factors, so that dimensions are only of relative importance (Belding, 1912; Loosanoff and Davis, 1950; Loosanoff et al., 1966).

Embryogenesis in *M. mercenaria* has not been described in any detail. I present next a summary of development of early bivalve stages generalized from accounts by Belding (1912), Loosanoff and Davis (1950), Quayle (1952), Ansell (1962), Raven (1966), D'Asaro (1967), Wada (1968), Camacho and Cabello (1974), Sastry (1979), Verdonk and van den Biggelaar (1983), Cooke (1986), and Bandel (1988). Elston (1980) described in excellent detail the functional anatomy, histology, and microstructure of the soft tissues of veligers of *Crassostrea virginica*, and Waller (1981), that of *Ostrea edulis*. Loosanoff et al. (1966), Raven (1966), Bayne (1971), Cragg and Crisp (1991), and others, have observed a high degree of

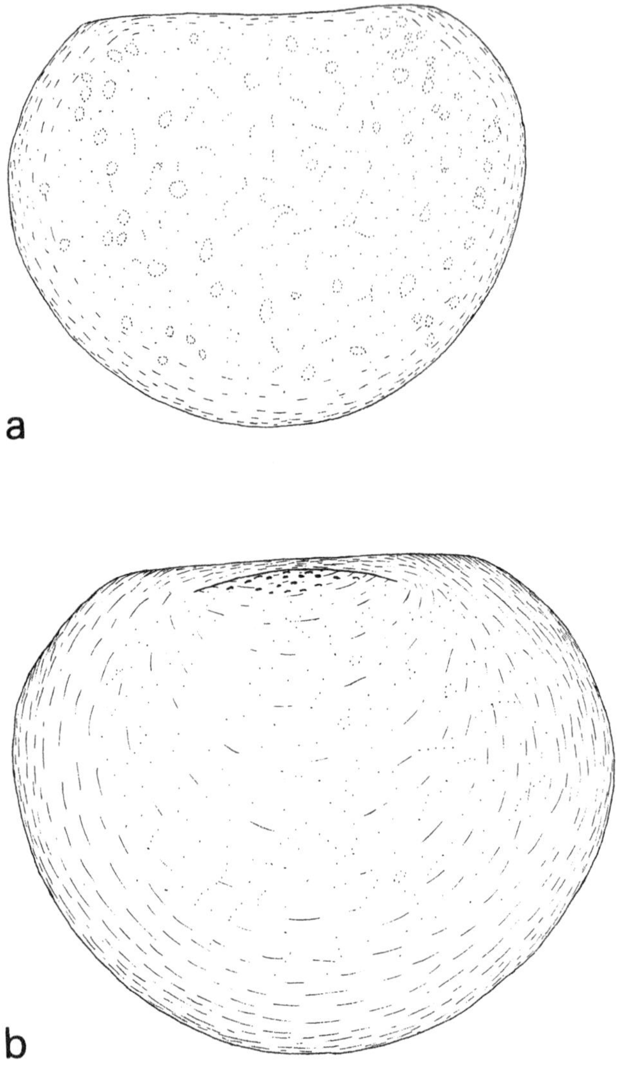

Fig. 3.2. Drawings of straight-hinged veligers of *Mercenaria mercenaria* reared in the Laboratory. (a) 100 μm long. (b) 120 μm long.

similarity in the organs and organ systems of different species of most bivalves. Thus the following summary should apply in a generic sense to the embryogenesis of *M. mercenaria*.

All bivalves exhibit a holoblastic, spiral mode of cleavage. In gastrulation, the outer micromeres (cells) divide rapidly and grow over the large inner endomeres, followed by invagination of the endomeres to form the blastopore. At first, the blastopore is wide, then deepens and narrows and is reduced to a longitudinal slit. The blastopore then closes, and the archenteron within forms as a closed sac. It is generally assumed that stomatoblasts (formative

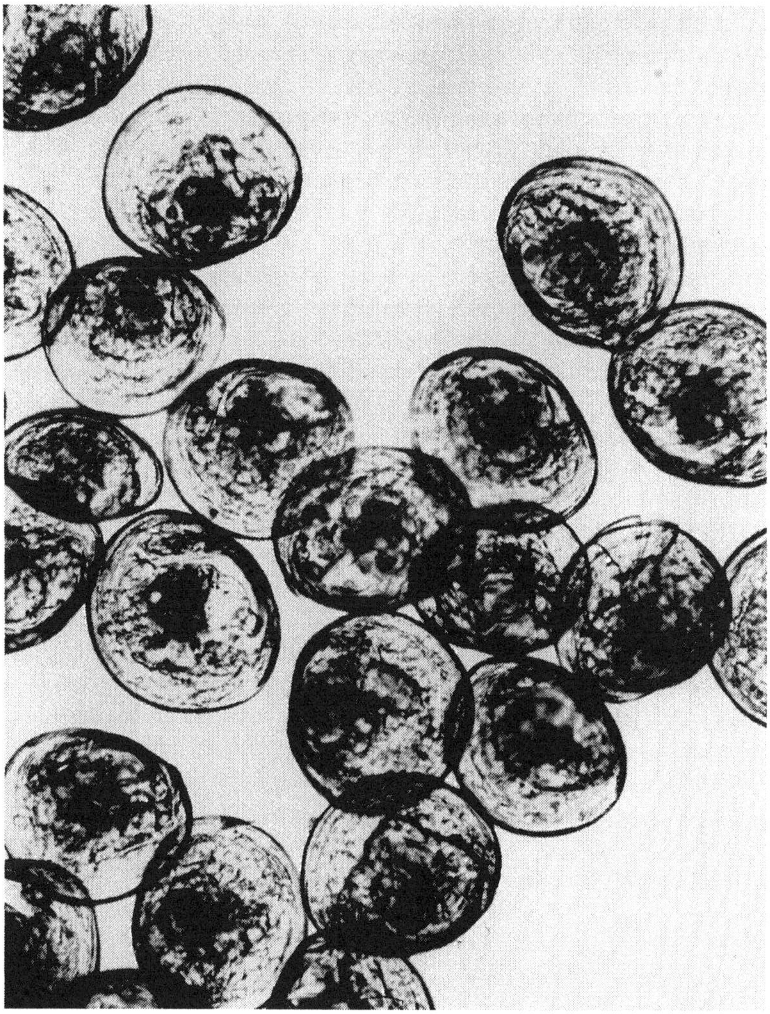

Fig. 3.3. Photomicrograph of late umboned veligers of *Mercenaria mercenaria*, living, cultured in the laboratory. Veligers approximately 180 μm long.

cells) lining the periphery of the blastopore invaginate to form the stomodaeum and later the mouth.

Following gastrulation, the blastopore, originally in the center of the vegetal hemisphere, is displaced anteriorly and ventrally. The anterior region is also shifted forward. Thus the main axis of the embryo from animal to vegetal pole becomes bent and the blastopore moves anterior to the prototroch, a band of cilia girdling the embryo. During this time, cells lining the periphery of the blastopore invaginate to form the stomodaeum.

Development in the trochophore is characterized by expansion of the apical ectoderm, differentiation of cilia, and formation of the stomodaeum and lumen of the gut. The prototroch consists of one to three rows of large cells bearing powerful cilia, and divides the trochophore body into the upper anterior (pretrochal, or head) region, and the lower posterior (posttrochal) region. Differentiation of the prototroch probably begins after gastrulation, and its change into the velum occurs during transformation of the trochophore into the veliger. This takes place by lateral outgrowth of the sides of the prototroch into the velar disc (= lobe, pad, disk). The mouth comes to lie midventrally in the velar disc. Cell rows of the prototroch telescope into each other, forming one row of ciliated cells in which the cilia increase in length to form the strong peripheral velar cilia. New rows of cilia are added here resulting in a total of three marginal ciliated bands girdling the fully developed velar disc. The velum is attached to the mantle by two pairs of velar retractor muscles. These muscles insert on the mantle immediately underlying the shell in the dorsal region and have multiple insertions in the velum. Numerous branching muscle fibers within the disc give it its marked mobility. The mouth, located in the ventral part of the velum, often described as a funnel-shaped structure, leads into the ciliated esophagus. A fully functional velum is present from the second day of the veliger stage until the structure degenerates during metamorphosis.

The stomodaeum, whose entrance becomes the mouth, originates as an invagination on the ventral side of the trochophore behind the prototroch. The narrowest part of the stomodaeum, of endodermal origin and often ciliated, differentiates into the foregut (the esophagus). Its inner end opens into the midgut, which becomes the stomach. The hindgut, eventually the intestine, arises as a posteriorly directed outgrowth from the midgut, and connects with ectodermal cells lying between the dorsal shell gland and the foot anlage, where the anus forms. The intestine, first a straight tube, exhibits considerable growth in length and becomes looped as development proceeds. The muscular coat and connective tissue of the gut arise from the mesoderm. The midgut endoderm bulges out as two lateral outgrowths, giving rise to the larval digestive gland. In some species, this disappears at the end of the embryonic stage, to be replaced rapidly by the definitive digestive gland, which forms by proliferation of cells from the wall of the stomach. The crystalline style sac differentiates early as a blind sac from the posterior part of the stomach. The digestive system becomes functional early in the veliger stage with the opening of the anus.

Each of the pair of protonephridia, of ectodermal origin, consists of two to three cells: a terminal cell with a 'flame' of cilia, a canalicular cell with an intracellular canaliculus, and possibly an aperture cell containing the nephridial pore. The terminal cell sends long protoplasmic extensions into the body cavity.

Besides isolated fibers from the stomodaeum toward the pretrochal region, the generally complicated larval musculature consists principally of retractors of the velum, larval adductors between the valves, and pedal retractors. The anterior adductor muscle develops in the early veliger stage; in the late veliger stage, spindle-shaped mesenchyme cells aggregate anterior to the anus and form the posterior adductor muscle. The larval musculature probably is derived mainly from mesenchyme (which arises from mesodermal bands) and the primary mesoderm. Paired retractors run anteroposteriorally through the body cavity of the larva and are symmetrical. At metamorphosis they become reduced. The adult shell ligament may originate by posteriorly directed growth from the larval shell ligament.

The pericardium and its accessory organs have a mesodermal origin. Cells of the anterior dorsal part of the left and right mesodermal bands form two vesicles, or coelomic cavities; these constitute the rudiments of the pericardium. These cavities increase in size on the two sides of the intestine, eventually coming in contact with each other and fusing. The ventricle is formed between this cavity and the intestine, the inner wall of the cavity becoming the ventricular wall. Invaginations in the pair of cavities form the anlagen of the two auricles and make connection with the ventricle. The rudiment of the kidney, adjacent to the pericardium, is a filiform tissue formed by mesodermal cells and contains a lumen. This soon opens into the pericardium, while the other end extends toward the body wall and eventually forms an opening to the outside. The pericardial rudiment is visible in the veliger stage, but pulsations of the heart are usually not detected until after settlement.

A ctenidium appears at the time of metamorphosis as a row of papillae on left and right sides of the pediveliger, the inner ctenidial lobes appearing earlier than the outer pair. Ctenidia do not function in food collection until after metamorphosis.

Differentiation of the reproductive system ordinarily lags far behind that of other organs. It maintains a close connection with the pericardium, heart, and kidney.

The foot originates as an ectodermal thickening on the ventral side behind the mouth. It is first indicated by an area of high columnar cells, and later by a ciliated outgrowth filled with mesenchyme.

The mantle develops during the trochophore stage as an expansion of ectoderm on the posterior surface of the larva, and forms later as a lateral fold on each side toward the anterior region. In the early veliger stage, the ventral edge of the mantle thickens forming two folds.

The velum of veligers of *M. mercenaria*, *V. pullastra*, and *V. striatula*, when unfolded and extended between the valves, takes the form of an elliptical disc with a border of long prominent cilia (Fig. 3.5), and occupies at least two-fifths of the volume of the shell cavity (Fig. 3.6). A long sensory apical flagellar tuft projects from the center of the cone-shaped apical plate, or sensory organ, and is present from the trochopore stage until metamorphosis. On the right and left of the apical plate lie the cephalic plates from which the cerebral ganglia develop. When the veliger is swimming, contact of the apical flagellae with an object stimulates retraction of the velum. Reextension between partly opened valves is preceded by reappearance of the apical flagellae. The tuft is characteristic of veligers of the molluscan family Veneridae, and is present, as well, in some other molluscan taxa. The velar disc has a thick, marginal cuboidal epithelium that folds and creases as the velum is retracted (Quayle, 1952; Ansell, 1962; D'Asaro, 1967; Sastry, 1979).

The veliger velum produces both locomotory and feeding currents. The velar margin is bordered by an outermost preoral band consisting of long compound cilia, or cirri, that produce both swimming and feeding currents; an innermost postoral band, with shorter finer cilia than the preoral ones, that beat toward the preoral cilia; and the food groove with small, fine cilia between the preoral and postoral bands, which transport particles toward the mouth. The apposed pre- and postoral ciliary bands capture and retain particles in the food groove (Strathmann and Leise, 1979; Strathmann, 1987; Gallagher, 1988; see also Maia, 1988; and Cragg, 1989).

Suspension feeding by veligers of *M. mercenaria* appears to include capture of planktonic cells by pre- and postoral cirri, their transport in mucus to the mouth in the food groove, concentration of cells into a bolus at the mouth, selection or rejection of cells for entry into the ciliated esophagus, and activation of a ciliated sphincter at the esophagastric junction that lets cells pass into the stomach or rejects them. Each ribbon-like cirrus, which consists of about 10 single cilia fused at the base, tapers toward the tip. Cirral spacing within rows normal to seawater flow is about 0.5 μm. Velar circumference, cirral length, beat frequency, and cirral tip velocity all increase by almost a factor of two during development of veligers. Thus in a veliger, about 100 μm long the circumference of the velar disc at the top of the recovery stroke is about 260 μm and length of frontal cilia is 32 μm; whereas in a veliger 235 μm long, these dimensions are 575 μm and 45 μm, respectively. Consequently, flow of water around velar cilia and clearance rate increase by a factor of two between 2 and 20 days of veliger development (Gallagher, 1988).

Physiological aspects of feeding by veligers of *M. mercenaria* are treated by Gallagher (1988) and Riisgard (1988), and general reviews are provided by Boidron–Métairon (1995) and Hart and Strathmann (1995). Galtsoff (1964) described a useful method for the examination of the velum of living larvae of oysters narcotized with menthol, or other narcotics, and made transparent with glycerol.

Metamorphosis, a marked morphological and physiological reorganization and complex change in life habit, begins a few hours to several days after the settling pediveliger stage is reached. It is during settlement that larval sense organs reach their peak of development (Crisp, 1974), and may be the most critical point in the life history of marine molluscs exhibiting the metamorphosal phenomenon (Bonar, 1976). Metamorphosis involves degeneration and loss of some of the organs specialized for larval life, rearrangement and renewal of others, and increased rate of development of potential adult organs (Quayle, 1952; Ansell, 1962; Raven, 1966; D'Asaro, 1967; Bonar, 1976; Cooke, 1986: Rowe and Ludwig, 1991). Major anatomical changes include: loss of velum, formation of labial palps and assumption of feeding by gills; reduction of larval musculature (less the adductor muscles), larval heart, and larval kidneys; and beginning of development of the definitive adult musculature, nervous system, intestinal tract, and heart and kidney.

After the larval velar feeding mechanism is lost, and before gill filaments and labial palps begin to function, there is a period of 1–3 days when the newly metamorphosed juvenile cannot feed and relies on stored nutrients for energy (Quayle, 1952; Bayne, 1976). The extent to which metamorphosis in *M. mercenaria* pediveligers can be delayed, has not been quantified (Crisp, 1984; Pechenik, 1990).

Of the major changes that take place during metamorphosis of most bivalve pediveligers, the most conspicuous is the shedding of the velum. The remaining apical plate sinks in close to the stomodaeum and connects secondarily with the epidermis above the mouth. The plate grows outward bilaterally and contributes to the formation of the labial palps. Loss of the velum is accompanied by rapid disintegration of the velar retractor muscles, which are replaced by pallial muscles.

In *M. mercenaria*, metamorphic loss of the velum involves the shedding of individual cells, which continue to swim in the seawater for some time (H.J. Turner, personal communication). Loss of the velum was also observed in *V. pullastra* and *V. striatula* by Quayle (1952) and Ansell (1962). In these two species, the greater part of the velum appears to disintegrate and is rapidly sloughed off, leaving behind only the velar apical region attached dorsal to the mouth where it forms the origin of the upper palp. This develops rapidly and forms a hood above and around the mouth, and serves to trap food particles from the inhalant stream of seawater passing over the ctenidia. At this stage the food-collecting mechanism of the ctenidia is lacking and the marginal groove is not yet formed.

3.4 ORGANOGENESIS

Information on the development of the organs and organ systems of *M. mercenaria* during veliger, metamorphic and early byssal plantigrade stages is fragmentary and scattered at best (Belding, 1912, 1931; Loosanoff, 1937; Loosanoff and Davis, 1950; Turner and George, 1955; Carriker, 1961). That on related venerid species is more comprehensive (Quayle, 1952 on *V. pullastra*; Ansell, 1962 on *V. striatula*). Ansell's (1962) paper and figures are thorough and especially helpful. The synoptic works of Raven (1966), Wada (1968), Sastry (1979), Moor (1982), and Fioroni (1982) serve as important sources of background information on bivalve organogenesis. The following generalized account of organogenesis of the early stages of *M. mercenaria* was synthesized from these various sources, and builds on the previous section on embryogenesis. The account has merit in view of the structural uniformity that exists among early stages of widely differing bivalve groups (Belding, 1912; Ansell, 1962; Bayne, 1971; Carriker, 1990), and reveals the paucity of knowledge on the morphogenesis of *M. mercenaria* and the many interesting facets yet in need of investigation.

3.4.1 Shell

The earliest valves of living straight-hinged veligers (prodissoconch I) are characterized by a stippled gray appearance with a slight tinge of translucent grayish-salmon color that fades to clear at the edges of the valves. Stippling is uneven and most pronounced in the center of each valve. Within the soft tissues, as seen through the translucent valves, globules are prominent, range in color from clear to dark gray, occur in a loosely scattered pattern, and vary in size from 1 to 4 μm (Fig. 3.4a,b).

Valves of straight-hinged veligers are slightly asymmetric, one end being slighted pointed (the anterior end) and the other rounded (posterior end). The asymmetry becomes more pronounced as individuals grow in size (Fig. 3.3). Because the hinge of straight-hinged veligers is relatively long in proportion to the length and height of the valves, and because of the D-shape of the valves in lateral view, this stage is often called the 'D-shaped' larva.

Valves of prodissoconch I veligers collected in the plankton in the field do not differ significantly in appearance from those reared in the laboratory. This probably comes as no surprise, and was verified in early laboratory culture of veligers (Carriker, 1961) to obviate the possibility that artificial culture might modify the form of the valves.

In prodissoconch II veligers about 120 μm in length, a smooth, pronounced arc appears just under the hinge line where umbones are forming (Fig. 3.2b). The light yellow color persists in the outer margins of the valves. This color is present in empty valves, and therefore is not pigment in underlying tissues. Tips of the umbones of a veliger about 130 μm long approach close to the hinge line. At a length of about 140 μm, the early umbonal veliger is characterized by umbones that rise slightly but uniformly above the hinge line (Fig. 3.3). The band of yellow shell around the valve margins persists.

Between a shell length of 140 and 200 μm, umbones of plantigrade II rise gradually but inconspicuously above the hinge line, giving the veliger a rounded profile (Fig. 3.3). Umbones in *M. mercenaria* never assume the prominence represented by those of late veliger stages of species in the genera *Crassostrea*, *Teredo*, and *Anomia* (Carriker, 1956). As prodissoconch II of *M. mercenaria* develops further, its straight hinge becomes eclipsed, the anterior end

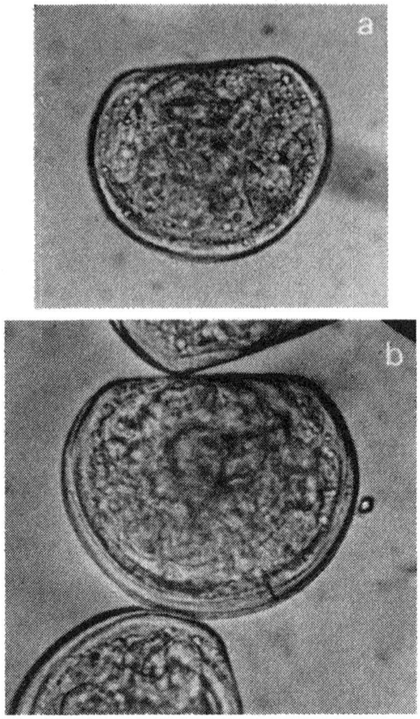

Fig. 3.4. Photomicrographs of early veligers of *Mercenaria mercenaria*, living, cultured in the laboratory. (a) straight-hinged veliger, 90 μm long. (b) straight-hinged veliger, 115 μm long.

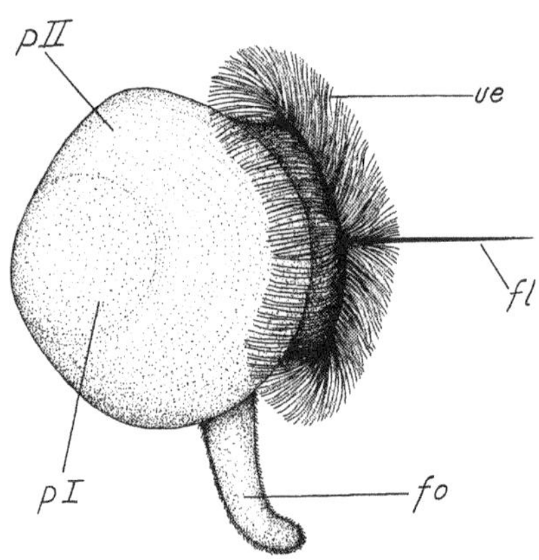

Fig. 3.5. Drawing of pediveliger of *Mercenaria mercenaria*, 200 μm in length. fl = apical flagellar tuft, fo = ciliated foot; ve = ciliated velum; pI = prodissoconch I; pII = prodissoconch II.

becomes slightly pointed, the posterior end retains its roundness, and shoulders off the umbones slope gently. Light yellow shell margins are still present in living veligers, even after they have fed on a diet of various microorganisms possessing different pigments.

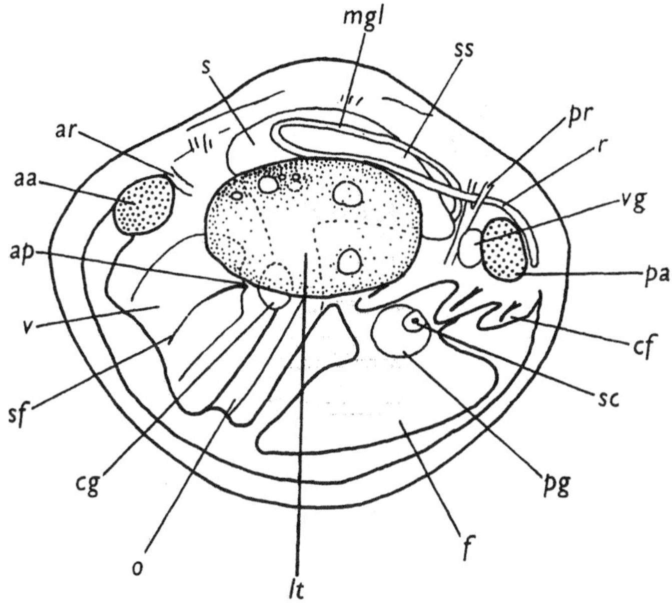

Fig. 3.6. Internal organs of the pediveliger of *Venus striatula*, about 190 μm long. aa = anterior adductor muscle; ap = apical plate; ar = anterior pedal retractor muscle; cf = ctenidial filament; cg = cerebral ganglion; f = foot; lt = left digestive tubule; mgl = mid-gut loop; o = esophagus; pa = posterior adductor muscle; pg = pedal ganglion; pr = posterior pedal retractor muscle; r = rectum; s = stomach (globular region); sc = statocyst; sf = anterior sensory flagellum; ss = style sac; v = velum; vg = visceral ganglion. Drawing from Ansell (1962).

Shell deposited at the rim of the valves of prodissoconch II of *M. mercenaria* is sculptured into minute commarginal growth striae. These are so fine that the shell surface appears smooth under low magnification.

With the genesis of the dissoconch valves, however, conspicuous primary commarginal ridges begin to appear on the external surface of the valves. These ridges, first reported by Belding (1912), consist of thin, more or less periodic elevations of the shell surface and run entirely around the valves from the hinge line (Fig. 3.7a). The elevations, coincident with the early life of newly settled byssal plantigrades, are small, set close together, and consist of soft, chalky-white shell material. On the average in faster growing plantigrades, the first ridge appears at a shell length of about 250 μm. In all, some 4–12 of these form on different individuals. What determines the total number of these ridges is uncertain. Because of their relatively soft composition, they soon erode, increasing the difficulty of counting them on plantigrades under 10 mm in length.

Deposition of conspicuous secondary shell ridges begins at a plantigrade shell length of about 1 mm, and coincides with early stages of development of the definitive exhalant siphon. Secondary ridges are set more widely apart, are many times higher, thicker, and harder than primary ridges (Fig. 3.7b), are translucent-white in color, and are secreted more or less periodically for the remainder of the life of individuals. The number of ridges on different byssal plantigrades of the same size is exceedingly variable. In one analysis, for example,

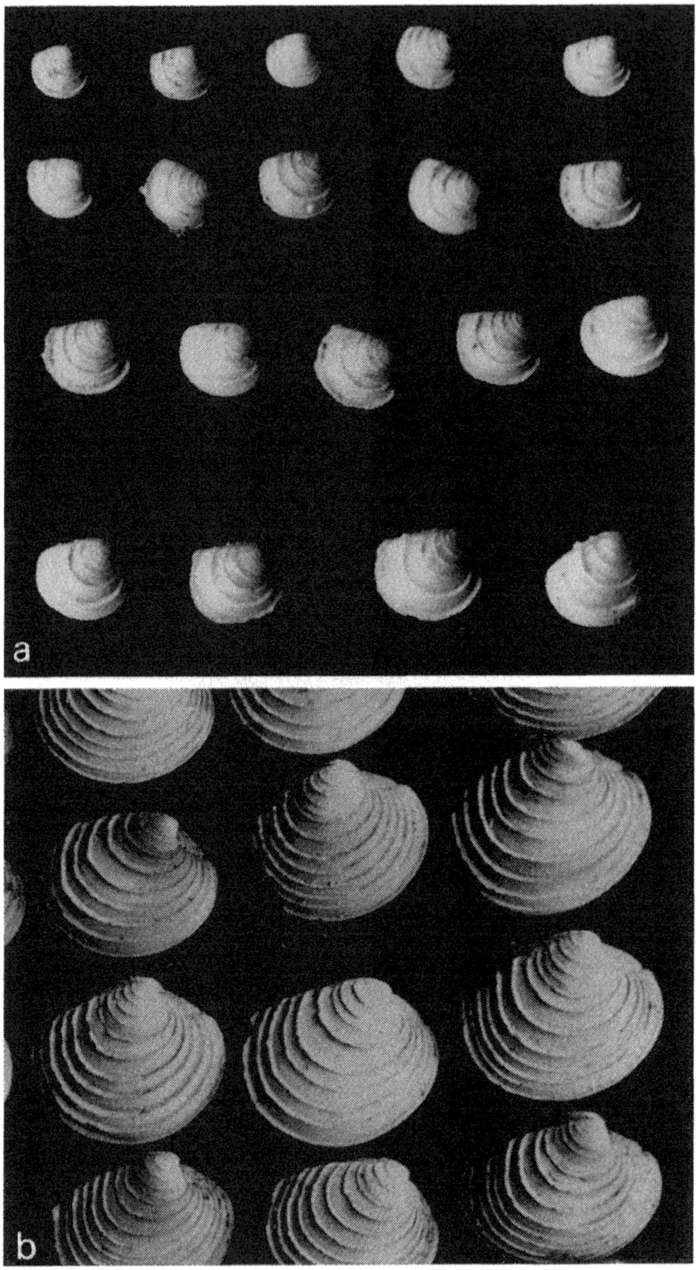

Fig. 3.7. Photomicrographs of shells of young *Mercenaria mercenaria* illustrating shell ridges. (a) byssal plantigrades, 1.7–3.2 mm long, showing primary ridges and first two secondary ridges (by T.C. Nelson and H.H. Haskin, personal communication). (b) juvenile plantigrades, 11–13 mm long, showing prominent secondary ridges (by T.C. Nelson, personal communication).

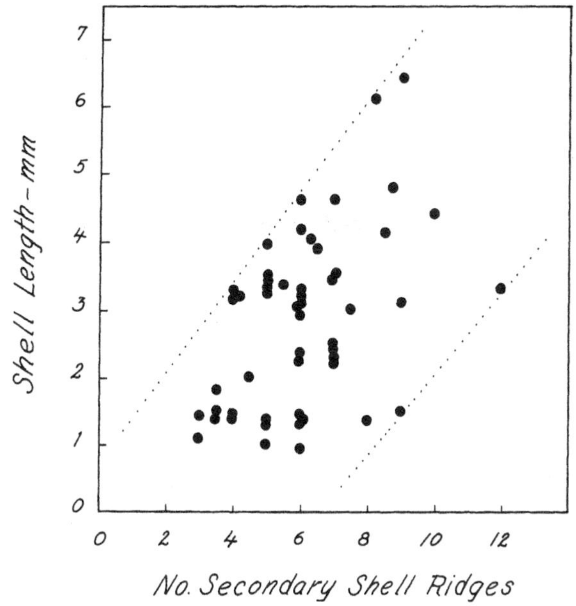

Fig. 3.8. Relationship of number of primary shell ridges and shell length in byssal plantigrades of *Mercenaria mercenaria*.

the number of ridges in individuals 3.0–3.5 mm long varied from 4 to 12 (Fig. 3.8). The range of number of ridges borne by laboratory-reared and native populations does not vary significantly (Belding, 1912; Carriker, 1961).

3.4.2 Mantle and Mantle Cavity

The mantle, a thin, sheet-like organ, formed of spindle-shaped cells, appears first in the trochophore stage as an expansion of the ectoderm in the posterior part of the larva; subsequently the mantle develops as a lateral lobe on left and right sides, extending to the anterior end of the larva (Wada, 1968). In *Venus striatula*, the mantle consists of a single layer of cells, except at the margins where it thickens to form two folds — as in the larva of *Ostrea edulis* (Cranfield, 1974, Fig. 3.9) — where the margin consists of three cell layers (Ansell, 1962). There is little differentiation of the mantle cells, which are sparsely ciliated except for the tract of long cilia near the ventral margin of the mantle lobes. This tract can be traced through development and forms the origin of the mantle rejection tract in the adult. The right and left mantle lobes are separated except for a short distance near the origin of the gills and dorsally between the adductor muscles where fusion with the visceral mass occurs.

In veligers, *Ostrea edulis* for example, mantle edges do not become retractable from the margin of the valves in response to irritation until growth of prodissoconch II is well along.

Formation of the mantle cavity begins with growth of prodissoconch II (Waller, 1981) as an invagination of the ventral body wall that penetrates beneath the valves in the area surrounding the foot rudiment (Raven, 1966). Eventual closure of the cavity occurs by outgrowth of the

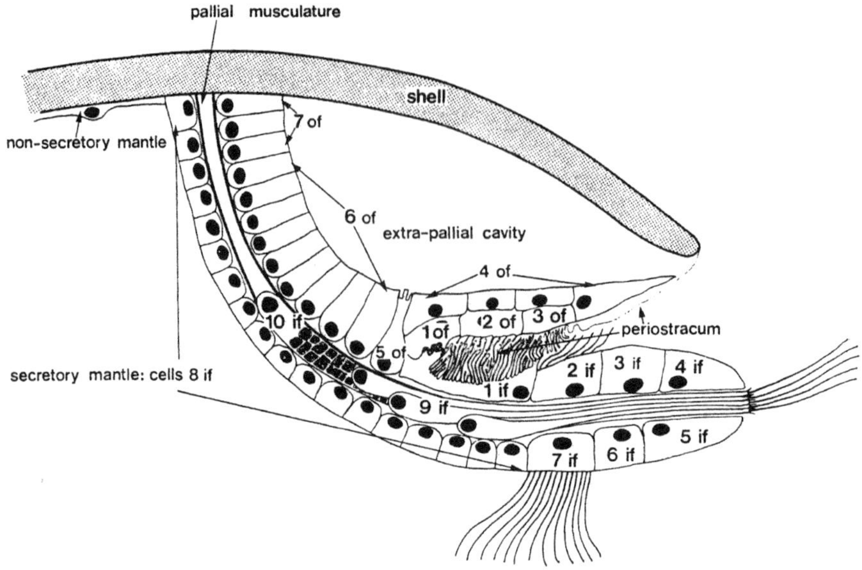

Fig. 3.9. Diagram of cells of folds of mantle of the pediveliger of *Ostrea edulis*. Outer fold of: 1, 2, 3, cells of periostracal groove; 4, 6, 7, columnar epithelial cells; 5, sensory cells. Inner fold of: 1, 2, cells of periostracal groove; 3–6, cells of outer part of inner fold; 7, 8, 9, sensory cells; 10, gland cells. From Cranfield (1974).

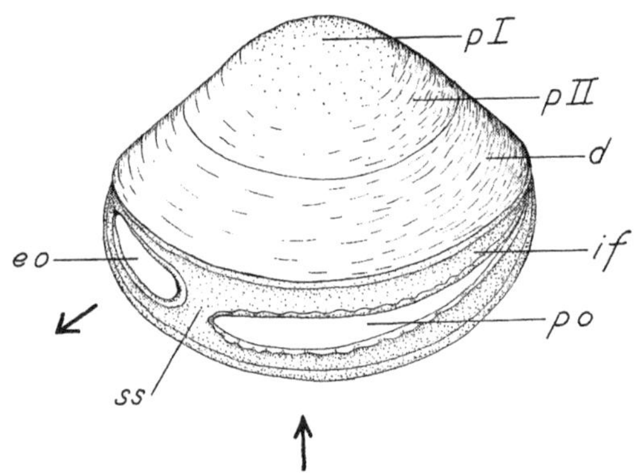

Fig. 3.10. Byssal plantigrade of *Mercenaria mercenaria*, 290 μm long. Fusion of mantle margins delimiting opening for exhalant siphon has begun; deposition of dissoconch shell has started, and two primary shell ridges are present. d = dissoconch; eo = exhalant opening, if = inner fold of mantle; po = pedal opening; pI = prodissoconch I; pII = prodissoconch II; ss = siphonal septum; arrows = direction of flow of seawater through the plantigrade.

mantle lobes (Fig. 3.10) (Moor, 1983). The tract of long cilia on the inner ventral side of the inner fold of the two mantle lobes may interlock during compression from the inner mantle folds to control flow of seawater into and out of the mantle cavity, and prevent entry of large suspended particles (Quayle, 1952).

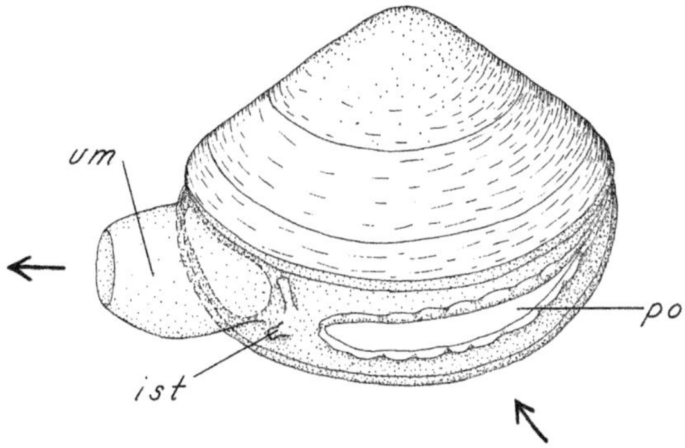

Fig. 3.11. Byssal plantigrade of *Mercenaria mercenaria*, 330 μm long. Primary exhalant siphon (vm) is forming out of exhalant opening, and two inhalant siphonal tentacles (ist), and two primary shell ridges are present; arrows, direction of flow of seawater.

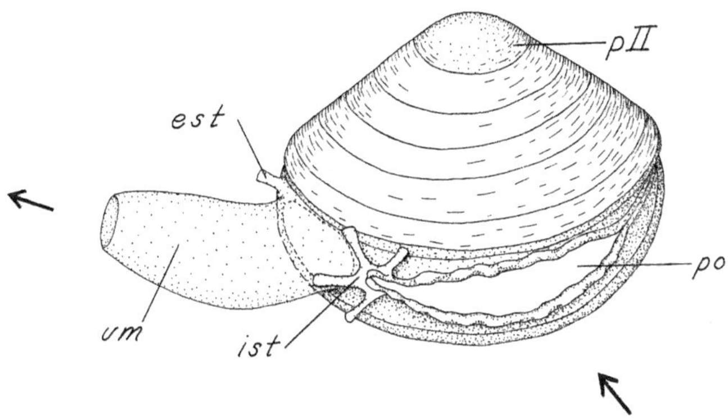

Fig. 3.12. Byssal plantigrade of *Mercenaria mercenaria*, 625 μm long. Primary exhalant siphon (vm) is further developed; four inhalant siphonal tentacles (ist), one exhalant siphonal tentacle (est) and 5 primary shell ridges are present; po = pedal opening; pII = prodissoconch II; arrows = direction of flow of seawater.

3.4.3 Mantle Fusion and Siphons

Minute suspended particles floating out of the mantle cavity through the primary exhalant siphon (Figs. 3.11–3.13), which now accommodates the excurrent flow of seawater from the mantle cavity, are clearly visible through the walls of the siphon. As the plantigrade opens its valves to pump seawater, the primary exhalant siphon billows outward rapidly, forced by pressure created within the mantle cavity and probably through ciliary activity, the distal tip unfurling from within (Figs. 3.11–3.14). When (a) hydrostatic pressure is momentarily cut off by constriction of the base of the primary exhalant siphon (see also Sellmer, 1959 and Ansell, 1962) and (b) mantle margins appose each other causing enlargement of the mantle cavity, the siphon collapses into a shrunken, wrinkled tube that dangles loosely externally. More frequently, probably upon further reduction of hydrostatic pressure, the siphon is inverted swiftly within the mantle cavity, distal tip first. Belding (1912) briefly described and illustrated the primary exhalant siphon in young *M. mercenaria*, and was probably the first

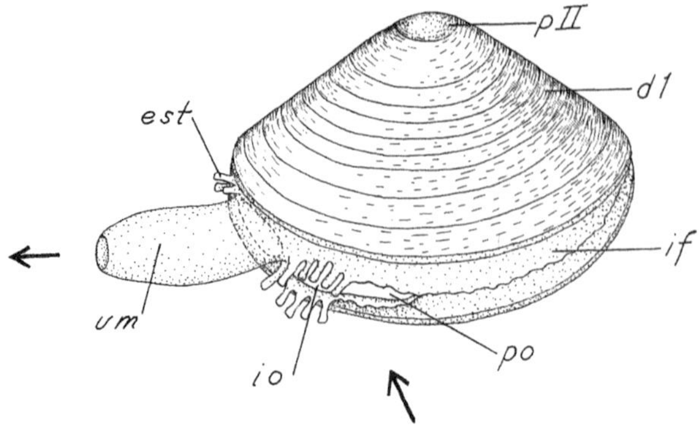

Fig. 3.13. Byssal plantigrade of *Mercenaria mercenaria*, 1.36 mm long. Primary exhalant siphon (vm) is further enlarged; 10 inhalant siphonal tentacles (io), three exhalant siphonal tentacles (est), and 10 primary shell ridges are present; d1, dissoconch with primary shell ridges; if = inner fold of mantle; io = inhalant opening; po = pedal opening; arrows = direction of flow of seawater.

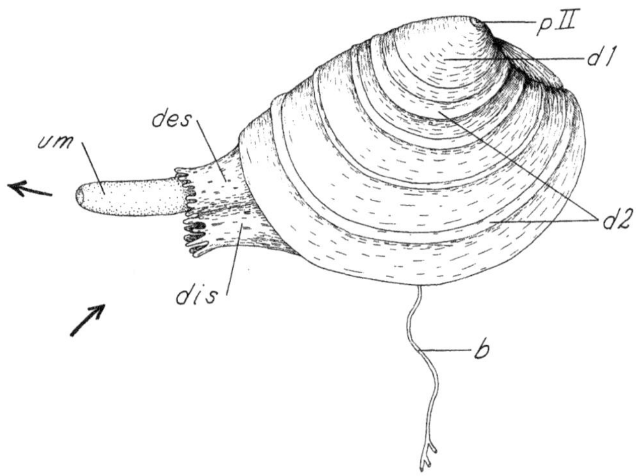

Fig. 3.14. Byssal plantigrade of *Mercenaria mercenaria* 4 mm long. Primary exhalant siphon is now borne on distal end of definitive exhalant siphon (vm, des), and the inhalant siphon (dis) has formed alongside and fused to the exhalant siphon. Numerous tentacles are present on the distal margins of both siphons. The byssus (b) extends ventrally between the valves, and four secondary shell ridges are present (d2). Prodissoconch (pII) and dissoconch (d1) valves with primary shell ridges are shown as smooth shell surface. Arrows, direction of flow of seawater.

to report it in this species. He noted that average time of expansion of the 'filmy telescopic attachment' is 4 s and time of contraction is 2–11 s.

During growth of byssal plantigrades to a shell length of about 1.5 mm, the primary exhalant siphon continues to enlarge (Table 3.2), though the ratio of shell length to length of the siphon in different individuals of the same size is quite variable. Thereafter as the bivalve grows to a length of approximately 7–20 mm, dimensions of the siphon decrease in proportion to the size of the bivalve. Eventually the siphon is reduced to a shallow flange at the tip of

TABLE 3.2

Ratio of length of shell to length of primary exhalant siphon in 10 young plantigrade *Mercenaria mercenaria* selected at random

Shell length (μm)	Length of primary exhalant siphon (μm)	Ratio: shell length to length of primary exhalant siphon
230	50	4.6
330	100	3.3
415	200	2.1
625	380	1.6
1,000	780	1.3
1,360	615	2.2
1,500	1,500	1.0
4,000	1,420	2.8
14,000	1,100	12.7

the definitive exhalant siphon just inside the circle of siphonal tentacles (Figs. 3.12–3.16). Especially during its more prominent stages, the primary exhalant siphon is characterized by extreme flexibility and by a relatively small distal orifice that reduces the diameter of the exhalant siphon to a fraction of that of the inhalant siphon (Figs. 3.13 and 3.14).

Shortly after genesis of the primary exhalant siphon, two small truncated tentacles, forerunners of the ring of sensory tentacles that will adorn the distal rim of the inhalant siphon, appear anterior and ventral to the base of the primary exhalant siphon (Fig. 3.11). Then two more tentacles form anterior to the first pair. Simultaneously the anterior part of the fused mantle splits at the site of the future definitive inhalant siphonal opening (Fig. 3.12); this permits conjoining of inhalant and exhalant siphonal walls as these siphons develop. At this stage, a single tentacle appears dorsal to the base of the primary exhalant siphon, the first of a ring of tentacles that will bound the distal opening of the definitive exhalant siphon. Tentacles continue to form anterior and ventral to the site of the first ones as definitive inhalant and exhalant siphons form (Fig. 3.13). By the stage at which two definitive siphons are visible beyond the margin of the valves, inhalant and exhalant tentacles form complete rings on the distal orifices of the siphons (Fig. 3.14), and the primary exhalant siphon, now projecting from

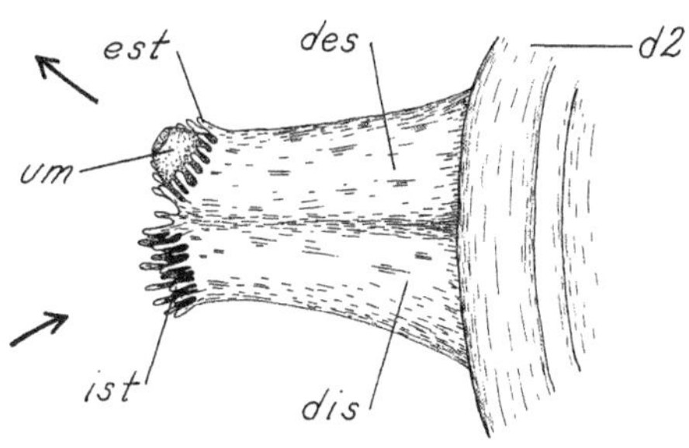

Fig. 3.15. Siphons of juvenile plantigrade 14 mm long, *Mercenaria mercenaria*. Siphons (des, dis) are fully formed, and the primary exhalant siphon (vm) has been reduced to a short funnel-shaped opening; est, ist, tentacles; d2 = dissoconch with secondary ridges; arrows = direction of flow of seawater.

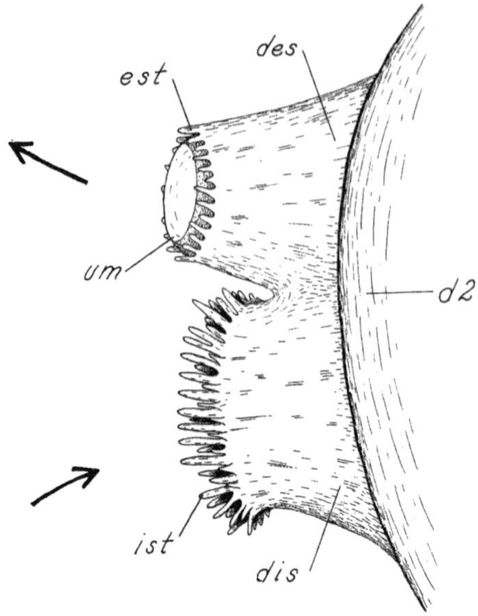

Fig. 3.16. Definitive siphons (des, dis) of adult *Mercenaria mercenaria*, shell 55 mm long. Primary exhalant siphon (vm) now reduced to a narrow shelf just inside the tentacular border (est) of the exhalant siphon (des); arrows = direction of flow of seawater.

inside the base of the border of tentacles, is borne on the distal end of the lengthening exhalant siphon. Belding (1912) reported that in a juvenile *M. mercenaria* 1 mm in length, there are 12 tentacles on the inhalant and 4 on the exhalant siphon (close to the stage shown in Fig. 3.13), and that tentacles are sometimes strongly pigmented.

Before formation of the inhalant siphon, seawater is drawn into the mantle cavity through the inhalant-pedal opening, and the pediveliger remains at the sediment–water interface with mantle margins in contact with ambient seawater. As siphons lengthen, the bivalve begins to burrow shallowly in the sediment, siphons extended upward toward the seawater. The plantigrade burrows deeper as definitive siphons lengthen, and shortly the entire shell is pulled within the sediment. Direct contact with overlying seawater is maintained by siphons extending to, or shortly above the seawater–sediment interface; or aqueous contact may be indirect through interstitial seawater when siphons are opened beneath the sediment.

Yonge (1948, 1957) noted that formation of siphons was a matter of major significance in the evolution of bivalves, enabling them to live sheltered within sedimentary substrata. This complex anatomical development has also permitted *M. mercenaria* to survive extraordinarily successfully in shallow coastal areas over long geological periods despite its numerous predators.

In *M. mercenaria*, siphons are of type B, according to Yonge's (1957) classification, in which combined siphons are formed as extensions of inner and middle folds of mantle margins. Like the primary exhalant siphon, siphonal tentacles are also derived from the inner fold of the mantle margins. Mantle fusion and siphonal formation in *M. mercenaria* resemble rather closely that described briefly by Quayle (1952) for *V. pullastra*, and in more detail for

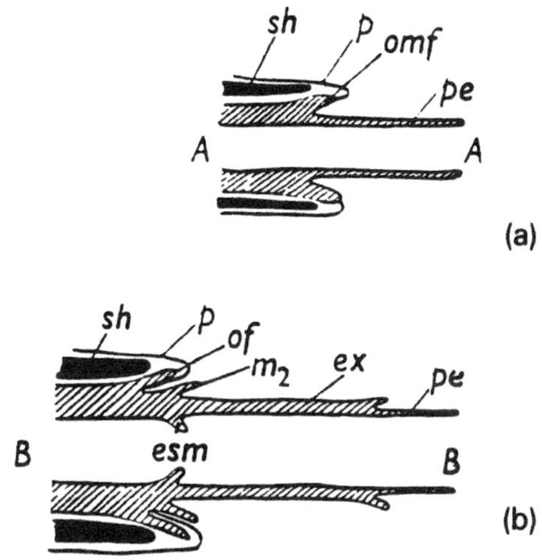

Fig. 3.17. Median longitudinal section of the exhalant siphon of *Venus striatula*; (a) young post larva; (b) adult. esm = exhalant siphonal membrane; ex = exhalant siphon; m_2 = middle mantle fold; of = outer mantle fold; omf = outer and middle mantle folds; p = periostracum; pe = primary exhalant siphon; sh = shell. From Ansell (1962).

V. striatula by Ansell (1962). In *V. pullastra*, and probably also in *M. mercenaria*, as siphons develop, pallial muscles at their base also enlarge. Muscles originate from the muscular lobe of the mantle near the siphonal septum, beginning as short muscle fibers that extend posteriorly into the mantle. As the bivalve grows, they extend and radiate outward until they cover an arc of nearly 180° (Quayle, 1952).

In *V. striatula*, inner folds of the mantle edge, which fuse to separate inhalant aperture and pedal gape, still consist of only two folds (Ansell, 1962). Later the outer fold gives rise to the outer and middle marginal folds of the adult (Fig. 3.17a,b). The outer surface of the outer fold secretes the outer calcareous layer of the adult shell, and the inner surface, the periostracum. The middle fold of the mantle edge of the adult later becomes duplicated, and also takes part in the further development of the siphons. A series of tentacles develops from the middle fold around both siphonal apertures, the middle folds fuse, and the definitive siphons form by a local growth of the fused inner and middle folds posteriorly. Mantle margins in *M. mercenaria* also first possess three, and later four marginal folds (Hillman, 1964).

Arrows in Figs. 3.10–3.16 illustrate the direction of flow of seawater into and out of the mantle cavity of *M. mercenaria*. In asiphonate byssal plantigrades restricted to the surface of sediments, ventral margins of mantle lobes separate slightly to permit ingress and egress of seawater, the region of inflow occupying more than half of the length of the gape. Initial fusion of mantle edges at the siphonal septum restricts egress of water to the posterior exhalant opening. As the inhalant siphon forms, it becomes the principal course of entry of incurrent seawater into the mantle cavity, and the pedal gape is used only infrequently.

In siphonate *M. mercenaria*, siphon tips (including the primary exhalant siphon) point away from each other (see especially Figs. 3.14–3.16), thus deflecting incurrent from excurrent seawater. As demonstrated by individuals 10–15 mm in length in seawater containing a suspension of carmine, the excurrent stream, which flows from the relatively small exhalant orifice, is characterized by a velocity and length of flow-path several times that of the incurrent flow through the larger inhalant opening. As a consequence of this differential flow, the seawater mass in the immediate vicinity of the siphon tips is moved away from the posterior end of the bivalve, forced by the relatively stronger flow of the excurrent stream. Thus mixing of incoming and effluent seawater is minimized. Consequently, in flowing seawater a siphonate bivalve probably draws in little, if any, of its own excurrent seawater. In relatively still seawater, however, probably more seawater is recycled.

Quayle (1952) assigned to the primary exhalant siphon in young sedentary *V. pullastra* the function of freeing the mantle cavity and gills of feces. In newly metamorphosed asiphonate *M. mercenaria*, however, no fouling of epithelia of the mantle chamber appears to occur, ciliary action and rapid closing of the valves likely serving to keep the cavity surfaces clean.

The principal function of the unique primary exhalant siphonal system in bivalves is thus to serve as a transitory exhalant conduit, deflecting incurrent and excurrent streams, and reducing the rate of recirculation of mantle seawater as the adult siphonal apparatus unfolds.

3.4.4 Alimentary Canal

In veligers of *V. striatula* (Ansell, 1962), food particles collected by velar cilia are carried into the mouth located immediately posterior to the velum (Fig. 3.18). The esophagus runs dorsally into the visceral mass, opening into the anterior ventral region of the stomach. This is a globular organ situated between the umbones in the dorsal region of the visceral mass. Posteriorly, the stomach is elongated to form the style sac (Fig. 3.19), which lies anterior to the posterior adductor muscle. From the posterior end of the right side of the stomach the midgut passes back around the end of the style sac, then forward to form an anterior loop to the left of the stomach. From there it runs over the posterior adductor muscle to open at

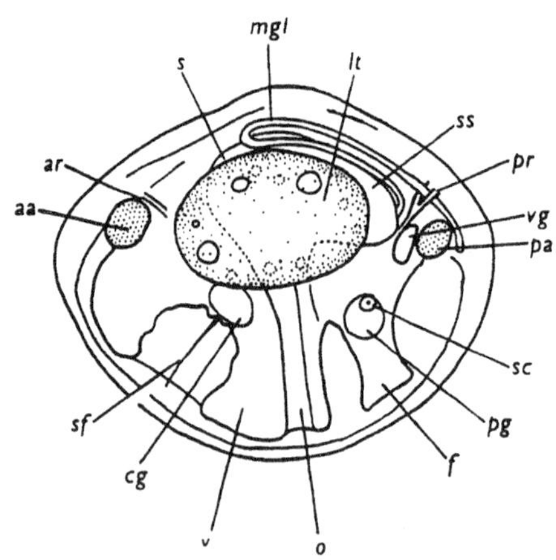

Fig. 3.18. Diagrammatic drawing of internal organs in the pediveliger of *Venus striatula*. aa = anterior adductor muscle; ar = anterior pedal retractor muscle; cg = cerebral ganglion; f = foot; lt = left digestive tubule; mgl = midgut loop; o = esophagus; pa = posterior adductor muscle; pg = pedal ganglion; pr = posterior pedal retractor; s = stomach; sc = statocyst, sf = anterior sensory flagellum; ss = style sac; v = velum; vg = visceral ganglion. From Ansell (1962).

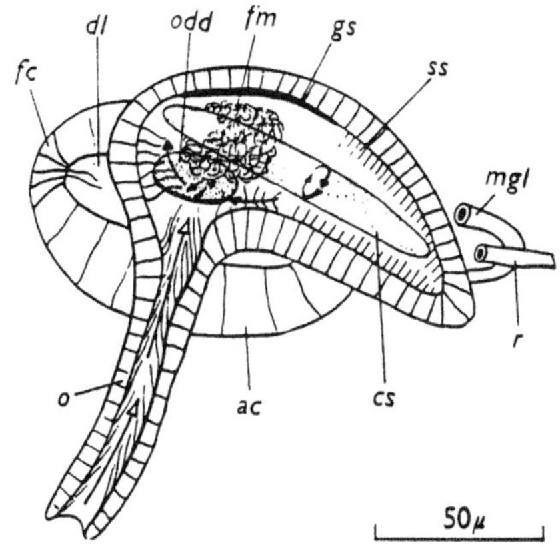

Fig. 3.19. Diagrammatic drawing of the internal structures of the stomach of the pediveliger of *Venus striatula*, illustrating the movement of food particles. ac = absorptive cells of digestive gland tubules; cs = crystalline style; dl = lumen of digestive gland; fc = flagellated cells of digested tubules; fm = food mass in stomach; gs = gastric shield; mgl = midgut loop; o = esophagus; odd = opening of digestive gland into stomach; r = rectum; ss = style sac. From Ansell (1962).

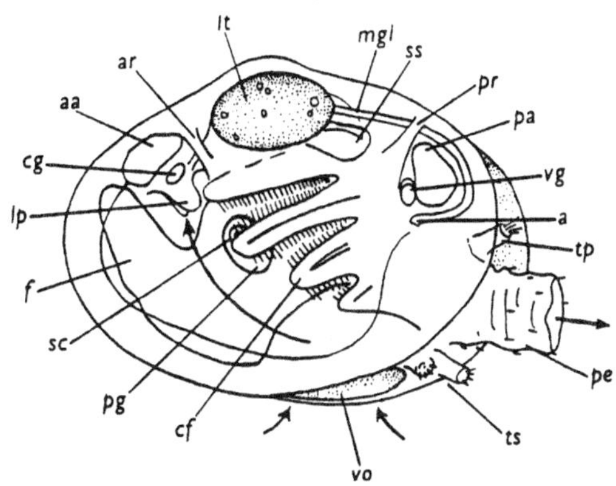

Fig. 3.20. Diagrammatic drawing of internal organs in a young post-larva of *Venus striatula*. Arrows indicate flow of seawater (ventrally and out siphon) and movement of particles (across foot). a = anus; aa = anterior adductor muscle; ar = anterior pedal retractor muscle; cf = ctenidial filament; cg = cerebral ganglion, f = foot; lp, lip, mouth; lt = left digestive tubule; mgl = midgut loop; pa = posterior adductor muscle; pe = primary exhalant siphon; pg = pedal ganglion; pr = posterior pedal retractor muscle; sc = statocyst; ss = style sac; tp = primary exhalant siphonal tentacle; ts = primary inhalant siphonal tentacles; vg = visceral ganglion; vo = ventral inhalant opening. From Ansell (1962).

the anus (Fig. 3.20). Tubules of the digestive gland lie to each side of the stomach, and open separately through the anterior wall on either side of the opening of the esophagus. Tubule lumina open directly into the lumen of the stomach, without primary and secondary ducts such as occur in the adult. The anatomy of the alimentary canal of veligers of *V. pullastra* (Quayle, 1952) is similar to that of *V. striatula*.

In veligers of *V. striatula*, the wall of the alimentary canal is one cell thick throughout.

The esophagus is a slender tube lined with ciliated cells, cilia near the opening into the stomach being especially long. The epithelium of the globular region of the stomach consists of columnar ciliated cells; in those over the roof and left side of the stomach, the free margin is modified to form the smooth gastric shield. The wall of the style sac consists of smooth epithelium bearing dense cilia. The crystalline style, lying within the style sac, projects forward into the globular region of the stomach. The midgut and rectum are slender tubes lined with smooth epithelium bearing scattered cilia and occasional mucous cells. The structure of the digestive glands of veliger larvae is similar to that of a single tubule of the adult glands, and is formed of flagellated and absorptive cells (Ansell, 1962). In pediveligers of *V. striatula*, phytoplanktonic cells passing from the esophagus into the stomach are trapped in mucus-entangled material around the head of the crystalline style. This material is rotated by action of cilia on the floor and anterior walls of the stomach and by clockwise movement of the crystalline style. The digestive glands on either side of the stomach contract alternately, drawing material into the lumen as they expand and forcing material out to rejoin the mass in the stomach as they contract. These movements of food particles can be seen in swimming veligers trapped in the surface film of seawater, or in pediveligers crawling over the substratum. Movements cease when the velum is retracted (Ansell, 1962).

With the loss of the velum in *V. pullastra*, the mouth and esophagus move dorsally and anteriorly, taking with them the cerebral ganglion and apical plate. The mouth comes to lie immediately in front of the anterior muscle, and the apical plate comes to rest directly above the mouth, forming the beginning of the upper palps. These now grow rapidly. In young plantigrades, the mouth is close to the adductor, but later moves a short distance back from it. Up to a shell length of 1 mm, the mouth opens directly downward, while the esophagus, now relatively shorter than in the veliger, bends around to enter the stomach. The mouth and esophagus are lined with tall columnar cells. The stomach enlarges rapidly and the posterior end tilts downward, bringing the style sac into a vertical position. After looping variously, the intestine becomes the rectum as it passes through the heart and pericardium and circles behind the posterior adductor muscle to which it is closely applied, then bends forward to end in the anus just below the visceral ganglion. The digestive gland increases in size forming numerous branching ducts that enter the anterior part of the stomach at the upper limit of the style sac (Quayle, 1952). In *M. mercenaria*, the intestine likewise increases in length by forming 'tortuous coils' in the visceral mass. After piercing the ventricle of the heart, it terminates behind the posterior adductor muscle (Belding, 1912).

According to Turner and George (1955), the digestive organs of the veliger of *M. mercenaria* become functional at the time that prodissoconch I is formed, usually 24–48 h after fertilization, depending on the ambient temperature. Development can proceed as far as the straight-hinged stage in filtered seawater without the addition of food. After prodissoconch I valves are formed, growth does not occur in the absence of food, and veligers will survive under conditions of starvation up to 12 days. Normal development occurs again when suitable food is provided. However, veligers that have developed to the early umbo stage on an optimal diet usually die when food is withheld.

In the Bivalvia, the marked structural changes that occur in the transition from a planktonic to a benthic life do not affect the alimentary tract, as the diet does not change fundamentally. The style sac is already present in the veliger and takes part in digestion of food; thus because the sac persists in the juvenile, it is typically a larvo–adult organ (Moor, 1983).

3.4.5 Ctenidia

In *M. mercenaria*, according to Belding (1912), a few simple V-shaped filaments, capable of extension and contraction, arise on each side of the foot, forming the inner gill. New buds repeat the process to form the outer gill. As the bivalve grows, the gills enlarge and the filaments increase in number.

Ansell (1962) provides a more detailed description of venerid ctenidial development in *V. striatula* (Fig. 3.21). Filaments of the inner lamella first appear when the veliger is 190–200 μm in length. In a fully developed veliger, active cilia are present between the filaments, but ctenidia do not take over the function of food collection until some time after settlement. After settlement, additional filaments are added to the ctenidium from the ctenidial anlage in the region of the siphonal septum. The first formed filament of the inner demibranch is attached to the visceral mass along the entire length and consists of a descending limb only. The next formed filaments at first develop the same way, but are attached at their origin, the ctenidial axis. Later the distal region of each of these filaments becomes attached to the visceral mass by a ciliary junction. With the filaments now fixed at each end, subsequent elongation results in the formation of descending and ascending parts; no bending is involved. Attachment of the distal region to the visceral mass takes place progressively earlier in later formed filaments, until both descending and ascending limbs develop together from the region of the ctenidial axis. The development of all filaments is thus essentially similar; in the earlier formed filaments the two limbs remain unequal in length, but the region where they occur becomes progressively less important as the overall size of the ctenidium increases. Development of the outer demibranch begins when the bivalve is about 1 mm long, by upward growth of filaments forming the supraxial extension. Details of this phase of development have yet to be reported. Quayle (1952) found that by the early juvenile stage, the number of filaments in the ctenidium of *V. pullastra* increases to about 12, and the adult type of ciliation develops.

Occasional small mucous cells are present between the cells bearing the frontal cilia and those supporting the laterofrontal cilia in *V. striatula* (Ansell, 1962). Large mucous cells occur in lines along the abfrontal region in each filament. At the free margin of each filament, a region remains undifferentiated; when the bivalve reaches a length of 2–3 mm, this region forms the marginal groove. The groove is preceded by an incipient oralwards current produced by cilia at the free margin. Ansell (1962) did not observe a continuous sheet of mucus on the surface of the ctenidium.

3.4.6 Kidney

In bivalves, the kidneys arise from the posterolateral parts of the common mesodermal cell mass. This rounds off forming the left and right nephric vesicles. These lie at first on either side of the hindgut. They then elongate into tubules, and shift beneath the pericardium, where they form a loop, consisting of an inner and outer arm connected dorsally. Both arms grow considerably in length. One of them fuses at its extremity with the ectoderm, and breaks through into the mantle cavity. The other arm grows forward, bends dorsally, and its tip opens into the pericardium. Here a funnel-shaped renopericardial duct, provided with a ciliary flame, is formed. The kidney may then be thrown into coils by considerable growth in length. In later stages, kidney cells may be laden with concretions (Raven, 1966).

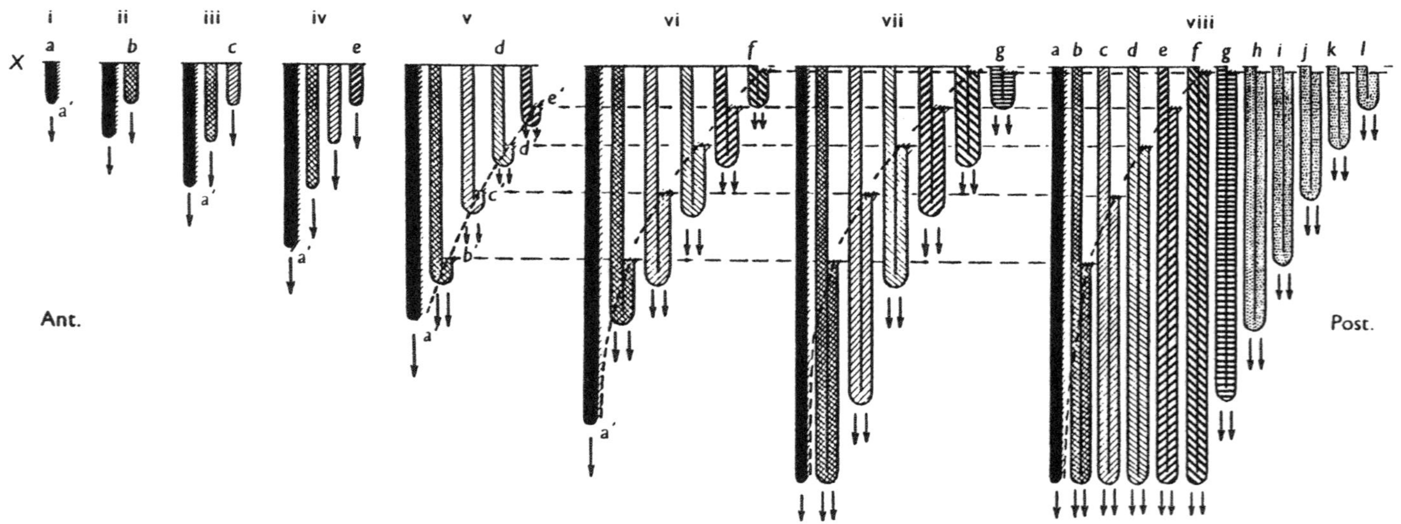

Fig. 3.21. Successive stages in the development of individual gill filaments in pediveliger-juvenile of *Venus striatula*; a–g under i–vii, individual filaments; a–g these filaments in juvenile plus new filaments h–y, under viii; a′ and a, first filament formed; b, second filament formed; direction of arrows, downward growth of filaments. From Ansell (1962).

In a plantigrade of *V. pullastra*, 280 μm long, the rudiment of the kidney occurs at the posterior end of the pericardial cavity against the posterior pedal retractor muscles. It consists of paired vesicles with groups of cells resembling the excretory cells of the mature kidney. In a plantigrade 0.5 mm long, the two kidneys contain a central lumen and consist of two tubes; the inner, more slender tube, is the renopericardial duct that leads from the pericardium into the dorsal part of the excretory section of the kidney; the excretory pore ends near the renopericardial opening, into what will later be the suprabranchial chamber. Right and left kidneys are united by a tube below the rectum; their walls later become greatly folded, increasing surface area (Quayle, 1952). Postlarval development of the kidney of *V. striatula* is similar to that of *V. pullastra* (Quayle, 1952; Ansell, 1962). There are no reports on the organogenesis of the kidney in veliger and byssal plantigrade *M. mercenaria*.

3.4.7 Heart and Vascular System

In most molluscs, the heart, pericardium, kidney, and often also the reproductive organs, arise from a common anlage. This cell group originates from the mesodermal bands of the embryo. In Bivalvia, part of the cells of this anlage may differentiate into primordial germ cells, and others to the left and right into the nephric vesicles. The remaining cells then spread medially and apply themselves against the hindgut, where they unite dorsally and ventrally forming a ring of cells around the gut, giving rise to the pericardium and heart. The pericardial cavity is present as a cleft in this cell mass. Between the inner wall of the pericardium and the gut a space becomes the ventricle of the heart. Adjacent to the ventricle there forms a mass of tissue within which a hemocoel becomes the lumen of the auricle (or atrium). At the juncture of the auricles and ventricle the auricular ventricular valves form. The heart next forms a transverse tube across the pericardial cavity, pierced by the hindgut; its lateral parts represent the auricles; its median part, the ventricle. Later the walls of the heart become multi-layered with muscle fibers and epithelia. Arteries and veins arise separately from the heart rudiments as cavities among the mesodermal cells. The anterior aorta, originating by joining together mesenchyme cells dorsal to the foregut, gives rise to arteries that supply the stomach, head region, liver, and foot. The posterior aorta, arising in the same way ventral to the hindgut, gives off the siphonal arteries that pass to the siphons. A venous sinus develops from the foot to the kidneys, and veins of the kidney empty into the dorsal part of the gill fold where the afferent gill vessel is formed. Similar vessels arise in the edge of the inner and outer gill lamellae. From cells of the walls of the auricle and pericardium a pericardial gland emerges (Raven, 1966).

In the pediveliger of *V. pullastra*, Quayle (1952) observed no anlage for the heart or the kidney, but these organs were unmistakable in the early juvenile. In a plantigrade 280 μm long, the thin-walled pericardium appears as a long, shallow cavity surrounding the rectum and extends from the posterior end of the shell ligament to the posterior pedal retractor muscles. The heart is a double membrane tightly enclosing the rectum, but with no cavity. At a shell length of 350 μm, the external heart membrane separates from the internal one to form a cavity surrounding the rectum. The heart is surrounded by the pericardium, bounded at the posterior by the kidney. By a shell length of 1 mm, muscle fibers begin to appear in the heart wall (Quayle, 1952). The post larval development of the heart and pericardium of *V. striatula* takes place as in *V. pullastra* (Quayle, 1952; Ansell, 1962).

3.4.8 Reproductive Organs

In bivalves, the gonadal primordium can be identified as primordial germ cells first begin to differentiate. These cells are large, round, with large clear vesicular nuclei, with one or two distinct nucleoli, peripherally arranged chromatin, and clear cytoplasm. As these cells continue to differentiate, they split off from the pericardial wall. At an early stage of gonadal development, mesodermal cells form a connective tissue envelope around the gonad. Later the gonad becomes lobate by local evaginations of its surface; secondary evaginations may give rise to acini, in which eventually ovo- and spermatogenesis take place. The rudiment of the primary gonoduct occurs at an early stage as a backwards directed ectodermal invagination in the posterior part of the mantle cavity, close behind the efferent duct of the kidney. Subsequently the gonoduct grows in, and connects at an early stage with the rudiment of the gonad (Raven, 1966).

Byssal plantigrades of *M. mercenaria* are distinctly hermaphroditic. The initial, or primary gonad, which is distinctly bisexual as individuals grow to a shell length of 4–6 mm, contains precursor gametes of both sexes; rapid proliferation of spermatogenic cells occurs before that of oogenic cells in individuals only a few millimeters long. In a sample of several hundred plantigrades 5–7 mm long, for example, functional spermatozoa were present in approximately 98% of the individuals. The remaining 2% developed directly into females, apparently without going through a functional male phase (Loosanoff, 1937).

Initial gonadal tissue first appearing in byssal plantigrades of *M. mercenaria*, 4–6 mm long, consists of a thin layer of cells between the muscular wall and the stomach, at the level of, or slightly below the heart. Gonadal follicles at first consist of a single layer of germinal epithelial cells that are irregular in shape and size, and possess large deeply staining nuclei. At first there is almost no follicular lumen, the walls of the follicles almost touching. As the plantigrade continues to grow, the follicles ramify through the loose connective tissue of the body. A few weeks later, the germinal epithelium begins to differentiate into oogonia, and the follicular lumen enlarges (Fig. 3.22). Rapid proliferation and specialization of cells follows in the six to eight follicles present at this stage. Different follicles of the same plantigrade exhibit widely different stages of development ranging from a few indifferent cells to male and female cells. In some cases, the follicle contains a few oogonia along its wall and a mass of spermatogenic cells in early stages of development in the lumen (Fig. 3.23). In others, gonads are distinctly bisexual, spermatozoa fill the lumen, and oocytes are present along the walls (Fig. 3.24). Phagocytic–nutritive cells, up to 12–14 μm in size, often occur in large numbers along the outer walls of gonadal follicles, and a few in their lumina (see also Wourms, 1987) (Fig. 3.25). Maximal density of phagocytic–nutritive cells occurs during the active stages of gametogenesis; they are present at all stages from indifferent to mature gonads. They appear to provide materials for the developing gametes, and to phagocytize degenerating cells (Loosanoff, 1937).

Plantigrades of *M. mercenaria* may function as males at the end of the first summer of life, retain this phase during the winter, and then become fully functional males at the age of about 1 year; other individuals may not become functional males until their second summer. About half of a population of plantigrades retains the male phase after the initial discharge of spermatozoa, whereas the other half transforms into fully functional females. The gonads of individuals that will become definitive females remain empty during the winter, and then

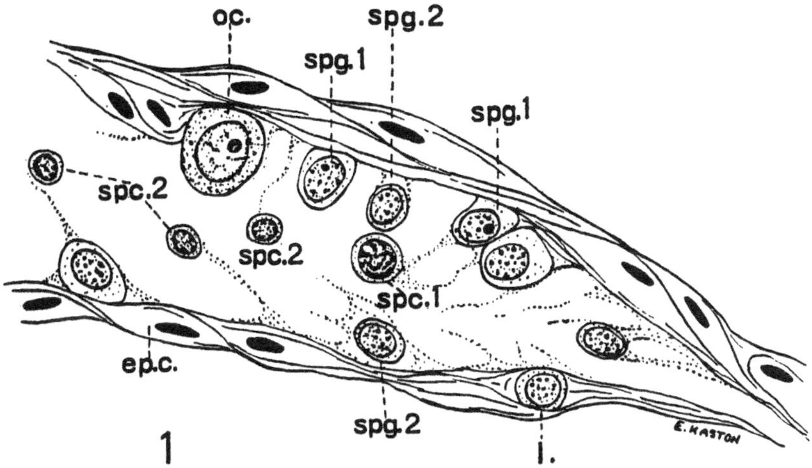

Fig. 3.22. Primitive bisexual gonad of a byssal plantigrade of *Mercenaria mercenaria* soon after formation of the gonadal lumen; i. = indifferent cell; epc. = follicular cells; oc. = young oocytes; spg.1 = primary spermatogonium; spg.2 = secondary spermatogonium; spc.1 = primary spermatocytes; spc.2 = secondary spermatocytes. From Loosanoff (1937).

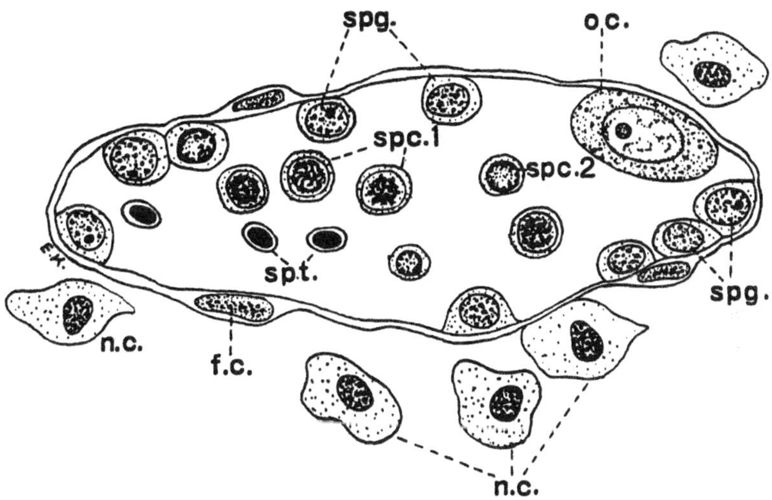

Fig. 3.23. Primary bisexual gonad in byssal plantigrade of *Mercenaria mercenaria*, 4–6 mm shell length, showing large oocyte (oc.), spermatogonia (spg.), primary spermatocytes (spc.1), secondary spermatocytes (spc.2), spermatids (spt.), follicular cells (fc.), and nutritive cells (nc). From Loosanoff (1937).

as the seawater warms in the spring, oogenesis takes place rapidly. Hence, in a population of sexually mature *M. mercenaria*, with only a few exceptions, male and female sexes are approximately equal and separate. Out of a population of 650 mature individuals examined, for example, only three were true functional hermaphrodites (Loosanoff, 1937). In *V. striatula*, the early development of the gonad appears to be similar to that of *M. mercenaria* (Ansell,

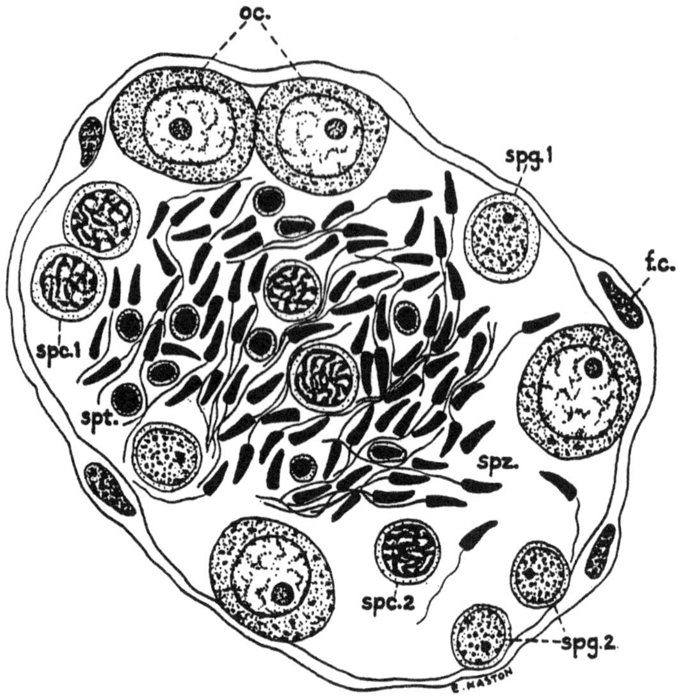

Fig. 3.24. Bisexual primary gonad of byssal plantigrade of *Mercenaria mercenaria* about 16 weeks old showing both male and female cells, but predominantly male cells. Primary spermatogonia (spg.1), secondary spermatogonia (spg 2), primary spermatocytes (spc.1), secondary spermatocytes (spc.2), spermatids (spt.), spermatozoa (spz.), oocytes (oc.), follicular cells (fc.). From Loosanoff (1937).

1961). General aspects of spermatogenesis and oogenesis in invertebrates are covered by Longo (1983), Franzén (1987) and Wourms (1987). Moor (1983) notes that when the gonadal primordium is formed relatively late, as in *V. striatula*, it proliferates from the pericardium.

3.4.9 Nervous System and Sensory Organs

The nervous system of pediveligers of *V. pullastra* and *V. striatula* is comparatively well developed. Ganglia are large and stain deeply with haemalum, eosin and alcian blue (Ansell, 1962). Cerebral ganglia are located anterior to the esophagus and posterior to the apical plate of the velum, one ganglion lying on each side of the apical groove and connected by the cerebral commissure. Pleural ganglia are close to the mantle against the thin roof of the velum and ventral to the cerebral ganglia to which they are linked by connectives. Visceral ganglia lie just anterior to the posterior adductor muscle. Fused pedal ganglia are situated in the proximal region of the foot (Quayle, 1952; Ansell, 1962).

As a rule, statocysts arise as ectodermal invaginations on the sides of pedal primordia near the boundary between foot and body. The cavity of the statocyst is narrow at first with columnar or cuboidal cells. Soon the cavity widens, cells of the wall flatten, and fine sensory

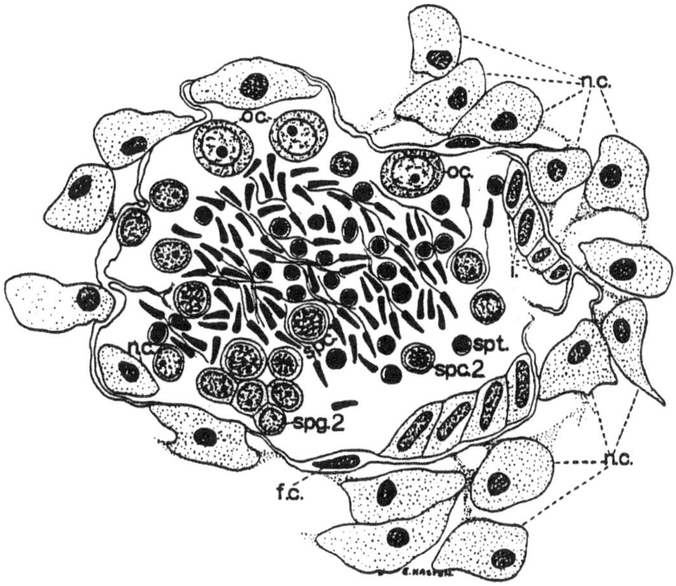

Fig. 3.25. Gonad of juvenile *Mercenaria mercenaria*, 10 mm shell length, surrounded by nutritive phagocytic cells (nc.), taken in late April–May. Remaining caption as in Fig. 3.24. From Loosanoff (1937).

cells appear on their surface. At an early stage, a large spherical statolith is formed in the lumen of the statocyst from organic cellular secretions. Each statocyst lies close to a pedal ganglion and is innervated from the cerebral ganglia. Cilia also line the funnel-shaped duct connecting with the exterior, and appear to direct an inward flow (Ansell, 1962; Raven, 1966; D'Asaro, 1967; Moor, 1983). These structures have not been examined histologically in *M. mercenaria*. There is, however, a detailed report on the ultrastructure of statocysts in the pediveliger of the bivalve *Pecten maximus* (Cragg and Nott, 1977).

With the loss, during metamorphosis, of the velum and movement of the mouth and esophagus dorsally and anteriorally, taking cerebral ganglia and the apical plates with them, the pleural ganglia become incorporated with the cerebral ganglia. These come to lie dorsal to the palps and posterior to the anterior adductor muscle. Statocysts and statoliths increase in size, and immediately below the visceral ganglia the epithelium differentiates into the osphradium (Quayle, 1952; Ansell, 1962).

3.4.10 Musculature

Organization of the musculature of the bivalve veliger is fundamentally the same in all species examined to date (Chanley and Andrews, 1971; Cragg, 1985). The careful study of the veliger of *P. maximus* (Cragg, 1985) contributes a general picture of the probable musculature of veligers of *M. mercenaria*. Differentiating muscle cells first appear as prodissoconch I of the larval shell is being secreted. At this time a functioning system of four pairs of velar retractors (1, 2, 3, 4, Fig. 3.26), three pairs of retractors attached to the posterior body wall (p1, p2, p3) and anterior adductor muscles (sm, st) form. Although branching

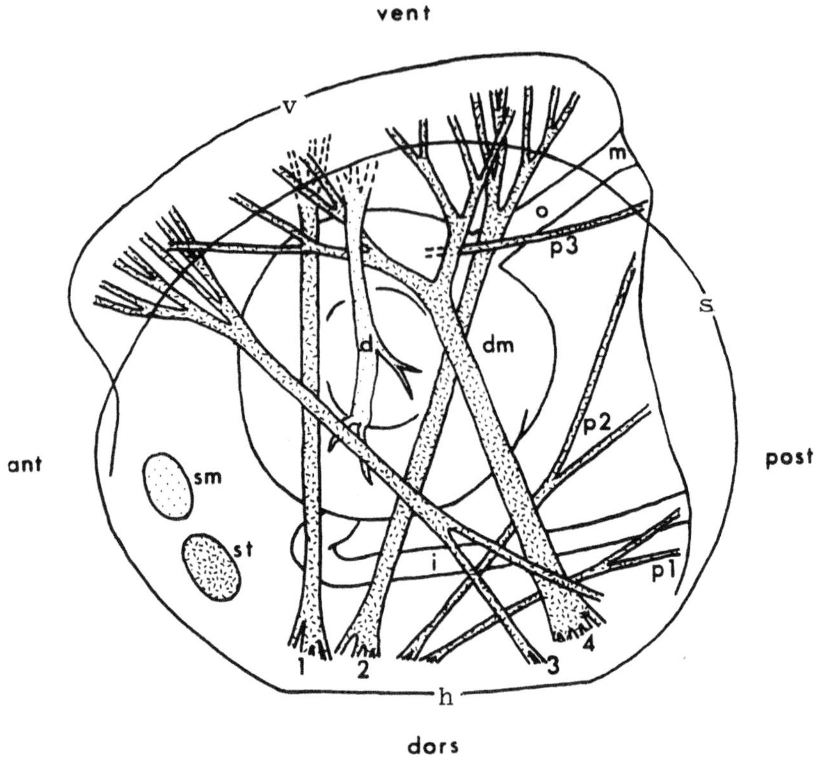

Fig. 3.26. Major muscles of the lefthand side of the veliger of *Pecten maximus*; dots, smooth muscle; dashes, striated muscle. dm = digestive gland; h = hinge; i = intestine; m = mouth; o = esophagus; p1, p2, p3, anterior retractor muscles; s = shell; sm = smooth, st = striated parts of anterior adductor muscle; v = velum; 1, 2, 3, 4, velar retractor muscles. From Cragg (1985).

of the retractor muscles becomes more profuse as the veliger grows, arrangement of velar retractors remains the same until they are lost at metamorphosis when the velum degenerates. A posterior adductor and pedal retractor muscles develop during the early pediveliger stage. In *M. mercenaria*, both adductor muscles differentiate before prodissoconch I is complete, the anterior adductor appearing before the posterior one; the two muscles remain in the same position as the veliger grows, enlarging as the animal increases in size (Belding, 1912).

Raven (1966) noted that in bivalves the posterior pedal retractor originates from paired groups of myocytes. The retractor forms strands that run ventrally on either side of the hindgut, enclosing the hindgut between them and uniting below it and dividing again in the foot into two branches that become fastened to the sides of the foot. The anterior pedal retractor develops similarly. Later, transverse musculature forms in the foot connecting its two sides.

In *P. maximum* (Cragg, 1985), velar and posterior retractors consist of striated muscle. The pattern of their distribution is symmetrical in the plane between the edges of the shell valves. Velar retractors generally branch profusely before inserting in the velum. On each side of the veliger there is a band of smooth muscle anchored dorsally to the digestive mass. The anterior adductor muscle consists of two distinct columns, a dorsal striated (st) one and a ventral smooth one (sm, Fig. 3.26). The posterior adductor of the pediveliger differs from the anterior one in having less clearly separated muscle columns.

Approximation of the two anterior columns, in accommodation to the opening and closing of the growing valves, probably occurs by addition of new muscle fibers to the enlarging margin of opposing myostracal muscle imprints in the valves. Whether imprints are composed of aragonite, as is the case in *Crassostrea virginica* (Carriker, 1996), has not been determined. In *C. virginica*, migration of the posterior adductor muscle similarly takes place by growth of new muscle fibers at the advancing edge, while obsolete fibers on the retreating margin are resorbed (Stenzel, 1971; Carriker, 1996).

Velar and posterior retractors originate from specialized parts of the epithelium lining the shell and insert in the epithelium bridging the rim of the two shell valves. This flexible part of the body wall is elaborated into the velum, mantle folds, boundary of the mantle cavity, and in the pediveliger, the surface of the foot. The body wall, in conjunction with the mantle of the shell, forms a body space containing hemolymph in which muscles, nerves, and digestive tract are immersed. The fluid constitutes a hydrostatic skeleton whose shape is determined by the combined activities of the retractor muscles, which control the rapid contractions of the velum. Complete retraction occurs spontaneously at intervals, or can be triggered by disturbing stimuli, and is preceded by abrupt cessation of beating of the swimming cilia, often followed by a momentary closing of the valves. Sequence of muscle action is probably coordinated by a system of nerves, which link sense organs and muscles, permitting the veliger to respond rapidly to external stimuli. There is no detailed anatomical evidence of a nervous system to the muscles, though cell profiles adjacent to muscle cell nuclei could be part of such a system (Cragg, 1985).

3.4.11 Foot and Byssal Glands

The foot of *M. mercenaria* develops functionally in late stage veligers as small as 166 μm in length, as a long, rounded, slender, heavily ciliated, highly supple, extensive muscular organ. In a byssal plantigrade 235 μm in length, the foot can be extended normally about 190 μm beyond the ventral margin of the valves; in individuals 415 μm in length, about 350 μm; and in individuals 7 mm in length, about 9 mm. Inside the mantle cavity the foot can be turned 180° and protruded rearward, and after the foot contacts the substratum the body realigns itself over the foot. On an open surface free of sediment, juveniles often extend the foot under themselves and turn themselves over. As individuals continue to grow, their foot becomes shorter in proportion to the size of the shell, and the base of the foot becomes more stocky.

In both pediveligers and byssal plantigrades of *M. mercenaria*, the foot is heavily ciliated. At the tip of the foot cilia are longer than on the remainder of the foot, enabling the foot tip to 'grip' the surface.

Ansell (1962) observed that in very young juveniles of *V. striatula* the foot may sometimes be extended through the primary exhalant siphon, an action that could serve to rid the interior siphonal walls of detritus. This behavior has not been observed in *M. mercenaria*.

The foot of pediveligers and byssal plantigrades of *M. mercenaria* have not been studied microstructurally nor histochemically. For such investigations in *Ostrea edulis* see Cranfield (1973a,b); in *P. maximus*, Gruffydd et al. (1975); and in *Mytilus edulis*, Lane and Nott (1975).

Because the foot of pediveligers and byssal plantigrades of *M. edulis* is highly active (Lane and Nott, 1975), as it is in *M. mercenaria*, it is probable the structure and function of the foot in both species are similar.

The foot of the pediveliger of *M. edulis* is histologically and histochemically complex (Lane and Nott, 1975), consisting of nine kinds of glands (Fig. 3.27). The pedal duct (d) opens at the heel of the foot and becomes a u-shaped, ciliated groove extending along the midventral surface of the foot to the pedal depression (pd) which widens. From the pedal depression toward the tip of the foot the groove takes the form in cross section, along the sole, of a shallow v-shaped concavity. The posterior duct arises from a pair of lateral pouches

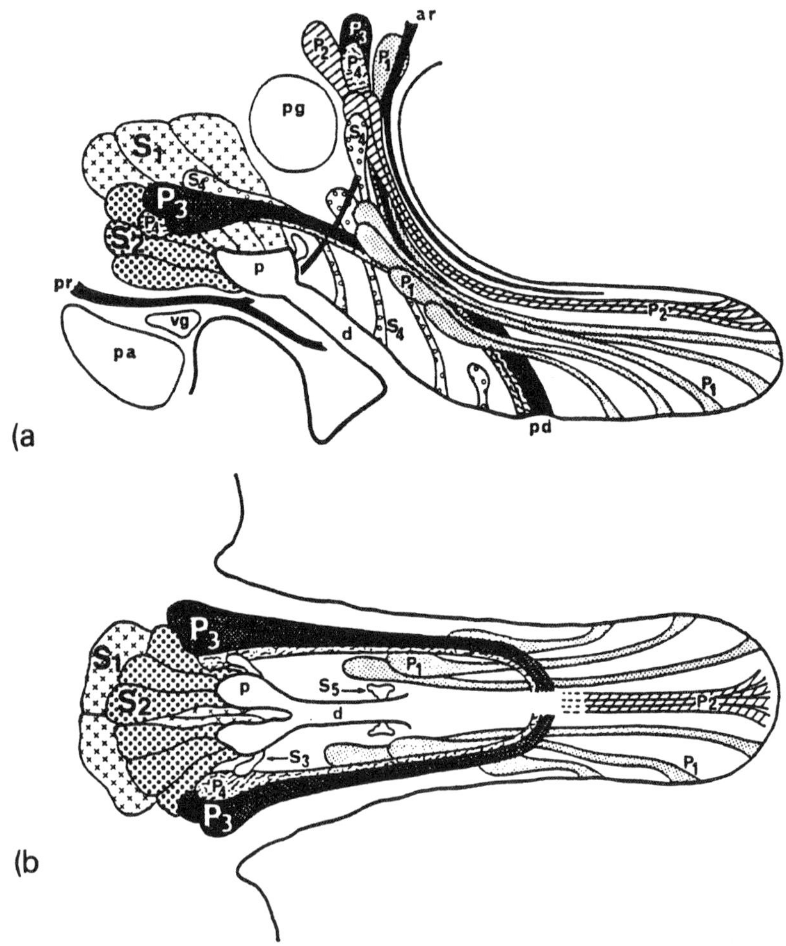

Fig. 3.27. Diagrams showing the location of glands in the foot of the pediveliger of *Mytilus edulis*. (a) sagittal section. (b) horizontal section. Cells of each of the 9 types of gland are indicated by P_1, P_2, P_3, P_4, S_1, S_2, S_3, S_4, S_5. ar = anterior retractor muscle; d = posterior duct; p = lateral pouch; pa = posterior adductor muscle; pd = pedal depression; pg = pedal ganglion; pr = posterior retractor muscle; vg = visceral ganglion. From Lane and Nott (1975).

(p) that are associated with the cells of the posterior glandular system. Apparently each type of gland has a specific function during crawling, production of the byssus and attachment plaque, and attachment. Lane and Nott (1975) suggested that secretions from glands P_3 and P_4 are moulded in the pedal depression (pd) to form the terminal attachment plaques of the byssus. Glands S_1, S_2, and S_3, which discharge into the lateral pouches (p), are involved in the formation of the primary byssus thread. The main structural component of the thread appears to be derived from the filamentous section of gland S_2. Collagenous secretion from gland S_4 passing into the groove and the duct, forms the secondary thread linking the primary thread to the terminal attachment plaque. The abundance and variety of these pedal glands indicate their important roles in the activity of the foot during crawling and byssal attachment. Whereas pediveligers of *M. edulis*, facilitated by the v-shaped pedal groove, settle primarily on filamentous substrata (Lutz and Kennish, 1992), pediveligers of *M. mercenaria*, with their flattened pedal sole, seek mostly flat surfaces. Also, the byssus of *M. mercenaria* is essentially a post-larval secreted structure, used only for initial settlement and subsequent burrowing; then after a brief, but vital appearance, it disappears (Yonge, 1962). Aspects of the chemistry of the byssus of adult bivalves are treated by Mercer (1972) and in *Mytilus edulis* by Waite (1992), and of the microstructure of the byssal forming system of *M. californianus* by Tamarin and Keller (1972), Tamarin et al. (1974, 1976) and Tamarin (1975). Although these reports reflect in a generic way the chemistry and microstructure of the byssus of young *M. mercenaria*, such a comparison has yet to be made.

Belding (1912) was the first to show in 1906 that *M. mercenaria* possesses a byssal gland and secretes a byssus (van der Feen, 1949). The gland, probably of ectodermal origin, occupies a major part of the heel of the foot. It arises as a pitlike invagination ventrally in the midline of the foot behind the pedal ganglia (Raven, 1966; Wada, 1968). The byssal groove is a longitudinal depression along the midventral region of the basal two thirds of the foot. This depression shallows as it passes distally from the heel, and the edges, which form prominent folds in the vicinity of the heel, also diminish in size distally. The gland becomes functional shortly after pediveligers evolve the plantigrade habit, and sometimes before the velum degenerates.

In *V. pullastra* (Quayle, 1952), the byssal gland also becomes functional at metamorphosis, when it connects with the byssal groove. Upon further development, the foot grows shorter and broader, retains its ciliation, and the byssal gland moves farther back into the foot and becomes smaller. In *M. mercenaria*, in which the byssus is such an important transitional structure, the byssus gland likely remains prominent throughout the byssal plantigrade stage to a length of the shell of about 5–9 mm. Resorption of the byssal gland after the byssal plantigrade stage has not been examined in this species.

3.5 ACKNOWLEDGMENTS

I wish to thank Michael Castagna, Albert Eble and John Kraeuter for especially helpful comments on the manuscript; Linda Leidy for typing the final draft of the typescript; Robert J. Bowden II for preparing photographic copies of the figures for publication; and the College of Marine Studies, University of Delaware, for facilities in which the literature search, synthesis, and writing for the chapter were done. The cost of preparation of the manuscript was supported in part by a grant from the Conchologists of America.

REFERENCES

Ansell, A.D., 1961. The functional morphology of the British species of Veneracea (Eulamellibranchia). J. Mar. Biol. Ass. U.K., 41: 489–515.

Ansell, A.D., 1962. The functional morphology of the larva, and the post-larval development of *Venus striatula* (da Costa). J. Mar. Biol. Ass. U.K., 42: 419–443.

Bandel, K., 1988. Stages in the ontogeny and a model of the evolution of bivalves (Mollusca). Paleontol. Z., 62: 217–254.

Bayne, B.L., 1971. Some morphological changes that occur at the metamorphosis of the larvae of *Mytilus edulis*. In: D.J. Crisp (Ed.), Fourth European Marine Biology Symposium, University Press, Cambridge, pp. 259–280.

Bayne, B.L. (Ed.), 1976. Marine Mussels: Their Ecology and Physiology. Cambridge University Press, Cambridge, 506 pp.

Belding, D.L., 1912. A report upon the quahog and oyster fisheries of Massachusetts, including the life history, growth and cultivation of the quahog (*Venus mercenaria*), and observations on the set of oyster spat in Wellfleet Bay. Commonwealth of Massachusetts, Wright and Potter, Boston, MA, 134 pp.

Belding, D.L., 1931. The quahaug fishery of Massachusetts. Mass. Dept. Conservation, Div. Fish Game, Mar. Fishery Serv., 2, 41 pp.

Boidron-Métairon, I.F., 1995. Larval nutrition. In: L. McEdward (Ed.), Ecology of Marine Invertebrate Larvae. CRC Press, Boca Raton, FL, pp. 223–248.

Bonar, D., 1976. Molluscan metamorphosis: a study in tissue transformation. Am. Zool., 16: 573–591.

Camacho, A.P. and Cabello, G.R., 1974. Desarrollo larvario de *Venerupis pullastra*. Bol. Inst. Esp. Oceanogr., 165: 1–36.

Carriker, M.R., 1956. Biology and propagation of young hard clams bivalve *Mercenaria mercenaria*. J. Elisha Mitchell Sci. Soc., 72: 57–60.

Carriker, M.R., 1961. Interrelation of functional morphology, behavior, and autecology in early stages of the bivalve *Mercenaria mercenaria*. J. Elisha Mitchell Sci. Soc., 77: 168–241.

Carriker, M.R., 1990. Functional significance of the pediveliger in bivalve development. In: B. Morton (Ed.), Proceedings of Memorial Symposium in Honour of Sir Charles Maurice Yonge, Edinburgh, 1986. Hong Kong Univ. Press, Hong Kong, pp. 267–282.

Carriker, M.R., 1996. The shell and ligament. In: V.S. Kennedy, R.I.E. Newell and A.E. Eble (Eds.), The Eastern Oyster: *Crassostrea virginica*. Maryland Sea Grant College, University of Maryland System, College Park, MD, pp. 75–168.

Chanley, P. and Andrews, J.D., 1971. Aids for identification of bivalve larvae of Virginia. Malacologia, 11: 45–119.

Cooke, C.A., 1986. Embryogenesis and morphology of larval structures in *Chlamys hastata*, with an examination of the effect of temperature on larval development and factors affecting settlement and metamorphosis. Master's Thesis, Univ. Victoria, Canada, 133 pp.

Cragg, S.M., 1985. The adductor and retractor muscles of the veliger of *Pecten maximus* (L.) (Bivalvia). J. Molluscan Stud., 51: 276–283.

Cragg, S.M., 1989. The ciliated rim of the velum of larvae of *Pecten maximus* (Bivalvia: Pectinidae). J. Molluscan Stud., 55: 497–508.

Cragg, S.M. and Crisp, D.J., 1991. The biology of scallop larvae. In: S.E. Shumway (Ed.), Scallops: Biology, Ecology and Aquaculture. Developments Aquaculture Fish. Vol. 21, Elsevier Science, Amsterdam, pp. 75–132.

Cragg, S.M. and Nott, J.A., 1977. The ultrastructure of the statocysts in pediveliger larvae of *Pecten maximus* (L.) (Bivalvia). J. Exp. Mar. Biol. Ecol., 27: 23–36.

Cranfield, H.J., 1973a. A study of the morphology, ultrastructure, and histochemistry of the foot of the pediveliger of *Ostrea edulis*. Mar. Biol., 22: 187–202.

Cranfield, H.J., 1973b. Observations on the function of the glands of the foot of the pediveliger of *Ostrea edulis* during settlement. Mar. Biol., 22: 211–223.

Cranfield, H.J., 1974. Observations on the mantle folds of the pediveliger of *Ostrea edulis* and their function during settlement. J. Mar. Biol. Ass. U.K., 54: 1–12.

Crisp, D.J., 1974. Factors influencing the settlement of marine invertebrate larvae. In: P.T. Grant and A.M. Mackie (Eds.), Chemoreception in Marine Organisms. Academic Press, London, pp. 177–265.

Crisp, D.J., 1984. Overview of research on marine invertebrate larvae, 1940–1980. In: J.D. Costlow and R.C. Tipper (Eds.), Marine Biodeterioration, an Interdisciplinary Study. Naval Inst. Press, Annapolis, MD, pp. 103–126.

D'Asaro, C.N., 1967. The morphology of larval and postlarval *Chione cancellata* Linne (Eulamellibranchia: Veneridae) reared in the laboratory. Bull. Mar. Sci., 17: 949–972.

Elston, R., 1980. Functional anatomy, histology and ultrastructure of the soft tissues of the larval American oyster, *Crassostrea virginica*. Proc. Natl. Shellfish. Ass., 70: 65–93.

Eversole, A.G., 1997. Gametogenesis of *Mercenaria mercenaria*, M. campechiensis and their hybrids. Nautilus, 110: 107–110.

van der Feen, P.J., 1949. Byssus. Basteria, 13: 66–71.

Fioroni, P., 1982. Larval organs, larvae, metamorphosis and types of development of Mollusca — a comprehensive review. Zool. Jb. Anat., 108: 375–420.

Franzén, A., 1987. Spermatogenesis. In: A.C. Giese, J.S. Pearse, V.B. Pearse (Eds.), Reproduction of Marine Invertebrates, Vol. 9, General Aspects, Seeking Unity in Diversity. Blackwell Scientific, Palo Alto, CA, pp. 1–47.

Gallagher, S.M., 1988. Visual observation of particle manipulation during feeding in larvae of a bivalve mollusc. Bull. Mar. Sci., 43: 344–365.

Galtsoff, P.S., 1964. The American oyster *Crassostrea virginica* Gmelin. Fishery Bull. Fish Wildl. Serv. U.S., 64: 1–480.

Goodsell, J.G. and Eversole, A.G., 1992. Prodissoconch I and II length in *Mercenaria* taxa. Nautilus, 106: 119–122.

Gruffydd, L.D., Lane, D.J.W. and Beaumont, A.R., 1975. The glands of the larval foot in *Pecten maximus* L. and possible homologues in other bivalves. J. Mar. Biol. Ass. U.K., 55: 463–476.

Hart, M.W. and Strathmann, R.R., 1995. Mechanisms and rates of suspension feeding. In: L. McEdward (Ed.), Ecology of Marine Invertebrate Larvae, CRC Press, Boca Raton, FL, pp. 193–221.

Hillman, R.E., 1964. The functional morphology of the fourth fold of the mantle of the northern quahog, *M. mercenaria* (L.). J. Elisha Mitchell Scient. Soc., 80: 8–12.

Keck, R., Maurer, D. and Malouf, R., 1974. Factors influencing the setting behavior of larval hard clams, *Mercenaria mercenaria*. Proc. Natl. Shellfish. Assoc., 64: 59–67.

Kraeuter, J., Castagna, M. and van Dessel, R., 1982. Egg size and larval survival of *Mercenaria mercenaria* (L.) and *Argopecten irradians* (Lamarck). J. Exp. Mar. Biol. Ecol., 56: 3–8.

Lane, D.J.W. and Nott, J.A., 1975. A study of the morphology, fine structure and histochemistry of the foot of the pediveliger of *Mytilus edulis* L. J. Mar. Biol. Assoc. U.K., 55: 477–495.

Levin, L.A. and Bridges, T.S., 1995. Pattern and diversity in reproduction and development. In: L. McEdward (Ed.), Ecology of Marine Invertebrate Larvae. CRC Press, Boca Raton, FL, pp. 1–48.

Linnaeus, C., 1758. Systema Naturae, 10th ed., I. Salvius, 824 pp.

Longo, F.J., 1983. Meiotic maturation and fertilization. In: N.H. Verdonk, J.A.M. van den Biggelaar, A.S. Tompa (Eds.), The Mollusca, Vol. 3, Development. Academic Press, New York, pp. 49–89.

Loosanoff, V.L., 1937. Development of the primary gonad and sexual phases in *Venus mercenaria* Linnaeus. Biol. Bull. Mar. Biol. Lab., Woods Hole, 72: 389–405.

Loosanoff, V.L. and Davis, H.C., 1950. Conditioning *V. mercenaria* for spawning in winter and breeding its larvae in the laboratory. Biol. Bull. Mar. Biol. Lab., Woods Hole, 98: 60–65.

Loosanoff, V.L., Miller, W.S. and Smith, P.B., 1951. Growth and setting of larvae of *Venus mercenaria* in relation to temperature. J. Mar. Res., 10: 59–81.

Loosanoff, V.L., Davis, H.C. and Chanley, P.E., 1966. Dimensions and shapes of larvae of some marine bivalve mollusks. Malacologia, 4: 351–435.

Lutz, R.A. and Kennish, M.J., 1992. Ecology and morphology of larval and early postlarval mussels. In: E. Gosling (Ed.), The Mussel *Mytilus*, Ecology, Physiology, Genetics, and Culture. Elsevier, New York, pp. 53–85.

Maia, B.C., 1988. Swimming responses of larvae of three mactrid bivalves to different salinity gradients. Masters Thesis, School of Marine Science, College of William and Mary, Virginia, 115 pp.

Menzel, W., 1989. The biology, fishery and culture of quahog clams, Mercenaria. Dev. Aquac. Fish. Sci., 19: 201–242.

Mercer, E.H., 1972. Byssus fiber — Mollusca. In: M. Florkin and B. Scheer (Eds.), Chemical Zoology, Vol. 7. Academic Press, New York, pp. 147–155.

Moor, B., 1983. Organogenesis. In: N.H. Verdont, J.A.M. van den Biggelaar and A.S. Tompa, The Mollusca, Vol. 3. Development, Academic Press, New York, pp. 123–177.

Pechenik, J.A., 1990. Delayed metamorphosis by larvae of benthic marine invertebrates: does it occur?. Ophelia, 32: 63–74.

Quayle, D.B., 1952. Structure and biology of the larva and spat of *Venerupis pullastra* (Montagu). Trans. R. Soc. Edinb., 62: 255–297.

Raven, C.P., 1966. Morphogenesis, the Analysis of Molluscan Development, Revised ed. Pergamon Press, New York, 365 pp.

Rice, M.A., 1992. The northern quahog: the biology of *Mercenaria mercenaria*. Rhode Island Sea Grant Publ. RUI-B-92-001, 59 pp.

Rieger, R.M., 1994. The biphasic life cycle — a central theme of metazoan evolution. Am. Zool., 34: 484–491.

Riisgard, H.U., 1988. Feeding rates in hard clam (*Mercenaria mercenaria*) veliger larvae as a function of algal (*Isochrysis galbana*) concentration. J. Shellfish Res., 7: 377–380.

Rosati, F. and Focarelli, R., 1996. Sperm–egg interaction in bivalve molluscs. Malacol. Rev., Suppl., 6: 63–69.

Rowe, L. and Ludwig, D., 1991. Size and timing of metamorphosis in complex life cycles: time constraints and variation. Ecology, 72: 413–426.

Russell-Hunter, W.D., 1979. A Life of Invertebrates. Macmillan, New York, 650 pp.

Sastry, A.N., 1979. Pelecypoda (excluding Ostreidae). In: A.C. Giese and J.S. Pearse (Eds.), Reproduction of Marine Invertebrates 5, Academic Press, New York, pp. 113–192.

Sellmer, C.P., 1959. Studies on the anatomy and life history of the gem clam, *Gemma gemma* (Totten). Doctoral Dissertation, Rutgers Univ., New Brunswick, NJ, 237 pp.

de Severeyn, Y.G., Severeyn, H.J. and Ewald, J.J., 1994. Early development of the estuarine mollusk *Polymesoda solida* (Philippi, 1846) (Bivalvia: Corbiculidae) in Lake Maracaibo, Venezuela. Am. Malacol. Bull., 11: 51–56.

Stasek, C.R., 1972. The molluscan framework. In: M. Florkin and B.T. Scheer (Eds.), Chemical Zoology 7, Academic Press, New York, pp. 1–44.

Stenzel, H.B., 1971. Oysters. Treatise on Invertebrate Paleontology, Part N, Vol. 3, Mollusca 6, Bivalvia, Geol. Soc. Am., Univ. Kansas, pp. N953–N1224.

Strathmann, R.R., 1987. Larval feeding. In: A.C. Giese, J.S. Pearse and V.B. Pearse (Eds.), Reproduction of Marine Invertebrates, Vol. 9, General Aspects: Seeking Unity in Diversity, Blackwell Scientific, Palo Alto, CA, pp. 465–550.

Strathmann, R.R. and Leise, E., 1979. On feeding mechanisms and clearance rates of molluscan veligers. Biol. Bull. Mar. Biol. Lab., Woods Hole, 157: 524–535.

Tamarin, A., 1975. An ultrastructural study of byssus stem formation in *Mytilus californianus*. J. Morphol., 145: 151–177.

Tamarin, A. and Keller, P.J., 1972. An ultrastructural study of the byssal thread forming system in *Mytilus*. J. Ultrastr. Res., 40: 401–416.

Tamarin, A., Lewis, P. and Askey, J., 1974. Specialized cilia of the byssus attachment plaque forming region in *Mytilus californianus*. J. Morphol., 142: 321–328.

Tamarin, A., Lewis, P. and Askey, J., 1976. The structure and formation of the byssus attachment plaque in *Mytilus*. J. Morphol., 149: 199–221.

Turner, H.J. and George, C.J., 1955. Some aspects of the behavior of the quahog, *Venus mercenaria*, during the early stages. Massachusetts Dept. Nat. Resources, Div. Mar. Fisheries, Investigations of the Shellfisheries of Massachusetts, 8th Rep., pp. 5–14.

Turner, H.J., Personal communication.

Verdonk, N.H. and van den Biggelaar, J.A.M., 1983. Early development and the formation of the germ layers. In: N.H. Verdonk, J.A.M. van den Biggelaar and A.S. Tompa (Eds.), The Mollusca, Vol. 3. Academic Press, New York, pp. 91–122.

Wada, S.K., 1968. Mollusca. In: M. Kumé and K. Dan (Eds.), Invertebrate Embryology. NOLIT Publ. House, Belgrade, Yugoslavia (Trans. by J.C. Dan, pp. 485–525, originally published by Bai Fu Kan Press, Tokyo, 1957).

Waite, H., 1992. The formation of mussel byssus: anatomy of a natural manufacturing process. In: S.T. Case (Ed.), Results and Problems in Cell Differentiation Biopolymers, Spring-Verlag, Berlin, pp. 27–54.

Waller, T.R., 1981. Functional morphology and development of veliger larvae of the European oyster, *Ostrea edulis* Linne. Smithsonian Contr. Zool., 328: 1–70.

Wells, H.W., 1957. Status of the name *Venus*. Ecology, 38: 160–161.

Wourms, J.P., 1987. Oogenesis. In: A.C. Giese, J.S. Pearse and V.B. Pearse (Eds.), Reproduction of Marine Invertebrates, Vol. 9, General Aspects: Seeking Unity in Diversity, Blackwell Scientific, Palo Alto, CA, pp. 49–178.

Yonge, C.M., 1948. Formation of siphons in Lamellibranchia. Nature, 161: 198–199.
Yonge, C.M., 1957. Mantle fusion in the Lamellibranchia. Pubbl. Staz. Zool. Napoli, 29: 151–171.
Yonge, C.M., 1959. Evolution within the bivalve Mollusca. Proc. XVth Int. Congr. Zool., London, 1958, 15: 367–370.
Yonge, C.M., 1962. On the primitive significance of the byssus in the Bivalvia and its effects in evolution. J. Mar. Biol. Ass. U.K., 42: 113–125.
Zinn, D.J., 1973. Quahog — queen of the mudflats. Maritimes, Nov.: 4–7.

Chapter 4

Anatomy and Histology of Mercenaria mercenaria

Albert F. Eble

4.1 INTRODUCTION

Research on *Mercenaria mercenaria* (Linnaeus 1758) really began with the pioneering work of Kellogg (1892) who was supported by the U.S. Commission of Fish and Fisheries and strongly encouraged by Professor W. Brooks of Johns Hopkins University. Professor Brooks suggested the use of histological sections to supplement the practice of dissections in order to more clearly and accurately describe bivalve anatomy. Kellogg continued his work on bivalve anatomy of commercially important species in addition to reporting on current practices of their fishery (Kellogg, 1910, 1915). Belding (1912, 1931) continued the studies on the fishery of *M. mercenaria* particularly with reference to life history studies. Ansell (1961) described the functional morphology of several British Veneracea and Jones (1979) wrote a comprehensive monograph of American Veneracea including *M. mercenaria*.

The literature on *M. mercenaria* continued to grow and was finally organized in the form of an annotated bibliography by McHugh et al. (1981). However, no definitive modern work on the comprehensive anatomy and histology of *M. mercenaria*, similar to that by Galtsoff (1964) and by Eble and Scro (1996) for the eastern oyster, *Crassostrea virginica*, has been published. This chapter was written to fill this void and to stimulate further research in the many glaring gaps in our knowledge of the gross anatomy, histology, histochemistry and fine structure of this commercially important species.

M. mercenaria (Eulamellibranchia, Veneracea) is an isomyarian bivalve with a well-developed foot and a short siphon. The two valves are equal in size and shape. Gills and labial palps are reduced in size compared to most bivalves. The mantle is attached to the valves at the pallial line; dorsal to the pallial line the mantle is soft and fleshy, but ventral to this line it is muscular and firm.

4.2 MANTLE

4.2.1 Anatomy

The mantle is bilobed and attaches to the ventral region of the shell at the pallial line that runs between the adductor muscles (Fig. 4.1). The connecting neck of tissue between the two lobes located in the mid-dorsal region is termed the mantle isthmus (Owen et al., 1953); this area represents the original mid-dorsal surface of the mantle. The mantle is also attached to the dorsal portion of each valve by a series of small muscular attachments. The mantle is pierced by the paired adductor and pedal retractor muscles. The margins of the right and left

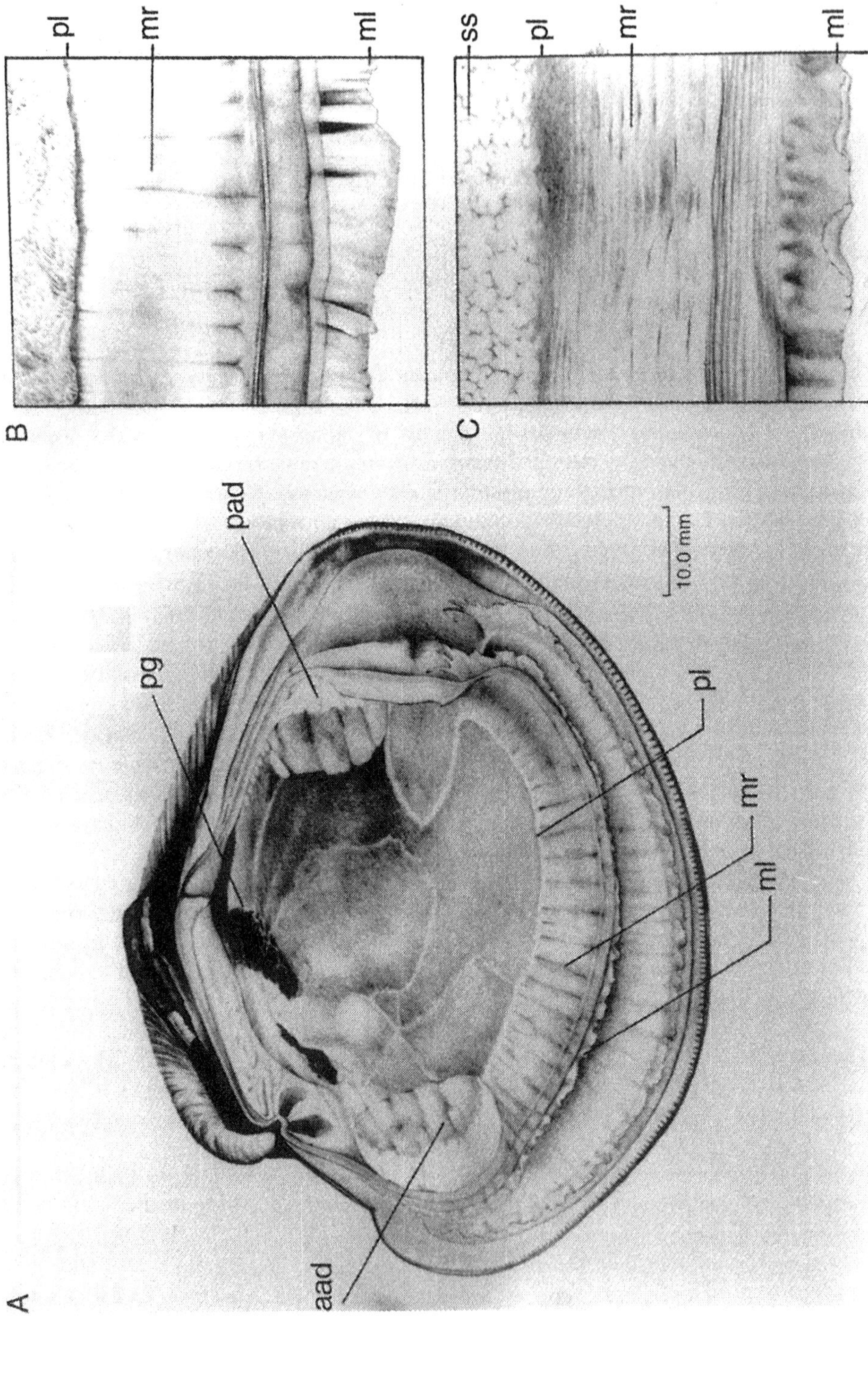
A
aad
pg
pad
pl
mr
ml
10.0 mm
B
pl
mr
ml
C
ss
pl
mr
ml

pallial lobes are in juxtaposition along most of the periphery with the exception of the pedal gape and region of the siphon.

Typically in the Bivalvia, the mantle margins are divided into three folds (Yonge, 1957). The outer fold secretes the two outer layers of the shell; the middle fold is sensory and usually is sub-divided into tentacles or some other anatomical arrangement which increases the sensory surface area; the inner fold is usually muscular and controls water movements into and out of the mantle cavity. Indeed, the inner fold is so prominent in the eastern oyster and scallop, forms that lack a siphon, that it is termed the pallial curtain (oyster) or velar fold (scallop) in recognition of its function (Galtsoff, 1964; Beninger and Le Pennec, 1991; Eble and Scro, 1996).

In the Veneridae, the mantle margin is divided into four folds: fold 1, a typical outer fold that secretes the outer calcareous layer of the shell from its outer surface and the periostracum from its inner surface; folds 2 and 3, a middle fold that is separated into a smaller outer fold and a larger inner fold — the outer fold (fold 2) is sensory, while the inner fold (fold 3) assumes the function of the typical third fold, i.e. controlling water flow into and out of the mantle cavity (Yonge, 1957); fold 4, a small flap-like fold that is directed dorsally that may be involved in removing foreign particles from the mantle (Hillman, 1964).

The broad outer face of each mantle lobe is divided into two distinct regions: from the mantle isthmus to the pallial line this surface is highly ruffled and the ruffles are tiered (Fig. 4.2) and newly secreted pieces of shell can be observed to be attached in this region; from the pallial line to the mantle edge the surface is smooth and dense bundles of longitudinal muscles radiate to the margins and penetrate the mantle folds (Fig. 4.2).

4.2.2 Histology

Serial sections of the mantle and all organs discussed in this chapter were treated with four basic staining protocols: (1) hematoxylin and eosin; (2) Mallory's trichrome; (3) Alcian blue, pH 2.6 — PAS (periodic acid Schiff); and (4) Alcian blue, pH 0.5.

The mantle consists of two epithelia, the outer or shell-secreting epithelium (shell epithelium) and the inner or mantle-cavity epithelium, separated by loose connective tissue permeated with many hemolymph sinuses (Fig. 4.3). My sections show the shell epithelium to be simple columnar, lacking cilia, with basally to centrally located nuclei about 6 μm in diameter with a central nucleolus. Cells average 7 × 28 μm and rest upon a basement membrane rich in neutral glycoproteins. The free surface of the shell epithelium presents a double line at the level of the light microscope which represents densely packed microvilli with a uniform diameter of 0.1 μm and an average length of 2–3 μm. Lateral cell membranes are convoluted and interdigitated; tight junctions are present towards the apex of cells and septate junctions appear immediately below them (Neff, 1972). Intercellular spaces vary from

Fig. 4.1. (A) Sketch of half-shell preparation in right valve to show general anatomy. Radial muscles of the mantle (mr) have their origin on the shell (pallial line, pl) and insert into the folds of the mantle (mf). aad = anterior adductor muscle; pad = posterior adductor muscle; pg = pericardial gland. (B) Sketch of mantle edge to show radial muscles (mr) originating on pallial line (pl) and inserting in pallial folds (mf). (C) Sketch of mantle edge to show shell-side surface (ss) and radial muscles originating on pallial line (pl) and inserting in pallial folds (mf).

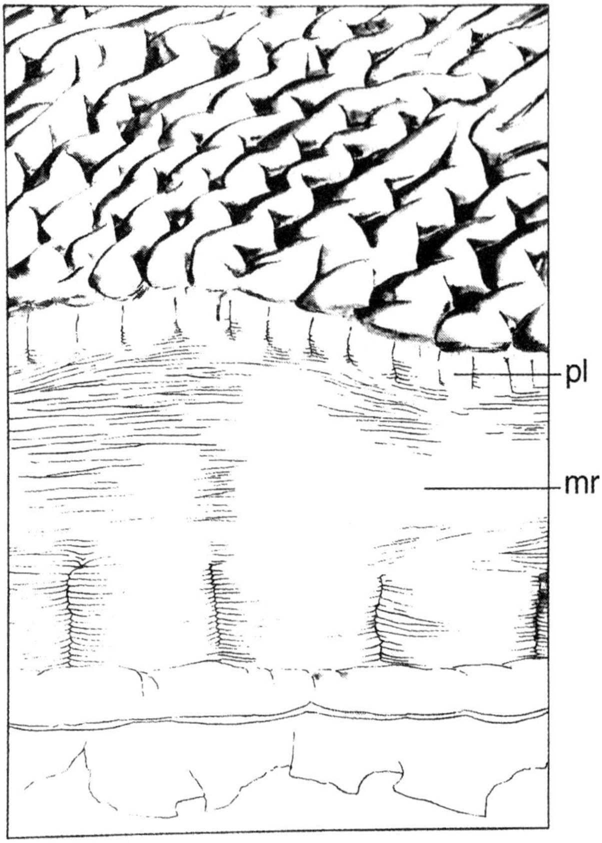

Fig. 4.2. Sketch of shell-side surface of mantle to show presence of many parallel ruffles. At pallial line (pl), the mantle becomes smooth because of the bundles of radial muscle (mr). 7.5×.

very narrow to large (1 μm). Basal lamina are convoluted, consist chiefly of acid glycoproteins and communicate directly with connective tissue spaces below.

Mitochondria in shell-secreting mantle epithelia are concentrated in the apical cytoplasm just below the microvilli; mitochondria are concentrated in the basal cytoplasm as well. Mitochondrial cristae are plate-like and the matrix is finely granular (Neff, 1972).

My sections show the mantle-cavity epithelium to be simple, low cuboidal, 12 × 16 μm, with centrally placed nuclei. The free surface is covered by a well-defined layer of material rich in acid glycosaminoglycans. Slender strands of longitudinal muscle are situated just under the epithelium; many hemolymph sinuses bathe the epithelium and muscle layers (Fig. 4.3). A loose connective tissue permeated with many hemolymph vessels and sinuses binds the two epithelia (Hillman, 1962). Hemolymph vessels and surrounding connective tissues are rich in neutral glycoproteins. Morrison (1993a) described the histology and histochemistry of the mantle epithelia of the eastern oyster, *Crassostrea virginica*, which is very similar to that of *M. mercenaria*.

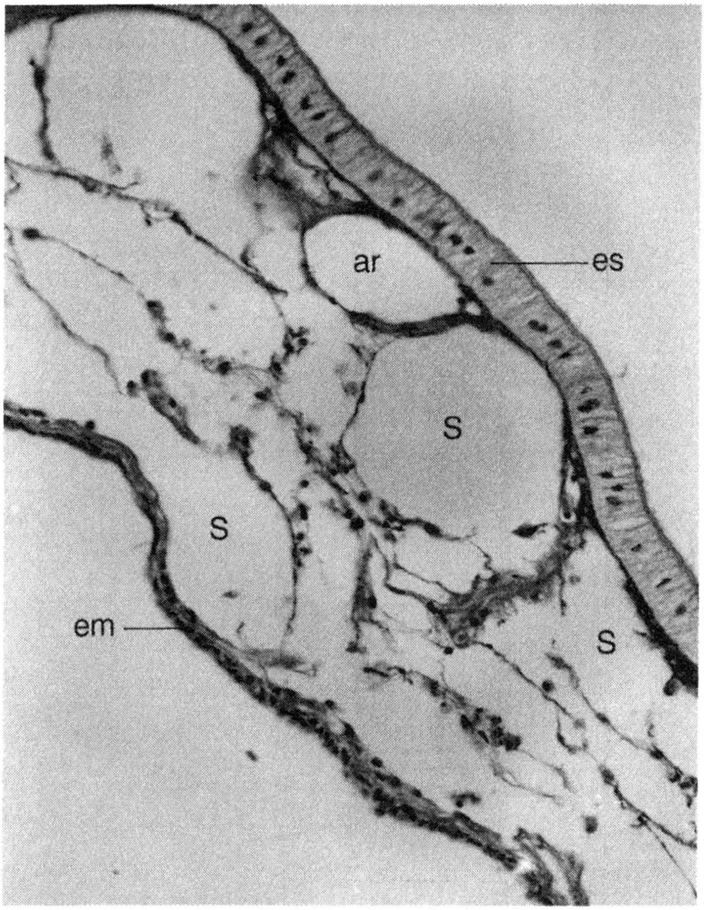

Fig. 4.3. Longitudinal section of mantle to show difference between shell-side (es) and mantle cavity-side (em) epithelia. A small artery (ar) and hemolymph sinuses (s) are obvious. Horizontal field width (HFW) = 405 μm.

Distal to the pallial line and continuing to the pallial folds, is found a thick longitudinal muscle under the shell epithelium (Fig. 4.4). The shell epithelium in this region changes in structure and is reduced to simple, low cuboidal cells. Close to the origin of the pallial folds, prominent subepithelial hemolymph sinuses force the shell epithelium into a series of prominent bulges (Fig. 4.5). An extensive discussion of the histology and histochemistry of the mantle can be found in Hillman (1962).

4.2.2.1 First pallial fold

The first pallial fold can be seen in Figs. 4.5 and 4.6. The outer, shell epithelium of this fold is convoluted and consists of simple, low columnar cells (6 μm [width] × 28 μm [height]). Small secretory cells are present in this epithelium as well as in subepithelial areas; my sections show that these cells contain acid glycosaminoglycans rich in sulfate groups (see also Hillman, 1968).

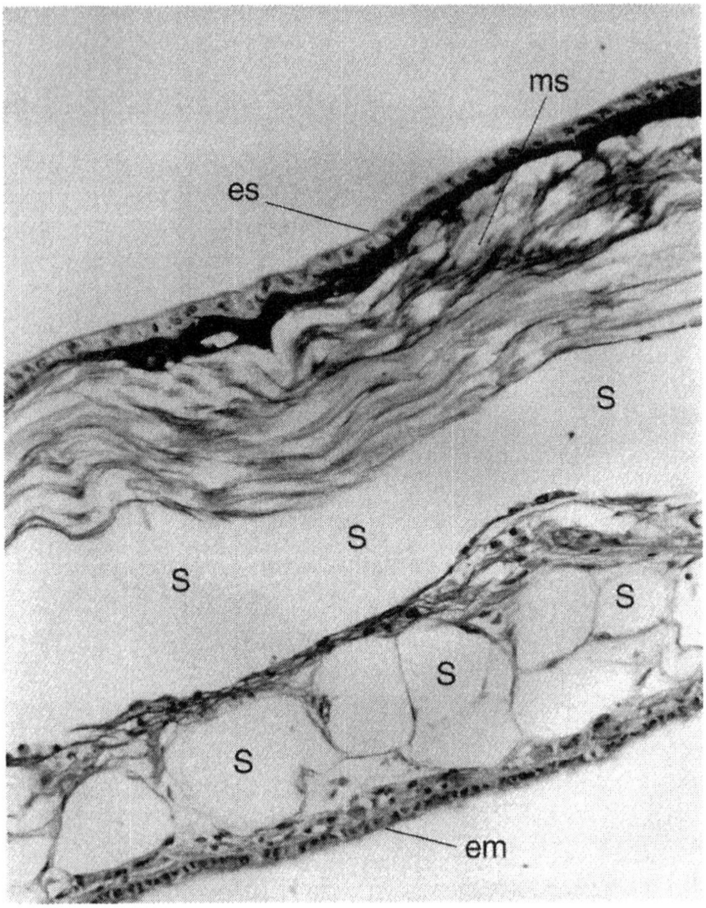

Fig. 4.4. Longitudinal section of mantle in zone of radial muscle (mr) which lies immediately under the shell-side epithelium (es). This epithelium becomes progressively less tall from the pallial line to the mantle folds. The epithelium bordering the mantle cavity (em) does not change in this area. Note the large hemolymph sinus (s) bathing the radial muscle and the complex of hemolymph sinuses (s) just under the epithelium bordering the mantle cavity. HFW = 405 μm.

The inner epithelium lacks convolutions but is similar to the outer epithelium; the fine structure has been described in detail by Neff (1972). Essentially, these cells have long, slender apical microvilli, basal nuclei and large accumulations of glycogen in the basal and middle regions of cells; the apical cytoplasm is filled with dilated cisternal profiles of rough endoplasmic reticula.

The proximal portion of the inner epithelium of the first pallial fold contains a cluster of cells that are simple, tall columnar (5 × 37 μm) and have a jagged free surface; they correspond in position to the periostracal gland of other bivalves (Figs. 4.6 and 4.7) and have been termed the basal bulb (Neff, 1972). The space between the inner portion of the first pallial fold and the outer surface of the second fold is the pallial groove. The most proximal

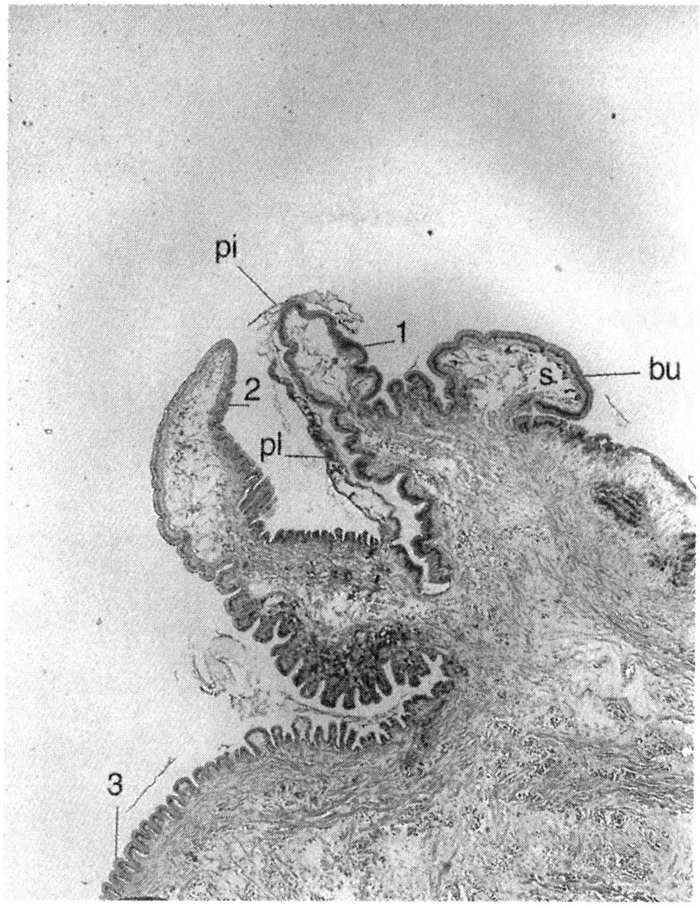

Fig. 4.5. Longitudinal section through first (1) and second (2) folds of mantle and part of third (3) fold. The periostracal fold (pl) and periostracum (pi) are clearly evident. Shell epithelium is thrown into a prominent bulge (bu) due to a large hemolymph sinus (s) near origin of first pallial fold. HFW = 810 μm.

cell of the basal bulb is the basal cell; the periostracum makes its first appearance in the intermembrane space next to the basal cell. The first layer of the periostracum will become the pellicle. The next layer of the periostracum to be secreted is the inner homogeneous layer which is made by the inner epithelium of the first fold that support the periostracal sheet and move it out of the periostracal groove as it is synthesized (Fig. 4.8). The free surface of the inner epithelium of the first pallial fold is covered with a layer of acid glycosaminoglycans rich in sulfate groups.

Slender strands of longitudinal muscle run in a subepithelial position the length of the first fold on both sides. A prominent hemolymph sinus is located in the middle of the fold; strands of muscle and connective tissue cross the hemolymph sinus at many levels (Figs. 4.5, 4.6 and 4.8).

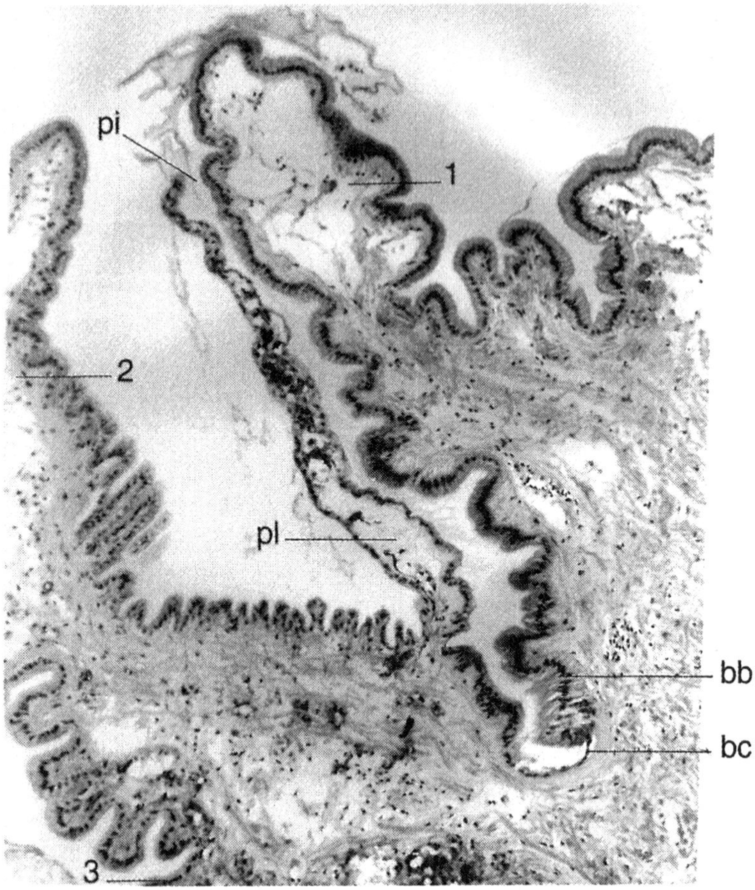

Fig. 4.6. Longitudinal section through first (1) and second (2) folds of mantle. Note origin of periostracal fold (pl) from basal portion of second fold. Basal bulb (bb) is located at base of first fold; the most proximal cell is termed the basal cell (bc). Periostracum (pi) can be seen originating at basal bulb and extending between the first and periostracal folds. HFW = 405 μm.

4.2.2.2 Second pallial fold

The second pallial fold is shown in Figs. 4.5–4.7. The short stretch of epithelium between the basal bulb and the beginning of the second fold is simple cuboidal (3 × 6 μm) that quickly grades into the outer epithelium of the second fold; the latter is simple, low columnar (5 × 12 μm). Cells contain a central nucleus, short, stout microvilli which are usually applied to the outer surface of the periostracum and bundles of tonofilaments oriented parallel to the apical surface (Neff, 1972). After running a short distance, the second fold branches into the periostracal fold, a slender sheet-like projection that runs parallel to and lies close to the inner surface of the first fold (Figs. 4.5–4.7). Hillman (1962) mentions this branch and refers to it as 'a small flap of epithelium'. I have termed it the peristracal fold because it is involved with the processing of the periostracum as described above (Figs. 4.5–4.8). The construction of the

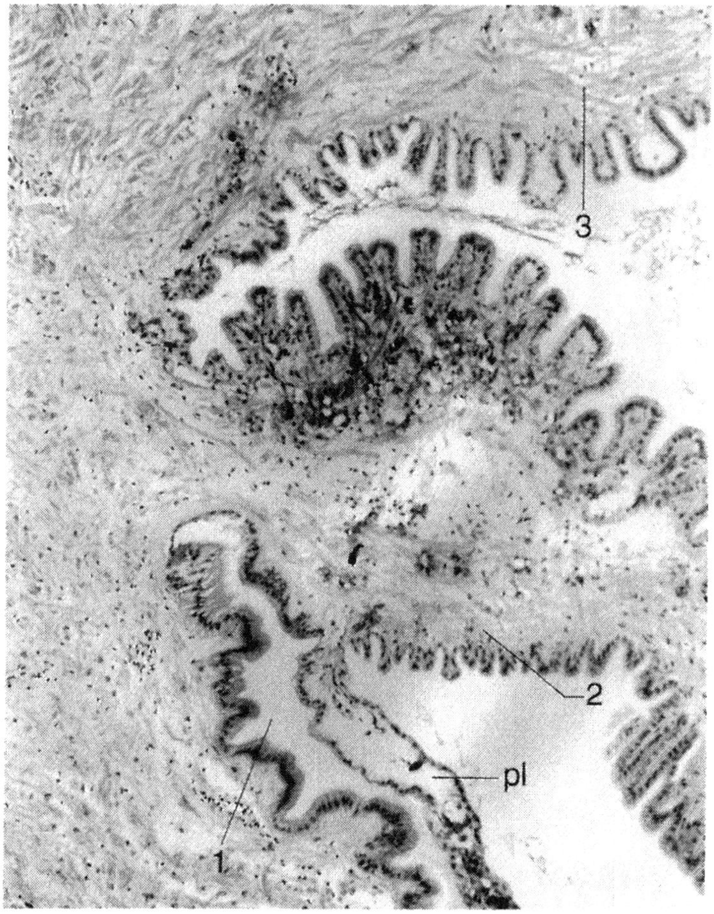

Fig. 4.7. Longitudinal section of the base of the first (1), second (2) and third folds (3). The origin of the periostracal fold (pl) and the highly folded epithelia, particularly the inner epithelium of the second folds, are evident. HFW = 436 μm.

periostracal fold is similar to that of the other folds: a low, cuboidal epithelium (3 × 6 μm) encloses a core of connective tissue and a large, central hemolymph sinus. Many prominent glands that secrete acid glycosaminoglycans rich in sulfate groups are located in the core of the periostracal fold.

Both outer (facing the first fold) and inner epithelia of the second fold are simple, low columnar (5 × 12 μm); the outer surface is highly convoluted (Figs. 4.5–4.8). A large hemolymph sinus, criss-crossed by many strands of muscle and connective tissue, occupies the middle of the fold (Figs. 4.5–4.8). Many subepithelial glands are located in the proximal area of the inner side; these glands secrete three different substances: acid glycosaminoglycans rich in sulfate groups, neutral glycoproteins and acid glycoproteins. Hillman (1969) describes two types of gland cells in this area: cell type I and type II. The former secretes acid glycoproteins, sulfated and carboxylated, while the latter secretes neutral 'mucosubstances'

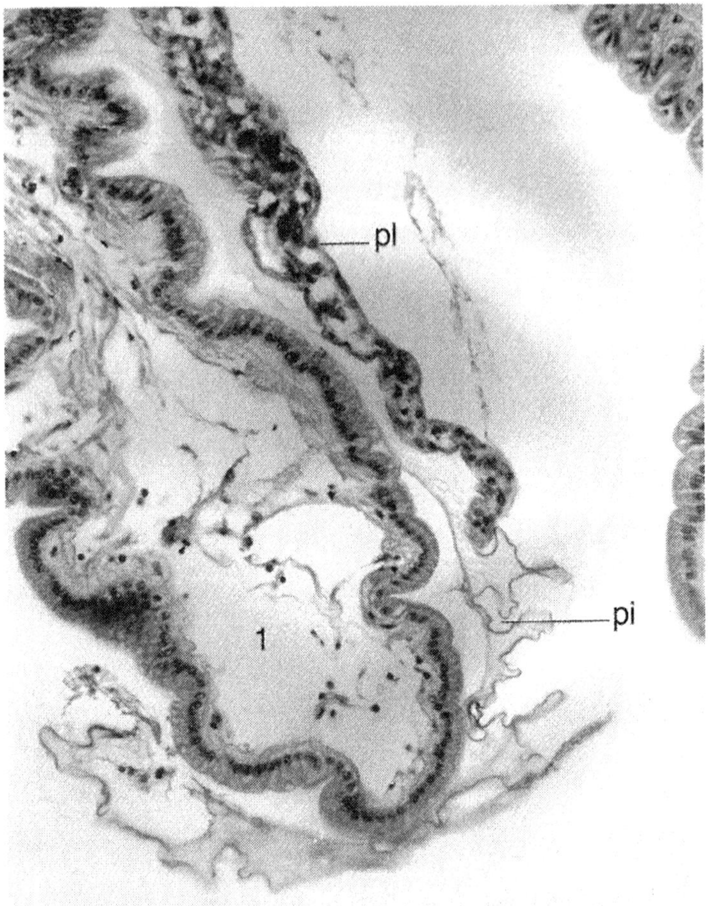

Fig. 4.8. Longitudinal section of first (1) and periostracal folds (pl). Note periostracum (pi) running between these folds until it reaches the tip of the periostracal fold at which point it is directed around the first fold. HFW = 220 μm.

(glycoproteins). Hillman (1969) speculated that one of these substances was probably similar to heparin that might aid in the processing of calcium, sodium and chloride by the clam; other investigators (Love and Frommhagen, 1953; Thomas, 1954; Burson et al., 1956) have reported heparin-like substances in bivalves, notably *Spisula solidissima*. Prominent layers of longitudinal muscle run in a subepithelial position the entire length of the fold; slips of muscle attach to the basement membrane of outer and inner epithelia.

4.2.2.3 Third pallial fold

The third pallial fold can be seen in Figs. 4.5, 4.7 and 4.9. This fold is the longest of the four pallial folds. The outer and inner epithelia of this fold are similar to that of the contiguous second fold. Glands that were described for the inner portion of the second fold continue into

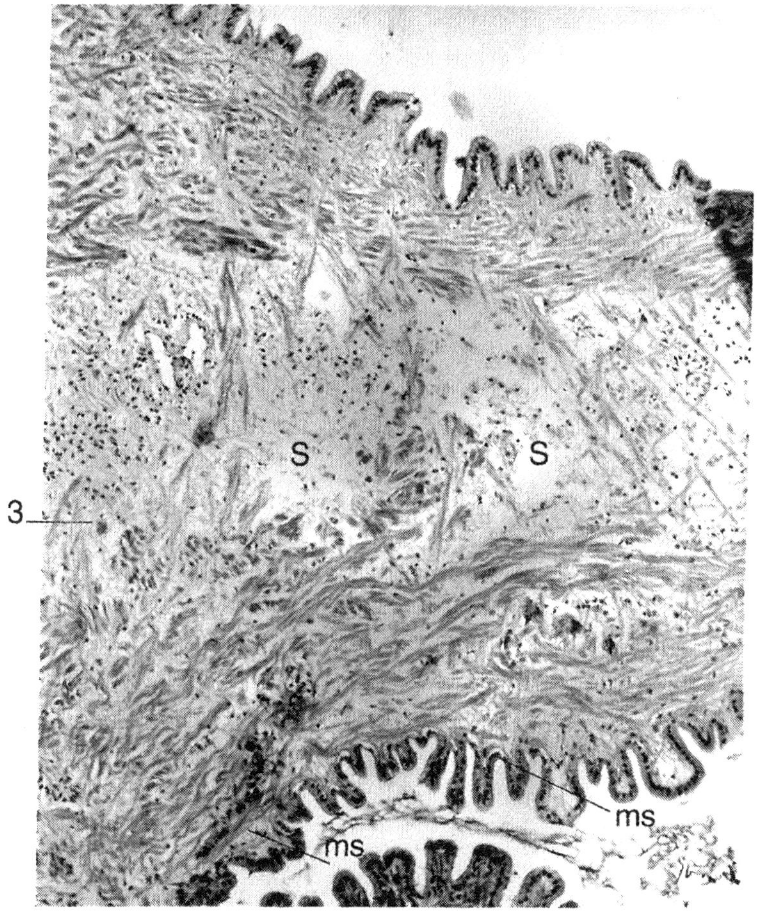

Fig. 4.9. Longitudinal section of third fold (3) showing a prominent hemolymph sinus (s) occupying the core. Subepithelial muscle (ms) is oriented longitudinally. HFW = 405 μm.

the outer portion of the third fold. On this fold, glands are located subepithelially and send long necks to the surface of the epithelium. Hillman (1968) reported that gland cell types I and II that he described for the second fold were also present in the third fold. Glands on the inner surface are scarce. Large bundles of muscle run longitudinally under both outer and inner epithelia the entire length of the fold. A prominent hemolymph sinus forms the core of the fold and strands of connective tissue and slips of muscle traverse the hemolymph sinus from base to tip of the fold (Fig. 4.5).

4.2.2.4 Fourth pallial fold

In many of my sections, representing dozens of animals, this fold is organized as a rounded projection followed by a pointed one (Figs. 4.10 and 4.11). In other species of bivalves, only

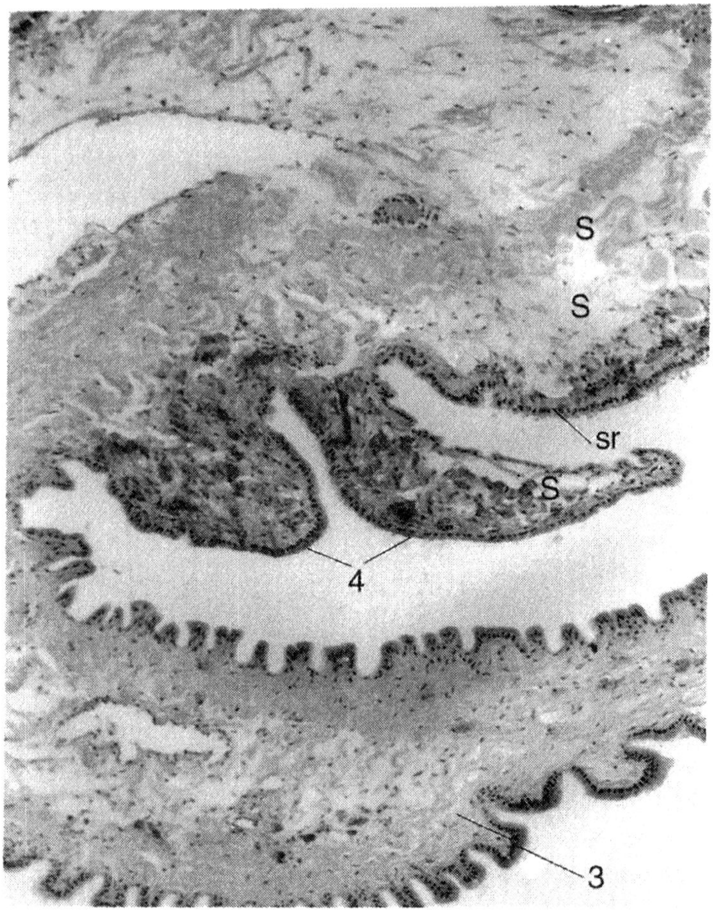

Fig. 4.10. Longitudinal section of third (3) and fourth (4) folds. Fourth fold consists of a rounded projection followed by a more pointed one and contains many unicellular gland cells. The secretory ridge (sr) lies next to the fourth fold. HFW = 810 μm.

the pointed fold is represented (Ansell, 1961; Hillman, 1962, 1964, 1968). The epithelium in both projections (when present) is simple cuboidal (12 × 4 μm). Many glands are located in the connective tissue supporting the epithelium; glands send long necks to the surface where they terminate in swellings interspersed between epithelial cells before discharging their contents to the surface. Gland secretions are of three types: acid (chiefly sulfated) glycosaminoglycans, neutral glycoproteins and acid glycoproteins. Hillman (1964) discussed the histology, histochemistry and possible functions of the fourth pallial fold of *Mercenaria mercenaria*. Glands are distributed in abundance throughout the rounded projection but restricted to the outer region of the pointed projection (Figs. 4.10 and 4.11). There is a weakly developed subepithelial musculature bathed by many hemolymph sinuses in both rounded and pointed lobes.

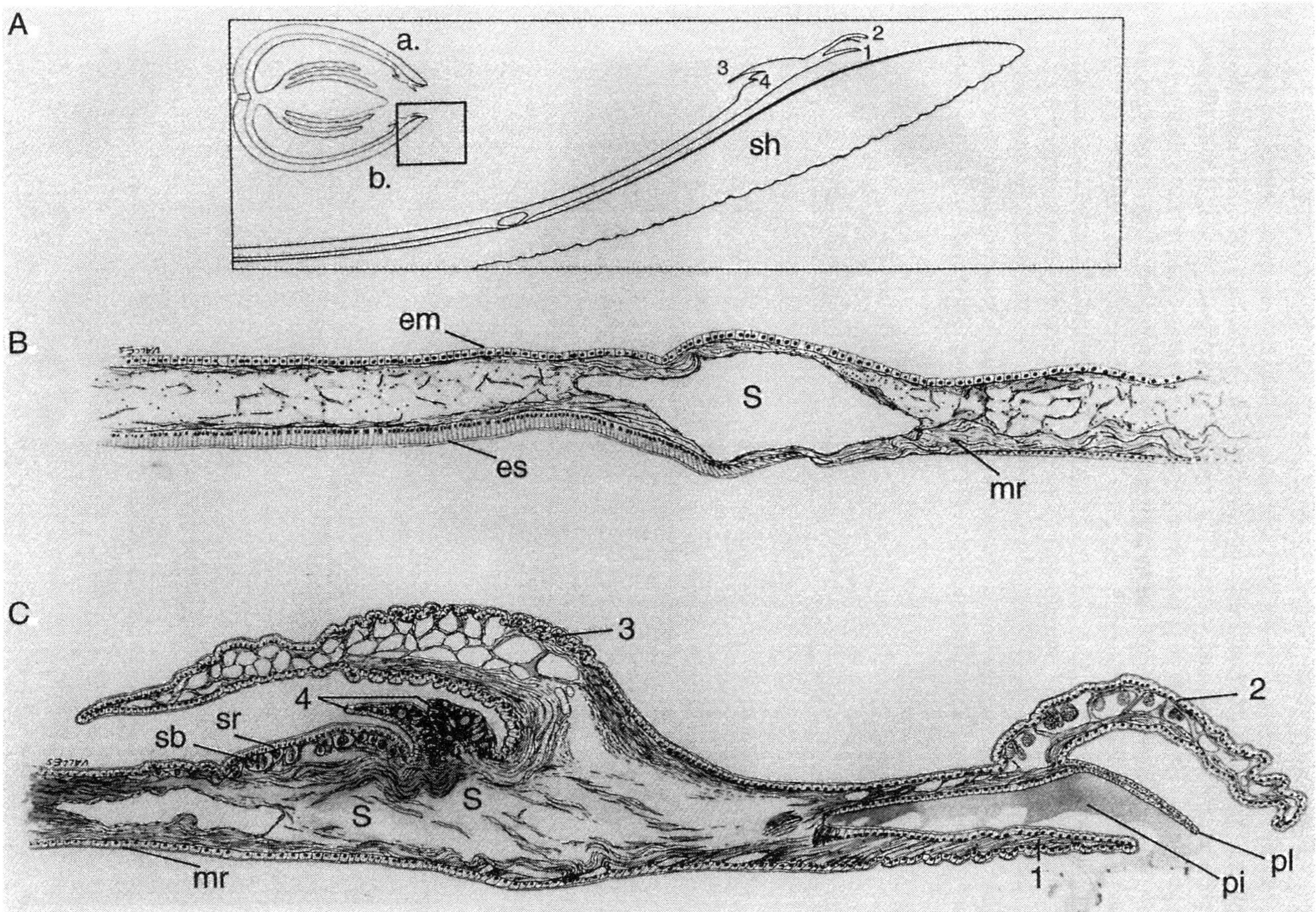

Fig. 4.11. Interpretive sketches of mantle: (A) Longitudinal section through mantle folds. b shows details of rectangle in a. 1, 2, 3 and 4 indicate folds of mantle; sh = shell. 15×. (B) Longitudinal section of mantle at origin of radial muscle (mr). em = mantle-cavity epithelium; es = shell epithelium; s = hemolymph sinus. 16.5×. (C) Longitudinal section of mantle terminating in lobes. 1 = first mantle fold; 2 = second mantle fold; 3 = third mantle fold; 4 = fourth mantle fold; mr = radial muscle; pi = periostracum; pl = periostracal fold; s = hemolymph sinus; sb = subepithetical glands; sr = secretory ridge. 43.5×.

4.2.2.5 Secretory ridge

The secretory ridge (Figs. 4.10–4.12) is that area of the mantle-cavity epithelium close to the origin of the fourth fold and partially covered by it (Hillman, 1962). The epithelium is low cuboidal (6 × 3 μm) and covered by mucous secretions rich in acid (sulfated) glycosaminoglycans. Hillman (1962) described this epithelium as ciliated but probably saw clumps of mucus adhering to the surface; I found no evidence of cilia. Two types of glands are present subepithelially (Hillman, 1962, 1964): a small gland that secretes acid (sulfated) glycosaminoglycans and a larger one that secretes neutral glycoproteins; both send ducts to the surface where they terminate in swellings interspersed between epithelial cells (Figs. 4.10–4.12). A third type of gland-like structure that secretes acid glycoproteins is present subepithelially along the entire secretory ridge. A large hemolymph sinus lies under the secretory ridge; deep to the hemolymph sinus is the large radial muscle of the mantle

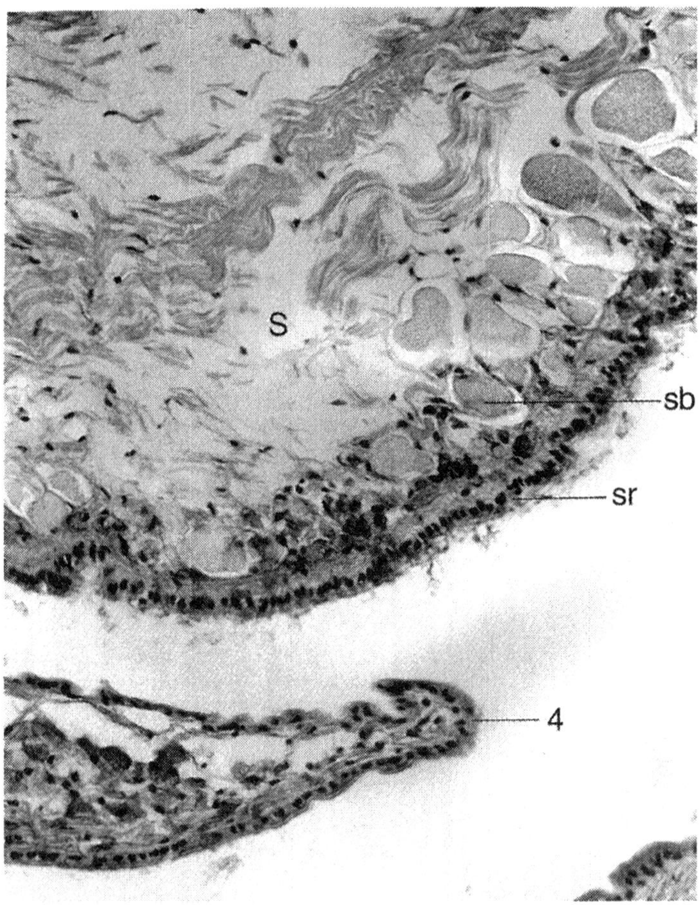

Fig. 4.12. Photomicrograph of a portion of the fourth mantle fold (4) and the secretory ridge (sr) containing prominent subepithelial glands (sb). s = hemolymph sinus. HFW = 406 μm.

(Figs. 4.10–4.12). Hillman (1964) stated that the fourth pallial fold and associated secretory ridge secrete copious amounts of mucus and mucoid substances which are probably involved in removing detritus and pseudofeces (rejecta) from the mantle cavity.

All four folds have a large, central hemolymph sinus which functions as a hydraulic skeleton: subepithelial muscles shorten the folds and force hemolymph out of them; folds can be extended again due to the action of invading hemolymph which maintains their shape and turgidity.

4.3 LABIAL PALPS

4.3.1 Anatomy

There are two pair of labial palps that accept potential food particles from the gills, subject them to further sorting and finally transfer them to the mouth. The palps are roughly triangular in shape; the base is fused with the mantle while the apex is greatly drawn out and extends towards the gills in the mantle cavity (Fig. 4.13). The palps are highly muscular and their complex movements can easily be observed with preparations on the half shell. Each palp has a smooth surface and a ridged surface; the two palps of a pair are arranged so that the surfaces with ridges are in apposition, while the smooth surfaces face away from each other (Fig. 4.14). The outer member of a palp pair lies lateral to the inner demibranch of that side while the inner palp lies medial to the inner demibranch; thus, the pair of palps form a sandwich around the anterior length of the inner demibranch (Fig. 4.14).

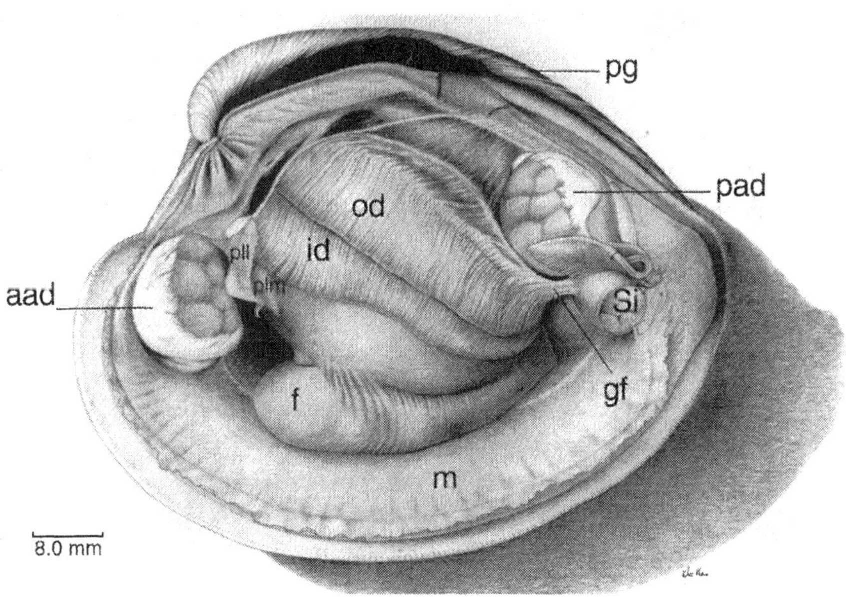

Fig. 4.13. Sketch of half-shell preparation in right valve. Note the inner demibranch of the left gill (id) inserted between the outer (od) and inner left (id) labial palps. aad = anterior adductor muscle; f = foot; gf = fusion of gill to siphon (see text); m = mantle; pad = posterior adductor muscle; si = siphon.

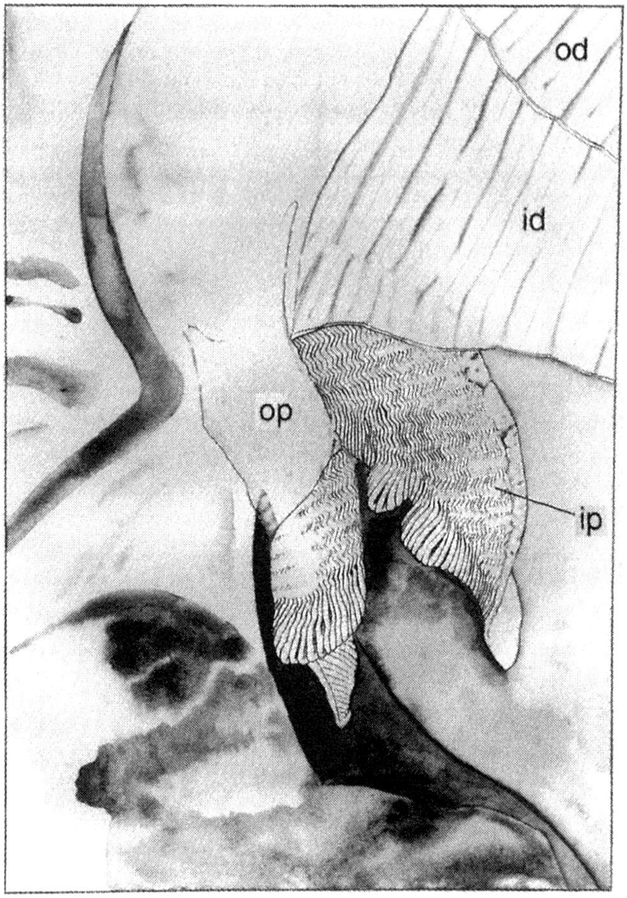

Fig. 4.14. Sketch of outer (op) and inner (ip) left labial palps. Note that ridged surfaces of a palp pair are in apposition while smooth surfaces face away from each other. id = inner demibranch of left gill; od = outer demibranch of left gill. 8×.

Mercenaria mercenaria belongs to Category II with respect to the association between gills and palps (Stacek, 1963): one in which the ventral regions of the anterior-most filaments of the inner demibranch are inserted into and fused to a distal oral groove (Kellogg, 1915). The following account of the sorting of particles on labial palps is based on my studies of clams in half-shell preparations using suspended carmine as tracer particles as well as the use of histological sections of palps. Finally, papers by Ansell (1961) and Foster-Smith (1978) were extensively utilized.

Ciliary tracts on the gills move particles ventrally to the median particle groove, which, in turn, conveys particles either to the distal palp oral groove and/or palp ridged surface where particles are further sorted. Large or densely concentrated particles are swept across the face of the ridged surface of the palp from the dorsal to the ventral edge where they are then transported to the palp tip (Fig. 4.15); the ventral edge of the palp in *M. mercenaria*,

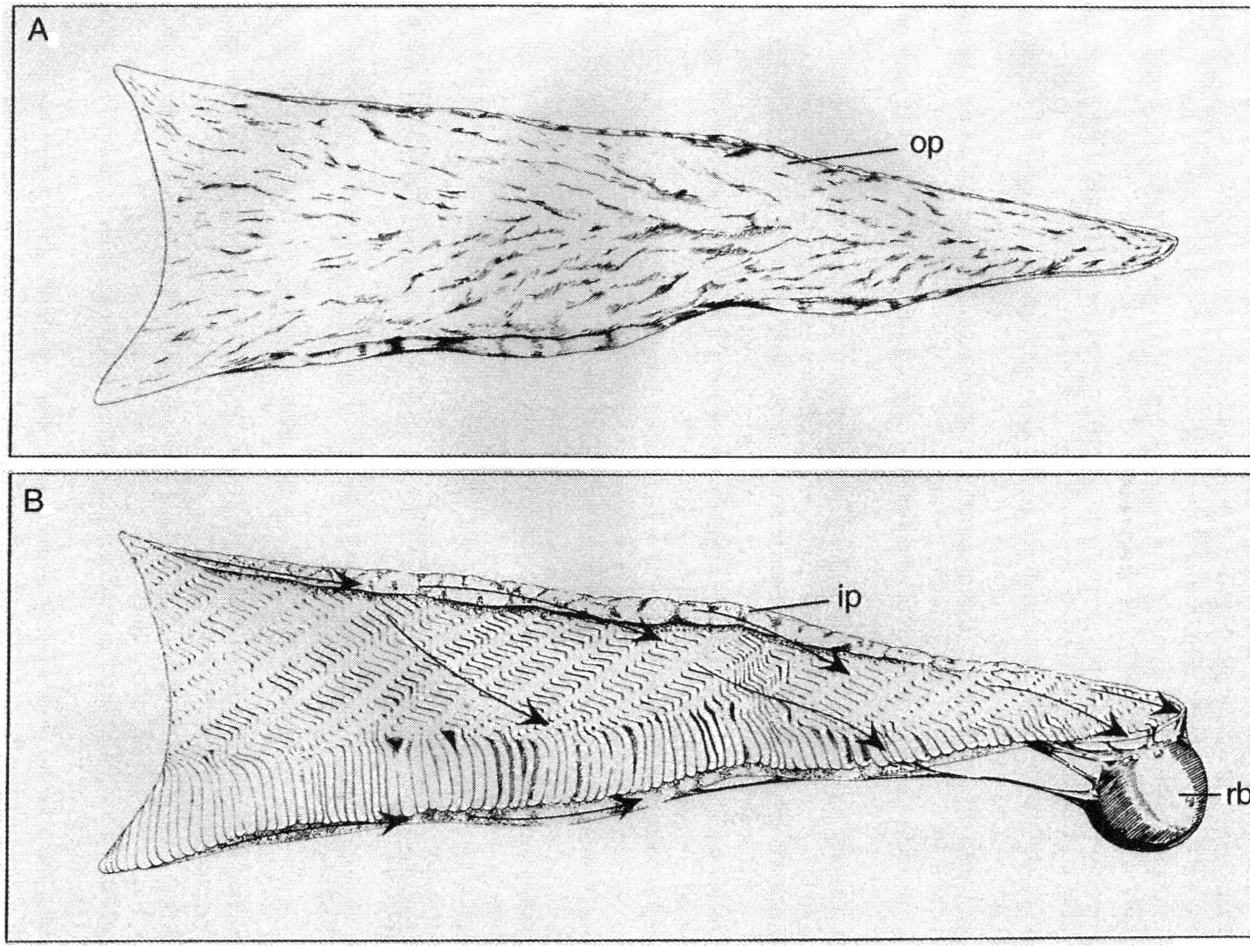

Fig. 4.15. Sketches showing outer (A) and inner (B) surfaces of left labial palps. Arrows show direction of rejection tracts; note ball of rejecta (rb) forming on tip of inner palp. ip = inner palp; op = outer palp. 16.5×.

as in most bivalves (Ansell, 1961; Foster-Smith, 1978), is a strong rejection tract. During these movements, particles are coated and mixed with a heavy mucus resulting in a growing ball of particles enmeshed in mucus that accumulates at the tip of the palp. Heavy loads of particles are held at the tips both by the mucous balls as well as the spiraling action of the distal portion of the palps which brings the strong rejection tracts on the ventral edge in contact with ridged surfaces (Fig. 4.15). Balls of particles wrapped in mucus are ultimately wiped from the tips of the palps onto the mantle by vigorous muscular activity of the palps and form rejecta (pseudofeces). Smaller particles are transferred from the anterior end of the inner demibranchs to the distal oral groove situated at the junction of the inner demibranch and mantle; particles in the distal oral groove are transferred to the lateral oral groove located at the junction of the inner and outer members of a palp pair and are ultimately conducted to the proximal oral groove which conveys particles to the mouth (Kellogg, 1915). Ciliary patterns on the palps as well as the position of the palp ridges, whether relaxed or contracted, determine the fate of particles. In the relaxed position, when ridges overlap, most particles move oralward over crests to the proximal food groove where they are transferred to the mouth. When ridges are erect, a condition caused by a high quantity of material reaching the palps, most particles move downwards into the deep rejection tracts (Figs. 4.16 and 4.18), where particles are moved towards the ventral edge of the palps and are rejected as mentioned above. Moderate amounts of material cause the ridges to undulate, which initiates resorting and rejecting behavior. Details of these movements to the mouth are complex and have not been studied in detail for *M. mercenaria*. Beninger has tried endoscopic observations with *M. mercenaria* but was unsuccessful due to failure of palp relaxation; he has been successful, however, with *Mya arenaria* and *Spisula solidissima* (personal communication). Ansell (1961) studied palp-sorting mechanisms for British species of Veneracea and Foster-Smith (1978) has summarized such data from 19 bivalves. Ward et al. (1991) used direct endoscopic observations to study feeding in several bivalve molluscs and Ward et al. (1994) reported on particle capture and sorting behavior by gills and palps in the eastern oyster, *Crassostrea virginica*. Newell and Langdon (1996) described mechanisms of feeding in *C. virginica*. These papers should be consulted before studies are attempted with the hard clam.

4.3.2 Histology

The smooth surface of each palp is lined with a low, simple cuboidal epithelium (6×6 μm); a prominent layer of longitudinal muscle (9–10 μm wide) is separated from the epithelium by a thin basement membrane (Fig. 4.16). The epithelium and muscle layers are bathed by an extensive system of hemolymph sinuses (Figs. 4.16–4.18). The smooth surface is not ciliated.

The ridged surface of each palp is complex and is divided into ridges and interridges. I have arbitrarily subdivided each ridge into four distinct epithelial zones. Zone I: (Figs. 4.16 and 4.18) situated on the aboral side has a simple, tall columnar (4×16 μm) epithelium with long cilia (this zone together with the adjacent interridge area forms a deep rejection groove for the transport of particles). Zone II: (Figs. 4.16 and 4.18) is also on the aboral surface and separated from Zone I by a deep groove (Figs. 4.16 and 4.18), the resorting tract; this zone is a simple, ciliated, columnar epithelium with three types of gland cells that secrete neutral glycoproteins, acid glycosaminoglycans rich in sulfate groups and acid glycoproteins

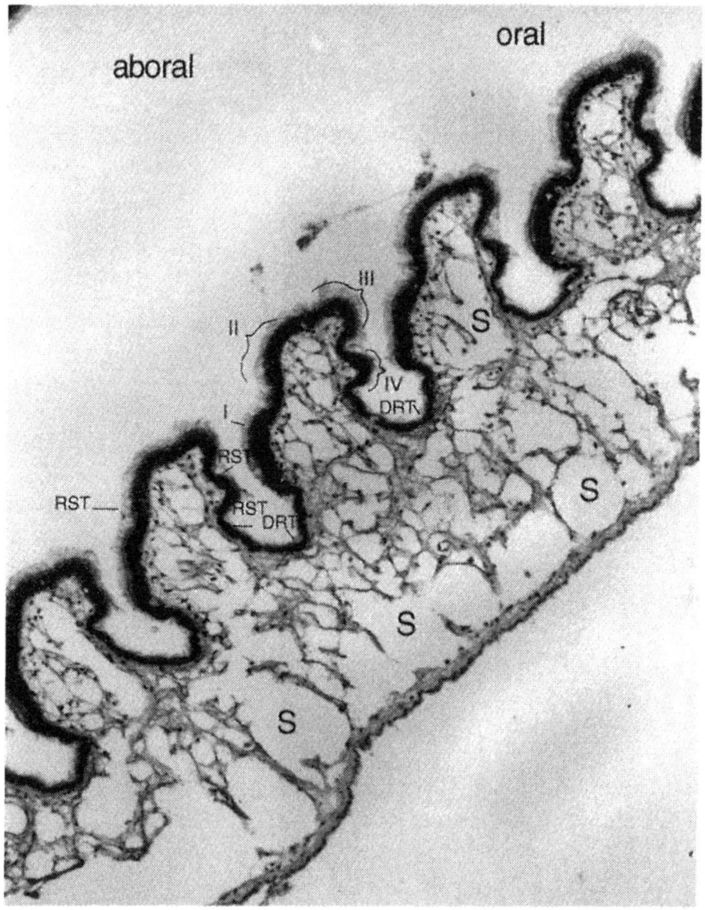

Fig. 4.16. Longitudinal section of labial palp to show ridged and smooth surfaces. Ridged epithelium bears prominent cilia in sorting tracts; smooth epithelium lacks cilia. Extensive hemolymph sinuses (s) pervade the entire organ. I = Zone I; II = Zone II; III = Zone III; IV = Zone IV (see text for details); DRT = deep rejection tract; RST = resorting tract. Oral surface (closer to mouth) lies to the right. HFW = 810 μm.

with both carboxyl and sulfate groups. Zone III: (Figs. 4.16 and 4.18) consists of a simple, columnar epithelium with tall cilia (9 μm) that forms the apex of each ridge, termed the crest rejection ridge; this is a powerful rejectory zone that moves particles rapidly towards the ventral edge of the palp. Zone IV: (Figs. 4.16 and 4.18) is a simple, ciliated, tall (4 × 12 μm) columnar epithelium with prominent grooves, the resorting tracts, (Foster-Smith, 1978) on each side.

The interridge epithelium is simple cuboidal (6 × 5 μm) with short (3 μm) cilia and forms the borders of the deep rejection tract (Fig. 4.18); gland cells are scarce in this area with the exception of those interridge epithelia near the tip of the palps where abundant gland cells are located that secrete neutral glycoproteins, acid glycoproteins and acid glycosaminoglycans, the latter two secretions rich in sulfate groups.

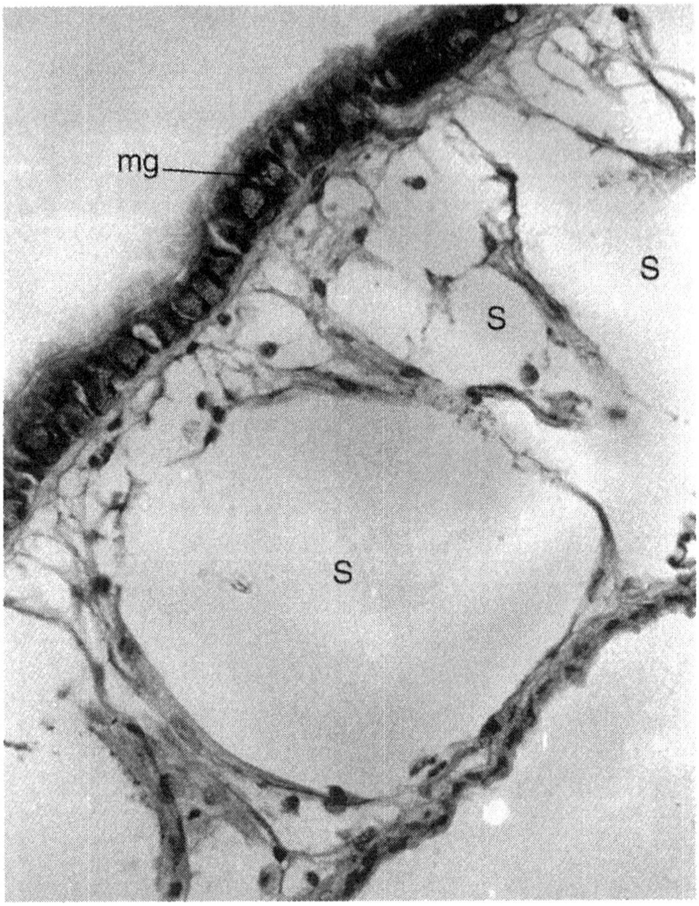

Fig. 4.17. Longitudinal section of ridged surface near ventral edge of labial palp. Several mm from dorsal and ventral margins, the ridged epithelium flattens out and becomes a ciliated, simple columnar epithelium that contains many unicellular gland cells (mg). Note the extensive hemolymph sinus system. s = hemolymph sinus. HFW = 405 μm.

Several millimeters from both dorsal and ventral edges of the palp, the ridged epithelium flattens out and becomes a ciliated, simple, tall columnar (19 μm) with many gland cells, particularly on the ventral side, that secrete acid glycosaminoglycans rich in sulfate groups (Fig. 4.17). These surfaces, particularly the ventral surface, constitute powerful rejection tracts.

A complex mixture of connective tissue and muscle supports the ridged epithelium, particularly under the interridge areas (Figs. 4.16 and 4.18). Each ridge is well supplied with a hemolymph sinus which, in addition to supplying nutrients for local tissues, acts as a hydraulic skeleton to maintain ridge turgidity (Figs. 4.16 and 4.18). The extensive system of hemolymph sinuses that bathes both smooth and ridged epithelia also ramifies throughout the palps (Figs. 4.16–4.18) and is responsible not only for supplying nutrients to local tissues but also for extending the palps following local contractions of longitudinal muscles; hemolymph also contributes to palp turgidity. Bundles of collagenous and reticular connective

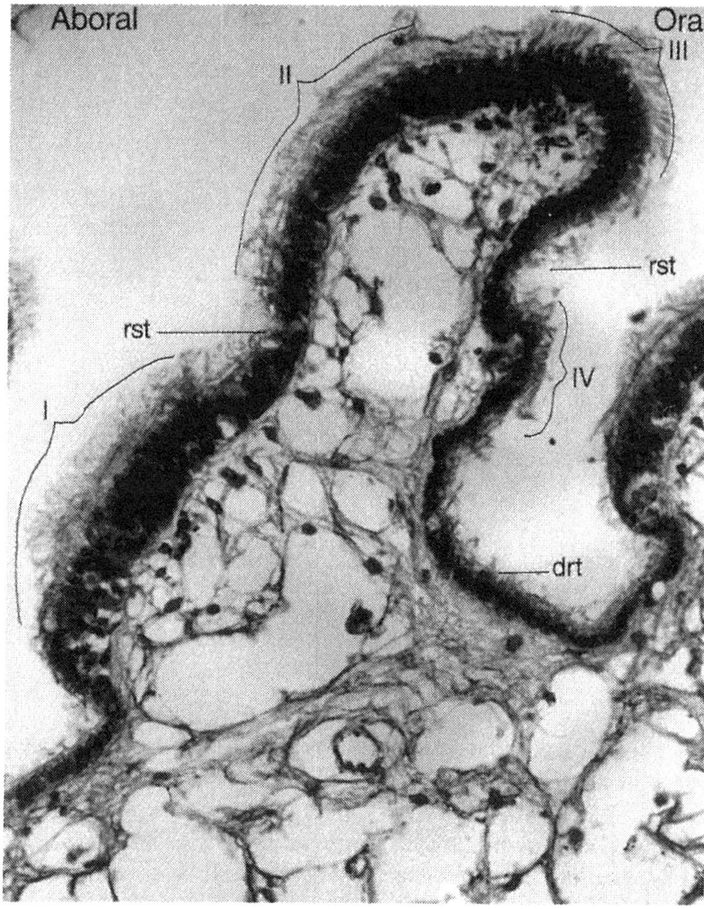

Fig. 4.18. Longitudinal section of ridged surface of labial palp. Extensive ciliation of sorting tracts is evident. Many hemolymph sinuses (s) are present in each ridge. I = Zone I; II = Zone II; III = Zone III; IV = Zone IV (see text for details); drt = deep rejection tract; rst = resorting tract. Oral = side of ridge facing mouth; aboral = side of ridge away from mouth. HFW = 405 μm.

tissue interlace throughout the palps, creating a three-dimensional framework through which hemolymph flows. Thus, palp form, extension after local contractions as well as changes in palp shape, are a function of a complex hydraulic system created by the muscles and connective tissue interacting with the vast hemolymph system.

4.4 GILLS

4.4.1 Anatomy

There are two gills, one on either side of the foot. Each gill consists of two demibranchs, an outer, shorter one and an inner longer member (Fig. 4.13). Demibranchs, or half-gills, in

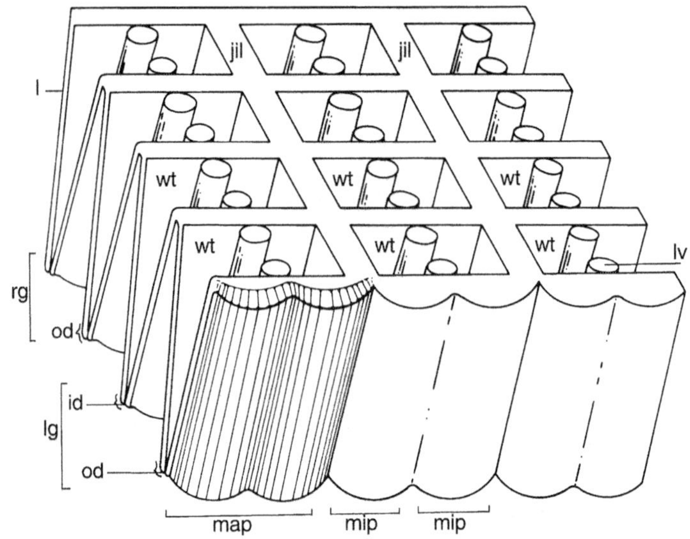

Fig. 4.19. Stereodiagram of both gills viewed from the region of the base of the gills, near the epibranchial chamber. Each major plica (map) has a shallow depression at its apex that divides it into two minor plicae (mip). Note the lamellar vessel (lv), a prominent hemolymph vessel that projects into the water tube (wt) at each junction of two minor plicae. Vascular interlamellar junctions not shown. id = inner demibranch; jil = interlamellar junction; lg = left gill; l = lamella of gill; od = outer demibranch; rg = right gill. 6.4×.

turn, are constructed on a basic plan: two plate-like lamellae consisting of hundreds of fused parallel filaments join to form a V-shaped structure; lamellae are thrown into gentle folds or pleats called plicae (Fig. 4.19).

Each plica (major plica) has a shallow depression at its apex that divides the major plica into two minor plicae; this anatomical arrangement greatly increases the surface area for filtration (Fig. 4.19). The two lamellae that constitute a demibranch are connected to each other at the free margin of the demibranch as well as at numerous interlamellar junctions that form organic connections between opposing lamellae (Fig. 4.19, Fig. 4.20A, B). A prominent hemolymph vessel, the lamellar vessel, is located at each junction of two minor plicae, in the middle of a major plica, and projects into the water tube (Figs. 4.19 and 4.20A, B); at regular intervals, opposing lamellar vessels enlarge and fuse to form a vascular interlamellar junction (Fig. 4.21). Vascular interlamellar junctions also connect bases of major plicae; two small auxiliary vessels, one on each side, spring from these junctions and project into the water tube on that side (Figs. 4.19 and 4.20). Thus, water tubes contain four hemolymph vessels: two lamellar vessels attached to apices of minor plicae and two auxiliary vessels attached to vascular interlamellar junctions that connect opposing major plicae in a demibranch. Accordingly, hemolymph can flow from one lamella to another in a demibranch via the hemolymph vessels located in vascular interlamellar junctions.

Filaments in a lamella are also connected with each other, much like rungs in a ladder; hemolymph sinuses are also located in these interfilamentar junctions (Fig. 4.21). Openings between interfilamentar junctions are the ostia through which water passes when the gill is pumping water (Fig. 4.21). Spaces created between opposing lamellae of a demibranch and

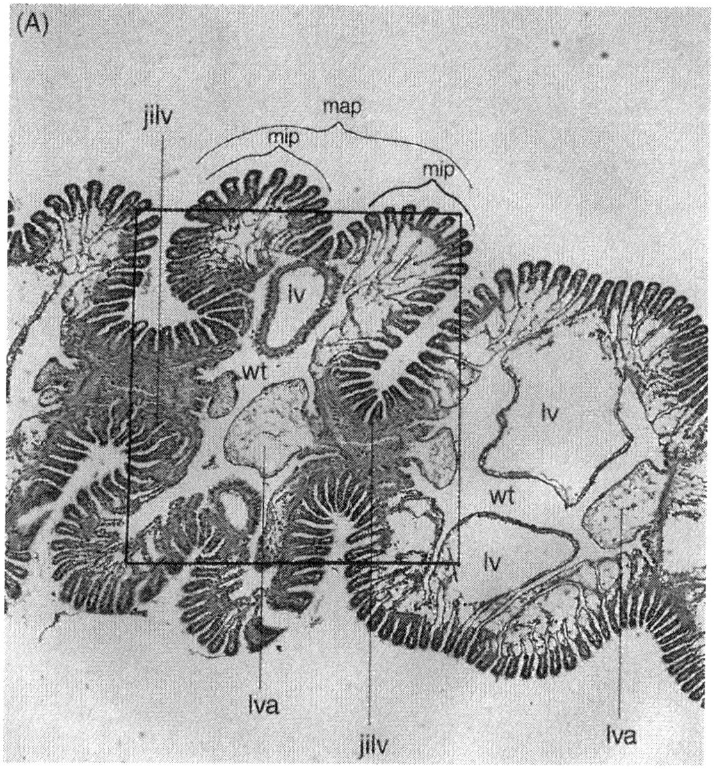

Fig. 4.20. (A) Frontal section of demibranch to show a shallow depression in apex of each major plica (map) that creates two minor plicae (mip). Lamellar vessels (lv) are attached where minor plicae join. Auxiliary lamellar vessels (lva) branch from vascular interlamellar junctions (jilv) that connect bases of major plicae. wt = water tube. HFW = 2.06 mm.

adjacent interlamellar junctions are termed water tubes (Figs. 4.20 and 4.21); these conduct water from the ostia to large channels located at the base of the demibranchs, the epibranchial chambers (Fig. 4.22).

Water tubes from both demibranchs of a gill empty into a common epibranchial chamber, one on each side of the foot; at their most posterior extremity, the gills extend past the foot at which point the water tubes of both gills empty into a single epibranchial chamber which, in turn, conducts water to the excurrent siphon (Fig. 4.22). Anteriorly, the gills fuse with the mantle: the outer, shorter demibranch ends at the distal oral groove while the inner demibranch terminates at the lateral oral groove. The base of the middle of the outer demibranch is fused to the mantle and courses over the lower portion of the pericardium; the base of the inner demibranch is fused to the basal region of the inner lamella of the outer demibranch, while the free margins of both demibranchs lie in the mantle cavity (Fig. 4.13). Posteriorly, the gills fuse with the dorsal edge of the partial flap of the siphon (Figs. 4.13 and 4.22).

Ciliary currents on both gills are essentially as diagrammed by Ansell (1961). The marginal groove, located at the apex of the larger, inner demibranch, conveys the bulk of the material collected by both demibranchs to the palps, since material in the marginal groove of the outer,

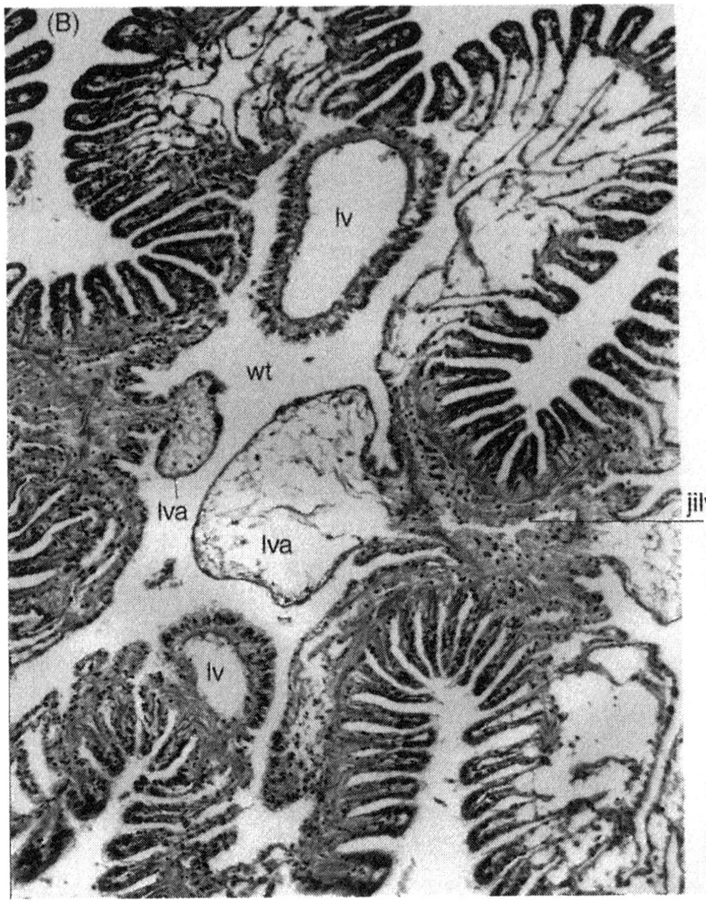

Fig. 4.20 (continued). (B) Frontal section of demibranch to show details of rectangle in A. Note that four hemolymph vessels project into each water tube: two lamellar vessels (lv) and two auxiliary lamellar vessels (lva). HFW = 880 μm.

smaller demibranch is transferred to the inner demibranch near the anterior extremity of each gill (Fig. 4.22). The dorsal groove, located between bases of demibranchs of each gill, also conveys material to the palps (Ansell, 1961).

4.4.2 Histology

Cells in the frontal region of each filament are arranged as a simple, columnar epithelium; cells are about 15 μm tall and bear short (3–4 μm) cilia (Fig. 4.23). Usually, two cells with moderate basophilic cytoplasm bear the long, laterofrontal cirri (compound cilia, 12 μm); two cells with eosinophilic cytoplasm bear the long (10 μm), lateral cilia (Fig. 4.23). Cells that bear lateral cilia appear to be arranged as a simple, low columnar, ciliated epithelium, but Fries and Grant (1991) show this area as two rows of cells in their electron micrographs. A

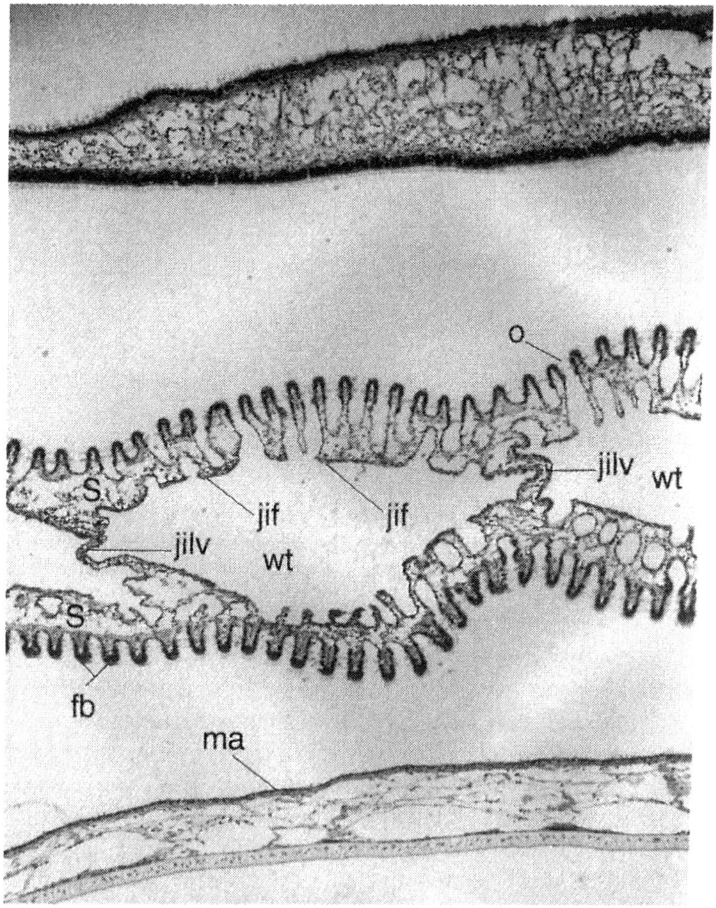

Fig. 4.21. Longitudinal section through demibranch to show two vascular interlamellar junctions (jilv) and several interfilamentar junctions (jif). fb = skeletal bars of filaments; ma = mantle; o = ostium; s = hemolymph sinus; wt = water tube. HFW = 2.06 mm.

single gland cell is frequently located just abfrontal to the cells that bear lateral cilia; the secretion stains positively for acid glycosaminoglycans rich in sulfate groups. The remaining abfrontal epithelium of the filament is a simple squamous that lacks cilia. Many lacunar cells (Atkins, 1937) span the large hemolymph sinus that occupies the abfrontal half of each filament (Fig. 4.23).

A pair of slender skeletal rods supports the frontal surface of filaments (Figs. 4.21 and 4.23); the older literature refers to them as chitinous but they are composed of collagen (Brown, 1952; Ruddall, 1955; Le Pennec et al., 1988). At levels of the interfilamentar junctions, skeletal rods elongate and serve to anchor the interfilamentar muscles that control the size of ostia (Figs. 4.21 and 4.23). Large hemolymph sinuses are present in interfilamentar junctions that are spanned by many lacunar cells (Figs. 4.21 and 4.23).

Interlamellar junctions are composed of connective tissue covered by pseudostratified,

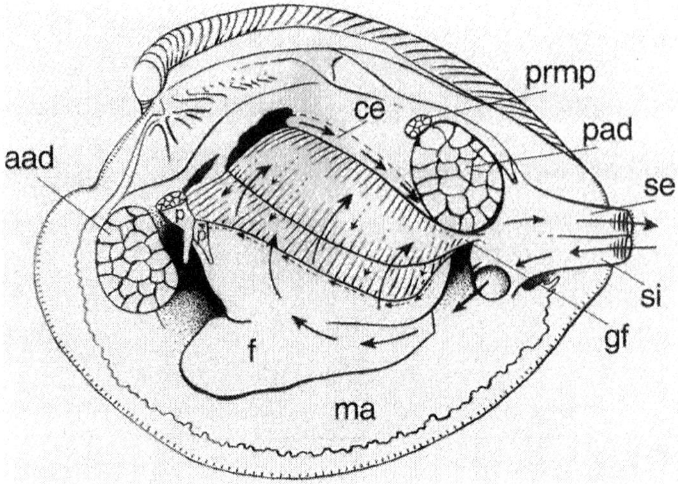

Fig. 4.22. Sketch of half-shell preparation in right valve to show direction of water currents (solid arrows) and ciliary pathways (dashed arrows) on gills leading to palps (p). aad = anterior adductor muscle; ce = epibranchial chamber; f = foot; gf = fusion of gill to part of siphon; ma = mantle; prmp = posterior pedal retractor muscle; se = excurrent siphon; si = incurrent siphon. 1.6×.

columnar epithelia (Fig. 4.21). These junctions occur at various levels from base to apex in the demibranch and have two functions: to maintain the integrity of the demibranch and to circulate hemolymph between the opposing lamellae. Lamellar hemolymph vessels, described above, are covered with pseudostratified, columnar epithelia since opposing lamellar hemolymph vessels, at various levels in the demibranch, meet to form vascular interlamellar junctions. Auxiliary interlamellar junctions are covered with simple, low cuboidal epithelia. Details of afferent and efferent hemolymph flow to and from filaments of a plica as well as hemolymph flow between opposing lamellae of a demibranch are not known but may be similar to that in the eastern oyster (Eble and Scro, 1996).

4.5 SIPHON

4.5.1 Anatomy

The siphon represents the third and fourth mantle folds fused and are of Type B according to the classification of Yonge (1957). Carriker (1961) has given a comprehensive description of the formation of the siphon in young *M. mercenaria*; he notes that mantle fusion and siphon formation resemble closely that described for the venerid clam, *Venerupis pullastra* (Quayle, 1951).

Both incurrent and excurrent chambers of the siphon are fringed with a single row of tentacles. Neither chamber has valves, unlike those reported for other venerids (Jones, 1979), but the excurrent chamber does possess the remnants of the valvular membrane, now termed the primary exhalent siphon (see Carriker, Chapter 3), a maroon-colored shallow flange just inside the circle of tentacles (Carriker, 1961). The primary exhalent siphon functions as the excurrent chamber in young clams and is replaced by the definitive organ as growth exceeds

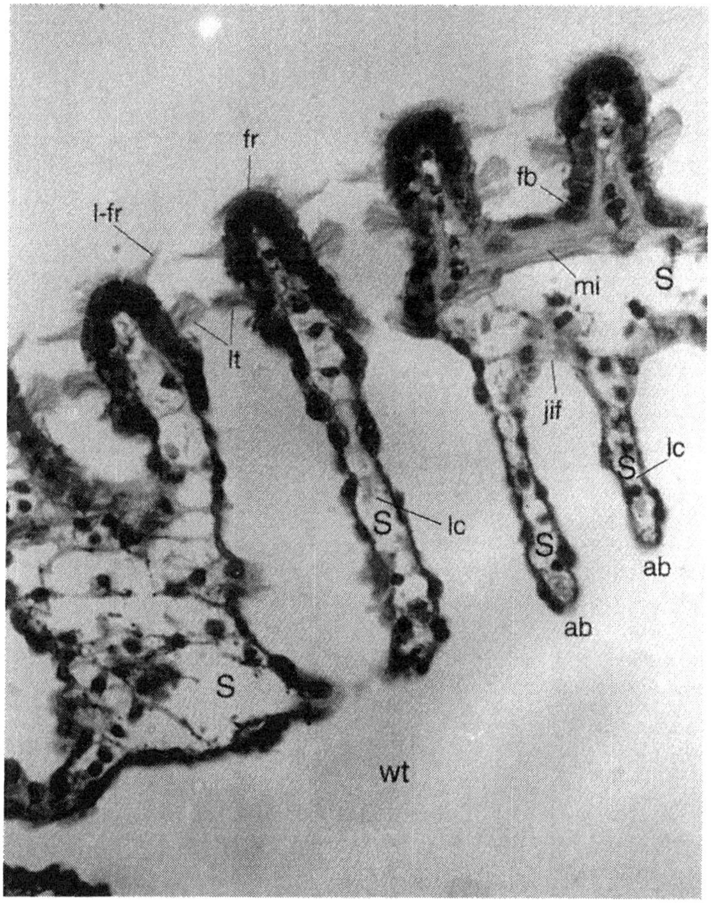

Fig. 4.23. Cross section of gill filaments to show principal ciliation: fr = frontal cilia; l-fr = laterofrontal cirri; lt = lateral cilia. Note slender lacunar cells (lc) spanning the hemolymph sinus (s) in each filament. At regular intervals, skeletal bars of filaments (fb) elongate to attach interfilamentar muscles (mi); this forces all hemolymph into hemolymph sinuses of the interfilamentar junctions (jif). ab = abfrontal surface of filament; wt = water tube. HFW = 55 μm.

about 20 mm; it narrows the opening of the excurrent chamber which has the effect of increasing the velocity of the water leaving this chamber, thus, minimizing the chances of the excurrent flow being reintrained with the incurrent water (Carriker, 1961). Externally, the siphon has a dark pigment fringing the orifices including the tentacles; this pigment is reduced to a series of irregular spots along the length of the cream-colored siphon. Internally, both chambers are lined with a bright-yellow pigment, the function of which is unknown. Pallial muscles in the region of the siphon become highly specialized and hypertrophied to form the siphonal retractor muscle; this muscle attaches to the shell along the indented pallial line in this region creating an area known as the pallial sinus (Figs. 4.1 and 4.24). Both valves contain much purplish pigment in the pallial sinus; this portion of the shell was polished, made into

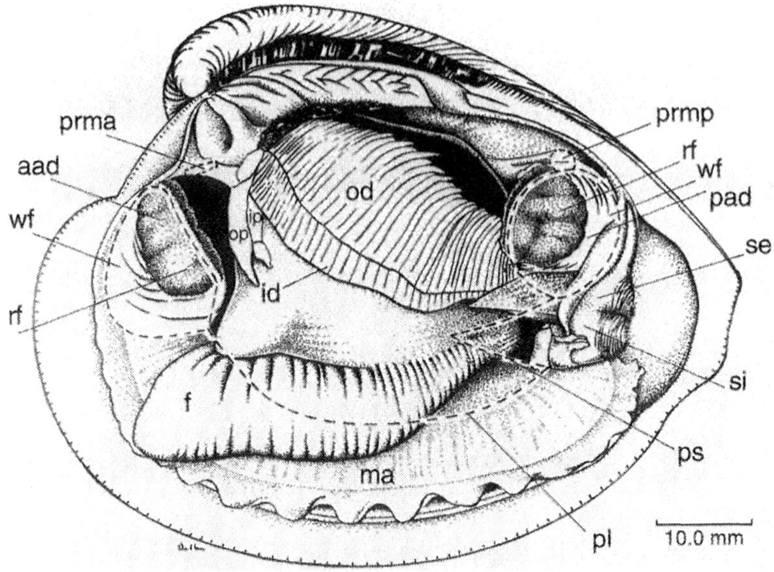

Fig. 4.24. Sketch of half-shell preparation in right valve. Muscle scars on valves are indicated by dashed lines. aad = anterior adductor muscle; pad = posterior adductor muscle; f = foot; id = inner demibranch of left gill; ma = mantle; od = outer demibranch of left gill; ip = inner left palp; op = outer left palp; pl = pallial line; prma = scar of anterior pedal retractor muscle; prmp = scar of posterior pedal retractor muscle; ps = pallial sinus; rf = red fibers of adductor muscles; se = excurrent siphon; si = incurrent siphon. wf = white fibers of adductor muscles.

beads and used as 'wampum' by certain indigenous tribes in the northeastern United States. Today, this beautiful region of the shell is made into costume jewelry.

The common epibranchial chamber at the posterior end of the gills empties directly into the excurrent chamber of the siphon as does the rectum (Fig. 4.22). The partial flap at the base of the incurrent chamber can apparently be raised to deflect water currents ventrally toward the ventral edge of the base of the incurrent siphon where rejecta from the mantle and palps finally accumulate (Fig. 4.25); once in suspension, this material can be forcefully ejected from the incurrent chamber by a sudden adduction of the valves (Kellogg, 1915; Jones, 1979).

4.5.2 Histology

The internal epithelium of both incurrent and excurrent siphonal chambers is simple, low columnar with basally situated nuclei. A prominent layer of circular muscle is organized just under the inner epithelium which has two functions: (1) acts to extend the siphon after the shell opens and; (2) acts as a sphincter at both incurrent and excurrent openings (Figs. 4.26 and 4.27). Many glands are located interspersed in the circular muscle with ducts that open onto the surface of the internal epithelium; gland secretions are rich in sulfated and carboxylated glycosaminoglycans (Fig. 4.26).

A prominent layer of circular muscle lies under the external epithelium, similar to the internal surface (Figs. 4.26 and 4.27). Large bundles of longitudinal muscle are organized

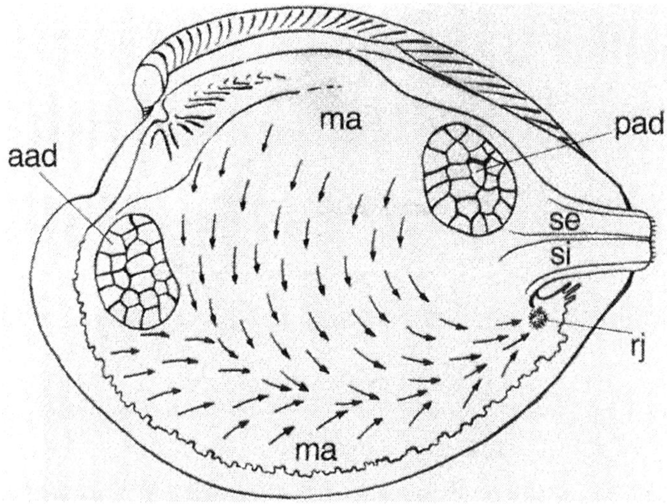

Fig. 4.25. Sketch of ciliary currents on right mantle. Note collection area for rejecta (rj) is situated at base of incurrent siphon (si). aad = anterior adductor muscle; pad = posterior adductor muscle; ma = mantle; se = excurrent siphon; si = incurrent siphon. Redrawn after Kellogg (1915). 1.1×.

immediately under the circular muscle; bundles are separated and organized into blocks of muscle by prominent connective tissue sheaths (Fig. 4.27). The histology of the siphon of *Mercenaria (Venus) striatula* (Ansell, 1961) is quite similar to that of *M. mercenaria*. The function of the longitudinal muscle is to shorten the siphon; contraction of longitudinal muscles occurs simultaneously with contraction of sphincter muscles resulting in a shortening of the siphon with a closing of the incurrent and excurrent apertures.

Siphonal tentacles are covered with two types of epithelia: the inner surface has an epithelium that resembles epithelia from the internal and external surfaces of the body of the siphon; the outer epithelium is highly folded and resembles taste-bud epithelia in mammals (Fig. 4.28). Many small nerve bundles lie in the proximity of this epithelium (Fig. 4.28) which apparently is highly sensory. A thin band of circular muscle enclosing gland cells lies under both internal and external epithelia; many more gland cells are located in the inner than in the outer side of the tentacles. Gland cells empty via large ducts directly onto the surface epithelium; gland secretions are glycosaminoglycans rich in carboxyl and sulfate groups. Large bundles of longitudinal muscle are aligned down the center of each tentacle and many prominent transverse muscle fibers are present at all levels in the tentacles from base to apex (Fig. 4.28).

4.6 FOOT

4.6.1 Anatomy

The foot in *M. mercenaria*, as in all the Veneridae, is large and wedge-shaped, being adapted for burrowing in soft substrata. The foot is flattened laterally, is highly muscular and extends nearly the entire ventral surface of the visceral mass. The anterior end is shaped like a

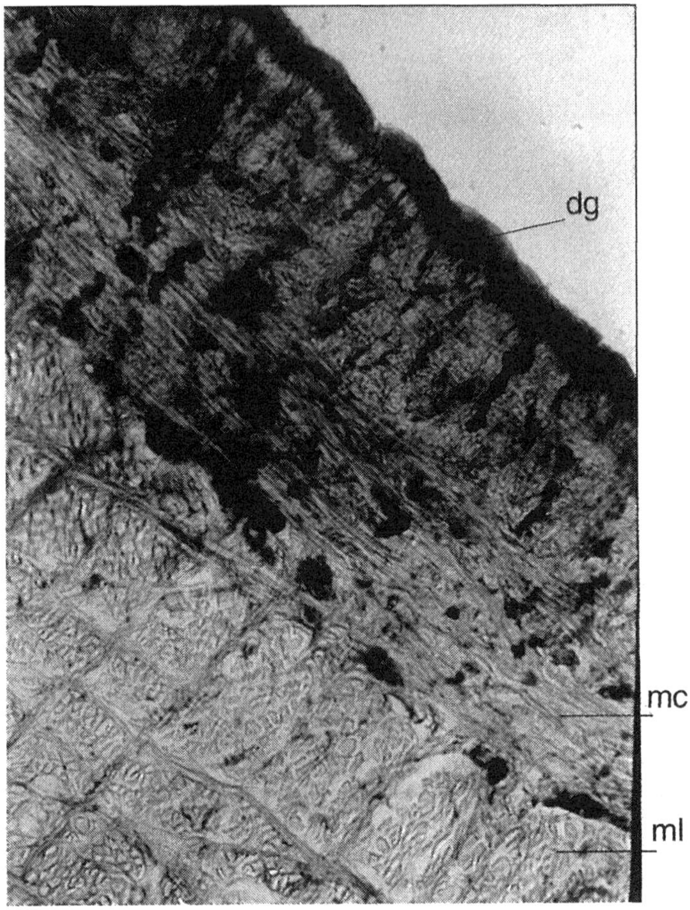

Fig. 4.26. Cross section of siphon to show profusion of unicellular glands interspersed throughout the circular muscle layer (mc) with ducts (dg) running to the epithelium. ml = longitudinal muscle. HFW = 405 μm.

plow and is greatly protrusible. The ventral surface of the foot terminates in a sharp ridge that is somewhat ruffled. To accommodate the large foot, there is a corresponding large pedal gape in the mantle margins.

The foot is retracted and ultimately withdrawn into the shell cavity by the action of the anterior and posterior pedal retractor muscles; each retractor muscle has its origin on the dorsal portion of the shell (Fig. 4.24) and inserts into the tissues of the foot especially at the ventral area where muscle fibers of both retractor muscles converge (Kellogg, 1892; Fig. 4.29). The foot is extended by the action of subepithelial circular muscles, described below, in addition to the pressure of hemolymph, which is forced into extensive hemolymph sinuses that bathe all muscle tissues (Fig. 4.30).

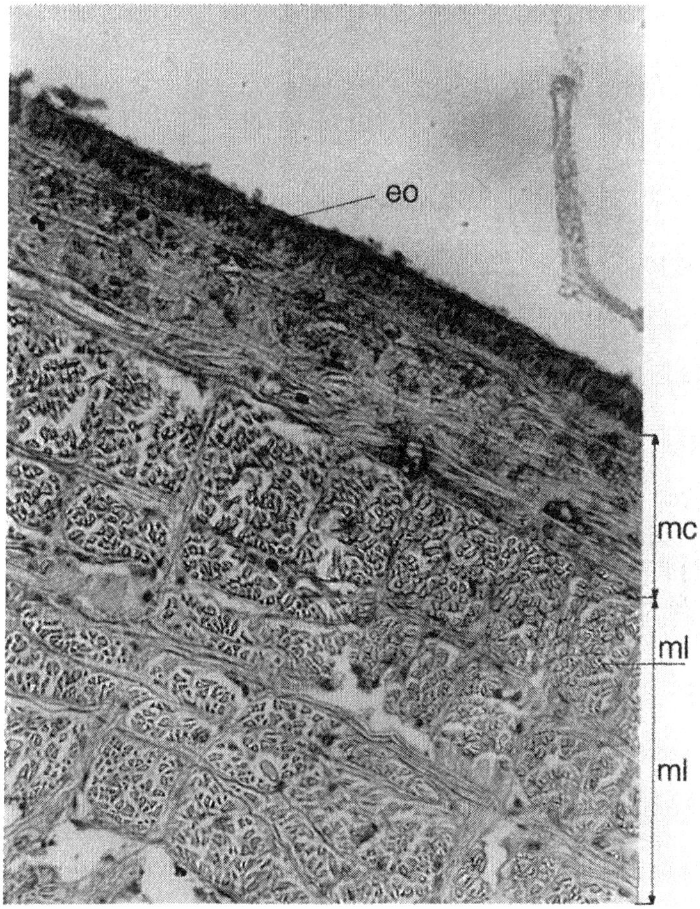

Fig. 4.27. Cross section of siphon to show arrangement of outer epithelium (eo), circular muscle (mc) and block-like bundles of longitudinal muscle (ml). HFW = 810 μm.

4.6.2 Histology

The foot is covered with a highly folded, simple columnar epithelium that is ciliated near the tip of the anterior extremity. Crests of folds are tall columnar (6 × 22 μm) while troughs are low columnar to cuboidal epithelium (6 × 7 μm; Figs. 4.30 and 4.31). The epithelium rests on a dense connective tissue support that contains many muscle fibers (Figs. 4.30 and 4.31). Prominent glands that secrete acid glycosaminoglycans rich in sulfate and carboxylate groups are located subepithelially (Fig. 4.31); long ducts convey secretions to the epithelium, which becomes coated with a prominent layer of acid glycosaminoglycans (Fig. 4.32).

Muscle bundles from the anterior and posterior retractor muscles branch repeatedly to send fibers into the periphery where they further branch into a fan-like network before inserting on the connective tissue that supports crests of the epithelium (Fig. 4.31). A layer of thick,

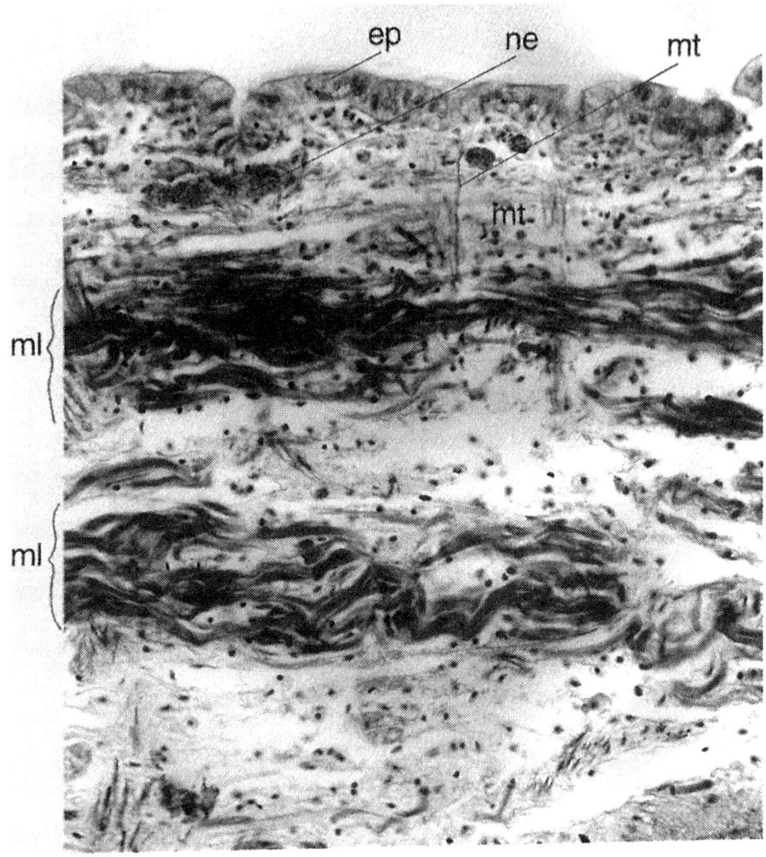

Fig. 4.28. Longitudinal section of outer surface of siphonal tentacle. Epithelium (ep) is similar to taste-bud epithelia of mammals. Small nerve bundles (ne) lie under the epithelium. Muscles are organized chiefly as longitudinal (ml) and transverse (mt) bundles. s = hemolymph sinus. HFW = 405 μm.

circular muscle bundles lies just peripheral to the centrally located, longitudinal muscles. All muscle bundles are enveloped by a fine, reticular-like connective tissue that contains acid glycosaminoglycans; this connective tissue borders all hemolymph sinuses prominent especially in the central domains of the foot (Fig. 4.30).

4.7 MUSCULAR SYSTEM

4.7.1 Anatomy

4.7.1.1 Foot

The muscular tissues intrinsic to the foot were described in the preceding Section 4.6.2. There are two pairs of extrinsic muscles: the left and right pedal anterior and posterior retractor muscles. The anterior pair originate on the left and right valves in a pit located

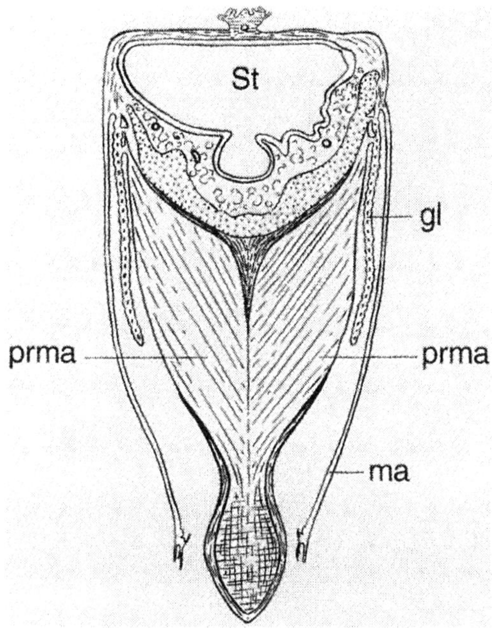

Fig. 4.29. Sketch of cross section of anterior region of visceral mass to show insertion of right and left anterior pedal retractor muscles (prma) in the foot (f). gl = gill; ma = mantle; st = stomach. Redrawn after Kellogg (1892). 5.6×.

under the anterior edge of the dental platform (Fig. 4.24; Kellogg, 1892; Jones, 1979); the muscle fibers run ventrally and medially in the foot and join in the anterior, mid-ventral region (Fig. 4.29). The pair of posterior pedal retractor muscles originate on the anterodorsal corner of the posterior adductor muscle (Fig. 4.24), course over the anterior edge of this muscle then enter the foot where they run ventrally and medially and eventually meet and intermingle with the fibers from the pair of anterior pedal retractor muscles (Fig. 4.29).

4.7.1.2 Adductor muscles

There are two adductor muscles, anterior and posterior (Fig. 4.24). Each adductor muscle consists of a small white portion and a large pinkish-to-red area; the former contains the so-called 'catch' fibers and the latter, the 'quick' fibers. In the anterior adductor muscle, the white fibers form the anterior region of the muscle with the pink, quick fibers constituting the remainder; white fibers are found in the posterior portion of the posterior adductor muscle, pink fibers forming the remainder. The pigment in pink portions of both adductor muscles was identified as hemoglobin (Fox, 1953).

4.7.2 Histology

Morrison et al. (1970) studied the histology of adductor muscles of bivalve molluscs from a variety of habitats and concluded that the classification system of Morton (1958) was essentially correct.

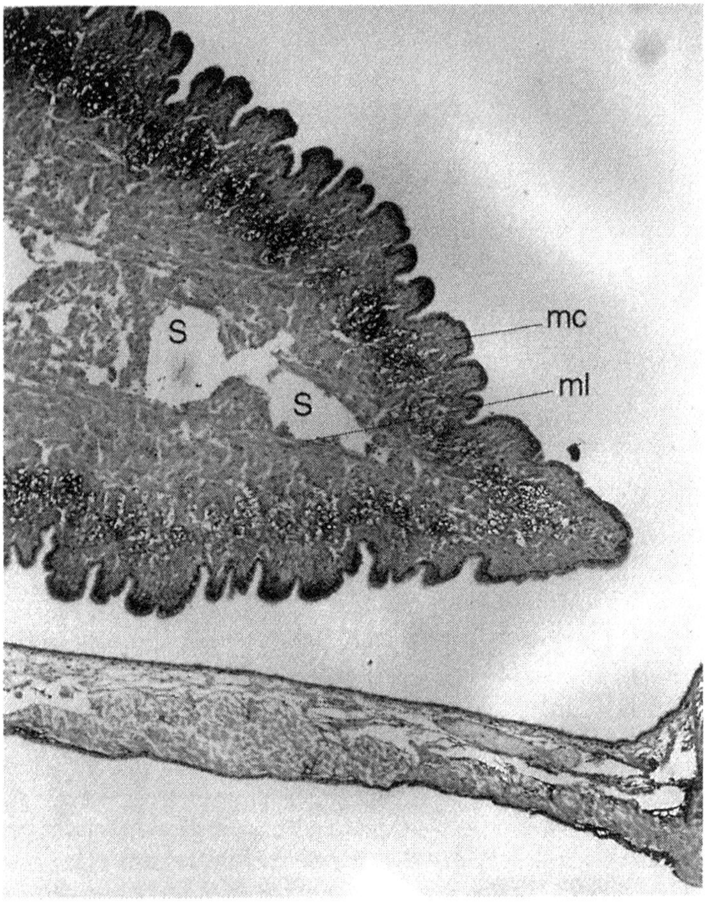

Fig. 4.30. Longitudinal section through anterior tip of foot. Note prominent hemolymph sinus (s) in core of foot. mc = circular layer of muscle; ml = longitudinal layer of muscle. HFW = 810 μm.

My studies show muscle fibers surrounded by a fine connective tissue, the endomysium (Figs. 4.33 and 4.34), rich in carboxylated glycosaminoglycans. Bowden (1958) describes the endomysium as delicate, argyrophilic reticular fibers that form a sheath around muscle fibers. Large bundles of adductor muscle fibers surrounded by connective tissue, the perimysium, stain identically to the endomysium, but can be separated from each other by prominent hemolymph sinuses (Fig. 4.35). In my studies, fibers from the white, 'catch' muscle averaged about 2.8 μm while fibers from the pink, 'fast' muscle averaged 1.2 μm in diameter. These figures are lower than those given by Bowden (1958) although he does not mention *Mercenaria* (= *Venus*) in his studies on dimensions of intact muscle fibers.

Muscle fibers of both white and pink muscles are of the smooth type. The same is true of both the extrinsic as well as intrinsic muscle fibers of the foot.

No work has been published on the fine structure of the adductor muscles of *M. mercenaria*; this work is badly needed. Workers are referred to Nicaise and Amsellem (1983) for a

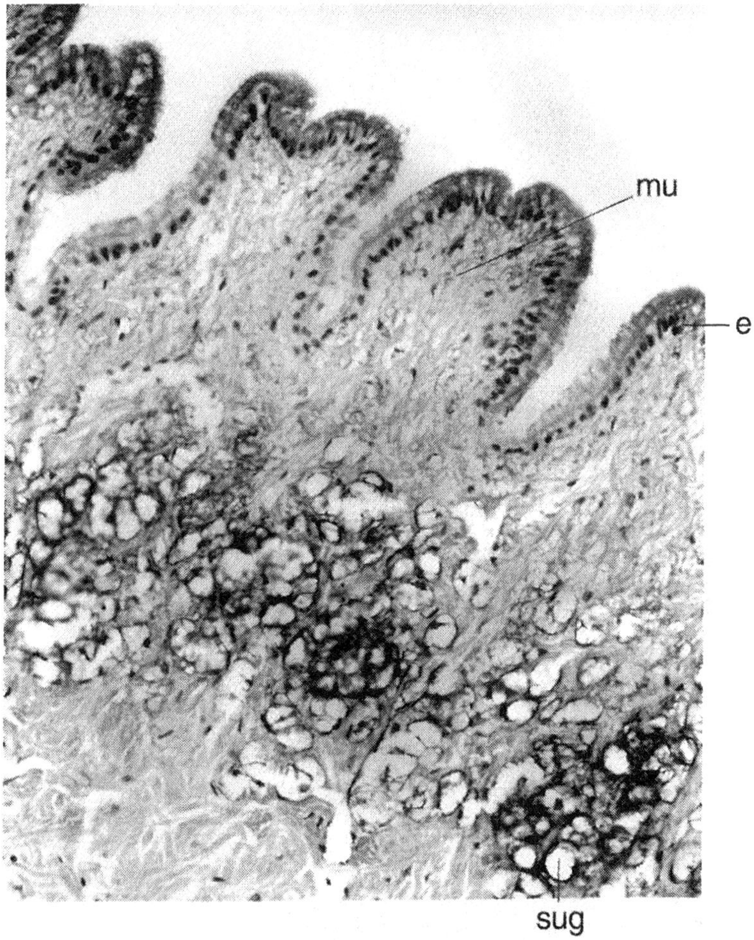

Fig. 4.31. Longitudinal section through foot near anterior extremity to show ciliated, simple columnar epithelium (e). Note prominent muscle bundles (mu) inserting on subepithelial connective tissue in a fan-like network. Subepithelial glands are evident (sug). HFW = 405 μm.

discussion on cytology of molluscan muscles, Muneoka and Twarog (1983) for an in-depth treatment of molluscan neuromuscular transmission and excitation–contraction coupling, Chantler (1983) for a comparison of biochemical and structural aspects of molluscan muscles, Chantler (1991) for a comprehensive discussion of scallop muscle, and Morrison (1993b, 1996) for a through treatment of the histology and fine structure of the adductor muscle of the eastern oyster, *Crassostrea virginica.*

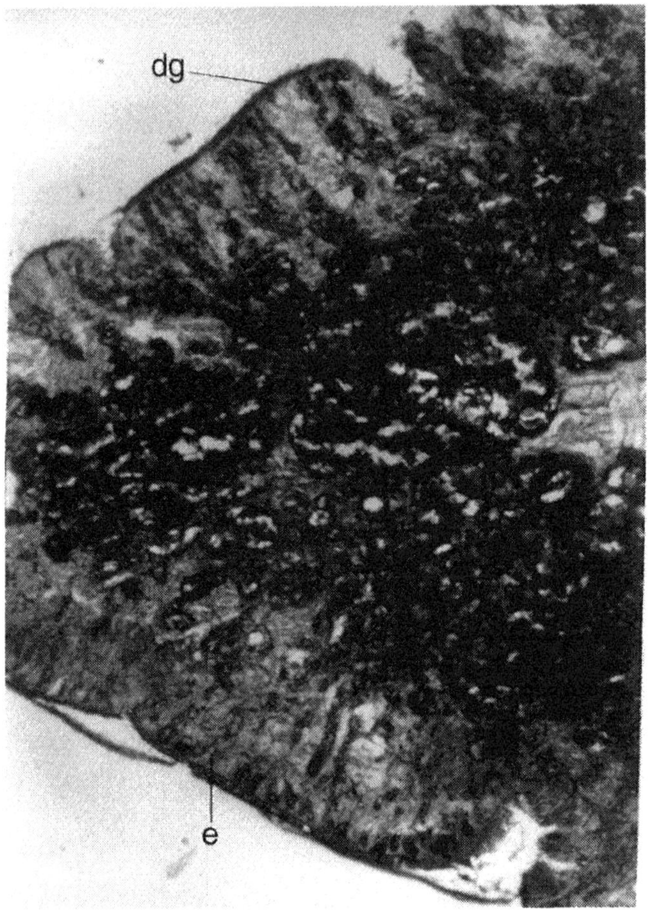

Fig. 4.32. Longitudinal section of foot near its anterior extremity to show unicellular subepithelial glands that secrete onto the epithelial surface via long ducts (dg). e = epithelium. HFW = 405 μm.

4.8 DIGESTIVE SYSTEM

4.8.1 Anatomy

4.8.1.1 Mouth

The mouth is situated in the median line in the isthmus of tissue that connects the left and right labial palps. The opening is usually oval shaped with the shortest radius on the dorsoventral axis; the tissue bordering the mouth is usually thrown into fine folds so that the mouth appears slightly 'puckered'.

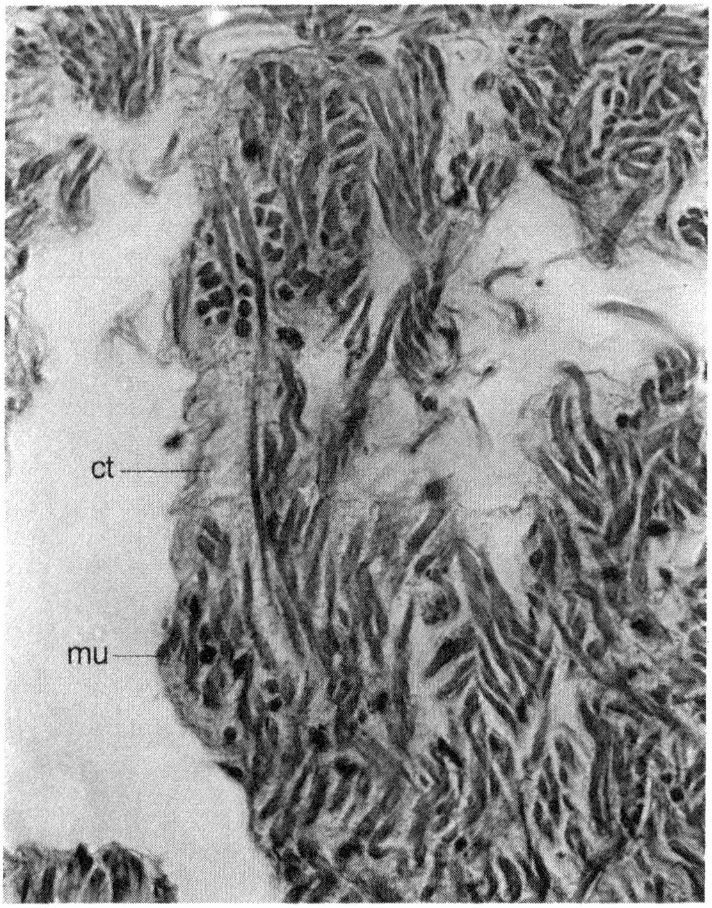

Fig. 4.33. Section through foot to show fine reticular connective tissue (ct) binding muscle bundles (mu). HFW = 44 μm.

4.8.1.2 Esophagus

The esophagus is a straight tube about 8–10 mm long that connects to the anterior ventral region of the stomach at approximately a 90° angle (Fig. 4.36). The lumen of the esophagus is lined with longitudinal rugae that are fine and closely spaced in *M. mercenaria* as well as in *Chione cancellata* and *C. undatella* (Jones, 1979).

4.8.1.3 Stomach

The stomach is roughly divided into a spacious, globular, anterior end and a narrow posterior region containing the style sac and intestine (Ansell, 1961). The anterior stomach receives the esophagus on its ventral surface. Slightly above the esophageal orifice the stomach wall evaginates into the right and left caeca from which primary ducts lead to the

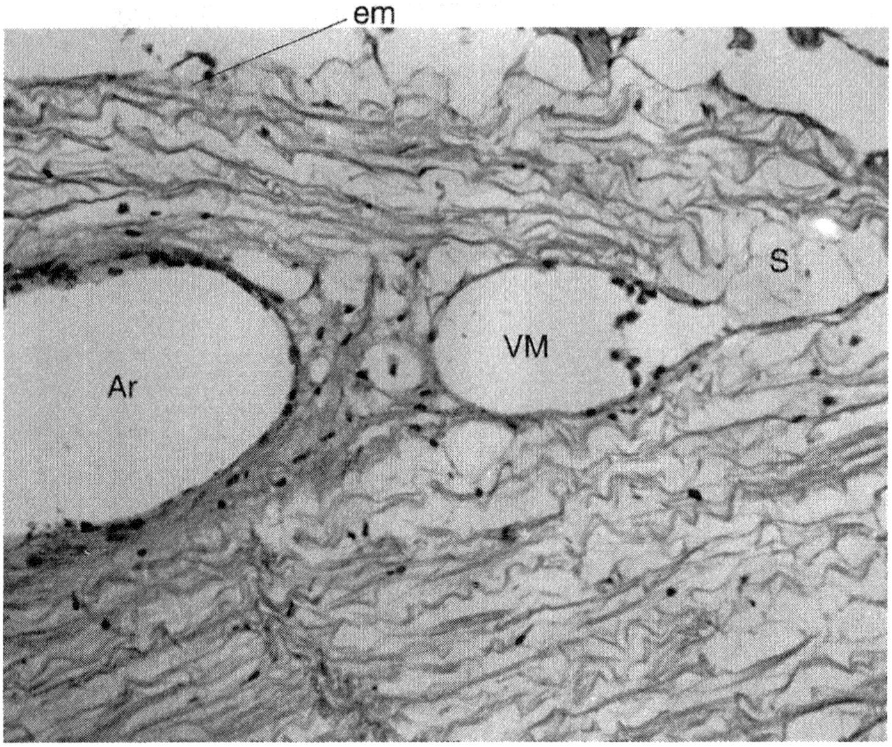

Fig. 4.34. Cross section through the adductor muscle to show an artery (ar), vein (vn) and hemolymph sinuses (s). em = endomysium. HFW = 44 μm.

paired digestive glands (Fig. 4.37). The external wall of the anterior stomach is highly folded except at the extreme anterior end where it forms a bulbous sac (Fig. 4.37).

The posterior stomach containing the style sac and intestine forms approximately a 90° angle with the anterior stomach and runs posteriad and ventrally towards the heel of the foot. The major and minor typhlosoles separate the style sac on the left from the intestine on the right (Figs. 4.37 and 4.38). Because of the angle the posterior stomach makes with the anterior stomach, the crystalline style is curved as it courses from the style sac through the anterior stomach where it abuts against the gastric shield embedded in the dorsal wall of the anterior stomach. The crystalline style may be easily dissected out intact but only from animals freshly removed from the field or aquaria. The gastric shield may be visualized by dissecting open the anterior stomach and gently removing the shield with a fine forceps. Ansell (1961) shows the gastric shield in situ while Jones (1979) illustrates it free from gastric tissues.

Details of food-sorting tracts and mechanisms in venerid stomachs are discussed and figured in Ansell (1961) and Jones (1979); the latter also illustrates the stomach in *M. mercenaria*. Reid (1965) discussed the structure and function of the stomach of 9 bivalves belonging to three orders of the Polysyringia, a new subclass of the Bivalvia suggested by Purchon (1960); he did not describe the stomach of *M. mercenaria*. Dinamani (1967) described the stomachs of 36 bivalves including some venerids, but not that of *M. mercenaria*.

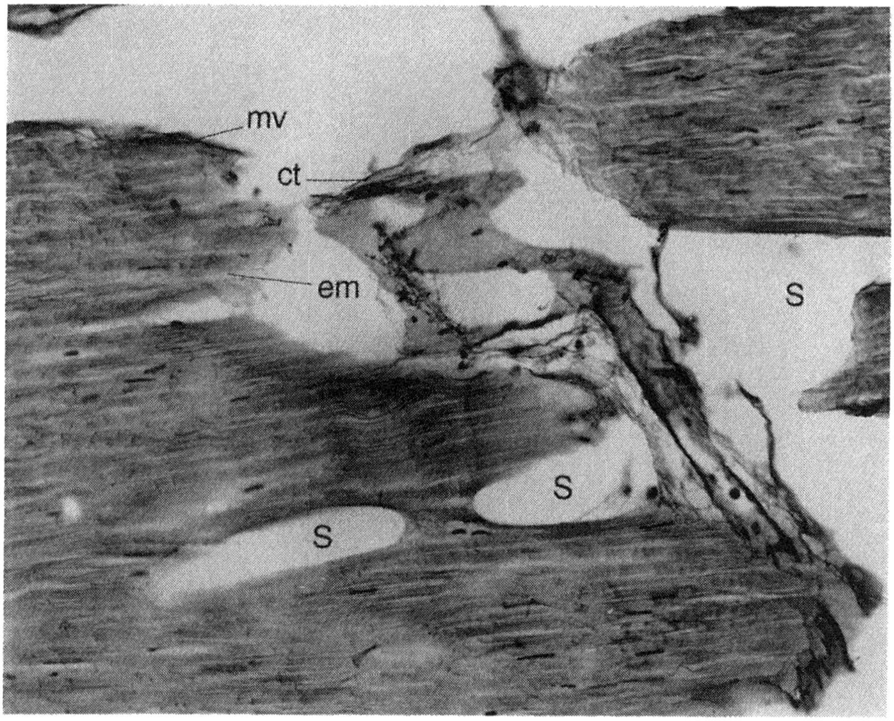

Fig. 4.35. Cross section through the adductor muscle to show fine, reticular connective tissue (ct), the perimysium, enveloping and spanning muscle bundles (mu). Note the extensive hemolymph sinus system (s). em = endomysium. HFW = 436 μm.

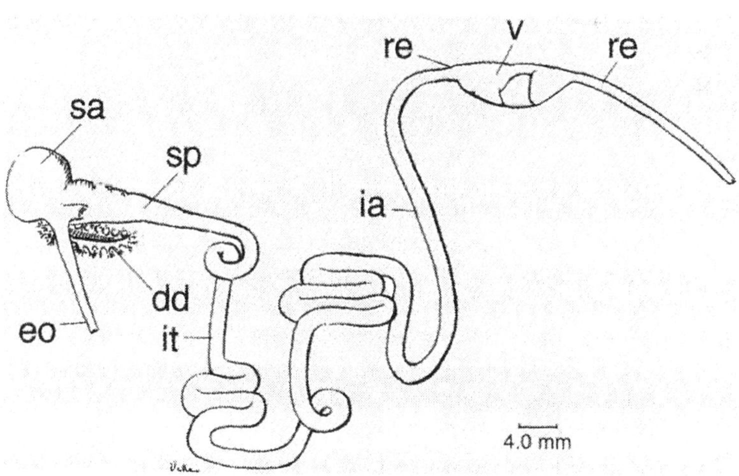

Fig. 4.36. Sketch of the digestive system from a vinyl acetate cast, left side. dd = digestive diverticula; eo = esophagus; ia = ascending intestine; it = anterior intestine; re = rectum; sa = anterior stomach; sp = posterior stomach; v = ventricle.

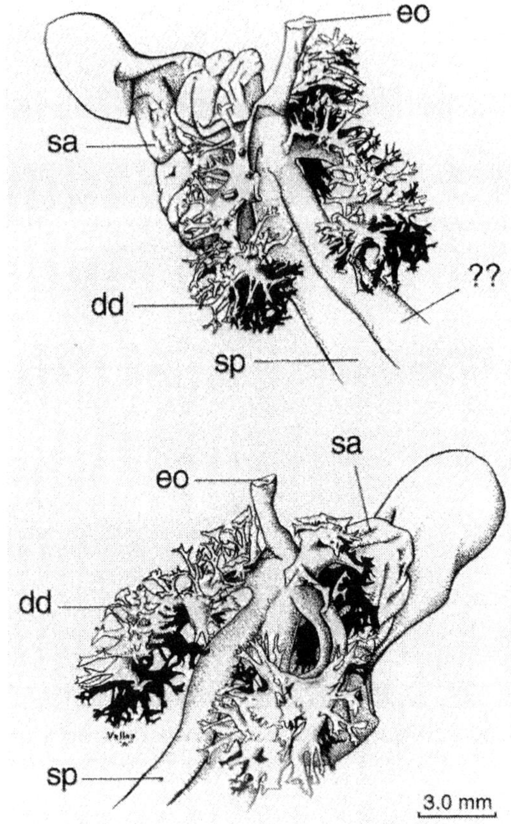

Fig. 4.37. Sketch of the esophagus (eo), anterior stomach (sa), posterior stomach (sp), and digestive gland (dd) from a vinyl acetate cast. Upper-right, lateral view; lower-left, ventral view.

4.8.1.4 Intestine

The intestine originates in and constitutes the right portion of the posterior stomach; this is obvious from vinyl-acetate casts made in our laboratory (Figs. 4.37 and 4.38). The intestine is divided into three areas based on location and histology: (1) anterior, (2) ascending and (3) rectum (Fig. 4.36). Immediately upon leaving the posterior stomach, the anterior intestine curves ventrally and courses in an anterior direction slightly to the left of midline in the ventral portion of the foot (Fig. 4.36, Fig. 4.39); at least two complete loops are formed in this region before the anterior intestine courses posteriad and slightly ventrally (Fig. 4.36). A prominent typhlosole fills most of the lumen of the anterior intestine that, perforce, is crescent shaped (Fig. 4.40). The anterior intestine grades into the ascending intestine by taking a right-angle turn dorsally (Fig. 4.36); shortly before this point, the typhlosole of the anterior intestine dramatically decreases in size and is absent in the ascending intestine. The latter ascends in the visceral mass until it reaches the level of the heart where it takes another right-angle turn

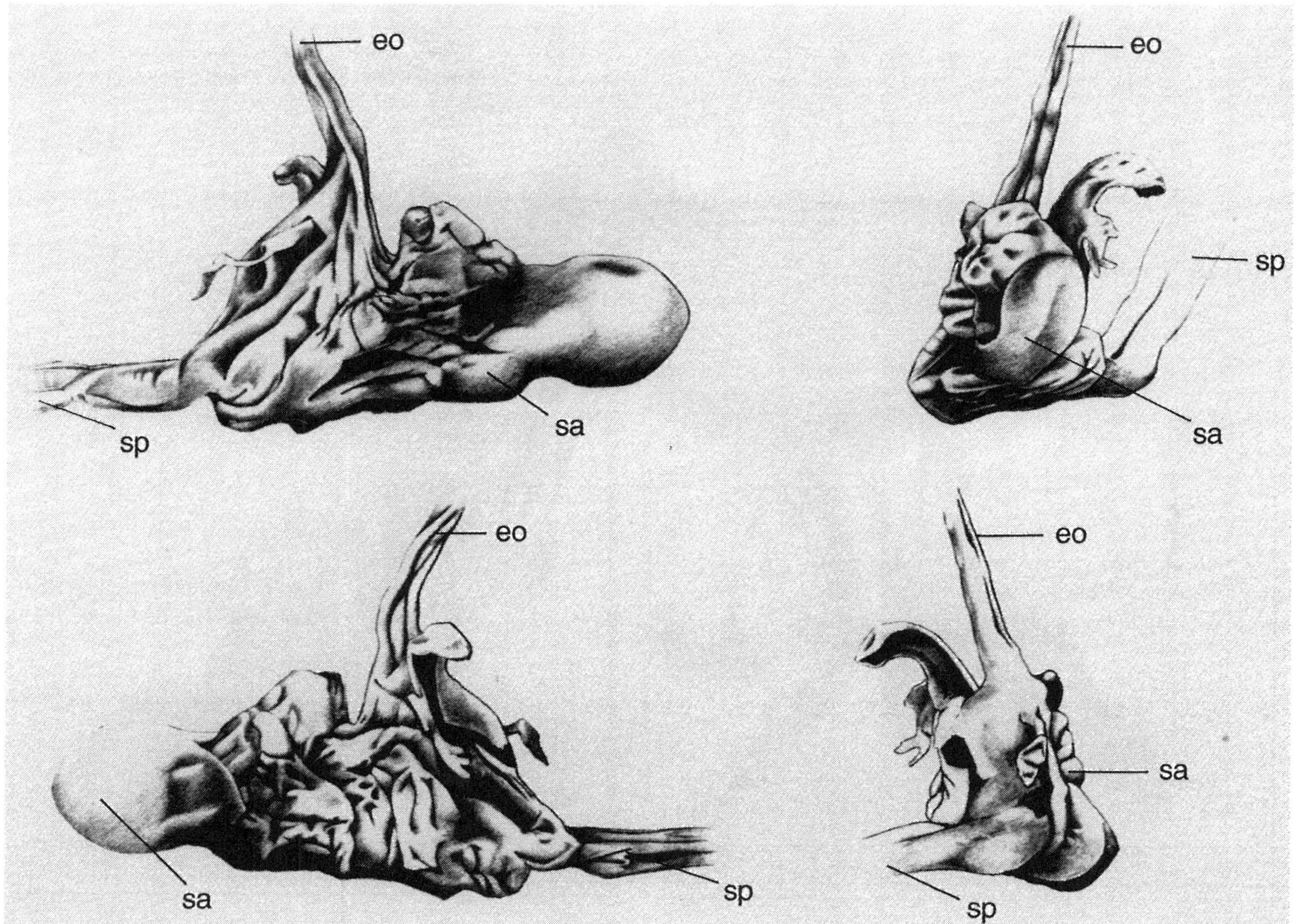

Fig. 4.38. Sketches of the esophagus (eo), anterior stomach (sa), and posterior stomach (sp) from a vinyl acetate cast; digestive gland not shown. Top left: left, lateral view. Top right: left anterior view. Bottom left: right, lateral view. Bottom right: posterior view. 16×.

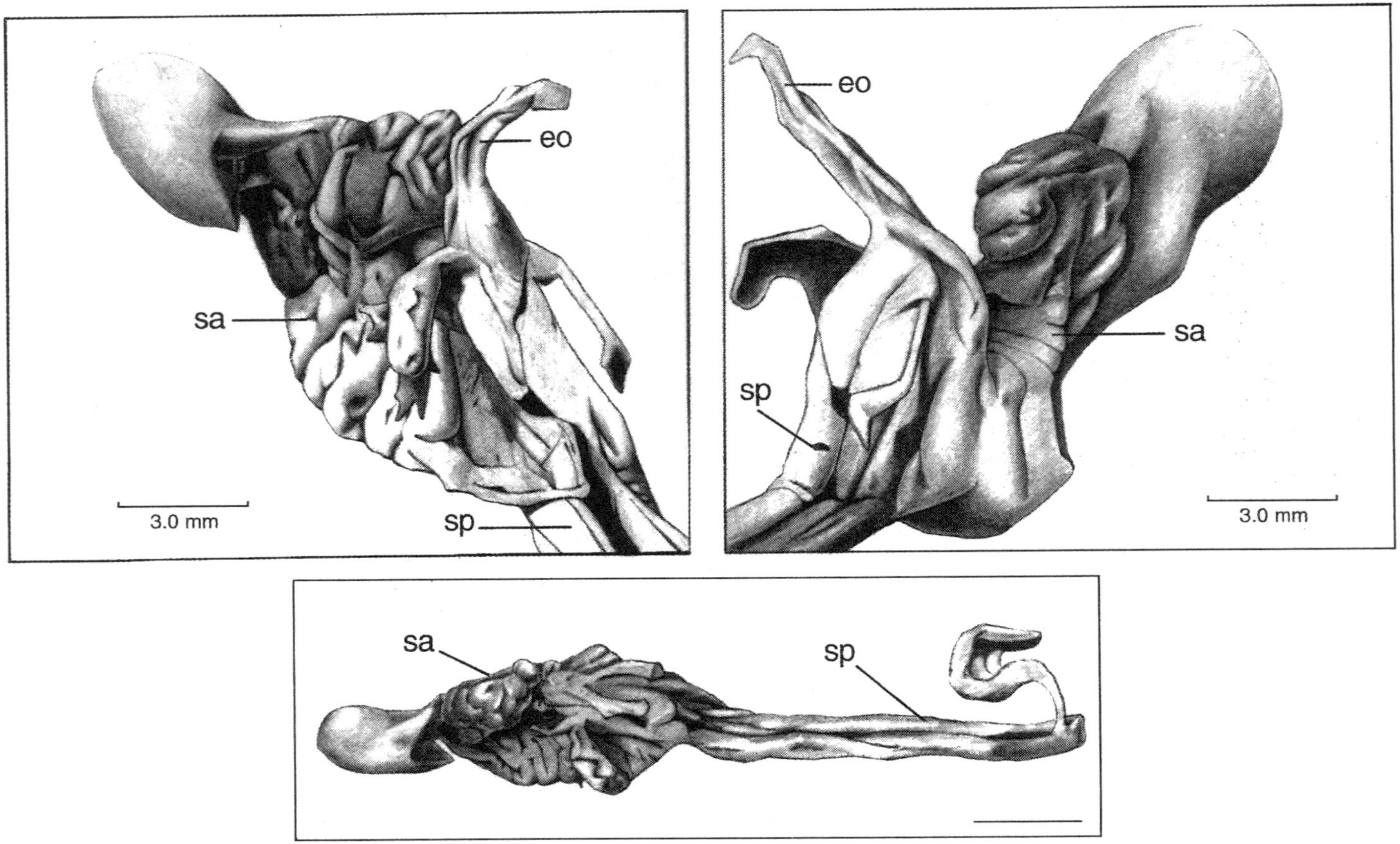

Fig. 4.39. Sketch of the esophagus (eo), anterior stomach (sa) and posterior stomach (sp) from a vinyl acetate cast; digestive gland not shown. Top left: right, ventral surface. Top right: left, ventral surface. Bottom: dorsal surface.

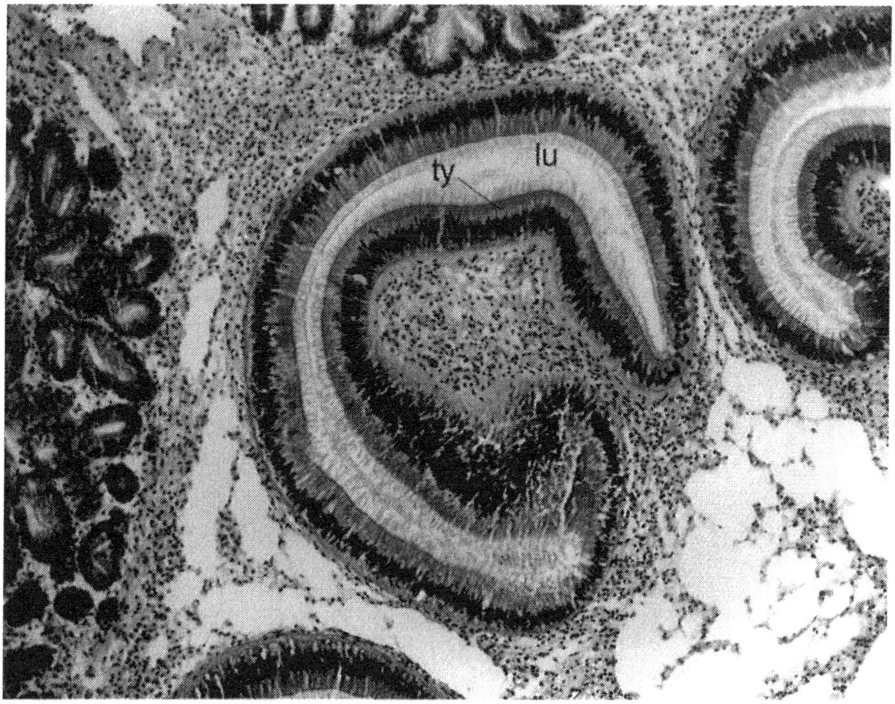

Fig. 4.40. Photomicrograph of cross section through the intestine. Note the prominent typhlosole (ty), the presence of which imparts a C-shape to the lumen (lu). HFW = 436 μm.

and enters the pericardial coelom in company with the anterior aorta at which point it becomes the rectum (Figs. 4.36 and 4.41).

4.8.1.5 Rectum

The rectum, with the anterior aorta attached to its dorsal surface, is found in the pericardial coclom (Fig. 4.41). The rectum is partially enveloped by the ventricle of the heart; posterior to the heart, the rectum is surrounded by the posterior aorta (Fig. 4.41). Approximately 8–10 mm from the heart the rectum is enveloped by the dorsal portion of the aortic bulb (Fig. 4.41). Posterior to the aortic bulb, the rectum leaves the pericardial coelom at its dorsoposterior corner and courses over and follows the contour of the posterior adductor muscle until it meets the excurrent siphon where it terminates as the anus (Fig. 4.22; Jegla and Greenberg, 1968a).

4.8.2 Histology

4.8.2.1 Esophagus

The lower esophagus, close to the mouth, has a narrow lumen in the shape of a cross caused by the presence of four large rugae. The internal epithelium is ciliated, pseudostratified

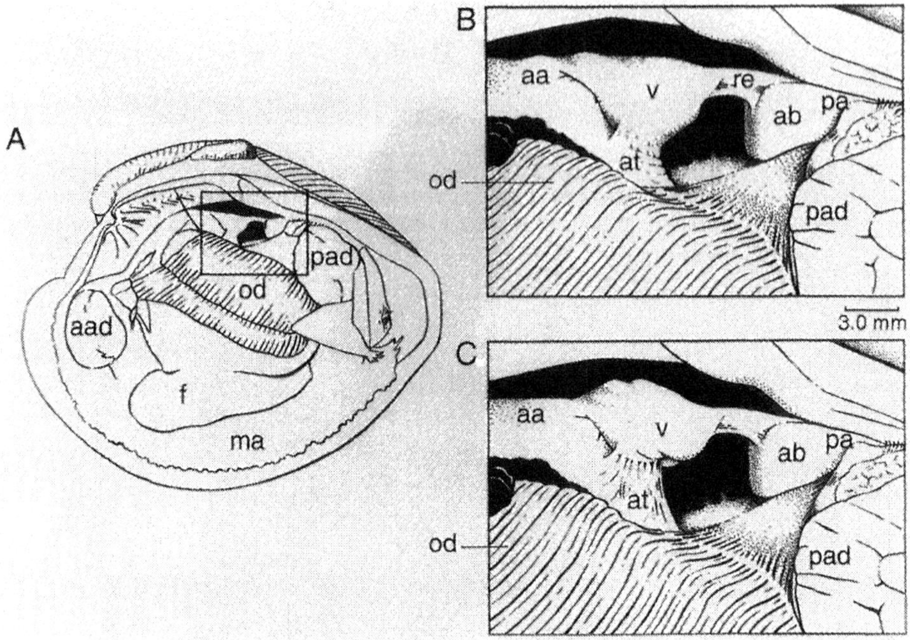

Fig. 4.41. (A) Sketch of half-shell preparation in right valve to show location of pericardial coelom (pc) in rectangle. aad = anterior aorta; f = foot; ma = mantle; od = outer demibranch of left gill; pad = posterior adductor muscle. 0.78×. (B,C) Sketches of details of pericardial coelom to show anterior aorta (aa), aortic bulb (ab), rectum (re), ventricle (v), left atrium (at), outer demibranch of left gill (od), posterior adductor muscle and posterior aorta (pa).

columnar about 45 μm high with prominent mucous glands situated at the end of each arm of the 'cross' (Fig. 4.42); mucous glands are a mixture of neutral glycoproteins and carboxylated glycosaminoglycans. Nuclei are centrally located, oval, have evenly distributed chromatin and average about 15 μm. Scattered brown or pore cells can be seen penetrating the epithelium by diapedesis en route to the lumen. Connective tissue supporting the epithelium is dense and consists of collagen and carboxylated glycosaminoglycans (Fig. 4.42).

The upper esophagus, close to the stomach, has 3–4 low rugae and the same type internal epithelium as the lower esophagus but is much higher (about 70 μm); there are many mucous glands containing acid glycoproteins. Nuclei are similar to those in the lower esophagus. Subepithelial connective tissue is similar to that described for the lower esophagus, but is thicker (Fig. 4.43).

4.8.2.2 Stomach

4.8.2.2.1 Anterior

The epithelium is simple, ciliated columnar, about 30 μm high, with many mucous glands containing carboxylated glycosaminoglycans. Nuclei are oval, about 10 μm in the long axis

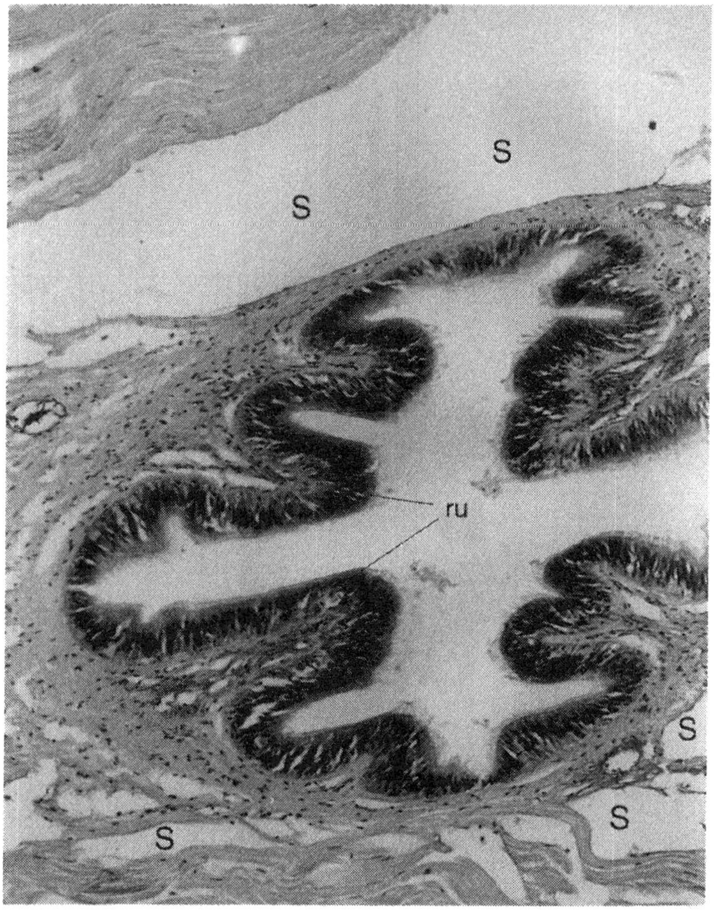

Fig. 4.42. Cross section through the anterior esophagus to show prominent rugae (ru) and large periesophageal hemolymph sinus (s). HFW = 405 μm.

and are centrally located; they have evenly scattered chromatin and do not contain nucleoli. Subepithelial connective tissue consists of dense inner collagenous fibers grading into an outer loose meshwork enclosing hemolymph spaces containing many brown or pore cells (Fig. 4.44).

4.8.2.2.2 Posterior

The posterior stomach consists of the style sac and intestine. The style sac epithelium is high (about 75–80 μm), pseudostratified, ciliated columnar; cilia are long (12–13 μm) and uniform — this is typical of style-sac epithelia of many bivalves (Eble and Scro, 1996). Nuclei are oval with evenly scattered chromatin containing one nucleolus; they are located in the middle of the pale-staining epithelium (Fig. 4.45). Subepithelial connective tissue is similar to

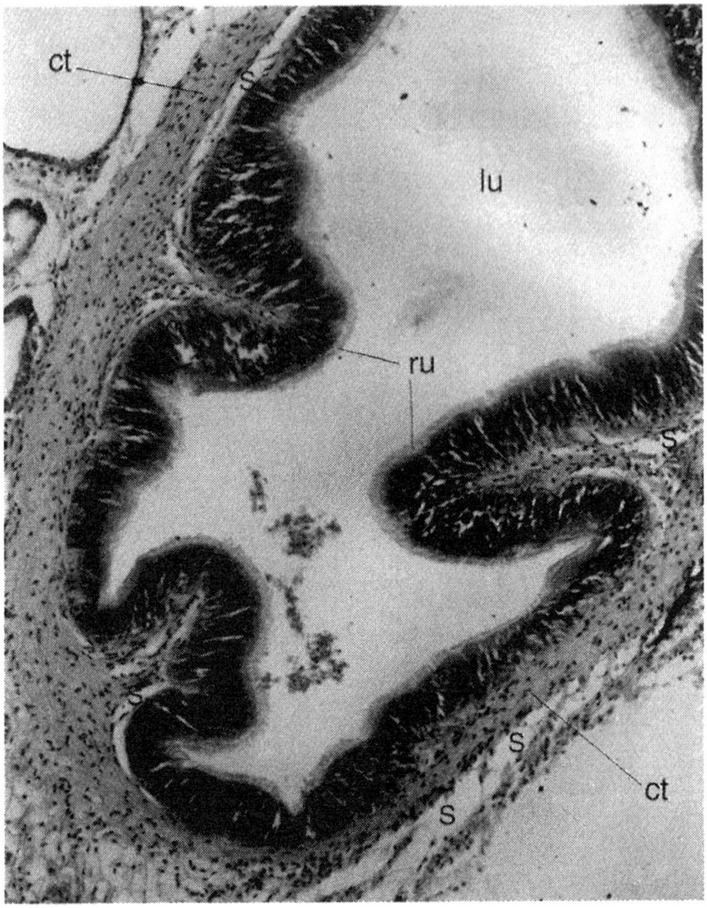

Fig. 4.43. Cross section through esophagus close to stomach. Subepithelial connective tissue (ct) becomes more prominent as esophagus approaches the stomach; rugae (ru) acquire large hemolymph sinuses (s). Compare with Figs. 4. and 4.42. lu = lumen of esophagus. HFW = 405 μm.

that described for the anterior stomach. The style sac is separated from the intestine by greater and lesser typhlosoles (Figs. 4.45 and 4.46). The greater typhlosole has a high (160–185 μm) ciliated, pseudostratified epithelium that stains darker than the epithelia of both the style sac and intestine which it separates (Fig. 4.46); nuclei are similar to those described for the style sac. The lesser typhlosole also has a ciliated, pseudostratified epithelium about 75–80 μm high. This epithelium appears to be contiguous with that of the intestine that has the same histological construction. Greater and lesser typhlosoles have a thin collagenous connective tissue supporting the basal lamina.

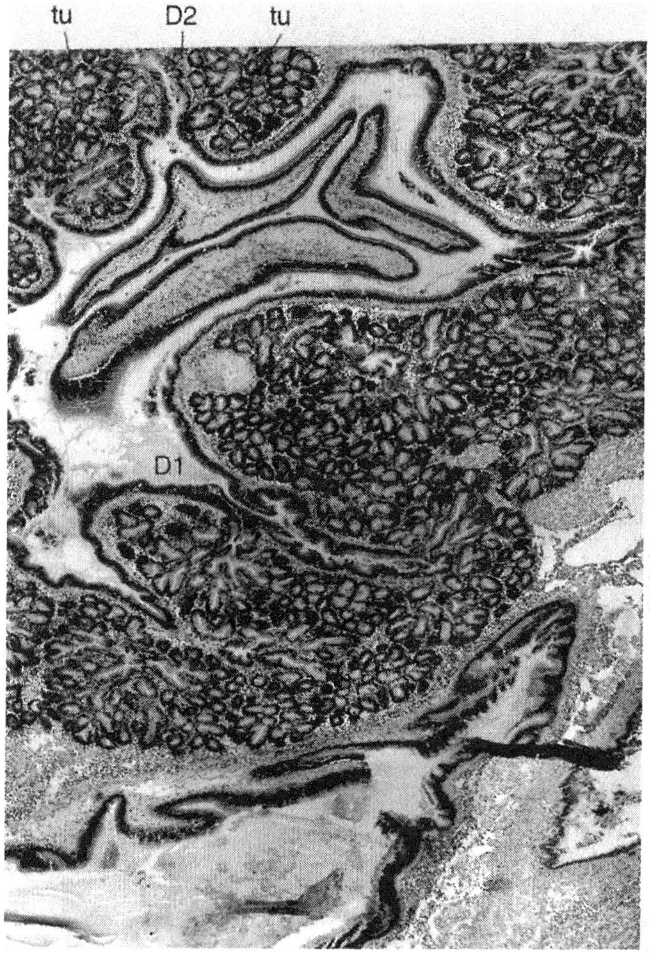

Fig. 4.44. Longitudinal section through the anterior stomach to show primary ducts (D1) leading to secondary ducts (D2) that lead to digestive tubules (tu). HFW = 2.06 mm.

4.8.2.3 Intestine

4.8.2.3.1 Anterior

The anterior intestine is characterized by a prominent typhlosole, the presence of which gives a C-shape to the lumen (Fig. 4.40). The epithelia on both typhlosole and intestine are high (50 μm), ciliated, pseudostratified columnar; many mucous glands containing carboxylated glycosaminoglycans are found in particularly high concentrations at junctions of typhlosole and intestine. Nuclei of both typhlosole and intestine are elongate and avenge about 10 μm; they possess evenly scattered chromatin and contain one nucleolus. Nuclei are present

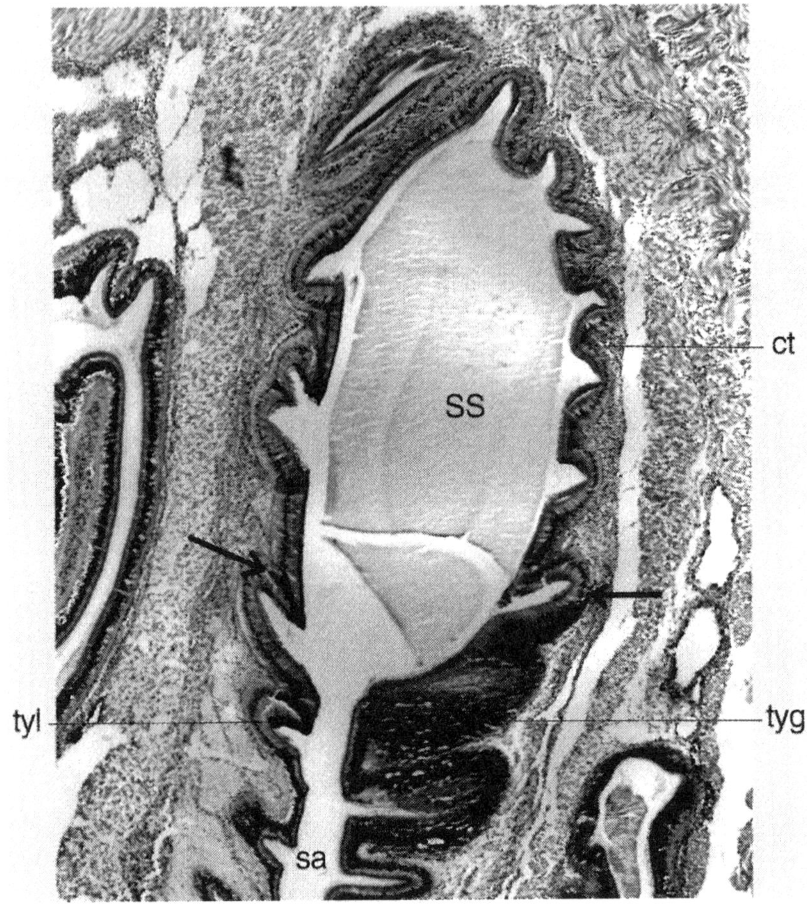

Fig. 4.45. Longitudinal section through junction of style sac (ss) and anterior stomach (sa); note change in epithelial structure at junction (arrows). ct = connective tissue; tyg = greater typhlosole; tyl = lesser typhlosole. HFW = 405 μm.

basally in the intestinal epithelium but in the middle of epithelial cells in the typhlosole. Many epithelial cells of both typhlosole and intestine appear to be budding vesicles into the lumen via apocrine secretion. Connective tissue supporting the intestine is prominent and consists of a mixture of collagen and carboxylated glycosaminoglycans.

4.8.2.3.2 Ascending

The ascending intestine is similar to the anterior intestine but it lacks the typhlosole; instead, the epithelium and underlying connective tissue are thrown into a series of folds.

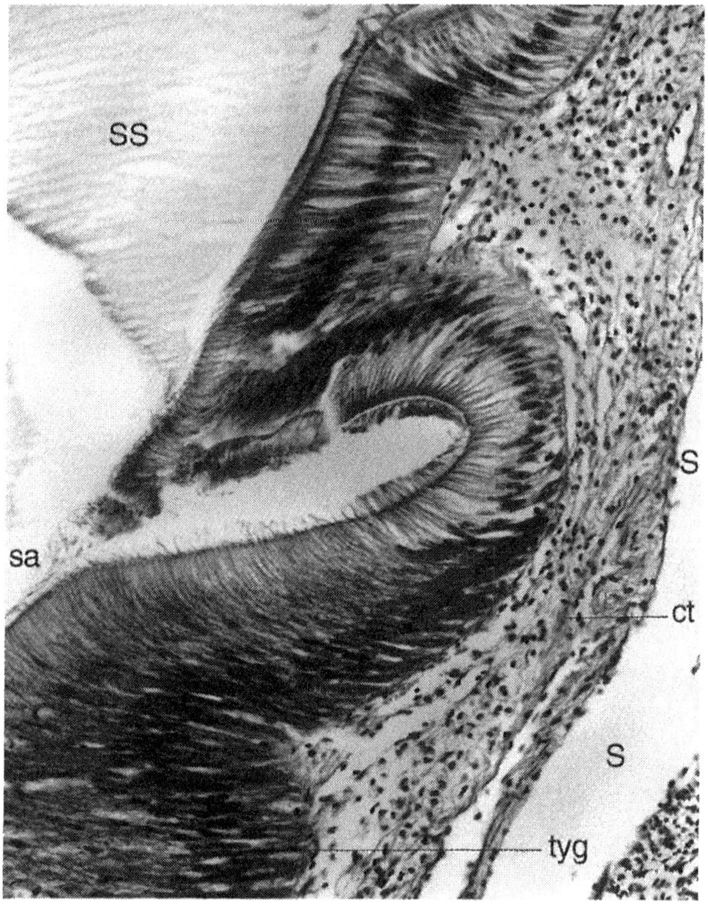

Fig. 4.46. Longitudinal section through junction of style sac (ss) and anterior stomach (sa) to show change in epithelial structure (arrow). ct = connective tissue; tyg = greater typhlosole; tyl = lesser typhlosole; s = hemolymph sinus. HFW = 405 μm.

4.8.2.4 Rectum

There is no typhlosole in the rectum but its connective tissue and epithelium are thrown into two large folds imparting a slit-like appearance to the lumen (Fig. 4.47). The epithelium is high (64 μm), ciliated, pseudostratified columnar with many mucous cells containing carboxylated glycosaminoglycans; many brown or pore cells appear to be penetrating the epithelium en route to the lumen (Fig. 4.48). One side of the rectum is thrown into 4–5 shallow folds or troughs covered with an epithelium with prominent cilia. Nuclei are located in the basal portion of the epithelium; they are narrow and elongate (12 × 3 μm), have evenly distributed chromatin and lack nucleoli (Fig. 4.47). Connective tissues supporting the rectum are arranged in inner and outer layers: the former is dense, about 110 μm thick and consists of a layer of neutral glycoproteins just under the epithelium, followed by a layer of carboxylated

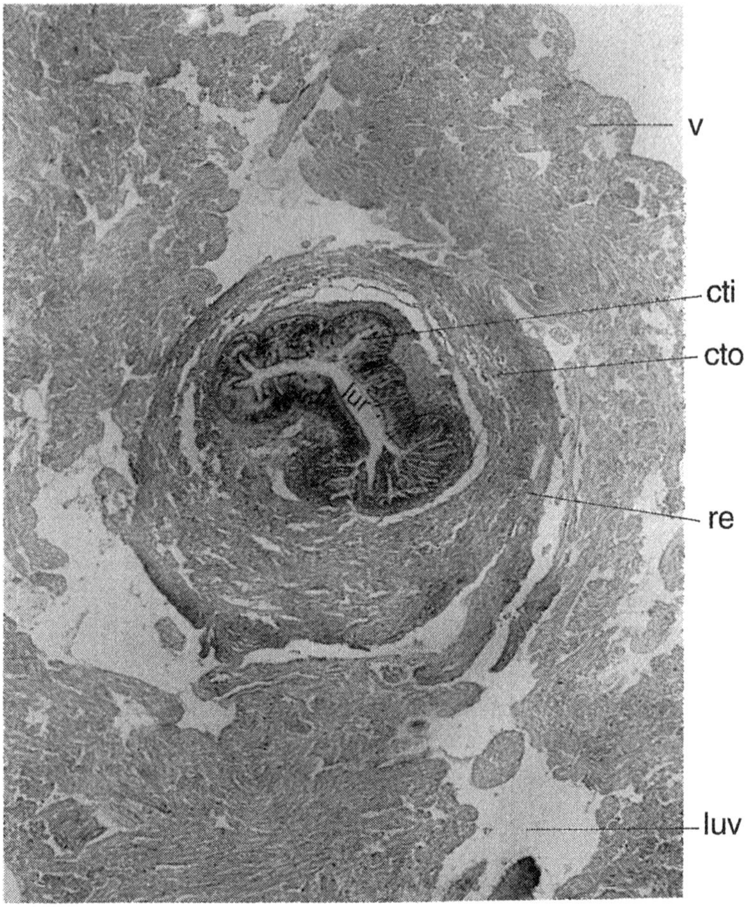

Fig. 4.47. Cross section through the ventricle (v) enclosing the rectum (re). Outer muscular tissues of rectum grade with and become contiguous with ventricular muscle fibers. cti = inner layer of connective tissue of rectum; cto = outer layer of connective tissue of rectum; lur = lumen of rectum; luv = lumen of ventricle. HFW = 2.18 mm.

glycosaminoglycans; the outer, connective tissue layer consists chiefly of collagenous fibers (Figs. 4.47 and 4.48). Many smooth muscle cells are interwoven throughout both layers of connective tissues: muscle cells are arranged in a circular pattern in the inner layer and in a longitudinal array in the outer layer; muscle cells are not organized into discreet sheets as they are in some bivalves (Figs. 4.47 and 4.48; Jegla and Greenberg, 1968b). Whether peristalsis of the rectum occurs has been discussed by Jegla and Greenberg (1968b) who concluded that it probably does not, although Greenberg and Jegla (1963) observed spontaneous contractions of in vitro preparations of rectums in *M. mercenaria*. A prominent hemolymph sinus is located between inner and outer connective tissue layers of the rectum (Figs. 4.47 and 4.48).

The histological appearance of the rectum in the aortic bulb is similar in all respects to that just described for the ventricle (Fig. 4.49). Usually many brown or pore cells are present penetrating the epithelium, becoming finally localized in the lumen (Fig. 4.49). Many rectal

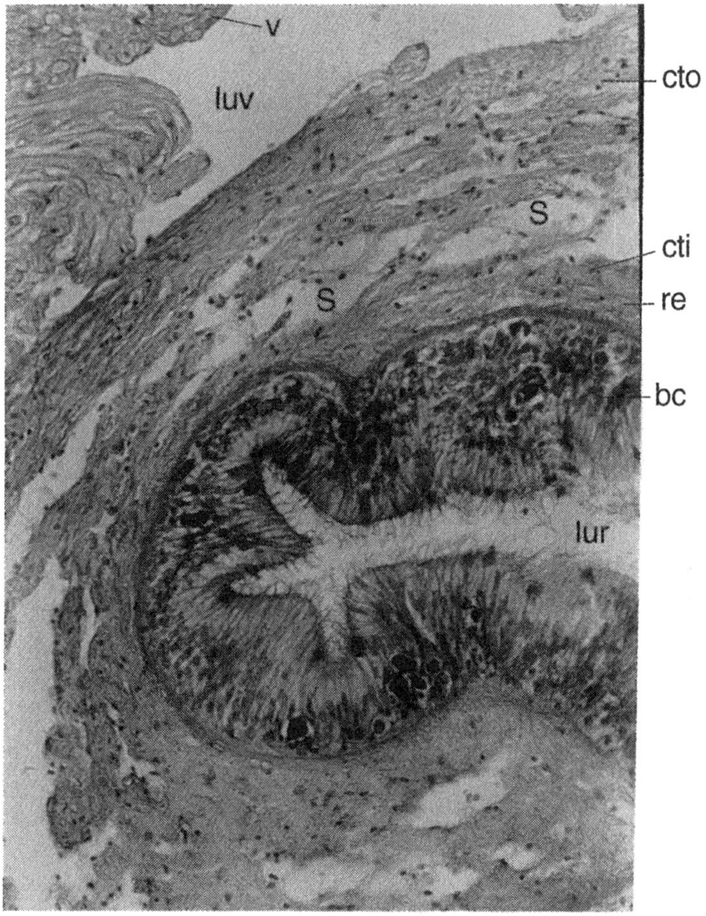

Fig. 4.48. Cross section of rectum (re) within ventricle (v). Many brown cells (bc) may be seen penetrating epithelium of rectum en route to lumen of rectum (lur). cti = inner layer of connective tissue of rectum; cto = outer layer of connective tissue of rectum; lur = lumen of rectum; luv = lumen of ventricle; s = hemolymph sinus; v = muscle fibers of ventricle. HFW = 405 μm.

epithelial cells lack cilia and show evidence of apocrine secretion; as such, the lumen is filled with a mixture of membranous materials, hemocytes and pore cells. Garofola and Eble (1995) reported the fecal route to be a major pathway in the elimination of the heavy metals, cadmium and zinc, especially during the first 7–10 days of exposure.

The rectum leaves the aortic bulb in company with the posterior aorta, which surrounds the rectum (Fig. 4.50). Here, the rectal epithelium is high (90 μm), ciliated, pseudostratified columnar showing evidence of much apocrine secretion; many brown or pore cells are in evidence en route to the lumen (Fig. 4.51). Connective tissues, including smooth muscle cells, are as described for the rectum both in the ventricle and aortic bulb. The rectum terminates in the anus which opens into the excurrent siphon (Fig. 4.22).

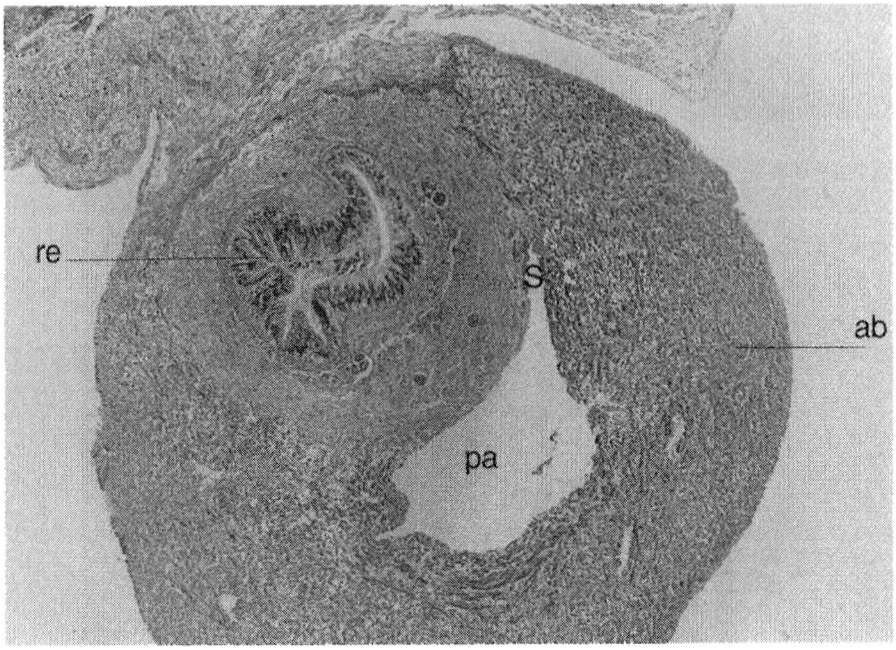

Fig. 4.49. Cross section through aortic bulb (ab) to show the rectum (re) embedded in dorsal half. Note many hemolymph sinuses (s) in the muscular, ventral half leading to the posterior aorta (pa). HFW = 2.18 mm.

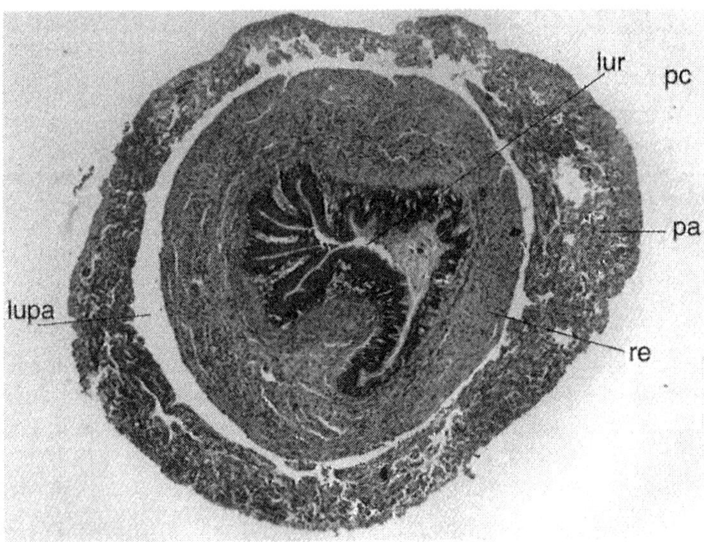

Fig. 4.50. Cross section through the posterior aorta (pa) enclosing the rectum (re). The space just outside the posterior aorta is the pericardial coelom (pc). lupa = lumen of posterior aorta; lur = lumen of rectum. HFW = 2.18 mm.

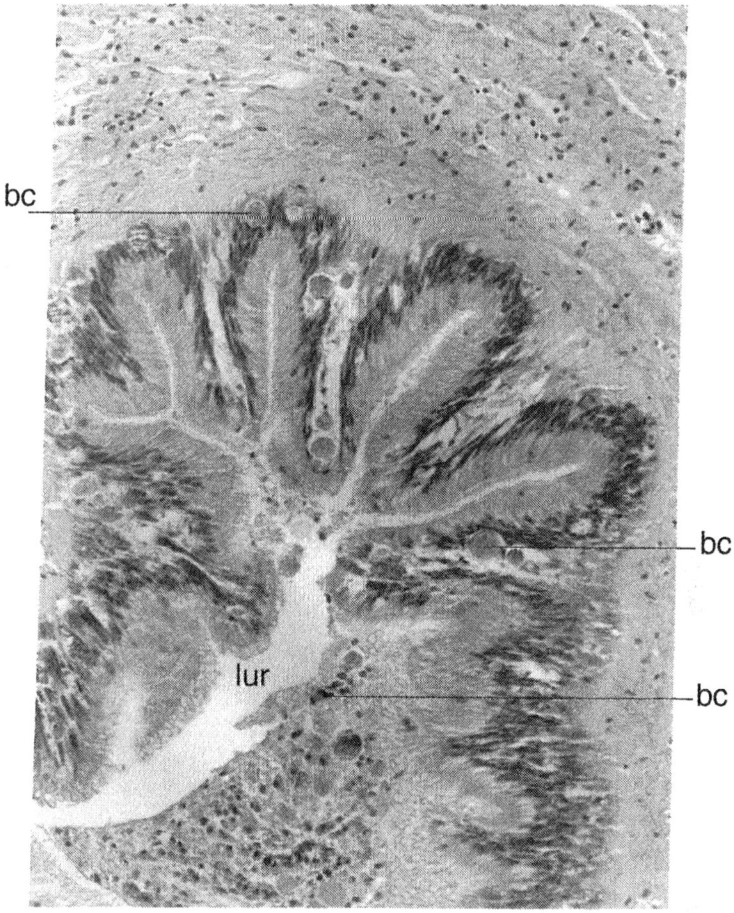

Fig. 4.51. Cross section through the rectum to show details of the epithelium. Note many brown cells (bc) penetrating the epithelium of the rectum and in the rectal lumen (lur). HFW = 405 μm.

4.9 DIGESTIVE GLAND

4.9.1 Anatomy

The digestive gland is a bilateral, compound, tubuloacinar gland that surrounds the stomach (Figs. 4.37 and 4.52). Primary ducts arise from various areas of the stomach: the left pouch, left and right caeca, and independently from the right side. Primary ducts are long and branch into short secondary ducts which, in turn, branch into pretubular ducts; the latter terminate in digestive tubules (Figs. 4.37 and 4.52).

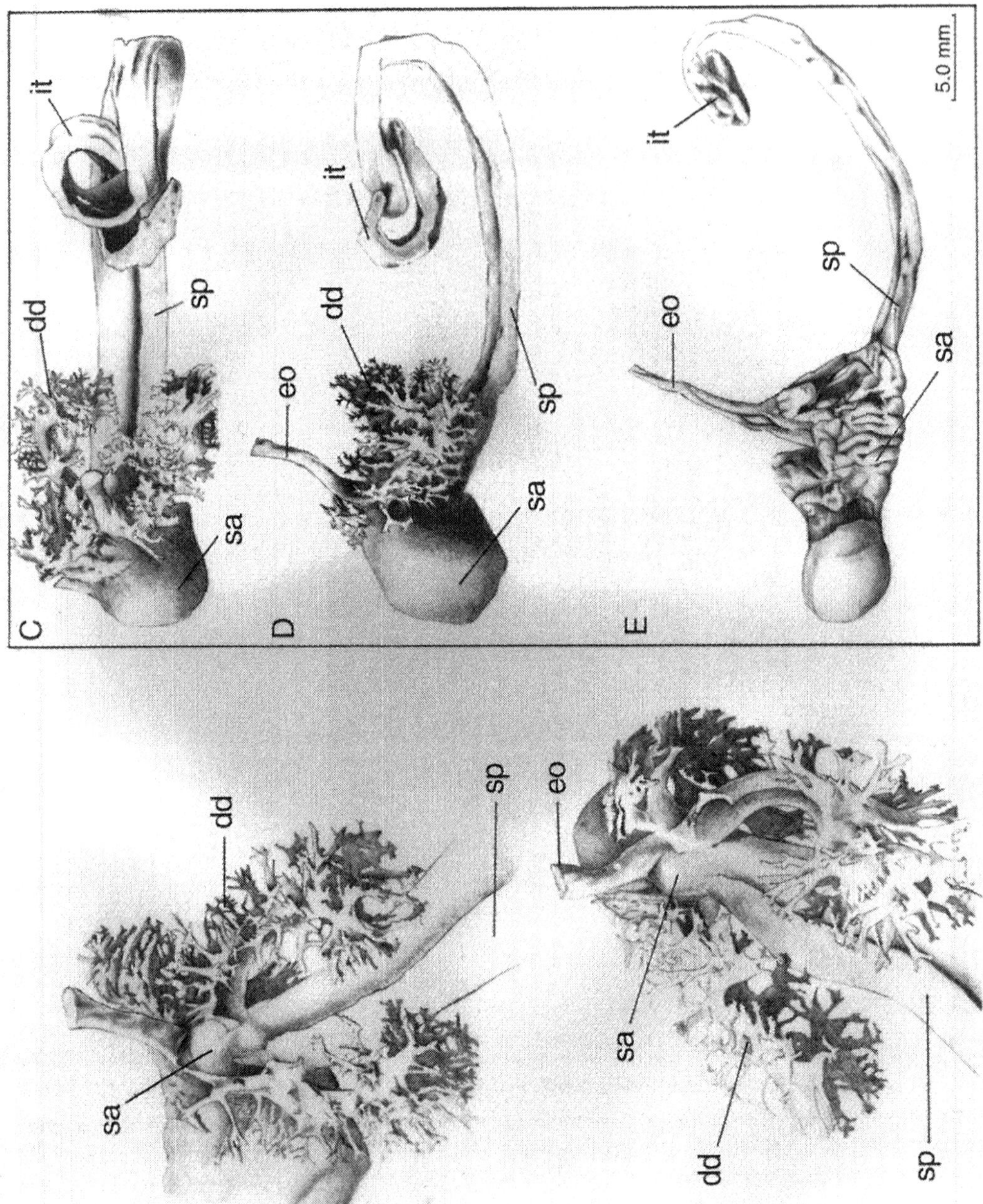
A
B
C
D
E
sa
dd
sp
eo
it
5.0 mm

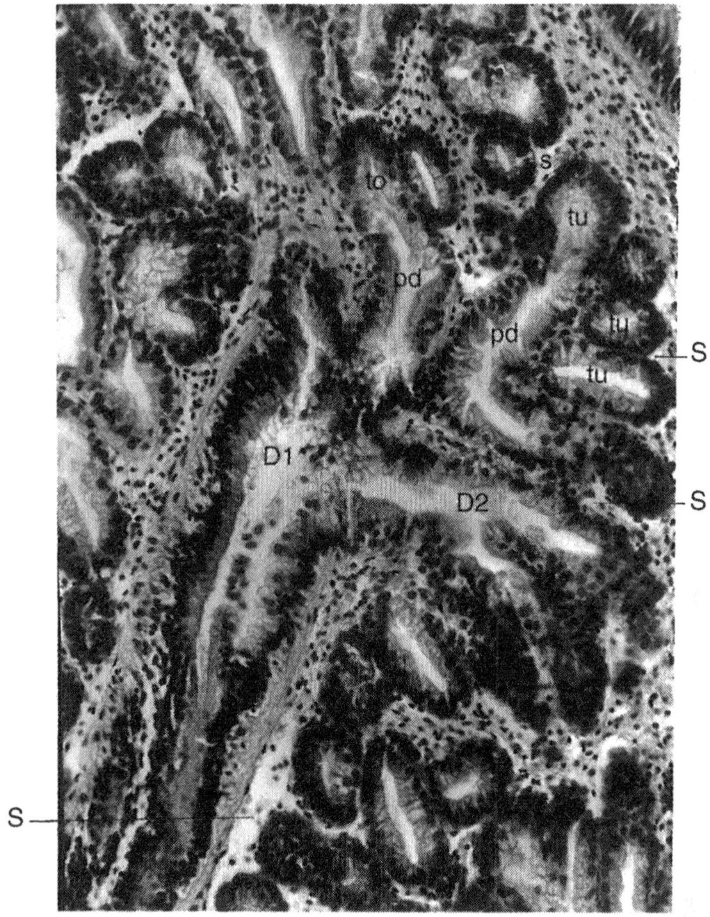

Fig. 4.53. Cross section through the digestive gland showing a primary duct (D1), a secondary duct (D2), a pretubular duct (pd) and digestive tubules (tu). s = hemolymph sinuses. HFW = 405 μm.

4.9.2 Histology

Primary ducts are similar to the stomach in histological structure: the epithelium is tall (40 μm), pseudostratified columnar; most cells display an active apocrine secretion hence many small vesicles occur in the lumen of the duct (Figs. 4.44 and 4.53). A ciliary tract is located on one surface of the epithelium; material in ciliary tracts is moved away from digestive tubules while partially digested food from the stomach is moved towards digestive

---

Fig. 4.52. Sketches of esophagus (eo) anterior stomach (sa), posterior stomach (sp), digestive gland (dd) and anterior intestine (it) from vinyl acetate casts. (A) Ventroposterior view. (B) Ventroposterior view. (C) Ventral view. (D) Ventrolateral view from right side. (E) Ventrolateral view from right side with digestive gland removed.

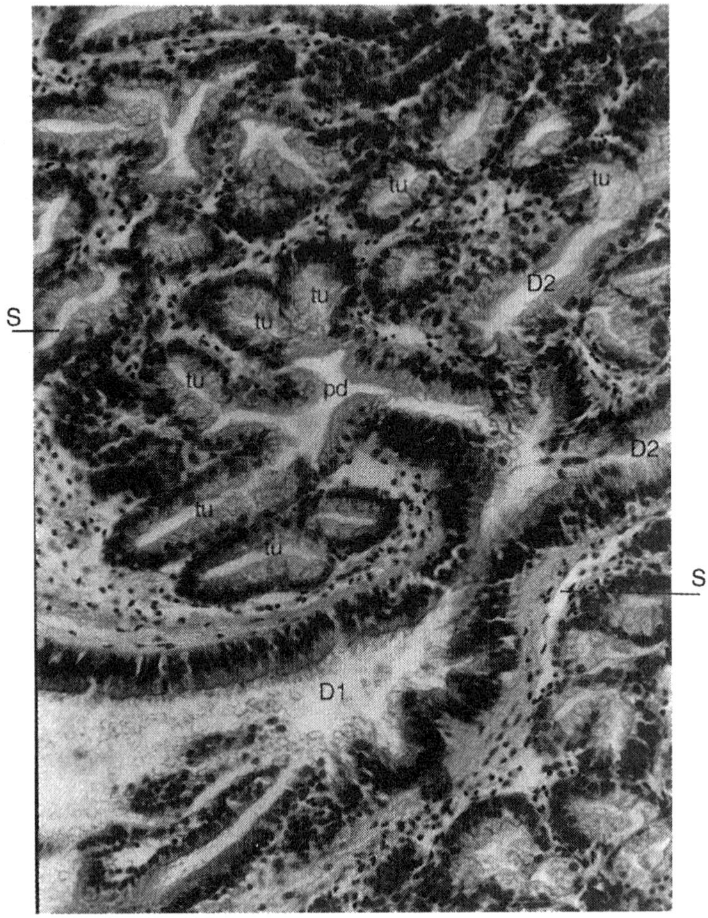

Fig. 4.54. Section through the digestive gland showing the primary duct (D1) and secondary duct (D2) leading into pretubular ducts (pd) which, in turn, lead into tubules (tu). s = hemolymph sinuses. HFW = 405 μm.

tubules by a countercurrent flow (Owen, 1955). Nuclei are basally situated, large (7–8 μm) and oval with evenly distributed chromatin and one nucleolus. The apex of epithelial cells is eosinophilic. Connective tissue surrounding primary ducts is thick (6–7 μm) and consists of a mixture of carboxylated glycoproteins and glycosaminoglycans. Some smooth muscle cells are scattered throughout this connective tissue. A large hemolymph sinus bathes the primary duct (Fig. 4.53); this condition is also present in the eastern oyster (Eble and Scro, 1996) and the scallop (Beninger and Le Pennec, 1991).

Secondary ducts have a pseudostratified columnar epithelium; nuclei are basally located and similar to those in primary ducts (Figs. 4.44 and 4.54). The cytoplasm of epithelial cells of secondary ducts is distinctive: the basal portion is basophilic and contains what seems to be fibrillar material; the apical area is large, clear and vacuolated (Fig. 4.54). Secondary ducts, like primary ducts, are bathed by a large hemolymph sinus (Figs. 4.53 and 4.54).

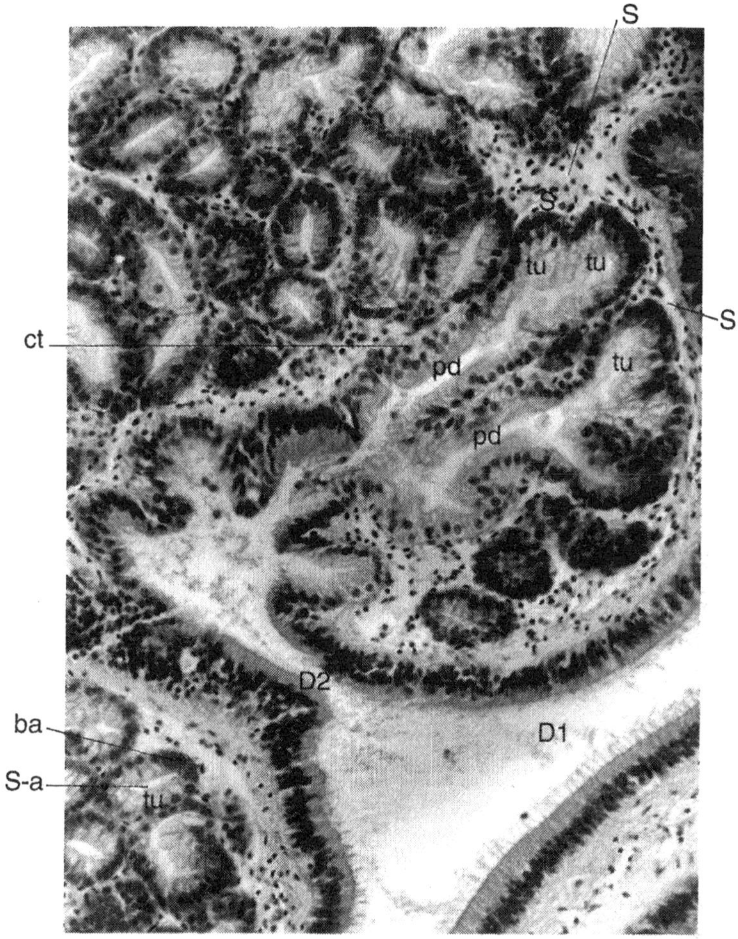

Fig. 4.55. Section through the digestive gland showing a primary duct (D1) and secondary duct (D2) leading into pretubular ducts (pd), which, in turn, lead into digestive tubules (tu). Digestive tubules are composed of secretory–absorptive cells (s-a) and basiphil cells (ba). ct = connective tissue; s = hemolymph sinuses. HFW = 405 μm.

There is an abrupt transition between secondary ducts and the next segment, the pretubular ducts. Connective tissues surrounding secondary ducts have the same composition as that of primary ducts but are greatly reduced: primary ducts have a connective tissue sheath that is 3 μm thick compared to only 1 μm thick for secondary ducts (Fig. 4.55). Pretubular ducts are small and have either H- or Y-shaped lumina in cross section (Fig. 4.55). The epithelium of pretubular ducts is simple columnar and its apical portion is eosinophilic; most cells have prominent, dense cilia but there is one small tract of cells that lack cilia (Fig. 4.55). Nuclei are similar to those described for secondary ducts.

Pretubular ducts connect with 3–4 digestive tubules (Figs. 4.53–4.55). Connective tissues supporting digestive tubules are reduced (1–2 μm) and appear to be composed chiefly of collagen. A prominent hemolymph sinus is present (Figs. 4.53–4.55).

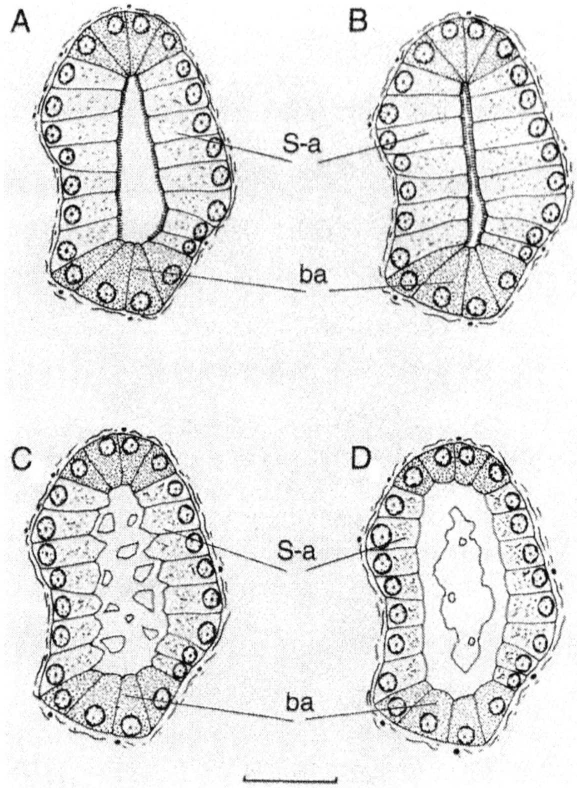

Fig. 4.56. Sketches of the four major phases of digestive-tubule cytology. (A) Type I = normal. (B) Type II = absorptive. (C) Type III = disintegrating. (D) Type IV = reconstituting. s-a = secretory–absorptive cells, ba = basiphil cells. Redrawn after Robinson and Langton (1980). Bar = 50 μm.

Digestive tubules are composed of at least two types of cells: (1) digestive or secretory-absorptive cells; and (2) basiphil cells (Figs. 4.53–4.55). Digestive cells cycle between a simple, tall (25 μm) columnar and a low (6 μm) cuboidal depending upon their physiological condition (Fig. 4.56); the cycle has been arbitrarily divided into four phases: Type I = normal; Type II = absorptive; Type III = disintegrating; Type IV = reconstituting (Platt, 1971; Langton, 1975; Robinson and Langton, 1980). Digestive cells possess microvilli, macrovesicles of the lysosomal-vacuolar system and a large (6 μm) centrally located nucleus with evenly distributed chromatin and one nucleolus. Digestive tubules, similar to pretubular ducts to which they connect, are enveloped by a thin connective tissue (1–2 μm) composed chiefly of collagen and are bathed by prominent peritubular hemolymph sinuses (Figs. 4.53–4.55).

Preliminary work in our laboratory suggests that there are two classes of basiphil cells in *M. mercenaria*, flagellated and non-flagellated. This finding agrees with Weinstein (1995) who described these cell types in the eastern oyster, *Crassostrea virginica*. Owen (1955) described and figured flagellated 'darkly staining cells' (= basiphils) of *Venerupis pullastra* and explained that in all Eulamellibranchia, basiphils extend the length of the tubules to meet at the apex. Flagellated basiphil cells (the older literature referred to these organelles as cilia) were first reported by Potts (1923) and later confirmed by Yonge (1926). Our work with *M. mercenaria* confirms the studies by Weinstein (1995) with *C. virginica*: non-flagellated basiphil cells have microvilli, abundant rough endoplasmic reticulum (RER), an extensive Golgi lying above the basally located nucleus and secretory vesicles lying close to the apical plasma membrane. Flagellated basiphil cells, on the contrary, possess a single, long flagellum in addition to microvilli; Golgi, RER and secretory vesicles are lacking in this cytotype.

It appears that the non-flagellated basiphil cell is specialized for production and secretion via exocytosis of extracellular enzymes. The flagellated basiphil cell has been considered to be a stem cell and mitotic figures have been reported, particularly in starved animals (Yonge, 1926; Owen, 1955). However, workers recently have begun to question whether these cells are truly stem cells because Pal et al. (1990) failed to find mitotic figures in *Meretrix meretrix*, and Weinstein (1995) did not find evidence of mitotic activity in these cells in *C. virginica*. Recent work in my laboratory, using 5-bromodeoxyuridine (BrdU), a thymidine analogue, as a probe for the S-phase of the cell cycle (Miller and Nowakowski, 1988), has confirmed that basiphil cells are stem cells. In all tubules studied from 25 animals, only basiphil cells showed uptake of BrdU, never secretory–absorptive cells. It was common to find several basiphil cells from the same area of a tubule in the S-phase of the cell cycle (Fig. 4.57). Much work remains to be done on the cell physiology and molecular biology of the bivalve digestive gland.

4.10 EXCRETORY SYSTEM

4.10.1 Anatomy

The excretory system consists of paired pericardial glands that open into the pericardial coelom; the renopericardial canal connects the pericardial coelom with the kidney, the lumen of which is a specialized region of the coelom, the renocoel. Morse (1987) reviewed the bivalve excretory system in general as well as the specific morphology and physiology of the pericardial gland and kidney in *M. mercenaria* (Fig. 4.58).

The pericardial glands lie in the mantle just anterior and slightly dorsal to the pericardial coelom (Figs. 4.1 and 4.13). Their color is dark brown to black and they stand out in stark contrast to the pale mantle tissue in which they are embedded. Meyhofer et al. (1985) ranked epithelial cells of the pericardial gland of bivalve molluscs as podocytes and they, together with their basal lamina, constitute the ultrafiltration barrier (Fig. 4.58).

The pericardial coelom is large and contains the heart and bulbus arteriosus; the rectum is enveloped by both the ventricle and the bulbus arteriosus as it traverses the pericardial coelom (Figs. 4.41, 4.47 and 4.49). The paired renopericardial canals are located in the floor of the pericardial coelom.

The paired kidneys lie just ventral to the pericardial coelom and extend from the anterodorsal corner of the posterior pedal retractor muscle to a point approximately under the ventricle (Fig. 4.59); in effect, the connective tissue that forms the floor of the pericardial coelom becomes the roof of the kidney and separates the pericardial coelom from the renocoel (Fig. 4.59). The kidneys are actually glandular coelomoducts that are differentiated into proximal and distal arms: proximal arms are bound to the roof of the renocoel and run through the renocoel; distal arms are connected to the floor of the kidney. Both proximal and

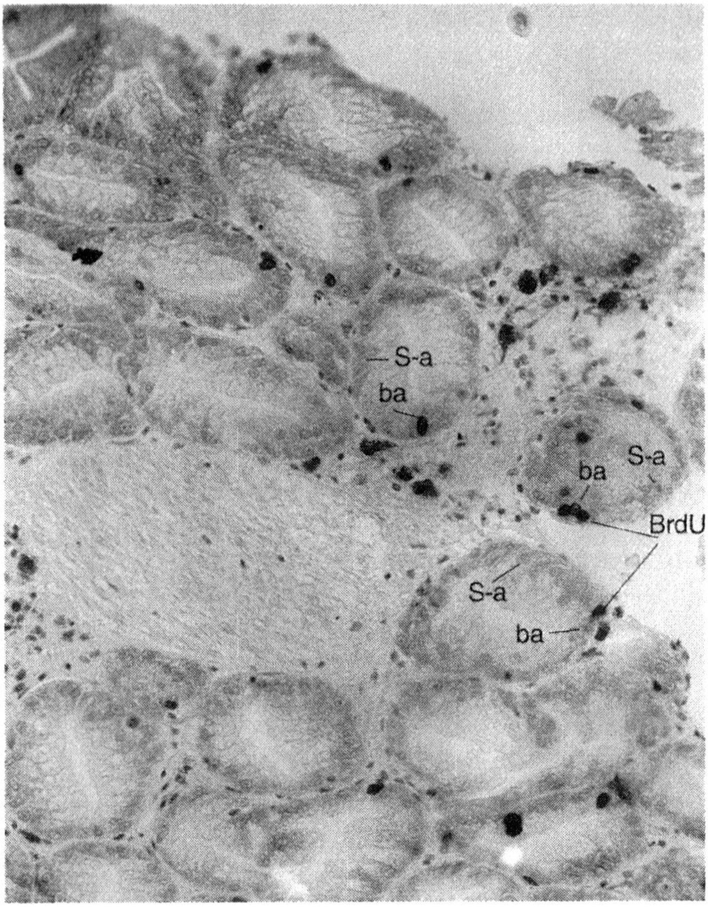

Fig. 4.57. Section through the digestive gland to show location of 5-bromodeoxyuridine probe (BrdU) in nuclei of basiphil cells (ba). Note absence of probe in secretory–absorptive cells (s-a). HFW = 405 μm.

distal chambers of the renocoel are divided into left and right compartments by means of a vertical septum (Fig. 4.59). Black concretions (0.5–2 mm in diameter) are usually scattered throughout both regions of the kidney. Tubules of the distal kidney empty into the epibranchial chamber via the renopore.

4.10.2 Histology

The pericardial gland is a compound tubular gland bathed by a large hemolymph sinus. Renal tubules are lined by a simple, low columnar epithelium; epithelial cells contain concretions in apical portions, microvilli and a basal nucleus with a distinctive distribution of chromatin: evenly distributed throughout the periphery of the nucleus leaving a clear eccentric zone (Fig. 4.60).

Epithelial cells of the pericardial gland are regarded as podocytes that separate two cavities,

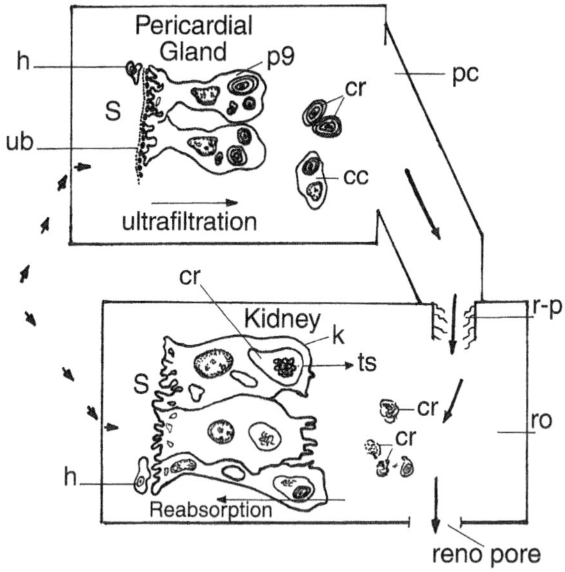

Fig. 4.58. Sketch to show relationships between pericardial gland and kidney to pericardial coelom (pc). Note that the pericardial coelom communicates with the lumen of the kidney, the renocoel (ro), another specialized compartment of the coelom, via the reno-pericardial canal (r-p). Long dark arrows show the generalized pathway of urine flow. Hatched arrows show pathways of ultrafiltration, reabsorption and tubular secretion (ts) in the bivalve excretory system. Short arrowheads show direction of flow of hemolymph. cc = cell with concretion in pericardial coelom; cr = concretion in pericardial coelom; h = hemocyte; s = hemolymph sinuses; ub = ultrafiltration barrier. Redrawn after Morse (1987).

the hemocoel and the pericardial coelom. Basal extensions of podocytes are the pedicels that attach to the diffuse basal lamina; this pedicel network together with the basal lamina represent the ultrafiltration barrier (Meyhofer et al., 1985). Distances between plasma membranes of adjacent pedicels, the slit widths, are variable and, in *Mytilus edulis*, ranged from 15 to 22 nm. Based on the variable size of slit widths and the fact that injected ferritin is retained in the hemolymph and did not pass the basal lamina in three species of lamellibranchs, *Mytilus edulis*, *Chlamys hastada* and *M. mercenaria*, Meyhofer and Morse (1996) concluded that the basal lamina is the principal ultrafilter in lamellibranchs. Since horseradish peroxidase (HRP) rapidly entered the ultrafiltrate, it was concluded by these authors that the range of particle size for the process of ultrafiltration in the above bivalves should be greater than 40 kDa (the molecular mass of HRP), but less than 400 kDa (the molecular mass of ferritin). In addition, a regular distribution of anionic sites on the basal lamina was demonstrated with ruthenium red; anionic sites on pedicels of podocytes were also present which may be important for maintenance of the integrity of ultrafiltration slits (Meyhofer and Morse, 1996).

Ducts of the pericardial coelom are filled with concretions and cellular debris suggesting a combination of apocrine and holocrine secretion by podocytes (Figs. 4.58 and 4.61); ducts lead directly into the pericardial coelom.

Andrews and Jennings (1993) sketched pericardial gland podocytes from various bivalves; they also discussed the functional significance of and mechanism of primary urine formation

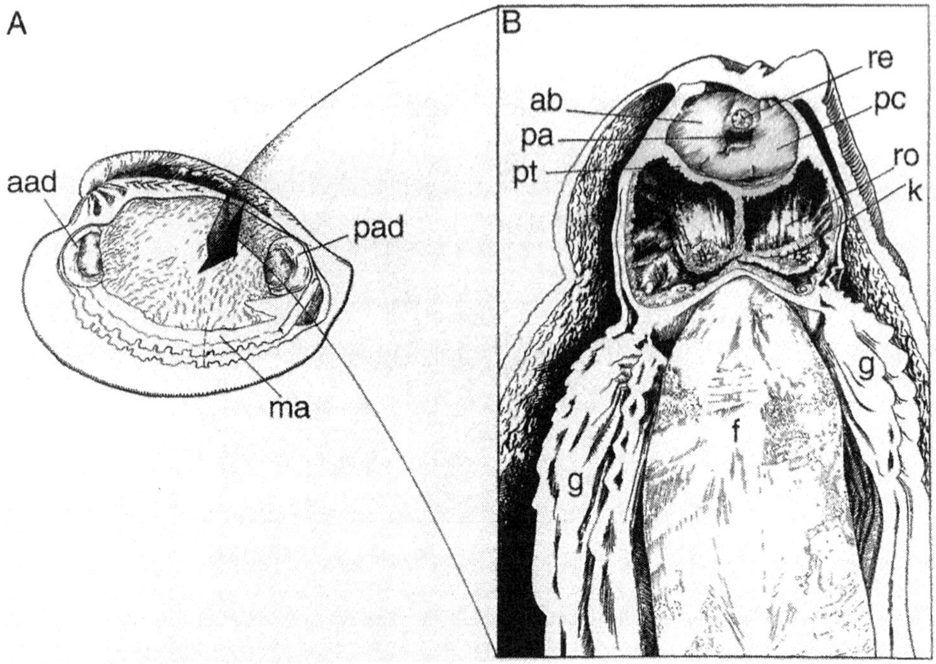

Fig. 4.59. (A) Sketch of half-shell preparation in right valve to show location of kidney; dashed line shows plane of cross section interpreted in B. aad = anterior adductor muscle; ma = mantle; pad = posterior adductor muscle. 0.52×. (B) Sketch of cross section to show location of kidney tubules (k) in renocoel (ro). Note that pericardial coelom (pc) and renocoel are separated only by a thin partition (pt). ab = aortic bulb; f = foot; g = gill; pa = posterior aorta; re = rectum. 9.36×.

by pericardial glands. Robinson and Morse (1994) analyzed the molecular size range of ultrafiltration in bivalves; they also found several proteins in the pericardial fluid that were absent in the hemolymph, indicating secretion by podocytes of the pericardial gland. Morse and Zardus (in press) discussed the comparative ultrastructure of podocytes and pericardial glands from a variety of bivalves including *M. mercenaria.*

The proximal kidney is composed of a simple, columnar epithelium with microvilli, a few cilia, deep, basal infoldings and a basal nucleus; most cells contain small, apically situated concretions. Large, concentrically layered concretions may be present in tubule lumina (Fig. 4.62). Morse (1987) regards this portion of the kidney as the major site of reabsorption because of the presence of both microvilli and basal infoldings.

The distal kidney also has a simple, columnar epithelium with scattered microvilli but the cells have a well-developed lysosomal-vacuolar system that is the source of many concretions found in the swollen apical ends of epithelial cells. Basal nuclei have the same unique cytology as that described for cells of the pericardial gland. Large concretions are frequently found in tubule lumina; similar to those found in the proximal kidney, concretions are usually concentrically layered (Fig. 4.63). The distal kidney always has a greater concentration of concretions in tubule lumina than does the proximal kidney. Renal tubules in the distal kidney

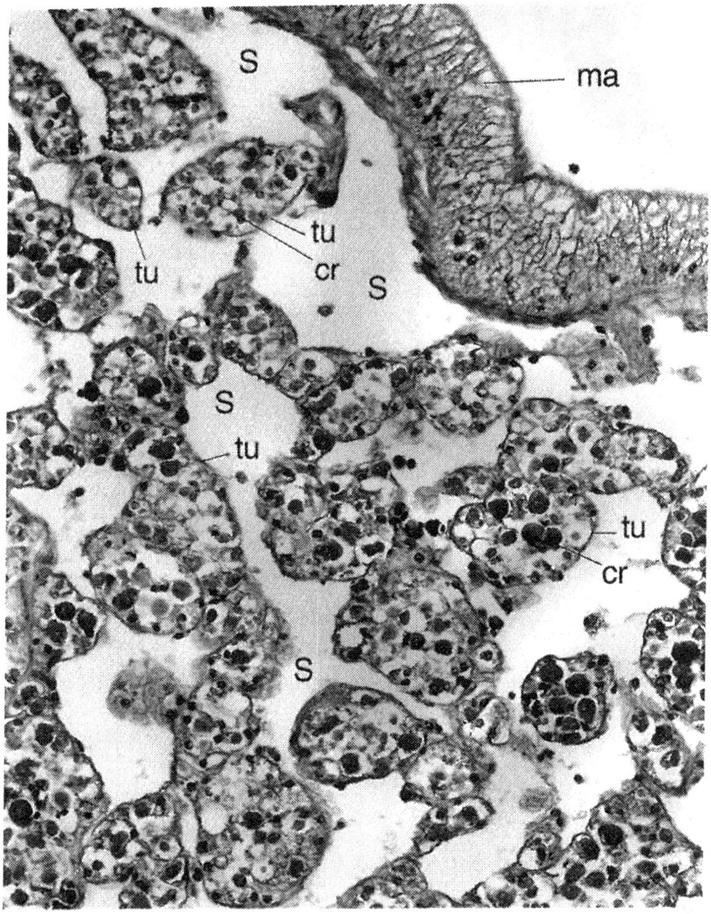

Fig. 4.60. Cross section through pericardial gland to show tubules (tu) lying in hemolymph sinuses (s). Note large concretions (cr) in tubule cells. ma = mantle. HFW = 405 μm.

are surrounded with a layer of collagenous connective tissue interspersed with many smooth muscle cells (Scheairs and Eble, 1995).

Sullivan et al. (1988a) isolated and characterized kidney granules (concretions) of *M. mercenaria*: 89% were less than 5.6 μm, but some were as large as 420 μm (extracellular granules); Ca was the most common metal followed by Mn, Zn and Fe; P was the predominant non-metal. It was concluded that P was complexed chiefly with Ca, but also with other metals.

Sullivan et al. (1988b) determined the subcellular distribution of metals in the kidney of *M. mercenaria*. Their results showed that Ba, Fe, Mn and Pb were concentrated chiefly in granules while Cd, Zn, Cu and Mg were associated with the soluble cytosolic fractions, probably complexed with metallothioneins.

Morse and Zardus (in press), discuss comparative renal ultrastructure in general and illustrate proximal and distal renal cells of *M. mercenaria* at both light and transmission electron microscope levels.

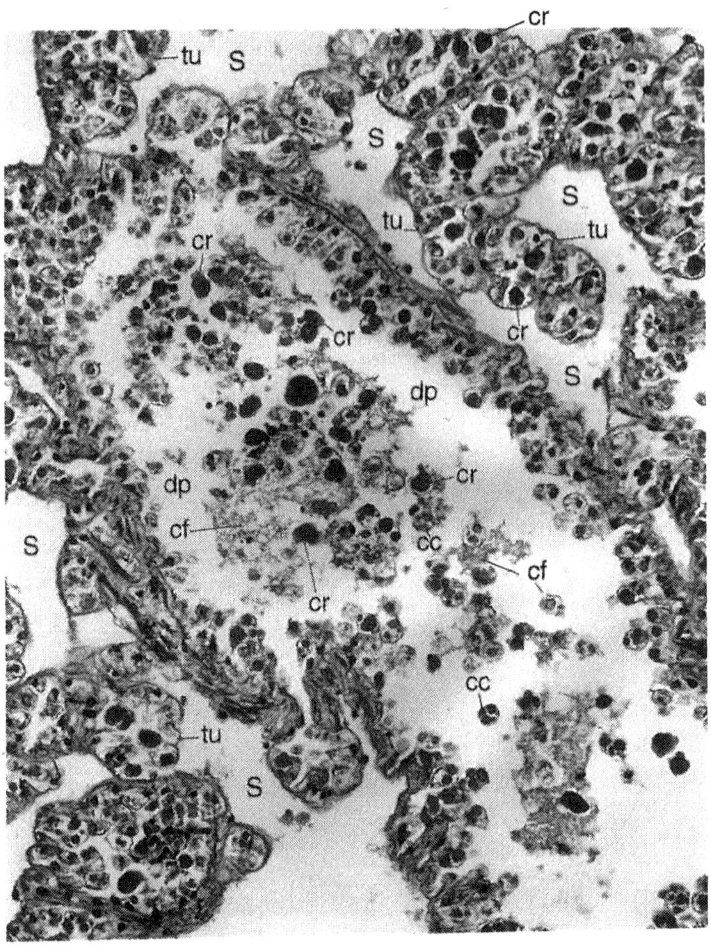

Fig. 4.61. Longitudinal section through a duct of the pericardial gland (dp) containing cell fragments (cf) and concretions (cr). cc = concretions in cell; s = hemolymph sinuses; tu = tubule of pericardial gland. HFW = 405 μm.

4.11 REPRODUCTIVE SYSTEM

Loosanoff (1936) was the first to study in detail changes in the sexual phases of *M. mercenaria* although Belding (1912) reported on spawning activities of this animal based upon macroscopic observations. Loosanoff (1937a) reported on seasonal changes in the gonads and was the first to study the details of the development of the primary gonad as well as the cytology of gametogenesis in *M. mercenaria* (Loosanoff, 1937b). Ansell (1961) reported on aspects of reproduction in the venerid clam, *Venus striatula* and several workers have described seasonal changes in the gonads of *M. mercenaria* from various localities (Porter, 1964; Keck et al., 1975; Knaub and Eversole, 1988).

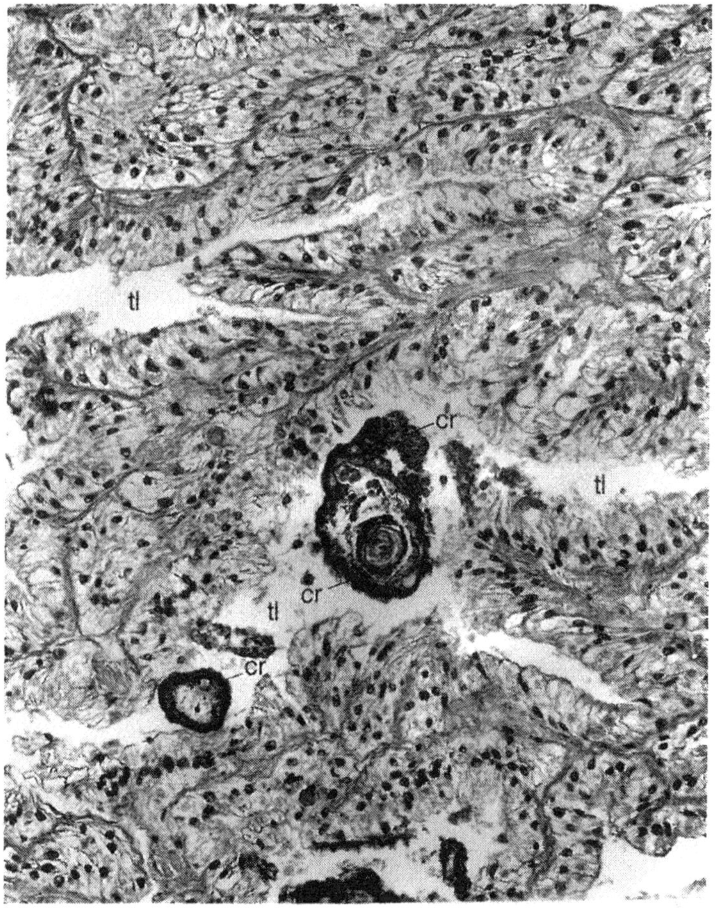

Fig. 4.62. Section through the proximal kidney showing large concretions (cr) in tubule lumina (tl). HFW = 405 μm.

4.11.1 Anatomy

M. mercenaria, like many bivalves, is considered protandric and usually spawns as a male the first year (this species is not strictly protandric, however, since Loosanoff (1937b) reported some animals spawning as females the first year); about half of the clams retained the male phase after spawning while the other half transformed into females (Loosanoff, 1937b). Gonadal follicles or acini first appear as a single layer of germinal epithelial cells between the body wall and the stomach; as the animal grows, acini begin to branch and germinal epithelia begin to differentiate into spermatogonia and oogonia (Loosanoff, 1937b). Growing acini vary in their state of differentiation: some only contain indifferent cells while others have developing male and female gonia. Gonads at these early stages are bisexual but because the proliferation of spermatogenic cells is so very rapid, the tissue acquires a male appearance. Young clams, 5–7 mm shell length, discharge sperm (Loosanoff, 1937b). Acini continue to grow and ramify throughout the visceral mass and foot. After the first year, about half of the

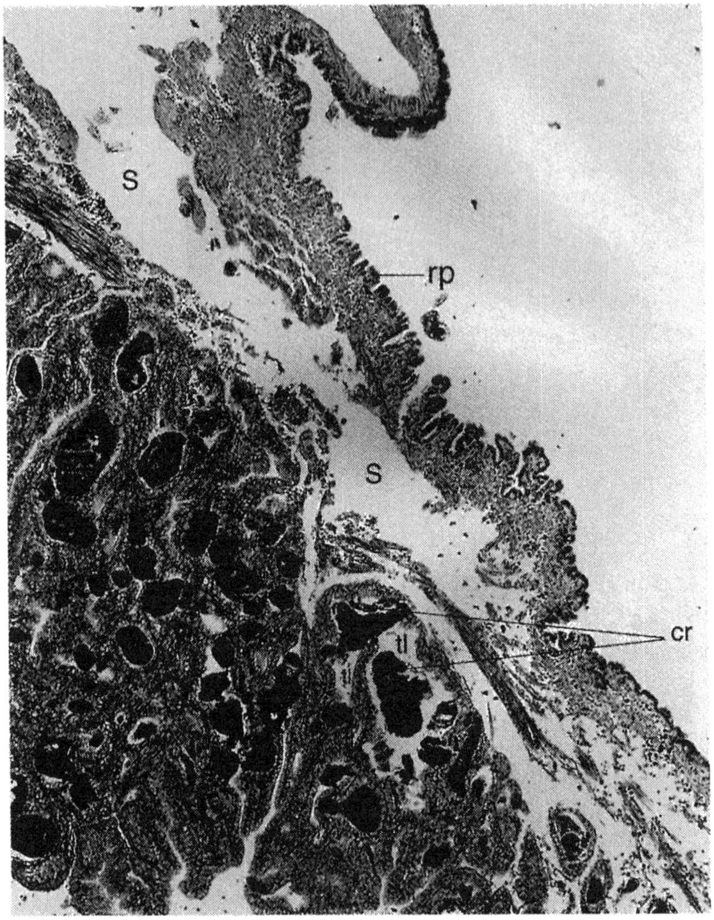

Fig. 4.63. Cross section through the distal kidney showing large concretions (cr) in tubule lumina (tl). rp = renal cortical epithelium. s = hemolymph sinuses. HFW = 405 μm.

males change sex and become female but females rarely change back into males (Loosanoff, 1937a). Eversole (Chapter 5) reports that males change sex in South Carolina, USA, between 20 and 35 mm shell length; he indicates there is no evidence that females change into males.

Spawning is usually completed in Long Island Sound (USA) at the end of August. By mid-October spermatogenesis and oogenesis are actively proceeding and by November gonadal follicles are fairly well filled with sperm (Loosanoff, 1937a). Young oocytes grow rapidly during October–November and attain sizes of 33–55 μm in diameter; acini ramify throughout the connective tissue of the visceral mass and into the foot. During winter months few changes take place in gonads. In the spring when water temperatures reach 15°C, spermatogenesis again becomes very active and the ripe condition is quickly reached. Spring temperatures also encourage rapid growth of oocytes that reach the maximum size of 66–70 μm in diameter (Loosanoff, 1937a). In Long Island Sound (USA), discharge of sperm occurs when water temperatures reach 23–25°C.

In northeastern waters of the United States, gonadal development takes place during falling water temperatures; cells produced during the autumn constitute the greatest portion of the next summer's spawning season and many oocytes ripen about 2–3 months before spawning (Loosanoff, 1937a). In southeastern waters (USA), gametogenesis occurs in all seasons.

4.11.2 Histology

A modified classification system based on Porter (1964) is used below to describe male and female follicles or acini in various phases of gonadal development including gametogenesis. Loosanoff (1937a,b) and Eversole (Chapter 5) should also be consulted by the serious reader; authors include lucid drawings and photomicrographs to supplement their descriptions. Photomicrographs presented here should be compared with their illustrations to gain a complete understanding of gonadal maturation.

4.11.2.1 Male

4.11.2.1.1 Immature

Acini at this stage are expanded and filled with follicle cells (Fig. 4.64). These cells are not discussed by Loosanoff (1937a,b); he draws them, but shows them as only contributing to the simple squamous lining of acini. Follicle cells were originally described and figured by Porter (1964) for *M. mercenaria*; Ansell (1961) briefly mentions them in connection with post-spawned female acini in *Venus striatula*. Follicle cells contain a centrally located, large (5.5 μm) nucleus that contains evenly distributed chromatin and an eccentric nucleolus. Fine strands of membrane radiate in web-like fashion from the nucleus to the plasma membrane leaving many vacuoles scattered throughout the cytoplasm (Figs. 4.64 and 4.65). These cells are presumed to be nutritive but their physiology has not yet been investigated.

Spermatogonia are present in the lining of acini and contribute to the cell population of the germinal epithelium (Figs. 4.64 and 4.65). Spermatogonia may easily be distinguished by the large nucleus (6.5 μm) with evenly spaced chromatin containing 1–2 nucleoli); this cell is small (8 μm) with a dense cytoplasm (Figs. 4.64 and 4.65). Some primary spermatocytes may be present as cords of cells interspersed within the follicle-cell matrix (Figs. 4.64 and 4.65).

Many nutritive cells are present in perigonadal hemolymph spaces (Figs. 4.64 and 4.65); similar cells form dense masses around gonadal arteries and occur in large numbers in hemolymph sinuses surrounding those portions of the stomach and style sac in proximity to gonadal follicles. Nutritive cells were first described by Loosanoff (1937a) who termed them phagocytic–nutritive cells. They are abundant throughout all stages of male follicle development and to a lesser extent in the maturation of female follicles (Fig. 4.66). Nutritive cells are 12–15 μm in diameter with the cytoplasm filled with fine-to-coarse yellowish granules that are composed of two different types of glycoproteins: a more common neutral glycoprotein and an acid glycoprotein rich in carboxyl groups; nuclei are elongate (5.5 μm), with evenly dispersed chromatin (Figs. 4.64 and 4.66). Nutritive cells in hemolymph sinuses surrounding gonadal acini lie within a connective tissue framework (Fig. 4.67) which consists of sulfated glycosaminoglycans. As male acini mature, granules within nutritive cells become less prominent and, in some cells, disappear entirely, suggestive of active use by developing

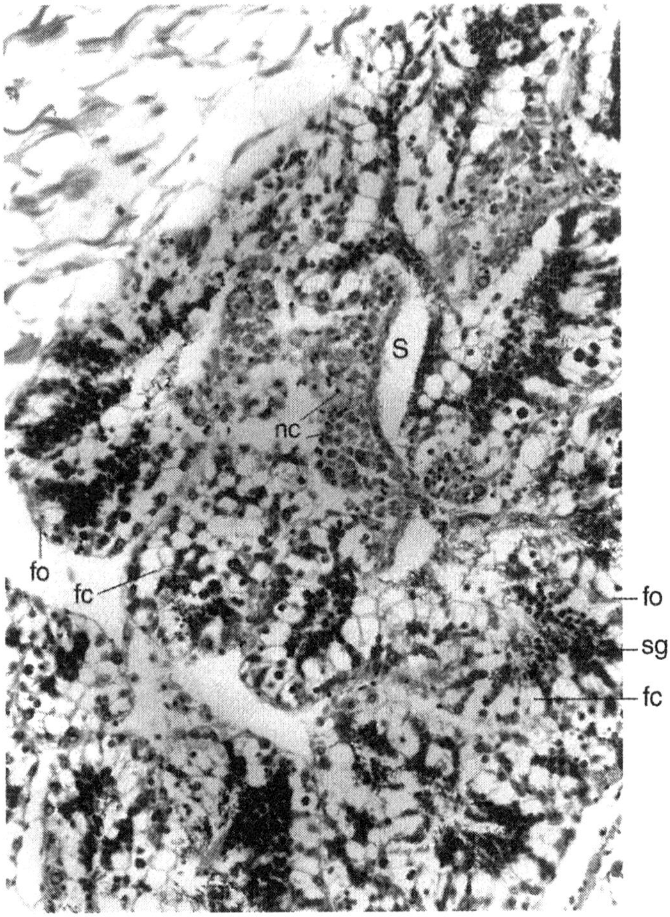

Fig. 4.64. Cross section through the immature testis to show follicles (fo) filled with follicle cells (fc) and some spermatogonia (sg). Many clusters of nutritive cells (nc) are in juxtaposition to follicles on one side and a hemolymph vessel (s) on the other. HFW = 405 μm.

gametes (Fig. 4.68). Much physiological research remains to be done on these interesting cells.

4.11.2.1.2 Mature

Follicle cells become restricted to the periphery of acini with dense zones of primary spermatocytes predominating. Just central to the zone of primary spermatocytes are islands of secondary spermatocytes with small central areas containing dense radiating bands of sperm (Fig. 4.68). Primary spermatocytes have large nuclei (4.7 μm) with evenly distributed chromatin and a faint, eccentric nucleolus; nuclei are usually in various stages of prophase I (Fig. 4.68). Secondary spermatocytes have smaller (3.4 μm), round nuclei with chromatin in various stages of prophase II (Fig. 4.68).

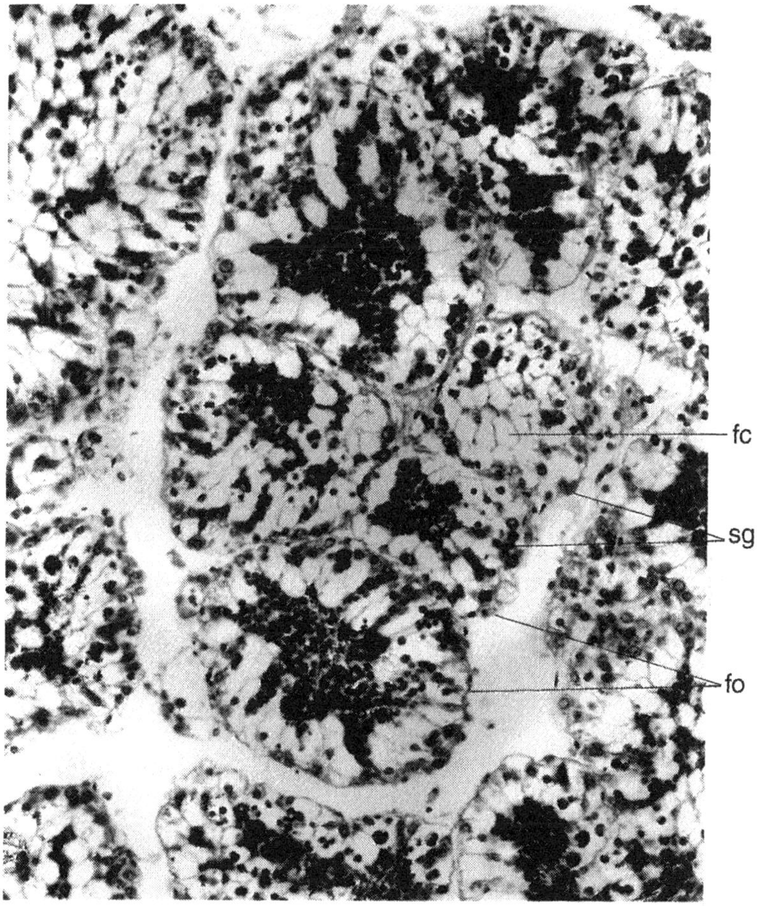

Fig. 4.65. Cross section through an immature testis to show follicle cells (fc) occupying most of the follicle (fo) with spermatogonia (sg) lining the periphery. HFW = 405 μm.

Sperm heads are large (5 μm) and scimitar shaped with the pointed end anterior; flagella are long (24–30 μm) and eosinophilic which enables them to stand out starkly from all the basophilic nuclei (Figs. 4.68 and 4.69).

4.11.2.1.3 Ripe

This stage is similar to the mature phase but dense bands of radiating sperm fill about one-half the area of acini; primary and secondary spermatocytes occupy a thin peripheral area (Fig. 4.69). In some acini, spermatids may be seen as small (2.5 μm) dark nuclei in juxtaposition to sperm heads.

It should be pointed out that the potential bisexual nature of the male gonad is visible in all stages described above; oogonia occur occasionally in male acini and may be seen as scattered cells in the germinal epithelium of acini (Fig. 4.68).

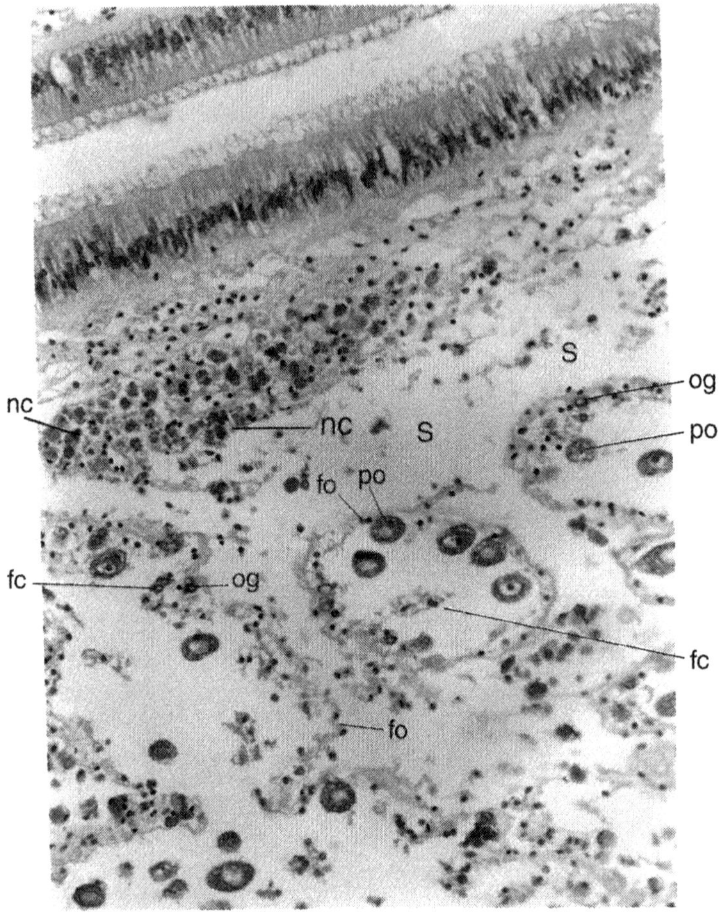

Fig. 4.66. Cross section through an immature ovary to show acini or follicles (fo) containing oogonia (og), young primary oocytes (po) and some follicle cells (fc). A cluster of nutritive cells (nc) is separated from three acini by a hemolymph sinus (s). HFW = 405 μm.

4.11.2.2 Female

4.11.2.2.1 Immature

Acini are usually filled with follicle cells; oogonia are numerous and distributed in the germinal epithelium of acini (Fig. 4.70) and, occasionally, developing primary oocytes occur. Oogonia are small cells (9–16 μm) with a granular cytoplasm and a large (5–10 μm) nucleus; chromatin is evenly distributed in small nuclei but becomes progressively associated with the nuclear envelope as nuclei enlarge leaving a central area with sparse chromatin. A prominent, eccentric nucleolus is found in all oogonia with nuclear sizes above 4.5 μm (Fig. 4.70).

Oocytes can be distinguished from oogonia by: (1) their granular, basophilic cytoplasm; and (2) large nucleus containing a prominent nucleolus; most of the chromatin is associated

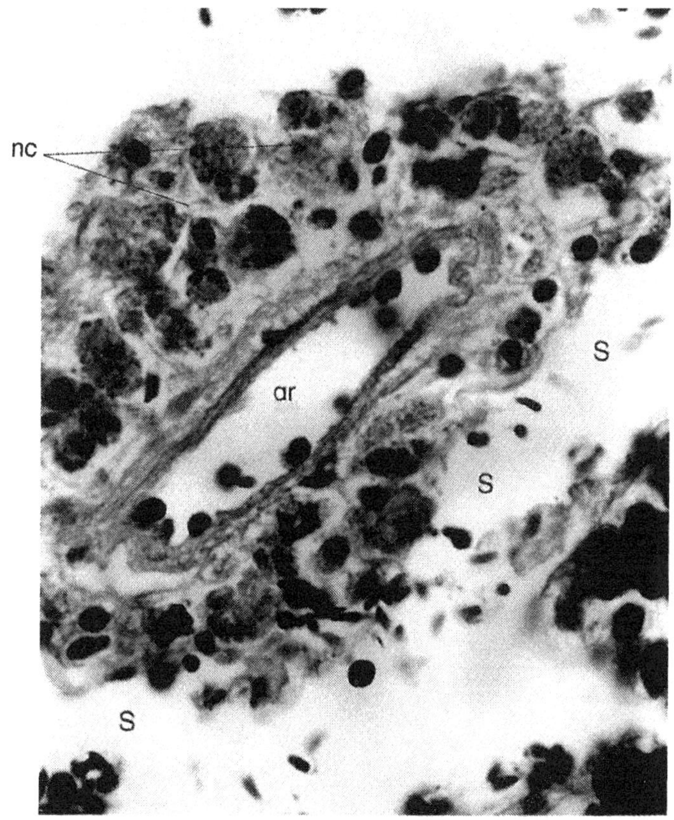

Fig. 4.67. Cross section through the testis to show an artery (ar) surrounded by nutritive cells (nc) that seem to lie within a connective tissue framework. Note large hemolymph sinuses (s) bathing nutritive cells. HFW = 95 μm.

with the nuclear envelope (Figs. 4.70–4.72). Oocytes at this stage are small (20–30 μm). Some nutritive cells may be seen in the peri-acinar hemolymph sinuses (Figs. 4.71–4.73).

4.11.2.2.2 Mature

Oogonia are present in the germinal epithelium of acini, scattered at bases of follicle cells; oocytes have started to grow and average 40–50 μm in diameter. All oocytes at this stage are connected to the acinar epithelium and, in some oocytes, this connection is starting to narrow down to a stalk (Fig. 4.71). Nutritive cells are still present in peri-acinar hemolymph sinuses (Fig. 4.71).

4.11.2.2.3 Ripe

Oogonia are scattered in the germinal epithelium of acini, but few follicle cells are present. Oocytes are quite large (50–70 μm) and fill the lumen of acini (Fig. 4.72). Many oocytes

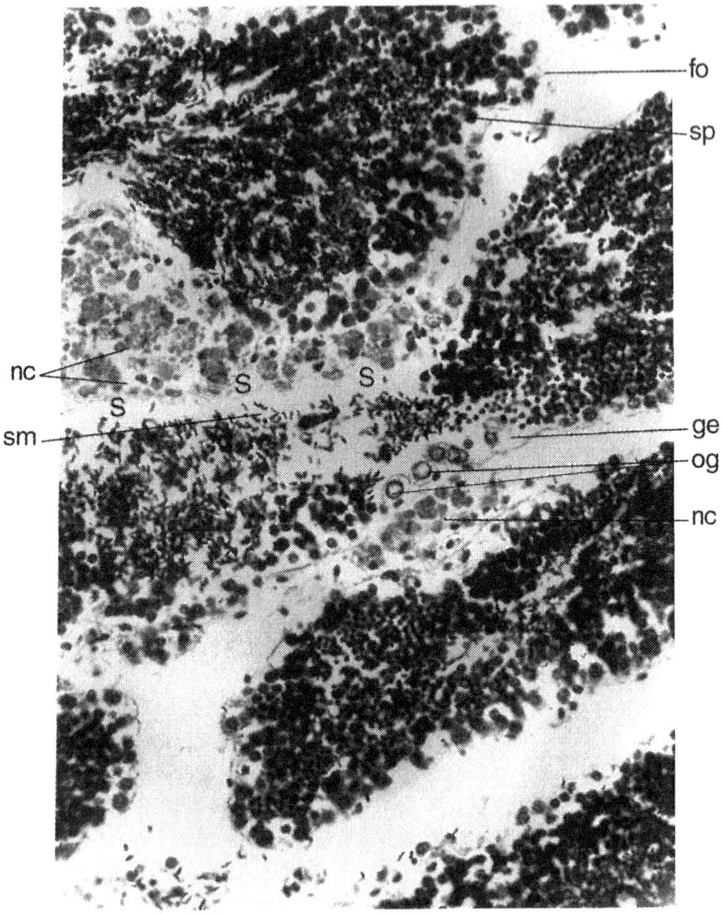

Fig. 4.68. Cross section through the testis to show follicles (fo) with developing spermatocytes (sp) and some sperm (sm). Clusters of nutritive cells (nc) can be seen surrounded by hemolymph sinuses (s) located between follicles. Note the four oogonia (og) in the germinal epithelium (ge) of the center follicle; oogonia are not uncommon in the testes, evidence of the bisexual condition of the gonad. HFW = 405 μm.

are still connected to the germinal epithelium of acini, but some seem to be floating free in the middle of the follicle (Fig. 4.72). Some nutritive cells are still present in the peri-acinar hemolymph sinuses, but their numbers steadily diminish as female acini mature and ripen (Fig. 4.72). Loosanoff (1937a) noticed this and thought that nutritive cells played less of a role in developing female than in male acini. Much work needs to be done on the physiology and seasonal occurrence of these cells for both sexes.

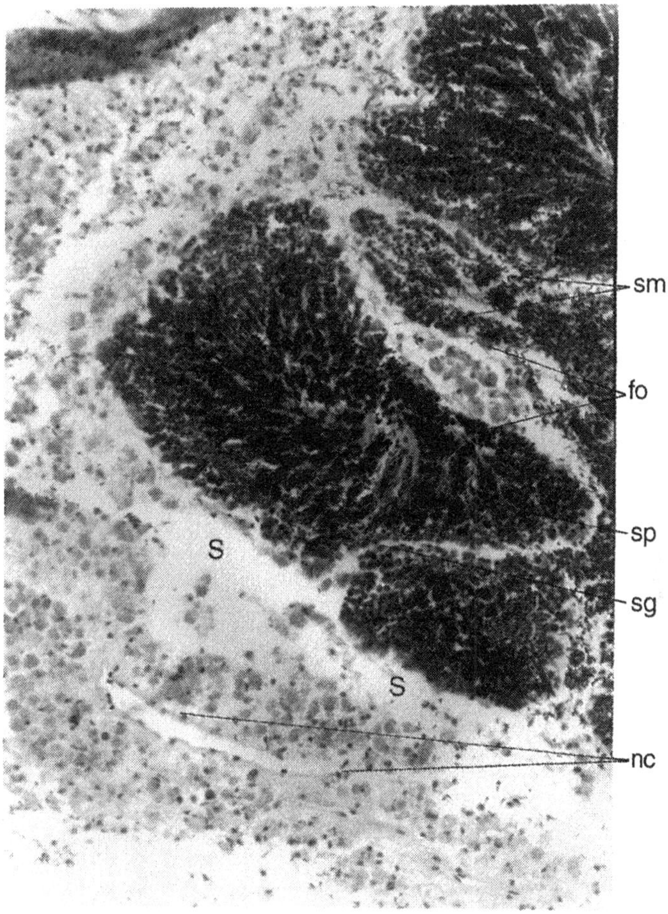

Fig. 4.69. Cross section through the ripe testis showing long columns of sperm (sm) in follicles (fo). Spermatogonia (sg) and primary spermatocytes (sp) occur in the germinal epithelium that lines the periphery of follicles. Note clusters of nutritive cells (nc) surrounded by hemolymph sinuses (s). HFW = 405 μm.

4.12 CIRCULATORY SYSTEM

4.12.1 Anatomy

4.12.1.1 Pericardial coelom

The heart lies in the pericardial coelom and consists of three chambers: two atria and a ventricle (Fig. 4.41). Atria are thin walled, colorless and receive hemolymph from gills, kidneys, the pericardial gland and large veins located just anteroventral to the pericardial coelom. Atrial muscle fibers are weakly developed imparting a membranous texture to the atria. The ventricle frequently has a pale or salmon-pink color, many well-developed muscle

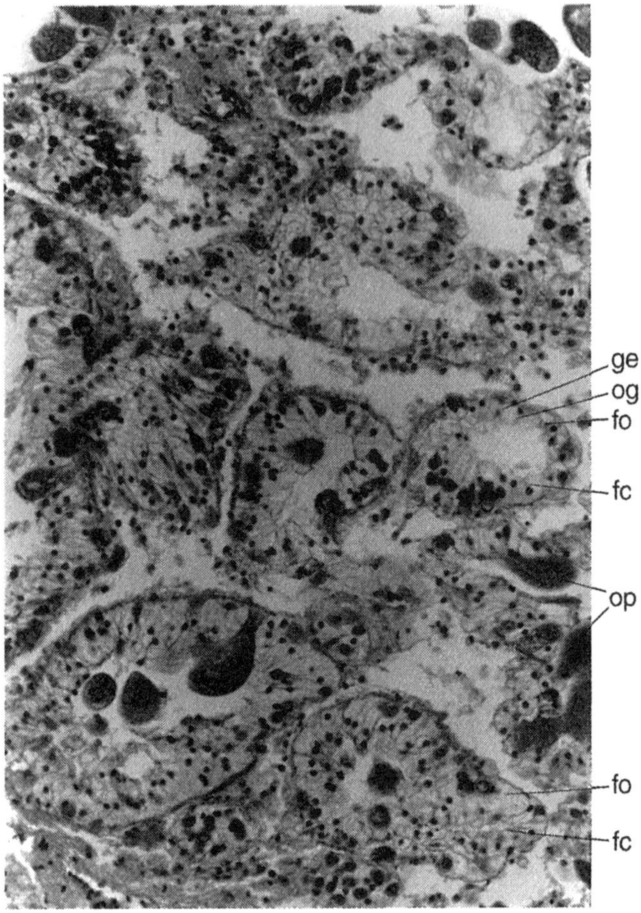

Fig. 4.70. Cross section through an immature ovary to show follicles (fo) filled with follicle cells (fc) and young primary oocytes (op). Oogonia (og) are present in germinal epithelia (ge) nestled between follicle cells. HFW = 405 μm.

bundles, and envelops the rectum (Fig. 4.41); White (1942) suggested two functions for this arrangement: contraction of the ventricle would facilitate movement of rectal contents and the rectum would provide a firm anchorage for the contracting ventricle. Two aortae leave the ventricle: an anterior aorta which is attached to the dorsal surface of the rectum and a posterior aorta which traverses the pericardial coelom in company with the rectum and enters the aortic bulb situated in the posterior region of the pericardial coelom (Fig. 4.41). The posterior aorta together with the rectum exit the posterior end of the aortic bulb and leave the pericardial coelom dorsal to the posterior adductor muscle (Fig. 4.41 Fig. 4.50).

The aortic bulb or bulbus arteriosus is a large, muscular organ about the same size as the ventricle (Fig. 4.41; Hersh, 1957). It is present in bivalves that have siphons, but unlike the heart, it does not pulsate (Hersh, 1957). The aortic bulb has a large, irregular lumen into which hemolymph is distributed as it enters from the posterior aorta; many hemolymph sinuses

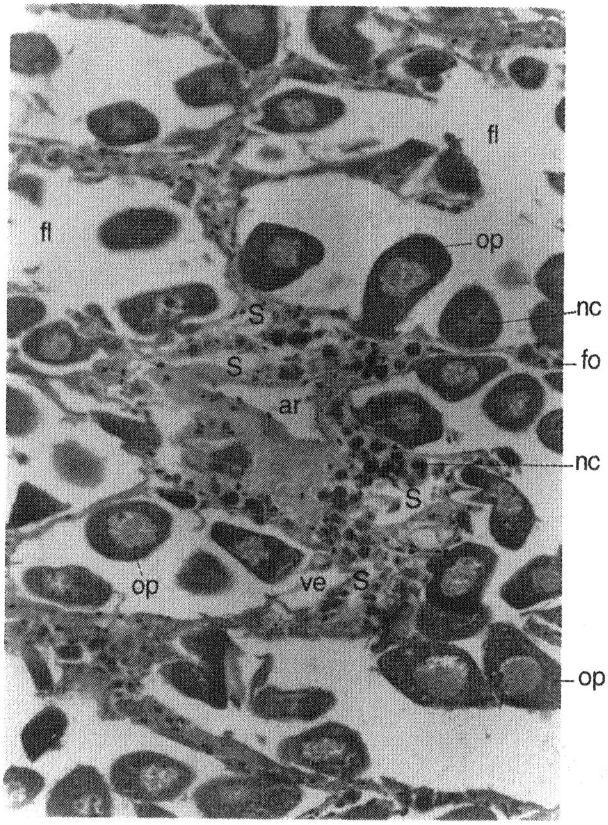

Fig. 4.71. Cross section through mature ovary to show follicles (fl) filled with developing primary oocytes (op) that are attached to the germinal epithelium (ge) by a stalk. Note clusters of nutritive cells (nc) surrounded by a perifollicular hemolymph sinus (s). ar = ovarian artery; ve = ovarian vein. HFW = 405 μm.

situated between muscle bundles drain into the central lumen and impart a 'spongy' texture to the interior. (Fig. 4.49). The aortic bulb acts as a temporary reservoir to store hemolymph when the siphons contract (Brand, 1972); this protects the ventricle from possible damage when sudden surges of hemolymph are forced backwards in the posterior aorta. Indeed, observations in my laboratory with clams prepared with a 'window' over the pericardial coelom (accomplished by striking one of the valves near the hinge with a heavy oyster knife and removing the pieces to expose the pericardial area) reveal that the aortic bulb is capable of considerable expansion. Bivalves with siphons invariably possess an aortic bulb and those that lack siphons do not have this organ (Eble and Scro, 1996).

4.12.1.2 Arterial system

Jones (1979) briefly discusses the arterial system in venerid clams, but gives few details. Nielsen (1963) and Joshi and Bal (1967) discuss aspects of the arterial system of *Katelysia marmarata*, a venerid clam that shares many features similar to those of *M. mercenaria*. The

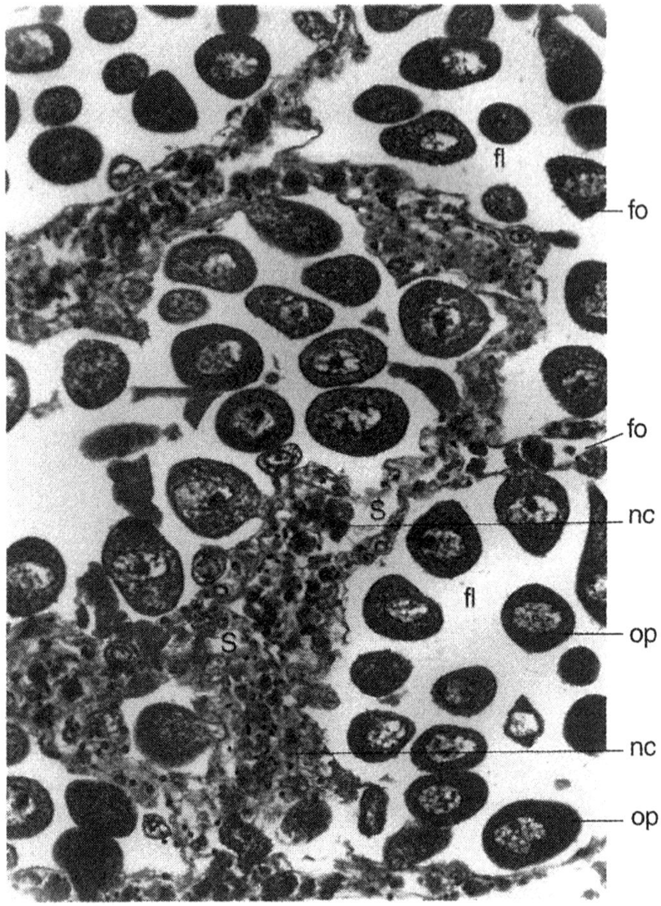

Fig. 4.72. Cross section through a ripe ovary showing follicles (fo) with fully formed primary oocytes (op) most of which are free floating in follicular lumina (fl). Note clusters of nutritive cells (nc) surrounded by perifollicular hemolymph sinuses (s). HFW = 405 μm.

work reported here was done in my laboratory using a red solution of vinyl acetate (Carolina Biological Supply House, Elon College, NC) that was injected into the anterior aorta and allowed to polymerize; blue vinyl acetate was used to inject veins. Clam tissues were then digested by strong alkaline hydrolysis (20% sodium hydroxide) leaving a cast of the arterial system (Shuster and Eble, 1961). The anterior aorta, situated on the dorsal surface of the rectum, arises from the anterodorsal region of the ventricle and plunges into the visceral mass (Fig. 4.73). Several pallial arteries and gonadal arteries spring from the aorta as soon as it enters the visceral mass (Fig. 4.73). After running a short distance in the dorsal portion of the visceral mass, the anterior aorta takes a right-angle turn and courses posteroventrally as the visceropedal artery (Joshi and Bal, 1967). The visceropedal artery descends into the visceral mass where it gives rise to many hepatic arteries that supply the digestive gland before dividing into two large branches: the more ventral branch (intestinal artery) supplies the coiled intestine while

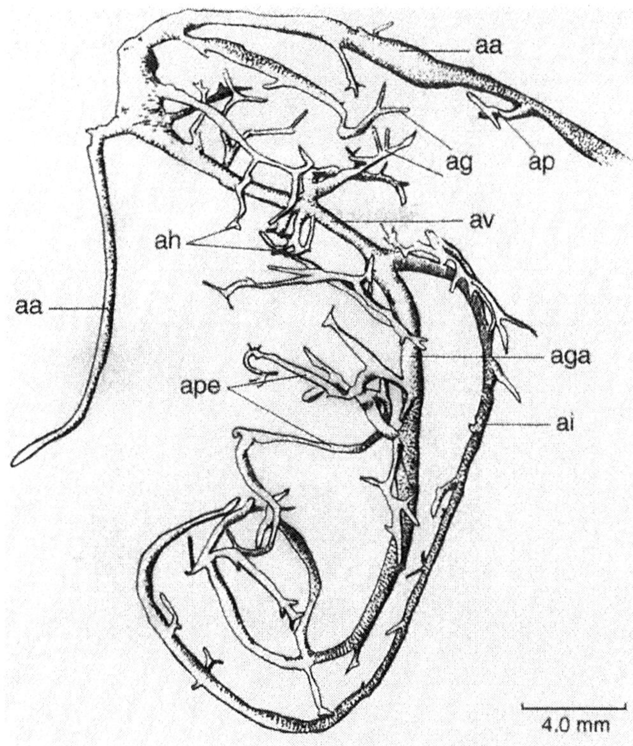

Fig. 4.73. Sketch of arterial system drawn from vinyl acetate cast. aa = anterior aorta; ag = gonadal artery; aga = gastric artery; ah = hepatic artery; ai = intestinal artery; ap = pallial artery; ape = pedal artery; av = visceropedal artery.

the more dorsal branch (gastric artery) supplies the stomach and style sac (Fig. 4.73). Both the intestinal and gastric arteries give off many pedal arteries that supply the foot (Fig. 4.73). At their anterior extremities, intestinal and gastric arteries recurve dorsally upon themselves and terminate by supplying many branches to gonadal and foot tissues (Fig. 4.73).

The anterior aorta continues anteriad as a slender vessel after giving off the visceropedal artery (Fig. 4.73). It runs under the anterior adductor muscle, which it supplies, and branches into the left and right anterior pallial arteries, which deliver hemolymph to the anterior region of the mantle (Joshi and Bal, 1967).

4.12.1.3 Venous system

The venous system originates as a dense network of highly branched venules which drain hemolymph sinuses in the anterior portion of the foot and visceral mass and which ultimately coalesce into several large visceral and hepatic veins situated on the left side of the animal (Fig. 4.74). These large veins ascend in the visceral mass in proximity to the descending anterior aorta and visceropedal artery and finally return hemolymph to the atria. The bulk of the venous system in the visceral mass and foot appears to take the form of a basket-like

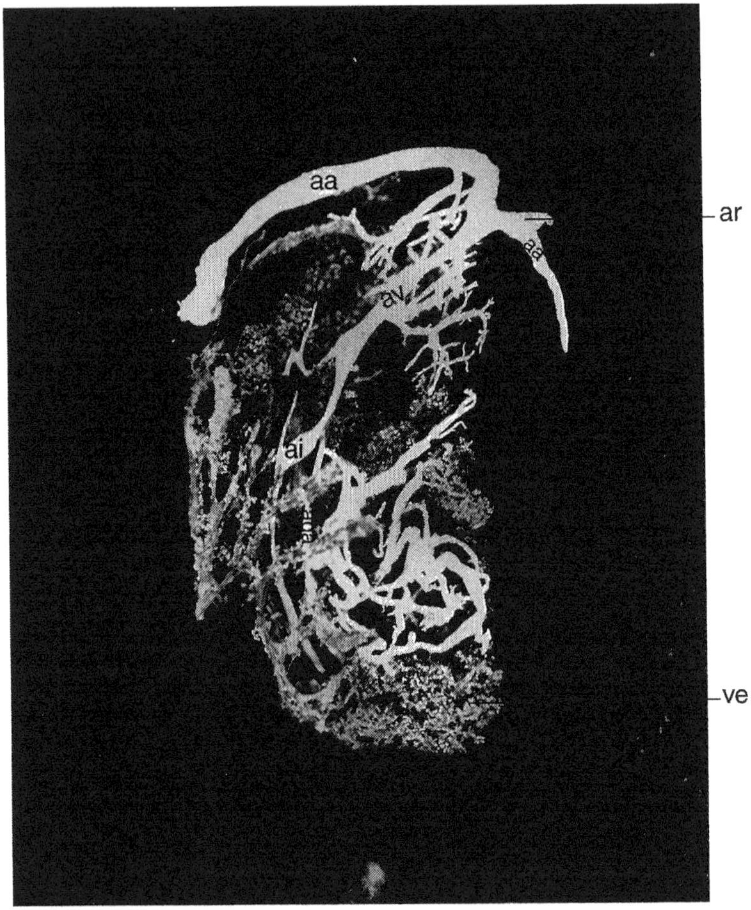

Fig. 4.74. Photograph of doubly injected specimen: red vinyl acetate was injected into the arterial system (ar) and blue vinyl acetate injected into the venous system (ve). Note the venous system forms a basket-like complex around the arterial system (compare with Fig. 4.73). aa = anterior aorta; aga = gastric artery; ai = intestinal artery; av = visceropedal artery. 1.6×.

meshwork lateral to the more centrally situated arterial system (Fig. 4.74). Much work remains to be done on both arterial and venous systems of *M. mercenaria*.

4.12.2 Histology

4.12.2.1 Heart

4.12.2.1.1 Atria

The epicardium has a simple columnar epithelium, which is thrown into a series of folds; a pronounced apical plasma membrane distinguishes the epithelium (Fig. 4.75). Nuclei are centrally located, average about 5.6 μm and have evenly distributed chromatin. A thin

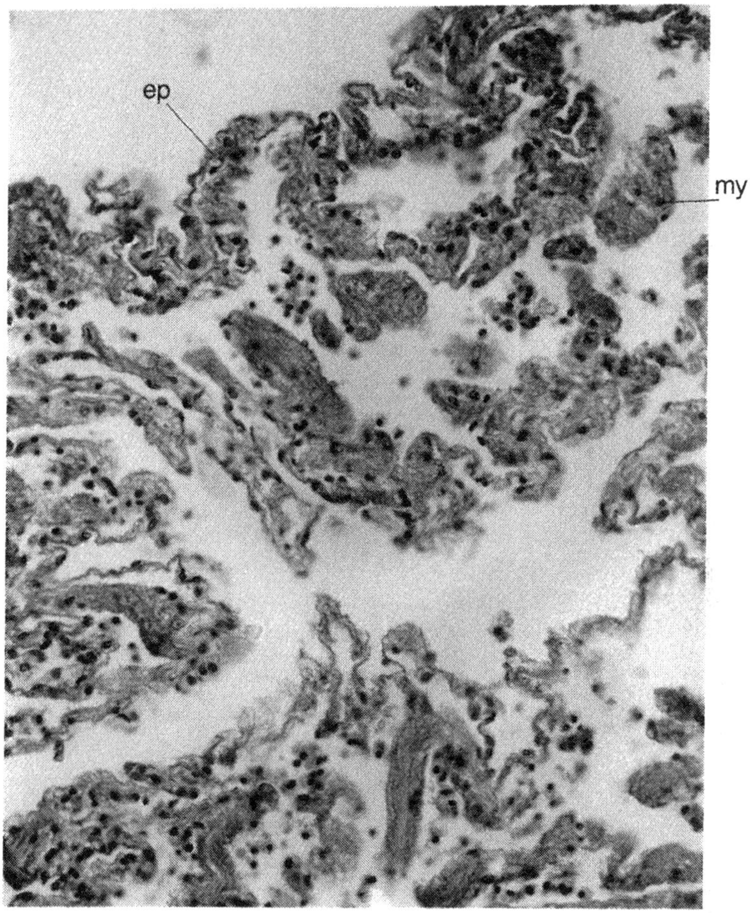

Fig. 4.75. Longitudinal section through the atrium to show the epicardium (ep) and myocardium (my). Note prominent cell membranes of epicardial cells. HFW = 405 μm.

connective tissue composed chiefly of glycosaminoglycans rich in carboxyl groups supports the epithelium.

The myocardium consists of scattered bundles of smooth muscle fibers (Fig. 4.75) bound by a delicate connective tissue made up of carboxylated glycosaminoglycans. Nuclei are oval to spherical, large (6.3 μm) with chromatin scattered in 2–3 clumps. Much glycogen is stored in muscle fibers of the myocardium, but little is found in the epicardium.

The presence of an endocardium is in doubt.

4.12.2.1.2 Ventricle

The epicardium has a simple, cuboidal epithelium supported by a thin connective tissue. Nuclei are oval to spherical, about 4.4 μm and have evenly distributed chromatin (Fig. 4.76).

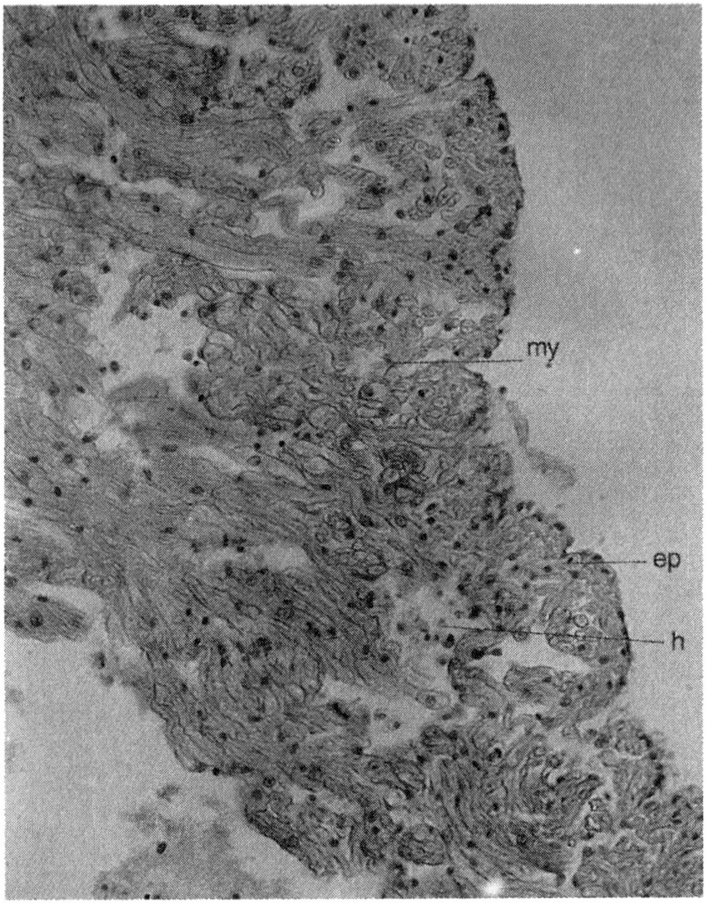

Fig. 4.76. Cross section through the ventricle to show the epicardium (ep) and myocardium (my). Compare the size and number of muscle bundles seen here with that in the atrium (Fig. 4.75). Note many hemocytes (h) in the lumen. HFW = 405 μm.

The myocardium has prominent muscle bundles consisting of large (5 μm in width) smooth muscle fibers. Nuclei are similar to myocardial atrial fibers both in size and distribution of chromatin (Figs. 4.47 and 4.76). Muscle fibers of the myocardium store glycogen and are bound together by a delicate connective tissue consisting of glycosaminoglycans rich in carboxyl groups.

Muscle fibers do not seem to possess an endocardium, but there is a suggestion of such a layer composed of a simple squamous epithelium with long, thin nuclei (1.25 × 7.5 μm) with evenly distributed chromatin situated at the interface between the rectum and myocardium of the ventricle.

Kelly and Hayes (1968) studied the ventricular musculature of *M. mercenaria* with light and electron microcopy. They determined that the muscle was of the smooth variety. Muscle fibers were rich in glycogen and we have confirmed that with studies in our laboratory; fibers were elongated and fusiform with a centrally placed nucleus surrounded by glycogen rosettes and mitochondria. The latter organelles were large with many cristae and were also found, together with much smooth endoplasmic reticulum, in the periphery of muscle cells. Most of the myofilaments were also located in the cell periphery; myofilaments were composed of actin and myosin (Kelly and Hayes, 1968). Hayes and Kelly (1968) studied the dense bodies of the heart of *M. mercenaria* and showed that they consist of aggregates of thin filaments embedded in an amorphous material; dense bodies were scattered throughout the cytoplasm of muscle cells, but were especially prominent in the periphery of cells. Furthermore, they demonstrated dense bodies in association with the cell membrane of muscle cells and termed the entire complex, attachment plaques; the latter were supported by a large network of collagen fibers. They postulated that the heart of *M. mercenaria* is yet another example of the collagen-net hypothesis for force transference as originally proposed by Mullins and Gunderoth (1965).

Smith (1985) studied the hearts of *M. mercenaria* and *Busycon canaliculatum* and concluded that these molluscs might functionally apply Starling's Law of the heart to accommodate increased output during exercise.

4.12.2.2 Aortic bulb

The aortic bulb is a highly muscular organ that consists of a dorsal portion containing the rectum and a ventral portion that receives hemolymph from the posterior aorta (Fig. 4.49). This ventral portion has a central lumen, roughly spherical in shape, that is surrounded by a complex network of muscle fibers enclosing many hemolymph sinuses all of which drain into the central lumen. Muscle fibers are of the smooth variety, large, and widely spaced. Nuclei are spherical, about 5.6 μm in diameter, and contain a prominent cluster of chromatin situated in the middle of the nucleus that is connected to the periphery by delicate 'spokes'. Muscle fibers are closely enveloped by a collagenous connective tissue imparting to the tissue a sponge-like texture. Since the function of the organ is to act as a temporary reservoir and absorb the 'pulse' of hemolymph from the contracting siphons, the structure is ideally suited to this function: hemolymph is forced into the central lumen from which it rapidly passes into the myriad of small connecting hemolymph sinuses (Fig. 4.49). To accommodate this surge of hemolymph, the aortic bulb increases in volume sometimes by a factor of 2 or 3; I have observed this many times! Muscle fibers stretched to accommodate this additional volume then contract to force hemolymph back into the posterior aorta.

Consistent with the construction of a hemolymph sinus system, there is no endothelium, and muscle fibers abut directly on the central lumen. A serosal layer is present lining the exterior of the aortic bulb, that consists of a highly folded, simple, columnar epithelium resting on a prominent collagenous connective-tissue layer; nuclei have a striking, tear-drop shape and usually possess a single nucleolus (Fig. 4.77). There appears to be much sloughing of serosal cells into the surrounding pericardial coelom which gives rise to a fine layer of cellular debris that clings to the outside of the serosal layer (Fig. 4.77).

4.12.2.3 Posterior aorta

The region of the posterior aorta which conducts hemolymph from the aortic bulb to the siphons and posterior adductor muscle surrounds the rectum and is composed of smooth, large

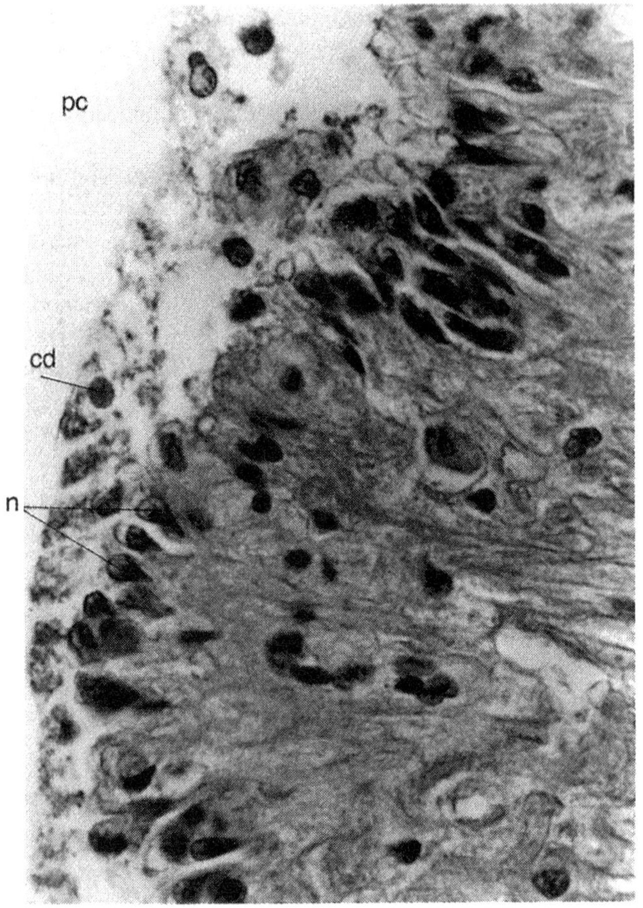

Fig. 4.77. Cross section through the serosal layer of the aortic bulb showing cone-shaped nuclei (n). Cellular debris (cd) just outside serosal layer suggests much sloughing of the serosal layer into the pericardial coelom (pc). HFW = 95 μm.

(3.4 μm in width) muscle fibers similar in general appearance, including nuclear cytology, to muscle fibers of the ventricle (Fig. 4.50). Muscle fibers are closely spaced and arranged in a net-like array although there is a predominance of circular fibers close to the central lumen and longitudinally oriented fibers close to the serosal surface. The serosal surface consists of a simple, squamous epithelium supported by a thin, collagenous connective tissue.

4.12.2.4 Arteries

Arteries possess an intima that is a simple, squamous endothelium resting on a thin connective tissue consisting largely of neutral glycoproteins. Nuclei of endothelial cells are long and thin (2.5×5.6 μm) and oriented with their long axes perpendicular to the longitudinal axis of the artery; chromatin is fine and evenly distributed (Fig. 4.67). The intima is reinforced

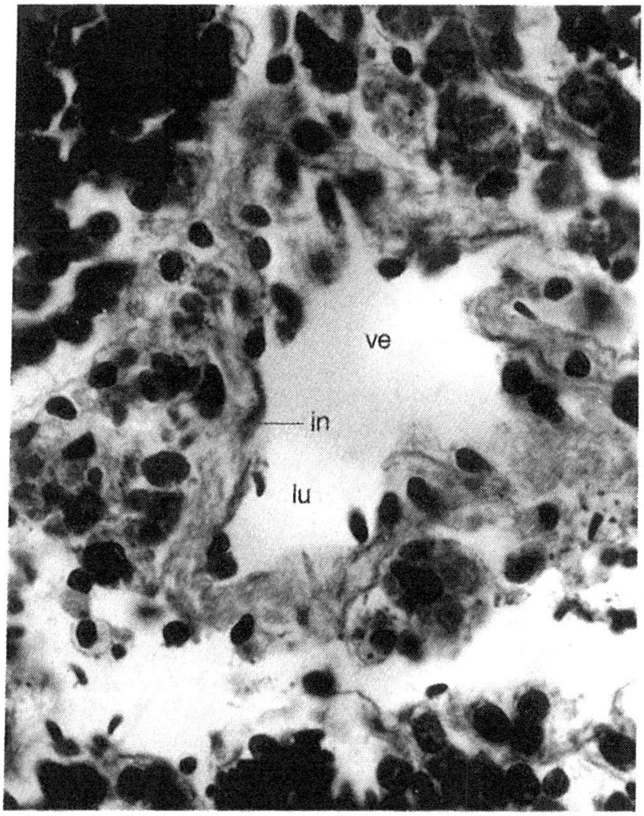

Fig. 4.78. Cross section through a vein (ve) to show thin intima (in) and irregular lumen (lu). HFW = 95 μm.

by connective-tissue fibers that arise in surrounding connective tissues and weave around and provide support; these connective-tissue fibers consist of acid glycoproteins (Fig. 4.67).

4.12.2.5 Veins

Veins are quite similar in histological structure to arteries, but have a larger, more irregular lumen as well as a less well-defined and thinner intima (Fig. 4.78).

4.12.3 Hemolymph (blood)

4.12.3.1 Hemocytes

Zachs and Welsh (1953) and Zachs (1955) discussed aspects of the cytochemistry of hemocytes in *M. mercenaria*; only one type of cell was mentioned, the granulocyte. Moore (1972) and Moore and Eble (1977) identified three cell types in *M. mercenaria*: an agranulocyte (5 μm), a small (25–30 μm) and a large granulocyte (40–45 μm); these authors also reported on the cytology and cytochemistry of these cell types. Foley and Cheng (1974) also identified three cell types in *M. mercenaria*: two different types of agranulocytes that they termed hyalinocytes and one type of granulocyte. They also reported on hemocyte numbers in clams from Buzzard's Bay, Massachusetts (1,954,900 cells/mL) and Great South Bay, New York (1,411,700 cells/mL). This agrees well with unpublished results from my laboratory which showed fed clams kept in laboratory aquaria at 28‰ and 21°C maintained hemocyte counts between 1.5 and 2.0 × 10^6/mL for a period of 8 weeks. Cheng and Foley (1975) reported on the ultrastructure of *M. mercenaria* hemocytes and identified a degranulated hemocyte as one they had previously classified as a fibrocyte; their classification system, then, included only two cell types, a hyalinocyte and a granulocyte.

Moore (1981) carried out a detailed study on *M. mercenaria* hemocytes including the cytology, cytochemistry and ultrastructure of the three cell types of Moore (1972) and Moore and Eble (1977); she also investigated aspects of the hemocyte cytoskeleton and studied the cell biology of phagocytosis at light- and electron-microscope levels. Moore (1981) concluded that only one hemocyte type is present in *M. mercenaria* which goes through three developmental phases (agranulocyte → large granulocyte → small granulocyte).

4.12.3.1.1 Cell types

Agranulocyte The agranulocyte (Figs. 4.79 and 4.80) is a small, round cell with a large nucleus that occupies most of the cytoplasm (Moore, 1972; Moore and Eble, 1977). The nucleus is spherical, has a dense rim of heterochromatin on the inside of the nuclear envelope with clumps scattered throughout the nucleoplasm; a nucleolus is occasionally observed (Moore, 1981). The thin rim of cytoplasm surrounding the nucleus contains mitochondria, smooth endoplasmic reticulum and glycogen. Moore (1981) found a developmental series of agranulocytes with progressively larger volumes of cytoplasm (Fig. 4.81). This cell comprises only 2% of the hemocyte population (Moore and Eble, 1977; Moore (1981).

Large granulocyte This cell type (Figs. 4.81 and 4.82) spreads thinly on glass slides and exhibits many filopodia (Moore and Eble, 1977). Cells appear not to move, but when viewed with phase-contrast, time-lapse cinematography and video microscopy, they show much membrane ruffling and cytoplasmic activity: a radial organization of microtubules emanating from the constantly shifting centrosome is obvious (Loy and Eble, 1974).

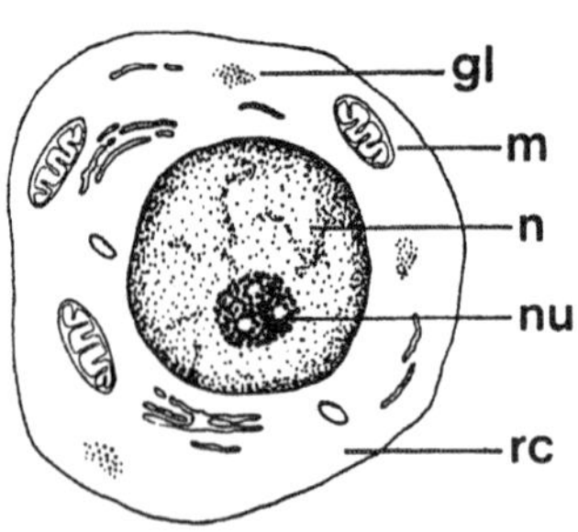

Fig. 4.79. Sketch of an agranulocyte to show glycogen (gl), mitochondria (m), nucleus (n), nucleolus (nu) and rim of cytoplasm (rc). From Moore (1981). HFW = 7875×.

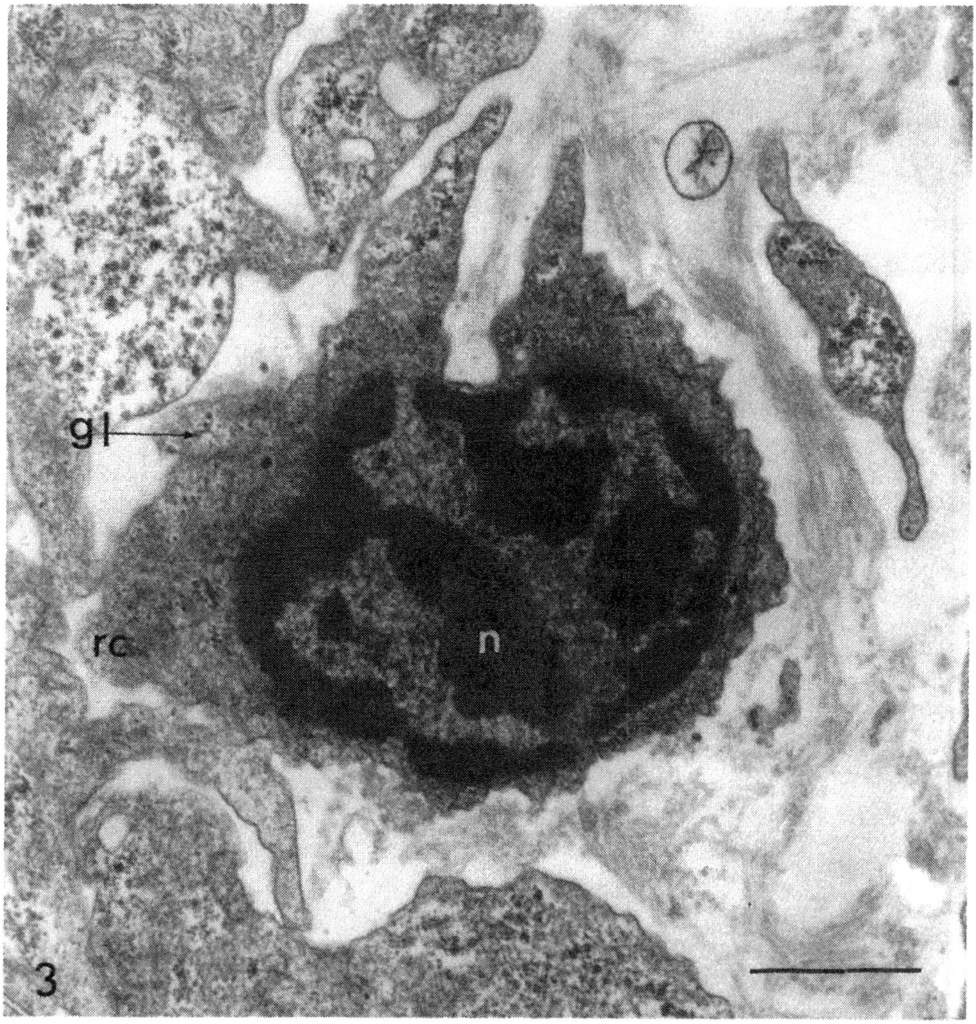

Fig. 4.80. Transmission electron micrograph of an agranulocyte showing a thin rim of cytoplasm (rc) containing glycogen granules (gl); the large nucleus (n) occupies most of the cell. From Moore (1981). Bar = 1 μm.

Nuclear cytology is similar to that of the agranulocyte but nucleoli are never observed (Moore and Eble, 1977; Moore, 1981). The most obvious cytoplasmic organelle is the membrane-bound, electron-dense vesicle; it is polymorphic and shows a wide range of sizes (Figs. 4.80–4.82). Vesicles vary from uniformly dense to containing either membrane-bound 'debris', lamellae or glycogen; further, smaller vesicles can be seen incorporated within larger vesicles and ultimately, these vesicles can be located within residual bodies (Moore, 1981). Before phagocytosis, these vesicles are usually homogeneously dense but quickly acquire debris, lamellae and finally, glycogen deposits following phagocytosis. This cell comprises 37% of the hemocyte population (Moore and Eble, 1977; Moore, 1981).

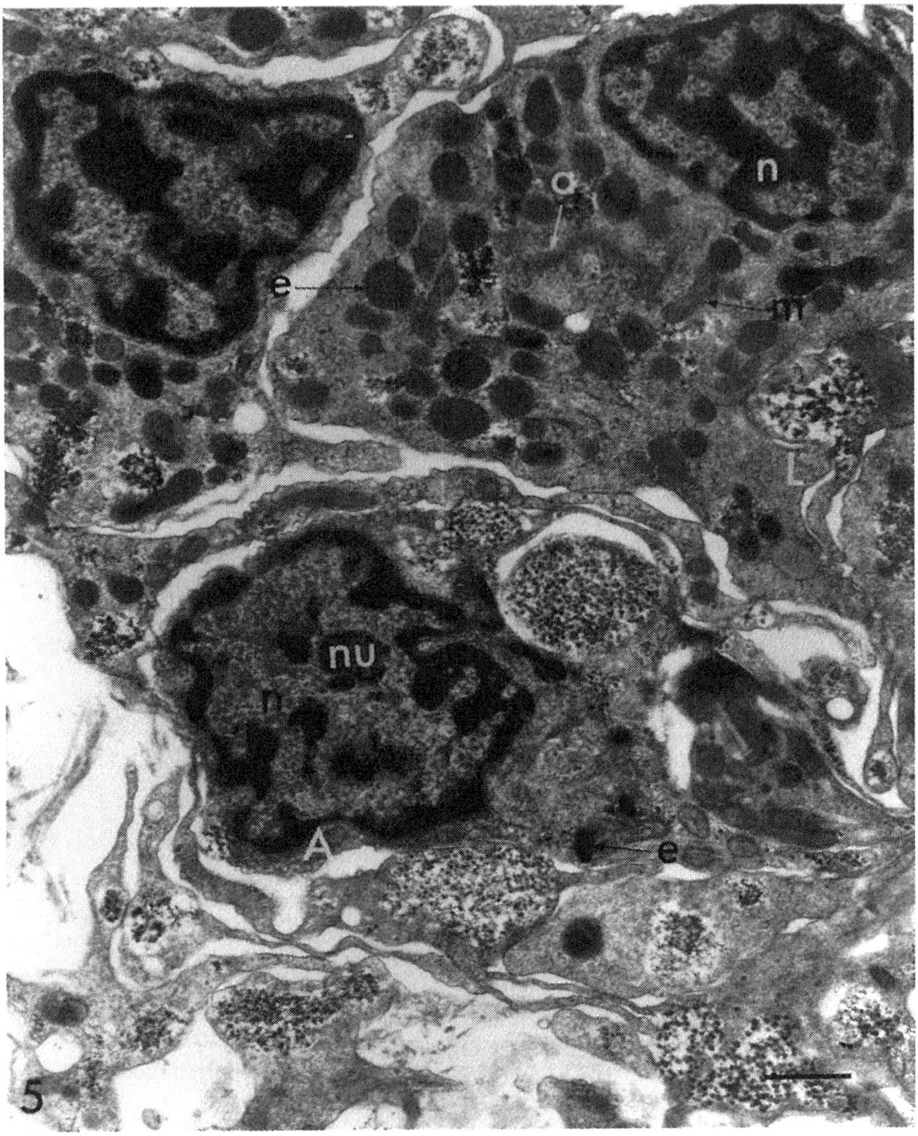

Fig. 4.81. Transmission electron micrograph of an agranulocyte (A) and a large granulocyte (L) to show Golgi apparatus (a), electron-dense blunt vesicle (e), mitochondrion (m), nucleus (n) and nucleolus (nu). From Moore (1981). Bar = 1 μm.

Small granulocyte Two features characterize this cell type: many large, electron-dense vesicles that fill the entire cell (Moore, 1972; Moore and Eble, 1977; Moore, 1981), and its obvious motility (Loy and Eble, 1974; Moore and Eble, 1977). Cells move rapidly and unidirectionally by extending lobopodia. When viewed with phase-contrast, time-lapse cinematography, extensive membrane ruffling and much movement of cytoplasmic organelles can be seen, sim-

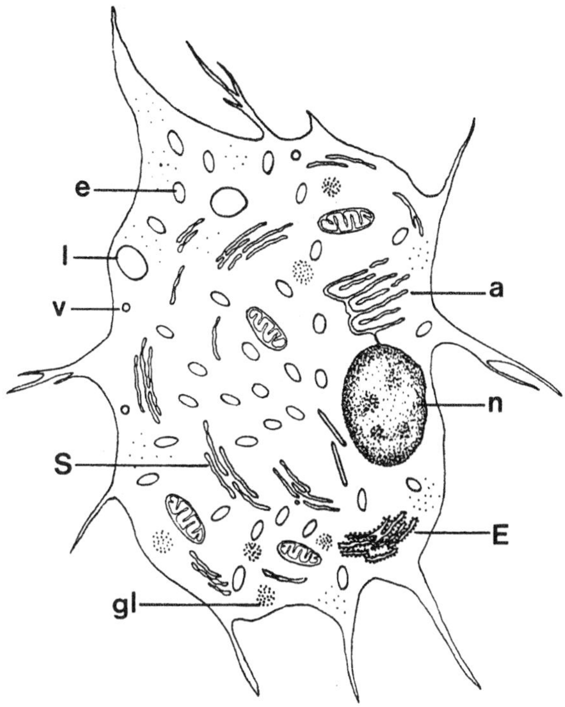

Fig. 4.82. Sketch of a large granulocyte to show Golgi apparatus (a), electron-dense blunt vesicle (e), rough endoplasmic reticulum (E), glycogen (gl), lipid droplet (l), nucleus (n), smooth endoplasmic reticulum (S) and electron-lucent vesicle (v). From Moore (1981). 2160×.

ilar to that described for the large granulocyte. Nuclear cytology is similar to that of both the agranulocyte and large granulocyte; nucleoli are occasionally observed (Moore and Eble, 1977; Moore, 1981). Cytoplasm is denser than that of the large granulocyte, but the same organelles, as described for the large granulocyte, are present, with the exception of the rough endoplasmic reticulum. Lipid droplets and glycogen deposits are similar to those found in large granulocytes (Moore, 1981). This cell type is the most numerous (61%) of the hemocyte population (Moore and Eble, 1977; Moore, 1981). See Figs. 4.83 and 4.84 for examples of a small granulocyte.

4.12.3.1.2 Phagocytosis

Cheng (1996) organizes phagocytosis into four phases: (1) attraction of phagocyte to nonself particles; (2) attachment of phagocyte to nonself particles; (3) internalization (endocytosis) of nonself particles into the phagocyte; and (4) intracellular degradation and distribution of molecules of nonself material. Cheng (1996) discusses in detail Phases 1 and 2 with special reference to the oyster, especially *Crassostrea virginica.* He also reviews Phase 3 and discusses three types of endocytic mechanisms. Phase-contrast, time-lapse cinematography (Loy and Eble, 1974) with boiled yeast cells show that *M. mercenaria* uses the Type 2 mechanism: formation of an invagination of the cell membrane with no involvement of filopodia.

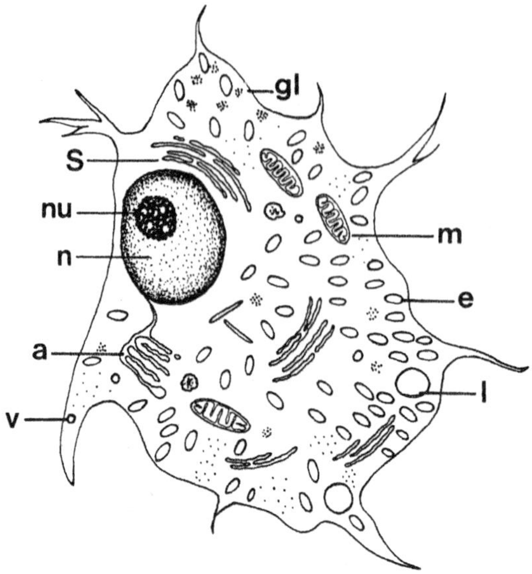

Fig. 4.83. Sketch of a small granulocyte to show the Golgi apparatus (a), an electron-dense blunt vesicle (e), glycogen (gl), a lipid droplet (l), a mitochondrion (m), nucleus (n), nucleolus (nu), smooth endoplasmic reticulum (S) and an electron-lucent vesicle (v). From Moore (1981). 2037.6×.

Phase 4 (intracellular degradation), has been the most thoroughly studied phase of phagocytosis in *M. mercenaria*. Before phagocytosis, cell organelles, especially electron-dense vesicles, are radially arranged around the centrosome; after internalization, these vesicles rapidly surround the newly formed phagosome (Loy and Eble, 1974; Moore and Eble, 1977; Moore, 1981; Moore and Gelder, 1983). Moore (1981) and Moore and Gelder (1983) presented algal cells (*Isochrysis galbana*) to *M. mercenaria* hemocytes in vitro. Internalization of algae and subsequent processing inside the phagosome were studied by transmission electron microscopy and microspectrofluorimetry. Before internalization, healthy, free-swimming *I. galbana*, as analyzed by microspectrofluorimetry, peaked at a mean wavelength of 682.4 nm with a relative fluorescence of about 30 Rf units (Fig. 4.85A); electron-dense vesicles showed a radial orientation around the centrosome in granulocytes (Fig. 4.85A). Shortly after internalization (5–10 min), electron-dense vesicles lost their radial orientation and surrounded the phagosome; *I. galbana* chloroplasts in the phagosome appeared less discrete and the red fluorescence began to diffuse throughout the algal cell; however, the emission peak stayed about the same (679.5 nm) while total fluorescence increased to about 220 Rf units (Fig. 4.85B). At this time, a secondary peak at 551 nm was observed indicating digestion of the algal cell within the phagosome. Furthermore, some of the green fluorescing chlorophyll could now be seen inside the closely adhering electron-dense vesicles (Fig. 4.85B). About 20–30 min after internalization, algae in phagosomes appeared shrunken and disorganized; fluorescence emission peaked at 547 nm with total fluorescence at only 9 Rf units (Fig. 4.85C). At this time, many of the green-fluorescing electron-dense vesicles were found distributed throughout the cytoplasm (Fig. 4.85C).

Thus, electron-dense vesicles participate actively in the degradation of ingested material

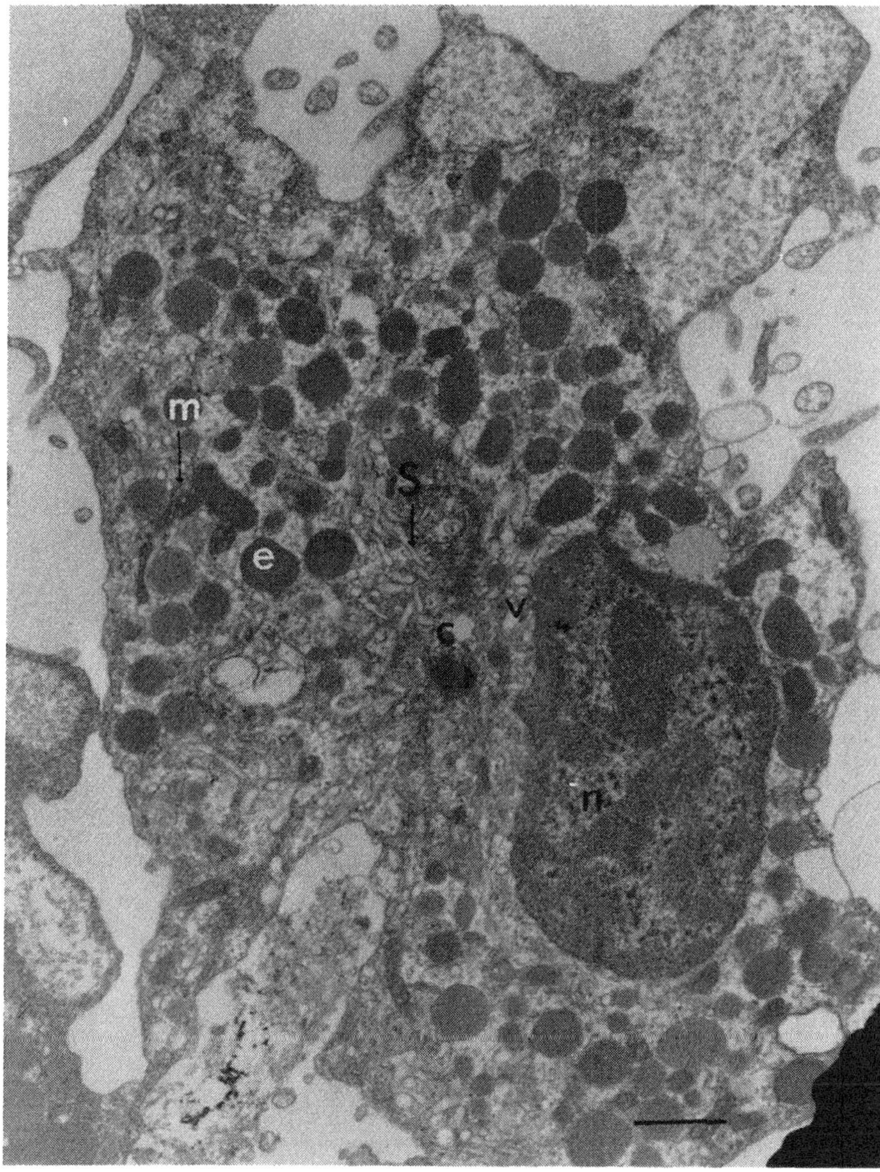

Fig. 4.84. Transmission electron micrograph of a small granulocyte to show centrosome (c), electron-dense blunt vesicle (e), mitochondrion (m), nucleus (n), nucleolus, smooth endoplasmic reticulum (S) and electron-lucent vesicle (v). From Moore (1981). Bar = 2 μm.

within phagosomes; further, synthesis of phagosomal contents into glycogen occurs by a phagosomal-electron-dense vesicle pathway (Fig. 4.86; Moore, 1981). In many instances, limiting membranes of electron-dense vesicles disappeared leaving packets of glycogen in electron-lucent vesicles (Fig. 4.86). Cheng and Cali (1974) also proposed a similar degrada-

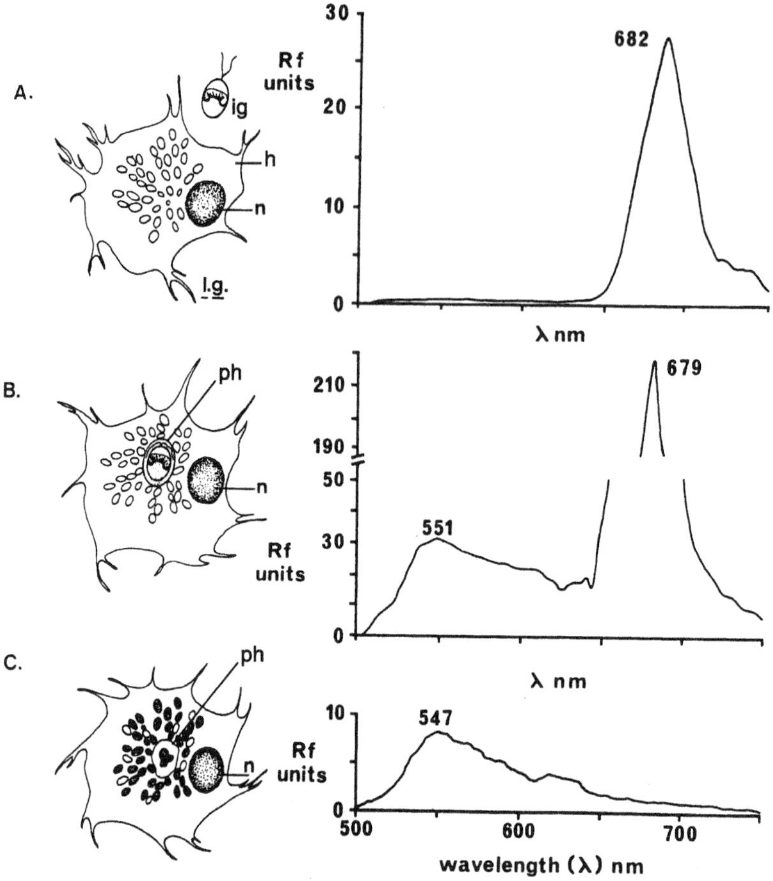

Fig. 4.85. The phagocytosis and intracellular degradation of *Isochrysis galbana* by *Mercenaria mercenaria* hemocytes. Left: Behavior of blunt vesicles and phagosome (ph) through stages of phagocytosis and degradation. (A) Hemocyte (h) and *I. galbana* (ig) before phagocytosis. n = nucleus. (B) Hemocyte and *I. galbana* in phagosome (ph) shortly after phagocytosis, demonstrating the initial breakdown of chlorophyll. n = nucleus. (C) Hemocyte and *I. galbana* in phagosome (ph) showing alga in an advanced stage of degeneration; note blunt vesicles contain chlorophyll degradation products (dark-colored vesicles). n = nucleus. Right: microspectrofluorimetric analysis of the emission from the chloroplast and blunt vesicles in phagocytosis and degradation stages corresponding to those diagrammed. (A) Emission analysis from *I. galbana* chloroplast before phagocytosis. (B) Emission analysis from *I. galbana* chloroplast shortly after phagocytosis. (C) Emission analysis from both the digested remnant of *I. galbana* and blunt vesicles (Note: both peaks were identical, therefore, only one peak is depicted). From Moore (1981).

tion-synthesis process in electron-lucent granules, which they termed 'secondary lysosomes', in hemocytes of *Crassostrea virginica* which had phagocytosed bacteria. These authors hypothesized that the newly synthesized glycogen was eventually released from secondary lysosomes into the cytoplasm and, finally, into the hemolymph. This hypothesis was later supported by Cheng and Rudo (1976) who traced the fate of injected ^{14}C-labelled *Bacillus megaterium* in *C. virginica*; they detected glycogen synthesized from bacterial by-products, initially in hemocytes and subsequently in hemolymph and finally in body tissues. Rodrick and Ulrich (1984) reported a significant increase in hemolymph glycogen in *C. virginica*, *Mercenaria campechiensis* and *Anadara ovalis* 1 h after a challenge with *Escherichia coli* and *Vibrio anguillarum*. Yoshino and Cheng (1976) localized acid phosphatase in some of the cytoplasmic vesicles in *M. mercenaria* using transmission electron microscopy, and reported these organelles as primary lysosomes. Moore and Gelder (1985) demonstrated four other lysosomal enzymes in primary lysosomes: β-glucuronidase, β-acetylglucosaminidase, acid β-galactosidase and esterases-A, -B and -C. Peroxidases including myeloperoxidase were localized in so-called 'blunt' and 'dot-like' granules (Moore and Eble, 1977; Moore and Gelder, 1985) seen with phase-contrast, light microscopy; potentiation of enzyme activity occurred in algal-challenged hemocytes (Gelder and Moore, 1986). Another lysosomal enzyme, aryl-sulfatase, was identified in primary lysosomes (Gelder and Moore, 1986); these authors also noted intense alkaline phosphatase reactions in the cytoplasm, probably glucose-6-phosphatase activity in the endoplasmic reticulum, immediately adjacent to phagosomes.

Cheng (1996) summarized the ultrastructure and functions of cytoplasmic granules of *C. virginica* and compared them with similar organelles from other bivalve species.

4.12.3.2 Hemolymph

Few studies have been reported on chemical parameters in the hemolymph of *M. mercenaria*. Unpublished results from my laboratory for fed clams kept in aquaria at 28‰ and 21°C were: protein, 395 μg/mL; glucose, 32 μg/mL.

Animals maintained these approximate levels for a period of 8 weeks.

Defense mechanisms by a variety of hemolymph factors have been reported in bivalve molluscs: lysins, chiefly lysozyme (McDade and Tripp, 1967a,b; Anderson, 1981), lectins (Arimoto and Tripp, 1977; Renwrantz, 1983, 1986; Tripp, 1992) and lysosomal hydrolases (see reviews by Bayne (1983); Cheng (1983a,b, 1983c, 1996); Chu (1988)). Granulocytes of *M. mercenaria* normally release lysozyme into the serum, but this process is considerably enhanced during phagocytosis (Cheng et al., 1975). Foley and Cheng (1977) showed that granulocytes of *M. mercenaria* exhibit degranulation during phagocytosis of bacteria; furthermore, subsequent to phagocytosis, cells showed a redistribution of granules. Mohandas et al. (1985) using scanning electron microscopy, demonstrated that lysosomes of granulocytes of *M. mercenaria* migrate to the surface of the cell, become coated with a portion of the cell membrane and are finally extruded as a double-membrane lysosome; the inner layer is the original lysosomal membrane and the outer layer is derived from the cell membrane. The fine structure of the double-membrane released lysosome was elaborated upon by Mohandas and Cheng (1985).

Tripp (1992) showed that red cells from six species of vertebrates (horse, burro, sheep, rabbit, cow and chicken), yeast cells, Gram-positive (*Staphylococcus aureus*) and Gram-negative (*Escherichia coli*) bacteria as well as latex spheres were phagocytosed equally well by hemocytes of *M. mercenaria* in artificial sea water and in homologous serum. He concluded that phagocytosis is a strong, but nonspecific process in hemocytes from this animal. Interestingly, Tripp (1992) also concluded that in *M. mercenaria*, soluble lectins in hemolymph do not serve as recognition molecules that promote phagocytosis of foreign particles; he further stated "It is still not clear what role lectins play in normal *M. mercenaria*, but they are not necessary for cellular defense." Much work remains to be done on the contribution of hemolymph factors to defense mechanisms in *M. mercenaria*.

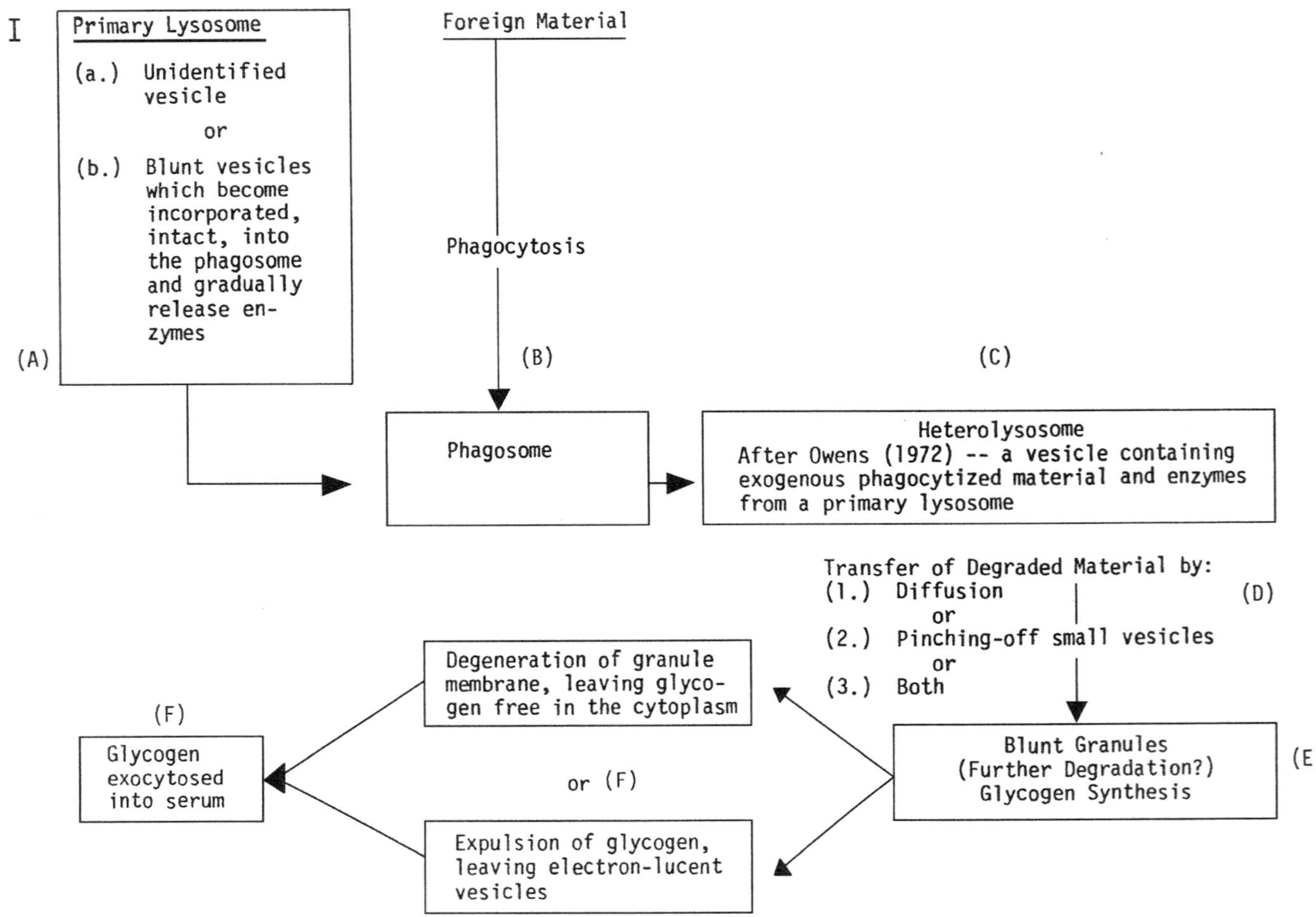
I
Primary Lysosome
(a.) Unidentified vesicle
or
(b.) Blunt vesicles which become incorporated, intact, into the phagosome and gradually release enzymes
(A)
Foreign Material
Phagocytosis
(B)
Phagosome
(C)
Heterolysosome
After Owens (1972) -- a vesicle containing exogenous phagocytized material and enzymes from a primary lysosome
Transfer of Degraded Material by:
(1.) Diffusion
or
(2.) Pinching-off small vesicles
or
(3.) Both
(D)
Blunt Granules
(Further Degradation?)
Glycogen Synthesis
(E)
Degeneration of granule membrane, leaving glycogen free in the cytoplasm
or (F)
Expulsion of glycogen, leaving electron-lucent vesicles
(F)
Glycogen exocytosed into serum

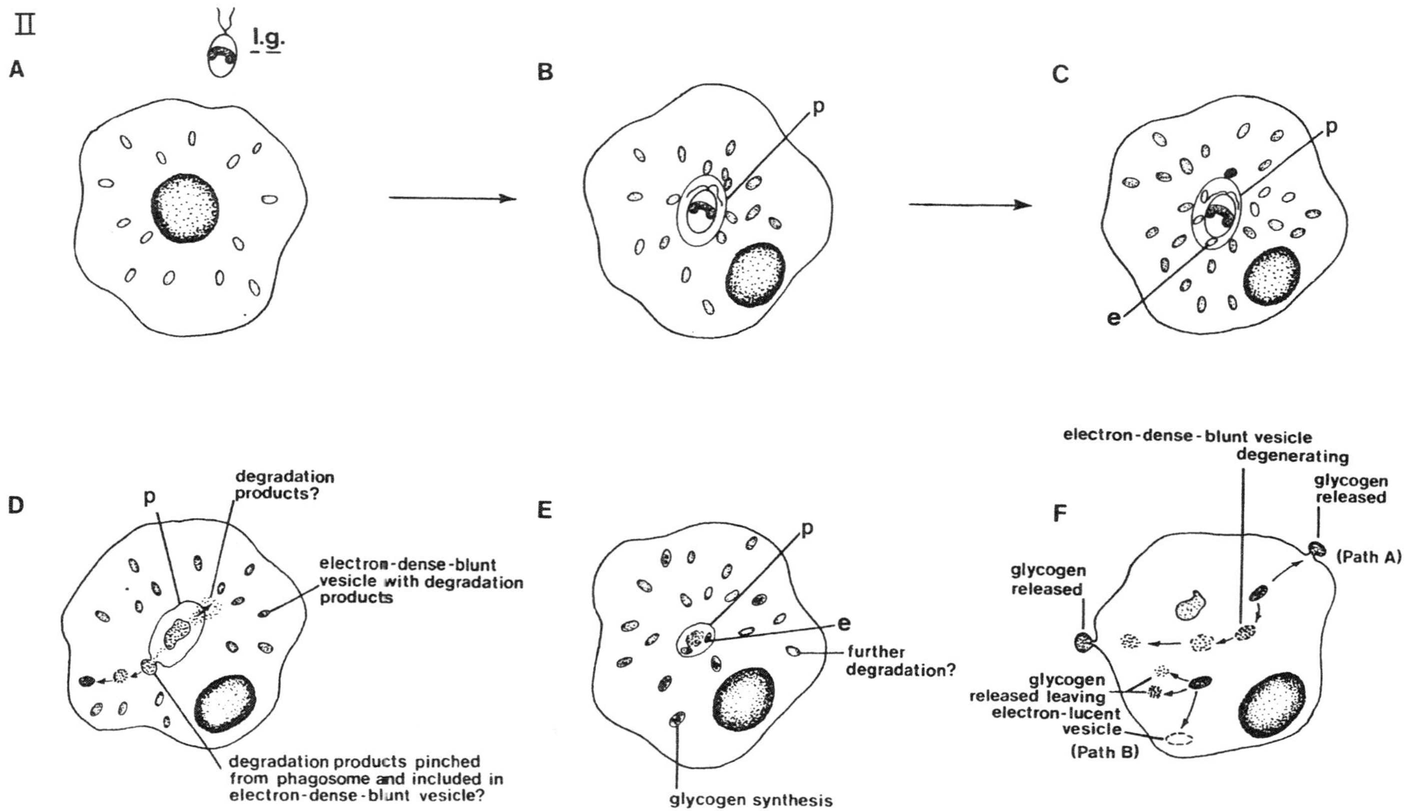

Fig. 4.86. Intracellular degradation–synthesis sequence for *Mercenaria mercenaria* hemocytes. (I) Flow chart indicating the phagocytic degradation–synthesis sequence. Letters refer to the corresponding diagram (II). (II) Diagrammatic representation of the sequence indicating alternative pathways. (A) Hemocyte and *I. galbana* (I.g.) prior to phagocytosis. (B) *I. galbana* incorporated into a phagosome (p). (C) The fusion of primary lysosomes with the phagosome (p); blunt vesicles (e) incorporated into the phagosome. (D) Degradation products from phagosome (p) transferred to blunt vesicles by diffusion or via minute vesicles. (E) Products transferred to blunt vesicles (e = dark-colored vesicles) either further degraded or synthesized into glycogen. (F) Glycogen released from blunt vesicles into the cytoplasm by: Path A = membrane surrounding blunt vesicle degenerates; Path B = glycogen released from blunt vesicles leaving electron-lucent vesicles behind. From Moore (1981).

4.13 NERVOUS SYSTEM

4.13.1 Anatomy

Ansell (1961) presents a brief account of the nervous system of *Venus striatula*, and included only a sketch of the posterior ganglia and nerves. Jones (1979) gives a thorough account of the nervous system of *M. mercenaria* including five detailed illustrations.

The nervous system consists of three pairs of principal ganglia: cerebral, visceral and pedal. Nerves and ganglia are pale yellow and enveloped by a thin connective-tissue sheath (Fig. 4.87; Jones, 1979).

Cerebral ganglia lie just anterior to the origins of the anterior pedal retractor muscles; they are joined by a short, thick supraesophageal commissure (Figs. 4.87–4.90). Fine nerves leave these ganglia and pass to labial palps, mantle, visceral mass, pedal retractor muscles and the anterior adductor muscle. Connectives also pass to the visceral and pedal ganglia (Figs. 4.87 and 4.88).

The closely fused visceral ganglia are embedded in connective tissue covering the anterior face of the posterior adductor muscle; connectives join with cerebral and pedal ganglia (Figs. 4.87 and 4.88). Nerves from visceral ganglia radiate to the mantle, gills, kidney and, in some venerids, to the atria and ventricle (Figs. 4.87 and 4.88; Carlson, 1905; Phillis, 1966);

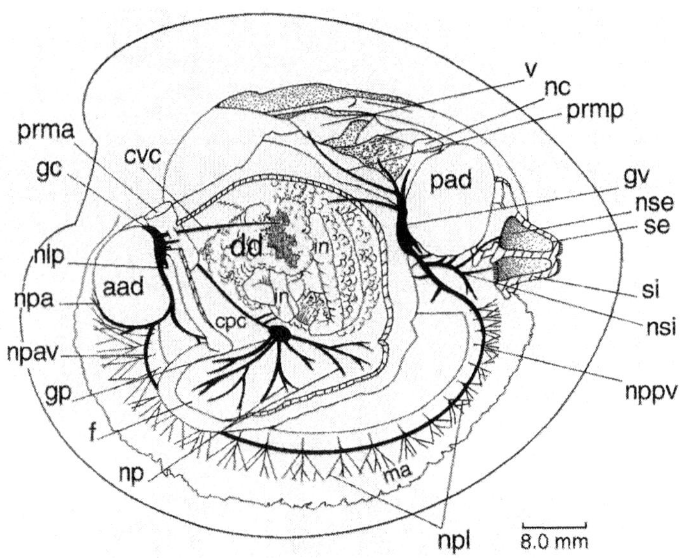

Fig. 4.87. Sketch of half-shell preparation in right valve to show major tracts and ganglia of the nervous system. aad = anterior adductor muscle; cpc = cerebropedal connective; cvc = cerebrovisceral connective; dd = digestive gland; f = foot; gc = cerebral ganglion; gp = pedal ganglion; gv = visceral ganglion; in = intestine; ma = mantle; nc = cardiac nerve; nlp = nerve to labial palp; nse = nerve to excurrent siphon; nsi = nerve to incurrent siphon; pad = posterior adductor muscle; np = pedal nerves; npa = anterior pallial nerve; npav = anteroventral pallial nerve; npl = nerves to pallial lobes; nppv = posteroventral pallial nerve; pad = posterior adductor muscle; prma = anterior pedal retractor muscle; prmp = posterior pedal retractor muscle; se = excurrent siphon; si = incurrent siphon; v = ventricle. Redrawn after Jones (1979). 1.68×.

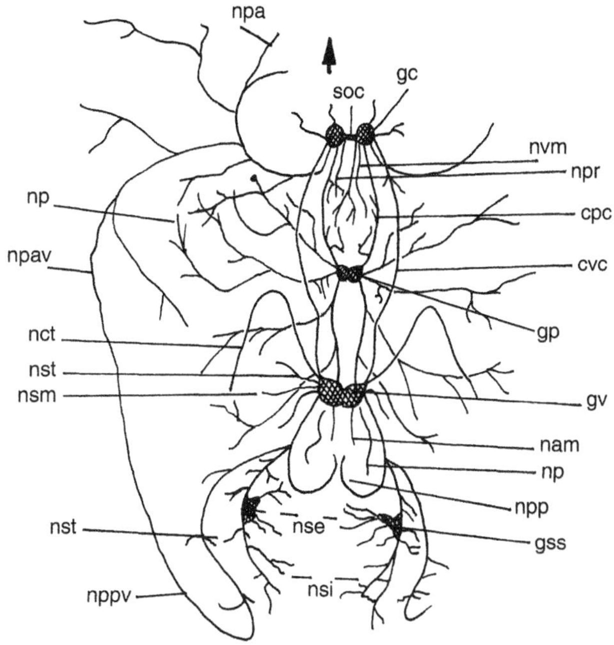

Fig. 4.88. Dorsal view of major tracts, branches and ganglia of the nervous system. cpc = cerebropedal connective; cvc = cerebrovisceral connective; gc = cerebral ganglion; gp = pedal ganglion; gss = subsidiary siphonal ganglion; nam = nerve to adductor muscle; nct = ctenidial nerve; np = pedal nerve; npa = anterior pallial nerve; npav = anteroventral pallial nerve; npp = posterior pallial nerve; nppv = posteroventral pallial nerve; npr = pedal retractor nerve; nse = nerve to incurrent siphon; nsi = nerve to incurrent siphon; nsm = nerve to shell surface of mantle; nsr = siphonal retractor nerve; nst = nerve to statocyst; nvm = nerve to anterodorsal visceral mass; gv = visceral ganglion; soc = supraesophageal commissure. Redrawn after Jones (1979). Arrow points towards anterior aspect. 4.2×.

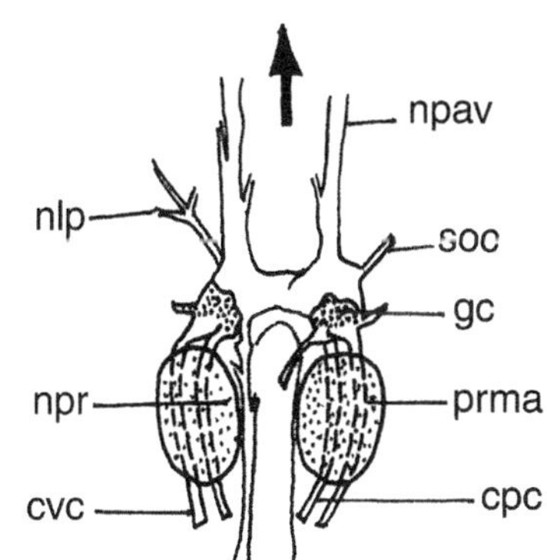

Fig. 4.89. Diagram of the anterodorsal aspect of the cerebral nervous system. cpc = cerebropedal connective; cvc = cerebrovisceral connective; eo = esophagus; gc = cerebral ganglion; nlp = nerve to labial palp; npav = anteroventral pallial nerve; npr = pedal retractor nerve; nvm = nerve to anterodorsal viscera; prma = anterior pedal retractor muscle; soc = supraesophageal commissure. Arrow points to anterior aspect. Redrawn after Jones (1979). 4.25×.

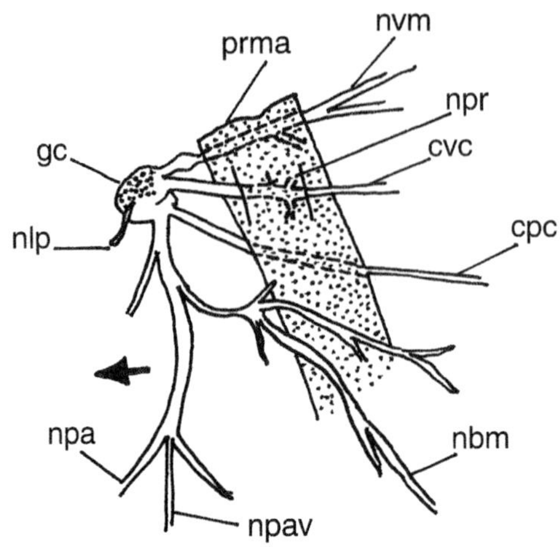

Fig. 4.90. Diagram of the left, lateral aspect of the cerebral nervous system. cpc = cerebropedal connective; cvc = cerebrovisceral connective; gc = cerebral ganglion; nbm = nerve to mantle (pallial cavity surface); nlp = nerve to labial palp; npa = anterior pallial nerve; npav = anteroventral pallial nerve; npr = pedal retractor nerve; nvm = nerve to anterodorsal visceral mass; prma = anterior pedal retractor muscle. Arrow points to anterior aspect. Redrawn after Jones (1979). 6.16×.

innervation of atria and ventricle by nerves from the visceral ganglia occurs in *M. mercenaria* (Fig. 4.87). Ono et al. (1992) showed large nerve fibers positive for 5-hydroxytryptamine, entering the atria as thick fascicles, continuing to the A-V (atrioventricular) valves and finally ending in the ventricle as nerve fascicles that repeatedly branched into small diameter fibers. Several nerves leave the dorsal aspect of the visceral ganglia to supply the posterior pedal retractor muscle and the posterior adductor muscle; large nerves course to the siphon where they terminate in subsidiary siphonal ganglia (Figs. 4.88 and 4.91) which, in turn, contain cell bodies of neurons that innervate the excurrent and incurrent siphons (Figs. 4.87, 4.88 and 4.91; Jones, 1979).

Cerebropedal connectives (Figs. 4.87 and 4.88) run from the posterior region of the cerebral ganglion ventrally through the anterior pedal retractor muscle, to which they send many fine nerves, and terminate in the anterior edge of pedal ganglia (Figs. 4.87 and 4.88; Jones, 1979). Pedal ganglia are fused and lie just anteroventral to the posterior stomach. Three pairs of nerves from pedal ganglia supply tissues of the foot (Figs. 4.87 and 4.88; Jones, 1979).

4.13.2 Histology

Ganglia may contain thousands of neuron cell bodies and communicate with one to several large nerve bundles (Fig. 4.92). The connective tissue sheath surrounding ganglia is thick (37 μm) but narrows to 3–5 μm surrounding nerves (Figs. 4.92 and 4.93). Connective-tissue nuclei are usually elongate and measure 3.1 × 6.2 μm (Fig. 4.93). Cell bodies of neurons in ganglia range in size from 5 to 19 μm (Fig. 4.93). Nuclei of cell bodies average 6.2 μm, but some are quite large (9.3 μm); nuclei have one eccentric nucleolus and a rim of heterochromatin on the inside of the nuclear envelope (Fig. 4.93).

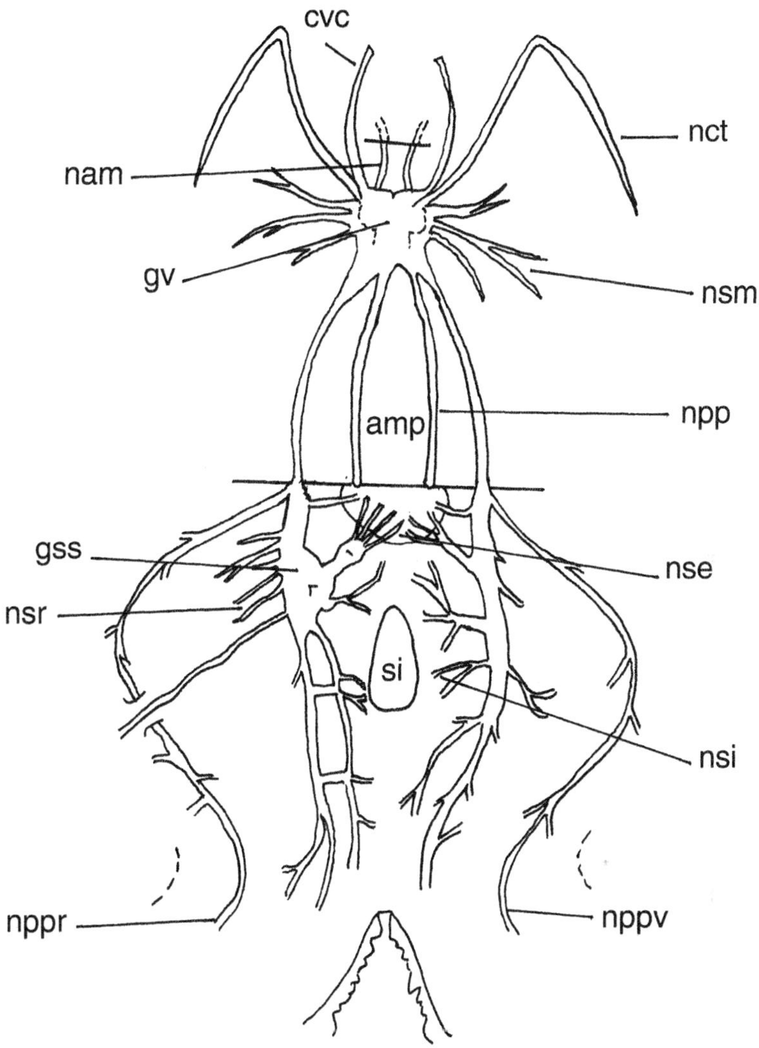

Fig. 4.91. Diagram of the anterior aspect of the visceral nervous system. amp = posterior adductor muscle; cvc = cerebrovisceral connective; gss = subsidiary siphonal ganglion; gv = visceral ganglion; nam = nerve to adductor muscle; nct = ctenidial nerve; npp = posterior pallial nerve; nppv = posteroventral pallial nerve; nse = nerve to excurrent siphon; nsi = nerve to incurrent siphon; nsm = nerve to shell surface of mantle; nsr = siphonal retractor nerve; se = excurrent siphon; si = incurrent siphon. Redrawn after Jones (1979). 8.5×.

4.14 SUMMARY

It has just been 100 years since the publication of the landmark work of Kellogg. Although much has been learned about the anatomy and histology of *Mercenaria mercenaria* since that seminal work, we have a long way to travel on that tortuous pathway known as research. A

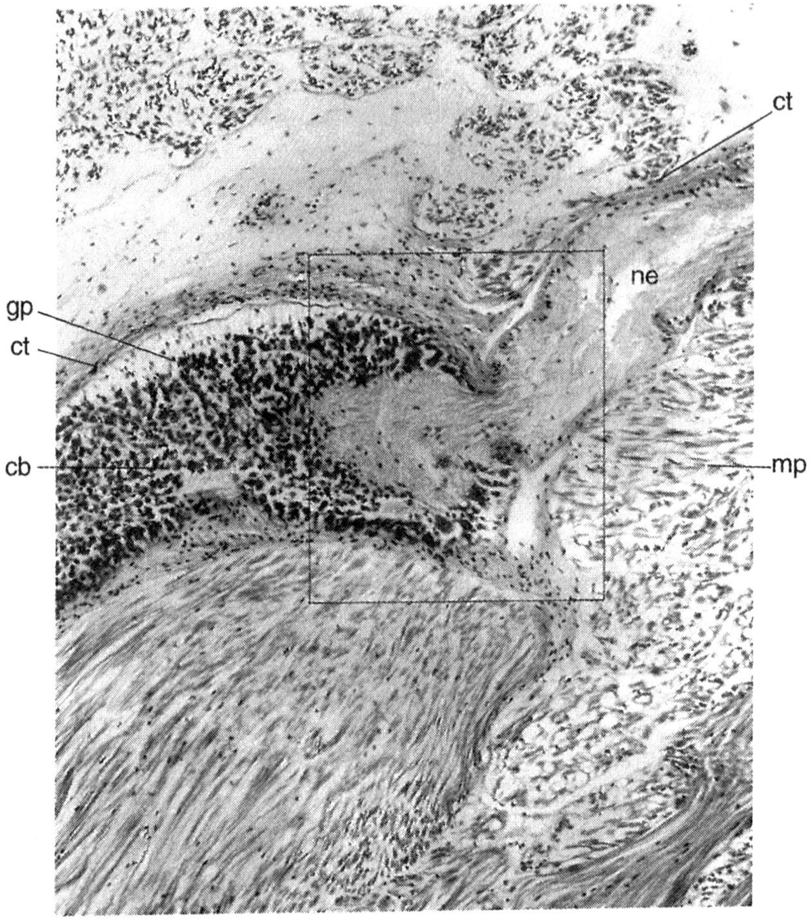

Fig. 4.92. Section through a portion of the pedal ganglion (gp) with emerging nerve (ne). cb = cell bodies of neurons; ct = connective tissue sheath; mp = pedal muscle. HFW = 2.06 mm.

case in point: although we now know much about the cytology, fine structure and functions of hemocytes, we have no knowledge about hematopoietic centers. Thanks to the pioneering work of Loosanoff and the many fine contributions of workers in the past 50 years, we understand the life cycle very well and can spawn animals upon demand throughout the year independent of latitude; yet we have no knowledge concerning the fine structure and functions of follicle and nutritive cells of gonadal acini.

In the past 15 years, however, substantial strides have been made in renal cytology, fine structure and general function largely due to the work of M.P. Morse and her students. It is expected that investigators will direct concentrated studies of this type to all organ systems and cell types. A *Mercenaria* genome project should not be in the too distant future.

A little more than 50 years ago, Thurlow Nelson stated that the eastern oyster, *Crassostrea virginica*, was the best known marine animal in the world; I expect that with continued and

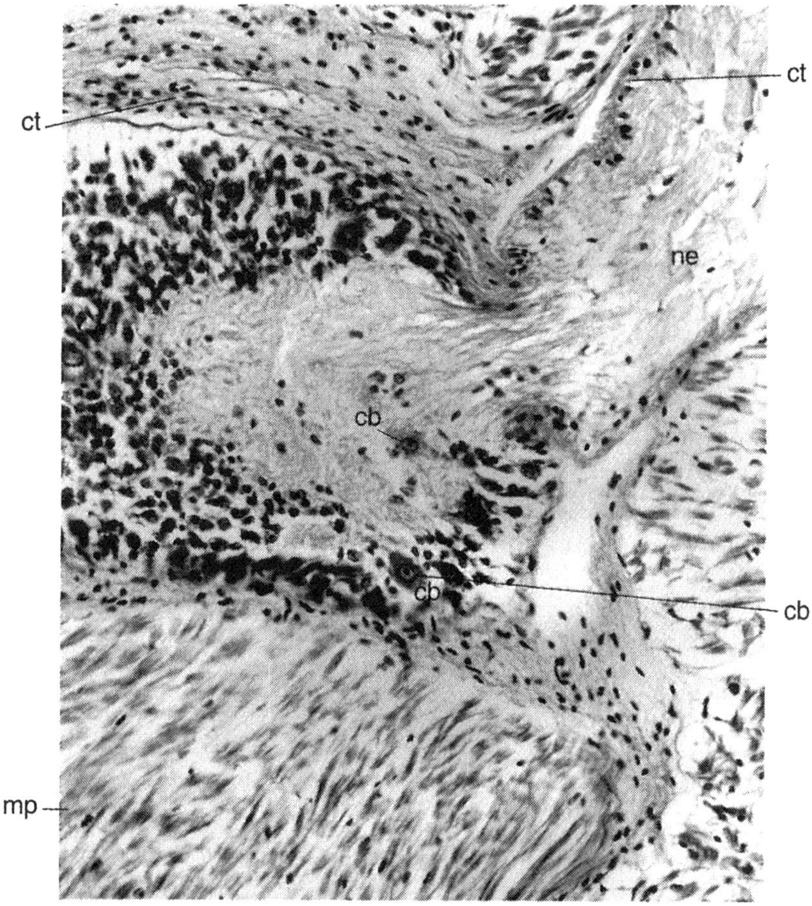

Fig. 4.93. Section through a portion of the pedal ganglion (gp) to show details of rectangle in Fig. 4.92. cb = cell bodies of neurons; ct = connective tissue sheath; mp = pedal muscle. HFW = 405 μm.

expanded research on this complex, hearty and commercially important animal, *Mercenaria mercenaria* may well become the marine animal about which we have the most scientific information.

4.15 ACKNOWLEDGMENTS

I wish to thank Melbourne R. Carriker for his critical review of the entire manuscript; as usual, his comments and suggestions proved invaluable. Peter Beninger also reviewed the manuscript and offered many helpful comments. Carol Moore reviewed the sections on hemocytes and hemolymph and graciously allowed me to use many figures and micrographs from her doctoral thesis. Arnold Eversole reviewed the section on reproduction and M. Patricia Morse reviewed the material on the kidney and excretory system. Robert E. Hillman reviewed the section on the mantle. I wish to thank them for their many helpful comments. The artistic talents of Chris Valles have made a substantial contribution to this work and to the field of bivalve biology. Lastly, I thank all my students who have helped with technical details.

REFERENCES

Anderson, R.S., 1981. Inducible hemolytic activity in *Mercenaria mercenaria* hemolymph. Dev. Comp. Immunol., 5: 575–585.

Andrews, E.B. and Jennings, K.H., 1993. The anatomical and ultrastructural basis of primary urine formation in bivalve molluscs. J. Moll. Stud., 59: 223–257.

Ansell, A.D., 1961. The functional morphology of the British species of Veneracea (Eulamellibranchia). J. Mar. Biol. Assoc. U.K., 41: 489–515.

Arimoto, R. and Tripp, M.R., 1977. Characterization of a bacterial agglutinin in the hemolymph of the hard clam, *Mercenaria mercenaria*. J. Invertebr. Pathol., 30: 406–413.

Atkins, D., 1937. On the ciliary mechanisms and interrelationships of lamellibranches. Part VII. Laterofrontal cilia of the gill filaments and their phylogenetic value. Q. J. Microsc. Sci., 80 (319): 345–436.

Bayne, C.J., 1983. Molluscan immunobiology. In: A.S.M. Saleuddin and K.M. Wilbur (Eds.), The Mollusca, Vol. 5, Physiology, Part 2. Academic Press, San Diego.

Belding, D.L., 1912. A report upon the quahog and oyster fisheries of Massachusetts, including the life history, growth and cultivation of the quahog (*Venus mercenaria*), and observations on the set of oyster in Wellfleet Bay. Commonwealth of Massachusetts, Wright and Potter, Boston.

Belding, D.L., 1931. The quahaug fishery of Massachusetts. Mass. Dept. Conservation, Div. Fish Game, Mar. Fishery Serv. 2, 41 pp.

Beninger, P.G. and Le Pennec, M., 1991. Functional anatomy of scallops. In: S.E. Shumway (Ed.), Scallops: Biology, Ecology and Aquaculture, Elsevier, Amsterdam.

Bowden, J., 1958. The structure and innervation of lamellibranch muscle. Int. Rev. Cytol., 7: 295–335.

Brand, A.R., 1972. The mechanism of blood circulation in *Anodonta anatina* (L.). J. Exp. Biol., 56: 361–379.

Brown, C.H., 1952. Some structural proteins of *Mytilus edulis*. Q. J. Microsc. Sci., 93: 487–502.

Burson, S.L., Fahrenbach, M.J., Frommhagen, L.H., Riccardi, R.A., Brown, R.A., Brockman, J.A., Lewry, H.V. and Stockstad, E.L.R., 1956. Isolation and purification of mactins, heparin-like anti-coagulants from mollusca. J. Am. Chem. Soc., 78: 5874–5878.

Carlson, A.J., 1905. Comparative physiology of the invertebrate heart. Biol. Bull., 8: 123–169.

Carriker, M.R., 1961. Interrelation of functional morphology, behavior, and autecology in early stages of the bivalve *Mercenaria mercenaria*. J. Elisha Mitchell Sci. Soc., 77: 168–241.

Chantler, P.D., 1983. Biochemical and structural aspects of molluscan muscle. In: A.S.M. Saleuddin and K.M. Wilbur (Eds.), *The Mollusca*, Vol. 4. Academic Press, New York, pp. 77–154.

Chantler, P.D., 1991. The structure and function of scallop adductor muscles. In: S.E. Shumway (Ed.), Scallops: Biology, Ecology and Aquaculture, Elsevier, New York.

Cheng, T.C., 1983a. Internal defense mechanisms of molluscs against invading microorganisms: personal reminiscences. Trans. Am. Microsc. Soc., 102: 185–193.

Cheng, T.C., 1983b. The role of lysosomes in molluscan inflammation. Am. Zool., 23: 129–144.

Cheng, T.C., 1983c. Triggering of immunologic defense mechanisms of molluscan shellfish by biotic and abiotic challenge and its applications. Mar. Technol. Soc., 17: 18–25.

Cheng, T.C., 1996. Hemocytes: forms and functions. In: V.S. Kennedy, R.A. Newell and A.F. Eble (Eds.), The Eastern Oyster. A Maryland Sea Grant Book, College Park, MD.

Cheng, T.C. and Cali, A., 1974. An electron microscope study of the fate of bacteria phagocytized by granulocytes of *Crassostrea virginica*. Contemp. Top. Immunobiol., 4: 25–35.

Cheng, T.C. and Foley, D., 1975. Hemolymph cells of the bivalve mollusc *Mercenaria mercenaria*: an electron microscopical study. J. Invertebr. Pathol., 26: 341–351.

Cheng, T.C. and Rudo, B.M., 1976. Distribution of glycogen resulting from degradation of ^{14}C-labelled bacteria in the American oyster *Crassostrea virginica*. J. Invertebr. Pathol., 27: 259–262.

Cheng, T.C., Rodrick, G.E., Foley, D.A. and Koehler, S.A., 1975. Release of lysozyme from hemolymph cells of *Mercenaria mercenaria* cells during phagocytosis. J. Invertebr. Pathol., 25: 261–265.

Chu, F.-L.E., 1988. Humoral defense mechanisms in bivalves. In: W.S. Fisher (Ed.), Disease Processes in Marine Bivalve Molluscs, Am. Fish. Soc. Spec. Publ. 18, Bethesda, MD.

Dinamani, P., 1967. Variation in the stomach structure of the Bivalvia. Malacologia, 5 (2): 225–268.

Eble, A.F., 1965. Some observations on the seasonal distribution of selected enzymes in the American oyster as revealed by enzyme histochemistry. Proc. Natl. Shellfish. Assoc., 56: 37–42.

Eble, A.F. and Scro, R., 1996. General Anatomy. In: V.S. Kennedy, R.E. Newell and A.F. Eble (Eds.), The Eastern Oyster. A Maryland Sea Grant Book, College Park, MD.

Foley, D.A. and Cheng, T.C., 1974. Morphology, hematologic parameters, and behavior of hemolymph cells of the quahaug clam *Mercenaria mercenaria*. Biol. Bull., 146: 343–356.

Foley, D.A. and Cheng, T.C., 1977. Degranulation and other changes of molluscan granulocytes associated with phagocytosis. J. Invertebr. Pathol., 29: 321–325.

Foster-Smith, R.L., 1978. The function of the pallial organs of bivalves in controlling ingestion. J. Molluscan Stud., 44: 83–99.

Fox, D.L., 1953. Animal Biochromes and Structural Colours. Cambridge University Press, London.

Fries, C.R. and Grant, D.M., 1991. Rickettsiae in gill epithelial cells of the hard clam, *Mercenaria mercenaria*. J. Invertebr. Pathol., 57: 166–171.

Galtsoff, P.S., 1964. The American oyster. Fish. Bull. Fish Wildlife Serv., 64: 1–480.

Garofola, N. and Eble, A.F., 1995. Analysis of ^{65}Zn uptake and depuration in the presence of background Cd^{2+} in the hard clam, *Mercenaria mercenaria*. Bull. NJ Acad. Sci., 40 (1): 30.

Gelder, S.R. and Moore, C.A., 1986. Cytochemical demonstration of several enzymes associated with phagosomal processing of foreign material within hemocytes of *Mercenaria mercenaria*. Trans. Am. Microsc. Soc., 105 (1): 51–58.

Greenberg, M.J. and Jegla, T.C., 1963. The action of 5-hydroxytryptamine and acetylcholine on the rectum of the venus clam, *Mercenaria mercenaria*. Comp. Biochem. Physiol., 14: 275–290.

Hayes, R.L. and Kelly, R.E., 1968. Dense bodies of the contractile system of cardiac muscle in *Venus mercenaria*. J. Morphol., 127: 151–162.

Hersh, E., 1957. Some observations on the bulbus arteriosus of *Venus mercenaria*. Biol. Rev., 19: 11–19.

Hillman, R.E., 1962. A histological and histochemical study of the mantle of the northern quahog, *Mercenaria mercenaria*. Ph.D. Dissertation, Univ. Delaware, Newark, DE.

Hillman, R.E., 1964. The functional morphology of the fourth fold of the mantle of the northern quahog, *Mercenaria mercenaria* (L.). J. Mitchell Soc., 80: 8–12.

Hillman, R.E., 1968. Histochemistry of mucosubstances in the mantle of the clam, *Mercenaria mercenaria*. I. A glycosaminoglycan in the first marginal fold. Trans. Am. Microsc. Soc., 87 (3): 361–367.

Hillman, R.E., 1969. Histochemistry of mucosubstances in the mantle of the clam, *Mercenaria mercenaria*. II. Mucosubstances in the second marginal fold. Trans. Am. Microsc. Soc., 88 (3): 120–125.

Jegla, T.C. and Greenberg, M.J., 1968a. Structure of the bivalve rectum. I. Morphology. Veliger, 10 (3): 253–263.

Jegla, T.C. and Greenberg, M.J., 1968b. Structure of the bivalve rectum. II. Notes on cell types and innervation. Veliger, 10 (4): 314–319.

Jones, C.C., 1979. Anatomy of *Chione cancellata* and some other chionines (Bivalvia: Veneridae). Malacologia, 19 (1): 157–199.

Joshi, M.C. and Bal, D.V., 1967. The circulatory system of clam *Katelysia marmorata* (Lam.). J. Univ. Bombay, 33 (3): 13–23.

Keck, R.T., Maurer, D. and Lind, H., 1975. A comparative study of the hard clam gonad developmental cycle. Biol. Bull., 148: 243–258.

Kellogg, J.L., 1892. A contribution to our knowledge of the morphology of lamellibranchiate mollusks. Bull. U.S. Fish. Comm., 10: 389–437.

Kellogg, J.L., 1910. Shellfish Industries. Henry Holt, New York.

Kellogg, J.L., 1915. Ciliary mechanisms of Lamellibranchs with descriptions of anatomy. J. Morphol., 26: 625–701.

Kelly, R.E. and Hayes, R.L., 1968. The ultrastructure of smooth cardiac muscle in the clam, *Venus mercenaria*. J. Morphol., 127: 163–176.

Knaub, R.S. and Eversole, A.G., 1988. Reproduction of different stocks of *Mercenaria mercenaria*. J. Shellfish Res., 7 (3): 371–376.

Langton, R.W., 1975. Synchrony in the digestive diverticula of *Mytilus edulis*. J. Mar. Biol. Assoc. U.K., 55: 221–229.

Le Pennec, M., Beninger, P.G. and Henry, A., 1988. New observations on the gills of *Placopectin magellanicus* (Mollusca: Bivalvia), and implications for nutrition. II. Internal anatomy and microanatomy. Mar. Biol., 98: 229–237.

Lill, S. and Eble, A.F., 1996. Histology, histochemistry and studies of proliferating cells in the kidney of the hard clam, *Mercenaria mercenaria*. Bull. N.J. Acad. Sci., 41: 43.

Loosanoff, V.L., 1936. Sexual phases in the quahog. Science, 83: 287–288.

Loosanoff, V.L., 1937a. Seasonal gonadal changes of adult clams, *Venus mercenaria*. Biol. Bull., 12: 406–416.

Loosanoff, V.L., 1937b. Development of the primary gonad and sexual phases in *Venus mercenaria*. Biol. Bull., 72: 389–405.

Love, R. and Frommhagen, I.H., 1953. Histochemical studies on the clam *Mactra solidissma*. Proc. Soc. Exp. Biol., 83: 838–844.

Loy, N. and Eble, A.F., 1974. Locomotion and phagocytic behavior of the amoebocytes of the hard clam *Mercenaria mercenaria* as revealed by time-lapse cinematography. Bull. N.J. Acad. Sci., 19: 27.

McDade, J.E. and Tripp, M.R., 1967a. Lysozyme in the hemolymph of the oyster, *Crassostrea virginica*. J. Invertbr. Pathol., 9: 531–535.

McDade, J.E. and Tripp, M.R., 1967b. Lysozyme in oyster mantle mucus. J. Invertbr. Pathol., 9: 581–582.

McHugh, J.L., Sumner, M.W., Flagg, P.J., Lipton, D.W. and Behrens, W.J., 1981. Annotated bibliography of the hard clam (*Mercenaria mercenaria*). U.S. Dept. Comm., Natl. Mar. Fish. Serv., 845 pp.

Meyhofer, E., Morse, M.P. and Robinson, W.E., 1985. Podocytes in bivalve molluscs: morphological evidence for ultrafiltration. J. Comp. Physiol. B, 156: 151–161.

Meyhofer, E. and Morse, M.P., 1996. Characterization of the bivalve ultrafiltration system in *Mytilus edulis*, *Chlamys hastata* and *Mercenaria mercenaria*. Invertebr. Biol., 115 (1): 20–29.

Miller, M.W. and Nowakowski, R.S., 1988. Use of bromodeoxyuridine-immunohistochemistry to examine the proliferation, migration and time of origin of cells in the central nervous system. Brain Res., 457: 44–52.

Mohandas, A. and Cheng, T.C., 1985. An electron microscope study of the structure of lysosomes released from *Mercenaria mercenaria* granulocytes. J. Invert. Pathol., 46: 332–334.

Mohandas, A., Cheng, T.C. and Cheng, J.B., 1985. Mechanism of lysosomal release from *Mercenaria mercenaria* granulocytes: a scanning electron microscope study. J. Invertebr. Pathol., 46: 189–197.

Moore, C.A., 1972. Cytology and cytochemistry of the amoebocytes of the hard clam *Mercenaria mercenaria*. Masters Thesis, Montclair State College, Montclair, NJ.

Moore, C.A., 1981. Cytological and physiological features of the hemocytes of *Mercenaria mercenaria* L. Ph.D. dissertation, Northeastern University, Boston.

Moore, C.A. and Eble, A.F., 1977. Cytochemical aspects of *Mercenaria mercenaria* hemocytes. Biol. Bull., 152: 105–119.

Moore, C.A. and Gelder, S.R., 1983. The role of the 'blunt' granules in hemocytes of *Mercenaria mercenaria*. J. Invertebr. Pathol., 41: 369–377.

Moore, C.A. and Gelder, S.A., 1985. Demonstration of lysosomal enzymes in hemocytes of *Mercenaria mercenaria* (Mollusca: Bivalvia). Trans. Am. Microsc. Soc., 104 (3): 242–249.

Morrison, C.M., 1993a. Histology and cell ultrastructure of the mantle and mantle lobes of the eastern oyster, *Crassostrea virginica* (Gmelin): a summary atlas. Am. Malacol. Bull., 10 (1): 1–24.

Morrison, C.M., 1993b. The histology and ultrastructure of the adductor muscle of the eastern oyster *Crassostrea virginica* (Gmelin). Am. Malacol. Bull., 10 (1): 25–38.

Morrison, C.M., 1996. Adductor and Mantle Musculature. In: V.S. Kennedy, R.G. Newell and A.F. Eble (Eds.), The Eastern Oyster. A Maryland Sea Grant Book, College Park, MD.

Morrison, C.M., Cameron, M.L. and Odense, P.H., 1970. Periodicities in the thick filaments of the opaque and translucent parts of the adductor of the oyster, *Crassostrea virginica* Can. J. Zool., 48: 608–609.

Morse, M.P., 1987. Comparative functional morphology of the bivalve excretory system. Am. Zool., 27: 737–746.

Morse, M.P. and Zardus, J.D. Bivalvia, In: Harrison (Ed.), Microscopic Anatomy of the Invertebrates, in press.

Morton, J.E., 1958. Molluscs. Hutchinson University Library, London.

Mullins, G.L. and Gunderoth, W.G., 1965. A collagen net hypothesis for force transference of smooth muscle. Nature (Lond.), 206: 592–594.

Muneoka, Y. and Twarog, B.M., 1983. Neuromuscular transmission and excitation–contraction coupling in molluscan muscle. In: A.S.M. Saleuddin and K.M. Wilbur (Eds.), The Mollusca, Vol. 4. Academic Press, New York, pp. 35–76.

Neff, J.M., 1972. Ultrastructural studies of periostracum formation in the Hard-shelled clam *Mercenaria mercenaria*. Tissue Cell, 4(2): 311–326.

Newell, R.I.E. and Langdon, C.I., 1996. Mechanisms and Physiology of Larval and Adult Feeding. In: V.S. Kennedy, R.E. Newell and A.F. Eble (Eds.), The Eastern Oyster. A Maryland Sea Grant Book, College Park, MD.

Nicaise, G. and Amsellem, J., 1983. Cytology of muscle and neuromuscular junction. In: A.S.M. Saleuddin and K.M. Wilbur (Eds.), The Mollusca, Vol. 4. Academic Press, New York, pp. 1–33.

Nielsen, B.J., 1963. Studies on the genus *Katelysia* Romer 1857 (Mollusca Lamellibraniata). Mem. Nat. Mus. Vict., 26: 219–251.

Ono, J.K., Hampton, J.D.R. and Koch, R.A., 1992. Immunohistochemical localization and radioenzymatic measurements of serotonin (5-hydroxytryptamine) in hearts of *Aplysia* and several bivalve mollusks. Cell Tissue Res., 269: 421–430.

Owen, G.E.R., 1955. Observations on the stomach and digestive diverticula of the Lamellibranchia. I. The Anisomyaria and Eulamellibranchia. Q. J. Microsc. Sci., 96 (4): 517–537.

Owen, G.E.R., Trueman, R. and Yonge, C.M., 1953. The Ligament in the Lamellibranchia. Nature, 171: 73–75.

Pal, S.G., Ghosh, B. and Modak, S., 1990. Fine structure of the digestive tubules of *Meretrix*. In: The Bivalvia-Proceedings of a Memorial Symposium in Honour of Sir Maurice Yonge, Edinburgh, 1986. University Press, Hongkong.

Phillis, J.W., 1966. Innervation and control of a molluscan (*Tapes*) heart. Comp. Biochem. Physiol., 17: 719–739.

Platt, A.M., 1971. Studies on the digestive diverticula of *Mytilus edulis* L. Ph.D. thesis, Queen's University, Belfast.

Porter, H.J., 1964. Seasonal gonadal changes of adult clams, *Mercenaria mercenaria* (L.), in North Carolina. Proc. Natl. Shellfish. Assoc., 55: 35–52.

Potts, F.A., 1923. The structure and function of the liver of *Teredo*, the shipworm. Proc. Cambridge Phil. Soc. (Biol. Sci.), 1: 1–17.

Purchon, R.D., 1960. The stomach in the Eulamellibranchia; stomach types IV and V. Proc. Zool. Soc. Lond., 135: 431–489.

Quayle, D.B., 1951. The rate of growth of *Venerupis pullastra* (Montagu) at Millport, Scotland. Proc. Roy. Soc. Edinb. B, 64: 384–406.

Reid, R.G.B., 1965. The structure and function of the stomach in bivalve molluscs. J. Zool., 147: 156–184.

Renwrantz, L., 1983. Involvement of agglutinins (lectins) in invertebrate defense reactions: the immuno-biological importance of carbohydrate-specific binding molecules. Dev. Comp. Immunol., 7: 603–608.

Renwrantz, L., 1986. Lectins in molluscs and arthropods: their occurrence, origin and roles in immunity. Symp. Zool. Soc. Lond., 56: 81–93.

Robinson, W.E. and Langton, R.W., 1980. Digestion in a subtidal population of *Mercenaria mercenaria* (Bivalvia). Mar. Biol., 58: 173–179.

Robinson, W.E. and Morse, M.P., 1994. Biochemical constituents of the blood plasma and pericardial fluids of several marine bivalve molluscs: implications for ultrafiltration. Comp. Biochem. Physiol. B, 107: 117–123.

Rodrick, G.E. and Ulrich, S.A., 1984. Microscopical studies on the hemocytes of bivalves and their phagocytic interaction with selected bacteria. Helgolander Meeresuntersuchungen, 37: 167–176.

Ruddall, K.M., 1955. The distribution of collagen and chitin. Symp. Soc. Exp. Biol., 9: 49–71.

Scheairs, D. and Eble, A.F., 1995. Cytology and cytochemistry of the kidney of the hard clam, *Mercenaria mercenaria*. Bull. N.J. Acad. Sci., 40: 38.

Shuster, C.N. and Eble, A.F., 1961. Techniques in visualization of organ systems in bivalve molluscs. Proc. Natl. Shellfish. Assoc., 52: 13–24.

Smith, P.J.S., 1985. Cardiac performance in response to loading pressures in *Busycon canaliculatum* (Gastropoda) and *Mercenaria mercenaria* (Bivalvia). J. Exp. Biol., 119: 301–320.

Stacek, C.R., 1963. Synopsis and discussion of the association of ctenidia and labial palps in the bivalved Mollusca. Veliger, 6: 91–97.

Sullivan, P.A., Robinson, W.E. and Morse, M.P., 1988a. Subcellular distribution of metals within the kidney of the bivalve *Mercenaria mercenaria*. Comp. Biochem. Physiol., 91C (2): 589–595.

Sullivan, P.A., Robinson, W.E. and Morse, M.P., 1988b. Isolation and characterization of granules from the kidney of the bivalve *Mercenaria mercenaria*. Mar. Biol., 99: 359–368.

Thomas, L., 1954. The localization of heparin-like blood anticoagulant substances in the tissues of *Spisula solidissima*. Biol. Bull., 106: 129–138.

Tripp, M.R., 1992. Phagocytosis by hemocytes of the hard clam, *Mercenaria mercenaria*. J. Invert. Pathol., 59: 222–227.

Ward, J.E., Beninger, P.G., MacDonald, B.A. and Thompson, R.J., 1991. Direct observations of feeding structures and mechanisms in bivalve molluscs using endoscopic examination and video image analysis. Mar. Biol., 111: 287–291.

Ward, E.J., Newell, R.I.E., Thompson, R.J. and MacDonald, B.A., 1994. Endoscopic observations of particle capture and transport in the eastern oyster, Crassostrea virginica. Gmelin. Biol. Bull., 186: 221–240.

Weinstein, J.E., 1995. Fine structure of the digestive tubule of the eastern oyster, *Crassostrea virginica* (Gmelin 1791). J. Shellfish. Res., 14 (1): 97–103.

White, K.M., 1942. The pericardial cavity and the pericardial gland of the Lamellibranchia. Proc. Malocol. Soc. Lond., 25 (2): 4–88.

Yonge, C.M., 1926. The digestive diverticula of the Lamellibranchs. Trans. Roy. Soc. Edinburgh, 54: 703–717.

Yonge, C.M., 1957. Mantle fusion in the Lamellibranchia. Pubbl. Staz. Zool. Napoli, 29: 150–171.

Yoshino, T.P. and Cheng, T.C., 1976. Fine structural localizations of acid phosphatase in granulocytes of the pelecypod *Mercenaria mercenaria*. Trans. Am. Microsc. Soc., 95: 215–220.

Zachs, S.I., 1955. The cytochemistry of the amoebocytes and intestinal epithelium of *Venus mercenaria* (Lamellibranchiata), with remarks on a pigment resembling ceroid. Q. J. Microsc. Sci., 96: 57–71.

Zachs, S.I. and Welsh, J.W., 1953. Cholinesterase and lipase in the amoebocytes, intestinal epithelium and heart muscle of the quahog *Venus mercenaria*. Biol. Bull., 105: 200–211.

Biology of the Hard Clam
J.N. Kraeuter and M. Castagna (Eds.),

Chapter 5

Reproduction in Mercenaria mercenaria

Arnold G. Eversole

5.1 INTRODUCTION

Bivalve reproduction has been the subject of much study. Sastry (1979) and Mackie (1984) gave accounts of marine and freshwater bivalve reproduction while Andrews (1979) and Eversole (1989) reviewed reproduction in oysters and clams, respectively. None of these authors restricted their discussions to one species of bivalve. In this chapter, an attempt was made to present the current state of our knowledge of reproduction in *Mercenaria*. Emphasis was placed on *M. mercenaria*; however, related research on the reproductive biology of *M. campechiensis*, *M. campechiensis texana* and hybrids was included to help complete the discussion of *Mercenaria* reproduction. Reference to *Mercenaria* throughout the chapter was made when more than one taxon or unidentified hybrids were discussed; otherwise, species were identified by name. Most of the generalities suggested for *M. mercenaria* in the chapter should also be applicable for the other *Mercenaria* taxa.

5.2 SEXUAL EXPRESSION

5.2.1 Sex Determination and Hermaphroditism

As a group, bivalves exhibit a wide range of sexuality from those species that are dioecious to those that are fully functional hermaphrodites. Although separate sexes predominate in the bivalves, hermaphroditism proves to be more varied and interesting. Coe (1943) proposed four hermaphroditic patterns in bivalves based on the sequence of reproductive events: (1) functional hermaphroditism (simultaneous production of eggs and sperm); (2) consecutive hermaphroditism (single sex change in life); (3) rhythmical consecutive hermaphroditism (repeated predictable sex changes); and (4) alternate hermaphroditism (erratic and less predictable sex changes). *Mercenaria mercenaria*, *M. campechiensis* and their hybrids exhibit consecutive hermaphroditism (Loosanoff, 1936, 1937a; Dalton and Menzel, 1983).

Initially, *M. mercenaria* goes through a juvenile sexual phase when clams are only a few months of age and 6–7 mm shell length (Loosanoff, 1936, 1937a). In a majority of the cases, the presence of "ripe" spermatozoa gives these clams a distinct male appearance; however, close examination reveals female sex cells and the bisexual nature of clams (Loosanoff, 1936, 1937a). Spermatogenesis dominates this sexual phase and clams that function as juvenile males.

Loosanoff (1936, 1937a) reported that 98% of the clams examined passed through a juvenile male phase with half undergoing a protandric sexual change. The remaining 2% developed directly into females without going through the juvenile male phase. Similar proportions of protandric development have been reported for cultured populations of *Mercenaria*. For example, in South Carolina, Eversole et al. (1980) observed over 90% of *M. mercenaria* less than 28 mm shell length to be males. Dalton and Menzel (1983) identified 99.7% of the juvenile *M. mercenaria*, *M. campechiensis* and reciprocal hybrids 7–14 mm shell length cultured in Florida to be males.

Juvenile sexuality as a stage preceding the development of an adult gonad was first described in *M. mercenaria* (Loosanoff, 1936). Since that time, juvenile sexuality has been observed in 21 species and 12 families of bivalves (Lucas, 1975). Juvenile sexuality phase in *Mercenaria* is followed by a male or female phase in the adult. Adult clams are strictly dioecious (Eversole, 1989).

Sexes of *Mercenaria* cannot be accurately distinguished by shell characteristics or tissue weight (Belding, 1931). Harkness (1980) used shell length, height, width, weight, internal shell volume, and tissue wet weight of 217 *M. mercenaria* (116 males and 101 females) collected from one population and discriminate analysis to test this hypothesis. Results of the analysis indicated that none of these six size parameters individually or in combination were good discriminators. Clams were only successfully assigned a sex 61–62% of the time using these parameters (Harkness, 1980). Reliable determination of sex entails spawning or microscopic examination of the gonad.

Hermaphroditism in adult *M. mercenaria* is rare and, according to Coe (1943), should be termed accidental or developmental functional hermaphroditism. Hermaphroditic clams were only observed in collections from natural populations at two localities: 3 of 650 clams from Connecticut (Loosanoff, 1937a) and 1 of 1066 clams from Chesapeake Bay (Otto, 1973). Three of these clams were bilaterally hermaphroditic (i.e., with mature spermatozoa and ova) while the fourth may be a case of an adult clam changing sex.

Knaub and Eversole (1988) detected three hermaphrodites when 45 *M. mercenaria* from a Massachusetts hatchery were forced repeatedly to spawn in water temperatures warmer than ambient conditions. The unusually high frequency (6.7%) of hermaphroditism may be attributed to stress (Eversole, 1989). No hermaphrodites were observed in the stock of *M. mercenaria* ($n = 45$) from South Carolina or in a cross of the two stocks ($n = 45$) during repeated spawning trials. The three hermaphrodites spawned first as females and then as males. Histological examination of the gonad of these hermaphrodites after 10 spawning trials revealed a few free ova and various stages of spermatogenesis in the follicles (Fig. 5.1). Sperm from the hermaphrodites successfully fertilized eggs and produced larvae. Although the production of eggs overlapped the production of spermatozoa in these hermaphrodites, we did not observe sperm and eggs in any one spawn.

Several hypotheses have been proposed to explain sexual determination in bivalves, but none has been substantiated (e.g., Coe, 1943; Sastry, 1979; Mackie, 1984). Some researchers support a genetic basis for sexual determination, but there is no evidence of sex-related chromosomes in *Mercenaria* (Menzel and Menzel, 1965; Menzel, 1968). Coe (1943) and others contend that genetic and environmental factors interact to determine sex in bivalves. The importance of physiological state and stress should not be overlooked as an underlying factor in sex determination.

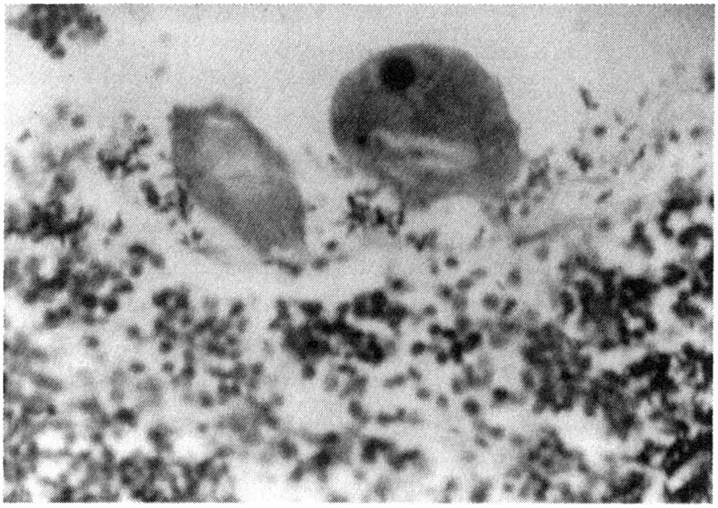

Fig. 5.1. Hermaphroditic gonad of a *M. mercenaria* that spawned initially as a female then as a male. Darkly stained bodies are spermatids and spermatocytes.

5.2.2 Sex Ratios

M. mercenaria are protandrous consecutive hermaphrodites, and at younger ages and smaller sizes males outnumber females (Loosanoff, 1937a). For example, 1.81 : 0.19 and 1.39 : 0.61 sex ratios were observed in 1-year-old *M. mercenaria* (Eversole et al., 1980; Walker, 1994). When Dalton and Menzel (1983) examined *M. campechiensis*, *M. mercenaria* and their hybrids within the range of 7–14 mm shell length, only two females were detected of 660 specimens. Loosanoff (1937a) reported that transformation of about 50% of the population to female occurred after the juvenile phase. Eversole et al. (1980) observed 1 : 1 sex ratios in 2- and 3-year-old *M. mercenaria*. Walker (1994) aged 2604 *M. mercenaria* and observed that the sex ratios of all but one of the 2-year-old and older age classes ($n = 37$) were equal. The sex ratio of the adult *Mercenaria* is approximately 1 : 1 over its geographical range (Table 5.1).

Sex ratio of adult *M. mercenaria* does not change from 1 : 1 with growth. A comparison of the number of males and females among 2635 clams in 18 sizes classes above a minimum breeding shell length (31–35 mm to 116–120 mm) revealed equal proportions in all but the 66–70 mm size class where males outnumbered females 1.00 to 0.64 (Walker, 1994). Equal sex ratios also occurred in the commercial littleneck (1.06 : 0.94) and chowder (1.12 : 0.88) sizes (Bricelj and Malouf, 1980; Pline, 1984).

Eversole (1989) indicated that stress may upset sex ratio equilibria in dioecious bivalve populations. For example, Knaub and Eversole (1988) observed a disparity in the sex ratio of *M. mercenaria* from Massachusetts induced to spawn in the warmer water temperatures of South Carolina. Of the 45 clams in the spawning trials, only 13 were males, 29 females and 3 were hermaphrodites. However, Walker (1994) failed to find a significant difference in the sex ratio of *M. mercenaria* with the increased stress of elevated intertidal level. The sex

TABLE 5.1
Sex ratio of adult *Mercenaria* (≥1 year of age) at different localities

Location	*N*	M : F	Source
Narragansett, RI	539	0.83 : 1.17	Barry and Yevich (1972)
Great South Bay, NY	327	1.07 : 0.93	Bricelj and Malouf (1980)
Raritan Bay, NY/NJ	358	1.06 : 0.94	Ropes (1987)
Core Sound, NC	264	0.81 : 1.19	Porter (1964)
Clark Sound, SC	204	0.95 : 1.05	Eversole (unpubl. data)
Folly River, SC	347 [a]	0.98 : 1.02	Eversole and Heffernan (1995)
Wassaw Sound, GA	886	1.05 : 0.95	Pline (1984)
Various locations, GA	2604	1.03 : 0.97	Walker (1994)
Indian River, FL	674 [a]	0.98 : 1.02	Hesselman et al. (1989)
Indian River, FL	1854 [a]	0.91 : 1.09	Bert et al. (1993)
Total	8056	0.98 : 1.02	

[a] These samples included *M. mercenaria*, *M. campechiensis* and their hybrids.

ratio data of clams collected from high, medium, and low intertidal stations in Georgia were 1.09 : 0.91, 1.01 : 0.99, and 1.03 : 0.97, respectively (Walker, 1994). Also, Pline (1984) observed a 1.01 : 0.99 sex ratio in adult clams after short-time exposure to low salinities (10–30‰).

5.3 GONAD DEVELOPMENT

5.3.1 Early Development

Primordial gonad of the juvenile phase first appears on the ventral side of the pericardium as paired follicles (Loosanoff, 1937a; Coe, 1943). Initially, the follicles are composed of a single layer of germinal epithelium and no lumen. Epithelial cells differentiate into gonia and a lumen forms as each primordium branches into the surrounding loose connective tissue. Rapid cell growth and differentiation follows. Examination of juvenile clams at this stage reveals a wide degree of development within an individual and among individuals. For example, a clam can have some follicles poorly developed and others with a few small oocytes along the follicle wall and spermatozoa in the lumen (Loosanoff, 1937a). Although female sex cells are present, the continued proliferation of the spermatogonia give the gonad a predominately male appearance. Loosanoff (1937a) observed that some of the follicles of juvenile clams collected from Long Island Sound in early fall appeared to have already discharged some spermatozoa. Little development was observed in the gonad during the juveniles' first winter. With spring and warmer water temperatures, rapid spermatogenesis was observed in many juveniles. Small oocytes were still present, but usually restricted to the follicular wall. Phagocytic–nutritive cells were found in great numbers around the follicles at this time. Presumably, these serve a nutritive function in gametogenesis and a phagocytic function after spawning (Loosanoff, 1937a). All summer, a branching system of follicles extends, in a posterior direction, surrounding the intestine.

M. mercenaria goes through a sex change after spermatozoa have been discharged in early fall of the second year in populations from more northern latitudes (Loosanoff, 1936, 1937a).

Those clams destined to be males will continue spermatogenesis in the fall and through winter at a reduced rate. Mature spermatozoa, which are present throughout this period, increase in the spring. The follicles of clams destined to become females remain in a distended state after discharge of sperm until spring. Oogenic activity begins slowly and continues through the summer in Long Island Sound (Loosanoff, 1937a). By summer's end, some follicles of a female contain large growing ova, whereas other follicles exist in various stages of early oogenesis. Obviously, it requires longer for a female clam to achieve fully functional status than a male clam; for this reason sex ratios favor males during early development.

5.3.2 Onset of Maturity

Theoretically, sexual maturity could be a function of size, age or the interaction of size and age of the clam (Eversole, 1989). Belding (1931) suggested that size, not age, was important in determining clam sexual maturity. *M. mercenaria* may have a minimum time to mature, but in most cases this is surpassed on the way to achieving a minimum maturation size. Minimum shell length of sexually mature *M. mercenaria* ranges from 20 to 35 mm (Belding, 1931; Loosanoff, 1936; Bricelj and Malouf, 1980; Eversole et al., 1980; Knaub and Eversole, 1988; Hesselman et al., 1989). It takes up to three years to achieve these sizes in the northern part of *M. mercenaria* distribution (Belding, 1931; Loosanoff, 1936; Stanley and DeWitt, 1983) and as little as one year in southern extremes of the distribution (Eversole et al., 1980; Hesselman et al., 1989; Walker, 1994). Environmental conditions such as water temperature, which affect the length of the growing season, also influence the time to reach sexual maturity. Apparently, sexual maturation size also involves some genetic component. Stocks of *M. mercenaria* from Massachusetts, South Carolina, and their hybrid matured at different sizes when cultured in South Carolina. The South Carolina stock matured at 20–25 mm shell length and the other stocks matured at 30–35 mm (Knaub and Eversole, 1988). The maturation size of northern field populations of *M. mercenaria* routinely averages 5–10 mm shell length larger than that of clams from southern populations (Belding, 1931; Bricelj and Malouf, 1980; Eversole et al., 1980; Stanley and DeWitt, 1983; Hesselman et al., 1989). Discounting variation among stocks and locations, *Mercenaria* accomplishes sexual maturity at about 25% of maximum shell length (Eversole, 1989).

5.3.3 Gametogenesis

5.3.3.1 Spermatogenesis

Spermatogenesis in *M. mercenaria* is known from the comprehensive studies of Loosanoff (1937a,b, 1937d). He reported that the progressive stages of sperm formation begins with primary spermatogonia lining the follicle walls as single cells (Fig. 5.2). At this stage, spermatogonia are extremely difficult to distinguish from primary oogonia. A large nucleus with a well-defined nucleolus and diffuse chromation occupies much of the primary spermatogonium cell. Secondary spermatogonia (7.7 μm in diameter) are smaller, frequently occurring in groups and farther from the follicle wall than primary spermatogonia (12 μm). The nuclear material is rounder and more concentrated in the secondary spermatogonia. This completes the mitotic divisions of spermatogenesis.

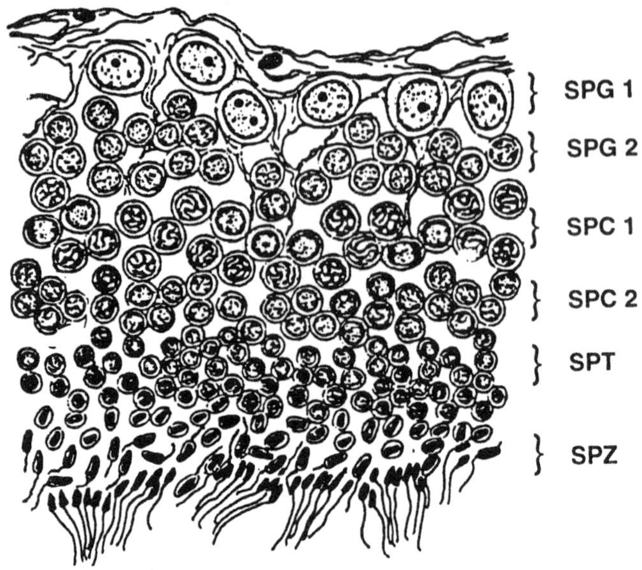

Fig. 5.2. Diagrammatic representation of spermatogenesis in *M. mercenaria*. Abbreviations: SPG 1, primary spermatogonia; SPG 2, secondary spermatogonia; SPC 1, primary spermatocyte; SPC 2, secondary spermatocyte; SPT, spermatid; SPZ, spermatozoa. Redrawn from Loosanoff (1937d).

Primary spermatocytes divide meiotically and give rise to secondary spermatocytes. Primary and secondary spermatocytes occur progressively farther from the follicle wall and closer to the lumen center (Fig. 5.2). Primary spermatocytes are relatively numerous during periods of rapid gametogenesis and somewhat rare at other times, indicating that this is a brief stage in gametogenesis. Primary spermatocytes (8.5 μm) are larger than secondary spermatocytes (5.8 μm), but occur in smaller groups. The second meiotic division yields spermatids that pass through several developmental stages before transforming into mature spermatozoa. In well-prepared sections fine protoplasmic strands can be seen extending from the follicle walls during all stages of to the early stages of spermateliosis. Loosanoff (1937b) hypothesized that these strands served a nutritive function.

Initially, spermatids (4.5 μm) and their nuclei are round, but with continued development the nuclei elongate and the quantity of cytoplasm gradually diminishes. Close examination will reveal the middle piece at the posterior end of the nucleus and the lightly stained acrosome forming at the anterior end. Concomitant with these changes is the formation and elongation of the tail. Average length of a mature spermatozoon head and tail is 3.8 μm and 30–35 μm, respectively (Loosanoff, 1937d). The lumen of ripe individuals are filled with radiating bands of mature spermatozoa arranged with heads facing the follicle wall and tails toward the lumen center. Except for brief periods after spawning, morphologically mature spermatozoa can be found in adult *M. mercenaria* all year (Loosanoff, 1937b).

Occasionally, atypical spermatogenesis results in congregations of darkly stained cells in *M. mercenaria* (Loosanoff, 1937b,d). These cells become pycnotic and are later cytolized. At present the nature of these cells is poorly understood; however, Coe and Turner (1938) noticed the ratio of atypical to normal cells in *Mya arenaria* increased during the harsh conditions of winter. Sperm balls that have been observed in other clams (Porter, 1974; Cain, 1975) have not been observed in *Mercenaria* taxa.

5.3.3.2 Oogenesis

There are no published detailed studies on oogenesis in *Mercenaria*. The description that follows was modified from published work with other bivalves (e.g., *Pinctada albina*; Tranter, 1958) and unpublished observations with *M. mercenaria* (Heffernan, personal communication, 1992). Oogonia line the follicle wall in clusters. The nuclear profile of primary oogonia is elongated, whereas secondary oogonia have a more rounded nucleus. It is not unusual to see a number of primary and secondary oogonia in a resting stage, not undergoing further division, during development.

Meiotic divisions of secondary oogonia produce primary oocytes. Margination of chromatin is detected in the nucleus of the small primary oocytes (approx. 15 μm). These cells remain in close contact with follicle cells that contain vesicles assumed to be nutrient stores. The nucleus and cytoplasm grow little during this previtellogenic phase of development.

Oocyte growth begins with the leptotene stage of meiotic prophase and when vitellogenesis is initiated. During vitellogenesis, the cytoplasm accumulates nutrients (e.g., lipid and glycogen) and the nucleus enlarges. Developing oocytes change shape and move away from

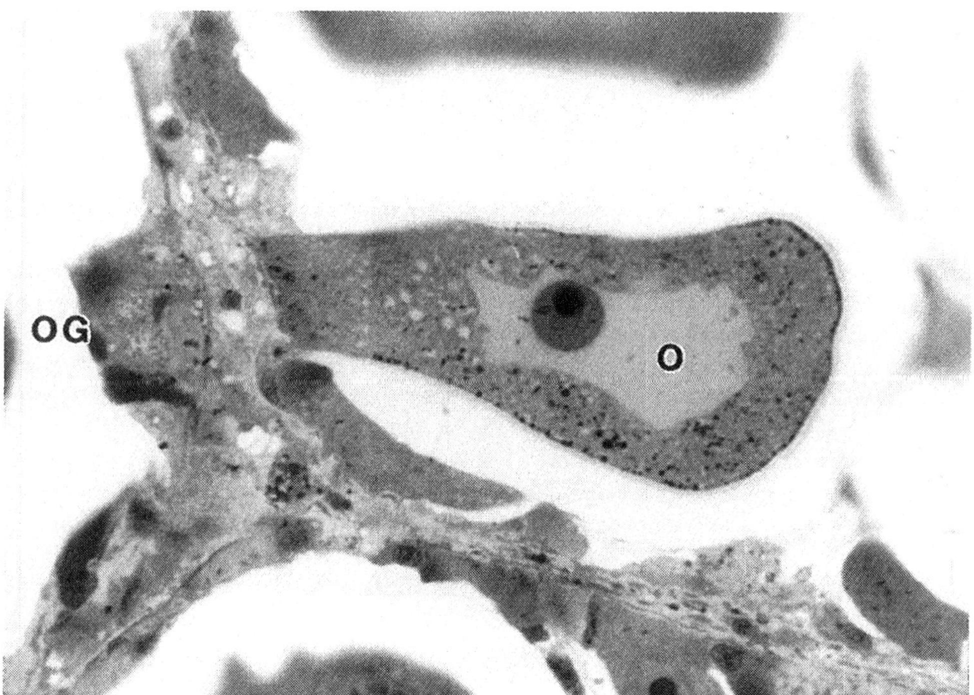

Fig. 5.3. Light micrograph of female gonad from *M. mercenaria*. Note the peduncle of the maturing oocyte (O) and an earlier developing oogonial stage (OG) surrounded by a follicle cell. Figure provided by P. Heffernan.

the follicle wall but maintain contact via a peduncle (Fig. 5.3). As vitellogenesis progresses, the oocytes enlarge and the vesicles in the cytoplasm of the follicle cells decrease in size. These follicle cells appear to play a nutritive role in vitellogenesis. A vitelline envelope surrounds individual oocytes as they increase in diameter from about 30 to 50 μm.

By late vitellogenesis, some oocytes are still connected by a thin peduncle while others are free within the lumen. Mature detached oocytes usually range from 50 to 70 μm in diameter. The vitelline coat is fully developed and a well-defined nucleus with nucleoli dominates the mature oocytes. Mature oocytes distend and fill much of the lumen in ripe females. The follicle cells, which are dominant in early gametogenesis, are greatly reduced in thickness and cellular content in ripe gonadal stages. The nutrient role of follicle cells in gametogenesis of *M. mercenaria* is supported by these observations and electron microscopy studies (P. Heffernan, personal communication, 1992).

It is not unusual to observe considerable overlap of gametogenic stages in individuals during oogenesis and spermatogenesis (Loosanoff, 1937b). For example, some follicles are out of phase with other follicles in the same gonad or various stages of gametogenesis are found in one follicle. Follicles with a series of stages of oogenesis or spermatogenesis are more frequently encountered later in gametogenesis. When space becomes available with the liberation of gametes, gametogenesis can resume and presumably, so can spawning. This is in contrast to those species (e.g., *Argopecten irradians*) which have only a few gametogenic stages present in mature gonads and exhibit more synchronous spawning (Sastry, 1979).

5.4 GAMETOGENIC CYCLE

5.4.1 Description

5.4.1.1 Qualitative methods

The gametogenic cycle of *Mercenaria* has been determined by a wide variety of techniques. These range from histological preparations of gonad tissue to the use of selected indices or indirect measures of gonadal condition. Although the gonad smear technique has been used with other economically important bivalves (Loosanoff, 1962), this technique has not been successful with *Mercenaria* to date.

Classical means of describing *Mercenaria* gametogenic cycle involve categorizing reproductive development into a series of subjective stages. Although the number and types of stages used by investigators have varied, inherent in nearly all the studies is a cycle that starts with a quiescent or recuperative stage after spawning. The recuperative stage is followed by an active developing stage, ripe stage, spawning and spent stages. The follicle wall of the recuperative stage is usually thicker than later stages, and few, if any, small gametogonia can be observed near the lumen periphery. Detection of sex is extremely difficult at this stage. Names for this stage in *Mercenaria* include: (1) indifferent (Keck et al., 1975; Kassner and Malouf, 1982; Dalton and Menzel, 1983); (2) inactive (Manzi et al., 1985); (3) in between (Porter, 1964); (4) recuperative (Pline, 1984); and (5) undifferentiated (Eversole et al., 1980).

A rapid proliferation of gametogonia and expansion of follicles characterizes those stages referred to as active and developing. Usually, follicles contain all stages of spermatogenesis from spermatogonia to spermatozoa or oogenesis from oogonia through mature oocytes. This

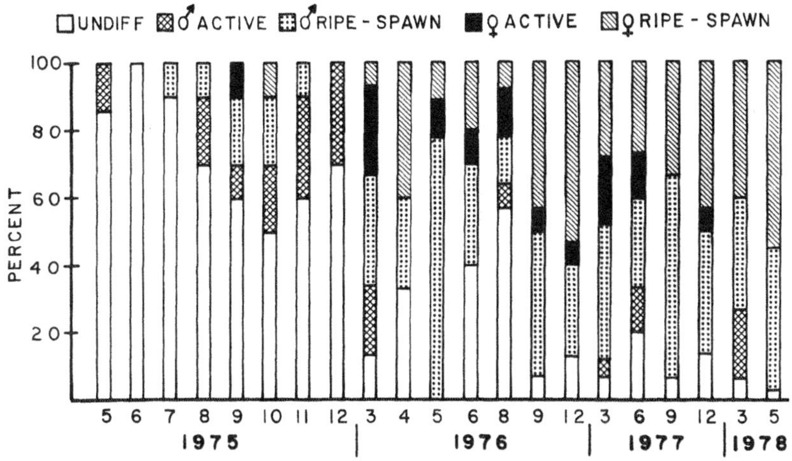

Fig. 5.4. Gametogenic stages of *M. mercenaria* stocked at 13 mm shell length in Clark Sound, South Carolina. The length of each shaded histogram represents the percentage frequency of clams in each stage and sex from May 1975 to May 1978 (Eversole et al., 1980).

is a continuous process, so the proportion of each gametogenic stage varies as the male and female ripen.

Follicles of ripe clams are characterized by a great number of large mature oocytes in females or a lumen filled with dense radiating bands of spermatozoa in males. These advanced stages of gametogenesis are more numerous than the earlier stages in development (e.g., spermatocytes or spermatids and small oocytes with peduncle attachment). The follicle wall is usually very thin and distended in ripe clams.

Spawning acts to empty the follicle of gametes, so the lumen has fewer sex cells than in ripe clams. Spermatozoa and large mature oocytes are found in the center of the follicle during spawning, whereas the gametogenic layer next to the follicle wall becomes extremely thin and more difficult to detect as spawning progresses. Spent or spawned-out clams represent the conclusion of spawning and the gametogenic cycle. At this time, the follicles are distended with only a few undischarged spermatozoa or large oocytes in the center of the follicle. Undischarged gametes remaining in the gonad of *M. mercenaria* after spawning may be extruded or cytolyzed (Loosanoff, 1937a,b; Bricelj and Malouf, 1980). The major difference between a spent and recuperating clam is the follicle wall, the latter having larger follicle cells evidencing resumption of gametogenic activity. Plotting the percentages of each gametogenic stage by sample date provides an impression of the gametogenic cycle (Fig. 5.4).

5.4.1.2 Quantitative methods

A gonadal index based on a method described by Kennedy (1977) has been successfully used to semi-quantify the gametogenic cycle of *M. mercenaria* (Kassner and Malouf, 1982; Heffernan et al., 1989). Each gametogenic stage can be assigned a value and an average value can be calculated for a sample date. We developed a numerical scale to facilitate comparisons of the gametogenic cycles among *M. mercenaria*, *M. campechiensis* and their

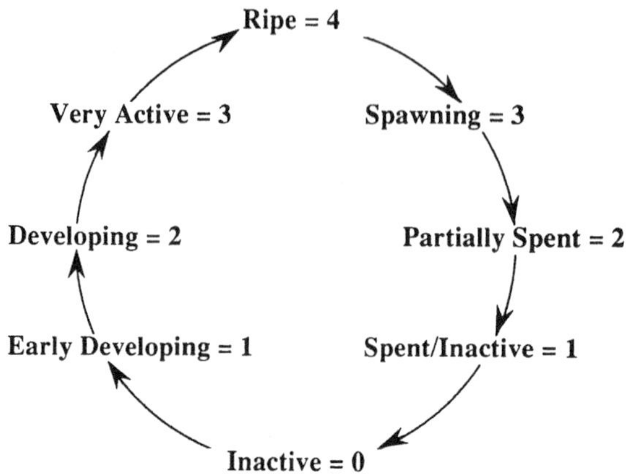

Fig. 5.5. Scoring system used to rank different stages of gametogenesis in *M. mercenaria*, *M. campechiensis* and their hybrids (Eversole, 1997).

hybrids (Fig. 5.5). A gonadal index can be determined for a sample date by multiplying the number of clams of a particular stage by the stage score, summing all the values and dividing by the sample size. An increase in the gonadal index corresponds to the proliferation of gametes, whereas a decrease represents spawning (Kassner and Malouf, 1982; Heffernan et al., 1989; Eversole, 1997). The magnitude of the changes in this index indicate the relative importance of gametogenic periods and spawning peaks.

Further attempts to improve the quantitative nature of gametogenic analysis involved measurements of particular aspects of gonad sections. Porter (1964) was the first to measure and enumerate small and large oocytes to describe the gametogenic cycle of *M. mercenaria*. Planimetry techniques have been used to estimate lumen area also to describe *M. mercenaria* gametogenic cycles (Keck et al., 1975; Eversole et al., 1980). Development of computer-aided image analysis made the assessment of reproduction condition less time consuming and more sensitive. Routine measurements include gonad area (% of field occupied by gonad), gamete area (% of follicle occupied by oocytes, spermatogenic stages and spermatozoa), oocyte number (per field or follicle) and mean diameter of nucleolated oocytes. Several investigators have observed strong agreement between qualitative methods of staging gonads and these quantitative methods (Keck et al., 1975; Eversole et al., 1980; Heffernan and Walker, 1989; Heffernan et al., 1989; Walker and Heffernan, 1994). Fig. 5.6 illustrates how percent lumen occupied by spermatogenic stages track the qualitative categories of gametogenesis over a 12-month period.

5.4.1.3 Indices

The gonadal–somatic index is one of the most widely used techniques for estimating gametogenic activity (Giese and Pearse, 1974). However, this method has not gained acceptance with *Mercenaria* because clam gonads are found within the visceral mass surrounding other tissues (e.g., digestive tract). Careful dissection will minimize the variability in separating gonad from somatic tissue. The other major limitation of the gonadal–somatic index is that this technique needs to be accompanied by microscopic examination to ensure that observed

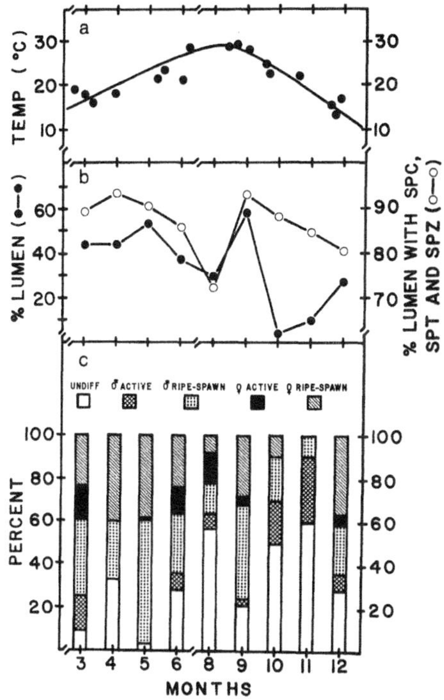

Fig. 5.6. Composite of the gametogenic stages of *M. mercenaria* in relation to water temperature and quantitative condition of the male gonad. (a) Water temperatures in Clark Sound, South Carolina. (b) Monthly means of percent follicle lumen (dots) and percent lumen with spermocytes (SPC), spermatids (SPT) and spermatozoa (SPZ) (circles). (c) The length of each shaded histogram represents the percentage frequency of clams in each gametogenic stage and sex (Eversole et al., 1980).

increases and decreases in the index are the consequence of changes in gametogenic tissue. Eversole et al. (1984) demonstrated that a gonadal–somatic index compared well with both a qualitative method of staging gonads and other quantitative methods in *M. mercenaria* (Fig. 5.7). Spawning was marked by decreases in gonadal–somatic indices, percent lumen and the proportion of ripe and spawning individuals with a corresponding increase in undifferentiated individuals. All three techniques reflected the same basic reproductive period with spawning peaks in May and June and in September and October in this *M. mercenaria* population (Fig. 5.7).

Indirect methods of assessing gametogenic cycles usually involve following seasonal changes in the condition of *Mercenaria* in relation to the reproductive cycle. Commonly, measures of tissue weights, water contents and biochemical constituents (e.g., protein, lipid and carbohydrates) are used to track seasonal changes in selected tissues or the total body of a standard-sized bivalve. Ansell and Lander (1967) observed cyclic changes in water content and total nitrogen content of soft tissue in relation to *M. mercenaria* reproductive cycle. However, Ansell et al. (1964b) failed to detect similar seasonal changes in adductor muscle, mantle, siphon, foot and other visceral tissue including the gonads and digestive gland of the clam. *M. mercenaria*, unlike other bivalves (e.g., *Tapes philippinarum*), rely more on the

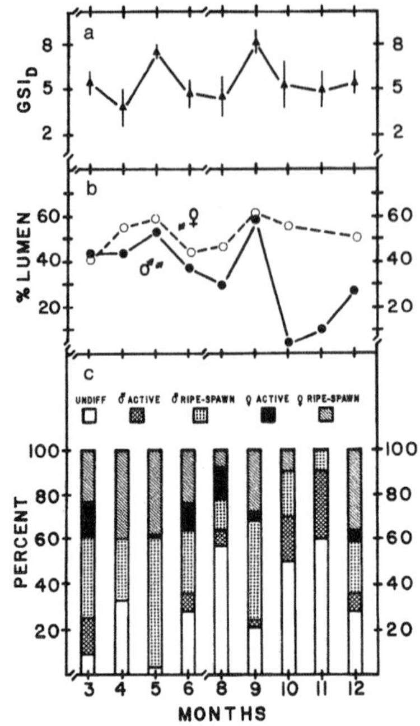

Fig. 5.7. Composite of the gametogenic stages of *M. mercenaria* in relation to gonadal–somatic index and quantitative condition of male and female gonads. (a) Triangles with vertical bars represent monthly means and standard errors of gonadal–somatic indices on a dry weight basis (GSI_D). (b) Circles and dots represent monthly means of percent lumen of male and female clams. Percent lumen was calculated by determining the area occupied by lumen (space within the follicle wall) in a standard area of gonad tissue. (c) The length of the shaded areas represents monthly frequencies of clams in each gametogenic stage and sex (Eversole et al., 1984).

immediate environment for resources needed for gametogenic activity and less on nutrient storage sites such as the foot and adductor muscles (Ansell and Loosmore, 1963; Bricelj and Malouf, 1980; Eversole, 1989). Coupling these measures of condition with qualitative methods of staging gonads and other quantitative methods (e.g., % follicle with spermatozoa) provides a clear picture of reproductive changes in *Mercenaria* and other bivalves.

5.4.2 Factors Influencing Gametogenic Cycles

5.4.2.1 Exogenus factors

The timing and duration of the gametogenic pattern appears to vary with a number of exogenous (environmental) factors, endogenous factors within an organism, and their interaction. Accordingly, the timing of spawning is linked to the gametogenic cycle because a certain degree of gonad ripeness must be achieved before *Mercenaria* can respond to stimuli and spawn. Belding (1931) indicated that, ultimately, the reproductive cycle of the clam is timed so environmental conditions are favorable for larval survival. This section will be devoted to summarizing those factors that influence the timing and duration of the reproductive pattern, whereas a later section will discuss those factors which prompt spawning.

Over its natural range, *M. mercenaria* exhibits either an annual or semiannual gametogenic cycle. These gametogenic cycles have been described for *M. mercenaria* in Connecticut (Loosanoff, 1937b), New York and New Jersey (Kassner and Malouf, 1982; Ropes, 1987), Delaware (Keck et al., 1975), North Carolina (Porter, 1964; Peterson and Fegley, 1986), South Carolina (Eversole et al., 1980; Manzi et al., 1985), Georgia (Pline, 1984; Heffernan et al., 1989) and Florida (Dalton and Menzel, 1983; Hesselman et al., 1989). Gametogenic cycles and spawning peaks appear to vary with latitude (Table 5.2). For example, gametogenesis

TABLE 5.2

Spawning times for populations of *Mercenaria* taxa by geographical locations and the temperatures (°C) during the first spawning peak per year. Solid line signifies periods of peak spawning with thicker line indicating the more intense spawning period

Location[1]	Temp. (°C)	Source[2]
Charles Is., CT	23 - 25	A
Gr. South Bay, NY[a]	20	B
Gr. South Bay, NY[b]	20	B
Raritan Bay, NY / NJ		C
Delaware Bay, DE	25 - 27	D
Chesapeake Bay, VA[3]		E
Core Sound, NC	27 - 30	F
Back Sound, NC		G
N. Santee Bay, SC	20	H
Clark Sound, SC	20 - 23	I
Wassaw Sound, GA	22 - 26	J
Wassaw Sound, GA[a]		K
Wassaw Sound, GA[b]		K
Alligator Harbor, FL[4]	16 - 20	L
Indian River, FL[5]	18 - 19	M

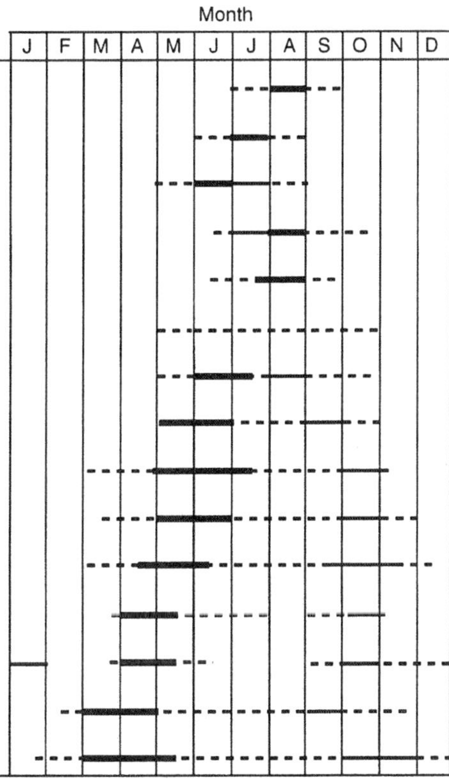

[1] Symbols [a] and [b] indicate different year gametogenic cycles from the same location.
[2] Sources: A: Loosanoff (1937b); B: Kassner and Malouf (1982); C: Ropes (1987); D: Keck et al. (1975); E: Porter (1964); F: Peterson and Fegley (1986); G: Manzi et al. (1985); H: Eversole et al. (1980); I: Pline (1984); J: Heffernan et al. (1989); K: Dalton and Menzel (1983); L: Hesselman et al. (1989).
[3] The spawning season was estimated from larval abundance at this location and from histological examination of gonads at the other locations.
[4] Spawning cycle of young male clams less than 2 years old.
[5] These samples include a complex of *M. Mercenaria, M. campechiensis* and their hybrids.

and spawning were initiated earlier and lasted longer in the more southerly latitudes, until a second developmental period and spawning period became possible. Populations in Connecticut, New York, New Jersey and Delaware exhibited an annual gametogenic cycle and one spawning peak (Loosanoff, 1937b; Keck et al., 1975; Kassner and Malouf, 1982). *Mercenaria* populations from North Carolina, South Carolina, Georgia, and Florida exhibited two periods of gametogenic activity and spawning (Porter, 1964; Eversole et al., 1980; Pline, 1984; Manzi et al., 1985; Peterson and Fegley, 1986; Hesselman et al., 1989; Heffernan et al., 1989). Measures of gametogenesis (e.g., % follicle occupied by oocytes) indicated that gametogenic activity and spawning in the spring were more intense than the gametogenesis leading up to spawning in the fall. Gametogenic cycles also varied among years within the same population of *M. mercenaria.* Heffernan et al. (1989) observed two periods of gametogenesis in 1984 and three periods in 1985 in the *M. mercenaria* sampled from Wassaw Sound, Georgia (Table 5.2).

Temperature is known to play a critical role in timing gametogenesis in bivalves (Giese and Pearse, 1974; Sastry, 1979). In *Mercenaria*, the gametogenic cycle and spawning correlate well with seasonal changes in temperature. Loosanoff (1937b) observed an activation of gametogenesis after completion of spawning in the fall when water temperatures were still relatively warm. Gametogenesis continued through the fall in spite of falling temperatures, finally stopping in midwinter as water temperatures approached 5°C. Although clams were feeding again at about 10°C, gonads remained quiescent until 15°C in late spring. Gametogenesis proceeded at a rapid rate producing mature gametes by early summer, well in advance of spawning that occurred at 23–25°C in late summer. An extension of the time critical temperatures occur for gametogenesis and spawning makes bimodal reproductive patterns possible. Protraction of the reproductive period will continue with decreasing latitudes until water temperatures exceed the optimum reproductive temperature for *Mercenaria.* In Florida, Hesselman et al. (1989) observed a period of gametogenic inactivity in late summer when temperatures exceeded 30°C.

Ansell et al. (1964b) showed that thermal effluent from a generating station influenced gonad ripening in *M. mercenaria.* The population closer to the generating station had an extended period of gametogenesis and two spawning peaks, a major one in early summer and minor peak in fall. In contrast, the population of clams removed from the immediate effects of the heated effluent had a more contracted gametogenic cycle and one spawning peak similar to the pattern described by Loosanoff (1937b). Loosanoff and Davis (1950, 1963) were successful in inducing gametogenesis in winter collections of *M. mercenaria* when clams outside the normal gametogenic cycle were conditioned at elevated water temperatures in the laboratory. As warmer water temperatures accelerate gametogenesis, colder temperatures delay gonad development in *M. mercenaria.* Loosanoff and Davis (1963) also observed that *M. mercenaria* collected from Long Island Sound in spring and transplanted to the colder waters of Maine delayed gametogenesis and spawning.

The influence of tidal level on the gametogenic cycle of *Mercenaria* has not been studied in detail until recently. Walker and Heffernan (1994) documented a difference in gonad development among clams planted at different tidal levels in Wassaw Sound, Georgia. Clams in the highest intertidal site (marsh) exhibited a significantly higher percentage of sexually undifferentiated individuals compared to those clams planted at a tidal level two hours above mean-low water (oyster zone), at mean low tide and just below the spring low-water mark. Sexually undifferentiated clams were observed at all the sample dates at the marsh site and at

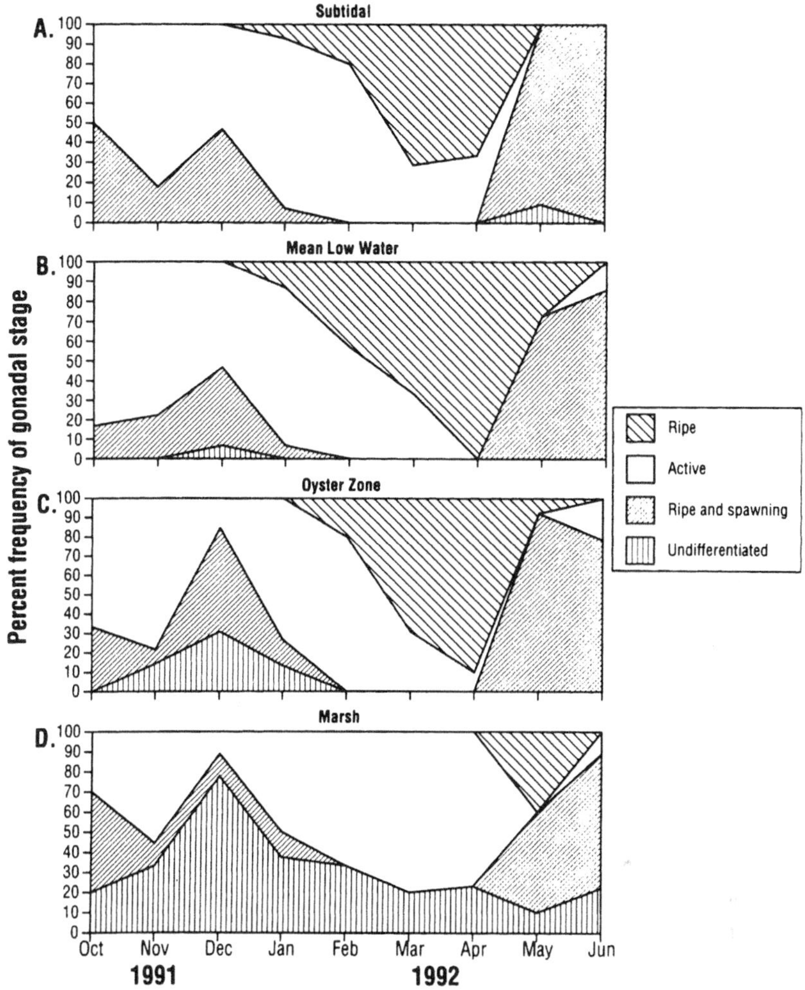

Fig. 5.8. Gametogenic stages of *M. mercenaria* cultured in a subtidal site (A), mean low water (B), two hours above mean water (C), and the marsh zone (D) in the Skidaway River, Wassaw Sound, Georgia. The height of each shaded area represents the percentage frequency of clams in each stage (Walker and Heffernan, 1994).

or less than three sample dates at the other three sites (Fig. 5.8). The major spawning period occurred from April to June at the three lower tidal sites and from May to June at the marsh site (Walker and Heffernan, 1994).

In a similar study, Eversole et al. (1980) failed to find a significant difference in the histological gametogenic staging of *M. mercenaria* planted in a subtidal site and intertidal site in Clark Sound, South Carolina. However, the overall proportion of gonad tissue devoted to spawning in the spring and fall peaks (average reproductive effort) measured as a gonadal–somatic index by Eversole et al. (1984) was significantly greater in the clams from the subtidal site than from the intertidal site (Fig. 5.9). Walker and Heffernan (1994) also observed that

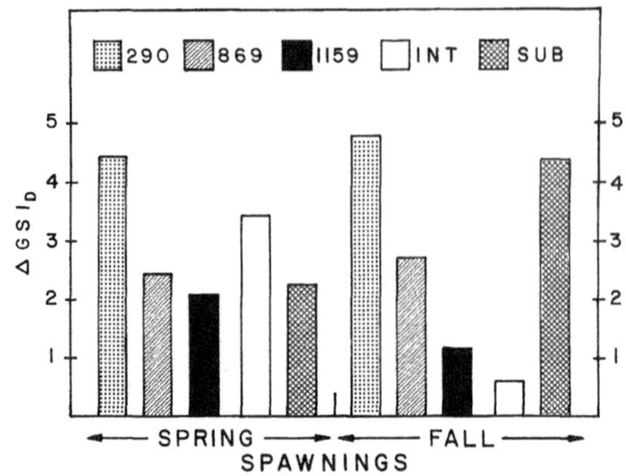

Fig. 5.9. Mean declines in gonadal–somatic index on a dry weight basis (ΔGSI_D) by density level (290, 869 and 1159 clams/m^2) and tidal location (intertidal [int] and subtidal [sub]) for the spring and fall spawning peaks of *M. mercenaria* cultured in Clark Sound, South Carolina (Eversole et al., 1984).

clams from the marsh (high intertidal) site had significantly lower reproductive potential (e.g., % gonad area, % gonad area occupied by spermatozoa or oocytes, mean number and diameter of oocytes) than clams from the three lower tidal sites. Although differences among the three other tidal sites in Georgia were not significant, an obvious trend of decreasing reproductive potential was observed with increased tidal exposure. *M. mercenaria* apparently employs a reproductive strategy of allowing gametogenesis and reproduction to proceed but at a reduced level in the higher intertidal sites. These two studies adequately demonstrate a tidal exposure effect on *M. mercenaria* gonad development; however, it is not clear if the clams are responding to differences in temperature, food abundance, time available for feeding or some other factor caused indirectly by tidal exposures.

Giese and Pearse (1974) suggested that seasonal fluctuations in the quantity and quality of food supplies may be a factor in regulating gametogenesis. Gonad development is an energy demanding process; approximately 30–50% of the total annual organic growth of *M. mercenaria* is allocated to reproduction (see Section 5.6). The relationship between gonad development and food supplies is not as fully understood in *Mercenaria* as it is with other species of bivalves (e.g., scallops; Sastry, 1979). The gametogenic cycle of *M. mercenaria* in Long Island Sound coincides with seasonal fluctuations of phytoplankton abundance (Loosanoff, 1937b). Although concurrent measures of phytoplankton abundance or chlorophyll *a* and gametogenic cycles of *Mercenaria* are rare, several examples support the link between the two. Heffernan et al. (1989) observed a trimodal reproductive cycle of clams in Wassaw Sound, Georgia, and reported that other studies detected periods of phytoplankton abundance in the Sound during the periods of peak spawning. Walker and Heffernan (1994) also observed that the gonad development of cultured *M. mercenaria* was retarded in trays fouled by sea squirts (*Mogula* sp.), and a recovery of gonad development occurred after the trays were cleaned. Presumably, the proliferation of gonad material cited in these two studies was related to increased food supplies. Gametogenesis appears to correlate well with phytoplankton abundance because the gametogenic activity of *Mercenaria* is more dependent on the existent intake of food rather than nutrient storage and its subsequent utilization for gametogenesis. Hatchery managers have known for years that adequate food levels need to be maintained in conditioning tanks to achieve gonad maturation and spawning out of season (Loosanoff and Davis, 1950).

Eldridge et al. (1979) observed a density-mediated growth effect in *M. mercenaria*; however, no differences were detected in either the histological stages or the quantitative measure of reproductive condition (e.g., % gonad occupied by oocytes) at 290 clams/m^2, 869 clams/m^2 or 1159 clams/m^2 (Eversole et al., 1980). When relative gonad mass (gonadal–somatic index on a dry weight basis) was determined, significant relationships were observed with shell length and density level (Eversole et al., 1984). Larger clams had proportionally more gonadal tissue than smaller clams and the gonadal–somatic index of clams at the lowest density was greater than the indices at the two higher densities. However, no differences in gonadal–somatic indices were observed with standard-sized clams because of the associated growth reductions at elevated densities. On the other hand, density-related reproductive effects were observed when average reproductive effort (the change in gonadal–somatic index) was compared among density treatments (Fig. 5.9). Average reproductive effort at 869 clams/m^2 and 1159 clams/m^2 was 54% and 33% of the reproductive effort of clams at 290/m^2, respectively (Eversole et al., 1984). Apparently *M. mercenaria* proceeded with gametogenesis but at a reduced rate at elevated densities.

Although there is no direct evidence, food could be the limiting resource and the mechanism by which density affects reproductive activity. Eversole et al. (1984) noted that both growth and the gonadal–somatic index were reduced to greater degree at the intertidal location where the time available for feeding was limited by the tidal cycle. The importance of the amount of food in the overlayer water and the time available for feeding to gametogenesis was reviewed above in this section.

There are a few studies reporting the effect of salinity on *Mercenaria* gametogenesis. Pline (1984) reported that spermatogenesis of clams subjected to 25‰ and 30‰ salinities appeared normal and healthy, whereas males exposed to 10, 15 and 20‰ salinities for eight weeks exhibited signs of stress. Gonads of these clams contained clumps of 8–16 or more spermatocytes resembling the sperm balls that have been observed in other bivalves (Shaw, 1965; Porter, 1974; Cain, 1975). The percentage of males with clumped spermatocytes reached 29% in 20‰ salinity, 63% in 15‰ salinity and 67% in 10‰ salinity (Pline, 1984). Sperm produced by the clams at 10‰ salinity exhibited low motility. No deleterious effects were detected in the gametogenesis of female clams subjected to low salinity for eight weeks. Pline (1984) also provided evidence that clumping of spermatocytes was a consequence of low salinity and not of starvation. He speculated further that clumping of spermatocytes was a subtle indication of gamete degeneration and reduced reproductive potential.

Reproductive activity may be affected by parasites and disease. The parasites and diseases of clams were recently reviewed by Gibbons and Blogoslowski (1989) and by Ford in Chapter 12 of this book. Trematodes were suggested as especially common parasites of the gonad of *M. mercenaria*. Eversole et al. (1980) found only one gonad out of 304 gonads of *M. mercenaria* heavily infested with digenetic trematodes (Fig. 5.10). In this case, the infestation appeared to interrupt gametogenesis and result in castration of the host.

Neoplasms have been found in the gonads of *Mercenaria* from Narragansett Bay, Rhode Island (Barry and Yevich, 1972), Clark Sound, South Carolina (Eversole and Heffernan, 1995) and Indian River, Florida (Hesselman et al., 1988; Bert et al., 1993). A gonadal

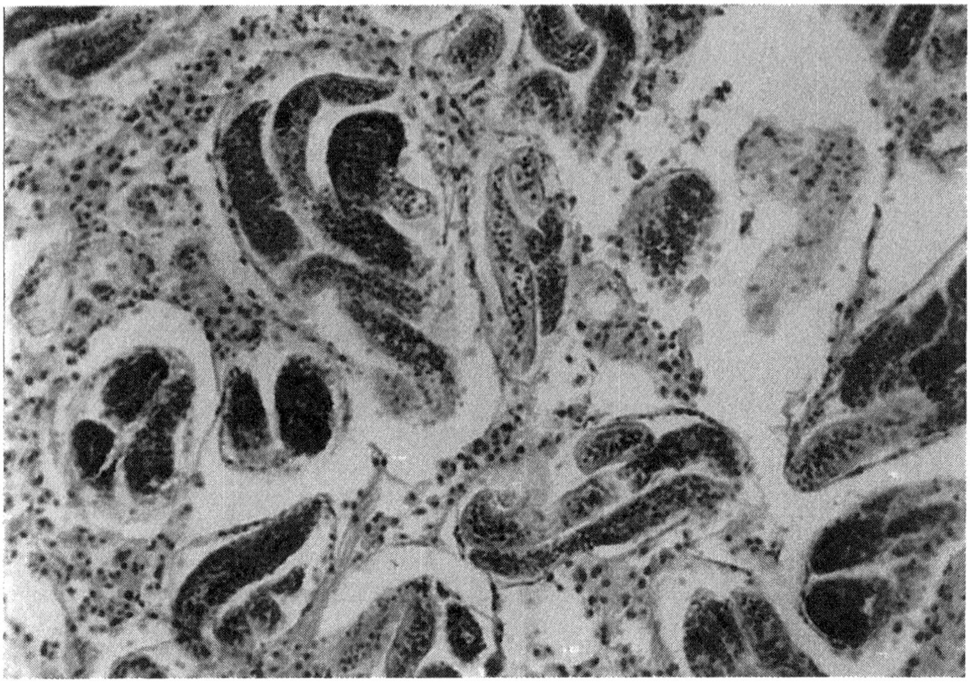

Fig. 5.10. Section of gonad infested with digenetic trematodes from *M. mercenaria* cultured in Clark Sound, South Carolina (Eversole et al., 1980).

neoplasia prevalence of 47% in Clark Sound was much higher than the prevalences at the other two localities. In the more advanced cases, the neoplastic cells filled the follicles, and gametogenesis did not appear to precede (Fig. 5.11). Recently, Barber (1996) observed that female *M. arenaria* with gonadal neoplasia from Whiting Bay, Maine, produced significantly fewer gametes than females without neoplasia. Neoplastic cell displacement in the gonadal follicles resulted in a 66% reduction in gametes. Concomitant with this gamete reduction was a disruption of normal oogenesis and an inhibition of spawning in the *M. arenaria* with gonadal neoplasia. A review of the comparative histopathology of gonadal neoplasia in other bivalves was given by Peters et al. (1994).

Peters et al. (1994) noted that gonadal neoplasms were found in many bivalves from degraded environments. Van Beneden et al. (1993) proposed a herbicide etiology for the gonadal neoplasia in *Mercenaria* collected from Indian River, Florida. *Mercenaria* collected from Florida exhibited abnormal gametogenesis and disruption in the more advanced cases of neoplasia (Hesselman et al., 1988). The direct effects of gonadal neoplasia and xenobiotic exposure on *Mercenaria* gametogenesis await confirmations.

There is no question that *Mercenaria* has suffered from exposure to xenobiotics and man's activities (e.g., dredgings); however, there are very few studies reporting these effects on gametogenesis. Ropes (1987) reported that the reproductive capacity of *M. mercenaria* in Raritan Bay was not affected by pollution. However, he noted that gametogenic cycle and spawning occurred 1 to 2 months later than was observed at the more northern locations

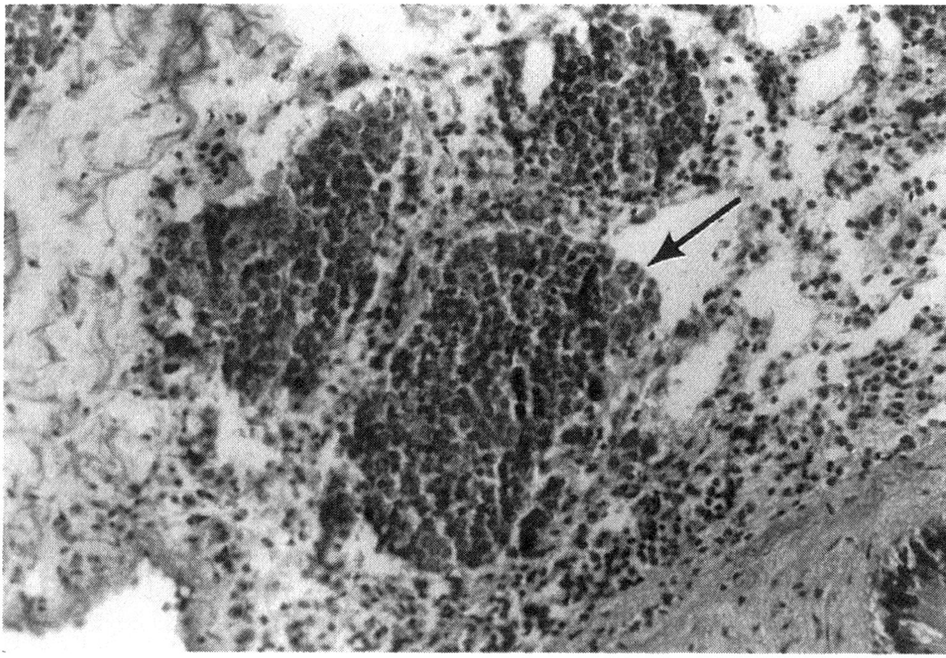

Fig. 5.11. Representative section of severe infestation of gonadal neoplasia. Neoplastic cells (arrow) occupy nearly all of the follicle of a *M. campechiensis* female × *M. mercenaria* male hybrid collected July 1988 from Folly River, South Carolina (Eversole and Heffernan, 1995).

(Table 5.2). It is possible that the timing and duration of these clams' gametogenic cycle may have been affected by sublethal exposure to organic and inorganic pollution in Raritan Bay. *Mercenaria* usually responds to an environmental insult by scaling back fecundity rather than interrupting gametogenesis. Noticeable effects of pollution-compromised environments are readily observed during embryogenesis and larval clam development (Calabrese, 1972; Stiles et al., 1991).

5.4.2.2 Endogenous factors

The gametogenic cycle varies between male and female *Mercenaria* (Eversole et al., 1980; Pline, 1984; Manzi et al., 1985; Heffernan et al., 1989; Hesselman et al., 1989). The male gametogenic cycle is more protracted than that of female clams. For example, Loosanoff (1937b) observed spermatogenesis and mature spermatozoa in all seasons of the year excluding a brief post-spawning period, whereas, ripe female clams in the Charles Island area of Long Island Sound were limited to May through September. As a consequence, males can spawn sooner and longer than females. Pline (1984) also reported that oogenesis followed a distinct pattern of events through the gametogenic cycle. Overall, spermatogenesis was less synchronous within the population and considerable asychrony also occurred among follicles of individual males. It was not unusual to observe ripe spermatozoa in a follicle adjacent to follicles containing less advanced spermatogenic stages. Male *M. mercenaria* exhibit greater reproductive plasticity as indicated by these observations and the quicker response to spawning stimulation. The proportions of body tissue devoted to reproduction appears to be similar in male and female clams (Eversole et al., 1984).

Differences in the gametogenic cycle were observed between chowder-sized clams (>7.8 cm shell length) and littleneck-sized clams (3.8–6.8 cm shell length) (Pline, 1984). In Wassaw Sound, Georgia, female chowders were ripe by March and spawned by April, whereas female littlenecks were actively developing in March and some females were still ripe and unspawned in May. The redevelopment period that followed the spring spawning in female littlenecks was more protracted than in female chowders. Spawning started sooner in the spring, lasted longer into the summer and occurred to a greater extent in the fall with female chowders. A similar gametogenic pattern to that of the female littlenecks and chowders was observed between male littleneck and chowder clams. The spawning intensity, indicated by a decline in oocytes per follicle area, was greater for the chowders than for littleneck clams (Pline, 1984).

The effect of age on gametogenesis in *M. mercenaria* has not been studied in great detail (Eversole et al., 1980, 1984; Walker, 1994). Few studies examined the influence of age on gametogenesis where the confounding effects of clam size were factored out. In one study Eversole et al. (1984) reported that relative gonad mass (gonadal–somatic index) varied significantly with both clam size and age. Using covariant analysis, Eversole et al. (1984) observed that older clams had proportionally more gonad tissue than younger clams of the same size. Similarly, large clams had proportionally larger gonads than smaller clams of the same age.

Dillon and Manzi (1989) documented the extensive hybridization of *M. mercenaria* and *M. campechiensis* in Indian River, Florida. In the Indian River, the gametogenic cycle of *Mercenaria* taxa was quite variable within and between monthly samples (Hesselman et al., 1989). Hesselman et al. (1989) speculated that some of the variation may have resulted from simultaneous collection of *Mercenaria* taxa that included *M. mercenaria*, *M. campechiensis* and their hybrids. Dalton and Menzel (1983) followed the gametogenic pattern of young male clams from laboratory *M. campechiensis*, *M. mercenaria* and reciprocal crosses. *M. campechiensis* exhibited a bimodal reproductive pattern with spawning peaks in the cooler months in Alligator Harbor, Florida. The spermatogenic cycle of the hybrid from the *M. campechiensis* female × *M. mercenaria* male cross was similar to the reproductive pattern of *M. campechiensis*, whereas the spermatogenesis of the hybrid from the reciprocal cross resembled *M. mercenaria*. Fig. 5.12 illustrates the gametogenic cycle of adult male and female progeny of *M. mercenaria*, *M. campechiensis* and reciprocal crosses cultured in Folly River, South Carolina (Eversole, 1997). Progeny from the two maternal *M. mercenaria* crosses exhibited similar bimodal gametogenic cycles with two spawning peaks, a major peak in spring followed by a smaller peak in early fall. Those clams from maternal *M. campechiensis* crosses spawned about 1 month later in the spring, and 2–3 months later in the fall/early winter. The period of gametogenic development between the year's first and second peaks was longer in crosses derived from maternal *M. campechiensis* than maternal *M. mercenaria* crosses. These data support the speculation of Dillon (1992) that the observed low rate of hybridization in South Carolina resulted from species-specific temporal differences in the gametogenic cycles.

The gametogenic cycle of *M. mercenaria* varies also in timing and duration throughout its geographical distribution (Table 5.2). Porter (1964) suggested that the differences in

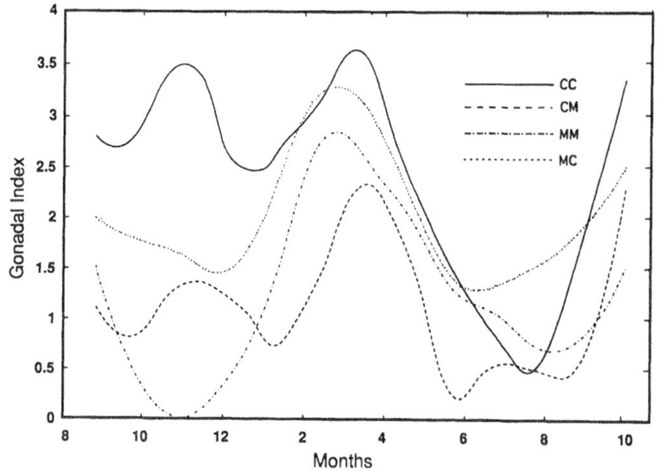

Fig. 5.12. Gonadal index for crosses of *M. mercenaria* female × *M. mercenaria* male (MM), *M. campechiensis* female × *M. campechiensis* male (CC), *M. mercenaria* female × *M. campechiensis* male (MC) and the reciprocal hybrid (CM). Fitted line by spline interpolation (Eversole, 1997).

gametogenic cycles were expressions of racial differences of the *M. mercenaria* populations along the east coast of the United States. Knaub and Eversole (1988) studied the reproductive cycle of a South Carolina wildstock (SCW), a hatchery stock from Massachusetts (ARC) and a cross between the two (ARC × SCW) cultured in South Carolina waters. Significant differences were found among stocks, months and interaction term in level of spawning condition (i.e., ripe and spawning clams vs. active clams). The ARC stock exhibited a trimodal reproductive pattern with a smaller peak in December and two major peaks in August and April (Fig. 5.13). The SCW clams spawned in July and March, whereas the reproductive pattern of ARC × SCW was somewhat intermediate of the parental stocks. Comparisons of laboratory spawning trials indicated these stocks of *M. mercenaria* differed in several other reproductive characteristics including fecundity, number of spawns/female, eggs/spawn, egg size and oocytes area/follicle area (Knaub and Eversole, 1988). These observations provide evidence that the reproductive characteristics of clams are genetically influenced and populations of *M. mercenaria* may respond differently to environmental cues.

Several methods have been explored in an attempt to develop faster growing *M. mercenaria* for aquaculture including hybridization, selection programs and induction of polyploidy. Triploid *M. mercenaria* have been produced with a treatment of cytochalasin B (Hidu et al., 1988; Buzzi, 1990). Clams treated by Buzzi (1990) were cultured to market size (45–50 mm shell length), and at that time, the triploids were significantly larger than diploids (Eversole et al., 1996). When 22 diploid and triploid clams each were conditioned and induced to spawn, none of the triploids spawned, whereas 82% of the diploids spawned and produced viable gametes. The gonads from diploid clams were ripe, male gonads contained mature spermatozoa and female gonads contained many free mature oocytes in the follicle (Fig. 5.14). Some of the oocytes were attached to the follicle wall indicating maturation was continuing

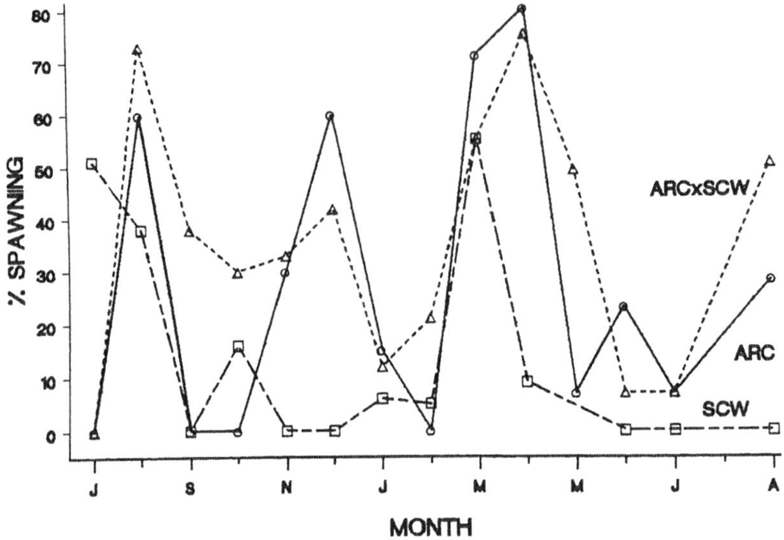

Fig. 5.13. Mean percentage of clam in spawning condition for stocks (ARC, hatchery stock from Massachusetts; SCW, South Carolina wildstock; ARC × SCW cross) of *M. mercenaria* cultured in South Carolina (Knaub and Eversole, 1988).

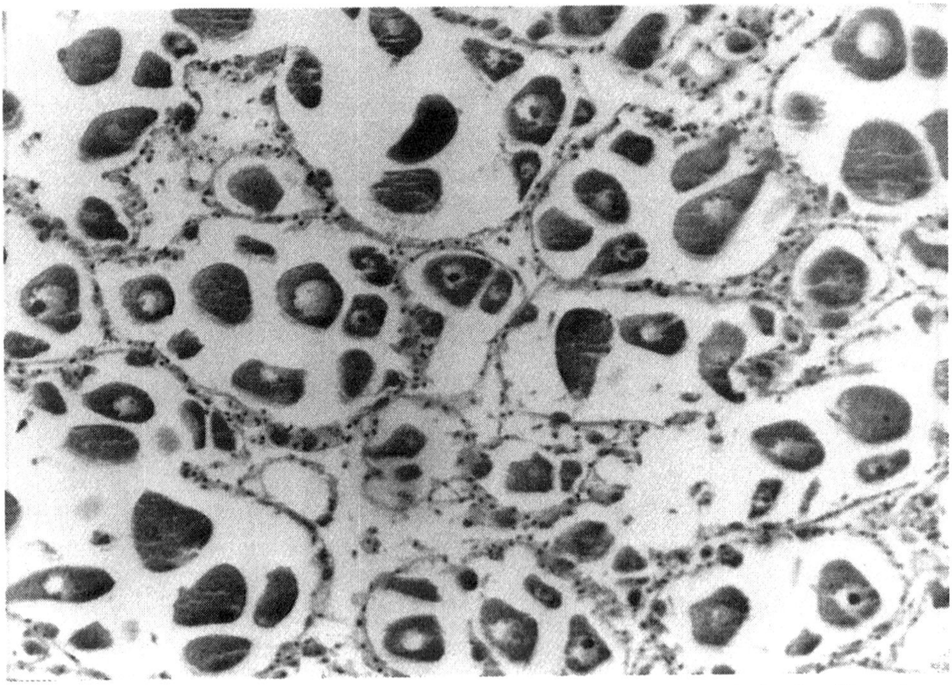

Fig. 5.14. Section of a gonad from a diploid female *M. mercenaria* cultured in South Carolina (Eversole et al., 1996).

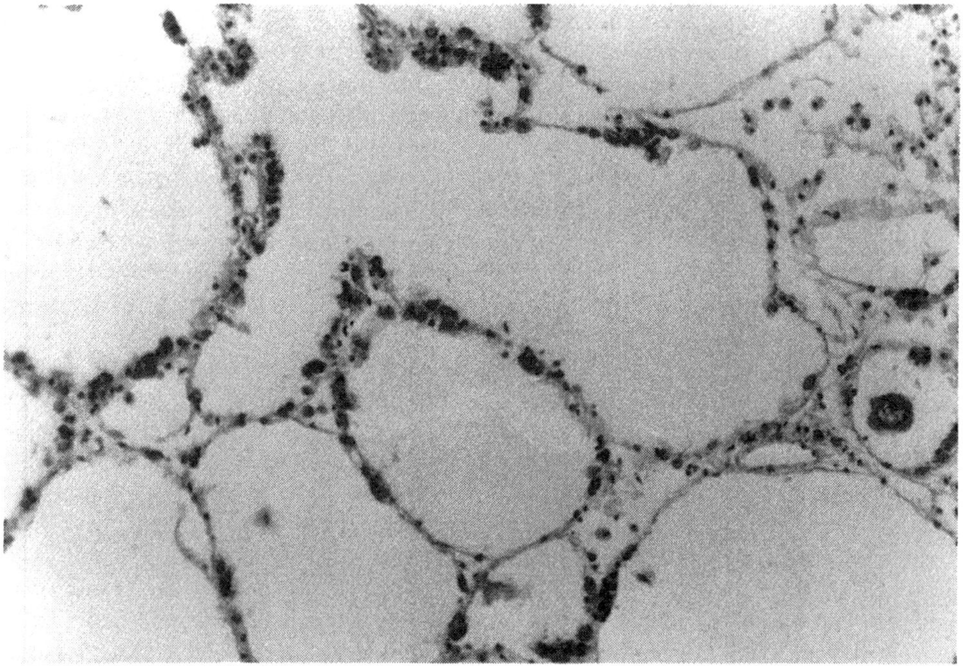

Fig. 5.15. Section of a gonad from a triploid female *M. mercenaria* cultured in South Carolina (Eversole et al., 1996).

in diploid clams. Triploid clams contained gonad tissue, but mature gametes were lacking. Free in the lumen of male triploids were clusters of darkly stained bodies that appeared to be degenerative sex cells. A few nucleated oocytes were observed in the gonad of female triploid clams (Fig. 5.15). These cells were generally smaller than the mature oocytes observed in diploid clams and somewhat degenerative in nature. The area of lumen occupied by spermatogenic stages in the diploid clam (85%) was significantly larger than the area occupied in triploid clams (20%). Eversole et al. (1996) also observed a significant difference in the area of lumen occupied by oocytes in diploid (34%) and triploid clams (15%). Although gonad tissues were present in triploid clams, reproductive potential was greatly reduced because gametogenesis was disrupted and viable gametes were not formed.

5.5 SPAWNING

5.5.1 Factors Influencing Spawning

5.5.1.1 Natural factors

A variety of environmental factors have been reported to stimulate spawning in bivalves (Loosanoff and Davis, 1963; Galtsoff, 1964; Sastry, 1979; Mackie, 1984; Eversole, 1989). Water temperature has been considered to be of paramount importance in synchronizing bivalve spawning to optimize fertilization and larval survival. Spawning in *M. mercenaria* was initially thought to be "triggered" by a particular maximum water temperature 24–25°C (Nelson, 1928; Belding, 1931). More recently, Keck et al. (1975) suggested that the rate of temperature change provides a stronger spawning cue than absolute temperature. Carriker (1961) observed that *M. mercenaria* generally spawned at or shortly after low tide when water temperatures changed. He suggested further that only slight changes in temperature stimulate spawning after the clam is competent to spawn. The warming and fluctuation of temperature associated with tidal cycle may have been responsible for inducing exposed beds of *M. mercenaria* to spawn as early as March in Wassaw Sound, Georgia (Pline, 1984). Spawning has also been reported to occur with declining temperatures in southern parts of *Mercenaria* distribution (Hesselman et al., 1989).

The temperature at which *M. mercenaria* spawns differs throughout its geographical distribution. In addition to those temperatures cited for initial spawning of *Mercenaria* in Table 5.2, spawning is also reported to occur at 18–19°C in Southampton waters, England (Raymont, 1972), at 21°C in Westport River, Massachusetts (Fiske et al., 1968), at 22–25°C in Long Beach, California (Crane et al., 1975), and at 24–26°C in Little Egg Harbor, New Jersey (Carriker, 1961). Differences in spawning temperatures were suggested as evidence for the existence of physiological race among populations of *M. mercenaria* (Porter, 1964; Keck et al., 1975). Knaub and Eversole (1988) provided experimental results to support the hypothesis that the timing of spawning involved a genetic component. In this study, the *M. mercenaria* from Massachusetts grown in South Carolina had a trimodal spawning pattern, and the native South Carolina clams had a bimodal pattern, whereas offspring from a cross of the two geographical stocks had a pattern intermediate of the two parental stocks (Fig. 5.13). Other differences in reproductive characteristics (e.g., fecundity, egg size) were observed when the clams from Massachusetts and South Carolina were spawned under controlled conditions (Knaub and Eversole, 1988). Loosanoff et al. (1951) also observed that there are physiologically different races of *M. mercenaria* based on larval setting and growth.

The restriction of spawning within a broad range of appropriate temperatures indicates that other factors such as the presence of food may also be important in stimulating spawning. Breese and Robinson (1981) stimulated *M. mercenaria* at 18–20°C to spawn by the addition of marine algae. The bulk of spawning in Lower Little Egg Harbor, New Jersey, took place during that segment of the tidal cycle which would result in most of the clam larvae being transported up-bay rather than toward the inlet and the open ocean (Carriker, 1961). Loosanoff (1937c) reported that the mere presence of gametes in the water stimulates *M. mercenaria* to spawn. A number of procedures to artificially induce clams to spawn were based on principles gathered from these field observations.

5.5.1.2 Induced methods

Practical methods used to artificially induce spawning in *Mercenaria* have been outlined by Loosanoff and Davis (1963), Castagna and Kraeuter (1981) and Hadley et al. (1997). Loosanoff (1937c) was the first to elucidate the use of temperature to induce *M. mercenaria* spawning. Presently, temperature cycling is the most frequently used method of artificially inducing *M. mercenaria* to spawn in hatcheries. Physiologically conditioned (ripe) clams are usually allowed to acclimate to ambient or slightly chilled seawater (20–22°C). After the clams begin siphoning, the water temperature is raised gradually to 26–28°C over a period of 30 min. Clams are left undisturbed at this temperature for about 30 min before recycling to 20–22°C. The elevated temperature should not exceed 30°C. Ripe clams usually spawn during the first temperature increase, whereas poorer conditioned clams require repeated temperature cycles. Goodsell (1991) observed that dry storage in a refrigerator (7–8°C) overnight facilitated the temperature spawning stimulus.

Loosanoff (1937c) noticed that the reaction time to elevated water temperatures was shorter for male than for female *M. mercenaria.* Of an equal number of male and female clams exposed to a thermostimulus, male spawners outnumbered females 220 to 132 (Ansell et al., 1964a). Castagna and Kraeuter (1981) also reported that moderately sized clams were easier to stimulate to spawn at high temperatures than large clams. Smaller clams have thinner shells and are presumably more sensitive to sudden temperature changes than the larger thicker-shelled clams.

Loosanoff (1937c) was the first to notice that factors in addition to temperature are important in inducing a spawning reaction in *M. mercenaria.* He noticed that the addition of small quantities of sperm and egg suspensions made from the gonadal material of ripe clams enhanced the effectiveness of thermal stimulation. Eggs from *M. mercenaria* will also induce spawning in other bivalves (Galtsoff, 1940). Sperm and eggs can easily be destroyed by pasteurization (i.e., heating to 60°C for 10 min) or freezing for use in controlled fertilizations. Sperm or egg infusions are usually added with an eyedropper to the incurrent siphon of a pumping clam during thermostimulation (Castagna and Kraeuter, 1981).

Serotonin, a molluscan neurotransmitter, injected in the anterior adductor muscle induced spawning in *M. mercenaria* (Gibbons and Castagna, 1984, 1985). Concentrations of 0.02–2.0 mM serotonin elicited a spawning response in male and female clams within 10–15 min (Gibbons and Castagna, 1985). Seven times more male clams than females spawned upon injection with serotonin. Cherrystones (36.4–41.2 mm shell width) responded more frequently to serotonin injections than littleneck (25.4–36.4 mm shell width) clams. Also, anterior muscle injections of serotonin proved to be more effective than either intragonadal injections or water-borne serotonin solutions in stimulating spawning.

Fertilizable eggs can be obtained from *M. mercenaria* by stripping ripe females and dissolving the germinal vesicles (Loosanoff and Davis, 1963). A 0.1 N solution of ammonia hydroxide will break the germinal vesicle of *M. mercenaria* eggs. Eggs exposed to ammonia hydroxide produce fewer larvae than those produced by thermostimulated spawns (Loosanoff and Davis, 1963).

Other substances and procedures induce spawning with various degrees of effectiveness in *M. mercenaria.* For example, phytoplankton (*Pseudoisochrysis paradoxa* and *Thallasiosera pseudonana*) at 2–2.5 million cells/ml induced spawning in *M. mercenaria* at 18–20°C (Breese and Robinson, 1981).

5.5.2 Behavior

M. mercenaria shed eggs and sperm into the surrounding water where fertilization and eventual larval development takes place. Gametes pass through the suprabranchial chamber and out the excurrent siphon (Loosanoff, 1937c). Males usually respond to spawning stimulation first (Loosanoff, 1937c; Ansell et al., 1964a). Sperm is emitted in a white thread-like stream. Macroscopic vision of the sperm stream dissipates about 3–5 cm from the siphon (Carriker, 1961). Eggs are also released in a steady stream, but the eggs disperse more slowly and have a granular appearance. Females occasionally discard clumps of eggs which accumulate near the siphon (Carriker, 1961).

M. mercenaria of both sexes appear relaxed during spawning; valves are open, the siphons are extended and the clams are vigorously pumping. The production of streams of gametes from siphons indicate that cilia are involved in the spawning act and not the adductor muscles (Loosanoff, 1937c). Contractions of the adductor muscle are occasionally observed near the end of spawning when remnants of gametes and feces are voided.

5.5.3 Gametes

5.5.3.1 Sperm

A mature sperm of *M. mercenaria* consists of an elongated conical shaped head, which appears slightly bent, and a long tail. Loosanoff (1937d) reported that the head of sperm is 3.8 μm long and 1.7 μm wide with a tail 30–35 μm in length. The structure of the sperm head consists of an acrosome in the anterior end and an aggregation of mitochondria surrounding a pair of centrioles near the posterior end of the head. The acrosome commonly ruptures when it comes in contact with an egg and a filament can be detected during fertilization (Lin, 1972). The tail is very thinly covered with cytoplasm. Lin (1972) counted nine microtubules in the tail, but admitted that the ultrastructure of *M. mercenaria* sperm needs more study.

Number of sperm in an infusion or spawn can be estimated using a spectrophotometer and the relationship developed by Bricelj (1979). The following relationship is effective up to 2×10^7 sperm/ml or an absorbance of 0.45. The regression equation is:

$$C = (45.4284 \times 10^6)A$$

where C is the concentration of sperm/ml and A the absorbance reading ($A = -\log$ (transmittance/100)) at 640 nm.

Data on the number of sperm released by any male in a single spawn or over a spawning season are lacking. Goodsell and Eversole (1991) presented preliminary observations that the number of sperm released was related to the season and relative condition of the clam. The average number of sperm produced by 91 male *M. mercenaria* thermostimulated to spawn was 7.08×10^9 with a range from 5.45×10^8 to 1.41×10^{10} sperm/spawn.

Gonadal–somatic indices of male and female *M. mercenaria* indicated that reproductive effort in terms of tissue was similar for the sexes (Eversole et al., 1984). Stanley and DeWitt (1983) reported that about 2000 spermatozoa are released for each egg. Assuming an average female clam produces 25 million eggs over a spawning season (Loosanoff and Davis, 1963), then average sperm production would equal at least 50 billion per spawning season.

5.5.3.2 Eggs

Mature eggs are usually spherical but may be somewhat irregular in shape when first discharged due to the pressure of the gonad (Belding, 1931). In water, eggs assume a normal spherical appearance and the surrounding gelatinous membrane increases in width to 2.4–3.2

times the diameter of the egg (Belding, 1931; Loosanoff and Davis, 1963). The large gelatinous membrane distinguishes clam eggs from eggs of other bivalves. This gelatinous membrane imparts some buoyancy (Belding, 1931) and continues to surround the embryo through late blastula and occasionally until the trochophore stage (Loosanoff and Davis, 1950).

Lee and Heffernan (1991) determined the major constituents of eggs isolated from *M. mercenaria* gonads. The dry weight was 51 ng/egg and protein (20 ng/egg), carbohydrate (4 ng/egg) and lipid (7 ng/egg) constituted 40%, 8% and 14% of dry weight, respectively. These relative amounts of protein, carbohydrate and lipid are similar to the relative amounts found in other bivalves (e.g., Bayne et al., 1975). Egg lipids were associated with lipid droplets (1.5 ng/egg), membranes (4.8 ng/egg) and lipoproteins (0.5 ng/egg). The numerous lipid droplets in *M. mercenaria* appear important in the survival and growth of larvae (Gallager and Mann, 1986). Lee and Heffernan (1991) determined that triglycerides constituted 98% of the lipid droplets in the clam eggs. These investigators implied that bivalve eggs with higher concentrations of triglycerides could survive longer after hatching before the initiation to feeding.

Fertilization and stocking protocols routinely require an estimate of the number of eggs in the spawn. Eggs are usually separated from the debris, feces and mucus accompanying spawning with a series of sieves. Collected eggs are redistributed in a known volume, subsampled and counted. Sedgewick–Rafter cells are frequently used in hatcheries to count eggs, whereas Coulter counters have gained acceptance in research laboratories. An explanation of step-wise procedures to collect and count gametes are presented in Castagna and Kraeuter (1981) and Hadley et al. (1997).

M. mercenaria spawn numerous small eggs that develop into planktonic larvae. Realized fecundity or spawned egg estimates are usually determined by counting the number of eggs released during induced spawning trials, whereas gonad mass determinations are used to estimate potential fecundity or reproductive effort (see Section 5.6). Egg counts proved a better estimator of fecundity because of the protracted nature of gametogenesis and spawning in *M. mercenaria*. Consequently, individual female clams should be induced to spawn over a time comparable to the normal spawning interval (Eversole, 1989).

Belding (1912) reported that a female *M. mercenaria* 63.5 mm in shell length produces an average of 2 million eggs, but this was unsubstantiated. Since then a series of experiments have been undertaken to determine the number of eggs produced by *M. mercenaria* (Table 5.3). Clams induced to spawn on 3-day, 7-day and 14 day intervals over 69-day, 98-day and 112-day experimental periods released from 8 to 39.5×10^6 eggs/female with an average of 24.6×10^6 eggs/female (Davis and Chanley, 1956). Estimates of the average fecundity of female clams collected from the waters of Southampton, England (Ansell, 1967), and induced to spawn over comparable times and intervals as Davis and Chanley (1956) were 7.11×10^6 eggs and 9.28×10^6 eggs (Table 5.3). These fecundity estimates were lower than Davis and Chanley (1956) but similar to the fecundities found by Bricelj and Malouf (1980) for larger clams from Great South Bay, New York. Although there are differences among the fecundity studies in Table 5.3, it is clear that *M. mercenaria* is more fecund than originally suggested by Belding (1912). The overall average number of eggs spawned/female was 7.39×10^6 and 45% of individual spawning events exceeded 2 million eggs in these studies.

Table 5.3 demonstrates the considerable variation in fecundity estimates that an individual investigator may encounter with *M. mercenaria*. Davis and Chanley (1956) observed one

TABLE 5.3

Number of laboratory spawning trials (total length/spawning interval in days), number of females spawned, shell length (≃ indicates estimate), mean and range (in parentheses) of spawns per female, eggs per spawn and eggs per female (NR equals not reported)

Location (stock)	Spawning trials (duration/intervals)	Females N	Shell length (mm)	Spawns/female	Eggs/spawn ($\times 10^6$)	Eggs/female ($\times 10^6$)
CT [a]	14 (69 d/3 d)	11	≃76–102	6.6 (3–10)	3.63 (0.3–17.7)	24.09 (17.1–37.4)
CT [a]	15 (98 d/7 d)	13	≃76–102	6.8 (2–11)	3.58 (<0.1–17.8)	24.25 (8.0–39.5)
CT [a]	9 (112 d/14 d)	14	≃76–102	5.4 (3–7)	4.75 (<0.1–24.3)	25.44 (11.2–36.5)
Eng [b]	22 (63 d/3 d)	18	NR	4.7 (1–12)	1.51 (<0.1–12.7)	7.11 (0.4–18.8)
Eng [b]	12 (77 d/7 d)	18	NR	3.1 (1–7)	3.04 (<0.1–23.7)	9.28 (0.6–29.9)
NY [c]	NR (NR/20–30 d)	3	<44	NR	NR	1.61 (0.9–2.4)
NY [c]	NR (NR/20–30 d)	37	44–68	NR	NR	2.76 (0.2–7.9)
NY [c]	NR (NR/20–30 d)	32	68–78	NR	NR	6.34 (0.3–16.8)
NY [c]	NR (NR/20–30 d)	16	>78	NR	NR	6.32 (0.6–16.2)
SC [d] (ARC)	10 (30 d/3 d)	29	52–74	2.6 (1–5)	0.52 (<0.1–3.1)	1.35 (<0.1–3.4)
SC [d] (ARC × SCW)	10 (30 d/3 d)	21	51–65	3.7 (1–6)	0.51 (<0.1–1.5)	1.88 (0.4–3.3)
SC [d] (SCW)	10 (30 d/3 d)	19	42–54	4.3 (1–10)	0.18 (<0.1–1.0)	0.77 (0.1–1.9)

Sources: [a] = Davis and Chanley, 1956; [b] = Ansell, 1967; [c] = Bricelj and Malouf, 1980; [d] = Knaub and Eversole, 1988 (ARC, clams from a Massachusetts hatchery; SCW, South Carolina wildstock; ARC x SCW, a cross between the two).

female release 24.3×10^6 eggs/spawn, whereas another spawning female released less than 0.1×10^6 eggs in a single spawn within the same experiment. Also, average fecundities varied from 8.0×10^6 to 39.5×10^6 eggs/female over a 98-day spawning period. Some of these discrepancies in the fecundity estimates may be explained by the timing of the spawning trials, clam size, clam condition and genetic stock of the clam. The average fecundities per female at the 3-day and 7-day intervals were similar, but increased slightly at the 14-day interval (Davis and Chanley, 1956). Ansell (1967) observed a similar trend when the spawning interval was increased from 3-day to 7-day intervals. Gonad proliferation and oocyte maturation continues during the spawning period (Loosanoff, 1937b). The total number of eggs released by a female was related to the size of the spawns and not the number of times a female spawned (Davis and Chanley, 1956; Ansell, 1967).

Bricelj and Malouf (1980) observed significant differences in total fecundity among three commercial sizes of *M. mercenaria* collected from two sites in Great South Bay, New York.

Chowders (>78 mm shell length) and cherrystones (68–78 mm) released significantly greater number of eggs than littlenecks (44–68 mm), but not more than each other (Table 5.3). For the two study sites, the overall correlation coefficients (r) were 0.50 and 0.39 between shell length and the number of eggs produced, indicating that between 25% and 15% of the variation in fecundity could be explained by clam size. Internal shell volume also explained a similar amount of variation in the total egg production (Davis and Chanley, 1956; Ansell, 1967; Bricelj and Malouf, 1980). The correlation coefficients ranged from 0.26 to 0.66, indicating 7–44% of variation in fecundity was attributed to internal shell volume of the clams used in these three studies.

Bricelj and Malouf (1980) proposed that the large variability in fecundity observed within a size class may be due to differences in the gametogenic condition or possible genetic differences among the individuals. *M. mercenaria* (76–102 mm shell length) collected from Long Island Sound in January released on average 24.25×10^6 eggs/female compared to 9.28×10^6 eggs spawned by females also collected in January from Southampton waters, England (Davis and Chanley, 1956; Ansell, 1967). Long Island Sound clams began gametogenesis in fall, whereas the majority of the gametogenesis activity of the Southampton clams did not occur until the following spring (Loosanoff, 1937b; Ansell et al., 1964b). Thus, the clams collected in January from Southampton water had experienced a shorter conditioning period than those clams from Long Island Sound. Ansell and Loosmore (1963) observed a direct relationship between condition and the percentage of *M. mercenaria* responding to spawning stimuli. In South Carolina, the number of *M. mercenaria* responding to spawning stimuli was greater for those clams collected and conditioned in spring and summer than in fall (Goodsell and Eversole, 1991). We also observed that fall-conditioned females released fewer eggs in a spawn (0.57×10^5) than either the clams conditioned in the spring (2.0×10^6) or in the summer (1.34×10^6). The periods of greater spawning success and fecundity coincided with the natural time of peak condition in South Carolina (Eversole et al., 1980).

Knaub and Eversole (1988) studied reproductive behavior of native South Carolina *M. mercenaria* (SCW), a hatchery stock from Massachusetts (ARC) and a cross between the two (ARC × SCW). The average number of eggs released per female over 10 spawning attempts was similar for ARC (1.35×10^6) and ARC × SCW (1.88×10^6) but different from the average number spawned by SCW clams (0.77×10^6). Similarities also existed between ARC and ARC × SCW clams in the mean number of eggs per spawn (Table 5.3). The ARC clams spawned less frequently but released more eggs per spawn than SCW clams. These observations and the gametogenic cycle evidence discussed in Section 5.4 support the hypothesis of Bricelj and Malouf (1980) that fecundity differences may be partially explained by genetic differences among test clams.

Bricelj and Malouf (1980) determined that spawned eggs of *M. mercenaria* from Great South Bay, New York, ranged from 44 to 100 μm in diameter (without the gelatinous envelope) and averaged 70.9 μm. The mean diameters of eggs spawned by littlenecks (44–68 mm shell length), cherrystones (68–78 mm) and chowders (>78 mm) were similar, but a significant decrease in egg size was observed between spawnings early in the season and spawnings later in the season (Bricelj and Malouf, 1980). The mean diameter of eggs recovered from 10 laboratory spawns of *M. mercenaria* from South Carolina (SCW), a hatchery stock from Massachusetts (ARC) and a cross of the two (ARC × SCW) decreased

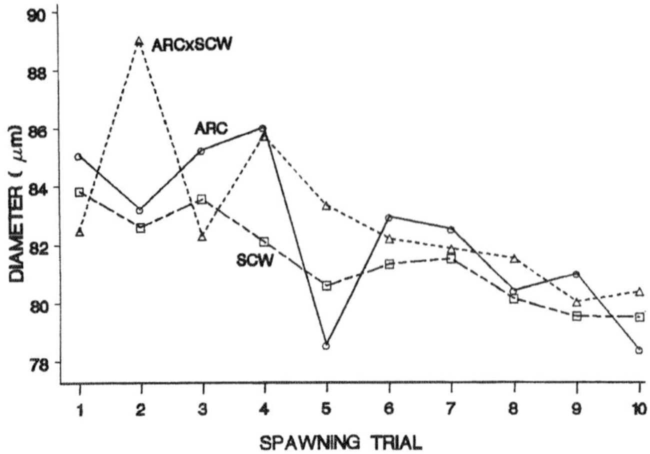

Fig. 5.16. Mean diameter of eggs released by *M. mercenaria* stocks (ARC, a hatchery stock from Massachusetts; SCW, South Carolina wildstock; and ARC × SCW cross) thermally induced to spawn 10 times (Knaub and Eversole, 1988).

significantly in successive spawning trials (Fig. 5.16). Knaub and Eversole (1988) attributed the observed decrease in egg diameter with continued spawning to incomplete gametogenic recovery and the stress associated with repeated spawning inductions. Bricelj and Malouf (1980) also considered elevated ambient water temperatures as an important factor in the observed decreases in egg size over time.

Lannan et al. (1980), working with oysters, observed an optimal conditioning window within the gametogenic cycle for high quality eggs. Gallager and Mann (1986) reported that egg lipid content was high (>18% of ash-free dry weight) in the initial spawns of laboratory-conditioned *M. mercenaria* but decreased in subsequent spawns. They suggested that this was a reflection of the stage of gametogenesis when clams were conditioned. Similar findings were observed when the egg diameters from spring, summer and fall laboratory conditioned and spawned *M. mercenaria*, *M. campechiensis* and *M. campechiensis texana* were compared (Goodsell and Eversole, unpublished data). All three of the *Mercenaria* produced their largest diameter eggs in spring (Fig. 5.17) which corresponds to the natural spawning peak (Dalton and Menzel, 1983; Eversole et al., 1984; Manzi et al., 1985; Heffernan et al., 1989).

Significant variations in egg size have been observed in the different *Mercenaria* taxa and stocks of *M. mercenaria*. Goodsell and Eversole (unpublished data) found that the mean diameter of eggs from a spring laboratory spawn of *M. mercenaria* ($\bar{x}$ = 79.1; 70–87 μm) was significantly smaller than that of eggs from *M. campechiensis* ($\bar{x}$ = 87.5; 70–115 μm) and *M. campechiensis texana* ($\bar{x}$ = 89.3; 83–95 μm). The overall difference in the mean egg diameters from 10 spawning trials of progeny from South Carolina wildstock ($\bar{x}$ = 81.3; 40–113 μm), a selected stock from Massachusetts ($\bar{x}$ = 80.6; 42–107 μm) and a cross of the two stocks of *M. mercenaria* ($\bar{x}$ = 82.1; 40–110 μm) were small but significantly different (Knaub and Eversole, 1988). It is also interesting to note that the distribution of egg diameters reflects taxonomic and stock relationships. For example, the size distributions of eggs released from *M. campechiensis* and *M. campechiensis texana* were similar to each other but different

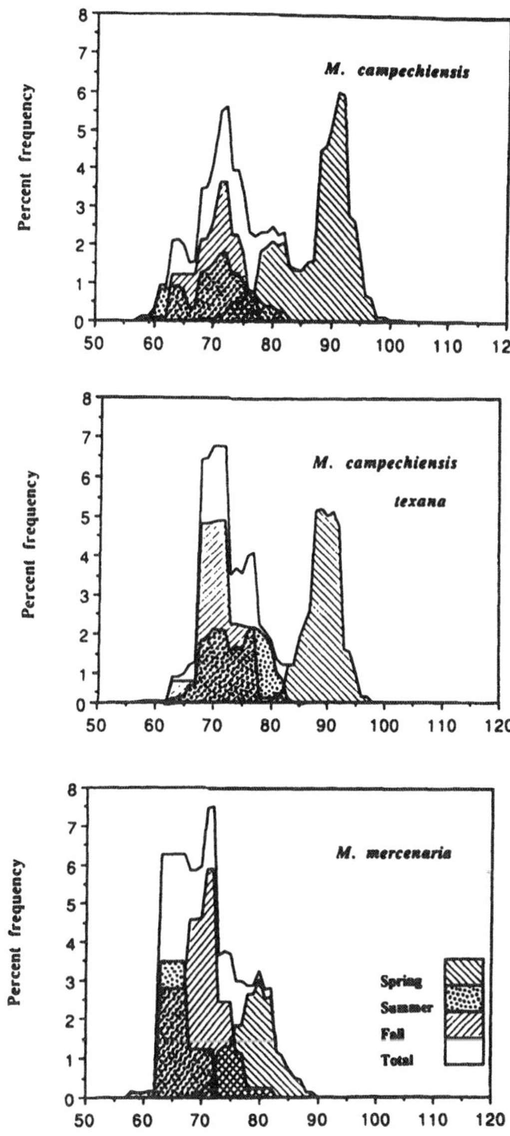

Fig. 5.17. Percentage frequency of diameters of eggs spawned by *M. campechiensis*, *M. campechiensis texana* and *M. mercenaria* in summer, fall and spring (Goodsell and Eversole, unpublished data).

from the distribution of eggs from *M. mercenaria* (Fig. 5.17). Egg diameter distributions from consecutive spawning trials with *M. mercenaria* from Massachusetts (ARC) were distinct from the egg size distributions of the wildstock clams from South Carolina (SCW) and the ARC × SCW cross which were quite similar to each other (Knorr, 1995).

Egg size distributions for *M. mercenaria* are reported to be unimodal (Walker, 1994; Goodsell and Eversole, unpublished data), bimodal (Bricelj and Malouf, 1980) and trimodal

(Gallager and Mann, 1986). Close examination of the egg size distribution from individual clams over several repeated spawns revealed that the number and placement of egg size peaks varied from one spawn to another. For example, one clam that spawned 10 consecutive times over 30 days had egg size distributions with as few as two peaks and as many as 4–5 peaks at any one spawning event (Knorr, 1995). Modal egg diameter for all 10 spawns was 80 μm and ranged from 78 μm to 91 μm in the individual spawning events.

Bricelj and Malouf (1980) suggested that reduced viability or growth rate of larvae may be related to the observed decrease in egg diameter later in the spawning season. When Kraeuter et al. (1982) cultured three sizes of fertilized *M. mercenaria* eggs, the largest eggs survived significantly better than the smallest eggs. The survival of fertilized *M. mercenaria* eggs to pediveliger was highly correlated with egg lipid content (Gallager and Mann, 1986). These investigators also observed that lipid content of eggs varied both with the time of year when conditioning protocol started and the length of the condition period as well as the length of the spawning period. Although the quality of *M. mercenaria* eggs varies significantly with levels of conditioning and stress of continued spawning, no differences in egg quality have been observed with different clam ages (Walker, 1994) or sizes of clams (Loosanoff and Davis, 1963).

5.5.3.3 Storage

Unfertilized gametes deteriorate fairly rapidly after spawning; however, sometimes it is necessary (e.g., controlled crosses) to store gametes. Short-term storage of both clam eggs and sperm is possible, but reduced viability is to be expected (Chanley, 1955). Freshly spawned clam eggs and sperm were challenged several times with gametes stored at 4°C and 22°C over 24 h to determine fertilization rate and survival to straight-hinge larvae (Goodsell and Eversole, 1990; Goodsell, 1991). Fertilization rate of stored eggs remained relatively high over time, but larval survival decreased from about 50% for eggs stored 1 h to 0% after 5 h storage. Sperm viability and larval survival decreased with storage time, and as expected, these decreases were slower at the colder temperature. Fertilization with sperm stored at 4°C for 4–5 h was over 80% and larval survival exceeded 40%, whereas the fertilization and larval survival rates for sperm stored at 22°C for the same time were 50% and <1%, respectively.

In a separate experiment, Bricelj (1979) observed that the concentration of stored sperm at 15°C for 4 h needed to be increased to 3100–5700 sperm : 1 egg to achieve a similar fertilization rate as that observed using an optimum sperm to egg ratio with fresh gametes. Increased sperm to egg ratio and the lower sperm storage temperature improved survival. Goodsell (1991) also reported that increased sperm to egg ratio increased the probability of producing larvae, but warned that high fertilization rates of stored gametes does not always translate into high larval survival rates.

Cryopreservation offers an opportunity to store gametes of *M. mercenaria* for extended periods with negligible losses. Most of the research effort in cryobiological technique development has involved the spermatozoa of oysters (e.g., Yankson and Moyse, 1991). Yankson and Moyse (1991) achieved a 71% fertilization rate and 55% survival rate to straight-hinge with fresh eggs and cryopreserved sperm of *Crassostrea tulipa*. Unfortunately, these techniques have not been perfected for gametes of *M. mercenaria*. Private companies (e.g., Celsys, Cambridge, England; TrochoFeed, Victoria, Canada) have had some success cryopreserving *M. mercenaria* and other bivalve larvae. To date, U.S. government funded research projects to determine the optimal concentrations of cryoprotectants and freezing protocol for *M. mercenaria* larvae have had a limited success.

5.5.4 Fertilization

M. mercenaria release gametes into the water where fertilization takes place. Spermatozoa are attracted to eggs (Belding, 1931). Upon contact, the acrosomal head region of the spermatozoa deteriorates releasing a filament that penetrates the egg (Lin, 1972). Shortly thereafter the egg surface changes which makes the egg refractory to additional spermatozoa. Polyspermy or multiple fertilization can occur in culture conditions when too many spermatozoa are added to an egg suspension. Polyspermy may result in abnormal larvae and poor survival (Bricelj, 1979). Bricelj (1979) reported an optimum ratio of 1800 spermatozoa to 1 egg for successful fertilization under culture conditions.

The membrane around the egg is visible after fertilization and often extra spermatozoa can be seen clinging to the egg membrane. Fertilized eggs develop rapidly; the first polar body should be visible within 15 min. At 27–30°C, cleavage begins at about 30 min and a ciliated gastrula can be seen spinning within the gelatinous envelope in 10 h (Belding, 1931). Larval development is discussed by Carriker in Chapter 3.

It is unknown what percentage of the eggs spawned in nature are fertilized. Coordination of spawning and the synchronous release of gametes increase the chance for successful fertilization. Those factors that time gametogenesis so the population can respond to a stimulus for synchronous release of gametes are discussed in Sections 5.4 and 5.5. To help ensure fertilization, numerous spermatozoa are produced and usually available when eggs are spawned. The optimum gamete ratios for nature are likely much higher than 1800 : 1 sperm to egg ratio reported for aquaculture situations (Bricelj and Malouf, 1980).

5.6 ECOLOGICAL CONSIDERATIONS

An r-strategist bivalve lives in shallow waters and relatively unstable habitats, matures at an early age, displays high fecundities, and experiences high-density-independent mortalities (Mackie, 1984). *M. mercenaria* appears to have selected an r-strategy. For example, *M. mercenaria* attains sexual maturity at a relatively young age (1–2 years) and small size (30–35 mm shell length) (Eversole, 1989). The fecundity of *M. mercenaria* is also quite high (see Section 5.5.3). Malinowski and Whitlatch (1988) reported that the age-specific reproduction contribution peaked early in life and remained relatively high throughout *M. mercenaria* adulthood. The reproductive contribution to future generations by these older clams was realized because of increased fecundity with growth and the absence of reproductive senility (Peterson, 1983, 1986; Malinowski and Whitlatch, 1988; Walker, 1994). These reproductive characteristics fit that of an r-strategist as do some of the other life history traits of *M. mercenaria* (e.g., high mortality rates).

The r-strategist usually allocates a relatively large proportion of available energy to reproduction. The energy remaining after respiration, termed the non-respired assimilated energy, is available for growth of the shell, soft body and gonad, and reproduction of the clam (Fig. 5.18). Energy is partitioned within the clam to organic growth or production, gametes released, gametes absorbed, and secretions and other losses. The relative amount of energy

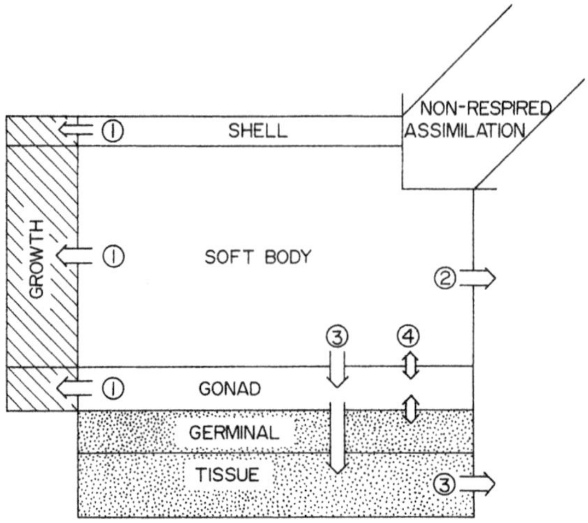

Fig. 5.18. Diagrammatic representation of non-respired energy flow through a clam. Open areas represent relative energy values for shell protein, gonads and soft parts of the clam whereas crosshatched areas represent growth during the year. Energy transfers include: (1) organic growth or production; (2) losses due to secretions, etc.; (3) gametes released; and (4) gametes resorbed (Eversole, 1989).

or mass that a clam devotes to reproduction has been calculated a number of ways: a ratio of the amount of energy or mass in the gonad, or released in the form of gametes in relation to either total annual growth (production) or the size or energy content of a standardized clam. The shell is usually not included in these calculations because a relatively small amount of organic matter is sequestered there (Eversole, 1989). These procedures for estimating reproductive effort in *M. mercenaria* have some weakness and difficulties, including (1) the technical difficulty associated with separating the gonad from the visceral mass, (2) the incomplete and protracted nature of spawning, and (3) that the spawning rate in one year is compared to the standing crop biomass of clams accumulated from several years of growth. Other definitions for bivalve reproductive effort are discussed in Lucas (1982) and Bayne and Newell (1983).

Hibbert (1977) estimated the amount of non-respired energy allocated to reproduction as the difference in soft body tissue weight (kcal) between summer periods with and without spawning. The population of *M. mercenaria* introduced into Hambit Spit, England, allocated 61 kcal m^{-2} $year^{-1}$ to reproduction and 72 kcal m^{-2} $year^{-1}$ to growth (Hibbert, 1977). In this case, reproduction represented 46% of the population's annual non-respired assimilation (133 kcal m^{-2} $year^{-1}$). Using a similar technique, Ansell and Lander (1967) estimated that spawning accounted for 1.0 g and 1.4 g of 3.75 g and 5.40 g of the tissue (wet weight basis) of a 40 mm shell length clam in Southampton waters, England, in 1962 and 1963, respectively. The average loss attributed to spawning was equivalent to 26.6% of the non-respired assimilated energy. These estimates of reproductive costs are conservative because they do not include gonad proliferation during spawning or spawning losses during times when weight is increasing. Although these may be underestimates, it is clear that a substantial percentage of the annualized non-respired assimilation is channeled into reproduction (27–46%). These percentages are probably fair estimates for both male and female *M. mercenaria* since no difference was detected in the relative proportion of gonad mass in male and female clams the same size and age (Eversole et al., 1984).

M. mercenaria allocates non-respired assimilated resources to growth of the shell, somatic mass and gonad mass. The relative contribution of non-respired assimilation can be inferred from gonad mass (Peterson, 1983) and the gonadal–somatic index (Eversole et al., 1984). Peterson (1983, 1986) reported that gonad mass increased significantly with shell length and internal shell volume. The increase of gonad mass was at a greater rate than the increase in body size indicating a greater reallocation to reproductive activities with increasing size (Peterson, 1983). In a separate study, Eversole et al. (1984) observed that increases in gonadal–somatic index accompanied increases in clam age independent of clam size, so that older clams had proportionately larger gonads than younger clams the same size. However, Peterson (1986) reported that clam size (shell length and internal shell volume) was a much better predictor of gonad mass than clam age. No evidence was detected of either absolute (zero) or partial (reduced gonad mass) reproductive senility with increasing size or advancing age (Peterson, 1983, 1986). One *M. mercenaria* aged at 46 years old showed no sign of senility (Peterson, 1986). Apparently, the gonad increases from the negligible amount ($<0.01\%$ of total tissue weight) found in juvenile (19 mm shell length) *M. mercenaria* (Peterson and Fegley, 1986) to a greater proportion of the soft body tissue in older and larger clams until death or until some balance is achieved between growth and reproduction.

As discussed in the above paragraph, *M. mercenaria* has the capacity to shift or reallocate resources between growth and reproduction through ontogeny. Peterson and Fegley (1986) observed a growth anomaly between juvenile and adult *M. mercenaria* in December and January that could not be explained by soma or gonad weights. They suggested that clams during these two months were storing resources in preparation for a burst of gametogenesis in spring. If this is correct, adult *M. mercenaria* reallocated resources between somal growth and reproduction. It is not unusual for bivalves to control metabolism and growth to meet gametogenesis requirements (Bayne, 1976). Eversole et al. (1984) also observed what appeared to be an adjustment between somal and gonadal mass with population density and tidal location. The gonadal–somatic index of clams cultured in a subtidal location and at the lowest density were significantly greater than indices of clams in the intertidal location and at the highest population density (Eversole et al., 1984). Although we know that *M. mercenaria* can reapportion items on the energetic balance sheet for reproduction, it is not clear what regulates the allocation of resources between growth and gametogenesis. It would also be interesting to know what effect reallocation will have on the physiological state of the clam when limited resources (energy) otherwise needed for growth and maintenance are diverted to reproduction.

5.7 SUMMARY

Most of the current information on the reproduction of *Mercenaria* is drawn from field observations and histological examination of collected specimens. In the late 1930s, Dr. Victor L. Loosanoff provided detailed descriptions of spermatogenesis, the juvenile male phase, gametogenesis and spawning in *M. mercenaria*. Since then a number of researchers have continued to study clams in an attempt to understand the gametogenic cycle, energetic partitioning and fecundity of *Mercenaria*. However, much remains to be studied with regard to gametogenesis (e.g., a description of oogenesis and the nutritive function of follicle cells, the role selected factors play in timing the gametogenic cycle, the effect of stressors on gametogenesis, and the specific dietary requirements necessary for gametogenesis), spawning (e.g., spawning cues and the synchrony of events), and reproductive effort (e.g., the development of easier and less time-consuming methods of estimating potential and realized fecundity).

With the increased interest in clam aquaculture, significant developments have been achieved with laboratory studies. Dr. Loosanoff and his colleagues at the National Marine Fisheries Laboratory in Milford, Connecticut, again provided the impetus by spawning *Mercenaria* and other bivalve taxa under controlled conditions. Manuals on techniques for conditioning and spawning (e.g., Castagna and Kraeuter, 1981; Hadley et al., 1997) are proof that hatchery techniques have advanced to the commercial application stage. An increased understanding of the conditioning requirements and gamete handling procedures (e.g., storage and cryopreservation) would enhance hatchery operations. The marriage of basic and applied research has expanded our knowledge and should provide the direction of future clam research.

5.8 ACKNOWLEDGMENTS

I wish to thank Dr. Randal L. Walker, Dr. Francis X. O'Beirn, Dr. Richard S. Knaub, and Ms. Nancy H. Hadley for helpful criticisms in earlier drafts of this chapter. Special thanks go to those graduate students, especially Dr. Richard S. Knaub, Dr. William K. Michener, and Ms. Joy G. Goodsell, who contributed significantly to our research efforts in *Mercenaria* reproduction. Dr. Randy Walker and Dr. Peter Heffernan were kind enough to provide figures for the chapter, and Chris Kempton rendered rough drawings into usable figures. Finally, I appreciate the patience and care of Ms. Jean Richardson in typing many revisions of this chapter. Funds for much of our research were provided by South Carolina Aquaculture Experiment Station and South Carolina Sea Grant Consortium. Technical Contribution No. 4139, published by permission of the Director, South Carolina Agricultural Experiment Station.

REFERENCES

Andrews, J.D., 1979. Pelecypoda: Ostreidae. In: A.C. Giese and J.S. Pearse (Eds.), Reproduction of Marine Invertebrates. Vol. V, Molluscs: Pelecypods and Lesser Classes. Academic Press, New York, pp. 293–341.

Ansell, A.D., 1967. Egg production of *Mercenaria mercenaria*. Limnol. Oceanogr., 12: 172–176.

Ansell, A.D. and Lander, K.F., 1967. Studies on the hard-shell clam, *Venus mercenaria*, in British waters. III. Further observations on the seasonal biochemical cycle and on spawning. J. Appl. Ecol., 4: 425–435.

Ansell, A.D. and Loosmore, F.A., 1963. Preliminary observations on the relationship between growth, spawning and condition in experimental colonies of *Venus mercenaria* L. J. Cons. Int. Explor. Mer, 28: 285–294.

Ansell, A.D., Lander, K.F., Coughlan, J. and Loosmore, F.A., 1964a. Studies on the hard-shell clam, *Venus mercenaria*, in British waters. I. Growth and reproduction in natural and experimental colonies. J. Appl. Ecol., 1: 63–82.

Ansell, A.D., Loosmore, F.A. and Lander, K.F., 1964b. Studies on the hard-shell clam, *Venus mercenaria*, in British waters. II. Seasonal cycle in condition and biochemical composition. J. Appl. Ecol., 1: 83–95.

Barber, B.J., 1996. Effects of gonadal neoplasms on oogenesis in softshell clams, *Mya arenaria*. J. Invertebr. Pathol., 67: 161–168.

Barry, M.M. and Yevich, P.P., 1972. Incidence of gonadal cancer in the quahaug *Mercenaria mercenaria*. Oncology, 26: 87–96.

Bayne, B.L., 1976. Aspects of reproduction in bivalve molluscs. In: M. Wiley (Ed.), Estuarine Processes. Academic Press, New York, pp. 432–448.

Bayne, B.L. and Newell, R.C., 1983. Physiological energetics of marine molluscs. In: A.S.M. Saleuddin and K.M. Wilbur (Eds.), The Mollusca, Vol. 4. Physiology. Part I. Academic Press, New York, pp. 407–515.

Bayne, B.L., Gabbott, P.A. and Widdows, J., 1975. Some effects of stress in the adult on eggs and larvae of *Mytilus edulis*. J. Mar. Biol. Assoc. UK, 55: 675–689.

Belding, D.L., 1912. A report upon the quahaug and oyster fisheries of Massachusetts, including the life history, growth and cultivation of the quahaug (*Venus mercenaria*), and observations on the set of the oyster spat in Wellfleet Bay. Wright and Potter Printing, Boston, MA, 134 pp.

Belding, D.L., 1931. The quahog fishery of Massachusetts. Commonw. Mass. Dep. Conserv., Div. Fish. Game, Mar. Ser. 2, 41 pp.

Bert, T.M., Hesselman, D.M., Arnold, W.S., Moore, W.S., Cruz-Lopez, H. and Marelli, D.C., 1993. High frequency of gonadal neoplasia in a hard clam (*Mercenaria* spp.) hybrid zone. Mar. Biol., 117: 97–104.

Breese, W.P. and Robinson, A., 1981. Razor clams, *Siliqua patula* (Dixon): gonadal development, induced spawning and larval rearing. Aquaculture, 22: 27–33.

Bricelj, V.M., 1979. Fecundity and related aspects of hard clam (*Mercenaria mercenaria*) reproduction in Great South Bay, New York. M.Sc. Thesis, State University of New York at Stoney Brook, Stoney Brook, 95 pp.

Bricelj, V.M. and Malouf, R.E., 1980. Aspects of reproduction of hard clams (*Mercenaria mercenaria*) in Great South Bay, New York. Proc. Natl. Shellfish. Assoc., 70: 216–229.

Buzzi, W.R., 1990. Effects of induced polyploidy on the growth and survival of juvenile hard clams *Mercenaria mercenaria*. M.Sc. Thesis, College of Charleston, South Carolina, 58 pp.

Cain, T.D., 1975. Reproduction and recruitment of the brackish water clam *Rangia cuneata* in the James River, Virginia. Fish. Bull., 73: 412–430.

Calabrese, A., 1972. How some pollutants affect embryos and larvae of American oyster and hard-shell clams. Mar. Fish. Rev., 34: 66–77.

Carriker, M.R., 1961. Interrelation of functional morphology, behavior, and autecology in early stages of the bivalve *Mercenaria mercenaria*. J. Elisha Mitchell Sci. Soc., 77: 168–241.

Castagna, M. and Kraeuter, J.N., 1981. Manual for growing the hard clam *Mercenaria mercenaria*. Va. Inst. Mar. Sci., Spec. Rep. 249, 110 pp.

Chanley, P.E., 1955. Possible causes of growth variations in clam larvae. Proc. Natl. Shellfish. Assoc., 45: 84–94.

Chanley, P. and Andrews, J.D., 1971. Aids for identification of bivalve larvae of Virginia. Malacologia, 11: 45–119.

Coe, W.R., 1943. Sexual differentiation in molluscs, I. Pelecypods. Q. Rev. Biol., 18: 154–164.

Coe, W.R. and Turner Jr., W.H., 1938. Development of the gonads and gametes in the soft-shell clam (*Mya arenaria*). J. Morphol., 62: 91–111.

Crane Jr., J.M., Allen, L.G. and Eisemann, C., 1975. Growth rate, distribution, and population density of the northern quahog *Mercenaria mercenaria* in Long Beach, California. Calif. Fish Game, 61: 68–81.

Dalton, R. and Menzel, W., 1983. Seasonal gonadal development of young laboratory-spawned southern (*Mercenaria campechiensis*) and northern (*Mercenaria mercenaria*) quahogs and their reciprocal hybrids in northern Florida. J. Shellfish Res., 3: 11–17.

Davis, H.C. and Chanley, P.E., 1956. Spawning and egg production of oysters and clams. Biol. Bull., 110: 117–128.

Dillon Jr., R.T., 1992. Minimal hybridization between populations of the hard clams, *Mercenaria mercenaria* and *Mercenaria campechiensis*, co-occurring in South Carolina. Bull. Mar. Sci., 50: 411–416.

Dillon Jr., R.J. and Manzi, J.J., 1989. Genetics and shell morphology in a hybrid zone between the hard clams, *Mercenaria mercenaria* and *M. campechiensis*. Mar. Biol., 100: 217–222.

Eldridge, P.J., Eversole, A.G. and Whetstone, J.M., 1979. Comparative survival and growth rates of hard clams, *Mercenaria mercenaria*, planted in trays subtidally and intertidally at varying densities in a South Carolina estuary. Proc. Natl. Shellfish. Assoc., 69: 30–39.

Eversole, A.G., 1989. Gametogenesis and spawning in North American clam populations: implications for culture. In: J.J. Manzi and M. Castagna (Eds.), Clam Mariculture in North America. Elsevier, Amsterdam, pp. 75–109.

Eversole, A.G., 1997. Gametogenesis of *Mercenaria mercenaria*, *M. campechiensis* and their hybrids. Nautilus, 110: 107–110.

Eversole, A.G. and Heffernan, P.B., 1995. Gonadal neoplasia in northern *Mercenaria mercenaria* (Linnaeus, 1758) and southern *M. campechiensis* (Gmelin, 1791) quahogs and their hybrids in South Carolina. J. Shellfish Res., 14: 33–39.

Eversole, A.G., Michener, W.K. and Eldridge, P.J., 1980. Reproductive cycle of *Mercenaria mercenaria* in a South Carolina estuary. Proc. Natl. Shellfish. Assoc., 70: 20–30.

Eversole, A.G., Michener, W.K. and Eldridge, P.J., 1984. Gonadal condition of hard clams in a South Carolina estuary. Proc. Annu. Conf. Southeast. Assoc. Fish Wildl. Agencies, 38: 495–505.

Eversole, A.G., Kempton, C.J., Hadley, N.H. and Buzzi, W.R., 1996. Comparison of growth, survival and reproductive success of diploid and triploid *Mercenaria mercenaria*. J. Shellfish Res., 15: 689–694.

Fiske, J.D., Curley, J.R. and Lawton, R.P., 1968. A study of the marine resources of the Westport River. Mass. Dep. Nat. Resour., Div. Mar. Fish., Monogr. 7, 52 pp.

Gallager, S.M. and Mann, R., 1986. Growth and survival of larvae of *Mercenaria mercenaria* (L.) and *Crassostrea virginica* (Gmelin) relative to broodstock conditioning and lipid content of eggs. Aquaculture, 56: 105–121.

Galtsoff, P.S., 1940. Physiological reproduction of *Ostrea virginica*. II. Stimulation of spawning in the male oyster. Biol. Bull., 78: 117–135.

Galtsoff, P.S., 1964. The American oyster *Crassostrea virginica* Gmelin. Fish. Bull., 64: 1–480.

Gibbons, M.C. and Blogoslowski, W.J., 1989. Predators, pests, parasites, and diseases. In: J.J. Manzi and M. Castagna (Eds.), Clam Mariculture in North America. Elsevier, Amsterdam, pp. 167–200.

Gibbons, M.C. and Castagna, M., 1984. Serotonin as an inducer of spawning in six bivalves species. Aquaculture, 40: 189–191.

Gibbons, M.C. and Castagna, M., 1985. Responses of the hard clam *Mercenaria mercenaria* (Linné) to induction of spawning by serotonin. J. Shellfish Res., 5: 65–67.

Giese, A.C. and Pearse, J.S. 1974. Introduction: general principles. In: A.C. Giese and J.S. Pearse (Eds.), Reproduction of Marine Invertebrates Vol. I. Academic Press, New York, pp. 1–49.

Goodsell, J.G., 1991. Egg size and shell morphological features of larval and early postlarval *Mercenaria*. M.Sc. Thesis, Clemson University, South Carolina, 61 pp.

Goodsell, J.G. and Eversole, A.G., 1990. Effects of gamete storage on fertilization in *Mercenaria mercenaria* (abstr.). J. Shellfish Res., 9: 432.

Goodsell, J.G. and Eversole, A.G., 1991. Seasonal variation in response to laboratory conditioning in *Mercenaria mercenaria* in South Carolina (abstr.). J. Shellfish Res., 10: 295.

Hadley, N.H., Manzi, J.J., Eversole, A.G., Dillon, R.J., Battey, C. and Peacock, N.M., 1997. A manual for the culture of the hard clam, *Mercenaria* spp., in South Carolina. S.C. Sea Grant Publ., 135 pp.

Harkness, L.L., 1980. Discriminate analysis: the classification of male and female clams. M.Sc. Thesis, Clemson University, South Carolina, 28 pp.

Heffernan, P.B., 1992. Marine Extension Service, University of Georgia, Savannah. Personal communication.

Heffernan, P.B. and Walker, R.L., 1989. Quantitative image analyses methods for use in histological studies of bivalve reproduction. J. Moll. Stud., 55: 135–137.

Heffernan, P.B., Walker, R.L. and Carr, J.L., 1989. Gametogenic cycle of three bivalves in Wassaw Sound, Georgia: I. *Mercenaria mercenaria* (Linnaeus, 1758). J. Shellfish Res., 8: 51–60.

Hesselman, D.M., Blake, N.J. and Peters, E.C., 1988. Gonadal neoplasms in hard shell clams *Mercenaria* spp., from the Indian River, Florida: occurrence, prevalence, and histopathology. J. Invertebr. Pathol., 52: 436–446.

Hesselman, D.M., Barber, B.J. and Blake, N.J., 1989. The reproductive cycle of adult hard clams, *Mercenaria* spp., from the Indian River lagoon, Florida. J. Shellfish Res., 8: 43–49.

Hibbert, C.J., 1977. Energy relations of the bivalve *Mercenaria mercenaria* on an intertidal mudflat. Mar. Biol., 44: 77–84.

Hidu, J., Mason, K.M., Shumway, S.E. and Allen, S.K., 1988. Induced triploidy in *Mercenaria mercenaria* L.: effects on performance in the juveniles (abstr.). J. Shellfish Res., 7: 202.

Kassner, J. and Malouf, R.E., 1982. An evaluation of 'spawners transplants' as a management tool in Long Island's hard clam fishery. J. Shellfish Res., 2: 165–172.

Keck, R.T., Maurer, D. and Lind, H., 1975. A comparative study of the hard clam gonad developmental cycle. Biol. Bull., 148: 243–258.

Kennedy, V.S., 1977. Reproduction in *Mytilus edulis aoteanus* and *Aulacomya maoriana* (Mollusca: Bivalvia) from Taylors Mistake, New Zealand. N. Z. J. Mar. Freshwater Res., 11: 255–267.

Knaub, R.S. and Eversole, A.G., 1988. Reproduction of different stocks of *Mercenaria mercenaria*. J. Shellfish Res., 7: 371–376.

Knorr, K.J., 1995. A statistical analysis of distribution of clam egg size. M.Sc. Thesis, Clemson University, Clemson, SC, 20 pp.

Kraeuter, J.N., Castagna, M. and van Dessel, R., 1982. Egg size and larval survival of *Mercenaria mercenaria* (L.) and *Argopectan irradians* (Lamarck). J. Exp. Mar. Biol. Ecol., 56: 1–8.

Lannan, J.E., Robinson, A. and Breese, W.P., 1980. Broodstock management of *Crassostrea gigas*. II. Broodstock conditioning to maximize larval survival. Aquaculture, 21: 337–345.

Lee, R.F. and Heffernan, P.B., 1991. Lipids and proteins in eggs of eastern oysters (*Crassostrea virginica* (Gmelin, 1791)) and northern quahogs (*Mercenaria mercenaria* (Linnaeus, 1758)). J. Shellfish Res., 10: 203–206.

Lin, C.C., 1972. Electron microscope study of the spermatozoa of the hard-shelled clam *Venus mercenaria* (Mollusca) by negative staining. Trans. Ky. Acad. Sci., 33: 57–63.

Loosanoff, V.L., 1936. Sexual phases in the quohog. Science, 83: 287–288.

Loosanoff, V.L., 1937a. Development of the primary gonad and sexual phases in *Venus mercenaria* Linnaeus. Biol. Bull., 72: 389–405.

Loosanoff, V.L., 1937b. Seasonal gonadal changes of adult clams, *Venus mercenaria* (L.). Biol. Bull., 72: 406–416.

Loosanoff, V.L., 1937c. Spawning of *Venus mercenaria*. Ecology, 18: 506–515.

Loosanoff, V.L., 1937d. Spermatogenesis in the hard-shell clam (*Venus mercenaria* Linnaeus). Yale J. Biol. Med., 9: 437–442.

Loosanoff, V.L., 1962. Gametogenesis and spawning of the European oyster, *Ostrea edulis*, in the waters of Maine. Biol. Bull., 122: 86–94.

Loosanoff, V.L. and Davis, H.C., 1950. Conditioning *V. mercenaria* for spawning in winter and breeding its larvae in the laboratory. Biol. Bull., 98: 60–65.

Loosanoff, V.L. and Davis, H.C., 1963. Rearing of bivalve molluscs. Adv. Mar. Biol., 1: 1–136.

Loosanoff, V.L., Miller, W.S. and Smith, P.B., 1951. Growth and settling of larvae of *Venus mercenaria* in relation to temperature. J. Mar. Res., 10: 59–81.

Lucas, A., 1975. Sex differentiation and juvenile sexuality in bivalves mollusc. Pubbl. Staz. Zool. Napoli, 39 (Suppl.): 532–541.

Lucas, A., 1982. Evaluation of reproductive effort in bivalve molluscs. Malacologia, 22: 183–187.

Mackie, G.L., 1984. Bivalves. In: A.S.M. Saleuddin and K.H. Wilbur (Eds.), The Mollusca, Vol. 7. Reproduction. Academic Press, New York, pp. 351–418.

Malinowski, S. and Whitlatch, R.B., 1988. A theoretical evaluation of shellfish resource management. J. Shellfish Res., 7: 95–100.

Manzi, J.J., Bobo, M.Y. and Burrell Jr., V.G., 1985. Gametogenesis in a population of the hard clam, *Mercenaria mercenaria*, in North Santee Bay, South Carolina. Veliger, 28: 186–194.

Menzel, R.W., 1968. Chromosome number in nine families of marine pelecypod molluscs. Nautilus 82: 45–50, 52–58.

Menzel, R.W. and Menzel, M.V., 1965. Chromosomes of two species of quahog clams and their hybrids. Biol. Bull., 129: 181–188.

Nelson, T.C., 1928. On the distribution of critical temperatures for spawning and for ciliary activity in bivalve molluscs. Science, 67: 220–221.

Otto, S.V., 1973. Hermaphroditism in two species of pelecypod mollusks. Proc. Natl. Shellfish. Assoc., 63: 96–98.

Peters, E.C., Yevich, P.P., Harshbarger, J.C. and Zaroogian, G.E., 1994. Comparative histopathology of gonadal neoplasms in marine bivalve molluscs. Dis. Aquat. Org., 20: 59–76.

Peterson, C.H., 1983. A concept of quantitative reproductive senility: application to the hard clam, *Mercenaria mercenaria* (L.)?. Oecologia, 58: 164–168.

Peterson, C.H., 1986. Quantitative allometry of gamete production by *Mercenaria mercenaria* into old age. Mar. Ecol. Prog. Ser., 29: 93–97.

Peterson, C.H. and Fegley, S.R., 1986. Seasonal allocation of resources to growth of shell, soma, and gonads in *Mercenaria mercenaria*. Biol. Bull., 171: 597–610.

Pline, M.J., 1984. Reproductive cycle and low salinity stress in adult *Mercenaria mercenaria* L. of Wassaw Sound, Georgia. M.Sc. Thesis, Georgia Institute of Technology, Atlanta, 74 pp.

Porter, H., 1964. Seasonal gonadal changes of adult clams. *Mercenaria mercenaria* (L.) in North Carolina. Proc. Natl. Shellfish. Assoc., 55: 35–52.

Porter, R.G., 1974. Reproductive cycle of the soft-shell clam, *Mya arenaria*, at Skagit Bay, Washington. Fish. Bull., 72: 648–656.

Raymont, J.E.G., 1972. Some aspects of pollution in Southampton water. Proc. R. Soc. London, B 180 (1061): 451–468.

Ropes, J., 1987. Age and growth, reproductive cycle, and histochemical tests for heavy metals in hard clams, *Mercenaria mercenaria*, from Raritan Bay 1974–75. Fish. Bull., 85: 653–662.

Sastry, A.N., 1979. Pelecypoda (excluding Ostreidae). In: A.C. Giese and J.S. Pearse (Eds.), Reproduction of Marine Invertebrates. Volume V, Molluscs: Pelecypods and Lesser Classes. Academic Press, New York, pp. 113–292.

Shaw, W.N., 1965. Seasonal gonadal cycle of the male soft-shell clam, *Mya arenaria*, in Maryland. U.S. Fish Wildl. Serv., Spec. Sci. Rep. 508, 5 pp.

Stanley, J.G. and DeWitt, R., 1983. Species profiles: life histories and environmental requirements of coastal fishes and invertebrates (North Atlantic) — hard clam. U.S. Fish Wildl. Serv., FWS/OBS-82/11.18. U.S. Army Corps of Engr., TP EL-82-4, 19 pp.

Stiles, S., Choromanski, J., Nelson, D., Miller, J., Greig, R. and Sennefelder, G., 1991. Early reproductive success of the hard clam (*Mercenaria mercenaria*) from five sites in Long Island Sound. Estuaries, 14: 332–342.

Tranter, D.J., 1958. Reproduction in Australian pearl oyster (Lamellibranchia). II. *Pinctada albina* (Lamarck): gametogenesis. Aust. J. Mar. Freshwater Res., 9: 144–158.

Van Beneden, R.J., Gardner, G.R., Blake, N.J. and Blair, D.G., 1993. Implications for the presence of transforming genes in gonadal tumors in two bivalve species. Cancer Res., 53: 2976–2979.

Walker, R.L., 1994. Reproductive ecology of the northern quahog, *Mercenaria mercenaria* (Linnaeus, 1758) in coastal Georgia and its resource management implications. Ph.D. Dissertation, University of Georgia, Athens, 184 pp.

Walker, R.L. and Heffernan, P.B., 1994. Temporal and spatial effects of tidal exposure on the gametogenic cycle of the northern quahog, *Mercenaria mercenaria* (Linnaeus, 1758), in coastal Georgia. J. Shellfish Res., 13: 479–486.

Yankson, K. and Moyse, J., 1991. Cryopreservation of the spermatozoa of *Crassostrea tulipa* and three other oysters. Aquaculture, 97: 259–267.

Biology of the Hard Clam
J.N. Kraeuter and M. Castagna (Eds.),

Chapter 6

Genetics of Hard Clams, Mercenaria mercenaria

Thomas J. Hilbish

6.1 INTRODUCTION

The term genetics has two principal uses. On one hand molecular genetics describes the structure, organization, expression, and regulation of genetic information in individual organisms while, alternatively, population genetics describes genetic variation among individuals and its organization within and among populations. In general, molecular genetics has been thoroughly developed for only a few model organisms (e.g. *Drosophila melanogaster*) while there is, at best, rudimentary information in the vast majority of other species. For example, in hard clams the karyotype has been described (Menzel and Menzel, 1965; Menzel, 1968) and a few studies have examined oncogene activation by pathological substances (Van Beneden et al., 1993). The only morphological genetic marker that has been described in *Mercenaria* is the 'notata' shell color pattern which may be due to segregation at a single gene (Chanley, 1961), although others have argued that the genetics of this system may be more complex (Newkirk, 1980; Humphrey and Walker, 1982).

The vast majority of information on the genetics of *Mercenaria* is at the population level. In this review I will focus on the population genetics of *Mercenaria* and will do so at four levels: quantitative genetic studies, population genetics of individual gene loci or molecular markers, studies of hybridization among species and genetic studies that attempt to infer the evolutionary history of *Mercenaria*.

6.2 QUANTITATIVE GENETICS

Studies of quantitative genetics examine the nature of phenotypic inheritance and employ statistical approaches to measure the level of phenotypic similarity among individuals with varying degrees of genetic relatedness. There have been two general categories of studies in *Mercenaria*: common garden experiments and sib-analysis.

6.2.1 Common Garden Experiments

Studies of this type obtain juvenile individuals from multiple populations and rear them in a single environment and compare the among-population variation in phenotype to the within-population variation. Common garden experiments assume that differences in phenotype among populations are due to differences in the genetic composition of the source populations. Common garden experiments were originally used in the analysis of phenotype variation among plant populations (e.g. Clausen et al., 1940) and can be a relatively fast and inexpensive method of evaluating genetic differentiation among populations. However, if significant amounts of among-population variation are due to either the maternal or early-juvenile environment these may be mis-interpreted as indications of genetic differences among populations.

A few efforts have been made to assess growth rate variation among populations of hard clams using common garden experiments. Menzel (1961, 1962, 1977, 1989) compared the growth of the progeny from six populations of *M. mercenaria*, collected from Maine to Florida. Adults were spawned, crossed, and their progeny reared in northwest Florida. Considerable growth rate variation appeared to exist among the source populations: progeny from populations in Maine took an average 730 days to reach a size of 50 mm, while progeny from the Florida population required only 570 days to attain the same size. There was a general latitudinal trend of increasing growth rate with a more southerly origin of the source population. These experiments, however, were conducted over several years; parents were spawned and the resulting progeny measured for growth rate during different seasons in separate years. This makes a direct comparison of the results impossible since the effects of the source population may be confused with season- or year-specific differences in growth environment.

Adamkewicz (1988) also explored growth rate variation among populations of *M. mercenaria* using common garden experiments. She constructed nine crosses within and among three populations of hard clams from Massachusetts, Virginia and South Carolina and then measured growth rate among progeny from each cross-transplanted to all three environments. Following transplantation there were significant differences in growth rate due to both the origins of the parents and the location where the clams were grown. The interpretation of these results was complicated by a significant interaction between the origin of the parents and the location where growth was tested. Despite this complication there was generally higher growth among clams with genes originating from southern populations.

The results of Adamkewicz (1988) are consistent with those of Menzel (1977) in that both indicate that southern populations of clams may exhibit greater growth rate when grown in the same environment than clams originating from more northern populations. Adamkewicz's results also illustrate that the magnitude of genetic variation in growth rate may depend upon the environment in which the clams are grown. Because of the confounding effects of interactions between genotype and environment in the study of Adamkewicz and of among-year effects, and spawning and planting time in the studies by Menzel, it is difficult to conclude that there is solid evidence for genetic differences in growth rate among populations of *M. mercenaria*. More common garden experiments conducted in multiple environments are necessary to assess the level of growth rate variation among populations of hard clams.

6.2.2 Sib-analysis

A second category of quantitative genetic approaches are analyses that use the relationship among relatives to determine the degree to which phenotypic variation is under genetic control. The vast majority of quantitative analysis of this type assess the level of genetic variation within populations and some times the role of environmental interactions in altering the expression of within-population genetic variation. It is possible to use the resemblance among relatives to address the level of differentiation among populations (e.g. Mousseau and Roff, 1989), but these types of studies require exceedingly large genetic designs and there has been no attempt to make this type of analysis in any bivalve mollusk.

Two basic approaches are usually taken to measure the association between phenotypic variance and genetic resemblance: sib-analysis and parent–offspring analysis. Sib-analysis determines the degree of phenotypic resemblance among siblings and/or half-siblings relative to unrelated individuals to measure the genetic components of variation in a quantitative character, such as growth rate, disease resistance, or shape. Parent–offspring analysis compares the phenotypic resemblance between parents and their offspring relative to unrelated individuals to make the same assessment. Parent–offspring analysis is impractical in hard clams because it requires that the same phenotype be measured in both the parents and progeny at the same phase of their life history. For species like *Mercenaria* such a study would require many years and accordingly no such efforts have been attempted in hard clams. Hard clams are, however, excellent subjects for sib-analysis. The development of aquaculture technology for *Mercenaria* makes possible large-scale, controlled breeding programs that are necessary to construct large numbers of full-sib families. External fertilization also makes possible the construction of half-sib families in which a male may be crossed to multiple females or vice versa.

The foundation of quantitative genetics is beyond the scope of this chapter. I will provide only a brief overview of quantitative inheritance here. An introductory description of both the theory and empirical basis of quantitative genetics is given in Falconer (1981). Quantitative genetics is a statistical description of the origins of phenotypic variation in a population. Fisher (1930) laid the theoretical foundation of quantitative genetics, by constructing linear models to subdivide phenotype variance into separate components. This allows quantitative genetics to be analyzed using analysis of variance techniques. Phenotypic variation in a population (V_P) can be grossly divided into two components: the genetic variance (V_G) and environmental variance (V_E). The genetic variance can be further subdivided into additive genetic (V_A), dominance genetic variance (V_D) and interaction variance (V_I) components. The environmental variance may also be subdivided to take into account sources of variation that can be identified (e.g. container, field) versus random environmental variation that can not be attributed to a specific source. One of the most important decompositions to be made in the environmental variance is to isolate maternal effects, since in many experimental designs these may be confounded with genetic components of the variance.

The additive genetic variance describes the degree to which individuals in the population can influence the appearance of their progeny through the alleles that they contribute. It is important to remember that in a sexual species parents pass their alleles on to their offspring and not their genotype. The additive genetic variance is a measure of the number of alleles available in a population that can influence the phenotypic value of the population. Accordingly, additive genetic variance is a function of the average phenotypic effect of the alleles in a population and the frequency of those alleles. For a population to have a high additive genetic variance for a trait it must contain alleles at one or more genes that effect the phenotype *and* these alleles must be common. Therefore, additive genetic variance is maximized when all of the alleles influencing a character are of equal abundance. For this reason additive genetic variance is a population-specific parameter and difference in additive genetic variance between populations is more likely due to changes in allele frequencies than to differences in the genes or alleles that contribute to the variance.

Dominance variance describes the departure from strictly additive effects between parents. Offspring is expected to have a mean phenotypic value that is the average of the values of the parents. The degree to which the offspring, on average, look like one parent more than the other contributes to the dominance variance. This, however, is an area where quantitative genetic and Mendelian genetic terminologies create confusion. Additive variance describes the variation in average effect of the parent on the appearance of the offspring while dominance variance describes any statistical departure from linearity in these relationships. Since dominance variance describes the *statistical* departure from strict linearity, it is possible for a phenotype to be controlled by genes that exhibit dominance in the Mendelian sense and yet most of the genetic variance is included in the additive component. Also, as noted above for V_A, the magnitude of V_D is a function of both gene action and the frequency of alleles. Likewise, while interaction variance is created by epistatic interaction among genes, a phenomenon we know to be widespread, this will usually be a very small component of the total genetic variance in a quantitative genetic model.

Additive genetic variance is usually the key genetic parameter that an investigator wishes to measure. In sexual species the genotype is not inherited; only alleles are transmitted from parent to offspring. Since dominance and epistatic effects are manifestations of the genotype, these components of quantitative variation cannot be transmitted from parent to offspring. Only the additive effects of the parent's genes are truly heritable. Therefore, the additive genetic variance describes the level of genetic variation that can be transmitted from one generation to the next. And, as Fisher (1930) demonstrated, it is the additive genetic variance upon which selection, either artificial or natural, can act. Additive genetic variance is therefore of keen interest to animal and plant breeders because it is this component of the total genetic variation that determines whether a trait will respond to artificial selection. Additive genetic variance is also of interest to evolutionary biologists because it was expected that traits closely related to fitness should have minimal additive variance and surprisingly this is not true (Mousseau and Roff, 1987). The maintenance of additive genetic variance for important fitness traits remains a major unresolved problem in evolutionary biology.

The relative magnitude of genetic variation in a character is often expressed as heritability. Narrow sense heritability (h^2) expresses the proportion of the total phenotypic variance in a trait that is attributed to additive genetic effects (V_A/V_P). Broad sense heritability is the proportion of the total phenotypic variance due to the total genetic variance (V_G/V_P). Heritabilities are convenient expressions that scale the genetic variance against the total phenotypic variance but they may also be misleading since two populations may have similar heritabilities and yet exhibit very disparate levels of phenotypic variation. It is possible for two populations, with similar heritability values, to exhibit very different responses to selection in the absolute value of a given trait. It is always preferable that investigators present estimates of both heritability and variance components when reporting the results of quantitative genetic studies.

6.2.2.1 Empirical studies

Hilbish and coworkers estimated the quantitative inheritance of numerous characters related to the growth and development of hard clams. They constructed a series of full- and half-sib families of *M. mercenaria* and progeny from these crosses were used to examine the additive genetic variance of growth and the physiological basis of growth rate variation. These experiments used hard clams from a natural population in Wellfleet Harbor, Massachusetts and employed a classic nesting mating design developed by Comstock and Robinson (1952) to partition variation in the different traits into genetic and non-genetic causes. In this design 31 male clams were mated to an average of 3 separate females, thereby generating 95 full-sib families and 31 half-sib families. One important advantage of hard clams (as well as other bivalve mollusks) is that successful matings generate large numbers of progeny that can be used in multiple experiments.

Rawson and Hilbish (1990) described the partitioning of growth rate variation among clams raised in Charleston, South Carolina at the site where the parents were spawned. They measured size after 9 months of age and specific growth rate over a 3-month period in the fall. Both experiments detected significant additive genetic variation in growth which corresponded to a narrow sense heritability of 0.72 (±0.32 S.E.) to 0.91 (±0.17) for size at 9 months and 0.37 (±0.13) for specific growth rate.

The results of Rawson and Hilbish (1990) indicated that appreciable levels of genetic variation for growth rate exist in natural populations of hard clams. Like most other bivalve species, *Mercenaria* has a life-history that includes a prolonged larval stage in which sibling larvae can potentially disperse across a variety of habitats. Rawson and Hilbish (1991) tested the hypothesis that genetic variation in growth rate may be differentially expressed in different environments. Differential expression of this type is termed genotype–environment interaction. Genotype–environment interaction may take the form of genotypes reversing their relative performance in alternate environments, indicating that there may be trade-offs among characters contributing to growth. Alternatively, genotype–environment interaction may result from variation among genotypes in some environments but not in others.

Rawson and Hilbish (1991) measured the level of genotype–environment interaction by transplanting juvenile clams from each of the full-sib families described above to five locations where specific growth rate was measured. Growth rate variation could be partitioned among six causal components in this design: location variance, additive genetic variance, non-additive genetic variance, interactions between location and the additive genetic variance and the non-additive genetic variance and environmental or error variance. Genotype by environment interaction accounted for over 37% of the non-error variance. Location and error were the two largest sources of variance. There was no evidence of a significant genetic variance (either additive or non-additive) across environments in this study (Table 6.1). Rawson and Hilbish (1991) found that genotype–environment interaction was at least partially due to a change in the amount of genetic variation expressed at each location, with significant additive genetic variation detected at Charleston and Georgetown, South Carolina and significant non-additive genetic variation expressed in Millsboro, Delaware. Reversals in relative family performance across locations were prevalent and there was no indication of a ubiquitously superior genotype.

These results are supported by those of Peterson and Beal (1989) who also found evidence of differential growth among clams sampled from different locations within an estuary in North Carolina. Growth differences persisted for at least two years suggesting that these responses may be genetic in origin. Adamkewicz (1988) also found that growth variation among clams from different source populations depended upon the environment in which they were reared.

These studies have several important implications. First, it is clear that significant genetic variation may be observed at any given location but this does not indicate the level of

TABLE 6.1

Sources of variation in specific growth rate of *Mercenaria mercenaria* tested in multiple environments (from Rawson and Hilbish, 1991)

Source of variation	Total variance (%)
Location	17.2
Additive genetic	−2.1
Non-additive genetic	4.7
Gentotype–environment interaction	12.6
Environmental (error)	67.6

Additive and non-additive genotype–environment interaction components have been pooled into a single term. Additive and non-additive genetic variances were significant at specific locations but when averaged across all locations neither genetic component was significant. Both additive and non-additive interactions between genotype and environment were highly significant sources of variance.

heritability for the same trait at other localities. Second, the absence of a generally superior genotype also suggests that it may be difficult for either natural or artificial selection to generate a high growth rate strain of clams that maintain their advantage in multiple environments. Third, these results suggest that larval dispersal may have an important role in determining the level of genetic variation present within natural populations. Dispersal has the effect of connecting different locations where genetic variation will be differentially expressed. When averaged across locations, Rawson and Hilbish (1991) found no evidence of additive genetic variation for growth. This result indicates that selection at the large scale may have already eroded genetic variation of this trait and that dispersal prevents effective selection at the local scale. Finally, Via and Lande (1985) have pointed out that genotype–environment interaction is the source of variation for the evolution of phenotypic plasticity. The presence of a high level of genotype–environment interaction in *M. mercenaria* suggests that this species retains ample variation for the evolution of growth plasticity.

Hilbish et al. (1993) examined growth rate variation in *M. mercenaria* during different periods of larval and juvenile development. They showed that there was significant additive genetic variation for larval hard clams at prodissoconch I, and 2 days and 10 days of age (Table 6.2). As described above, hard clams from these same families exhibited significant heritability for shell length at 9 months of age. Hilbish et al. (1993) then measured the genetic covariation for shell length between successive stages of development. They found that shell length at prodissoconch I and 2 days of age were genetically correlated, which is not surprising since the prodissoconch makes up the majority of the shell of a 2-days-old larva. There was, however, no evidence of significant genetic covariation among other stages of development (Table 6.2). Shell length of 2-days-old larvae did not genetically covary with shell length of 10-days-old larvae, which, in turn, did not covary with the size of 9-months-old juveniles.

Genetic covariance measures the degree to which two characters are related at the genetic level. Covariances of zero indicate that two traits are influenced by separate sets of genes. The study of Hilbish et al. (1993) therefore indicates that shell growth during one stage of development is not strongly related to the genetic variation in growth during other ontogenetic

TABLE 6.2

Mean shell length and estimated components of variance ($\times 10^{-3}$) and narrow sense heritability of shell length for prodissoconch I, 2-days-old, 10-days-old and 9-months-old *Mercenaria mercenaria* (from Hilbish et al., 1993)

	Prodissoconch I	2 days	10 days	9 months
Mean shell length (μm)	90	106.9	165.8	2950
V_A	2.16^{**}	3.04^{**}	28.2^{*}	118.4^{**}
V_{NA}	−0.25	−1.15	16.1	8.6
V_E	1.83	0.93	−9.7	12.1
h^2	0.58 (0.16)	1.08 (0.29)	0.82 (0.28)	0.85 (0.22)
COV_A	1.78^{*}	1.88	3.56	
r_g	0.74 (0.12)	0.20 (0.18)	0.06 (0.21)	

The subscripts A, NA, and E indicate components of the total phenotypic variance due to additive and non-additive genetic variance and environmental variance, respectively. h^2 indicates the narrow sense heritability for each trait ± one standard error given in parentheses. Genetic covariances (COV_A) and genetic correlations (r_g) are given for each sequential set of characters (e.g. shell length at prodissoconch I vs. 2-days-old larvae). Significance of COV_A is given by an asterix and standard errors of r_g are given in parentheses.
$^{*}P < 0.05$; $^{**}P < 0.01$.

periods. Hilbish et al. (1993) suggested that the genetic uncoupling of growth variation during development might be due to metamorphic events that change the nature of the feeding apparatus. They also suggested that the high level of genetic variation for larval growth indicated that larvae may not be under as strong selection for maximizing growth as had been previously supposed.

Hilbish et al. (1993) expected to observe substantial non-additive genetic variation in the size of prodissoconch I larvae since egg size (a maternal character) and the size of prodissoconch I are known to be positively correlated among bivalve species (Ockelmann, 1965; Jablonski and Lutz, 1980, 1983). This was not the case, however; estimates of non-additive variance for the size of prodissoconch I did not differ from zero. Therefore, correlations between egg and prodissoconch I that are valid for individuals among populations may not hold for individuals within populations.

6.2.2.2 *Cautionary notes*

The studies of growth rate variation in *M. mercenaria* also provide several cautionary tales for future studies of quantitative genetics in bivalve molluscs. First is the consideration of sample size. At first glance the crosses constructed by Hilbish and coworkers for the study of quantitative variation in growth rate above seem to contain large sample sizes. This impression, however, is illusionary. As originally constructed there were 95 full-sib families and 31 half-sib families. Usually 10 or more progenies were measured from each full-sib family. For example, 1728 juvenile clams were measured to obtain the estimate of additive genetic variance for specific growth rate in the study by Rawson and Hilbish (1991). It is important to note, however, that additive genetic variance is the variance in breeding values for a given trait. In experiments like these, the breeding value of each male is twice the deviation of the phenotypic value of his progeny from the average phenotypic value of all the sires (Falconer, 1981). The number in progeny measured per full-sib family and the number of females mated to each male only influence the *precision* with which the breeding value for each male is measured. Since, at most, there were only 31 males used in the experiment, there were only 31 breeding values used for estimating the additive genetic variance. It is inevitable that any population parameter estimated with a sample size of only 31 will have a large sampling variance. It is indeed the case that sampling variances around quantitative genetic parameters are notoriously large. Large experimental designs are necessary for accurately estimating quantitative genetic parameters.

One consequence of the necessity for large experimental designs is that the logistics for constructing all of the necessary crosses can become very cumbersome. It required 8 weeks to construct all of the crosses used in the studies by Hilbish and coworkers (Rawson and Hilbish, 1991; Hilbish et al., 1993). It is possible that culture conditions may change during this time, and this may influence the phenotype being measured. All of the crosses comprising the full-sib families sired by an individual male are likely to be made at one time (when the male spawns). If there is a temporal effect on the phenotype of the progeny then half-sib progeny share more than a common father, they also share a common environment. In the studies on *Mercenaria* the data sets were examined to determine whether there was a significant effect of the 'date-of-birth' on each trait being analyzed and there was no evidence of such an effect. Hilbish and coworkers also re-spawned a few males so that it was possible to compare the breeding value of a given male measured at different times during the construction of the families. Again, there was no evidence of a temporal effect on the breeding value. It is important to take such precautions and to attempt to insure a constant rearing environment; otherwise temporal effects can be misinterpreted as additive genetic effects on the phenotype.

Container effects can also be a significant problem in quantitative genetic experiments using bivalves. By necessity, full-sibling larvae must be reared together in a common container. If there is a significant effect of the larval or nursery rearing containers on the trait then in a full-sib/half-sib design this will inflate the estimate of the non-additive genetic variation. Rawson and Hilbish (1991) observed a large non-additive effect on shell-length at 9 months of age. This was attributed to culture density. In their experiment these effects could be statistically removed. If it is possible to split each full-sib family into separate containers for larval and nursery culture then it is possible to factor out potential container effects. Splitting the brood, however, has severe consequences on the logistics of the experiment. Even rearing two cultures of each full-sib family necessitates doubling the culture facilities, personnel and the number of animals measured. If these resources were available it is probably better to expand the number of half-sib families (breeding values) than it is to increase the number of replicates of each family.

The development of new statistical approaches for analyzing quantitative genetic designs using maximum-likelihood approaches (Shaw, 1987) may alleviate some of the difficulties described above. Many of the logistical difficulties encountered were due to the attempt to maintain a roughly balanced design where three females were mated to each male. If clams spawned but did not maintain this ratio then many potential parents were wasted. Maximum-likelihood techniques do not require balanced designs and can incorporate any number of genetic relationships among relatives which can reduce the wastage of parents and the amount of time necessary to complete crosses. For example, if during a given attempt four females and two males spawn, this would have produced only a single half-sib family using the design described above. Instead, all possible pair-wise matings could be made producing eight full-sib families and six half-sib families, two of which share sires and four of which share dams. Maximum-likelihood analysis can incorporate many such 'mini-partial diallele' crosses to estimate the quantitative genetic parameters of interest. The advantages of maximum-likelihood techniques should be seriously considered in future quantitative genetic studies of *Mercenaria* or other bivalves.

6.2.3 Selection Studies

An alternative approach to sib-analysis for assessing genetic variation in a trait is to determine the response of the phenotype to selection. The rate at which a population responds to a known selective pressure is directly dependent upon the heritability of the trait. Selective breeding, however, only provides an estimate of heritable variation for the trait under selection in the environment where the experiment was conducted. Selective breeding is also very difficult in species with long maturation periods like *Mercenaria*. Nonetheless, Hadley et al. (1991) successfully completed two generations of selective breeding in *M. mercenaria* and reported a realized heritability of 0.4 for growth rate to market size. This result is consistent with the estimates of heritability made by Rawson and Hilbish (1990) for full- and half-sib families of hard clams grown in the same habitat.

6.2.4 Summary

Quantitative genetic analysis of growth rate variation in *M. mercenaria* indicates that there are high levels of genetic variation resident in natural populations. Some reports (Hadley et al., 1991) indicate that it may be possible to exploit this variation to improve growth rate in selected strains. However, the lack of genetic covariation of growth rate across environments suggests that it may be difficult to establish selectively bred strains that have ubiquitously enhanced growth rate. Genotype–environment interactions coupled with larval dispersal may act to maintain the high levels of genetic variation for growth rate observed in these studies.

6.3 POPULATION GENETICS

The development of methods for the electrophoretic separation and staining of specific enzyme-coding genes (allozymes) in the mid-1960s (Lewontin, 1974) fueled the analysis of organization of genetic information within and among populations of a huge variety of organisms. Allozyme loci have many important advantages as genetic systems for study: they usually have Mendelian inheritance, they are co-dominant so that frequencies of homozygotes and heterozygotes may be directly determined, many loci can be resolved using standard techniques that are often applicable across a wide diversity of taxa, and they are relatively inexpensive and fast which allows large numbers of animals to be assayed. Despite the recent proliferation of molecular approaches to population analysis, and some evidence that allozymes may yield different answers than other techniques (Karl and Avise, 1992; Pogson et al., 1995), allozyme approaches remain a mainstay in the arsenal of techniques for resolving genetic variation and population structure. A conflict between the results of allozyme and alternative approaches is not an indictment of either technique but rather a strong indication that different portions of the genome respond differently to those evolutionary forces that structure populations. Indeed, such conflicts can provide new insight into the evolutionary dynamics of a species.

6.3.1 Population Structure

As with other taxa, allozyme approaches have been used to assess the level of population differentiation among populations of *Mercenaria*. *M. mercenaria*, in particular, has an exceptionally large geographic range, from Nova Scotia, Canada to Florida, that includes at least three major geographic provinces. The basic question is whether the geographic range of *M. mercenaria* is subdivided into genetically distinct populations. Planktotrophic larval dispersal argues for large-distance gene flow, which should promote panmixia, while the distribution of populations across major zoogeographic provinces with different thermal environments should promote local adaptation and genetic divergence.

Populations of *M. mercenaria* from Massachusetts to Florida are not strongly differentiated suggesting that they may be interconnected by gene flow. Humphrey (1981) and Humphrey and Crenshaw (1989) assayed eleven gene loci in populations of *M. mercenaria* and found minimal divergence among populations (Fig. 6.1). Nei's genetic distance (calculated pair-wise among locations) ranged from 0.005 to 0.02 which indicates little to no genetic differentiation. These results are supported by studies of allozyme variation by Pesch (1972, 1974) and by Dillon and Manzi (1987, 1992). These results also point out a distinction between ecological

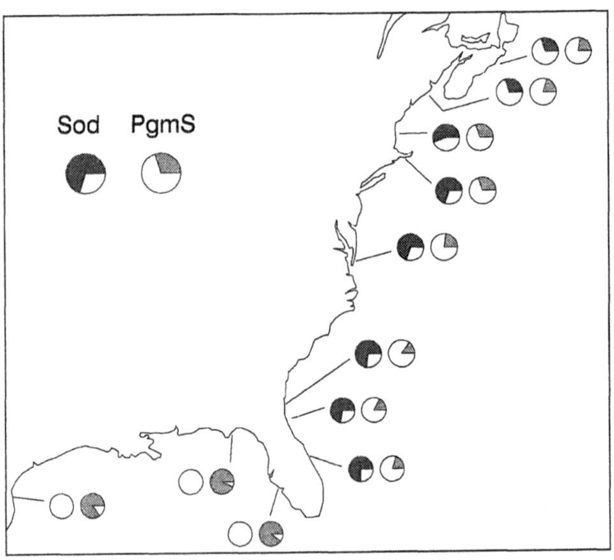

Fig. 6.1. Allele frequencies at the superoxide dismutase (Sod) and phosphoglucomutase-slow (PgmS) loci in *M. mercenaria* (Atlantic coast), *M. campechiensis* (eastern Gulf of Mexico) and *M. texana* (western Gulf of Mexico). The solid portion of the pie chart for Sod and the open portion for PgmS indicate the frequency of the most common allele in *M. mercenaria*, while the remainder of each pie chart indicates the frequency of all other alleles combined. Allele frequency data are compiled from Humphrey and Crenshaw (1989) and Dillon and Manzi (1989a,b, 1992). PgmS corresponds to Pgm-slow in Dillon and Manzi (1989a,b, 1992) and Pgm-2 in Humphrey and Crenshaw (1989). Sod corresponds to the tetrazolium oxidase (TO) locus of Humphrey and Crenshaw (1989).

and genetic populations. Assemblages of clams found at different locations appear to be strongly linked by gene flow and accordingly the genetic population is likely to be very large in both population size and geographic area. Nonetheless, collections of clams at different locations are likely to be effected by different ecological forces that influence their local abundance, demography and reproductive success (e.g. Peterson and Beal, 1989). Therefore, when considered at the ecological level clam populations are probably much more localized than they are when considered from a genetic perspective. In my view this illustrates the difficulty in using genetic differentiation as an exclusive criterion for defining populations or stocks for management purposes. A genetically homogeneous region may contain many ecologically independent populations of clams.

North of Massachusetts populations of *M. mercenaria* become increasingly fragmented. Pesch (1972, 1974) found moderate evidence of genetic differentiation between populations north and south of Massachusetts. Dillon and Manzi (1992) studied differentiation in this region in greater detail. They found that clam populations in Maine, New Brunswick and Nova Scotia were differentiated from those from Massachusetts at every gene locus examined (seven loci all together). Differentiation was particularly marked at Mpi and Sod. Dillon and Manzi (1992) also found that northern populations had significantly lower heterozygosity than southern populations, primarily due to the loss of rare alleles. They argued that northern populations may be more frequently subject to population bottlenecks than populations to the south. It is also possible that northern populations are more frequently extinguished and then recolonized which also accelerates genetic drift.

The population structure of *M. campechiensis* has been studied to a much lesser degree than that of *M. mercenaria* but the results appear to be similar. Populations on the Atlantic coast and in the Gulf of Mexico have similar allozyme frequencies (Pesch, 1972, 1974; Humphrey, 1981; Dillon and Manzi, 1989a,b). These results have been confirmed in a more recent and larger geographic survey of genetic variation in *Mercenaria* (T.M. Bert, pers. commun.).

The absence of significant population structure, inferred from allozyme analysis, across the majority of the range of *M. mercenaria* has been confirmed by Brown and Wolfinbarger (1989) who used restriction analysis to examine geographic patterns of mitochondrial DNA (mtDNA) variation. Even though mtDNA techniques often provide greater resolution of genetic structure than allozyme methods (Avise, 1994), Brown and Wolfinbarger (1989) found genetic homogeneity in *M. mercenaria* populations from New Jersey to North Carolina.

6.3.2 Variation within Populations

In *Mercenaria*, Adamkewicz et al. (1984a) have demonstrated Mendelian inheritance for phosphoglucose dehydrogenase (Pgd), phosphoglucomutase (Pgm), leucine aminopeptidase (Lap) and phosphoglucoisomerase (Pgi). They also showed that these loci were not significantly linked. Dillon (1985) compared electrophoretic techniques for different allozyme loci in *Mercenaria* and a gastropod.

Intrapopulation variation at allozyme loci in *Mercenaria* is well within the range observed for most outcrossing species (Nevo, 1978). Humphrey (1981) estimated the probability that an individual will be heterozygous at a randomly chosen allozyme locus is 0.13 to 0.15.

Many populations of marine bivalves exhibit a deficiency of heterozygotes at some loci relative to the expectations of Hardy–Weinberg equilibrium (Zouros and Foltz, 1984;

Gaffney et al., 1990). Reports for hard clams vary. Humphrey (1981) and Humphrey and Crenshaw (1989) report that four of the seven loci they studied exhibited large and significant deficiencies of heterozygotes. Slattery et al. (1993) also found a deficiency of heterozygotes in a population of *M. mercenaria* sampled in New Jersey. Dillon and Manzi (1992), however, report that the eleven gene loci they examined exhibited genotype frequencies that were in close agreement with Hardy–Weinberg equilibrium. There are many possible explanations for heterozygote deficiency in natural populations. One of these is the mixing of zygotes produced in populations that differ in allele frequency (the Wahlund effect). This possibility can be excluded for hard clams. There is insufficient allele frequency variation among populations of *M. mercenaria* throughout their geographic range to explain heterozygote deficiencies of the magnitude observed by Humphrey (1981). It is also unlikely that the mixing of larvae between *M. mercenaria* and *M. campechiensis* can account for observed heterozygote deficiencies since some of the observed departures occur at loci that are not differentiated between the two species. Slattery et al. (1993) were also able to exclude inbreeding as an explanation of heterozygote deficiency in *M. mercenaria.*

Adamkewicz et al. (1984a) report significant departures from Hardy–Weinberg equilibrium at three of five allozyme loci examined in nursery cultures of *M. mercenaria.* Since Wahlund effects cannot occur within controlled matings, this explanation is again excluded and Adamkewicz et al. (1984a) favor genotype-specific selection as an explanation for the observed departure of genotype frequencies from random expectation. Presently there is no satisfactory explanation for the presence of heterozygote deficiencies in natural populations of bivalves and given the conflicting reports between Humphrey (1981) and Dillon and Manzi (1992) it is not clear whether there is a deficiency to be explained in *Mercenaria.*

Pesch (1972, 1974) reports that Ldh exhibited an excess of heterozygotes in all populations of *Mercenaria.* Populations north of Massachusetts had higher heterozygosity, and Canadian populations contained only heterozygotes. This result conflicts with those of Dillon and Manzi (1992) who found lower heterozygosity among northern populations of *M. mercenaria.* The results of Pesch for Ldh should be considered with some skepticism, however, since it is extremely unlikely that any population of clams will contain only heterozygotes. It is much more likely that there are two isomers of Ldh that are either encoded by two genes or by a single gene with tissue- or environment-dependent interconversion. If expression of these isomers changes with environment then the frequency of two bands staining for LDH activity may increase in more northern populations until the two isomers are continuously expressed.

6.3.3 Allozyme Effects on Phenotypic Variation

A positive association between individual multi-locus heterozygosity at allozyme loci and growth or size has been reported for many species and populations of marine bivalves (Mitton and Grant, 1984; Zouros and Foltz, 1987). Attempts to make such an association in *Mercenaria* have used hatchery-reared clams produced from mass-spawns (Adamkewicz et al., 1984b), crosses within and between hatchery selected strains (Dillon and Manzi, 1988), and natural populations (Slattery et al., 1993). In none of these experiments was a relationship between heterozygosity and growth or size detected.

Adamkewicz et al. (1984b) and Dillon and Manzi (1988) both detected associations between specific genotypes and growth among the progeny of hatchery crosses. Adamkewicz et al. (1984b) concluded that specific genes or linked loci effected growth rate while Dillon and Manzi (1988) argued that their results were more consistent with individual gene loci that mark the parental origin of large segments of the genome which influence growth. Gaffney and Scott (1984), however, have argued that hatchery crosses often contain high levels of linkage disequilibrium among genes. This may obscure true associations between genotype at specific loci and the phenotype.

Koehn et al. (1980) investigated the effects of genotype at the Lap locus on aminopeptidase activity. They found genotype-specific enzyme activity in oceanic populations of *M. mercenaria* and *Mytilus edulis*. Overall variation in activity decreased in lower-salinity habitats in Long Island Sound and differences among genotypes disappeared in both species. They argue that parallel variation of enzyme activity among genotypes in different species suggests that the observed diversity of enzymatic phenotypes is maintained by natural selection.

6.3.4 Summary

In most respects the population genetics of hard clams is unremarkable. *M. mercenaria* appears to be homogeneous over the majority of its geographic range. In the northern portion of its range populations show evidence of both differentiation and loss of genetic variation suggesting that populations in this region may be subject to drift. It is possible that northern populations were initially established with relatively small numbers of colonists or are subject to an ongoing process of extinction and recolonization. Studies differ on whether they report a heterozygote deficiency within populations of *M. mercenaria* and there is no evidence of a relationship between heterozygosity and growth or size. The population genetics of *M. campechiensis* has not been as well studied, but all existing evidence suggests that this species is also homogeneous throughout its geographic range.

6.4 HYBRIDIZATION

Hybridization between taxa is most often studied to resolve questions of taxonomic status. However, hybridization is also an important evolutionary process that potentially acts as a source of genetic variation within species and often leads to complex interactions between gene flow and natural selection (Barton and Hewitt, 1985; Harrison, 1993). Hybridization also can offer insight into the genetics of the variation between species. Hybridization between differentiated taxa is widespread and therefore is a significant phenomenon in its own right. In hard clams hybridization occurs in two contexts: infrequent hybridization that occurs throughout the overlapping ranges of two species and frequent hybridization that occurs in the Indian River, Florida.

6.4.1 Infrequent Hybridization

M. mercenaria and *M. campechiensis* hybridize in culture and produce fully viable and fertile offspring. F_1 and F_2 hybrids are also interfertile and may be successfully backcrossed (Loosanoff, 1954; Chestnut et al., 1956; Haven and Andrews, 1956; Menzel, 1962, 1977, 1989; Menzel and Menzel, 1965). In nature the ranges of the two species overlap broadly but are rarely truly sympatric (Menzel, 1989). In the Atlantic, *M. campechiensis* occurs offshore as far north as Cape May, New Jersey, while *M. mercenaria* occurs in more shallow-water habitats from Florida to Canada. Along this parapatric distribution hybridization occurs infrequently. Dillon (1992) estimated in one estuarine site that about 0.1% of the clams were *M. campechiensis* and an even smaller proportion (≈0.06%) were hybrids; over 99% of the clams were *M. mercenaria*. Levels of hybridization in offshore environments are not known but Brown and Wolfinbarger (1989) found evidence of hybridization in Oregon Inlet, North Carolina; mtDNA haplotype patterns in this region indicated a mixture of the two species and the high level of heteroplasmy suggests the possibility of hybridization. Despite evidence of a relatively high level of sympatry in Oregon Inlet the results of Brown and Wolfinbarger (1989) are generally consistent with those of Dillon (1992); the two species typically do not co-occur and natural hybridization is rare.

6.4.2 The Indian River

An important exception to this conclusion are clam populations that occupy the Indian River lagoon system on the east coast of central Florida. In this area *M. mercenaria* and *M. campechiensis* are sympatric and hybridization occurs frequently (Dillon and Manzi, 1989a; Bert and Arnold, 1995). The frequency of hybrids varies between 20% and 50% in clam populations in the Indian River system. The level of hybridization depends upon habitat (Arnold et al., 1996) and increases northward in the lagoon. Arnold et al. (1996) demonstrated that hybrid and parental genotypes varied in growth rate and that variation among genotypes depended upon habitat. *M. campechiensis* occurred in three out of six regions of the lagoon and had higher growth rates than *M. mercenaria* and hybrid genotypes. When *M. campechiensis* was absent, hybrids grew faster than *M. mercenaria* in shallow-water, vegetated habitats and the reverse occurred in deeper-water areas that were devoid of vegetation. Bert et al. (1993) also found that hybrid clams are more likely to suffer from gonadal neoplasia than are parental clams. Since this condition often results in sterility it is yet another selective force determining the characteristics of the hard clam hybrid zone in the Indian River.

Bert and Arnold (1995) made a detailed examination of the population genetics of clam populations in the Indian River to determine the strength and form of selection among hybrid and parental genotypes. The purpose of this study was to test two competing hypotheses for the maintenance of hybrid zones that differ in whether selection is exogenous (i.e. environmental) or endogenous (i.e. genetic background). They found that selection occurred among genotypes in the Indian River clam populations, that it was strongest in the northern regions of the lagoon, and that hybrids were generally less fit than either parental type but that selection also affected some portions of the genome differently than others. In particular, while selection was generally against hybrid genotypes, alleles derived from *M. mercenaria* at the linkage group marked by the amino acid transferase (Aat) locus appeared to be selectively favored.

The results of the analysis by Bert et al. (1993), Bert and Arnold (1995) and Arnold et al. (1996) indicate that the Indian River hybrid zone is influenced by a complex array of both endogenous and exogenous selective mechanisms. The general inferiority of hybrid genotypes is consistent with the endogenous selection model, but the dependence of selection upon the environment, habitat-dependent growth variation among genotypes, and evidence that some linkage groups may be differentially subject to selection, all support the exogenous selection model.

6.4.3 Summary

Studies of hybridization in hard clams suggest that divergence in habitat affinity is an important mechanism of reproductive isolation in this group. In open coastal regions *M. mercenaria* and *M. campechiensis* reside in physically separated habitats and seldom interbreed. The Indian River is, however, an enclosed system that contains a vast array of habitat types. Within this system the two species dominate in different habitats but sufficient intermediate habitats occur and the two species are in close enough proximity that interbreeding does occur and hybrids are found at high frequency. Much remains to be determined about the circumstances that promote hybridization within this system and maintains reproductive isolation of these species outside of the Indian River.

6.5 EVOLUTIONARY GENETICS

Genetic methods can also be used to infer the evolutionary history of individual species or groups of species. Low levels of genetic variation within populations often suggest a history of small population size, and a high degree of endemism of alleles among populations suggests limited migration among populations. The magnitude of genetic divergence among populations, subspecies and species has also been used to estimate the time since these units shared a common ancestor. More recently, population samples of DNA sequences have been used to infer the phylogenetic history of divergence of populations within species and among species (Avise et al., 1987; Avise, 1994).

6.5.1 The Evolutionary Relationship of *M. mercenaria* and *M. campechiensis*

The evolutionary relationships among different taxa within *Mercenaria* have been the subject of much debate and remain unclear. Fossil evidence indicates that *M. mercenaria* and *M. campechiensis* originated more than 12 million years ago in the Miocene (Stenzel, 1955). Allozyme studies, however, show relatively low genetic divergence between the two species indicating a considerably more recent time of separation (Humphrey, 1981; Dillon and Manzi, 1989a,b). Studies of DNA sequence divergence in the mitochondrial genomes of these species endorse the results of the allozyme analyses. O'Foighil et al. (1996) found 2.8–3.5% sequence divergence between *M. mercenaria* and *M. campechiensis* at the 16S mitochondrial RNA gene. If the 16S mtRNA gene evolves at rates similar to those reported for the same gene in the mussel *Mytilus* (Rawson and Hilbish, 1995) this level of divergences suggests that *Mercenaria mercenaria* and *M. campechiensis* have been evolving as separate lineages for 3–4 million years.

There is a clear discrepancy between the divergence times inferred from genetic approaches relative to those estimated from the fossil record. It is possible that the two 'species' have maintained an extensive period of hybridization that has led to homogenization of their genomes. It seems unlikely, however, that over 75% of the evolutionary history has been erased by extensive introgression. Alternatively, it is possible that one (or both) of the species identified in the fossil record morphologically resemble the modern forms of either *M. mercenaria* or *M. campechiensis* but are in fact not the ancestor of the modern species. In this case, the two species may appear to have an evolutionary history longer than their actual time of divergence. In either case it is apparent that significant disparities exist between the genetic and fossil histories of these species.

6.5.2 What is *M. texana*?

The relationship between *M. mercenaria*, *M. campechiensis* and *M. texana* has also been actively debated. The latter taxon is morphologically distinct from both *M. mercenaria* and *M. campechiensis*, although several authors have concluded that *M. texana* is more closely related to *M. mercenaria* (Menzel, 1989). Dillon and Manzi (1989a), however, concluded that *M. texana* is a subspecies of *M. campechiensis* since they share nearly identical patterns of allozyme allele frequencies.

Recent studies by O'Foighil et al. (1996) indicate that *M. texana* has a complex evolutionary history. O'Foighil et al. (1996) sequenced 444 nucleotides of the mitochondrial 16S ribosomal gene for four representative populations of hard clams. Their analysis consistently resolved three well supported clades (Fig. 6.2). As noted above, *M. mercenaria* and *M. campechiensis* contained distinct mtDNA lineages. Populations of *M. texana* contain two distinct mtDNA

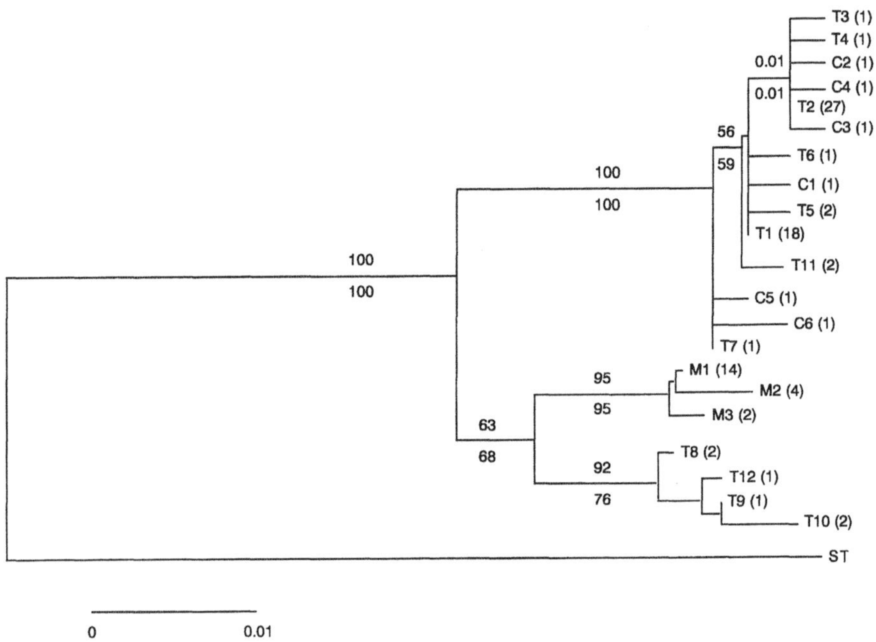

Fig. 6.2. Phylogenetic tree of North American *Mercenaria* mitochondrial lineages (redrawn from O'Foighil et al., 1996). The three taxa of *Mercenaria* are indicated by capital letters (M = *M. mercenaria*, collected from South Carolina; C = *M. campenchiensis*, collected from Florida and Mexico; T = *M. texana* collected from Texas) and the subsequent number designates individual haplotypes. Thus, M2 represents the second unique haplotype observed in *M. mercenaria*. The numbers in parentheses indicate the number of individuals that were observed to contain each haplotype. The tree was constructed using a neighbor-joining method. Confidence limits of major nodes are presented if they were supported at levels >50% and were estimated by bootstrapping (500 iterations). The Japanese congener *M. stimpsoni* (ST) was used as an outgroup.

lineages: about 87% of the individuals had a mtDNA haplotypes indistinguishable from those found in *M. campechiensis* while 13% contained mtDNA haplotypes from a third clade that was more closely related to that found in *M. mercenaria*. O'Foighil et al. (1996) concluded that this third clade diverged from the *M. mercenaria*-specific clade about 2 million years ago. Evidence of *M. texana* populations harboring two distinct mtDNA lineages was also found by Brown and Wolfinbarger (1989).

O'Foighil et al. (1996) concluded that the evolutionary history of *M. texana* probably includes a secondary contact between *M. campechiensis* and a third taxon of *Mercenaria*, followed by hybridization and strongly biased mtDNA and nuclear genetic introgression from *M. campechiensis*. The consequences of this process is that the third taxon is now only represented by relict mtDNA lineages preserved in *M. texana* and nuclear genes that encode unique external shell sculpture features. The majority of both the nuclear and mtDNA genomes of *M. texana* are clearly derived from *M. campechiensis*. Taxonomic issues regarding the status of *M. texana* have always been problematic and will likely continue to be so since the evolutionary history of this taxon contains a dichotomous phylogenetic signal stemming from an ancient hybridization process that has differentially influenced allozyme, shell morphology and mtDNA phenotypes.

6.5.3 Summary

The evolutionary history of hard clams has clearly been complex and many aspects remain unclear. The population genetics of hard clams in many ways is unremarkable. These species are wide spread and exhibit relatively little inter-specific divergence, as might be expected in a species that has a high dispersal larval stage. On the other hand, hard clams also exhibit patterns of divergence that indicate that their evolutionary history includes episodes of hybridization that has led to differential patterns of selection and introgression between taxa.

6.6 ACKNOWLEDGMENTS

This work was supported by the National Science Foundation grants BSR-8615052, DEB-9208014, OCE-9203320 and DEB-9509742, and the South Carolina Sea Grant Consortium.

REFERENCES

Adamkewicz, L., 1988. Geographical effects on growth rate in the hard clam *Mercenaria mercenaria*. J. Shellfish Res., 7: 146.

Adamkewicz, L., Taub, S.R. and Wall, J.R., 1984a. Genetics of the clam *Mercenaria mercenaria*, I. Mendelian inheritance of allozyme variation. Biochem. Genet., 22: 215–219.

Adamkewicz, L., Taub, S.R. and Wall, J.R., 1984b. Genetics of the clam *Mercenaria mercenaria*, II. Size and genotype. Malacologia, 25: 525–533.

Arnold, W.S., Bert, T.M., Marelli, D.C., and Cruz-Lopez, H., 1996. Genotype-specific growth of hard clams (genus *Mercenaria*) in a hybrid zone: variation among habitats. Mar. Biol., 125: 129–139.

Avise, J.C., 1994. Molecular Markers, Natural History and Evolution. Chapman and Hall, New York.

Avise, J.C., Arnold, J., Ball, R.M., Bermingham, E., Lamb, T., Neigel, J.E., Reeb, C.A. and Saunders, N.C., 1987. Intraspecific phylogeography: the mitochondrial bridge between population genetics and systematics. Annu. Rev. Ecol. Syst., 18: 489–522.

Barton, N.H. and Hewitt, G.M., 1985. Analysis of hybrid zones. Annu. Rev. Ecol. Syst., 16: 113–148.

Bert, T.M. and Arnold, W.S., 1995. An empirical test of predictions of two competing models for the maintenance and fate of hybrid zones: both models are supported in a hard-clam hybrid zone. Evolution, 49: 276–289.

Bert, T.M., Hesselman, D.M., Arnold, W.S., Moore, W.S., Cruz-Lopez, H. and Marelli, D.C., 1993. High frequency of gonadal neoplasia in a hard clam (*Mercenaria* spp.) hybrid zone. Mar. Biol., 117: 94–104.

Brown, B.L., and Wolfinbarger, L., 1989. Mitochondrial restriction enzyme screening and phylogenetic relatedness in the hard shell clam genus *Mercenaria*. Part II. population variation. Virginia Sea Grant Report VSG-89-02.

Chanley, P.E., 1961. Inheritance of shell marking and growth in the hard clam, *Mercenaria mercenaria*. Proc. Natl. Shellfish. Assoc., 50: 163–169.

Chestnut, A.F., Fahy, W.E. and Porter, H.J., 1956. Growth of young *Venus mercenaria, Venus campechiensis* and their hybrids. Proc. Nat. Shellfish. Assoc., 47: 50–56.

Clausen, J., Keck, D.D., and Hiesey, W.M., 1940. Experimental studies on the nature of species, I. Effect of varied environments on western North American plants. Carnegie Inst. Wash. Publ., 520.

Comstock, R.E., and Robinson, H.F., 1952. Estimation of averages dominance of genes. In: J.W. Gowen (Ed.), Heterosis. State College Press, Ames, Iowa, pp. 494–516.

Dillon Jr., R.T., 1985. Correspondence between the buffer systems suitable for electrophoretic resolution of bivalve and gastropod isozymes. Comp. Biochem. Physiol., 82B: 643–645.

Dillon Jr., R.T., 1992. Minimal hybridization between populations of the hard clams, *Mercenaria mercenaria* and *Mercenaria campechiensis*, co-occurring in South Carolina. Bull. Mar. Sci., 50: 411–416.

Dillon Jr., R.T. and Manzi, J.J., 1987. Hard clam *Mercenaria mercenaria*, broodstocks: genetic drift and loss of rare alleles without reduction in heterozygosity. Aquaculture, 60: 99–105.

Dillon Jr., R.T. and Manzi, J.J., 1988. Enzyme heterozygosity and growth rates in nursery populations of *Mercenaria mercenaria* (L.). J. Exp. Mar. Biol. Ecol., 116: 79–86.

Dillon Jr., R.T. and Manzi, J.J., 1989a. Genetics and shell morphology of hard clams (genus *Mercenaria*) from Laguna Madre, Texas. Nautilus, 103: 73–77.

Dillon Jr., R.T. and Manzi, J.J., 1989b. Genetics and shell morphology in a hybrid zone between the hard clams *Mercenaria mercenaria* and *M. campechiensis*. Mar. Biol., 100: 217–222.

Dillon Jr., R.T. and Manzi, J.J., 1992. Population genetics of the hard clam, *Mercenaria mercenaria*, at the northern limit of its range. Can. J. Fish. Aquat. Sci., 49: 2574–2578.

Falconer, D.S., 1981. Introduction to Quantitative Genetics. Longman Press, New York.

Fisher, R.A., 1930. The Genetical Theory of Natural Selection. Clarendon Press, Oxford, 340 pp.

Gaffney, P.M. and Scott, T.M., 1984. Genetic heterozygosity and production traits in natural and hatchery populations of bivalves. Aquaculture, 42: 289–302.

Gaffney, P.M., Scott, T.M., Koehn, R.K. and Diehl, W.J., 1990. Interrelationships of heterozygosity, growth rate, and heterozygote deficiencies in the coot clam, *Mulinia lateralis*. Genetics, 124: 687–699.

Hadley, N.H., Dillon Jr., R.T. and Manzi, J.J., 1991. Realized heritability of growth rate in the hard clam *Mercenaria mercenaria*. Aquaculture, 93: 109–119.

Harrison, R.G., 1993. Hybrids and hybrid zones: historical perspective. In: R.G. Harrison (Ed.), Hybrid Zones and the Evolutionary Process. Oxford Press, Oxford, pp. 3–12.

Haven, D.S. and Andrews, J.D., 1956. Survival and growth of *Venus mercenaria, Venus campechiensis* and their hybrids in suspended trays. Proc. Natl. Shellfish. Assoc., 47: 43–48.

Hilbish, T.J., Winn, E.P. and Rawson, P.D., 1993. Genetic variation and covariation during larval and juvenile growth in *Mercenaria mercenaria*. Mar. Biol., 115: 97–104.

Humphrey, C.M., 1981. Ecological genetics of the hard clams (*Mercenaria mercenaria* Linné and *Mercenaria campechiensis* Gmelin): electrophoretic estimates of enzyme variation and the use of shell morphology as a species indicator. Ph.D. Dissertation, University of Georgia, Athens, 93 pp.

Humphrey, C.M., and Crenshaw Jr., J.W., 1989. Clam genetics. In: J.J. Manzi, and M. Castagna (Eds.), Clam Mariculture in North America. Elsevier, Amsterdam, pp. 323–356.

Humphrey, C.M. and Walker, R.L., 1982. The occurrence of *Mercenaria mercenaria* form notata in Georgia and South Carolina: calculation of phenotypic and genotypic frequencies. Malacologia, 23: 75–80.

Jablonski, D., and Lutz, R.A., 1980. Molluscan larval shell morphology: ecological and paleontological applications. In: D.C. Rhoads and R.A. Lutz (Eds.), Skeletal Growth of Aquatic Organisms. Plenum Press, New York, pp. 323–377.

Jablonski, D. and Lutz, R.A., 1983. Larval ecology of marine benthic invertebrates: paleobiological implications. Biol. Rev., 58: 21–89.

Karl, S.A. and Avise, J.C., 1992. Balancing selection at allozyme loci in oysters: implications from nuclear RFLPs. Science, 256: 100–102.

Koehn, R.A., Hall, J.G. and Zera, A.J., 1980. Parallel variation of genotype-dependent aminopeptidase, I. Activity between *Mytilus edulis* and *Mercenaria mercenaria*. Mar. Biol. Lett., 1: 245–253.

Lewontin, R.C., 1974. The Genetic Basis of Evolutionary Change. Columbia University Press, New York, 346 pp.

Loosanoff, V.L., 1954. New advances in the study of bivalve larvae. Am. Sci., 42: 607–624.

Menzel, R.W., 1961. Seasonal growth of northern quahog *Mercenaria mercenaria* and the southern quahog *M. campechiensis* in Alligator Harbor, Florida. Proc. Natl. Shellfish. Assoc., 52: 37–46.

Menzel, R.W., 1962. Seasonal growth of the northern and southern quahogs *Mercenaria mercenaria* and *M. campechiensis* and their hybrids in Florida. Proc. Natl. Shellfish. Assoc., 53: 111–118.

Menzel, R.W., 1968. Cytotaxonomy of species of clams (*Mercenaria*) and oysters (*Crassostrea*). Symp. Mollusca, Mar. Biol. Assoc. India, Part I, pp. 75–84.

Menzel, R.W., 1977. Selection and hybridization in Quahog clams (*Mercenaria* spp.). In: J.W. Avault Jr. (Ed.), Proceedings of the Eighth Annual Meeting of the World Mariculture Society. Louisiana State University Division of Continuing Education, Baton Rouge, LA, pp. 507–521.

Menzel, W., 1989. The biology, fishery and culture of quahog clams, *Mercenaria*. In: J.J. Manzi and M. Castagna (Eds.), Clam Mariculture in North America. Elsevier, Amsterdam, pp. 201–242.

Menzel, R.W. and Menzel, M.Y., 1965. Chromosomes of two species of quahogs and their hybrids. Biol. Bull., 129: 181–188.

Mitton, J.B. and Grant, M.C., 1984. Associations among protein heterozygosity, growth rate, and developmental homeostasis. Annu. Rev. Ecol. Syst., 15: 479–499.

Mousseau, T.M. and Roff, D.A., 1987. Natural selection and the heritability of fitness components. Heredity, London, 59: 181–197.

Mousseau, T.M. and Roff, D.A., 1989. Adaptation to seasonality in a cricket: patterns of phenotypic and genotypic variation in body size and diapause expression along a cline in season length. Evolution, 43: 1483–1496.

Nevo, E., 1978. Genetic variation in natural populations: patterns and theory. Theor. Pop. Biol., 13: 121–177.

Newkirk, G.F., 1980. Review of the genetics and the potential for selective breeding of commercially important bivalves. Aquaculture, 19: 209–228.

Ockelmann, K.W., 1965. Development types in marine bivalves and their distribution along the Atlantic coast of Europe. Proc. 1st Eur. Malac. Congr., 1962, pp. 25–35.

O'Foighil, D., Hilbish, T.J., and Showman, R.M., 1996. Mitochondrial gene variation in *Mercenaria* clam sibling species reveals a relict secondary contact zone in the western Gulf of Mexico. Mar. Biol., 126: 675–683.

Pesch, G., 1972. Isozymes of lactate dehydrogenase in the hard clam, *Mercenaria mercenaria*. Comp. Biochem. Physiol., 43B: 33–38.

Pesch, G., 1974. Protein polymorphisms in the hard clams *Mercenaria mercenaria* and *Mercenaria campechiensis*. Biol. Bull., 146: 393–403.

Peterson, C.H. and Beal, B.F., 1989. Bivalve growth and higher order interactions: importance of density, site, and time. Ecology, 70: 1390–1404.

Pogson, G.H., Mesa, K.A. and Boutilier, R.G., 1995. Genetic population structure and gene flow in the Atlantic cod *Gadus morhua* — a comparison of allozyme and nuclear RFLP loci. Genetics, 139: 375–385.

Rawson, P.D. and Hilbish, T.J., 1990. Heritability of juvenile growth for the hard clam *Mercenaria mercenaria*. Mar. Biol., 105: 429–436.

Rawson, P.D. and Hilbish, T.J., 1991. Genotype–environment interaction for juvenile growth in the hard clam *Mercenaria mercenaria* (L.). Evolution, 45: 1924–1935.

Rawson, P.D. and Hilbish, T.J., 1995. Evolutionary relationships among the male and female mitochondrial DNA lineages in the *Mytilus edulis* species complex. Mol. Biol. Evol., 12: 893–901.

Shaw, R.G., 1987. Maximum likelihood approaches applied to quantitative genetics in natural populations. Genetics, 99: 323–335.

Slattery, J.P., Lutz, R.A. and Vrijenhoek, R.C., 1993. Repeatability of correlations heterozygosity, growth, and survival in a natural population of the hard clam *Mercenaria mercenaria* L. J. Exp. Mar. Biol. Ecol., 165: 209–224.

Stenzel, H.B., 1955. Ancestors of the Quahog. J. Sediment. Petrol., 25: 145.

Van Beneden, R.J., Gardner, G.R., Blake, N.J. and Blair, D.G., 1993. Implications for the presence of transforming genes in gonadal tumors in two bivalve mollusk species. Cancer Res., 53: 2976–2979.

Via, S. and Lande, R., 1985. Genotype–environment interaction and the evolution of phenotypic plasticity. Evolution, 39: 505–522.

Zouros, E. and Foltz, D.W., 1984. Possible explanations of heterozygote deficiency in bivalve molluscs. Malacologia, 25: 583–591.

Zouros, E. and Foltz, D.W., 1987. The use of allelic isozyme variation for the study of heterosis. Isozymes: Current Topics Biol. Med. Res., 13: 1–59.

Section 2

Environmental Biology

Biology of the Hard Clam
J.N. Kraeuter and M. Castagna (Eds.),

Chapter 7

Functional Morphology and Behavior of Shelled Veligers and Early Juveniles

Melbourne R. Carriker

7.1 SHELLED VELIGER

7.1.1 Straight-Hinged Veliger

A glance under low microscopic magnification of an actively swimming, straight-hinged veliger, its valves slightly parted, velum fully extended, rapidly beating cilia projecting widely beyond the edge of the glistening valves, brings into view an exquisite, vibrating icon. Clean, translucent, aragonitic valves shimmeringly mirror light as the mite spirals slowly in no particular direction, its velar retractor muscles twitching from time to time in response to a velar collision with some tiny planktonic particle. Not to have seen a living, vibrant molluscan veliger is to have missed one of marine biology's larval treasures.

In culture jars, early straight-hinged veligers of *Mercenaria mercenaria* tend to concentrate in surface layers of seawater. In nature this behavior could have survival value by positioning them above near-bottom strata with their potential toxic conditions and ever present benthic predators. Turner and George (1955) determined that at temperatures of 22°C straight-hinged veligers of *M. mercenaria* swim upward at a rate of 1.17–1.3 mm s^{-1}, and show no detectable reaction to light. In a glass tube in which layers of diluted seawater are placed one above the other, they are indifferent to a considerable range of salinity gradients. They swim through sharp transitions separating layers differing by as much as 5 ppt with no loss in velocity until the gradient between 20 and 15 ppt is traversed; at this stratum, swimming velocity decreases and veligers swim slowly in circular patterns just above the 20–15 ppt interface. Swarming at the surface could be disadvantageous when upper levels of highly stratified estuaries move seaward more rapidly than the lower layers, movements being accentuated by wind blowing in the same direction.

7.1.2 Umbonal Veliger

Infrequently, the margin of one of the valves of prodissoconch II of veligers collected in the plankton has been chipped and replaced by newly deposited shell. Mending probably takes place in much the same way that adults renew fringes of shell lost through unsuccessful attempts of predators to penetrate them (Carriker, 1996). How planktonic veliger valves are thus broken is not known.

Whereas most tissues of healthy living umbonal veligers of *M. mercenaria* are translucent-grayish in color, the digestive gland is often brightly pigmented, the color varying somewhat with the quality and quantity of the diet. Actively feeding veligers cultured in the laboratory, at one time or another, exhibit digestive glands ranging in color from light yellow to orange-brown to bright orange to green, the color of the gland corresponding more or less with that of the phytoplankton ingested (see also Loosanoff and Davis, 1951). At a microscopic magnification of about 150×, freshly consumed food particles averaging 1 to 2 μm in size are visible in the stomach of veligers. These food particles rapidly rotate in little clusters, propelled by active gastric cilia. The digestive gland of wild veligers likewise varies in color. Starved individuals in the laboratory possess noticeably less pigment. The color of the digestive gland is not a reliable diagnostic taxonomic character of bivalve species, whereas that of the valves of some other species often is (Chanley and Andrews, 1971). Uptake, lysis, and digestion or rejection of algal cells can also be observed microscopically through the translucent veliger valves by autofluorescence of chlorophyll *a* and its derivatives, as was done by Babinchak and Ukeles (1979) with grazing veligers of *Crassostrea virginica*.

Umbonal veligers tend to remain more or less uniformly distributed in culture jars, swimming in slow, stately, circling patterns, the velum fully extended, and valves slightly parted. When pipetted into a watch glass in a little seawater, some veligers remain on the bottom with valves momentarily closed; others immediately dart away, a few so swiftly they are difficult to follow at a magnification of 35×. As they spiral off the bottom, some individuals spread their valves, quickly extend the velum, and swim hinge first in helical paths — they may then suddenly reverse direction and advance velum first. Still others may proceed velum first directly off the bottom, sailing majestically surfaceward. The speed of swimming can thus vary widely; the stimulus(i) that prompts the changes has not been determined. In whatever direction, these ciliary movements are gracefully performed.

Even crowded in densities of 50 per cubic centimeter, veligers seldom collide with each other, currents provided by the swimming organ seeming to act like the 'slip stream of an aeroplane' (Turner and George, 1955). In a more dense culture, though, collisions can be frequent. Some of the victims fall slowly toward the bottom, others open and flip away before reaching the bottom, and those striking the bottom dart swiftly off after turning rapidly in place.

Mild pipette-driven currents of seawater and gentle knocks on walls of culture containers do not cause veligers to halt swimming, whereas a drop of seawater falling on the surface near them is sufficient stimulus to provoke closure of the valves and sinking for a short distance before swimming is resumed.

When swirled into a heap in the center of a small concave dish, swimming veligers generally continue moving about with the velum outstretched, and soon disperse into the seawater. In nature, veligers, seemingly fragile in the laboratory, never rest on the bottom of coastal waters; they pass the entire planktonic veliger stage of eight days or more suspended in the seawater. Underfed or diseased individuals do fall to the bottom in culture jars, for a time making sporadic efforts to swim. Presumably the same occurs in nature, except that predators may soon consume them.

In coastal waters horizontal distribution of veligers is markedly uneven and can be widespread. Swarms vary in distribution, density, and size relative to the intensity and time of spawning of their benthic parents. The single highest concentration of early veligers sampled by Carriker (1961) was 67,200 per 100 l of seawater in Little Egg Harbor, New Jersey. Veligers tend to remain well above the bottom (and bottom-dwelling predators) during daylight hours, concentrating more or less in the middle strata. During hours of darkness, however, they are generally more broadly distributed vertically, late umbonal stages especially spreading closer to the bottom. Whether vertical distribution of *M. mercenaria* veligers is influenced by vertical salinity gradients has not been determined. Turner and George (1955) did report, though, that veligers are indifferent to a wide range of salinities (for response of larvae of other bivalve species, see Chia et al., 1984; Maia, 1988). Turbulence from wave action and tidal currents, as during maximal tidal flow and wind velocities, may stimulate larval swarms of *M. mercenaria* to rise in the water column during daylight hours. This response needs to be confirmed.

While early veligers may remain in the swarms originating from parents at the time of spawning, in time, because of hydrographic mixing processes, later-stage veligers, unable to counteract horizontal dispersion, tend to become widely dispersed (Young, 1995). Mortality rate during planktonic existence can be extremely high (Rumrill, 1990). In one closely monitored study, a single widespread, dense swarm was reduced in concentration from about 800 straight-hinged veligers to two late umbonal veligers per 100 l of seawater in 8 days (Carriker, 1961; see also Levin, 1990).

7.1.3 Pediveligers

Toward the end of their planktonic cycle, pediveligers develop a strong, highly ciliated, mobile, muscular foot. For an unspecified period (one or more days, Turner and George, 1955), varying with individuals and probably with availability of settling surfaces (Burke, 1983), they alternate between a swimming and crawling mode. In both the laboratory and the field, they start settling some 8 days after fertilization of the ovum (Carriker, 1961) at temperatures ranging from 23 to 26°C. At higher temperatures, approximately 30°C, minimal settling time is about 6 days (Loosanoff et al., 1951).

A population of pediveligers from a single pair of parents, and grown under identical conditions, can display wide variation in rate of growth and duration of the planktonic phase (Loosanoff et al., 1951; Carriker, 1961). In laboratory cultures, for example, maximal range of length at which pediveligers settle is most commonly 200–210 μm, though individuals as small as 100 μm have been known to settle (Loosanoff et al., 1951). Larger pediveligers do not always settle before the smaller ones; nor are their shape and size altered by development at different temperatures, a matter of some taxonomic significance (Loosanoff and Davis, 1963).

Thorson (1957) conjectured many years ago that the capacity of larvae of many bottom invertebrates to respond to substrata is possible because of the presence of the 'swimming–crawling' stage, a normal developmental feature of most bivalves and gastropods (Carriker, 1990). By pediveligers of *M. mercenaria*, the search for a favorable settling surface involves principally the velum and the foot, which are liberally supplied with nerve endings and come most intimately in contact with potential settling surfaces.

A searching pediveliger of *M. mercenaria* can be extremely active. As it approaches the sediment–seawater interface, it extends the foot beyond parted valves, alights, and energetically creeps smoothly over the bottom, moving back and forth in an apparent exploratory pattern. If the surface is not attractive, possibly because of particle size, chemical composition, or other negative stimuli, it suddenly extends the velum, spirals away to a new location, stops there, and continues its zigzag creeping trail. This pattern of swimming and alighting continues until an attractive settling site is encountered. Its capacity to swim and be transported by currents, unquestionably extends widely the area that it can explore; just how broadly, is not known (Turner and George, 1955; Wilson, 1990). The swimming rate of the pediveligers of *M. mercenaria* has not been recorded. Mann et al. (1991) in laboratory studies at 23°C and a salinity range of 20 to 30 ppt, determined that pediveligers of the mactrid *Spisula solidissima* swim upward (net vertical movement) at a rate ranging from 0.30 to 0.40 mm s^{-1}. Umbo larvae, consistent with their lesser weight, swim at a rate slightly greater than that of pediveligers, 0.22 to 0.49 mm s^{-1}. It is likely that the swimming rate of *M. mercenaria* pediveligers is consistent with that given for *S. solidissima*, though during their sudden spurts of swimming the rate is undoubtedly faster.

Pediveligers searching for a settling spot are attracted to specific types and contours of substrata. This can be demonstrated by observing their response to different kinds and conformations of surfaces: glass, bleached shell of *Crassostrea virginica* and *M. mercenaria*, glass dipped in melted paraffin and sprinkled with fine bay-bottom sand, leached oyster and clam shell excavated by the boring sponge *Cliona celata*, leached pitted slag, fine detritus (of microscopic fecal pellets, unconsolidated organic fragments, silt from the bay bottom), and fine and coarse bay-bottom sands. In a laboratory experiment, one set of these substratal types was placed on the upper surface of tilted, slowly vertically moving, aerating glass plunger-plates in three culture jars containing a moderately dense population of umbo veligers; a second set of substrata was located on the bottom of the jars (Carriker, 1961; see also Turner and George, 1955).

After leaving the plankton, most pediveligers bysally attached on or near the bottom of the jars where the only motion was that of slowly circulating seawater; none affixed to substrata on the plunger plates, perhaps because of vibration. Although a few plantigrades were present on the underside of some slightly elevated bivalve shells, most had affixed to upper surfaces of the different substratal types. Quayle (1952) also observed that most non-cementing species of bivalves prefer to settle on upper surfaces.

Although some *M. mercenaria* settled in the sandy substrata (see also Crisp, 1974 and Keck et al., 1974) and in open pits in the slag and shell, by far densest sets occurred in pits filled, and on hard surfaces covered, with fine detritus. In some of the detritus-filled pits, plantigrades were so tightly packed that others could not have entered (see also Crisp, 1984). Aggregations of fine detritus on open surfaces also supported plantigrades and in densities equivalent to those in the detritus-filled pits. This same response was noted by Turner and George (1955) by byssal plantigrades of *M. mercenaria* and by Quayle (1952) by *Venerupis pullastra*. These settling responses suggest that pediveligers can select microhabitats in spatial scales ranging from millimeters to tens of centimeters, based on the characteristics of sediment other than grain size (Butman, 1987). Crisp (1974) likewise observed that larvae of many invertebrate marine species settle in concavities, and emphasized the importance of sedimentary particle size in larval choices.

Turner and George (1955) provided pediveligers of *M. mercenaria* a choice of clean sieved sand of varying grain sizes, but the pediveligers expressed no preference for any particular size. This lack of response was probably because the sand was freed of organic matter for ease in sieving. In still, four-hour experiments in which pediveligers of *M. mercenaria* were given a choice between two extreme types of sediment (natural, organic-rich mud, and an abiotic mixture of spherical glass beads), selection occurred in only 5 of the 23 experiments, and these five selections were for mud rather than beads (Bachelet et al., 1992; see also Jonsson et al., 1991). It is possible that four hours was insufficient time within which selection could take place.

The fact that pediveligers of *M. mercenaria* in the plunger-plate jars actively chose favorable substrata in the relatively still seawater (Carriker, 1961), in contrast to the low response by pediveligers in experiments in flowing seawater in a flume (Butman et al., 1988), may be explained, in part, by the absence of strong currents in the plunger-plate jars. However, Butman et al. (1988) noted that when pediveligers were given a longer time, they could select between two different types of sediment in flowing seawater.

It has been shown that biologically modified substrata can enhance settlement by *M. mercenaria*. Keck et al. (1974), a case in point, in preliminary laboratory experiments observed that settling pediveligers prefer sediment treated with the mantle cavity fluid of adults to untreated sediment. Also, Ahn et al. (1993) demonstrated that dense assemblages of the bivalve *Gemma gemma* enhance settlement of *M. mercenaria* pediveligers among them.

Gotelli (1990) cautioned that larval preference for substrata may interact with the gregarious settlement response, gregariousness reinforcing the underlying settlement response. In the experiments cited in this section, the possible effect of gregariousness was not considered; if present, it simply would have exaggerated the magnitude of substratal choices.

7.2 PLANTIGRADE STAGES

7.2.1 Byssal Plantigrade

Foot and locomotion. Pedal locomotion by pediveligers and newly settled plantigrades consists primarily of ciliary gliding. Cilia on the ventral surface of the foot beat toward the heel of the foot, smoothly moving the bivalve forward. A similar type of locomotion was described by Quayle (1952) for pediveligers of *V. pullastra* (see also Chia et al., 1984). As development proceeds, ciliary locomotion is partly superseded by muscular crawling. For a brief time after loss of the velum, *M. mercenaria* plantigrades can propel themselves through the water by a combination of ciliary action and pedal kicking (Belding, 1912; Loosanoff and Davis, 1950). Crawling is effective for movement through sediments, and is retained for the remainder of the bivalve's life.

Muscular crawling by byssal plantigrades, about 0.4 mm in length, on hard open surfaces is characteristic of all sizes of this stage, and agrees generally with the description given by Belding (1912) for crawling by individuals 2 to 3 mm in length. In brief, from the resting position on one valve, a plantigrade extends the foot anteriorly between the margins of its valves and makes contact with the substratum, the forward tip, or toe, of the foot acting as a searching and guiding organ (Figs. 7.1 and 7.2). The middle part of the foot is then appressed to the substratum, flattened to one and a half times its normal width; the area of contact passes distally until the whole distal half of the foot is appressed and effects purchase with the substratum. During this maneuver the plantigrade draws its valves upright, pulling them forward over the foot by contraction of the pedal muscles. If the foot when first extended, strikes an object, it withdraws within the mantle cavity and extends anew in a different direction, moving by means of both ciliary and muscular–vascular action. Occasionally, the shell may fall to one side, but is soon righted. With each forward prolongation of the foot, the plantigrade opens its valves slightly, the mantle margins of the pedal gape maintaining

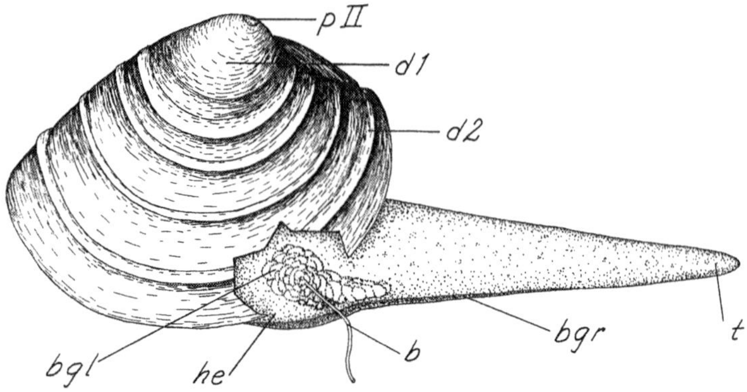

Fig. 7.1. Drawing of byssal plantigrade of *Mercenaria mercenaria*, shell 4 mm long. Part of left valve cut away to show byssal gland in heel (he) of foot; b = byssus; bgl = byssal gland; bgr = byssal groove; d1 = dissoconch primary shell ridges; d2 = dissoconch secondary shell ridges; pII = prodissoconch II; t = toe.

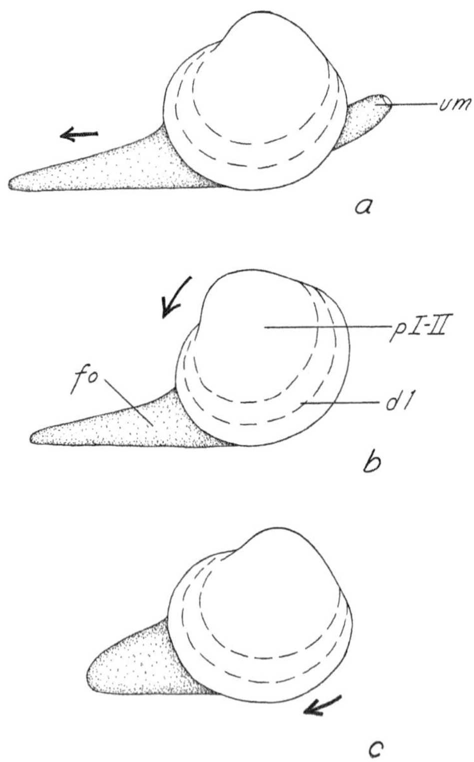

Fig. 7.2. Diagrams of mode of locomotion in a byssal plantigrade of *Mercenaria mercenaria*, shell 415 μm long; d1 = dissoconch primary shell ridges; fo = foot; pI and pII = prodissoconchs I and II; vm = primary exhalant siphon.

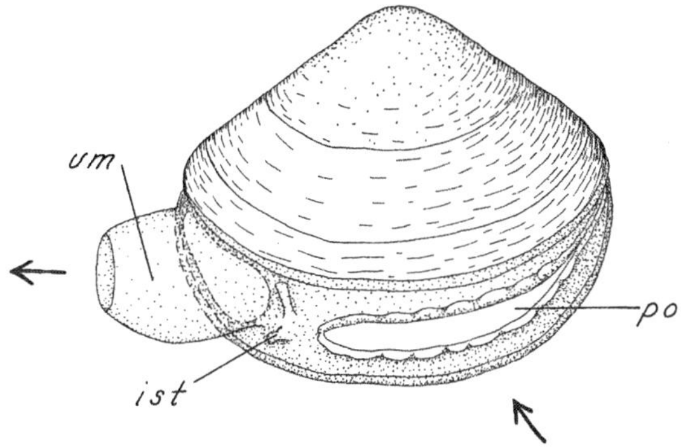

Fig. 7.3. Byssal plantigrade of *Mercenaria mercenaria*, 330 μm long. Primary exhalant siphon (vm) is forming out of exhalant opening, and two inhalant siphonal tentacles (ist), and two primary shell ridges are present; arrows indicate direction of flow of seawater.

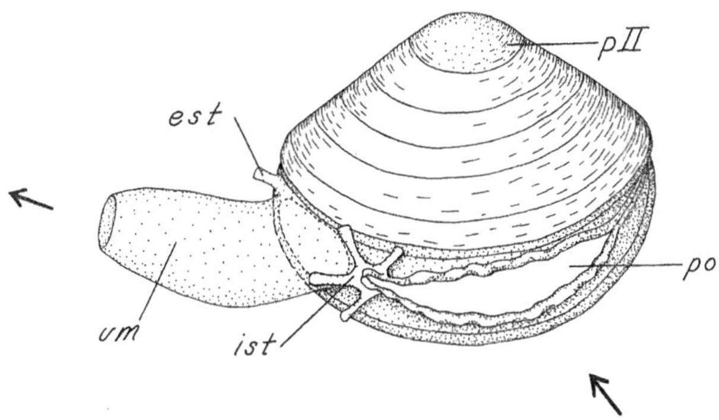

Fig. 7.4. Byssal plantigrade of *Mercenaria mercenaria*, 625 μm long. Primary exhalant siphon (vm) is further developed; four inhalant siphonal tentacles (ist), one exhalant siphonal tentacle (est) and five primary shell ridges are present; po = pedal opening; pII = prodissoconch II; arrows indicate direction of flow of seawater.

a close seal around the foot thereby excluding sedimentary grains as the foot is withdrawn; simultaneously, the bivalve everts the primary exhalant siphon (Fig. 7.2a, Figs. 7.3–7.5). Upon initial contraction of pedal muscles, the valves close gently upon the foot and the primary exhalant siphon is withdrawn completely (Fig. 7.2b); the anterior part of the valves is then pulled downward toward the substratum (Fig. 7.2b), and immediately following, the ventral region of the shell is drawn forward (Fig. 7.2c) over the foot as the foot contracts. The overall action is of a bivalve moving forward by rocking its shell over the foot, the ventral rim of the valves sliding over the bottom much as a sled slides on its runners. The bottom of the foot can adhere to firm substrata with considerable tenacity.

7.2.2 Byssus and Byssal Attachment

The pediveliger initiates attachment to firm substrata by means of the translucent, elastic, byssal thread (Van der Feen, 1949). Byssal fluid passes from the byssal glands through the byssal duct to the midventral base of the foot where it flows into the pedal byssal groove.

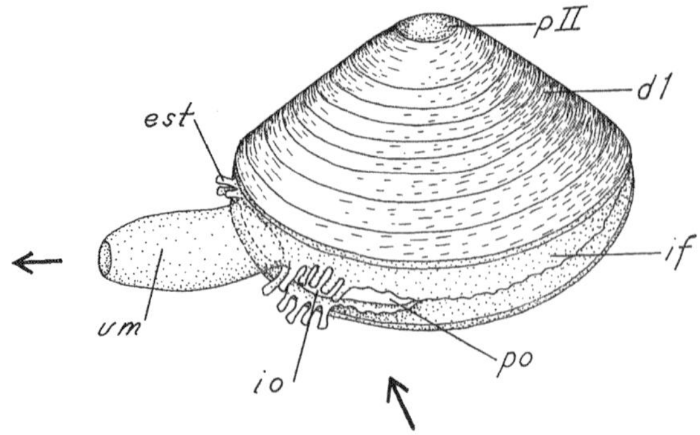

Fig. 7.5. Byssal plantigrade of *Mercenaria mercenaria,* 1.36 mm long. Primary exhalant siphon (vm) is further enlarged; ten inhalant siphonal tentacles (ist), three exhalant siphonal tentacles (est), and ten primary shell ridges are present; dl = dissoconch with primary shell ridges; if = inner fold of mantle; io = inhalant opening; po = pedal opening; arrows indicate direction of flow of seawater.

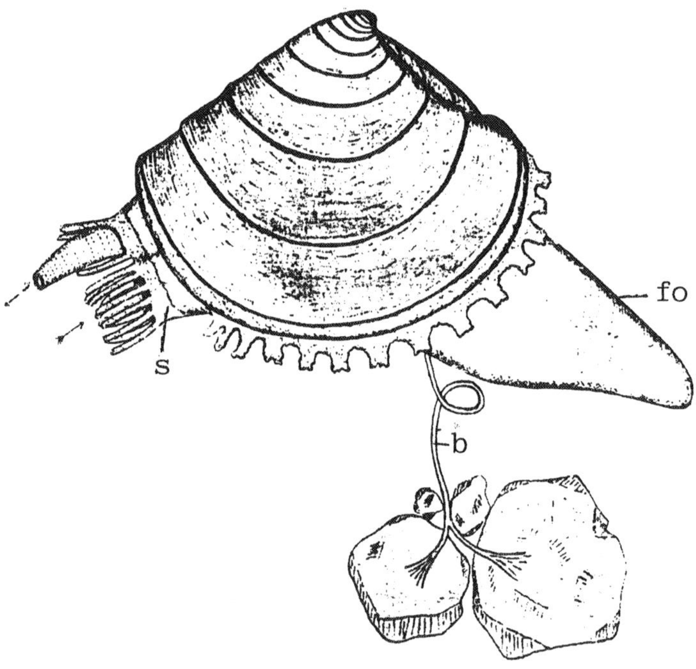

Fig. 7.6. Byssal plantigrade of *Mercenaria mercenaria,* shell length 1 mm, attached by sand grains by the byssus (b). Foot (fo) and siphons (s) are extended, as is the primary exhalant siphon from the definitive exhalant siphon. Arrows indicate direction of flow of seawater (from Belding, 1912).

Early plantigrades effect initial fixation to flat surfaces or grains of sediment by means of a very fine transparent byssus scarcely visible at a magnification of 70× (Fig. 7.6). As the bivalve grows, the byssal gland, duct, and groove likewise enlarge. In individuals 270 fm in length the byssus has a maximum diameter of 5 μm and a length of about 0.3 mm, though

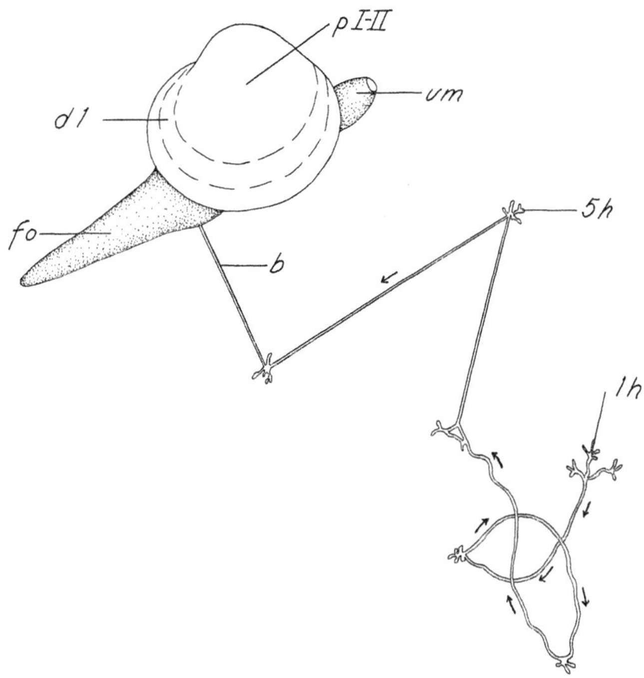

Fig. 7.7. Diagram of the byssal trail laid down by a byssal plantigrade of *Mercenaria mercenaria*, shell length 415 μm, in shallow sand in a finger bowl of seawater; b = byssus; d1 = dissoconch primary shell ridges; fo = foot; PI–II = prodissoconchs I and II; vm = primary exhalant siphon; 1h = first holdfast; 5h = fifth holdfast of byssus.

the length can be extremely variable; in an individual 400 μm in length, the byssus is 8 μm in diameter; in one 6 mm in length, 30 μm in diameter and up to 4 mm long to the first holdfast, and may extend through a number of holdfasts to a total length of 2 cm (Fig. 7.7). By the time a plantigrade has grown to a shell length of 0.5 mm, the byssus is a sturdy, elastic, anchor line; and at shell lengths ranging from 0.5 to 1 mm, plantigrades adhere so firmly to hard surfaces by means of the byssus that a strong stream of seawater directed at them from a bulb pipette does not dislodge them. The longest byssus encountered proportionate to the length of the shell was 5.5 mm in an individual 1.4 mm long (Carriker, 1961). The average shell length at which plantigrades discontinue use of the byssal gland has not been determined, but a gross estimate is about 7 mm. Belding (1912) reported finding a byssus in an individual 9 mm long. He noted that two or three branches extend from the distal part of the thread, and that their distal ends divide again into strands like "the delta of a river, each similar to strands of prepared gelatin and apparently fastening by little suckers or stickers to the substratum".

Byssal formation and attachment are readily observed through a binocular microscope secured in an inverted position on a cantilevered support lighted from beneath by a strong spotlight (Carriker, 1961). Plantigrades ranging in length from 0.5 to 7 mm are placed in a shallow dish with a small quantity of fine bay sand and clean seawater for viewing from beneath under the microscope. Light does not appear to disturb the bivalves so long as they remain buried in the sediment. This device permits an observer to follow the movements of plantigrades at various magnifications from underneath as they crawl about and effect byssal attachment. Carriker (1961) made his observations on plantigrades of several representative populations of active, fast-growing *M. mercenaria* reared in the laboratory.

These plantigrades will anchor the byssus to an open surface of the dish, but attach more quickly and frequently when buried completely under sediment. While affixing the byssus, a plantigrade extends the foot slowly, pushing it through the sand, the toe held closely against the glass surface. At full extension of the foot, the toe is pressed firmly against the surface for a period varying from 0.5 to 3 s. During this activity, edges of the byssal groove meet midventrally, converting the byssal groove into a tube, and minute peristaltic waves pass along the foot pressing byssal fluid along the tube. Formation and fixation of the byssus is extremely rapid. After attachment of the byssal holdfast the foot is withdrawn slowly (at the same rate as when crawling), paying the byssus out behind it. The holdfast consists of a hardened drop of byssal fluid securely glued to the glass surface. Byssus and holdfast harden quickly upon contact with seawater, as indicated by the fact that attached plantigrades hold against a stream of seawater from a small pipette within less than a minute after attachment. Following fixation, the plantigrade may withdraw the foot completely within the mantle cavity and close the valves upon the byssus, or may hold valves apart, and when in sediment, siphon interstitial seawater from among the sedimentary particles; or if under sediment, it may crawl to the surface to siphon, still anchored by its byssus; or if on an open surface, it may secrete a longer byssus with a number of holdfasts (Fig. 7.7) before becoming quiescent.

The byssal holdfast is usually characterized by a distinct digitate blob of byssal secretion. If the plantigrade is among sedimentary grains, the holdfast contains sand grains trapped in the fluid as it was molded against the substratum. The relatively small size of the holdfast (Fig. 7.7) suggests the high degree of constriction exerted on the hardening byssal fluid by the pedal tip at the moment of attachment. The complexity of representative byssal trails (Fig. 7.7) suggests the considerable versatility and maneuverability of the byssal groove, foot, and toe during placement of successive holdfasts.

In deep sediment byssal plantigrades affix the byssus to several grains of sediment a short distance under the sediment–seawater interface (Fig. 7.6) and then crawl surfaceward, taking a position in the sediment, the depth depending on the length of the siphons.

Removal of sedimentary grains from around byssal plantigrades stimulates them promptly to release the byssus at its insertion in the byssal duct, move to another site, and there reaffix. While discarding the byssus, young plantigrades crawl forward seemingly effortlessly, leaving behind the byssus still connected to its holdfast. Older plantigrades, those approaching a length of 300 μm, may have to make two or three attempts, circling about, resting occasionally, pulling tightly against the holdfast, before the byssus is dislodged from the byssal duct. A still older individual, 2 mm or more in length, may turn about on its foot and tug at the holdfast for a minute or so before freeing itself. One such individual crawled slowly and with apparent effort from the holdfast for about 5 min, stretching the byssus to a length of 5.5 mm before it gave way from the foot, contracting onto the substratum like a released stretched rubber band. Whether the byssus is held in the byssus duct by a sphincter muscle, or by hardening of byssal branches in the branching ducts of the byssal gland itself, is not known; in view of the apparent effort involved in release of the byssus by older plantigrades, the latter may be the explanation.

TABLE 7.1

Rate of byssal attachment by *Mercenaria mercenaria* in the laboratory

Interval plantigrades were undisturbed in sand (m)	Percent plantigrades attaching during this interval
5	5
10	7
15	17
20	19
30	48
40	32
50	42
60	39
120	32
180	60

The interval between disturbance and release of the byssus and reattachment of different individual byssal plantigrades varies from a few minutes to several hours. In a laboratory experiment, 20 plantigrades were placed in each of 10 different petri dishes with a little heap of fine bay sand in fresh sand-filtered seawater (salinity 35.5 ppt) (Carriker, 1961). Individuals and sand were swirled into the center of each dish, and dishes were set aside under glass covers. No artificial aeration or food was provided during the observations; room temperature ranged from 23°C in the morning to 30°C in the late afternoon. At intervals plantigrades that had affixed in each dish were exposed by gentle flushing aside of sand grains with a stream of seawater. At first, reattachment was made by only a few individuals, whereas after 3 h, more than half had attached (Table 7.1).

In a second laboratory experiment under similar conditions of salinity and water temperatures, 150 plantigrades similar to those reported in Table 7.1, were left uninterrupted for 18 h. Of these 91% made byssal attachment. Probably under natural conditions, especially in flowing seawater, plantigrades will affix much more rapidly than in the laboratory. Rate of attachment in a range of current velocities has not been studied.

Byssal attachment is a common and important means by which many species of young bivalves anchor themselves to hard surfaces and thereby remain close to the general area of settlement (for example, *Mytilus edulis*, Kellogg, 1892; Bayne, 1976; *Mya arenaria*, Kellogg, 1901; Ryder, 1989; *Hiatella gallicana*, Hunter, 1949; *Venerupis pullastra*, Quayle, 1952; *Lasaea rubura*, Morton, 1960 and others). According to Yonge (1962), the byssus is widespread, if not universally present, in post-larvae, and when present in the adult, represents the persistence of a post-larval organ, these adults thus being neotenous or paedomorphic.

7.2.3 Response to Contact

In the laboratory in depths of fine sediment that only slightly exceed the length of the byssus and plantigrade combined, the plantigrade affixes the byssus to the underlying solid surface and then remains suspended in the overlying sediment near the water–sediment interface. In deeper sediments the majority of plantigrades attach to the sides of the culture container slightly below the surface of the sediment, a minority to grains of sand in the sediment. The exterior surface of the shell of plantigrades reared in the absence of sediment soon becomes coated with accumulating organic detritus, whereas that of plantigrades with sediment to burrow into remain clean and smooth from movement within the abrasion sediment.

Without exception byssal plantigrades placed in clean glass dishes devoid of sediment persistently try to 'dig' the distal tip of the foot into the glass surface. In an experiment a small population of plantigrades 0.2 to 0.4 mm in length was held without sediment for 8 days (Carriker, 1961). These bivalves were then transferred to a petri dish in which the bottom was covered unevenly with a thin layer of clean fine bay sand, particles ranging in diameter from 10 to 120 μm. Plantigrades soon began crawling actively about, snail-like, through the sand (and conspicuously more vigorously than on the clean glass surface before), until deeper layers of sand surrounded them. At that point they crept further until valves were completely covered with sediment, and then ceased crawling and attached. These same plantigrades, when put in a dish with larger particles of sand (about 400 μm in diameter), crawled less actively and attached less readily than when among the fine grains. Thus grains of particle sizes smaller than the plantigrades appear to stimulate affixation more than larger particles. Attractiveness to byssal plantigrades of grains of different sizes and kinds of organic and inorganic sediment needs to be examined. Where thigmotactic sensory receptors are located in young *M. mercenaria* also needs to be determined.

On an open hard surface free of loose sediment, plantigrades aggregate and appear to receive contact stimulation from each other's shells, attaching and reattaching many times, forming a loose tangled mat of interwoven byssal threads. Some individuals even attach by means of forked byssi. How this might be done is hard to visualize, and has not been checked. In a laboratory experiment within an hour after some 50,000 plantigrades ranging in length from 0.2 to 0.5 mm were placed in shallow seawater in a clean glass container, the majority of individuals had aggregated into a mass of grape-like clusters, affixing to the bottom and to each other's shells (Carriker, 1961). How much of this behavior was a gregarious response, and how much a preference for contact, is not clear; perhaps both operated, the two reinforcing each other (Gotelli, 1990). Hunter (1949) reported a similarly strong thigmotactic response by byssal plantigrades of the rock-boring bivalve *Hiatella gallicana*. A search in the field for heavy sets of *M. mercenaria* revealed a large population of plantigrades settling in fine sediments in and among empty oyster valves. Upturned cupped valves lying on the surface of the bottom on an intertidal oyster bar and containing a mixture of fine organic detritus and sand–silt, when flushed clean of sediment by gentle washing with seawater, exposed dense numbers of plantigrades ranging in length from 1 to 4 mm. These were still attached to the inner surface of the oyster shells, even in valves covered by a centimeter or so of sediment. Plantigrades were not present on the surface of the sediment, nor on naturally clean shells nearby. In contrast sediment beneath and around the perimeter of the oyster valves contained only a small fraction of the number of plantigrades occupying the cupped valves. Several weeks later, oyster valves covered with sediment contained only an occasional plantigrade, whereas many were located in the sediment outside around the valves. This distribution suggests that upon reaching a certain size or age, plantigrades crawl out of the cupped valves into surrounding sediment (Carriker, 1961).

In nature in unstable sand, or sandy mud devoid of hard surfaces, in swiftly flowing seawater, plantigrades of *M. mercenaria* often secrete multiple attachments to several grains of sand (Fig. 7.6). Belding (1912) observed the same behavior in the young of this species.

Settlement of *M. mercenaria* frequently occurs in eddies or on the sides of channels in areas of swift currents (Belding, 1912; Carriker, 1961; Eckman, 1983). Orton (1937) reported that the bivalves *Cardium edule* become concentrated and settle out of the plankton in localities where 'slack and eddy waters' occur. Turner (1953; see also Butman, 1987; Butman et al., 1988) hypothesized that distribution of bivalves may be influenced by the same hydrographic features that affect zonation and sorting of inert sedimentary particles; that is, small byssal bivalves being only slightly less dense than accompanying sand grains, become concentrated in the corresponding sedimentary zone (see also Eckman, 1983, and Jonsson et al., 1991).

There is little likelihood of burial or breakage of byssal plantigrades of *M. mercenaria* in nature when waters and sediments become turbulent (Turner and George, 1955). As newly settled plantigrades, sand, and seawater are stirred with sufficient violence to move the sediment, the plantigrades (slightly less dense than sand particles) tend to remain above the sand in the agitated seawater. As motion of the seawater subsides, the sand settles out first, leaving the plantigrades on the surface. Byssal plantigrades can then attach the byssus to clumps of algal detritus that may be present and be rafted away on the next flowing tide, coming to rest on the bottom elsewhere on the following slack tide.

Abundance of siphonate byssal plantigrades of *M. mercenaria* along channels and inlets of swift tidal flows demonstrates that they are exceptionally well adapted to life in shifting sediments, and able to survive highly turbulent conditions. Earliest plantigrade stages, however, before siphons have developed, would appear to be more vulnerable; just how vulnerable has yet to be determined (Young and Chia, 1987).

7.2.4 Response to Light

Light reflected from the northern sky directed onto byssal plantigrades of *M. mercenaria* (0.2–1.3 mm long) on a surface devoid of sediment stimulates them to move, and they crawl randomly about, their behavior suggesting a strong negative photokinetic response. Measuring the size of young bivalves illuminated on a binocular microscope stage thus presents difficulties. Locomotory activity commences about 30 s after they are first illuminated. Those buried in clean fine sand also become active under illumination, the most exposed being the most active. Those buried more deeply, particularly under dark grains, are less, or not at all, affected, apparently the sediment shielding them from the light.

Quantitative response to light of about 500 active, rapidly growing byssal plantigrades ranging in length from 0.4 to 1.3 mm proved instructive. These bivalves were heaped several layers deep within a circle 2 cm in diameter in the center of each of two 6-cm petri dishes, in seawater 5 mm deep (35 ppt, 27°C). The control dish was set aside in dim laboratory light (no artificial lights) on a heavily overcast day. The experimental dish was placed on the glass stage of a binocular microscope and a beam of light from a microscope lamp, cooled through a water cell, was directed underneath the plantigrades by a flat substage mirror; approximately 700 foot-candles of light struck the bivalves. A mask with an opening 5.5 cm in diameter under the petri dish provided a narrow shaded zone around the edge of the dish. Light was most intense in the 2-cm circle in the center of the 5.5-cm lighted circle and graded into dim light peripherally. Almost as soon as light shone on the plantigrades, they began crawling outward from the illuminated center. In 10 min, half of the bivalves had left the center of 2 cm circle; in 20 min, most had departed and concentrated in the dimmer peripheral light;

a few crawled to the outer dark confines of the dish. In the control dish, at the end of the 20 min, most plantigrades remained aggregated in the center of the dish; only a few had moved outward about 1.5 cm from the original 2-cm circle of bivalve concentration. Plantigrades of different sizes were about equally distributed throughout the zone of dispersion, indicating that all sizes tended to move away from the bright light at about the same rate (Carriker, 1961).

The experiment was repeated with plantigrades ranging in length from 1.0 to 2.4 mm, 25 in each of the control and experimental dishes. In 20 min all bivalves, but one, crawled out of the lighted 2-cm circle, whereas those in the control remained clustered within the original 2-cm circle. Degree of dispersion was less marked in the larger than in the smaller bivalves.

Hence, when out of sediment, plantigrades of *M. mercenaria* crawl away from light of moderate intensity into areas of dimmer light. For siphonate plantigrades this means burying in the sediment, when this is present, where they can be shaded by sedimentary particles. The young of *Hiatella gallicana* are also negatively phototactic (Hunter, 1949). Whether byssal plantigrades in nature respond to bright daylight by burying more deeply in the sediment, as at low tide in intertidal areas, is not known. Negative phototaxis thus reinforces positive geotaxis enhancing survival, not only of young but also of older individuals.

7.2.5 Response to Flow of Seawater

Byssal plantigrades of *M. mercenaria* often open the valves and commence crawling when a stream of seawater is directed over them from a bulb pipette in a dish devoid of sediment. The possibility of a rheotactic response by byssal plantigrades was tested on 55 individuals ranging in length from 1.0 to 2.4 mm, placed 60 cm downstream from the inflow in a shallow trough 280 cm long and 10 cm wide. Seawater pumped directly from a nearby sound (35 ppt, 28°C) flowed at the rate of 2.5 cm/s in the trough and covered the bivalves several times their length. Light from laboratory windows (25 foot-candles) illuminated the trough at right angles to the flow of seawater. In 4 h, five bivalves had moved 1 to 4 cm from the original aggregation, two upstream, and three downstream. Consequently, at least under these conditions, there was no response to flow. In nature young bivalves generally experience alternating ebbing and flooding currents, rather than continuous unidirectional flows, so their lack of response to flow in the trough could have been anticipated.

What minimal locomotory activity was displayed was probably a response to absence of sediment; had sediment been present, in all likelihood there would have been no movement (Carriker, 1961).

7.2.6 Siphons, Foot, Valves, and Burrowing

The behavior of byssal plantigrades of *M. mercenaria* normally living within sediment can be observed clearly under low microscopic magnification in small sediment-filled glass cells. Open rectangular receptacles constructed of clear glass plate, slightly wider internally than the width of young plantigrades, and filled with fine muddy sand and seawater, are secured openside up against the stage of a binocular dissecting microscope held in a horizontal position. Plantigrades, confined loosely between the interior sides of the cells, move up and down readily, and can be followed closely, illuminated by dim natural light from a laboratory window. Seawater (ambient coastal salinities and summer temperatures) in the cells is changed frequently, or preferably run continuously (Carriker, 1961).

During burrowing, the plantigrade extends its foot downward out of the mantle cavity, lengthening it into a long, slender, tapering organ that worms its way among the sedimentary grains. At maximal extension of the foot the distal tip is distended, providing a partial anchor in the sediment against which the remainder of the foot pulls as it contracts and draws the anterior end of the plantigrade downward. Immediately after the plantigrade draws its posterior end downward and forward, the direction of movement of the valves more or less parallels the direction of the comarginal ridges on the valves (Cenni et al., 1989). In this direction the ridges appear to offer minimal resistance to movement of the valves in the sediment (Stanley, 1981); yet, after the plantigrade becomes stationary, the ridges, because of their relative size and curvature, themselves seem to aid in holding the valves in position. The seesaw, or rocking movement, of the valves continues until the plantigrade has completely buried itself. In loose sediment it can bury itself in a half dozen seesaws in less than a minute.

Burrowing activities of a byssal plantigrade of *M. mercenaria* are similar to those of an adult, which with siphons closed squirts seawater ventrally from the mantle cavity and simultaneously extends the foot forming a bulb on the end for anchorage; then the anterior pedal retractor contracts rotating the shell forward, followed by pulling by the posterior retractor, which rocks the shell backward (Trueman, 1966; Trueman and Ansell, 1969). Stanley (1975) confirmed that adult *M. mercenaria* 'walk' downward into the sediment by forward-and-back rocking movements of the shell. The blunt anterior end of the shell tends to reduce upward slippage of the anterior end, and the axis of backward rotation of the shell comes to lie anterior to the axis of forward rotation with each alternating contraction of the pedal retractor muscles. To what extent the bluntness of the anterior end of byssal plantigrade shells plays a part in burrowing as it is more compressed than that of adults remains to be determined.

When moving to the sediment–seawater interface a byssal plantigrade of *M. mercenaria* pushes itself upward with its foot, moving slowly and intermittently, more or less posterior end first, and omitting the seesaw movements of the valves. If siphons point in some other direction than surfaceward, the plantigrade reorients itself; what sensory cues make this possible, is not known. The foot is extended slowly pressing firmly against sedimentary grains; then it is withdrawn to variable lengths, rests for several seconds, extends the foot anew, and pushes the body further upward. Plantigrades can emerge from a depth of loose sediment at least five times their shell length.

During movements within the sediment a byssal plantigrade continues to pump seawater through its mantle cavity, though at a reduced rate that seems to correspond to the porosity of the surrounding sediment. Tentacles on the inhalant siphon fold inward forming a partial screen across the opening that admits seawater and small particles but excludes larger sedimentary grains and detritus. If the screen becomes blocked by large particles the plantigrade appresses its valves slightly, but forcefully, emitting spurts of seawater principally out of the inhalant siphon. The force of these jets can be considerable. A juvenile 3 cm in length buried in sediment with siphons extended just out of shallow water, for example, can jet a stream of seawater through an aerial trajectory some 75 cm from the inhalant siphon. For plantigrades submerged in sediment these jets clean the inhalant siphonal screen, and also tend to clear a small relatively sediment-free space in front of both siphons into which the siphons can then extend. The primary-definitive exhalant siphon does not clear a path for the bivalve among the sedimentary particles by discharging seawater, but in a limited way, as permitted by confinement imposed by surrounding sediment, does divert excurrent streams of seawater to one side. As a plantigrade nears the surface both inhalant and exhalant siphons may be extended to the surface and begin siphoning actively. Occasionally, only a single current stream of seawater may be seen bubbling from a juvenile at the sediment surface, indicating that incurrent seawater is being drawn interstitially (T.C. Nelson, pers. commun.). Burrowing movements by *M. mercenaria* resemble those described by Quayle (1952) for *V. pullastra*, except that the latter uses its pedal heel in digging.

M. mercenaria can extend siphons beyond the margin of its valves to a length at least twice the length of the valves. Measurement of siphonal lengths in juveniles ranging in length from 5 to 35 mm under conditions of low oxygen tension in containers, disclosed a range of ratios of length of siphons to length of shell of 1.0 to 2.3. Because of their capacity to pump seawater interstitially, however, clams can burrow into sediment more deeply than the length of fully extended siphons permits. At summer temperatures, a case in point, young individuals can bury below the sediment–seawater interface as much as five times the length of the shell. Deep burrowing becomes possible as soon as the circle of tentacles on siphonal tips becomes fully formed and functional, and the individual's geotactic response has reversed. Plantigrades in the length range of 2–13 mm burrow more rapidly and in larger numbers per unit time than do large individuals 14–22 mm in length (Baptist, 1955). This is possible because of the greater overall activity and superior burrowing ability of small individuals, which possess a relatively larger foot than older stages, a juvenile advantage that undoubtedly enhances survival of the species.

7.3 SOME AFTERTHOUGHTS

An American species, represented by fossil shells more than 25 million years old and identical to their modern counterparts (Abelson, 1956), *M. mercenaria* probably evolved along the western North Atlantic. Its recent range in lower estuaries, embayments, and ocean inlets in a narrow strip of eastern North America from Nova Scotia to Yucatan, has since been extended by man to western Europe and the western United States (Ansell, 1968). Within its indigenous coastal ecotone its veliger larvae can be widely dispersed at the risk of loss to the sea (Shanks, 1995). And some veligers are swept into the ocean by outflowing tidal currents (Carriker, 1961); the puzzling question is why all are not thus lost, and generally why sufficiently large, seeding swarms remain to maintain resident populations. A partial answer is intrinsic in the complex behavioral–embryogenic–organogenic development of the species: adaptation to bodies of seawater possessing attenuated exchange with the ocean, a relatively brief holoplanktonic veliger larva, a benthic-searching pediveliger, a byssally settling–anchoring plantigrade, rapid siphonate development, and facile burrowing in loose sediment. Early ontogenetic characteristics also account for the conspicuous biologic success of the species, notably evidenced by the dense native populations widespread before the current chemo-anthropogenic blitzkrieg!

Still, the early life history does not explain why populations of *M. mercenaria* occur primarily in sandy to sand–mud sediments in relatively clean seawater (Carriker, 1961). *Crassostrea virginica*, in contrast, has also occupied this same coastal rim, its ancestors possibly from before the Jurassic, but distributed mainly in the inner, brackish, more heavily sedimented reaches of estuaries (Shrock and Twenhofel, 1953; Carriker, 1967; Stenzel, 1971). Whether during early geologic periods ecologic ranges of both species coincided, and then in time diverged, can only be conjectured. Today, in fact, distribution of veligers of both species may still overlap in some coastal waters. An experimental study of differences in the vertical distribution of veligers of the two species relative to up-and-down estuarine tidal movements (Carriker, 1961, 1967) and to chemo-environmental tolerances (Castagna and Chanley, 1973; Carriker, 1986), especially of pediveligers and early byssal plantigrades, might plumb an answer to this distributional dissimilarity.

There is, moreover, the unexplained, generally dissimilar biogeographic distribution of the species *M. mercenaria* and its close relative *M. campechiensis* (J.N. Kraeuter, pers. commun.). Ranges diverge even though the two species hybridize and their early ontogenetic stages are quite similar and of equal duration (Loosanoff, 1959; Menzel, 1971). *M. mercenaria* occurs in mouths of estuaries, bays, and inlets from the Gulf of St. Lawrence to Florida, and sparsely in some places in the northern Gulf of Mexico. *M. campechiensis*, though, ranges from New Jersey to the Gulf of Mexico and the West Indies in shallow oceanic bottoms off the coast in its northern range, and inshore and sometimes with *M. mercenaria* in its southern range. An explanation could be sought in a comparative ecophysiological–behavioral study of the responses of early ontogenetic stages of the two species and their hybrids. Environmental factors that could be considered, singly and synergistically, include salinity, summer and winter temperatures, stability of the bottom sediment, pressure, and predator avoidance. Stages to utilize should be veligers, pediveligers, and byssal plantigrades, as these more sensitively serve as barometers of the extent of survival than older plantigrades.

The quite extraordinary anatomic–physiologic–behavioral transformation of *M. mercenaria* from veliger to plantigrade infallibly sets the stage for the discrete metamorphic shift from the planktonic to the benthic niche (Rowe and Ludwig, 1991). Singularly perilous is the ontogenetic transition from epifaunal to the infaunal habit. Two instinctive seemingly incompatible needs arise in the shift: maintaining contact with the seawater for feeding and respiration, and retiring within the sediment to avoid disturbance and achieve protection from predators. These conflicting needs are resolved by rapid development of the siphonal complex: early formation of the primary and definitive siphons, accompanied by rapid growth of the inner ctenidial demibranch and upper palp, together facilitating burrowing and respirational seawater exchange and feeding (Ansell, 1962).

Once in place within its sedimentary microcosm, an early byssal plantigrade should have gained some security. Ordinarily this would be the case, except that occasionally it can be displaced by violent storm currents, forced to feed in dense clouds of suspended inorganic particles raised by storm waves and currents, accidently buried in shifting sediments, and engulfed by waiting predators. Notwithstanding this catalog of environmental assaults, *M. mercenaria* has flourished well. Although its early life history stages may appear overly fragile and vulnerable to biological, chemical, and physical onslaughts, it is undeniably a hardy species, uncommonly well adapted to its coastal milieu.

Aspects of the biology of early stages of *M. mercenaria* bring up some provocative experimental questions, answers to which might be of consequence in the mariculture of the species. Could, for example, the duration of the veliger stage be abbreviated by genetic manipulation from its present minimal eight days to a few days? Shortening the planktonic phase would facilitate larval hatchery management. Size of early veligers is closely related to size of ova (Goodsell and Eversole, 1992; Hadley, 1993). Could selection likewise reduce the length of time for development to the pediveliger stage? Byssal plantigrades are raised successfully in the absence of sediment (Loosanoff and Davis, 1963; Castagna, 1984). But is sediment actually of benefit in the development and functioning of plantigrades? Does the presence of appropriate sediment contribute to a faster rate of growth, a healthier physiological condition, greater resistance to disease, longer life, than the absence of sediment? Restlessness of plantigrades when not surrounded by sediment suggests an inherent behavioral pattern for burrowing long imprinted in the species. Perhaps too deeply intrinsic to be altered? It goes without saying that raising young plantigrades in the absence of sediment dramatically simplifies hatchery management. There is, further, the burdensome matter of the extremely heavy mineralized valves of *M. mercenaria* (Newcombe et al., 1938). Growth of populations not only exacts a drain on the calcium and other minerals in culture seawater, but excessive weight of valves hampers handling, harvesting, and shipping to market. A proportionate reduction in both weight and volume of the valves, without unduly sacrificing shell strength, would decidedly benefit commerce. Shell modification possibly could be brought about by genetic manipulation, commencing with selection among early byssal plantigrades. There is substantial genetic variation for shell growth in this species, but apparently this variation is not stable during development (Hilbish et al., 1993).

A lackluster afterthought question: why modify the species? Its present biological attributes signal it a valuable choice for continued cultivation in hatcheries and increasingly in commercial grow-out systems (Menzel, 1989). And not to be ignored is the undeniable fact that with its culinary appeal, *M. mercenaria* is already in notable favor with gourmets — commercially speaking, a matter of quintessential importance!

7.4 ACKNOWLEDGMENTS

I wish to thank Michael Castagana, Albert Eble and John Kraeuter for especially helpful comments on the manuscript, Linda Leidy for typing the final draft of the typescript, Robert J. Bowden II for preparing photographic copies of the figures for publication, and the College of Marine Studies, University of Delaware, for facilities in which the literature search, synthesis, and writing for the chapter were done. The cost of preparation of the manuscript was supported in part by a grant from the Conchologists of America.

REFERENCES

Abelson, P.H., 1956. Paleobiochemistry. Sci. Am., 195: 83–92.

Ahn, I.-Y., Malouf, R. and Lopez, G., 1993. Enhanced larval settlement of the hard clam *Mercenaria mercenaria* by the gem clam *Gemma gemma*. Mar. Ecol. Progr. Serv., 99: 51–59.

Ansell, A.D., 1962. The functional morphology of the larva, and the post-larval development of *Venus striatula* (da Costa). J. Mar. Biol. Assoc. U.K., 42: 419–443.

Ansell, A.D., 1968. The rate of growth of the hard clam *Mercenaria mercenaria* (L.) throughout the geographical range. J. Cons. Perm. Int. Explor. Mer, 31: 364–409.

Babinchak, J. and Ukeles, R., 1979. Epifluorescence microscopy, a technique for the study of feeding in *Crassostrea virginica* veliger larvae. Mar. Biol., 51: 69–76.

Bachelet, G., Butman, C.A., Webb, S.M., Starczak, V.R. and Snelgrove, P.V.R., 1992. Non-selective settlement of *Mercenaria mercenaria* L. larvae in short term, still water, laboratory experiments. J. Exp. Mar. Biol. Ecol., 161: 241–280.

Baptist, J.P., 1955. Burrowing ability of juvenile clams. Spec. Sci. Rep., Fisheries No. 140, Fish. Wildl. Serv., 13 pp.

Bayne, B.L. (Ed.), 1976. Marine Mussels: Their Ecology and Physiology. Cambridge University Press, Cambridge, 506 pp.

Belding, D.L., 1912. A report upon the quahog and oyster fisheries of Massachusetts, including the life history, growth and cultivation of the quahog (*Venus mercenaria*), and observations on the set of oyster spat in Wellfleet Bay. Commonwealth of Massachusetts, Wright and Potter Printing Co., Boston, 134 pp.

Burke, R.D., 1983. The induction of metamorphosis of marine invertebrate larvae: stimulus and response. Can. J. Zool., 61: 1701–1719.

Butman, C.A., 1987. Larval settlement of soft-sediment invertebrates: the spatial scales of pattern explained by active habitat selection and the emerging role of hydrodynamical processes. Oceanogr. Mar. Biol. Annu. Rev., 25: 113–165.

Butman, C.A., Grassle, J.P. and Webb, C.M., 1988. Substrate choices made by marine larvae settling in still water and in a flume flow. Nature, 333 (6175): 771–773.

Carriker, M.R., 1961. Interrelation of functional morphology, behavior, and autecology in early stages of the bivalve *Mercenaria mercenaria*. J. Elisha Mitchell Sci. Soc., 77: 168–241.

Carriker, M.R., 1967. Ecology of estuarine benthic invertebrates: a perspective. In: G.H. Lauff (Ed.), Estuaries. Am. Assoc. Adv. Sci., Publ. 83, Washington, DC, pp. 442–487.

Carriker, M.R., 1986. Influence of suspended particles on biology of oyster larvae in estuaries. Am. Malacol. Bull., Spec. Ed., 3: 41–49.

Carriker, M.R., 1990. Functional significance of the pediveliger in bivalve development. In: B. Morton (Ed.), Proceedings of Memorial Symposium in Honour of Sir Charles Maurice Yonge, Edinburgh, 1986. Hong Kong Univ. Press, Hong Kong, pp. 267–282.

Carriker, M.R., 1996. The shell and ligament. In: V.S. Kennedy, R.I.E. Newell and A.E. Eble (Eds.), The Eastern Oyster: *Crassostrea virginica*. Maryland Sea Grant College, University of Maryland System, College Park, MD, pp. 75–168.

Castagna, M., 1984. Methods of growing *Mercenaria mercenaria* from postlarval- to preferred-size seed for field planting. Aquaculture, 39: 355–359.

Castagna, M. and Chanley, P., 1973. Salinity tolerance of some marine bivalves from inshore and estuarine environments in Virginia waters on the western mid-Atlantic Coast. Malacologia, 12: 47–96.

Cenni, S., Cerrato, R.M. and Siddall, S.E., 1989. Periodicity of growth lines in larval and postlarval shells of *Mercenaria mercenaria*. J. Shellfish Res., 8: 444–445.

Chanley, P. and Andrews, J.D., 1971. Aids for identification of bivalve larvae of Virginia. Malacologia, 11: 45–119.

Chia, F.-S., Buckland-Nicks, J. and Young, C.M., 1984. Locomotion of marine invertebrate larvae: a review. Can. J. Zool., 62: 1205–1222.

Crisp, D.J., 1974. Factors influencing the settlement of marine invertebrate larvae. In: P.T. Grant and A.M. Mackie (Eds.), Chemoreception in Marine Organisms. Academic Press, London, pp. 177–265.

Crisp, D.J., 1984. Overview of research on marine invertebrate larvae, 1940–1980. In: J.D. Costlow and R.C. Tipper (Eds.), Marine Biodeterioration, an Interdisciplinary Study. Naval Inst. Press, Annapolis, MD, pp. 103–126.

Eckman, J.E., 1983. Hydrodynamic processes affecting benthic recruitment. Limnol. Oceanogr., 28: 241–257.

Goodsell, J.G. and Eversole, A.G., 1992. Prodissoconch I and II length in *Mercenaria* taxa. Nautilus, 106: 119–122.

Gotelli, N.J., 1990. Stochastic models of gregarious larval settlement. Ophelia, 32: 95–108.

Hadley, N.H., 1993. Effects of hard clam hatchery management practices on productivity and on broodstock quality. World Aquacult., 24: 30–31.

Hilbish, T.J., Winn, E.P. and Rawson, P.D., 1993. Genetic variation and covariation during larval and juvenile growth in *Mercenaria mercenaria*. Mar. Biol., 115: 97–104.

Hunter, W.R., 1949. The structure and behavior of *Hiatella gallicana* (Lamarck) and *H. arctica* (L.), with special reference to the boring habit. Proc. R. Soc. Edinburgh, B 63: 271–289.

Jonsson, P.R., André, C. and Lindegarth, M., 1991. Swimming behavior of marine bivalve larvae in a flume boundary-layer flow: evidence for near bottom confinement. Mar. Ecol. Progr. Serv., 79: 67–76.

Keck, R., Maurer, D. and Malouf, R., 1974. Factors influencing the setting behavior of larval hard clams, *Mercenaria mercenaria*. Proc. Natl. Shellfish. Assoc., 64: 59–67.

Kellogg, J.L., 1892. A contribution to our knowledge of lamellibranchiate mollusks. Bull. U.S. Fish. Comm., 10: 389–436.

Kellogg, J.L., 1901. Clam and scallop industries of New York State. Bull. N.Y. Sta. Mus., 8 (43): 605–630.

Levin, L.A., 1990. A review of methods for labelling and tracking marine invertebrate larvae. Ophelia, 32: 115–144.

Loosanoff, V.L., 1959. The size and shape of metamorphosing larvae of *Venus (Mercenaria) mercenaria* grown at different temperatures. Biol. Bull. Mar. Biol. Lab., Woods Hole, 117: 308–318.

Loosanoff, V.L. and Davis, H.C., 1950. Conditioning *V. mercenaria* for spawning in winter and breeding its larvae in the laboratory. Biol. Bull. Mar. Biol. Lab., Woods Hole, 98: 60–65.

Loosanoff, V.L. and Davis, H.C., 1951. Delaying spawning of lamellibranchs by low temperature, Sears Found. J. Mar. Res., 10: 197–202.

Loosanoff, V.L. and Davis, H.C., 1963. Rearing of bivalve mollusks. Adv. Mar. Biol., 1: 1–136.

Loosanoff, V.L., Miller, W.S. and Smith, P.B., 1951. Growth and setting of larvae of *Venus mercenaria* in relation to temperature. J. Mar. Res., 10: 59–81.

Maia, B.C., 1988. Swimming responses of larvae of three mactrid bivalves to different salinity gradients. Masters Thesis, School of Marine Science, College of William and Mary, Williamsburg, VA, 115 pp.

Mann, R., Campos, B.M. and Luckenbach, M.W., 1991. Swimming rate and responses of larvae of three mactrid bivalves to salinity discontinuities. Mar. Ecol. Progr. Serv., 68: 257–269.

Menzel, W., 1971. The mariculture potential of clam farming. Am. Fish Farmer, 2 (8): 8–14.

Menzel, W., 1989. The biology, fishery and culture of quahog clams, *Mercenaria*. Dev. Aquacult. Fish. Sci., 19: 201–242.

Morton, J.E., 1960. The responses and orientation of the bivalve *Lasaea rubra* Montagu. J. Mar. Biol. Assoc. U.K., 39: 5–26.

Newcombe, C.L., Thompson, S.J. and Kessler, H., 1938. Variations in growth indices of *Venus mercenaria* L. from widely separated environments of the Atlantic coast. Can. J. Res., 16: 1–5.

Orton, J.H., 1937. Some interrelations between bivalve spatfalls, hydrography and fisheries. Nature, 140: 505–506.

Quayle, D.B., 1952. Structure and biology of the larva and spat of *Venerupis pullastra* (Montagu). Trans. R. Soc. Edinburgh, 62: 255–297.

Rowe, L. and Ludwig, D., 1991. Size and timing of metamorphosis in complex life cycles: time constraints and variation. Ecology, 72: 419–426.

Rumrill, S.S., 1990. Natural mortality of marine invertebrate larvae. Ophelia, 32: 163–198.

Ryder, J.A., 1989. The byssus of the young of the common clam (*Mya arenaria* L.). Am. Nat., 23: 65–67.

Shanks, A.L., 1995. Mechanisms of cross-shelf dispersal of larval invertebrates and fish. In: L. McEdward (Ed.), Ecology of Marine Invertebrate Larvae. CRC Press, Boca Raton, FL, pp. 323–367.

Shrock, R.R. and Twenhofel, W.H., 1953. Principles of Invertebrate Paleontology. McGraw-Hill, New York, 816 pp.

Stanley, S.M., 1975. Why clams have the shape they have: an experimental analysis of burrowing. Paleobiology, 1: 48–58.

Stanley, S.M., 1981. Infaunal survival: alternative functions of shell ornamentation in the Bivalvia (Mollusca). Paleobiology, 7: 384–393.

Stenzel, H.B., 1971. Oysters. Treatise on Invertebrate Paleontology, Part N, Vol. 3, Mollusca 6, Bivalvia. Geol. Soc. Am., Univ. Kansas, pp. N953–N1224.

Thorson, G., 1957. Bottom communities. Treatise on Marine Ecology and Paleoecology, Vol. 1. Geol. Soc. Am. Mem. 67, pp. 461–534.

Trueman, E.R., 1966. Fluid dynamics of burrowing. Science, 152: 523–525.

Trueman, E.R. and Ansell, A.D., 1969. The mechanism of burrowing into soft substrata by marine animals. Oceanogr. Mar. Biol. Annu. Rev., 7: 315–366.

Turner, H.J., 1953. A review of the biology of some commercial molluscs of the east coast of North America. Investigations of the shellfisheries of Massachusetts, 6th Rep., Massachusetts Dept. Nat. Resources, Div. Marine Fisheries, pp. 39–74.

Turner, H.J. and George, C.J., 1955. Some aspects of the behavior of the quahog, *Venus mercenaria*, during the early stages. Massachusetts Dept. Nat. Resources, Div. Marine Fisheries, Investigations of the Shellfisheries of Massachusetts, 8th Rep., pp. 5–14.

Van der Feen, P.J., 1949. Byssus. Basteria, 13: 66–71.

Wilson, F.S., 1990. Temporal and spatial patterns of settlement: a field study of molluscs in Bogue Sound, North Carolina. J. Exp. Mar. Biol. Ecol., 139: 201–220.

Yonge, C.M., 1962. On the primitive significance of the byssus in the Bivalvia and its effects in evolution. J. Mar. Biol. Assoc. U.K., 42: 113–125.

Young, C.M., 1995. Behavior and locomotion during the dispersal phase of larval life. In: L. McEdward (Ed.), Ecology of Marine Invertebrate Larvae. CRC Press, Boca Raton, FL, pp. 249–277.

Young, C.M. and Chia, F.-S., 1987. Abundance and distribution of pelagic larvae as influenced by predation, behavior, and hydrographic factors. In: A.C. Giese, J.S. Pearse and V.B. Pearse (Eds.), Reproduction of Marine Invertebrates, 9. General Aspects: Seeking Unity in Diversity. Blackwell, Palo Alto, CA, pp. 385–463.

Biology of the Hard Clam
J.N. Kraeuter and M. Castagna (Eds.),

Chapter 8

Physiological Ecology of Mercenaria mercenaria

Raymond E. Grizzle, V. Monica Bricelj and Sandra E. Shumway

8.1 INTRODUCTION

8.1.1 Overview

This chapter reviews the literature on physiological ecology, with emphasis on feeding, nutrition, growth, and production. It also includes a review of whole-organism behavior relevant to feeding. A consideration of energy flow through the organism provides the organizing framework for most topics discussed. Although there is no discussion of energy flow from populations of *Mercenaria mercenaria* to consumers at other trophic levels, the effects of the clam's individual physiological processes at population and higher ecological levels are discussed where appropriate. Some of the subjects discussed here will be covered in other chapters of this volume. In most cases, when a topic that is covered in more detail in another chapter is addressed here, discussion is kept to a minimum. For example, studies concerned with reproduction are covered in more detail elsewhere, even though they are relevant to the hard clam's overall energetics. In some cases, however, because the emphasis here is different from that in other chapters (e.g., environmental factors affecting feeding and growth), there will be some overlap.

Section 8.2 introduces the major terms that deal with energetics. The intent here is to define terms that will be discussed in detail in following sections, and show the overall relationships between the various anatomical structures and associated physiological processes. Section 8.3 reviews aspects of energy acquisition, which include feeding and various post-ingestion processes. Energy expenditures such as biodeposition, excretion, and other metabolic "costs" such as respiration are covered in Section 8.4.

The largest section by far is on nutrition, growth, and production. The vast majority of the research on hard clams has been concerned with growth and factors that influence it. This section focuses mainly on extrinsic (environmental) factors affecting growth. Although we provide a brief review of genetic factors influencing growth, this area is covered more fully in a separate chapter. This section integrates some of the topics covered earlier, and emphasizes the fact that feeding responses largely explain the patterns of individual growth responses.

The last section concludes the chapter with a review of whole-organism behavior and provides a fluid mechanical perspective on the feeding environment experienced by clams in nature. Feeding typically occurs very close to the bottom, well within the benthic boundary layer, because of the hard clam's infaunal life habit and the relatively short length of its siphons (a few centimeters when fully extended in the adult). Moreover, because the hard clam lives in relatively shallow waters that are affected by tidal and wind-induced currents, it generally feeds within flowing water. It is thus necessary to have some understanding of the fluid environment in which the hard clam lives to gain an ecologically realistic understanding of its behavior and physiology. The section ends with a review of mathematical modeling of processes relevant to feeding and growth.

8.2 ENERGETICS AND PHYSIOLOGY

The total energy budget for heterotrophic organisms from a mass-balance perspective is:

$$I = G + M + E + F \tag{8.1}$$

where I = ingestion (or consumption) rate, G = growth rate (= somatic growth + reproductive growth), M = total metabolic rate, including aerobic and anaerobic respiration, E = excretion rate, and F = fecal production (or egestion).

Feeding physiology and nutrition are mainly concerned with "energy acquisition" processes (components I and G, respectively, in the above equation), and are the focus of Sections 8.3 and 8.5. "Energy expenditures" (M, E, and F) are dealt with in Section 8.4.

Fig. 8.1 is a schematic of energy flow through an individual suspension-feeding bivalve like *Mercenaria mercenaria*, showing the anatomical structures associated with major physiological processes (see Chapter 4 for anatomical details) and components of the above energy budget equation.

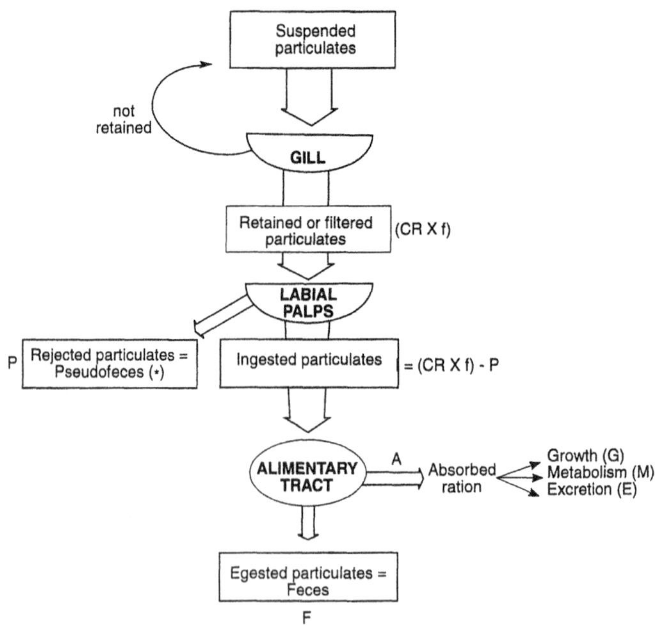

Fig. 8.1. Flow diagram of the components of energy balance and pathways of particulates (food) in a typical suspension-feeding bivalve like the hard clam (modified from Bayne and Newell, 1983 by Malouf and Bricelj, 1989). See text for definitions of symbols. *Pseudofeces are only produced above a threshold concentration of particulates (see text).

Suspension-feeding, lamellibranch bivalves including the hard clam, are particularly complex in their feeding physiology because it involves sophisticated particle processing (i.e., particle retention, transport, and selectivity) that precedes actual ingestion. In brief, water currents are produced by ciliary action on the gills which function in gas exchange and food capture. Suspended particles are removed from the pumped water stream onto the gills, transported oralward along dorsal and ventral gill tracts, and transferred to the labial palps. Sorting of particles occurs on the gills and/or palps such that particles are either transferred to the mouth for ingestion, or rejected in pseudofeces through the inhalant siphonal opening.

Individual components of the energy budget are defined below.

PR = pumping rate = ventilation rate = volume of water flowing through the gills per unit time (in L h^{-1})

CR = clearance rate (sometimes referred to as filtration rate) = volume of water filtered completely free ("cleared") of particles per unit time (in L h^{-1})

When all suspended particles are removed from the water flowing through the gills (particle retention efficiency = 100%), CR = PR. When no pseudofeces are produced (all particles are ingested):

$$I = \text{ingestion rate (in mg h}^{-1}\text{)} = \text{CR} \times f \tag{8.2}$$

where f = particle concentration (in mg L^{-1}).

When pseudofeces are produced:

$$I = (\text{CR} \times f) - P \tag{8.3}$$

where P = rate of pseudofeces production (in mg h^{-1}).

The energy budget equation above essentially partitions the various fates of ingested particles in the organism. The growth term (G) represents the sum of somatic and reproductive growth, both of which depend on the amount of food energy absorbed in the gut. The rate of absorption (A) is related to ingestion as follows:

$$A = \text{absorption rate} = G + M + E = I - F = I \times (\text{AE}/100) \tag{8.4}$$

where AE = absorption efficiency = $A/I \times 100$.

Absorption efficiency is generally calculated from the ratio of ash (inorganic matter or non-absorbed component of the diet) to ash-free dry weight (AFDW = organic matter or absorbed dietary component) of a representative sample of the feces and ingested particulates, using a generalized form of the equation by Conover (1966):

$$\text{AE (\%)} = 1 - \left[(\text{Ash}/\text{AFDW}_{\text{food}}) \,/\, (\text{Ash}/\text{AFDW}_{\text{feces}})\right] \times 100 \tag{8.5}$$

At concentrations below the threshold for production of pseudofeces, or when pseudofeces are produced but ingestion selectivity does not occur (there is no discrimination between particles ingested and those rejected in pseudofeces), the inorganic to organic ratio of ingested particulates can be determined directly from that of suspended particles (assuming 100% particle retention efficiency). More sensitive dual radiotracer techniques can be used instead of Conover's gravimetric method, in which an inert radiolabel (^{51}Cr or ^{241}Am) acts as a tracer for the inorganic component of the diet, and ^{14}C tracks organic carbon (Bricelj and Malouf, 1984; Bricelj et al., 1984a and Wang and Fisher, 1996 applied this method to suspension-feeding bivalves). The various assumptions, relative merits, and conditions for suitable application of each of these methods have been reviewed elsewhere (Calow and Fletcher, 1972).

The term "assimilation rate" refers to the sum of G and M. Somatic and reproductive growth (G) is dependent upon the efficiency with which the absorbed food energy is converted to biomass. Growth efficiency (K) can be expressed in two ways:

$$K_1 = \text{gross growth efficiency} = G/I \times 100 \tag{8.6}$$

$$K_2 = \text{net growth efficiency} = G/A \times 100 \tag{8.7}$$

8.3 ENERGY ACQUISITION: FEEDING PHYSIOLOGY

Energy acquisition includes processes that range from pumping water through the mantle cavity to removing, sorting, rejecting, and ingesting suspended food particles, as well as post-ingestive processes (sorting in the gut, digestion, and absorption). These processes have been examined in *M. mercenaria*, although not to the extent they have in other bivalves such as the blue mussel, *Mytilus edulis*. This section centers on what is known about the hard clam, but it also deals with other more extensively studied bivalves. This is done because there are generalities in the response patterns among species that allow development of a general theory of suspension-feeding behavior for bivalves of relevance to the hard clam.

Previous work on bivalve feeding physiology has provided a basic understanding of many of the relevant processes (for reviews see: Bayne and Newell, 1983; Malouf and Bricelj, 1989; Bayne and Hawkins, 1992; Bayne, 1993, 1998; Navarro and Iglesias, 1993). However, the following statements by Bayne (1993, pp. 1–2) point to the complex nature of the feeding process, and the preliminary nature of our overall understanding at this time:

> "In spite of considerable experimental and observational study over many years, controversies and uncertainties still exist concerning fundamental features of feeding behavior in suspension-feeding bivalves... those (models) which are to be most effective will recognize the existence of both physical and physiological influences on feeding behavior, set in the context of response to the considerable spatial and temporal variability within the natural food environment ... When these processes are viewed in their totality, it is the coupling of physiological and morphological processes that can be seen to comprise and to control integrated feeding behavior."

8.3.1 Basic Anatomy and Physiology of the Feeding Process

Hard clams are capable of obtaining some nutrition from dissolved organic matter (Rice and Stephens, 1988; see below), but their primary food source is suspended particles (phytoplankton, heterotrophic microorganisms, and detritus), as is the case for suspension-feeding bivalves generally (Bayne and Hawkins, 1992). Studies on bivalve feeding physiology have concentrated on understanding the mechanisms involved in particle capture and processing. Studies on several species, including hard clams, have elucidated the following generalized process.

Specialized cilia on the gill surface create currents that move water into and out of the mantle cavity (Fig. 8.1; see Chapter 4 for anatomy and other details). In infaunal, siphonate bivalves, water enters and exits through extensible siphons. The pumped water flows across the mantle surface, through the gill filaments where gas exchange occurs, and where particles are removed from suspension. Some sorting and selection of particles typically occurs on the labial palps before ingestion. In some bivalves (e.g., *Crassostrea* spp.) sorting may also occur on the gills (Ward et al., 1994; 1997). Sorting involves discrimination of the retained particles into either particles that will be transferred to the mouth for ingestion, or pseudofeces that will be rejected from the mantle cavity. In the hard clam, pseudofeces accumulate below the labial palps and are sporadically ejected through the incurrent siphon by rapid contraction of the adductor muscles and siphons. Pseudofeces production provides a mechanism to reject excess material filtered by the gills and thereby control the total amount of food ingested. Additionally, particle selection associated with pseudofeces formation provides the potential to modify food composition before ingestion. For example, ingested seston may be enriched organically compared to the filtered seston by pre-ingestive rejection in the pseudofeces of less nutritious inorganic particles Detailed reviews on feeding processes and controversies surrounding feeding mechanisms in suspension-feeding bivalves (other than hard clams) are provided elsewhere (Jørgensen, 1990, 1996; Ward, 1996; Ward et al., 1993a,b; Beninger and St-Jean, 1997; Beninger et al., 1997, etc.).

8.3.2 Particle Retention

Most post-settlement through adult stages of suspension-feeding bivalves, including *Mercenaria mercenaria* (see Fig. 1 in Riisgård, 1991), and the venerid clam *Venerupis pullastra* (Møhlenberg and Riisgård, 1978), are able to retain particles >3 to 4 μm in diameter with ~100% efficiency. This is in contrast to scallops (Pectinidae) which only retain particles >5 to 6 μm with maximum efficiency (Fig. 8.2). Retention efficiency (RE) is generally determined by simultaneous measurement of clearance rates of different size particles, and expressed relative to that of larger particles assumed to be captured with 100% efficiency. The efficiency of particle capture typically declines steeply with decreasing particle size, and for particles <3 μm it varies considerably among species. This variability is generally attributed to differences in the complexity of gill cirri, i.e., to the presence and magnitude of development of the gill latero-

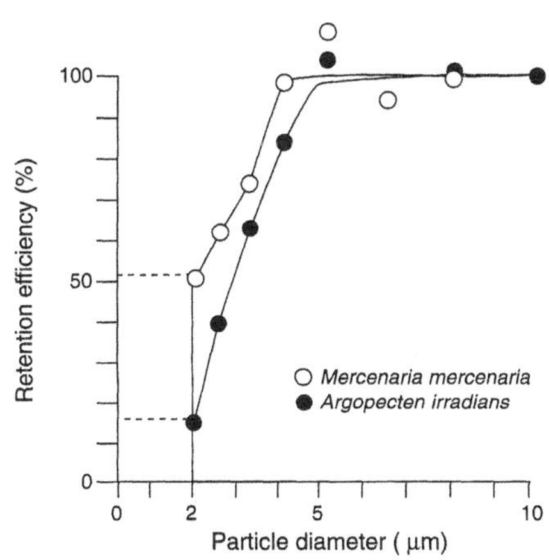

Fig. 8.2. *Mercenaria mercenaria* and *Argopecten irradians*. Particle retention efficiency as a function of particle diameter for two adult bivalves: the hard clam compared to a pectinid, the bay scallop (from Riisgård, 1991).

frontal cirri (Owen and McCrae, 1976; Silverman et al., 1996). Changes in the beat frequency of cirri may also cause some intraspecific variation in the size at which 100% RE occurs.

The mechanism of particle retention by gill filaments, however, is not fully understood. Several mechanisms have been postulated, including mechanical sieving by latero-frontal cirri in which RE is controlled only by the spacing between gill cirri [a mechanism argued to be incompatible with low Reynolds number fluid mechanics of bivalve suspension-feeding (Jørgensen, 1981)], hydrodynamic entrainment (Jørgensen, 1990), and non-sieve-like capture mechanisms such as direct interception of particles (reviewed by Ward, 1996). Indeed, particle retention may be a dynamic process, which depends on the integrated activity of the entire gill-ciliated surface and its associated musculature, and on a combination of mechanisms.

Although changes in RE in response to environmental parameters have not been determined for *M. mercenaria*, Stenton-Dozey and Brown (1992) found that in another venerid clam, *Venerupis corrugata*, overall clearance rate, and RE varied in relation to particle availability over the tidal cycle. Thus, RE of large particles (9 to 13 μm in diameter) increased significantly at high tide relative to low tide, presumably in response to the increased availability of particles in that size range. Most studies have focused on the retention of small particles to establish a lower size threshold for particle capture. Additional work is required to determine the upper size limit of effective particle capture.

Adult *M. mercenaria*, which have large latero-frontal cirri, retain all particles >4 μm with maximum efficiency, and retain 2 μm particles with relatively high (50%) efficiency compared to bay scallops (15% RE for 2 μm particles), which lack latero-frontal cirri (Fig. 8.2). In contrast, high-speed video-microscopy and direct counts by epifluorescence microscopy have demonstrated that both *Argopecten irradians* and *M. mercenaria* larvae capture 1 and 4 μm particles with equal efficiency, and can therefore effectively use picoplankton-sized particles (0.2 to 3 μm) as food (Gallager et al., 1989, 1994). Discrepancies with previous work which showed poor retention of small particles by bivalve larvae (Riisgård et al., 1980) were attributed to contamination by larval fecal pellets at the lower size range of particle size spectra obtained with an electronic particle counter.

8.3.3 Particle Selection and Pseudofeces Production

Pre-ingestive particle selection is associated with pseudofeces production in bivalves. As the filtered particles are processed on the gills and labial palps, they are either ingested or rejected as pseudofeces. In the process of pseudofeces formation, a variable amount of selection can occur. Comparison of the composition of the pseudofeces relative to the ambient seston provides an estimation of selection efficiency (Kiørboe and Møhlenberg, 1981). It is well established that bivalves can selectively ingest algae and reject sediment particles in pseudofeces from a mixed suspension, but this ability varies greatly among species, and among populations adapted to different turbidities (Kiørboe and Møhlenberg, 1981; see brief review by Bayne, 1993). High selection efficiency appears to be correlated with the ability to tolerate environments with high suspended sediment loads. It has also been associated with the relative size of feeding organs, i.e., large labial palps relative to the size of the gills (Kiørboe and Møhlenberg, 1981) and with the distribution and type of secretion of gill mucocytes (not yet studied in *M. mercenaria*) (Beninger et al., 1993). Selection efficiency of *M. mercenaria*, measured by comparing the ratio of chlorophyll *a* to dry weight of pseudofeces and seston of animals fed a mixed algal–sediment suspension, was somewhat lower than that of *M. edulis* (6.2 and 8.7, respectively) (Bricelj, 1984). It is moderately high, however, when compared to selection efficiencies similarly determined by Kiørboe and Møhlenberg (1981), which ranged from 2.9 to 15.8 for ten species of bivalves. Enrichment of pseudofeces in coarse mineral particles relative to seston in *M. mercenaria* further suggests that hard clams are able to discriminate among sediment particles on the basis of size or a parameter which covaried with size (Bricelj, 1984). Overall, ingestion selectivity allowed hard clams to lose a maximum of only 18% of the algae cleared from a mixed algal–silt suspension, in pseudofeces (Fig. 8.3B). Comparison of pseudofeces and mixed algal suspensions using flow-cytometric techniques demonstrated that bivalves can also discriminate among algal species of comparable size (Shumway et al., 1985), but this work has not been extended to hard clams.

Pseudofeces production in bivalves typically only occurs above a certain threshold seston

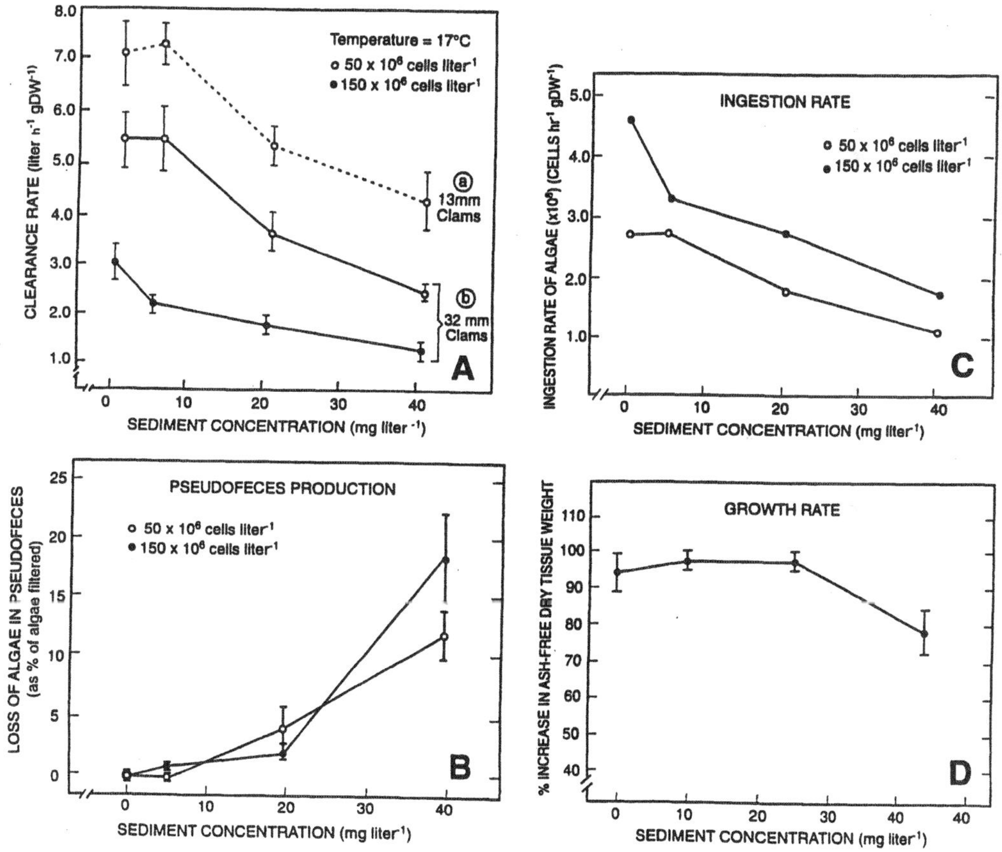

Fig. 8.3. *M. mercenaria*. Effects of sestonic sediment concentration on: (A) weight-specific clearance rate; (B) percent loss of algae in pseudofeces; (C) algal ingestion rate at algal concentrations of 50×10^6 and 150×10^6 *Pseudoisochrysis paradoxa* cells L^{-1}; (D) 3-week growth rate of hard clams fed 50×10^6 *Pseudoisochrysis paradoxa* cells L^{-1} supplemented with different sediment concentrations (all at 21°C; from Bricelj and Malouf, 1984; Bricelj et al., 1984b).

concentration that allows approximately maximal ingestion rate. In their review, Bayne and Newell (1983) showed that this threshold lies between 1 and 5 mg dry weight L^{-1} total seston for several species, a range that is well below the seston concentrations typically encountered in some coastal waters where hard clams live. This threshold is difficult to determine in clams such as *M. mercenaria*, which eject pseudofeces intermittently via contraction of the inhalant siphon, because it will depend on some arbitrary selection of the time interval between ejections. However, in hard clams, Bricelj and Malouf (1984) observed that pseudofeces production, in terms of bulk dry weight, was inconspicuous below ca. 10 mg dry weight L^{-1}.

The threshold of pseudofeces production in bivalves fed pure algal diets has also been expressed in terms of cell packed volume (volume of cells per volume of seawater) to standardize for differences in algal size. Based on data available for *M. edulis*, Malouf and Bricelj (1989) suggested a range of 2 to 20 μl of algae L^{-1} for this threshold, although a higher value (ca. 95 μl L^{-1}) was reported for small juvenile *M. mercenaria* (Wikfors et al., 1992). It is important to note, however, that some species, such as the hard clam, appear to regulate their ingestion rate at high seston concentrations primarily via reduction in clearance rates, rather than through copious production of pseudofeces (Fig. 8.3A), as documented in mussels, *M. edulis*, and oysters, *Crassostrea virginica* (Tenore and Dunstan, 1973; also see Section 8.4.2 below). As pointed out by Malouf and Bricelj (1989), in the former species, high selection efficiency in the absence of high rates of pseudofeces production will likely provide a relatively inefficient mechanism for the removal of unwanted particles.

The mechanisms of particle sorting on the labial palps and the basis for particle discrimination (size, chemical composition, etc.) in bivalves remain debatable. Reduced viscosity of mucus at the palps appears to allow for dispersion and sorting among individual particles and suggests that sorting is not necessarily incompatible with mucus feeding (Newell and Jordan, 1983; Ward et al., 1994). While the labial palps are generally considered the predominant site of particle selection, sorting also takes place during transport on the gills in some species. Video-endoscopy has shown that in bivalves with complex pseudolamellibranch, heterorhabdic gills, such as the oysters *Crassostrea* spp., algae are transported preferentially along the dorsal food tracts. In contrast, non-nutritious, detrital particles move along the ventral tracts, where they are more likely to be rejected in pseudofeces (Ward et al., 1993a,b; Levinton et al., 1996; Newell and Langdon, 1996). This *in vivo* technique of observation has not yet been applied to the study of *M. mercenaria*, which has eulamellibranchs gills, but sorting on the gills was not observed in other eulamellibranchs (Beninger et al., 1997).

Control of gut passage time can provide a mechanism for post-ingestive selection of particles and maximization of energy gain. Using dual radiotracer techniques, Bricelj et al. (1984a) demonstrated that juvenile hard clams had shorter gut residence times for algae that were inefficiently absorbed (chlorophytes and cyanobacteria) than for *Pseudoisochrysis paradoxa*, a species absorbed with high efficiency. This was presumably because the former were transferred directly to the hind gut, while the latter was diverted into the digestive gland.

8.3.4 Feeding Rate

Of the three measures of suspension-feeding (clearance, pumping, and ingestion), clearance rate is the most commonly measured. Pumping rates of bivalves are less frequently measured, largely due to the difficulties in directly measuring flow rate through the mantle cavity.

Ingestion rates are also difficult to measure directly when pseudofeces are produced. Hence, much of the bivalve literature deals with clearance rate because it can be more easily estimated. This can be done in a static system by measuring the decrease in seston concentration in the experimental chamber (Coughlan and Ansell, 1964), or more realistically in a flow-through system (for discussion of appropriate temporal scales for such measurements see McClatchie, 1992). If 100% retention rate is assumed for seston in the water flowing through the mantle cavity, clearance rate in a flow-through system can be calculated by the following equation (Hildreth and Crisp, 1976):

$$\text{Clearance rate} = F\left[(C_i - C_o)/C_o\right] \quad (8.8)$$

where F = flow rate in the experimental chamber, C_i = seston concentration of inflow water, and C_o = seston concentration of outflow water.

Physiological rates, including feeding rate, are known to be strongly influenced by body size. The relationship between feeding rate, measured by pumping or clearance rate, and body size is described by the generalized allometric equation:

$$\text{Feeding rate} = a\,(\text{body size})^b \quad (8.9)$$

where a and b are fitted parameters (reviewed by Winter, 1978 and Bayne and Newell, 1983).

Tables 8.1 and 8.2 summarize data from studies on the relationship between feeding rate and body size in *M. mercenaria*. Data obtained for juveniles are presented separately in Table 8.2, because smaller individuals are often excluded from allometric studies. Bayne and Newell (1983) noted that the weight exponent, b, tends to be lower when bivalves are fed natural seston, than when they are fed cultured algae. Thus, the weight exponents calculated by Hibbert (1977b) using natural particulates are low (0.27 to 0.31 in Table 8.1) relative to the expected value of 0.75.

In general, there is a negative relationship between CR and body size when the former is expressed on a per unit weight basis (= $CR/W = CR_W$, referred to as the weight-specific

TABLE 8.1

Relationship between clearance rate and body size of *Mercenaria mercenaria* (modified from Malouf and Bricelj, 1989)

Reference	Experimental suspension	T (°C)	W (g)	a	b	L (mm)	a'	b'
Coughlan and Ansell (1964)	dye	18–20	0.33–4.81	2.595 [a]	0.730 [a]	30–83	0.0026	1.809
Hibbert (1977a,b) [b]	natural seston	12	0.68–7.29	0.896	0.288	43–88	0.026	0.917
		17		1.171	0.281		0.044	0.865
		20		1.196	0.310		0.032	0.954
		25		1.511	0.271		0.063	0.834
Doering and Oviatt (1986)	natural seston	13.5–21				32–107	0.033	0.967

Clearance rate (L ind^{-1} h^{-1}) = $a(W)^b$; W = tissue dry weight (unless noted) in g; clearance rate (L ind^{-1} h^{-1}) = $a'(L)^{b'}$; L = shell length in mm.

[a] Calculated by Winter (1978).

[b] W is ash-free dry weight. Ash content of adult *M. mercenaria* soft tissues = 10% (Ansell et al., 1964). Relationship between shell length and weight in November used for 12°C experiment, and that in August used for higher temperatures (Hibbert, 1977a).

TABLE 8.2

Average clearance rates relative to body size of juvenile *Mercenaria mercenaria* (from Malouf and Bricelj, 1989)

Reference	Experimental suspension	T (°C)	Shell length (mm)	Soft tissue dry wt. (mg)	CR_W ($L\ h^{-1}\ g^{-1}$)	CR ($mL\ ind^{-1}\ h^{-1}$)
Bricelj (1984)	*Pseudoisochrysis paradoxa* (50×10^6 cells L^{-1})	17	13	20	7.162	136.6
Bricelj (unpubl.)	natural seston	12	10	16	0.801	–
		20	10–15	14–64	2.480	–
		27	11–15	15–61	4.282	–
Walne (1972)[a]	*Isochrysis galbana* (50×10^6 cells L^{-1})	20–22	4.5	0.89	5.381	4.8
	Dunaliella tertiolecta (50×10^6 cells L^{-1})	20–22	4.5	0.89	3.700	3.3
	Phaeodactylum tricornutum (50×10^6 cells L^{-1})	20–22	4.5	0.89	17.040	15.2

[a] Calculated from Walne's data using: W (mg) $= 0.0102\ L$ (mm)$^{2.973}$ ($r^2 = 0.89$); ash content of tissues = 17 to 23%; regression calculated for clams 7 to 14 mm in shell length (Bricelj, unpubl. data).

clearance rate). CR_W thus only adequately corrects for differences in body size when the size range of individuals is relatively small (Table 8.2). Hence, Bayne and Newell (1983) recommended that for comparative purposes, physiological rates (e.g., CR) be corrected for differences in weight as follows:

$$CR_s = (W_s/W_e)^b \times CR_e \tag{8.10}$$

where CR_s = CR of a standard-sized animal, CR_e and W_e are the uncorrected clearance rate, and weight, respectively, of the experimental animal, and b is the weight exponent of the allometric equation. When the standard size is selected as that of a 1 g animal, the above equation becomes:

$$CR_{(1\,g)} = CR_e/(W_e)^b \tag{8.11}$$

In general, weight-standardized clearance rates of *M. mercenaria* were found to be comparable to those of other bivalves, including oysters, cockles and mussels, although they were generally lower than those of scallops (Malouf and Bricelj, 1989).

Infaunal taxa like *M. mercenaria* are known to respond to a variety of physical and chemical changes in their environment. Such behavioral responses, which are on time scales of seconds to minutes, affect feeding rate because they involve closure and/or retraction of the siphons. Hence, the feeding process is easily altered and can be widely variable over even short time intervals, especially in larger/older individuals. Bivalve pumping rates have been shown to be affected by both environmental and physiological factors, and feedback between the two (for reviews see: Bayne and Newell, 1983; Bayne and Hawkins, 1992; Bayne, 1993). Although the general anatomy and overall functioning of the gill pump is reasonably well known (see Chapter 4), there is only a preliminary understanding of the environmental and physiological factors that control it. The gill pump certainly should not be viewed as simply being "on" or "off". Rather, it should be viewed as a complex organ with numerous control

TABLE 8.3

Studies on the effects of environmental factors on three measures of individual feeding response of *Mercenaria mercenaria*

Reference	Feeding response			Environmental factor					
	pumping ($L\ h^{-1}$)	clearance ($L\ h^{-1}$)	ingestion ($g\ h^{-1}$)	temp.	seston conc.	seston comp.	O_2 conc.	water flow	other
Rice and Smith (1958)		×			×	×			
Hamwi (1969)	×			×		×	×		× [a]
Hamwi and Haskin (1969) [b]	×						×		
Walne (1972)		×			×	×		× [c]	
Tenore and Dunstan (1973)		× [d]			×				
Walsh (1974)	×						×		
Epifanio and Srna (1975)		×							× [e]
Van Winkle (1975)	×						×		
Hibbert (1977b)	×			×					
Bricelj and Malouf (1984)		×	×		×	×			
Turner (1990)					×			×	
Lesser and Shumway (1993)		×							× [f]

An "×" indicates that feeding response or environmental factor was studied. Selected results from the listed studies are summarized graphically in Figs. 8.3–8.9.

[a] Hamwi reported on the effect of salinity and salinity × temperature interactions.

[b] Verduin (1969) disputed Hamwi and Haskin's conclusion that pumping rate is regulated by oxygen need; Van Winkle (1975) discusses the problems involved in relating oxygen requirements to pumping rates.

[c] Walne's conclusions on the effects of water flow were disputed by Hildreth and Crisp (1976) who concluded (based on a re-analysis of Walne's data) that water flow does not affect feeding rate (but see discussion in text of more recent work).

[d] Tenore and Dunstan (1973) reported their data as the amount of carbon removed per unit dry-weight of animal.

[e] Epifanio and Srna (1975) studied the effects of ammonia, nitrite, nitrate, and phosphate ions.

[f] Lesser and Shumway (1993) studied the effects of several toxic dinoflagellates.

mechanisms, probably routinely capable of a wide range of pumping rates (Bayne, 1993). Table 8.3 lists the major environmental factors affecting feeding rates in hard clams.

8.3.4.1 Temperature and salinity

Of the five environmental factors that have been studied and can be expressed quantitatively as continuous independent variables (temperature, seston concentration, oxygen concentration, water flow, and salinity), only temperature (Fig. 8.4) and salinity (Fig. 8.5) have been studied across a range that approaches typical variability in nature. Both show an inverted parabolic relationship to the dependent variable, feeding rate. This pattern is typical of how organisms respond generally to changes in environmental factors. There is a range across which the relationship is positive, then some optimum is reached, after which further increases in the independent variable cause decreased response levels. This pattern is sometimes explained using the "law of tolerance" which states that there are upper and lower limits of tolerance by organisms to most abiotic environmental factors (Smith, 1996). Somewhere between the extremes is a range of optimal conditions. Various combinations of temperature and salinity show a similar pattern (Fig. 8.6).

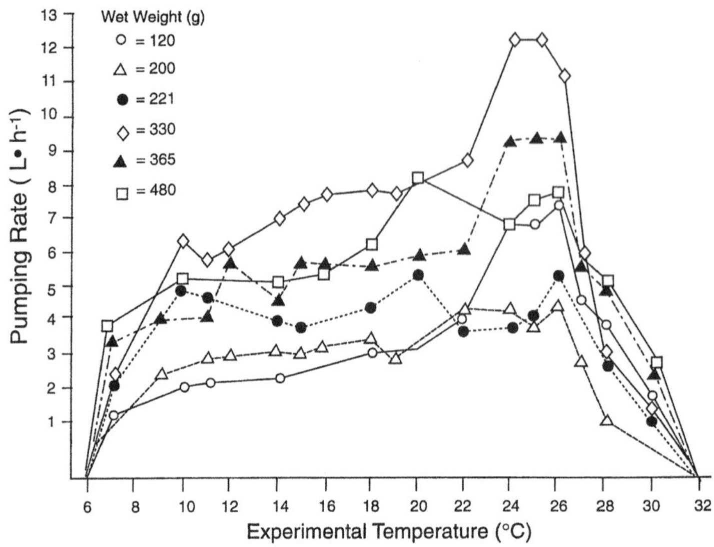

Fig. 8.4. *M. mercenaria.* Pumping rate at different water temperatures for six different size classes (total body wet weight), held at 23 to 27 ppt salinity (from Hamwi, 1969).

Inhibited pumping rates at temperature and salinity extremes (Figs. 8.4–8.6) may in part be due to discontinuous feeding. Based on continuous records of shell movements, Loosanoff (1939) determined that adult *M. mercenaria* show discontinuous feeding activity at lower temperatures. Hard clams remained open 69 to 90% of the time (average 19.5 h day^{-1}) within a temperature range of 11° to 18°C. This percentage decreased rapidly at lower temperatures, such that all clams remained closed at 2° to 3°C. Activity, measured as the percent of the test population with extended and open siphons, was found by Van Winkle et al. (1976) to be a function of salinity as well as temperature. Based on this criterion, ~20% of the hard clams were active at salinities below 14 to 21 ppt depending on the temperature.

Feeding responses to salinity and temperature have been explained mechanistically at the organismic level by other studies on tolerance limits of *M. mercenaria.* Salinities below 15 ppt generally have a negative effect on hard clams, inhibiting, in addition to feeding rate, burrowing, growth, and long-term survival of both juveniles and adults (Chanley, 1958; Castagna and Chanley, 1973). The upper salinity range has not been well studied for juveniles and adults, but salinities above 32 ppt are detrimental to developing eggs and larvae (Davis, 1958; Feng, 1969). There is a similar trend for temperature. Below 6°C, hard clams cease pumping (Ansell, 1964). Growth of juveniles and adults does not occur, and there are various detrimental effects on eggs and larvae, outside a range of 9° to 31°C (Ansell, 1968; Bardach et al., 1972). Salinity and temperature interactive effects on various physiological responses also have been studied (Davis and Calabrese, 1964; Feng, 1969; Hamwi, 1969; Lough, 1975). For both temperature and salinity, regardless of the response being studied, the univariate response pattern across a wide range of the independent variable typically approximates an inverted parabola.

Metabolic rates and essentially all physiological processes of poikilotherms like bivalves are strongly controlled by ambient temperature. The relationship between temperature and

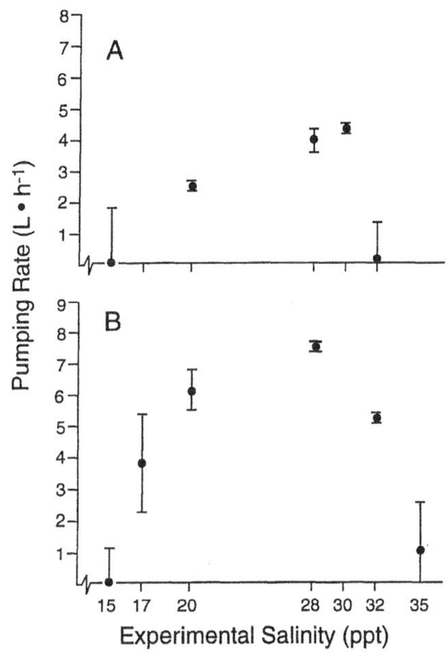

Fig. 8.5. *M. mercenaria.* Pumping rate at different water salinities. (A) Non-acclimated. (B) Acclimated for 4 to 7 days at the experimental salinity (from Hamwi, 1969).

various physiological rates has been described using temperature coefficients. The ratio of a rate function at one temperature to that at 10°C lower provides an index (Q_{10}) of the dependence of the rate on temperature. Table 8.4 lists the few studies that provide Q_{10} values for feeding rates in *M. mercenaria.* The published Q_{10} values are quite similar when

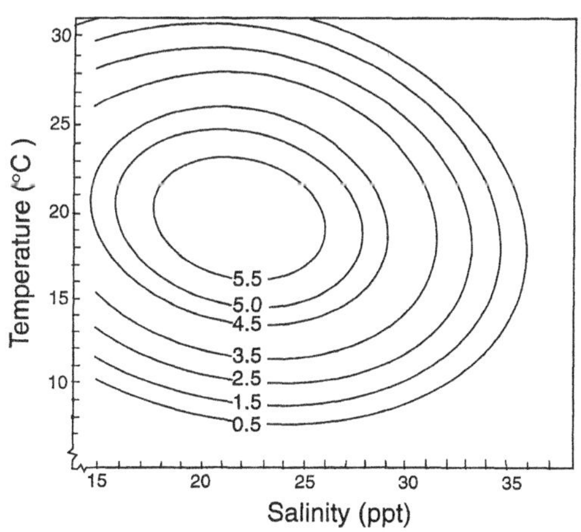

Fig. 8.6. *M. mercenaria.* Interaction effects of temperature and salinity on pumping rate (contours show rate in L h^{-1}) of hard clams with mean total wet weight of ~200 g. A total of 63 temperature/salinity combinations were tested, with 632 total observations (from Hamwi, 1969).

TABLE 8.4

Temperature coefficients (Q_{10})[a] for feeding (clearance or pumping) rates of *Mercenaria mercenaria*

Reference used for calculation	Data source	Temperature range (°C)	Q_{10}
Hibbert (1977b)	Ambient, seasonal	12–17	1.81
		17–25	1.36
Doering and Oviatt (1986)	Ambient, seasonal	13–21	1.77
Hamwi (1969) (his Table 11A)	Laboratory, acclimation	10–20	4.58
		20–25	1.01

[a] Prosser and Brown (1961): $Q_{10} = (K_1/K_2)^{[10/(T_1-T_2)]}$ where K_1 and K_2 = clearance rates at temperatures T_1 and T_2.

calculated over a comparable temperature range. Nonetheless, there is apparently considerable variation in Q_{10} values among bivalve taxa. Some of the variation may be attributed to different study techniques (for comparative review see Malouf and Bricelj, 1989) and to the temperature range over which the coefficient is calculated.

It is important to note that the Q_{10} values shown in Table 8.4 reflect the hard clam's degree of acclimatization (where animals are exposed to gradual, seasonal changes in temperature), or acclimation rather than their acute response to sudden temperature changes. Complete acclimatization or acclimation (where animals are exposed to constant experimental temperatures) to temperature is indicated by a $Q_{10} = 1$. In contrast to *Mytilus edulis*, in which clearance rate is independent of seasonal or acclimation temperatures between 10° and 20°C (e.g., Bayne et al., 1977 and Widdows, 1978), *Mercenaria* does not fully acclimate its feeding rate to temperature, as indicated by $Q_{10} > 1$ (Table 8.4).

Predictive equations of clearance rate (CR) in relation to both body size (L = shell length) and temperature (T, °C) for *M. mercenaria* were determined in two studies as follows (Hibbert, 1977b):

$$\text{CR (L clam}^{-1}\text{ h}^{-1}) = \left[L^{0.892}\text{ (mm)}\right]/10^{a} \tag{8.12}$$

where $\log a = -0.005T + 0.241$, and (Doering and Oviatt, 1986):

$$\text{CR (mL clam}^{-1}\text{ min}^{-1}) = \left[L^{0.96}\text{ (cm)}\right] \times (T^{0.95}) \times 0.339 \tag{8.13}$$

The above, two-variable model was found to explain more of the variation in the observed CR (40%) than either of the single variable models. Pumping rate in relation to temperature for hard clams of different sizes was also determined by Hamwi (1969) (Fig. 8.4). In this study, which tested a wider temperature range (7° to 30°C), pumping rate was maximized at 24° to 26°C, and declined sharply above 27°C in all size classes.

8.3.4.2 *Seston concentration and composition*

M. mercenaria occupies shallow estuarine environments that have widely fluctuating sediment loads, especially near the seabed. This has generated interest in the effect of resuspended sediments on feeding and growth. Hamwi (1969) showed that pumping rate was inversely related to added silt concentrations in the range of 100 to 2000 mg L^{-1}. Similarly, experiments by Bricelj and Malouf (1984) using mixed suspensions of algae and silt, showed a negative relationship between clearance rate of *M. mercenaria* and suspended sediment concentrations ranging from 0 to 44 mg L^{-1} (Fig. 8.3A). On average, CR declined by 0.08 L h^{-1} g^{-1} (1.3%) for every mg L^{-1} increase in sediment load. Due to this effect and the increased loss of algae in pseudofeces (Fig. 8.3B), ingestion of algae from a mixed suspension declined with increasing sediment load. It is noteworthy that rejection of algae in pseudofeces did not become significant (12 to 18% of algae filtered) until sediment concentrations attained levels of 40 mg dry weight L^{-1}. Thus, as indicated earlier, *M. mercenaria* regulates ingestion at high seston concentrations primarily through reduction in CR, and to a much lesser degree via increased pseudofeces production (reviewed by Malouf and Bricelj, 1989). This is consistent with the observation that pseudofeces production, as measured by the number of ejections $clam^{-1}$ h^{-1}, increased by a factor of only 2.6 when wave-induced resuspended sediment concentrations in flume experiments increased from 6.5 to 100 mg dry weight L^{-1} (Turner and Miller, 1991). Although these authors did not measure pumping or clearance rates, they noted that juvenile *Mercenaria* did not exhibit intermittent closure of valves and appeared to feed continuously, although the siphons were more constricted at high sediment concentrations.

Feeding and biodeposition rates as a function of food concentration (a mixed algal suspension dominated by diatoms) were compared for *Mytilus edulis*, *Crassostrea virginica* and *Mercenaria mercenaria* by Tenore and Dunstan (1973). The relative feeding rate of hard clams, defined as % of carbon removed from the culture water presented to the bivalves in a flow-through system, increased with food concentration up to ~450 to 650 g C L^{-1}, a concentration typical of estuarine environments, and declined thereafter (Fig. 8.7). At all food levels, hard clams exhibited lower relative rates than mussels and oysters. Absolute feeding rates (C removed or filtered per unit time) increased linearly with increasing food concentration in all species, but were consistently lower in hard clams than in mussels or oysters. A logarithmic increase in biodeposition rate (feces + pseudofeces production) with increasing food concentration, mainly reflecting an increase in pseudofeces production, was also common to the three bivalves (see Section 8.4). However, *Mercenaria* was characterized

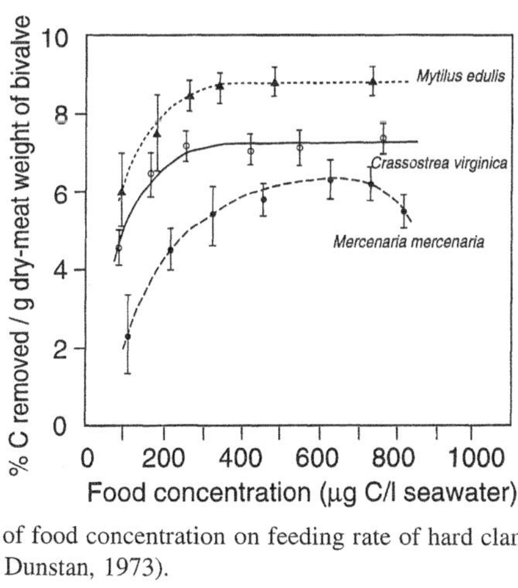

Fig. 8.7. *M. mercenaria*. Effects of food concentration on feeding rate of hard clams compared to blue mussels and Eastern oyster (from Tenore and Dunstan, 1973).

by the lowest biodeposition rate. These results suggested that hard clams are not as well adapted for feeding at high food concentrations as oysters and mussels.

Limited data are available on the functional response of CR vs algal cell concentration in *Mercenaria mercenaria*. For example, a 55% reduction in clearance rate (but increase in ingestion rate) was described in juvenile hard clams with an increase in *Pseudoisochrysis paradoxa* concentration from 20 to 150 cells μL^{-1} (Bricelj, 1984). Studies of other venerid clams, such as *Venerupis pullastra* (Foster-Smith, 1975; Beiras et al., 1993) and *Tapes philippinarum* (Coutteau et al., 1994) showed an inverse relationship between CR and algal cell density above a threshold cell concentration. The latter varies with the algal species used.

There is some evidence that food quality (i.e., the type of algal species) can influence clearance rate in *M. mercenaria*. For example, Rice and Smith (1958) showed that clearance rates were higher in suspensions of the diatom *Nitzschia closterium* than in those of *Nannochloris* and *Chlorella*, two taxa that are of less nutritional value (see below). Walne (1970) also found considerable (5-fold) differences in CR of juvenile clams with three different algal species (Table 8.2). Most studies, however, have focused on the effects of different algae on growth rate (see below), without identifying the physiological mechanisms responsible for a differential growth response.

Iglesias et al. (1992) suggested that the mechanism regulating ingestion with increasing seston levels (control of CR or pseudofecal production) was largely a function of seston quality, rather than being algal species-specific. In their study, diets with a varying proportion of silt and algae, administered at different concentrations, allowed independent variation of food quantity and quality. Cockles, *Cerastoderma edule*, fed a higher-quality diet (greater proportion of algae in the suspension), were found to regulate ingestion rate mainly by adjusting CR. In contrast, when fed low-quality diets, cockles regulated ingestion rate primarily by pseudofecal production, which allows energetic benefits to be achieved through pre-ingestive selection. Copious pseudofecal production has never been described in hard clams, even under conditions of low food quality (Bricelj, 1984), whereas Navarro et al. (1992) reported that cockles rejected large amounts of pseudofeces when exposed to increasing seston sediment concentrations (up to 30 mg L^{-1}).

8.3.4.3 Dissolved oxygen concentration

The relationship between feeding (pumping) rate and oxygen concentration has been studied primarily from the perspective of oxygen uptake or oxygen requirements, as opposed to how oxygen concentration affects pumping (see relevant references cited in Table 8.3; also see Loveland and Chu, 1969 and review by Van Winkle, 1975). Hamwi (1969), however, varied oxygen concentration and measured pumping rate in the hard clam. He found a strong, positive linear relationship between oxygen concentrations from <1 to 5 mg O_2 L^{-1} and pumping rate (Fig. 8.8). Moreover, Hamwi found that individuals that had been held in cold storage for 24 to 48 h, and presumably incurred an oxygen debt, had significantly and substantially higher pumping rates (for a few hours) than the control animals held in aerated, warmer water (Fig. 8.9).

Hamwi's (1969) experiments show that oxygen concentration can strongly affect pumping rate of the hard clam, but only at concentrations below about 5 mg L^{-1}. He did not, however, test the effects of higher oxygen concentrations. Because supersaturated ($>100\%$) oxygen has been shown to inhibit clam growth (Malouf et al., 1972; Bisker and Castagna, 1985),

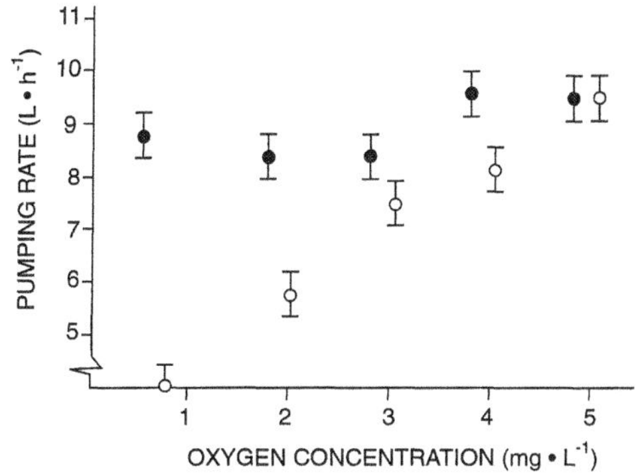

Fig. 8.8. *M. mercenaria.* Effects of oxygen concentration on pumping rates of adult hard clams. Solid circles show pumping rates of control clams held at ~100% oxygen saturation; open circles show rate after transfer from control to experimental concentration (from Hamwi, 1969).

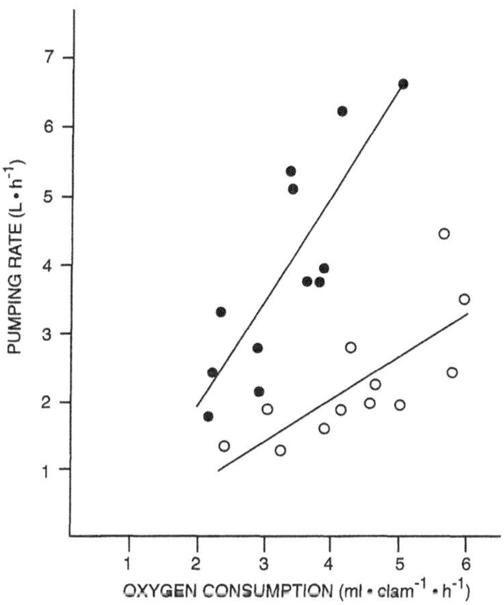

Fig. 8.9. *M. mercenaria.* Oxygen consumption vs. pumping rates of adult hard clams. Solid circles are individuals held at <5°C (had ceased pumping) for 24 to 48 h and transferred to 24°C water; open circles are individuals held at 24°C (from Hamwi, 1969).

such conditions might also inhibit pumping rate. If this is true, then the relationship between oxygen concentration and feeding rate would also be an inverted parabola.

8.3.4.4 Water flow

Although water flow strongly affects growth of bivalves, including *M. mercenaria* (see Section 8.5 below), there has been relatively little research on its effect on individual feeding responses. The general assumption for both feeding and growth is that water flow is mainly relevant when it prevents seston depletion by a population of bivalves (e.g., Bayne and Hawkins, 1992, p. 268; for additional discussion see: Butman et al., 1994; Newell and Langdon, 1996; Wildish and Kristmanson, 1997).

The few studies on the effects of water flow on feeding rates of *M. mercenaria* are inconclusive. Walne (1972) reported a positive linear relationship between flow and clearance rate for several bivalves including *M. mercenaria*, but his results were disputed by Hildreth and Crisp (1976) and Møhlenberg and Riisgård (1979) who ascribed the positive relationship to experimental artifacts involving re-pumping of exhalant water at the lower flow speeds. Both of these subsequent studies (not done on *M. mercenaria*) suggested that clearance rate and water flow rate are independent above some minimum critical flow rate. In the only published paper where inhalant pumping rates of hard clams were directly measured by measuring flow speed into the inhalant siphon, Turner (1992) showed no differences in inhalant speeds at ambient water flow speeds of 20 and 30 cm s^{-1}. However, these flow speeds are greater than the range where most variation in feeding rates might be expected.

In contrast, the effect of water flow on feeding rates of other suspension-feeding bivalves has received considerable attention (Wildish et al., 1987, 1992; Wildish and Miyares, 1990; Wildish and Saulnier, 1993; Lenihan et al., 1996). Wildish and Kristmanson (1997) recently reviewed and re-interpreted much of their earlier work. Fig. 8.10, based on experiments with the giant scallop, reflects their interpretation of what is known at this time for a non-siphonate species. In particular, it shows that there is an interaction between the effect of water currents and seston concentration. Inspected in univariate fashion, both variables (flow speed and seston concentration) have an inverted hyperbolic shape, but the response of each is dependent on the other. Wildish and Kristmanson (1997, pp. 139–142) discuss the hypothetical causes for the overall pattern. Briefly, at low ambient flow speeds seston

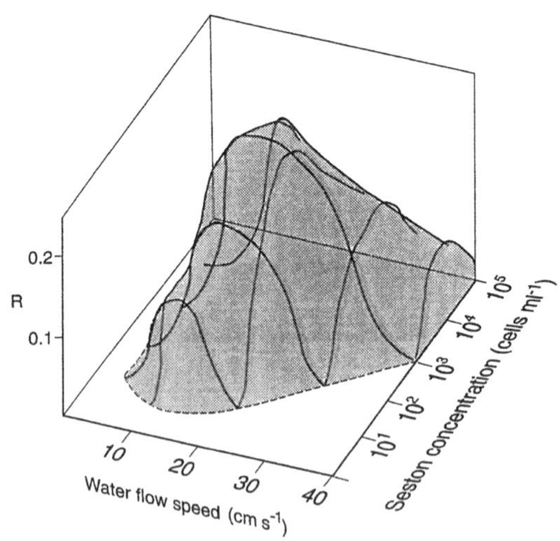

Fig. 8.10. Giant scallop (*Placopecten magellanicus*) feeding rates at different combinations of water flow speed and algal concentration (1×10^x cells mL^{-1} of *Chroomonas salina*). 'R' is relative feeding rate (from Wildish and Kristmanson, 1993).

depletion may occur in the immediate vicinity of the bivalve and previously cleared water may be re-pumped. The associated response is reduced pumping rates. There is a broad range of ambient flow speeds (and seston concentrations; Fig. 8.7) where pumping rate is near maximal, and reduced pumping occurs at high flow speeds. Direct hydrodynamical inhibition due to pressure differences between inhalant and exhalant regions and/or drag forces may occur at high flows, and cause closing of the valves.

Little work has been done on *M. mercenaria*, but it seems reasonable to conclude that water flow does affect feeding rate (see more discussion below). More comprehensive studies are needed to address the effects of multiple factors, particularly factor interactions, on feeding responses of all bivalve taxa. Such approaches are especially needed for infaunal, siphonate species such as hard clams, because they may differ considerably from scallops in their responses to changes in water flow (Grizzle et al., 1992).

8.3.4.5 Noxious algae and other toxic substances

Several studies have shown that metabolites produced by algal species associated with toxic and "noxious" blooms strongly affect feeding rates and other physiological processes of hard clams. Coastal waters in many areas are sporadically affected by blooms of microalgae that affect marine life, including bivalves, in significant and sometimes catastrophic ways (see Shumway, 1990, for review of effects on shellfish generally, and Bricelj and Shumway, 1998, for effects of dinoflagellates producing paralytic shellfish poisoning [PSP] toxins; also see Section 8.5 below). Several algal taxa associated with noxious blooms inhibit feeding of hard clams. Lesser and Shumway (1993) showed that clearance rates of the hard clam (and several other bivalves) were significantly reduced when fed unialgal cultures of the toxic dinoflagellates *Alexandrium tamarense* and *Gyrodinium aureolum* compared to clearance rates when fed *Isochrysis* spp. However, results of this study may be confounded by biomass-dependent effects, since *Isochrysis* and the toxic dinoflagellates tested varied greatly in cell size, and were offered at comparable densities but widely different biomass levels.

Mercenaria mercenaria closed their valves and stopped feeding when exposed to a unialgal suspension of a highly toxic, PSP-producing dinoflagellate (*Alexandrium fundyense*, strain GtCA29, toxicity = 96 pg saxitoxin equivalents [STXeq] $cell^{-1}$; Bricelj et al., 1991, or strain Gt429; Shumway and Cucci, 1987). Clams resumed feeding and maintained a constant ingestion rate over 2 weeks of exposure when this isolate was offered in combination with a non-toxic diet (*Thalassiosira weissflogii* at 30% of total algal biovolume; Bricelj et al., 1991). However, clearance rate on this mixture was significantly (5×) lower than that on a monospecific suspension of the low-toxicity strain, *A. tamarense*, GtLI22, 6 pg STXeq $cell^{-1}$ (Lee, 1993). Weight-standardized clearance rate on strain GtLI22 at 16°C was in turn comparable to that of published values for *Mercenaria* fed non-toxic diets routinely used in commercial production, which ranged from 2.6 to 3.4 L h^{-1} for a clam 1 g in dry tissue weight at 17–19°C (reviewed by Lee, 1993). These results indicate that both specific toxicity and relative abundance of *Alexandrium* cells in a mixed phytoplankton assemblage, are important determinants of feeding rate and thus of the maximum PSP toxin levels accumulated in clam tissues. They are also consistent with the observation that hard clams attain relatively low (10^3 μg STXeq 100 g^{-1}) or undetectable toxicity levels during severe PSP outbreaks involving highly toxic *Alexandrium* spp. (Twarog and Yamaguchi, 1972; Bricelj and Shumway, 1998), but attained up to 10^4 μg STXeq 100 g^{-1} when fed a non-toxic/toxic algal mixture in the laboratory (Bricelj et al., 1991). The mechanism responsible for feeding rate inhibition, e.g., pre-ingestive, chemosensory detection of toxic cells or physiological incapacitation following initial toxin incorporation, remains to be determined.

Some dinoflagellates not known to produce toxins that affect humans (e.g., *Prorocentrum* spp.) can also have detrimental effects on growth of hard clams (Wikfors and Smolowitz, 1993; see Section 8.5). Although feeding rates were not reported in this study, the results suggest that *Prorocentrum minimum* may interfere with the uptake and/or absorption of other nutritious microalgae present in a mixed diet, while *Prorocentrum micans* may inhibit feeding and/or be nutritionally deficient.

Brown tides of the picoplanktonic alga *Aureococcus anophagefferens* (Pelagophyceae), which occur in shallow, mid-Atlantic US estuaries, cause severe inhibition of clearance rates in *M. mercenaria*, although the magnitude of inhibition (14-fold at 10^6 *Aureococcus* cells mL^{-1}, relative to an equivolume suspension of *Isochrysis galbana*, T-ISO) is less than that observed in *Mytilus edulis* (Tracey, 1988). This agrees with the finding that juvenile mussels experienced significantly greater growth inhibition than *M. mercenaria* during a moderate brown tide event, when field concentrations reached 3×10^5 *Aureococcus* cells mL^{-1} (Bricelj and Lonsdale, 1997). As reviewed by these authors, low retention efficiency due to the alga's small size (ca. 2 μm) only partially explains these results. Gainey and Shumway (1991) showed that direct contact with *Aureococcus* cells or amylase digests of these cells, caused inhibition of ciliary beat of excised gill tissues in several bivalve species, including *M. mercenaria*. They concluded that grazing inhibition is likely associated with a bioactive compound residing in the exocellular polysaccharide-like layer of *Aureococcus* cells. This compound remains uncharacterized, but mimicked the action of the neurotransmitter dopamine.

Roegner and Mann (1990) reviewed the literature on toxicological studies on the hard clam. Most studies characterized conditions of lethality for various contaminants. Some studied sublethal responses, but little has been done on feeding responses. Epifanio and Srna (1975) showed that clearance rates of both juvenile and adult hard clams were strongly inhibited by high concentrations of ammonia ($\geq 4 \times 10^{-4}$ M) nitrite ($\geq 4 \times 10^{-2}$ M), and nitrate ($\geq 1.6 \times 10^{-1}$ M); phosphate ions at all concentrations tested did not affect feeding rates. Several organic contaminants and metals have been shown to inhibit growth (Roegner and Mann, 1990), but their effects on feeding are not known.

In conclusion, seston characteristics are presently considered one of the most important factors (other than temperature) controlling feeding behavior in bivalves (Bayne and Hawkins, 1992; Willows, 1992; Bayne, 1993, 1998). As the above discussion shows, other factors can also strongly affect feeding. The challenge is to determine how these factors, both intrinsic and extrinsic, interact in various combinations to control feeding in nature.

8.3.5 Post-Ingestion Processes: Digestion and Absorption

After entering the mouth, food is transported through the short esophagus along ciliated tracts to the stomach where most extracellular digestion occurs. The stomach, which is also ciliated, is connected to a sac containing the crystalline style, a gelatinous rod-like structure which releases digestive enzymes, and presumably aids physically in the breakdown of food material. The food empties from the stomach either directly into the intestine or to the digestive gland. The digestive gland is the site of most intracellular digestion, which is associated with digestive tubules lining the gland, as well as initial absorption of nutrients. Material leaving the digestive gland also enters the intestine. Apparently little further digestion occurs in the intestine but some absorption may occur before discharge of the undigested material through the rectum and anus.

In several bivalve taxa that live in the intertidal zone, and some subtidal taxa, the overall digestion process is discontinuous, typically cyclic and correlated with the tidal cycle (Morton, 1973). Hard clams live both intertidally and subtidally, and thus might be expected also to have discontinuous digestion. Studies by Robinson and colleagues, however, have not shown a cyclic process.

Extracellular and intracellular digestive processes of a subtidal population of adult *M. mercenaria* were studied by Robinson and Langton (1980), and Robinson et al. (1981). These studies found no significant variation in crystalline style length or in the secretory activity of the style sac epithelium, indicative of extracellular digestion in the stomach, in response to changes in tidal phase and seston concentrations, although clams starved in the laboratory showed a complete absence of the crystalline style. Secretion and dissolution of the crystalline style occurred asynchronously within the natural population. Furthermore, histological examination of the cell morphology of digestive gland tubules showed no clear evidence of rhythmic cyclic or discontinuous intracellular digestive activity. A 2 to 3 h time lag was observed, however, between the time of peak food availability and peak levels of digestive tubules in the absorptive stage within the population, suggesting that there is some correlation between food availability and digestive phase. However, there was pronounced variability in tubule morphology within a single animal. Overall, these results led the authors to conclude that intracellular digestion in subtidal *Mercenaria* was a relatively continuous process, and that hard clams, as well as suspension-feeding bivalves generally, should be viewed as opportunistic feeders, in which the timing of digestion is primarily controlled by food availability.

Natural seston which comprises the diet of suspension-feeding bivalves contains microalgae with amylase and laminarin as energy storage products, cell walls containing cellulose (e.g., chlorophytes), as well as cellulose-rich detrital particles derived from the degradation of vascular plants. Limited information is available on digestive enzyme activity in *M. mercenaria*. In common with other suspension-feeding bivalves such as *Mytilus edulis*, *Mya arenaria* and *Argopecten irradians*, hard clams have the potential for enzymatic degradation of polysaccharides (starch, cellulose and laminarin) associated with their diet (Brock et al., 1986; Brock and Brock, 1989). *In vitro* experiments detected significant amylase and laminarase activity in both the crystalline style and digestive gland of adult clams (Brock and Brock, 1989). As expected, digestive gland enzymatic activity increased with temperature, and for *Mercenaria* it peaked at 24°C [Q_{10} of 1.94 and 2.02 for amylase (calculated between 4° and 24°C) and laminarase (calculated between 8° and 28°C) respectively] (Brock et al., 1986). The digestive gland cellulase from *Mercenaria*, which peaked in activity at 28°C, differed from that of other bivalves tested, in that it was only active at higher temperatures (22° to 34°C). The ecological significance of this finding has not been determined.

Chloropigments (chlorophyll and its degradation products) have been used in the past as natural tracers, instead of radioactive tracers, to study feeding and digestive processes in bivalves (Bricelj, 1984; Hawkins et al., 1986). Ingestion of algae by bivalves, results in at least partial degradation of chlorophyll (Chl) *a* to phaeopigments due to low pH and/or enzymatic action during gut passage. In juvenile *M. mercenaria*, phaeophorbide *a* is the dominant component (92 to 99% by weight) of total phaeopigments (phaeophorbide *a* and phaeophytin *a*, determined by high-performance liquid chromatography) present in feces (Bricelj, 1984). Molar conversion of Chl *a* to a single degradation product allows conversion of total chloropigments in feces to units of Chl *a* equivalents. Collection of feces must be carried out in the dark, however, since Moreth and Yentsch (1970) showed evidence of photo-oxidation of chlorophyll *a* in hard clam feces. In bivalves, chloropigments cannot be used as the inert (non-absorbable) tracer to replace ash in the calculation of absorption efficiency (AE) with Conover's (1966) ratio, since Hawkins et al. (1986) demonstrated that total chloropigments were absorbed by mussels with 46 to 80% efficiency. Total chloropigments, measured fluorometrically, can be used however, as an indicator of the absorbable component in Conover's AE equation. Both Hawkins et al. (1986) and Navarro and Thompson (1994) found that comparable AE values were obtained with chloropigments or ash-free dry weight as indicators of the absorbable fraction in mussels (*Mytilus edulis* and *Modiolus modiolus*). The use of chloropigments to determine AE assumes that the measured loss of chloropigments represents absorption during gut passage. However, it has been shown in copepods that significant and variable amounts of chloropigments may also be lost via breakdown to non-fluorescing products (e.g., Head and Harris, 1992). Additional work is therefore needed to gain a better understanding of chlorophyll pigment conservation and loss in bivalves.

The contribution of Chl *a* to total chloropigments in *M. mercenaria* feces was found to increase at high algal densities (Bricelj, 1984), when gut passage time presumably decreases, and an increasing number of undigested, intact algal cells are eliminated in feces. Thus, Chl *a* comprised 9 to 14% of total fecal chloropigments at low to moderate algal densities ($\leq 50 \times 10^6$ *Pseudoisochrysis paradoxa* cells L^{-1}), but increased to 52% at 150×10^6 cells L^{-1}. It was therefore suggested that this percentage may provide a relative index of the absorption efficiency of algal diets.

Low absorption efficiency of some algal species can explain their inability to support good growth of bivalves. Further discussion of the absorption of various algal diets by *M. mercenaria* is provided in Section 8.5.

8.4 ENERGY EXPENDITURES: BIODEPOSITION, EXCRETION, AND OTHER METABOLIC PROCESSES

8.4.1 Overview

Energy gains and losses for the individual bivalve are shown in Fig. 8.1 above. Energy can be lost prior to ingestion as pseudofeces (P) or following ingestion as feces (F). Both contribute to the production of biodeposits. In turn, energy absorbed from the ingested food can be lost as excretion (E), metabolic heat losses generally (M), and reproductive output (included in the growth term, G). Hibbert (1977b) estimated these energy components for a wild population of *M. mercenaria* (Fig. 8.11). Of the mean total annual seston removal rate of 1292 kcal m^{-2}, only about 5% went to population production, based on annual production estimates for several year classes. Most energy loss (59% of consumption) occurred as feces and pseudofeces (measured together as total biodeposition). Tenore et al. (1973) found that

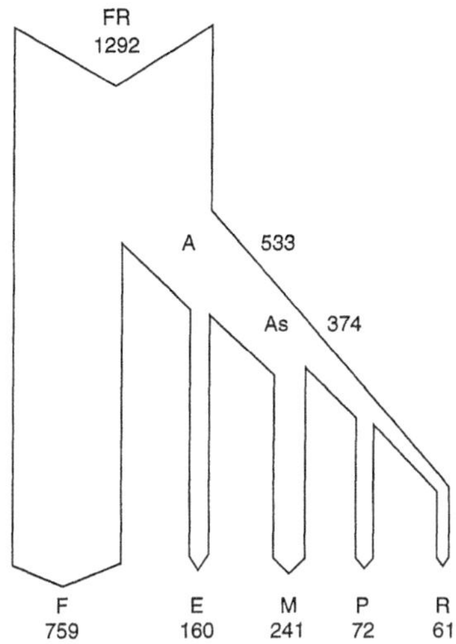

Fig. 8.11. *M. mercenaria.* Energy budget (in kcal m^{-2} $year^{-1}$) for an intertidal population of hard clams; 1292 = total seston removed during feeding; F = egestion or biodeposition (includes feces and pseudofeces); E = excretion (not measured directly, calculated by difference); M = respiration or metabolic heat losses; R = reproductive output; A = absorption ($= I - F$); A_s = assimilation ($= P + R + M + E$); and P = population production (summation of individual growth and losses due to mortality) (modified from Hibbert, 1977b).

24% of the energy from food filtered went to growth (population production) in a short-term study using hard clams in flow-through tanks supplied with mixed algal cultures as food.

Bivalves are often major primary consumers in shallow coastal waters. A considerable body of research has focused on assessing their role in energy fluxes or material cycling and nutrient regeneration, particularly from the perspective of benthic–pelagic coupling (for reviews see Dame, 1993, 1996). The following subsections review the major topic areas with emphasis on *M. mercenaria.*

8.4.2 Biodeposition and Benthic–Pelagic Coupling

Suspended particulates removed ("filtered") from the water column are rejected before ingestion as pseudofeces, or if ingested and not absorbed, are eventually eliminated as feces (collectively termed biodeposits). In siphonate bivalves like *M. mercenaria* pseudofeces are eliminated sporadically through the inhalant siphon while feces are discharged via the exhalant siphon. Biodeposits may be swept away in swift water currents, but they typically settle nearby and accumulate on the bottom. In this way, suspension-feeding bivalves are a direct link between the water column and benthos. Bivalve biodeposits consist of a wide variety of elements (carbon, nitrogen, phosphorus, etc.), but much of the research on bivalve biodeposition has concentrated on carbon.

In nature, infaunal bivalves like hard clams typically do not occur at densities sufficient to produce excessive sediment buildup in their immediate vicinity, as do reef-building bivalves like oysters and mussels. Nonetheless, because they feed on particles suspended in the water column infaunal bivalves can play an important role in transferring materials from the

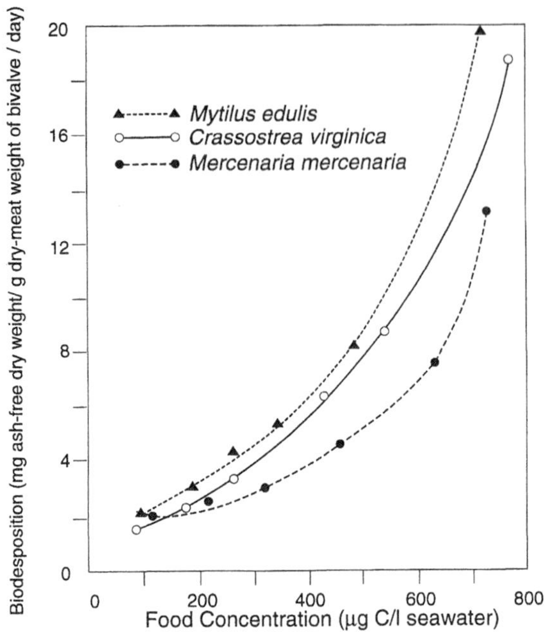

Fig. 8.12. *M. edulis*, *C. virginica*, and *M. mercenaria*. Biodeposition rates at different food (mixed algae) concentrations (from Tenore and Dunstan, 1973).

water column to the benthos, particularly in shallow, well-mixed coastal waters (Haven and Morales-Alamo, 1972). Tenore and Dunstan (1973) quantified carbon transfer to the bottom by comparing biodeposition rates over a range of algal cell concentrations for three common bivalves, *Mytilus edulis*, *Crassostrea virginica*, and *M. mercenaria*. They found logarithmic increases in biodeposition rate with increasing food concentration for all three bivalves (Fig. 8.12) mostly because of increases in pseudofeces production. *M. mercenaria* had the lowest biodeposition rates of the three.

Based on mesocosm experiments representative of conditions experienced in the natural environment, and using radioactive carbon labeling, Doering et al. (1986) and Doering and Oviatt (1986) found that adult hard clams at a density of 16 individuals m^{-2} increased net organic carbon sedimentation (including incorporation of label in clam tissues) by 58% relative to controls with no hard clams. Much of this increase could be accounted for by clam filtration. However, in contrast to the findings by Hibbert (1977b), very little of the sedimented carbon was biodeposited. Most of it was absorbed and then either respired or incorporated into clam tissues.

In addition to biodeposited solids that contribute to material cycling, bivalves also excrete organic and inorganic nitrogenous metabolites (see Section 8.4.4 below). The result is a direct return of various dissolved substances to the water column. Substantial attention has been paid to how hard clams affect elemental fluxes, particularly of nitrogen and phosphorus.

Most relevant studies have been done using either laboratory tanks or mesocosms. Tenore et al. (1973) used flow-through tanks (with the bivalves held in trays with no sediment

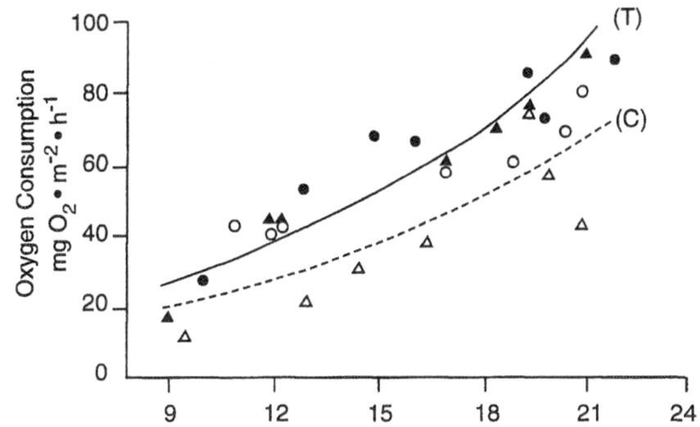

Fig. 8.13. *M. mercenaria*. Effects of temperature (°C) on oxygen consumption in mesocosms with hard clams at 16 individuals m^{-2} ('T', solid circles and triangles), and mesocosms with no hard clams present ('C', open circles and triangles) (from Doering et al., 1987).

substrate) to estimate the effects of three bivalve taxa (including *M. mercenaria*) on nitrogen and phosphorus flux rates. Hard clams increased total N concentrations in the outflow water by 26% and PO_4 by 13%. Doering et al. (1986) found that mesocosms with hard clams (including juveniles and adults) at a density of 16 individuals m^{-2} had increased rates of remineralization of dissolved inorganic nitrogen (DIN) as a result of enhanced bottom deposition of organic matter compared to mesocosms without hard clams. In a companion study, Doering et al. (1987) quantified the effects of hard clams on flux rates of oxygen, carbon dioxide, silicate, DIN, and phosphate. Hard clams significantly increased oxygen consumption (Fig. 8.13), but had no significant effect on PO_4 or CO_2 fluxes. Hard clams also increased the DIN flux from sediments, with most of the increase attributed to release of ammonia. This contributed to the increase in water column primary production in treatments with hard clams (Doering et al., 1986). No reduction in phytoplankton biomass via clam filtration was detected at the hard clam densities used in the experiments.

Two field studies on benthic communities dominated by *M. mercenaria* indicated its potential importance to elemental cycling. Using a bottom chamber, Hale (1975) measured oxygen uptake and nitrogen (NH_4 and NO_3) and PO_4 fluxes for three different benthic communities, including one dominated by *M. mercenaria* (no hard clam density data were presented). There were no significant differences in oxygen fluxes among the three benthic communities, but the hard clam-dominated community had significantly different NO_3 and PO_4 fluxes compared to at least one of the other communities. There was also a strong positive correlation between most flux rates and temperature from about 2° to 25°C. Somewhat in contrast to Hale's findings, Murphy and Kremer (1985) estimated that over their 2-year study a dense assemblage of hard clams (mean of 143 individuals m^{-2} and including a small proportion of other suspension-feeding bivalves) contributed 68% of the oxygen uptake and ammonium release for the total community. Thus, although it is typically found that microbial processes dominate fluxes of dissolved materials to and from the benthos, bivalves can be important when occurring at high densities.

In sum, studies on biodeposition and material fluxes indicate that hard clams play a potentially important role in elemental cycling. Such knowledge has important applied use, particularly for water quality management. Most studies on bivalves like hard clams and oysters have been aimed at aquaculture and/or fisheries-related topics, but their potential role in ameliorating the effects of eutrophication and other water pollution-related changes has been recognized (e.g., Officer et al., 1982; Newell, 1988; Ulanowicz and Tuttle, 1992).

8.4.3 Metabolic Rate

Oxygen consumption rate (V_{O_2}, mL O_2 h^{-1} $clam^{-1}$) of *M. mercenaria* increases with body size (total wet wt., W, for hard clams ranging from 5 to 284 g) according to the following allometric relationship, at 25°C and 22 ppt (Loveland and Chu, 1969):

$$\log_{10} V_{O_2} = -0.344 \log_{10} W - 1.023 \tag{8.14}$$

Unlike *Mytilus edulis*, which can fully acclimate its routine rate of oxygen consumption over a wide temperature range (5° to 20°C), *M. mercenaria* shows a marked increase in V_{O_2} (as well as feeding rate) with increasing acclimation or seasonal temperature between 10° and 27°C (Fig. 8.14; Q_{10} values shown in Table 8.5). The relationship between V_{O_2}, shell length (L, mm) and temperature (T, °C) ranging from 12° to 25°C was described by Hibbert (1977b):

$$\log_{10} V_{O_2} = \left(1.016 \log_{10} L\right) - A \tag{8.15}$$

where $\log_{10} A = -0.012T + 0.478$. If food supply is not limiting, this increased energy expenditure with increasing temperature can be offset by an increase in pumping rate (Fig. 8.14). In a study in which juvenile hard clams were exposed to ambient seston in a temperature-controlled, flow-through system, both V_{O_2} and clearance rate (CR) increased with temperature (Fig. 8.14C). Overall, the increase in CR was greater than that in V_{O_2}, resulting in a net energy gain between 12° and 27°C. It is noteworthy, however, that while CR increased at a higher rate with temperature than V_{O_2} between 12° and 20°C, leading to a marked increase in growth rate, V_{O_2} increased somewhat faster than CR between 20° and 27°C, thus explaining the observed decline in growth rate above 20°C. Similarly, although growth was not measured in his study, Hamwi (1969) showed that CR increased faster than V_{O_2} between 10° and 20°C, but the opposite was true between 20° and 25°C (Fig. 8.14A), and both declined rapidly at temperatures $\geq$30°C.

TABLE 8.5

Temperature coefficients (Q_{10})[a] for oxygen consumption of *Mercenaria mercenaria*

Reference	Temperature range (°C)	Q_{10}
Hamwi (1969)[b]	10–20	2.76
	20–25	1.56
	25–30	0.29
Bricelj (unpubl.)[b]	12–20	2.38
	20–27	2.89
Hibbert (1977b)[c]	15–20	2.54
	20–26	3.38

[a] Q_{10} as in Table 8.4, where K_1 and K_2 = oxygen consumption rates at temperatures T_1 and T_2.
[b] Acclimation temperature.
[c] Seasonal or acclimatization temperature.

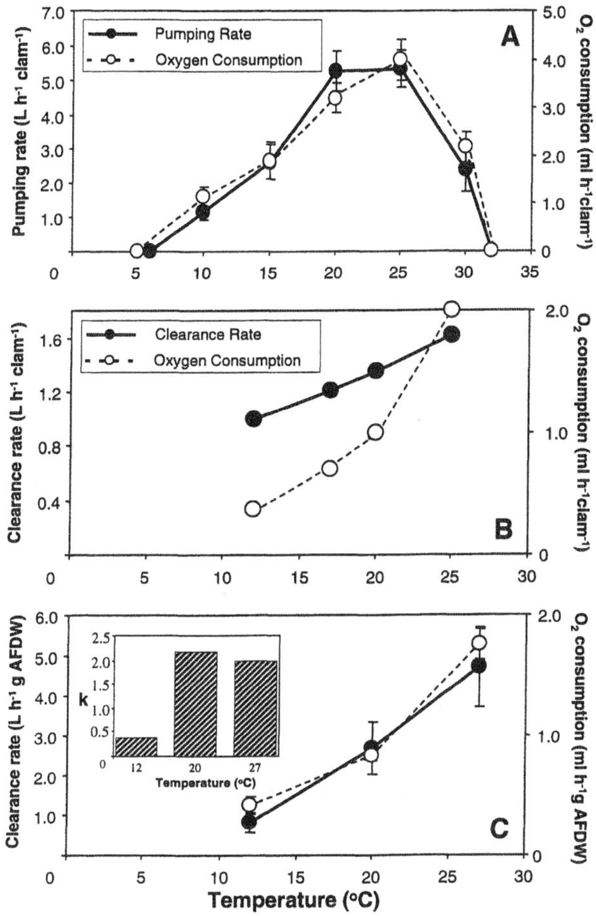

Fig. 8.14. *M. mercenaria.* Relationship between acclimation temperature (A and C) or acclimatization temperature (B) and feeding rate (clearance or pumping) and oxygen consumption rate. Error bars = ±1 SD. (A) Data from Hamwi (1969) for hard clams averaging 200 g in total body wet weight, at a salinity of 20–21 ppt. (B) Data from Hibbert (1977b) for hard clams averaging 50 mm shell length (SL) fed natural seston. (C) Weight-specific CR and V_{O_2} from Bricelj (unpublished data) for juvenile hard clams 10 to 15 mm SL fed natural seston from Great South Bay, New York; values represent seasonal means calculated over 4 to 8 sampling periods throughout the fall (mid-October to mid-December); insert shows instantaneous growth coefficients (*k*, % daily increase in soft-tissue AFDW).

Although no work to this effect has been done on *M. mercenaria*, oxygen consumption of other clam species also has been shown to vary with diet quality. This was demonstrated for juvenile *Tapes semidecusatta*, in which V_{O_2} was lower when clams were fed *Chroomonas salina*, a relatively poor diet, than when fed algal diets that yielded high growth rates such as *Isochrysis galbana* (T-ISO), and the diatoms *Skeletonema costatum* and *Chaetoceros calcitrans* (Laing et al., 1987). These differences among diets became more pronounced at higher temperatures.

Hypoxia or anoxia can be induced in bivalves either by oxygen depletion in the surrounding water or by shell closure. These effects may occur during air exposure or sporadically in submerged animals. *M. mercenaria* can regulate its rate of oxygen consumption with decreasing external oxygen concentrations down to 5 mg L^{-1}, by increasing the efficiency of oxygen removal from the water column (Hamwi, 1969). Suppression of V_{O_2} below this critical oxygen level allows hard clams to conserve energy during environmental hypoxia.

Bivalve molluscs are also capable of respiring anaerobically during environmental hypoxia and shell closure. Furthermore, Hammen (1980) measured a higher ratio (27% increase) of the rate of heat production (a measure of total metabolic rate) to V_{O_2} (a measure of aerobic metabolism), relative to that expected based on carbohydrate oxidation, in a single specimen of *M. mercenaria* held in aerated seawater, suggesting that anaerobic metabolism can contribute significantly to total metabolism even under aerobic conditions. This confirmed results of an earlier study (Hammen, 1979). Detailed partitioning of total metabolic rate into aerobic and anaerobic components using simultaneous calorimetry and respirometry as a function of emersion time and oxygen concentration, has been conducted for other bivalves such as *Mytilus edulis* (e.g., Wang and Widdows, 1993), but not for *M. mercenaria.* Hard clams, however, have a relatively high ratio of cytoplasmic activities of the glycolytic enzymes malate dehydrogenase and lactate dehydrogenase (MDH/LDH). This ratio has been suggested as an index of tolerance to hypoxia for bivalves (Schapiro and Bobkova, 1976), and indicates that *M. mercenaria* is euryoxic. Unlike other bivalves such as scallops, softshell clams, and ribbed mussels, hard clams are capable of complete and sustained valve closure [for up to 18 days at low temperatures (−1° to 6°C), Loosanoff, 1939] and can therefore survive extended periods of exposure. This long "shelf life" is a major benefit for commercialization of the species.

In *M. mercenaria,* as in many other marine bivalves, succinate and alanine are the major end-products of anaerobic metabolism (Korycan and Storey, 1983). However, in contrast to *M. edulis* and *C. virginica,* hard clams produced succinate in greater amounts than alanine in all tissues. During anoxia (96 h) aspartate and glycogen are used as anaerobic substrates (Korycan and Storey, 1983). Recovery from anoxia is rapid: tissue levels of alanine, aspartate, glucose and ATP are restored to control levels within 24 h of return to aerated seawater and succinate levels are re-established within 48 h.

Anaerobic metabolism in *M. mercenaria* has also been studied in the context of shell formation and dissolution. Upon valve closure, hard clams experience a rapid decrease in the O_2 tension of the extrapallial fluids, and a buildup of succinate, the main end product of anaerobiosis (Crenshaw and Neff, 1969; Crenshaw, 1980). This acidic product, which accumulates in high levels in the mantle fluid as well as tissues, is neutralized by the dissolution of shell calcium carbonate. Thus, the alternation of aerobic and anaerobic respiration leading to shell dissolution and deposition, has been proposed as a mechanism of subdaily or daily shell growth-line formation (Lutz and Rhoads, 1977, 1980).

8.4.4 Excretion

As indicated earlier, excretion of nitrogenous compounds by bivalve molluscs is of ecological significance in that it can contribute to nutrient fluxes from sediment to the overlying water column and can thus stimulate primary production in shallow estuaries. Such an effect was demonstrated in mesocosms stocked with and without hard clams by Doering et al. (1986).

The relationship between ammonia excretion (V_{NH_4}, μmoles day^{-1}) and body size (W, g dry wt. of soft tissues) for *M. mercenaria* (at 20°C and 30 ppt) was described by Srna and Baggaley (1976):

$$\log_{10} V_{NH_4} = 1.33 + 0.94 \log_{10} W \qquad (8.16)$$

This study showed that excretion rate by hard clams was greater and showed more scatter around the fitted regression line (correlation coefficient, $r = 0.67$) than that of *Crassostrea virginica* ($r = 0.84$). The magnitude of V_{NH_4} in both species was sensitive to temperature perturbations. However, only ammonia excretion rate was measured, and bivalves can excrete organic nitrogenous compounds (e.g., amino acids). Hammen (1968) measured the relative contributions of various nitrogenous excretory products in several bivalve species; NH_4, amino acids, and uric acid contributed 66%, 30%, and 4%, respectively, to the total nitrogen excreted by *M. mercenaria*. There was no measurable contribution of urea; the relative proportion of different nitrogenous products excreted by bivalves and the rate of total nitrogen excretion are influenced by environmental factors and physiological condition. For example, excretion rate in *Mytilus edulis* increases in stressed animals subjected to starvation (due to increased protein catabolism after depletion of glycogen reserves), or hypoosmotic (low salinity) shock (Bayne et al., 1976; Deaton et al., 1984). Much of our detailed knowledge of nitrogen metabolism is derived from studies of *M. edulis*, which is more euryhaline than *M. mercenaria*.

8.4.5 Intracellular Osmotic and Volume Regulation

In osmoconforming bivalves, such as *M. mercenaria*, the hemolymph and pallial fluids are isosmotic with the external seawater medium, as long as the shell valves remain open (Anderson and Prosser, 1953). In marine bivalves, intracellular cell volume regulation following prolonged exposure to a lowered or increased salinity, is achieved by regulation of the concentration of intracellular solutes, primarily nonessential free amino acids (FAA). Thus, the intracellular pool of FAA tends to rise in response to an extended increase in salinity. Increased amino acid and NH_4 excretion (considered to be derived from the deamination of FAA) is observed following a decrease in salinity. DuPaul and Webb (1974) showed that *M. mercenaria* subjected to a sudden increase in salinity responded by a rapid initial increase in the gill concentration of the amino acid taurine, and a slower, delayed increase in alanine. The response differed from that of *Mya arenaria* and *Spisula solidissima*, in which alanine was the main contributor to the initial increase in the FAA pool. Hence, taurine is a major component of the FAA pool in the hard clam. Rice and Stephens (1988) demonstrated a net loss of intracellular FAA, especially taurine, and of radiolabeled alanine to the external seawater medium one week following exposure of *M. mercenaria* to a decrease in salinity from 34 to 17 ppt. This prolonged loss led them to suggest that the hard clam is poorly adapted to a hyposaline environment. In contrast, net uptake of FAA and ^{14}C-alanine by hard clams occurred when they were exposed to an increase in salinity. It should be noted that the hard clams in this study were artificially irrigated and not allowed to close their valves. The ratio of taurine to glycine also has been shown to provide an index of stress by hydrocarbon pollution or laboratory confinement (Jeffries, 1972).

Shumway et al. (1977) argued that an estuarine species such as *M. mercenaria* is more likely to experience a fluctuating salinity regime rather than a prolonged and continued reduction in salinity in its natural habitat. When the osmotic concentration of seawater drops below about 50% that of full-strength seawater, the hard clam typically responds by tightly closing its valves, thus behaviorally regulating hemolymph osmotic pressure. Shumway et al. (1977) found that hard clams exposed to short-term salinity fluctuations in the laboratory, showed an increased FAA concentration in adductor muscle tissue during abrupt or gradual decreases in salinity, rather than the decrease expected if FAA were being extruded to the hemolymph to achieve intracellular isosmotic regulation. The increase in the FAA pool generally followed shell closure and was not associated with increased NH_4 concentration in tissues. Significant fluctuations in the tissue water content of hard clams exposed to a fluctuating salinity regime only occurred in specimens in which the shell valves were wedged open to prevent isolation from the external medium (Shumway et al., 1977). They concluded that shell closure in *M. mercenaria* serves as a protective mechanism allowing temporary isolation from the external medium, and therefore precludes the need for intracellular volume regulation by solute extrusion.

8.5 NUTRITION, GROWTH, AND PRODUCTION

From an energetics perspective, overall growth (G) represents the net energy gain for the individual. Growth is typically partitioned into reproductive tissues and other tissues (= somatic growth). This section deals with nutrition and overall somatic growth, and the result at the population level, namely production. Partitioning of growth between reproductive tissues and soma, as well as more detail on partitioning among somatic tissues, is dealt with below.

Most studies at the organismic level on *M. mercenaria* have dealt with somatic growth, and/or production which refers to the total increase in biomass for a population and takes into consideration both growth and survival. These topics are of concern to fisheries scientists because age–size relationships define when a given organism reaches market size, and the overall production of the population reflects potential harvest rates. Similarly, a major concern of aquaculturists is the production rate of their operation. The term "somatic growth" may be used to refer to growth of various body parts; however, studies on *M. mercenaria* and other bivalves mainly have been directed at shell growth and/or overall "soft tissue" growth. In this section, unless otherwise noted, the term "growth" will be used indiscriminately to refer to either or both components because the overall response pattern for both is similar. The two are not, however, necessarily strongly coupled on some temporal scales (see below). After a brief discussion of techniques for measuring growth and production, this section is divided into eight subsections, each dealing with major aspects of growth or production.

8.5.1 Growth and Production Measurement Techniques

The size of a hard clam is typically measured in one of three ways: (1) linear shell dimensions, usually height (maximum dorso-ventral dimension; see Chapter 2 for shell terminology) and/or length (maximum antero-posterior dimension); (2) total body volume; and (3) wet or dry weight of shell or soft tissue (see Crisp [1984] for review of measurement techniques, and Malouf and Bricelj [1989] for brief discussion of the advantages and disadvantages of the various size measurements for estimating growth). Shell size, however, is the most commonly used measure in growth studies of juvenile hard clams because it can be quickly measured, and shell growth generally tracks soft tissue growth. Shell size is also used to determine market size in most areas. Volume is also commonly used in aquacultural operations, and total wet biomass in studies involving small spat.

Regardless of the size measurement(s) used, growth (= change in size) is expressed in either absolute or relative terms. Absolute growth refers to change in size, with the change usually expressed over a given time period, so it becomes an absolute growth rate (GR; e.g., in mm week^{-1}) using the following equation:

$$GR = (L_2 - L_1)/(t_2 - t_1) \tag{8.17}$$

where L_1 and L_2 = initial and final length (or other size measure), and $t_2 - t_1$ = elapsed time (typically in days to months).

Absolute growth measures are sometimes converted to relative growth in order to take the initial size of the bivalve into consideration. For example, an absolute shell length growth rate of 5 mm in one week represents a 10% (1.4% per day assuming equal daily growth increments) change in size for a hard clam that was 50 mm in size, but only a 5% (0.7% per day) change for a 100 mm hard clam. Thus, relative growth rate declines with increasing age/size.

Growth of small juveniles that are experiencing exponential growth (see next subsection) is sometimes expressed as an "instantaneous growth rate" using the following equation:

$$k = [\ln(L_2) - \ln(L_1)]\,/(t_2 - t_1) \tag{8.18}$$

where L_1, L_2, t_2 and t_1 are defined as above, and time usually is expressed in days. The growth coefficient k multiplied by 100 is used to express growth in "% per day".

Applying this equation to the example given above yields an instantaneous growth rate (k) of 0.014 for the 50 mm hard clam, and 0.007 for the 100 mm hard clam. Such measures using a single number to describe growth are most useful for comparisons of bivalves of similar initial size because the rate of growth typically changes dramatically over the full life cycle. Studies of long-term growth patterns require more complicated approaches, as discussed in the next subsection.

Production can be defined as the change in biomass of a population. It is typically expressed per unit time as a rate, which is termed "productivity" (Pr). Production (or productivity) can be measured using two different approaches (Crisp, 1984). The first is simply a summation of all the individual growth rates of the members of a population over a specified time interval, with the rate expressed per unit bottom area for benthic organisms like *M. mercenaria* (e.g., in mg m^{-2} year^{-1}). The second approach is more complicated and less direct in that it ignores individual growth and instead considers the fate of the total biomass produced by the population over a specified time interval, according to the following:

$$\mathrm{Pr} = [(B_2 - B_1) + M]\,/\Delta t \tag{8.19}$$

where B_1 = initial standing crop or biomass, B_2 = final standing crop or biomass, M = loss of biomass from the population (due to predation, etc.), and Δt = time interval.

Both approaches have been used in production studies on hard clams. The first approach is easiest and is typically used when good estimates of growth are available and there has been negligible loss of individuals from the population. If individual growth rates are not known, the second approach must be used.

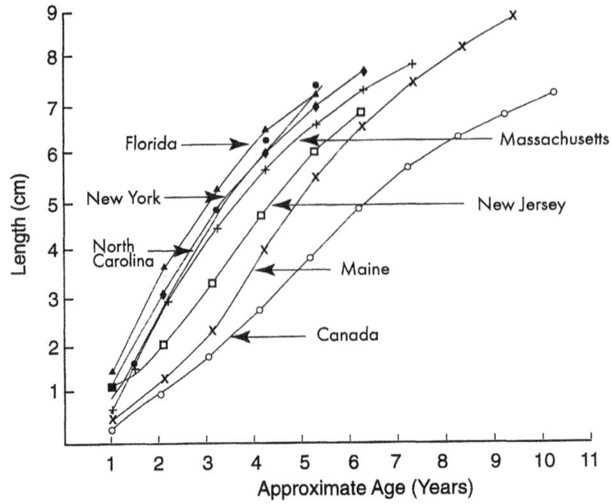

Fig. 8.15. *M. mercenaria.* Growth curves for representative 'best' sites for North America: ▲ Florida; • New York; ♦ North Carolina; + Massachusetts; □ New Jersey; × Maine; ○ Prince Edward Island, Canada (from Ansell, 1968; see his Fig. 2 and Table 2 for data sources).

8.5.2 Ontogenetic Growth

Ansell (1968) summarized early research on the individual growth response pattern over the full lifetime (= ontogenetic growth) of *M. mercenaria* across its geographical range. His review emphasized the effect of temperature (see below), while recognizing that other environmental factors affect growth and that there is wide variability in growth rates in any given locale. Fig. 8.15 is a plot of the "best" sites for growth reviewed by Ansell and covering the hard clam's latitudinal range in North America. These early studies showed fastest growth in Florida and slowest in Canada, but otherwise no clear latitudinal gradient.

Ontogenetic growth curves published since Ansell's (1968) review, however, suggest a latitudinal gradient (Fig. 8.16). These studies described and modeled growth using various approaches. For example, Kennish and Loveland (1980) compared logistic, Gompertz, and monomolecular models. The von Bertalanffy model, however, has probably become the most widely used in studies on bivalve fisheries generally (Seed, 1980; Arnold et al., 1991; Allison, 1994). Von Bertalanffy models were only published in three of the six studies summarized in Fig. 8.16, but these did show a gradient of mean Brody growth coefficients (k) with greatly decreasing values from Florida to Canada.

Ontogenetic models typically are based on year-to-year shell growth increments, so they do not include a seasonality component, even though seasonal growth variations are dramatic in temperate areas. For example, growth ceases during the colder months from Virginia north along the Atlantic coast (Fig. 8.17). In contrast, individuals from Florida grow all year, though not at uniform rates. Hard clams from northern latitudes may grow faster during the warmer months compared to those in southern latitudes (Fig. 8.17), but all-year growth in the southern areas results in faster annual rates for the first five or so years (Fig. 8.16). More recent bivalve

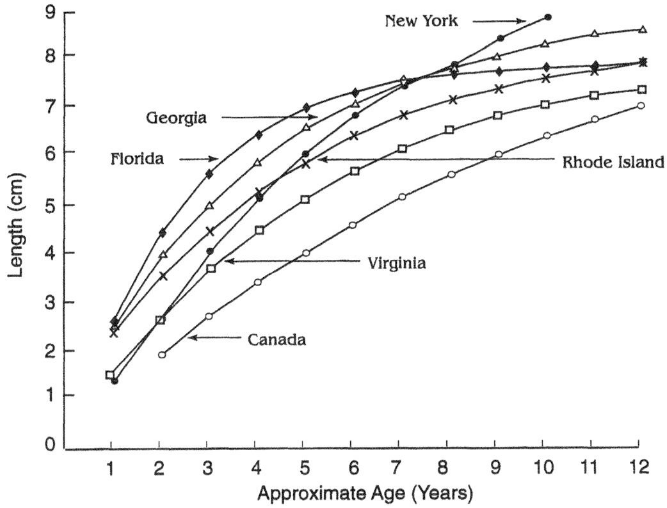

Fig. 8.16. *M. mercenaria*. Growth curves for representative sites for North America, from studies subsequent to Ansell (1968), for natural populations unless indicated: ♦ Florida, Matanzas River, Atlantic, calculated from von Bertalanffy equation in Jones et al. (1990) using a H/L conversion factor of 0.91; △ Georgia, Little Tybee Island, growth described by power function in Fig. 4 in Walker and Tenore (1984); × Rhode Island, Narragansett Bay, calculated from von Bertalanffy equation in Jones et al. (1989) using their mean H/L ratio of 0.91; • New York, Great South Bay, calculated from data in Buckner (1984) using his height–length conversion equation; □ Virginia, York River, hard clams in experimental plots, curve calculated using Ford–Walford plot (Loesch and Haven, 1973); ○ Canada, Prince Edward Island, Hillsborough River, intertidal population (Landry et al., 1993).

ontogenetic modeling incorporates seasonal variations (Allison, 1994), but such approaches have not been used for hard clams.

Most studies on growth rates have emphasized the first few years of growth when growth rates are potentially maximal. Table 8.6 summarizes the range of reported maximal growth rates in shell length (SL) for juvenile hard clams exposed to natural seston in enclosures in the field, or in laboratory raceways or upwellers. All these studies were done during the growing season for each area, so they probably represent maximal growth rates. Hence, there is no latitudinal gradient evident. Maximal juvenile growth rates range from 0.42 to 1.11 mm SL week^{-1}, with a mean of 0. 74 mm SL week^{-1}. These can be compared to those realized when hard clams are fed controlled algal diets (Table 8.6B) which average 0.59 mm SL week^{-1}.

A major concern has been the time required to reach market size, which varies greatly across the geographic range of *M. mercenaria*, as is suggested by the ontogenetic growth variations discussed above. Table 8.7 summarizes selected studies providing information on time to market size, which is assumed for comparative purposes to be 25.4 mm shell thickness (corresponding to a shell length of about 48 mm; see morphometrics discussion below). The range is from 1.9 years for a Florida population to 13.0 years in Canada. The same overall north-to-south trend of decreasing time to market size is seen here as mentioned above for ontogenetic growth rates, although there is wide variability in the mid-Atlantic areas. Moreover, it is typically the case that variability in growth rates within an estuary is greater than among latitudes over a broad range (e.g., two-fold variation among sites in Great South

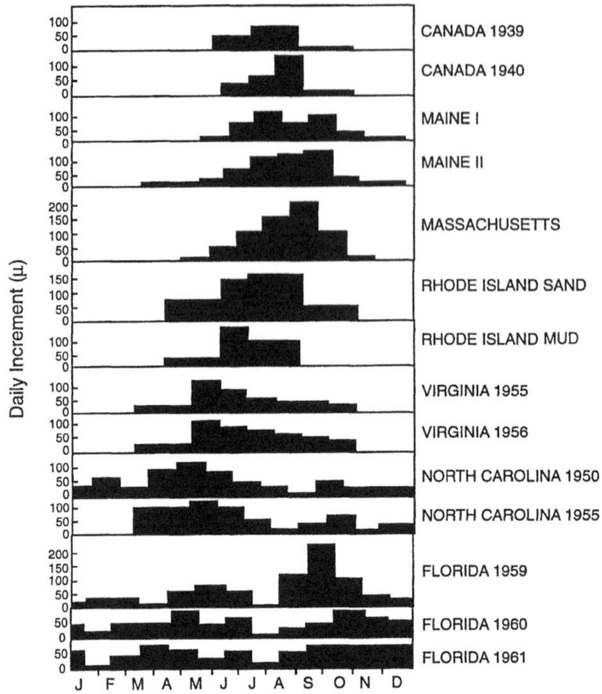

Fig. 8.17. *M. mercenaria.* Average monthly shell growth rates at North American sites (from Ansell, 1968; data sources as in Fig. 8.15).

Bay, New York, Greene, 1978; three-fold variation in Cape Lookout, North Carolina areas, Peterson and Beal, 1989; and two-fold variation at Florida sites, Arnold et al., 1991). Such variability suggests that in addition to temperature, which is presumed to be the major cause of latitudinal gradients, other environmental factors are of major importance in controlling growth (see below).

8.5.3 Morphometrics of Shell Growth and Age–Size Relationships

Overall shell morphology is mainly discussed in Chapter 2. This subsection briefly deals with changes in shell morphometrics (i.e., external size relationships), especially as related to ontogenetic growth patterns, and important age–size relationships.

Stanley (1970) and Seed (1980) review the literature on shell growth patterns for bivalves generally. For *M. mercenaria*, various shell size dimensions (L = length, H = height, and W = width or thickness) are approximately related by the ratios (Stanley and DeWitt, 1983): $L/H = 1.25$; $H/W = 1.52$; $L/W = 1.90$; ratio of shell volume to internal volume = 0.60. Such ratios vary from area to area and among habitats in a given area.

Growth curves shown in Figs. 8.15 and 8.16 reflect information on maximum size and age. Studies in most areas indicate that hard clams can exceed 20 years in age, with maximum reported ages of 28 years (Florida; Jones et al., 1990), 31 and 46 years (North Carolina;

TABLE 8.6

Growth rates of juvenile *Mercenaria mercenaria* exposed to (A) natural seston and (B) cultured algae

Site; period	Shell length (L_o–L_f)	Temp. (°C)	Growth rate (mm week^{-1})	Source
(A) Natural seston				
Northumberland Straight, NB,	3.4–6.8	17–22	0.57 [a]	C. Gionet, Shippagan Hatchery records
Canada (upwellers)	2.9–5.4	17–22	0.42 [b]	(pers. commun., 1996)
Napeague Harbor, NY;				
July–Aug.	10.3–14.2	22–28	0.96	Bricelj and Borrero, unpubl.
Great South Bay, NY;				
Oct.–Nov. (raceways)	10.5–15.4	27	0.54	Bricelj, unpubl.
Fishers Island, NY;				
Aug. (upwellers)	4.6–5.7	22	1.05	Applemans (1989)
Shinnecock Bay [c], NY;				
July–Oct.	7.9–15.4	–	0.62	Flagg and Malouf (1983)
Folly River, SC;				
Feb.–Aug.	3.9–16.9	8–32	0.48	Hadley and Manzi (1984)
May (raceways)		21–26	1.11 [d]	
Clark Sound [c], SC;				
May–Dec.	13.0–26.9	–	0.45	Eldridge et al. (1979)
Wassaw Sound [c], GA;				
(intertidal)	6.1–28.3	–	1.08	Walker (1984)
Alligator Harbor, FL;				
April	5.4–9.0	17–26	0.84	Menzel (1963)
			Mean = 0.74	

Diet	Shell length (L_o–L_f)	Temp. (°C)	Growth rate (mm week^{-1})	Source
(B) Cultured algae				
Pseudoisochrysis paradoxa	8.0–15.5	21	0.63	Bricelj et al. (1984b)
Isochrysis galbana	6.1–24.3	20	0.58	Epifanio (1979a)
Thalassiosira pseudonana	10.0–23.2	20	0.67	Epifanio (1979b)
Mixed algae	12.0–31.0	20	0.46	Epifanio et al. (1975)
			Mean = 0.59	

L_o and L_f = initial and final shell length, respectively, in mm). Maximum (subtidal and density-independent) growth rates were selected where several conditions were tested. Growth rates were determined in enclosures in the natural environment unless otherwise noted (from Bricelj, 1993).

[a] Notata strain, hatchery-produced juveniles.

[b] Wild strain, hatchery-produced juveniles from wild broodstock.

[c] Clams grown in substrate.

[d] Maximum seasonal growth rate.

Peterson et al., 1984; Peterson, 1983, 1986), 36 years (New Jersey; Lutz and Haskin, 1985), and 40 years (Rhode Island; Jones et al., 1989). A comparison of these maximum reported ages with latitudinal trends in growth rates (see above) indicates an overall negative relationship between growth rate and longevity (also see Jones et al., 1990). Maximum mean shell length is about 100 mm in most areas, though some individuals may reach 150 mm (Carriker, 1959; Morris, 1973).

TABLE 8.7

Average time (in years) to attain legal market size (= 25.4 mm in shell thickness; see text) of *Mercenaria mercenaria* natural populations along the species' latitudinal range, from north to south

Time (years)	Location	Source
8.0–13.0	Prince Edward Island, Canada	Fig. 3 in Landry et al. (1993)
6.0	Prince Edward Island, Canada	Fig. 5 in Ansell (1968)
4.4	Maine	Fig. 5 in Ansell (1968)
3.2	Monomoy Point, Massachusetts	Fig. 5 in Ansell (1968)
4.0 (3.0–4.8)	Narragansett Bay, Rhode Island	Jones et al. (1989) [a]
3.5 (3.0–4.0)	Great South Bay, New York	App. 4 in Buckner (1984)
3.5 (2.5–5.0)	Great South Bay, New York	Greene (1978)
3.0	Barnegat Bay, New Jersey	Kennish and Loveland (1980) [a]
4.3 (3.8–4.6)		From Table 5, Kennish (1980)
4.4	York River, Virginia	From Fig. 3 in Loesch and Haven (1973)
2.4	Core Sound, NC	Peterson et al. (1983) [b]
3.0–4.0	Wassaw Sound, Georgia (intertidal)	Walker and Tenore (1984)
2.0	Kings Bay, southern GA	Jones et al. (1990) [c]
2.2–2.3	Indian River, Atlantic coast of Florida	Jones et al. (1990) [c]
2.1 (1.9–2.5)		Arnold et al. (1991)
2.6	Gulf Coast, Florida	Fig. 5 in Ansell (1968)

Range is shown in parentheses; unless indicated, time to market size is calculated from fitted von Bertalanffy, Gompertz or logarithmic growth equations (from Bricelj, 1993).

[a] Shell height (H) converted to length using an H/L ratio = 0.933.

[b] Assuming that age in years = number of annual bands.

[c] Using a H/L conversion factor = 0.91.

Ontogenetic growth patterns generally can be thought of as the net response over the long-term to a host of underlying factors affecting both feeding and growth. There has been a plethora of research on the various factors affecting growth in an attempt to gain a mechanistic understanding of the growth response. These controlling factors can be either "extrinsic" or "intrinsic" (see Rice and Pechenik, 1992, for a recent review of the factors affecting growth of larval, juvenile, and adult hard clams). In the next subsection, each environmental (extrinsic) factor known to affect growth rate is discussed individually, concluding with a discussion of research aimed particularly at characterizing the effects of various combinations of factors. The second subsection below deals with how intrinsic factors like genetics affect growth.

8.5.4 Environmental Factors Affecting Growth

The literature on the effects of environmental factors on growth of hard clams is by far the most voluminous of all the topics reviewed in this chapter. Pioneering studies by Kellogg (1903, 1910) and Belding (1909, 1912) identified most of the major environmental influences on growth. Subsequent research has resulted in at least a preliminary quantitative understanding for some factors, but a holistic, ecologically realistic, quantitative understanding of how combinations of environmental factors affect growth of *M. mercenaria* is still lacking. Growth response patterns will be compared to feeding responses discussed in Section 8.3 because feeding responses are generally thought to be the major underlying physiological control on growth (Bayne and Newell, 1983). The intent is to identify known underlying mechanisms for the growth responses as well as to identify areas where more research is needed.

8.5.4.1 Temperature

Ansell (1968) attributed much of the geographic variation in growth to variations in water temperature (see discussion above). Maximal shell growth occurs at 20° to 24°C with decreasing growth at lower and higher temperatures and cessation of growth below 7°C and above 31°C (Fig. 8.18). This is the same general relationship between temperature and pumping rate reported by Hamwi (1969) (see Fig. 8.4), supporting the notion that feeding rate is the major determinant of growth (Bayne and Newell, 1983). Ansell described the overall response pattern as that of a "limiting geometric catenary curve", which is essentially an inverted parabola, the general biotic response pattern to environmental factors as discussed in Section 8.3.4.1.

One of the major results of temperature-related variations in shell growth are shell banding patterns, particularly internal patterns that have been used in a number of growth studies.

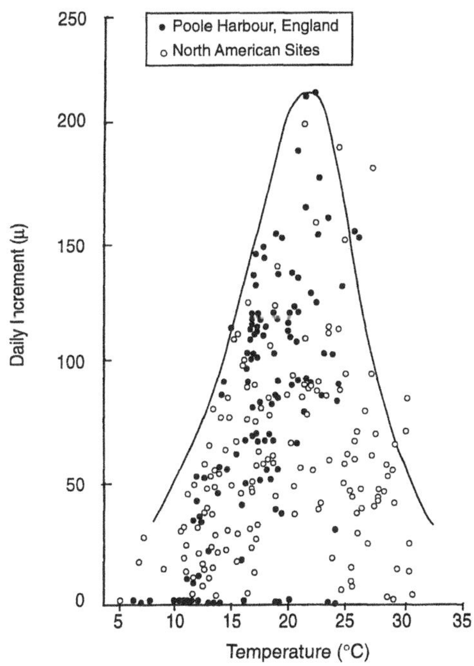

Fig. 8.18. *M. mercenaria.* General relationship between water temperature and growth rate (from Ansell, 1968).

Geographic variations in annual internal banding patterns of bivalves are mainly attributed to differences in ambient water temperatures (Lutz and Rhoads, 1980). Arnold et al. (1991) summarized studies on the relationship between water temperature and internal banding patterns of hard clams. In the cut, polished shell, clusters of daily slow-growth increments appear as a translucent, dark band. In contrast, clusters of fast-growth increments appear as an opaque, light-colored band. Slow-growth bands occur when water temperatures are inhibitory for growth, which corresponds to excessively low or high temperatures (Fig. 8.18). Temperatures between 15° and 25°C are optimal and result in fast-growth bands. Slow-growth bands typically occur in both summer and winter in some temperate areas, and fast-growth bands occur in fall and spring, following seasonal variations in ambient water temperature (e.g., Fritz and Haven, 1983; Grizzle and Lutz, 1988). In the southeastern US, slow-growth bands occur in summer and fall, corresponding to high water temperatures, and fast-growth bands occur during winter and spring when water temperatures are optimal for growth (see review in Arnold et al., 1991).

8.5.4.2 Salinity

Hard clams are typically only abundant in coastal waters with salinities that range from about 20 to 30 ppt, suggesting that they do not do well in reduced salinities. Studies on larvae (Davis, 1958; Carriker, 1961; Davis and Calabrese, 1964; Castagna and Chanley, 1973; Castagna and Kraeuter, 1981) and juveniles and adults indicate that growth rates decrease with decreasing salinities, although adults tolerate lower salinities than larvae (Fig. 2.1 in Malouf and Bricelj, 1989). Chanley (1958) reported progressively decreased growth of juveniles as salinity was lowered from 27 to 15 ppt. Castagna and Chanley (1973) found that growth of adults was decreased at salinities <17.5 ppt. Godwin (1968) recommended salinities of at least 25 ppt for maximum growth, based on field studies at six sites with transplanted hard clams. Davis (1958) reported optimal growth of larvae at 26 to 27 ppt. Reduced salinities clearly cause reduced growth.

Although we are aware of no studies on the effects of hypersaline waters on growth, presumably salinities above 30 ppt would be inhibitory on growth. If it can be assumed that the feeding response largely controls growth, this would almost certainly be the case (note in Fig. 8.5 that salinities below 20 and above 30 ppt strongly inhibit pumping rates). Hence, the overall pattern for the salinity vs. growth response probably also resembles an inverted parabola.

8.5.4.3 Food quantity and quality: a nutritional perspective

Suspension-feeding bivalves like *M. mercenaria* derive most of their nutrition from particulates removed from suspension, but some studies have shown that dissolved organic matter (DOM) can also partially meet their nutritional demands, especially in the larval stages. Rice and Stephens (1988; see their brief review of earlier literature) showed that adult *M. mercenaria* are capable of uptake of alanine and other free amino acids directly from ambient water at concentrations typically found in nature. Moreover, these amino acids were rapidly translocated to internal tissues and incorporated into macromolecules in these tissues. Thus, hard clams are capable of deriving some amount of nutrition from dissolved organic substances. However, under most conditions suspension-feeding bivalves generally (including hard clams) meet most of their nutritional needs from the wide variety of particulate matter typically consumed.

Both the quantity (concentration) and quality of suspended particulates (seston) affect growth of bivalves. Ideally both should be considered together in nutrition and growth studies, but because there is a wide diversity of natural particulates potentially used as food, gross estimates of seston quality and quantity together (e.g., total particulate organic matter) are typically reported. The generalized bivalve feeding response to seston concentration shows an inverted parabolic shape (Winter, 1978; Bayne, 1993). Increasing food concentrations, up to some optimal concentration, result in potentially increased ingestion rates and thus actual food intake available for digestion. Above this concentration, pseudofeces production costs energetically (see Section 8.2 above) and other inhibitory effects related to digestion physiology (see below) result in decreasing growth as food concentration increases. At least two studies have shown such a relationship for seston concentration and growth of hard clams.

Juvenile *M. mercenaria* showed a four-fold increase in growth rates when food (the picoplanktonic alga *Micromonas pusilla*) concentrations were increased from 50 to 500 cells μL^{-1}, then a similar magnitude of decline as concentrations were increased from 500 to 2000 cells μL^{-1} (Walne, 1970) (Fig. 8.19A). Coutteau et al. (1994) found that daily seston ration expressed as "% dry weight of food per wet weight of hard clams per day" also showed an apparently inverted parabolic-shaped relationship to growth of juvenile hard clams, with a steeper slope at low than at high food levels, though extremely high food concentrations were not tested (Fig. 8.19B). A gradual decline in growth rate at algal concentrations exceeding 50 cells *Isochrysis* μL was also described in juvenile *Mercenaria campechiensis* (Goldstein and Roels, 1980). Although the overall response pattern, and the underlying physiological mechanisms for it, seem well established at least for pure algal diets, more work is needed to adequately characterize how seston composition (food quality) is related to hard clam growth.

Studies discussed above show that *M. mercenaria* is capable of both pre- and post-ingestive selection of food items. When this is considered in light of the large spatial and temporal variability in seston concentration and composition typically occurring in nature, nutrition and growth studies can indeed be complicated. A major problem is determining the food value of different seston components. A few bivalve taxa, including hard clams, have been shown to meet some of their nutritional needs from detrital particles (see below). Although the relative importance of algal vs. non-algal organic particles in bivalve nutrition has not been extensively investigated, algal cells undoubtedly are important food items much of the time in nature. It is certain, based on well-established culture techniques, that some algal diets provide essentially all nutritional needs of many bivalves, including hard clams (Manzi and Castagna, 1989). Nonetheless, because of the wide variability in both the quantity and composition of even the algal component of natural seston, determining the overall nutritional value of the seston at any given time remains problematic. Researchers have addressed this problem in several ways: by assessing the food value of various algal taxa based on growth responses to carefully controlled diets, by determining what non-algal organic particles support growth, and by comparative assessments of experimental artificial and natural diets.

It is well-established that there is marked variation among algal species in their nutritional value, and thus their ability to support bivalve growth. This research primarily has been aimed at the overall growth response, but underlying processes that affect growth (e.g., variations in absorption efficiency) have also been studied.

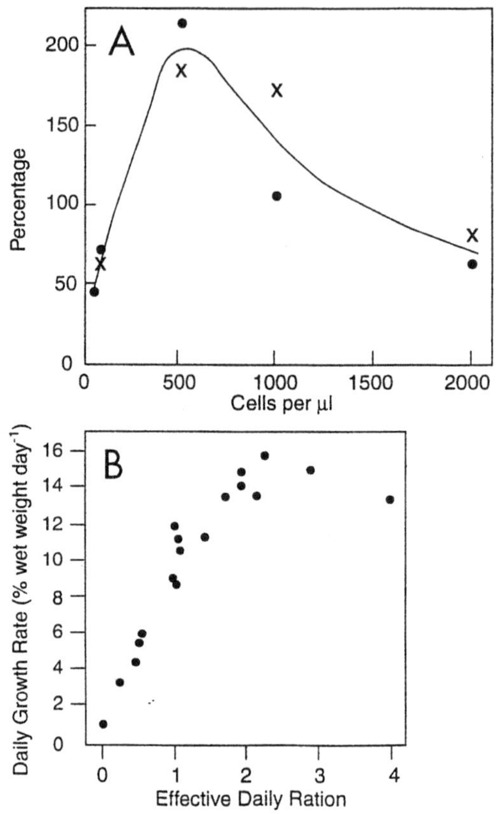

Fig. 8.19. *M. mercenaria.* Growth in relation to algal food quantity. (A) Relative growth rates (Percentage = % of the overall mean growth in shell length) of juvenile hard clams fed *Micromonas pusilla* for 21 days. Two different experiments indicated by dots and crosses (from Walne, 1970). (B) Daily growth rates of juvenile hard clams fed different effective daily rations (% dry weight of food per wet weight of clams) a 50 : 50% mixture of *Chaetoceros gracilis* and *Isochrysis galbana* (from Coutteau et al., 1994).

Table 8.8 compares the major studies on algal diets and growth of *M. mercenaria*. Walne (1970) conducted an extensive series of experiments on four bivalve species, including *M. mercenaria* and showed dramatic differences in food value for the algal species tested. Epifanio (1979b) tested the effects of fifteen different combinations of four algal taxa on growth of *M. mercenaria* (and *Crassostrea virginica*). His study was aimed in particular at assessing the effects of mixed vs. unialgal diets. In general, the more diverse diets produced the greatest growth, with particular algal species (*Isochrysis galbana*, *Thalassiosira pseudonana*) strongly associated with enhanced growth (see similar findings by Hartman et al., 1973). Epifanio (1979b) found no correlation between gross biochemical composition of the algal diets and growth (see more discussion below).

Wikfors et al. (1992) provide one of the most comprehensive studies to date on the nutritional value of algal taxa commonly used in the culture of hard clams (Table 8.8). Wikfors et al. also used multiple regression analysis in an attempt to relate the chemical

TABLE 8.8

Growth rate of juvenile *Mercenaria mercenaria* fed unialgal or mixed experimental diets, expressed as a proportion of the growth obtained with *Isochrysis galbana* (relative growth = 1.00 in all 3 studies), in order to provide a ranking of the food value of various diets (see sources for additional strain designations)

Modified from Wikfors et al. (1992)			Epifanio (1975), modified from Walne (1970)		Modified from Epifanio (1979b)	
diet	#[a]	relative growth	diet	relative growth	diet	relative growth
Isochrysis galbana (t-Iso)	(1)	1.00	*Skeletonema costatum*	3.30	B + C + D	2.01
Isochrysis galbana (Parke)	(2)	0.68	*Pyramimonas grossii*	1.19	A + B + C + D	1.88
Chaetoceros calcitrans	(3)	0.42	*Cricosphaera carterae*	0.70	B + D	1.14
Pavlova lutherii	(4)	0.42	*Tetraselmis suecica*	1.11	C + D	1.69
Tetraselmis maculata	(5)	0.39	*Isochrysis galbana*	1.00	A + C + D	1.59
Phaeodactylum tricornutum	(6)	0.37	*Nannochloris atomus*	0.92	A + B + C	1.41
Nitzschia closterium	(9)	0.30	*Dicrateria inornata*	0.67	*Isochrysis galbana* (Parke) = C	1.00
Rhodomonas sp.	(10)	0.27	*Olisthodiscus* sp.	0.75	*Thalassiosira pseudonana* (3H) = D	1.19
Dunaliella tertiolecta	(12)	0.23	*Micromonas pusilla*	0.74	A + C	1.17
Chlorella autotrophica	(16)	0.13	*Monochrysis lutherii*	0.59	A + B + D	0.98
Unfed 1	(18)	0.03	*Phaeodactylum tricornutum*	0.44	*Platymonas* (= *Tetraselmis*) *suecica* = B	0.87
Thalassiosira pseudonana (3H)	(19)	0.02	*Chlorella stigmatophora*	0.31	A + B	0.66
Unfed 2	(21)	–0.14	*Chlamydomonas coccoides*	0.19	A + D	0.48
			Dunaliella tertiolecta	0.14	*Carteria chuii* = A	0.33

[a] Numbers correspond to those plotted in Fig. 8.20.

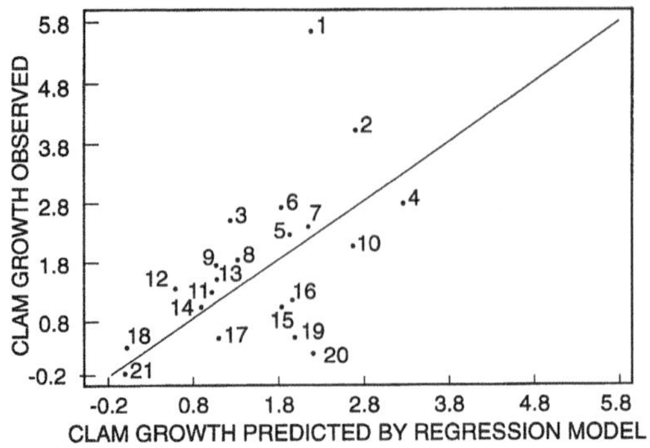

Fig. 8.20. *M. mercenaria.* Growth rates (change in total wet weight) of juvenile hard clams fed 19 algal diets (with two unfed treatments included); line is best-fit multiple regression model relating algal rations of protein, lipid, and carbohydrate to growth ($R^2 = 0.688$, $p < 0.0001$). Same diets as shown in Table 8 (from Wikfors et al., 1992). Numbers refer to diets listed in Table 8.8. 7, 11, 13 = *C. calcitrans*; 8, 20 = *T. maculata*; 14, 15, 17 = *N. closterium*, grown at nutrient concentrations that yielded less than maximum clam growth on that strain.

composition of the 19 algal cultures tested to hard clam growth responses (Fig. 8.20). Differences in protein, lipid, and carbohydrate content accounted for 69% of the variance in growth, with protein and lipid content by far being most strongly correlated with hard clam growth. They discussed these findings in light of experiments by Epifanio (1979b) that showed no correlation between algal chemical composition and hard clam growth rates, suggesting that previous studies were more limited in the range of biochemical constituents tested, and that a multiple regression analysis might have shown correlations not revealed in the univariate analyses used previously [also see Laing et al. (1987) and Bass et al. (1990) for related discussion].

Published growth rates, however, typically cannot be directly compared because different growth measures were used in the different studies as well as individuals of different initial sizes. Nonetheless, several algal taxa consistently support good growth of hard clams, and others do not. For example, the flagellate *Isochrysis galbana* supports fast growth. This alga has been used routinely in bivalve culture for many years, and is sometimes used as a standard against which other diets are compared, as in Table 8.8. The diatom *Skeletonema costatum* gave fastest growth in the studies by Walne (1970), and Laing et al. (1987) ranked it equal to *I. galbana*. At the other end of the spectrum, the chlorophyte *Dunaliella tertiolecta* was reported to only support minimal growth by Walne (1970) and Wikfors et al. (1992). One difference to note is the disparity in relative growth rates for *Nannochloris atomus* reported by Walne (1970) and Bass et al. (1990; see further discussion below). In Walne's studies, it ranked 5th out of 14 algal taxa tested. In contrast, Bass et al. (1990) found that hard clams did not grow when fed only *N. atomus*.

What explains the differences in food value among algal taxa? Studies on *M. mercenaria* by Bass et al. (1990, based on Bass, 1983) were aimed at assessing the nutritional value of various "small forms" (<1 to 4 μm; picoplankton) of chlorophytes and cyanobacteria that have been associated with coastal eutrophication and poor growth in bivalves. They also compared absorption efficiencies of *Pseudoisochrysis paradoxa* (known to support good growth in hard clams) and the "small forms". Mean absorption efficiencies of the *P. paradoxa* cultures were

TABLE 8.9

Percent absorption efficiency (AE) of organic matter from algal diets by juvenile *Mercenaria mercenaria*

Algal species (clone)	% AE	Source
Nannochloris atomus	23.8	Bass et al. (1990) [a]
Nannochloropsis sp.	17.6	Bass et al. (1990) [a]
Synechococcus bacillaris	31.1	Bass et al. (1990) [a]
Synechococcus sp. (ASN C-3)	29.4	Bass et al. (1990) [a]
Pseudoisochrysis paradoxa	80.3–86.5	Bass et al. (1990) [a]
	81.9	Bricelj (1984) [b,c]
Alexandrium tamarense (GtLI22)	61.8	Bricelj et al. (1991) [b]

[a] AE measured using $^{51}Cr:^{14}C$ dual radiotracer method.
[b] AE measured using Conover's ash-ratio method.
[c] Note that the AE of organic N and organic C averaged 86.8 and 80.6, respectively.

80.3% and 86.5%, reflecting this alga's high food value, compared to mean efficiencies of 17.6% and 31.1% for the small forms (Table 8.9). Thus, differences in absorption efficiencies probably explain at least part of the differences in growth rates reported among algal taxa. Poor absorption of these chlorophytes and cyanobacteria was associated with the presence of indigestible cell walls, containing sporopollenin, a highly refractory, polymerized carotenoid typically found in spores of higher plants (Bass et al., 1990). Low absorption efficiency was also correlated with very short residence times for these species, compared to *Isochrysis galbana*, suggesting that *M. mercenaria* regulates gut passage time in response to food quality (Bricelj et al., 1984a).

Few measurements have been made on absorption efficiencies of algal diets by *M. mercenaria* (summarized in Table 8.9). Absorption values for a wide range of algal species have been obtained for other bivalves, such as oysters and mussels, but it may not be possible to extrapolate these data to *M. mercenaria*. Absorption efficiency in hard clams is known to decline with increasing cell density [e.g., decreased from 82% at 50 $\times$ 10^6 cells L^{-1} to 62% at 250 $\times$ 10^6 cells L^{-1} of *Pseudoisochrysis paradoxa* (Bricelj, 1984)]. The values reported in Table 8.9 represent maximum values, since they were obtained at low to moderate cell densities. It is also noteworthy that similar absorption efficiency values were obtained (for *Pseudoisochrysis paradoxa*) using the dual radiotracer technique and Conover's ash-ratio method, suggesting that mucus and metabolic losses associated with the breakdown and exocytosis of the cellular contents of digestive cells did not contribute significantly to the organic content of clam feces.

In conclusion, planktonic algal taxa clearly vary widely in their relative food values, and commonly used gross chemical measures of seston food quantity/quality like chlorophyll or total organic content should not be expected to consistently reflect relative nutritional value or to correlate with growth performance. The multiple regression analysis by Wikfors et al. (1992) has important implications for designing artificial diets as well as indicating what kinds of measurements of natural seston might yield an estimate of its overall nutritional value.

It was first hypothesized by Petersen and Jensen (1911) that suspension-feeding bivalves derive nutrition from detritus. Subsequent studies have conclusively demonstrated that at least some bivalve species meet part of their nutritional needs from detrital particles (e.g., Newell and Langdon, 1986; Crosby et al., 1989; Langdon and Newell, 1990). Using a modification of the $^{14}C:^{51}Cr$ dual radiotracer method, Bricelj (1984) showed that juvenile *M. mercenaria* were able to absorb sedimentary organics with 22% efficiency. The organics derived from suspended silt allowed hard clams to compensate for the dilution of algae present in a mixed suspension, and maintain a constant rate of absorption of organic matter up to silt additions of 20 mg dry weight L^{-1}. This study also indicated that low seston sediment concentrations (ca. 5 mg L^{-1}, 11% organic content) might provide a supplementary food source during periods of low phytoplankton concentrations ($\leq$100 μg C L^{-1}). No growth enhancement was observed, however, when low sediment levels were added to an algal diet equivalent to 300 μg C L^{-1} (Fig. 8.3D).

Several studies have compared the nutritional value for *M. mercenaria* of various artificial diets relative to known algal diets as well as natural seston. The major impetus for such studies has been the development of artificial diets for bivalve aquaculture in order to minimize the need for maintaining algal cultures (for review see Coutteau and Sorgeloos, 1993). Three major kinds of artificial or manipulated diets have been tested with bivalves generally: centrifuged wet algal pastes or dried algae, yeast-based, and synthesized, multi-component, microencapsulated diets. Experiments involving all three have been carried out with hard clams.

Laing and Verdugo (1991) reported similar growth rates for juvenile *M. mercenaria* (and four other bivalve taxa) fed spray-dried *Tetraselmis suecica* compared to live diets of the same alga. Growth was enhanced when the dried *T. suecica* diet was supplemented with live cells of the same alga as well as other algal species. Spray-dried *T. suecica* (70% of total ration) combined with live *Skeletonema costatum* or *Isochrysis galbana* has also been successfully used to condition adult hard clams yielding the same results as mixed live diets.

Yeast cells offer the advantages that they are high in protein content, suitable in size, highly stable in suspension, and are not expensive to produce. Poor food value of 100% yeast diets relative to algal diets has been attributed to the poor digestibility of the yeast cell wall and/or deficiency in essential polyunsaturated fatty acids. Epifanio (1979a) found that mixed diets of live algae (*Thalassiosira pseudonana*) and up to 50% spray-dried yeast yielded growth of juvenile *M. mercenaria* equal to growth of the controls fed 100% algae. Hurd et al. (1989) tested freeze-dried algal diets and yeast-based diets with and without supplements on growth of juvenile hard clams. Freeze-dried *Isochrysis galbana* combined with dried yeast supported moderate growth over the 3-week experimental period. The same alga/yeast diet supplemented with either corn oil or cod liver oil gave significantly increased growth in both shell and tissue. Recent use of enzymatically or chemically treated yeast to increase its digestibility has considerably improved on earlier results (e.g., Epifanio, 1979a). Manipulated yeast has now been tested as a diet for juvenile clams on a commercial scale (*Tapes philippinarum* in Spain and *M. mercenaria* in the US; Coutteau et al., 1994). However, satisfactory results are generally only obtained when yeast is used as a partial substitute (50 to 80% of the live algal diet).

Synthesized microencapsulated diets have also been tested for their ability to support growth of *M. mercenaria*. Laing (1987) fed hard clam and oyster spat microencapsulated diets compared to no food and live *Chaetoceros calcitrans*. Three oyster species and two clam species (*M. mercenaria* and *T. semidecussata*) were tested, and the data were presented only on the basis of either "oysters" or "clams". The two synthesized diets alone resulted in significantly reduced growth of clams (54% and 64%) compared to the live algal diet. But similar growth rates were achieved by a wide range of combinations of synthesized and live algae, including a substitution of up to 85% of the algae by synthesized diet. The synthesized diets were more inhibitory for the oyster spat than for clams.

These experiments demonstrate the ability of a variety of experimental diets based on dried food organisms and synthesized microparticulates to support growth of hard clams. Such diets generally only yield growth rates comparable to live algae when used as a partial dietary supplement. These findings have obvious implications for aquaculture. They also point to the feasibility of further development of such approaches as substitutes for cultured as well as natural seston in feeding and growth experiments generally (e.g., Kreeger et al., 1996).

8.5.4.4 Water flow

Hard clams live in shallow coastal waters that are typically affected by tidal and/or wind-generated water movements. Early field studies (Kellogg, 1903; Belding, 1912; Kerswill, 1941, 1949; Haskin, 1952; Carriker, 1959) indicated that water flow enhanced growth rates of *M. mercenaria*. Subsequent research has elucidated a more complicated relationship between current speed and hard clam growth, particularly with respect to how flow and seston may interact.

A problem with determining how current speed affects growth of bivalves centers around determining if one is measuring an individual response or a population-level response. For example, most field studies are based on sampling of individuals from a population, and information on how the members of the bivalve population (or other benthic organisms not sampled) in the immediate area collectively may have affected important environmental variables like seston concentration, are not known. Hence, in such studies, the effect of currents on growth of individuals might be better interpreted as a secondary effect of the population on seston depletion or some other factor (see additional discussion in: Grizzle and Lutz, 1989; Bayne and Hawkins, 1992; Judge et al., 1992; Lenihan et al., 1996).

Aquaculture-related studies are typically concerned with population-level phenomena like production. For example, studies on the effects of water flow on growth of hard clams typically consider flow rate with respect to bivalve densities as well as food concentration (e.g., Eldridge et al., 1979; Manzi et al., 1984, 1986; Malinowski and Siddall, 1989; Eversole et al., 1990). These studies provide useful information relevant to culture protocols, but not for how water flow *per se* is directly related to individual growth (see Grizzle et al., 1992, for a summary of the effects of water currents on individual growth of bivalves generally, emphasizing avenues for future research; also see Wildish and Kristmanson, 1993, 1997).

A descriptive-correlative field study at eight sites found that mean near-bottom (within 16 cm of the sediment surface) current speed approximated an inverted parabolic-shaped relationship to hard clam growth (Grizzle, 1988; cf. Fig. 6 in Grizzle and Lutz, 1989). The average current speed at these sites ranged from 5.3 to 13.8 cm s^{-1}, with maximal values of 11.2 to 40.0 cm s^{-1}, respectively. In a field study involving experimental alteration of tidal currents at three sites, Judge et al. (1992) found that mean near-bottom current speeds from 4 to 11.3 cm s^{-1} (maximal 13 to 27.4 cm s^{-1}) did not affect growth of hard clams. Hard clams had substantial growth at all sites. Two flume experiments by Grizzle et al. (1992) indicated that hard clam growth was positively related to increasing flow speeds between 0 and 9 cm s^{-1}.

Current speeds above 10 cm s^{-1} and perhaps up to 30 cm s^{-1} appear to produce maximal growth. Speeds >30 cm s^{-1} are probably inhibitory. The overall pattern is the same as that discussed above for feeding responses (Fig. 8.10); however, the underlying causal mechanism(s) for the effects of water currents on both feeding and growth are not fully understood.

Grizzle et al. (1992) offered the "inhalant pumping speed" hypothesis (IPS) to account for the positive relation between growth and water current speeds up to the inhalant pumping speed. IPS is based on a conceptual model (see Section 8.6 below) which indicates that at ambient current speeds well below the speed of the inhalant stream, water is entrained from areas above the level of the siphon. From an energetics perspective, this means that the bivalve is "actively" pulling in water for its feeding and ventilation currents. At ambient current speeds equal to the inhalant pumping speed, however, the ambient currents are flowing with no pressure differential into the inhalant siphon. This may mean less energy expenditures for the bivalve, and thus maximal growth (assuming other factors like food concentration are equal). The negative relation between growth and higher current speeds is probably due to other mechanisms. At speeds sufficient to entrain bottom sediments, bedload transport would interfere with feeding (Turner and Miller, 1991). Frictional drag forces on the siphons at high current speeds might also be a factor (Eckman et al., 1989).

8.5.4.5 Sediment characteristics

Several descriptive-correlative field studies have indicated that hard clams grow fastest in sandy sediments and slowest in muds (e.g., Carriker, 1959; Kennish and Olsson, 1975; Greene, 1979). Such studies, however, are typically plagued by the possibility that other environmental factors associated with sediment type (e.g., water currents) may have contributed to or been the actual cause of the reported growth differences. Sediment characteristics and water current regimes are typically correlated, with muddy sediments accumulating in areas with slowest currents. Because sluggish water currents can inhibit hard clam growth (see above), experimental approaches have been used in an attempt to elucidate sediment effects.

Belding (1912) reported a 24% decrease in shell volume growth for hard clams transplanted into boxes containing mud compared to coarse sand. Pratt (1953) found ~20% growth inhibition for hard clams placed into experimental boxes containing sandy mud compared to hard clams in adjacent boxes containing sand. Based on a series of experiments involving boxes containing different sediment types and placed at eight sites, Pratt and Campbell (1956) showed a strong negative correlation between percent silt–clay (mud) in the sediment and hard clam growth. Rhoads and Pannella (1970), using mud-filled and sand-filled trays at two sites showed growth differences of 19% and 30% between hard clams in sand compared to mud, with fastest growth in sand at both sites.

In contrast to the above studies, and also based on a transplant experiment, Kerswill (1941, discussed in Pratt, 1953) found no significant sediment effects. In a reciprocal transplant experiment designed to test both sediment and site effects, Grizzle and Morin (1989) found only a marginally significant ($p = 0.13$) sediment effect, with growth inhibition of 6% in mud compared to sand (Fig. 8.21). The site effect (which was primarily explained as differences in water currents and/or seston fluxes; see more below) was significant, with an 11% difference between the slow-growth and fast-growth sites. These two studies (also see Hibbert, 1976)

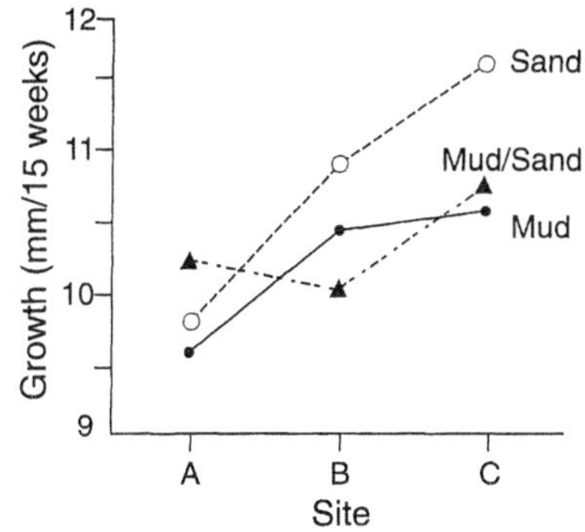

Fig. 8.21. *M. mercenaria*. Shell growth at three reciprocal transplant sites by sediment type indicating a differential sediment effect (from Grizzle and Morin, 1989).

suggest that a sediment effect may not always be strong, and other environmental factors may mask it.

Thus, it seems reasonable to conclude that sedimentary characteristics can affect growth of *M. mercenaria*, even though the effects of other environmental factors may be stronger. The most-reported relationship is that increased silt–clay (mud) content causes decreased growth. If the relationship is really one of cause-and-effect, what might be the underlying mechanisms?

Using laboratory and field studies, Pratt and Campbell (1956) identified at least three potential causal mechanisms for the inhibitory effects of fine-grained sediments. First, they observed that hard clams held in aquaria in muddy sediments remained near the sediment surface and maintained an open burrow. In contrast, hard clams in sand often remained completely covered, pumping inhalant water through the overlying sediment. Mud content was negatively correlated with sediment permeability. Subsequent field studies showed that the average depth of burial of hard clams in experimental sediment boxes was positively correlated with sediment permeability. Hence, they hypothesized that reduced permeability of mud affects burrowing and feeding behavior, with the net result being increased energy expenditures and decreased growth.

In another set of aquarium experiments, Pratt and Campbell (1956) found that sediment mud content was positively correlated with pseudofeces production rate by the hard clams. Fine-grained sediments are apparently more easily entrained in the feeding currents than sand and at least in part must be expelled along with other non-food items. The result is an increase in food dilution by sediment particles and perhaps an increase in expended energy for food sorting, and thus decreased growth for hard clams living in mud sediments compared to sand. Bricelj et al. (1984b) found significant growth inhibition (16%) of hard clams after suspended sediment (silt) concentrations reached 44 mg DW L^{-1}. Reduced growth in mud may also result from increased bioturbation due to a general abundance of deposit feeders in muddy substrates (Murphy, 1985). Turner and Miller (1991) found up to 38% growth inhibition of *M. mercenaria* at suspended sediment concentrations of 193 mg DW L^{-1} resulting from wave action.

A third hypothesis proposed (but not tested) by Pratt and Campbell (1956) is that muddy sediments may at times produce hypoxic or anoxic conditions and accumulate toxic substances (e.g., hydrogen sulfide) to which the hard clam is potentially exposed during pumping when its valves are gaping. Sandy sediments are less likely to become anoxic.

In conclusion, sediment characteristics can significantly affect growth of hard clams, but the actual mechanisms involved remain to be elucidated. Pratt and Campbell's (1956) hypotheses would be a good starting point for future work. Future experiments (particularly in the field) need to be carefully designed so as to control other factors that affect growth, or adequately characterize them so they can be statistically separated. A major shortcoming of most research in this area has been the use of only gross characterizations of the sediment. It will be necessary to characterize more than grain size distribution to adequately assess sedimentary effects on growth.

8.5.4.6 Noxious algae and other factors

In many coastal areas blooms of various algal taxa have harmful effects on grazers, including bivalves. Much of the concern has been from a public health perspective in that humans consuming bivalves that have fed on the toxic algae may contract amnesic, diarrhetic, and paralytic shellfish poisoning. Some microalgae also have detrimental effects on bivalves (for review see Shumway, 1990). Algal species known to affect growth of hard clams adversely include: the chlorophyte *Nannochloris atomus*, *Aureococcus anophagefferens* (Pelagophyceae), and the dinoflagellate *Alexandrium fundyense*, the causative agents of "green", "brown", and "red" tides, respectively. The first two are picoplanktonic algae ($\sim$2 μm diameter) which because of their small size are expected to be poorly retained by the hard clam's feeding apparatus. However, growth inhibition by these algae has been largely attributed to factors other than their small size.

Summer blooms of *N. atomus* were documented in New York's southern bays on Long Island in the 1950s. Laboratory studies subsequently demonstrated that monospecific cultures of this alga do not support growth of either larval (Tiu et al., 1989) or juvenile (Bass et al., 1990) hard clams, and cause growth inhibition when fed in combination with other nutritive algal taxa. Bricelj et al. (1984a) showed that the hard clam has a short gut retention and low absorption efficiency of ingested organics from this alga.

Aureococcus anophagefferens first occurred in Narragansett Bay, Rhode Island (Sieburth et al., 1988) and in eastern and southern Long Island bays in 1985 (Cosper et al., 1987), and has reappeared in New York embayments since (see review by Bricelj and Lonsdale, 1997). It causes severe inhibition of filtration rates in the hard clam (Tracey, 1988), and inhibition of ciliary beat in excised gill tissue (Gainey and Shumway, 1991). The expected growth inhibition was documented during a moderate brown tide outbreak. In this instance, up to 30% reduction in tissue growth rates of juvenile hard clams suspended off-bottom in several Long Island bays occurred at concentrations of *A. anophagefferens* ranging between 1.7 and 3.2×10^8 cells L^{-1} (Bricelj and Lonsdale, 1997).

Some dinoflagellates not known to produce toxins that affect mammals (e.g., *Prorocentrum* spp.) can also have detrimental effects on growth of hard clams (Wikfors and Smolowitz, 1993). For example, persistent blooms of *Prorocentrum* spp. (*P. micans*, *P. minimum* and *P. redfeldii*) have been associated with reduced growth of *M. mercenaria* in Long Island Sound. In controlled laboratory experiments, post-set hard clams showed poor growth on a unialgal suspension of *P. micans* (strain CCMP693), but grew at rates only slightly below those of

Isochrysis galbana (T-ISO) controls when offered a T-ISO/*P. micans* mixture (Wikfors and Smolowitz, 1993). In contrast, a mixed diet of T-ISO and *P. minimum* (strain CCMP 1329, EXUV) produced no detectable growth of clams. Although feeding rates were not reported in this study, these results suggest that *P. minimum* may interfere with the uptake and/or absorption of other nutritious microalgae present in a mixed diet, while *P. micans* may inhibit feeding and/or be nutritionally deficient. The presence of many intact, undigested *P. minimum* cells in feces suggested that this alga may be poorly absorbed by hard clams. The lack of growth even when the diet was supplemented with T-ISO, a known good food source, suggests that this dinoflagellate may also be toxic. High-toxicity strains of *Alexandrium* spp. are known to inhibit feeding rates of hard clams (see above); therefore, although growth responses of PSP-producing dinoflagellates have not been determined for *M. mercenaria*, it is likely that they also inhibit growth.

A final factor to consider in this subsection is gas-bubble disease caused by exposure to water supersaturated with gases. It results from formation of gas emboli in blood when hyperbaric pressures are uncompensated (for review see Wietkamp and Katz, 1980). Malouf et al. (1972) produced supersaturated conditions by rapidly heating cool ambient seawater, and found that it was harmful to hard clams. Bisker and Castagna (1985) found that total gas saturation of 115% inhibited growth of juvenile hard clams. Such conditions may occur in hatcheries where compressed air is injected into the culture water, or ambient seawater is heated.

8.5.4.7 Miscellaneous environmental factors

Kerswill (1949) found that *M. mercenaria* in plots within dense eelgrass beds showed growth inhibition of 60% compared to hard clams not in the vicinity of eelgrass. Those in less dense eelgrass grew about 50% slower. Subsequent research has shown variable effects of seagrasses (and other macrophytes), sometimes inhibiting growth (Peterson and Beal, 1989), sometimes increasing growth (Peterson et al., 1984; Arnold et al., 1991; Irlandi and Peterson, 1991; Slattery et al., 1991), and in some cases having no effect (Peterson and Beal, 1989). Macrophytes can affect several environmental factors (e.g., water flow, sediment characteristics) that probably more directly affect growth. Hence, Irlandi and Peterson (1991) considered the effect of seagrasses as most likely being indirect (see more discussion of multiple factors below). Positive effects of seagrasses on growth have been attributed to increased near-bottom food supply resulting from enhanced particle settlement (Peterson et al., 1984), resuspension of benthic or epiphytic algae produced within the seagrass bed (Judge et al., 1993), reduced siphon nipping activity by fish (see below), and reduced sediment resuspension (Irlandi and Peterson, 1991). The major hypothetical negative effect on growth is decreased water flow in seagrass beds which results in inhibited food supply and/or a buildup of muddy sediments (Kerswill, 1949).

Several researchers have investigated the effects of bottom elevation, particularly intertidal vs. subtidal position, on hard clam growth. Nearly all studies have demonstrated that hard clams grow fastest subtidally (Belding, 1912; Newcombe, 1935; Walker and Heffernan, 1990), though the effect is sometimes not great (Eldridge et al., 1979), or is influenced by other factors (Rhoads and Pannella, 1970). The elevation effect usually has been interpreted as the result of differences in feeding time, though other factors could be involved.

8.5.4.8 Biotic interactions

Some fishes and birds feed on the siphons of infaunal bivalves like *M. mercenaria*, removing only a part of the siphon and not killing the bivalve (see brief review by Coen and Heck, 1991). For example, Festa (1975) found that hard clam siphons represented 13% of the diet of winter flounder (*Pseudopleuronectes americanus*) in a New Jersey study. This form of sublethal predation, referred to as siphon "nipping", "cropping", or "browsing", has been shown to reduce growth rates of some bivalves (e.g., Peterson and Quammen, 1982; De Vlas, 1985). The effect of siphon nipping on hard clams has been simulated by excising standardized portions of the siphon. This procedure significantly reduced growth, and hard clams from natural populations in three areas (Massachusetts, New Jersey, and Alabama/Florida) showed evidence of incidences of siphon nipping ranging from 4 to 25% (Coen and Heck, 1991). They hypothesized that siphon nipping could inhibit growth via three mechanisms: energy losses needed for regeneration of the siphon, decreased feeding efficiency and particle selectivity due to loss of siphon tentacles, and further loss of siphonal tissue to predators due to diminished light-sensing capabilities because of loss of photoreceptors on the siphon tentacles.

Irlandi and Peterson (1991) found that hard clams collected from unvegetated sand flats had lower siphon weights relative to other soft tissues than hard clams collected concurrently in a nearby seagrass bed. They interpreted this to be evidence of siphon nipping, and as a partial explanation for reduced growth rates of hard clams on sand flats compared to those in seagrasses. They also demonstrated in a laboratory experiment that the presence of a whelk, *Busycon carica*, reduced the amount of time spent feeding by hard clams. Such predator avoidance behavior could also lead to reduced growth rates.

Several studies have been carried out to determine the effects of hard clam population densities on growth. Such studies are perhaps best interpreted as population-level phenomena involving competition for food, as discussed above in the section on water currents. Interference or other forms of competition potentially involving as few as two individuals may occur, but to our knowledge, no such studies have been done on hard clams. Some population-level studies show a strong density effect (Eldridge et al., 1979; Hadley and Manzi, 1984; Walker, 1984; Eversole et al., 1990), while others show no effect (Manzi et al., 1980; Widman and Goldberg, 1990). Density-dependent reductions in growth rate up to 33% were reported for juveniles 1.7 cm in initial SL at 3027 hard clams m^{-2} (Walker, 1984). Density effects generally are determined by food supply rates, hard clam feeding rates, and other more direct influences on individual growth (see discussion in Hadley and Manzi, 1984). Hence, any density effect would have to be site-specific, or characterized in terms of other environmental and experimental conditions (see Table 3 in Bricelj, 1993).

8.5.4.9 Combinations of environmental factors: multiple causes

The studies reviewed above have resulted in a quantitative understanding of the effects of several environmental factors on growth of *M. mercenaria*. Even so, an ecologically realistic, quantitative understanding of how different combinations of factors control growth has not been attained. Hard clams in nature are exposed to constantly changing environmental conditions. To develop general, predictive models, the relative importance of the major factors affecting growth as well as the effects of different combinations of factors, must be determined.

Some 19th century biologists were already concerned with "... multiple variables in the full context of their interactions in nature..." (McIntosh, 1985, p. 25). Although much new knowledge has been gained on how environmental factors affect growth of hard clams, Belding's (1912) general assessment is still valid today (except for his exclusion of temperature and food quality and quantity):

> "The chief natural agents affecting growth [of *M. mercenaria*] are current, tide, soil, depth, and salinity of the water, arranged in order of individual importance, yet so closely interwoven that their separate actions cannot always be clearly demonstrated. Their various combinations form a favorable or unfavorable environment for the quahaug."

A quantitative understanding of such multiple causes will only be attained using multi-factor, experimental approaches, in both the field and laboratory. Unfortunately, only a few experimental studies have been particularly aimed at such a goal, and most have involved larval growth and development (e.g., Davis and Calabrese, 1964; Lough, 1975; Calabrese et al., 1977).

Based on a field study of *M. mercenaria*, Peterson et al. (1984) invoked a "hydrodynamic hypothesis" to explain how current speed and food concentration are potentially inter-related as they affect growth of suspension-feeding bivalves generally, particularly from the perspective of how aquatic macrophytes may influence growth (see above). For example, the positive relationship between water currents and growth may be offset if food concentration is higher in areas of sluggish flow such as seagrass beds compared to higher flow areas.

Peterson and Beal (1989) experimentally investigated the effects of five factors: year, hard clam density within enclosed plots, the presence of enclosure walls, the source of hard clams, and site (as well as various two-way interactions of the five) on growth of hard clams. They found that most factors significantly affected growth under some conditions, and some interaction effects were significant. A simple additive model adequately described how site effects combined with various experimental effects to produce a net effect on growth (see more discussion in modeling subsection below).

Based on a reciprocal transplant experiment, Grizzle and Morin (1989) estimated the relative effects of site and sediment type on growth of hard clams. They found a significant site effect ($p = 0.006$), a marginally significant sediment effect ($p = 0.13$) and no interaction effect ($p = 0.26$) (Fig. 8.21). They hypothesized that the site effect resulted from differences in near-bottom water currents and/or seston fluxes. Following these experimental findings, Grizzle and Lutz (1989) used multiple regression analysis to relate different combinations of water currents, seston fluxes, and sediment characteristics to hard clam growth at eight sites. Fastest growth was best correlated with sandy sediments and intermediate water flow speeds and/or seston fluxes; growth rates decreased as water speeds increased or decreased and as percent silt–clay in the sediments increased (see further discussion below).

The notion that horizontal seston fluxes, as opposed to either seston concentration or water currents alone, may affect individual growth rates was first proposed by Grizzle (1988), Grizzle and Morin (1989), and Grizzle and Lutz (1989) in their field studies on *M. mercenaria*. It has been experimentally tested, however, only on two other species, and in both cases water speed or seston concentration was better correlated with growth than seston fluxes (Cahalan et al., 1989; Lenihan et al., 1996).

Much more work is needed on how water currents and seston affect growth, particularly from a fluid mechanical perspective. The model presented below could provide an organizing framework for the design of further studies. Irlandi and Peterson (1991) provide a particularly useful and comprehensive qualitative assessment of how growth of *M. mercenaria* might be affected by multiple environmental factors in both direct and indirect fashion [also see Goldberg and Wikfors (1992) and discussion of aquatic macrophytes above].

8.5.5 Genetic Factors Affecting Growth

Most studies show wide variability in growth rates among individuals to any given set of environmental conditions. It is generally assumed that part of the variability is attributable to differences in genetic makeup of the individuals comprising the study population. Unless it is the actual subject of study, genetic variability is typically assumed to be a part of the overall variability that statistical analyses factor in when assigning error ranges or probabilities. In aquaculture operations, however, knowledge of the genetic component of variability in growth responses can potentially be important because it allows selection for faster growth in breeding and seed stock production. Such research on most bivalve species (including *M. mercenaria*) has been, however, only rudimentary (see review by Humphrey and Crenshaw, 1989). To date, most studies on genetics of hard clams emphasize either aquacultural concerns (for a brief review see Manzi et al., 1991) or ecological and evolutionary processes (see Slattery et al., 1993).

Multiple-locus heterozygosity (i.e., increased genetic diversity) and growth rate have been positively correlated in a number of studies on bivalve populations in nature (Gaffney and Scott, 1984; Garton et al., 1984; Koehn and Gaffney, 1984), though not involving the hard clam. Menzel (1962) reported increased growth rates (and wider environmental tolerance) of putative hybrids from crosses between *M. mercenaria* and *M. campechiensis*, with the result presumably caused by increased heterozygosity of the hybrids. We are aware of no subsequent hybridization efforts involving hard clams. Crossbreeding studies involving separate, inbred lines of hard clams have been conducted in an attempt to increase heterozygosity, and thereby develop faster-growing offspring, and there have been some attempts to develop fast-growth stocks via artificial selection.

Selective breeding for increasing growth rates of cultured bivalves generally has been recommended and pursued with some success for decades (see brief review by Hadley et al., 1991), but results involving *M. mercenaria* thus far have been mixed. Heffernan et al. (1991) selected wild stock hard clams over two generations for fast-growing individuals, finding no long-term benefit from selection. The F_2 progeny from the fast-growth line actually grew significantly slower than the controls. In contrast, Rawson and Hilbish (1990) found substantial heritability (defined as the proportion of the total phenotype variance in a trait [e.g., growth rate] that is attributable to additive genetic effects) of enhanced growth rates via artificial selection in two separate experiments using wild hard clams. They found that culture density strongly affected the outcome of one experiment, and concluded that selection has the potential to cause rapid change in growth characteristics. Likewise, Hadley et al. (1991) showed substantial heritability and significantly increased growth in two of three selection experiments using wildstock *M. mercenaria* grown over 2 years under different environmental conditions. They emphasized the importance of control populations and environmental conditions in interpreting the results of such experiments.

Dillon and Manzi (1987) described two stocks of *M. mercenaria* that had been selected over three or four generations for fast growth. Crosses of individuals from the two stocks showed no relationship between overall heterozygosity measured at seven enzyme loci (Dillon and Manzi, 1988). They discussed several possible reasons for why overall heterozygosity may not be a good indicator of growth. Similarly, Manzi et al. (1991) found no correlation between heterozygosity and growth rates involving several crosses between two hatchery stocks of hard clams reared under hatchery conditions and grown in the intertidal zone in cages for 2 years.

Genetic changes over time resulting from differential selection is the core theoretical mechanism for evolutionary change. Hence, ecologically oriented studies (whether in the laboratory or nature) on hard clam genetics typically have emphasized evolutionary aspects. Adamkewicz et al. (1984) identified two genes (*Lap* and *Pgm-3* loci) associated with shell size, with one fitting a model based on additive effects only and another having both additive and interaction effects for some genotypes. They found no correlation, however, between heterozygosity *per se* and growth (shell size). Adamkewicz et al. interpreted their data as strong evidence for selection at the *Lap* locus, and discussed possible mechanisms in nature for selection. Similarly, Slattery et al. (1991) found no correlation between multi-locus heterozygosity and growth. They found that 98% of the genetic variability in *M. mercenaria* was within populations, with only 1.7% occurring across three widely separated localities. In a follow-up study, Slattery et al. (1993) found a weak correlation between heterozygosity and age. They also showed the importance of sample size, particularly with respect to the representation of young age classes, in detecting such correlations. Both papers (1991, 1993) discuss potential environmental conditions (i.e., selective pressures) that may affect genetic characteristics of populations.

Recent research on hard clams has shown the presence of genotype-environment interactions in determining various traits, including growth rate. In a study involving five sites ranging geographically from New Hampshire to South Carolina, Rawson and Hilbish (1991) attempted to partition variations in growth into genetic and environmental causes. They found both genetic and environmental effects, with the additive genetic × location effect to be predominant. Such interactions were at least partly due to a change in the amount of genetic variation expressed at each site. They concluded that genotype–environment interactions should act to constrain the evolution of juvenile growth rate, preserving heritable variations and leading to phenotypic plasticity for growth. Hilbish et al. (1993) also showed that genetic variation for shell growth is not stable over time, with shell growth at one ontogenetic stage not being similar to growth at another. They discussed potential reasons for why natural selection may not affect growth rates.

From a physiological ecology view there is much yet to be learned of the relationship between genetics and environmental conditions and their effects on growth rates. A more thorough discussion of *Mercenaria* genetics is provided in Chapter 6.

8.5.6 Production in Wild and Cultured Populations

Fisheries scientists, aquaculturists, and ecologists are concerned with production, defined as the increase in biomass per unit area per unit time for a population. Surprisingly, few studies have been done on hard clam production rates. As the above review indicates, most

TABLE 8.10

Production rates (P) and production/biomass ratios (P/B) of *Mercenaria mercenaria* with associated environmental conditions

Location	Environmental conditions	Density (# m^{-2})	Mean biomass (g m^{-2} $year^{-1}$)	Mean production (g m^{-2} $year^{-1}$)	P/B
Southampton, England [a]	high intertidal; muddy-sand to sandy-mud	0.25–5.75	7.73 (ash-free wt.)	3.99 (ash-free wt.)	0.52
	mid intertidal; muddy-sand to sandy-mud	0.29–12.57	50.04	14.00	0.28
	low intertidal; mud	0.25–3.33	36.54	6.19	0.17
Georgia, USA [b]	intertidal; mud to sand; tidal creeks, shell deposits, oyster reefs	18 (SD = 5)	53.17 (ash-free wt.)	7.66 (ash-free wt.)	0.14
		12 (SD = 6)	11.41	2.66	0.23
		49 (SD = 14)	119.76	6.01	0.05

[a] Hibbert (1976).
[b] Walker and Tenore (1984).

effort has gone into studies of individual growth. Population production rate is essentially the summation of all individual growth rates of the members of a population, so the two are obviously related. Mean growth measurements typically form the basis of production estimates, but other approaches are sometimes used (see above). Perhaps the major reason why production has not been emphasized, even in aquaculture-related research, is that time to market is generally considered more important in controlling profits. Studies of population production under various environmental conditions indicate the range of production rates for *M. mercenaria*, some of the factors affecting it, and how it may be related to individual growth rates (Table 8.10).

Walker and Tenore (1984) used their data to show the relationship between growth, production, and production/biomass ratios over the lifespan of the hard clam (Fig. 8.22). Eversole et al. (1990) also presented extensive data on production rates for hard clams planted

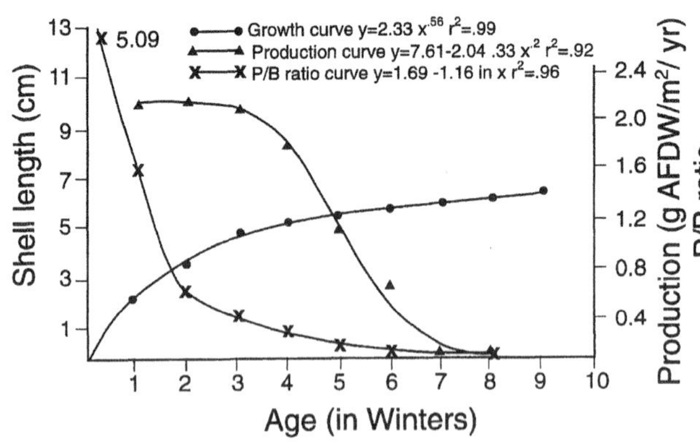

Fig. 8.22. *M. mercenaria*. Production/biomass ratios, growth rate, and production rate by age (from Walker and Tenore, 1984).

at different densities into trays placed intertidally. Their data, however, were expressed in wet weights so no direct comparisons can be made with previous studies. One interesting finding from their study is that production rates were highest at the higher densities (869 and 1159 hard clams m^{-2}) compared to the lowest density (290 hard clams m^{-2}), even though mean individual growth was higher in the latter.

8.5.7 Partitioning Between Shell Growth and Growth of Soft Tissues

As previously discussed, the growth term (G) in Section 8.2 refers to all additions of biomass to the bivalve. The major categories of growth include shell, somatic tissues, and reproductive tissues. In most studies on growth, however, the term is used to refer in a collective sense to shell growth and/or growth of all soft tissue because in the long term both occur at similar rates. On some time scales, especially seasonal, the various kinds of growth rates can be de-coupled, as illustrated for *Mytilus edulis* by Hilbish (1986). This subsection deals with such differential growth, which is essentially a partitioning of energy among shell, somatic tissues, and gonadal tissues.

Hibbert (1977a) developed monthly linear regression equations relating shell length of *M. mercenaria* to soft tissue ("flesh") dry weight for populations in Southampton Water, England. Over a 15-month period, both were highly correlated, with r values ranging from 0.990 to 0.998. There was more pronounced growth of soft tissue compared to shell growth during early and late summer, presumably due to gametogenic activities. Hibbert (1977a) found maximal caloric value of soft tissue in late summer. Mitchell (1974) showed that summer peaks in soft-tissue caloric content resulted from increases in lipid levels and/or protein, though there were differences in timing and occurrence of seasonal peaks among the studies.

Peterson and Fegley (1986) compared monthly, size-corrected growth rates in soft tissue (gonads and soma) and shell volume (based on external shell dimensions) for juvenile and adult hard clams in North Carolina. Both age classes had seasonal maxima in April–May, July–August, and November (Fig. 8.23). The juveniles had much greater shell growth rates relative to the adult hard clams during December and January. Growth of soft tissues did not show the same pattern as that of shell growth. Growth of somatic tissue of juveniles showed no seasonal trend, except for the occurrence of negative growth during September (Fig. 8.24). Adult somatic growth showed no seasonal trends, but there was wide variability from positive to negative growth rates. Gonadal tissue had two seasonal peaks: April and August. Differences in volumetric shell growth of the two size classes (most pronounced in winter) were attributed to differential resource allocation: greater diversion of energy towards the accumulation of energy reserves (growth of somatic mass) in preparation for spring gametogenesis, in adults relative to pre-reproductive juveniles. The possibility that juveniles and adults showed differential utilization of winter food sources, or differential tolerance to low water temperature stress, could not be discounted.

8.5.8 Changes in Condition and Biochemical Composition

Changes in the biochemical composition of *M. mercenaria* during early development (before and after larval settlement) were described by Mann and Gallager (1984). Metamorphosis was associated with a rapid decline in lipid and carbohydrate reserves, which extended

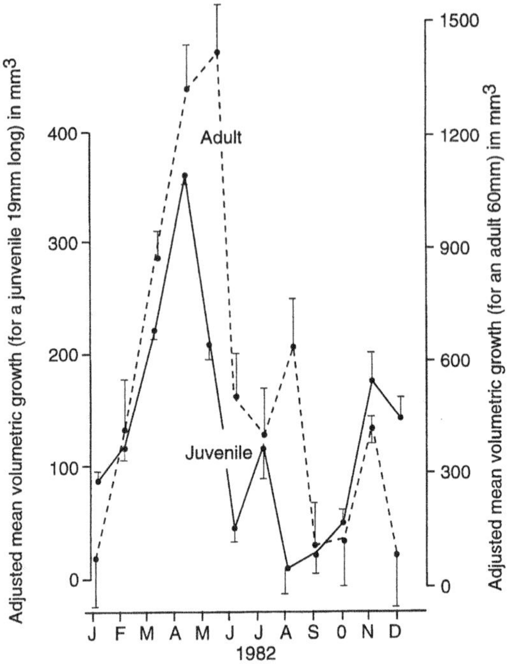

Fig. 8.23. *M. mercenaria.* Mean volumetric growth (±SE) each month, adjusted to a hard clam of constant size: 19 mm shell length for juveniles and 60 mm for adults (from Peterson and Fegley, 1986).

through the first 8 days following settlement. Lipid made the greatest contribution to the total caloric loss. The ontogenetic transition from lipid–protein-based metabolism in bivalve larvae to carbohydrate–protein-based metabolism after settlement is of particular interest because it may be related to important functional changes. For example, a shift from the use of lipids in planktonic larvae, to that of carbohydrate (glycogen) as a catabolic substrate in spat, may explain the ability of post-metamorphic stages to undergo anaerobiosis and tolerate hypoxia/anoxia. Mann and Gallager (1984) showed that in hard clams this biochemical transition occurs about 40 days following settlement, but the physiological implications of this transition were not described in their study.

Seasonal changes in condition (soft-tissue wet or dry weight as a % of total body weight) and biochemical composition of adult *M. mercenaria* have been described for a naturalized population in British waters affected by thermal discharges (Ansell and Loosmore, 1963; Ansell et al., 1964; Ansell and Lander, 1967). Condition index showed a positive, linear relationship with spawning potential, as measured by the ability of hard clams to respond to laboratory induction of spawning prior to the occurrence of spawning in nature. Changes in condition were therefore mainly related to the reproductive cycle. Gonadal growth of field populations was most consistently accompanied by an increase in condition and protein (nitrogen) content, which reached a maximum just prior to spawning and declined during spawning (Ansell and Lander, 1967). In some bivalve species, such as *Argopecten irradians* (Epp et al., 1988) and *Mytilus edulis* (Bayne, 1976), gonadal development takes place largely

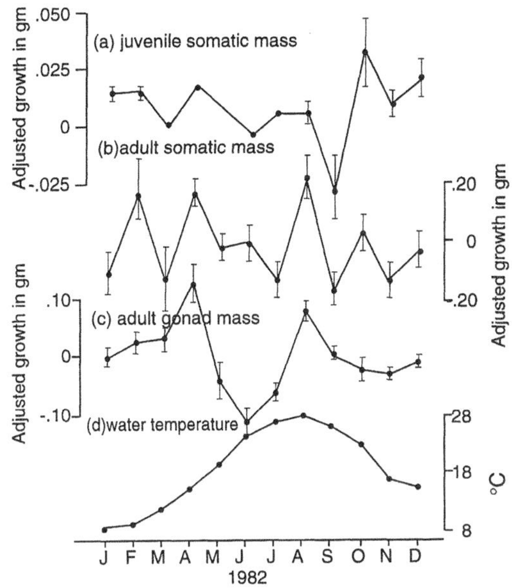

Fig. 8.24. *M. mercenaria*. Monthly changes in average adjusted growth: (a) total biomass (all of which was soma) for juveniles of 19 mm shell length; (b) somatic biomass for adults of 60 mm shell length; (c) adult gonad biomass; (d) water temperature averaged from daily maxima and minima (from Peterson and Fegley, 1986).

at the expense of endogenous reserves stored in somatic tissues during periods of high food availability, although this may vary with environmental conditions. This results in alternating cycles of energy storage in somatic tissue, and its utilization for reproductive growth. In contrast, it appears that in *M. mercenaria* little storage of reserves occurs, and gonadal development occurs sporadically largely in response to the external food supply (Ansell and Loosmore, 1963). Additionally, hard clams show no apparent antagonism between gonadal development and somatic growth.

8.5.9 Reproductive Output Relative to Body Size and Age

One of the major components of energy partitioning during growth and development is growth of reproductive tissues. The latter is typically considered to be an energy loss because the added tissue is temporary and represents a loss in energy that could be diverted to somatic tissues. Reproduction is covered in Chapter 5. This subsection is a brief review of how reproductive output relates to the overall growth process.

Several studies have shown that gonadal growth and reproductive (gamete) output generally increase with age and size, but the relationships are not simple. Bricelj and Malouf (1980) found a significant linear correlation between shell length and egg production as determined from laboratory-induced spawning for female hard clams, but there was wide variability among individuals, particularly within larger size classes. These findings agreed well with earlier studies (Davis and Chanley, 1956; Ansell, 1967) which also showed wide variability in egg output relative to hard clam size. Peterson (1986) used multiple regression to relate log-shell

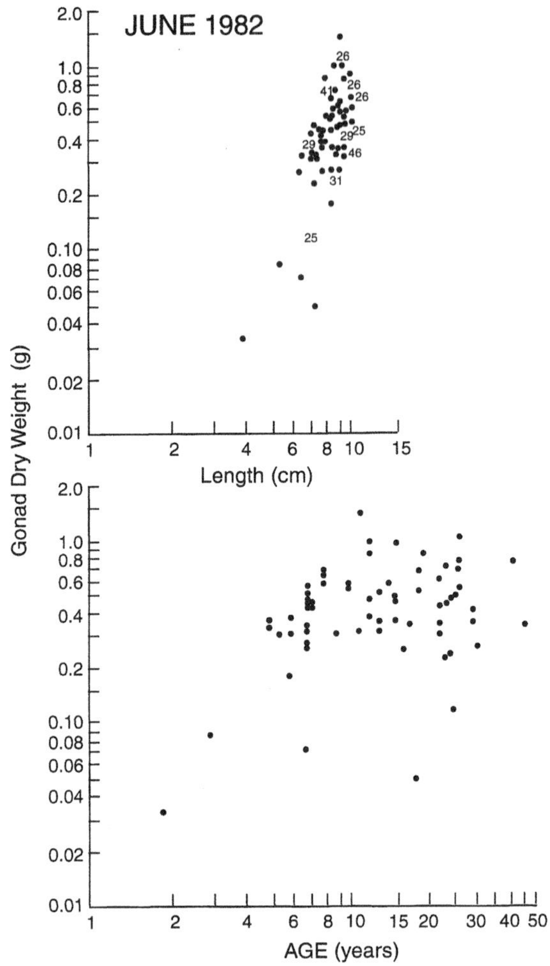

Fig. 8.25. *M. mercenaria.* Gonadal biomass as a function of shell length and age (from Peterson, 1986). In the top graph numbers (age) are plotted instead of points for all clams ≥25 years old. Length is shown to be a better predictor of gonadal mass than age.

length and hard clam age to log-transformed gonadal dry weights prior to spawning; he found that only shell length was strongly and highly significantly related to gonad weight (Fig. 8.25) and that age did not explain a significant amount of the residual variance in this relationship. His study included individuals up to 46 years old, and confirmed his earlier (Peterson, 1983) contention of a lack of quantitative reproductive senility in hard clams. This occurs when older age classes achieve a lower gonadal mass immediately prior to spawning than expected from the allometric (power) curve relating gonad mass to body size (shell length) in younger adults. Furthermore, the exponents of the linear regression relating log gonad mass to log length, and gonad mass to shell cavity volume, did not differ significantly from those (3.0 to 1.0, respectively) expected under conditions of isometric growth (Peterson, 1986).

Ansell and Lander (1967) estimated that in a hard clam 4 cm in SL, gonad production accounts for 40 to 60% of the total organic production per year, on a dry weight basis, and that 20 to 25% of total organic production is released as spawned gametes each year. However, these figures may not provide an accurate estimate of total reproductive effort, since they do not take into account that gametogenesis and spawning can occur in synchrony in this species. Reproductive effort (the proportion of total production allocated in reproduction) in bivalves also typically increases with age/size. Whereas the relationship between reproductive output (fecundity) and age/size has been documented for *M. mercenaria* (see discussion above), the relationship between total reproductive effort and age/size has not.

8.6 WHOLE-ORGANISM BEHAVIOR, FLUID MECHANICS, AND MODELING (with Larry Sanford)

The literature on whole-organism behavior of *M. mercenaria* has not been reviewed previously. Here, we summarize from a fluid mechanical perspective what is known concerning the behavior of juvenile and adult hard clams that is relevant to feeding and growth. After a brief consideration of some basic fluid mechanical principles, a new mathematical framework is presented which characterizes the feeding response under a range of hydrodynamic conditions. We then review bivalve feeding studies in general that are relevant to the model, and hard clam behavioral studies interpreted in the context of the model. The section concludes with a review of mathematical modeling generally that has involved hard clams.

8.6.1 Basic Fluid Mechanical Principles and Ecological Implications

A fluid (air, water) moving across a "rough" solid (soil, sediment) is affected by friction so that very close to the fluid–solid interface the velocity of the fluid is zero. The drag on the moving fluid caused by friction slows the fluid so that a velocity gradient is set up where horizontal velocities increase with increasing distance from the bottom. This region of differential velocities is defined as a "boundary layer" (Fig. 8.26; for introductions to boundary layer theory from a benthic ecology perspective see: Nowell and Jumars, 1984, 1987; Muschenheim et al., 1986). In benthic boundary layers there is also typically a gradient of seston concentrations with highest concentrations nearest the bed. There also can be substantial differences in seston quality in the boundary layer occurring on a scale of a few centimeters. Therefore, movements of hard clams vertically just a few centimeters can result in dramatic differences in environmental conditions affecting food availability and feeding processes. Hence, it is necessary to have an understanding of the fluid mechanical environment in which the hard clam lives to have an ecologically realistic understanding of processes that affect feeding and growth.

8.6.2 A Fluid Mechanical Perspective on Hard Clam Feeding

Fig. 8.26 presents a mathematical framework that specifies the factors affecting feeding in a moving fluid. No formal mathematical modeling has been done, but the framework provides the conceptual basis for a fluid mechanical perspective. Hence, the framework may be considered a conceptual model. It shows how two flow fields, horizontal ambient flow and

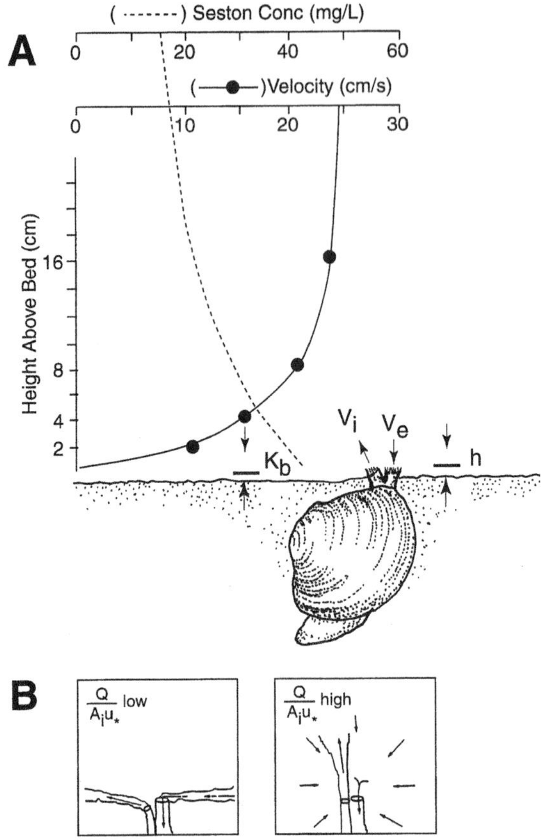

Fig. 8.26. (A) Schematic of spatial relations between typical near-bottom flow conditions and seston concentrations and an adult quahog (see vertical axis for scale). (B) Range of conditions for how ambient horizontal flows affect the "feeding zone" of an infaunal bivalve like the quahog. When $Q/A_i u_*$ is low (ambient horizontal flow speed exceeds inhalant pumping speed) the quahog's feeding zone only extends slightly above the height of the siphon (h). When $Q/A_i u_*$ is high (no-flow ambient conditions) the feeding zone is hemispherical in shape extending several centimeters away from the tip of the inhalant siphon (modified from Grizzle, 1988; see text for detail).

the bivalve's inhalant and exhalant currents, interact. We suggest how the model might be used to characterize the potential "feeding zone" (volume of the water column from which water is pumped into the mantle cavity) of an actively pumping suspension feeder like the hard clam for a range of ambient water flow conditions. But it might equally well be used to characterize or analyze other aspects of hard clam feeding.

Referring to Fig. 8.26, consider an infaunal, siphonate bivalve like the hard clam with its inhalant siphon extended a distance (h) above the bottom. The inhalant siphon has an opening cross-sectional area (A_i) and it is pumping at a rate (Q), so that the velocity of water (v_i) into the siphon $= Q/A_i$. The inhalant flux of seston (F_i) is a product of the concentration of seston at the siphon opening (c_i) and the pumping rate (Q). The exhalant siphon has a cross-sectional opening area (A_e) so that the velocity of water leaving the exhalant siphon

$(v_e) = Q/A_e$. For bivalves generally, $A_e < A_i$ and $v_e > v_i$, ensuring that the animal does not re-filter its own exhalant water (LaBarbera, 1984, and references therein). The boundary layer flow is assumed to be in a steady-state condition. It is characterized near the bed by the bed roughness height upstream of the animal (k_b), the boundary shear velocity (u_*), and the kinematic viscosity of the water (ν). The seston has a representative settling velocity (w_s) and an upstream seston concentration at height h of $c(h)$. The upstream flow profile and upstream concentration profile are completely specified (at least in theory) through application of appropriate boundary layer and suspended sediment transport theories.

It is important to distinguish between cases with an individual bivalve in an otherwise physically controlled boundary layer and cases in which a population of bivalves plays an active role in the boundary layer dynamics. In the former case, the seston concentration generally increases towards the bed, whereas in the latter case a seston deficit may exist near the bed due to the collective feeding of the bivalves. In other words, the situation described by the model is for near-bottom conditions for an individual bivalve.

The variables identified above may be regarded as completely characterizing the complex interactions between flow, seston, and an actively pumping hard clam. This allows the problem to be analyzed using dimensional analysis techniques. These can substantially simplify development and testing of hypotheses, and can aid in interpretation of data. Bivalve feeding is expressed formally as a function of 10 variables, $f(A_i, A_e, Q, h, c_i, c(h), w_s, u_*, k_b, \nu) = 0$. These ten variables can be expressed in terms of three dimensions (length, time, and seston mass), and dimensional analysis may be used to reduce the problem to a dependence on $10 - 3 = 7$ non-dimensional variables. For the present case, we may write:

$$\frac{c_i}{c(h)} = g\frac{u_* k_b}{\nu}, \frac{w_s}{\kappa u_*}, \frac{h}{k_b}, \frac{A_i}{A_e}, \frac{Q}{A_i u_*}, \frac{Q}{A_e^{1/2}\nu} \tag{8.20}$$

The dependent variable on the left-hand-side of this equation is the ratio of the seston concentration at the opening of the inhalant siphon (c_i) relative to the seston concentration upstream at the same height [$c(h)$]. The independent variables on the right-hand-side of the equation are, in sequence, the roughness Reynolds number ($u_* k_b/\nu$), the seston Rouse number ($w_s/\kappa u_*$), the relative height of the siphon opening (h/k_b), the relative areas of the two siphons (A_i/A_e), the dimensionless inhalant pumping speed ($Q/A_i u_*$), and the exhalant pumping Reynolds number ($Q/A_e^{1/2}\nu$).

The present non-dimensional description of the bivalve–fluid–particle interaction problem focuses on the relationship between the seston entering the inhalant siphon and the seston approaching the inhalant siphon at some distance upstream. The two are not necessarily the same, as illustrated by the following example.

Consider a case in which the physical characteristics of the upstream boundary, the settling speed of the particles, the extension of the siphon, and the siphon opening characteristics are fixed, but either the boundary layer flow speed or the bivalve's pumping rate may vary. If the dimensionless inhalant pumping speed $Q/A_i u_*$ is very low (ambient horizontal flow speed is greater than inhalant pumping speed), the boundary layer flow dominates and the feeding zone (the region of fluid effectively sampled by the bivalve) is a horizontal zone at height (h) upstream of the siphon opening (Fig. 8.26B). In this case, c_i should be virtually the same as $c(h)$. On the other hand, if $Q/A_i u_*$ is very high (no-flow ambient conditions), the flow induced by pumping dominates and the feeding zone approximates a hemisphere with its center at the inhalant siphon opening. In this case, c_i represents the average of the seston concentration throughout the feeding zone, which may be quite different from $c(h)$.

In conclusion, the conceptual model (Fig. 8.26) shows that the "feeding zone" of infaunal bivalves like hard clams is usually restricted to the bottom few centimeters above the seabed. This region, however, is strongly influenced by ambient water flow speeds. The model, and previous research, also indicate that the near-bottom fluid mechanical environment is complex and dynamic. Hence, future studies should consider this complexity, and the perspective offered here can function as a guide. We are aware of no research based on our mathematical framework, but there have been several relevant studies that warrant discussion.

8.6.3 Relevant Research on the Fluid Mechanics of Bivalve Feeding Generally

Based upon flow visualization techniques using upstream dye releases, the infaunal cockle, *Clinocardium nutalli*, was found to pull water from only about 1 to 2 cm above its siphon and 3 to 4 cm laterally at low ambient speeds, but adequate data to compare ambient and inhalant speeds were not given (Ertman and Jumars, 1988). Monismith et al. (1990) used physical models of paired bivalve siphons to study exhalant siphonal flows under various flow conditions in a flume. Though not directly concerned with feeding zones or seston fluxes, inspection of their published photographs shows that the exhalant dye plumes under various flow speeds compare well with the predicted exhalant flow regions in our Fig. 8.26B (also see O'Riordan et al., 1995). At ambient flows of 12 cm s^{-1}, the exhalant siphonal flow (about 10 cm s^{-1}) was deflected into the ambient flow after the exhalant flow had risen a few centimeters. In contrast, at ambient flows greatly exceeding the exhalant speed, the exhalant plume was deflected into the ambient stream a few millimeters above the siphon height.

A combined field (near-bottom sampling of water currents and seston characteristics) and laboratory (measurement of feeding rates) approach showed that the feeding zone of a bed of blue mussels (*Mytilus edulis*) extended on average only about 3.5 cm above the tops of the bivalves (Muschenheim and Newell, 1992). Preliminary studies (R.E. Grizzle, unpublished data) that were part of the research reported in Grizzle et al. (1992), indicated that dye releases upstream of actively pumping individual mussels, oysters, and hard clams generally followed the model predictions for ambient current speeds at and below 5 cm s^{-1}. The feeding zone never extended vertically more than about 2 cm above the height of the inhalant siphon.

These studies indicate the general utility of the framework presented here. Several additional studies demonstrate the importance of characterizing small-scale, near-bottom conditions.

Near-bottom water flows and seston (derived from particulates scraped from the upper 3 mm of an intertidal sand flat) fluxes were characterized at 1-cm vertical intervals in a flume by Muschenheim (1987). Under shear velocities (u_* in Fig. 8.26) averaging 0.22 cm s^{-1}, inorganic particulates had flux maxima nearest the bottom (as illustrated in Fig. 8.26 to generally be the case). Lighter (lower w_s) organic-rich particulates, however, had maximal fluxes up to several centimeters above the bottom. Hence, changes in feeding height of only a few centimeters could strongly affect the quality of available food for infaunal bivalves.

The mean orientation of the siphons of the infaunal softshell clam, *Mya arenaria*, were found to be perpendicular to the predominant current direction, and variability in orientation was related to the variability in ambient current direction (Vincent et al., 1988). It was suggested that such orientation optimizes energy acquisition by minimizing re-pumping of exhalant waters. Likewise, the feeding behavior of three species of *Macoma* was found to be strongly affected by ambient, near-bottom water flow speed as well as bedload sediment transport (Levinton, 1991). These infaunal bivalves switched back and forth from deposit to suspension feeding largely depending upon ambient water flow conditions. In still water, feeding was active with the inhalant siphons typically extended up to 2 cm from the burrow hole laterally and up to 1 cm vertically into the water column. As water speed increased, the feeding radius decreased. Cessation of feeding, or feeding only within the confines of the burrow hole occurred at current speeds that caused bedload transport.

8.6.4 Hard Clam Behavior and Feeding

In one of the earliest studies on hard clam behavior and water flow, Carriker (1961) found no trend in horizontal movements of byssal plantigrade stages (1.0 to 2.4 mm shell length) in relation to water currents up to 2.5 cm s^{-1}. Subsequent research has not emphasized horizontal movements, which are apparently only substantial during early life history stages (Ahn, 1990; Ahn et al., 1993). In contrast, vertical movements (at spatial scales that the model indicates could be quite important with respect to feeding) of both juveniles and adults have been shown to be strongly affected by various environmental factors.

Pratt and Campbell (1956) carried out laboratory and field studies on behavior of hard clams that were aimed at identifying the causes for growth inhibition in mud. They found that burial depth and feeding behaviors generally were related to sediment characteristics, particularly sediment permeability. For example, clams in mud (low permeability) typically maintained an open burrow and fed near the sediment surface. Those in sand (high permeability) often were completely buried (up to 3 cm below the sediment surface) and fed through the sand layer (see further discussion below). Their study demonstrated the potential importance of considering bottom sediment characteristics in feeding research (see more discussion above).

Several recent studies indicate the importance of considering hydrodynamic conditions at small vertical spatial scales for feeding and growth studies. Judge et al. (1993) developed a protocol for sampling seston in the field in near-bottom waters under flowing conditions. Their method simulated the inhalant siphon size and pumping speed of *M. mercenaria*, and collected seston in a fluid mechanically realistic manner, as opposed to isokinetic sampling (where sampler pumping speed is the same as ambient flows) as is sometimes recommended (for a similar approach also see Grizzle et al., 1992). They generally found the highest chlorophyll *a* concentrations at 1 cm above the bottom. Turner and Miller (1991) characterized near-bottom boundary layer conditions during simulated storm events in a water tunnel that produced oscillatory flows. They found that suspended sediment concentrations greatly increased in the lower 4 cm nearest the bottom during "storm" flows. Under such conditions, juvenile *M. mercenaria* maintained their siphons at a lower height, produced more pseudofeces, and had a 35% growth reduction compared to non-storm conditions. The hard clams did not, however, appear to alter their siphon orientation in response to changes in water flow speeds. Turner (1990) modeled these findings from the perspective of energy optimization theory (e.g., Lehman, 1976; Bayne, 1981) indicating that overall behavioral changes represent a balancing of the energetic costs of increased pseudofeces against net energy gains when the hard clam is feeding in a turbid environment.

The above studies show that hard clams have complex feeding behaviors that can be interpreted in the context of fluid mechanical considerations, particularly food availability. Predation risks, however, must also be considered. Several studies have shown that the presence of potential predators elicits vertical movements of up to several centimeters, and predation risk is decreased at greater burial depths. For example, Doering (1976, 1982) showed that the predatory seastar, *Asterias forbesi*, caused hard clams to increase their burial depth to as much as 4.5 cm below the sediment surface. He experimentally demonstrated that such behavior reduces predation risks. Similarly, Roberts et al. (1989) used field experiments to show that hard clams buried 2 cm or more had reduced risk of predation by gulls. They also showed that hard clams migrated vertically in synchrony with tidal stage, and that the likely causal mechanism was changes in water pressure associated with changing water depths. Clams buried deepest at low tide when gulls would be feeding on the tide flats and shallowest at high tide. Burrowing activity generally is also affected by temperature (Savage, 1976), and siphon extension (which can affect fish predation on the siphons themselves; see above) is affected by temperature and salinity (Van Winkle et al., 1976).

8.6.5 Conclusions Concerning our Feeding Model

As already noted, the conceptual framework presented here (Fig. 8.26) has not yet been applied to feeding studies. The above review shows that *M. mercenaria* exhibits behaviors important for feeding that can be interpreted from the fluid mechanical perspective of the model. It also suggests that small-scale vertical movements such as those involved in predator avoidance are probably not uncommon. This indicates that gaining a comprehensive understanding of how whole-organism behavior is related to feeding (and growth) will be difficult. As reviewed above, however, much is already known about how factors in the model such as seston concentration and water flow affect feeding (and growth) of *M. mercenaria* and other suspension-feeding bivalves. So, what is the significance of the model?

It is presented primarily to provide a conceptual basis for use in the design of future research relevant to feeding of suspension-feeding, infaunal bivalves. Because most bivalves live in environments that are hydrodynamically widely variable, fluid mechanical considerations must be given detailed attention if an ecologically realistic understanding of feeding responses to environmental factors is to be achieved. The model provides a preliminary, comprehensive fluid mechanical perspective identifying the factors that must be considered in order to develop a comprehensive theory for the feeding ecology of the individual bivalve (as opposed to population-level modeling). A major intent is to show how changes at spatial scales smaller than typically have been characterized in past research may need to be considered in the future.

8.6.6 Other Models

This final subsection is a brief review of mathematical modeling approaches concerned with other than whole-organism behavior and that tie together some of the topics previously discussed. Production of a bivalve population is dependent upon growth rates of the individuals making up the population. Individual growth is determined by feeding and nutrition. Mechanistic modeling efforts aimed at either growth or production would have to include processes acting at the next lowest hierarchical level. This means that both physiological ("intrinsic" or "endogenous") and environmental ("extrinsic") factors have to be included. To date, no such models have emerged. Nonetheless, theoretical models that consider multiple factors at one level (e.g., environmental factors affecting growth), and models that have bridged the gap from individual to population levels, have been developed.

Theoretical models attempt to organize information so that the relationships between various components of the system being studied are shown. They range from simple "box" models that consist of particular spatial arrangements of various boxes each representing some factor or process, to complex series of differential equations that are intended to be predictive. Models are in one sense attempts to demonstrate a functional or mechanistic understanding of the process or processes being studied. Suspension-feeding bivalves have been included in a wide variety of modeling efforts (for comprehensive reviews see chapters in Dame, 1993) because they are often abundant members of the benthos in coastal waters and support important fisheries and aquaculture operations in many areas. When a detailed enough understanding is achieved, models can be developed that have substantial utility for aquaculture, water quality management, and other activities. The intent here is not to review all of this literature, but rather to briefly discuss attempts (particularly those that have included *M. mercenaria*) at modeling feeding, nutrition, growth, and/or production that are aimed at mechanistic understandings of the processes themselves. The intent is also to argue for more use of models in guiding research.

Three hierarchical levels of modeling are considered below: physiological modeling of feeding and nutrition, individual growth models, and ecological modeling of population-level phenomena based explicitly on individual-level responses.

8.6.6.1 Physiological models

Until recently, models of individual bivalve feeding and nutrition were primarily conceptual in nature and typically based on responses to a single environmental variable, or narrowly focused on a particular aspect of the feeding process. For example, Figs. 8.4–8.10 can be thought of as conceptual models of how various environmental variables affect feeding rates. The relationships in these feeding studies are valid, but not ecologically realistic because in nature the bivalve is typically responding to concurrent changes in several variables that may affect feeding. Nonetheless, such data could be the basis for predictive mathematical models.

Willows (1992) described the first attempt at a comprehensive model of feeding and nutrition. His model was based on optimization of energy and it attempted to tie together much of what is known about bivalve feeding and nutrition. The quantitative relationships were based mainly on work on *Mytilus edulis* because it is the most-studied bivalve species. His approach considered the bivalve gill pump and its operation as responsive to various environmental and physiological factors. In contrast to such a view, Jørgensen et al. (1986) proposed a solely hydrodynamically based model (see Bayne, 1993, 1998, for critiques of this perspective; for further development of the hydrodynamical theory of particle capture generally also see Shimeta and Jumars, 1991). Although research on *M. mercenaria* has not contributed to development of these models, many of the relationships shown would be relevant for future work on hard clams.

8.6.6.2 Individual growth models

Individual growth of hard clams has been well-studied, but quantitative models of the relationships involved have not been developed. Peterson and Beal (1989) used ANOVA techniques to model the relative effects of year, density, enclosure walls, source, and site, as well as various two-way interactions on hard clam growth. They concluded that a simple additive model adequately described how site effects combined with various experimental effects. Grizzle and Lutz (1989) used multiple regression to model the combined effects of water currents, seston, and sediment characteristics (see section 8.5.4.9 for more discussion of both these studies).

8.6.6.3 Population-level models

In nature, bivalves occur in populations. Many population models, however, are based on individual-level responses, particularly feeding rates. This is the modeling perspective that has been most developed, and researchers have approached the problem from several directions. Dame (1993) provides a major review and synthesis on the ecological functions of bivalves in coastal ecosystems, and includes important contributions on modeling approaches (e.g., Herman, 1993; Newell and Shumway, 1993; Wildish and Kristmanson, 1993; Grant et al., 1993). Ecological models dealing with bivalves are proliferating rapidly, but few have been based on research on *M. mercenaria.*

Doering and Oviatt (1986) briefly reviewed the literature on pumping rates of infaunal bivalves (mostly *M. mercenaria*) and developed a multiple regression model using water temperature and animal size (weight and shell length) to predict filtration rates of hard clams in model mesocosms. They also compared models based on feeding studies using algal monocultures with those using natural seston. Only those based on natural seston yielded estimates similar to observed rates in the mesocosms. Doering and Oviatt (1986) emphasized the need to use natural seston in deriving feeding (and biodeposition) rates for models.

In conclusion, research on the hard clam has not thus far emphasized development of mathematical models. The literature on mathematical modeling of a variety of physiological and ecological processes of bivalves generally, however, is proliferating rapidly, and this research has included both basic and applied topics. Aquaculture-related models (e.g., Newell and Shumway, 1993; Grant et al., 1993; Raillard and Ménesguen, 1994) and models aimed at elucidating the role that both infaunal and epifaunal bivalves play in the control of eutrophication and other water quality changes (e.g., Gerritsen et al., 1994) are increasingly receiving attention. The hard clam is widely distributed, supports important commercial fisheries, and is cultured in many areas. It is an ideal candidate for model-based research on these and other topics.

8.7 ACKNOWLEDGMENTS

Support for this review was provided by Taylor University, the Institute for Marine Biosciences, National Research Council of Canada, Southampton College, Long Island University, and Jackson Estuarine Laboratory of the University of New Hampshire. The fluid mechanical model presented in Section 8.6 was provided by Larry Sanford. We thank Jim Rollins for preparation of most of the figures and Pam Shephard-Lupo for library research. The manuscript was greatly improved by the comments of five reviewers.

REFERENCES

Adamkewicz, L., Tabu, S.R. and Wall, J.R., 1984. Genetics of the clam *Mercenaria mercenaria.* II. Size and genotype. Malacologia, 25: 525–533.

Ahn, I.-Y., 1990. Effects of the gem clam, *Gemma gemma*, on the settlement and the post-settlement migration, growth and survival of the hard clam, *Mercenaria mercenaria*. Ph.D. Dissertation, State University of New York at Stony Brook, 131 pp.

Ahn, I.-Y., Lopez, G.L. and Malouf, 1993. Effects of the gem clam *Gemma gemma* on early post-settlement emigration, growth and survival of the hard clam *Mercenaria mercenaria*. Mar. Ecol. Prog. Ser., 99: 61–70.

Allison, E.H., 1994. Seasonal growth models for great scallops (*Pecten maximum* L.) and queen scallops (*Aequipecten opercularis* L.). J. Shellfish Res., 13: 555–564.

Anderson, J.D. and Prosser, C.L., 1953. Osmoregulation capacity in populations occurring in different salinities (abstr.). Biol. Bull., 105: 369.

Ansell, A.D., 1964. Some parameters of growth of mature *Venus mercenaria* L. J. Cons. Int. Explor. Mer, 29: 214–220.

Ansell, A.D., 1967. Egg production of *Mercenaria mercenaria*. Limnol. Oceanogr., 12: 172–176.

Ansell, A.D., 1968. The rate of growth of the hard clam *Mercenaria mercenaria* (L.) throughout the geographical range. J. Cons. Perm. Int. Explor. Mer, 31: 364–409.

Ansell, A.D. and Lander, K.F., 1967. Studies on the hard-shell clam, *Venus mercenaria*, in British waters. III. Further observations on the seasonal biochemical cycle and on spawning. J. Appl. Ecol., 4: 425–435.

Ansell, A.D. and Loosmore, F.A., 1963. Preliminary observations on the relationship between growth, spawning and condition in experimental colonies of *Venus mercenaria* L. J. Cons., 28(1): 285–294.

Ansell, A.D., Loosmore, F.A. and Lander, K.F., 1964. Studies on the hard-shell clam, *Venus mercenaria*, in British waters. II. Seasonal cycle in condition and biochemical composition. J. Appl. Ecol., 1: 83–95.

Applemans, N.L., 1989. Effects of variations in chlorophyll-*a* and temperature on growth of hard clams, *Mercenaria mercenaria*, L. in an upflow culture system. Ph.D. Dissertation, State University of New York at Stony Brook, 91 pp.

Arnold, W.S., Marelli, D.C., Bert, T.M., Jones, D.S. and Quitmyer, I.R., 1991. Habitat-specific growth of hard clams *Mercenaria mercenaria* (L.) from the Indian River, Florida. J. Exp. Mar. Biol. Ecol., 147: 245–265.

Bardach, J.E., Ryther, J.H. and McLarney, W.O., 1972. Aquaculture. The Farming and Husbandry of Freshwater and Marine Organisms. Wiley–Interscience, New York, 868 pp.

Bass, A.E., 1983. Growth of hard clams, *Mercenaria mercenaria*, feeding on chlorophyte and cyanobacterial picoplankton. M.S. Thesis, State University of New York at Stony Brook, 66 pp.

Bass, A.E., Malouf, R.E. and Shumway, S., 1990. Growth of northern quahogs (*Mercenaria mercenaria* (Linnaeus, 1758)) fed on picoplankton. J. Shellfish Res., 9: 299–307.

Bayne, B.L., 1976. Aspects of reproduction in bivalve molluscs. In: M. Wiley (Ed.), Estuarine Processes. Vol. 1. Uses, Stresses and Adaptation to the Estuary. Academic Press, New York, pp. 432–448.

Bayne, B.L., 1981. Theory and observation: benthic predator–prey relationships. In: A. Academic Press, New York, pp. 573–606.

Bayne, B.L., 1993. Feeding physiology of bivalves: time-dependence and compensation for changes in food availability. In: R.F. Dame (Ed.), Bivalve Filter Feeders in Estuarine and Coastal Ecosystem Processes. Springer, New York, pp. 1–23.

Bayne, B.L., 1998. The physiology of suspension feeding by bivalve molluscs: an introduction to the Plymouth 'TROPHEE' workshop. J. Exp. Mar. Biol. Ecol., 219: 1–19.

Bayne, B.L. and Hawkins, A.J.S., 1992. Ecological and physiological aspects of herbivory in benthic suspension-feeding molluscs. In: D.J. John, S.J. Hawkins and J.H. Price (Eds.), Plant–Animal Interactions in the Marine Benthos. Systematics Assoc. Spec. Vol. 46, pp. 265–288.

Bayne, B.L. and Newell, R.C., 1983. Physiological energetics of marine molluscs. In: K.M. Wilbur and A.S. Saleuddin (Eds.), The Mollusca. Vol. 4. Academic Press, New York, pp. 407–515.

Bayne, B.L., Widdows, J. and Thompson, R.J., 1976. Physiology II. In: B.L. Bayne (Ed.), Marine Mussels: their Ecology and Physiology. Int. Biol. Program Vol. 10. Cambridge University Press, Cambridge, pp. 207–260.

Bayne, B.L., Widdows, J. and Worrall, C.W., 1977. Some temperature relationships in the physiology of two ecologically distinct bivalve populations. In: F.J. Vemberg, A. Calabrese, F.P. Thurberg and W. Vernberg (Eds.), Physiological Responses of Marine Biota to Pollutants. Academic Press, New York, pp. 379–400.

Beiras, R., Pérez-Camacho, A. and Albentosa, M., 1993. Influence of food concentration on energy balance and growth performance of *Venerupis pullastra* seed reared in an open-flow system. Aquaculture, 116: 353–365.

Belding, D.L., 1909. A Report upon the Mollusk Fisheries of Massachusetts. Wright and Potter Printing Co., Boston, MA, 243 pp.

Belding, D.L., 1912. A Report upon the Quahog and Oyster Fisheries of Massachusetts. Wright and Potter Printing Co., Boston, MA, 134 pp.

Beninger, P.G. and St-Jean, S.D., 1997. The role of mucus in particle processing by suspension-feeding marine bivalves: unifying principles. Mar. Biol., 129: 389–397.

Beninger, P.G., St-Jean, S., Poussart, Y. and Ward, J.E., 1993. Gill function and mucocyte distribution in *Placopecten magellanicus* and *Mytilus edulis* (Mollusca: Bivalvia): the role of mucus in particle transport. Mar. Ecol. Prog. Ser., 98: 275–282.

Beninger, P.G., Dufour, S.C. and Bourque, J., 1997. Particle processing mechanisms of the eulamellibranch bivalves *Spisula solidissima* and *Mya arenaria*. Mar. Ecol. Prog. Ser., 150: 157–169.

Bisker, R. and Castagna, M., 1985. The effect of various levels of air-supersaturated seawater on *Mercenaria mercenaria* (Linné), *Mulinia lateralis* (Say), and *Mya arenaria* Linné, with reference to gas-bubble disease. J. Shellfish Res., 5: 97–102.

Bricelj, V.M., 1984. Effects of suspended sediments on the feeding physiology and growth of the hard clam, *Mercenaria mercenaria* L. Ph.D. Dissertation, State University of New York at stony Brook, 157 pp.

Bricelj, V.M., 1993. Aspects of the biology of the northern quahog, *Mercenaria mercenaria*, with emphasis on growth and survival during early life history. In: M.A. Rice and D. Grossman-Garber (Eds.), Proceedings of the Second Rhode Island Shellfish Industry Conference, Rhode Island Sea Grant, University of Rhode Island, Kingston, pp. 29–46.

Bricelj, V.M. and Lonsdale, D.J., 1997. *Aureococcus anophagefferens*: Causes and ecological consequences of brown tides in U.S. mid-Atlantic coastal water. Limnol. Oceanogr., 42: 1023–1038.

Bricelj, V.M. and Malouf, R.E., 1980. Aspects of reproduction of hard clams (*Mercenaria mercenaria*) in Great South Bay, New York. Proc. Natl. Shellfish. Assoc., 70: 216–229.

Bricelj, V.A. and Malouf, R.E., 1984. Influence of algal and suspended sediment concentrations on the feeding physiology of the hard clam *Mercenaria mercenaria*. Mar. Biol., 84: 155–165.

Bricelj, V.M. and Shumway, S.E., 1998. Paralytic shellfish toxins in bivalve molluscs: occurrence, transfer kinetics and biotransformation. Rev. in Fish. Sci., 6(4): 315–383.

Bricelj, V.M., Bass, A.E. and Lopez, G.R., 1984a. Absorption and gut passage time of microalgae in a suspension feeder: an evaluation of the $^{51}Cr:^{14}C$ twin tracer technique. Mar. Ecol. Prog. Ser., 17: 57–63.

Bricelj, V.A., Malouf, R.E. and de Quillfeldt, C., 1984b. Growth of juvenile *Mercenaria mercenaria* and the effect of resuspended bottom sediments. Mar. Biol., 84: 167–173.

Bricelj, V.M., Lee, J.H. and Cembella, A.D., 1991. Influence of dinoflagellate cell toxicity on uptake and loss of paralytic shellfish toxins in the northern quahog *Mercenaria mercenaria*. Mar. Ecol. Prog. Ser., 74: 33–46.

Brock, V. and Brock, A., 1989. Kinetics of amylase, cellulase, and laminarinase in four commercially important bivalves. In: R.Z. Klekowski, E. Styczynaska-Jurewicz and L. Falkowski (Eds.), Proceedings of the 21st European Mar. Biol. Symp., Gdansk, Polish Academy of Sciences, pp. 31–40.

Brock, V., Kennedy, V.S. and Brock, A., 1986. Temperature dependency of carbohydrase activity in the hepatopancreas of thirteen estuarine and coastal bivalve species from the North American east coast. J. Exp. Mar. Biol. Ecol., 103: 87–101.

Buckner, S.C., 1984. Aspects of the population dynamics of the hard clam, *Mercenaria mercenaria* L., in Great South Bay, New York. Ph.D. Dissertation, State University of New York at Stony Brook, 217 pp.

Butman, C.A., Fréchette, M. and Geyer, W.R., 1994. Flume experiments on food supply to the blue mussel *Mytilus edulis* L. as a function of boundary-layer flow. Limnol. Oceanogr., 39: 1755–1768.

Cahalan, J.A., Siddall, S.E. and Luckenback, M.W., 1989. Effects of flow velocity, food concentration and particle flux on growth rates of juvenile bay scallops *Argopecten irradians*. J. Exp. Mar. Biol. Ecol., 129: 45–60.

Calabrese, A., Thurberg, F.P. and Gould, E., 1977. Effects of cadmium, mercury, and silver on marine animals. Mar. Fish. Rev., 39: 5–11.

Calow, P. and Fletcher, C.R., 1972. A new radiotracer technique involving ^{14}C and ^{51}Cr for estimating the assimilation efficiencies of aquatic, primary consumers. Oecologia, 9: 155–170.

Carriker, M.R., 1959. The role of physical and biological factors in the culture of *Crassostrea* and *Mercenaria* in a salt-water pond. Ecol. Monogr., 29: 219–266.

Carriker, M.R., 1961. Interrelation of functional morphology, behavior, and autecology in early stages of the bivalve *Mercenaria mercenaria*. J. Elisha Mitchell Sci. Soc., 77: 168–241.

Castagna, M. and Chanley, P., 1973. Salinity tolerance of some marine bivalves from inshore estuarine environments in Virginia waters on the western mid-Atlantic coast. Malacologia, 12: 47–96.

Castagna, M. and Kraeuter, J.N., 1981. Manual for Growing the Hard Clam, *Mercenaria mercenaria*. Special Report in Applied Science and Ocean Engineering No. 249. Virginia Institute of Marine Science, Gloucester Point, VA, 110 pp.

Chanley, P.E., 1958. Survival of some juvenile bivalves in water of low salinity. Proc. Natl. Shellfish. Assoc., 48: 52–65.

Coen, L.D. and Heck Jr., K.L., 1991. The interacting effects of siphon nipping and habitat on bivalve (*Mercenaria mercenaria* (L.)) growth in a subtropical seagrass (*Halodule wrightii* Aschers) meadow. J. Exp. Mar. Biol. Ecol., 145: 1–13.

Conover, R.J., 1966. Assimilation of organic matter by zooplankton. Limnol. Oceanogr., 11 (3): 346–354.

Cosper, E.M., Carpenter, E.J. and Cottrell, M., 1987. Recurrent and persistent brown tide blooms perturb coastal marine ecosystems. Estuaries, 10: 284–290.

Coughlan, J. and Ansell, A.D., 1964. A direct method for determining pumping rate of siphonate bivalves. J. Cons., 29: 205–213.

Coutteau, P. and Sorgeloos, P., 1993. Substitute diets for live algae in the intensive rearing of bivalve mollusks — a state of the art report. World Aquacult., 24: 46–51.

Coutteau, P., Hadley, N.H., Manzi, J.J. and Sorgeloos, P., 1994. Effect of algal ration and substitution of algae by manipulated yeast diets on the growth of juvenile *Mercenaria mercenaria*. Aquaculture, 120: 135–150.

Crenshaw, M.A., 1980. Mechanisms of shell formation and dissolution. In: D.C. Rhoads and R.A. Lutz (Eds.), Skeletal Growth of Aquatic Organisms: Biological Records of Environmental Change. Plenum Press, New York, pp. 115–132.

Crenshaw, M.A. and Neff, J.M., 1969. Decalcification at the mantle–shell interface in mollusks. Am. Zool., 9: 881–885.

Crisp, D.J., 1984. Energy flow measurements. In: N.A. Holme and A.D. McIntyre (Eds.), Methods for the Study of Marine Benthos. Blackwell, Oxford, pp. 284–372.

Crosby, M.P., Langdon, C.J. and Newell, R.I.E., 1989. Importance of refractory plant material to the carbon budget of the oyster *Crassostrea virginica*. Mar. Biol., 100: 343–352.

Dame, R.F. (Ed.), 1993. Bivalve Filter Feeders in Estuarine and Coastal Ecosystem Processes. Springer, New York.

Dame, R.F., 1996. Ecology of Marine Bivalves: An Ecosystem Approach. CRC Press, Boca Raton, FL.

Davis, H.C., 1954. On food and feeding of larvae of the American oyster, *C. virginica*. Proc. Natl. Shellfish. Assoc. for 1952: 54–69.

Davis, H.C., 1958. Survival and growth of clam and oyster larvae at different salinities. Biol. Bull., 114: 296–307.

Davis, H.C. and Calabrese, A., 1964. Combined effects of temperature and salinity on development of eggs and growth of larvae of *M. mercenaria* and *C. virginica*. Fish. Bull., 63: 643–655.

Davis, H.C. and Chanley, P.E., 1956. Spawing and egg production of oysters and clams. Biol. Bull., 110: 117–128.

Deaton, L.E., Hilbish, T.J. and Koehn, R.K., 1984. Protein as a source of amino nitrogen during hyperosmotic volume regulation in the mussel *Mytilus edulis*. Physiol. Zool., 57 (6): 609–619.

De Vlas, J., 1985. Secondary production by siphon regeneration in a tidal flat population of *Macoma balthica*. Neth. J. Sea Res., 19: 147–164.

Dillon Jr., R.T. and Manzi, J.J., 1987. Hard clam, *Mercenaria mercenaria*, broodstocks: genetic drift and loss of rare alleles without reduction in heterozygosity. Aquaculture, 60: 99–105.

Dillon Jr., R.T. and Manzi, J.J., 1988. Enzyme heterozygosity and growth rate in nursery populations of *Mercenaria mercenaria* (L.). J. Exp. Mar. Biol. Ecol., 166: 79–86.

Doering, P.H., 1976. A burrowing response of *Mercenaria mercenaria* (Linnaeus, 1758) elicited by *Asterias forbesi* (Desor, 1848). Veliger, 19: 167–175.

Doering, P.H., 1982. Reduction of sea star predation by the burrowing response of the hard clam *Mercenaria mercenaria* (Mollusca: Bivalvia). Estuaries, 5: 310–315.

Doering, P.H. and Oviatt, C.A., 1986. Application of filtration rate models to field populations of bivalves: an assessment using experimental mesocosms. Mar. Ecol. Prog. Ser., 31: 265–275.

Doering, P.H., Oviatt, C.A. and Kelly, J.R., 1986. The effects of the filter-feeding clam *Mercenaria mercenaria* on carbon cycling in experimental marine mesocosms. J. Mar. Res., 44: 839–861.

Doering, P.H., Kelly, J.R., Oviatt, C.A. and Sowers, T., 1987. Effect of the hard clam *Mercenaria mercenaria* on benthic fluxes of inorganic nutrients and gases. Mar. Biol., 94: 377–383.

DuPaul, W.D. and Webb, K.L., 1974. Salinity-induced changes in the alanine and aspartic aminotransferase activity in three marine bivalve molluscs. Arch. Int. Physiol. Biochim., 82: 817–822.

Eckman, J.E., Peterson, C.H. and Cahalan, J.A., 1989. Effects of flow speed, turbulence, and orientation on growth of juvenile bay scallops *Argopecten irradians concentricus* (Say). J. Exp. Mar. Biol. Ecol., 132: 123–140.

Eldridge, P.J. and Eversole, A.G., 1982. Compensatory growth and mortality of the hard clam, *Mercenaria mercenaria* (Linnaeus, 1758). Veliger, 24: 276–278.

Eldridge, P.J., Eversole, A.G. and Whetstone, J.M., 1979. Comparative survival and growth rates of hard clams *Mercenaria mercenaria* planted in trays subtidally and intertidally at varying densities in a South Carolina estuary. Proc. Natl. Shellfish. Assoc., 69: 30–39.

Epifanio, C.E., 1975. Culture of bivalve mollusks in recirculating systems: nutritional requirements. In: K.S. Price, Jr., W.N. Shaw and K.S. Danberg (Eds.), Proceedings of the 1st International Conference on Aquaculture Nutrition, Delaware Sea Grant Publ. DEL-SG-17-76, pp. 173–194.

Epifanio, C.E., 1979a. Comparison of yeast and algal diets for bivalve molluscs. Aquaculture, 16: 187–192.

Epifanio, C.E., 1979b. Growth in bivalve molluscs: nutritional effects of two or more species of algae in diets fed to the American oyster *Crassostrea virginica* (Gmelin) and the hard clam *Mercenaria mercenaria* (L.). Aquaculture, 18: 1–12.

Epifanio, C.E. and Srna, R.F., 1975. Toxicity of ammonia, nitrite ion, nitrate ion, and orthophosphate to *Mercenaria mercenaria* and *Crassostrea virginica*. Mar. Biol., 33: 241–246.

Epifanio, C.E., Logan, C.M. and Turk, C., 1975. Culture of six species of bivalves in a recirculating seawater system. In: G. Persoone and E. Jaspers (Eds.), Proceedings of the 10th European Symposium on Marine Biology, Ostend, Vol. 1, Mariculture. Universa Press, Ostend, pp. 97–108.

Epp, J., Bricelj, V.M. and Malouf, R.E., 1988. Seasonal partitioning and utilization of energy reserves in two age classes of the bay scallop, *Argopecten irradians irradians*. J. Exp. Mar. Biol. Ecol., 121: 113–136.

Ertman, S.C. and Jumars, P.A., 1988. Effects of siphonal currents on the settlement of inert particles and larvae. J. Mar. Res., 46: 797–813.

Eversole, A.G., Goodsell, J.G. and Eldridge, P.J., 1990. Biomass, production and turnover of northern quahogs, *Mercenaria mercenaria* (Linnaeus, 1758), at different densities and tidal locations. J. Shellfish Res., 9: 309–314.

Feng, S.Y., 1969. Responses of some estuarine bivalves to temperature and salinity changes. In: Proc. Conf. Shellfish Culture, Nassau–Suffolk Regional Marine Resources Council, pp. 49–57.

Festa, P.J., 1975. Observations on the predator–prey relationships of winter flounder, *Pseudopleuronectes americanus* and the hard clam, *Mercenaria mercenaria*, in Little Egg Harbor, New Jersey. M.S. Thesis, 126 pp.

Flagg, P.J. and Malouf, R.E., 1983. Experimental plantings of juveniles of the hard clam *Mercenaria mercenaria* (Linné) in the waters of Long Island New York. J. Shellfish Res., 3: 19–27.

Foster-Smith, R.L., 1975. The effect of concentration of suspension on the filtration rates and pseudofaecal production for *Mytilus edulis* L., *Cerastoderma edule* (L.) and *Venerupis pullastra* (Montagu). J. Exp. Mar. Biol. Ecol., 17: 1–22.

Fritz, L.W. and Haven, D.S., 1983. Hard clam, *Mercenaria mercenaria*: shell growth patterns in Chesapeake Bay. Fish. Bull., 81: 697–708.

Gaffney, P.M. and Scott, T.M., 1984. Genetic heterozygosity and production traits in natural and hatchery populations of bivalves. Aquaculture, 42: 289–302.

Gainey Jr., L.F. and Shumway, S.E., 1991. The physiological effect of *Aureococcus anophagefferens* ("brown tide") on the lateral cilia of bivalve mollusks. Biol. Bull., 181: 298–306.

Gallager, S.M., Stoecker, D.K. and Bricelj, V.M., 1989. Effects of the brown tide on growth, feeding physiology et al.feeding physiology and locomotory behavior of scallop larvae (*Argopecten irradians*). In: E.M. Cosper,

V.M. Bricelj and E.J. Carpenter (Eds.), Novel Phytoplankton Blooms. Causes and Impacts of Recurrent Brown Tides and Other Unusual Blooms. Springer, New York, pp. 511–542.

Gallager, S.M., Waterbury, J.B. and Stoecker, D.K., 1994. Efficient grazing and utilization of the marine cyanobacterium *Synechococcus* sp. by larvae of the bivalve *Mercenaria mercenaria*. Mar. Biol., 119: 251–259.

Garton, D.W., Koehn, R.K. and Scott, T.M., 1984. Multiple locus heterozygosity and the physiological energetics of growth in the coot clam, *Mulinia lateralis*, from a natural population. Genetics, 108: 445–455.

Gerritsen, J., Holland, A.F. and Irvine, D.E., 1994. Suspension-feeding bivalves and the fate of primary production: an estuarine model applied to Chesapeake Bay. Estuaries, 17: 403–416.

Godwin, W.F., 1968. The distribution and density of the hard-clam, *Mercenaria mercenaria*, on the Georgia coast. Georgia Game and Fish Commission, Marine Fisheries Division, Brunswick, GA. Cont. Ser., 10, 30 pp.

Goldberg, R. and Wikfors, G.H., 1992. Growth of hard clams in Long Island Sound: sorting out the determining factors. Environ. Manage., 16: 521–529.

Goldstein, B.B. and Roels, O.A., 1980. The effect of feed density on the growth of juvenile *Mercenaria campechiensis*, the southern hard clam. Proc. World Maricul. Soc., 11: 192–201.

Grant, J., Dowd, M., Thompson, K., Emerson, C. and Hatcher, A., 1993. Perspectives on field studies and related biological models of bivalve growth and carrying capacity. In: R.F. Dame (Ed.), Bivalve Filter Feeders in Estuarine and Coastal Ecosystem Processes. Springer, New York, pp. 373–420.

Greene, G.T., 1978. Population structure, growth and mortality of hard clams at selected locations in Great South Bay, New York. Ph.D. Dissertation, State University of New York at Stony Brook, 199 pp.

Greene, G.T., 1979. Growth of clams (*Mercenaria mercenaria*) in Great South Bay, New York (abstr.). Proc. Natl. Shellfish. Assoc., 69: 194–195.

Grizzle, R.E., 1988. The relative effects of tidal currents, seston and bottom sediments on individual growth of *Mercenaria mercenaria* (Linné): Feeding ecology and aquacultural implications. Ph.D. Dissertation, Rutgers University, New Brunswick, NJ, 293 pp.

Grizzle, R.E. and Lutz, R.A., 1988. Descriptions of macroscopic banding patterns in sectioned polished shells of *Mercenaria mercenaria* from southern New Jersey. J. Shellfish Res., 7: 367–370.

Grizzle, R.E. and Lutz, R.A., 1989. A statistical model relating horizontal seston fluxes and bottom sediment characteristics to growth of *Mercenaria mercenaria*. Mar. Biol., 102: 95–105.

Grizzle, R.E. and Morin, P.J., 1989. Effect of tidal currents, seston, and bottom sediments on growth of *Mercenaria mercenaria*: Results of a field experiment. Mar. Biol., 102: 85–93.

Grizzle, R.E., Langan, R. and Howell, W.H., 1992. Growth responses of suspension-feeding bivalve molluscs to changes in water flow: Differences between siphonate and nonsiphonate taxa. J. Exp. Mar. Biol. Ecol., 162: 213–228.

Gustafsen, A.H., 1955. Growth studies in the quahog, *Venus mercenaria*. Proc. Natl. Shellfish. Assoc., 45: 140–150.

Hadley, N.H. and Manzi, J.J., 1984. Growth of seed clams, *Mercenaria mercenaria*, at various densities in a commercial scale nursery system. Aquaculture, 36: 369–378.

Hadley, N.H., Dillon Jr., R.T. and Manzi, J.J., 1991. Realized heritability of growth rate in the hard clam *Mercenaria mercenaria*. Aquaculture, 93: 109–119.

Hale, S.S., 1975. The role of benthic communities in the nitrogen and phosphorus cycles of an estuary. University of Rhode Island Marine Reprint 57, pp. 289–308.

Hammen, C.S., 1968. Aminotransferase activities and amino acid excretion of bivalve molluscs and brachiopods. Comp. Biochem. Physiol., 26: 697–705.

Hammen, C.S., 1979. Metabolic rates of marine bivalve molluscs determined by calorimetry. Comp. Biochem. Physiol., 62: 955–959.

Hammen, C.S., 1980. Total energy metabolism of marine bivalve mollusks in anaerobic and aerobic states. Comp. Biochem. Physiol., 67A: 617–621.

Hamwi, A., 1969. Oxygen consumption and pumping rate of the hard clam *Mercenaria mercenaria* L. Ph.D. Dissertation, Rutgers University, New Brunswick, NJ, 177 pp.

Hamwi, A. and Haskin, H.H., 1969. Oxygen consumption and pumping rates in the hard clam *Mercenaria mercenaria*: a direct method. Science, 163: 823–824.

Hartman, M., Epifanio, C.E., Pruder, G. and Srna, R., 1973. Farming the artificial sea: growth of clams in a recirculating seawater system. Proc. 26th Annu. Session Gulf Caribb. Fish. Inst., pp. 59–74.

Haskin, H.H., 1952. Further growth studies on the quahog, *Venus mercenaria*. Proc. Natl. Shellfish. Assoc., 42: 181–187.

Haven, D. and Morales-Alamo, R., 1972. Biodeposition as a factor of fine suspended solids in estuaries. Geol. Soc. Am. Mem., 133: 121–130.

Hawkins, A.J.S., Bayne, B.L., Mantoura, R.F.C. and Lewellyn, C.A., 1986. Chlorophyll degradation and absorption throughout the digestive system of the blue mussel, *Mytilus edulis* L. J. Exp. Mar. Biol. Ecol., 96: 213–223.

Head, E.J.H. and Harris, L.R., 1992. Chlorophyll and carotenoid transformation and destruction by *Calanus* spp. grazing on diatoms. Mar. Ecol. Prog. Ser., 86: 229–238.

Hilbish, T.J., 1986. Growth trajectories of shell and soft tissue in bivalves: seasonal variation in *Mytilus edulis* L. J. Exp. Mar. Biol. Ecol., 96: 103–113.

Heffernan, P.B., Walker, R.L. and Crenshaw Jr., J.W., 1991. Negative larval response to selection for increased growth rate in northern quahogs *Mercenaria mercenaria* (Linnaeus, 1758). J. Shellfish Res., 10: 199–202.

Herman, P.M.J., 1993. A set of models to investigate the role of benthic suspension feeders in estuarine ecosystems. In: R.F. Dame (Ed.), Bivalve Filter Feeders in Estuarine and Coastal Ecosystem Processes. Springer, New York, pp. 421–454.

Hibbert, C.J., 1976. Biomass and production of a bivalve community on an intertidal mud-flat. J. Exp. Mar. Biol. Ecol., 25: 249–261.

Hibbert, C.J., 1977a. Growth and survivorship in a tidal-flat population of the bivalve *Mercenaria mercenaria* from Southampton water. Mar. Biol., 44: 71–76.

Hibbert, C.J., 1977b. Energy relations of the bivalve *Mercenaria mercenaria* on an intertidal mudflat. Mar. Biol., 44: 77–84.

Hilbish, T.J., Winn, E.P. and Rawson, P.D., 1993. Genetic variation and covariation during larval and juvenile growth in *Mercenaria mercenaria*. Mar. Biol., 115: 97–104.

Hildreth, D.I. and Crisp, D.J., 1976. A corrected formula for calculation of filtration rate of bivalve molluscs in an experimental flowing system. J. Mar. Biol. Assoc. U.K., 56: 111–120.

Humphrey, C.M. and Crenshaw Jr., J.W., 1989. Clam genetics. In: J.J. Manzi and M. Castagna (Eds.), Clam Mariculture in North America. Elsevier, Amsterdam, pp. 323–356.

Hurd, T., Gallagher, M.L. and Horton, W.J., 1989. Effects of fatty acid supplementation of a minimal diet on growth of the hard clam (*Mercenaria mercenaria*). J. Elisha Mitchell Sci. Soc., 105: 88–92.

Iglesias, J.I.P., Urrutia, M.B., Navarro, E., Alvarez-Jorna, P., Larretxea, X., Bougrier, S. and Heral, M., 1992. Variability of feeding processes in the cockle, *Cerastoderma edule* (L.) in response to changes in seston concentration and composition. J. Exp. Mar. Biol. Ecol., 197: 121–143.

Irlandi, E.A. and Peterson, C.H., 1991. Modification of animal habitat by large plants: mechanisms by which seagrasses influence clam growth. Oecologia, 87: 307–318.

Jeffries, H.P., 1972. A stress syndrome in the hard clam, *Mercenaria mercenaria*. J. Invertebr. Pathol., 20: 242–251.

Jones, D.S., Arthur, M.A. and Allard, D.J., 1989. Sclerochronological records of temperature and growth from shells of *Mercenaria mercenaria* from Narragansett Bay, Rhode Island. Mar. Biol., 102: 225–234.

Jones, D.S., Quitmyer, I.R., Arnold, W.S. and Marelli, D.C., 1990. Annual shell banding, age, and growth rate of hard clams (*Mercenaria* spp.) from Florida. J. Shellfish Res., 9: 215–225.

Jørgensen, C.B., 1981. A hydromechanical principle for particle retention in *Mytilus edulis* and other ciliary suspension feeders. Mar. Biol., 61: 277–282.

Jørgensen, C.B., 1990. Bivalve Filter Feeding: Hydrodynamics, Bioenergetics, Physiology and Ecology. Olsen and Olsen Press, Fredensborg.

Jørgensen, C.B., 1996. Bivalve filter feeding revisited. Mar. Ecol. Prog. Ser., 142: 287–302.

Jørgensen, C.B., Møhlenberg, F. and Sten-Knudsen, O., 1986. Nature of relation between ventilation and oxygen consumption in filter feeders. Mar. Ecol. Prog. Ser., 29: 73–88.

Judge, M.L., Coen, L.D. and Heck Jr., K.L., 1992. The effect of long-term alteration of in situ currents on the growth of *Mercenaria mercenaria* in the northern Gulf of Mexico. Limnol. Oceanogr., 37: 1550–1559.

Judge, M.L., Coen, L.D. and Heck Jr., K.L., 1993. Does *Mercenaria mercenaria* encounter elevated food levels in seagrass beds? Results from a novel technique to collect suspended food resources. Mar. Ecol. Prog. Ser., 92: 141–150.

Kellogg, J.L., 1903. Feeding habits and growth of *V. mercenaria*. N.Y. State Mus. Bull. 71, 27 pp.

Kellogg, J.L., 1910. The hard clam. In: Shell-fish Industries. American Nature Series, Henry Holt and Co., New York, pp. 321–332.

Kennish, M.J., 1980. Shell microgrowth analysis: *Mercenaria mercenaria* as a type example for research in population dynamics. In: D.C. Roads and R.A. Lutz (Eds.), Skeletal Growth of Aquatic Organisms. Plenum Press, New York, pp. 255–294.

Kennish, M.J. and Loveland, R.E., 1980. Growth models of the northern quahog, *Mercenaria mercenaria* (Linné). Proc. Natl. Shellfish. Assoc., 70: 230–239.

Kennish, M.J. and Olsson, R.K., 1975. Effects of thermal discharges on microstructural growth of *Mercenaria mercenaria*. Environ. Geol., 1: 41–64.

Kerswill, C.J., 1941. Some environmental factors limiting growth and distribution of the quahog *Venus mercenaria* L. Ph.D. Dissertation, University of Toronto, 104 pp.

Kerswill, C.J., 1949. Effects of water circulation on the growth of quahogs and oysters. J. Fish. Res. Bd. Can., 7: 545–551.

Kiørboe, T. and Møhlenberg, F., 1981. Particle selection in suspension-feeding bivalves. Mar. Ecol. Prog. Ser., 5: 291–296.

Koehn, R.K. and Gaffney, P.M., 1984. Genetic heterozygosity and growth rate in *Mytilus edulis*. Mar. Biol., 82: 1–7.

Korycan, S.A. and Storey, K.B., 1983. Organ-specific metabolism during anoxia and recovery from anoxia in the cherrystone clam, *Mercenaria mercenaria*. Can. J. Zool., 61: 2674–2681.

Kreeger, D.A., Hawkins, A.J.S. and Bayne, B.L., 1996. Use of dual-labeled microcapsules to discern the physiological fates of assimilated carbohydrate, protein carbon, and protein nitrogen in suspension-feeding organisms. Limnol. Oceanogr., 41: 208–215.

LaBarbera, M., 1984. Feeding currents and particle capture mechanisms in suspension feeding animals. Am. Zool., 24: 71–84.

Laing, I., 1987. The use of artificial diets in rearing bivalve spat. Aquaculture, 65: 243–249.

Laing, I. and Verdugo, C.G., 1991. Nutritional value of spray-dried *Tetraselmis suecica* for juvenile bivalves. Aquaculture, 92: 207–218.

Laing, I., Utting, S.D. and Kilada, R.W.S., 1987. Interactive effect of diet and temperature on the growth of juvenile clams. J. Exp. Mar. Biol. Ecol., 133: 23–28.

Landry, T., Sephton, T.W. and Jones, D.A., 1993. Growth and mortality of northern quahog, *Mercenaria mercenaria* (Linneaeus, 1758) in Prince Edward Island. J. Shellfish Res., 12: 321–327.

Langdon, C.J. and Newell, R.I.E., 1990. Utilization of detritus and bacteria as food sources by two bivalve suspension-feeders, the oyster *Crassostrea virginica* and the mussel *Geukensia demissa*. Mar. Ecol. Prog. Ser. 299–310.

Lee, J., 1993. The kinetics of PSP toxin transfer from the toxic dinoflagellate *Alexandrium* spp. to two bivalve mollusc species, *Mytillus edulis* and *Mercenaria mercenaria*. M.S. Thesis, Marine Science Research Center, State University of New York at Stony Brook, 169 pp.

Lehman, J.T., 1976. The filter feeder as an optimal forager and the predicted shapes of feeding curves. Limnol. Oceanogr., 21: 501–516.

Lenihan, H.S., Peterson, C.H. and Allen, J.M., 1996. Does flow speed also have a direct effect on growth of active suspension-feeders: An experimental test on oysters. Limnol. Oceanogr., 41: 1359–1366.

Lesser, M.P. and Shumway, S.E., 1993. Effects of toxic dinoflagellates on clearance rates and survival in juvenile bivalve molluscs. J. Shellfish Res., 12: 377–381.

Levinton, J.S., 1991. Variable feeding behavior in three species of *Macoma* (Bivalvia: Tellinacea) as a response to water flow and sediment transport. Mar. Biol., 110: 375–383.

Levinton, J.S., Ward, J.E. and Thompson, R.J., 1996. Biodynamics of particle processing in bivalve molluscs: models, data, and future directions. Invertebr. Biol., 115 (3): 232–242.

Loesch, J.G. and Haven, D.S., 1973. Estimated growth functions and size–age relationships of the hard clam, *Mercenaria mercenaria* in the York River, Virginia. Veliger, 16: 76–81.

Loosanoff, V.L., 1937. Spawning of *Venus mercenaria* (L.). Ecology, 18: 506–515.

Loosanoff, V.L., 1939. Effect of temperature upon shell movements of clams, *Venus mercenaria* (L.). Biol. Bull., 76: 171–182.

Lough, R.G., 1975. A reevaluation of the combined effects of temperature and salinity on survival and growth of bivalve larvae using response surface techniques. Fish. Bull., 73: 86–94.

Loveland, R.E. and Chu, D.S.K., 1969. Oxygen consumption and water movements in *Mercenaria mercenaria*. Comp. Biochem. Physiol., 29: 173–184.

Lutz, R.A. and Haskin, H.H., 1985. Some observations on the longevity of the hard clam *Mercenaria mercenaria* (Linné). J. Shellfish Res., 5: 39.

Lutz, R.A. and Rhoads, D.C., 1977. Anaerobiosis and a theory of growth line formation. Science, 198: 1222–1227.

Lutz, R.A. and Rhoads, D.C., 1980. Growth patterns within the molluscan shell: an overview. In: D.C. Rhoads and R.A. Lutz (Eds.), Skeletal Growth of Aquatic Organisms. Biological Record of Environmental Change. Plenum Press, New York, pp. 203–254.

Malinowski, S.M. and Siddall, S.E., 1989. Passive water reuse in a commercial-scale hard clam, *Mercenaria mercenaria*, upflow nursery system. J. Shellfish Res., 8: 241–248.

Malouf, R.E. and Bricelj, V.M., 1989. Comparative biology of clams: environmental tolerances, feeding, and growth. In: J.J. Manzi and M. Castagna (Eds.), Clam Mariculture in North America. Elsevier, Amsterdam, pp. 23–73.

Malouf, R.E., Keck, R., Maurer, D. and Epifanio, C., 1972. Occurrence of gas-bubble disease in three species of bivalve molluscs. J. Fish. Res. Bd. Can., 29: 588–589.

Mann, R. and Gallager, S.M., 1984. Physiological consequences of metamorphosis in marine bivalve larvae. Woods Hole Oceanographic Institution, Woods Hole, MA, Sea Grant Annu. Rep. 1983–84, pp. 9–12.

Manzi, J.J. and Castagna, M., 1989. Introduction. In: J.J. Manzi and M. Castagna (Eds.), Clam Mariculture in North America. Elsevier, Amsterdam, pp. 1–21.

Manzi, J.J., Burrell, V.G. and Carson, W.Z., 1980. A mariculture demonstration project for an alternative hard clam fishery in South Carolina: preliminary results. Proc. World Maricult. Soc., 11: 79–89.

Manzi, J.J., Hadley, N.H., Battey, C., Haggerty, R., Hamilton, R. and Carter, M., 1984. Culture of the northern clam *Mercenaria mercenaria* (Linné) in a commercial-scale, upflow, nursery system. J. Shellfish Res., 4: 119–124.

Manzi, J.J., Hadley, N.H. and Maddox, M.B., 1986. Seed clam, *Mercenaria mercenaria* culture in an experimental-scale upflow nursery system. South Carolina Marine Resources Contrib. 206.

Manzi, J.J., Hadley, N.H. and Dillon Jr., R.T., 1991. Hard clam, *Mercenaria mercenaria* broodstocks: growth of selected hatchery stocks and their reciprocal crosses. Aquaculture, 94: 17–26.

McClatchie, S., 1992. Time-series measurement of grazing rates of zooplankton and bivalves. J. Plankton Res., 14: 183–200.

McIntosh, R.P., 1985. The Background of Ecology. Cambridge University Press, Cambridge, 383 pp.

Menzel, R.W., 1962. Seasonal growth of northern and southern quahogs, *Mercenaria mercenaria* and *M. campechiensis*, and their hybrids in Florida. Proc. Natl. Shellfish. Assoc., 52: 111–119.

Menzel, R.W., 1963. Seasonal growth of the northern quahog *Mercenaria mercenaria* and the southern quahog *M. campechiensis* in Alligator Harbor, Florida. Proc. Natl. Shellfish. Assoc., 53: 37–46.

Mitchell, R., 1974. Aspects of the ecology of the lamellibranch *Mercenaria mercenaria* (L.) in British waters. Hydrobiol. Bull., 8: 124–138.

Møhlenberg, F. and Riisgård, H.U., 1978. Efficiency of particle retention in 13 species of suspension feeding bivalves. Ophelia, 17: 239–246.

Møhlenberg, F. and Riisgård, H.U., 1979. Filtration rate, using a new indirect technique, in thirteen species of suspension-feeding bivalves. Mar. Biol., 54: 143–147.

Monismith, S.G., Koseff, J.R., Thompson, J.K., O'Riordan, C.A. and Nepf, H.M., 1990. A study of model bivalve siphonal currents. Limnol. Oceanogr., 35: 680–696.

Moreth, C.M. and Yentsch, C.S., 1970. The role of chlorophyllase and light in the decomposition of chlorophyll from marine phytoplankton. J. Exp. Mar. Biol. Ecol., 4: 238–249.

Morris, P.C., 1973. A Field Guide to the Shells of the Atlantic and Gulf Coasts and the West Indies. Houghton Mifflin Co., Boston, MA, 3rd ed., 330 pp.

Morton, B., 1973. A new theory of feeding and digestion in the filter feeding lamellibranchia. Malacologia, 14: 63–79.

Murphy, R.C., 1985. Factors affecting the distribution of the introduced bivalve, *Mercenaria mercenaria*, in a California lagoon — The importance of bioturbation. J. Mar. Res., 43: 673–692.

Murphy, R.C. and Kremer, J.N., 1985. Bivalve contribution to benthic metabolism in a California lagoon. Estuaries, 8: 330–341.

Muschenheim, D.K., 1987. The dynamics of near-bed seston flux and suspension-feeding benthos. J. Mar. Res., 45: 473–496.

Muschenheim, D.K. and Newell, C.R., 1992. Utilization of seston flux over a mussel bed and estimation of an effective "feeding zone". Mar. Ecol. Prog. Ser., 85: 131–136.

Muschenheim, D.K., Grant, J. and Mills, E.L., 1986. Flumes for benthic ecologists: theory, construction and practice. Mar. Ecol. Prog. Ser., 28: 185–196.

Navarro, E. and Iglesias, J.I.P., 1993. Infaunal filter-feeding bivalves and the physiological response to short-term fluctuations in food availability and composition. In: R.F. Dame (Ed.), Bivalve Filter Feeders in Estuarine and Coastal Ecosystem Processes. Springer, New York, pp. 25–56.

Navarro, J.M. and Thompson, R.J., 1994. Comparison and evaluation of different techniques for measuring absorption efficiency in a suspension-feeding bivalve. Limnol. Oceanogr., 39: 159–164.

Navarro, E., Iglesias, J.I.P. and Ortega, M., 1992. Natural sediment as a food source for the cockle *Cerastoderma edule* (L.): effect of variable particle concentration on feeding, digestion and the scope for growth. J. Exp. Mar. Biol. Ecol., 156: 69–87.

Newcombe, C.L., 1935. Certain environmental factors of a sand beach in the St. Andrews region, New Brunswick, with a preliminary designation of the intertidal communities. J. Ecol., 23: 334–355.

Newell, R.I.E., 1988. Ecological changes in Chesapeake Bay: Are they the result of overharvesting the American oyster, *Crassostrea virginica*? In: M.P. Lynch and E.C. Krome (Eds.), Understanding the Estuary: Advances in Chesapeake Bay Research. Chesapeake Research Consortium, Solomons, MD, pp. 536–546.

Newell, R.I.E. and Jordan, S.J., 1983. Preferential ingestion of organic material by the American oyster *Crassostrea virginica*. Mar. Ecol. Prog. Ser., 13: 47–537.

Newell, R.I.E. and Langdon, C.R., 1986. Digestion and absorption of refractory carbon from the plant *Spartina alterniflora* (Loisel) by the oyster *Crassostrea virginica* (Gmelin). Mar. Ecol. Prog. Ser., 34: 105–115.

Newell, C.R. and Shumway, S.E., 1993. Grazing of natural particulates by bivalve molluscs: a spatial and temporal perspective. In: R.F. Dame (Ed.), Bivalve Filter Feeders in Estuarine and Coastal Ecosystem Processes. Springer, New York, pp. 85–145.

Newell, R.I.E. and Langdon, C.R., 1996. Mechanisms and physiology of larval and adult feeding. In: V.S. Kennedy, R.I.E. Newell and A.F. Eble (Eds.), The Eastern Oyster *Crassostrea virginica*. Maryland Sea Grant, College Park, MD, pp. 185–229.

Nowell, A.R.M. and Jumars, P.A., 1984. Flow environments of aquatic benthos. Annu. Rev. Ecol. Syst., 15: 303–328.

Nowell, A.R.M. and Jumars, P.A., 1987. Flumes: theoretical and experimental considerations for simulation of benthic environments. Oceanogr. Mar. Biol. Annu. Rev., 25: 91–112.

Officer, C.B., Monismith, S.G. and Koseff, J.R., 1982. Benthic filter feeding: A natural eutrophication control. Mar. Ecol. Prog. Ser., 9: 203–210.

O'Riordan, C.A., Monismith, S.G. and Koseff, J.R., 1995. The effect of bivalve excurrent jet dynamics on mass transfer in a benthic boundary layer. Limnol. Oceanogr., 40: 330–344.

Owen, G. and McCrae, J.M., 1976. Further studies on the laterofrontal tracts of bivalves. R. Soc. London B, 194: 527–544.

Petersen, C.G.J. and Jensen, P.B., 1911. Valuation of the sea. I. animal life of the sea bottom, its food and quantity. Beret. Dansk Biol. Stat., 20: 1–76.

Peterson, C.H., 1983. A concept of quantitative reproductive senility: application to the hard clam, *Mercenaria mercenaria* (L)? . Oecologia, 58: 164–168.

Peterson, C.H., 1986. Quantitative allometry of gamete production by *Mercenaria mercenaria* into old age. Mar. Ecol. Prog. Ser., 29: 93–97.

Peterson, C.H. and Beal, B.F., 1989. Bivalve growth and higher order interactions: importance of density, site, and time. Ecology, 70: 1390–1404.

Peterson, C.H. and Fegley, S.R., 1986. Seasonal allocation of resources to growth of shell, soma, and gonads in *Mercenaria mercenaria*. Biol. Bull., 171: 597–610.

Peterson, C.H. and Quammen, M.L., 1982. Siphon nipping: its importance to small fishes and its impact on growth of the bivalve *Protothaca staminea* (Conrad). J. Exp. Mar. Biol. Ecol., 63: 249–268.

Peterson, C.H., Duncan, P.B., Summerson, H.C. and Safrit Jr., G.W., 1983. A mark-recapture test of annual periodicity of internal growth band deposition in shells of hard clams, *Mercenaria mercenaria*, from a population along the southeastern United States. Fish. Bull., 81: 765–779.

Peterson, C.H., Summerson, H.C. and Duncan, P.B., 1984. The influence of seagrass cover on population structure and individual growth rate of a suspension-feeding bivalve, *Mercenaria mercenaria*. J. Mar. Res., 42: 123–138.

Pratt, D.M., 1953. Abundance and growth of *Venus mercenaria* and *Callocardia morrhua* in relation to the character of the bottom sediments. J. Mar. Res., 12: 60–74.

Pratt, D.M. and Campbell, D.A., 1956. Environmental factors affecting growth in *Venus mercenaria*. Limnol. Oceanogr., 1: 2–17.

Prosser, C.L. and Brown, F.A., 1961. Comparative Animal Physiology. W.B. Saunders, Philadelphia, PA, 2nd ed., 688 pp.

Raillard, O. and Ménesguen, A., 1994. An ecosystem box model for estimating the carrying capacity of a macrotidal shellfish system. Mar. Ecol. Prog. Ser., 115: 117–130.

Rawson, P.D. and Hilbish, T.J., 1990. Heritability of juvenile growth for the hard clam *Mercenaria mercenaria*. Mar. Biol., 105: 429–436.

Rawson, P.D. and Hilbish, T.J., 1991. Genotype-environment interaction for juvenile growth in the hard clam *Mercenaria mercenaria* (L.). Evolution, 45: 1924–1935.

Rhoads, D.C. and Pannella, G., 1970. The use of molluscan shell growth patterns in ecology and paleoecology. Lethaia, 3: 143–161.

Rice, M.A. and Pechenik, J.A., 1992. A review of the factors influencing the growth of the northern quahog, *Mercenaria mercenaria* (Linnaeus, 1758). J. Shellfish Res., 11: 279–287.

Rice, M.A. and Stephens, G.C., 1988. Influx, net flux and transepithelial flux of amino acids in the hardshell clam *Mercenaria mercenaria* (Linné): influence of salinity. Comp. Biochem. Physiol., 89A: 631–636.

Rice, T.R. and Smith, R.J., 1958. Filtering rates of the hard clam (*Venus mercenaria*) determined with radioactive phytoplankton. Fish. Bull., 58: 71–82.

Riisgård, H.U., 1991. Filtration rate and growth in the blue mussel, *Mytilus edulis* Linnaeus, 1758: Dependence on algal concentration. J. Shellfish Res., 10: 29–35.

Riisgård, H.U., Randløv, A. and Sand Kristensen, P., 1980. Rates of water processing, oxygen consumption and efficiency of particle retention in veligers and young post-metamorphic *Mytilus edulis*. Ophelia, 19: 37–47.

Roberts, D., Rittschof, D., Gerhart, D.J., Schmidt, A.R. and Hill, L.G., 1989. Vertical migration of the clam *Mercenaria mercenaria* (L.) (Mollusca: Bivalvia): environmental correlates and ecological significance. J. Exp. Mar. Biol. Ecol., 126: 271–280.

Robinson, W.E. and Langton, R.W., 1980. Digestion in a subtidal population of *Mercenaria mercenaria* (Bivalvia). Mar. Biol., 58: 173–179.

Robinson, W.E., Pennington, M.R. and Langton, R.W., 1981. Variability of tubule types within the digestive glands of *Mercenaria mercenaria* (L.), *Ostrea edulis* (L.), and *Mytilus edulis* (L.). J. Exp. Mar. Biol. Ecol., 54: 265–276.

Roegner, G.C. and Mann, R., 1990. Habitat requirements for the hard clam, *Mercenaria mercenaria*, in Chesapeake Bay. Virginia Institute of Marine Science, Gloucester Point, VA, Spec. Sci. Rep. 126, 31 pp.

Savage, N.D., 1976. Burrowing activity in *Mercenaria mercenaria* (L.) and *Spisula solidissima* (Dillwyn) as a function of temperature and dissolved oxygen. Mar. Behav. Physiol., 3: 221–234.

Schapiro, A.Z. and Bobkova, A.N., 1976. The role of malate dehydrogenase during adaptation to hypoxia in invertebrates. J. Evol. Biochem. Physiol. (Engl. Transl. Zb. Evol. Biokhim. Fiziol. 11(5) 1975: 478–479.

Seed, R., 1980. Shell growth and form in the Bivalvia. In: D.C. Rhoads and R.A. Lutz (Eds.), Skeletal Growth of Aquatic Organisms. Biological Records of Environmental Change. Plenum Press, New York, pp. 23–67.

Shimeta, J. and Jumars, P.A., 1991. Physical mechanisms and rates of particle capture by suspension-feeders. Oceanogr. Mar. Biol. Annu. Rev., 29: 191–257.

Shumway, S.E., 1990. A review of the effects of algal blooms on shellfish and aquaculture. J. World Aquacult. Soc., 21: 65–104.

Shumway, S.E. and Cucci, T.L., 1987. The effects of the toxic dinoflagellate *Protogonyaulax tamarensis* on the feeding and behavior of bivalve molluscs. Aquat. Toxicol., 10: 9–27.

Shumway, S.E., Gabbott, P.A. and Youngson, A., 1977. The effect of fluctuating salinity on the concentrations of free amino acids and ninhydrin-positive substances in the adductor muscles of eight species of bivalve molluscs. J. Exp. Mar. Biol. Ecol., 29: 131–150.

Shumway, S.E., Cucci, T.L., Newell, R.C. and Yentsch, C.M., 1985. Particle selection, ingestion, and absorption in filter-feeding bivalves. J. Exp. Mar. Biol. Ecol., 91: 77–92.

Sieburth, J.M., Johnson, P.W. and Hargraves, P.E., 1988. Ultrastructure and ecology of *Aureococcus anaphagefferens* gen. et sp. nov. (Chrysophyceae): the dominant picoplankter during a bloom in Narragansett Bay, Rhode Island, Summer 1985. J. Phycol., 24: 416–425.

Silverman, H., Lynn, J.W. and Dietz, T.H., 1996. Particle capture by the gills of *Dreissena polymorpha*: structure and function of latero-frontal cirri. Biol. Bull., 191: 42–54.

Slattery, J.P., Vrijenhoek, R.C. and Lutz, R.A., 1991. Heterozygosity, growth, and survival of the hard clam *Mercenaria mercenaria*, in seagrass vs. sandflat habitats. Mar. Biol., 111: 335–342.

Slattery, J.P., Lutz, R.A. and Vrijenhoek, R.C., 1993. Repeatability of correlations between heterozygosity, growth and survival in a natural population of the hard clam *Mercenaria mercenaria* L. J. Exp. Mar. Biol. Ecol., 165: 209–224.

Smith, R.L., 1996. Ecology and Field Biology. Harper and Collins, New York.

Smolowitz, R. and Shumway, S.E., 1997. Possible cytotoxic effects of the dinoflagellate, *Gyrodinium aureolum*, on juvenile bivalve molluscs. Aquacult. Int., 5: 291–300.

Srna, R.F. and Baggaley, A., 1976. Rate of excretion of ammonia by the hard clam *Mercenaria mercenaria* and the American oyster *Crassostrea virginica*. Mar. Biol., 36: 251–258.

Stanley, J.G. and DeWitt, R., 1983. Species Profiles: Life Histories and Environmental Requirements of Coastal Fishes and Invertebrates (North America) — Hard Clam. U.S. Fish Wildl. Serv. FWS/OBS-82/11.18, 19 pp.

Stanley, S.M., 1970. Relation of shell form to life habits of the Bivalvia (Mollusca). Geol. Soc. Am. Mem. 125, 296 pp.

Stenton-Dozey, J.M.E. and Brown, A.C., 1992. Clearance and retention efficiency of natural suspended particles by the rock-pool bivalve *Venerupis corrugatus* in relation to tidal availability. Mar. Ecol. Prog. Ser., 82: 175–186.

Tenore, K.R. and Dunstan, W.M., 1973. Comparison of feeding and biodeposition of three bivalves at different food levels. Mar. Biol., 21: 190–195.

Tenore, K.R., Goldman, J.C. and Clarner, J.P., 1973. The food chain dynamics of the oyster, clam, and mussel in an aquaculture food chain. J. Exp. Mar. Biol. Ecol., 12: 157–165.

Thompson, R.J. and Bayne, B.L., 1974. Some relationships between growth, metabolism and food in the mussel, *Mytilus edulis*. Mar. Biol., 27: 317–326.

Tiu, T.A., Vaughan, D., Chiles, T. and Bird, K., 1989. Food value of eurytopic microalgae to larvae of *Cyrtopleura costata* (Linnaeus, 1758), *Crassostrea virginica* (Gmelin, 1791) and *Mercenaria mercenaria* (Linnaeus, 1758). J. Shellfish Res., 8: 399–405.

Tracey, G.A., 1988. Feeding reduction, reproductive failure, and mortality in *Mytilus edulis* during the 1985 "brown tide" in Narragansett Bay, Rhode Island. Mar. Ecol. Prog. Ser., 50: 73–81.

Turner, E.J., 1990. Suspension feeding in oscillatory flows with resuspended bottom sediments. Ph.D. Dissertation, University of Delaware, Lewes, Delaware.

Turner, E.J., 1992. A new method for directly measuring bivalve pumping velocity (abstr.). J. Shellfish Res., 11: 275.

Turner, E.J. and Miller, D.C., 1991. Behavior and growth of *Mercenaria mercenaria* during simulated storm events. Mar. Biol., 111: 55–64.

Twarog, B.M. and Yamaguchi, H., 1972. Resistance of paralytic shellfish toxins in bivalve molluscs. In: V.R. LoCicero (Ed.), Proceedings of the 1st International Conference on Toxic Dinoflagellate Blooms. Massachusetts Science and Technology Fund, Wakefield, MA, pp. 382–393.

Ulanowicz, R.E. and Tuttle, J.H., 1992. The trophic consequences of oyster stock rehabilitation in Chesapeake Bay. Estuaries, 15: 298–306.

Van Winkle, W., 1975. Problems in establishing the relationship between pumping rate and oxygen consumption rate in the hard clam, *Mercenaria mercenaria*. Comp. Biochem. Physiol., 50A: 657–660.

Van Winkle, W., Feng, S.Y. and Haskin, H.H., 1976. Effect of temperature and salinity on extension of siphons by *Mercenaria mercenaria*. J. Fish. Res. Bd. Can., 33: 1540–1546.

Verduin, J., 1969. Hard clam pumping rates: energy requirements. Science, 166: 1309–1310.

Vincent, B., Desrosiers, G. and Gratton, Y., 1988. Orientation of the infaunal bivalve *Mya arenaria* L. in relation to local current direction on a tidal flat. J. Exp. Mar. Biol. Ecol., 124: 205–214.

Walker, R.L., 1984. Effects of density and sampling time on the growth of the hard clam *Mercenaria mercenaria*. Nautilus, 98: 114–119.

Walker, R.L. and Heffernan, P.B., 1990. Intertidal growth and survival of northern quahogs *Mercenaria mercenaria* (Linnaeus 1758) and Atlantic surf clams *Spisula solidissima* (Dillwyn 1817). J. World Aquacult. Soc., 21: 307–313.

Walker, R.L. and Tenore, K.R., 1984. The distribution and production of the hard clam, *Mercenaria mercenaria* in Wassaw Sound, Georgia. Estuaries, 7: 19–27.

Walne, P.R., 1970. Studies on the food value of nineteen genera of algae to juvenile bivalves of the genera *Ostrea*, *Crassostrea*, *Mercenaria* and *Mytilus*. Fishery Investigations, Series II, Vol. 26, 62 pp.

Walne, P.R., 1972. The influence of current speed, body size and water temperature on the filtration rate of five species of bivalves. J. Mar. Biol. Assoc. U.K., 52: 345–374.

Walsh, D., 1974. The responses of the bivalve *Mercenaria mercenaria* to declining oxygen tensions. M.S. Thesis, College of William and Mary, Williamsburg, VA.

Wang, W.-X. and Fisher, N.S., 1996. Assimilation of trace metals and carbon by the mussel *Mytilus edulis*: Effects of food composition. Limnol. Oceanogr., 41 (2): 197–207.

Wang, W.X. and Widdows, J., 1993. Metabolic responses of the common mussel, *Mytilus edulis*, to hypoxia and anoxia. Mar. Ecol. Prog. Ser., 95: 205–214.

Ward, J.E., 1996. Biodynamics of suspension-feeding in adult bivalve molluscs: particle capture, processing, and fate. Invertebr. Biol., 115 (3): 218–231.

Ward, J.E., MacDonald, B.A. and Thompson, R.J., 1993a. Mechanisms of suspension feeding in bivalves: Resolution of current controversies by means of endoscopy. Limnol. Oceanogr., 38 (2): 265–272.

Ward, J.E., MacDonald, B.A., Thompson, R.J. and Beninger, P.G., 1993b. The role of mucus in bivalve feeding — A reply to the comment by Jørgensen. Limnol. Oceanogr., 38: 467.

Ward, J.E., Newell R.I.E., Thompson, R.J. and MacDonald, B.A., 1994. *In vivo* studies of suspension-feeding processes in the eastern oyster, *Crassostrea virginica* Gmelin. Biol. Bull., 186: 221–240.

Ward, J.E., Levinton, J.S., Shumway, E.E. and Cucci, T.L., 1997. Linking individual function to ecosystem control: identification of the gills as the locus of particle selection in a bivalve mollusk. Nature, 390: 131–132.

Widdows, J., 1978. Combined effects of body size, food concentrations and season on the physiology of *Mytilus edulis*. J. Mar. Biol. Assoc. U.K., 58: 109–124.

Widman, J.C. and Goldberg, R., 1990. Growth and survival of hard clams at various densities in three Long Island Sound locations (abstr.). J. Shellfish Res., 9: 488.

Wietkamp, D.E. and Katz, M., 1980. A review of dissolved oxygen gas supersaturation literature. Trans. Am. Fish. Soc., 109: 659–702.

Wikfors, G.H. and Smolowitz, R.M., 1993. Detrimental effects of a *Prorocentrum* isolate upon hard clams and bay scallops in laboratory feeding studies. In: T.J. Smayda and Y. Shimizu (Eds.), Toxic Phytoplankton Blooms in the Sea. pp. 447–452.

Wikfors, G.H., Ferris, G.E. and Smith, B.C., 1992. The relationship between gross biochemical composition of cultured algal foods and growth of the hard clam, *Mercenaria mercenaria* (L.). Aquaculture, 108: 135–154.

Wildish, D.J. and Kristmanson, D.D., 1993. Hydrodynamic control of bivalve filter feeders: A conceptual view. In: R.F. Dame (Ed.), Bivalve Filter Feeders in Estuarine and Coastal Ecosystem Processes. Springer, New York, pp. 299–324.

Wildish, D.J. and Kristmanson, D.D., 1997. Benthic Suspension Feeders and Flow. Cambridge University Press, Cambridge.

Wildish, D.J. and Miyares, M.P., 1990. Filtration rate of blue mussels as a function of flow velocity: preliminary experiments. J. Exp. Mar. Biol. Ecol., 142: 213–219.

Wildish, D.J. and Saulnier, A.M., 1993. Hydrodynamic control of filtration in *Placopecten magellanicus*. J. Exp. Mar. Biol. Ecol., 174: 65–82.

Wildish, D.J., Kristmanson, D.D., Hoar, R.L., DeCoste, A.M., McCormick, S.D. and White, A.W., 1987. Giant scallop feeding and growth responses to flow. J. Exp. Mar. Biol. Ecol., 113: 207–220.

Wildish, D.J., Kristmanson, D.D. and Saulnier, A.M., 1992. Interactive effect of velocity and seston concentration on giant scallop feeding inhibition. J. Exp. Mar. Biol. Ecol., 155: 161–168.

Willows, R.I., 1992. Optimal digestive investment: A model for filter feeders experiencing variable diets. Limnol. Oceanogr., 37: 829–847.

Winter, J.E., 1978. A review on the knowledge of suspension-feeding in lamellibranchiate bivalves, with special reference to artificial aquaculture systems. Aquaculture, 13: 1–33.

Biology of the Hard Clam
J.N. Kraeuter and M. Castagna (Eds.),

Chapter 9

Demography and Dynamics of Hard Clam Populations

Stephen R. Fegley

9.1 INTRODUCTION

Hard clams, *Mercenaria mercenaria* L., are frequently one of the most abundant large infaunal suspension feeders in soft-substrates throughout their range. Hard clam populations have supported fisheries, especially in southern New England and the mid Atlantic states, since the 19th century. Despite the ecological and economical importance that can be attached to hard clam populations, little information has been gathered that provides insight into their dynamics. Neither resource managers nor shellfish biologists can identify, confidently, which populations are endangered or which ones are most important for local recruitment. No one knows what the minimum viable population size is for hard clams or what the demographics of an unharvested hard clam population are. No one can identify the relative importance of the various factors that affect hard clam numerical abundances at different stages of their life-history. By drawing together the available information on hard clam population demography and dynamics it may be easier to identify those areas we need to know more about in order to better understand changes in hard clam abundance.

One problem I encountered in reviewing the available information on hard clam populations is the absence of consistency among investigators in sampling methodology (Table 9.1). The wide ranges of techniques and sampling efficiencies used in the various studies limit the extent to which we can compare the respective results and generalize conclusions. The limits imposed by methodological diversity must be considered when examining the following information.

9.2 ABUNDANCE

9.2.1 Population Density

Adult *Mercenaria* population densities vary over two orders of magnitude (Table 9.2). Rarely, very high densities (>500 individuals m^{-2}) have been observed over areas of several m^2 (Dow and Wallace, 1955; Crane et al., 1975). More typically, *Mercenaria* population densities range from 1 to 15 individuals m^{-2}; over 80% of reported mean densities fall within that range (Fig. 9.1). Population density frequency distributions tend to be log-normal with overall mean densities occurring between 4 and 8 individuals m^{-2} (Fig. 9.1). No apparent regional trends in clam density exist. Population densities are similar in the northeast US, southeast US, and in areas where *Mercenaria* have been introduced, such as California and England (Table 9.2).

The roles of environmental factors, such as sediment composition (sand versus muds), sediment organic content, and local flow regimes on population densities are complex and,

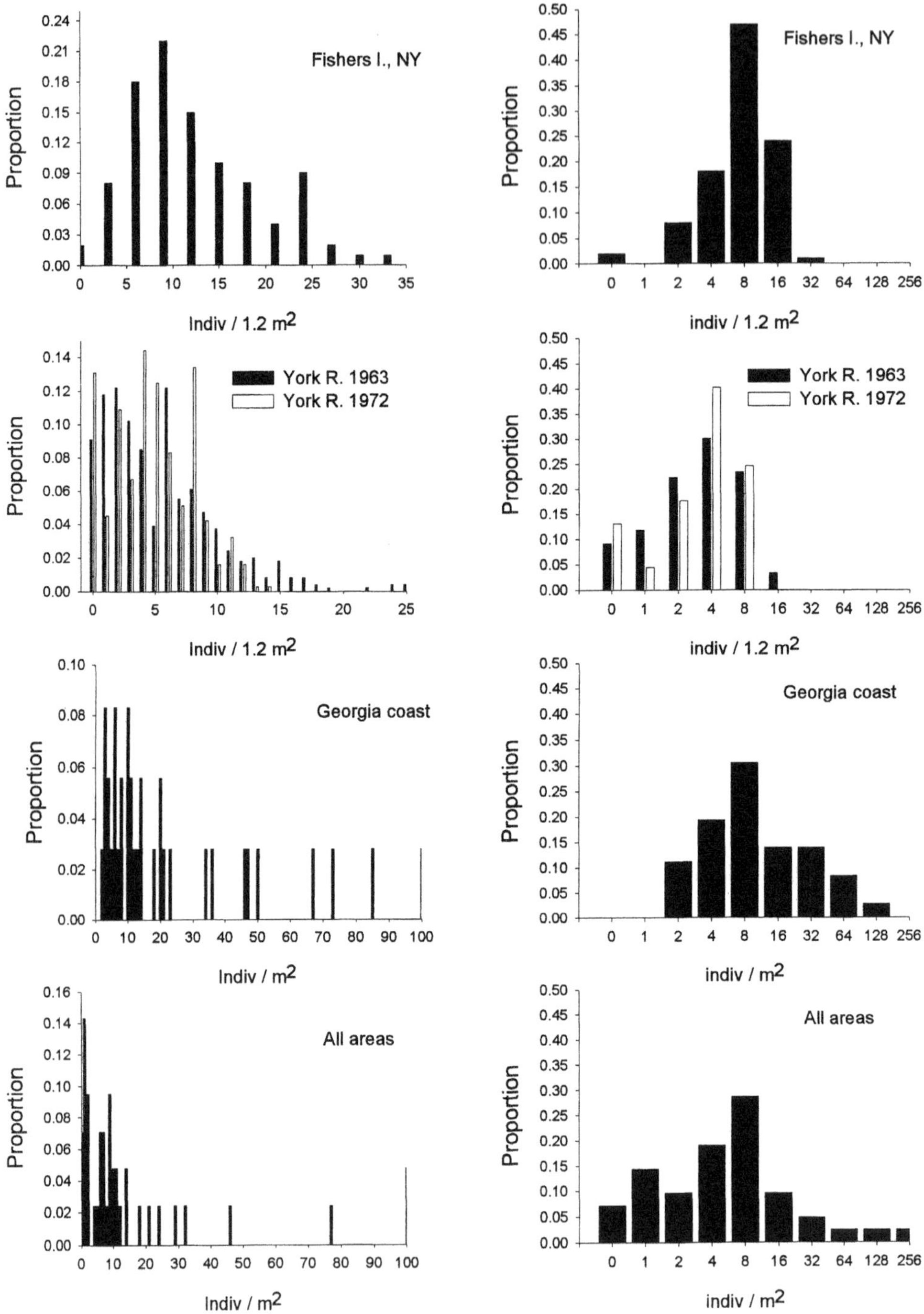
Fishers I., NY
Proportion
Indiv / 1.2 m2
indiv / 1.2 m2
York R. 1963
York R. 1972
Georgia coast
Indiv / m2
indiv / m2
All areas

TABLE 9.1

Methods used by investigators to sample hard clam populations

Author	Method used	Smallest size
1. Beal (1983)	finger plowing and sieving 1 m^2 quadrats	6.4 mm
2. Buckner (1984)	clam shell bucket grabs	6.4 mm
3. Carriker (1961)	various methods	variable
4. Connell (1983)	excavation of 0.5 m^2 area	1.5 mm
5. Craig and Bright (1986)	'treading' 5 m^2 areas	?
6. Crane et al. (1975)	excavation of 1.0 m^2 quadrats	6.4 mm
7. Dame (1979)	excavation of 0.0675 m^2 quadrats	1 mm
8. Dow and Wallace (1955)	excavation of 0.186 m^2 plots	?
9. Greene (1978)	'standard' clam tong grabs	15 mm
10. Hibbert (1976)	excavation of 0.25 m^2 quadrats	10 mm
11. Joseph (1989)	hydraulic clam dredge	30 mm
12. Landry et al. (1993)	suction dredge of 1 m^2 quadrats	?
13. Loesch (1977)	patent tong grabs	?
14. Loesch and Haven (1973)	hydraulic escalator	no sieving
15. MacKenzie (1977)	excavation of 0.3 m^2 quadrats or 0.5 m^2 grabs	4 mm
16. Malinowski (1985)	excavation of 0.33 m^2 quadrats and raking	2 or 24 mm
17. Mitchell (1974)	excavation of 0.25 m^2 quadrats	6 mm
18. Peterson (1982)	finger plowing and sieving 1 m^2 quadrats	1 mm
19. Peterson et al. (1984)	excavation of 0.25 m^2 quadrats	3 mm
20. Peterson et al. (1987)	excavation of 0.25 m^2 quadrats	3 mm
21. Rhodes et al. (1977)	hydraulic escalator commercial catch data	no sieving
22. Rice et al. (1989)	excavation of 0.25 m^2 quadrats	5 mm
23. Richardson and Walker (1991)	0.5 m^2 Baird grabs	?
24. Russell (1972)	rocking-chair dredge	no sieving
25. Saila et al. (1967)	construction bucket grabs	6.4 mm
26. Slattery et al. (1991)	hand digging	no sieving
27. Slattery et al. (1993)	hand digging	no sieving
28. Stickney and Stringer (1957)	clam shell bucket grabs	12 mm
29. Walker (1987)	hand digging 0.44 m^2 quadrats	no sieving
30. Walker and Rawson (1985)	raked 0.3 m^2 quadrats or 0.2 m^2 patent tong grabs	?
31. Walker and Tenore (1984)	raking and excavation of 1.0 m^2 quadrats	5 mm
32. Wells (1957)	clam rakes or oyster tongs in 9.3 m^2 quadrats	20–50 mm

The smallest size individual that could be captured is based on sieve size or dimensions of the capture device used. In each case, the smallest sized clam that could be captured reliably was almost certainly larger. Question marks indicate the author did not mention whether sieving occurred. Numbers preceding the author(s) identify sources in subsequent tables and figures.

Fig. 9.1. Relative frequency distributions of hard clam densities. Distributions on the left side of the figure are repeated to the right using an exponential scale to ease comparisons among the studies. The Fisher's Island distribution comes from a single population in Long Island Sound. The York River, VA distributions come from the same location sampled at two times. The Georgia coast distribution has samples from many different habitats and populations. The final distribution pools the densities recorded in all studies not already presented in the top three distributions. All sources for the data are in Table 9.1.

TABLE 9.2

Population densities of large (at least 0.3 cm in shell length and generally much greater) hard clams from different regions

Location		n	Mean (± 1 SE)	Individual sample	Sediment
Canada	Prince Edward Island[12]		4.6–16.4		sandy and muddy
Maine	Maquoit Bay[8]		856		
Rhode Island	Greenwich Bay[28]		2		mud
	Greenwich Bay[28]		2		sand
	Greenwich Bay[28]		6		silty sand
	Greenwich Bay[28]		12		silty sand and shell
	Narragansett Bay[22]	30	190.4 (28.8)	32–500	sand
	Narragansett Bay[22]	30	77.6 (7.8)	8–184	muddy sand
	Narragansett Bay[22]	30	46 (4.7)	0–100	muddy sand and shell
Connecticut	Long Island Sound[15]		1		
New York	Fisher's Island[16]	80	12.3 (0.8)	0–33	coarse sand, some seagrass
	Northport Harbor[15]		7		
	Great South Bay[15]		18		
	Great South Bay[9]	28	24.1 (3.4)	5–81	
	Great South Bay[2]		5.1–21.4		
New Jersey	Horseshoe Cove[15]		14		
	Little Egg Harbor[3]			<1–538	
Maryland	Chincoteague Bay[32]	23	0.4 (0.6)		mud
	Chincoteague Bay[32]	33	1.0 (0.2)		sand–mud mix
Maryland	Chincoteague Bay[32]	29	1.1 (0.2)		sand
	Chincoteague Bay[32]	8	3.3 (0.8)		shell
Virginia	York River[13]	494	4.5 (0.17)		
	York River[13]	311	3.9 (0.15)		
North Carolina	Bogue Sound[18]	9	0.4 (0.4)		unvegetated fine sand
	Bogue Sound[18]	20	4.2 (1.0)		sparsely grassy fine sand
	Bogue Sound[18]	28	11.3 (1.7)		densely grassy fine sand
	Back Sound[19]	36	1.0 (2.2)–2.0 (3.4)		sand flat
	Back Sound[19]	36	7.3 (5.9)–10.2 (8.7)		seagrass bed
	Back Sound[19]		0.2–14.1		
South Carolina	Santee River[21]		<1–27		
	North Inlet[7]		5–6		
Georgia	several areas[29]	3	11.5 (8.3)–13.5 (18.4)		mud
	several areas[29]	3	<1–67.7 (52.1)		shell/mud
	several areas[29]	3	<1–36.0 (45.0)		shell/sand
	Christmas Creek[29]	3	151.0 (10.2)		mud/shell
	several areas[29]	3	14.0 (10.2)–73.5 (27.2)		shell
	Wassaw Sound[31]	?	36 (6)		
	Wassaw Sound[31]	?	3 (2)		mud
	Wassaw Sound[31]	?	13 (6)		sand
	Wassaw Sound[31]	?	19 (9)		sandy-mud
	Wassaw Sound[31]	?	26 (9)		shelly sediments
	Wassaw Sound[31]	?	31 (6)		among shell on oyster bars
	Wassaw Sound[31]	?	0.53 (0.03)		among oysters on oyster bars
	several areas[30]		<1–91		
Texas [a]	Christmas Bay[5]	45	0.20 (0.07)–3.68 (0.37)	0–5.6	mud mixed with shell
California	Colorado Lagoon[6]		<1–556		

TABLE 9.2 (CONTINUED)

Location		n	Mean (± 1 SE)	Individual sample	Sediment
England	Southampton[10]		6.25–32.58		
	Southampton (<1 year old)[17]		0–202	475	sands to muds
	Southampton (>1 year old)[17]		<1–162		sands to muds

All values are no. of individuals m^{-2}. Standard errors associated with means are presented if any measure of variance was reported by the authors. Sample sizes (n) are presented for studies reporting variation. Sediment type is presented if known. If no sediment information is given, either the author does not specify sediment types for the samples taken or a wide range of sediment types were combined when taking the samples or calculating the means. Superscripts after the locations refer to sources identified in Table 9.1. Unless indicated otherwise, all locations are in the United States.

[a] All individuals collected in this study were *Mercenaria mercenaria texana*.

in some cases, not clear. Many studies have sampled with insufficient attention to changes in sediment structure, depth, and habitat characteristics and consequently confound our interpretation of these factors that could influence local densities. In addition, some studies report mean densities without any indication of variation in density or the total area sampled. This makes it difficult to compare results among different studies and to discern important demographic information, such as within site spatial variability in density. Finally, the great variation in sampling techniques used to assess hard clam abundances (Table 9.1) prevents rigorous comparison of the results among the different studies. Those populations from which smaller clams are captured more efficiently will appear to have higher densities, when they may not.

Experimental studies have begun to reveal the nature of the relationship between sediment type and hard clam abundance. More heterogeneous substrates (those with shell or seagrass present) often support higher densities than nearby homogeneous substrates (Table 9.2, the most notable exception occurs in Narragansett Bay where the highest mean density occurred in the most homogeneous sediments; Rice et al., 1989). Experimental studies have indicated that shell and seagrass root mats decrease the ability of predators to discover and manipulate clams (Castagna and Kraeuter, 1977; Peterson, 1982; Irlandi and Peterson, 1991). Peterson et al. (1995) planted marked, hatchery-reared juvenile clams (16–22 mm in length). They found that survival was significantly higher in sediments with oyster shell hash or seagrasses present than more homogeneous sediments. Beal (1994) who provided evidence that grassbeds do not protect hard clams <5 mm in length. Heterogeneous sediments may produce refuges from predation for clams but other factors may play a role. Lack of information on site-specific settlement limits our ability to make conclusions about the importance of juvenile supply in determining density. Carriker (1961) found that pediveligers preferentially attach to hard substrate when they first settle in the laboratory and he had greatest success in finding byssally attached clams in the field by inspecting shells that were lightly covered with sediments. This suggests that heterogeneous sediments may receive higher initial densities of clam recruits than more homogeneous sediments. Peterson (1986a) examined the relative importances of differential settlement densities and post-settlement survival for clams in grass beds versus unvegetated areas and found evidence that both differential settlement patterns and post-settlement mortality were important in determining higher adult clam densities in grassbeds.

Wells (1957), who also found that shelly sediments where characterized by higher clam densities, detected a positive relationship between hard clam density and rapid current speeds in Chincoteague Bay, Maryland. Locations with characteristically low current speeds (<0.05 m s^{-1}) had mean densities less than 8 individuals m^{-2}. Sites with higher current speeds (0.1–0.5 m s^{-1}) had mean densities over twice as great. In an experimental study, Beal (1983) demonstrated very high mortality rates of hard clams (10–80 mm in length) placed in a high current speed, sandy site. Within 3 days, 100% of the clams were consumed by whelks. Clearly the association between current speeds and clam survival is not clear. Although several studies have demonstrated a positive relationship between hard clam growth and higher current speeds (Kerswill, 1949; Haskin, 1952; Grizzle and Morin, 1989), I have not found other studies reporting a relationship between local current speeds and hard clam density.

Not surprisingly, harvest activity can decrease hard clam density. Peterson et al. (1983, 1987) demonstrated that both hand-operated and mechanical gear could significantly decrease clam abundance in discrete areas. Walker (1987) reports that a 3-day episode of illegal clam harvesting decreased the mean abundance of clams from 90 to 20 m^{-2} in an area approximately 90 m^2.

Several investigators have sampled areas with different harvest histories (Greene, 1978; Buckner, 1984; Rice et al., 1989) or have sampled a specific area before and after harvest (Russell, 1972; Greene, 1978) and generally found decreased abundances where harvest intensity was higher. However, in many cases, harvesting does not always have dramatic effects on population density nor are its effects clear. Greene (1978) estimated hard clam densities at 29 sites located in New York's Great South Bay. The mean (±1 S.E) density across all sites was 24.1 (3.3) clams m^{-2}. Sites in uncertified waters (n = 6) where harvest was excluded (for over three decades in some cases) had a mean (±1 S.E.) of 43.5 (8.6) clams m^{-2} while lightly clammed (n = 5) and heavily clammed (n = 4) sites had mean (±1 S.E.) densities of 13.0 (2.8) and 16.2 (2.2) clams m^{-2}, respectively. Despite a trend of higher clam densities in areas closed to harvest, interpreting these results is difficult. Greene (1978) provides harvest information on only 15 of the 29 sites he sampled. Discussion in the text and examination of figures showing the location of the sample sites suggest that some of the sites for which he does not provide harvest histories were also closed to harvesting. These sites generally have low abundances and would decrease the difference in mean clam densities between unharvested and harvested sites if they were included in the calculations. In addition, Greene (1978) states that many of the sites that are open to harvest are not clammed frequently because natural densities are perennially low even in the presence of low harvest pressure. Working in the same area several years later, Buckner (1984) found higher densities in uncertified waters in each of two consecutive years. In 1978 (n = 349 sites) mean population density in uncertified waters was 21.4 (95% confidence limits of 14.0 and 32.1) clams m^{-2} and in certified waters was 7.9 (95% c.l. of 7.2 and 8.6) clams m^{-2} while in 1979 (n = 354 sites) mean density in uncertified waters was 11.3 (95% c.l. of 8.3 and 15.1) clams m^{-2} and in certified waters was 5.1 (95% c.l. of 4.7 and 5.6) clams m^{-2}. These data appear to show a strong effect due to harvesting on clam abundances. However, the mean percentage decrease in densities from 1978 to 1979 in both uncertified and certified waters was virtually the same (−30.84% and −31.78%, respectively). The absence of an added relative decrease in density where harvest occurred over locations where it did not occur indicates that harvest was less important in determining clam densities between years than some other factor(s). In Georgia, which has a modest commercial hard clam fishery, the range of clam densities in areas experiencing intense harvest is well within the range of clam densities seen in habitats experiencing little to no harvest activity (Walker, 1987). While comparing two adjacent sites in Narragansett Bay, Rice et al. (1989) found the mean density of clams to be over 2× greater (190 individuals m^{-2} versus 78 individuals m^{-2}) at the site closed to harvest. Size frequency distributions showed that the clams occurring at the site open to harvest were predominately smaller than the legal size limit while the great majority of clams at the closed site were above the legal size limit in size. The authors suggest that the differences in density and size structure were a function of differential harvest effort between the two sites. A third site sampled in the study, which was also closed to harvest, showed a size frequency slightly dominated by larger clams but with a population density of 46 individuals m^{-2}, almost half the density seen at the site open to harvest. The third site differed from the other two sites in several environmental characteristics (e.g., more heterogeneous sediments, higher ambient current speeds). This suggests that factors other than harvest intensity can affect clam population densities. More rigorous testing must be done to determine the relative importance of harvest activity in determining hard clam densities and to reveal the interactions of harvest activity with other factors that influence hard clam abundances.

Cohorts, consisting primarily of small (<1 cm in length), juvenile clams, have been discovered infrequently. In part, this is a function of sampling methodology. Few studies have used methods or sieve sizes that capture small clams efficiently (Table 9.1). When juvenile clams have been sampled the densities of individuals can exceed those generally associated with larger adult clams. For example, Carriker (1961) found aggregations of juveniles near shells, pilings, and other structures that protruded above the sediment surface (the two highest densities he cites had 125 and 54 byssal plantigrades m^{-2}). Connell (1983) found over 800 juveniles m^{-2} in Shark River, New Jersey. A density of 475 0- and 1-year old clams m^{-2} was sampled in a shell-gravel ridge in Southampton Water (England) by Mitchell (1974).

Observed field population densities of hard clams lie well below many of the population densities created in experimental or aquacultural settings. Peterson and Beal (1989) increased the density of clams (10–80 mm in length) 8× over mean natural densities in enclosures at several locations in North Carolina. Despite this substantial increase in clam density only modest, and generally non-significant effects were seen in clam growth and either positive or no effects on survival over a 2-year interval. No negative effects of increased density were seen in clam recruitment. In eastern Long Island Sound Malinowski (1985) created densities of 12, 33, 81, and 201 individuals m^{-2} in field enclosures using clams 5–21 mm in length. He found that after 1 year survivorship was not affected by density and only at the highest density was there any significant depression in meat weight. Overall, growth was poor in all treatments, a condition that should have enhanced any competitive interactions associated with clam density. The densities of hard clams used in aquacultural field grow-out to market size typically exceed natural densities by an order of magnitude or more without substantial increases in mortality or decreases in growth rates (Castagna, 1984; Summerson et al., 1995). Clams can show crowding effects on growth. Eldridge et al. (1979) detected reduced growth rates in clams held at very high densities (869 and 1158 individuals m^{-2}) and demonstrated

TABLE 9.3

Proportions of total area surveyed containing the indicated densities of hard clams

Location	Area (km^2)	Density categories							Total no. of individuals $\times 10^6$
		<2	2–10	10–20	>20				
Greenwich Bay, RI[28]	11.2	0.433	0.473	0.061	0.033				50–70
		2.4–12	12.1–23.9	24–35.9	36–60	>60			
Providence River, RF[25]	17.5	0.315	0.204	0.177	0.116	0.091			400–425
		<5	5–10	10–20	20–30	30–40	40–50	>50	
Great South Bay, NY[2]									
1978	60.7	0.456	0.316	0.16	0.054	0.009	0	0.004	410–470
1979	60.7	0.692	0.219	0.072	0.014	0.001	0	0.001	210–290
		none	>2.15	2.15–5.3	>5.3		leased		
Raritan Bay (NJ waters)[11]	82.8	0.213	0.37	0.273	0.144		0		280–290
Shark River, NJ[11]	5.5	0	0.461	0.318	0.066		0		10–20
Lower Barnegat Bay, NJ[11]	84.8	0.039	0.607	0.317	0.036		0		190–200
Little Egg Harbor, NJ[11]	41.5	0	0.466	0.313	0.114		0.107		130–140
Great Bay, NJ[11]	34.1	0.062	0.726	0.088	0.04		0.084		50–60

Total area surveyed is indicated behind each location. All densities are individuals m^{-2}. Proportions and areas were determined by digitizing figures ('distribution maps') found in the respective publications. Approximate population size, calculated as the sum of the products of each density category and its respective area, are presented in the last column. Superscripts after the state abbreviations refer to sources identified in Table 9.1.

later that these clams exhibited compensatory growth when densities were reduced to 290 individuals m^{-2} (Eldridge and Eversole, 1982). The absence of strong intraspecific effects on growth and survival of clams at the artificial densities used in most manipulated systems suggests that intraspecific competition is not a significant factor determining natural population densities in juvenile and adult clams. In most areas, adult hard clam populations are well below local carrying capacities (Malinowski, 1985).

9.2.2 Population Size

Attempts have been made to estimate the actual sizes of *Mercenaria* populations by surveying all areas suspected of having clams. Although potentially useful, surveys frequently suffer from several shortcomings. As an example, in 1983 the State of New Jersey Bureau of Shellfisheries began a survey of commercial shellfish resources occurring in all New Jersey estuarine waters. The project took several years to sample slightly over half of the intended survey area (Saila et al., 1965 and Loesch, 1974 have advocated using sequential sampling to increase efficiency and decrease cost associated with clam resource surveys). By 1989, the survey program was discontinued due to expiration of funds (Joseph, 1989). The failure to complete the survey limits the utility of the data that was collected; no information is available on a large proportion of the resource. Even if the project had been completed, the pace of progress was so slow that comparing information from the beginning of the survey to that at the middle or end of the project would have been confounded by the period of time that elapsed. Finally, surveys frequently sample uniform, regular arrays of sample locations to ease logistics in the field. Taking samples in this manner decreases the range of statistical procedures that can be used confidently to analyze the data (Cochran, 1977), although the non-uniform dispersion of clams in the field make it very unlikely that any bias is introduced by using uniform sampling in this case.

Surveys can provide information on large-scale (several km) spatial patterns of shellfish populations. The New Jersey shellfish survey (Joseph, 1989) found the majority of the bottom consisted of either moderate (2.15–5.3 individuals m^{-2}) or low densities ($0 < 2.15$ individuals m^{-2}) of clams. The high clam or no clam categories were usually much less than 20% of the area (Table 9.3). Several embayments (Shark River, Lower Barnegat Bay, and Little Egg Harbor) had at least some clams in all samples taken. However, the relative areas of different clam densities varied among the embayments. For example, Great Bay consisted largely of low densities. Raritan Bay had the most even distribution of clam densities. Stickney and Stringer (1957) determined the distributions of several dominant infauna in Greenwich Bay, Rhode Island using a 'clamshell' construction bucket (essentially a large bottom grab). Over 43% of the bottom consisted of hard clam densities <2 m^{-2} (Table 9.3). Only about 10% of the bottom had high densities (10 m^{-2}) of clams. Highest abundances were correlated with very heterogeneous sandy sediments containing shell. Saila et al. (1967) surveyed approximately 17.5 km^2 of bottom in the Providence River, Rhode Island. They found much higher densities of clams covering much larger areas than occurred in New Jersey or Greenwich Bay (Table 9.3). Saila et al. (1967) relate the distribution of hard clam densities to a range of environmental factors. Although they found some strong correlations among sediment size (>2 mm in diameter) and organic carbon content with hard clam density it is not clear what direct role, if any, these factors play in determining the distribution of clams in the Providence River. Finally, Buckner (1984) provides clam survey information from Great South Bay, New York between the town of Babylon and Nicoll Pt. in two successive years. The general patterns he found are similar to the other studies. Most of the bottom has no or low clam densities and higher densities of clams tend to be aggregated. However, comparing the results of the two successive years reveals large changes in the distribution of hard clam densities (Table 9.3). These changes occurred even in areas that were not subject to harvesting. Sampling error probably accounts for some of this change. Even though his surveys have >300 samples, distances between the individual samples are 10s to 100s to 1000s of meters apart. Consequently, these surveys were probably not sampling intensively enough to provide reliable estimates for patterns of spatial distribution over the large area and with the level of spatial variability present.

Indeed, some of the survey studies (Stickney and Stringer, 1957; Saila et al., 1967) found that the highest density strata of clams were usually surrounded by the second highest density strata of clams which were in turn surrounded by the third highest density strata, etc. creating successive 'halos' of decreasing clam density. It is not clear whether this pattern of nested densities is real (it would be expected if the distributions were contagious at this scale) or results from artifacts such as: (1) the resolution of the sampling grid; or (2) how the strata are created. Not all survey studies found similar spatial patterns (Buckner, 1984; Joseph, 1989). Greater spatial resolution of hard clam areal distribution will be required to determine whether density decreases change abruptly or not in these areas.

Surveys can also be used to estimate total clam population size by summing the products of clam density strata and their respective areas. This assumes: (1) the average density within each designated area equals the midpoint of the category range for that area; (2) the areas are estimated well; and (3) the regular sampling arrays have not produced biased data. I provide estimates of the ranges of hard clam population size for each survey mentioned above (Table 9.3). The estimates range from 10×10^6 clams to over 400×10^6 clams.

Attempts have also been made to assess population size using modifications of the Leslie–DeLury method (Leslie and Davis, 1939; DeLury, 1947). The method depends on observing the extinction of catch size with continued effort and requires that a substantial proportion of the population be removed before a reasonably accurate estimate can be made (Ricker, 1958). The Leslie–DeLury method uses less time than surveys but provides less information because the mechanical harvesters, which are generally used, homogenize much of the spatial information on clam distribution. In addition, the interaction of mechanical gear with changes in the nature of the substrate and with different-sized clams introduces error into the estimations because the catch efficiency of the gear varies with substrate type (Loesch and Haven, 1973). Finally, this method has obvious disadvantages if the resource being estimated is intended for subsequent commercial harvest. For example, using an hydraulic escalator dredge, Loesch and Haven (1973) produced reasonably accurate estimates of clam abundances in areas that were subsequently completely harvested to determine the actual amount of clams present (their estimates ranged from 0 to 7.6% of the amounts of clams subsequently harvested). Although they do not report the proportion of clams harvested while they were sampling, the time spent sampling ranged from 25 to 82% of the time required to completely harvest the area. Hence, the Leslie–DeLury method information about what was present in the bottom and can best be used only as an after-harvest assessment of hard clam population sizes if accurate records are kept during the harvest season.

Rhodes et al. (1977) used harvest and effort data from a hydraulic escalator fishery to estimate hard clam abundances in several coastal South Carolina embayments. In an area of ~104 ha in the South Santee River they estimated that 6,442,042 harvestable-sized clams (equals 6.2 clams m^{-2}) were present (before harvesting). In two locations in the North Santee River, one of 53 ha and one of 159 ha, they estimated 5,007,000 and 10,685,417 harvestable-sized clams, respectively (equals 9.5 and 6.7 clams m^{-2}, respectively).

In the lower West Passage of Narragansett Bay, RI, Russell (1972) used both industry catch per unit effort data and dredge samples collected in stratified random surveys conducted before and after harvest to provide independent estimates of the size of the local hard clam population. The independent estimates of the size of the harvestable population (with 95% confidence ranges) were 18,147 (12,444–23,852) bu from the before-harvest sampling and 21,880 (18,170–28,510) bu based on the fishery catch data. Russell concludes that stratified sampling provided a reasonable and efficient means of estimating the size of the catchable population (the actual harvest fell within the confidence limits of his survey). The survey technique may have been better than he realized. The estimated population size (and 95% confidence range) from the after-harvest sampling was 7,235 (5,068–9,402) bu. The difference between the before- and after-harvest estimates from the stratified random samplings is 10,912 bu, which is strikingly close to the 10,925 bu that the fishery harvested from that area in the interval between the two surveys. It is unfortunate that information on the means and variances of the number of clams in their bushels was not gathered. Without such information, we cannot extract information about the actual size, density, and age-structure of the population.

Despite the differences in approach, it seems clear that hard clam populations typically number in the 10s of millions of individuals in many coastal embayments. This is true even in regions that perennially experience harvest activity.

9.2.3 Population Dispersion

Hard clam populations display aggregated dispersion patterns over spatial scales extending from less than a meter to 100s of meters. Frequency distributions of hard clam densities are generally best fit by either log-normal or negative binomial distributions, indicating that the original samples come from aggregated dispersions (Loesch, 1977; Buckner, 1984). Variance to mean ratios of clam density typically are much greater than one (Fig. 9.2). At large scales (100s of m), clams also occur in aggregated patterns (Wells, 1957; Saila et al., 1967; Russell, 1972; Crane et al., 1975; Joseph, 1989).

The aggregated dispersion of clams has important implications, not only for selecting sampling methodologies that should incorporate this fact in their design, but also for understanding the biology of clams. Recent studies have demonstrated the importance of relatively close proximity of adults to successful reproduction among some marine species that spawn in the water column (Levitan et al., 1992). Under conditions where large numbers of widely dispersed spawners occur, lower fertilization rates are likely. Given the limited mobility of adult clams (see below) and the probable tendency for harvest practices to decrease the number and sizes of aggregations of hard clams, reduced spawning efficiency could be an important factor affecting recruitment of hard clam populations.

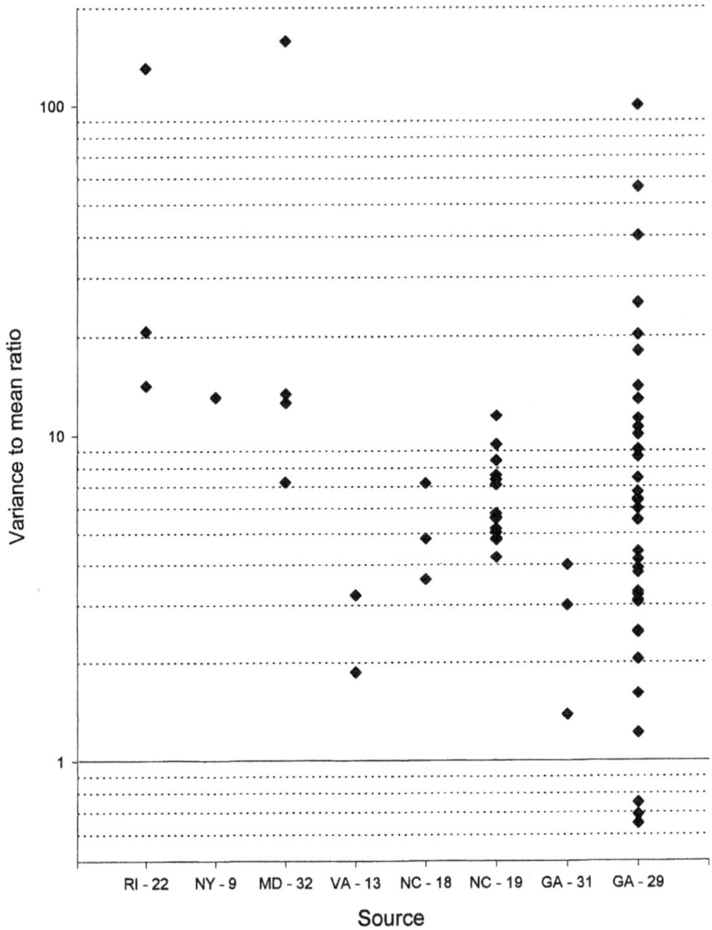

Fig. 9.2. Variance to mean ratios of hard clam densities. The state in which the samples were taken are indicated on the abscissa along with the numbers indicating the source of the data (as used in Table 9.1).

9.2.4 Temporal Changes in Abundance

Few data exist of within- and between-year temporal changes in hard clam abundances at specific locations. Commercial landings data have been used as indicators of change in the harvestable population. Unfortunately, using landings data as more than a very general indicator of hard clam population dynamics is confounded by several factors. First, landings are subject to variable effort between regions and between years within a region. Changes in the local per unit price of clams, availability of competing fisheries (in some areas, such as North Carolina, many clammers participate in several coastal fisheries, personal observation), and other seasonal employment can change the number of harvesters. Landings data in Georgia do not reflect very much information about hard clam populations there because the local fishery has very few participants; only a very small fraction of the available clam population receives any harvest effort (Walker, 1987). Second, management determines the sizes of clams that may be harvested and regulates where clamming may occur. Size regulations are not constant between states and, with respect to clamflat closures, are not constant within states. As the area of harvestable waters increases or decreases due to changes in water quality, the proportion of the hard clam population subject to harvest changes in response. Comparing landings data between states or within states at different times is complicated by these factors. Third, under reporting of landings is apparently frequent and widespread (McCay, 1988). Tax revenues and fees are generally assessed as a function of the size of individual harvests which produces a financial incentive for the harvester (and the shipper) to engage in inaccurate reporting. Fourth, estimates of recreational landings are rare. Fifth, clammers do not sample for clams, they hunt them. They spend most of their time where clams are abundant and avoid areas where clams are scarce or hard to collect. Areas that have moderate abundances of large clams (>8 cm in length) will not be visited frequently by clammers because of the very low market value of these clams. Sixth, landings are generally measured in total volume or mass harvested; information is lacking on size structure and relative abundances of different age groups in the catch. Seventh, by design, landings data do not provide information on the abundances of clams below legal sizes. The absence of information on the abundance of small clams prevents any use of landings data in projecting clam population sizes in the future. Finally, landings are reported for relatively large regions (states or counties usually), thereby pooling information from many separate locations within the region. This prevents detecting variations in abundance at spatial and temporal levels that correspond to the scales of processes determining local population success (e.g., spawning, age-specific survival, settlement, etc.). In conclusion, landings data can provide only the crudest indication of actual clam population changes and then only when independent information is available over the same time interval on some of the factors mentioned above (e.g., proportion of flats available, actual numbers of harvesters). At best, we may be able to discern order of magnitude, long-term changes in hard clam populations from landings data.

No consistent pattern exists in landings among the four states with the historically most productive hard clam fisheries (Fig. 9.3). Since the 1940s, New York and Rhode Island landings have demonstrated large amplitude (and nearly complementary) fluctuations. New Jersey and Massachusetts landings declined generally over the same interval with relatively small, short-term increases and relatively long-term periods of nearly constant annual production. With few exceptions, annual decreases and increases in landings are very smooth; the state fisheries (and their respective clam populations?) do not experience dramatic collapses or expansions in size. It would be informative and fascinating to have independent, quantitative estimates of hard clam abundances in any of these regions over the same time intervals.

A few data sets exist that sample hard clam populations sequentially in order to determine the importance of harvesting to local clam abundance. Comparison of separate patent tong surveys indicates that the hard clam population in the York River, VA was less dense and more dispersed in 1972 than in 1963 (Fig. 9.1; Loesch, 1977). Given the absence of any other information on the clam population before and after these samplings (and the long interval of time between samplings) we cannot conclude that the changes were a function solely of harvest over natural processes. Greene (1978) sampled near Water Island, NY before (May) and after (October) the local harvest season. The October sample had less than half as many clams and the greatest relative decrease occurred in the number of clams in the 35–40 mm

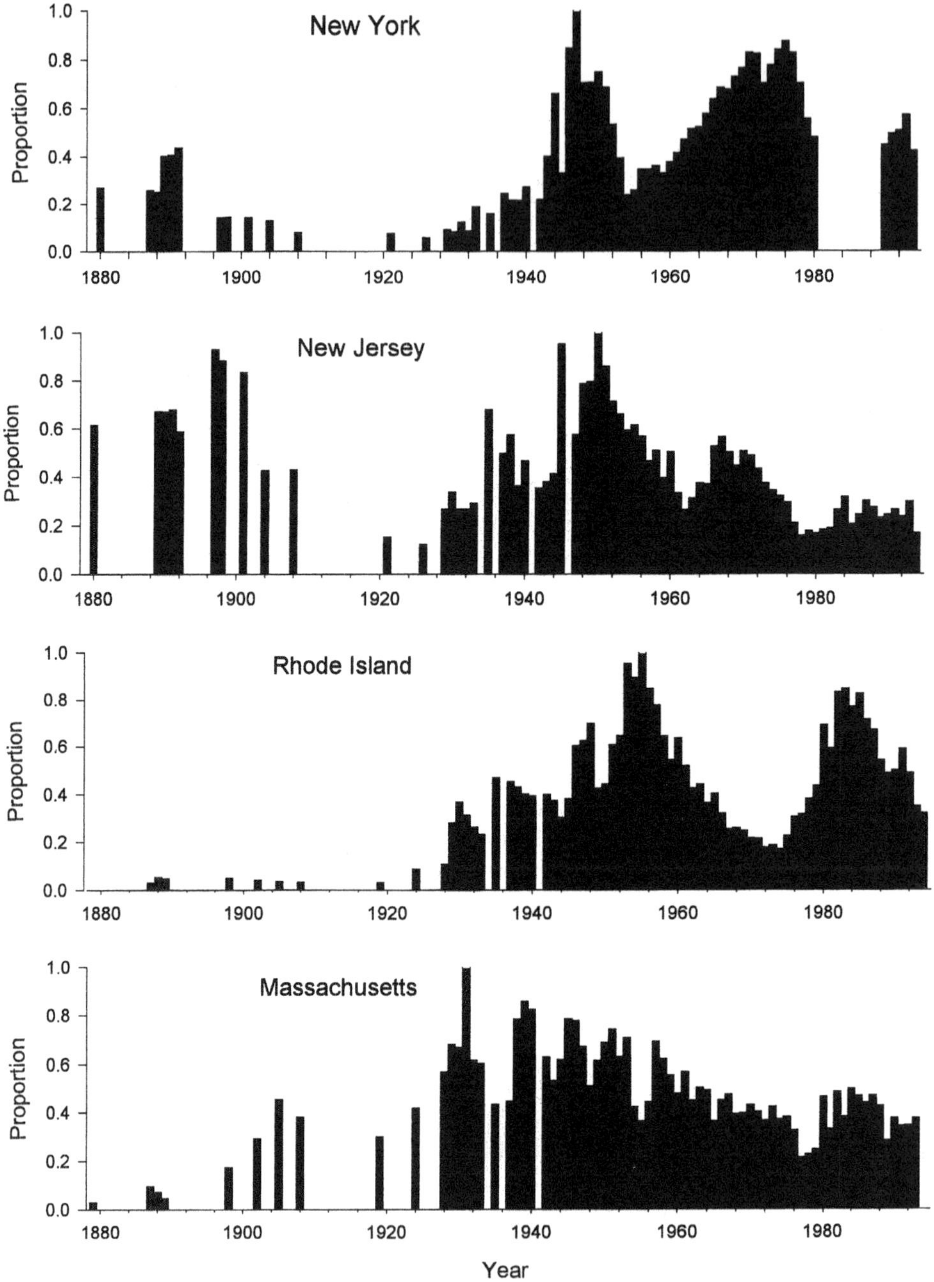

Fig. 9.3. Annual hard clam landings from 1879 to 1994 from the historically most productive states. To ease comparison of patterns among states, the landings are presented as a proportion of the maximum annual harvest reported for each state. The year and mass of maximum harvest for each state are: New York, 1947, 4,686.1 metric tons; New Jersey, 1950, 2306.5 metric tons; Rhode Island, 1955, 2277.0 metric tons; Massachusetts, 1931, 1328.1 metric tons. The absence of a bar indicates that data are not available for that year (Lyles, 1969; National Marine Fisheries Service, 1996).

length range. A significant decrease in mean abundance (measured in number of 80 lb bushels collected with a dredge) is apparent in the data collected by Russell (1972) before and after commercial harvest in Narragansett Bay (mean (SD) = 1.80 (0.89) bushels tow^{-1} before versus 0.54 (0.28) bushels tow^{-1} after harvest: t-test is significant; $P = 0.007$). It is important to note that Russell did not sample juvenile clams. We know what changes occurred in the harvestable component of the clam population, but we do not know what changes occurred in the population as a whole.

9.3 AGE AND SIZE STRUCTURE

9.3.1 Size Frequency Distributions

Hard clam populations exhibit wide ranges of different-sized individuals (Table 9.4 and Fig. 9.4). Many clam populations north of Georgia are dominated by large individuals (12 of 30 means, 11 of 30 medians, and 14 of 30 modes are equal to or greater than 6 cm in length; Table 9.4). Of the 26 size frequency distributions available from clam populations in Georgia waters, 16 of them have a median size greater than or equal to 7 cm in length. Few of these Georgia populations experience harvesting (Walker, 1987). North Carolina and Prince Edward Island clam populations showed even size frequencies or greater frequencies of smaller clams (3 cm in length or less: Table 9.4 and Fig. 9.4). Most of the size frequency distributions are unimodal and, if significant skews are present, are skewed to the left (22 of 28 instances where skew was significant it is negative) indicating that populations are generally dominated by large individuals. Kennish (1978) found similar results for (height-based) size frequency distributions of both living and dead (valve) assemblages of hard clams at several sites in Barnegat Bay, NJ. It is important to remember, however, that most of the sampling techniques used to collect clams in these studies did not collect clams less than 2 cm in length with consistency, if at all.

Comparing size frequency distributions among different studies is limited because the various investigators have used different sampling methods with varying capture efficiencies (of small clams especially) and because the data are often presented using different size class intervals. However, consistent characteristic size frequency distributions associated with specific habitat types (mud or sand, presence or absence of grass or shell) or regions do not appear except in areas with differential harvest activity. Greene (1978), Buckner (1984), Rice et al. (1989), and Walker (1989) argue that populations experiencing little or no harvest activity will tend to have greater proportions of large clams than populations that are harvested frequently. They reason as follows, individual hard clams: (1) can live for several decades (see Chapter 3; however, examination of dead assemblages of clams in New Jersey showed that the majority of clams live less than 9 years, Kennish, 1980); (2) display asymptotic growth while attaining most of their size in the first few years of life (see Chapter 5); and (3) experience reduced predation rates due to a size refuge (see Chapter 7). If recruitment into adult size classes is infrequently successful (an implicit assumption in their argument), an accumulation of large individuals within an area would be expected in the absence of harvest activity. In contrast, Malinowski (1985) suggests that clam recruitment is consistent from year to year, but only with very low numbers of individuals entering adult stages in each year. Where harvest occurs, large individuals are frequently at greater risk than small to medium-sized individuals

TABLE 9.4

Descriptive statistics of *Mercenaria mercenaria* population size frequency distributions (based on shell lengths)

Location	Habitat	Size class				Mean	SD	Skewness	Kurtosis	*n*
		Interval (mm)	Range	Mode	Median					
PEI, Canada										
West R.[12]	mud	10	5–85	25.0	45.0	43.3	22.0	ns		46
Hillsborough R.[12]	mud	10	5–85	50.0	35.0	50.0	23.4	ns	ns	58
Pownal Bay[12]	sand	10	5–65	15.0	5.0	21.6	17.4	+ [b]	− a	164
Massachusetts										
Nauset Marsh[26]	sand	10	15–85	55.0	55.0	52.2	11.6	− [b]	ns	106
	grass	10	25–95	65.0	65.0	61.3	16.0	ns	ns	106
Rhode Island										
Greenwich Cove[22]	sand	3	5–98	62.0	62.0	62.0	16.2	− [c]	+ [c]	1426
Greenwich Bay[22]	sand (open to harvest)	3	5–74	14.0	26.0	30.4	14.5	+ [a]	− [c]	590
	sand (closed to harvest)	3	5–98	80.0	62.0	59.5	21.2	− [c]	ns	349
New York										
Fisher's Island[16]	sand	10	30–110	80.0	80.0	76.2	15.6	− [c]	+ [c]	328
Great South Bay[9]	uncertified	2	26–94	66.0	60.0	59.9	10.6	− [c]	ns	1354
	light clamming intensity	2	30–104	70.0	74.0	71.8	12.7	− [c]	ns	939
	heavy clamming intensity	2	12–78	52.0	48.0	46.6	11.1	− [c]	+ [c]	921
Great South Bay[2]	certified 1978	1.85	6.4–106.6	49.1	49.1	51.1	19.7	+ [c]	− [c]	4497
	uncertified 1978	1.85	5.9–92	62.1	52.7	47.2	20.0	− [c]	− [c]	1127
	certified 1979	1.85	3.8–107.8	45.4	52.9	54.3	18.7	ns	− [c]	2858
	uncertified 1979	1.85	8.4–95.7	44.4	53.9	52.3	16.1	− [c]	ns	913
Great South Bay[15]	fine sand	5	5–60	45.0	45.0	37.3	17.7	− [b]	ns	50
New Jersey										
Horseshoe Cove[15]	sand and 'stones'	5	5–100	5.0	35.0	31.8	24.7	+ [a]	ns	113
Shark R.[27], 1986	sand	10	35–95	65.0	65.0	65.7	14.0	ns	ns	91
Shark R.[27], 1987	sand	10	35–95	65.0	65.0	68.7	12.4	ns	ns	98
Shark R.[27], 1988	sand	10	15–95	65.0	65.0	62.0	15.7	− [c]	+ [b]	181
Marshelder I.[26]	sand	10	35–105	75.0	75.0	76.4	11.1	− [b]	ns	112
	grass	10	45–105	75.0	75.0	74.3	11.1	− [a]	ns	101
North Carolina										
Bogue Sd.[18], 10/77	fine sand, grass	20	10–110	30.0	50.0	49.3	25.1	+ [c]	− [a]	405
Bogue Sd.[18], 6/78	fine sand, grass	20	10–110	30.0	50.0	49.1	26.4	+ [a]	ns	87
Bogue Sd.[18], 11/78	fine sand, grass	20	10–110	30.0	50.0	49.0	25.7	ns	ns	77
Back Sound[19]	sand	10	5–95	5, 15	35.0	37.7	27.3	ns	ns	85
	grass	10	5–95	65.0	55.0	50.7	22.3	ns	− [c]	488

TABLE 9.4 (CONTINUED)

Location	Habitat	Size class				Mean	SD	Skewness	Kurtosis	n
		Interval (mm)	Range	Mode	Median					
Shackleford Bank [26]	sand	10	15–75	45.0	45.0	43.4	11.4	ns	ns	107
	grass	10	25–95	65.0	55.0	55.8	17.4	ns	ns	107
Georgia										
Several areas[31]	intertidal flats	17.5	9.5–69.5	39.5	39.5	37.7	16.5	ns	ns	131
	creek bottoms	17.5	24.5–95	69.5	69.5	67.0	13.5	–[b]	ns	222
North Cabbage[31]	shell	5	3–73	43.0	38.0	38.4	17.0	ns	ns	131
Wassaw I.[31]	shell	5	23–88	78.0	73.0	68.2	13.0	–[c]	+[b]	222
Cabbage I.[29]	intertidal shell/sand	5	17.5–97.5	57.5	55.0	53.6	15.6	ns	ns	122
Cabbage I.[29]	creek mud	5	77.5–112.5	102.5	102.5	97.0	10.0	ns	ns	33
Wassaw I.[29]	feeder crk, shell/mud	5	27.5–112.5	72.5	72.5	71.6	15.8	ns	+[a]	290
Cabbage I.[29]	creek mud	5	77.5–112.5	102.5	102.5	97.0	10.0	ns	ns	33
Wassaw I.[29]	feeder crk, shell/mud	5	27.5–112.5	72.5	72.5	71.6	15.8	ns	+[a]	290
Skidaway I.[29]	feeder crk, shell/mud	5	57.5–107.5	77.5	82.5	82.6	8.6	ns	ns	47
Little Tybee I.[31]	sand	5	33–88	78.0	73.0	70.8	13.0	–[b]	ns	100
Little Tybee I.[29]	feeder crk, sand/shell	5	22.5–92.5	72.5	67.5	64.6	12.3	–[c]	ns	218
	creek shell	5	42.5–97.5	62.5	62.5	63.3	13.7	ns	ns	100
	creek shell/sand	5	67.5–107.5	77.5	82.5	82.9	9.7	ns	ns	68
Little Tybee I.[29]	creek shell/mud	5	67.5–102.5	82.5	82.5	85.2	10.6	ns	ns	39
St. Catherine's I.[29]	creek shell/mud	5	37.5–82.5	62.5	60.0	58.6	11.5	ns	ns	54
	creek mud	5	42.5–92.5	72.5	72.5	70.8	10.2	ns		49
	intertidal shell/sand	5	82.5–102.5	97.5	97.5	94.8	6.1			11
Sapelo Sound[29]	creek shell/mud	5	32.5–97.5	77.5	77.5	72.2	13.2	–[c]	+[c]	224
	feeder crk shell/mud	5	17.5–102.5	77.5	67.5	67.2	15.1	–[c]	ns	216
	intertidal flat shell	5	47.5–67.5	57.5	57.5	57.0	5.5	ns		31
	intertidal shell/mud	5	37.5–72.5	57.5	57.5	56.1	7.6	ns		43
Christmas Creek[29]	feeder crk shell/mud	5	42.5–102.5	82.5	77.5	76.2	13.1	ns	ns	173
	feeder creek shell	5	37.5–92.5	82.5	77.5	75.6	11.5	–[b]	ns	93
	feeder crk mud/shell	5	57.5–97.5	82.5	82.5	81.6	7.4	–[b]	ns	132
	feeder crk shell to mud	5	47.5–97.5	72.5	72.5	71.4	8.8	ns	ns	100
Cumberland I.[29]	intertidal sand/grass	5	32.5–82.5	67.5	57.5	58.5	12.7	ns		35
Crooked River[29]	sand	5	62.5–107.5	87.5	92.5	89.4	8.2	–[b]	ns	59

The significance of skewness and kurtosis was determined following the procedure in Sokal and Rohlf (1995) for distributions with $n > 150$ and following the procedure in Snedecor and Cochran (1967) for distributions based on $n < 150$. A blank cell occurs where n (number of individuals measured) was too small to produce a testable estimate of skewness or kurtosis. ns = not significant. Superscripts after the locations refer to sources identified in Table 9.1.
[a] $P < 0.05$. [b] $P < 0.01$. [c] $P < 0.001$.

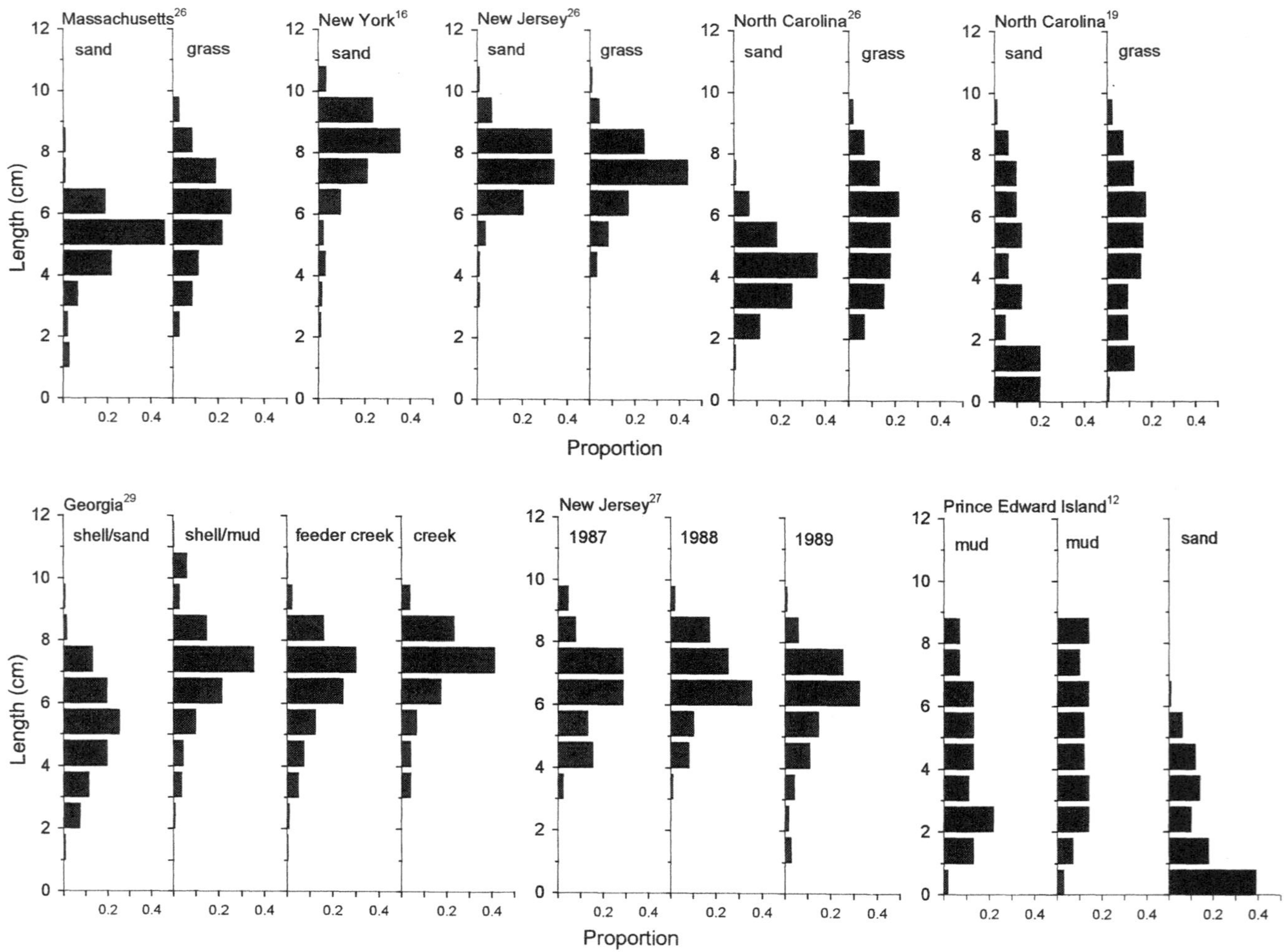

Fig. 9.4. Selected size frequency distributions of hard clam populations from several different locations and habitats. Sources are indicated by the superscripts which refer to Table 9.1.

because most hard clam harvest methods are biased towards discovery of larger individuals. Harvested populations would be expected to experience higher removal rates of large clams, despite their lower market value, and experience a shift in the relative abundance of sizes towards smaller, medium sized individuals.

Non-harvest factors can lead to shifts in population size structure. MacKenzie (1977), using information on size frequency distributions of live clams and dead valves, found few individuals greater than 65 mm in length, presumably due to harvesting. However, he also estimated the relative abundances of predators and found evidence suggesting that small clam predators were more numerous than those of large clams. The absence of small clams was assumed to be a consequence of the relatively high numbers of small clam predators. In a field study, Peterson (1982) found a shift towards smaller size classes of clams in experimental plots exposed to predation by whelks (*Busycon* sp.) versus plots in grassbeds (*Halodule wrightii*) where whelk predation was ineffective. The sizes of the clams consumed by the whelks were well within the range of commercial harvest sizes (4–10 cm in length). In a subsequent North Carolina study, examination of clam size frequency distributions between vegetated (*Zostera marina*) and unvegetated sites also showed greater proportions of larger clams in the grassbed (Peterson et al., 1984, Fig. 9.4). This pattern is not consistent across regions. In grassbeds in Nauset Marsh, MA and Marshelder Island, NJ larger clams are not disproportionately abundant compared to those occurring in nearby unvegetated sediments (Slattery et al., 1991) suggesting that different sediment–predator interactions occur in different areas of *Mercenaria*'s range.

Beal (1983) sampled a range of unvegetated substrates in North Carolina and found that mud sites were dominated by large clams, sand sites were highly variable, and a shelly site (located near the mud sites) consisted mostly of smaller clams (Table 9.5). Although the numbers of individual clams and replicates are small, these results indicate those differential patterns of recruitment, survival, and growth may occur in different substrates. It would be interesting to see the results of a single study following recruitment, survival, and growth conducted across an array of sediment types in several different regions within *Mercenaria*'s range. Sediment type–hydrodynamic interactions have been shown to have effects on hard

TABLE 9.5

Size frequency distributions of hard clams (>6.4 mm in length) collected from several different habitats in the Cape Lookout region of North Carolina (Beal, 1983)

Size class (mm)	Mud		Sand			Shell
	Intertidal	Subtidal	Site a	Site b	Site c	
≤29.9	0.0	0.05	0.60	0.30	0.17	0.36
30.0–49.9	0.14	0.05	0.20	0.25	0.03	0.26
50.0–69.9	0.14	0.35	0.00	0.40	0.41	0.27
≥70.0	0.71	0.55	0.20	0.05	0.38	0.11
Total no. of individuals	7	20	5	20	29	249
Individuals m^{-2}	0.6	1.7	0.4	1.7	2.4	20.8

The three sand sites occurred at (a) Cape Carteret, (b) North River, and (c) Core Sound, NC. A fourth sand site in Middle Marsh, NC contained no hard clams. The mud and shell sites were located in Middle Marsh.

clam growth (Ashley and Grizzle, 1988). Experiments looking at several factors would help explain much of the variation that can be seen in the results of studies exploring clam–sediment interactions from different locations.

9.3.2 Age Frequency Distributions

Ages of clams can be determined by direct examination of annual growth lines (Chapter 2 and references there), detection of year class 'peaks' in size frequency distributions (Greene, 1978; Cerrato, 1980), or indirectly from growth models (Buckner, 1984). As both environmental and genetic factors affect growth of same-aged individuals (Chapter 5) there is greater confidence in age frequency distributions developed via direct age determination. Several studies support this conclusion. Peterson et al. (1984) demonstrated age-specific differences in growth rates between habitats. A large amount of variation in age of similar-sized clams was detected by Rice et al. (1989) for clams collected in the same habitat. Buckner (1984) found substantial overlap in size ranges of clams of different ages. Walker and Tenore (1984) were unable to confidently identify cohorts of clams from size frequency distributions alone; there was too much overlap of sizes among *Mercenaria* of very different ages. Finally, Ansell (1968) concluded from his extensive review of individual growth in *Mercenaria* that variation in growth within regions was high, throughout its range. Because of the ambiguity in interpretation introduced by variability in size at age observed in *Mercenaria*, I will focus only on age frequency distributions determined directly from sectioned shells.

Age frequency distributions show great variety among sites within regions and between regions (Table 9.6, Fig. 9.5). Some of the observed differences between regions and habitats results from the different sampling methods used to collect the clams. However, some differences exist even where sampling methodology does not confound the results.

Peterson et al. (1984) found a significant difference in age–frequency distributions between clams collected in a North Carolina *Zostera marina* grassbed and a nearby sandflat (Fig. 9.5). In both habitats, the mean age did not differ and both frequency distributions were dominated by clams less than 4 years old. The significant difference occurred in the relative frequency of clams less than 1 year old. Only 22% of the clams in the grassbed were less than 1 year old while 40% of the sandflat clams were less than 1 year old. Slattery et al. (1991) found a pattern similar to Peterson et al. (1984) when comparing clam age–frequency distributions between grass and sandflat habitats in North Carolina but found very different patterns in age structure between grass and sandflat habitats in Massachusetts and New Jersey. In North Carolina (in the same general region sampled by Peterson et al. (1984) but 7 years later), clam populations in both habitats were dominated by clams less than 5 years old. Between habitats, younger clams (1 and 2 years old) were relatively more abundant in the sandflat and older clams (3 years old and greater) were more abundant in the grassbed. In New Jersey, the overall distributions were similar between habitats, but differed from North Carolina in that the majority of clams were between 6 and 12 years old. The between habitat difference was small as sandflat clams were marginally older than grassbed clams. In Massachusetts, clams collected from the grassbed tended to be older than sandflat clams, but the shapes of the age–frequency distributions between the habitats are very different. The age–frequency distribution from the grassbed is bimodal with one mode occurring around 4- and 5-year-old clams and the second, and larger mode occurring around 9–12-year-old clams. In the sandflat,

TABLE 9.6

Descriptive statistics of *Mercenaria mercenaria* population age frequency distributions

Location	Habitat	Age class			Mean	SD	Skewness	Kurtosis	n
		Range	Mode	Median					
Massachusetts									
Nauset Marsh[27]	sand	0–14	4	4	4.6	2.4	$+^{b}$	$+^{b}$	106
	grass	1–14	11	9	8.2	3.7	ns	ns	107
New York									
Fisher's Island[16]	sand	1–15	11	11	10.3	3.8	$-^{a}$	ns	106
New Jersey									
Shark R.[27] 1986	sand	2–18	6	6	6.5	3.5	$+^{b}$	ns	90
Shark R.[27] 1987	sand	1–18	9	8	8.7	4.3	$+^{b}$	ns	96
Shark R.[27] 1988	sand	0–18	8	7	6.9	3.5	$+^{b}$	ns	181
Marshelder I.[26]	sand	2–20	9	9	10.3	4.2	$+^{b}$	ns	102
	grass	2–20	8	8	9.7	4.7	$+^{b}$	ns	107
North Carolina									
Shackleford Bank[26]	sand	0–11	1	2	2.3	2.0	$+^{b}$	$+^{b}$	107
	grass	0–13	1	3	3.8	2.6	$+^{b}$	ns	108
Georgia									
Cabbage I.[29]	intertidal shell/sand	1–9	3	4	4.0	2.0	$+^{b}$	ns	124
	creek mud	4–40	24	26	25.3	8.9	ns		37
Wassaw I.[29]	feeder crk, shell/mud	1–30	4	8	8.2	4.8	$+^{b}$	ns	222
Skidaway I.[29]	feeder crk, shell/mud	6–27	14	14	13.6	4.2	$+^{b}$		45
Little Tybee I.[29]	feeder crk, sand/shell	1–34	4	5	6.5	5.0	$+^{b}$	$+^{b}$	216
	creek shell	1–14	4	4	4.4	2.5	$+^{b}$	ns	100
	creek shell/sand	3–23	8	9	10.8	5.3	$+^{b}$	ns	68
	creek shell/mud	4–35	14	13	13.7	6.9	$+^{b}$		39
St. Catherine's I.[29]	creek shell/mud	2–19	3	4	5.0	3.6	$+^{b}$	$+^{b}$	54
	creek mud	2–13	7	7	6.6	2.1	ns	ns	53
	intertidal shell/sand	9–32	18	20	20.3	6.2			11
Sapelo Sound[29]	creek shell/mud	1–36	5	9	11.3	7.8	$+^{b}$	ns	223
	feeder crk shell/mud	0–36	2	5	9.0	8.8	$+^{b}$	ns	214
	intertidal flat shell	3–38	6	15	17.6	11.3	ns		28
	intertidal shell/mud	1–7	2	3	3.2	1.2	$+^{b}$		43
Christmas Creek[29]	feeder crk shell/mud	1–36	4	13	13.2	8.8	$+^{b}$	ns	175
	feeder creek shell	1–33	23	19.5	17.3	8.1	$-^{a}$	ns	94
	feeder crk mud/shell	2–35	5	11	11.7	7.2	$+^{a}$	ns	99
	feeder crk shell to mud	6–35	19	22	22.4	5.1	ns	ns	133
Cumberland I.[29]	intertidal sand/grass	1–18	5	5	6.9	4.5	$+^{a}$		35
Crooked River[29]	river sand	7–28	10	14	13.9	4.6	ns	ns	58
England									
Southampton[23], 1983		4–18	6	9	8.9	3.3	$+^{a}$	ns	94
Southampton[23], 1988		3–22	3	11	10.1	6.9	ns		40

The significance of skewness and kurtosis was determined following the procedure in Sokal and Rohlf (1995) for distributions with $n > 150$ and following the procedure in Snedecor and Cochran (1967) for distributions based on $n < 150$. A blank cell occurs where n (number of individuals aged) was too small to produce a testable estimate of skewness or kurtosis. ns = not significant. Superscripts after the locations refer to sources identified in Table 9.1.
^{a} $P < 0.05$. ^{b} $P < 0.01$.

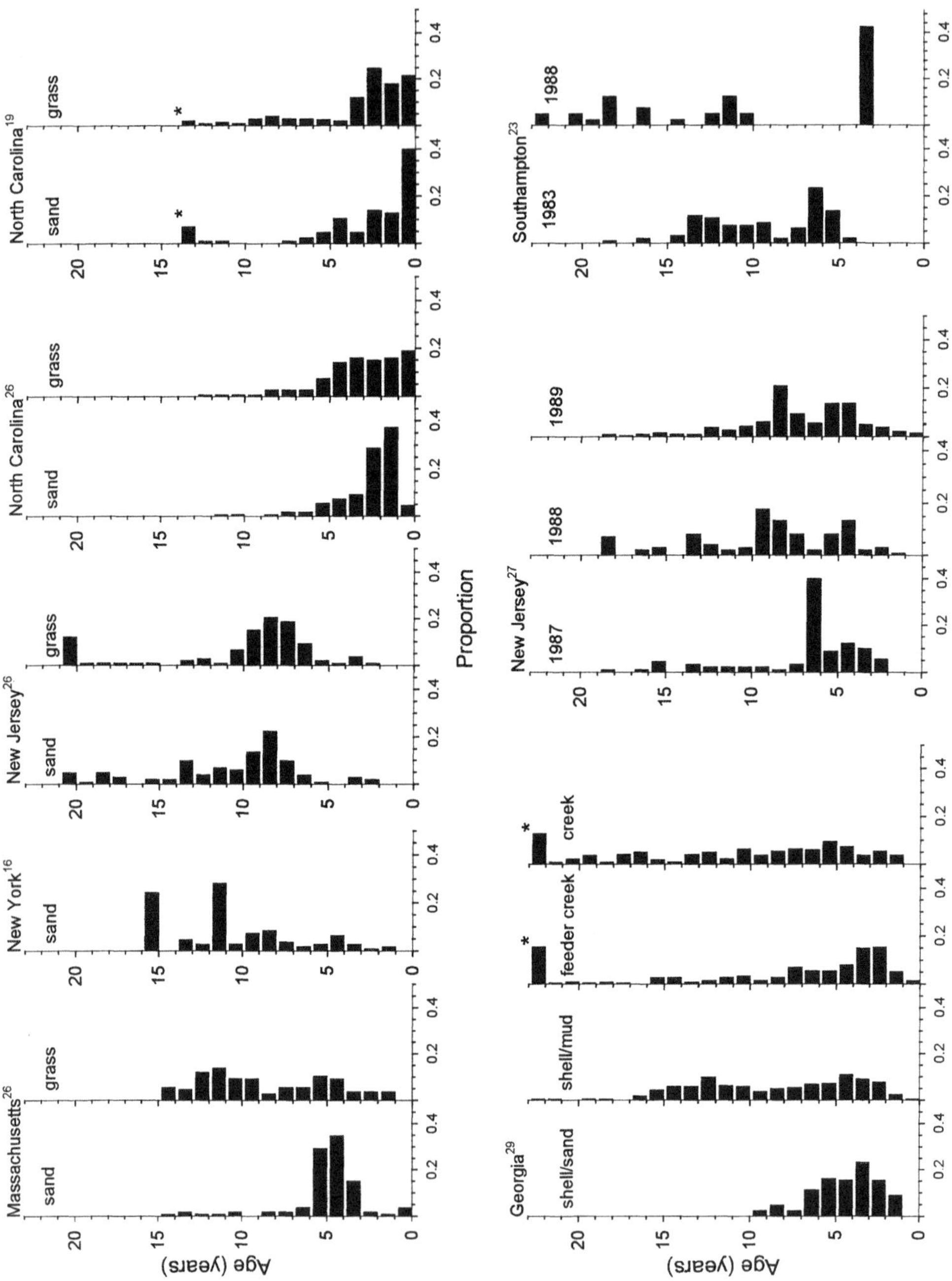

Massachusetts26
sand
grass
New York16
sand
New Jersey26
sand
grass
North Carolina26
sand
grass
North Carolina19
sand
grass
Georgia29
shell/sand
shell/mud
feeder creek
creek
New Jersey27
1987
1988
1989
Southampton23
1983
1988
Age (years)
Proportion

a single, very dominant mode occurs around clams 3–5 years of age. In Long Island Sound the age–frequency distribution of clams from a sandflat differs from those described above (Malinowski, 1985; Fig. 9.5). Older clams were most abundant with just two cohorts, 12- and 16-year-old clams contributing most of the individuals. Sandflats and grassbeds do not appear to exert similar effects on the survival of hard clam cohorts across all regions.

In coastal Georgia, fairly even representation of age classes occurs in the clam populations (Fig. 9.5). Only sediments consisting of shell/sand mixtures contained populations dominated by relatively young clams. Most of the sites sampled in Georgia (especially in the feeder creeks) have experienced virtually no clamming (Walker, 1989). The relatively high proportions of very old (>20 years) clams may represent how clam populations would appear elsewhere if harvesting had not occurred (although this could be just a characteristic of southern hard clam populations).

In all of the studies mentioned above, most age classes were represented by at least a few individuals. Richardson and Walker (1991) found a clam population in Southampton Water (Great Britain, where *Mercenaria* is an introduced species) that showed a different pattern. In 1983, the population consisted largely of individuals 5–15 years old with none younger. However, in 1988 a dominant 3-year-old age class was present (Fig. 9.5). This dominant age class was the only successful recruitment in a 9-year interval (the clams in 1988 that are 10 years and older are recruitment classes that were represented in the 1983 population). They relate the years of apparent good recruitment to selected environmental factors (mean seawater temperature and river flow rate of a nearby river), but the relationship among these factors is not strong and alternative explanations are possible. The pattern of age–class dominance followed by many years of complete recruitment failure is not observed in hard clam populations from Massachusetts to Georgia. This suggests that although the Southampton population of *Mercenaria* has persisted for almost half a century, that area may be marginal habitat for the species.

Interestingly, Mitchell (1974) presents spatial information on hard clam population age structure over 16 sites within Southampton Water (England). Some year classes (the 1965 and 1968 cohorts) are relatively abundant at most of the sites (10 out of 16 for both cohorts) throughout the entire sampling area. In contrast, the 1964 cohort is abundant at only two of sites. Sites that are approximately 1 km apart can have different age structures while those that are >5 km apart can be similar. These data indicate that factors that affect age structure vary spatially and temporally.

The importance of adequate sampling to determine age structure is illustrated by results presented in Slattery et al. (1993). They collected clams from a single 20 × 30 m area located in the Shark River, NJ in three successive years (91 clams in 1986, 98 clams in 1987, and 181 clams in 1988; Fig. 9.5). Over a third of the clams collected in 1986 were from one age class, the 1980 set. This age class was almost four times larger than the second largest age class. In 1987, 1980 clams represented only 6% of the clams sampled and the age class fell from first

Fig. 9.5. Selected age frequency distributions of hard clam populations from several different locations and habitats. The presence of an asterisk by a histogram bar indicates that the frequency for that category includes all older categories as well. Sources are indicated by the superscripts which refer to Table 9.1.

to fifth in relative size. In 1988, when almost twice as many clams were collected, 1980 clams comprised the largest age class and accounted for over a third of all clams again. As discussed below, clams do exhibit some ability to move, but the limited range of movement of large clams and the size of the area sampled in the study suggests that between year differences in the study result most likely from sampling error. Although a collection of 100 individuals would normally be considered an adequate sample size, it is clear that variability in hard clam distributions within habitats demands greater sample sizes to be assured of adequate representation of the sample results.

9.4 SEX RATIO

Hard clam population sex ratios have been reported rarely. Bricelj and Malouf (1980) found non-significant (compared to 1 : 1, small excesses of males collected from two Great South Bay, New York sites. A subsample of older clams from one of the sites also showed no significant difference between the number of female and male clams suggesting that no differential, gender-specific mortality occurred in this population.

9.5 DISPERSAL AND MOVEMENT

9.5.1 Vertical Movement

Clams exhibit vertical movements over the tidal cycle. In North Carolina, Roberts et al. (1989) placed clams (2.0–5.4 cm in width) in regular arrays in three different substrates in the field (sand, mud, and oyster rubble). By inserting a metal rod vertically into the sediments until it encountered the clam, they measured the depth beneath the sediment surface at successive intervals through the tidal cycle. They found the clams moved vertically, regardless of substrate type. The clams were closest to the sediment surface at high to early ebb tide and were deepest at low to early flood tide. The amount of mean vertical movement over the tidal cycle was 1.5–2.0 cm.

Doering (1976, 1982) demonstrated that clams move in response to the presence of asteroid seastars in laboratory experiments. He found that clams moved 1–2 cm deeper in sandy sediments when either seastars or effluent that had bathed seastars were present.

9.5.2 Lateral Movement

Chestnut (1952) mapped the locations of individually marked clams placed in two substrates (sand, exposed to some wave action, and mud, in a sheltered location) at monthly intervals. More clams were displaced in the sand substrate than the mud, smaller clams were displaced more frequently and further than large clams, and the frequency and distance of displacement varied monthly (most displacement occurred from April through July). It is unclear how much of the displacement was active movement versus physical dispersal followed by reburrowing by the clams. In the exposed site, the most frequent direction of movement was in the same direction as the local wave action. However, clams did move in the protected site and some of the clams moving in the exposed site were quite large (one 5.3-cm-long clam moved a total distance of 2.1 m over the year-long study).

Roberts et al. (1989) detected relatively little lateral migration in their study of clam vertical movements. Most clams showed no lateral movement over the relatively short periods (one to two tidal cycles) of observation. However, a few clams were observed to move over longer periods; one clam moved a distance of 15 cm over a 6-month time period.

Passive dispersal of even relatively large (several cm in length) clams occurs. Dow and Wallace (1955) surveyed the periphery of a large aggregation of clams (individuals were predominately 3–4 cm in length, Wallace personal communication, 1996) in Maquoit Bay, Maine before and after several winter storms. Initially the aggregation covered an area of 1.3×10^4 m^2 and had a mean density of 856 individuals m^{-2} of clams ranging from 2.7 to 5.3 cm in length. After the storms the aggregation covered an area of 2.7×10^4 m^2 with the center of the aggregation displaced approximately 33 m to the northwest. Clams were displaced over all of the pre-storm, aggregation borders except on the southwest side. The greatest amount of displacement occurred on the northwest side (the strongest winds recorded during the winter storms were from the southeast). Densities of clams were reduced throughout the aggregation not only by dispersal, but also by mortality presumably caused by direct exposure to freezing temperatures as the clams were uncovered from the sediments by the storm.

Other studies provide evidence that adult, large clams move laterally. Prezant et al. (1990, and personal communication, 1996) has documented sufficiently high rates of dispersal among juvenile and adult hard clams in Georgia waters to suggest that the process may be important in repopulating sites locally. Experimental studies that place marked or measured clams in enclosures have found that the clams can be eroded away by storms (Greene, 1978), move outside enclosures (Malinowski, 1985) or move within enclosures to locations where they were not planted (frequently clams were found adjacent to enclosure walls despite being consistently planted away from enclosure walls in many experiments, personal observation).

9.6 POPULATION CHARACTERISTICS OF EARLY LIFE-HISTORY STAGES

9.6.1 Spawning and Fertilization

No one has examined spawning of hard clams in the field. Logistical and technical problems associated with quantifying spawning have prevented this. Much information is available on spawning in laboratory and commercial settings (Chapters 5, 7 and 15).

To summarize:

(1) adult clams produce gametes throughout their life with no apparent decrease in fecundity with age (Peterson, 1983 Peterson, 1986b);

(2) clams produce large numbers of gametes (number of ova 1×10^6 to $>20 \times 10^6$ ova per female per spawning event; Davis and Chanley, 1956; Ansell, 1967; Bricelj and Malouf, 1980);

(3) larger females produce greater numbers of ova (Bricelj and Malouf, 1980; Peterson, 1983, 1986a,b);

(4) only 15–20% of competent individuals spawn in a given spawning event (Ansell, 1967; Bricelj and Malouf, 1980);

(5) individual females require several spawning events over several weeks to release all of their ova (Loosanoff, 1937; Davis and Chanley, 1956).

In the laboratory, hard clam gametogenesis shows few effects to variations of a wide range of factors such as day length, type of plankton used for adult food, tidal rhythms, etc. (Loosanoff, 1953). Unfortunately, we do not know to what degree gametogenesis and spawning are influenced by environmental factors in the field. Kassner and Malouf (1982) detected very different patterns of gametogenesis in native clams (from Great South Bay, New York) in successive years. They found by histological examination of gonads that spawning began 3 weeks earlier and lasted twice as long in 1979 as it did in 1978 in the same population of clams. Differences do occur between laboratory and field spawning performance. By examining the density of ripe oocytes remaining in the ovaries of clams spawned in the laboratory to those of clams collected in the field at the end of the spawning season Bricelj and Malouf (1980) found that spawning was more complete in the natural habitat. A total of 15% of laboratory clams possessed spent gonads (<2 oocytes/0.391 mm^2 of gonad) versus 46% of the field collected clams. Laboratory clams that retained oocytes had a mean ($\pm$1 SD) of 20.8 (11.71) oocytes/0.391 mm^2 of gonad versus 11.0 (14.73) oocytes/0.391 mm^2 of gonad in clams with unspent ovaries from the field. Other factors, such as dispersion and age structure, may influence spawning success as well. Older clams produce more gametes, but may be less likely to release them (personal observation of hatchery spawning). If spawning is not well synchronized, the high fecundity of the adult clams may be wasted. Further effort in determining age-specific reproductive success under natural conditions is needed.

Kraeuter et al. (1982) demonstrated that larval survival (up to 48 h after fertilization) and post-set juvenile survival (up to 1 month after setting) are significantly increased with increased initial egg size. This result, which has an obvious importance in a hatchery setting, also suggests that determination of egg sizes released in the field may be important to estimating reproductive success of hard clams.

Carriker (1961) estimated the frequency of spawning in Little Egg Harbor by back-calculating from the appearance of early-stage larvae in the water column. In four successive summers (1948 through 1951), he found evidence for 21, 9, 23, and 14 separate spawnings, respectively. The earliest spawn he detected occurred on 19 June in both 1949 and 1951. The latest spawn occurred on 1 September 1948. Although the sizes of the larval swarms from the different spawns differed greatly (see below), it is clear that *Mercenaria* populations spawn over an interval of several months in the field. Both individual variation in gonadal development and environmental variation (water depth, water temperature, etc.) throughout the population probably contribute to the pattern of dispersed spawnings.

Stiles et al. (1991) detected substantial changes in the proportions of normally developing embryos and larval mortality (up to the straight-hinge larval stage) from spawnings derived from clams collected at different locations and maintained in water from different sites. Given their design, it is unclear how much of the variation in embryo development and survival is a function of either adult history, the water used to incubate the embryos, or other factors. However, the demonstration of large variation in developmental success argues that we cannot ignore the possibility that this is an important part of hard clam success in the field.

9.6.2 Larval Clams

Only Landers (1954), Carriker (1961), and Mitchell (1974) have intensively sampled clam larvae in the field. Landers sampled two embayments (Wickford Harbor and Greenwich

Bay) in western Narragansett Bay, RI over successive summers (1950–1952). He reports only average weekly abundances of the smallest (individual larvae up to 135 μm in diameter, which is not defined) and largest (individual larvae over 186 μm in diameter) *Mercenaria*. Larval abundances vary between years and within and between sites (Fig. 9.6). In Wickford Harbor the densest swarms of early-stage larvae occurred from mid to late June (weeks 22–24). In 1950, early-stage larvae were abundant for most of the summer, but not in 1951 to 1952. In Greenwich Bay, early-stage larval swarms were generally less dense and the time when the highest abundances occurred is highly variable. Within a year, there was little temporal correspondence between densities of early-stage larvae on opposite sides of Greenwich Bay (a distance of approximately 4.5 km).

Late-stage larvae displayed some of the patterns seen in the early-stage larvae. The densest swarms occurred early in the summer. Wickford Harbor densities exceeded those of Greenwich Bay by about 5×. In contrast, the late-stage larvae show more similarity in temporal patterns of abundance between years in Wickford Harbor and more similarity in spatial abundance patterns within Greenwich Bay than occurred with the early-stage larvae.

Correspondence between patterns of abundance of early-stage and late-stage larvae is variable. Significant correlations occur between early-stage and late-stage abundances at a 1-week interval in 1950, a 2-week interval in 1951, and 3–4-week intervals in 1952 (Table 9.7). Greenwich Bay data show either non-significant or meaningless correlation between early- and late-stage larvae at both sites in both years sampled. The lack of consistent correlation between early- and late-stage larval abundances prevents estimating larval survivorship or identifying factors affecting larval survival using these data.

Carriker (1961) sampled several permanent stations in Little Egg Harbor, NJ in four successive summers (1948–1951). Larval densities in Little Egg Harbor ranged from similar to much higher than those seen by Landers in Rhode Island (Fig. 9.7). Some patterns are apparent. The largest and densest swarms of larvae always occurred in July. In 1949, 1950, and 1951, the densest swarms were early in the month. The densest swarms were sampled in late July in 1948 but sampling began later that year than in subsequent years and some early larval swarms could have been missed. Mean densities vary between years, as do the frequencies of larval swarms in the water column (Fig. 9.7).

Carriker (1961) provides estimates of larval clam survivorship in the field. Based on his measurements of local current patterns and drift buoy studies, he detected little net water movement away from the central basin of lower Little Egg Harbor. Concluding that larval swarms occurring in the central basin would be retained with little attenuation by currents, he used counts of larvae from samples taken on successive days, combined with measurements of individual clam size and ontological features as indicators of age, to estimate survivorship. Caution should be extended to these estimates. It is essential that the densities of late-stage larvae be compared to the densities of those same larvae when they were younger. Carriker recognized spatial (horizontal and vertical) variation in density of the larvae within Little Egg Harbor and suggests that recognition of the separate swarms was possible although no data is provided to support this claim. The range of larval lengths in samples taken sequentially increased fairly linearly over time, supporting his assumption that he was sampling the same larval swarm over successive sampling periods. Assuming that discrimination of several different swarms was possible the method of sampling he used prevents estimating density variations within a swarm. Clearly, samples taken at different times from the same swarm, but

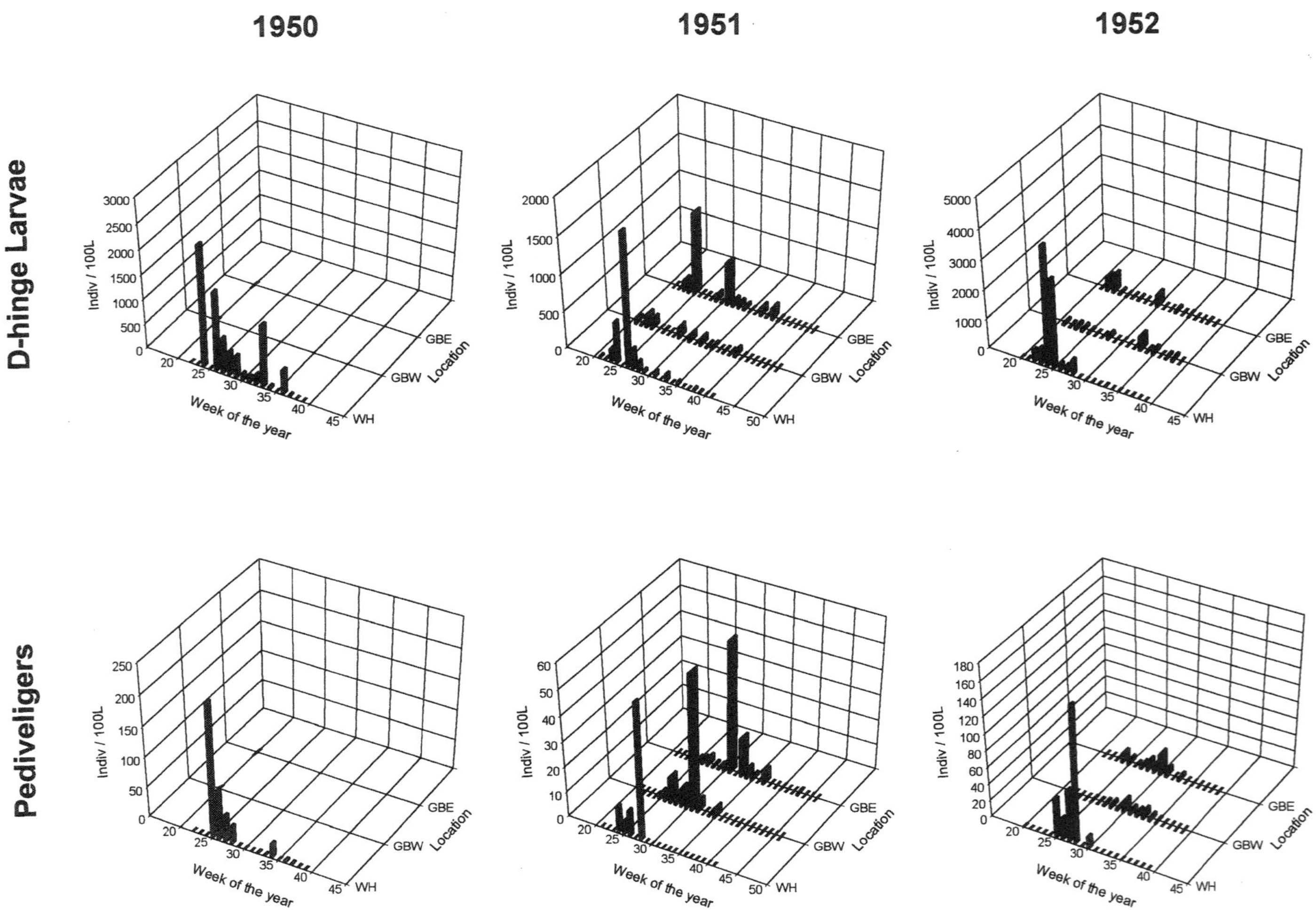

Fig. 9.6. Temporal patterns of mean abundance for early- and late-stage hard clam larvae in Wickford Harbor (WH), east Greenwich Bay (GBE), and west Greenwich Bay (GBW), RI (Landers, 1954). No data were collected in the Greenwich Bay locations in 1950. The size of the y-axis varies between years and whether early- or late-stage larvae are plotted. The week of the year in which the swarms were sampled is indicated on the x-axis.

TABLE 9.7

Correlations between early- and late-stage hard clam larvae with varying time lags at different locations in different years in Narragansett Bay (Landers, 1954)

Location	Time lag (weeks)						
	0	1	2	3	4	5	6
Wickford Harbor							
1950	ns	0.753 (0.243)	ns	ns	ns	ns	ns
1951	ns	ns	0.947 (0.229)	ns	ns	ns	ns
1952	ns	ns	ns	0.776 (0.243)	0.674 (0.250)	ns	ns
Greenwich Bay							
West 1951	ns	ns	ns	ns	ns	ns	ns
West 1952	ns	ns	ns	ns	ns	ns	ns
East 1951	0.401 (0.196)	ns	ns	ns	ns	ns	0.811 (0.224)
East 1952	ns	ns	ns	ns	ns	ns	ns
West vs east 1951	ns	ns	0.486 (0.204)	ns	ns	ns	ns
West vs east 1952	ns	ns	ns	ns	ns	ns	ns
East vs west 1951	0.410 (0.196)	ns	ns	ns	ns	ns	0.804 (0.224)
East vs west 1952	ns	ns	ns	ns	ns	ns	ns

Where correlations were significant ($P < 0.05$) the magnitude of the correlation is presented along with its standard error. In the bottom four rows, abundances of early- and late-stage larvae are compared between the western and eastern parts of the bay within each year.

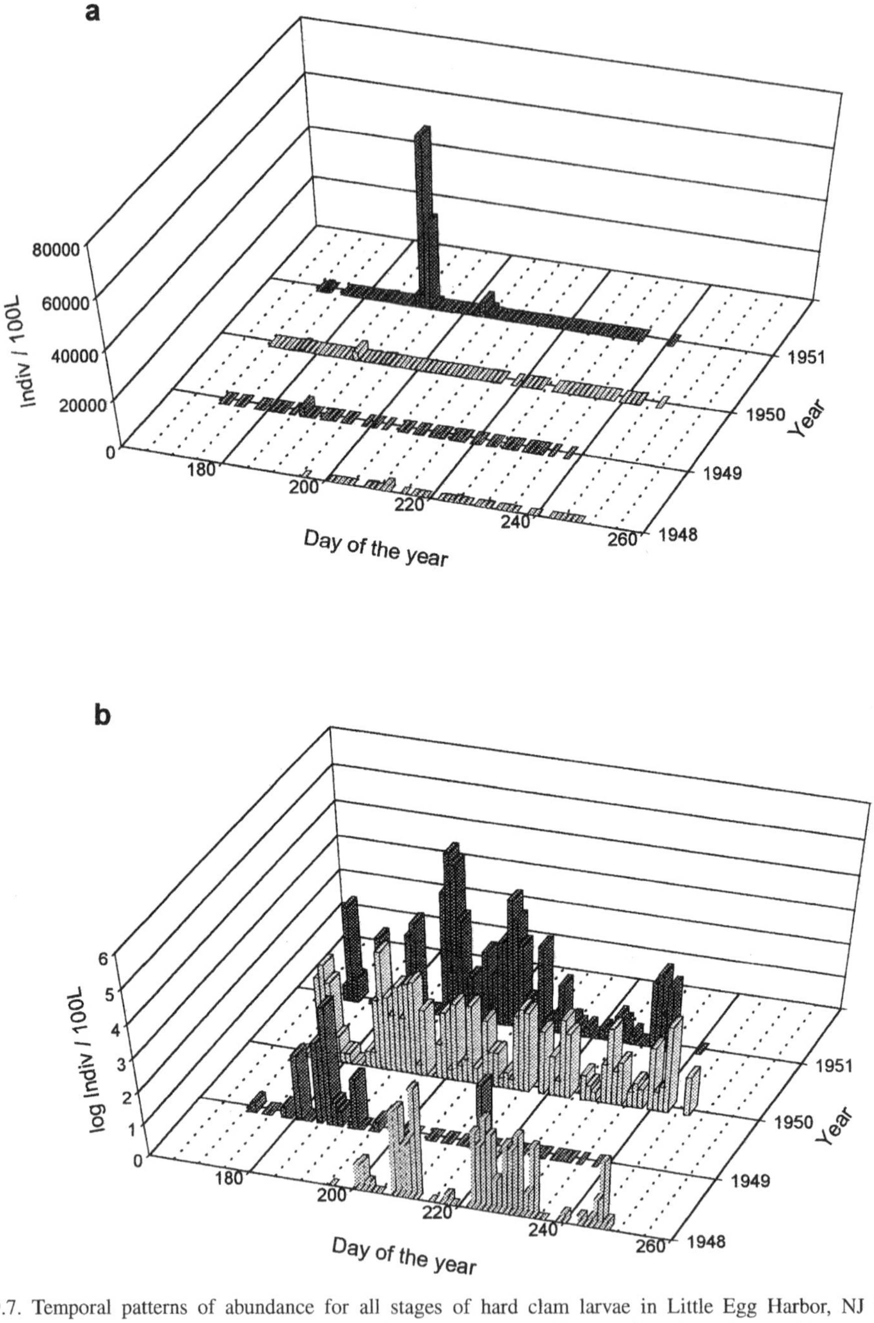

Fig. 9.7. Temporal patterns of abundance for all stages of hard clam larvae in Little Egg Harbor, NJ in four successive summers (Carriker, 1961). The upper figure is repeated with the values logged to enable comparisons of the densities among years (the extraordinarily large swarm observed in 1951 overwhelms all other variations in abundances using a linear z-axis). The day of the year is indicated on the x-axis.

TABLE 9.8

In situ estimates of larval clam survivorship in Little Egg Harbor, NJ (Carriker, 1961)

Date	Interval (days)	Density (# 1OO L^{-1})		Survivorship (%)
		Early stage	Late stage	
July 1950	6	112	8	7.1
August 1948	8	240	20	8.3
June 1950 [a]	5	800	2	0.25
July 1951	5	1,630	1	0.06
July 1949	–	2,100	–	0.0
July 1948	–	2,780	–	0.0
July 1950	4	3,300	36	0.11
July 1951	5	6,000	1	0.02
July 1951	–	34,000	–	0.0
July 1951	–	67,000	–	0.0

The interval in days is the time between when the early- and late-stage samples were collected. If no late-stage larvae were found for a particular larval swarm it is indicated by a '–' in the interval and late-stage density columns. Survivorship is estimated by dividing the number of late-stage larvae by the respective number of early-stage larvae.
[a] Daily estimates of larval clam density (in individual 1OO L^{-1}) from a single, isolated spawning event were made for this June 1950 swarm: June 3, 800; June 4, 255; June 5, 60; June 6, 10; June 7, 3; June 8, 2.

in areas of different density would lead to false indications. Nevertheless, these represent the only and best estimates we have of in situ clam larval survival.

Survivorship was 0% on four of ten occasions when larval swarms that were discrete enough to be followed occurred (Table 9.8). Larger swarms tended to have lower survival rates than smaller swarms. In part, this pattern could result from mathematical relationships; a few late-stage clams sampled from an initially smaller swarm of early-stage larvae will represent a relatively larger proportion than if the same number of late-stage larvae are found from a large early-stage larval swarm. However, it is interesting that Carriker was more likely to identify survivors from smaller than larger, early-stage swarms. This suggests that larval clam survival may be a function of initial larval density. More data of this type will be required to test this hypothesis.

Carriker (1961) also presents daily survivorship data on a single swarm (Table 9.8). Most mortality (or loss) occurs over a 1-day interval, presumably 2–3 days after spawning occurred (based on growth estimates, the entire larval period was about 8 days). The density of larvae decreased exponentially (density of larvae $= 2724 * d^{-3.63}$ where time (d) is measured in days, $r^2 = 0.978$, regression significance $P < 0.001$). No information is available on the mortality source. Despite earlier comments to the contrary, some of the loss could be due to export from Little Egg Harbor. Carriker (1961) found that freshwater input from rain into Little Egg Harbor increased flushing rates. In addition, he sampled ebb and flood waters near the oceanic inlet of Little Egg Harbor and, although he sometimes found low abundances of clam larvae being exported from the bay during ebb tide, he never found larvae entering the bay during flood tide. Transport into coastal waters appears to result in net losses in larval clam abundance within the embayment. It may be worthwhile investigating the association of larval retention and success in embayments as a function of the occurrence and magnitude of freshwater input into the system.

Mitchell (1974) sampled clam larval abundances at two locations in Southampton Water in England. He does not provide information on the abundances of the different larval stages and probably undersampled the earliest (D-hinge) stage because he used a 92.5 × 92.5 μm mesh. Most larvae were collected in late June and early July, most densities were less than 400 individuals 100 L^{-1}, and the infrequent very dense swarms generally consisted of 1000–2000 individuals 100 L^{-1}. Larval patterns in England are not very different from those observed by Landers (1954) and Carriker (1961).

Mitchell (1974) also sampled, in a single day, clam larval abundances at nine different locations spanning an area of approximately 2 × 18 km (this area enclosed most of the adult population of clams). Highest abundances (29–138 individuals 100 L^{-1}) were concentrated in the same region where the highest adult abundances occurred in the benthos. The correspondence of larval and adult abundances suggested to Mitchell that the local distribution of adult clams might be determined primarily by the dispersion of larvae produced by the densest beds of adult clams.

All of the larval data indicate that substantial losses occur in the larval stage of *Mercenaria*. Entire spawns can disappear and generally much less than 1% of the early-stage larvae survive to a size competent to settle into the benthos.

9.6.3 Post-Settlement Juvenile Clams

No field data are available on the abundances and fate of juvenile clams <1 mm in length. The absence of information on very small clams in the benthos is surprising and disturbing. Clams settle from the plankton and metamorphose when they are 200–230 μm in length (Carriker, 1961), usually in association with solid objects such as shell or pilings that protrude into the water column. Based on Carrikers (1961) laboratory observations and results from his clam setting indicators, recently metamorphosed clams require several weeks to reach 1 mm or more in length. This time interval is more than sufficient for biotic and abiotic factors to affect local patterns of juvenile clam survival and distribution. Existing evidence is conflicting. Watzin (1986) demonstrated in an experimental field study that increases in abundances of some meiofaunal turbellarian species and, separately, of nematodes and copepods had variable effects on juvenile venerid bivalve abundances. Clam settlement and/or short-term (2 weeks) survival were significantly affected positively, negatively, or not at all depending on month. No differences were detected in juvenile *Mercenaria* (200–600 μm in length) survival or growth in the presence and absence of nematode and copepod assemblages in laboratory experiments (Zobrist and Coull, 1994). Competitive interactions can occur. Ahn et al. (1993b) demonstrated that high densities of *Gemma gemma* (in the laboratory) decreased juvenile *Mercenaria* growth and increased juvenile *Mercenaria* emigration (where the experimental design permitted movement to occur). Passive dispersal of juvenile hard clams is likely as well. Juvenile *Mercenaria* siphons do not form completely until the clams reach shell lengths of 1 mm or more (Carriker, 1961). Prior to that time the clams must remain at the sediment surface to respire and feed (Carriker, 1961). Even after both inhalant and exhalant siphons have formed, the lengths of the siphon is short relative the shell length and *Mercenaria* stay relatively close (several mm) to the surface (Carriker, 1961; Stanley, 1970). Proximity to the surface increases the probability of passive dispersal via eroding currents, despite the ability of small clams to form temporary byssal threads for attachment (Carriker, 1961). Considerable redistribution of clam juveniles by hydrodynamic conditions may occur during this stage of the life history.

Hydrodynamic conditions play an important role in the initial settlement of clams. Carriker (1961) suggested that objects extending into the water column act as baffles, slow water currents and lead to greater local settlement of competent *Mercenaria* larvae. This produces high local abundances of juvenile clams. Peterson (1986a) found evidence that seagrasses enhanced early *Mercenaria* settlement by examining the relative abundances of 5–25-mm-long individuals in a neighboring grassbed and sandflat. More directly, Wilson (1990) used arrays of settlement traps to capture settling clams and found that submerged grass blades did increase initial bivalve settlement from the water column relative to unvegetated areas. Biotic interactions may affect initial settlement as well. Increased densities of *Gemma gemma* in sediment-filled laboratory containers increases settlement of pediveliger hard clams into those containers (Ahn et al., 1993a).

Some field data are available on juvenile clams greater than 1 mm and less than 20 mm in length. Carriker (1961) discovered several sets of 1–4 mm clams in Little Egg Harbor, NJ. The variation in size among the juveniles was sufficiently great to suggest to Carriker (1961) that the juvenile clams at each site came from several independent settlement events. The sets did not coincide well with adult spatial distributions. The clams occurred in greater numbers where coarse materials, oyster valves primarily, were mixed with fine sediments. The highest density observed was 125 m^{-2}; the next highest was 54 m^{-2}. At the former site, a 50% decrease in density was seen over a 2-week period. Shell fragments and empty valves were collected with the clams indicating that some of the losses were due to mortality, but emigration could have been occurring as well.

Connell (1983) followed a cohort of juvenile clams over a 13-month interval in the Shark River, NJ beginning in September of 1979. The initial mean density of 0-year class clams was over 800 m^{-2}. A 55% reduction in clam density was observed over the interval from October to December. A further 25% reduction in abundance occurred over the winter. Connell suggests that predation, by different suites of predators was primarily responsible for the observed mortality. During the fall, crab predation was most important. In the winter, predation by ducks appeared most important (feeding pits produced by the ducks had lower abundances of juvenile clams than nearby, undisturbed sediments).

Recruitment into the smallest size category he sampled (individuals 2–3 mm in length) occurred in September and October suggesting that larval settlement from the plankton occurred during early to mid summer. Observed recruitment in the second fall of his study was much smaller than the recruitment in the preceding fall. In October 1979 the mean density of individuals <5 mm in length was 440 m^{-2} while in October 1980 the mean density of individuals in this size range was 2 m^{-2} (the mean density of the 1979 cohort was 51 m^{-2} at this time). It is unfortunate that he did not sample after this date to determine whether recruitment was truly less in the second year or just delayed (the latter seems unlikely based on the observations of Landers (1954) and Carriker (1961) on temporal patterns of clam larval abundances).

Malinowski (1985) conducted experimental field studies determining survivorship of juvenile clams in two locations of eastern Long Island Sound. In one experiment, he planted three sizes (5, 10, and 15–21 mm in length) of hatchery-reared *Mercenaria* (at a density of 100 individuals m^{-2} in separate 0.25 m^2 quadrats) in May. In September, he also planted

1–2 mm clams, at the same density, in quadrats. In November, he harvested the clams in all enclosures. There was a size × site interaction in the proportion of clams remaining. Only 10–20% of the smallest clams (1–2, 5, and 10 mm) were still in the enclosures in November at both sites. With the larger clams (15–21 mm) site was important. A mean of about 18% survived at the Fishers Island site while a mean of 80% of the individuals survived at the Poquonock River site.

Malinowski (1985) also established enclosures of clams (of the same size ranges except for the 1–2 mm size range) that he sampled monthly over the May to November interval mentioned above. He detected a significant size × month × site interaction. As before, larger clams survived relatively better in the Poquonock River than Fishers Island. In the Poquonock River survivorship was slightly higher for all size groups in the fall than in the summer months. Monthly survivorship was variable among size groups at Fishers Island. In this experiment, he also manipulated clam densities (100, 600, and 1200 individuals m^{-2} in separate 0.25 m^2 enclosures) and found a strong negative relationship between initial clam density and survivorship. Interestingly, the final density of clams was 20–60 individuals m^{-2} (5–15 per enclosure), regardless of the initial stocking density.

During an experimental study examining the effects of site and clam density to clam survivorship and growth in coastal North Carolina, Beal (1983) measured native, juvenile clam recruitment into his experimental plots. He detected no significant differences in recruitment between plots with different densities of adult clams (10 or 80 individuals m^{-2}). There was considerable variation in the number of recruits among sites with the highest density (means >6 individuals m^{-2}) of recruits occurring in sediments containing shell and the least (means <1 individual m^{-2}) in the sandy and muddy sites. Grassbeds had intermediate mean densities (2–3 individuals m^{-2}) of hard clam recruits.

The experimental studies indicate that juvenile clam survivorship varies depending on a range of site-specific factors. Consequently, hard clam recruitment among habitats may be quite variable even if settlement were fairly uniform and high throughout the region.

9.7 POPULATION DYNAMICS

Life-history tables have been constructed for hard clams by both Kennish (1978) and Buckner (1984). Both life tables begin with clams 1–2 years of age. Kennish (1978) collected empty clam valves from several locations in Barnegat Bay, New Jersey and aged them to estimate age-specific mortality. Buckner (1984) sacrificed live clams collected from locations open and closed to harvest in Great South Bay, New York. Kennish's use of an assemblage of dead shells produces some problems with interpretation of the life table because it is not clear how well collections of dead shells represent the population structure of live clams that produced the assemblage. Kennish (1978) argues that there was little evidence of dispersion of clam valves in Barnegat Bay, but admits that valves of smaller clams were clearly lacking in assemblages of dead valves. In addition, clam transplant experiments that he performed in the same locations where he collected the dead valves, large proportions of the marked, transplanted clams (and their shells) were never found. Microscopic examination of growth breaks and rings in dead shells can be useful. For example, Kennish (1978) found, via examination of the growth breaks in dead clam valves, that clams were about twice as likely to die in the summer and winter than in the spring or fall of the year. In addition, summer and winter mortality appeared to be spread across each season; there was no evidence of widespread, short-term, synoptic mortality. These results suggest we should test to see whether clam mortality is season dependent.

There are similarities between Kennish's and Buckner's clam life-history tables. Both found that life expectancy of 1-year-old clams was 3–6 years. Buckner estimated life expectancy to be 3–3.5 years for clams experiencing both natural and harvest mortality and 6–6.5 years for clams experiencing just natural mortality. Kennish estimated life expectancy to be around 5 years for clams in Barnegat Bay. Both studies found that age-specific mortalities were less than 10% for younger clams (2–3 years old in Great South Bay and 3–4 years old in Barnegat Bay). The studies differed in their estimates of mortality rates for older clams. Great South Bay clams did not experience (natural + harvest) mortality rates over 50% until they were 9–10 years old while Barnegat Bay clams reached mortality rates this high when they were around 6 years old. Neither study had age-specific fecundity data available for the local clam populations so they could not estimate population growth rates from their life-history data.

Only Malinowski and Whitlatch (1988) have attempted to model hard clam population dynamics. Using a Leslie matrix population model, they determined that changes in juvenile survivorship affected population growth rate the most. They concluded that management efforts aimed at enhancing juvenile recruitment and survival would be the most effective means to provide sustainable harvests of adult clams for fisheries.

Although their conclusion may be correct (as they point out, MacKenzie (1977) independently came to the same conclusion by observing the effects of predators on juvenile hard clams over a range of field sites), critical assessment of this point is not possible. First, due to the paucity of relevant information, Malinowski and Whitlatch (1988) were forced to use data for the population parameters in their model that were collected from hard clam populations located in very different locations (New Jersey, Virginia, and England) sampled at different times. It is not clear that these parameters can be used jointly. Second, they could not incorporate larval or early post-set survival (once again, because virtually no information is available) into the model. They used a constant, calculated larval survivorship of 1.1×10^{-6} (using an equation for estimating the minimum survivorship necessary to maintain a stable population, Vaughn and Saila, 1976). This prevented them from assessing the relative importance of variation in larval life-history stage to subsequent population changes. Third, the Leslie matrix model is time invariant (although time varying and stochastic forms of the Leslie model exist, Caswell, 1989). The assumption that age-specific mortalities and fecundities do not change in hard clams is unlikely to be true. It is not clear what the effect of variation in demographic factors would be on hard clam population dynamics.

9.8 SUMMARY

It is astonishing that so little information exists on hard clam populations given that this species has supported long-standing, economically important fisheries. A common problem that limits our ability to compare studies and make generalizations about regional or habitat-specific patterns in hard clam population structure is inadequate sampling. Because sampling effort is low (relative to the observed variability), we have little ability to detect differences or to be confident that observed patterns are real.

It is tempting to use the data presented here to summarize hard clam populations by speculating on applicable models for their dynamics. Yet the gaps in information, diversity in quality of what is available, and the absence of appropriate between-region comparisons clearly indicate that such an effort is unwarranted. I have even avoided making comparisons to populations of other bivalve species because the limited information on *Mercenaria* prevents assessing which demographic characteristics can be compared reasonably to other species and which ones cannot.

Instead of speculating, I will close by suggesting areas of inquiry that would help us understand hard clam population dynamics better as revealed by consideration of the data we have available. What is the role of adult dispersion patterns in hard clam spawning success? Does gamete viability change with age of the adult clam? What are the realized age-specific fecundities of clams? What are the relative importances of flushing rates and larval predation in determining larval survival? What factors influence settlement success? What proportion of the clams in a new cohort are progeny of the adult clams in the local population? What are the spatial and temporal patterns of abundance of pediveligers and juvenile clams <1 mm in length? What role does lateral advection of juveniles play in local hard clam recruitment? What effects do the various harvest methods have on juvenile hard clam settlement and survival? What are the temporal patterns of adult hard clam abundance? Finally, how do aspects of hard clam population biology change regionally? The last question requires that future investigations either be conducted jointly with common techniques or that region-specific projects make serious attempts to relate their results to similar projects done elsewhere.

Wherever *Mercenaria* have been abundant they have been harvested. This places some importance in studying hard clam populations located in areas that have been closed to harvest continuously for several decades. All other existing populations are either at non-equilibrium conditions (due to harvesting) in acceptable hard clam habitats or at equilibrium conditions in marginal hard clam habitats. The data that we gather from most clam populations will have to be viewed in this context. But the effort is worthwhile. In the absence of robust information on hard clam populations enlightened management of this species and understanding their role in estuarine communities will be impossible. Only when unconfounded, long-term abundance data, synoptic life-history information, and experimental studies of factors influencing hard clam demographic parameters have been conducted will it be possible to understand hard clam populations as a comprehensive whole.

9.9 ACKNOWLEDGMENTS

I thank B. Barber for use of his digitizing equipment. B. Beal, M. Castagna, J. Fegley, J. Kraeuter, and J. Loesch provided useful reviews of an early version of this chapter. Numerous conversations over the past few years with J. Kraeuter helped shape my perspectives of hard clam populations.

REFERENCES

Ahn, I., Lopez, G. and Malouf, R., 1993a. Enhanced larval settlement of the hard *clam Mercenaria mercenaria* by the clam *Gemma gemma*. Mar. Ecol. Prog. Ser., 99: 51–59.

Ahn, I., Lopez, G. and Malouf, R., 1993b. Effects of the gem clam *Gemma gemma* on early post-settlement emigration, growth and survival of the hard clam *Mercenaria mercenaria*. Mar. Ecol. Prog. Ser., 99: 61–70.

Ansell, A.D., 1967. Egg production of *Mercenaria mercenaria*. Limnol. Oceanogr., 12: 172–176.

Ansell, A.D., 1968. The rate of growth of the hard clam *Mercenaria mercenaria* (L.) throughout the geographical range. J. Cons. Perm. Int. Explor. Mar., 31: 364–409.

Ashley, G.M. and Grizzle, R.E., 1988. Interactions between hydrodynamics, benthos, and sedimentation in a tide-dominated coastal lagoon. Mar. Geol., 82: 61–81.

Beal, B.F., 1983. Effects of environment, intraspecific density, predation by snapping shrimp and other consumers on the population biology of *Mercenaria mercenaria* near Beaufort, North Carolina. Masters thesis. Univ. of North Carolina at Chapel Hill, 105 pp.

Beal, B.F., 1994. Biotic and abiotic factors influencing growth and survival of wild and cultured individuals of the soft-shell clam, *Mya arenaria* L., in Eastern Maine. Ph.D. diss. Univ. of Maine at Orono, 482 pp.

Bricelj, V.M. and Malouf, R.E., 1980. Aspects of reproduction of hard clams (*Mercenaria mercenaria*) in Great South Bay, New York. Proc. Natl. Shellf. Assoc., 70: 216–229.

Buckner, S.C., 1984. Aspects of the population dynamics of the hard clam, *Mercenaria mercenaria* L., in Great South Bay, New York. Ph.D Diss., SUNY at Stony Brook, 217 pp.

Carriker, M.R., 1961. Interrelation of functional morphology, behavior, and autoecology in early stages of the bivalve, *Mercenaria mercenaria*. J. Elisha Mitchell Sci. Soc., 77: 168–241.

Castagna, M., 1984. Methods of growing *Mercenaria mercenaria* from post-larval to preferred-size seed for field planting. Aquaculture, 39: 355–359.

Castagna, M. and Kraeuter, J.N., 1977. *Mercenaria* culture using stone aggregate for predator protection. Proc. Natl. Shellf. Assoc., 67: 1–6.

Caswell, H., 1989. Matrix Population Models: Construction, Analysis, and Interpretation. Sinauer, Sunderland, MA, 328 pp.

Cerrato, R.M., 1980. Demographic analysis of bivalve populations. In: Rhoads, D.C. and Lutz, R.A. (Eds) Skeletal Growth of Aquatic Organisms, Plenum Press, NY, pp. 417–465.

Chestnut, A.F., 1952. Growth rates and movements of hard clams, *Venus mercenaria*. Proc. Gulf Carib. Fish. Inst. 4th Ann. Sess., pp. 49–59.

Cochran, W.G., 1977. Sampling techniques. John Wiley and Sons, New York, NY, 3rd ed.

Connell Jr., R., 1983. Seasonal mortality and population dynamics of juvenile hard clams, *Mercenaria mercenaria* L., on an intertidal mudflat in New Jersey. M.S. Thesis, Rutgers Univ..

Craig, M.A. and Bright, T.J., 1986. Abundance, age distributions and growth of the Texas hard clam, *Mercenaria mercenaria* texana, in Texas bays. Contrib. Mar. Sci., 29: 59–72.

Crane Jr., J.M., Allen, L.G. and Eisemann, C., 1975. Growth rate, distribution, and population density of the northern quahog, *Mercenaria mercenaria* in Long Beach, California. Calif. Fish Game, 61: 68–81.

Dame, R.F., 1979. The abundance, diversity and biomass of macrobenthos on North Inlet, South Carolina intertidal oyster reefs. Proc. Natl. Shellf. Assoc., 69: 6–10.

Davis, H.C. and Chanley, P.E., 1956. Spawning and egg production of oysters and clams. Biol. Bull., 110: 117–128.

DeLury, D.B., 1947. On the estimation of biological populations. Biometrics, 3: 145–167.

Doering, P.H., 1976. A burrowing response on *Mercenaria mercenaria* (Linnaeus, 1758) elicited by *Asterias forbesi* (Desor, 1848). Veliger, 19: 167–175.

Doering, P.H., 1982. Reduction of sea star predation by the burrowing response of the hard clam *Mercenaria mercenaria* (Mollusca: Bivalvia). Estuaries, 5: 310–315.

Dow, R.L. and Wallace, D.E., 1955. Natural redistribution of a quahog population. Science, 122: 641–642.

Eldridge, P.J., Eversole, A.G. and Whetstone, J.M., 1979. Comparative survival and growth rates of hard clams, *Mercenaria mercenaria*, planted in trays subtidally and intertidally at varying densities in a South Carolina estuary. Proc. Natl. Shellf. Assoc., 69: 30–39.

Eldridge, P.J. and Eversole, A.G., 1982. Compensatory growth and mortality of the hard clam, *Mercenaria mercenaria* (Linnaeus, 1758). Veliger, 24: 276–278.

Greene, G.T., 1978. Population structure, growth and mortality of hard clams at selected locations in Great South Bay, New York, Master Thesis, SUNY, Stony Brook, 199 pp.

Grizzle, R.E. and Morin, P.J., 1989. Effect of tidal currents, seston, and bottom sediments on growth of *Mercenaria mercenaria*: results of a field experiment. Mar. Biol., 102: 85–93.

Haskin, H.H., 1952. Further growth studies on the quahog, *Venus mercenaria*. Proc. Natl. Shellf. Assoc., 42: 181–187.

Hibbert, C.J., 1976. Biomass and production of a bivalve community on an intertidal mudflat. J. Exp. Mar. Biol. Ecol., 25: 249–261.

Irlandi, E. and Peterson, C.H., 1991. Modification of animal habitat by large plants: mechanisms by which seagrasses influence clam growth. Oecologia, 87: 307–318.

Joseph, J., 1989. Inventory of New Jersey's estuarine shellfish resources. NJDEP Completion Report to NMFS, NOAA, Dept. of Commerce. Unpubl. Report, 75 pp.

Kassner, J. and Malouf, R.E., 1982. An evaluation of 'spawner transplants' as a management tool in Long Island's hard clam fishery. J. Shellf. Res., 2: 165–172.

Kraeuter, J.N., Castagna, M. and van Dessel, R., 1982. Egg size and larval survival of *Mercenaria mercenaria* (L.) and Argopecten irradians (Lamarck). J. Exp. Mar. Biol. Ecol., 56: 3–8.

Kennish, M.J., 1978. Effects of thermal discharges on mortality of *Mercenaria mercenaria* in Barnegat Bay, New Jersey. Environ. Geol., 2: 223–254.

Kennish, M.J., 1980. Shell micrograph analysis: *Mercenaria mercenaria* as a type example for research in population dynamics. In: D.C. Rhoads and R.A. Lutz (Eds.), Skeletal Growth of Aquatic Organisms. Plenum Press, New York, NY, pp. 255–294.

Kerswill, C.J., 1949. Effects of water circulation on the growth of quahogs and oysters. J. Fish. Res. Bd. Can., 7: 545–551.

Landers, W.S., 1954. Seasonal abundance of clam larvae in Rhode Island waters, 1950–1952. U.S. Fish and Wildlife Serv. Spec. Rep., Fisheries No. 117.

Landry, T., Sephton, T.W. and Jones, D.A., 1993. Growth and mortality of northern quahog, (Linnaeus, 1758) *Mercenaria mercenaria* in Prince Edward Island. J. Shellf. Res., 12: 321–327.

Leslie, P.H. and Davis, D.H.S., 1939. An attempt to determine the absolute number of rats on a given area. J. Anim. Ecol., 8: 94–113.

Levitan, D.R., Sewell, M.A. and Chia, F., 1992. How distribution and abundance influence fertilization success in the sea urchin *Strongylocentrotus franciscanus*. Ecology, 73: 248–254.

Loesch, J.G., 1974. A sequential sampling plan for hard clams in lower Chesapeake Bay. Chesapeake Sci., 15: 134–139.

Loesch, J.G., 1977. A comparison of frequency distributions of hard clam, patent-tong catches. Chesapeake Sci., 18: 79–80.

Loesch, J.G. and Haven, D.S., 1973. Estimates of hard clam abundance from hydraulic escalator samples by the Leslie method. Chesapeake Sci., 14: 215–216.

Loosanoff, V.L., 1937. Spawning of *Venus mercenaria* (L.). Ecology, 18: 506–515.

Loosanoff, V.L., 1953. Lack of relation between age of oysters or clams and quality of their spawn. Fish. Biol. Lab. Milford CT Bull., 17: 1–2.

Lyles, C.H., 1969. Historical catch statistics (shellfish). U.S. Fish and Wildlife Spec. Publ. C.F.S. , 5007: 1–116.

MacKenzie Jr., C.L., 1977. Predation on hard clam (*Mercenaria mercenaria*) populations. Trans. Am. Fish. Soc., 106: 530–537.

Malinowski, S.R., 1985. The population ecology of the hard clam, *Mercenaria mercenaria*, in Eastern Long Island Sound. Ph.D. dissertation. Univ. of Connecticut, 101 pp.

Malinowski, S. and Whitlatch, R.B., 1988. A theoretical evaluation of shellfish resource management. J. Shellf. Res., 7: 95–100.

McCay, B.J., 1988. Muddling through the clam beds: cooperative management of New Jersey's hard clam spawner sanctuaries. J. Shellf. Res., 7: 327–340.

Mitchell, R., 1974. Aspects of the ecology of the lamellibranch *Mercenaria mercenaria* in British waters. Hydrobiol. Bull., 8: 124–138.

National Marine Fisheries Service, 1996. NMFS Fisheries Statistics Homepage, http://remora.ssp.nmfs.gov/ (10 Sept., 1996).

Peterson, C.H., 1982. Clam predation by whelks (*Busycon* spp.): experimental tests of the importance of prey size, prey density, and seagrass cover. Mar. Biol., 66: 159–170.

Peterson, C.H., 1983. A concept of quantitative reproductive senility: application to the hard clam, *Mercenaria mercenaria* (L)? Oecologia, 58: 164–168.

Peterson, C.H., 1986a. Enhancement of *Mercenaria mercenaria* densities in seagrass beds: is pattern fixed during settlement season or altered by subsequent differential survival? Limnol. Oceanogr., 31: 200–205.

Peterson, C.H., 1986b. Quantitative allometry of gamete production by *Mercenaria mercenaria* into old age. Mar. Ecol. Prog. Ser., 29: 93–97.

Peterson, C.H. and Beal, B.F., 1989. Bivalve growth and higher order interactions: importance of density, site, and time. Ecology, 70: 1390–1404.

Peterson, C.H., Summerson, H.C. and Fegley, S.R., 1983. Relative efficiency of two clam rakes and their contrasting impacts on seagrass biomass. Fish. Bull., 81: 429–434.

Peterson, C.H., Summerson, H.C. and Duncan, P.B., 1984. The influence of seagrass cover on population structure and individual growth rate of a suspension feeding bivalve, *Mercenaria mercenaria*. J. Mar. Res., 42: 123–138.

Peterson, C.H., Summerson, H.C. and Fegley, S.R., 1987. Ecological consequences of mechanical harvesting of clams. Fish. Bull., 85: 281–298.

Peterson, C.H., Summerson, H.C. and Huber, J., 1995. Replenishment of hard clam stocks using hatchery seed: combined importance of bottom type, seed size, planting season, and density. J. Shellf. Res., 14: 293–301.

Prezant, R.S., Rollins, H.B. and Toll, R.B., 1990. Dispersal of adult hard clams as an adjunct to larval recruitment. Am. Zool., 30: 89.

Rhodes, R.J., Keith, W.J., Eldridge, P.J. and Burell Jr., V.G., 1977. An empirical evaluation of the Leslie–DeLury method applied to estimating hard clam, *Mercenaria mercenaria*, abundance in the Santee River estuary, South Carolina. Proc. Natl. Shellf. Assoc., 67: 44–52.

Rice, M.A., Hickox, C. and Zehra, I., 1989. Effects of intensive fishing effort on the population structure of quahogs, *Mercenaria mercenaria* (Linnaeus 1758) in Narragansett Bay. J. Shellf. Res., 8: 345–354.

Richardson, C.A. and Walker, P., 1991. The age structure of a population of the hard-shell clam, *Mercenaria mercenaria* from Southampton Water, England, derived form acetate peel replicas of shell sections. ICES J. Mar. Sci., 48: 229–236.

Ricker, W.E., 1958. Handbook of computations for biological statistics of fish populations. Bull. Fish. Res. Bd. Canada, 119: 1–300.

Roberts, D., Rittschof, D., Gerhat, D.J., Schmidt, A.R. and Hill, L.G., 1989. Vertical migration of the clam *Mercenaria mercenaria* (L.) (Mollusca: Bivalvia): environmental correlates and ecological significance. J. Exp. Mar. Biol. Ecol., 126: 271–280.

Russell Jr., H.J., 1972. Use of a commercial dredge to estimate a hard shell clam population by stratified random sampling. J. Fish. Res. Board Can., 29: 1731–1735.

Saila, S.B., Flowers, J.M. and Campbell, R., 1965. Applications of sequential sampling to marine resource surveys. Ocean Sci. Ocean Eng., 2: 782–802.

Saila, S.B., Flowers, J.M. and Cannario, M.R., 1967. Factors affecting the relative abundance of *Mercenaria mercenaria* in the Providence River, Rhode Island. Proc. Natl. Shellf. Assoc ., 57: 83–89.

Slattery, J.P., Vrijenhoek, R.C. and Lutz, R.A., 1991. Heterozygosity, growth, and survival of the hard clam, *Mercenaria mercenaria*, in seagrass vs sandflat habitats. Mar. Biol., 111: 335–342.

Slattery, J.P., Lutz, R.A. and Vrijenhoek, R.C., 1993. Repeatability of correlations between heterozygosity, growth and survival in a natural population of the hard clam *Mercenaria mercenaria* L. J. Exp. Mar. Biol. Ecol., 165: 209–224.

Snedecor, G.W. and Cochran, W.G., 1967. Statistical Methods. The Iowa State Univ. Press, Ames, IA. 6th ed., 593 pp.

Sokal, R.R. and Rohlf, F.J., 1995. Biometry. W.H. Freeman, New York, NY. 3rd ed., 887 pp.

Stanley, S.M., 1970. Relation of shell form to life habits of the bivalvia (Mollusca). Geol. Soc. Am. Mem., 125: 1–296.

Stickney, A.P. and Stringer, L.D., 1957. A study of the invertebrate bottom fauna of Greenwich Bay, Rhode Island. Ecology, 38: 111–121.

Stiles, S., Choromanski, J., Nelson, D., Miller, J., Grieg, R. and Sennefelder, G., 1991. Early reproductive success of the hard clam (*Mercenaria mercenaria*) from five sites in Long Island Sound. Estuaries, 14: 332–342.

Summerson, H.C., Peterson, C.H. and Hooper, M., 1995. Aquacultural production of northern quahogs, *Mercenaria mercenaria* (Linnaeus, 1758): high water temperatures in the nursery and growth penalties of predator control by gravel. J. Shellf. Res., 14: 25–31.

Vaughn, D.S. and Saila, S.B., 1976. A method for determining mortality rates using the Leslie matrix. Trans. Am. Fish. Soc., 105: 380–383.

Walker, R.L., 1987. Hard clam *Mercenaria mercenaria* (Linne') populations of coastal Georgia. Ga. Mar. Sci. Center, Tech. Rep. 87-1: 1–73.

Walker, R.L., 1989. Exploited and unexploited hard clam, *Mercenaria mercenaria* (L.), populations in coastal Georgia. Contrib. Mar. Sci., 31: 61–75.

Walker, R.L. and Rawson, M.V., 1985. Subtidal hard clam, *Mercenaria mercenaria* (Linne'), resources in coastal Georgia. Ga. Mar. Sci. Center, Tech. Rep. 85-1: 1–164.

Walker, R.L. and Tenore, K.R., 1984. The distribution and production of the hard clam, *Mercenaria mercenaria*, in Wassaw Sound, Georgia. Estuaries, 7: 19–27.

Wells, H.W., 1957. Abundance of the hard clam *Mercenaria mercenaria* in relation to environmental factors. Ecology, 38: 123–128.

Watzin, M.C., 1986. Larval settlement into marine soft-sediment systems: interactions with the meiofauna. J. Exp. Mar. Biol. Ecol., 98: 65–113.

Wilson, F.S., 1990. Temporal and spatial patterns of settlement: a field study of molluscs in Bogue Sound, North Carolina. J. Exp. Mar. Biol. Ecol., 139: 201–220.

Zobrist, E.C. and Coull, B.C., 1994. Meiofaunal effects on growth and survivorship of the polychaete Streblospio benedicti Webster and the bivalve *Mercenaria mercenaria* (L.). J. Exp. Mar. Biol. Ecol., 175: 167–179.

Biology of the Hard Clam
J.N. Kraeuter and M. Castagna (Eds.),

Chapter 10

Integrating Nutritional Physiology and Ecology to Explain Interactions between Physics and Biology in Mercenaria mercenaria

Charles H. Peterson

10.1 INTRODUCTION

To presume to review all the interactions between nutritional physiology and ecology of a species is madness. There are countless numbers of such interactions, even important ones, that enter into the biology of every organism. Thus, some explanation for the topic of this contribution is required. Here, I do not attempt to catalogue all significant physiological–ecological interactions that affect the biology of hard clams, a universe limited solely by the scope of measurements made. Instead, I will argue from first principles and from the limited suite of studies conducted on hard clams and analogous bivalves how some particular interactions that bridge these traditionally separate levels of biological organization are especially important. I focus explicitly on examination of those important processes where understanding the interactions between physical and biological factors requires integration of nutritional physiology and ecology.

The specific interactions that I consider fall into a small number of categories. First, I address the role of what has been termed trophic group amensalism (Rhoads and Young, 1970), the process whereby deposit feeders exert negative influence(s) on the abundance of suspension feeders through their modification of the physical environment. Second, I review the complex set of interactions within the seagrass habitat between physical and biological factors that affect the growth of hard clams. Third, I explore some important interactions between the physiological state of a hard clam and its ecology, explicitly addressing ecological implications of interactions among multiple physiological stressors.

10.2 TROPHIC GROUP AMENSALISM

The trophic group amensalism hypothesis was first proposed by Rhoads and Young (1970) to explain an apparent segregation of deposit feeders and suspension feeders in soft-sediment communities. Deposit feeders prefer finer sediments because they are typically richer in organic content, implying greater microbial food supply rates (Sanders, 1958). In contrast, according to Sanders' arguments, suspension feeders prosper where water currents are strong enough to enhance flux of suspended foods, which implies occupation of coarser sediments. Nevertheless, actively pumping suspension feeders should be able to feed at adequate rates even where currents are slow and sediments are fine. Any depression of abundance of active suspension feeders in the presence of deposit-feeding communities in fine sediments would represent a surprising phenomenon, unexplained by direct effects of flow speed. Trophic group amensalism offers an explanation for a general depression of suspension feeders in fine sediments (Rhoads and Young, 1970). Deposit feeders actively burrow through the sediments during foraging, which has the effect of increasing the water content of the sediments and thereby rendering them more readily eroded and resuspended under any given flow regime. When this resuspension of sediments occurs, it presumably interferes with feeding of suspension feeders by clogging gills and other filtration devices. Furthermore, the destabilization of the sediments during burrowing by deposit feeders can bury and kill settlers, especially the relatively immobile post-larvae of colonizing suspension feeders. Thus, trophic group amensalism describes a process of indirect competition between feeding types (Peterson, 1980), such that deposit feeders inhibit suspension feeders in their vicinity by altering the local physical environment in a fashion that renders it unsuitable for occupation by most suspension feeders.

Although the hypothesis of trophic group amensalism addresses patterns of abundance, Rhoads and Young (1970) tested the concept by evaluating growth rather than abundance of a suspension feeder, specifically *Mercenaria mercenaria*. They used trays holding hard clams to assess whether growth rates were enhanced by elevating clams above a muddy bottom populated by deposit feeders. When growth rates proved higher with increasing elevation above the bottom, they interpreted the result as evidence of suppression of growth nearest the seafloor by the turbidity induced indirectly by deposit-feeder activity. This study represented one of the first uses of experimental hypothesis testing in soft-bottom benthic ecology, a Rubicon which, once crossed, has led to substantial improvement in the evaluation of processes in this environment. Nevertheless, we now realize that increased elevation off the bottom also implies enhanced flow speeds (e.g., Nowell and Jumars, 1984) and enhanced food quality through reduction in inorganic materials (Muschenheim, 1987), independent of the presence of deposit feeders below. Thus, the experimental demonstration of faster hard clam growth with increasing height above a muddy bottom containing deposit feeders does not necessarily implicate deposit feeders and cannot be used to show that growth rates of *Mercenaria* are reduced by indirect effects of deposit feeder activity (Dayton and Oliver, 1980). In fact, one group of hard clams in Rhoads and Young's (1970) study maintained over sandy sediments could not be shown to grow significantly faster than those clams over the fine sediments. Furthermore, growth responses cannot be readily linked to abundance patterns, which are the focus of the trophic group amensalism hypothesis.

Conceivably, the consequences of lower growth and reduced fitness in finer sediments, due to both increased time spent at sizes most vulnerable to predation and lower reproductive output of the smaller clams (Peterson, 1983a, 1986b), could represent an effective selection force in suspension feeders for evolution of behavioral avoidance of fine sediments or behavioral preference for coarser sediments during larval settlement. Such settlement behaviors could thereby lead to segregation of suspension feeders and deposit feeders, but this hypothetical linkage lacks a general test. For *M. mercenaria*, larvae appear to show little selectivity for particle size during settlement even in still water conditions (Bachelet et al., 1992). Laboratory experiments reveal that *M. mercenaria* settles on a wide range of substrata (Loosanoff and Davis, 1950), with conflicting demonstrations of preference for large (Carriker, 1961) and small (Keck et al., 1974) sedimentary substrata. In a flume study with slow, turbulent flows, Butman et al. (1988) showed no active selection by settling *Mercenaria* larvae. In laboratory experiments, Ahn et al. (1993a) demonstrated preferential larval settlement of *M. mercenaria* around dense patches of another suspension-feeding bivalve, *Gemma gemma*, and around sediments enriched in *Gemma* shells. This settlement behavior could conceivably have evolved to enable *Mercenaria* to select habitats suitable for suspension-feeder occupation by using an abundant surface-dwelling suspension feeder and its shells as a settlement cue; however, Ahn et al. (1993b) published results of experiments showing that in muddy sand substrata the presence of dense *Gemma* reduced the subsequent growth and survival of post-settlement *Mercenaria*. Data on *Mercenaria* larval settlement in the field are essentially non-existent. Peterson (1986a) compared densities of 0-year-class and older age groups of *Mercenaria* in unvegetated sand flats to those in nearby siltier sediments in seagrass beds. This contrast revealed that although densities are twice as high inside seagrass beds at the conclusion of settlement season, the disparity grows tremendously as the clams reach older age classes. Consequently, while differential larval or post-larval settlement may conceivably play a role in establishing an apparent association of recruits with finer (not coarser) sediments, later post-settlement differences in survival are even more important. Wilson (1990) used buried sediment traps inside seagrass beds and on adjacent sand flats to sample settling larvae and post-larvae, showing that bivalves as a group, presumably including many *M. mercenaria*, were settling at higher numbers into the siltier sediments of the seagrass bed. Such a pattern could be generated by either active larval habitat selection or passive transport and erosion/deposition. In either case, these observations fail to support the hypothesis that *Mercenaria* settles preferentially in coarser sediments. However, these studies confound the changes in sediment characteristics with the presence/absence of seagrass, so they do not represent pure tests of sediment grade alone.

Trophic group amensalism has strong appeal and continues to be invoked as an explanation because of its reliance on processes based on first principles in physiology and ecology of invertebrates, sedimentology, and fluid dynamics (e.g., Taghon et al., 1980). Levinton (1977) suggested that the low abundance of *M. mercenaria* in the eelgrass habitat of Quisset Harbor, Massachusetts, could be explained by the trophic group amensalism induced by *Nucula proxima*. *Nucula* is an active surface deposit feeder that congregates around the siphon holes of any *Mercenaria*, thereby intensifying local sediment resuspension in the very near vicinity of feeding hard clams. Myers (1977) broadened the trophic group amensalism concept by arguing that even in sandy sediments actively burrowing benthic deposit feeders can suppress growth and survival of sedentary suspension feeders, especially bivalve molluscs. He supported this idea by provision of data showing reduced abundance of *M. mercenaria* and other sedentary suspension feeders in the presence of the active deposit-feeding holothurian *Leptosynapta tenuis*. Myers hypothesized that when suspension-feeding bivalves are exposed to sediment instability they suffer reduced growth because of feeding interruptions and energetic costs of re-establishing contact with the overlying water column and they experience enhanced mortality because of more frequent exposure to surface predators.

The concept of trophic group amensalism has been expanded further by Brenchley (1981) to include all the effects exerted by mobile species on more sedentary species in soft sediments. She employed laboratory experiments to show that sediment instability could be induced by burrowing benthic animals of all sorts, not solely deposit feeders, and that sediment instability inhibited feeding by relatively immobile benthic invertebrates, including suspension feeders such as *Mercenaria*. Field sampling and experiments provide support for Brenchley's (1981) expanded concept that active burrowers inhibit sedentary benthic animals: Peterson (1977), Murphy (1985), and Posey (1986) each demonstrated negative relationships between the burrowing, but suspension-feeding, ghost shrimp *Callianassa* and suspension-feeding bivalves. In Murphy's (1985) study, the bivalve that was negatively associated with the ghost shrimp was *M. mercenaria*. Using field experiments, Hunt et al. (1987) showed that the mobile surface deposit feeder, *Ilyanassa obsoleta*, inhibits successful recruitment of bivalves, whereas the sedentary suspension feeder, *M. mercenaria*, does not. In this example of depressed recruitment of bivalves by active deposit feeders, there is potential for predation by the deposit feeders to represent the effective mechanism, rather than physical disturbance of the sediments. Trophic group amensalism in the form of the broadened mobile vs. sessile distinction also plays an important part in the Woodin (1976) general model of how adult–larval interactions structure soft-sediment communities at high density.

Recent reviews of animal–sediment relationships by Posey (1990), Snelgrove and Butman (1994), and Wildish and Kristmanson (1997) challenge the notion that deposit feeders segregate from suspension feeders in the strict fashion that the trophic group amensalism hypothesis was designed to explain. Application of improved understanding of the trophic behavior of benthic invertebrates to reinterpret earlier data reveals that suspension and deposit feeders exhibit widespread co-occurrence. More critically, perpetuation of the concept of trophic group amensalism is due largely to ad hoc modifications and qualifications of the original idea so as to embrace apparent contradictions. The mobility group hypothesis also does not appear to make accurate predictions except when the mobile animal is substantially larger than the sedentary species that suffers from the sediment destabilization (Posey, 1987). This recent and compelling reconsideration of the status of the trophic group amensalism hypothesis casts serious doubt on its applicability to hard clams. Snelgrove and Butman (1994) argue that more complete inclusion of principles of fluid dynamics will be needed to develop predictive theory to replace the insufficient concept of trophic group amensalism. There is more compelling evidence to support the less complex trophic group mutual exclusion hypothesis of Wildish (1977). This hypothesis explains patterns of abundance and production of suspension feeders and deposit feeders as largely independent responses to flow, rather than driven by indirect biological interactions. Furthermore, the application by Micheli (1997) of foraging theory to explain patterns of blue crab predation on *Mercenaria* as a function of habitat characteristics implies that individually based behavioral models for key predators, such as the blue crab, are also critically needed to explain and predict associations of *M. mercenaria* abundance with sedimentary and other habitat variables in nature. Thus, a more complete predictive model relating the abundance of *M. mercenaria* to sedimentary habitats will surely involve interactions among important factors, as does the trophic group amensalism concept, but will incorporate mechanistic understanding of how post-larval survival is related to sediment dynamics, how fluid dynamics affect feeding success, and how predator behavior responds to changes in habitat.

10.3 INDIVIDUAL GROWTH AS A FUNCTION OF VEGETATION

Like the trophic group amensalism concept, which attempts to define how interactions between the sedimentary habitat and deposit-feeding animals influence the success of suspension feeders such as *M. mercenaria*, a substantial area of research on factors controlling growth in hard clams has also involved interactions between physics and physiological ecology.

Specifically, the question of how growth of *Mercenaria* is altered by the presence of emergent seagrasses has required understanding of how fluid dynamics interact with emergent structural habitat to alter the ability of these suspension-feeding bivalves to achieve net growth. This area of research thus requires interdisciplinary study to combine boundary layer hydrodynamics with feeding physiology of bivalves.

The earliest study (Kerswill, 1949) of how *Mercenaria* growth is affected by seagrass habitat showed reduced growth inside seagrass beds, which was interpreted as a simple consequence of the slower current flows underneath the canopy of vegetation (Fig. 10.1:

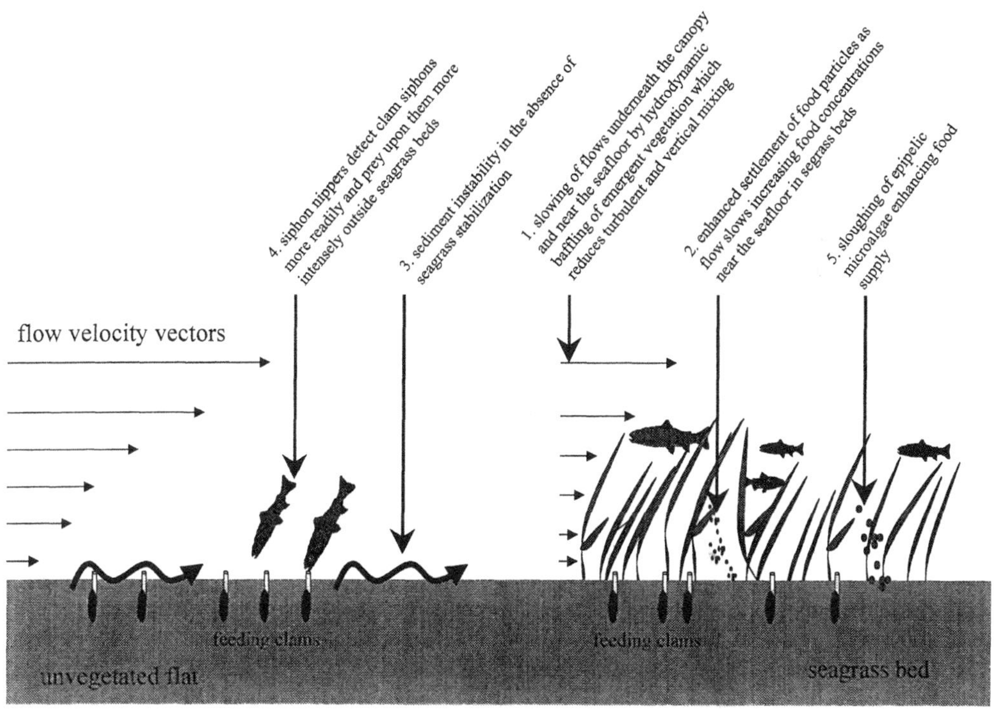

Fig. 10.1. Cartoon depiction of how the presence of seagrass influences factors that may cause net growth rate of individual hard clams to differ from in- to outside seagrass beds. (1) Emergent seagrass blades baffle the current flow such that underneath the canopy, flow velocity is slower than outside on the unvegetated flat at any given elevation. (This can reduce hard clam growth if food concentration is low enough that turbulence and vertical mixing over the flat is important for preventing near-bed food depletion.) (2) As baffling acts to reduce flow velocities and reduce turbulent vertical mixing underneath the canopy, settlement of food particles toward the seafloor will be enhanced. (This can increase food concentration inside the seagrass bed near the seafloor and thus enhance hard clam growth rates.) (3) The seagrass plants help bind the sediments and reduce flow velocities and bottom shear stress, thereby stabilizing the sediments. (This can increase hard clam growth rates inside seagrass beds because greater sediment stability reduces the need for clams to re-establish position in the sediments and reduces the degree of feeding inhibition caused by mobilized and resuspended inorganic sediments.) (4) Seagrass represents a barrier to vision to help hide clam siphons from siphon-nipping fishes. (This can increase net growth of hard clams in seagrass beds because loss of siphons requires energy for regeneration and may reduce feeding efficiency.) (5) Epipelic microalgae grow on seagrass blades and slough off in flows. (This can enhance the supply and concentration of food particles for hard clams inside seagrass beds.)

process 1). Subsequent tests failed to confirm this pattern and, in stark contrast, showed dramatically enhanced growth of hard clams inside seagrass bed habitat (Peterson et al., 1984; Irlandi and Peterson, 1991). Peterson et al. (1984) suggested that the higher growth inside seagrass beds might be a consequence of baffling of near-bed current flows by the emergent blades of grass. This creates deceleration of flows, deposition of fine particles, and potential enhancement of the rate of deposition of particulate foods, which enhances food concentrations near the seabed where feeding takes place. The viability of this explanation was tested in a kelp bed by Eckman et al. (1989a), who indeed showed greater deposition of particulates inside the canopy of an understory kelp bed as a response to the longer residence time of slower moving water under the canopy. This hydrodynamic baffling explanation requires that the enhanced vertical flux of food particles inside the seagrass canopy more than counteract the effects of flow reduction (Fonseca et al., 1982; Eckman, 1987; Gambi et al., 1990) and turbulence reduction (Ackerman and Okubo, 1993; Leonard and Luther, 1995) underneath the canopy. Both of these mechanisms serve to increase the potential for food depletion. Depending upon variables such as food concentration, suspension-feeder density, absolute flow velocity, and distance into the bed, this balance between opposing effects of the seagrass baffle on food supply rate could tilt in either direction. Indeed, Peterson and Beal (1989) demonstrated in three separate localities examples of all three possible outcomes, one with faster, one with slower, and one with equal rates of *M. mercenaria* growth inside the seagrass as compared to the adjacent unvegetated habitat.

Irlandi and Peterson (1991) proposed several alternative, but not mutually exclusive, hypotheses to explain the faster growth of *Mercenaria* under seagrass canopies in their study (Fig. 10.1: processes 2–5). In addition to (1) the baffling hypothesis (slowing of flow creating greater vertical flux of food particles), they suggested (2) that the seagrass bed may serve as a partial refuge for hard clams from siphon-nipping predators (thereby enhancing net growth), (3) that the presence of the seagrass may help stabilize the sediments (thus reducing the negative influence of intense sediment transport on the unvegetated sand flat), and (4) that epiphytic algae sloughing off the seagrass leaves may be enhancing food concentrations inside the seagrass bed. Support for the siphon-nipping explanation comes from observations of greater frequencies of siphon damage among *Mercenaria* found on unvegetated bottom than inside shoalgrass beds (Coen and Heck, 1991) and experimental manipulations of siphon-nipping fish by Irlandi (1994). Circumstantial support for the sediment stability hypothesis is provided by Irlandi and Peterson's (1991) demonstration that hard clam growth was slower at the edge of the seagrass bed that received faster flow velocities. Indirect support for the hypothesis that epiphytic algae enhance food abundance inside seagrass beds comes from a study of ribbed mussel (*Geukensia demissa*) growth at different densities, tidal elevations, and positions within a *Spartina alterniflora* marsh (Lin, 1989). Ribbed mussels at one study site grew much faster in the marsh interior position, which could be explained by a subsidy of epiphytic algae associated with the *Spartina* culms. Judge et al. (1992) provided direct support for the epiphytic algae hypothesis by demonstrating in a study of *M. mercenaria* that up to 90% of the chlorophyll-bearing particulates sampled below 5 cm inside a seagrass bed was comprised of pennate diatoms. This revealed how important epipelic microbes can be as hard clam food supplies.

The absence of a single definitive explanation for observed patterns of *Mercenaria* growth in- and outside seagrass habitat arises (1) from the reality of the complex situation in which multiple processes contribute, and (2) from the failure to develop a complete mechanistic understanding of the factors that determine growth in this species. In principle, a passive suspension feeder should grow in proportion to food flux, whereas an actively pumping suspension feeder should respond to food concentration rather than flux, other factors remaining the same (Eckman et al., 1989b). The importance of food concentration to growth in *Mercenaria* (e.g., Walne, 1972) and other suspension-feeding bivalves, especially mussels (Bayne et al., 1976; Widdows et al., 1979), seems reasonably well established by both first principles and experiment. However, the role of flow velocity is much less clear. For example, Judge et al. (1992) constructed field flumes to test whether growth responded to flow velocity, holding water mass and presumably available food constant for *M. mercenaria* and found that a 40% reduction and a 65% increase in flow velocity did not change observed growth rates. Nevertheless, several laboratory experiments with hard clams have revealed positive effects of flow velocity on *Mercenaria* growth (e.g., Manzi et al., 1986; Grizzle et al., 1992).

The positive relationships between flow velocity and growth in *Mercenaria* and other active suspension-feeding bivalves can best be explained by a combination of direct, fine-scale effects at low flows and indirect effects on food concentration over a wide range of flows (Lenihan et al., 1996), illustrated in Fig. 10.2. At very low flow velocities (from 0 to a few centimeters per second), increases in flow may enhance growth (1) by creating increased shear that disrupts the food-depleted envelope of water in the immediate vicinity of the animal (O'Riordan et al., 1993; Wildish and Salnier, 1993; Lenihan et al., 1996), and (2) by providing a pressure gradient that, depending upon orientation of the animal to the flow, can enhance the efficiency of water passage through the feeding system of the animal (Jørgensen et al., 1986; Wildish et al., 1987; Eckman et al., 1989b; Lenihan et al., 1996; Wildish and Kristmanson, 1997). Over a wide range of flows, increased flow velocity acts indirectly on growth of active suspension feeders by increasing food concentrations in the vicinity of food-gathering organs. Flow counteracts depletion on larger spatial scales in two ways. First, faster flows move parcels of water more quickly past beds of feeding suspension feeders and thereby reduce the amount of re-filtration of previously processed water by animals downstream (O'Riordan et al., 1993). This effect probably explains the impacts of enhancing flow velocity in aquaculture settings for hard clams (e.g., Manzi et al., 1986). Second, and even more importantly, increased flow creates greater vertical mixing of the water column, which allows net vertical transport of suspended foods to replace those depleted near the seafloor by feeding suspension feeders (Wildish and Kristmanson, 1979, 1985). This successful application of boundary layer hydrodynamics to feeding biology of suspension feeders represented a true breakthrough in understanding environmental forcing of an important biological transfer process of importance to all suspension feeders (e.g., Cloern, 1991), including *M. mercenaria.* The effect of enhanced mixing is perhaps the major cause for the large-scale association of greater biomass and production of suspension feeders with faster current flows (e.g., Wildish and Peer, 1983; Wildish and Kristmanson, 1997). The enhanced depletion of food nearer the seafloor (e.g., Fréchette et al., 1989) and the poorer food quality closer to the bottom (Muschenheim, 1987) probably explain the frequently observed behavior of many infaunal bivalves, including *M. mercenaria*, to extend their siphon tips upwards above the seafloor.

This explanation for the role of flow velocity in enhancing growth of active suspension feeders on the seafloor assumes that localized food depletion can be achieved by beds of suspension feeders. Such depletion has indeed been demonstrated multiple times using a

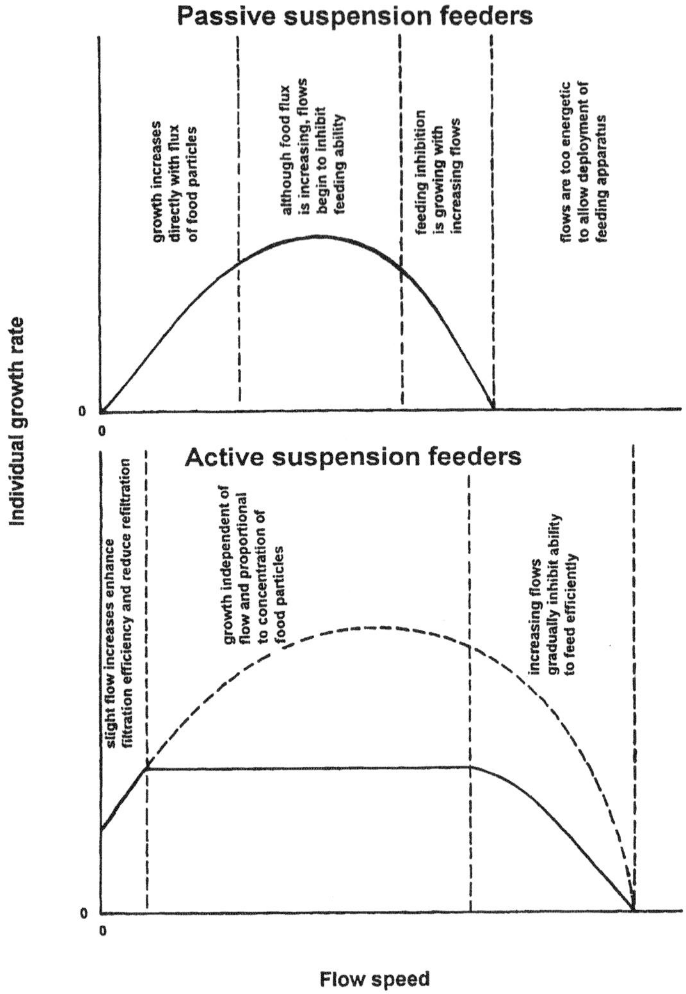

Fig. 10.2. Graphical depiction of the complex relationships between flow speed as it encounters a feeding suspension feeder and individual growth rate, assuming no change in concentration of food particles in the flow. The diagram compares the response of a passive benthic suspension feeder to an active suspension feeder like a hard clam. Note. (1) Growth falls to near zero in still waters for a passive suspension feeder but an active suspension feeder can still achieve net growth because it can pump a feeding current. (2) Growth increases with flow speed for active suspension feeders only over a narrow range of flows above zero, where such slow flows may enhance pumping efficiency and also remove depleted waters in the immediate vicinity of a feeding individual, whereas growth is proportional to flow speed for passive suspension feeders over a wider range of flows until the flows become so energetic as to begin to inhibit effective deployment of the feeding apparatus. (3) Growth of active suspension feeders is independent of flow speed per se over a wide range of flows, as indicated by the unbroken line in the middle section of the curve (Note that in nature, as flows increase through this range, growth rates will generally increase but because of the indirect effect of enhanced turbulent vertical mixing, which enhances food concentrations in the near-bed benthic boundary layer by increasing the rate of mixing of overlying food-rich waters into the depleted bottom waters, as indicated by the dashed line.) (4) High flow speeds are ultimately inhibitory to benthic suspension feeders of both types, but complete inhibition occurs at lower flow speeds for passive suspension feeders because of their need to expose their feeding apparatus directly to the damaging shear forces and turbulence of the flow.

variety of bivalve molluscs, both indirectly through manipulations of bivalve density to test its effect on growth (Peterson, 1982; Peterson and Black, 1987), and directly by measuring chlorophyll concentrations in the water column (Fréchette and Bourget, 1985; Fréchette et al., 1989; Peterson and Black, 1991). These studies include tests of *M. mercenaria*, which revealed reduced growth as density was increased from 290 to 869 to 1159 per m^2 in aquaculture trays (Eldridge et al., 1979) and as density was varied eight-fold from 10 to 80 in 1 square meter plots of natural bottom habitat (Peterson and Beal, 1989). Nevertheless, it is likely that hard clams also occur in situations of low density and high food concentrations where depletion would not occur, thereby rendering growth constant over a wide range of flow velocities (Judge et al., 1992). Evaluation of the degree to which food is limiting seems an important step in resolving the contradictory evidence on the effects of seagrass on hard clam growth.

At sufficiently high flows, suspension-feeder growth typically exhibits suppression (Wildish and Kristmanson, 1997). Suppression results when flow damages or interferes with the feeding apparatus (e.g., Okamura, 1987) or when flow creates need for energetically demanding re-burrowing activity to retain a position at the sediment–water interface (Myers, 1977). Bottom shear stress can even reach high enough levels to cause erosion of clams and transport away from energetic sites, so effects on growth become moot (Emerson and Grant, 1991). Because of this inhibitory effect of high flow velocities on suspension feeders, Wildish and Kristmanson (1997) suggested that a general curve relating growth of active suspension feeders like *M. mercenaria* would be characterized (1) by a short rising segment, where fine-scale effects of increased flow enhance filtration efficiency and reduce re-filtration, (2) by a long flat segment in which growth and flow are independent, and (3) by a declining segment of the curve where feeding inhibition is induced. Flume observations by Grizzle et al. (1992) demonstrated the first two segments of this curve for *M. mercenaria*, but flow velocities did not range high enough to demonstrate the inhibitory phase.

Some papers have suggested that it is appropriate to conclude that growth of active suspension feeders responds to food flux (e.g., Grizzle and Morin, 1989). Food flux clearly is the appropriate variable with which to predict growth in passive suspension feeders (Muschenheim, 1987), at least until flow rates are reached that endanger or inhibit feeding organs. However, for active suspension feeders, this relationship is only fortuitous when observed, in that growth is actually responding independently to the two components of food flux, food concentration and flow speed (Fréchette et al., 1993; Lenihan et al., 1996; Wildish and Kristmanson, 1997). Over the range of slow flow velocities where growth responds positively, food flux and growth should be positively correlated. Over the range of flows in which growth is independent of flow, concentration alone dictates growth: depending on how concentration varies relative to flow, their product may or may not correlate with concentration, and thus may or may not explain growth. In the portion of the curve where increasing flow inhibits growth, food flux and growth should be totally decoupled. Combining these two independent factors into a single variable is not only misleading by displacing focus away from the mechanisms but also generally incorrect. For these reasons, further pursuit of the explanation for variable effects of seagrass habitat on hard clam growth should treat separately both flow effects and food concentration effects.

Although the debate in the literature about the impacts of seagrass on *Mercenaria* growth has focused on flow speed, food concentration, sediment disturbance, and siphon nippers, food quality represents another factor influencing growth of suspension feeders (e.g., Campbell and Newell, 1998) that may be altered by the presence of seagrass habitat. Specifically, the amount of inorganic sediment suspended along with edible food particles can vary among habitats in nature and influence hard clam growth (see: Fegley et al., 1992; Judge et al., 1993; Bock and Miller, 1994). There has been a substantial effort among feeding physiologists to characterize the influence of suspended silts on *Mercenaria* growth in laboratory settings, but relatively little analysis of seston quality in the field as a function of habitat. Relatively high concentrations of suspended inorganic particles tends to reduce growth in *M. mercenaria* (Bricelj and Malouf, 1984; Bricelj et al., 1984; Murphy, 1985; Huntington and Miller, 1989; Turner and Miller, 1991), although addition of small amounts of silt has no effect or even stimulates growth (e.g., Davis, 1960). Grizzle and Lutz (1989) showed reductions in growth of *Mercenaria* in the field as suspended sediment concentrations increased over a range of 51–111 mg per l. Integrating an understanding of effects of seagrass habitat on seston quality as well as quantity would seem to represent a useful next step in solving the mystery of how hard clam growth is affected by seagrass habitat. These results then need to be considered in a broader context of the fluid dynamics of boundary layer flows in the presence of suspension-feeding bivalves, which can act simultaneously as both sinks for particulates and as sources of fluid jets. The insightful models of these interactions (Fréchette et al., 1989; Monismith et al., 1990; O'Riordan et al., 1993) have not attempted to include effects of emergent seagrass blades. Mechanistic inclusion of fluid dynamic modeling with our understanding of the physiology of suspension feeding has led already to new appreciation of how physical processes influence hard clam growth. This approach represents the most likely avenue for explaining the remaining mysteries of how seagrass habitat acts to modify growth of suspension feeders. One major limitation to progress in explaining patterns of growth in hard clams is the continuing uncertainty over what exactly represents food among the diverse suite of suspended organic particulates and how to weight different types of food particles by their relative value for inducing growth. This problem might best be solved by measuring growth of hard clams under various controlled diets.

10.4 INTERACTIONS BETWEEN MULTIPLE PHYSIOLOGICAL STRESSORS

The principle that organisms respond in their physiological condition to interactions among multiple stressors is well accepted among physiologists. The most common manifestation of this principle in studies of marine invertebrates is the repeated demonstration that salinity tolerance limits depend on temperature and temperature tolerance limits depend on salinity (Kinne, 1964). As a consequence of the interaction between these two basic environmental variables in their effects on marine invertebrates, it is impossible to specify fixed temperature and salinity tolerances for any species: tolerance is most accurately depicted within a temperature–salinity bivariate plot. An analogous concept applies to human physiology: epidemiologists recognize that the likelihood of contraction of disease upon exposure to an appropriate transmission vector is influenced by physiological condition, such that pre-existing illness and even exhaustion predisposes one towards contraction of another disease.

The implications of interactions among multiple physiological stressors to the ecology of marine organisms, including hard clams, are only beginning to be investigated (see Grosholz, 1992). Many of the factors that represent stressors to hard clams and other marine invertebrates are physical or chemical variables in the environment. For example, sedimentation can bury and kill suspension-feeding bivalves (e.g., Kranz, 1974), excessively high flow speeds can inhibit feeding (Wildish and Kristmanson, 1997), low-salinity events during major storms and freezing temperatures can cause mass mortality (Greene and Becker, 1977), and low oxygen and high sulfide concentrations can cause mortality of sedentary benthic invertebrates. However, biological factors can also represent stressors of importance to hard clams. For example, locally high population density can cause food depletion, reduce growth, and lower the physiological condition of *M. mercenaria* (Nakaoka, 1999). Disease and parasitism commonly have serious physiological manifestations. Consequently, there is substantial, and largely unexplored, potential for physical and biological factors to interact to dictate both physiological state of hard clams and their ecological responses.

Perhaps the most explicit example of how biological and physical factors can interact to affect the physiology and ecology of a suspension-feeding bivalve is found in a study by Peterson and Black (1988). They showed for two species of suspension-feeding bivalves that a history of being maintained at high density caused not only lower growth but also subsequent higher mortality when exposed to the stress of burial by sediments. This represents an interaction between a biological factor, the historic intraspecific crowding, and a physical stressor, sedimentation, in effecting observed mortality rate. The results of this experiment help provide an explanation for why an intense storm that deposited about 10 cm of fine sediments over suspension-feeding bivalves caused higher mortality rates where local densities were higher (Peterson, 1985). It seems reasonable to expect that *M. mercenaria* would exhibit a similar interaction between intense crowding (and consequent reduced physiological condition) and susceptibility to another stress, such as sedimentation. While analogous data for hard clams are unavailable, the rate of infection by a protozoan parasitic oyster disease and the intensity of infection have both been shown to respond to change in oyster reef habitat. Near crests of tall, undegraded reefs, where flow speeds are highest, oyster growth is maximal, condition index is highest, and disease infection rates and intensities are lowest (Lenihan et al., 1999). Similar interactions among physiological stresses are doubtless important to the ecology of hard clams.

In addition to the role of intraspecific population density and the role of parasites and disease organisms in creating physiological stresses of ecological significance, several other biological interactions between hard clams and other types of species are potentially important. Locally high densities of other suspension feeders may reduce food concentration through interspecific competition. This potential for interspecific competition has not been subjected to test using *M. mercenaria*, although tests of the importance of interspecific competition for food among other species of co-occurring suspension-feeding bivalves have been conducted and have not demonstrated large impacts (e.g.: Peterson, 1982; Peterson and Black, 1993). *Mercenaria* can also be affected physiologically by the presence of epibionts and shell borers. For example, Peterson (1983b) demonstrated small reductions in growth of the suspension-feeding bivalve *Chione undatella*, when its shell was occupied by a suspension-feeding slipper limpet. *M. mercenaria* is also host for another species of *Crepidula (fornicata)*, but no study has evaluated the consequences to *Mercenaria*. *Mercenaria* shells are frequently bored by the polychaete, *Polydora*. This pest probably has much more serious physiological and ecological consequences than slipper limpets because the burrow can ultimately penetrate the shell. Nevertheless, studies of interactions between *Polydora* and other stressors in *Mercenaria* are lacking. Finally, siphon nippers are likely to have substantial impact on the physiological condition of hard clams, given the growing evidence of how commonly this process of cropping is directed towards hard clams (e.g.: Coen and Heck, 1991; Irlandi, 1994). Many species of juvenile demersal fishes pass through a developmental period when the bulk of their diet is comprised of bivalve siphons (Peterson and Skilleter, 1994).

10.5 FUTURE DIRECTIONS

The integration of feeding physiology with ecology has tremendous potential for future advances in understanding how physical and biological processes interact to influence the biology of *M. mercenaria*. There has been much more progress made using this approach on two other groups of suspension-feeding bivalves, marine mussels and scallops, than on hard clams. A long tradition of successful integration of feeding physiology and ecology of blue mussels has been established through the vision of Bayne and colleagues (e.g.: Bayne et al., 1976, 1993; Widdows et al., 1979; Thompson, 1984; see also Kiørboe et al., 1981). Similar research efforts at the interface of feeding physiology and ecology have characterized research on scallops (Vahl, 1980; Bricelj et al., 1987; Shumway et al., 1987; Wildish et al., 1987; MacDonald and Ward, 1994). Surprisingly, far fewer such studies have been conducted on hard clams and other infaunal bivalves (Iglesias et al., 1992; Navarro and Iglesias, 1993). Because hard clams occupy a soft-sediment habitat, where flows are more readily characterized and even modeled than in the topographically complex rocky shore or in the intertidal zone where wave action also substantially complicates the fluid dynamics (Denny, 1988), there is reason to believe that understanding how physical–biological interactions influence hard clam biology might be more feasible than for blue mussels. On the other hand, because hard clams occupy a habitat comprised of unconsolidated sediments, whereas blue mussels attach to hard substrata, it seems likely that issues of habitat stability and inorganic contamination of food resources will represent a greater complication in understanding physical–biological coupling in hard clams. Scallops have proven to be a popular study organism because they can be more easily observed because of their epibiotic living position, but this too means that they are not influenced in quite the same ways by sediment stability as an infaunal organism like the hard clam.

Some new technology has promise for future advance in assessing how physical and biological interactions combine to influence hard clam biology. The use of flow chambers in the field (Ward and MacDonald, 1996) and flumes and wave tanks in the laboratory (Butman et al., 1988; Miller et al., 1992) to replace unrealistic observation in still water allows evaluation of how feeding physiology is influenced by flow regime. One of the most exciting technological developments that is improving our understanding of feeding physiology and its relationships to physical and biological variables is the use of the endoscope (Ward et al., 1991). This tool adapted from medical science allows direct observation of particle sorting processes, which is providing new insights and challenging old dogma in bivalve feeding physiology (Ward et al., 1993). When this device is used to study feeding in hard clams, we can expect to be able to improve our mechanistic understanding of the process of particle sorting and thus the costs and benefits of variation of food quality and quantity, thereby resolving some unsolved mysteries. The physics of near-bed flow and particle transport are now being much more accurately and carefully described, in part because physical oceanographers have begun to study coastal processes despite their complexities, and, in part, because true collaborations between physical and biological oceanographers are growing in appeal (e.g.: Fréchette et al., 1989; O'Riordan et al., 1993). These partnerships of biologists and fluid dynamicists have involved ecologists for some time, but the active collaboration of fluid dynamicists with feeding physiologists is much more in its infancy. Expanded research of this type holds much promise for improved physiological and ecological appreciation of how hard clams function both in nature and in aquaculture settings. Finally, the integration of fluid dynamics, ecological measures, and feeding physiological rates into process models that can be tested against field data (e.g.: Campbell and Newell, 1998; Grant and Bacher, 1998; Hawkins et al., 1998) represents a major advance in evaluating and improving our understanding of how feeding physiology dictates growth and production of suspension-feeding bivalves, such as *M. mercenaria*. The use of modeling is driven in large part by the needs in aquaculture for assessing carrying capacity in the field to determine appropriate stocking densities. Advances in integrating physics, feeding physiology, and ecological processes thus have enormous practical as well as intellectual value.

REFERENCES

Ackerman, J.D. and Okubo, A., 1993. Reduced mixing in a marine macrophyte canopy. Funct. Ecol., 7: 305–309.

Ahn, I.-Y., Malouf, R. and Lopez, G., 1993a. Enhanced larval settlement of the hard clam. *Mercenaria mercenaria* by the gem clam *Gemma gemma*. Mar. Ecol. Prog. Ser., 99: 51–59.

Ahn, I.-Y., Lopez, G. and Malouf, R., 1993b. Effects of the gem clam *Gemma gemma* on early post-settlement emigration, growth and survival of the hard clam *Mercenaria mercenaria*. Mar. Ecol. Prog. Ser., 99: 61–70.

Bachelet, G., Butman, C.A., Webb, C.M., Starczak, V.R. and Snelgrove, P.V.R., 1992. Non-selective settlement of *Mercenaria mercenaria* (L.) larvae in short-term, still-water, laboratory experiments. J. Exp. Mar. Biol. Ecol., 161: 241–280.

Bayne, B.L., Widdows, J., and Thompson, R.J., 1976. Physiological integrations. In: B.L. Bayne (Ed.), Marine Mussels: Their Ecology and Physiology. Cambridge University Press, Cambridge, pp. 261–299.

Bayne, B.L., Iglesias, J.I.P., Hawkins, A.J.S., Navarro, E., Héral, M. and Deslous-Paoli, J.M., 1993. Feeding behavior of the mussel, *Mytilus edulis*: responses to variations in quantity and organic content of the seston. J. Mar. Biol. Assoc. U.K., 73: 813–829.

Bock, M.J. and Miller, D.C., 1994. Seston variability and daily growth in *Mercenaria mercenaria* on an intertidal sandflat. Mar. Ecol. Prog. Ser., 114: 117–127.

Brenchley, G.A., 1981. Disturbance and community structure: an experimental study of bioturbation in marine soft-bottom environments. J. Mar. Res., 39: 767–790.

Bricelj, V.M. and Malouf, R.E., 1984. Influence of algal and suspended sediment concentrations on the feeding physiology of the hard clam *Mercenaria mercenaria*. Mar. Biol., 84: 155–165.

Bricelj, V.M., Malouf, R.E. and de Quillfeldt, C., 1984. Growth of juvenile *Mercenaria mercenaria* and the effect of suspended bottom sediments. Mar. Biol., 84: 167–173.

Bricelj, V.M., Epp, J. and Malouf, R.E., 1987. Comparative physiology of young and old cohorts of the bay scallop, *Argopecten irradians irradians* (Lamarck): mortality, growth and oxygen consumption. J. Exp. Mar. Biol. Ecol., 112: 73–91.

Butman, C.A., Grassle, J.P. and Webb, C.M., 1988. Substrate choices made by marine larvae settling in still water and in a flume flow. Nature, 333: 771–773.

Campbell, D.E. and Newell, C.R., 1998. MUSMOD, a production model for bottom culture of the blue mussel, *Mytilus edulis* L. J. Exp. Mar. Biol. Ecol., 219: 171–203.

Carriker, M.R., 1961. Interrelation of functional morphology, behavior, and autecology in early stages of the bivalve *Mercenaria mercenaria*. J. Elisha Mitchell Sci. Soc., 77: 168–241.

Cloern, J.E., 1991. Tidal stirring and phytoplankton bloom dynamics in an estuary. J. Mar. Res., 49: 203–221.

Coen, L.D. and Heck, K.L., 1991. The interacting effects of siphon nipping and habitat on the bivalve (*Mercenaria mercenaria* (L.)) growth in a subtropical seagrass (*Halodule wrightii* Aschers) meadow. J. Exp. Mar. Biol. Ecol., 145: 1–13.

Davis, H.C., 1960. Effects of turbidity-producing materials in sea water on eggs and larvae of the clam (*Venus mercenaria*). Biol. Bull., 118: 48–54.

Dayton, P.K., and Oliver, J.S., 1980. An evaluation of experimental analyses of population and community patterns in benthic marine environments. In: K.R. Tenore and B.C. Coull (Eds.), Marine Benthic Dynamics. University of South Carolina Press, Columbia, SC, pp. 93–120.

Denny, M.W., 1988. Biology and the Mechanics of the Wave-Swept Environment. Princeton University Press, Princeton, NJ.

Eckman, J.E., 1987. The role of hydrodynamics in recruitment, growth and survival of *Argopecten irradians* (L.) and *Anomia simplex* (D'Orbigny) within eelgrass meadows. J. Exp. Mar. Biol. Ecol., 106: 165–191.

Eckman, J.E., Duggins, D.O. and Sewell, A.T., 1989a. Ecology of understory kelp environments. I. Effects of kelps on flow and particle transport near the bottom. J. Exp. Mar. Biol. Ecol., 129: 173–188.

Eckman, J.E., Peterson, C.H. and Calahan, J.A., 1989b. Effects of flow speed, turbulence, and orientation on growth of juvenile bay scallops *Argopecten irradians concentricus* (Say). J. Exp. Mar. Biol. Ecol., 132: 123–140.

Eldridge, P.J., Eversole, A.G. and Whetstone, J.M., 1979. Comparative survival and growth rates of hard clams, *Mercenaria mercenaria*, planted in trays subtidally and intertidally at varying densities in a South Carolina estuary. Proc. Natl. Shellfish. Assoc., 69: 30–39.

Emerson, C.W. and Grant, J., 1991. The control of soft-shell clam (*Mya arenaria*) recruitment on intertidal sand flats by bedload sediment transport. Limnol. Oceanogr., 36: 1288–1300.

Fegley, S.R., MacDonald, B.A. and Jacobsen, T.R., 1992. Short-term variation in the quantity and quality of seston available to benthic suspension-feeders. Estuarine Coastal Shelf Sci., 34: 393–412.

Fonseca, M.S., Fisher, J.S., Zieman, J.C. and Thayer, G.W., 1982. Influence of the seagrass, *Zostera marina* L., on current flow. Estuarine Coastal Shelf Sci., 15: 351–362.

Fréchette, M. and Bourget, E., 1985. Energy flow between the pelagic and benthic zones: factors controlling particulate organic matter available to an intertidal mussel bed. Can. J. Fish. Aquat. Sci., 42: 1166–1170.

Fréchette, M., Butman, C.A. and Geyer, W.R., 1989. The importance of boundary layer flows in supplying phytoplankton to the benthic suspension feeder, *Mytilus edulis* L. Limnol. Oceanogr., 34: 19–36.

Fréchette, M., Lefaivre, D., and Butman, C.A., 1993. Bivalve feeding and the benthic boundary layer. In: R. Dame (Ed.), Bivalve Filter Feeders in Estuarine and Coastal Ecosystem Processes. NATO ASI Ser. Vol. 33, Springer, Berlin, pp. 325–369.

Gambi, M.C., Newell, A.R.M. and Jumars, P.A., 1990. Flume observations on flow dynamics in *Zostera marina* (eelgrass) beds. Mar. Ecol. Prog. Ser., 61: 159–169.

Grant, J. and Bacher, C., 1998. Comparative models of mussel bioenergetics and their variation at field culture sites. J. Exp. Mar. Biol. Ecol., 219: 21–44.

Greene, G.T., and Becker, D.S., 1977. Winter kill of hard clams in Great South Bay, N.Y., 1976–77. Marine Sciences Research Center, State University of New York, Stony Brook, NY, Spec. Rep. 9, 77-5, pp. 1–23.

Grizzle, R.E. and Lutz, R.A., 1989. A statistical model relating horizontal seston fluxes and bottom sediment characteristics to growth of *Mercenaria mercenaria*. Mar. Biol., 102: 95–105.

Grizzle, R.E. and Morin, P.J., 1989. Effects of tidal currents, seston, and bottom sediments on growth of *Mercenaria mercenaria*: results of a field experiment. Mar. Biol., 102: 85–93.

Grizzle, R.E., Langan, R. and Howell, W.H., 1992. Growth responses of suspension-feeding bivalve molluscs to changes in water flow: differences between siphonate and nonsiphonate taxa. J. Exp. Mar. Biol. Ecol., 162: 213–228.

Grosholz, E.D., 1992. Interactions among intraspecific, interspecific, and apparent competition with host–pathogen population dynamics. Ecology, 73: 507–514.

Hawkins, A.J.S., Bayne, B.L., Bougrier, S., Héral, M., Iglesias, J.I.P., Navarro, E., Smith, R.F.M. and Urrutia, M.B., 1998. Some general relationships in comparing the feeding physiology of suspension-feeding bivalve molluscs. J. Exp. Mar. Biol. Ecol., 219: 87–103.

Hunt, J.H., Ambrose Jr., W.G. and Peterson, C.H., 1987. Effects of the gastropod, *Ilyanassa obsoleta* (Say), and the bivalve, *Mercenaria mercenaria* (L.), on larval settlement and juvenile recruitment of infauna. J. Exp. Mar. Biol. Ecol., 108: 129–140.

Huntington, K.M. and Miller, D.C., 1989. Effects of suspended sediment, hypoxia, and hyperoxia on larval *Mercenaria mercenaria* (Linnaeus, 1758). J. Shellfish Res., 8: 37–42.

Iglesias, J.I.P., Navarro, E., Alverez Jorna, P. and Armentaria, I., 1992. Feeding, particle selection, and absorption in cockles *Cerastoderma edule* (L.) exposed to variable conditions of food concentration and quality. J. Exp. Mar. Biol. Ecol., 162: 177–198.

Irlandi, E.A., 1994. Large- and small-scale effects of habitat structure on rates of predation: how percent coverage of seagrass affects rates of predation and siphon nipping on an infaunal bivalve. Oecologia, 98: 176–183.

Irlandi, E.A. and Peterson, C.H., 1991. Modification of animal habitat by large plants: mechanisms by which seagrasses influence clam growth. Oecologia, 87: 307–318.

Jørgensen, C.B., Famme, P., Kristensen, H.S., Larsen, P.S., Møhlenberg, F. and Riisgard, H.V., 1986. The bivalve pump. Mar. Ecol. Prog. Ser., 34: 69–77.

Judge, M.L., Coen, L.D. and Heck Jr., K.L., 1992. The effect of long-term alteration of in situ water currents on the growth of the hard clam *Mercenaria mercenaria* in the northern Gulf of Mexico. Limnol. Oceanogr., 37: 1550–1559.

Judge, M.L., Coen, L.D. and Heck Jr., K.L., 1993. Does *Mercenaria mercenaria* encounter elevated food levels in seagrass beds? Results from a novel technique to collect suspended food resources. Mar. Ecol. Prog. Ser., 92: 141–150.

Keck, R., Maurer, D. and Malouf, R., 1974. Factors influencing the settlement behavior of larval hard clams, *Mercenaria mercenaria*. Proc. Natl. Shellfish. Assoc., 64: 59–67.

Kerswill, C.J., 1949. Effects of water circulation on the growth of quahaugs and oysters. J. Fish. Res. Bd. Can., 7: 545–551.

Kinne, O., 1964. The effects of temperature and salinity on marine and brackish water animals. II. Salinity and temperature–salinity relations. Oceanogr. Mar. Biol. Annu. Rev., 2: 281–339.

Kiørboe, T., Møhlenberg, F. and Nøhr, O., 1981. Effect of suspended bottom material on growth and energetics in *Mytilus edulis*. Mar. Biol., 61: 283–288.

Kranz, P.M., 1974. The anastrophic burial of bivalves and its paleoecological significance. J. Geol., 82: 237–265.

Lenihan, H.S., Peterson, C.H. and Allen, J.M., 1996. Does flow speed also have a direct effect on growth of active suspension feeders: an experimental test on oysters, *Crassostrea virginica* (Gmelin). Limnol. Oceaonogr., 41: 1359–1366.

Lenihan, H.S., Micheli, F., Shelton, S.W., and Peterson, C.H., 1999. Experimental demonstration on restored reefs that enhanced flow speeds reduce parasitic infection in oysters. Limnol. Oceanogr., 44: in press.

Leonard, L.A. and Luther, M.E., 1995. Flow hydrodynamics in marsh canopies. Limnol. Oceanogr., 40: 1474–1484.

Levinton, J.S., 1977. Ecology of shallow water deposit-feeding communities, Quisset Harbor, Massachusetts. In: B.C. Coull (Ed.), Ecology of Marine Benthos. Belle Baruch Library in Marine Science 6. University of South Carolina Press, Columbia, SC, pp. 191–227.

Lin, J., 1989. Importance in location in the salt marsh and clump size on growth of ribbed mussels. J. Exp. Mar. Biol. Ecol., 128: 75–86.

Loosanoff, V.L. and Davis, H.C., 1950. Rearing of bivalve molluscs. Adv. Mar. Biol., 1: 1–136.

MacDonald, B.A. and Ward, J.E., 1994. Variation in food quality and particle selectivity in the sea scallop *Placopecten magellanicus* (Mollusca: Bivalvia). Mar. Ecol. Prog. Ser., 108: 251–264.

Manzi, J.J., Hadley, N.H. and Maddox, M.B., 1986. Seed clam, *Mercenaria mercenaria*, culture in an experimental-scale upflow nursery system. Aquaculture, 54: 301–311.

Micheli, F., 1997. Effects of predators' foraging behavior on patterns of prey mortality in marine soft bottoms. Ecol. Monogr., 67: 203–224.

Miller, D.C., Bock, M.J. and Turner, E.J., 1992. Deposit and suspension feeding in oscillatory flows and sediment fluxes. J. Mar. Res., 50: 489–520.

Monismith, S.G., Koseff, J.R., Thompson, J.K., O'Riordan, C.A. and Nepf, H.M., 1990. A study of model bivalve siphonal currents. Limnol. Oceanogr., 35: 680–696.

Murphy, R.C., 1985. Factors affecting the distribution of the introduced bivalve, *Mercenaria mercenaria*, in a California lagoon. The importance of bioturbation. J. Mar. Res., 43: 673–692.

Muschenheim, D.K., 1987. The dynamics of near-bed seston flux and suspension-feeding benthos. J. Mar. Res., 45: 473–496.

Myers, A.C., 1977. Sediment processing in a marine subtidal sandy bottom community. II. Biological consequences. J. Mar. Res., 35: 633–647.

Nakaoka, M., 1999. Non-lethal effects of predators on prey populations: predator-mediated change in bivalve growth. Ecology: in review.
Navarro, E.I., and Iglesias, J.I.P., 1993. Infaunal filter-feeding bivalves and the physiological response to short-term fluctuations in food availability and composition. In: R.F. Dame (Ed.), Bivalve Filter Feeders in Estuarine and Coastal Ecosystem Processes. Springer, Berlin, pp. 25–56.
Nowell, A.R.M. and Jumars, P.A., 1984. Flow environments of aquatic benthos. Annu. Rev. Ecol. Syst., 15: 303–328.
Okamura, B., 1987. Particle size and flow velocity induce an inferred switch in bryozoan suspension-feeding behaviour. Biol. Bull., 173: 222–229.
O'Riordan, C.A., Monismith, S.G. and Koseff, J.R., 1993. A study of concentration boundary layer formation over a bed of model bivalves. Limnol. Oceanogr., 38: 1712–1729.
Peterson, C.H., 1977. Competitive organization of the soft-bottom macrobenthic communities of southern California lagoons. Mar. Biol., 43: 343–359.
Peterson, C.H., 1980. Approaches to the study of competition in benthic communities in soft sediments. In: V. Kennedy (Ed.), Estuarine Perspectives. Academic Press, New York, pp. 291–302.
Peterson, C.H., 1982. The importance of predation and intra- and interspecific competition in the population biology of two infaunal suspension-feeding bivalves, *Protothaca staminea* and *Chione undatella*. Ecol. Monogr., 52: 437–475.
Peterson, C.H., 1983a. A concept of quantitative reproductive senility: application to the hard clam, *Mercenaria mercenaria* (L.)?. Oecologia, 58: 164–168.
Peterson, C.H., 1983b. Interactions between two infaunal bivalves, *Chione undatella* (Sowerby) and *Protothaca staminea* (Conrad), and two potential enemies, *Crepidula onyx* Sowerby and *Cancer anthonyi* (Rathbun). J. Exp. Mar. Biol. Ecol., 68: 145–158.
Peterson, C.H., 1985. Patterns of lagoonal bivalve mortality after heavy sedimentation and their paleoecological significance. Paleobiology, 11: 139–153.
Peterson, C.H., 1986a. Enhancement of *Mercenaria mercenaria* densities in seagrass beds: is pattern fixed during settlement season or altered by subsequent differential survival?. Limnol. Oceanogr., 31: 200–205.
Peterson, C.H., 1986b. Quantitative allometry of gamete production by *Mercenaria mercenaria* into old age. Mar. Ecol. Prog. Ser., 29: 93–97.
Peterson, C.H. and Beal, B.F., 1989. Bivalve growth and higher order interactions: importance of density, site, and time. Ecology, 70: 1390–1404.
Peterson, C.H. and Black, R., 1987. Resource depletion by active suspension feeders on tidal flats: influence of local density and tidal elevation. Limnol. Oceanogr., 32: 143–166.
Peterson, C.H. and Black, R., 1988. Density-dependent mortality caused by physical stress interacting with biotic history. Am. Nat., 131: 257–270.
Peterson, C.H. and Black, R., 1991. Preliminary evidence for sestonic food depletion in incoming tide over a broad tidal sand flat. Estuarine Coastal Shelf Sci., 32: 405–414.
Peterson, C.H. and Black, R., 1993. Experimental tests of the advantages and disadvantages of high density for two coexisting cockles in a Southern Ocean lagoon. J. Anim. Ecol., 62: 614–633.
Peterson, C.H. and Skilleter, G.A., 1994. Control of foraging behavior of individuals within an ecosystem context: the clam *Macoma balthica*, flow environment, and siphon-cropping fishes. Oecologia, 100: 256–267.
Peterson, C.H., Summerson, H.C. and Duncan, P.B., 1984. The influence of seagrass cover on population structure and individual growth rate of a suspension-feeding bivalve, *Mercenaria mercenaria*. J. Mar. Res., 42: 123–138.
Posey, M.H., 1986. Changes in the benthic community associated with dense beds of a burrowing deposit feeder, *Callianassa californiensis*. Mar. Ecol. Prog. Ser., 31: 15–22.
Posey, M.H., 1987. Influence of relative mobilities on the composition of benthic communities. Mar. Ecol. Prog. Ser., 39: 99–104.
Posey, M.H., 1990. Functional approaches to soft-sediment communities: how useful are they?. Rev. Aquat. Sci., 2: 343–356.
Rhoads, D.C. and Young, D.K., 1970. The influence of deposit-feeding organisms on sediment stability and community trophic structure. J. Mar. Res., 28: 150–178.
Sanders, H.L., 1958. Benthic studies in Buzzards Bay. I. Animal–sediment relationships. Limnol. Oceanogr., 3: 245–258.

Shumway, S.E., Selvin, R. and Schick, D.F., 1987. Food resources related to habitat in the scallop *Placopecten magellanicus* (Gmelin, 1791): a qualitative study. J. Shellfish Res., 6: 89–95.

Snelgrove, P.V. and Butman, C.A., 1994. Animal–sediment relationships revisited: cause versus effects. Oceanogr. Mar. Biol. Annu. Rev., 32: 111–177.

Taghon, G.L., Newell, A.R.M. and Jumars, P.J., 1980. Induction of suspension feeding in spionid polychaetes by high particle fluxes. Science, 210: 562–564.

Thompson, R.J., 1984. The reproductive cycle and physiological ecology of the mussel *Mytilus edulis* in a subarctic, non-estuarine environment. Mar. Biol., 79: 277–288.

Turner, E.J. and Miller, D.C., 1991. Behavior and growth of *Mercenaria mercenaria* during simulated storm events. Mar. Biol., 111: 55–64.

Vahl, O., 1980. Seasonal variation in seston and in the growth rate of the Iceland scallop, *Chlamys islandica* (O.F. Müller) from Balsfjord, 70°N. J. Exp. Mar. Biol. Ecol., 48: 195–204.

Walne, P.R., 1972. The influence of current speed, body size, and water temperature on the filtration rate of five species of bivalves. J. Mar. Biol. Assoc. U.K., 52: 345–374.

Ward, J.E. and MacDonald, B.A., 1996. Pre-ingestive feeding behaviors of two sub-tropical bivalves (*Pinctada imbricata* and *Arca zebra*): responses to an acute increase in suspended sediment concentration. Bull. Mar. Sci., 59: 417–432.

Ward, J.E., Beninger, P.G., MacDonald, B.A. and Thompson, R.J., 1991. Direct observations of feeding structures and mechanisms in bivalve molluscs using endoscopic examination and video image analysis. Mar. Biol., 111: 287–291.

Ward, J.E., MacDonald, B.A., Thompson, R.J. and Beninger, P.G., 1993. Mechanisms of suspension feeding in bivalves: resolution of current controversies by means of endoscopy. Limnol. Oceanogr., 38: 265–272.

Widdows, J., Fieth, P. and Worrall, C.M., 1979. Relationships between seston, available food and feeding activity in the common mussel *Mytilus edulis*. Mar. Biol., 50: 195–207.

Wildish, D.J., 1977. Factors controlling marine and estuarine sublittoral macrofauna. Helgol. Wiss. Meeresunters., 30: 445–454.

Wildish, D.J. and Kristmanson, D.D., 1979. Tidal energy and sublittoral macrobenthic animals in estuaries. J. Fish. Res. Bd. Can., 36: 1197–1206.

Wildish, D.J. and Kristmanson, D.D., 1985. Control of suspension-feeding bivalve production by current speed. Helgol. Wiss. Meeresunters., 39: 237–243.

Wildish, D.J., and Kristmanson, D.D., 1997. Benthic Suspension Feeders and Flow. Cambridge University Press, Cambridge.

Wildish, D.J. and Peer, D., 1983. Tidal current speed and production of benthic macrofauna in the lower Bay of Fundy. Can. J. Fish. Aquat. Sci., 40 (Suppl. 1): 309–321.

Wildish, D.J. and Salnier, A.M., 1993. Hydrodynamic control of filtration in the giant scallop. J. Exp. Mar. Biol. Ecol., 174: 65–82.

Wildish, D.J., Kristmanson, D.D., Hoar, R.L., DeCoste, A.M., McCormick, S.D. and White, A.W., 1987. Giant scallop feeding and growth responses to flow. J. Exp. Mar. Biol. Ecol., 113: 207–220.

Wilson, F.S., 1990. Temporal and spatial patterns of settlement: a field study of molluscs in Bogue Sound, North Carolina. J. Exp. Mar. Biol. Ecol., 139: 201–220.

Woodin, S.A., 1976. Adult–larval interactions in dense infaunal assemblages: patterns of abundance. J. Mar. Res., 34: 25–41.

Biology of the Hard Clam
J.N. Kraeuter and M. Castagna (Eds.),

Chapter 11

Predators and Predation

John N. Kraeuter

11.1 INTRODUCTION

Most field predation studies on hard clams, *Mercenaria mercenaria*, have focused on elucidating information for aquaculture activities and involve manipulative experiments with some form of protection. The control conditions for these studies can be used to examine some ecological relationships, but the plots may have been seeded at relatively high densities. In addition, seed that are placed in the field are much larger than newly settled hard clam recruits. Studies on hard clam larvae and their predators are almost non-existent, and few studies examine predation on newly set individuals. To expand the information available on predators and predation, I have chosen to include some predation studies on other bivalves, such as oysters and mussels, and infaunal species, such as cockles, soft-shell, Manila and bent-nose clams, where these species are preyed upon by the same or similar predators as the hard clam.

The small size of bivalve set makes them relatively difficult to study in the field, and such studies on *Mercenaria mercenaria* have been particularly difficult. Spat of *Cerastoderma edule*, *Tapes philippinarum*, *Mya arenaria*, *Ensis directus* and other infaunal species are abundant, at least locally, but it is rare that concentrations of hard clam seed are reported. This rarity often creates the impression that recruitment has not taken place, but in subsequent years, the year class that was thought to be missing becomes apparent. For instance, Hibbert (1975) working on the introduced population in Southampton, England could not find spat of *Mercenaria* in spite of finding larvae in the plankton. He notes that Mitchell (1974) recorded similar results in 1969 and 1970 as did Rodhouse (1973), but Rodhouse (1973) found adult populations ascribable to sets in 1969 and 1970, and these correspond to the time when Mitchell (1974) did not find the set. These data and other anecdotal examples in the US suggest a form of sampling bias against newly set hard clams.

There is more information on the fate of seed clams once they have reached 1–3 mm shell length. As the clams reach 20–25 mm shell length, losses to predators appear to decline. There have been few studies to indicate rates of predatory loss to populations of clams >40 mm shell length. The exceptions to this are the papers of Peterson (1982a) who reported on whelk predation on natural hard clam populations, and Kraeuter and Castagna (1980) who documented cow-nose ray predation on aquacultured hard clams.

The interactions between sediment type, clam size, predator species, predator size and factors, such as temperature, make all but the most generalized information about consumption rates difficult to interpret. These scale difficulties are increased as the size of the predator increases to the point that containers are often many times smaller than the area normally occupied by the predator. The question of how faithfully data collected by placing an adult blue crab or busyconnine whelk in a container of less than several square meters can be extrapolated to field conditions is difficult to determine. Furthermore, interactions between various predators and prey have not been examined to any appreciable degree.

Several lists of hard clam predators have recently been published (Gibbons and Blogoslawski, 1989; Rice, 1992). It is likely that these lists of predators represent only a small portion of species that could be consumers of hard clams. This is particularly true of larval and newly set clams.

I have chosen to consider hard clam predators on a taxon-by-taxon basis, with consideration of predation on larvae and newly set first, followed by data on larger seed and adults. In some instances, where there were a large number of studies, I have provided an introductory paragraph depicting information on the abundance of the predator and general information on its prey. The information in this general paragraph is then followed by data on scallops, mussels, cockles and finally clams. In these cases, I have placed the data on the hard clam at the end of the section. Where information on predator density in the field seemed to be important to the evaluation of laboratory studies, I have included a brief description of data from selected studies, but I have made no attempt to make an exhaustive search of the literature for each predator. Within each section of the text, I have established a rate of predation for a class of predators on a number of bivalve species. Most often, I have calculated the data to indicate the number of prey eaten per day. Given the differing methods, density, substrates, sizes of predators and prey and other factors, I did not feel that extrapolating from these data, without giving the reader some background information, was justifiable. In many instances, I have used European literature on mussels, cockles, soft-shell clams or other species as a basis from which the reader can evaluate the potential effects of the predator taxa on hard clams. For species with substantial data I have provided a brief summary at the end of the section and have pointed out where there are potential discrepancies between studies. Lastly, I have tried to combined the information into a brief discussion of predators as guilds, latitudinal differences in predators, and attempted to evaluate the *relative* importance of predator taxa and to hard clam survival. The reader is also referred to the information on higher order predator–prey–environment growth interactions provided by Peterson (Chapter 10).

For convenience, I have provided sizes for predators or prey within parentheses after the species or common names. Prey sizes refer to length, except where noted. Most predator sizes refer to length, except for Reptantia where the size is reported as carapace width.

11.2 PROTOZOA

Loosanoff (1959) found that the ciliate, *Condylostoma*, a large heterotrich, was able to ingest up to 6 larval clams in laboratory cultures. Whether other protozoa, such as foraminifera, are able to consume newly set bivalves has not been determined, but Buchanan and Hedley (1960) indicated that these protozoa can be locally important components of the benthic community, and ingest amphipods, cumaceans, copepods, nematodes and newly set urchins, *Echinocardium flavescens*. It is likely that large protozoa can ingest newly set hard clams, but studies are not available to suggest whether or not they are important in clam recruitment.

11.3 CNIDARIA

Hydrozoa

There do not appear to be any studies that have examined the effects of various hydrozoa on survival of hard clam larvae, but McCormick (1969) and Zelickman et al. (1969) have shown that hydromedusae ingest bivalve larvae in waters off Oregon, USA and in the Barents Sea, respectively.

Scyphozoa

The medusa stage of the seanettle jellyfish, *Chrysaora quinquecirrha*, ingest the larvae of the eastern oyster, *Crassostrea virginica* (Purcell et al., 1991), but in most instances these are egested alive. Larvae of eastern oysters, blue mussels, *Mytilus edulis*, and coot clams, *Mulinia lateralis*, placed on the oral arms of the medusae were rejected. Ephyrae of the same seanettle also ingested, but did not digest, veliger larvae. The benthic stage of this scyphozoan (scyphistoma) consumed about one veliger day^{-1}, but had no overall effect on larval settlement (Purcell et al., 1991). These authors concluded that the net effect of *C. quinquecirrha* might be to enhance survival of bivalve larvae because the scyphozoan reduced the numbers of ctenophores, *Mnemiopsis leidyi*, an important predator of bivalve larvae.

Anthozoa

MacKenzie (1977a) reported that the anemone *Diadumene leucolena* could be an important predator on eastern oyster larvae. This was substantiated by Steinberg and Kennedy (1979) who found predation rates up to 4.9 oyster larvae min^{-1} (larval density 4 ml^{-1}) and 1.7 larvae min^{-1} (larval density 1 ml^{-1}). How this relates to daily consumption is difficult to evaluate, but rates of hundreds of larvae ingested per day seem possible.

Posey and Hines (1991) placed 25 individuals of the burrowing anemone *Nematostella vectensis* in 50 cm^2 cups and the placed them in field plots. These results were compromised because of low numbers of recruits, but in laboratory studies with larval *Macoma mitchelli* exposed to predation for 12-h periods, the anemone ingested significant numbers of clam pediveliger larvae. Forty five live clams recovered in the control containers and only 20 in those containing anemones (Posey and Hines, 1991). No information exists to indicate whether these species or other anemones commonly found on clam flats are predators on larval hard clams, but Sellmer (1967) reported finding 50–100 *Gemma gemma* in the gastrovascular cavity of the anemone *Paractis rapiformis* in a sand substrate on Union Beach, New Jersey. In general, anemone densities are lower in areas of soft substrates than on rocky shores, but forms such as *Edwardsia elegans*, *Haloclava productus*, *Sagartia luciae*, and the cerianthid *Ceriantheopsis americanus* are often present, and can be locally abundant in shallow water soft substrates. How much effect the presence of these forms has on hard clam larvae or recruiting juveniles is not known.

Ctenophora

Ctenophores have been reported to prey on bivalve larvae, but their overall importance in the larval survival of hard clams has not been clearly quantified. Nelson (1925a,b) reported that eastern oyster sets were diminished when the ctenophore *Mnemiopsis leidyi* was abundant, and he found 126 oyster larvae in a comb jelly (3 cm). Field collections provided further evidence of the localized importance of this predator, in that 75% of the *Mnemiopsis* collected by Nelson had eaten bivalve larvae. Carriker (1961) suggested that ctenophores might be important predators on hard clams in Little Egg Harbor Bay, NJ, but no studies were conducted. Quayle (1964) reported that *Pleurobranchia bachei* consumed bivalve larvae in British Columbia. Purcell et al. (1991), following Loosanoff (1966), suggested that ctenophores can be predators on oyster larvae, and calculated that this predator may consume between 0.2 and 1.7% daily of the oyster veligers in the central portion of Chesapeake Bay. The numbers consumed in the Chesapeake tributaries were estimated to be minimal because the jellyfish *C. quinquecirrha* reduced the numbers of ctenophores in these systems.

Quaglietta (1987) sampled ctenophore populations in Great South Bay, New York and conducted laboratory studies on ingestion of both copepods and hard clam larvae. Ingestion of adult copepods was independent of prey density in the range of prey densities studied (50–800 L^{-1}), and clearance rate of the water increased with ctenophore size to about 28 L day^{-1} for a 50 mm *M. leidyi*. There was only a slight increase in the volume of water cleared of hard clam larvae with the increasing size of the ctenophore; the maximum was about 9 L day^{-1} for a 50-mm *M. leidyi*. When both copepods and hard clam larvae were placed in containers with ctenophores, there was a slight preference for copepods. While the clearance rate for copepods in the copepod + clam trials remained the same as with copepods alone, the rate of clearance for hard clam larvae increased significantly relative to the clam-only test. Presumably, the additional ingestion of hard clam larvae was caused by the increased filtration rate due to the presence of copepods. If the laboratory rates of the mixed species experiment are combined, the average clearance rate was 3.38 L day^{-1} ml^{-1} of ctenophore. Based on this and the estimated *M. leidyi* abundances in August and September, Quaglietta (1987) calculated that an average of 11% (1985) and 36% (1986) of the water could be cleared of prey daily during peak feeding. *M. mercenaria* larvae were in the water column in Great South Bay during this time and there was a strong correlation between the appearance of ctenophores and the reduction in bivalve larvae during 1986; however, Quaglietta (1987) did not report if the field collected ctenophores had consumed hard clam larvae. If these data are compared with the estimated impact of ctenophores on oyster larvae in Chesapeake Bay (Purcell et al., 1991), the losses *M. mercenaria* larvae in Great South Bay, New York could be proportionally greater because the estimated percentage of the water volume cleared by the ctenophores in New York is greater. This may be offset by the lack of concordance between the ctenophore and clam larval peaks in Great South Bay.

11.4 PLATYHELMINTHES

Gallani et al. (1980) reported that the flatworm *Stylochus mediterraneus* preferentially consumed small (<25 mm) mussels *Mytilus galloprovincialis* at rates of 0.07–0.33 mussels $worm^{-1}$ day^{-1}. These authors also provide a literature review of the prey species of polyclad flatworms.

Predation of oyster spat by the flatworm *Stylochus ellipticus*, has been inferred from presence data (Webster and Medford, 1959) and has been documented (Provenzano, 1959; Landers and Rhodes, 1970; Newell and Kennedy, 1992). Landers and Rhodes (1970) found that adult *Stylochus ellipticus* could open and eat *Crassostrea virginica* spat as large as 61 mm long, and Newell and Kennedy (1992) (vide Kennedy, 1995) reported that microscopic juvenile *S. ellipticus* could cause extensive mortality to newly set oysters. Landers and Rhodes (1970) examined the influences of salinity, temperature and source of the predator on the rate of predation and the type of food preferred by *S. ellipticus*. In general, *S. ellipticus* predation rates were in the range of 0.05–0.1 oyster spat worm^{-1} day^{-1} at temperatures 10°C and above (Landers and Rhodes, 1970). Newell and Kennedy (1992) reported that on some natural oyster bars in Chesapeake Bay, most oyster spat <2 mm may be consumed by juvenile *S. ellipticus*.

Kato (1944) found that *Stylochus uniporus* consumed *Tapes philippinarum*. MacGinitie and MacGinitie (1949) observed *Stylochus californicus* consuming the bivalves *Barnea pacifica* and *Zirfaea gabbi*. Thorson (1966) reported that, in experiments done by Christensen, the turbellarian *Discocelides longi* (5–6 mm) consumed an average of 2.75 spat of newly set (1–2 mm) *Spisula elliptica* per day. Littlewood and Marsbe (1990) found that *Stylochus frontalis* was a major predator on cultivated *Crassostrea rhizophorae*, but that these infestations could be controlled by dipping the oysters in brine solutions.

Turbellarians are common in benthic communities, and many are known to be predators (Jennings, 1957). Watzin (1983) manipulated the densities of meiofaunal turbellarians and other meiofauna to determine if there was an effect on recruiting macrofauna. Recruits of some taxa were adversely affected, but there was no statistical reduction in the recruitment of bivalves (Watzin, 1983). While there are a significant number of flatworms inhabiting both intertidal and subtidal sediments, there does not appear to be any information on the intensity or importance of larger flatworms as predators of setting hard clams.

11.5 NEMERTEA

There are no studies documenting nemerteans consuming larval, newly set seed, or adult hard clams, but one report suggests juvenile hard clams may be a prey.

Cerebratulus lacteus

A number of reports describe the predatory mechanisms of the heteronemertean *Cerebratulus lacteus* on various species of adult infaunal bivalves (Atlantic jackknife clams, *Ensis directus*, McDermott, 1976; soft-shell clams, *Mya arenaria*, Kalin, 1984; Rowell and Woo, 1990). *Cerebratulus lacteus* preys on jackknife clams by enveloping the entire lower portion of the clam (shell and all) with its mouth. This aggression may cause the clam to exit the sediment in an effort to escape (McDermott, 1976). When this predation occurs at low tide, jackknife clams on intertidal flats may be 2/3 exposed to the air and thus could be subject to additional predation from birds and/or terrestrial predators. Schneider (1982) described a similar response by *E. directus* trying to escape attacks by the gastropod *Neverita duplicata* on intertidal flats. In this instance, some jackknife clams were observed at night completely exposed on the surface of an intertidal flat while others were protruding from the sediment. Those completely exposed would flip themselves across the flat. As opposed to the siphon-first exposure noticed by McDermott (1976), jackknife clams attempting to escape *N. duplicata* would emerge foot first and travel up to several meters before digging back into the flat (Schneider, 1982). No information exists on the predatory effects and the escape actions of the clams in the subtidal or in intertidal areas when they are covered with water, but mechanisms that would cause clams to leave the substrate would make them more susceptible to predatory fish. Similar attack and escape behavior was not noted for hard clams even though they can be locally common on the flats where McDermott (1976) reported nemertean attacks on *Ensis directus*. Whether this indicates a lack of predation on the hard clams on these flats because of its more robust shell shape, or its ability to completely close or the abundance of alternate prey is not known, but Landry et al. (1993) (see below) suggest that hard clams may be a prey for this worm.

Kalin (1984) found that *C. lacteus* preyed on soft-shell clams by entering through the siphon and then consuming the soft parts. There was no indication whether the *Mya arenaria* were alive when this predation took place. Rowell and Woo (1990) were able to show experimentally that *C. lacteus* actively seeks and preys on soft-shell clams. Fifty or 100 clams (14–32 mm) were planted in a series of plots. After 55 days, these plots had experienced 100% mortality and most of the clams had articulated shells with only about 2% bearing drill holes. In laboratory experiments, replicated groups of 14 and 10 clams (15–74 mm) from two sites were placed in buckets filled with sediments and four *C. lacteus*. After 35 days, only 17% of the clams were surviving and after 47 days there were no surviving clams (approximately 0.06 clams nemertean^{-1} day^{-1}). In a field experiment, 25–32 mm clams were placed in replicated field plots at 50 animals 0.25 m^{-2} . After 30 days, the mortality ranged from 12 to 20% and seven and three *C. lacteus* were found, but by day 60, mortality was 92–96% and the numbers of worms had increased to 15 and 10. Samples outside these plots yielded less than 1 worm 0.25 m^{-2}. If the maximum numbers of worms (15) are assumed to have been in the plots for the entire 60 days the predation rate would have been 0.055 clams worm^{-1} day^{-1}.

No published studies have directly documented *C. lacteus* to be a predator of hard clams of any size, but Landry et al. (1993) suggested that the elevated abundance of these nemertean worms may have been the cause of high mortality levels in hard clams transplanted into the West River, Prince Edward Island. The authors reported native populations of hard clams were regularly found partly exposed on this site, and when these clams were removed from the substrate nemertean worms were often observed. Although no predation studies were conducted, the high abundance of *C. lacteus* at this site relative to Pownal Bay, Prince Edward Island was thought to be partly responsible for the decreased clam survival at the West River site (Landry et al., 1993). Landry (personal communication) has found hard clams with the proboscis of *C. lacteus* inserted between the valves. When these clams were opened, the meats were covered with slime and appeared to be partly digested.

11.6 ANNELIDA

In spite of the common name of a clam worm for some annelid species, surprisingly little information exists on the consumption of larvae, newly set, or post set clams by annelids. Lebour (1922) observed that the larvae of the polychaete *Magelona papillicornis* prey exclusively on bivalve larvae. MacGinitie (1938) noted that *Chaetopterus variopedatus* could consume bivalve veligers, and Mileikovsky (1959) and Daro and Polk (1973) both reported that polychaete larvae fed on bivalve larvae. Mileikovsky (1959) was able to document that both *Mya arenaria* and *Macoma balthica* veligers were ingested by the larvae of *Nephtys ciliata*. Breese and Phibbs (1972) reported that *Polydora ligni* would consume up to 20 veligers of *Tapes semidecussata* and *Crassostrea gigas*. MacKenzie (1981) found that oyster larvae captured by *Polydora ligni* were usually released from the mucous unharmed. I have not found reports of larval polychaetes feeding on hard clam larvae.

Nereis spp.

Olafsson (1989) experimentally manipulated the densities of adult *Nereis diversicolor* (0–40 per core, = 0–5100 m^{-2}) and then introduced juvenile (0.25–0.3 mm) *Macoma balthica* to the cores. The experiment was allowed to run for 15 days and clam survival in cores containing *Nereis diversicolor* was reduced by 2/3 relative to the control. Because seawater was allowed to flow through the cores, some additional recruitment of clams took place, but final densities in cores for 5, 10, 20, and 40 adult worms had 38, 38, 46, and 33 clams compared with 106 clams in the control core. These results seem to indicate that 5 worms exerted the same predation pressure as the higher worm densities. Interference competition among the worms may have been an important experimental variable. If the clam survival data at the lowest worm density are utilized and subtracted from the control the consumption rate of these worms must have been 0.91 clams $worm^{-1}$ day^{-1}. Whether similar rates of consumption can be extrapolated to field conditions and further extrapolated to predation on hard clams is not known.

Reise (1979) enclosed the polychaete *Nereis diversicolor* in various densities in field experiments and found that some species of macrofauna, including bivalve spat, decreased in abundance while other species were unaffected. He calculated that this species caused a 70% reduction in cockles, *Cerastoderma edule* on a tidal flat in the Wadden Sea (Reise, 1979). *Nereis succinea* has been reported to cause a 40% reduction in *Macoma* spp. in Chesapeake Bay (Ambrose, 1991) citing Haddon and Hines (unpublished).

Ambrose (1984a,b,c) conducted field experiments in which the densities of the annelids *Glycera dibranchiata* and *Nereis virens* were manipulated in buckets, and recruitment of other infauna were enumerated after 8, 16 or 24 weeks. In general, two factors appeared to be of major importance: (1) time the test sediments were placed out; and (2) presence of *N. virens*. Bivalve recruitment was reduced in *N. virens* treatments relative to both control and *G. dibranchiata* treatments. Apparently, predation by *G. dibranchiata* was an important mechanism controlling *N. virens*, and this latter species was significantly more important than *G. dibranchiata* as a predator or disruptor of recruitment processes on other invertebrates. Because *N. virens* makes fecal mounds and actively feeds on the surface by extending its body from the burrow, separating the effects of predation and surface disruption on recruitment of other species was not attempted. The only bivalve present in significant numbers was *Macoma balthica*, but *Mya arenaria* can be abundant in this area. Analysis of prey and prey parts from the fecal material of both *N. virens* and *G. dibranchiata* revealed the presence of both species of bivalves in the former and only *Macoma* in the latter (Ambrose, 1984c). In general, *G. dibranchiata* alone had little effect on infaunal recruitment or abundance. The importance of infaunal/infaunal interactions, in the presence and absence of epifaunal predators, such as crabs, fish and birds that may prey either on the infaunal predator or its prey, has not been investigated. In a two-predator system, such as that described by Ambrose (1984c), a predator that consumed proportionally more *G. dibranchiata* could affect an indirect change in the bivalve populations by increasing the survival of *N. virens*.

The field data of Ambrose (1984a,b,c) are supported by the laboratory studies of Dean (1981) who placed *Nereis virens* in aquaria with *Mya arenaria*, *Ensis directus* and *Hiatella arctica*. Two experiments were conducted in aquaria supplied with flowing seawater and bottoms covered with sandy sediments, one in light and one in darkness. Bivalves included (length in mm): *Mya arenaria* (5–12, 7–20), *Ensis directus* (19–26, 22–39) and *Hiatella arctica* (6–10, 11–17) for experiments 1 and 2, respectively. The average initial weight of the worms was 3.3 g in experiment 1 and 4.3–5 g in experiment 2. In the first experiment (55 days), the four worms increased in weight to an average of 10.85 g and all the five *H. arctica* and five *E. directus* plus 44 of the original 50 *M. arenaria* were gone (average consumption of 0.25 clams $worm^{-1}$ day^{-1}). Crushed shells were found throughout the aquarium. A control aquarium of the same species without worms experienced no mortality. In the second experiment (28 days), more aquaria were used and two aquaria with worms, but no clams, were controls. The worms in the controls lost weight while those with the clams gained weight. None of the *M. arenaria* or *H. arctica* were consumed in this experiment, but 13 of the 15 *E. directus* were consumed (0.15 clams $worm^{-1}$ day^{-1}). Both the worms and the clams were slightly larger in this experiment, and it was conducted in 24 h of light. Whether this altered the feeding of the worms is unknown. One *E. directus* and one *M. arenaria* died in the control tanks. This study clearly demonstrated that *N. virens* can consume small clams. None of the control or experimental tanks with worms had any alternative source of food, such as soft-bodied invertebrates or algae, so it is difficult to relate these results to feeding preferences in the field. It is likely that even smaller clams would be consumed at a higher rate, but how the results from these three bivalve species with thin shells and/or gaping valves can be related to hard clams with their thick shell and tightly closing valves cannot be evaluated.

Landers (1967) reported that the polychaete *Polydora ciliata* infested the shells of 3–16 mm clam seed being held in trays in the laboratory. High percentages of the clams that had shells infested with this worm died. Subsequent experiments indicated that as long as clams are in sediments these infestations do not develop. Taylor and Saloman (1972) reported on shell blisters caused by *Nereis arenaceodentata* in the southern hard clam.

While there are a number of nereid polychaete species present throughout the range of the hard clam there do not appear to be any studies on their effects or those of other annelids on mortality of larval, juvenile or adult hard clams or on hard clam populations.

11.7 MOLLUSCA

11.7.1 Bivalvia

While bivalves are not commonly thought of as predators on other molluscs, there are a significant number of studies that indicate that, in the course of their filter feeding activities, bivalves may ingest their own larvae or the larvae of other filter feeders. Species such as *Pecten opercularis* and *Pecten varius* have both been reported to filter veliger larvae from the water column (Blegvad, 1914; Hunt, 1925). Nelson (1921) was one of the first to report bivalve larvae in the stomach of other bivalves, finding 63 umbo stage larvae of *Crassostrea virginica* in the stomach of an adult eastern oyster and 71 live eastern oyster larvae still alive in the feces of another adult oyster. MacKenzie (1981) reported that pediveligers of *Crassostrea virginica* were able to escape from the pseudofeces of adult eastern oysters, blue mussels (*Mytilus edulis*), coot clams (*Mulinia lateralis*) and hard clams after the mucous had been dispersed by a current from an eyedropper. Both blue mussels and hard clams ingested oyster larvae and excreted them alive in the fecal strings. The larvae were unable to escape from these strings (MacKenzie, 1981).

Tamburri and Zimmer-Faust (1996) found that *C. virginica* adults consumed larvae of annelids, molluscs and crustaceans at nearly equal rates. These rates were equivalent to >75% of the larvae being ingested and digested. Additional larvae perished in pseudofecal strings. When phytoplankton were fed along with the larvae, the quantity of mucous in the pseudofecal strings increased, and the mortality of larvae incorporated into these strings increased. More than 80% of larval (172 μm) *Mercenaria mercenaria* were ingested by the oysters (Tamburri and Zimmer-Faust, 1996). Only one larval species, the bryozoan *Bugula neritina*, was not ingested at the same rate, and this negative selection was believed to be due to the surface chemistry of this species.

Other oysters such as the Pacific oyster, *Crassostrea gigas* (Quayle, 1988) and the European flat oyster, *Ostrea edulis* (Korringa, 1941) have been shown to ingest their own larvae. Kristensen (1957) and Andre et al. (1993) reported that cockles were capable of filtering significant numbers of their own larvae from suspension. Kristensen (1957) found that only the smallest cockles tested (600–900 μm) were inhaled and not ingested, but expelled in the pseudofeces with a coating of mucus. Of the 14 larvae expelled, only two were alive a day later, and none had been able to extract itself from the mucus.

Kurkowski (1981) examined, through laboratory experiments, the effects of larval size, adult size and temperature on the interactions between adult and larval hard clams (Table 11.1). Small larvae (98 μm) and larger larvae (183 μm) were more affected by the presence of adult clams than those of intermediate size (155 μm). Smaller larvae were apparently killed by the filtration of the adults and their shells were reduced to fragments. None of the shell fragments or intact small larvae were found in feces. Mortality of larger larvae was due to their being trapped in the pseudofeces of the adult. Few intermediate sized larvae suffered either fate. Additional experiments indicated that larvae smaller than 163 μm were not likely to stimulate the production of pseudofeces. In concordance with the reports of Nelson (1921) and MacKenzie (1981), these experiments indicated that the pseudofecal strings of *M. mercenaria* retained larvae even after the strings had broken into clumps. All larvae in

TABLE 11.1

Effect of filtration by *Mercenaria mercenaria* adults on larval survival under various conditions

A		B		C	
Larval size (μm)	Mean Survival (%)	Temperature (°C)	Mean survival (%)	Adult length (mm)	Mean survival (%)
98	23	17	26	19	6.4
155	96	21	25	52	21
183	71	25	18	75	51
		29	19		

Data are adjusted for control mortality. A = larval size effects; B = temperature effects; C = Adult size effects. After Kurkowski, 1981.

the pseudofeces died within a few days (Kurkowski, 1981). As temperature increased from 17 to 29°C, larvae surviving adult filtration decreased from 26 to 19%. This was ascribed to the increased filtration rate of the adults with increasing temperature (Kurkowski, 1981). Lastly, Kurkowski (1981) examined the effects of clam size on larval survival. Pumping rates of clams (19 and 75 mm) were estimated and the clams arrayed to approximate a pumping rate of 13.28 L h^{-1}. This amounted to twenty-seven 19-mm clams, four 52-mm clams, and two 75-mm clams per test container. Larval clam survival increased with increasing size of the adults. Kurkowski (1981) utilized data from other studies to estimate the effect of filtration of larval hard clams by dense populations of adults. Utilizing a density of 81 clams m^{-2} and larval densities of 670 L^{-1} (98–100 μm) he calculated 285,000 could be filtered m^{-2} h^{-1} (3518 $clam^{-1}$ h^{-1}), while for densities of larger larvae (0.2 L^{-1}) only 85 m^{-2} h^{-1} would be destroyed.

11.7.1.1 Population and community effects

While it is clear from the above studies that some bivalves including the hard clam are capable of ingesting their own larvae or larvae of other species, there are conflicting reports in the literature concerning the potential of bivalve filter feeders to affect recruitment to their own populations. Studies of Best (1978), Williams (1980), Peterson (1982b), Jensen (1985), Moller (1986), Andre and Rosenberg (1991) and Andre et al. (1993) demonstrated a negative effect of suspension feeders on densities of other infaunal species, including congenerics, but Young and Young (1978), Maurer (1983), Hunt et al. (1987), Black and Peterson (1988), Ertman and Jumars (1988) and Young (1989) did not. Hines et al. (1989) and Gallagher et al. (1983) have shown both positive and negative effects of bivalves on other species. Ahn et al. (1993b) reported a positive effect of one suspension feeder on the recruitment of another. These studies, with the exception of Hunt et al. (1987) are evaluated below. The data from Hunt et al. (1987) are presented in the section on snail predators.

Nuculoida
Nuculidae

Nucula proxima
Levinton (1977) presented evidence that the protobranch *Nucula proxima* would congregate around the siphons of the hard clam. This crowding inhibited the effectiveness of hard clam filter feeding and was hypothesized as a mechanism by which deposit feeding clams could selectively inhibit suspension feeders.

Mytiloida
Mytilidae

Mytilus edulis
Commito (1987) reported that filter feeding by blue mussels, *Mytilus edulis*, did not appear to affect recruitment to the mussel bed even though Thorson (1946), Mironov (1948), Bayne (1976) and Cowden et al. (1984) have reported that this species and other mussels ingest larvae. The latter authors suggested that many mussels may recruit to nearby surfaces and then migrate, when they are larger, to the area densely covered by mussels rather than recruiting directly to the mussel bed. Mileikovsky (1974) reported that some larvae can be ingested, pass through the

consuming organism and emerge alive. Similar data on long-term survival by bivalves ingested by other organisms were provided by Thorson (1966) and Purcell et al. (1991). Cowden et al. (1984) conducted laboratory studies in which they placed 100 polychaete, crustacean, echinoid, and asteroid larvae in bowls with one *M. edulis* (32–41 mm) for 1.5 h. Fewer than 50% of the introduced larvae survived for the test period, and for most larvae less than 20% survived. These would be equivalent to consumption rates of >800–1280 larvae day^{-1}.

Veneroida

Cardiidae

Cerastoderma edule

Jensen (1985) found that cockles, *Cerastoderma edule*, negatively affected recruitment of *Macoma balthica* on tidal flats in the Danish Wadden Sea. Field experiments at 0, 0.25, 0.5, 1.0 and 2.0 times the natural density (3000 m^{-2}) of 2-year-old cockles showed a highly significant linear reduction in *M. balthica* recruitment with increasing numbers (surface area occupied) of cockles. Jensen (1985) attributed this reduction to the consumption of *Macoma balthica* larvae by the cockles.

Andre and Rosenberg (1991) provided data that support the importance of adults in controlling the intensity of larval settlement and additional data that refute the hypothesis. Field data were collected based on quantitative sampling of areas in which there were high and low densities of adult cockles (*Cerastoderma edule*). Bivalve juveniles were separated according to species (*C. edule* or *M. arenaria*) unless they were <0.4 mm in which case all bivalve spat were combined. As a group, the data clearly showed that, during the first sampling the station with high adult density of *C. edule* (mean = 1188 m^{-2}) had statistically fewer total bivalve juveniles than the station with low density adults (mean = 24 m^{-2}). Data on specific taxa did not reveal differences. The following month, only juvenile *C. edule* were significantly reduced at the high adult density station while the densities of other taxa, small bivalves and the combined bivalves (small bivalves + *M. arenaria* + *C. edule)* were not statistically different. In a third sampling, both *Mya* and *C. edule* juveniles were significantly reduced at the station with high abundance of adult cockles (Andre and Rosenberg, 1991). Manipulative field experiments in which both adult *M. arenaria* and *C. edule* densities were controlled (0, 100 and 400 m^{-2}) showed similar disparities in results (Andre and Rosenberg, 1991). On the first sampling date, newly set cockles were significantly fewer in the highest density of adult cockles, but newly settled *M. arenaria* and small (<0.4 mm) bivalves were not. Sampling at later dates found no effects of adult density on any of the three juvenile bivalve parameters; however, at all of the sampling dates and all experimental replicates there were fewer small bivalves and total juvenile bivalves in the high density of *Mya*. These data were combined and the resulting analysis showed a clear significant difference ascribable to the highest density of adult clams. These data indicate that is it possible to demonstrate an interaction between adults and settlement, but suggest that in areas with high continuous settlement, the overall effect may not be of major importance to population recruitment.

Andre et al. (1993) examined the effects of feeding by the adults of the cockle, *Cerastoderma edule*, on various components of cockle larval behavior. The authors then utilized a computer simulation model to examine the predation risk experienced by the larvae as they passed over a bed of adults. The flume studies indicated that the larval settlement density could be reduced by up to 33% within the 5 cm^2 area surrounding an adult. Fluorescent pigments used to track the larvae. Larval shells were found in intestines of the adults, confirming that the larvae had been ingested. As expected, the amount of time the larvae spent over the bed of the adults and the density of those adults had significant effects on larval survival. Larvae drifted through a dense (380 m^{-2}) bed of adults in about 1 min. Computer models of this process suggested that mean survival time of larvae passing over a 400 adult m^{-2} bed of cockles was about the same as indicated by the tests, but even at 50 adults m^{-2}, mean survival times for larvae were less than 15 min (Andre et al., 1993).

Based on field and laboratory studies it is apparent that *C. edule* is capable of ingesting a wide variety of larvae, and when adults are present in high density, this bivalve can affect recruitment of many species. In direct contrast to the hard clam, the setting density of cockles is very high, and because of this high density a small portion of the larvae escape predation by the adults. The data suggest that, even with potentially high losses due to adult filtering activities, enough larvae survive to establish annual recruitment pulses.

Tellinidae

Macoma balthica

Olafsson (1989) examined the effects of this deposit feeding bivalve on other bivalves. He reported that deposit- and filter-feeding populations of *Macoma balthica* had no direct effect on recruitment of *Macoma balthica*. The importance of the feeding behavior of *Macoma balthica* on recruitment of its own juveniles at two field sites was examined by constructing cages in which adult populations of clams were manipulated to replicate the range of natural density at the site and populations 4× that found at the site. The muddy sand (deposit feeding) site had population densities ranging from 250 to 4000 clams m^{-2} while the sand (filter feeding) site had bivalve densities of 100–1000 m^{-2}. No effects were found on the numbers of clams recruited, but the mean size of the juveniles was significantly less on the muddy sand site in cages with the highest densities of adults. Reductions in size of the recruits were also reported on the sand site, but the effect was less. Laboratory experiments confirmed that adult density had no effect on juvenile recruitment (Olafsson, 1989).

Macoma balthica and *Mya arenaria*

Hines et al. (1989) found both negative and positive effects of the bivalves *Macoma balthica* and *Mya arenaria* on recruitment of bivalves and other infaunal organisms. This effect appeared to vary not only with bivalve density, but also with the intensity of macrofaunal recruitment. Effects that were significant during a year with high recruitment were not found during periods of lower recruitment (Hines et al., 1989). In the year of high recruitment, bivalves had statistically significant effects on their own recruitment and on the recruitment of other species of invertebrates. The nature of these effects (positive or negative) varied depending on the density of the clam species and the other invertebrate species involved. No general patterns, based on trophic group or life history pattern, could be observed (Hines et al., 1989).

Veneridae

Tapes philippinarum

Williams (1980) experimentally examined the effects of the manila clam, *Tapes philippinarum*, on recruitment of its own spat. Clam densities were manipulated in the field to yield from 0 to 480 individuals m^{-2}. While large numbers of spat were found in the densest adult populations, a negative correlation was found between the highest adult densities and the number of clam spat recruited (Williams, 1980). The adult manila clam is smaller than adult hard clams, but is found in densities that far exceed those typical for the latter species. Spat numbers reported for manila clams in the highest density treatment were 18,200 m^{-2}. These densities far exceed those reported for setting for hard clams and thus it would be difficult to demonstrate a relationship between adult density and recruitment for hard clams in nature.

Protothaca staminea and *Chione undatella*

Peterson (1982b) experimentally manipulated the density of two suspension feeding bivalves, *Protothaca staminea* and *Chione undatella*, in 1-m^2 plots to examine their effects on recruitment. Densities were adjusted from 0.5× to 8× background levels on sand sites and 0.5× to 4× background on mud sites. No recruitment effects due to increased density of adults were noted on sand sediments, but in muds high *P. staminea* density decreased recruitment by about 40%. *C. undatella* recruitment was low in all cases and no density effects were observed on the recruitment of this species, but in one test the presence of *P. staminea* adults reduced *C. undatella* recruitment. Growth rate of the adults was reduced by higher density treatments. The lack of density effects on *P. staminea* in sandy substrate were interpreted to be due to post-recruitment emigration of the smaller clams.

Katelysia scalarina and *Katelysia rhytiphora*

Black and Peterson (1988) manipulated the density of the suspension feeding bivalves *Katelysia scalarina* and *Katelysia rhytiphora* in 1 m^2 plots by placing bivalves, either as single species or as combined species, in densities ranging from 20 to 320 m^{-2}. Subsequent core sampling did not find any significant differences in abundance or number of species of small infaunal invertebrates at either site or with any bivalve density, including effects on recruitment of bivalves.

Callista impar, Anomalocardia squamosa and *Circe lenticularis*

Black and Peterson (1988) manipulated the density of the suspension feeding bivalves *Callista impar*, *Anomalocardia squamosa* and *Circe lenticularis* in plots either as single species or in two-species combinations at either 50 or 100 m^{-2}. Subsequent core sampling did not find any significant differences in abundance or number of species of small infaunal invertebrates at either site or with any bivalve density, including effects on recruitment of bivalves.

Gemma gemma

Ahn et al. (1993a,b) examined the effects of dense assemblages of the gem clam, *Gemma gemma* on recruitment of hard clam larvae. In general, settlement of hard clam larvae was enhanced in both sand and mud by the presence of the gem clams (Ahn et al., 1993b). Ahn et al. (1993a) followed the post-settlement processes and found hard clam post-set emigration increased with increasing gem clam density and with decreased food. Increased gem clam density also increased hard clam growth in sand, but survival was not affected. Increased gem clam density inhibited hard clam growth and reduced recruit survival in sandy mud sediments. Levinton (1977) reported similar effects for *G. gemma* on adult hard clams. Ahn et al. (1993a) found that the absence of gem clams allowed the greatest hard clam growth in sandy mud sediments. These authors interpreted the results to mean that when small hard clams were undisturbed and stayed near the surface they fed more actively. Gem clams disturbed small hard clams and thus caused their emigration. Sandy muds have a small oxidized layer over an anoxic zone and this kept the small clams near the surface to feed. In sand, hard clams buried deeper, and could not feed effectively. Thus the lack of gem clams (disturbance) and the factors concentrating the hard clams near the surface (anoxic sediments) increased the ability of post-set hard clams to feed in sandy mud sediments. In sands, clam seed buried deeper and thus feeding was interrupted and growth was less. Predation on hard clam larvae was apparently not a factor.

Mercenaria mercenaria

The study of Best (1978) on the effects of *M. mercenaria* on infaunal community structure showed that recruitment of other species was reduced. Infaunal species with pelagic larvae experienced greater reductions than those which produced benthic larvae, and the higher the density (0, 5, 18 *Mercenaria* m^{-2}), the greater the reduction. Infaunal species that brood their young were not affected by the density of the hard clams. Similar data indicating no effect of filter feeding bivalves on the density of invertebrates that brood their young was provided by Commito (1987) who examined the effects of *Mytilus edulis* on recruitment. The Best (1978) study was for a relatively short period of time (3.5 months) and did not show any effects of hard clams on hard clam recruitment, but overall recruitment of clams was low, and the lack of effect may be the result of low recruitment or low clam density similar to the effects described by Hines et al. (1989).

Young and Young (1978) placed hard clams in 4 m areas at densities of 50 m^{-2} in caged and uncaged test plots in Indian River estuary, Florida. These plots were located within a *Halodule wrightii* seagrass meadow. Other plots included controls, caged fish, caged crabs, caged clipped seagrass, and caged and uncaged plots to which fertilizer had been added. These authors did not report any significant difference on infaunal macroinvertebrate densities, including recruitment of hard clams, that could be ascribed to the presence of clams or manipulations of clam density.

Maurer (1983) placed young hard clams (82.5–330 m^{-2}) in trays of sand that had been defaunated. The trays were all placed in a large tank which was continuously supplied with seawater for the duration (May 4 to October 9) of the experiment. Maurer (1983) then examined the recruitment of other benthic species to these trays. The data were condensed to: mean number of species; mean numbers of individuals; mean wet weight biomass; mean species richness; and mean dominance index per trial. These were then used to examine subsequent recruitment of benthic species to these trays. The study concluded that these populations levels of the hard clam did not cause significant differences in any of the five variables examined. No data were provided that examined the effects of clam density on individual species. It is difficult to compare these data to those of other authors who focused on consumption of individuals rather than on community level effects, but this study is consistent with that of Hines et al. (1989) and others.

Myoida

Myidae

Mya arenaria

Ertman and Jumars (1988) examined the flow of water around Nuttalls cockle, *Clinocardium nuttalli*, in laboratory flumes and conducted field experiments on the effects of individual

Mya arenaria on the recruitment of the polychaete *Hobsonia florida* and oligochaetes on a tidal flat. This study was designed to carefully measure recruitment within short distances from the siphon of the bivalve. They found no effect from the filtering activity of either bivalve on particles or recruitment beyond 2 cm from the siphon. The major effect was an increase in the variability of deposition downstream from the siphon due to the interaction of the excurrent flow from the bivalve with the ambient flow across the siphons. These authors concluded that it will be difficult to predict, based on areal density alone, what constitutes a dense assemblage. Calculations based on their data, without considering factors such as enhancement of particle deposition and depletion of larvae by the filtering of upstream bivalves, indicate that even with densities as high as 1000 *M. arenaria* m^{-2} there would be room for recruitment. The importance of the scale of the recruitment processes and the scale of the effects to be expected has also been emphasized by Peterson and Beal (1989).

Data provided by Moller (1986) established the importance of two mechanisms by which bivalves can cause a reduction in recruitment. Increasing densities of *Mya arenaria* from 0 to 200 m^{-2} in both caged and uncaged areas inhibited recruitment of the spat of *Mya arenaria*, *Cerastoderma edule*, *Spisula subtruncata* and *Mytilus edulis* (Fig. 11.1). Fewer spat were found in uncaged areas, supporting those who believe that the losses of spat due to predators and losses to adult *M. arenaria* are important recruitment parameters. Other studies indicated that in years when the numbers of adult bivalves were high, spat recruitment was low (Moller, 1986). The recruitment biomass of *M. arenaria* exceeded that of the cockles in all years from 1977 to 1983, except 1982, when a warmer than normal August caused rapid cockle growth (Moller, 1986). In spite of the spat recruitment of *M. arenaria* being higher than that of cockles in all years, this increased growth rate allowed the cockles to become too large for most of predators (Moller and Rosenberg, 1983). The resulting increased cockle density allowed them to completely cover the upper 5 cm of sediment. This coverage, and increased cockle biomass, was correlated with a decrease in the abundance and biomass of *M. arenaria* (Moller, 1986). The disturbance caused by the movement of the cockles (Jensen, 1985) was thought to be responsible for the severe reduction in *M. arenaria* (from 80 g ash free dry weight (AFDW) m^{-2} during the period 1979–1982 to only a few g AFDW m^{-2} in 1984) (Moller, 1986). In this instance, one adult bivalve population was able to effectively reduce recruitment of the other, through both predatory and amensalistic activities, until conditions changed to those that favored the other species.

11.7.1.2 Summary of bivalve predation

Observations by fishermen that, in areas of high adult densities of hard clams there appears to be low recruitment, are supported by the data of Greene (1978) and Fitch (1965) who showed that dense beds of large clams have relatively low recruitment. Johnson (1994) has shown that, under unexploited conditions, many ecosystems are dominated by species that have large individual biomass and a long life span. He provided examples from fish in arctic lakes, European virgin forests, giant tortoises on an island, swan mussels in an arctic lake and other areas. This condition appears to be true of hard clams as well, but undisturbed populations are difficult to find.

Whether the dominance of a population by older individuals and low recruitment is due to predation of larvae by the adults or some form of competitive exclusion as was seen

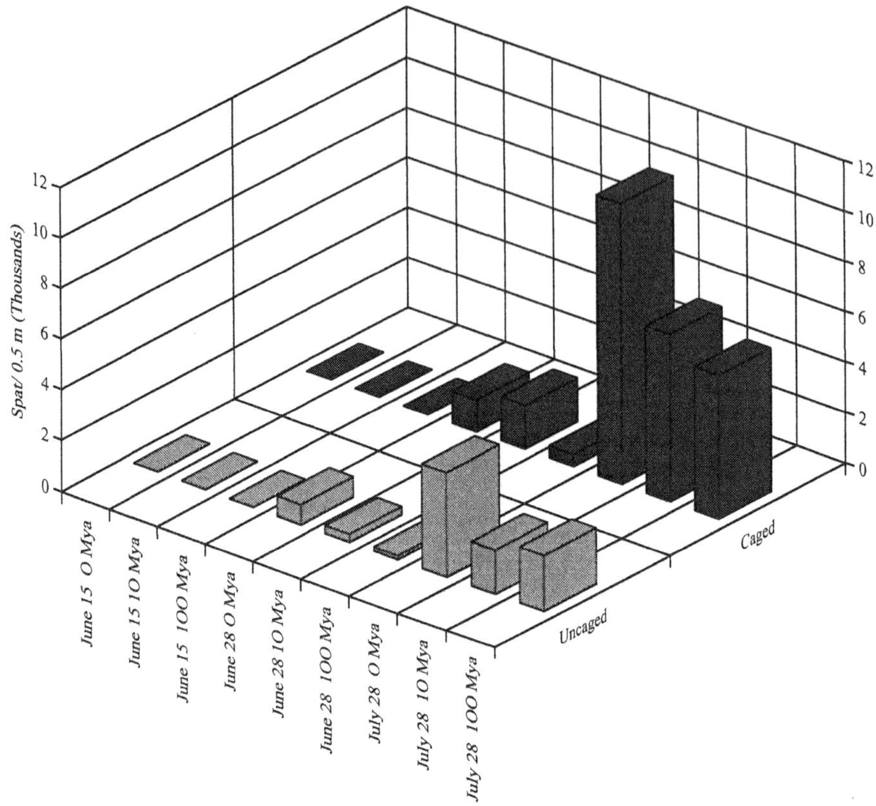

Fig. 11.1. Influence of three densities of adult soft shell clam, *Mya arenaria*, on recruitment of *Mya arenaria* and *Cerastoderma edule* on three consecutive sampling dates. Sampling dates are presented in a left to right sequence with densities of 100, 10 and 0 adults. Except for June 15, one set of adult densities was covered with a cage to prevent additional predation (caged) while another was uncaged. Data from Moller (1986).

with *Cerastoderma edule*, *Mya arenaria*, *Gemma gemma* and *Protothaca staminea* remains an area to be clarified by additional experiments. The data provided by Cowden et al. (1984) for *Mytilus edulis* feeding on larvae from a variety of taxa and Kurkowski (1981) for the hard clam feeding on its own larvae (900–1280 larvae mussel^{-1} day^{-1} and 285,000 clam^{-1} day^{-1}, respectively) suggest these effects could be locally important when high densities of adults are present and larval supply is low. It is clear that sediment type or factors related to sediment type such as current speed, food supply, the depth of the anoxic zone or other factors associated with sediment chemistry can have a demonstrable density-mediated recruitment effect. The above data suggest there are mechanisms to support the observations of reduced recruitment in the presence of adult bivalves, but as noted by Ertman and Jumars (1988) the extrapolation from areal density to potential effects on recruitment involves assumptions about a number of critically important variables, such as larval supply, spatial scales, and fluid dynamics. A number of recent papers on bivalve feeding (see also Chapter 8) may help elucidating this potential, but none of these test the effects on larvae directly (Frechette et al., 1989; Monismith et al., 1990; O'Riordan et al., 1995; Peterson and Black, 1987). The data of Moller (1986) and Hines et al. (1989) point out the difficulty in ascribing effects in field studies where interannual changes in recruitment of many species may affect the potential for other species to recruit. Studies in northern Europe, where large numbers of larvae, post-set and high adult densities are common, seem to be where interactions between adults and recruiting larvae have been shown to be significant. In areas with significantly lower numbers of recruiting larvae and low adult densities, interactions between adults and larvae have not proven to be significant. This latter condition appears to be the case for most hard clam populations, and those few studies that have examined the effect of hard clam adults on recruitment processes (Best, 1978; Young and Young, 1978; Maurer, 1983) indicate little or no effect.

11.7.2 Gastropoda

11.7.2.1 Predation on newly set clams

The only report of filter feeding gastropods taking in bivalve larvae is that of MacKenzie (1981) who placed oyster larvae in containers with *Crepidula plana* and *Crepidula fornicata*. Both species removed larvae from the water, but these were egested alive in the pseudofeces.

While many different snails have been reported to be predators on hard clams, only a few of these species interactions have been the subject of research that extends beyond simple observation. Predatory effects of snails on newly set clams have received little attention. Hunt et al. (1987) examined the effects the mud snail, *Ilyanassa obsoleta*, and of *Mercenaria mercenaria* on the recruitment of benthos by manipulating the densities of these two species. Mud snails reduced the density of the infauna by as much as 45%, but had no effect on recruitment of any bivalve species. Mud snails and clams were combined in field plots in densities (all 0.1 m^{-2}) of 0 clams and 0 snails, 15 clams and no snails, no clams and 50 snails, and 15 clams and 50 snails. The combined effects of mud snails and hard clams could be explained as simple function of the numbers of mud snails. The addition of hard clams did not change recruitment of any species. While not explicitly stated in Hunt et al. (1987), these data support those of Carriker (1961) and Turner and George (1955) who suggested that mud snails did not consume hard clams larger than 0.5 mm long. No data are available to suggest whether mud snail effects on communities are due to predation, disturbance, or a combination of these factors. Hunt et al. (1987) point out that the effects of suspension feeders and deposit feeders may be important at substantially different scales. Effects of deposit feeders, such as the mud snail, on recruitment would be in the immediate area, while suspension feeders like the hard clam may deplete larvae over large area. Thus, effects of suspension feeding on recruitment would be diffuse. These interactions are covered more thoroughly by Peterson (Chapter 10).

Carriker (1957) is the only individual who has reported on gastropod predation of newly settled hard clams. His laboratory tests found that newly hatched *Urosalpinx cinerea* could consume from 1 to 19 hard clams day^{-1}. These studies were conducted in 35–36 ppt seawater at temperatures from 22 to 31°C. The drills measured 0.9–2.0 mm long and clams ranged from 220 to 950 μm. The greatest mortalities occurred when the drills preyed on 0.22–0.29 mm newly set hard clams in 1-day tests. Whether such rates occur in the field or over long periods of time was not assessed.

11.7.2.2 Predation seed and larger clams

A number of studies have found that once hard clams reach seed size, predatory snails can consume considerable numbers. Gibbons and Blogoslawski (1989) have listed most of these species, but, as noted below, some of these species do not occupy the same habitat as hard clams except for rare instances.

Mesogastropoda
Naticidae

Most naticid snails feed exclusively on live molluscan prey by boring holes in the shell in much the same manner as muricids. The architecture of the drill hole (countersunk or not countersunk) allows discrimination of the predator type. In general, naticids inhabit sandy sediments, while muricids prefer substrates that include some hard bottom, such as shell or rock. The strong association of naticids with sediment was clearly shown in a study of predation on an isolated population of the mussel, *Choromytilus meridionalis*, by the naticid *Natica tecta*. Surveys found a low density of snails (12 m^{-2}) on sand, 69 m^{-2} on sand where a mussel bed was partly buried, and none on adjacent rocks covered with the mussel prey (Griffiths, 1981). As with other studies, large snails consumed larger mussels, but even snails 30 mm long could not consume 55 mm mussels. Rates of consumption were obtained from laboratory studies, and these rates predicted the mussel population would be completely consumed in <1 year, but growth to a refuge size prevented complete eradication of mussels in the sand (Griffiths, 1981).

Naticid predators such as *Natica maculosa* and *Neverita didyma* were reported to be among the most important predators on the manila clam *Tapes philippinarum* (= *Venerupis japonica*) in Japan (Cahn, 1951). Chew (1989) reported that small *Polinices lewisii* (5–6 mm shell diameter) fed on manila clams as small as 3 mm. Larger *P. lewisii* (25–100 mm) fed on larger manila clams. There are a substantial number of reports from field studies, such as that of Stickney and Stringer (1957), that indicate naticid gastropods prey on hard clams, but many do not provide size data for predator or prey or other pertinent information. Ansell (1960) reported that *Natica alderi* fed on a wide variety of bivalves near Millport, UK, including the genera *Nucula*, *Thyasira*, *Montacuta*, *Cyprina*, *Dosinia*, *Venus*, *Venerupis*, *Tellina*, *Garia*, *Cultellus*, *Spisula*, *Corbula*, and *Thracia*. The largest bivalve bored by this snail measured 22.3 mm shell length. Laboratory experiments with several species of bivalves indicated that predators with spire heights of 4–5 mm were able to drill prey about the same length as their spire height. These animals were unable to drill through an individual *Venus* that was 10 mm in shell length. Ansell (1960) estimated that 15%, 5% and 1–2% of the first, second and third year, respectively, of a large set of *Venus striatula* was consumed by this predator.

Ansell (1982) examined predation by juveniles of *Polinices catena* on the infaunal bivalves *Tellina tenuis* and *Venerupis decussata*. Although there were effects due to temperature, prey size and prey density, the rate of predation rarely exceeded one prey per gastropod per day, perhaps due to the amount of time the predator must spend handling the prey. When small prey must be utilized, the reduced food available per unit of work required to obtain it can severely limit predator growth rates. Increase in prey size resulted in a decrease in predation rate, and different species of prey were preyed on at different rates. For a similar shell size of prey, the growth of the predator was greater when it consumed *T. tenuis* than when it consumed *V. decussata* because the former has more meat per unit shell length. Ansell and Mace (1978) provided a list of dry tissue weight for four species of bivalves that could serve as prey for *Polinices alderi* in the North Atlantic and Mediterranean. At both sites the biomass yield of *T. tenuis* was greater than *Venus gallina*, *Spisula subtruncata*, or *Donax trunculus*, indicating that more energy would be available from this species (Ansell and Mace, 1978). As the predator grew, it preferentially consumed larger prey. If prey size was maintained near the optimum size, the number of prey consumed per unit time remained relatively constant and the consumption rate could be described by a simple function of temperature and reproductive index (i.e., number of egg collars produced). Maximal prey consumption was about 0.5 clams snail^{-1} day^{-1}. This was equivalent to 15% of the body weight per day. In the absence of reproduction (no egg collars being laid) the consumption was reduced to 1.7% (at 10°C) to 4.8% (at 25°C) of snail body weight per day on *T. tenuis*. Feeding rates on *Mya arenaria* peaked at 0.6 prey day^{-1} at high temperatures and on a yearly basis the snails consumed about 1% body weight per day.

Rogers and Rogers (1989) placed large (>40 mm shell length *Tapes philippinarum*) in cages 4.7 × 4.7 × 4.7 cm and smaller *T. philippinarum* in cages 4.7 × 2.4 × 1.6 cm buried in an intertidal flat. One *Polinices lewisii* was placed in each cage. These snails were able to consume only clams with shell length smaller than themselves, and on average, consumed about 0.14 clams snail^{-1} day^{-1}.

Euspira (= Lunatia)

Franz (1977) examined the effects of *Euspira heros* predation on the surf clam *Spisula solidissima* off Long Island. By collecting clams from the beach he was able to show that *E. heros* selected clams in the smaller size classes with peak predation occurring in prey measuring 47–75 mm length (age 3–4 years) (Franz, 1977). Unfortunately, as with most similar field studies, there are no data on the size of the predators, rates of predation or on the overall importance of these factors on the prey population. Dietl and Alexander (1997) examined the same site as Franz (1977), but also surveyed a wider spectrum of shoreline. They reported that the greatest concentration of naticid boreholes was found in 20–59 mm *S. solidissima*. Large boreholes were in larger clams and these appeared to be caused by *E. heros*. Laboratory studies have brought some naticids, such as *Euspira heros* and *Euspira triseriata* (Berg and Porter, 1974; Berg, 1975), into contact with *M. mercenaria*, but these two snail species are not commonly found in abundance near commercial hard clam beds except in rare cases in the most northern part of the hard clam's range. The data are, however, useful in establishing the potential effect of predation by naticids in general on hard clams. These data become particularly useful when species known to be abundant in hard clam beds, such as *Neverita duplicata*, are incorporated into an experimental design that allows direct comparisons of rates to be made (Hanks, 1952; Berg and Porter, 1974; Berg, 1975).

Commito (1982) compared the effects of *E. heros* predation on population dynamics of *Mya arenaria* and *Macoma balthica* on an intertidal flat in Maine. Most of the snails on the flat were >20 mm diameter, and as with the studies of Franz (1977), *E. heros* only preyed on smaller clams. Predation was higher on *M. arenaria* than *M. balthica*, presumably because the latter were able to reach a spatial (depth) refuge faster. Whether consumption of small bivalves was a function of size or the inability of the predator to reach deeply burrowed individuals was not examined.

Medcof and Thurber (1958) planted 38 mm long *Mya arenaria* at a density of 172 m^{-2} and found that *E. heros* migrating into the plots caused 79% mortality within 12 days. The investigators attempted to remove drills by picking them off the surface of some plots and were able to reduce mortality to 66%. They also reported that these drills killed more than half the *M. arenaria* without boring their shells. The authors counted drills in the experimental plots, but noted that even in areas where they attempted to remove the predators, they were unsuccessful in eradicating the snails. Average density of drills in an area 7 m^2 was estimated by collecting them from the surface and then removing sediment from the selected areas and passing it through a sieve. Medcof and Thurber (1958) estimated that between 300 and 500 drills were entering the clam plots each day (32–54 m^{-2}). Since the number of drills remained the same, an equal number must have been leaving the plots. Assuming an average density of 40 drills m^{-2}, and the consumption of 139 of the 172 clams in 12 days; the daily consumption rate per drill was 0.29 clams $drill^{-1}$ day^{-1}.

Neverita spp.

Rodrigues et al. (1987) examined the mechanisms use by the gastropod *Neverita didyma* to prey on *Tapes phillipinarum*. Prey were apportioned into four size classes (15–20, 20–25, 25–30, and 30–35 mm shell length and placed into 38 × 26 × 16 cm containers containing 5 cm of fine sand. Predators were separated into organisms with operculum lengths of 20–25, 25–30, 30–35 and 35–40 mm. In general, smaller predators consumed smaller prey and larger predators selected against the smallest prey. This process resulted in the smallest predators consuming more of the smaller prey. The authors concluded that mechanical aspects of prey selection were more important to this predator than optimum foraging for energy. This supports the work of Boggs et al. (1984) who reached the conclusion that naticid predators were unable to select individual prey, and Ansell (1982) who found that handling time was a very important component in the feeding of *Polinices lewisii*.

Examination of weekly consumption of manila clams by this species indicated that there was a significant correlation between feeding during 1 week and feeding the following week. The more food consumed in the first week; the less consumed the following week.

Rodrigues (1986) continued these studies and reported that both predator and prey exhibited more activity at night than during the day. The clams exhibited two escape responses as the predator approached; burying deeper into the sediment or leaping to another place. The latter response appeared to be more effective. Observations indicated that the snails often crawled over prey before selecting one to attack. Once the prey had been selected, the predator would wait beside it until some factor stimulated it to suddenly attack. Rates of consumption were slightly higher when a pair of predators was present (average about 0.75 clams $snail^{-1}$ day^{-1}), but at satiation the rate, was about two prey $week^{-1}$ (0.29 clams $snail^{-1}$ day^{-1}).

Neverita (= Polinices) duplicata

Based on observations of shells, Turner (1948) concluded that boring snails were responsible for about 3% of the losses of soft-shell clams planted on intertidal flats, and that crabs consumed approximately 38% of the clams. Field studies (Turner, 1949) revealed that the moon snail, *Neverita duplicata*, consumed approximately 10 clams $month^{-2}$ (0.33 clams $snail^{-1}$ day^{-1}) during the summer. Additional laboratory studies by Sawyer (1950) and Hanks (1952) found that there was a strong relationship between the consumption of clams and

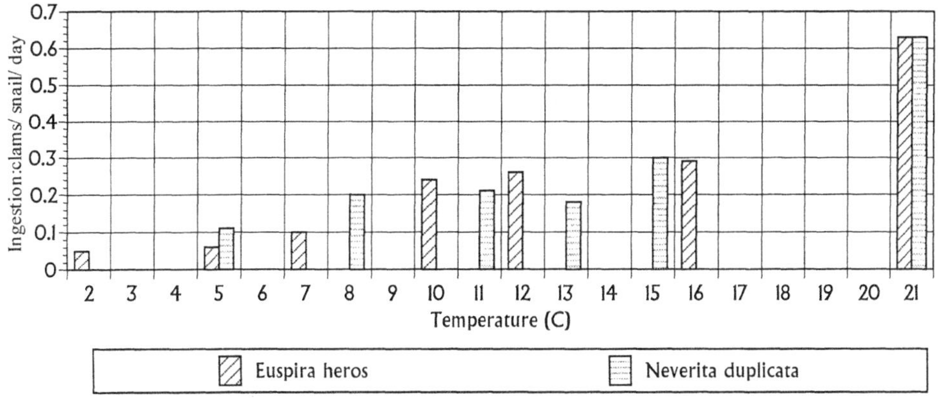

Fig. 11.2. Influence of temperature on the feeding rate of two species of naticid snails *Neverita (= Polinices) duplicata* and *Euspira (= Polinices) heros* on *Mya arenaria*. Data from Hanks (1952).

temperature and salinity. Sawyer (1950) found that at 21°C, 10 *N. duplicata* ate 202 of 300 clams in a tray of sand in 30 days (0.67 clams $drill^{-1}$ day^{-1}), while at 10°C, they consumed only 1/5 as many clams (0.13 clams $drill^{-1}$ day^{-1}). Hanks (1952) compared the feeding rate of *N. duplicata* and *Euspira heros* at temperatures and salinities ranging from 2 to 21°C and 6 to 32 ppt, respectively (Figs. 11.2 and 11.3). All temperature studies were conducted at 32 ppt and all salinity studies were done at temperatures between 14 and 18°C (Hanks, 1952). The feeding rates of the two species at constant salinity declined in concert until water temperatures were maintained below 10°C, whereupon *N. duplicata* feeding rate began to decline more rapidly. Only *E. heros* fed at 2°C. At constant temperature, *N. duplicata* continued to feed until salinity was reduced below 8 ppt, while *E. heros* feeding was greatly reduced at 14 ppt and no feeding was observed at 10 ppt (Hanks, 1952).

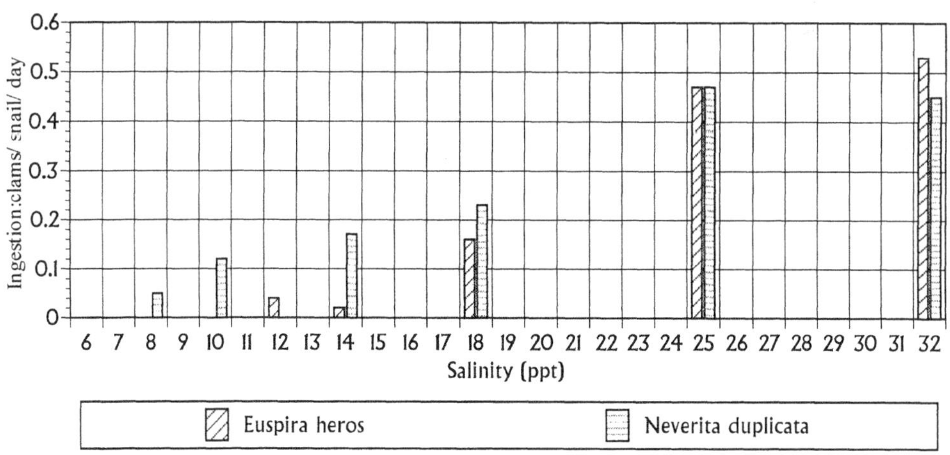

Fig. 11.3. The influence of salinity on the feeding rate of two species of naticid snails *Neverita (= Polinices) duplicata* and *Euspira (= Polinices) heros* on *Mya arenaria*. Data from Hanks (1952).

Edwards (1974) collected prey species of *N. duplicata* on Barnstable Harbor, MA flats. In 1969 he collected 37 bored valves (mean ± SD, 11 ± 7 mm) of hard clams from these flats. From 1968 to 1972, he observed seven attacks on hard clams by *N. duplicata*, and reported that the mean size and standard deviation of the predatory snails was 31 ± 8 mm, and they preyed on hard clams of 17 ± 6 mm. These observations reflected trends found for other prey species during his study in that larger snails preyed on larger prey. Snails less than 20 mm diameter did not prey on organisms larger than 14 mm while snails 40 mm in diameter preyed on organisms ranging from about 10 to 45 mm (Edwards, 1974). In other studies Edwards and Huebner (1977) and Huebner and Edwards (1981) reported that feeding by *N. duplicata* varied with temperature and ceased at about 5°C. Gross growth efficiency of the snails was inversely related to snail size and these predators were relatively efficient in converting energy gains to production (respiration was only 44% of consumed energy). Because of the nearly 4-month period of no activity in winter, these snails consumed only about 1% of their own body weight day^{-1}. *Neverita duplicata* densities in the Barnstable Harbor area typically range from 0.4 to 0.7 m^{-2}, but in selected areas density can be as high as 2.2 m^{-2} (Edwards, 1974). Other workers found densities of 0.6 m^{-2} for this species on nearby flats (Russell-Hunter and Grant, 1966).

Wiltse (1978, 1980a,b) studied predation on the Barnstable flats and noted that *Neverita duplicata* preferred thin-shelled molluscs (chiefly *Mya arenaria*), although on plots where *N. duplicata* were excluded by fences that allowed epifaunal organisms to access the plots, hard clams increased in abundance nearly four times. Importantly, Wiltse (1980a) noted that sizes of most molluscan species on these flats were <10 mm thus indicating that they were young of the year. Adult hard clams were not present. *Neverita duplicata* ranged from 3 to 40 mm in diameter and snails <19 mm averaged 0.68–6.64 m^{-2} (Wiltse, 1980b). Wiltse (1980b) reported a strong preference for prey of a certain size selection for *N. duplicata*, but only one snail (size class 17.9–23.4 mm) preyed on the small thick-shelled gem clams, *Gemma gemma*, during laboratory tests. These studies revealed that larger snails fed regularly on *Mya arenaria* even when they ignored smaller prey (Wiltse, 1980b).

Greene (1978) reported that whelks (*Busycon carica* and *Busycotypus canaliculatus*) were more abundant than *N. duplicata* in Great South Bay, New York. Moon snails were found on 5 of the 25 stations surveyed, and had an average density of 0.04 m^{-2}. This is nearly 10 times less than densities reported in Barnstable Harbor, MA, but may reflect a bias due to larger sampling gear used, and the generally subtidal habitat surveyed during the New York study. Most of the New York moon snail populations were in high salinity water near Fire Island inlet. Three of the stations nearest the inlet had snails that ranged from 40 to 55 mm long, while those from the inner bay stations were larger. So few snails were found overall that the data are difficult to interpret.

Neverita duplicatus feeding on *M. arenaria*, could have consumed 52 clams year^{-1} based on a yearly average feeding rate of 0.27 clams snail^{-1} day^{-1} for clams 15–45 mm (Edwards and Huebner, 1977) and a snail density of 0.53 m^{-2}. Carriker (1951) found that *N. duplicata* did not prey on hard clams in field plots when other species (eastern oysters, ribbed mussels, *Geukensia demissa*) were present, but with those species absent the moon snails consumed an average of 0.025 clams snail^{-1} day^{-1}. These rates appear to be low, but may reflect the size of the clams used, because exact sizes were not specified. Wiltse (1980b) reported snails (22–35 mm) fed on *Mya arenaria* (11–25 mm) at a maximal rate of 3.2 clams snail^{-1} day^{-1}

at 22–24°C even when they ignored the smaller prey (Wiltse, 1980b). The rate at which *N. duplicata* consumed the smaller gem clams declined from about 5 clams snail^{-1} day^{-1} as the snails grew (Wiltse, 1980b), and few snails >10 mm fed on gem clams. Wiltse (1980b) concluded that *N. duplicata* is probably not an important predator on *Gemma gemma* because it rarely consumes more than 15% of the clam population. How this size/species preference would affect hard clams that set near a dense population of another species such as *G. gemma*, has not been determined.

Laboratory studies on predation rate of moon snails (46–50 mm) on hard clams (20–65 mm) at 20°C yielded rates of 0.1–0.15 clams snail^{-1} day^{-1} (Greene, 1978). Feeding moon snails pulled the clams into the sediment so that both the predator and prey remained completely beneath the sediment surface. These snails were provided with clams ranging in size from 20 to 65 mm long and ate clams ranging from 23 to 55 mm. They did not feed on clams >55 mm even when the snails were starved for 2 weeks. Based on laboratory predation rates and field population counts, Greene (1978) estimated that moon snails would consume 2.2% of the hard clam population in the four warmest months of the year, and that this predation would be directed toward the 21% of the clam population that was less than 60 mm in shell length. Extrapolation of these numbers suggests that moon snails would eat approximately 10% of the available clams in four months.

Haskin (1951) reported on drill predation of different sizes of hard clam seed planted on the New Jersey side of Delaware Bay. From April to July *N. duplicata* consumed 28.8% of the clams placed on the flat, with an additional 35.6% unaccounted for (Fig. 11.4). These may have been lost or consumed by other predators, but unaccounted for losses were higher in the smaller size categories. Loss due to snail predation was similar in all weight classes from 1.7 to 5.7 g. There appeared to be a slight elevation in predation at the intermediate size clams

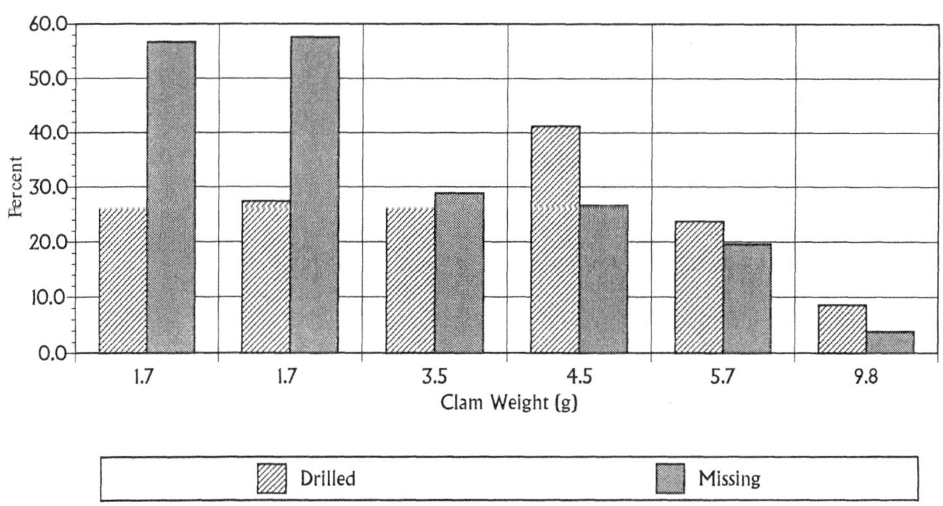

Fig. 11.4. Loss of hard clam, *Mercenaria mercenaria*, seed from plantings of 5 weight classes (two separate 1.7 g plantrings) in Delaware Bay, New Jersey due to predation by *Neverita (= Polinices) duplicata*. Missing clams were, in part, due to crab predation. Data from Haskin (1951).

(4.5 g) and a slight reduction in the largest size (9.8 g). Haskin (1951) averaged all data and estimated that *Neverita duplicata* consumed an average of 13–16 clams snail^{-1} month^{-1}. This is roughly equivalent to 0.5 clams snail^{-1} day^{-1} peak rates of *N. duplicata* feeding on small *Mya arenaria* on Massachusetts flats (Edwards and Huebner, 1977), but considerably less than the maximal 3.2 *Mya* snail^{-1} day^{-1} reported by Wiltse (1980b). The rates of predation recorded by Wiltse (1978, 1980a,b) are higher than those given by Greene (1978) for the same species preying on juvenile hard clams and for other species of *Neverita* examined elsewhere.

Summary Naticidae

Because there are so few studies that provide information on size ranges of predators, prey and their densities, it is difficult to determine the broad-scale effects of naticids on hard clams. Most studies on naticids have been done in more northern waters where numbers of recruiting and adult *Mya arenaria* far exceed those of *Mercenaria mercenaria.* None of the studies report observations similar to those of Schneider (1982) where *Ensis directus* actively avoided *N. duplicata* by exiting the substrate and leaping across the bottom. Adult hard clams have not been shown to exhibit such escape responses, but smaller clams actively move across intertidal flats. The data provided by Greene (1978) appear to be the only ones that combine the needed information. All data suggest that moon snails could be important predators in high salinity sandy areas, and this appears to be confirmed by the Delaware Bay data for seed planted on intertidal areas (Haskin, 1951). This report did not mention the presence of *Gemma gemma,* but these small clams are common faunal constituents of Delaware Bay sand flats. Whether the naticid populations shifted from preying on the latter species to the newly planted hard clams cannot be determined from the data. There are a number of studies that have examined moon snail predation on the genus *Spisula* (mostly by examining shell material), but none that provided rates of consumption. Hughes (1985) presented evidence that *Natica unifasciata* will pursue and attack smaller conspecifics and other gastropods. How this behavior affects predation on bivalve populations has not been measured. In any case, most moon snail predation seems to be on the order of 0.5–1 prey snail^{-1} day^{-1}, which is relatively low when compared to the effects of crustaceans (see below).

The study of Kitchell et al. (1981) is the only one that has attempted to model many of the aspects of the predator/prey interaction in snails. This effort examined prey selection by *Neverita duplicata* (25–60 mm) on three species of bivalves: *Mya arenaria, Mytilus edulis* and *Mercenaria mercenaria.* All tests were conducted in 38 × 8 × 10 cm chambers in 3–4 cm of fine sand. Temperatures were 19–22°C and salinity ranged from 28 to 34 ppt. Examination of energetic value (kJ) available per unit of prey length indicated that energy content of hard clams exceeded that of soft-shell clams which, in turn, exceeded that of mussels across all prey lengths evaluated. The energetic content had no effect on the selection process of the snails. The mean drilling rate for this species was 0.0223 mm h^{-1} regardless of the prey, but the ratio of drilling time to ingestion time was significantly different for different species (soft-shell clam 1.0, hard clam 2.4). Predators <30 mm were able to penetrate slightly greater prey : predator size ratio hard clams, but this ratio remained about the same between 30 and 50 mm prey size. In opposition to the nearly level curve for the hard clam, for soft-shell clams the prey : predator size ratio declined sharply to about 45 mm, indicating that larger predators were competing with smaller predators for the same size prey. More large hard clams with multiple boreholes were successfully drilled by smaller predators than those with only one borehole. Based on these data, the (Kitchell et al., 1981) developed a model that suggests that naticid gastropods appear to be selecting prey species and sizes that maximize net energy gain to the predator per unit of drilling time. A comparison of these rates of drilling with those available for muricids (Kitchell et al., 1981) indicated that naticids appear to gain more energy per unit of drilling and ingestion time. Kitchell et al. (1981) concluded that the soft-shell clam was preferred over the hard clam chiefly because the net energy gain was greater. There are negative interactions between moon snails and whelks that may account for some of the lack of information on *N. duplicata* preying on hard clams in the mid Atlantic (see below).

Neogastropoda

Muricidae

Muricids are important predators on many bivalve species. Marsh (1986) reported that *Nucella emarginata* and *Nucella canaliculata* were responsible for significant (up to 85%) losses of mussels, *Mytilus* spp., settling in experimental plots established on the Oregon coast. In some cases, these snails were more important than birds in controlling mussel populations. Bayne and Scullard (1978) demonstrated that, in general, larger *Thais lapillis* consumed larger *Mytilus edulis*, but there was a wide range of mussels selected by similar sized snails. Ingestion rates were dependent on temperature, and ranged from 0.25 to 0.38 mussels snail^{-1} day^{-1} (9–20°C).

Cahn (1951) cites a number of muricid species that are predators on clam genera as *Tapes* (= *Venerupis*) and *Meretrix* in Japan. Wells (1958b) reported that *Muricanthus fulvescens* could consume 3.5 hard clams per week (0.5 clams snail^{-1} day^{-1}), and utilized both grinding and chemical secretions to open the clams. This species is not usually found in the same habitats as hard clams. *Ocenebra erinacea* was reported to feed on *Mercenaria mercenaria*, *Cerastoderma edule*, and *Venerupis* in laboratory trials, but it is seldom found in direct contact with these species in typical field situations (Humphrey, 1990). Peterson and Black (1995) reported that the muricid *Bedeva paivae* was a significant predator on the bivalves *Katelysia* spp. By placing 36 snails (25–30 mm) in a single 39 × 29 × 15 aquarium with no sediment and 12 small, 12 large *Katelysia scalarina* and 12 *Katelysia rhytiphora* they were able to show that the snail preyed on the smallest prey (6–17 mm) *Katelysia scalarina* first, then proceeded to prey on larger (31–38 mm) *K. scalarina* and finally on the larger (39–46 mm) *Katelysia rhytiphora* (Peterson and Black, 1995). The rate of consumption in this situation (23–25°C, 35 ppt) was 0.011 bivalves snail^{-1} day^{-1} over the 49-day experiment (Peterson and Black, 1995).

Stramonita (= Thais) haemastoma

Combinations of environmental variables are known to have important effects on the activity of invertebrate predators. Although the southern oyster drill, *Stramonita (= Thais) haemastoma*, is not typically found in hard clam habitats (it may be if the clams are closely associated with oyster beds), the data on the interactive effects of environmental variables are the primary reason for their inclusion in this section. Garton and Stickle (1980) placed southern oyster drills (45 mm) in 38-L aquaria with cultchless eastern oyster spat (20–30 mm) at different combinations of temperature and salinity. All studies were done by dividing the aquaria into 8 compartments, and the numbers of spat consumed were counted each day and replaced. All combinations of 7.5, 10, 15, 20, 25, 30, and 35 ppt salinity and 10, 20 and 30°C were tested. There was no feeding at 10°C, and feeding was maximal at 30°C and 20 ppt salinity. The feeding rate at 30°C ranged from 0.4 spat $drill^{-1}$ day^{-1} to a maximum of 2.4 at 30°C and 20 ppt and then tapered off to 1.4 spat $drill^{-1}$ day^{-1} at 35 ppt. At 20°C, feeding rates were lower than at 30°C, and there was no significant salinity effect between 10 and 30 ppt. When the snails were subject to changing salinity (30–10 and 10–30 ppt) over a period of 24 h, consumption rates were not different from those held under constant salinity (Garton and Stickle, 1980).

Urosalpinx cinerea and *Eupleura caudata*

Most feeding studies have used fixed salinity, temperature or other environmental variables as a basis for comparisons, but in most estuarine habitats salinity and temperature can fluctuate greatly over a tidal cycle. Zachary and Haven (1973) evaluated the effects of fluctuating salinity on the survival of the oyster drill *Urosalpinx cinerea.* They reported results based on a progression in snail activity that ranged from attachment to a substrate, movement, feeding, and finally egg deposition and mortality. The major difference between constant and fluctuating salinity was in the time to the onset of mortality. Under fixed salinity, mortality was highest during the first 2 weeks, but when salinity was allowed to fluctuate, it took 10 days before the onset of mortality suggesting that stress was greater under the constant condition. How this stress may affect the results of feeding trials is unknown.

Most of the work identifying muricids as clam predators has been done on the oyster drills *Urosalpinx cinerea* and *Eupleura caudata*, and by far more data are available for the former (Carriker, 1955). Both of these species open clams by drilling holes through the shell and then use their radula to rasp out the meat. The removal of the flesh was described by Carriker (1957) who observed small drills feeding on newly set hard clams. Once most of the body flesh was consumed, the adductor muscles were rasped away. These mechanisms are the same as those reported for the same predator species preying on oysters. More information is available on the predation of oysters by muricids than for the same predators preying on the hard clam.

Hancock (1954) found *U. cinerea* density to range from 6 to 11 m^{-2} in oyster beds of the River Crouch in England. He noted that this species had been reported to feed on a number of native and imported molluscs including *Ostrea edulis* and its spat, *Mytilus edulis, Cerastoderma edule*, *Paphia aurea* and *Crepidula fornicata.* He reported that, as soon as they hatched from the egg capsules, juvenile drills were able to consume oyster spat and barnacles. Drills 2–3 mm long consumed oyster spat 4–5 mm in diameter. Field studies found that 58% of the oyster set for 1 year had been consumed by drills. Further sampling suggested that 86 oyster spat m^{-2} were consumed by drills each year.

Oyster beds in Long Island Sound had densities of *U. cinerea* ranging from 0.5 to 25.2 m^{-2} and *E. caudata* ranging from 0.3 to 6.3 m^{-2} (MacKenzie, 1981). Greene (1978) reported that oyster drills were present in Great South Bay, New York, but did not provide data on their numbers or predation rates because the gear being utilized (tongs and clam rake) excluded most small organisms including the drills. WAPORA (1982) found that *E. caudata* was more abundant in Great South Bay than *U. cinerea* by 5.3 : 1. In the WAPORA study, numbers of *Eupleura* ranged from 2 to 50 m^{-2}, but densities as high as 372 m^{-2} were found. Similar data for *Urosalpinx* were 2–10 m^{-2} and they were usually found in eelgrass beds. Reflecting the species respective preferences for hard substrate and soft substrate, when *Urosalpinx* was abundant *Eupleura* was not (WAPORA, 1982). The highest densities of *Eupleura* were found on mud bottom, and it was usually absent from eelgrass beds (WAPORA, 1982). The average densities reported by WAPORA (1982) for oyster drills were similar to those provided by MacKenzie (1970b) for stations from Massachusetts to New York (3.3–40.8 m^{-2}). These data on relative abundance of the two species were in marked contrast to the data reviewed by Carriker (1955) where *Eupleura* were typically found to be less than 10% of the oyster drill population. It is important to note that most of the data cited by Carriker (1955) focused on oyster beds and thus would be expected to be biased toward *Urosalpinx*, the species that prefers hard substrate. Little data are available on densities of these snail populations from other locations. WAPORA (1982) reported *Eupleura* ranged in length to 34 mm. These data were similar to those reported by Galtsoff et al. (1937) of 19–45 mm and the maximum height data on *Urosalpinx* (27–61 mm) compiled by Carriker (1955).

Cole (1942), who had enclosed drills a cage with 1-year-old oyster spat, found that the drills consumed 59 oysters in 5 months (0.38 oysters $drill^{-1}$ day^{-1}). These are similar to the predation rates documented by Carriker (1955) for *Urosalpinx* feeding on eastern oyster spat (0.01–4.8 spat drilled day^{-1}). The numbers of spat consumed generally increased as the size of the spat decreased in these laboratory studies. MacKenzie (1981) followed the feeding rates of drills in field studies and reported that drills consumed 0.13 oysters $drill^{-1}$ $week^{-1}$ in May–June up to 0.7 oysters $drill^{-1}$ $week^{-1}$ in June to July (MacKenzie, 1981). These rates are equivalent to 0.02–0.1 oyster $drill^{-1}$ day^{-1}. By August, drills on this oyster bed had killed 50% of the oysters and 33% of the spat, but data from other beds, with differing conditions and population densities, ranged from 0 to a high of 6.4% (MacKenzie, 1981).

Hanks (1957) reported on the predation rate of *Urosalpinx cinerea* on eastern oysters and blue mussels under controlled temperatures from 5 to 30°C. All experiments were conducted at 25 ppt salinity in 7.1 × 7.9 × 1.2 cm enamel trays to which 20 drills (20–25 mm) and 30–40 oyster spat or 40 mussels were added. In general the drills preferred oyster spat (10–30 mm) to mussels (20–30 mm) across all temperatures. Over a period of 34–102 days (dependent on the experiment), the highest rates of predation were recorded at 25°C on oyster spat (0.2 spat $snail^{-1}$ day^{-1}) and mussels (0.12 mussels $snail^{-1}$ day^{-1}). The lowest feeding temperature was about 7.5°C and the rate of consumption of both prey rose until the temperature exceeded 25°C. Between 25 and 30°C, feeding rates on both species of prey dropped.

Manzi (1970a) examined predation rates of *E. caudata* and *U. cinerea* on eastern oyster spat over a range of temperature and salinity combinations. Maximum average predation rates for *U. cinerea* and *E. caudata* were 0.17 and 0.14 bivalves $snail^{-1}$ day^{-1}, respectively. Both snail species ate the most spat at highest temperature and salinity combinations (Fig. 11.5). Studies of the effect of temperature on the rate at which *Eupleura* consumed 24.8 mm length oyster spat and 16.4 mm coot clams, *Mulinia lateralis*, revealed that more clams were consumed at all temperatures (Manzi, 1970a) (Fig. 11.6). Highest rates of consumption, 0.22 bivalves $snail^{-1}$ day^{-1}, were reached at intermediate temperatures (Fig. 11.6). These rates of predation were similar both in numbers eaten per day and the range of numbers eaten to those for the muricid *Nucella lapillis* feeding on mussels (0.25–0.38 mussels $Nucella^{-1}$ day^{-1}) in England (Bayne and Scullard, 1978), and to the 0.33, 0.31 and 0.25 organisms $snail^{-1}$ day^{-1} for recurved mussels, *Ischadium recurvum*, eastern oysters and Atlantic rangia clams, *Rangia cuneata*, respectively, being eaten by *Stramonita (Thais) haemastoma* (Brown and Richardson, 1987) in Louisiana. The rates were mostly based on laboratory studies, and taken

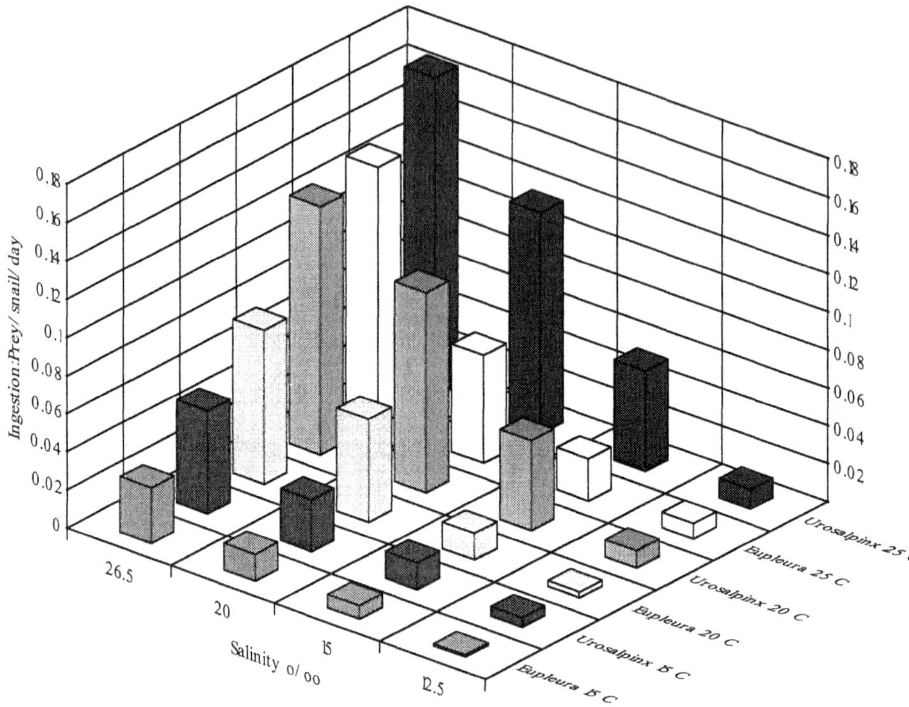

Fig. 11.5. Combined influence of temperature and salinity on consumption rate of two species of muricid snails *Eupleura caudata* and *Urosalpinx cinerea* feeding on the spat of *Crassostrea virginica*. 15 and 20°C refer to experimental temperatures. Data from Manzi (1970a).

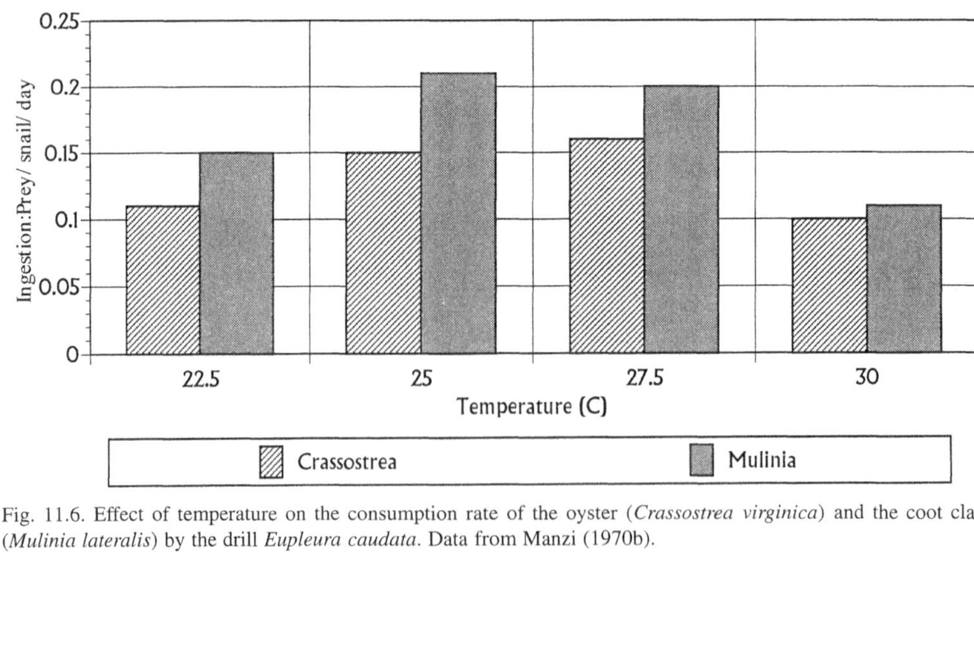

Fig. 11.6. Effect of temperature on the consumption rate of the oyster (*Crassostrea virginica*) and the coot clam (*Mulinia lateralis*) by the drill *Eupleura caudata*. Data from Manzi (1970b).

as a set, they clearly indicated the effects of size (relationship between predator and prey), temperature, season, reproductive condition, salinity and other variables on predation rates.

Stickney and Stringer (1957) reported that 34–36% of all juvenile hard clams in Greenwich Harbor, RI were drilled, and they indicated that *E. caudata*, because of its widespread distribution and tendency to congregate around newly set bivalves, was the most important snail predator. They also noted that drilling of hard clam seed was most evident in the 2–7 mm size classes, and clams >15 mm were seldom drilled.

MacKenzie (1977a) examined the effects of hard clam predators at three sites (Milford, CT; Great South Bay, NY; and Horseshoe Cove, NJ). The dominant muricid snail predators in these studies were the drills *Urosalpinx cinerea* and *Eupleura caudata*. By examining shell left by these predators in these three locations he inferred that in Great South Bay and Horseshoe Cove 60% of the dead hard clams were in the 4.5–10 mm length group. MacKenzie (1977a) estimated that drills killed most hard clams when the prey were between 4.5 and 20 mm long. These estimates were based on residual shell materials extracted from field-collected sediments. While such studies can provide an estimate of the importance of drills, crabs or other crushing predators do not leave large quantities of shell, particularly from smaller sized clams, and some species may move the prey before consuming it. Estimating relative percentage mortality in this fashion biases the data and may inflate the importance of drill predation relative to predation caused by crabs.

Melongenidae

Busycon carica and *Busycotypus (= Busycon) canaliculatus*

The term whelk in this review is taken to represent *Busycon carica* and *Busycotypus canaliculatus*, but other large snails, such as the Buccinidae (*Buccinum undatum*) and Strombidae (*Strombus* spp.), have also been referred to as 'whelks' in the literature. Alternatively, some authors use the term 'conch' for the Strombidae and Busyconidae. *Buccinum undatum* is known to consume infaunal bivalves, such as *Cerastoderma edule* in Europe (Nielsen, 1975; Hylleberg et al., 1978), so it can be considered to be a potential predator, but its range and that of the hard clam overlap only in rare instances. The buccinids *Cominella eburena* and *Cominella tasmanica* were both reported to consume bivalve prey on Australian tidal flats (Peterson and Black, 1995).

Prescott (1990) placed knobbed whelks, *Busycon carica* (170–250 mm height) in 75×75×35 cm aquaria containing 15 cm of sediment. These were supplied with running seawater (18°C) and if the single scallops (50–80 mm) were consumed, they were replaced. The experiment lasted for 1 month and 16 scallops were consumed by 2 whelks, 0.27 scallops whelk^{-1} day^{-1}.

There are a number of studies that provide evidence of whelk predation on hard clams. These studies date from the observations of Colton (1908) and Warren (1916) to the report by Stickney and Stringer (1957) who mentioned that channeled whelks were potentially important predators of hard clams in Rhode Island. Many studies combined observations of two or more species of whelks and often did not differentiate between the species when predation rates were discussed. Other reports such as those of Chestnut (1952) indicated the relative importance of whelk predation (69% of mortality vs 31% due to sea gulls) in North Carolina, and the field experimental studies of Peterson (1982a) attempt to place whelk predation in the context of the environment and other predators.

Greene (1978) reported that, based on abundance, *Neverita duplicata*, and the channeled whelk, *B. canaliculatus*, and knobbed whelk, *Busycon carica*, were the only gastropod species that appeared to be able to create a significant impact on adult hard clam populations in Great South Bay. His data were compiled using gear that would have excluded many smaller snails, such as the muricids, which he indicated could consume up to 10% of the seed clam population per year. *Busycon carica*, like moon snails are thought to feed only on live molluscan prey while *B. canaliculatus* forages more widely and will act as a scavenger. Whelks in Great South Bay, New York reached densities as high as 0.64 m^{-2} with average densities of 0.15 m^{-2} (Greene, 1978).

Flagg and Malouf (1983) reported that *B. carica* was a numerically dominant hard clam predator in Napeague Harbor, Long Island. Whelk density during their experiments (May to September) ranged from 0.3 to 7 whelks m^{-2}. Average density on three sites in the harbor ranged from 1.4 to 3.7 whelks m^{-2}.

Visual estimates of whelks inhabiting intertidal flats behind a barrier island in Virginia were made by searching the area at low tide (Kraeuter and Castagna, unpublished data). The average number of whelks found per survey was 225, with the highest numbers recorded being 641 *B. carica* and 17 *B. canaliculatus*. The searched area was approximately 423,000 m^2 and included many areas in the high intertidal where these species were rarely found. This large area made the 0.0005–0.0015 whelks m^{-2} seem low when compared with other studies, but most whelks were concentrated on a small portion of the flats. Mark/recapture studies conducted on these flats (unpublished) suggested that the population may be on the order of 10,000–12,000 individuals (0.028 m^{-2}), but these data also indicated significant immigration and emigration from the study area.

In contrast to the intertidal flats of Virginia, most of the whelks found in the WAPORA survey of Great South Bay were channeled whelks, and trapping studies conducted in Nantucket Sound also sampled channeled whelks (Davis and Sisson, 1988). The latter effort utilized commercial whelk traps and diver transects to evaluate whelk density. This extensive effort reported similar results for the two methods and an average density of 0.0048 whelks m^{-2} and a high density of 0.0078 whelks m^{-2}.

WAPORA (1982) reported that whelk densities based on two sampling methods commercial clam tongs and suction dredging. Sampling with tongs found whelk densities up to 0.83 m^2, similar to the data of Greene (1978), while maximum density for the suction dredge was 20 m^{-2}. Typical densities for whelks ranged from 2 to 4 m^{-2} (WAPORA, 1982). These were much higher than reports from other sites, but are nearly identical to those reported by Flagg and Malouf (1983) in Napeague Harbor, Long Island. Peterson (1982a) reported that densities of *B. carica* ranged from 0.0523 to 0.128 m^{-2}, *B. canaliculatus* 0.035–0.154 m^{-2} and *Busycon contrarium* 0.077 m^{-2}. Thus densities of all species combined ranged from 0.088 to 0.359 m^{-2} in a Bogue Sound, North Carolina seagrass habitat (Peterson, 1982a). Slightly farther south in Wassaw Sound, Georgia, Walker (1988) reported whelk densities of 0.024 m^{-2} in the same areas where hard clam densities averaged $>$1–4 m^{-2}. Highest clam densities at the Georgia sites were 100 m^{-2}, and, in general, more whelks were found in areas with higher clam densities (Walker, 1988). On the Gulf Coast, Menzel and Nichy (1958) and Kent (1983), both working in Alligator Harbor, Florida found *B. contrarium* at 0.04 m^{-2}; and whelks to 0.0016 m^{-2}, respectively.

A number of authors have noted interactions between various whelk species, and differences in prey selection. Kent (1983) noted significant differences in the interactions of *Busycotypus spiratus*, thin shelled, and *Busycon contrarium*, thicker shelled, whelks and their prey. While both species preferred active bivalves (scallops) over passive bivalves, the thinner shelled whelk species was more aggressive and more efficient at capture of active bivalves than the thicker shelled whelk. In contrast, the thicker shelled predator was more efficient when attacking thicker shelled passive prey, such as hard clams. These results were similar to those of Paine (1962) who maintained *B. spiratus* and *Busycon contrarium* in aquaria with nine species of pelecypods, and found that *B. spiratus* selectively preyed on thinner shelled or gaping species, such as *Ensis minor*, *Tagelus divisus* and *Mactra fragilis*.

B. contrarium the heavier shelled snail, preyed on such tightly closed forms as *Lucina floridana*, *Chione cancellata*, *Cardita floridana*, *Macrocallista nimbosa*, *Noetia ponderosa* and *Modiolus americanus* (Paine, 1962). Nichy and Menzel (1960) reported that 60% or more of the mortality on uncaged oysters in Alligator Harbor, Florida was due to *B. contrarium* predation. Peterson et al. (1989) and Prescott (1990) both reported that the knobbed whelk may be an important predator on bay scallops in the sounds of North Carolina, but field evidence was limited to some observations and tethering experiments.

Although there are limited data, the importance of snails preying on other snails and these interactive effects on hard clams should be examined more carefully. It is a common observation that *Neverita duplicata* is generally rare on flats that have high densities of *Busycon*. Magalhaes (1948) reported that a *B. canaliculatus* was found eating a *N. duplicatus* in a bucket while these were being returned from the North Carolina flats. Paine (1962) observed a *B. spiratus* that had eaten a *N. duplicatus* on the Alligator Harbor flats. Whether maintaining the thinner shelled species of *Busycotypus* on intertidal sand flats with hard clams decreases overall predation by moon snails and benefits the clam population will depend on the size of the clams, and the density and predation rates by both species. Based on consumption rates and population densities, it appears that moon snails could be a more important predator on clam seed than the thin shelled species of whelks, but the thicker shelled whelks are probably a more important consumer of adult hard clams (Peterson, 1982a) than moon snails.

Magalhaes (1948) did not examine the consumption rate of whelks in her studies at Beaufort, North Carolina, but listed the size range of the various prey species eaten by the three whelk species common to the area. The size range for hard clams (11.6–90.2 mm) and the size range for other bivalve species with over 10 specimens found eaten were *C. cancellata* (14.8–41.1 mm), *Doscinia discus* (45.3–77.7 mm), *Geukensia demissa* (34.1–79.2 mm).

Carriker (1951) provided a description of the wedging and chipping mechanism used by whelks to open hard clams. Again, many of the tests in this study simply indicate whelks and did not differentiate, but in one test (131 days at 20–24°C — clams ranging from 20 to 75 mm) a large *B. carica* consumed an average of 0.86 hard clams day^{-1}. The clams eaten in this study ranged from 33 to 56 mm. Carriker (1951), based on laboratory predation studies and field cages, suggested that in the 5 warm summer months channeled whelks at a density of 0.1 m^{-2} could have consumed 700 clams or about 0.05 clams snail^{-1} day^{-1}.

Greene (1978) reported predation rates by *Busycon* (20°C, 120–170 mm length) to be 0.111–0.115 hard clams snail^{-1} day^{-1} with a mean of 0.112 clams snail^{-1} day^{-1}, but the species was not identified. Based on the description of the predation in the text it appears that these rates refer to knobbed whelks. Paine (1962) did not directly compare the size of prey and size of predator in his experiments, but if all his data are combined, daily predation rates were 0.69 and 0.68 prey whelk^{-1} day^{-1} for *Busycotypus spiratus* and *Busycon contrarium*, respectively.

In laboratory studies, Peterson (1982a) reported that during a 10-month experiment, 3 *B. carica* consumed 35 hard clams and 25 *Chione cancellata* (0.067 clams snail^{-1} day^{-1}) while similar data for *B. contrarium* and *B. canaliculatus* were 32 hard clams and 19 *C. cancellata* (0.057 snail^{-1} day^{-1}) and 3 hard clams and 15 *C. cancellata* (0.02 snail^{-1} day^{-1}), respectively. Peterson (1982a) also observed that when the data from the above experiments were combined there was an obvious seasonal component to whelk predation. During the six colder months, 24 hard clams and 16 *C. cancellata* were damaged by whelks. Similar data for the four warmer months were 45 hard clams and 42 *C. cancellata*. Thus if predation rates had been computed on the basis of summer data the combined daily rate for all species would have been over twice as high as indicated above.

The data by Peterson (1982a) and the data by Paine (1962) on other whelks suggest that channeled whelks are less likely to consume hard clams than the heavier shelled species, such as the knobbed whelk. Personal observations on tidal flat populations in Virginia confirm these data. In over 4 years of observation, channeled whelks were never found consuming an intact hard clam (Kraeuter, unpublished). The only caveats to these observations are that there were relatively few channeled whelks on the flat and all searches were conducted during daylight hours. My data from Virginia (Fig. 11.7) suggest that knobbed whelks preferentially prey on larger hard clams (see also the data on oystercatchers on this same flat, below). This was also shown by the experiments of Peterson (1982a) for hard clams and *Chione cancellata*. What is also interesting is that the rate of predation on the smaller size *C. cancellata* was slightly less than for hard clams, but the smallest *C. cancellata* were consumed at a rate greater than hard clams of equivalent size (Peterson, 1982a). This suggests that in addition to a size preference, there are additional factors in prey selection. Peterson's experiments also indicated that there

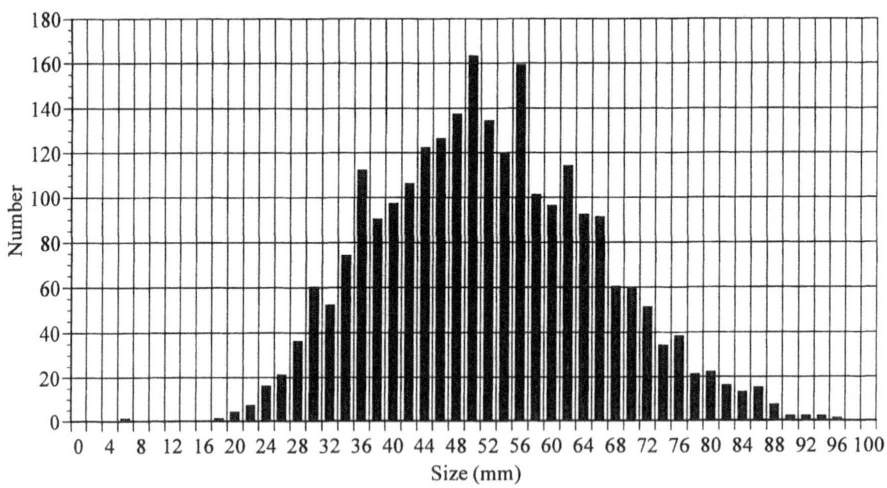

Fig. 11.7. Size–frequency distribution of shell lengths of the hard clam, *Mercenaria mercenaria*, collected from an intertidal flat, Cedar Island, Virginia. Mortality was caused by predation by whelks, mostly *Busycon carica*. Data were collected over several years.

was no density-dependent effect (the rate of predation did not increase with increasing clam density).

Irlandi and Peterson (1991) reported on the whelk predation in 1-m^2 experimental plots in which 49 clams m^{-2} (16.4–51.8 mm) were placed inside and outside grass beds. Two major classes of predators accounted for most of the losses of adult hard clams during the study. Crab predation was responsible for 61% and whelk predation for 38% of the losses in seagrass dominated areas while on sand flats the respective percentages were 77 and 22%. In a second experiment, crabs preyed on 56% and whelks 44% of the hard clams in grass beds compared to 86 and 14% and 84 and 15% on pruned sea grasses and sand flats, respectively. In a laboratory study, a trough, supplied with flowing seawater, was filled with sediment and divided into two sections to which 16 clams were added. The presence on one *Busycon carica* decreased the time hard clams spent feeding from 89 to 44% in the side to which the whelk was added compared with an 82–66% reduction in the control side. These data exclude those clams directly disturbed by physical contact with the predator.

In addition to the lack of density dependence, there are several other important generalizations can be found in Peterson (1982a): (1) whelks selected larger sized individuals of both prey species; (2) whelks were able to consume the largest sizes of clams present (there was no size refuge for hard clams); (3) sea grasses protected the clams, and the increased numbers of clams found in this habitat was due to protection, not increased recruitment.

Nassariidae

Ilyanassa obsoleta

One of the most abundant snails on clam flats is the mud snail *Ilyanassa obsoleta*. Carriker (1961) placed hard clams (3–15 mm) into a sand filled tray with adult mud snails and found no predation after 11 days. These studies further support those on newly settled clams (see above) that suggested that mud snails are not predators of hard clams.

Fasciolariidae

Fasciolaria lilium (= Fasciolaria hunteria)

Wells (1958a) reported that the banded tulip snail *Fasciolaria lilium* would feed on hard clams. In a series of experiments conducted in Beaufort, North Carolina, he demonstrated that *F. lilium* generally preferred oysters to other pelecypods, and that gastropods, such as the oyster drill, were preferred over pelecypods. *Fasciolaria lilium* attacked hard clams by wedging its shell between the clam's valves in much the same way that has been described by Carriker (1951) for *Busycon*. Wells (1958a) erected 0.37-m^2 cages in the field and 20 *F. lilium* (79–99 mm) were placed with 100 eastern oysters (48–118 mm), 100 ribbed mussels (51–94 mm), 30 clams (48–75 mm) and 12 bay scallops (42–58 mm). A similar control cage was maintained. After 51 days a net total (test minus control) of 67 oysters, 28 mussels, four scallops and one clam were consumed by this predatory snail. These data yielded an average of 0.098 pelecypods *Fasciolaria*$^{-1}$ day^{-1}, a rate that was nearly equal to the rates found in laboratory studies. Wells (1958a) pointed out that *F. lilium* was more common on shell and oyster bottoms than in most hard clam habitats and that the snail was unable to dig clams out of the bottom. These data suggest that in its native habitat it is probably not an important hard clam predator.

Prescott (1990) also experiment with the predation of banded tulip snails (60–80 mm) by placing them in 75 × 75 × 35 cm aquaria containing 15 cm of sediment. These aquaria were supplied with running seawater 16.5°C and one scallop was placed in each tank with a single predator. All scallops (50–80 mm) were replaced when consumed. Experiments lasted 2 months and the three snails tested consumed 9 scallops (test minus control), and the consumption rate was approximately 0.05 scallops $snail^{-1}$ day^{-1}.

11.7.3 Summary of molluscan predation

It is clear that filter feeding bivalves can ingest large numbers of larvae, but the effects of this ingestion at the population level have been more difficult to evaluate. In areas where there are high densities of adult filter feeders, reductions in recruitment have been documented. In areas with low adult density, recruitment effects have not been noted. In spite of the clear laboratory evidence that adult hard clams can damage their own larvae, none of the field studies in which hard clams density was manipulated found an effect due to the presence of adults.

As with other taxa, there are clearly important size-related interactions between snail predators and prey, but additional interactions with temperature, salinity, sediment type and predator reproductive state may define the rate of consumption. How these factors interact with predator and prey behavior has not been sufficiently researched. Where studies have concentrated on comparing size and density of predators and prey, there is remarkable similarity based on maximal numbers of prey species consumed by snail predators per day at about 0.1–0.9. In spite of the similarities noted, and the predation rates that have been derived from the data (Table 11.2), the reader is cautioned from extrapolating this information too far. Menge (1978) compared the rates of predation by the muricid *Nucella lapillis* feeding on barnacles and mussels in field plots. He noted that the factors normally considered to affect feeding rates of the snails, such as prey abundance, prey productivity and the presence of other predators were relatively unimportant (but, see Vadas et al., 1994 for a laboratory study that emphasized the importance of snail age, starvation and chemical cues from predators). Canopy forming algae, desiccation, snail phenotype and history had

TABLE 11.2

Summary of feeding rates of molluscan predators on a variety of bivalves

Predators	Size eaten	Consumption ($predator^{-1}$ day^{-1})		Density of predator
		Clams	Other species	
Bivalves	98–200 μm	25–84,400		80 m^{-2}
Muricidae	0.9–2 mm to 30 mm	1–19 0.22–0.5	0.01–4.8 spat 0.02–0.38	2–50 m^{-2}, 2–10 m^{-2}
Naticidae	<10 mm to 65 mm	5 0.1–3.2		0.04 m^{-2}, 0.53 m^{-2}, 0.68 to 6.64 m^{-2}
Melongenidae	to 100+ mm	0.1–0.86		2–4 m^{-2}, 0.64 m^{-2}, 0.15 m^{-2}, 0.4 m^{-2}
Fasciolariidae			0.05–0.1	

Data are combined from the references in the text.

significant effects (Menge, 1978). Wave shock was not significant, but the studies excluded the stormiest part of the year. Importantly, many of the interactions between the variables were statistically significant. The study clearly pointed out that many of the interactions between the main variables tested (tidal height, canopy cover, month and snail phenotype) were not additive, but some combinations were more important than others. For instance, month and canopy cover interacted so that warm weather combined with no canopy reduced predation intensity more than would be expected by their effects examined separately (Menge, 1978). The study demonstrated that individual predators should not be considered to be equivalent, and that higher order interactions between tested variables may greatly influence the outcome of even simple systems. Such interactions as those reported by Ambrose (1984a) for two predatory polychaetes, the evidence of channeled whelk feeding on moon snails (Magalhaes, 1948), and the observations that one of these snails preys on hard clams <20 mm while the other preys on larger hard clams clearly point out that knowledge of the particular assemblage occupying hard clam habitat is essential to interpreting even the best designed field studies.

Murdoch (1971) noted that snails tend to reach their maximum feeding rate quickly as prey density changes. This is because snails feed until satiated, and tend to eat all of an individual prey if possible. This process may be substantially different from the strategy employed by crabs and other active predators. The computer model of Murdoch (1971) indicated that, with snail predators and high prey density, the percentage mortality decreased (negative density dependence).

Studies such as that by Murdoch (1971) need to be viewed in context of the apparent ingestion conditioning associated with snail predation. Food conditioning studies conducted on muricids, such as *Nucella lapillis* (Hughes and Dunkin, 1984), and *Urosalpinx cinerea* (Wood, 1968), found that the foraging behavior of snails was strongly dependent on the prey species they were used to consuming. This preferential feeding can be reversed, but it takes time. Thus although there may be a negative density dependence, theory would suggest that snail predators will concentrate on a prey species even though the population density of the prey is very low. This may not be true of other predators.

Virtually nothing is known about the influence of various molluscan predators on recently set hard clams. The influence of combinations of predators, substrate type and exposure on recruitment of hard clams clearly needs additional work. There is a clear indication that most molluscan predators consume juvenile hard clams, and the rates of consumption are generally less than one clam per predator per day. At least one group (*Busycon*) appears to specialize on adult clams. The only other taxa reported to consume adult hard clams are vertebrates.

11.8 ARTHROPODA

The classic work Belding (1912) on hard clams did not list arthropods as serious predators, and only mentioned crustaceans as probable enemies of young clams. Until the recent emphasis on planting seed clams for aquaculture, the omission of crabs as one of the more serious predators on small clams has also been common among clam harvesters. In the early literature, boring snails and starfish were usually cited as the most important predators because the shells remain intact after the attack, and may accumulate from year to year. Crab predation typically leaves little or no visible evidence and thus may easily be underestimated. Other than barnacles, there do not appear to have been reports of arthropod predation on larval clams, but

Thorson (1966) cites laboratory studies by Christensen in which harpacticoid copepods ate newly set lamellibranch spat. Reed (1969) and Roberts (1974) both attempted to use bivalve larvae as a laboratory food for studies on the larval stages of *Cancer magister* and *Pagurus longicarpus*, respectively, but with little success.

Chelicerata
Merostomata
Xiphosura

Limulus polyphemus

Horseshoe crabs, *Limulus polyphemus*, have been reported as a predator on soft-shell clams, *Mya arenaria* (Turner et al., 1948a; Schuster, 1950; Smith and Chin, 1951; Smith, 1953; Smith et al., 1955). Turner et al. (1948a) placed a 5.6-cm carapace width horseshoe crab in a tray containing sand and 100 soft-shell clams slightly less than 13 mm long. After 72 h only, one clam remained (33 clams *L. polyphemus*$^{-1}$ day^{-1}). Smith (1953) planted soft-shell clams averaging 39 mm long in plots near Newburyport, MA, and reported that horseshoe crabs focused their digging in areas where clams were concentrated. In addition to consuming the soft-shell clams on the flats, *Macoma* sp. and *G. gemma* were also crushed by the crabs. Large plots (10 × 20 feet) were planted with 7600 *Mya arenaria* in June and by November only 8% were recovered (Smith and Chin, 1951). These authors indicated that most of the loss took place during the first 2 weeks when 31 horseshoe crabs were found in the three plots. If these data are utilized (10.1 crabs plot^{-1}, mortality of 6992 clams, in 14 days), predation rate on 39 mm *M. arenaria* would be 160 clams horseshoe crab^{-1} day^{-1}. Subsequent studies indicated that green crabs, *Carcinus maenas*, were also a major source of predation in the area, and thus the rates of clam loss cannot be completely ascribed to horseshoe crabs (Smith and Chin, 1951). Additional studies indicated that large (46 mm) clams could be protected by a fence that kept out horseshoe but not green crabs. In one experiment, survival of the protected clams was 95% after 8 months, but only 50% in an unprotected control plot. Smaller clams (16 mm) did not survive well in either the protected or unprotected sites. This may indicate green crab predation on these smaller clams because green crabs could enter and leave the plots through the mesh, while the horseshoe crabs could not.

Schuster (1950) listed *Ensis directus*, *Mya arenaria*, *Macoma* sp. and *Gemma gemma* as food items for horseshoe crabs in Massachusetts. Botton and Haskin (1984) reported on the stomach contents of *L. polyphemus* collected in surf clam dredges on the New Jersey continental shelf. Bivalves such as *Spisula solidissima*, *Tellina* sp., and *Siliqua costata* were listed as important food items, but on one station blue mussels were the predominant food. These authors note that in laboratory studies (Botton, 1982) a male crab (203 mm prosomal width) ate surf clams measuring 40.6, 43.8 and 42.4 mm. A female crab (279 mm) consumed a 46- and a 36.2-mm surf clam. In most of these cases, shell was not ingested, thus making identification of *S. solidissima* in the diet of wild caught animals difficult.

Stickney and Stringer (1957) mention that horseshoe crabs were an important predator on clams in Rhode Island. Botton (1982, 1984), examined the feeding of horseshoe crabs on *Mulinia lateralis*, *Mya arenaria*, *Spisula solidissima*, *G. gemma* and *M. mercenaria* under laboratory conditions. In general, the horseshoe crabs preferred those species with softer shells. When presented with thicker shelled species such as the hard clam (8–18.2 mm), the crabs preferentially consumed small individuals, but there was evidence that an attempt had been made to eat the largest clams. In most cases, those consumed were the smallest presented. Horseshoe crabs consumed up to 39 hard clam seed *L. polyphemus*$^{-1}$ day^{-1}. When thinner shelled species such as *M. lateralis* were tested, larger clams (>10 mm) as opposed to smaller (4–10 mm) clams were selected. In studies with gem clams, Botton (1982) reported that up to 1190 clams (2–4 mm) were consumed per day, but field sampling revealed an average of only 4.1 gem clams in the crab's gut. It is clear that horseshoe crabs could be a significant predator on hard clam seed until the clams reach about 15 mm.

Kraeuter and Fegley (1994) reported that horseshoe crabs can disrupt sediments to a maximal depth of 17.7 cm. The entire surficial sediment of an intertidal flat in Delaware Bay could have been completely mixed to an average depth of 11.1 cm in the 7-week spawning period. Large (>65 mm) hard clams were often pushed from the sediments by the horseshoe crabs burrowing activities. These clams were often picked up by humans and may have been more vulnerable to gulls (see Section 11.11).

Mandibulata
Crustacea
Cirripedia

Balanus spp.

Barnacles have been shown to ingest a wide variety of invertebrate larvae and inhibit recruitment in some epifaunal species, but these effects were not important in all cases (Young and Gotelli, 1988). MacKenzie (1981) found that *Balanus eberneus* removed oyster larvae from the water column, and that the larvae entered the barnacle's mantle cavity, but then were released unharmed. *Balanus improvisus*, a common estuarine barnacle, reduced the density of both umbonal and pediveliger larvae of the eastern oyster, *Crassostrea virginica*, in laboratory experiments (Steinberg and Kennedy, 1979). These authors reported >30 semidigested larvae in a large barnacle, but only one in smaller specimens. Barnacles are not common in typical soft-bottom hard clam habitats and what effect, if any, barnacle populations on oyster reefs, rocky shores, docks and other hard substrate may have on hard clam larvae would be difficult to evaluate.

Malacostraca
Stomatopoda

Squilla empusa

Laboratory experiments indicated that adults of the mantis shrimp, *Squilla empusa*, were able to break the shells and consume 10–20 mm hard clam seed as well as 10-mm-long *Mya arenaria* (Bisker, personal communication as cited in Gibbons and Blogoslawski, 1989). Whether mantis shrimp are a serious predator in the field is unknown, in part, because they burrow deeply and are not adequately enumerated with typical benthic sampling devices. They are present in many hard clam habitats.

Amphipoda

Monoporeia (= Ponotporeia) affinis

Segerstrale (1962) noted that literature reports indicated an inverse correlation between populations of the amphipod *Monoporeia affinis* and aggregations of *Macoma balthica*. This amphipod can occur at densities up to 12,000 m^{-2} (Segerstrale, 1962). Ankar (1980) reported over 80,000 subadults m^{-2}, and (Elmgren et al., 1986) found 1400 1-year-old individuals m^{-2}. Initial experiments with newly set clams were not done because *M. balthica* were not available, but Segerstrale (1962) substituted newly set *Mytilus edulis* (300–375 μm) for the smaller *M. balthica* post set (250–300 μm). Within 6 weeks the experimental containers, with 35 amphipods and 100 mussels, had no living *M. edulis* (0.07 mussels amphipod^{-1} day^{-1}). Unfortunately, mortalities in the control experiments were high, and only 31 of the initial 100 mussels survived. If the experimental containers are corrected for the control mortalities, the rate of predation drops to 0.02 mussels amphipod^{-1} day $^{-1}$. Elmgren et al. (1986) placed 0, 10, 20, and 40 *M. affinis* in aquaria with 405 *M. balthica* to evaluate the interaction between these two species. Increased amphipod populations led to increased mortality of the bivalves (number alive, 370, 329, 296, and 245, for 0, 10, 20 and 40 amphipods, respectively). Additional experiments found that the cause of the mortality was physical damage to the bivalve shell (predation), and not burial by amphipod activities. Increasing the depth of the sediment from 10 to 80 mm increased bivalve spat survival. Rates of predation for the 19-day initial experiment (adjusted for control losses) ranged from 0.23 bivalves amphipod^{-1} day^{-1} when 10 amphipods were present to 0.17 bivalves amphipod^{-1} day^{-1} when 40 amphipods were present. These studies indicated that some species of amphipods were capable of causing mortality of bivalve set, but no reports are available to indicate whether amphipod species abundant in hard clam habitats are significant source of predation on newly set individuals.

Observations from individuals engaged in hard clam aquaculture in New Jersey indicated that the tube building activities of amphipods, genus *Ampelisca*, caused suffocation of clams when the seed were confined beneath predator protection netting (Crema, Mathis, Maxwell and Zodl, personal communication). These same individuals indicated that clams in natural sediments migrate into the tube masses of the amphipods and thus become easier to harvest. Whether this latter activity affects the vulnerability of the clams to predators, other than humans, in the natural situation has not been investigated.

Isopoda

Saduria entomon

Little information exists on the feeding habits of isopods, but most have been considered to be scavengers or selective deposit feeders. Sandberg and Bonsdorff (1990) and later Ejdung and Bonsdorff (1992) demonstrated that the isopod *Saduria entomon* was a significant predator on small bivalves and selected specific sizes of newly recruited *Macoma balthica*. The latter authors placed single individuals of *S. entomon* in aquaria (81 or 144 cm^2) containing 2–3 cm of sediment and supplied with 5–6 ppt seawater at 8–13°C. Groups of 25–200 *M. balthica* were added to either a control or experimental container and allowed to establish themselves for about 1 h. The isopods were about 30 mm total length. Five experiments with 5–8 replicates per experiment were conducted with clam seed of a variety of sizes (0.3, 0.8, 1.2, 1.9 and 3.2 mm for experiments 1–5, respectively). Survival ranged from

102 to 94% in controls and 99 to 26% in the presence of isopods. Survival was highest in the experimental units with the smallest isopods (99%) and remained near 50% for 0.8, 1.2 and 1.9 mm clams, but decreased to 26% when the seed were 3.2 mm. Experiments ranged from 6 to 17 h, and daily consumption rates for the 5 experiments (test–control) averaged 0, 267, 392, 96 and 88 clams isopod^{-1} day^{-1}, respectively, for the five clam sizes. Field experiments with boxes of sediment, protected with screen mesh, placed in the field to allow for benthic recolonization showed that boxes containing isopods had similar numbers of species and populations densities except for *M. balthica* which exhibited a statistically significant 50% reduction in numbers (Ejdung and Bonsdorff, 1992).

Decapoda
Natantia

Palaemonetes vulgaris

MacKenzie and Stehlik (1988) conducted laboratory experiments in 1 L dishes containing 50 1-mm hard clams, but no sediments. Duplicate tests 24 h long with one adult *Palaemonetes vulgaris* resulted in all 50 clams being consumed (50 seed shrimp^{-1} day^{-1}). It seems, as indicated below, that other species of this genus could be equally voracious on small hard clams.

Palaemonetes pugio

Posey and Hines (1991) examined the effects of trophic interactions between grass shrimp, *Palaemonetes pugio*, killifish, *Fundulus heteroclitus*, and a burrowing anemone, *Nematostella vectensis*, and their combined effects on the benthic population of a shallow water mesohaline system in Chesapeake Bay. In general, these experiments indicated that both the shrimp and the anemone were predators on small clams. Laboratory tests were conducted in 18 cm diameter by 12 cm deep dishes with either fine sand, silt, or no sediment, and field tests were conducted in 0.25 m^2 cages covered with 3-mm mesh. Field studies utilized a number of shrimp and fish per cage, but laboratory tests were conducted for 48 h with a single shrimp (35–50 mm). In the laboratory, shrimp consumed *Macoma mitchelli* up to about 1.35 mm shell length. Up to 80% of the 0.33 and 0.63 mm *M. mitchelli* were consumed within 48 h, but there was no sediment effect. Additional studies found that grass shrimp also consumed significant numbers of small *Mya arenaria* (all <0.5 mm, 32% of 1–2 mm, and 5% of >2 mm clams). Nearly 100% of the 0.34 mm length *Mulinia lateralis* were consumed in similar tests. In field studies, comparisons of infauna in cages indicated that, when shrimp were added, clam numbers were reduced only during the recruitment period. Numbers of *M. mitchelli* were reduced by over 60% in one test on sand substrate and over 80% in a deeper water silt bottom. The presence of killifish also affected predation rates. When fish and shrimp were maintained together, clam survival was intermediate to areas where there were no predators. When only shrimp were present, clam losses were significantly higher.

Uguccioni and Posey (1992) conducted laboratory predation experiments in which *P. pugio* was offered hard clams (0.4, 0.6, 0.8 and 1.0 mm). Three substrates were tested in 12 cm diameter by 12 cm deep buckets, no sediment, sand and silt. All sediments were 1 cm deep, and all treatments allowed a single shrimp to prey on 50 clams for 48 h. For both the smallest seed sizes, predation was highest in the no sediment container, but there were no differences with sediment type until clams reached 1 mm and at this point survival in sand was slightly greater than in silt or no sediment. This suggests that the clams may be reaching a size that depth (either directly or in the mechanical difficulty of penetrating the sediment to find the clams) is beginning to offer a spatial refuge. In general, predation was greatest on smaller sizes (0.4 and 0.6 mm) and no different from controls (0.8–1.0 mm). Rates, estimated from the histograms, indicate that the 0.4 mm seed in no sediment controls were consumed at about 23 clams shrimp^{-1} day^{-1} and 0.6 mm at about 20 clams shrimp^{-1} day^{-1}, but above this size the rate was very low and no different from losses in the controls. When sediment was added, recovery of the small seed in control containers was reduced and although predation was high it is difficult to determine the rates.

Alpheus spp.

Beal (1983a,b) reported that snapping shrimp *Alpheus normanni* and *Alpheus heterochaelis* could crush and consume hard clam seed up to 15 mm in length. These shrimp were abundant in areas of seagrass and in substrates covered by shell. *Alpheus heterochaelis* was more common in mud areas with clumps of oysters while *A. normanni* was more common in seagrass areas. Beal (1983a) reported densities of 6.1 *A. normanni* and 1.1 *A. heterochaelis* m^{-2} in an eelgrass bed in North Carolina. Laboratory studies (24–27°C; 32–34‰) in which 40 clams and one shrimp were placed in 0.56 m^2 bowls with a mesh top for 7 days indicated an average daily consumption of hard clam seed smaller than 15 mm of 0.72 clams shrimp^{-1} day^{-1}. Shell damage due to chipping or crushing by these alpheid shrimp was indistinguishable from that caused by crabs. All experiments were conducted in finger bowls to which shell materials (all materials <3 mm were removed by sieving) had been added. Marked clams were separated into four size classes (6.0–8.0, 8.1–10.0, 10.1–15.0 and 15.1–20 mm) and placed 10 mm deep in the shell material. The shrimp held the clams with the minor chela and used the major chela to fracture the clams shell. Once the shell had been fractured, the minor chela was used to remove the flesh. Clams in the smallest size category were preferred (49% of all mortalities were in this class), but two clams in the largest size group were eaten. Evidence for predation in the field under natural conditions was not reported, but Beal (1983a) found crushed clams and snapping shrimp in cages that had been placed in the field to protect juvenile clams.

Whetstone (1978) examined the intestinal contents of two *Alpheus* species collected from trays containing hard clam seed and did not find any clam shells. This result was clearly different from those reported by Beal (1983a) and may be ascribed to differences in the substrate incorporated into the experimental design (Beal used shell while Whetstone provided sediments for the clams) or to differences in collection and preservation techniques.

Crangon septemspinosa

A large number of studies have reported that crangonid shrimp are carnivores. There has been a substantial amount of work on *Crangon crangon* in Europe, but there is much less information on the common sand shrimp, *Crangon septemspinosa*, of the US east coast. Haefner (1972) and Modlin (1980) examined the biology of *C. septemspinosa* in Maine and Connecticut, respectively. Adults of this species apparently remain on the continental shelf for most of the warmer months and utilize the estuary as a nursery. Large adults disappeared from the shore and the Mystic River estuary, CT when water temperature exceeded 20°C, but reappeared in the fall when temperature dropped below 20°C. Peak abundance of juveniles in deep water occurred in June and July (120–156 m^{-2}). Along the shore, peak numbers of juveniles were 20–38 m^{-2} in July and August (Modlin, 1980). Feeding studies on both species have reported significant quantities of sand, algae, detritus and mud in the digestive tracts (Lloyd and Young, 1947; Allen, 1960; Price, 1962; Kosaka, 1970; Wilcox and Jeffries, 1974). Much of the material consumed by crangonids is quickly changed into an unidentifiable form by trituration, and thus animal remains are difficult to separate from other materials. These reports suggest that these shrimps are capable of consuming a wide variety of materials, but are opportunistic carnivores. Of the 16 foods tested in the laboratory, *Mercenaria* flesh promoted the best growth (Wilcox and Jeffries, 1974).

Janssen and Kuipers (1980) examined the density of *Crangon crangon* on tidal flats of the Wadden Sea, and found that when shrimp reached 30–40 mm they shifted from inhabiting the intertidal zone toward the subtidal. Shrimp <30 mm remained in the intertidal area even at low tide, and, typically, larger numbers were found farther from the channels at both low and high water. At 100 m from the channel, there were about 10 shrimp m^{-2} while at 600 m from the channel the average was about 32 shrimp m^{-2} (Janssen and Kuipers, 1980).

Evans (1983) examined the effect of *Crangon crangon* and other epifaunal predators on the production of a soft-bottom community in Sweden. He reported that three species of bivalves were present in the macrofauna of this area (*Mya arenaria*, *Cerastoderma edule* and *Tellina tenuis*), and that molluscan remains were often found in the shrimps stomach through the three summer months. In many instances, molluscs were >30% of the stomach contents. In spite of these results, his calculations indicated that the shrimp plus two fish species were only able to crop 12–17% of the total macro- and meiofaunal production. It should be emphasized that these studies examined intact fauna and did not account for potential increased production if predators had been isolated from sections of the flat. In this respect, it is difficult to compare these results with studies such as those of Virnstein (1977) who used cages to exclude predators, and found increased bivalve numbers.

Pihl and Rosenberg (1984) examined the food and feeding of *Crangon crangon* in Sweden. They reported that small shrimp consumed meiofauna, but larger shrimp consumed 0-year class *Spisula subtruncata*, *Tellina* sp., *Mya arenaria* and *Cerastoderma edule* as well as *Nereis* spp., *Corophium volutator*, mysids and newly set *Carcinus maenas*. More bivalves were consumed on clay–silt bottom areas than in areas with other substrate types. Over a 3-year period, *C. crangon* consumed (in terms of ash free dry weight biomass) between 2–35%, and 4–68% of the 0-year-class *M. arenaria* and *C. edule* production, respectively, at the Gullmarsvik station (Pihl and Rosenberg, 1984). *C. crangon* with carapace lengths of 2 mm were able to consume *M. arenaria* smaller than 1 mm while 8–11 mm carapace length shrimp were able to eat *Mya* slightly >2 mm (Pihl and Rosenberg, 1984). Thus the high biomass consumption by this shrimp was important to the population dynamics of the prey because the shrimp consumed soft-shell clams and cockles <3 and <2 mm, respectively (Moller and Rosenberg, 1983). The numbers of *C. crangon* in these areas ranges from 50 to 95 m^{-2} (Pihl and Rosenberg, 1982). At these densities, the predators would be capable of thoroughly searching the bottom for newly set bivalves on a daily basis.

These data were in accordance with the studies of Reise (1977a, 1985) who experimented with *C. crangon* in aquaria to which fine and coarse sand substrates had been added. Six shrimp (3 cm total length) were introduced to the sediments stocked with 300 >2 mm cockles, *C. edule*, and allowed to feed for 39 h. After this period only 22 cockles survived in fine sediments while in coarse sediments 88 survived (28.5 and 22 cockles shrimp^{-1} day^{-1}, respectively).

In the field portion of this study, Reise (1977b) placed cages made of 20-, 5-, 2-, 1- and 0.5-mm mesh on the tidal flats and found that the 20-mm mesh provided no protection. Control and 20 mm macrofaunal populations were about 800 cm^{-2} while the populations under the 5-, 2-, and 1-mm mesh were 2700, 3000, and 3100 individuals cm^{-2}, respectively. The 0.5-mm mesh allowed a population of 6000 infauna cm^{-2} to develop. Bivalves that matured in the cages included *Spisula subtruncata*, *Mactra corallina*, *Venerupis pullastra* and *Abra alba*. These do not normally occur in the intertidal zone and their presence was attributed to the lack of predation by *Crangon crangon* and two other species of predator, the crab *Carcinus maenas* and the gobiid fish *Pomatoschistus microps*.

Newell (1983) reported that *Crangon* spp. consumed >100 post set *Mya arenaria* in a 3-h. period. This was a relatively short-term experiment, but suggests that consumption rates of >1000 newly set clams day^{-1} are possible.

Raffaelli et al. (1989) placed *C. crangon* in cages on mud flats in Scotland, and although both *Mytilus edulis* and *Macoma balthica* were consumed, the researchers did not find significant reductions in densities of the infaunal community between control and experimental plots (0.25 m^{-2}) to which 5 or 15 shrimp had been added. Mattila et al. (1990) conducted a similar caging study in Sweden with 0.093 m^2 cages to which 0, 1, 2, 4 or 6 *C. crangon* were added. These densities represent 0, 10.3, 20.6, 41.1 and 61.7 individuals m^{-2}. The total abundance of infauna, number of taxa, Spionidae, Oligochaeta, *Nereis diversicolor*, and *Macoma balthica* were found to be significantly reduced in the cages with the shrimp, but *Mya arenaria* were not. There was a statistically non-significant decrease in the length of *M. balthica* with increasing shrimp density. The authors attributed the lack of significance to the high variance in the data, but in every case the size of the clams decreased with increasing shrimp density (mean clam lengths 0.73, 0.71, 0.61, 0.50, and 0.48 mm for the 0, 1, 2, 4, and 6 *C. crangon* treatments, respectively).

There does not appear to be any information that connects consumption of hard clams with *Crangon septemspinosa* in the field. MacKenzie and Stehlik (1988) conducted laboratory experiments in 1-L dishes containing 50 1-mm-long *Mercenaria mercenaria*. No sediments were placed in the dishes. Duplicate tests 24 h. long with one adult *Crangon septemspinosa* resulted in all 50 clams being consumed. Thus minimal rates of consumption for unprotected seed were 50 clams shrimp^{-1} day^{-1}. Gibbons (in Gibbons and Blogoslawski, 1989) reported that *C. septemspinosa* can crush and consume post set (0.25–0.5 mm) hard clams.

Summary Natantia

In general, shrimp appear to be capable of consuming significant numbers of post set and small bivalve seed. The alpheids can consume larger seed than the other natant species. While large populations of shrimp occur in all hard clam habitats, there is no data to indicate what effect these taxa may have on population dynamics of hard clam. The high local populations of some of these species suggest their predatory effects could be locally important. In addition, some species appear to make extensive migrations so that large numbers of individuals are present only for a portion of the year. The importance of a species such as the sand shrimp to populations of bivalves that do not reach a refuge size by fall when the adult shrimp reappear in the estuary has not been evaluated. Finally, there do not appear to be any studies that indicate the importance of penaeid shrimp to bivalve recruitment, but Leber (1985) examined the effects of *Penaeus duorarum* on benthic communities in vegetated and unvegetated habitats. He found significantly lower numbers of *Tellina* sp. in those caged areas containing the shrimp than in control plots. The large numbers of these shrimp (up to 97 m^2 (Leber, 1985)) in the southeastern estuary systems suggest they could be important predators on newly settling bivalves including the hard clam.

Reptantia

Homarus americanus

Elner and Jamieson (1979) observed that adult lobsters held individually in 0.18-m^2 aquaria would crush the shell of sea scallops *Placopecten magellanicus* up to 10 cm shell height with their mouthparts, and shells of scallops from >10 to 70 mm with their chelae. Juvenile lobsters used their chela to crush scallop prey. Lobsters of 13–14 cm carapace length preferred scallop prey in the 30–40 mm size class while smaller lobsters (7.0–8.0 cm carapace length) preferred scallops of the smallest size tested (20–30 mm). Daily feeding rates of lobsters fed 40–50-mm-height scallops ranged from 3 to 10 scallops $lobster^{-1}$ day^{-1}. These authors also found that small size lobsters consumed an average of 5–8 scallops $lobster^{-1}$ day^{-1} while large lobsters consumed between 7 and 13 scallops $lobster^{-1}$ day^{-1}. The largest scallops (76 mm shell height) consumed were opened by a 14.7 cm carapace male lobster.

Malinowski (1985) reported that lobsters consumed hard clams in the laboratory, and that field observations and laboratory experiments confirm that they are capable of crushing clams 15–21 mm. Mortality rates of larger clams at field sites in New York were observed to be greatly reduced when lobsters were in the soft shell state (June–July and September). No data were provided on rates of consumption.

Lobsters and hard clams ranges overlap slightly (chiefly north of Long Island), but the data suggest that in certain locations, lobsters could be a serious predator of large hard clam seed.

Pagurus spp.

Thorson (1966) conducted laboratory experiments at 17.5°C with *Spisula subtruncata* and hermit crabs, *Pagurus bernhardus*, placed in aquaria containing sand from the nearby flats. He reported that the crabs (carapace length = 14.7 mm) were able to locate and consume 51 *Spisula subtruncata* spat (0.5–3.15 mm) in 1 day. His data support the observations of Hunt (1925) who indicated that the same hermit crab species could be an important predator on a number of bivalves.

Morgan et al. (1980) listed both *Pagurus longicarpus* and *Pagurus pollicaris* as potential predators on newly released bay scallop seed in Connecticut, and WAPORA (1982) noted the same two species were locally abundant in high salinity waters of Great South Bay, NY. No data were provided on actual abundance of the predators or their importance to hard clam abundance. Malinowski (1985) noted that *Pagurus* spp. were abundant at both the Poquonock River and Fishers Island, New York sites where predators caused significant loss of seed clams, but the effects of the hermit crabs on hard clams were not specifically documented. MacKenzie and Stehlik (1988) conducted laboratory experiments in 1-L dishes containing 50 1-mm-long *Mercenaria mercenaria* and one *Pagurus longicarpus*. No sediments were placed in the dishes. Duplicate 24-h-long tests resulted in all 50 juvenile clams in each dish being consumed, indicating a consumption rate of >50 clams $crab^{-1}$ day^{-1}.

TABLE 11.3

Number of two sizes of hard clams ingested by individual *Pagurus longicarpus* in 1 day at four temperatures

Temperature (°C)	Clam size (mm)	
	1	3
10	88.4	17.9
15	150	18.7
20	225.8	19.7
25	239.6	23.3

After Gibbons, 1984.

Gibbons (1984) placed 9–11 mm carapace length individual crabs in 0.03 m^2 culture bowls with prey species and reported that gravel reduced predation by *Pagurus longicarpus* on hard clam seed when these rates were compared to sand and bare substrates. The greatest predation rate observed in these laboratory studies was 143.5 clams $crab^{-1}$ day^{-1} at 18.5°C. Other studies to determine energetics of the crab in which individual crabs (9–11 mm carapace width) were fed clam seed (1–3 mm) ad libitum, indicated that 1 mm seed could be consumed at rates of nearly 240 clams $crab^{-1}$ day^{-1} at 25°C (Gibbons, 1984) (Table 11.3, Fig. 11.8). Predation by this species ceased at 4.5°C in the fall and resumed at 6°C in the spring.

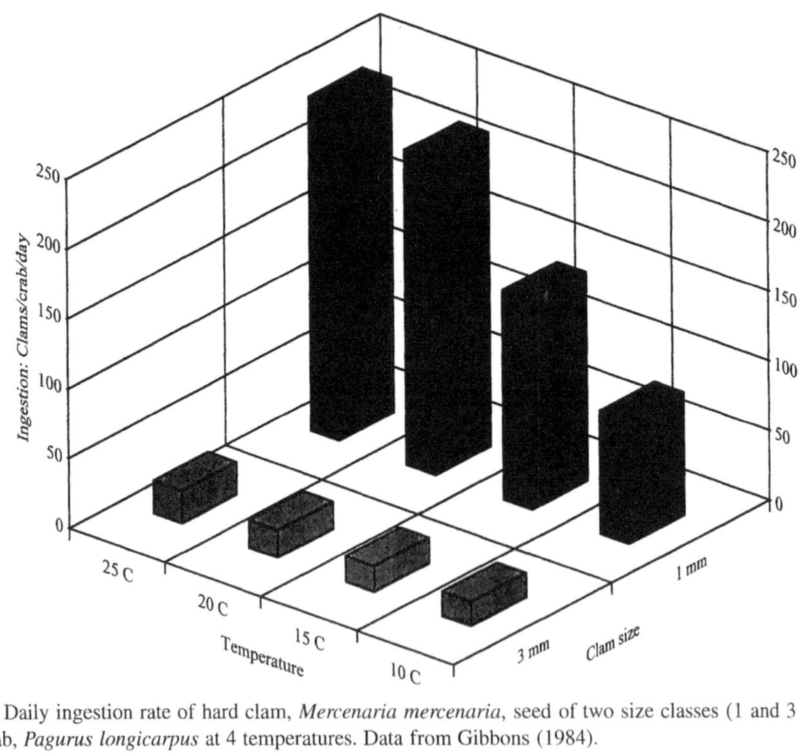

Fig. 11.8. Daily ingestion rate of hard clam, *Mercenaria mercenaria*, seed of two size classes (1 and 3 mm) by the hermit crab, *Pagurus longicarpus* at 4 temperatures. Data from Gibbons (1984).

Libinia spp.

There do not appear to be any reports directly linking spider crabs to predation on hard clam seed. Stickney and Stringer (1957) mention *Libinia emarginata* in a list of hard clam predators. Pohle et al. (1991) found that *Libinia dubia* (42–53 mm carapace width) was an important predator of bay scallop seed. In laboratory studies conducted in 0.25 m^2 tanks, the rate of predation by the spider crabs on the bay scallops attached near the base of the eelgrass blades was as great as that of scallops consumed by the mud crab, *Dyspanopeus sayi*, but spider crabs did not consume scallops near the tips of the blades. Turner (1950) reported that spider crabs made broad conical excavations in sand flats, and that crabs found in these pits had soft-shell clam shells and necks in their stomachs. Neither Greene (1978) nor WAPORA (1982) mention spider crabs as predators of the hard clam in Great South Bay, NY. Ropes (1988) listed pelecypods as an important food item in field collected *Libinia emarginata*, but the only species found in the gut was the blue mussel. Whether this or other species of spider crabs consume hard clams cannot be determined from the literature, but the data presented in Pohle et al. (1991) on scallops and Turner (1950) for soft-shell clams, and the anecdotal report by Stickney and Stringer (1957) all suggest that these crabs could prey on small clam seed, particularly in eelgrass beds.

Cancer spp.

Although *Cancer* species have not been reported to be abundant in hard clam habitats, most clam surveys in areas south of Long Island have focused efforts in summer months and the rock crab, *Cancer irroratus*, is known to migrate into these inshore waters in the winter. Morgan et al. (1980) listed both *Cancer irroratus* and the jonah crab, *Cancer borealis*, as predators causing losses of bay scallop seed that had been released into Connecticut waters. Elner (1981) reported that the diet of the rock crab was similar to that of *Carcinus maenas*.

Elner and Jamieson (1979) compared the maximum size consumed and consumption rates of two sizes of *Cancer irroratus* on sea scallops. All experiments were conducted in 0.18-m^2 aquaria with individual predators. Scallops were divided into groups of 10 mm height intervals from 20–30 to 60–70 mm. The largest scallop (72 mm) was consumed by a 130-mm crab. Larger crabs (120–130 mm) preferentially consumed scallops in the 40–50-mm size class. Smaller crabs (90–100 mm) consumed scallops of the smallest size at the greatest rate, between 2 and 7 scallops $crab^{-1}$ day^{-1}, while larger crabs consumed between 4 and 8 scallops $crab^{-1}$ day^{-1}. Barbeau and Scheibling (1994a,b) reported that when *C. irroratus* (45–120 mm) were presented with multiple size *P. magellanicus* seed, the crabs selected larger scallops. In experiments with large (19–23 mm shell height) scallops, the crabs consumed more at 15°C than at either 4 or 8°C. Predation rates at both lower temperatures were the same. The increased consumption was due mostly to the shorter handling times at the higher temperature. Field studies of tethered scallops (Barbeau et al., 1994) indicated that crab predation rate increased with scallop density, but the density of crabs did not. Water temperature was an important variable influencing predation rates, but varied with site. At one site, crab predation on scallops increased with temperature, while at another site predation rate was independent of temperature (Barbeau et al., 1994).

Lake et al. (1987) reported that *Cancer pagurus* consumed more scallops, *Pecten maximus*, than the other three species of crab tested. Individual crabs placed in 0.08 m^2 circular pipes at 10.6–13.8°C and fed scallops (40 or 50 mm) for 5 days revealed no differences in a consumption rate between male and female crabs (104 mm). These crabs consumed 3.1 and 1.1 scallops crab^{-1} day^{-1} for 40 and 50 mm scallops, respectively. Additional studies with a variety of crab and scallop sizes indicated that smaller scallops were preferred over larger scallops. Predation was greatly reduced when 70 mm scallops were tested. Furthermore, larger crabs ate more scallops of all sizes than smaller crabs. When ten, 106 mm crabs were placed in a larger tank (10.5 m^2) and supplied with scallops (45 mm) for 6 days, the predation rate was considerably less (0.6 scallops crab^{-1} day^{-1}) than indicated by the laboratory study.

Dumbauld et al. (1993) examined the effect of piles of intertidal oyster shell on the recruitment of the dungeness crab, *Cancer magister*, and found shell enhanced settlement, but numbers of recruits subsequently declined to 10 crabs m^{-2}. Crab populations on shell bottom and in eelgrass beds were higher than on adjacent sand and mud flats. High densities (155–298 crabs m^{-2}) of *Cancer magister* juveniles were reported by Fernandez et al. (1993) in oyster shell habitats, but they also noted that the first cohort of crabs reduced the density of subsequent cohorts of the same species.

Gotshall (1977) working in Humbolt Bay, California and Feder and Paul (1980) working in Cook Inlet, Alaska found that bivalves, chiefly *Siliqua patula* were of major importance in the diet of dungeness crabs. Adult dungeness crabs in Alaska also fed heavily on *Mactromeris polynyma*. A dietary shift of dungeness crabs from one prey source to another with age was reported by Bernard (1979) who found that bivalves (*Siliqua patula* and *Tellina carpenteri*) were important for small crabs, but later in life *Crangon* sp. became the dominant prey. Stevens et al. (1982) examined the gut contents of freshly caught *Cancer magister* from Grays Harbor, Washington. Small bivalves including *Cryptomya californica*, *Macoma* sp. and *Tellina* sp. were important in the crab's diet for the first year, but decreased in importance during year 2 when other crustaceans, chiefly *Crangon* spp. (*Crangon franciscorum*, *Crangon nigricauda*, and *Crangon stylirostris*) became the dominant prey items. Once the crabs reached year 3, their chief forage became fish of a variety of species. Many of these studies indicated significant cannibalism of the young crabs by older individuals.

Pearson et al. (1981) studied the effect of oiled sediment on the rate of predation by *Cancer magister* on *Protothaca staminea*. Twelve clams were placed in containers with sediment and allowed to burrow in shallow (5 cm) and deep (10 cm) sand with and without oil. One crab (155 mm) was placed in each container and allowed to consume clams for 19 days at 13°C. In general, more clams were consumed in shallow sediments and more were consumed when oil was present than when it was absent. Deep sand sediment without oil had the lowest predation rate (1.7 clams crab^{-1} day^{-1}). Clams in shallow sand and in the oiled sand of both depths were consumed at rates between 3 and 4 clams crab^{-1} day^{-1}. Within the size range of clams tested in the laboratory (26–35, 36–45, 46–55 and 56–65 mm), a higher percentage of clams in the smallest class was consumed. These findings did not translate to the field control sites where all sizes of clams were consumed at an equal rate. Oiled sediments in the field had higher percentages of small clams eaten. Consumption in the field ranged from 0.88 clams crab^{-1} day^{-1} for the control site in a 29-day experiment to 4.34 clams crab^{-1} day^{-1} in a 13-day experiment in oiled sediment. Clam density in the experiments ranged from 40 to 48 clams m^{-2}, and are typical of those found in Sequim Bay, Washington.

Asson-Batres (1986) reported that juvenile dungeness crabs were considered to be a major cause of loss of the recruits of the bivalve *Transennella tantilla*. The feeding behavior of the juveniles (10–30 mm) was evaluated by placing 5 crabs in tanks that contained 10 cm of sand

and 150 bivalves. After 26 days, no bivalves were alive in the tanks with crabs, and only 118 of the 150 clams were alive in the controls. If the control mortality is subtracted from the initial stocking, the crab's rate of consumption must have been at least 0.91 clams crab^{-1} day^{-1}. Sufficient *T. tantilla* were not available for additional experiments and the crabs were fed on *Clinocardium nuttalli* to establish the maximum daily intake of clam flesh. The crabs could not break the *C. nuttalli* shells so they were fed broken cockles. Rates on broken clams consumed in 82 h ranged from 3.8 to 16.9 cockles (1.1–4.9 cockles crab^{-1} day^{-1}) with larger crabs (24 mm) consuming the greatest number of cockles and the greatest amount of tissue (682 mg).

Prey selection by the dungeness crab on various sizes of the bivalve *Protothaca staminea* was evaluated by Juanes and Hartwick (1990) using 190-L aquaria without substrate. Adult male crabs (160–185 mm) were placed in the containers with clams of six size classes (15–20, 20–25, 25–30, 30–35, 35–40 and 40–45 mm). The authors found that the crabs selected more clams from the smallest size offered, and that crabs with broken or worn claws were significantly less effective in opening clams. They noted that some crabs suffered claw damage when trying to open larger clams. This suggests that crabs reduce the probability of damage by selecting prey that requires less strength to open. The general applicability of this extrapolation is suspect because Smith and Palmer (1994) evaluated the response of the claw size and strength of crab *Cancer productus* when it was fed shelled and unshelled prey. Strength of the crab's claws increased after molting when they were provided hard shelled prey, and the increase was due to strengthening on the hard diet not a weakening due to the soft diet. In addition to the strengthening, claw size increased.

Predation by *Cancer productus* on a variety of clam species has been extensively documented in a series of studies (Anderson et al., 1982; Boulding, 1984; Boulding and Hay, 1984; Boulding and LaBarbera, 1986; Peterson, 1982b). Peterson (1982b) generally found low levels of predator loss for *Protothaca staminea*, but noted that *C. productus* can crush and consume the clams. Boulding (1984) documented that shell meristics, such as shape, thickness and gape, all greatly affected the ability of *C. productus* to consume a variety of clam species. Clam species studied ranged from relatively globose, tightly closing, shallow burrowing species, such as *Clinocardium nuttalli*, *Tapes japonica*, *Protothaca staminea* (both a thick and thin shelled morphotype), *Macoma inquinata*, *Macoma nasuta*, and the deeper dwelling slightly to widely gaping clams *Saxidomus giganteus*, *Mya arenaria* and *Tresus capax*. None of the gaping clams had a size refuge from crab predation. In general, larger size crabs could prey on larger sized clams, but all sized crabs preferentially fed on clams smaller than the maximum size they could consume.

Studies on the thick and thin shelled *P. staminea* revealed that the thin morphotype shell failed at significantly lower compressive load than the thicker shelled morphotype (Boulding and LaBarbera, 1986). This failure was reflected in the rate at which crabs were able to break the shell. Shell thickness was only a partial refuge from predation, because continued pulses of loading eventually fractured the shell or the predator chipped away at the shell edges (Boulding and LaBarbera, 1986). Since the hard clam is similar in general shell structure to *Protothaca* and *Transennella* it seems reasonable to assume that crab predation by some *Cancer* sp. on *Mercenaria mercenaria* could mirror that described by Boulding (1984). These data need to be extrapolated with caution, because Peterson (1983) was able to show significantly a different predation rate by *Cancer anthonyi* on two species of bivalves, *Chione undatella* and *Protothaca staminea*. In field experiments, in 1-m^2 cage enclosures in seagrass beds, the deeper borrowing *P. staminea* was consumed to a greater extent than the shallower burrowing thicker shelled, *C. undatella*, but epibiotic molluscs were preferred over either infaunal species (Peterson, 1983).

Boulding and Hay (1984) conducted field studies in which prey were placed in 0.25 m^2 cages with either 5 or 20 prey specimens of one of three sizes (30–35, 40–45 or 50–55 mm). They found that *C. productus* caused greater mortality in areas of high clam density than in areas of low clam density. There was no difference in the predation rate between clams in the two smaller size classes. Laboratory studies with crabs ranging in size from 111 to 170 mm and clams from 31 to 60 mm found no differences in predation due to size of either species. Predation rate averaged 3 clams $crab^{-1}$ day^{-1} (Boulding and Hay, 1984). Chew (1989) reported that *Cancer productus* was considered to be a serious predator on intertidal Manila clam populations in Washington.

There is evidence that rock crabs can consume hard clams, but the effects in the field have apparently not been studied. Spear (1955), in a cryptic note reported that: "*Cancer* crabs readily eat *Venus*." MacKenzie (1977a, 1981) reported that in Connecticut, *C. irroratus* juveniles range from 3.6 to 57 crabs m^{-2}, and adults from 0.7 to 1.1 crabs m^{-2}. Feeding studies in which hard clams were placed in trays with no substrate revealed that juvenile rock crabs could consume up to 100 clams day^{-1}. Adult crabs were able to consume clams up to 15 mm long (MacKenzie, 1977a). Greene (1978) listed *Cancer irroratus* as a predator young clams, but did not provide any evidence for their importance in regulating the abundance of clams in Great South Bay, NY. WAPORA (1982) reported that this species ranged in abundance up to 0.75 crabs m^{-2}, and that almost all specimens were collected at stations where salinity was at least 30 ppt. Highest crab abundance was at stations with abundant eelgrass. Within this habitat most crabs appeared to be feeding on fauna from the eelgrass blades (WAPORA, 1982). Malinowski (1985) reported that *Cancer irroratus* was common where he found the highest predation on larger hard clam seed (15–21 mm), but no direct evidence was presented to indicate these crabs were the source of the seed losses.

Summary Cancer

Data from the Pacific northwest indicate that this genus can be a significant predator on bivalves, and it seems possible that from Chesapeake Bay northward, *Cancer irroratus* could be a significant predator on hard clams in cooler high salinity water and during the winter, but data to support such a prediction are lacking. There do not appear to be any data that would indicate if the diet of this species shifts with age as was reported for *Cancer magister* in the Pacific northwest.

Carcinus maenas

Carcinus maenas, commonly known as the green crab in the United States or the shore crab in Europe, has been reported to be a serious predator of bivalves (Turner et al., 1948a; Glude, 1955; Hanks, 1961; Ropes, 1968; Dare and Edwards, 1976; Elner, 1981; Scherer and Reise, 1981; Pihl and Rosenberg, 1982; Dare et al., 1983; Sanchez-Salizar et al., 1987b). Ropes (1968) reported that green crabs were omnivorous, but at least a third of their food was composed of bivalve molluscs of the genera *Mytilus*, *Gemma* and *Mya*. He noted that feeding activity was greater at night, and that crabs smaller than 30 mm did not consume significant quantities of molluscs. Like other crabs, there was evidence of cannibalism. Elner (1981) reported that the diet of green crabs was similar to that of rock and jonah crabs (*Cancer irroratus* and *Cancer borealis*, respectively) and lobsters (*Homarus americanus*). Stomach analysis revealed that a number of species of bivalves were the most important food items. The most frequently encountered species were *Mytilus edulis* and *Mya arenaria*, but both *Ensis directus* and *Macoma balthica* were also present. *Mercenaria mercenaria* was not listed in this study, but it was not present in the study area.

Scherer and Reise (1981) working in the intertidal zone of the North Sea near Sylt reported densities of adult *C. maenas* to be 0.01, 0.055, and 0.11 m^{-2} in a *Corophium* bed, a seagrass bed and on the sand flat, respectively. Maximum setting of the 0-year-class crabs reached 2000 m^{-2}, and they were most abundant in seagrass beds (to 500 m^{-2}) and mussel reefs. Fewer 0-year-class animals were found on the sand flat, and their density distribution was irregular. Average density of 0-year-class individuals on these flats was estimated to be about 125 m^{-2}. In addition to estimating the abundance of the crabs, these authors constructed cages for small crabs (83 cm^2) and adult crabs 0.25 m^{-2} to which 100 small (2–4 mm) and 25 adult, male or female, crabs were added, respectively. After six tidal cycles (small) or 7 days (adult) the benthos inside the cages was sampled and enumerated. Small crab numbers were significantly reduced by cannibalism, but their feeding and disturbance activity greatly reduced both meiofauna and macrofauna inside the cages with crabs relative to the controls. Most of the predation by 0-year-class crabs was confined to the surface sediments, but populations of benthos were reduced to a depth of 5 cm. There were significant reductions in macrobenthos in the cages with adult green crabs, and female crabs consumed more polychaetes while male crabs consumed more juvenile molluscs (*Macoma balthica* and *Cerastoderma edule*). Cages were constructed on the sand flat, mud flat and in a seagrass bed, and in general, the crabs preyed on the most abundant species in a particular habitat, but in all cases, a significant portion of the diet was soft-bodied prey. These studies reveal the complexity of the feeding habits of the green crab, and suggest caution when attempting to extrapolate laboratory studies in which the crabs are offered only one or two species of prey to field situations. Even this study, while oriented toward the field, should be viewed with caution because the density of crabs in the enclosures was 100–1000 times higher than the density reported from field samples of crab density.

In Sweden, Pihl and Rosenberg (1982) reported densities the 0- and 1-year classes of *Carcinus maenas* to be 60 m^{-2}. Pihl (1985) found that the diet of these crabs varied from year to year, but in some years *Mya arenaria*, *Mytilus edulis* and *Cerastoderma edule* were the most important components. Green crabs of the 0–1-year class consumed 24 and 20% of the soft-shelled clam and cockle production, respectively, on these Swedish tidal flats.

ap Rheinallt and Hughes (1985) examined predation on *Mytilus edulis*, *Carcinus maenas* and *Littorina rudis* by the velvet crab, *Liocarcinus puber*. The velvet crab will consume a variety of prey, but often is a significant predator on other crab species. These studies attempted to quantify handling time and other processes associated with feeding. Handling time included: picking-up, breaking, and ingestion times with the latter generally being the longest process. The time spent breaking the largest prey often exceeded ingestion time. The authors noted that because ingestion was related to mouth part and stomach size and because ingestion was usually the longest handling time process, total handling time was more closely related to carapace width than chela height. This is similar to the results of Haddon and Wear

(1987) who found that the numbers of cockles eaten was best explained by the size of the crab's foregut, but differs from the studies of Elner (1980) who explained a significant amount of the predation by green crabs on blue mussels based on chela size. These studies emphasize the importance of considering the effects of cannibalistic crustacean predators or crustaceans consuming other crustacean species when trying to evaluate the effect a predator may have on sessile species such as bivalves.

Recently, Grosholz and Ruiz (1996) reviewed the history of invasion of new habitat by the green crab and found that across all habitats, the diet remained nearly the same with molluscs being the most important dietary items in Europe, eastern North America, western North America and South Africa. These authors suggest that the invasion of the green crab into Bodega Harbor, California has caused a significant decline in molluscs. This crab is abundant in the cold waters north of Cape Cod and is found as far south as Delaware Bay. It may cause significant loss of hard clams throughout its range, but would be most important from Long Island northward.

Lake et al. (1987) placed individual green crabs of two sizes at 10.6–13.8°C in 0.02 m^2 circular pipes and fed them *Pecten maximus* (40 or 50 mm shell height) for 5 days. Predation rates of 0.1 and 0 scallops $crab^{-1}$ day^{-1} for 45 mm carapace width crabs on smaller and larger scallops, respectively, were observed. Similar data for the larger 75 mm carapace width crabs was 0.6 and 0.3 scallops $crab^{-1}$ day^{-1}.

Experimental plantings of bay scallops experienced large losses due to predation by a number of crustacean predators, and Morgan et al. (1980) indicated that green crabs were locally important in these losses. Pohle et al. (1991) also found that *C. maenas* was an important predator on juvenile bay scallops. This was particularly true for the scallops that were attached near the base of eelgrass blades. As opposed to the mud crab, *Dyspanopeus sayi*, green crabs were not effective predators on scallops that were attached higher on eelgrass blades.

Dare and Edwards (1976) reported that unprotected mussels relayed into plots in the Menai Straits, Wales suffered 70–85% mortality during the first year due to green crabs. Similar plots that were protected by a cage covered with 20-mm plastic mesh suffered only 17–41% losses. The same unprotected plots suffered an additional 22–57% mortality in year 2. The second year loss was attributed to storms and bird predation. The caged mussel plots suffered similar losses during year two, but this was attributed to overcrowding and smothering by biodeposits.

Davies et al. (1980) reported that in the Tal-y-foel (UK) experimental area, green crab (>20 mm) densities in the intertidal zone at high tide averaged 2–4 m^{-2}. Interestingly, because the experimental mussel plots were placed in the intertidal zone, green crabs that entered (or were placed in) the fenced areas climbed over the fence to retreat with the ebbing tide. The fences were effective in preventing most crabs from entering the plots on the flooding tide, and thus the behavior of the crabs made the plots nearly self-cleaning in spite of the presence of an abundant food resource.

Elner (1980) investigated the differences in foraging rates of green crabs and was able to explain preferences of the crabs for certain mussel sizes by comparing the size, number and energy content of the mussels with the size (height) of the crabs' master chela. Experiments were conducted in aquaria about 0.1 m^2 in size and individual crabs were placed into each tank. Crabs with smaller chela fed at rates that would be considered to be sub optimal by an optimal foraging model. There were also differences in consumption rates based on the sex of the crab. Females fed on smaller mussels than males with similar carapace widths. Most

TABLE 11.4

Daily consumption of six size classes of blue mussels by six sizes of green crabs in laboratory experiments (22°C)

Crab size (mm)	Mussel size (mm)					
	15	20	25	30	35	40
25	7.0	2.0	0.7	–	–	–
35	15.5	7.7	–	1.2	–	–
44	–	9.6	3.0	0.2	–	–
55	–	21.3	12.6	1.5	1.0	–
65	–	19.0	12.4	3.5	1.1	–
75	–	32.5	13.3	9.1	2.4	1.1

After Dare et al., 1983.

of this variation could be explained by differences in chela size. Feeding rates as high as 28 mussels crab^{-1} day^{-1} were reported.

Dare et al. (1983) reported the sizes of Pacific oysters and blue mussels eaten by green crabs. Laboratory experiments were conducted with individual crabs in 0.05 m^2 aquaria. Generally, fewer oysters were eaten than mussels, and the maximum size of any prey species that was consumed increased with the size of the crab. The largest crabs (75 mm carapace width) could consume the largest oyster (55–60 mm shell length) offered. In contrast, the smallest crab (25 mm) could only consume small oysters (18–22 mm). Average consumption rates of oysters did not exceed 3 oysters crab^{-1} day^{-1}. The smallest crabs (25 mm) were able to consume 7 mussels (15 mm) crab^{-1} day^{-1} at 22°C while the same size crab could only consume 2 mussels (20 mm) in a day. Larger crabs (75 mm carapace width) consumed an average of 32.5 mussels (20 mm) crab^{-1} day^{-1} at 22°C and 1.1 mussels (40 mm) crab^{-1} day^{-1} at 22°C (Table 11.4). An experiment in an outdoor tank (14 m^2) with 100 crabs was conducted to determine if the predation rate was similar to that obtained under the laboratory conditions; in general, predators consumed mussels smaller than 35 mm. After 32 days, the mortality of 40–45 mm mussels was 25–30% while there were no small mussels (<20 mm) left. A third series of studies, on an intertidal mudflat for 23 days, also yielded lower predation rates than in the laboratory. Mussels >35 mm had only 10% mortality in the field studies (Dare et al., 1983).

Jubb et al. (1983) placed individual green crabs in aquaria of about 0.1 m^2 and found that starved green crabs fed on the first mussels encountered regardless of size, but after the first few were eaten (about 30 min) they began to reject certain prey. Satiation was reached in about 3 h. Crabs generally ate mussels in proportion to the rate at which they were picked up, but medium mussels were preferred.

Cohen et al. (1995) placed 10 female (55–60 mm) green crabs in 25 × 25 cm tanks with a variety of prey species at 15°C. The prey species, *Mytilus* sp., and the clams, *Tapes philippinarum* and *Potamocorbula amurensis* were between 10 and 20 mm shell length. Two series of experiments were conducted, one with no substrate and a second in which sterilized sand was added and the prey allowed to bury or, in the case of the mussels, allowed to form a clump on the surface. All prey were presented in combinations for a 2-h test period. Without sediment, the mussel and *P. amurensis* were consumed at the same rate, but both were selected more frequently than *T. philippinarum*. In the presence of sediment, the crabs selected *P. amurensis* over the other species. The rates of consumption for the clam species were as high or higher in sediment than without it, but more mussels were consumed when sediment was not present. The highest rate of consumption for mussels, *P. amurensis*, and *T. philippinarum* was 7.6, 12.2 and 2.5 bivalves crab^{-1} day^{-1}, respectively. If the rates when both infaunal species were tested simultaneously in sediment are combined, the crabs consumed 14.2 bivalves crab^{-1} day^{-1}.

Jensen and Jensen (1985) placed juvenile (7.3 and 4.9 mm average size) *C. maenas* in cages (0.01 m^2) with juvenile cockles (mean 4 $\pm$ 3.0–4.9 mm SD) and after 44 h the larger crabs had consumed an average of 2.2 cockles crab^{-1} day^{-1}. Data were not presented for the smaller crabs. Generally, in order to consume a prey, the crabs had to have a carapace width 16% larger than the length of the cockle (Jensen and Jensen, 1985). In laboratory experiments conducted in small (0.005 m^2) aquaria with individual crabs, green crabs selected *C. edule* over *Macoma balthica* or a number of annelid species. Combined predation rates indicated that green crabs consumed an average of 6 juvenile cockles crab^{-1} day^{-1}, and this rate, when combined with crab density, was enough to reduce field populations of the prey from 33,000 to 7400 in 1 month (Jensen and Jensen, 1985).

Raffaelli et al. (1989) placed green crabs in cages on mud flats in Scotland, and although cockles, mussels and *Macoma balthica* were consumed, there were no significant reductions in density of the infaunal community between control and experimental plots (0.25 m^2) to which 3 or 10 crabs had been added.

Sanchez-Salizar et al. (1987b) examined predation on *Cerastoderma edule* by *Carcinus maenas*. Laboratory studies were conducted in sandy substrate in which three sizes of crabs (35, 52 and 68 mm) were placed in 0.15-m^2 tanks with cockles ranging from 2 to 35 mm. These, and other studies at three temperatures (6, 9.5 and 15.5°C), demonstrated that the crabs selected cockles 33% smaller than the largest size they could consume, and the cockles were also smaller than the size predicted by optimal foraging calculations. Green crabs searched for prey by probing the sediment surface with their chelae and dactyls. Once detected, the prey would be grasped by the chelae and opened using one of three methods: simple crushing by the master chela, continued reorientation in the master chelae with repeated attempts to crush the shell, and insertion of the tips of the chela between the edges of the cockle shell. Greater numbers of cockles were eaten as temperature increased. The largest crabs consumed an average of 13 cockles (~7 mm) day^{-1} while the smallest crabs consumed an average of 7 cockles day^{-1} (~4.5 mm). As in the studies of Elner and Hughes (1978), Griffiths and Seiderer (1980), Seed (1980, 1982) and Boulding (1984), crabs of larger sizes did not select either the smallest or largest cockles but preferred intermediate sizes. This may reflect chela size as suggested by Elner (1980), and recently emphasized by Mascaro and Seed (2000a). These latter investigators examined prey selection by green crabs (40–55 and 55–70 mm) offered *Mytilus edulis*, *Ostrea edulis*, *Crassostrea gigas* and *Cerastoderma edule* (all 5–40 mm). Studies were conducted in 30 $\times$ 20 aquaria with running seawater (12–17°C). As with other studies, smaller crabs generally selected smaller prey, 5–15 mm mussels, 5–10 mm cockles, while larger crabs selected 15–25 mm mussels and 10–20 mm cockles. There was no size preference for either species of oyster. The crabs appeared to be selecting prey based on the minimum shell dimension and this was related to the maximal cross-sectional dimension of the chela. Further studies (Mascaro and Seed, 2000b), presented paired combinations of the same prey to green crabs suggested that the crabs were selecting on the basis of prey shape and volume. This was confirmed by the use of models shaped to resemble the bivalves (Mascaro and Seed, 2000b).

Field studies associated with Sanchez-Salizar et al. (1987b) found that up to 96% of the cockle spat low on the shore failed to survive their first summer, while higher on the shore 47% of the spat survive, (Sanchez-Salizar et al., 1987a). Peak densities were approximately 240 cockles m^{-2}, and an average of 84 crabs migrated into the intertidal per meter of shore line per tidal cycle on these flats. Green crabs fed selectively on cockles <15 mm long, and most predation was on cockles <10 mm. Oystercatchers, *Haematopus ostralegus*, selected cockles >20 mm long on these same flats. By fall, most of the intertidal cockle population was from the new year class. During the ensuing winter mortality was independent of tidal location. In the following summer, the mortality of smallest cockles was highest in the lower shore (96%) and least on the upper shore (47%). The mortality was chiefly from predation by green crabs (Sanchez-Salizar et al., 1987a). Crab predation was limited by temperature and crabs did not appear in the intertidal zone below 8°C. As cockles increased in size the mortality pattern reversed and highest losses (88% annually) occurred higher on the shore. This was interpreted to be due to oystercatcher predation. Based on this study the authors estimated that crabs remove 25 times the number and 2 times the cockle biomass of oystercatchers. Given the high density of spat in this population, the crabs remove 236×10^3 individuals (2.4 kg dry weight flesh) per linear meter of shore per year.

Smith and Chin (1951) reported that green crabs were a significant source of losses to soft-shell clam plantings. One crab 64 mm wide was observed feeding on a freshly excavated 50 mm *Mya arenaria.* Glude (1955) reported that most losses of soft-shell clams planted at several localities in New England were due to predation by green crabs. He attempted to provide background data on year-to-year clam and crab abundances based on anecdotal reports by fishermen. In general, these reports suggested that clam recruitment was poor in those years with high abundances of green crabs. Smith et al. (1955) found that green crabs were a major source of mortality of recruiting soft-shell clams in Massachusetts. Survival of seed planted inside plots protected with wire mesh ranged from 12 to 69% after 2 years. Invariably, small seed planted without protection perished within a short time.

Hanks (1961) placed fish treated with the pesticide lindane on lines suspended over intertidal flats in Maine to determine if this would reduce green crab predation on *Mya arenaria.* In years of high green crab population the treatment resulted in the 3–10 mm soft-shell clams from 16 m^{-2} in the control site compared to 194 m^{-2} in the treated areas. Catches of crabs in traps were used to evaluate the crab population density. In years of high abundance, over 700 crabs $trap^{-1}$ $month^{-1}$ were found in the control areas while the treatment reduced crab catches to less than 100 $trap^{-1}$ $month^{-1}$. A cold winter was responsible for reducing crab catches at a site in Kittery, Maine from >700 $trap^{-1}$ $month^{-1}$ in 1958 to <30 $trap^{-1}$ $month^{-1}$ in 1959. At this level of crab abundance, the treatments had no effect on clam recruitment.

Sellmer (1967) working on Union Beach, New Jersey reported that a 4 cm carapace width *Carcinus maenas* consumed 50 adult 4 mm gem clams (*Gemma gemma*) in 30 min. How this rate would compare to a daily rate is difficult to assess, but the shell of 4 mm gem clams would present as formidable a test of chela strength as 4 mm hard clams. Ropes (1988) also examined field collected green crabs and found they had consumed gem clams.

Spencer et al. (1992) conducted a series of experiments to determine the effectiveness of

TABLE 11.5

Average number of hard clam seed of different sizes consumed per day by size of *Carcinus maenas*

Carapace width (mm)	Seed size (mm)							
	4	5	7	9	10	11	13	20
32.5	–	1.41	0.22	0	–	–	–	–
32.9	2.98	1.51	0.28	–	–	–	–	–
37.1	–	3.24	1.76	0.41	–	–	–	–
38.0	–	–	1.97	0.23	–	0.01	–	–
45.5	–	–	3.72	–	–	–	0.2	0
46.2	–	–	3.13	–	0.69	–	0.01	–

The – indicates no data collected. After Walne and Dean, 1972.

various protective netting types for prevention of green crab predation on Manila clam seed. In general all clams that were left unprotected, either in summer or winter, succumbed to predation. In a preliminary laboratory study, 15 mm clam seed were introduced into sand filled boxes (38 × 30 × 7 cm) were placed in a 122 × 50 × 15 cm fiberglass tank filled with 16–17°C seawater. One hundred clam seed were placed in each box and seven green crabs were added to the tank. Within 4 days, all the seed in the unprotected box had been eaten (3.57 clams crab^{-1} day^{-1}); in addition 5 clams in one of the netted boxes had been consumed raising the consumption rate to 3.75 clams crab^{-1} day^{-1}. Field studies yielded similar results, but consumption rates cannot be derived because the numbers of crabs could not be quantified.

Spear (1955) noted that green crabs preferred *Mya arenaria*, *Macoma balthica*, *Gemma gemma* and then *Mercenaria mercenaria* and that smaller size hard clams (6–10 mm) were eaten before larger hard clam seed (20–25 mm). The relationship between the size of the predator and size of the prey was more thoroughly evaluated by Walne and Dean (1972). They placed 5 individuals of three sizes of hard clams in trays (0.02 or 0.05 m^2) containing individual or multiple *Carcinus maenas*. All experiments were conducted at 20–21°C and lasted 7 days, numbers of bivalves consumed daily were recorded, and prey individuals were replaced as they were consumed. No substrate was provided in the initial trials and the smallest clams offered were preferred by each size group of crabs (Table 11.5).

Additional experiments were conducted to examine the effects of substrate (sand or mud) and numbers of crabs (5 or 10 per container) on predation rate (Table 11.6, Fig. 11.9). Control containers without substrate were compared to the initial studies where one crab was introduced into each container. Daily rates of predation for equivalent seed and crab carapace width combinations (compare rates at carapace width 32.6 mm and seed size 6.5 mm at five crabs per container (Table 11.6) with those 32.5 mm wide and either 5 or 7 mm seed at one crab per container (Table 11.5) were substantially higher when more than one crab was present. The data indicate that seed could be consumed at rates of up to 7.8 clams crab^{-1} day^{-1}, but most rates were near the 4 clams crab^{-1} day^{-1} indicated by Hibbert (1975).

Results presented by Walne and Dean (1972) concerning substrate type (sand vs mud) were opposite what most field data indicate. In general, field and experimental data indicate that most crab species are more effective predators in mud when compared to sand (Pratt, 1953; Gibbons, 1984; Lipcius and Hines, 1986; Eggleston et al., 1992), but this may depend on the species of clam, its burial depth, and siphonal configuration and siphonal length (Lipcius and

TABLE 11.6

Predation on the hard clam by *Carcinus maenas*

Carapace width (mm)	Seed size (mm)	Crabs per container					
		5			10		
		None	Sand	Mud	None	Sand	Mud
22.6	3	–	–	–	2.51	2.61	1.77
24.5	4.3	–	–	–	7.77	6.47	5.38
32.6	6.5	6.71	3.77	1.77	–	–	–

Average number of hard clam seed of three size classes in different substrates (none, size and mud) consumed per day by *Carcinus maenas* at two densities. The – indicates no data collected. After Walne and Dean, 1972.

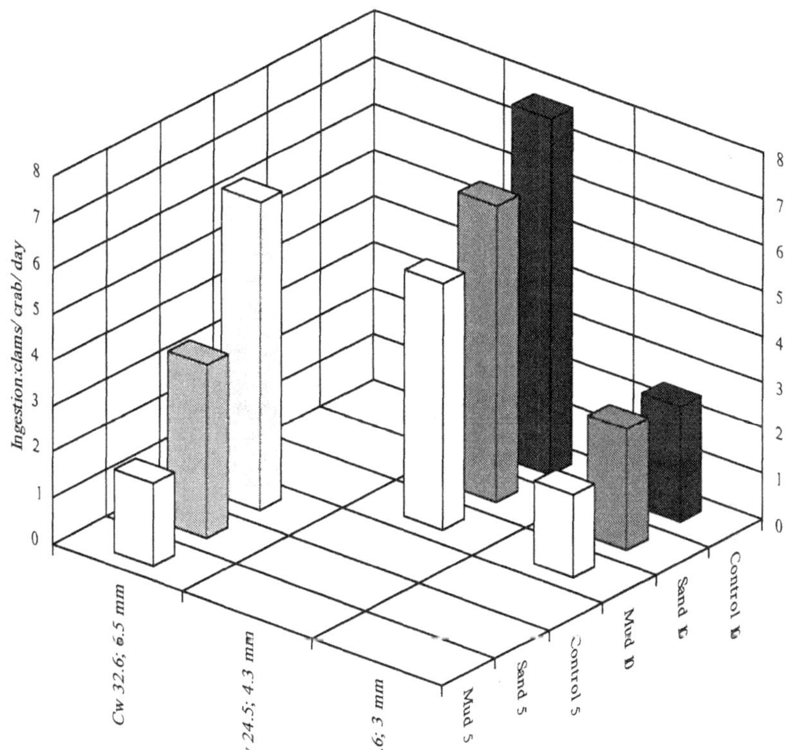

Fig. 11.9. Daily ingestion rate of hard clam, *Mercenaria mercenaria*, seed on three substrates by the green crab, *Carcinus maenas* (control = no substrate; Cw = carapace width). Data from Walne and Dean (1972).

Hines, 1986). Data on *Carcinus maenas* suggest this is not always true. In every experiment, more clams were eaten from sand sediments when compared to mud (Table 11.6). The reasons for these results were not explained. In addition to differences in predation rate attributed to seed size, carapace size, substrate type and density of predators, Walne and Dean (1972) demonstrated that the source of predator and prey can also affect the experimental results.

Data for the predation of *Carcinus maenas* on *Mytilus edulis* clearly showed that crabs collected from some areas were more voracious that those from other areas, and that the source of the prey also affected the rate of predation. Unfortunately, the reasons for these differences were not examined, nor are any data provided on the densities of predator or prey in the field. No information is available on predation rates in the field.

Malinowski (1985) reported that *Carcinus maenas* was abundant at Fishers Island, NY and Poquonock River, NY. He cited laboratory studies in which green crabs (50 mm) were allowed to feed on hard clams (no sediment) as large as 12 mm, and that a single crab could consume up to three 5-mm seed min^{-1}. Once, 200 5-mm seed were consumed by 1 crab h^{-1}. It is unlikely that this high rate of consumption would continue for 24 h, but at 3 clams min^{-1} (180 h^{-1}) the daily total would be 4320 clams, a factor of 1000 greater than reported by Walne and Dean (1972). A portion of this discrepancy is certainly due to the shorter time period of the observations, but some of the difference may be due to the larger crab size in the Malinowski observations. Malinowski (1985) planted 4.7, 9.8, 15, 17.6, and 20.4 mm seed clams in 0.25 m^2 plots at two sites. His studies indicated that density had that greatest effect on clam survivorship. Lower densities (100 clams m^2) survived at a 1.6 times greater rate (53%) than the same sizes planted at 600 m^2 (33%) and more than 4 times as well as clams at 1200 m^2 (13%). Based on these data, more clams survived (198 m^{-2}) at intermediate densities than at either the low (53 m^{-2}) or the high (156 m^{-2}) density plantings. Other studies in which 1–2 mm seed were planted in similar fashion indicated that smaller seed were consumed at equal rates on all sites. Larger seed showed significantly different survival levels at different sites. This was attributed almost exclusively to crustacean predation, and size-specific differences between sites were attributed to the different suites of predators. Daily observations suggested that most of the mortality in plots took place within a relatively short time (one to several days). Once high density patches of hard clams (5–10 mm seed) were located, the predators continued to forage until densities were reduced to 28–52 clams m^{-2}. This is considerably higher than the density of hard clam seed that is reported from most natural populations, but the result is opposite of those who have found high density refuges for a number of bivalve species.

Davies et al. (1980) provided a table listing the average size and weight at which three species of bivalves reach a size refuge from green crab predation. The shell length for blue mussels, *Mytilus edulis*, and Pacific oysters, *Crassostrea gigas*, were nearly the same at 40–45 and 45–50 mm, respectively. Hard clams, *Mercenaria mercenaria*, reached a refuge size at 25–30 mm shell length. Wet weight of the animals at the size refuge was identical for the blue mussel and hard clam (7–10 g) and 8–10 g for the Pacific oyster (Davies et al., 1980).

Summary Carcinus

Throughout its range this species has been considered a major bivalve predator. Along the western Atlantic, this reputation has been based on the reduction in recruitment of the soft-shell clam on New England tidal flats. Extensive studies in Europe have documented the importance of the juveniles of this species to recruitment and establishment of bivalve populations. These data seem to provide the best documentation of the importance of the foraging behavior of any species of crab predator on bivalve populations. Field and laboratory studies both in Europe and North America have clearly shown that predation by juvenile green crabs can have substantial population level effects on recruiting cockles and soft-shell clams. Laboratory studies have shown that this species is capable of consuming small heavy shelled clams, such as the gem clam. In laboratory studies, short-term rates of consumption on juvenile hard clams were similar to those found for gem clams. Populations of green crabs are not abundant south of Long Island, so any major effects of this introduced species on hard clam populations would be confined to cooler waters north of Long Island.

Ovalipes ocellatus

Crabs of the genus *Ovalipes* are significant predators on a variety of species including bivalve molluscs. *Ovalipes ocellatus*, the lady crab, is irregularly abundant in high salinity water on clean sand bottoms near ocean inlets from Cape Cod south along the US east coast. Stickney and Stringer (1957) mentioned *O. ocellatus* as an important clam predator in Rhode Island. Densities in areas of Great South Bay near Fire Island Inlet ranged to 2 m^{-2} (WAPORA, 1982). Flagg and Malouf (1983) reported that lady crabs were locally abundant in New York and that they were responsible for heavy predation on planted hard clam seed, but gave no information on predator density. Malinowski (1985) reported that this species was likely to be present during surveys of his Fishers Island, NY site, but was only occasionally seen at the Poquanock River, NY site. No data were given on the importance of this species in preying on hard clams at either site.

Davidson (1986) examined how *Ovalipes cantharus* opened blue mussels and concluded that predation rate varied with predator size, prey size, prey vulnerability and prey availability. This species had a variety of methods for opening the mussels: crushing (small mussels <10 mm length), anterior and posterior crushing (10–35 mm mussels) with the former more common, wedging, inserting the crusher chela until the cutter chela could reach an adductor muscle, and chipping. The latter two techniques were used only on mussels too large for crushing. Laboratory experiments indicated that, as with other crabs, *O. cantharus* opened smaller mussels at a greater rate than predicted by optimum foraging models based on energy maximization. This species, and *Callinectes sapidus*, achieved greatest energy return (profitability) from the smallest prey sizes, while *Carcinus maenas* and *Liocarcinus puber* both had maximum profitability with medium size prey. While *O. cantharus* preferred smaller prey, when the percentage of larger prey was increased, the increased encounter rate enabled crabs to reduce the time it took to open and consume the prey.

Similar data were reported by Seed (1990) for *Thalamita danae* preying on green-lipped mussels, *Perna viridis*. Due to chela morphology, the breaking time required for these crabs to open mussels increased rapidly with mussel size. There was a size refuge of about 50 mm, above which few mussels were consumed. The time crabs required to open mussels also decreased as crab size increased, and small mussels (10–15 mm) were consumed by all crabs while larger mussels (20–25 mm) were opened by 30 and 54% of medium (46–49 mm carapace width) and large (58–61 mm) crabs, respectively (Seed, 1990). Laboratory studies at 22.5°C (aquaria size not reported) in which groups of mussels of different sizes were used yielded rates of 0.19–0.50 mussels $crab^{-1}$ day^{-1}, but an unlimited diet of small mussels (replacement of those eaten) yielded rates of 3.8 mussels $crab^{-1}$ day^{-1} over an experimental period of 6–9 days. Over shorter periods, daily consumption rates equivalent to 20 mussels $crab^{-1}$ day^{-1} were observed (Seed, 1990).

Du Preez (1984) considered *Ovalipes punctatus* to be a significant predator on the gastropods *Bullia* spp. and the bivalves *Donax* spp. Prey species were detected by the dactyls of the crabs as the predator moved across the sand surface. Once detected, the prey were either crushed, or the valves were pried apart. The size of the *Donax* spp. crushed increased with the size of the crab, but there was no relationship between the size of the crab and its ability to open the clams by prying. The data indicated that clams that were pried open ranged between 57 and 64 mm shell length, and these were opened by crabs ranging from 36 to 65 mm carapace width. Laboratory studies in 50 L aquaria for 24 h examined the numbers of clams eaten by 5 size classes of crabs (40–65 mm). All size classes of crabs except the largest (60–65 mm) consumed between 13 and 17 *Donax serra* crab^{-1} day^{-1}. At peak consumption the largest size crabs ate about 7 of the 40–45 mm size class clams crab^{-1} day^{-1}. The smallest two size classes of crabs consumed clams 25–35 mm at the highest rate, but 50–55 and 55–60 mm size class crabs consumed the greatest numbers of clams from the 35 to 40 mm size range. Du Preez (1984) also examined the abundance of prey species at two locations, and while prey species were similar (the bivalves, *Donax serra* and *Donax sordidus*; and the gastropods, *Bullia rhodostoma*, and *Bullia pura*) the most abundant prey species differed on the two beaches. Laboratory studies on prey selection indicated that the crabs preferred to prey on the species that was *least* abundant on the beach from which the crabs were collected. This may suggest prey depletion, or a predation refuge.

Haddon and Wear (1987) examined the feeding of *Ovalipes cantharus* on three species of prey: two bivalves, *Austrovenus stutchburyi*, the cockle, and *Paphies subtriangulata*, the tuatua; and one crustacean, *Callianassa filholi*. These authors found no significant difference between the amount of food consumed in a 24-h period between crabs starved for 28 days and those freshly caught. In spite of the starvation, individual crab consumption rates were highly variable and were not standardized by up to 21 days of starvation. Feeding was greatly reduced below 10°C. When comparisons were made between freshly caught crabs and those that had been fed to satiation for 25 days, field caught crabs ate nearly twice as many cockles, but after 13 days of feeding the numbers of cockles eaten by field caught crabs had decreased from a mean of about 36 cockles crab^{-1} day^{-1} to a mean of 18 cockles crab^{-1} day^{-1}. Those that had been fed in the laboratory did not change appreciably from the 10 cockles crab^{-1} day^{-1} on the first day of the experiment. As with other studies, Haddon and Wear (1987) reported that the numbers of cockles eaten crab^{-1} day^{-1} were greater for larger sized crabs, but they were able to attribute this difference to the difference in the size of the crab foregut. Crabs of 95 mm had foregut volumes averaging 8.47 ml, while the foregut volumes of 107 mm crab averaged 11.57 ml. This difference closely corresponded to the amount of meat derived from the 42 and 60 cockles crab^{-1} day^{-1} consumed by this species (about 5 cockles ml^{-1} of foregut).

Experiments with *Ovalipes catharus* and the bivalve, *Paphia ventricosa*, the toheroa, were conducted to determine the effects of density and burial depth as refuges from predation (Haddon et al., 1987). All experiments were conducted in outdoor pools with clams 31–40 mm long at densities similar to that found in intertidal flats in New Zealand. All experimental units in the pool were available to each of the 15 crabs that were free to range throughout the pool. Using densities of 500, 1253 and 2000 toheroa m^{-2}, Haddon et al. (1987) found that although numbers of clams eaten from each density were statistically equal (43, 54 and 57, respectively), the proportion of prey eaten (57, 59, and 17%, respectively) was reduced by increasing density. Because all *Paphia ventricosa*, were available for 6 nights, the average mortality rate over all experimental units was 19.7 clams crab^{-1} day^{-1}. Toheroa were placed into containers with 26 cm of sand substrate and allowed to burrow to depths of 0, 5, 10 and 26 cm by restricting the burial depth by screens (Haddon et al., 1987). Although the numbers of clams eaten at 0, 5 and 10 cm were not significantly different, the standard error of these estimates increased with increasing depth of burial. The standard error decreased when clams were at the deepest burial depth and significantly fewer were eaten. Over all four experimental depths, predation averaged 25.3 clams crab^{-1} day^{-1}, but the average number eaten in the deepest experimental depth was 35 clams in the 3-day period vs an average of 64 for the other three depths. The authors noted that when the toheroa were placed in the tanks they were commonly found near the sand surface, but when they were disturbed, they burrowed more deeply into the substrate.

On the Gulf of Mexico coast of Florida, Cake (1970) reported that *Ovalipes guadulpensis* could consume *Macrocallista nimbosa*.

Turner (1948) reported that *Ovalipes ocellatus* may prey on *Mya arenaria* in Massachusetts. Sellmer (1967) reported that a population of adult gem clams, *Gemma gemma*, at Union Beach, New Jersey was reduced from 1000 m^2 to virtually 0 in 2 weeks by an invasion of lady crabs, *Ovalipes ocellatus*. Examination of the stomach contents of five of these crabs confirmed that all had been eating the clams. Up to that time, the salinity was apparently too low for the crabs to invade, but reduced precipitation allowed salinity levels to increase and the crabs to invade (Sellmer, 1967). How often similar short-term invasions of this species occur in coastal bays that support hard clam populations is unknown. Ropes (1988) examined lady crab stomach contents from Rhode Island and reported that gem clams and blue mussels were significant portions of diet.

Gibbons (1984) examined the effects of *Ovalipes ocellatus* (40.2–47.8 mm) on hard clam seed (3, 5 and 7 mm) placed in 0.03-m^2 culture containers with a substrate of sand, crushed aggregate and no substrate with temperatures that approximated the normal cycle in Long Island Sound, New York. As expected, temperature exerted a significant effect on predation, with feeding activity declining until activity reached 0 at about 3°C. In the spring, feeding activity was first noted at about 5°C. Gravel inhibited predation at most temperatures. Highest predation rates were reported in the bare trays (586 clams crab^{-1} day^{-1}) at 18°C for the smallest clams. The highest average rates of consumption for small clams in bare trays, sand and gravel were about 380, 145 and 20 clams crab^{-1} day^{-1}, respectively, for fall temperature conditions. Smaller clams were eaten at greater rates than larger clams at all temperatures (Table 11.7, Fig. 11.10).

Sponaugle and Lawton (1990) studied the predation of *O. ocellatus* (65–80 mm) on *M. mercenaria* seed (15–20 mm) at 5 different densities (24, 48, 72, 96, 120 clams m^{-2}) in sediments composed of sand and sand plus shell. All studies were conducted in 0.5-m^2 aquaria and included a video recording of crab behavior in order to determine handling time and other variables associated with crab feeding. In every case, more clams were consumed in sand than when sand and shell were combined. Consumption rates in sand ranged from 7 to 22 clams crab^{-1} day^{-1}, while those in sand and shell ranged from 0.5 to 17.5 clams crab^{-1} day^{-1}. Generally, higher rates of consumption were reported at higher clam densities in both sand and sand/shell mixtures. Additional experiments with 24 clams m^{-2} with sand, sand and gravel, and sand and shell mixtures revealed consumption rates of 6, 2.5 and 0.1 clams crab^{-1} day^{-1}, respectively. These rates were nearly the same as those for the same density

TABLE 11.7

Number of hard clams of three sizes ingested by individual *Ovalipes ocellatus* per day at four temperatures

Temperature (°C)	Clam size (mm)		
	3	5	7
10	98.4	48.6	14.2
15	120.3	67.5	20.9
20	137.9	71.4	30.4
25	196.9	97.9	39.6

After Gibbons, 1984.

and substrate combination from the first series of experiments. These data suggested a low density refuge for hard clam seed, particularly in substrates where sand was mixed with some coarser material, such as shell. The taped recordings of behavior indicated that crabs spent more time foraging in sand and shell substrates, but consumed fewer prey. This appeared to be due to increased handling of shell or gravel particles. Recordings taken when densities of prey were examined indicated that, while a successful encounter with a prey often increased more searching in the same area, there were no significant trends in the amount of time spent foraging in the different densities of prey. Definitive field studies with this species of predator and hard clam prey in substrates of sandy mud and shell have yet to be conducted.

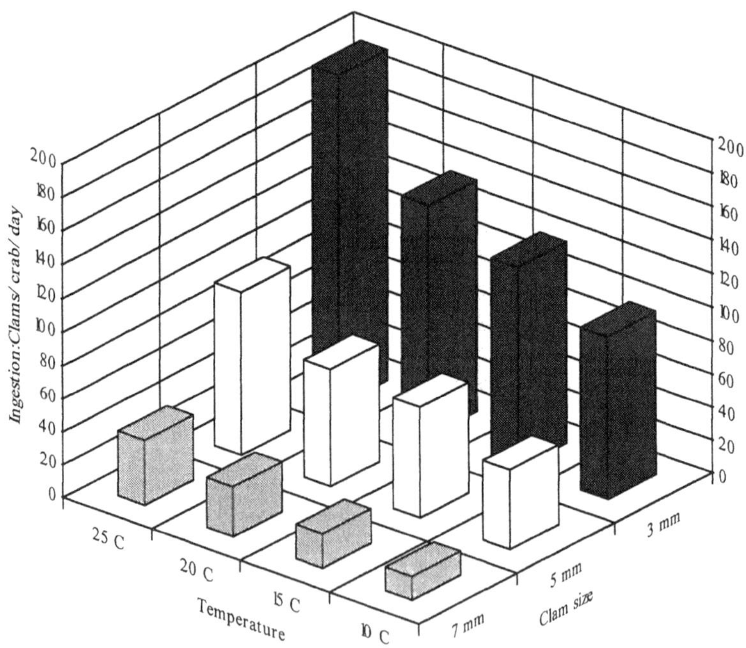

Fig. 11.10. Daily ingestion rate of hard clam, *Mercenaria mercenaria*, seed of three size classes (3, 5 and 7 mm) by the lady (calico) crab, *Ovalipes ocellatus* at 4 temperatures. Data from Gibbons (1984).

Callinectes sapidus

The numbers of blue crabs estimated by Hines et al. (1990) based on drop net studies in a subestuary in mid Chesapeake Bay (total area not provided) were nearly identical (0.08 crabs m^{-2}) to those reported the following year by van Montfrans et al. (1991). van Montfrans et al. (1991) estimated the population of blue crabs in a 10,000 m^2 creek/marsh system in the lower portion of Chesapeake Bay to be between 0.15 and 0.08 individuals m^{-2}. Farther south, in Georgia salt marsh systems, Fitz and Wiegert (1991b) estimated the blue crab population to be between 40 and 50 crabs ha^{-1} (0.005 crabs m^{-2}), considerably lower than that reported for the Chesapeake Bay. By tagging the crabs, these authors also found that most recaptures occurred within 10 day of the initial capture, after that most crabs had emigrated from the study area. These data are important because most laboratory experiments on blue crab feeding have been conducted at much higher crab densities. It is important to note that most of the crab population data and the experimental data on bivalve consumption refer to larger blue crabs. Few studies have been conducted to evaluate the density of the more numerous juvenile crabs in large areas. Pile et al. (1996) evaluated the recruitment process of blue crabs and reported that this species has higher survival in grass beds through the fifth instar, but following this molt, the crabs begin to migrate from the grass bed to the open sand and mud habitats. How this migration may affect the predation on hard clams that set within or outside the grass beds has not been investigated.

Few studies have reported on the diet of blue crabs in its natural habitat, but Laughlin (1982) showed that bivalves were the dominant prey items (35.7%) for all sizes of these predators in a northern Florida estuary. The next largest category of food was crustacean (25.3%, including 9% blue crab) followed by fish (11.4%). The dominant bivalves in the diet were reported to be *Rangia cuneata* (45%), *Brachidontes* spp. (37.5%) and *Crassostrea virginica* (12.5%), but the percentage of the diet composed of individual species generally reflected the abundance of the food source. *Mercenaria mercenaria* were not found in the food items in this study, but the dominant prey suggest the study was conducted in the oligohaline portion of the bay where few hard clams would be found.

Alexander (1986) reported on the diet of blue crabs in Galveston Bay. A variety of habitats were sampled and the diet of the crabs reflected the habitat from which they were collected. In general, 42% of the stomachs contained molluscs, 29% vascular plants, 26% fish, 25% benthic algae and 21% crustacean parts. None of the molluscs were identified to species.

Ropes (1988) reported on the stomach contents of blue crabs collected in the Pettaquamscutt River, Rhode Island. As in other locations, these crabs consumed a wide variety of prey including annelids, molluscs, crustaceans and fish. The molluscs consumed included blue mussels, gem clams, hard clams and soft-shell clams. Of the five crab species examined, the blue crab was the only species shown to consume hard clams.

An 8-year study of monthly samples (April to December) in a subestuary of Chesapeake Bay (Hines et al., 1990) found that blue crabs consumed whole clams (30–55%), chiefly *Macoma balthica*, blue crabs and fish. Early in the season, the crabs consumed a significant number of amphipods, but when clams became more abundant due to recruitment, the crabs' diets shifted to this preferred prey. Clam densities were typically 200 clams m^{-2} in the early summer, but a large 1981 spring set increased clam density to 2700 m^{-2}. This was subsequently reduced to 600–700 clams m^{-2} by the fall. Spring recruitment in 1982, 1983 and 1984 of 800–1000 clams m^{-2} increased the prey numbers until predators became active. Poor recruitment after 1984 caused clam density to fall back to the base of about 200 m^{-2}. This decline coincided with increased crab abundance during 1984 and 1985. Cage studies associated with the sampling of the predators indicated that, when cages were present, higher density of the clams *Mya arenaria*, *Mulinia lateralis* and *Macoma mitchelli* were found at all stations.

Virnstein (1977, 1979) examined the effects of blue crab predation on the density of infaunal organisms in Chesapeake Bay and determined that excluding predators caused a significant increase in three bivalve species; *Mulinia lateralis*, *Lyonsia hyalina*, and *Mya arenaria*. If crabs were maintained at densities of 1 m^{-2}, numbers of infaunal species were greatly reduced, and bivalves were nearly eliminated. If the density of blue crabs from van Montfrans et al. (1991) and Hines et al. (1990) extrapolated to these cage studies, the experimental predator densities were approximately 50 times greater than the naturally occurring populations. Virnstein (1977, 1979) reported that fish predation on juvenile bivalves, crab predation on small individuals and ray predation on larger clams may be the chief cause of the lack of infaunal bivalve dominance in Chesapeake Bay. In cages that excluded larger predators, *Mulinia lateralis* density increased to nearly 15,000 m^{-2} and were several layers thick. As clams in this dense assemblage grew, they forced other individuals outside the protective mesh where they were rapidly consumed. One small crab (60–80 mm) entered a 0.25-m^{-2} exclusion cage that contained 3675 *M. lateralis* (14,700 m^{-2}). At the end of 2 months the crab was 130 mm wide and all clams were gone. If we assume the crab was responsible for all mortality of the clams, this is equivalent to a minimum of 59 clams $crab^{-1}$ day^{-1}. In a separate control cage in which there were no predators observed, the number of clams was reduced to 1987 (7950 m^{-2}) 18.2-mm-long clams chiefly due to the individuals on the fringes being forced from the cage. Many clams were lying on the surface and unable to enter the sediment. If this is considered to be a base level for the consumption then the blue crab was responsible for a loss of 32 clams $crab^{-1}$ day^{-1}. Four days after the cage was removed from these clams, none remained alive, and only empty and broken shells could be found. Cage removal in colder months (November and February) when blue crabs and other predators were relatively inactive did not result in severe clam losses. These data are similar to those reported for *Mytilus edulis* populations in Barnegat Bay, NJ (Peterson, 1975, Peterson, 1979).

Results similar to those of Virnstein (1977) were also reported by Reise (1977b) on a European tidal flat and Holland et al. (1980) in the meshohaline portion of Chesapeake Bay. These latter investigators covered sand and muddy sand sediments with cages made of 12-mm mesh and reported that infauna that recruited in summer were the species that were most affected by predation. Those affected by this process included juvenile *Macoma balthica* and *Mya arenaria*, while adult soft-shell clams were not affected. Other species not affected by the cages were relatively small forms or deep burrowers. This suggests both a size and depth refuge, or that the small burrowers were influenced more by other infaunal than the epifaunal predators this study was designed to exclude. Reise (1977b) examined the effect of varying the mesh size on cages and found that while 20-mm mesh provided little or no protection to recruiting infauna, 5-, 2- and 1-mm meshes increased the total infauna from 800 to 2700, 3000 and 3100 cm^{-2}, respectively. Use of a 0.5-mm mesh further increased the faunal abundance to 6000 individuals cm^{-2}. A number of bivalve species not normally found on these flats recruited to the cages.

Morgan et al. (1980) indicated that blue crabs were one of 5 crab predators suspected in causing significant losses of hatchery reared bay scallops released into Connecticut waters.

The combination of these predators reduced the numbers of scallops <22 mm by 50% in less than 1 day. For larger scallop seed (>24 mm) the time to 50% mortality was about 6 days.

Seed (1980) conducted feeding experiments on three size classes of blue crabs (60–70, 100–110 and 150–160 mm carapace width) eating ribbed mussels, *Geukensia demissa*. All experiments were conducted in 0.16-m^2 aquaria. The largest crabs were able to crush all sizes of mussels except those larger than 80 mm. Medium size and small crabs consumed mussels up to 55 and 48 mm, respectively. Blue crabs 65 mm carapace width consumed an average of 18.8 mussels (20–30 mm) day^{-1}, and, when offered a variety of sizes, the crabs selected smaller mussels. Comparisons between predation on similar sized *G. demissa* and *Brachidontes exustus* revealed that the crabs did not select one species over the other, but when offered ribbed mussels and eastern oysters the crabs usually selected mussels. This latter tendency was less noticeable when small oysters were placed with the mussels. Other studies (Seed, 1982) indicated that blue crabs shift from crushing ribbed mussels to chipping the edges at 30, 20, and 11 mm mussel length for crabs of 117, 80 and 52 mm carapace width, respectively. Hughes and Seed (1981) and Seed (1982) also reported that crabs select smaller mussels and then proceed to larger sizes, but that the largest mussels were seldom selected. The exception to this generality was no difference in selection for mussels <25 mm (Hughes and Seed, 1981). The highest rate of consumption in the studies (Seed, 1982) was >50 mussels day^{-1}. Hughes and Seed (1981) noted that in some experiments the crabs appeared to be avoiding the smallest mussels, but subsequent studies found this was an artifact resulting from the crabs being unable to find the smaller prey in the shell debris of the container.

Lin (1991) manipulated the clump structure of ribbed mussels, *Geukensia demissa*, in laboratory studies to determine their vulnerability to blue crab (110–120 mm) predation. Mussel clumps were constructed by binding them together with nylon threads. This permitted development of clumps that were bound together with different strengths, and allowed arrangements of multiple mussels of the same size or alternating sizes in which small ones could be protected by larger mussels. All studies were conducted in 0.49-m^2 aquaria for 4 h. Some clumps were offered completely exposed, while others were partly buried in sediments. Field studies indicated that all juvenile mussels (<30 mm) were attached to clumps of mussels and completely buried in sediments. Laboratory studies confirmed that when mussels were exposed, there was no preference for size (30–40 vs 50–60 mm). When large mussels were exposed and small were buried, the crabs selected the larger mussels, and when small mussels were exposed and the larger buried the crabs selected the smaller mussels. In general, crabs consumed smaller mussels in preference to larger ones. When small mussels were attached to mussels 80–90 mm long they suffered little or no mortality, but similar experiments with small mussels and 50–60 mm long adults revealed that both the adults and juveniles were eaten at the same rate.

Menzel and Hopkins (1956) reported no differences in mortality of large oysters between those caged with blue crabs and those without crabs, but when spat were included in the cage the mortality rate increased to 2.15 oyster spat $crab^{-1}$ day^{-1}, and 19 oyster spat $crab^{-1}$ day^{-1}. Subsequently, Menzel and Nichy (1958) kept blue crabs in individual cages with 50–60-mm oysters, and reported that none of the oysters were consumed and that the crabs died. When oyster size was reduced to 40 mm, a blue crab consumed 15 oysters in 5 weeks before it too died (0.4 oysters $crab^{-1}$ day^{-1}).

Krantz and Chamberlin (1978) suggested that blue crabs were important predators on eastern oyster spat. In their study, a late September planting of 125,000 cultchless oyster spat

(6, 12, 20 and 25 mm) was completely eradicated by spring, and the 6 mm seed were all consumed within 2 weeks. The following year, 3–40 mm oyster seed were placed in open (no top) and closed trays near Chesapeake Bay oyster beds and within 1 month 99.7% of the unprotected spat were dead. In laboratory experiments, blue crabs (100–150 mm) were able to consume cultchless oysters up to 40 mm, while smaller crabs (65–80 mm) could consume oysters up to 25 mm.

Bisker and Castagna (1987) examined the predation of *Callinectes sapidus* on individual eastern oyster spat. Oysters set on shell were divided into 4 size classes (mean size 3.4, 7.6, 13.9, and 24.4 mm) and crabs were divided into 6 size classes (mean size 9.3, 17.2, 24.5, 31.7, 39.2 and 85.5 mm). Fifty oyster spat were placed in each 0.05-m^2 container with one crab, and experiments were conducted for 48 h. Temperature ranged from 20 to 28°C and salinity was 24–35 ppt. Blue crabs in the smallest size category could not consume oysters larger than the 13.9 mm size class, and the larger the crab the more spat consumed in 24 h (Fig. 11.11). Predation, when averaged for all sizes of crabs within a spat size, decreased with increasing spat size with the exception of the largest size crabs. These showed a peak predation rate for the 7.6-mm oyster seed. In comparisons with the mud crab, the authors found that within similar size (carapace width) categories, mud crabs could consume larger spat than blue crabs, and that the rate of predation was greater for mud crabs than for blue crabs. Mud crabs were found to have a ratio of 1.03 (length of prey/carapace width of predator) indicating that this

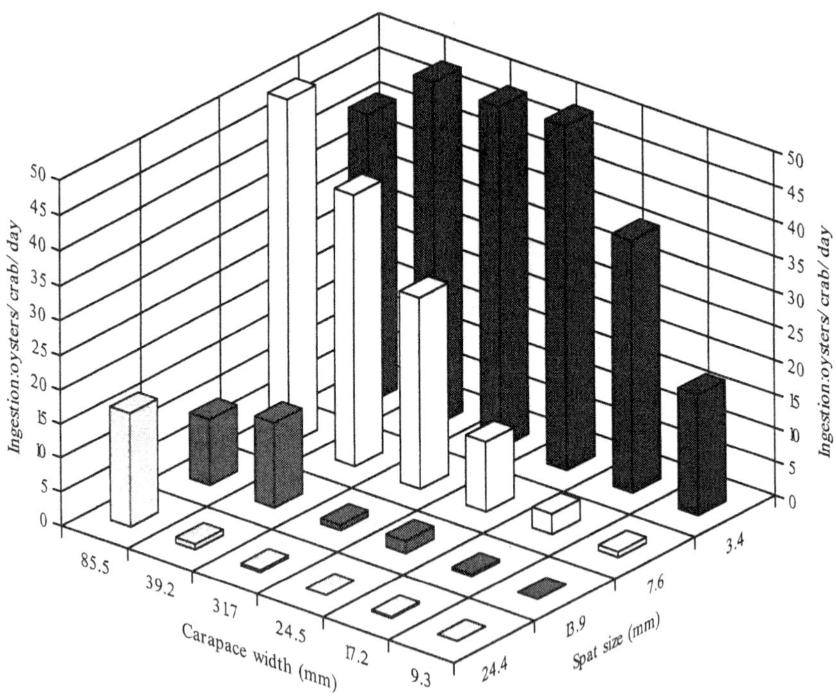

Fig. 11.11. Daily ingestion rate of 4 sizes of oyster *Crassostrea virginica*, spat (3.4, 7.6, 13.9 and 24.4 mm) by 6 sizes of blue crab, *Callinectes sapidus*. Data from Bisker and Castagna (1987).

species could consume oysters as long as its carapace width. In contrast, blue crabs could only consume oysters with ratios of 0.4 to 0.42.

All the above rates are lower than the 142 oysters $crab^{-1}$ day^{-1} reported by Eggleston (1990a) for blue crabs feeding on oyster spat (on cultch) averaging 15 mm shell height. Similar data for larger oyster seed (25 and 35 mm) were 27 and 7 oysters $crab^{-1}$ day^{-1}, respectively. These studies were done in tanks with oyster shell bottom cover, and with oyster density of 5, 10, 20, 30, 40 and 50 oysters per 0.33-m^2 tank for the larger seed and 10, 20, 40, 80, 120, 250, and 350 oysters per tank for the smallest seed. The higher densities of smaller seed were required to assure that maximal predation rates could be measured. The largest blue crabs could not consume oyster seed of 45 mm shell height. As expected, the numbers of oysters consumed were a function of prey density and shell height. The greatest predation took place on the smallest prey size at the highest densities. The amount of time required to consume an oyster (handling time) was related to prey size, but not density. The crabs spent more time foraging as the prey density increased, up to a point above which it remained the same, but the persistence (time spent trying to open an oyster) decreased with increasing density. This combination of increasing encounter rates and decreasing persistence led to an increased success rate for each attack. Presumably this was due to an increased chance of finding a weaker oyster spat. These studies suggested that blue crab predation can lead to local extinction of oyster spat and that no low density refuge exists.

Most studies with blue crabs have focused on predation by male crabs. Eggleston (1990c) examined the predatory rate of both male and female blue crabs on 6 sizes of oysters. Crabs were grouped into two carapace width categories (60–85 and 135–165 mm) and oysters on cultch were divided into 25 and 35 mm shell height groups. The experimental design was similar to those of Eggleston (1990a) with oysters arrayed in 6 densities. The consumption rate for large crabs of both sexes increased with oyster density. Daily consumption rates were 1.54 large oysters at 5 oysters 0.3 m^{-2} and 7.06 were consumed at 40 oysters 0.3 m^{-2}. Similar data for large blue crabs preying on small oysters ranged from 4.0 oysters $crab^{-1}$ day^{-1} at the lowest density to 26.5 oysters $crab^{-1}$ day^{-1} at the highest density. There were no differences with crab sex in the proportional mortality (oysters eaten/oysters present) of small oysters being preyed on by large crabs. Proportionally, more oysters were eaten at the lowest densities. These data suggest that small oysters do not have a prey size refuge from blue crab predation. There were significant differences between crab sex and the proportion of oysters eaten with respect to large oyster seed. Large female crabs consumed proportionally fewer seed at lower densities than male crabs. Female crabs ate proportionally more seed at density 5 than density 10, indicating the possibility of a low density refuge for small seed from female blue crab predation. This refuge was estimated to be about 16.7 oysters m^{-2} (Eggleston, 1990c).

Eggleston (1990b) suggested that an additional low density refuge for small oysters may exist for 25 mm seed (the only size tested) at low temperatures (13–14°C). These studies used only large male crabs in an experimental design similar to those described above. Three temperatures 13–14, 19–20, and 26–27°C were used and density was the same as in the above experiments. In general, consumption rates increased with temperature and oyster density. Significantly more oysters were consumed at the upper 5 prey densities than at the lowest density, irrespective of temperature. At the lower two temperatures, proportional mortality was significantly different, but in both cases it was higher at intermediate densities. At the lowest density and lowest temperature combination the oysters achieved a total prey refuge.

Interestingly, proportional mortality was significantly higher at lowest prey density at the highest temperatures.

Blundon and Kennedy (1982a) examined the effects of burial depth of 30–40 mm long *Mya arenaria* (5, 10, 15 and 20 cm sand depth) on predation by blue crabs (121–180 mm). These depths approximated those at which the authors found the soft-shell clam in the field. Ten clams were introduced into each 0.18 m^2 experimental unit plus one blue crab that was allowed to forage for 48 h. Average predation rate was statistically the same for the 5- and 10-cm-deep clams (4.9 and 4.5 clams $crab^{-1}$ day^{-1}, respectively), and the 15- and 20-cm-deep clams (2.3 and 0.8 clams $crab^{-1}$ day^{-1}, respectively), but the authors noted that in spite of these differences there was no spatial refuge that provided complete protection. Similarly, artificial vegetative cover designed to mimic submerged aquatic plants reduced predation, but conveyed no complete spatial refuge. In these latter experiments, *Mya arenaria* (10–20 mm) were allowed to bury up to 7 cm deep in sediment with either no grass mimic (plastic strips), sparse grass or dense grass mimics. Medium size blue crabs (61–120 mm) were introduced for 24 h, and then the survival of the clams was recorded. Average consumption was 9.7, 3.8 and 0.8 clams $crab^{-1}$ day^{-1} for the control, sparse and dense 'grass' experimental conditions, respectively. These data were compared to field studies that indicated that bivalve populations rapidly increased in the late fall, winter and spring months, but declined rapidly in late spring and summer. Four species of bivalves *Mya arenaria*, *Mulinia lateralis*, *Macoma balthica*, and *Macoma mitchelli* reached peak average densities of approximately 500, 80, 200 and 200 m^{-2}, respectively, only to diminish by a factor of 10 later in the year. Blundon and Kennedy (1982a) suggested that these precipitous declines were due to predation by the blue crab and possibly other benthic feeding species. It is important to remember that in these experiments blue crab density was 69 times as high as the approximately 0.08 crabs m^{-2} reported in field studies, and thus inferring that no spatial refuge would be present may be misleading.

Blundon and Kennedy (1982b) reported that the crushing force of blue crab (100–165 mm) chela ranged from approximately 60 to nearly 200 N. This was sufficient to crush shells of all sizes of the mussels *Mytilopsis leucophaeta* and *Ischadium recurvus*, and the clams *Mulinia lateralis*, *Macoma mitchelli*, *Macoma balthica*, *Mya arenaria*, and *Tagelus plebeius* tested. The largest specimens of *Rangia cuneata* (>40 mm long) had shell strength exceeding the range exhibited by the blue crab. These data indicate that blue crabs must use other methods of opening bivalves if the shell strength exceed 200 N, but prey shell shape may constrain this mechanical advantage below that maximal point. Blundon and Kennedy (1982b) did not find any preference of the crabs for *Macoma mitchelli*, *Macoma balthica* or *Mya arenaria*. No selection studies were conducted for other species. Studies with various sized *Mya arenaria* showed that larger crabs exhibited a statistically significant preference for larger clams and that small crabs preyed on small or large clams equally, but had reduced predation on intermediate size clams (Fig. 11.12). Average consumption rates varied from 0.72 to 3.83 clams day^{-1}.

Ebersole and Kennedy (1994) found that blue crabs preferred the smallest (10–20 mm) adult Atlantic rangia, *Rangia cuneata*, offered, and this selection was in accordance with an optimal energy foraging model. This shallow burrowing clam occupies the oligohaline portions of the estuary and had the strongest shell of the eight species tested by Blundon and Kennedy (1982b). *Rangia cuneata* were divided into 6 size classes (10–20, 20–30, 30–40, 40–50, 50–60 and 60–70 mm), and placed into 12 tanks (0.09 m^2) with individual crabs

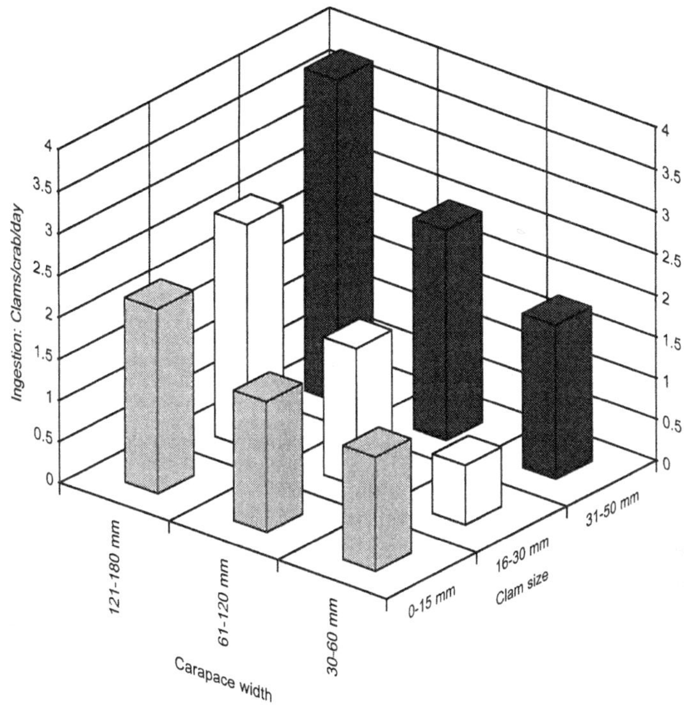

Fig. 11.12. Daily ingestion rate of four size classes (0–15, 16–30, and 31–50 mm) of soft shell clams, *Mya arenaria*, by three size classes of blue crab, *Callinectes sapidus*. Data from Blundon and Kennedy (1982b).

(71–166 mm). Two sets of experiments were conducted: a 5-day experiment in which clams were replaced, and a 14-day experiment where clams were not replaced. No substrate was used, and all studies were conducted in 6–12-ppt seawater. Clams larger than 30–40 mm were attacked, but it often took many days before they were consumed. In the longer term non-replacement experiments, smaller clams were consumed first and larger clams later. When clams were replaced with similar sized individuals, smaller clams were continuously selected. The 10–20-mm class was consumed at a rate of slightly greater than 4 clams $crab^{-1}$ day^{-1} and this was significantly different from the slightly more than 3 clams $crab^{-1}$ day^{-1} consumed in the 20–30-mm clam size class. Sizes larger than 20–30 mm were consumed at a very low rate (<0.3 clams $crab^{-1}$ day^{-1}). There was no significant relationship between crab carapace width and prey size.

In a companion series of experiments, Ebersole and Kennedy (1995) examined the feeding of the blue crab on three species of bivalves: *Rangia cuneata* (tightly closed heavy shell, shallow burrower); *Mya arenaria* (gaping at either end light shell, deep burrower); and recurved mussel *Ischadium recurvum* (completely closed, except at byssal notch, ribbed moderately strong shell, epifaunal clump former). Laboratory experiments, in which two of the three species were placed in the same 0.05-m^2 aquarium and subject to blue crab predation, were designed to determine the effects of prey preference, prey location, prey refuge, and prey density. Profitability models correctly predicted that the soft-shell clam (50–70 mm) would be selected when compared with *Rangia cuneata* (30–40 mm), but when the mussel (10–60 mm) and the soft-shell clam, or the mussel and the rangia clam were compared the profitability model lost predictive capability. Deep sand (20 cm) provided a better refuge for the soft-shell clam than for rangia clams, but in both cases more soft-shell clams were consumed (26 vs 6% deep and 59 vs 2% shallow) in shallow sand (3 cm). Soft-shell clams were consumed by crushing the shell while rangia were opened by chipping the edges and consuming the meat. Deep sand also provided a better refuge for the soft-shell clam than clustering did for the mussel. Prey density affected prey consumption, but not prey preference.

Lipcius and Hines (1986) noted that populations of *Mya arenaria* persisted in sandy sediment in Chesapeake Bay in spite of blue crab predation. Experimental tanks (0.36 m^{-2}) were filled with 25 cm of sand and/or mud sediments, and soft-shell clams (48–60 mm) at six densities: 2, 4, 8, 16 and 32 clams per tank (5.6–89 clams m^{-2}). All studies were conducted at 25°C and 10–13 ppt salinity. Individual adult blue crabs (130–140 mm carapace width) were introduced into the tanks (35× reported field densities), and allowed to feed for 3 days. Preliminary experiments indicated that satiation was 2.75 clams $crab^{-1}$ day^{-1}. Consumption rates were significantly higher in mud than sand (1.5 vs 0.83 clams $crab^{-1}$ day^{-1}) and more clams were consumed at the highest three densities than at the two lowest, regardless of sediment type (1.4, 1.6 and 1.9 clams $crab^{-1}$ day^{-1} for 8, 16 and 32 clams per tank, respectively, vs 0.4 and 0.6 clams $crab^{-1}$ day^{-1} for 2 and 4 clams per tank, respectively). In spite of numbers of clams consumed increasing with clam density, the proportion of the clam population consumed decreased. When the data were expressed as numbers eaten/numbers present, a negative density dependency resulted. This indicated that very high densities can provide a refuge similar to that reported by Haddon et al. (1987), presumably because the clams act as their own coarse material. At the lowest density, 60% of the clams were consumed in the 3-day period, while at the highest density only 17.8% of the clams were consumed. Even though more clams were consumed at the highest density (1.9 vs 0.4), only 69% of the satiation level was being eaten. Why satiation feeding was not achieved is unknown. This suggests that some form of high density refuge may have been present. Proportions of clams eaten by sediment type were significantly different. More clams were consumed in mud (60–90%) than in sand (17–54%). This was even more pronounced at low clam densities with 17–35% being eaten in sand while 80–90% were consumed in mud. Lipcius and Hines (1986) concluded that a low density refuge in sand was probably due to reduced sediment penetrability reducing the crab's ability to sense the clam. The importance of the difference in the predation at low densities on the soft-shell clam in sand and mud was used to explain the persistence of these populations in the mesohaline portions of Chesapeake Bay.

Mansour and Lipcius (1991) examined the relationship between predator and prey densities, and the potential for interaction between these two factors. Blue crabs were placed in 1-m^2 tanks that had been filled with 15 cm of natural sediments and *Macoma balthica* (27–33 mm). All experiments were conducted for 3 days at 22°C with clam densities of 4 or 16 m^{-2} and either 1, 2 or 4 crabs m^{-2} (12.5–50 times the crab field population density). Crabs ranged in size from 115 to 160 mm. Highest consumption rates (about 4.7 clams $crab^{-1}$ day^{-1}) occurred when crab density was low and clam density was high. Density-dependent foraging was found in all treatments. When additional crabs were added to the tank the numbers of clams consumed in the highest prey density decreased to less than 1 clam $crab^{-1}$ day^{-1}, indicating some interference between the predators, but the interaction between clam density and crab density was not statistically significant. This inference should not be surprising since Fernandez et al. (1993) have shown that early recruits of the dungeness crab were capable of effectively controlling subsequent settlement of later recruiting *C. magister* by cannibalism. In addition, food habit studies by Laughlin (1982) and Hines et al. (1990) and others suggested that blue crabs were cannibalistic, and that other blue crabs were a significant portion of their diet. A low density refuge (about 1–4 clams m^{-2}) was observed even when crab density was high (Mansour and Lipcius, 1991).

Eggleston et al. (1992) compared the survival of densities of 1, 2, 4, 8, 16 and 24 *Macoma balthica* per 0.23-m^2 tank (4.3–102.6 clams m^{-2}) in both 24-cm-deep sand and mud sediments. Individual blue crabs (135–165 mm) were introduced into the tank with the *M. balthica* (25–35 mm) and allowed to forage for 2 days. This experimental design was similar to that used by Lipcius and Hines (1986) for studies on the soft-shell clam. Salinity ranged from 18 to 20 ppt and temperature from 24 to 28°C. More clams were consumed at higher densities regardless of sediment type. Consumption rates ranged from 0.215 to 6.9 clams $crab^{-1}$ day^{-1} in sand to 0.0 to 6.25 clams $crab^{-1}$ day^{-1} in mud. The lowest consumption was found in tanks with lowest prey density, and the highest rates occurred in tanks with high prey density. When predation was placed on a proportional basis (number of clams eaten/number of clams present) no high density refuge was indicated. In sand, the clams achieved a partial refuge from predation at a density of 1 clam per tank, but at low densities, clams in mud were not consumed (a complete refuge). These data suggested that low densities of *Macoma balthica* can survive in both sand and mud even with heavy crab predation, but that densities should be higher in the muddy sediment. When comparing these data with similar data, and an earlier study by Lipcius and Hines (1986) for the soft-shell clam, the authors provided an explanation for the persistence of soft-shell clams in sandy sediments, and the survival of *Macoma balthica* in both sand and mud sediments. The interactive effects of burial depth and prey density only partly explain this differential survival. The authors suggested that the lower numbers of *M. balthica* in sand were due to it being found in shallower depths in the coarser grained sediments. The differential survival of the two species may have been due the ability of the crab to detect the larger siphons of the soft-shell clam, but confirmation of this speculation requires additional experimentation (Eggleston et al., 1992).

Sellmer (1967) listed the blue crab as an important predator on the gem clam, *Gemma gemma*, but did not provide data on consumption rates or stomach analysis.

Stickney and Stringer (1957) reported that blue crabs were important predators on hard clams in Rhode Island, and Haven and Andrews (1957), working on a site relatively close to that investigated by Virnstein (1977), found that hard clams (25–33 mm) planted on plots in Chesapeake Bay were heavily preyed upon. They did not find the predator, but based on the size of the clams, the seasonal nature of the losses, and the shell remains, the authors suggested that *C. sapidus* was the most likely candidate. Carriker (1959) observed blue crabs digging pits as deep as 10 cm to excavate and consume various species of bivalves. He placed a group of 3453 hard clams (2–40 mm long) in a plot in the field. Within 6 days, all but 18 of the clams had been consumed. Carriker (1959) suggested that blue crabs were responsible for most of these losses, and showed that a blue crab (50 mm) consumed 615 clams (2–16 mm) in 2 days (308 clams $crab^{-1}$ day^{-1}).

Menzel et al. (1976) and Menzel and Sims (1964) reported that blue crabs were responsible most of the mortality experienced by hard clam seed planted in unprotected plots in Florida.

In these studies, clams were protected with wire mesh cages, but even the cages were not sufficient to protect 10 mm seed from 100% losses. Hard clam seed (33–44 mm) yielded 82–95% survival when planted within the cages, but nearly all the seed were lost when planted outside the cages.

Walker (1983) reported that juvenile blue crabs were a significant source of the mortality of hard clam seed placed in experimental plots in Georgia. First year mortality for Georgia, Massachusetts and Virginia clam stocks planted in these waters was 71, 86 and 69%, respectively.

Arnold (1983, 1984) investigated the interactions between seed size *Mercenaria mercenaria*, blue crab size and substrate type. In paired comparisons of substrate type, crabs preferred sandy mud, mud, and sand over crushed oyster shell, large (average diameter 5 cm) gravel, and fine gravel (mean diameter 2 cm). The fine gravel was the least preferred. In all crab size classes investigated (small = ≤75 mm, medium = 75–125 mm, large = ≥125 mm), the crabs preferred sandy mud and mud over other types of sediments, but there were significant interactions between crab size and sediment type. Arnold's tanks were nearly 3 m^2, and only one crab was placed in each tank. These densities were still >4 times the average blue crab densities reported from field studies in Virginia and Maryland. As opposed to the substrate preference studies (Arnold, 1984) when clam consumption studies were conducted, no interaction was found between crab size and substrate type (Fig. 11.13). The description of these studies suggests that sediments were not mixed with the shell or gravel, but the latter were simply added on top of the sand, this is an important distinction when comparing the laboratory studies with those in the field. Clam size was the most significant factor in the rate of predation. Only the largest crabs could consume 25-mm clams and no 50-mm clams were consumed (Fig. 11.14). As with other studies, smaller crabs generally consumed smaller seed, but larger crabs did not consume as many small (5 mm) seed as the medium size crabs. The addition of oyster shell or fine gravel significantly reduced predation rates (Fig. 11.13). Maximum densities of hard clams in Georgia Sounds found in sand/mud or sand were 16 m^{-2} and 12 m^{-2} , respectively, while areas with shell deposits had densities of 26 m^{-2} or higher (Arnold, 1984; Walker et al., 1980). Arnold (1984) also noted that although predation rates between oyster shell and fine gravel were not statistically different, the rates were consistently higher in the crushed shell. Interestingly, large crabs consume 25 mm long clams only on bare substrate, fine gravel and crushed shell. The importance of the experimental set-up (no apparent mixing of the shell or gravel in the sediments) was apparent when the author noted that the largest clams were not consumed in sand or sandy mud substrates, but were consumed in bare tanks or in those with oyster shell. This may reflect the fact that larger clams were placed in the substrate rather than digging themselves in, or other factors relating to the behavior of crabs over shell substrates. Rates of predation computed from Arnold (1983, 1984) based on seed size and substrate, and seed size and crab size are provided in Tables 11.8 and 11.9.

Malinowski and Whitlatch (1984) found 4 times greater survival of hard clam seed (5 and 10 mm) planted at 100 m^{-2} relative to those planted at 1200 m^{-2}. They suggested this indicates low density refuge.

Sponaugle and Lawton (1990) studied the predation of 100–130 mm carapace width *C. sapidus* on *M. mercenaria* seed (15–20 mm length) in 0.5-m^2 tanks at a density of 24 clams m^{-2} with sediments composed of sand, sand and shell, and sand and gravel. All studies were conducted at 22–25°C, 28–33 ppt seawater, and included a video recording of crab behavior

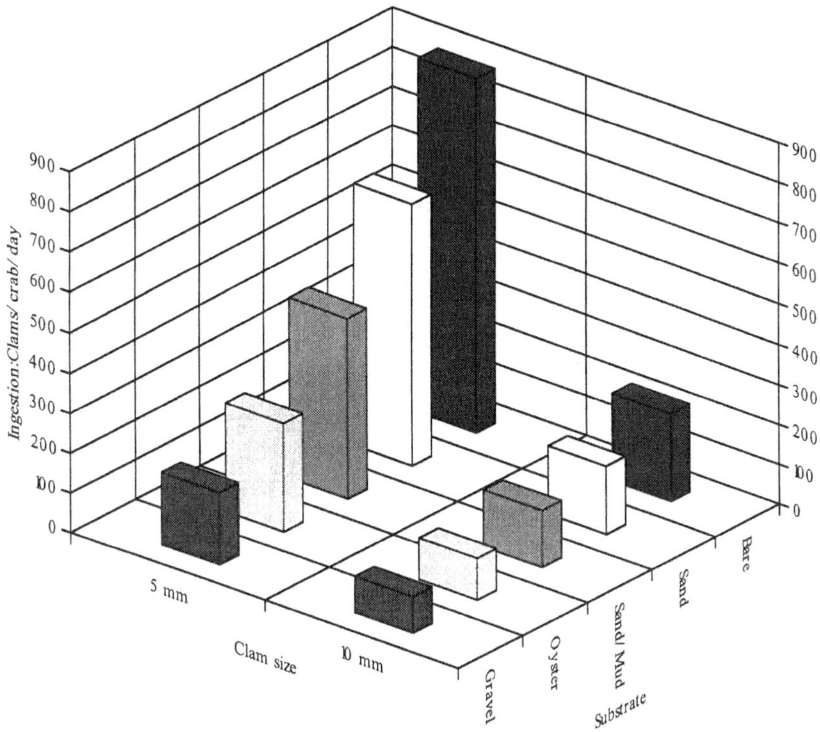

Fig. 11.13. Daily ingestion rate of two sizes (5 and 10 mm) of hard clam *Mercenaria mercenaria*, seed by the blue crab, *Callinectes sapidus*, on five types of substrate. Data from Arnold (1984).

in order to determine handling time and other variables associated with crab feeding. More clams were consumed in sand than in sand and shell or sand and gravel. Consumption rates were approximately 5, 2.5 and 1 clams crab^{-1} day^{-1}, respectively. These rates were much lower than those reported for similar studies by Arnold (1984).

Peterson (1990) examined selectivity of three sizes of blue crabs preying on six sizes of hard clams. The experimental units were 0.24-m^2 aquaria without substrate. Crab carapace width classes were <75, 75–125, and >125 mm, the same as utilized by Arnold (1983, 1984). Seed clams were graded into classes 5–10, 10–15, 15–20, 20–25, 25–30 and 30–35 mm

TABLE 11.8

Numbers of hard clam seed consumed daily by blue crabs, *Callinectes sapidus*, in various test substrates

Seed size (mm)	Sediment type				
	Bare	Sand	Sand/mud	Oyster shell	Fine gravel
5	885	650	450	270	180
10	220	170	145	105	90

Modified from Arnold (1983, 1984).

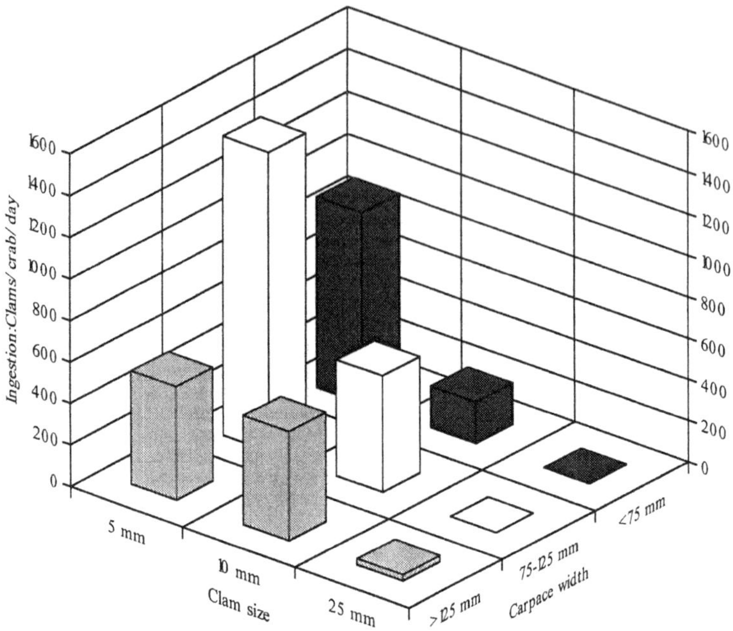

Fig. 11.14. Six-hour ingestion rate of three sizes (5, 10 and 25 mm) of hard clam *Mercenaria mercenaria*, seed by three size classes of blue crab, *Callinectes sapidus*. Data from Arnold (1984).

(Table 11.10). The smallest crabs consumed more clams in the smallest size category and only consumed clams up to 15 mm. Intermediate size crabs consumed all sizes of clams up to 20–25 mm, but ate proportionally more in the smallest two seed size classes. The largest crabs consumed all seed sizes, but highest consumption occurred in the 10–15, 15–20 and 20–25 mm seed. Consumption rates were 8–10 seed crab^{-1} in 8 h (24–30 crab^{-1} day^{-1}) for the large and intermediate size crabs at the preferred seed size. Smaller crabs consumed 33% of the smallest seed available at a rate of 3.3 seed crab^{-1} in 8 h (9.9 seed crab^{-1} day^{-1}). Additional experiments with substrate using similar seed and crab sizes found no differences in 2-h consumption tests when 6 cm of muddy substrate was compared with the no substrate. The largest crabs in this test (145–165 mm) had significantly lower consumption of small

TABLE 11.9

Numbers of hard clam seed consumed by blue crabs, *Callinectes sapidus*, of various sizes

Seed size(mm)		Clams consumed in 6 h			Clams consumed in 24 h		
	Carapace width (cm):	<7.5	7.5–12.5	>12.5	<7.5	7.5–12.5	>12.5
5		230	360	135	920	1440	545
10		50	140	130	200	560	520
25		0	0	7	0	0	30

Data are based numbers of clams consumed in 6 h (after Arnold, 1983, 1984), and those same data extrapolated to 24-h rates.

TABLE 11.10

Numbers of hard clam seed consumed in 8 h by blue crabs, *Callinectes sapidus*, of various sizes

Seed size (mm)		Clams consumed in 8 h			Clams consumed in 24 h		
	Carapace width (cm):	<7.5	7.5–12.5	>12.5	<7.5	7.5–12.5	>12.5
5–10		3.3	8.6	6.7	9.9	25.8	20.1
10–15		0.8	9.5	8.0	2.4	28.5	24.0
15–20			5.5	7.7		16.5	23.1
20–25			2.2	7.4		6.6	22.2
25–30				3.8			11.4
30–35				1.5			4.5
Total		4.1	25.8	35.1	12.4	77.4	105.3

After Peterson, 1990; and those same data extrapolated to 24-h rates.

seed (5–10 mm seed) than the next 3 largest seed size classes. Smaller crabs (118–135 mm) consumed these smaller seed at rates similar to the other sizes, but did not consume 27.5 mm or larger seed (Peterson, 1990).

Micheli (1995) examined the ability of blue crabs to select various size hard clams, and whether the selection process was ingrained or modified through experience. Individual male *C. sapidus* (124–158 mm) were placed in 0.1375-m^2 tanks containing 5 cm of fine sand as a base, and supplied with ambient running seawater. Water temperature ranged from 20°C for the first experiment to 15°C during the last. Ten hard clams in each of three size classes (15–20, 21–25 and 26–30 mm) were placed in each tank, and experiments were conducted for 24 h. Every 3 h, clams that had been consumed were replaced. Two sets of experiments were conducted, crabs starved 1 day and crabs starved 3 days. There was no statistically significant effect of hunger level on size selectivity, but there was a general trend for crabs starved for longer periods to consume more clams per unit time. Numbers of clams consumed decreased with an increase in clam size with nearly 8 clams (15–20 mm) being consumed per day by animals starved for 3 days vs about 0.4 clams (26–30 mm) consumed per day by crabs starved for 1 day.

Additional experiments were conducted to determine the basis for clam selection. Sham clams of two size classes (15–20 and 26–30 mm) were constructed by filling the cavity with sand and gluing the clam shut (Micheli, 1995). These were offered in combinations of live and sham, small and large clams. The crabs consumed the greatest percentage of large items when the sham clams were placed in combination with small live clams. Crabs crushed more sham clams of both sizes than live ones, and this was apparently because the force required to break the large sham clams was less than that for live large clams, but the force to break large sham clams was not different from that required for small live and small sham clams. Conditioning crabs to different prey sizes significantly affected their choice 24 h later. Thus the prey size, ease of breakage, and the experience of the individual crab all significantly affected predation rate.

Studies by Kraeuter and Castagna (1977, 1985a,b) utilized crab traps, in conjunction with a number of other predator control techniques, such as gravel beds and various mesh, to remove large blue crabs from the area of aquaculture plots planted with hard clam seed. These studies clearly indicate the importance of blue crabs as predators on hard clam seed, but also showed the importance of other predators, such as mud crabs in regulating overall survival of small clams. In virtually every case, interactive effects of protective devices were important components of the study, and combined predator protection of two or more devices conferred greater survival than would have been expected based on a simple summation of the protection of the individual devices. Many of these devices were designed to affect either blue crabs or mud crabs, and given the studies above these additive effects should not be surprising.

Micheli (1997) examined the effects of blue crab predation on hard clams tethered in various habitats in the shallow waters of North Carolina. Predation on the clams varied seasonally by habitat, and this was explained by an interaction between blue crabs and various bird predators. In fall predation on clams was greater in subtidal sand and just inside salt marshes than on intertidal sand flats. In summer predation on clams was the same in all three habitat types. Crabs consumed more clams when gulls and terns were excluded from sand flats than when similar conditions were tested in salt marshes. In the fall, *Larus argentatus* and *Laurus delawarensis* were more abundant than in summer, and in the latter period crab density in the intertidal sand flats was 1.5–3 times higher than in the fall. This increase in crab density was considered to be the cause of the seasonal habitat specific change in predation on hard clams. These changes resulted from modification to the crabs behavior not direct predation on the crabs by the gulls because no gulls were ever observed to feed on the crabs. This is certainly not always the case in the general area of these studies because Prescott (1990), also working near Beaufort, NC, reported that blue crabs were a significant part of the diet of gulls. In any case, the crabs shifted their distribution pattern to the safer habitats, and this resulted in fewer predator/prey interactions with clams in the intertidal sand areas in fall.

Summary Callinectes

Blue crabs are significant predators on most bivalve species. There is clearly a relationship between the size of the predator, size and species of the prey, prey and predator density and the type of substrate that dictates the rate of consumption. Many of the laboratory studies done with blue crabs were at densities much higher (4–90×) than reported field predator density. How this affects the interpretation of high and low density refuges in field situations is difficult to ascertain. Rates of consumption are clearly affected by the type of substrate, substrate depth, and the species involved, and seasonal specific habitat predation can be the result of predators, such as gulls, altering crab distribution. Consumption rates reported by Lipcius and Hines (1986) as satiation of 134–140 mm crabs feeding on 48–60 mm *Mya arenaria* at 2.75 clams crab^{-1} day^{-1} are consistent with the data of Blundon and Kennedy (1982b) who report 121–180 mm crabs feeding on 31–50 mm soft-shell clams consumed an average of up to 3.83 clams day^{-1}. Other studies by Blundon and Kennedy (1982a) indicate consumption rates of 9.7 *Mya arenaria* day^{-1}, 3 times greater than satiation reported by Lipcius and Hines (1986). These differences may be attributed to slightly different clam and/or crab sizes or experimental protocols. Given these data, and the presumed voraciousness of the blue crab, it is remarkable that Prescott (1990), based on studies in aquaria, did not find this crab was a significant predator on adult bay scallops. Even more remarkable is that the rates of consumption for hard clam seed are greater than those reported for any other species. Only data for consumption of oyster spat (Bisker and Castagna, 1987; Eggleston, 1990a) by blue crabs only begins to approach the rates reported for the hard clam seed. How hard clam populations are affected by other *Callinectes* sp. or other swimming crabs has apparently not been investigated. This could be locally important in waters south of Cape Hatteras. Field studies that evaluate the importance of blue crab predation on hard clam populations apparently have not been conducted so there is no way to determine if the predation rates or the sizes consumed determined by laboratory studies are typical.

Xanthidae

Panopeus spp., *Rithropanopeus harrissi* and *Eurypanopeus depressus*

There are a substantial number of studies on the ecology, natural history and taxonomy of the genus *Panopeus*, and a number of these report bivalves to be an important part of the crabs' diet. Williams (1983) revised the taxonomy of *Panopeus herbstii* s.l. and separated this taxon into six species. The ranges of these species overlap those of the hard clam, and studies done in the southeastern United States prior to the taxonomic revision may have included at least two of the species. Generally, studies north of Cape Hatteras, North Carolina can be considered to have been on *P. herbstii*, but any report of this species from the Gulf of Mexico could be attributed to *Panopeus simpsoni* (oyster beds) or *Panopeus obesus* (salt marshes). Similarly, studies on *Panopeus herbstii* s.s in the Carolinian Province could be attributed to either *P. herbstii* s.s. (oyster bars) or *Panopeus obesus* (salt marshes). Unless the author of a particular paper provides a taxonomic reference to one of these species, I have considered all of them to be *P. herbstii* s.l. and will rely on the reader to separate the species based on location. The other two genera are included in this section because they were included in a number of the *Panopeus* s.l. studies. In general, they appear to be less important predators of hard clams than *Panopeus* s.l.

Meyer (1994) reported that *Panopeus herbstii* and *Eurypanopeus depressus* abundances were highly correlated with shell cover and cluster volume of oyster shell in tidal marsh creeks. These analyses indicated that *P. herbstii* exploited the sediment and oyster shell strata as opposed to the shell cluster habitat utilized by *E. depressus*. Numbers per unit area varied with shell cover and season, but *P. herbstii* was considerably more abundant, with peak density exceeding 140 m^{-2} while peak *E. depressus* density was 28 m^{-2} (Meyer, 1994). Grant and McDonald (1979) examined the resistance of *E. depressus* to desiccation noted that its habitat preference (greater intertidal density in summer at 55 m^{-2}) did not correspond to its low desiccation resistance. This discrepancy was explained by its preference for microhabitats in the oyster reef (Grant and McDonald, 1979). Population densities of *P. herbstii* ranged from 30 to 103 individuals m^{-2} on oyster reefs in South Carolina (Dame and Vernberg, 1982) with smaller individuals predominating throughout the year. Dame and Patten (1981) suggested that, although the energy flow through this species may be small relative to the amount processed by a typical oyster reef, its predatory behavior may be one of the dominating influences in the structure of oyster reefs. All of these studies are supported by the data reported by McDonald (1982) that indicate that *Panopeus herbstii* and *Eurypanopeus depressus* have different life history and feeding strategies. The former is larger, faster growing, longer lived and feeds primarily on bivalves and barnacles while the latter matures earlier, produces more broods and has a more omnivorous diet feeding on detritus, algae scraped from the substrate, and other crustaceans (McDonald, 1982).

Lin (1990) examined the effects of mud crabs, *Panopeus herbstii*, on ribbed mussels, *Geukensia demissa*, in laboratory and marsh field experiments. Laboratory studies were conducted in 0.56-m^2 aquaria and 0.16-m^2 exclosures in the marsh. Even though the crabs fed only when they were submerged, tidal elevation had no effect on predation intensity. Juvenile mussels (<30 mm) attached to adult mussels had less mortality than those attached to oysters. In addition, juvenile mussels buried in the mud were consumed at rates similar to those that were not buried. When mussels reached a size of 50–60 mm height, predation rate dropped dramatically indicating a size refuge. Laboratory experiments with male *P. herbstii* (30–40 mm) and juvenile mussels (15–25 mm) attached to clumps of adult mussels (50–60 mm) were conducted at 23°C for 4 h. Crabs consumed 1.9 juvenile mussels in the 4-h period (11.4 mussels crab^{-1} day^{-1}). Crab density at the two sites studied by Lin (1990) was nearly an order of magnitude less than reported by Lee and Keib (1994) (see below), with an average of 4.6 and 1.6 crabs m^{-2}. Winter studies on the effects of marsh location on mussel survival (54 days) revealed a loss of 58 mussels with no difference between locations. If we assume equal numbers of mussels were eaten at each site, and the average crab densities from above, the rate of predation is between 0.34 and 0.12 mussels crab^{-1} day^{-1}. A separate study the following year revealed difference in predation rate between the sites and higher rates of predation, but daily consumption estimates cannot be made from the data provided (Lin, 1990).

Lee and Kneib (1994) emphasized the importance of utilizing structural elements found in typical field conditions when attempting to evaluate predation and predation rates. They examined the effects of *Panopeus herbstii* on the ribbed mussel *Geukensia demissa* when the latter were incorporated into an oyster reef matrix. Lee and Kneib (1994) introduced male *P. herbstii* into aquaria of 0.085 m^2 containing mussels (10 to 19 to 50 to 59 mm in 10-mm increments) that were attached to clumps of *Crassostrea virginica*. These clumps were combined so that 10 mussels were on the outside of the clump and 10 were on the interior. The position of the mussels within the clump greatly influenced the predation rate, with daily consumption rates of 0.6 ± 0.85 mussels day^{-1} for the prey on the interior of the clumps versus 2.3 ± 1.67 mussels day^{-1} for prey on the exterior of the clumps. More mussels of intermediate size (30–50 mm) were consumed than those larger or smaller size, and size selection was greater for exposed versus concealed prey (Table 11.11).

Mean density of *Panopeus herbstii* in the field was 68 ± 4.6 m^{-2} while mussel density was 212 ± 168 m^{-2} with mussels <15 mm in length comprising 69% of the population; mussels in the intermediate sizes were under represented in the field collections (Lee and Kneib, 1994).

TABLE 11.11

Size of ribbed mussels (*Geukensia demissa*) placed in experimental groups of oysters and consumed in 22 trials by the mud crabs *P. herbstii* (29.2–47.8 mm carapace width)

Size of mussel (mm range)	Number consumed (mussels day^{-1})	
	Exposed	Concealed
11–20	6	1
21–30	10	3
31–40	17	3
41–50	13	4
51–60	4	3

After Lee and Kneib, 1994. Concealed = mussels on the interior of the oyster clump. Exposed = mussels on the exterior of the oyster clump.

TABLE 11.12

Numbers of ribbed mussels (*Geukensia demissa*) of four sizes placed in trays and consumed by four sizes of the mud crab *P. herbstii*

Size of crab carapace width (mm)	Prey size (length, mm)	Number consumed (mean # mussels day^{-1})
8–12	5–10	2.00
19	10–20	5.86
24–25	10–20	5.14
35–38	10–20	13.71

After Seed, 1980.

These data differ somewhat from those of Seed (1980) who placed *Panopeus herbstii* (42–43 mm) in 0.08-m^2 aquaria and found that the crabs selected mussels 10–30 mm rather than those smaller or larger (Table 11.12). The reason for this shift in size selection was not apparent, but once the preferred sizes were consumed the crabs shifted to smaller and then to larger prey (Seed, 1980).

It is apparent from these relationships that there must be some size and/or spatial refuge for the mussels because, at the reported densities of crabs in the field, there were only enough mussels in the reefs to satisfy the crab population for less than 1 week.

McDermott (1960) reported on laboratory experiments in which *Panopeus herbstii* were maintained in 0.39-m^2 aquaria with 1–2-year-old oysters, *Crassostrea virginica*. In general, the crabs (22–35.5 mm) consumed smaller oysters (range offered 17–71 mm) in preference to larger ones. The predation rate varied from 0.15 to 0.7 oysters $crab^{-1}$ day^{-1} (McDermott, 1960). Additional studies were undertaken to estimate the rate of predation by *P. herbstii* and *Eurypanopeus depressus* on oyster spat attached to clam shells. *Panopeus herbstii* (23.4–32 mm) consumed 2 spat $crab^{-1}$ day^{-1} while *E. depressus* (16.4–22.9 mm) consumed between 0.6 and 1.6 oyster spat $crab^{-1}$ day^{-1} in laboratory studies and 3.9 spat $crab^{-1}$ day^{-1} in field studies.

Bisker and Castagna (1987) estimated population densities (13–103 crabs m^{-2}) on oyster reefs in Virginia, and examined the predation of *Panopeus herbstii* on individual oyster spat. Oysters set on shell were divided into 4 size classes (means 3.7, 8.1, 15.1 and 24.6 mm) and crabs were divided into 5 size classes (means 7.1, 13.0, 17.8, 25.2 and 34.4 mm). Fifty oyster spat were placed in each 0.05 m^2 container with one crab and experiments were conducted for 48 h (20–28°C, 24–35 ppt). Mud crabs in the smallest size category could not consume oysters larger than the 8.1 mm size class, and the larger the crab the more spat were consumed in 24 h (Fig. 11.15). Predation, when averaged for all sizes of crabs within a spat size, decreased with increasing spat size (Bisker and Castagna, 1987). In comparisons with the blue crab (see above) Bisker and Castagna (1987) found that within similar size (carapace width) categories, mud crabs could consume larger spat than blue crabs, and that the rate of predation was greater for mud crabs than for blue crabs. Mud crabs were found to have a predator ratio of 1.03 (length of prey/carapace width of predator) indicating that this species could consume oysters as long as its carapace width. In contrast, blue crabs could only consume oysters with predator ratios of 0.4–0.42. As with other studies, if field density of predators is coupled with laboratory-derived predation rates, it becomes obvious that the crabs are capable of consuming all of the spat in a very short time. Other factors must intercede to reduce these high predation rates.

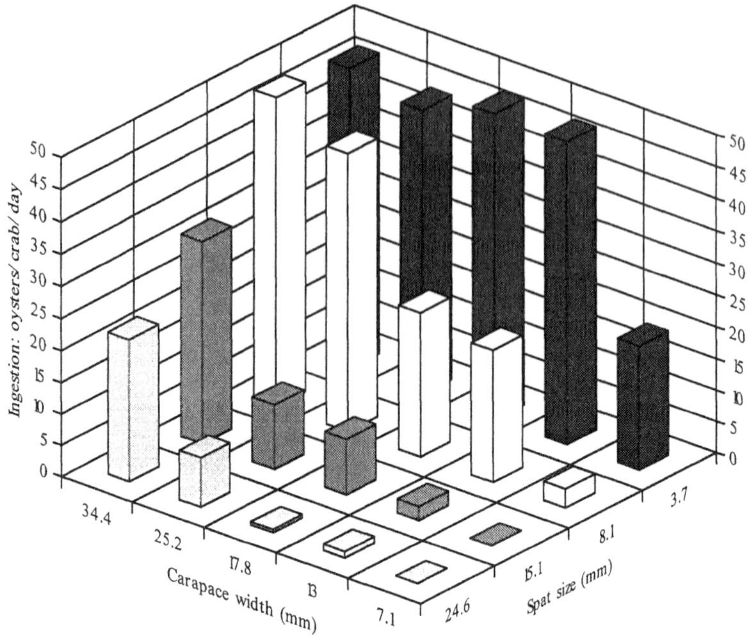

Fig. 11.15. Daily ingestion rate of 4 sizes of oyster, *Crassostrea virginica*, spat (3.7, 8.1, 15.1 and 24.6 mm) by 5 sizes of mud crab, *Panopeus herbstii*. Data from Bisker and Castagna (1987).

Sellmer (1967) reported that *Panopeus herbstii* consumed the gem clam, but did not provide evidence from stomach analyses or consumption rates.

Whetstone (1978) examined the gut contents of a number of crustacean species found in field trays of hard clam seed and reported that *Panopeus herbstii* was a dominant predator on these seed. Laboratory experiments were conducted with 10.5 and 20 mm hard clam seed and four *P. herbstii* (one <15, two 20.0–29.9 and one >30 mm) placed in trays at 10, 17 and 24°C (Whetstone, 1978; Whetstone and Eversole, 1981). Because of high variance in the data, no temperature related differences were found between the feeding rates of the crabs on the 20 mm seed (Table 11.13). Similarly there were no differences in predation losses

TABLE 11.13

Predation of *Panopeus herbstii* on four sizes of hard clam juveniles at three temperatures

Seed size (mm)	Temperature (°C)	Mean mortality	95% Confidence limits	Clams crab^{-1} day^{-1}
10	10	4.790	2.4–7.2	0.30
	17	34.57	15.2–54.0	2.16
	24	58.51	47.5–69.5	3.65
20	10	3.72	1.3–6.1	0.23
	17	8.51	3.9–13.1	0.53
	24	7.45	4.2–10.7	0.47

After Whetstone, 1978.

between the smaller seed at 10ºC and any of the larger seed at any temperature. Smaller seed at 17 and 24ºC experienced significantly greater predation mortality than any of the other tested combinations (Whetstone, 1978). The rate of loss ranged from 0.23 clams crab^{-1} day^{-1} for the 20 mm seed at 10ºC to 3.65 clams crab^{-1} day^{-1} for 10 mm seed at 24ºC (Whetstone, 1978). Experiments that related clam seed size and crab carapace width indicted that *Panopeus herbstii* could consume hard clams with lengths of 65% of the crabs' carapace width. This is considerably different from the predator ratio of 1.03 (length of prey/carapace width of predator) for the same species (see Bisker and Castagna, 1987) preying on oysters. The different shape and shell thickness for a given length of hard clams relative to oysters could explain the difference in the relationships.

Field studies (Eldridge et al., 1976, 1979; Whetstone and Eversole, 1978; Eldridge and Eversole, 1982) examined the relative importance of predators on clam seed placed in trays with natural sediment. Trays containing 13 mm seed clams were placed in the intertidal (10 trays) and subtidal (10 trays) areas of Clark Sound, South Carolina at three densities (290, 869 and 1159 clams m^{-2}, or 200, 600 and 800 clams per tray). The design included lining the trays with mesh to retain sediments and covering the trays with protective mesh. The trays were sampled monthly to determine the presence of various predators. Clams that had been consumed were replaced. *Panopeus herbstii* was the most abundant organism found in the trays (74% of all species and 84% of all crustaceans). Stomach analysis revealed that pieces of clam shells were more frequent in larger crabs, and that larger *P. herbstii* were more abundant in late summer (August/September) than at other times of the year. Other species that were found with clam shell in their stomach were *Callinectes sapidus*, *Menippe mercenaria*, and *Eurypanopeus depressus*, and in general, larger crabs of these taxa also had a higher percentage of clam shell in their stomach. As the clams grew the percentage of *Panopeus herbstii* containing clam shell dropped and by the second summer of the study only large (>25 mm carapace width) crabs had clams in their stomachs. When clams had grown to an average size of 15.5 mm the numbers of *P. herbstii* containing shell bits dropped by 50% and when clams reached 38 mm the percentage was <5% (Whetstone and Eversole, 1978). Losses of clams were also less in the winter than in the warmer months when mud crabs were more active, but this simple temperature/activity relationship was compounded by a decrease in the average size of crabs in the fall and winter. Even with the decrease in average size, absolute numbers of large crabs appeared to remain nearly the same throughout the year (Whetstone, 1978), although there were large fluctuations in the numbers of large crabs recovered from the trays. More mud crabs (60% of the total collected) were found in subtidal trays, but on 4 of the 14 collection dates the numbers collected were greater in the intertidal trays (Whetstone, 1978). Because clams that had been consumed in these experiments were replaced with like size clams during each interval, an estimate of the effect of constant density was obtained (Eldridge et al., 1979). Survival of 16–17-mm seed was nearly 95% and reached 99% when average size reached 21–22 mm. Most of the mortality (95%) occurred in the first 4 months when the clams grew from 13 to 16–17 mm. Clams at higher densities grew more slowly than those at lower densities. These results appear to be equivalent to the studies on *Paphia ventricosa* (Haddon et al., 1987), which suggest a high density refuge as well as the low density refuges, indicated by experiments of Lipcius and Hines (1986). Survival was lower in the intertidal zone, but growth did not vary with location.

Dyspanopeus sayi (= Neopanope texana, = Neopanope sayi)

Numbers of *Dyspanopeus sayi* in field populations are seldom reported, but it is considered to be one of the most numerous crabs in waters north of Delaware Bay, and is among the most numerous in waters south of Delaware Bay. WAPORA (1982) reported that highest densities of mud crabs (apparently the genera *Dyspanopeus* and *Panopeus* combined) in Great South Bay, NY were found on shell and gravel bottoms, and could reach densities of 102 m^{-2}. In areas with predominately muddy sediments these crabs were nearly absent (WAPORA, 1982), and bay-wide densities averaged about 4.4 m^{-2}. MacKenzie (1977b, 1981) reported that densities of mud crabs on shellfish beds in Long Island Sound, Connecticut ranged from 7.5 to 53.8 m^{-2}. Flagg and Malouf (1983) reported *D. sayi* at densities of 0.1–1.7 m^{-2} in Napeague Harbor, New York, but trays placed in the field containing coarse gravel and clam seed collected 306 mud crabs m^{-2}. Strieb et al. (1995) reported mud crab densities up to 225 crabs m^{-2} in the eelgrass beds of Long Island Sound. Typical densities of older crabs ranged from 7 to 36 crabs m^{-2}, while in unvegetated habitats the average density was 0.5 m^{-2}.

D. sayi has been recognized as a predator on oysters (McDermott and Fowler, 1951), bay scallops (Pohle et al., 1991; Strieb et al., 1995), gem clams (Sellmer, 1967) and hard clams (Landers, 1954, Landers, 1955).

Strieb et al. (1995) reported that *D. sayi* can consume bay scallops at a size that approximates the crabs' carapace width, and that a size refuge for the scallops exists above about 28 mm shell height. This size is approximately the size at which scallops drop from the eelgrass canopy and begin their benthic existence.

Landers (1954, 1955) conducted a number of laboratory experiments on *Dyspanopeus sayi* in which 50 hard clams (10 mm average length) were placed in tanks (no size given) supplied with running seawater and sand. Five *Dyspanopeus* were added to the tank, and 30 days later 33 hard clams had been consumed (0.22 clams $crab^{-1}$ day^{-1}). In a second series of tests, 0-year class clams (37 individuals 1–7 mm) and yearling clams (25 individuals, 15–20 mm) were placed in separate sand bottom tanks. Three *D. sayi* (15–20 mm) were placed in each tank and allowed to feed for 18 days. All 0-year class clams (0.69 clams $crab^{-1}$ day^{-1}), but only 4 larger clams (0.074 clams $crab^{-1}$ day^{-1}) were consumed. All the smaller clams had been crushed while all the larger clams had been opened by chipping around the edges. Clam seed (11 individuals 1–9 mm and 11 individuals 14–33 mm) were also placed in separate tanks with no sediment (20–23°C) and two crabs. Within 24 h, all the smaller seed had been consumed (5.5 clams $crab^{-1}$ day^{-1}), but none of the larger seed. Water temperatures were then lowered to 11–12.5°C and 30 clam seed were added to the tank. Clams eaten during the preceding 24 h were enumerated. During the first 24 h, 11 clams were consumed, then 8 clams. After 5 days, all 30 seed had been crushed (3 clams $crab^{-1}$ day^{-1}). Seed clams (1–9 mm) planted in a mesh covered box in Wickford Harbor, RI, and allowed to remain for 1 year yielded only 12.9% survival. Identifiable remains indicated that 0.65% had been drilled and 1.95% exhibited crab damage. The cause of the losses of the remaining 1690 clams (84.5%) could not be determined, but since crabs typically crush small clams, it is plausible that the majority of these losses were the result of crab predators (Landers, 1954).

MacKenzie (1977b) found that *D. sayi* consumed up to 14 hard clams $crab^{-1}$ h^{-1} when 5 mm clams were exposed in a bare tray for 8 h, but when trays were filled with sand and the test was conducted for 14 days the rate dropped to 1.6 clams $crab^{-1}$ day^{-1}.

Gibbons (1984) placed individual *D. sayi* (19.6–21.3 mm) in 0.03 m^2 culture bowls and

TABLE 11.14

Numbers of three sizes of hard clam ingested by individual *Dyspanopeus sayi* per day at four temperatures

Temperature (°C)	Clam size (mm)		
	3	5	7
10	44.7	23.1	10.5
15	56	32.1	14.9
20	71.2	38.6	19
25	114.6	49.6	21.5

After Gibbons, 1984.

observed the mortality of hard clam seed (3, 5 and 7 mm) placed in trays containing sand, crushed aggregate and no substrate. Temperature approximated the normal cycle in Long Island Sound, NY, and, as expected, exerted a significant effect on predation. Contrary to studies by Day (1987) (see below), Gibbons (1984) found that gravel inhibited predation at most temperatures. Highest predation rates were reported in bare trays (136 clams crab^{-1} day^{-1}) at 18.5°C, and feeding ceased when water temperatures declined to 2–3°C (Gibbons, 1984). Smaller clams were eaten at greater rates than larger clams at all temperatures (Table 11.14, Fig. 11.16). Additional studies that were designed to examine the bioenergetics

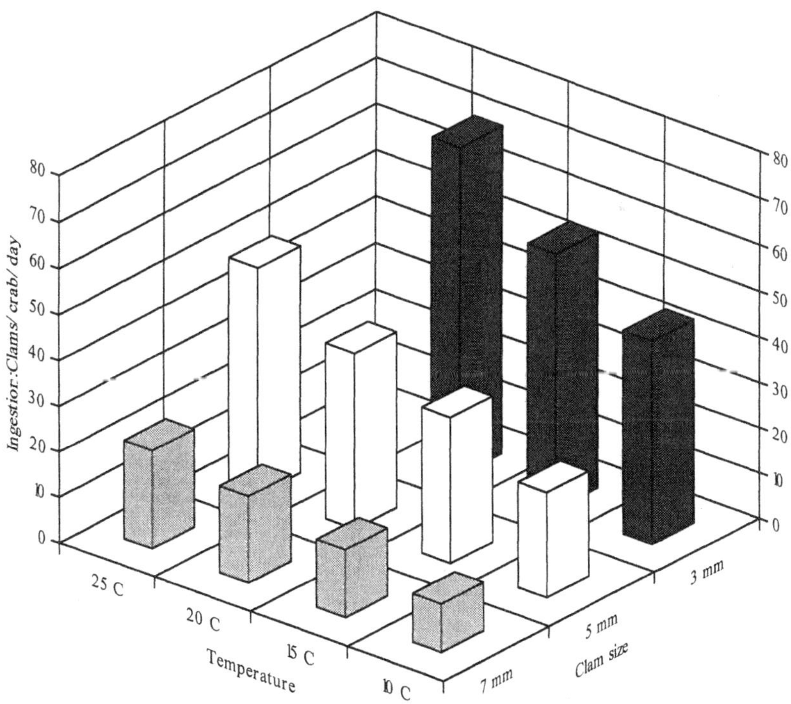

Fig. 11.16. Daily ingestion rate of hard clam, *Mercenaria mercenaria*, seed of three size classes (3, 5 and 7 mm) by the mud crab, *Dyspanopeus (= Neopanope) sayi* at 4 temperatures. Data from Gibbons (1984).

of the crab revealed predation rates as high as 115 clams crab^{-1} day^{-1} at 25°C (Gibbons, 1984). In these studies, individual crabs (19.6–21.3 mm) were placed in containers and fed clam seed (3, 5 and 7 mm) ad libitum. The rapid drop in consumption rates with increased seed size is evident from these data.

Day (1987) conducted laboratory studies on *D. sayi* to determine the effectiveness of gravel as protection for juvenile hard clams. All studies were conducted for 22 h at 28–33 ppt salinity and 22–25°C in 0.025-m^2 buckets or 0.52-m^2 aquaria. Crabs ranged from 22 to 23.5 mm in carapace width and clam seed ranged from 8.5 to 11 mm. Predation was greater in substrates with gravel, and clams had a more difficult time establishing themselves in these substrates. Predation rates ranged from 2.25 clams crab^{-1} day^{-1} on sand substrate to 11.2 clams crab^{-1} day^{-1} on small gravel. Predation rates on large gravel at 7.1 clams crab^{-1} day^{-1} and sand + small gravel at 7.9 clams crab^{-1} day^{-1} were intermediate between sand and small gravel (Day, 1987).

Day (1987) reported that when toad fish, *Opsanus tau*, were placed in 0.52-m^2 tanks containing clam seed, sand and *D. sayi* or clam seed, or small gravel and *D. sayi*, more clams survived in tanks containing fish. Only one crab was eaten during these experiments, but the crabs changed behavior by spending more of their time buried in the substrate after the fish were introduced. Predation rates were 0.1–1.0 clams crab^{-1} day^{-1} on sand and small gravel, respectively, with fish present, and 4.9 and 11 clams crab^{-1} day^{-1} on sand and gravel, respectively, without fish. These data support the experiments of Gibbons and Castagna (1985) who reported that crab predation on clam seed in field plots containing gravel was significantly reduced when toad fish were introduced into the cages.

Summary mud crabs (Xanthidae)

Studies have indicated that mud crabs can be a significant predator on a number of bivalves, but few studies have examined field populations. The few field surveys indicate that the abundance of these crabs can be very high when compared to more aggressive species, such as the blue crab. Evidence on the importance of juvenile stages of *P. herbstii* and other mud crab species on juvenile bivalve populations is lacking. Flagg and Malouf (1983) noted that *D. sayi* abundance increased as gravel grain size increased. This observation prompted Day and Lawton (1988) to undertake substrate preference investigations of three species of mud crabs (*Dyspanopeus sayi*, *Panopeus herbstii*, and *Eurypanopeus depressus*) on mud, sand, small gravel, large gravel and broken oyster shell. All substrates were provided in binary combinations. Substrate selection and activity of the crabs were recorded at intervals to examine activity periodicity. In general, all species preferred shell over other habitats and sand was the least preferred.

These authors emphasized the importance of testing each species over a complete day/night cycle. Nocturnal activity altered the choices of both *D. sayi* and *E. depressus*. These substrate selection studies were then used as a basis to examine the effectiveness of utilizing crushed stone or shell as protection for juvenile hard clams in aquaculture (Day, 1987), and more clams were consumed when gravel was present. This observation led Day (1987) to conclude, that when gravel is present, mud crab predation on hard clam seed will increase. These results do not match field observations from New England through Florida that prove that when shell or other large particles were mixed with sandy mud or mud substrates, more hard clams were present or that crab predation was reduced (Pratt, 1953; Stickney and Stringer, 1957;

Wells, 1957; Saila et al., 1967; Parker, 1975; Kraeuter and Castagna, 1977, 1985a,b; Arnold, 1983, 1984; Beal, 1983a; Gibbons, 1984; Walker and Tenore, 1984; Craig and Bright, 1986; Sponaugle and Lawton, 1990; Peterson et al., 1995).

Beal (1983a) found that in North Carolina, areas of shell had higher natural clam densities than any other area including nearby grass beds, and that recruitment was highest in shell areas. In addition, when gravel was used to cover mud substrates and when mesh screen was applied, survival of clam seed increased (Kraeuter and Castagna, 1989). Day (1987) concluded that gravel mixed with sand substrate may offer protection against large predators, but would be ineffective against *D. sayi*, an observation that seems to explain the results of Flagg and Malouf (1983) that gravel, in particular gravel larger than 25 mm did not protect small clam seed. The importance of gravel size was also noted by Arnold (1984) who reported that blue crab predation was less in substrates of fine gravel than in coarse gravel.

Day (1987) did note that clams were not able to burrow as easily in the gravel substrates. This inability to burrow may have made the clams more vulnerable to crab predation. In part, the conflicting results may be due to the fact that many laboratory studies appear to simply add gravel, shell or other coarse material on top of an existing substrate. This does not replicate field conditions where fine sediments and organic matter often are quickly mixed into the sedimentary fabric, filling the voids typically occupied by small crabs. The presence or absence of significant quantities of suspended sediments may also explain why gravel or shell does not always work when added to coarse substrates. In her discussion, Day (1987) made the assumption that mud crabs were not present in the area seeded by Kraeuter and Castagna (1977) or other areas where shell has worked (Arnold, 1983, 1984). Mud crabs including *D. sayi* are abundant in the shallow waters of coastal Virginia seaside bays, and were considered to be a prime source of predation by Castagna and Kraeuter (1977), although their abundance was not quantified. It seems likely that the studies by Day (1987) may explain the short-term losses often experienced when clams are initially seeded in aquaculture plots without proper protection. It is clear from the differences between the laboratory studies, distributional observations, and the irregular results of clam seeding in aquaculture that we are far from understanding the role of crab predation on hard clams in the field, and it is equally clear that this species and other xanthids are important predators on small seed.

Menippe mercenaria

Prescott (1990) examined the rate of bay scallop predation by *Menippe mercenaria* by placing one scallop (50–80 mm) with each predator (60–90 mm) in a 10 gallon aquaria supplied with running seawater (18°C). Experiments lasted 1 month and 19 scallops (test minus control) were consumed by 3 crabs for a rate of 0.21 scallops crab^{-1} day^{-1}.

Menzel and Hopkins (1955) and Menzel and Nichy (1958) reported that the stone crab, *Menippe mercenaria*, was capable of consuming large adult oysters. Samples on oyster beds indicated that there were about 0.8 stone crabs m^{-2}, and these could consume about 219 oysters crab^{-1} year^{-1} (0.6 oyster crab^{-1} day^{-1}) (Menzel and Hopkins, 1955). Subsequent studies, conducted with an 80 mm crab caged with 50–60 mm oysters for July and August, found that 237 oysters were consumed. The maximum rate of consumption was 60 week^{-1} (8.6 oysters crab^{-1} day^{-1}) with the average daily rate over the same period of 3.7 oysters crab^{-1} day^{-1}.

Whetstone and Eversole (1978) reported finding clam shell in the stomachs of *Menippe mercenaria* in South Carolina, but it represented only 8 specimens (0.41%) of the 1956 individuals of all potentially predatory species sampled. This species, as other xanthids, is capable of consuming hard clams, but no direct information exists to indicate if it is a serious predator in the field. Any predatory activities by this species on hard clams would decline with northward latitude along the eastern seaboard of the US. Stone crabs are only occasionally reported North of Cape Hatteras, NC.

Hemigrapsus spp.

Peterson (1982b) found that while *Hemigrapsus oregonensis* was a significant source of post mortem movement of *Protothaca staminea* shells, the crab did not appear to prey on the live organisms. The crabs would carry the dead clams to the top of cages that had been constructed to contain various concentrations of the clams, and consume the flesh without leaving any marks on the shell. Chew (1989) indicated that *Hemigrapsus* spp. were considered to be a minor predator on intertidal Manila clam beds in Washington. The recent arrival of *Hemigrapsus sanguineus* (McDermott, 1998) and its range expansion throughout the mid-Atlantic will bring this invader into contact with the hard clam. Whether this intertidal rocky substrate species will have significant interactions with hard clam populations is unknown, but the intertidal nature of this invading species and its preference for rocky shores should limit potential interactions.

Uca spp.

Fiddler crabs, *Uca* spp. normally feed only in the high intertidal zone of areas near marsh grasses. Because the habitats of fiddler crabs only partly overlap that of most commercially valuable clam species, little information exists on the potential for these crabs to consume clams. Turner et al. (1948b) reported observing fiddler crabs consuming small *Mya arenaria* on intertidal flats in Massachusetts.

Summary Reptantia

There are a significant number of studies indicating that crabs select individual bivalves much smaller than the maximum size they could open (Griffiths and Seiderer, 1980; Sanchez-Salizar et al., 1987a; Seed, 1990). Juanes (1992) examined a large number of published studies on crustacean predators and noted that smaller prey were preferred over larger prey across many predator and prey species combinations. Juanes (1992) hypothesized that this pattern may be due to an increasing probability of damage to the predator as it attacks the hard shelled prey. This damage may be mechanical, such as chipping of a claw, or damage due to an attack by another predator while the prey is being handled. Whether these considerations are significant from the point of view of the prey remain to be investigated. At least one study (Smith and Palmer, 1994) has shown that some species of crabs can change their claw size and strength between molts in direct response to prey with harder shells. In general, studies suggest that if the prey is immobile it is better to be large or thick shelled or to have some form of spatial refuge from predation. Many clam species respond to the presence of predators by retracting their siphons, retracting into pre-established burrows and by digging deeper into the substrate (Hughes, 1970; Blundon and Kennedy, 1982a; Elmgren et al., 1986, Haddon et al., 1987), but Nakaoka (1996) was unable to confirm such protection for *Yoldia notabilis* when it was subject to predation by *Paradorippe granualta*. In general, size was the most important variable affecting the outcome of this predator–prey interaction (Nakaoka, 1996).

Some studies have shown that the crabs may be maximizing energy intake (Elner and Hughes, 1978), and/or minimizing handling time (Hughes and Seed, 1981; ap Rheinallt and Hughes, 1985), thus anything the bivalve can do to affect these two factors should reduce predation. Studies based on optimal foraging theory appear to predict outcomes for some predator/prey combinations, but not for others (Ebersole and Kennedy, 1994, 1995). Kneib (1995) provides a review of crab predation that emphasizes similar concerns and emphasizes the complexity associated with a fundamental assessment of foraging in decapods. Studies which simply present food to decapods do not indicate how intense predation will be in nature. This may be due to the numerous other factors, in addition to the size of the crab and size of the prey, such as the crab's sex, chela morphology, prey species, prey shell structure, presence of alternate prey, substrate type, presence of seagrasses or coarse hard particles on the bottom, depth of the prey in the substrate, presence or absence of other predators, spatial and temporal scales of both predator and prey etc. All these add variability to predator choice or predation rates, and these effects seem to be particularly important in defining predation by highly motile, highly tactile predators, such as crabs, when compared to other forms of predation. A recent evaluation of prey selection (Mascaro and Seed, 2000a,b) found that a particular size green crab selected bivalves based on the minimum shell dimension. This dimension was related to the maximum cross section of the chela. This relationship was generally related to profitability and handling time, but the shape of the bivalve greatly influenced the selection process.

There is a significant amount of evidence indicating that cannibalism may be an important source of mortality in many crab species (Gotshall, 1977; Botsford and Wickham, 1978; Hines et al., 1990; Fitz and Wiegert, 1991a; Fernandez et al., 1993). This evidence suggests that crab species and populations may be self-regulatory and interference among and between crab species and individuals may be a significant factor in their predation rates. Recently, Higgins et al. (1997) have shown that such density-dependent behavior as cannibalism is important in regulating the populations of the Dungeness crab in the Pacific northwest. This study indicated these biological processes, when coupled with even small environmental effects on crab populations, tended to increase population fluctuations. How such effects would translate into the effects on prey populations would add another layer of stochastic processes and complexity.

In direct contrast to many other studies, Raffaelli et al. (1989) were unable to show large effects on the density of benthic fauna from either *Carcinus maenas* or *Crangon crangon*. These authors attributed this lack of effect to the lower density of predators used in their studies. Raffaelli et al. (1989) used cages with predator densities of 30 and 140 m^{-2} or roughly what was found from field sampling. In contrast, other studies have utilized *Carcinus* and *Crangon* at higher densities. Whether the results achieved by the various authors can be ascribed to differences in density, the way in which the prey response was measured or other methodological artifacts (see Micheli, 1996, for an experimental example), all data based on high densities of predators and/or prey in enclosures should be extrapolated with considerable caution.

A significant number of studies on crabs initiate predation after the predator has been starved for 12–48 h. Haddon and Wear (1987) were unable to find significant differences between the appetite of crabs that were starved for 24 h and those that were freshly caught. In addition, individual appetite was highly variable and starvation, even for 21 days, did not allow standardization of the feeding studies. They were able to link the upper limit of food intake with the foregut volume, and that volume with carapace width. This suggests a mechanism by which predation by various crab species could be compared and a means of investigating the commonly described relationship between the prey size consumed and carapace width. This relationship would not explain the disparities between consumption rates and satiation levels that have been reported. With all these variables, it has been difficult to make general conclusions that will predict predation rates based simply on density of predator and prey. Given this difficulty, it is somewhat surprising that the studies of Abolmasova (1970), which showed an exponential relationship between daily food intake and body weight of 4 species of crabs, have not been more widely evaluated. This study reported that green crabs, *Carcinus maenas*, (51–60 mm) consumed 1.42% wet weight food/wet weight crab daily, while smaller crabs (31–40 mm) consumed 3.02% of their body weight daily. Generally, with all species studied, the smaller crabs consumed higher percentages of their body weights daily. Similar relationships are used to allocate food to fish in aquaculture production, and it has proven to be one of the more effective means of food and growth management. Typically smaller fish receive a higher percentage of their body weight as food per day than adult forms. For most species of fish, the typical adult ration approximates 3% body weight per day (wet fish weight to dry food weight).

Based on the rates of consumption, there is little doubt that crabs are among the most significant predators of post set bivalves of all types. Few studies have compared rates of predation by different phyla in field situations. Irlandi and Peterson (1991) described the relative predation ascribable to crabs and whelks on hard clams (16–52 mm) placed in eelgrass beds and on sand flats in North Carolina (Table 11.15). Survival in these experimental plots was 73% (grass), 20% (sand), and 86% (grass), 29% (sand) for experiments 1 and 2, respectively (Table 11.15). Crabs were probably unable to damage the largest clams in this experiment so the size of the prey may have skewed the observed relative effects of predation toward emphasizing the snails. If greater numbers of small clams had been included, the balance undoubtedly would have been skewed toward greater effects of crab predation in both substrate types. Conversely, a population of adult hard clams would be more sensitive to whelk predation relative to incursions by crabs. In addition to predation, Irlandi and Peterson (1991) reported that siphons of clams on the sand sediments were significantly smaller than those from clams collected in grass beds, indicating higher predation rates on sandy areas relative to grass beds, and a similar result was obtained by Coen and Heck (1991). These sublethal effects are covered by Peterson (Chapter 10).

TABLE 11.15

Predation on hard clams (16–52 mm) in two habitats due to crabs and whelks in a comparative study conducted in North Carolina

	Seagrass		Sand	
	Crab	Whelk	Crab	Whelk
Experiment 1	61	38	77	22
Experiment 2	56	44	84	15

After Irlandi and Peterson, 1991. Data are presented as % of total predation observed by collection of shells.

While comparisons of predation may offer information on the relative intensity of two or more predators to consume a prey species, other studies have shown that some predators may provide the opportunity for smaller predators to prey on species or sizes of prey that would not otherwise be available. Large crustaceans have been shown to be major disturbance agents that can allow other crustaceans or fish to take advantage of prey resources. Auster and Crockett (1984) described the foraging of several crustaceans associated with the development of lateral burrows. These crustaceans (*Cancer borealis*, *Carcinus maenas* and *Pagurus longicarpus*) were observed to dig vertical pits in the sediment and then extend these laterally to obtain prey. The activity attracted other predators such as *Crangon septemspinosus*, *Pleuronectes americanus*, *Pagurus pollicaris*, *Panopeus* spp., and *Fundulus* spp. These other predators took advantage of the disturbed sediment to forage for prey the digging predator missed. Many of these predators were observed to feed on small bivalves.

Peterson (Chapter 10) has reviewed the interactive effects of various environmental variables, such as substrate, vegetative cover, and density on hard clam growth. For many years, it has been observed that more clams are found in areas with significant amounts of shell in the bottom (Pratt and Campbell, 1956). This was noted and modified by Castagna and Kraeuter (1977, 1981) and Kraeuter and Castagna (1977, 1985a,b) who were able to increase survivorship of high densities of small seed planted in gravel substrates coupled with other protective devices. Similarly, Peterson et al. (1995) have recently demonstrated the importance of the interaction between seed size and substrate type in the survival of low density plantings of hard clams in North Carolina.

In contrast to the low density refuge examined by Peterson et al. (1995), high density can also be a predator refuge for hard clams (Eldridge et al., 1976, 1979), but Boulding and Hay (1984) reported that high density patches of *Protothaca staminea* experienced greater predation than low density. Haddon et al. (1987) reported a negative density dependence for toheroa being preyed upon by *Ovalipes cantharus*. This negative density-dependent influence is similar to the data of Lipcius and Hines (1986) for blue crabs feeding on *Mya arenaria*. Malinowski (1985) reported that density and size refuges may differ greatly from location to location depending on the guild of predators present. Beal (1983a) reported the presence of a size refuge from chipping predators for hard clams in all habitats, and muddy sites were a spatial refuge from rasping predators. When all data for all sites were compared, clams placed at 8 times natural density (80 clams m^2) had higher survival than those at natural density (Beal, 1983a). It is clear that there are numerous ways in which hard clams are able to survive in the presence of intense predation, and that our ability to envision the interactions of the crustaceans with the environmental parameters and prey responses will require building models, but once the models are completed they will require rigorous and imaginative field testing.

Finally, it is clear from the rates of predation, that crustaceans are among the most voracious predators on juvenile bivalves in general and hard clams in particular (Table 11.16). Robles (1987) demonstrated that the lack of a population of large mussels in the high intertidal zone of a protected shoreline in California was due to predation by spiny lobsters, *Panulirus interruptus*. Due to its nocturnal habits this lobster was not normally seen on the site, but night diving, and cage studies found that they were relatively abundant at night (night maximum mean density 0.08 m^{-2}) and foraged preferentially on larger mussels (Robles, 1987). These data, suggest that large crustaceans such as lobsters can be a significant factor in influencing

TABLE 11.16

Summary of feeding rates of arthropod predators on a variety of bivalves

Predator	Size eaten	Consumption (predator^{-1} day^{-1})		Predator density
		Clams	Other species	
Limulus polyphemus	2–46 mm	30–1200		0.5 m^{-2}
Pontoporeia affinis	300–375 μm		0.02	1400 adults 80,000 juveniles m^{-2}
Shrimp	<0.5–15 mm	0.7 to >50	>1000	20–100 m^{-2}
Hermit crabs	to 3 mm	50–240		
Cancer spp.	to 70 mm	1.7–100	0.6–8	10 adults m^{-2}, juveniles 300 m^{-2}
Carcinus maenas	4–60 mm	14–4000	32–5	125 adults m^{-2}, juveniles 2000 m^{-2}
Ovalipes spp.	3–65 mm	17–586	4–20	2 m^{-2}
Callinectes sapidus	2–55 mm	0.3–1440	>50	0.1 m^{-2}
Xanthids	3–60 mm	0.5–136	17	140 adults m^{-2}, juveniles 225 m^{-2}

Data are combined from references cited in the text.

bivalve distributions, even when prey appear to be scarce. In general, most shrimp and hermit crabs appear to be potentially more important predators on hard clams <1 mm in size, while crabs can cause significant seed mortality until hard clams reach 20–25 mm shell length.

11.9 ECHINODERMATA

Asteroida

Although starfish have been reported to be predators of hard clams, there do not appear to be any data on the effects of asteroids or asteroid larvae on clam larvae. Belding (1912) indicated that *Asterias forbesi* could be a predator on hard clams, but noted that, because their habitats did not usually overlap, under most conditions the starfish would not consume many clams. This lack of overlap becomes more evident south of Long Island. Young starfish were reported to be an important predator on newly set oysters (Loosanoff, 1961; MacKenzie, 1970b). Carriker (1961) mentions the possibility of starfish predation on newly set hard clams, but notes that no data were available. There do not appear to be any reports of starfish consuming newly set hard clams in the field.

Lavoie (1956) and Burnett (1960) described how starfish open clams and the latter author reported the effects of *Asterias forbesi* on *Mercenaria mercenaria*. In general, once a clam was removed from the substrate the asteroid pried open the shell with pressure from its tubed feet, and once a slight gap existed between the valves, the starfish inserted its stomach (Burnett, 1960). The release of enzymes digested the clam meat, and the resulting material was ingested. Feder (1955) reported that the starfish *Pisaster ochraceus* only required an opening of 0.2 mm to insert its stomach and begin digestion of bivalve molluscs. Mauzey et al. (1968) described a unique circumstance in which the starfish *Orthasterias koehleri* used its tubed feet to pull chips of shell off the bivalve *Humilaria kennerleyi* until a small opening is made in the valves and the bivalve can be digested. In contrast, many *Astropecten* spp. ingest the entire prey organism shell and all, and evidence presented by Ribi and Jost (1978) suggests that *Astropecten* spp. may surround and force their prey open by depriving the latter of oxygen. This may explain the low food intake (0.33–1.3% of dry tissue weight day^{-1}) of the Astropectinidae (Ribi and Jost, 1978). The highest consumption rates of the Astropectinidae are about 50% that of *Asterias amurensis* (2.7% of dry tissue weight day^{-1}, Hatanaka and Kosaka, 1959) and *Pisaster ochraceus* (3.2% of dry tissue weight day^{-1}, Feder, 1956) based on a similar relationship between dry tissue weight of prey to dry tissue weight of predator. The latter species open their prey in much the same way as *A. forbesi*.

A number of starfish genera are known to excavate prey from sediments (Belding, 1912; Lavoie, 1956; Burnett, 1960; Anger et al., 1977; Doering, 1981a,b, 1982, *Asterias*; Mauzey et al., 1968, *Luidia, Pisaster, Evasterias, Orthasteria* and *Pycnopodia*; Sloan and Robinson, 1983, *Luidia, Mediaster, Pisaster* and *Pycnopodia*; Smith, 1961; Toba et al., 1992, *Pisaster*; Quayle and Bourne, 1972, *Pycnopodia* and *Evasterias*; Ribi et al., 1977; Ribi and Jost, 1978, *Astropecten*). Most of these species are not found within the range of the hard clam, but illustrate the potential of asteroids to consume even large bivalves. Sloan and Robinson (1983) are the only authors to suggest that a starfish, in this case *Mediaster aequalis*, may graze on newly settled bivalve larvae.

Anger et al. (1977) found densities of *Asterias rubens* from 2 to 31 m^{-2} on fine sediments and that young starfish fed on different prey than adults. The small starfish on fine sediments preferred to consume the snail *Hydrobia ulvae*, while the larger individuals fed on the clam *Macoma balthica*. This species of starfish exhibited a diurnal feeding pattern with activity being less during the night. Larger *A. rubens* selected and ate larger *M. balthica*. A starfish 7 cm in diameter consumed an average of 0.65 *Macoma balthica* (13.5 mm average) day^{-1} for the 51-day test period. In situ studies indicated that the amount of time required to eat small clams was almost independent of the size of the predator, but increasingly positive correlations were found between feeding time, larger prey and starfish size. These data are supported by data from Paine (1976) who examined the interactions of the starfish *Pisaster ochraceus* and its prey species on the rocky intertidal of Washington, USA. In general, larger starfish consumed large prey and avoided small prey. Starfish 4 cm in diameter did not attack *Mytilus* sp. larger than about 1.2 cm long, but starfish larger than 28 cm diameter did not ingest these smaller prey. Starfish 40 cm and greater selected mussels as small as 5 cm and as large as 21 cm (Paine, 1976). Similarly, McClintok and Robnett (1986) reported that *Pisaster ochraceus* selected mussels relative to the size of the starfish. When asteroids were arrayed in 3 size classes (70–80, 81–90 and 91–110 mm) and fed mussels averaging (20, 35, 55, and 85 mm), the smallest size class selected 35 mm mussels while 81–90 mm asteroids selected 35–55 mm mussels and the largest starfish selected 55 mm mussels. Shells of some mussels were filed to determine if the size selection was based on ease of opening. No change in the pattern of selection was found. This was interpreted to be due to the attachment strength of the larger mussels.

The presence of more than one asteroid attacking a prey can also alter predation rate. Doering (1981a) made observations on *A. forbesi* preying on hard clams in Rhode Island. These data amplify the description of Belding (1912) on the technique used by this species to remove clams from the sediment. In laboratory studies, clams were excavated from the sediments either by extending one arm into the sediment or by 'humping' the clam out with all the starfish arms. In the latter method, the starfish centered itself over the clam and then formed a pit by moving sand outward with its tubed feet. Once the clam had been contacted, the starfish raised its center disk (humping) to extract the clam from the substrate. Using these and other data, Doering (1981a) was able to relate the size of the clam opened to the size of the starfish. When single starfish opened clams the relationship was: Clam width (cm) = 0.218 starfish radius (cm) + 1.73. When a group of starfish combined their efforts the relationship was: Clam width (cm) = 0.134 starfish radius (cm) + 1.96, indicating that larger clams were able to be consumed by starfish acting in unison. Doering (1981a) found that single starfish opened clams averaging 40.8 mm wide while starfish operating in unison opened clams averaging 46.5 mm wide. It is difficult to extrapolate such data too far because Palumbi and Freed (1988) provided clear evidence that *Pisaster ochraceus* exhibited antagonistic behavior toward one another. When placed in close proximity the starfish rapidly moved away from each other and within 24 h density approached that which existed before the individuals were aggregated (Palumbi and Freed, 1998).

There is also evidence that adult starfish reduce population density in younger individuals of the same species (Guillou, 1990). In addition, starfish themselves form feeding guilds, with some species preying predominately on bivalves, while others such as *Luidia* sp. feed preferentially on other starfish (Guillou, 1990). Thus, simple presence of starfish cannot be interpreted to be a sign of potential predation on bivalves.

Mauzey et al. (1968) reported the diets of 18 species of starfish in Puget Sound. Of the 18 species, only 8 consumed bivalves, and one of the 8 appeared to prefer snails to clams. One species of starfish preferred the clam *Saxidomus giganteus* to all other prey, but instead of switching to other clam species, laboratory and field data indicated that a sea urchin was the preferred alternate prey. Data on the other 18 species indicated that some were very specialized feeders while still others change prey preferences with location, season or prey abundance. Laboratory data on predator/prey selection did not always mirror field observations and predator avoidance behavior was indicated to be a potentially important factor in interpretation of distributions in the field (Mauzey et al., 1968).

In a similar study, Sloan and Robinson (1983) evaluated the feeding of four species of starfish in Puget Sound. These species, *Luidia foliolata*, *Mediaster aequalis*, *Pisaster brevispinus* and *Pycnopodia helianthoides*, all consumed bivalves. The density of the dominant species in the area, *M. aequalis*, was 0.27 m^{-2}, but this species was found to consume the least numbers of bivalves. Laboratory studies on the escape response (leaping) of 8 size classes of *C. nuttalli* to each of the asteroids found differences in percent prey responding between asteroids and, in some cases, prey size (Sloan and Robinson, 1983). *L. foliolata* elicited the least prey response from larger cockles, but a greater response from small prey. This is in accordance with the prey preferences of this species. *P. brevispinus* and *P. helianthoides* elicited the greatest prey escape response. When sand was incorporated and the cockles were allowed to dig into the substrate *L. foliolata* consumed a greater proportion of small cockles while *P. brevispinus* and *P. helianthoides* consumed more large prey. The sand provided a partial shelter for small cockles from *P. brevispinus* and for both small and large *P. helianthoides* when compared with the no sand control (Sloan and Robinson, 1983). The studies of Mauzey et al. (1968) and Sloan and Robinson (1983) clearly illustrate, as with most other forms of predator/prey interactions, it is difficult to generalize asteroid predator/prey relationships. It is important to evaluate the species of starfish and the potential for interactions among the starfish predators as well as interactions with size and behavior of their prey.

Astropecten

As opposed to the eversion of the stomach as a means of feeding as done by *Asterias* sp., astropectinids typically feed by completely ingesting the prey and then digesting the flesh. Ribi et al. (1977) reported prey species size selection by two coexisting species of starfish. In this study, similar prey were eaten, but different sizes were consumed by different species. The larger predator had arm lengths of 107–190 mm while arm lengths ranged from 65 to 103 mm for smaller species. Both species consumed bivalves, but the larger (*Astropecten aranciacus*) selected the echinoid *Echinocardium mediterraneum* 56% of the time with the highest ranking bivalve being eaten only 27% of the time (Ribi et al., 1977).

Christensen (1962) examined the stomach contents of *Astropecten irregularis* and found large numbers of *Spisula subtruncata* and *Venus gallina* spat. Thorson (1966), citing Christensen (1962), provided data on an *Astropecten irregularis* with more than 400 bivalve spat in its stomach. While Christensen's data suggest that both *Spisula subtruncata* and *Venus gallina* were ingested, *Astropecten irregularis* were preferentially attracted to the higher pumping rate of *Spisula subtruncata*. The rate at which the starfish was able to kill and digest *Spisula subtruncata* was also greater because *Venus gallina* was able to remain closed for longer periods of time. Data indicated *V. gallina* could survive within the starfish stomach for up to 18 days (Thorson, 1966). Whether *Astropecten* spp. or other starfish are important predators of hard clams in the high salinity southeastern estuaries of the US has not been investigated, and this may reflect the relatively few starfish present in most soft-bottom estuarine situations. An alternate view of the potential of asteroids to affect bivalves is presented by VanBlaricom (1982) working in an exposed area off the California coast. This study documented the control of crab (*Cancer gracilis* and *Portunus xantusii*) recruitment by *Astropecten verrilli* (VanBlaricom, 1982). Similarly, Auster and DeGoursey (1994) reported that blue crabs, *Callinectes sapidus*, overwintering in a sandy silt bottom in Connecticut were attacked by *A. forbesi*. These attacks became more successful as temperature decreased. On one occasion in February, these authors found up to 81% of the crabs were being attacked by one or more starfish. Thus the presence of a starfish may control predators, and, because the rate at which crabs can consume bivalve seed is much greater than bivalve consumption by starfish, the latter may enhance the chances of bivalve survival.

The only report of *Astropecten* spp. feeding on *Mercenaria* spp. appears to be that of Wells et al. (1961), and that is for one *Mercenaria campechiensis* found in one of 124 *Astropecten articulatus* collected off Ocracoke Inlet, North Carolina.

Asterias spp.

In colder water, the genus *Asterias* is commonly reported as a bivalve predator. Loosanoff (1964) described the timing and intensity of starfish set in Long Island Sound for the period 1937–1961, based on starfish counted from bags of 20 oyster shells collected twice a week. These data indicate sets nearly every summer with nearly 3 orders of magnitude variation in numbers of set between years. Unfortunately, these data cannot be related to numbers of starfish on the bottom. MacKenzie (1977a) found *Asterias forbesi* in low densities (less than 1 m^{-2}) in Lower Norwalk Harbor, Connecticut. Menge (1979) examined the potential for competition between *Asterias vulgaris* and *A. forbesi* in New England. While these studies focused on rocky habitats with *Mytilus edulis* being the principal prey, the data from one site can be utilized to compare the density of these two predators on subtidal gravel and sand substrate (Menge, 1979). Over 40 starfish m^{-2} were found on hard substrate, but the average combined density on the sand and gravel was 2.6 m^{-2}. WAPORA (1982) reported that although starfish were present in Great South Bay, South Oyster Bay and Hempstead Bay, New York, they were not abundant. *Asterias* spp. were found at only 9 locations of the 409 stations and numbers based on the 9 stations ranged from 0.3 to 12 m^{-2}. Greene (1978) found starfish only on one station in Great South Bay. In recent years, it appears that starfish have been abundant only near inlets and in deep channels (WAPORA, 1982), but the broad fluctuations in abundance described by Loosanoff (1964) suggest this may be a period similar to the middle to late 1940s when few starfish were present. The lack of large recruitment pulses may also reflect the preference of *Asterias* spp. for high salinity cool water. At temperatures above 23.5°C *A. forbesi* will die (MacKenzie, 1969). These latter data suggest a strong latitudinal gradient for *Asterias forbesi* with little chance for interactions between *A. forbesi* and hard clams south of Long Island except near inlets and in years when a large fall set of starfish takes place.

Thorson (1958) and Gulliksen and Skjaeveland (1973) reported that *Asterias rubens* consumed 15–30% of its body weight day^{-1} on a wet weight to wet weight basis. Hancock (1955) offered *A. rubens* a variety of prey species and found that adults of this asteroid preferred mussels and *Crepidula fornicata* to flat oysters. On oyster beds, large numbers of starfish were most closely associated with concentrations of *C. fornicata*. Loosanoff (1961) found that a medium *Asterias forbesi* could consume up to five 1-year-old oysters in a day. Hylleberg et al. (1978) reported that *Asterias rubens* preyed on cockles, and found that there was a positive relationship between the size of the starfish and the size of cockles at the same station. There were no data to indicate whether this was due to migration of the starfish or other factors, but Hylleberg et al. (1978) believed that this predator was responsible for destroying most of the cockles at one of their subtidal stations.

Barbeau and Scheibling (1994a,b) found that 80–100 mm diameter *A. vulgaris* were able to capture and consume more small (5–9 mm height) sea scallops, *Placopecten magellanicus* at 15°C than at 4 or 8°C. The capture and consumption rates at the latter two temperatures were not significantly different from each other. When multiple sizes of scallops were presented to the starfish, the predator consumed more small scallops than medium or large ones. The reason for the consumption of smaller size prey was related to the increased vulnerability of the smaller scallops relative to medium and large ones (Barbeau and Scheibling, 1994a). In the field trials based on the above laboratory study, Barbeau et al. (1994) found that predation by starfish was not affected by scallop density.

MacKenzie (1969) reported the rates of *Asterias forbesi* predation on oysters at constant temperatures from 5 to 25°C. The numbers of oysters consumed reached a peak at 20°C at 0.178 oysters $starfish^{-1}$ day^{-1} and then declined to 0.1 oysters $starfish^{-1}$ day^{-1} at 22.5°C and further to 0.039 oysters $starfish^{-1}$ day^{-1} at 25°C. MacKenzie (1969) noted that starfish lost weight at the highest temperature and experienced significant mortality during the experiment. Additional studies of oysters in trays suspended in Milford, CT harbor found a temperature-mediated bimodal peak in starfish predation. There was a period of low feeding rate in the coldest months, an abrupt increase as temperature rose in the spring, a summer decrease, and an increase in the fall (MacKenzie, 1969). Average densities of juvenile starfish in Long Island Sound were 1–6 m^{-2}, but ranged up to 45 m^{-2} (MacKenzie, 1981). Each juvenile starfish consumed between 0.6 and 2.1 oyster spat day^{-1} (MacKenzie, 1981). Adult starfish were observed to move in and out of the oyster beds. In one instance, density of starfish increased to 75 m^{-2} and this resulted in 83% of the spat being consumed in two weeks (MacKenzie, 1981).

A strong bimodal temperature-mediated feeding pattern for *Asterias forbesi* was also reported in Rhode Island with greatest activity in the late spring and again in the fall (Doering, 1981a). As in the studies of MacKenzie (1969) on oysters, the peak feeding activity on hard clams occurred in the late spring (18.5–19.5°C, MacKenzie, 1969; 5–10°C, Doering, 1981a) and fall (3.6–11°C, MacKenzie, 1969; 5–10°C, Doering, 1981a). Predation on the hard clam was greater in the spring than in the fall, even when the temperature was the same. Doering (1982) attributed this increased predation by the starfish to the greater activity on the part of the clams in the fall when they attempted to avoid predation by burrowing deeper (Doering, 1982). Rates of predation on 34 buried clams (mean 55–57 mm) by 10 starfish (mean radius 91–97 mm) ranged from 0.007 clams $starfish^{-1}$ day^{-1} at 3.7°C and 0.017 clams $starfish^{-1}$ day^{-1} at 16°C, but peaked at 0.03 clams $starfish^{-1}$ day^{-1} at 9°C in the first set of experiments. Highest rates of predation were reported during the second experiment at 4.9°C when 0.043 clams were eaten $starfish^{-1}$ day^{-1} (Doering, 1982). Deeper sand (10 vs 7 cm) reduced predation, and clams responded to water that had passed across starfish by digging deeper.

Whether the factor of 3–5 times higher predation rate of *Asterias forbesi* preying on oysters relative to starfish preying on hard clams was due to species preferences, infaunal vs epifaunal habitat of the prey, or other factors is not known. Kim (1969) examined the time it took the starfish *Asterias amurensis* to open five species of bivalves that were glued to a substrate and attached to a kymograph. The starfish was able to open the scallop *Patinopecten yessonsis* within 2.5 h, *Tapes phillipinarum* in 3.5 h, *Scapharca broughtonii* in 4.9 h, *Crassostrea virginica* in 6.9 h and *Mytilus edulis* in 8.8 h. Thus epifaunal prey that under field conditions would be able to avoid predation by escape responses were the easiest to open. Infaunal species that have some protection from sediments were the next easiest to open and the vulnerable fixed epifauna were the most difficult to open (Kim, 1969). How any of these data relate to predation in the field has apparently not been examined. It is clear that predation by *Asterias forbesi* was limited by high water temperatures and low salinity. Unless high densities of starfish are present, their predation on hard clam populations must be small. Due to high summer temperatures, *Asterias* spp. activity south of Long Island is generally restricted to offshore waters near deep high salinity inlet areas.

Other Echinodermata

While few ophiuroids and echinoids are found in habitats occupied by *Mercenaria mercenaria*, they are capable of ingesting newly set bivalves. In a classic benthic study, Blegvad (1914) reported finding 1 mm long *Spisula subtruncata* that had been ingested by *Echinocardium cordatum* and that the brittle star *Ophiura texturata* was capable of ingesting bivalves. Thorson (1966) reported that newly set *Spisula subtruncata* were nearly all consumed by *Ophiura texturata* and *Natica alderi* that set nearly simultaneously, but after the clam set. Holothurians such as *Thyone briareus* and *Leptosynapta tenuis* can be found in hard clam habitats, but these forms are considered to be filter feeders or deposit feeders. Deposit feeding holothurians can ingest significant quantities of sediments, and have been shown to enhance populations some species of deposit feeding bivalves (Rhoads and Young, 1971), but there does not appear to be any record of them consuming bivalves.

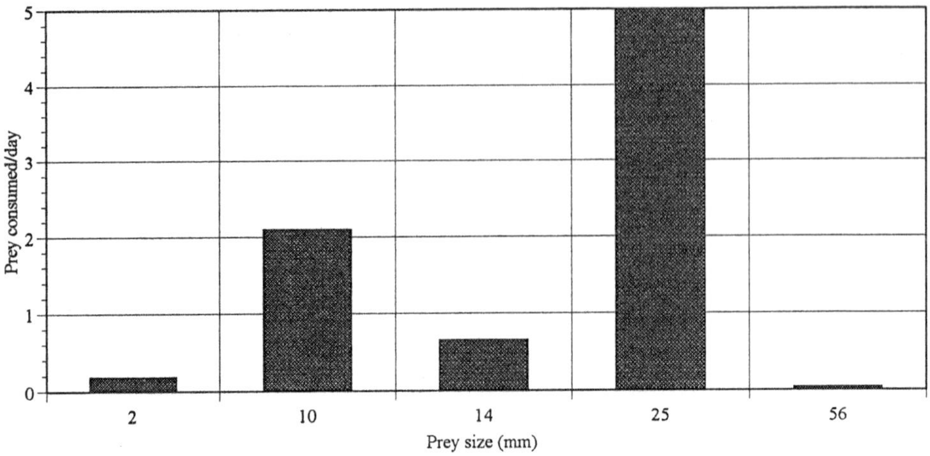

Fig. 11.17. Numbers of various size prey consumed daily by a variety of echinoderm predators.

Because of the predatory activities of starfish, echinoderms have been considered to be serious predators on a variety of bivalves. In general, the destructive nature of this predation has generally been investigated only in areas where large populations of starfish were apparent. In general, bivalve consumption by starfish is relatively low (<5 day^{-1}) (Fig. 11.17) when compared with crustaceans, but is relatively high when compared with typical rates of predation by snails (see data on these taxa above). Data on other echinoderms are insufficient to determine their potential effects.

11.10 CHORDATA

Ascidiacea

As with coelenterates and ctenophores, there is no information on the predation of *Mercenaria mercenaria* larvae by the filter feeding ascidians. While ascidians are not abundant on intertidal sand and mud flats, they can occupy significant areas that have some hard substrate cover such as large gravel or shell. Osman et al. (1989) reported that solitary ascidians were able to capture and ingest oyster larvae. *Mogula manhattensis* is a common pest in hard clam aquaculture in that large numbers of post set clams can be lost when the clams are used as substrate by the setting ascidians. This is an artifact of the culture methods which prevent the clams from being covered by sediments, and it is unlikely that similar conditions are found in nature. Young (1989) examined the effects of ascidians on epifaunal settlement and provided evidence that while ascidians ingest and consume a variety of larvae in laboratory studies, consumption in the field was not high enough to affect recruitment to nearby substrates.

Cowden et al. (1984) conducted laboratory studies in which they placed 100 each of polychaete, gastropod, crustacean, echinoid, and asteroid larvae in bowls with one *Styela gibbsii* (25–38 mm) for 3 h. There was a range of effects on the larvae, with fewer echinoderm larvae being eaten than annelids or gastropods. Fewer than 40% of the gastropod and annelid larvae survived for the test period, while for most echinoderm larvae >60% survived. These rates would be equivalent to a loss of >480 gastropod or polychaete larvae day^{-1}.

11.10.1 Vertebrata

11.10.1.1 Pisces

There is little information on the effects of predatory fish on the planktonic stages of the hard clam. Belding (1912) did not report data on fish predation on the hard clam, but noted that fish may be important predators of larvae and newly set. The literature on fish and fish larval predation on zooplankton focuses on copepods or other arthropods as the primary source of food. A recent review (MacKenzie et al., 1990) examined laboratory-derived ingestion rates for larval fish and extrapolated these data to field conditions. Within this review there was a listing of microzooplankton densities at 46 marine and estuarine sites, and while no bivalve larvae were explicitly included, the upper size limit for data inclusion was 200 μm (MacKenzie et al., 1990); a size spanning the typical range for bivalve larvae. How this relates to the potential for fish larvae to be important predators on bivalve larvae is unknown, but larval fish can be selective, and consumption rates have been found to be independent of food density (MacKenzie et al., 1990). Cahn (1951) reported that gobies, thred herring, and flounder consume pelagic larvae and newly set *Mactra sulcataria. Acanthnogobious flavimanus* and *Taenisides rubicundus* were significant predators on the bivalve *Sinonovacula constricta*, but this study did not provide data on the severity of the predation or rates of prey consumption (Cahn, 1951). While bivalve larvae were not mentioned as an item in the diet of the menhaden, *Brevoortia tyrannus*, Durbin and Durbin (1975, 1998) have shown that the minimum size zooplankton consumed by adult and juvenile menhaden was 13 and 7–9 μm, respectively. These schooling fish may have large localized effects on both phytoplankton and zooplankton (Durbin and Durbin, 1998). These authors equated the effect in Narragansett Bay, Rhode Island to be nearly equal or greater than the effects of the ctenophore *Mnemiopsis leidyi*, and thus it is entirely plausible that the effects on bivalve larvae could be similar to that of the ctenophore. Another fish species that has been shown to have substantial effects on zooplankton is the bay anchovy, *Anchoa mitchilli.* Baird and Ulanowicz (1989) reported that their model suggested that the bay anchovy could consume 70–90% of the zooplankton consumed by all fish species in the middle reaches of Chesapeake Bay. There does not appear to be any reason bivalve larvae would be excluded from the diet of such a predator.

It is illustrative to note that in a recent workshop (Dame, 1993) invertebrate and avian predators of bivalves each received a chapter, but no mention was made of the effects of fish on bivalve filter feeders or their ecosystems. Carriker (1961) attempted an experiment to determine if puffers, *Sphaeroides maculatus*, fed on hard clam seed, but the fish did not feed. T.C. Nelson (see Carriker, 1961) reported to Carriker that puffers washed young clams from the bottom, and Bigelow and Schroeder (1953) indicated that puffers consume small molluscs in their diet, but there was nothing specific written with respect to hard clams. Belding (1912) mentioned that young Atlantic jackknife clams, *Ensis directus*, and egg cockles, *Laevicardium mortoni*, were often found in flounder stomachs and implied that hard clams could also be consumed, but did not mention predation of adult clams by fish.

In addition to flatfish, Toba et al. (1992) indicated that pile perch, *Rhachochilis vaca*, may consume Manila clams, but no data were provided. What effect schools of filter feeding fish, such as the menhaden, *Brevoortia tyrannus*, may have on swarms of bivalve larvae is apparently unknown.

A large number of studies have reported on the effects of siphon nipping on hard clams and other bivalves, and because these processes do not generally result in mortality, these can be considered a higher order effect and are covered by Peterson (Chapter 10).

Chondrichthyes

Reports of predation by rays and skates on bivalves by fishermen or aquaculturists are common (Tiller et al., 1952; Merriner and Smith, 1979; Kraeuter and Castagna, 1980), but the scientific literature rarely indicates the importance of this phenomenon. This may be because the predation events of schooling species, such as the cow-nosed ray, *Rhinoptera bonasus*, are often temporally far apart (Orth, 1975, Orth, 1977; Kraeuter and Castagna, 1980), making it difficult to predict when the event will take place and even more difficult to design experiments of sufficient scale to evaluate the effects of these predators. Reports indicate that rays can be significant predators on oysters (Menzel and Hopkins, 1956; Krantz and Chamberlin, 1978), but neither of these studies quantified the rates or intensity of the predation. The mechanism by which rays excavate sediments was described by Howard et al. (1977) and Gregory et al. (1979). Circular to semicircular pits were excavated by flapping of the wings and/or by water jetted downward either through the mouth or gill clefts. Wide surface pits were reported to be caused by the flapping of the wings (Howard et al., 1977; Gregory et al., 1979), but the latter authors believed that the deeper depressions near where the mouth contacted the sediment were caused by water jets. This latter mechanism was presumed to be responsible for the pits made when rays feed on deeper dwelling organisms.

Thrush et al. (1989) ascribed some of the spatial distribution patterns of pits on intertidal sandflats to the feeding activities of rays (species not identified). Rays disrupted about 50% of the surface area of the flat each month and produced pits 30 cm in diameter and 20 cm deep (Thrush et al., 1989). This depth is sufficient to excavate many bivalves.

Howard et al. (1977) reported that the stomach contents of *Dasyatis sabina*, *Dasyatis sayi* and *Dasyatis americana* collected from intertidal flats in Georgia contained *Tagelus* spp., *Barnea truncata* and *Mya arenaria*. Although *Mercenaria* spp. were locally abundant in these waters, it was not included in the list of species in the stomach analysis of over 300 ray specimens. Menzel et al. (1976) reported feeding pits of sting rays *Dasyatis* spp. and butterfly rays *Gymnura micrura* were common in areas planted with 7–10 mm long hard clam seed. Survival of seed in areas not protected by wire cages was poor, and these authors assumed that the clams were being consumed by the fish and other predators, but no data were provided.

Talent (1982) compared the food habits of four shallow water elasmobranchs from Elkhorn Slough, California, and found that three of the four (*Mustelus californicus*, *Mustelus henlei*, and *Rhinobatos productus*) fed primarily on crustaceans, but the bat ray *Mylobatis californica* consumed clams and echiuroids. This study examined 310 bat ray stomachs and found that smaller fish fed almost entirely on gaper clams *Tresus nuttallii*, with smaller rays feeding on smaller clams. Larger rays consumed clams and echiuroids (Talent, 1982).

In New Zealand, eagle rays *Myliobatis tenuicaudatus* were found to eat bivalves such as *Macomona liliana*, *Chione stutchburyi* and *Amphidesma australis*, but no quantitative data were provided (Gregory et al., 1979). Orth (1975, 1977), working in Chesapeake Bay has shown that schools of cow-nosed rays, *Rhinoptera bonasus*, can nearly eradicate an eelgrass bed when they disrupt the roots and rhizomes to feed on *Mya arenaria*. In this study, 60–1000 juvenile *Mya arenaria* m^{-2} were found prior to the area being invaded by *R. bonasus*, and none were found after. Stomach analysis of 9 *R. bonasus* specimens from the study area revealed that most of the contents were *Mya* tissue and smaller amounts of *Zostera marina* roots and pieces of *Mya arenaria* shell (Orth, 1975). Kraeuter and Castagna (1980) reported the loss of 85–90% of hard clams in unprotected aquaculture plots due to ray predation. The size of clams destroyed averaged 39 mm hinge to lip. All clams were crushed and large pits were left throughout the graveled experimental area. Because the plots were destroyed within a 2-week period, a school *R. bonasus* was presumed to be the most likely predator. Similar predation in the same area on unprotected plots of hard clams indicated that once clams reach 60–65 mm hinge to lip the predator tends to push them aside and concentrate on smaller individuals (unpublished observations). Similarly, nearby small seed (<15 mm) were not consumed to any great extent.

Osteichthyes

Acipenseriformes

Acipenseridae

Acipenser oxyrhynchus

Johnson et al. (1997) examined the stomach contents of Atlantic sturgeon, *Acipenser oxyrinchus* caught in the Atlantic Ocean off the New Jersey coast by commercial fishermen. Molluscs were a relatively minor part of the diet, but hard clams were 5.9% of the stomach contents of fish collected in the fall of 1992. Since hard clams are rare or non-existent on the New Jersey continental shelf, this suggests the fish were feeding in the bays behind the barrier islands and had recently moved into the ocean.

Cyprinodontiformes

Cyprinodontidae

Fundulus heteroclitus

Kelso (1979) examined the effects of predation by the mummichog *Fundulus heteroclitus* on the soft-shell clam in shallow water areas near Essex, Massachusetts. *F. heteroclitus* were collected with seines and gut contents were examined for the presence of *Mya arenaria* during the spring, summer and early fall for 3 years. Predation was limited to clams smaller than 12 mm and the smallest clams reported to be eaten were 2 mm. Clams larger than 12 mm were never found whole in the stomachs of the mummichogs. Whether newly set soft-shell clams were preyed upon was not mentioned, and the reason for larger clams not being eaten was not examined. Kelso (1979) did speculate that by the time clams reached 15 mm they may have reached a depth refuge from the fish. Female fish of a given size ate substantially more clams than did male fish. Rates of predation averaged 14 clams $fish^{-1}$ day^{-1}, and predation was distinctly bimodal with respect to season. Highest rates were in May/June and again in mid-August through mid-September. Based on fish abundance data, Kelso (1979) estimated that in the May/June period, fish >50 mm (6 fish m^{-2}) eating 7 clams day^{-1} would consume 546,000 soft-shell clams per kilometer of low tide shore per day. No estimates of clam density were provided.

Kneib and Stiven (1982) examined the effects of *Fundulus heteroclitus* on the infaunal community of salt marshes in North Carolina. Cages were constructed on the marsh and size and densities of *F. heteroclitus* were manipulated. The only clam reported in these studies was the venerid *Gemma gemma*. These authors found that the numbers of gem clams were highest in the presence of larger fish and lowest in the controls or experimental plots containing smaller fish. These effects were enhanced at high fish density (range 1–4 m^{-2}). They hypothesized that these results were due to caging of mummichogs with another predator/disturber *Palaemonetes pugio*. The *F. heteroclitus* controlled the *P. pugio* abundance and thus reduced predation on the infaunal species in the plots.

Posey and Hines (1991) examined the effects of *Fundulus heteroclitus* and its interactions with *P. pugio* in a shallow water mesohaline benthic habitat in Chesapeake Bay and reported that the presence of mummichogs in tanks with the grass shrimp reduced predation on the coot clam *Mulinia lateralis*. This effect was caused by the shrimp population shifting to shallower water when the fish were present. The fish did not prey on the clams even though the clams were 0.75 mm long. These effects were small in comparison to the predator-free controls, but field studies indicated about a 25% greater survival of clams in cages where fish and shrimp were combined compared to cages with shrimp alone.

Whether *Fundulus heteroclitus* or other species such as *Fundulus majalis* or *Cyprinidon variegatus* are important predators of newly set clams or are important in reducing predation on *Mercenaria mercenaria* has not been determined, but Bigelow and Schroeder (1953) list molluscs as food for these species.

Gadiformes

Gadidae

Melanogrammus aeglefinus

Tyler (1972), working in the Passamaquoddy Bay region of Canada, reported that the haddock, *Melanogrammus aeglefinus*, was one of three species of fish found to consume bivalves. *Yoldia* sp. was the only species of bivalve found in the stomachs of this fish species and it was not considered to be an important component of the diet. The ranges of haddock and the hard clam do not overlap, except in rare instances, so the effects are not important. The potential for other gadids to interact with the hard clam would also appear to be minimal.

Perciformes

Sciaenidae

Leisostomus xanthurus

Virnstein (1977, 1979) reported that the spot *Leiostomus xanthurus* reduced abundance of infauna in soft sediments of mesohaline Chesapeake Bay nearly as much as the blue crab. He noted that *Mulinia lateralis* were among the most common stomach contents in spot collected during the summer. In the stomachs he examined, there were an average of 11.2 coot clams most of which were juveniles (1–3 mm), but some were adult size (5–10 mm). In addition, two other species of bivalves *Lyonsia hyalina* and *Mya arenaria* were eliminated from cages that were disturbed by mixing the sediments with his fingers, or contained either blue crabs or spot. When predators were kept from influencing infaunal densities, the numbers of clams reached over 14,000 m^{-2} and as the clams grew they forced some individuals beyond the protective mesh, where they were quickly eaten. These studies indicated that predation maintained the community below carrying capacity for many infaunal organisms, and the combined effects of fish and crabs might be the reason that dense assemblages of infaunal bivalves were not present in Chesapeake Bay. *Mercenaria mercenaria* was abundant enough to be harvested commercially in deeper water just beyond the area studied by Virnstein (1977, 1979) and it was common in eelgrass areas near where the caging studies were conducted, but it was not considered to be a dominant in the area. The above studies found *Mya arenaria* (1.2 mm average) recruited to >65,000 m^{-2} in May, and a second fall set reached 3000 (2.0 mm) individuals m^{-2}. Large sets of shallow dwelling bivalves *Mulinia lateralis* and *Lyonsia hyalina* were reported and populations increased in cages where fish and other large predators were excluded. Large *Mya arenaria* were present at low densities throughout the study area, but hard clams were not mentioned.

In the Maryland portion of Chesapeake Bay, Hines et al. (1990) examined the stomach contents of *C. sapidus*, *L. xanthurus*, hogchoker, *Trinectes maculatus*, and croker, *Micropogonias undulatus* caught in trawls on sand and mud bottoms. The crab, hogchoker and spot all consumed clams, but these two fish fed primarily on *Macoma balthica* siphons while croaker fed primarily on amphipods. This study did not find that fish consumed significant numbers of juvenile bivalves.

Peterson and Skilleter (1994) and Skilleter and Peterson (1994) have shown that siphon cropping by spot can cause significant loss of siphonal material in the clam *Macoma balthica*. These studies also showed a significant interaction between the depositing feeding *M. balthica*, spot and the suspension feeding clam *Rangia cuneata*. The siphon loss of the deposit feeder was greater in the presence of the suspension feeder than when the deposit feeder was present alone.

Pogonias chromis

Anecdotal reports suggest that the black drum, *Pogonias chromis*, can be a predator on many shellfish. For instance, Nelson (1903) reported that drum crushed and consumed eastern oysters, in beds planted by oystermen, near Tuckerton, NJ. At least one time, the oystermen resorted to exploding dynamite in the water to kill and/or frighten away the fish. Matthiessen (1971) and Smith and Merriner (1978) both reported that drum consume oysters, but no reports specifically link this species with significant losses of hard clams or clam seed.

Sparidae

Archosargus probatocephalus

Sheepshead, *Archosargus probatocephalus*, were important predators of oysters in the Gulf of Mexico, but no data were given on rates of consumption or sizes of oysters consumed (Menzel and Hopkins, 1956). These fish tend to graze on hard substrates such as pilings, but if seed clams were available it seems unlikely that they would be consumed by this species.

Gilt-head sea bream, *Sparus auratus* and the whitehead sea bream, *Diplodus sargus* have both been reported to consume mussels (IFREMER, 1988; Mason, 1976; Spencer, 1991).

Labridae

Tautoga onitis

Greene (1978) stated that flounder and tautog *Tautoga onitis* "usually feed only on clams <10 mm long", but did not provide evidence for this statement. Rheault (personal communication) found that large tautog removed from an aquaculture pond on Fishers Island, invariably had guts packed with *Mya arenaria* (6–10 mm) and some fragments of oyster spat.

Zoarcidae

Macrozoarces americanus

Tyler (1972) reported on the feeding of 13 species of demersal fish in the Passamaquoddy Bay region of Canada including ocean pout *Macrozoarces americanus*. The bivalves *Yoldia* sp. *Nuculana* sp. *Musculus niger*, *Arctica islandica*, and *Venericardia borealis* were reported to be principal prey for the ocean pout. The range of the ocean pout and the hard clam do not overlap except in rare instances in New England so it is unlikely to be a significant hard clam predator.

Pleuronectiformes

Bothidae, Pleuronectidae and Soleidae

Most studies have lumped together many species of flat fish, and to avoid redundant statements under each of the family taxa I have chosen to combine the information in much the same way as the various authors.

De Vlas (1979) found that consumption of entire bivalves, bivalve siphons, and bivalve feet were important components of the diet for plaice, *Pleuronectes platessa*, and flounder, *Pleuronectes flesus*, in the Wadden Sea. While siphon and foot parts were very important components of the diets of both species, juvenile *Macoma balthica*, *Cerastoderma edule* and *Mya arenaria* were all consumed by these fish. The numbers of individuals consumed were directly related to the density of the molluscan species on individual transects. As many as 1000 *C. edule* were consumed annually from 1 m^2 of sea bottom by plaice, and as many as 241 were consumed by flounder. No data are given on the average abundance of these organisms, but up to 4.2 g dry weight of whole molluscs m^{-2} (range 0–4.2 g dry weight m^{-2}), and a maximum of an additional 1.3 g dry weight m^{-2} of bivalve parts were consumed annually when data on all three species of prey were combined (De Vlas, 1979). Other authors have also reported that *Macoma balthica* is an important food for plaice (Braber and de Groot, 1973; Kuipers, 1977).

Kristensen (1957) reported that a number of authors had determined that small cockles were eaten by both plaice (Redeke, 1906; Smidt, 1951) and flounder (Redeke, 1906; Van Breemen and Redeke, 1907; Blegvad, 1917), but neither fish appeared to eat adult bivalves. Cockles were also believed to be consumed by flounder (Hancock, 1970). Hylleberg et al. (1978) examined the stomach contents of turbot, *Psetta maxima*; dab, *Limanda limanda*; plaice, *P. platessa* and flounder, *P. flesus*, and found that only the flounder consumed cockles. Braber and de Groot (1973) examined the diet of the flat fish turbot, plaice, dab, sole, *Solea solea*, and brill, *Scophthalmus rhombus*. Although molluscs were found in three of the five species, only plaice and dab consumed significant quantities of molluscs. Bivalves were found in both these species, but five species of bivalves were found in the plaice stomachs. Cockles removed from the stomach of these fish averaged 10.1–12.9 mm, and once the cockles became too large, none was consumed. Comparisons of predation on newly set *Cerastoderma edule* and *Mya arenaria* by *Crangon crangon*, *Carcinus maenas* and *Platichthys flesus* revealed that crabs and shrimp were more important predators of bivalves than the 0 age group flounder (Jensen and Jensen, 1985).

Similarly, Evans (1983) examined the effect of *Crangon crangon* and *Pleuronectes platessa* on the production of a soft-bottom community in Sweden. He reported that three species of bivalves were present in the macrofauna of this area (*Mya arenaria*, *Cerastoderma edule* and *Tellina tenuis*), and that molluscan remains were found in shrimp stomachs on all three occasions. The same prey were present in most of the size classes of shrimp sampled. In many instances, molluscs were >30% of the shrimp stomach contents, but were a significant part of the food only in larger size plaice and even then they rarely comprised more than 20% of the stomach contents. In spite of these results, calculations indicated that the shrimp plus the fish were able to crop 12–17% of the total macro- and meiofaunal production in this area. It should be emphasized that these studies examined intact fauna and did not account for potential increased production (such as noted in many cage studies) if predators had been isolated from sections of the flat.

Moller and Rosenberg (1983) reported that *P. flesus* consumed about 2.2 g of juvenile *Mya arenaria* per night (about 600 individuals) during July, but that this was reduced to about 0.9 g wet weight (70 individuals) by September. Conversely, the number of *C. edule* consumed increased in September to 1.5 g (200 individuals) and dropped to 2 g (50 individuals) in October. Maximum size soft-shell clams and cockles eaten were 12 and 10 mm, respectively (Moller and Rosenberg, 1983). Pihl (1985) found that 0 age group *Pleuronectes platessa* consumed 1% of the *Mya arenaria* and 4% of the *Cerastoderma edule* annual production on a tidal flat in Sweden. He noted that this was an unusually high percentage of the production, but siphons were not important in the diet in this area because most bivalves do not survive the winter on these flats. Thus only newly set bivalves were available for the predators (Pihl, 1985). If the data above are reduced based on an 8-h feeding cycle, then a maximum of 75 soft-shell clams fish^{-1} h^{-1} and 25 cockles fish^{-1} h^{-1} were consumed.

Ansell and Gibson (1990) examined the effects of predation by juvenile plaice, *Pleuronectes platessa*; flounder, *Platichthys flesus*, and dab, *Limanda limanda*, on a sandy intertidal flat in Scotland. The dominant food on these flats was *Donax vittatus*, and it comprised over 93% of the food for age 1 groups of all species. Whole bivalves comprised only a small part of the fish's diet, and never exceeded 6.1% in any species.

Toba et al. (1992) indicated that in Washington state, Manila clams <20 mm are eaten by rock sole *Lepidopsetta bilineata*, English sole *Parpophrys vetulus* and starry flounder *Platichthys stellatus*, but did not indicate whether these were major predators. Chew (1989) listed the same species of fish, but added the pile perch, *Rhacochilis vacca* to the list of Manila clam predators and noted that clams up to 10 mm were commonly found in the gut of this species and rock sole.

Jewett and Feder (1980) found that the diet of the starry flounder, *Platichthys stellatus*, in Norton Sound, Alaska was dominated (72.8%) by *Yoldia hyperborea*, *Serripes groenlandicus*, and *Musculus niger*. In Port Clarance, Alaska, *Y. hyperborea* was 41% of the diet, but the starry flounder in the Chukchi Sea did not appear to be feeding on molluscs at all (Jewett and Feder, 1980).

Tyler (1972), working in the Passamaquoddy Bay region of Canada, reported that American plaice, *Hippoglossoides platessoides*, was one of three species of fish found to consume bivalves. *Yoldia* sp. was the only species of bivalve found in the stomachs of the plaice and it was not considered to be an important component of the fish's diet.

MacKenzie (1970a) reported that summer flounder, *Paralichthys dentatus*, could damage oyster populations in Connecticut. Greene (1978) reported that in Great South Bay, NY both winter flounder *Pseudopleuronectes americanus* and summer flounder both consumed juvenile clams under a year old.

Mitchell (1974) cites a personal communication that hard clam shells have been found in the stomach of plaice and founders in England, and (Platt, 1969) noted that a hard clam (3 mm) was reported from a dragonet, *Callionymus lyra*, in England.

Frame (1974) reported that diet of young winter flounder, *Pseudopleuronectes americanus*, in the Weweantic River, Massachusetts was composed of fish and a wide variety of benthic invertebrates including two hard clams. Other bivalves such as *Laevicardium mortoni*, *Tellina agilis*, *Nucula proxima*, and *Yoldia limatula* were important diet items, as were unidentified bivalve siphons (Frame, 1974). Unfortunately, no size information was given on the hard clams that were consumed, but based on the other prey species, they were probably <25 mm.

11.10.1.1.1 Summary Pices

MacKenzie (1977a, 1979) indicated that other than the information on ray predation, there were no documented studies to indicate that any fish species consume hard clams, although, fishermen had reported finding small hard clams in the stomachs of northern puffer *Sphaeroides maculatus*, summer flounder, and tautog (MacKenzie, 1977a, 1979). There are a large number of fish species that potentially could consume hard clam seed and small adults (see Gibbons and Blogoslawski, 1989), but little or no information exists on the significance of their effects. There are differences between the rates of predation on *Mercenaria mercenaria* in unvegetated and vegetated substrates, and presumably this is partly the result of fish predation.

Virnstein (1977), Reise (1977a), Young and Young (1978) Kneib and Stiven (1982) and Posey and Hines (1991) all found decreased infaunal abundance associated with increased numbers of small decapod predators. Whether fish predation on these decapod species enhances hard clam abundance, or the same species of fish also consume hard clam seed is not known, but evidence from effects on cockles and soft-shell clams suggest such interactions are possible. Within an aquaculture setting, toadfish, *Opsanus tau*, has been found to enhance survival of hard clam seed, presumably by reducing crustacean predators (Gibbons and Castagna, 1985), although field trials in large plots have not shown similar effects (Barnett et al., 1987). The experiments of Kneib and Stiven (1982) and Posey and Hines (1991) also suggest that small fish such as mummichogs can effect similar changes in benthos by controlling crustacean predation, but this may be by displacement of one predator population or predatory activity. Because many of the predators are omnivorous, the outcome of various predator/predator and predator/prey interactions are difficult to evaluate without experimental evidence. On the other hand, Martin et al. (1989) found that blue crabs and spot, *Leiostomus xanthurus*, although they share many prey species, enhanced each other's survival rather than exhibiting interference competition. There is the additional possibility that hard clam seed may behave like *Gemma gemma* and be able to pass through many fish predators unharmed, but data to support this assertion are lacking. Consumption of small adult clams by fish, such as black drum and cow-nosed rays, has been documented, but the rates of consumption, and effects on hard clam populations have not been investigated. Lastly is it clear that many flat fish consume bivalves, or bivalve parts and may have important effects on bivalve populations. How much these species affect hard clams remains to be investigated.

11.10.1.2 Aves

Other than occasional references to hard clams being in gut contents, the effects of bird predation on *Mercenaria mercenaria* populations do not appear to have been documented. Most information about bird predation on bivalves is based on data from intertidal flats, and the most thoroughly documented data comes from studies conducted in northern Europe. Birds that utilize bivalves for food typically search the intertidal zone, but a number of duck species dive to considerable depths to consume bivalve prey. Anecdotal reports on predation by eider ducks on clam seed concentrated by aquaculture are passed back and forth by culturists, but no definitive studies have been published. Similar reports on disturbance or predation of aquaculture beds by black ducks, brandt and other species are also common, but data are not available. Britton (1991) considered the oystercatcher, *Haematopus ostralegus*, to be a major predator on Manila clams, *Tapes philippinarum*, being cultured in Ireland, but no data were presented. Spencer (1991) reported that oystercatchers and eiders (*Somateria mollissima*) were the most important avian predators on molluscs in Europe, and listed a number of control measures for aquaculture sites. Meire (1993) provided a list of species of birds that feed on bivalves in northern Europe. This included 12 ducks chiefly scaups, *Aythya* spp.; eiders, *Somateria* spp.; scoters, *Melanitta* spp. and goldeneyes, *Bucephala islandica*. The most important wading predators included the oystercatcher, knot *Calidris canutus*, bartailed godwit *Limosa lapponica* and the curlew *Numenius arquata*. Gulls, *Larus* spp. can be locally important predators of intertidal bivalves (Meire, 1993). Cahn (1951) lists birds that were considered to be significant predators by clam culturists in Japan, and the genera *Anas* and *Aythya* were significant predators to cultured *Venerupis* and *Anadara*. An unusual situation where the eastern gray heron, *Ardea einerea jouyi* and egret, *Egretta* sp. preyed on the razor clam, *Sinonovacula constricta* was described. All these bird genera are present in North America and most could be expected to prey on clam populations.

Anseriformes

Carriker (1959) in his study of hard clams noted that "ducks and geese" were abundant on Gardiners Island, New York flats and caused extensive sediment disturbance. He did not, however, determine if these birds consumed hard clams or clam seed. Martin and Uhler (1939) reported that ducks and geese consumed a wide variety of bivalves, and Meire (1993) has reported that, in addition to *Venerupis* sp. bivalves eaten include *Arctica islandica*, *Astarte* spp., *Cardium lamarckii*, *Cerastoderma edule*, *Chlamys* spp., *Crassostrea* sp., *Donax* sp., *Ensis* sp., *Hiatella* sp., *Leda* sp., *Macoma balthica*, *Mactra* spp., *Modiolaria* spp., *Modiolus modiolus*, *Mya* spp., *Mytilus edulis*, *Nucula* spp., *Ostrea edulis*, *Pecten* spp., *Scrobicularia plana*, *Solen* sp., *Spisula* spp. and *Venus* spp.

MacKay (1891, 1892) reported large flocks of oldsquaw, *Clangula hymenalis*, and scoters,

Melanitta nigra, *Melanitta deglandi*, and *Melanitta perspicillata* in a number of sites in New England. He noted that the oldsquaw consumed a variety of food including *Siliqua costata*, *Astarte castanea*, mussels and "a little shellfish very small, resembling a diminutive quahog (*Venus mercenaria*), but not one" (MacKay, 1892). This was probably a gem clam, but indicates this species could easily consume small hard clams. Scoters were reported to consume *Modiolus modiolus*, *Spisula solidissima* and *Argopecten irradians* and *Siliqua costata* (MacKay, 1891).

Yocom and Keller (1961) surveyed the duck populations in Humboldt Bay, California in the fall and winter of 1956/57, and compiled data on gut contents of these birds from 1953–1959. There were 19 species of waterfowl seen, but the widegon, *Mareca americana*, was the dominant species with over 47% of the birds, and eelgrass was its principal food. Food habits of the 11 most common species were reported, and clams were second in importance to vegetation. Clam remains were found in mallard, *Anas platyrhynchos*; canvasback, *Aythya valisineria*; lesser scaup, *Aythya affinis*; greater scaup, *Aythya marila*; buffelhead, *Bucephala albeola* and white-winged scoter, *Melanitta deglandi*, but were not an important item in the diet of mallards. By frequency of occurrence, clams appeared to be particularly important for the canvasback, greater scaup, buffelhead and white-winged scoter, but were more important in volume for the canvasback, lesser scaup, greater scaup and buffelhead (Yocom and Keller, 1961).

Cantin et al. (1974) examined the food of common eiders (*Somateria molissima dresseri*) feeding along the shores of the St. Lawrence estuary in summer. This study emphasized the diets of the adult females and juveniles feeding in the intertidal zone. Most birds fed at low tide and preferred snails (*Littorina* spp.). From June to September the female and juvenile eiders removed between 9.78% and 32.8% of the standing stock biomass of the snails, depending on location. Juvenile eiders were selective, taking the smallest snails first, and as the birds grew, they consumed larger snails. Two bivalve species, *Mya arenaria* and *Mytilus edulis*, made up 8 and 29%, respectively, of the adult diet, but only 0.4–9% of the duckling's food. The St. Lawrence study is similar to those conducted in Europe in that the bird population (25,000 pairs) was capable of removing a significant portion of the annual production of molluscs in the intertidal zone. Nehls (1989) estimated that most mussels taken by eiders were 20–30 mm, and the bird population (62,000) consumed 56,000 metric tons of mussels per year. Raffaelli et al. (1990) found that eider ducks (*Somateria mollissima*) removed over 35% by number of the mussels from unprotected plots. Unprotected sections of the bed had 4360 mussels m^{-2} removed in 60 days, and most of the losses were in the 10–25-mm size range. Individual birds fed on these mussel beds each low tide and the authors estimated that in the 4 h a day they spend feeding each bird could remove 176 mussels (176 mussels $bird^{-1}$ day^{-1}).

Goudie and Ankney (1986) compared the fall and winter diets and energy requirements of for species of sea ducks at Cape St. Mary's, Newfoundland. The smaller ducks, harlequin ducks, *Histrionicus histrionicus*, and oldsquaws, *Clangula hymenalis*, consumed more amphipods and isopods than did the black scoter, *Melanitta nigra*, and the eider, which fed predominately on mussels and sea urchins. All species fed on mussels and other molluscs to some extent.

Nehls (1989) reported that up to 151,000 common eiders could be present in the Schleswig Holstein area of the Wadden Sea during migrations, and that 30,000–40,000 may stay over the winter. Cockles, *Cerastoderma edule* comprised about 75% of the diet for these birds during the migratory population peaks. Although these birds can dive to 10 m to feed, most cockles were removed from the intertidal flats by puddling. The preferred size was thought to be those around 30 mm shell length. Although individual consumption rates were not computed, calculations indicated this species was consuming 12.5% of the total cockle + mussel biomass or 56,575 metric tons of bivalves per year.

Baird and Milne (1981) found that 60% of the eider's diet was mussels, but they also consumed both shore crabs, *Carcinus maenas*, and periwinkles *Littorina* spp. In the Dutch Wadden Sea, Swennen (1976) found that mussels and cockles were more than 80% of the eider's diet.

Data from shallow burrowing heavy shelled genera, such as *Cerastoderma*, *Venerupis* and *Arctica*, could be extrapolated to examine the potential effects of diving bird predators on *Mercenaria mercenaria*. These heavy shelled genera were prey for scaup, eiders and scoters (Cahn, 1951; Glude, 1964; Bourne, 1984; Meissner and Brager, 1990; Toba et al., 1992). The latter two bird taxa consumed *Arctica islandica* mean size 23 and 15 mm, respectively (Meissner and Brager, 1990). Bardach et al. (1972) reported that Japanese culturists consider scoters to be important predators of Manila clams. Toba et al. (1992) indicated that three scoter species (*Melanitta fusca delgandi*, white-winged scoter; *Melanitta nigra americana*, black scoter; and *Melanitta perspicillata*, surf scoter) were more significant in the destruction of natural beds of Manila clams in the state of Washington than were gulls and crows. The supporting data for this statement appeared to be from Glude (1964) and Bourne (1984). Glude (1964) found that white-winged and surf scoters were more important bivalve predators in Dabob Bay, Washington than black scoters. Predation was heaviest on smaller clams (6–19 mm), but some ducks fed on 25 mm Manila clams (Glude, 1964). Cahn (1951) citing Saito (1936) reported ducks consumed up to 52 small (10–20 mm) Manila clams bird^{-1} day^{-1}. Thompson (1995) speculated that size selective feeding by scoters may be a reason for the disappearance of Manila clams in intertidal sediments during the winter months. Clams >20 mm were more abundant in plots that received gravel or gravel + shell cover than in nearby controls where the ducks could feed more easily (Thompson, 1995).

Historically, diving ducks wintering in the Chesapeake Bay system fed mainly on submerged aquatic grasses, but Perry and Deller (1996) reported that there has been a shift in their feeding over time. Due to a decline in submerged aquatic vegetation, some of the bird species now have higher percentages of aquatic invertebrates in their gut contents. Perry and Uhler (1988) found that in the 1970–1979 decade, the diet of the canvasback, *Aythya valisineria*, was primarily molluscan. *Macoma balthica* was the predominant prey, but other species found in the gut and gizzard included *Mya arenaria*, *Macoma mitchelli*, *Congeria leucophaeta*, and *Rangia cuneata*. Plants, which formerly occupied >75% of the diet were reduced to 6% of the gullet and 4% of the gizzard contents (Perry and Uhler, 1988). Because there can be between 700,000 and 1,100,000 wintering waterfowl in the Chesapeake Bay system (Perry and Deller, 1996), a slight shift in feeding could have substantial consequences to local invertebrate populations. How the fluctuations in the submerged grasses, chiefly *Zostera marina*, in hard clam habits have affected the feeding of ducks and geese within the shallow bays along the mid Atlantic coast has apparently not been assessed.

Greene (1978) listed canvasback, black duck *Anas rubripes*, and scaup *Marila marila* as potential predators of juvenile (<1 year) hard clams in Great South Bay, NY. There do not appear to be any studies on the potential effects of ducks or geese on hard clam populations.

Charadriiformes
Haematopodidae

The effects of oystercatchers on bivalves have been studied more than most bird species. Dewar (1922) reported that the European oystercatcher, *Haematopus ostralegus*, was an important predator on *Ostrea edulis*, but Meire (1993) noted that the birds did not appear to be feeding on the imported species *Crassostrea gigas*.

Dare and Mercer (1973) examined the food of oystercatchers in Morecambe Bay, and reported the birds fed on *Mytilus edulis*, *Cerstoderma edule*, *Macoma balthica*, *Tellina tenuis*, *Littorina* spp., *Carcinus maenas*, *Crangon crangon*, *Ligia oceanica*, and, during periods of food scarcity, earthworms and insects from nearby fields. Most birds concentrated their feeding on either sandflats or mussel beds. Individual birds frequenting the sandflats preferred either *Macoma balthica* or cockles depending on the availability of prey. Those feeding on the mussel beds ate mostly mussels, but a few birds (1%) consumed periwinkles (Dare and Mercer, 1973). Fields were an alternate source of food during times when the preferred food on the tidal flats was sparse. Even though few annelids were consumed on the tidal flats, oligochaetes were the dominant prey species in the fields (Dare and Mercer, 1973).

Bryant (1979) also reported on the foraging of wading birds including the oystercatcher at 14 sites along a salinity gradient in the Forth Estuary, England. This effort examined the relationship between the invertebrate prey density, the area, configuration and exposure of the flats and the time the birds spent feeding. In general, there were statistically significant associations between feeding hours km^{-1} and numbers of birds km^{-2} and the density of at least one of the dominant prey species. The dominant prey of the oystercatcher was *Mytilus edulis*, but there were also significant positive correlations with density of *Cerastoderma edule* and *Macoma balthica*.

Drinnan (1958) examined the effects of oystercatchers on mussels at two sites on the river Conway. The birds fed at low water during both night and day periods. During a 7-h low water period, an average oystercatcher would consume 186 (37.5 mm) to 574 (25.7 mm) mussels. This consumption rate was approximately equal to the bird consuming a wet weight of mussels equivalent to its own body weight each day. At one site, the bird predation was severe enough that the author believed they caused a significant reduction in the larger size mussels on the bed. There was a clear preference for mussels >20 mm even though these were a small part of the population. As larger sizes became very scarce the oystercatchers began to attack mussels <20 mm, but still were consuming the largest mussels in the population (Drinnan, 1958).

Cayford and Goss-Custard (1990) found that oystercatchers selected larger mussels (40–45 mm) in all seasons but the spring when the average size dropped to 25–35 mm. The authors could not explain this shift on the basis of mussel density, or with an optimal diet model.

Zwarts and Drent (1981) reported that in the Dutch Wadden Sea nearly 40% of the mussels >40 mm were consumed by oystercatchers. Dare and Edwards (1976) reported that European oystercatchers were observed to regularly feed on second year mussel plots that were established to study losses and mussel production in the Menai Straits. As with most northern European sites oystercatchers were observed only during fall and winter.

Experiments with captive oystercatchers showed that the rate of consumption of mussels between 25 and 70 mm was dependent on the size of the mussel (Hulscher, 1974). More small mussels were consumed than larger ones, but the volume of mussel flesh consumed was not significantly different. The birds fed continually for 24 h did not consume any more food than those fed for 7.5 h per day (simulated tidal feeding conditions). Comparisons between the volume of flesh consumed when two bivalves (cockles and mussels) were present indicated that the birds always consumed a greater volume of cockle. When these data were analyzed on an ash-free dry weight basis, there were no significant differences in the weight consumed, indicating that the birds took in the same amount of food (ash-free dry weight) per day irrespective of the prey species (34.9 g AFDW of bivalve flesh for birds averaging 464.9 g body weight) (Hulscher, 1974).

High oystercatcher densities on mussel beds have been shown to lead to the dominant birds stealing mussels from subdominant ones (Ens and Goss-Custard, 1984). After additional studies, Horwood and Goss-Custard (1977) developed a model of the Burry Inlet cockle fishery studied by Davidson (1967). They concluded that the fishing mortality may have been underestimated and that the bird's rate of consumption must be reduced when cockle populations drop below 50–100 m^{-2} or they could extirpate the prey in years of low cockle recruitment. Conversely, Sutherland (1982b) reported that oystercatchers preying on cockles had greatest success in terms of profitability at intermediate prey densities of 25–150 cockles m^{-2}. Mortality of prey was greatest in the area of highest profitability for the birds and less both below and above that point (12 and 600 cockles m^{-2}) (Sutherland, 1982b).

Wanink and Zwarts (1985) examined the response of oystercatchers to *Scrobicularia plana* that had been experimentally placed at various densities and depths in the substrate. At low prey densities, the birds took all the prey they encountered, but at high prey densities, the birds became more selective by decreasing their probing depths (Wanink and Zwarts, 1985).

Davidson (1967) examined the effects of the oystercatcher on cockle stocks. On average, the birds fed approximately 7 h each day and consumed between 0.73 and 2 cockles $bird^{-1}$ min^{-1} or 306–420 cockles $bird^{-1}$ day^{-1}. Brown and O'Connor (1974) estimated that oystercatchers consumed 827 cockles day^{-1}, but they assumed similar feeding during night and day. Sutherland (1982c) found that oystercatchers were equally active at night, but the prey were generally smaller due to the lack of visual cues. In three winters (1961/62, 1962/63 and 1964/65), Davidson (1967) found the oystercatcher population consumed between 555,000,000 and 1,237,000,000 year 2 and older cockles (Davidson, 1967). The fishery removed between 47,000,000 and 128,000,000 cockles during the same time (Davidson, 1967).

Hulscher (1982) found that oystercatchers could select those *Macoma balthica* that were less infected with trematodes even though the infected clams were more visible on the tidal flats. Parasitized larger clams were rejected more often than parasitized small clams. This is in contrast to the studies of Bartoli (1974) (as reported in Thomas et al., 1997) in which *Venerupis aurea* infected with a trematode were more vulnerable to predation by *H. ostralegus*, the definitive host for the parasite. In this instance, the parasite causes the clam to reverse its position in the substrate and this behavioral alteration places the siphons farther from the surface, and the clam must move toward the surface to feed. The result is that the ventral margin of the shell is now near the surface, and this plus the shallower burial increases the clam's vulnerability to the oystercatcher.

Hulscher (1982) compared the food requirements for individual oystercatchers with the quantity of food which could be collected during daytime and night time tides. He found that feeding was less efficient at night, and the birds must feed during that time in order to meet metabolic requirements. Density of prey was important in the decision process of the birds. An experiment was conducted that placed *Macoma balthica* (11–22 mm) and *Cerastoderma edule* (25–35 mm) in a test plot at densities of 305 and 450 m^{-2}, respectively, and a second plot at 140 and 62 m^{-2}, respectively. At the higher density, 12 *M. balthica* were eaten for every cockle, while at the lower density, only 1.4 *M. balthica* were consumed for every cockle. In addition, the same bird species required *Macoma balthica* populations greater than 50 and 200 m^{-2} in June and March, respectively, to provide enough food (41–55 g ash free dry weight $bird^{-1}$ day^{-1}) for the birds (Hulscher, 1982). Estimated feeding rates at two sites ranged from 0.7 to 1.93 *Macoma balthica* $bird^{-1}$ min^{-1}. If oystercatchers consumed only *M. balthica*, they would have to eat about 1000 day^{-1} to meet metabolic requirements. Based on the annual average number of birds in the Dutch Wadden Sea (130,000), Hulscher (1982) concluded that densities of the clam were insufficient to support the birds. Only by using a combination of *Macoma balthica, Mytilus edulis* and *Cerastoderma edule* (depending on the varying densities of each species) was there sufficient food for the oystercatchers.

Heppleston (1971) found that the oystercatcher population partitioned the resources in the Ythan estuary. On low tides in winter, 63% of the population fed on *Mytilus edulis*, 26% on *Macoma balthica* and other mudflat fauna, and 11% in fields where they consumed earthworms and other fauna. The numbers of birds feeding in fields at high tide increased to 90% in December. The shift to feeding in the fields was thought to be due to the birds not being able to obtain enough food from the estuarine fauna. In a 9-h low tide feeding period an oystercatcher could consume between 270 and 400 mussels (Heppleston, 1971).

Even the lowest density of cockles being preyed on in Europe represents a high population level for hard clams, but oystercatchers have been reported to feed on hard clams. Tomkins (1947) cites Baldwin (1946) who describes a dead oystercatcher that had a clam attached to its bill. Because of the lack of data on bird predation on *Mercenaria mercenaria*, unpublished data on oystercatcher predation collected during an examination of feeding and growth of the knobbed whelk *Busycon carica* in Virginia (Kraeuter et al., 1989; Castagna and Kraeuter, 1994) are presented below.

From 1977 to 1979 an intensive field program was undertaken to study the growth rate, egg laying and movements of the whelk, *Busycon carica*. In the course of that study, the predatory effects of *B. carica* on bivalves occupying the intertidal flats of Cedar Island, Virginia was examined. Throughout the year, the intertidal flats were surveyed during low tide and all fresh bivalve shells with two valves still attached were collected. The length of these valves was recorded, and the type of predator was assessed by examination of shell damage. The organism responsible for the damage was determined by observing the predators in action and then recovering the shells from these activities.

Oystercatchers, *Haematopus palliatus*, were observed feeding on hard clams during early spring and throughout fall and early winter of several years. There was usually one, but at times two or three pairs of these birds on the beach during these periods. The chief prey species for these pairs were hard clams, stout tagelus, *Tagelus plebeius*, and Atlantic jackknife (razor), *Ensis directus*. The oystercatchers concentrated their foraging on the intertidal sand flats and did not actively search on nearby mud flats or intertidal oyster reefs. This is in direct contrast to the observations of Tomkins (1947) who noted that in Georgia oystercatchers concentrated feeding activity on oyster beds.

Fig. 11.18. The shell of a hard clam, *Mercenaria mercenaria*, collected by the author on a tidal flat, Cedar Island, Virginia, showing the distinctive v notch indicating predation by the oystercatcher, *Hematopus ostralegus*.

In Virginia, oystercatchers feeding on hard clams would thrust their bill into the sand, make several sideways jerking movements of their head and extract the clam from the sediments, and make several jabbing motions to open the clam. This is similar to the methods described for the European oystercatcher feeding on cockles. The cockles were opened when the bird severed the adductor muscle. Those opening the hard clams must have used a similar technique. A distinctive mark (Fig. 11.18) was present on the posterior region near the siphonal opening of clams the oystercatchers consumed.

A nearby (less than 0.8 km) population of hard clams was actively preyed upon by herring gulls (*Larus argentatus*). These birds picked up large hard clams from an intertidal mud flat and dropped them on intertidal oyster bars until the clam shell broke. While these clams may not have been removed from the same flat as that being used by the oystercatchers and whelks, they are included to indicate the size electivity of the bird and snail species.

Limited data were collected on the hard clam populations on the flat frequented by the oystercatchers. Data on the size frequency distribution of *Mercenaria mercenaria* from a preliminary study based on 30 random stratified 1/3 m^2 suction dredge samples from this flat are provided (Fig. 11.19). These data clearly show a dominance of smaller clams with a peak at 20 mm. This mirrors the pattern described by Sanchez-Salizar et al. (1987a,b) for the distribution of cockles in Wales. Cockles recruited throughout the intertidal and subtidal zone, but predation on smaller cockles (from crabs) increased with increasing tidal coverage. Crabs typically consumed cockles <15 mm, and most of these prey would be <10 mm

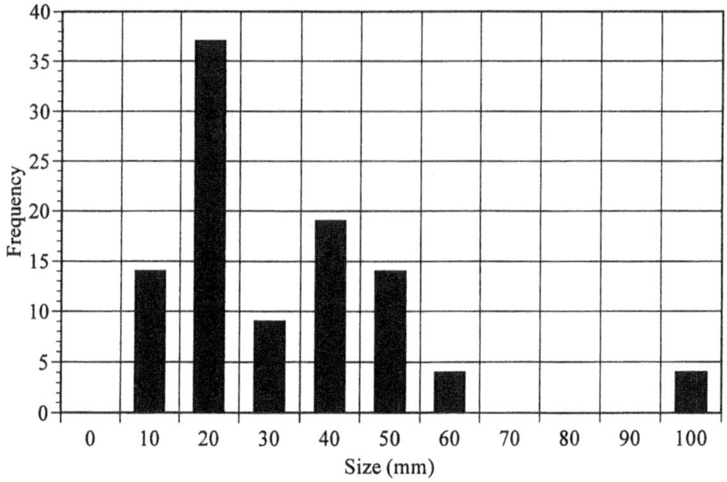

Fig. 11.19. Size–frequency distribution of hard clam *Mercenaria mercenaria* length collected in 0.25 m^2 suction dredge samples on an intertidal flat in Cedar Island, Virginia, April and May, 1978.

(Sanchez-Salizar et al., 1987b). Cockles >20 mm were removed from the intertidal zone by oystercatchers. The oystercatchers preferred cockles >30 mm and the mean length of cockles consumed was about 34 mm (Sanchez-Salizar et al., 1987a). The data from hard clam shells collected in Virginia (Fig. 11.20) clearly indicate that oystercatchers were selecting small (mean length 38.6 mm, SD = 10.2, n = 523) individuals while gulls were selecting only larger clams (mean length 73.8 mm, SD = 7.2, n = 78). In contrast, whelks preyed on a wide range of sizes with the predominant size being intermediate between that selected by

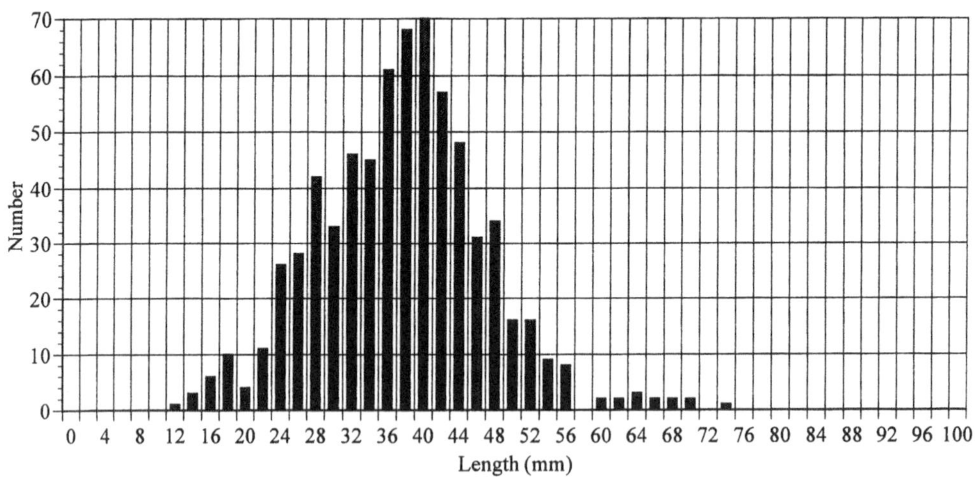

Fig. 11.20. Size–frequency distribution of hard clam *Mercenaria mercenaria* length collected from an intertidal flat in Cedar Island, Virginia over several years. All clams had the distinctive v shape notch indicating predation by the oystercatcher, *Hematopus ostralegus*.

the two bird species (mean length 51.4 mm, SD = 13.7, n = 2386). The maximum size eaten by oystercatchers was 78.2 mm and the herring gull was 89.8 mm. Less than 1% of the oystercatcher prey items were smaller than 15 mm and less than 5% were above 55 mm. Hulscher (1982) also found that oystercatchers preferred larger clams. In his studies, oystercatchers selected *Macoma balthica* between 15 and 20 mm, and it was rare that individuals <11 mm were eaten. It should be noted that there were only a few individuals of this species >20 mm (Hulscher, 1982). Norton-Griffiths (1967) found that oystercatchers' selection of mussels depended on the size distribution of the mussels on the bed. There was a clear preference for mussels >16 mm, but beyond that point, mussel position on the bed became an important variable in the bird's selection. In general, oystercatchers consumed a higher percentage of mussels once the prey reached 25 mm shell length (Norton-Griffiths, 1967), but beyond this lower size selection, the size range of mussels consumed mirrored that on the bed. The largest mussels on an individual bed (approximately 55–65 and 34–41 mm) were consumed at a lower rate relative to their abundance than mussels in slightly smaller size classes. The reasons for this selection were not explained.

In Virginia, larger clams were consumed by whelks, oystercatchers and gulls, the latter preyed on clams >58 mm, but all three species foraged on intertidal flats and formed a guild that preyed on nearly the entire size spectrum of clams available (Fig. 11.21). The latter probably reflects the inability of the gulls to break smaller clams on the oyster reef. Selection of larger size prey by birds that drop the prey to break the shell is generally in accordance with optimal foraging models (Kent, 1981; Richardson and Verbeek, 1986, 1987; Ward, 1991; Zack, 1979 and Zwarts and Drent, 1981), but these predictions seem to hold only for particular sizes. Ward (1991) found that oystercatchers preferred *Donax serra* 31–44 mm shell height, but this preference varied by site. At some sites, the preference range was 21–30 mm shell height, while at others, it was 51–60 mm shell height (Ward, 1991). Gulls preferred clams 51 mm and larger (Ward, 1991). It is difficult for oystercatchers to utilize larger clams because of their inability to penetrate the larger clams generally thicker shell at a point where the adductor muscle could be severed. The size range of hard clams eaten by oystercatchers was slightly larger than reported for cockles (15–40 mm) (Drinnan, 1957; Hancock and Urquhart, 1965; Davidson, 1967; Brown and O'Connor, 1974; Goss-Custard et al., 1977a; Sutherland, 1982c; Sanchez-Salizar et al., 1987a), mostly because larger clams were available. Oystercatchers in Wales fed on limpets *Patella* spp. (12–59 mm) while gulls feeding on the same limpets fed predominately on smaller individuals (5–38 mm) (Harris, 1965). While no data were given on the size distribution of the limpet populations, Harris (1965) found that the size curve for those preyed upon by oystercatchers peaked at 31 mm, while that for gulls peaked at approximately 15 mm. Thus, oystercatchers appear to have a restricted size range on which they can feed and this is species specific. The truncation of the upper end of the size range in the cockles relative to the data on hard clams is due to the smaller size of the cockles.

Predation on hard clams, such as described for an intertidal flat in Virginia, is important only in areas with extensive intertidal areas suitable for predatory birds, but in such areas birds can consume significant numbers of clams. Dankers (1993) has estimated that oystercatchers consume 20% of the cockles in the Dutch Wadden Sea each year, and this consumption is directed at the larger, more economically valuable sizes. Davidson (1967) estimated that the European oystercatcher can consume up to 381 cockles day^{-1}. The size distribution of hard

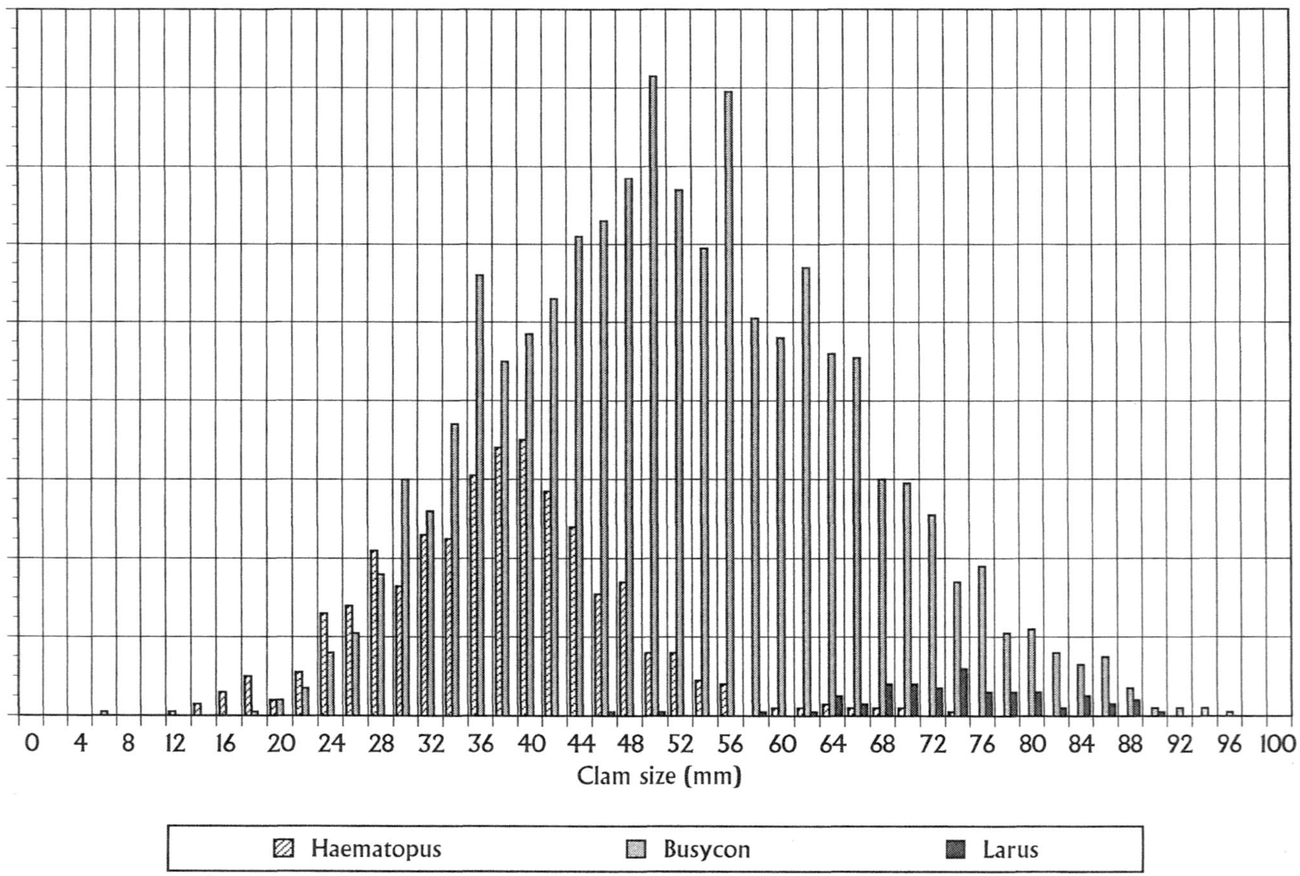

Fig. 11.21. Size–frequency distribution of hard clam *Mercenaria mercenaria* length collected from intertidal flats in Virginia over several years. Clams had been preyed upon by the whelk, *Busycon carica*; the oystercatcher, *Hematopus ostralegus*; and gulls — chiefly herring gulls, *Larus argentatus*, but occasionally by the black backed gull, *Larus fuscus*.

clams taken by oystercatchers on the Virginia intertidal flats clearly fit the pattern found for the European oystercatcher, and were the most economically valuable sizes. The lack of studies makes it impossible to assess impacts of these birds on *Mercenaria mercenaria* populations.

Charadriidae

Studies on shore birds have generally focused on one portion of their range. Recher (1966) observed the migrations of shorebirds in both California and New Jersey. The majority of the observations were conducted on the Palo Alto, California tidal flats. Gizzard contents were examined for the semipalmated plover, *Charadrius semipalmatus*; black-bellied plover, *Pluvialis squatarola*; avocet, *Recurvirostra americana*; short-billed dowitcher, *Limnodromus griseus*; least sandpiper, *Calidris minutilla*; western sandpiper, *Calidris mauri*; dunlin, *Calidris alpina*; knot, *Calidris canutus*; marbled godwit, *Limosa fedoa* and the willet, *Catoptrophorus semipalmatus*. *Gemma gemma* were found in the gizzards of all species and comprised over 40% of the items found in the avocet, knot and willet. *Modiolus demissa* was consumed by the willet, and *Mya arenaria* and *Macoma inconspicua* were found in the gizzards of the black-bellied plover, dowitcher, dunlin, marbled godwit and willet (Recher, 1966). Holmes and Pitelka (1968) examined food overlap among 4 species of sandpiper during the breeding season in Alaska. Although there was some broad habitat separation, the diet of all species was predominately insects or developmental stages of insects. Conversely, Baker and Baker (1973) examined the feeding behavior of six species of wading birds in their wintering and breeding habitats. They found that on the breeding grounds in summer, the food density was higher and foraging behavior more selective. This combination permitted the authors to statistically differentiate the summer microhabitat and food of each species, but in winter, with low food resources, there was little differentiation between the feeding of the species. These two studies emphasize the difficulty of extrapolating from the food taken and the feeding behavior by these birds in one area or season to an entirely different area or season. The studies examined below indicate the types of food taken by various species. The potential effects of any of these species on a relatively low density infaunal organism such as the hard clam are open to further study.

Bengtson and Svensson (1968) examined the diets of dunlins and *Calidris minuta* as they foraged on tidal flats in Sweden. Both species were selective with the former feeding on the polychaete *Nereis diversicolor* and the latter on insects. A number of crustaceans and molluscs were very abundant on these flats, but very few were consumed by the birds. Feeding habits and bill length explained a portion of the food differentiation, but neither of these hypotheses were able to account for the lack of crustaceans in the diets (Bengtson and Svensson, 1968). Reduction in the numbers of *Nereis diversicolor* may have a positive effect on the recruitment of small bivalves (see Section 11.6).

There are few published reports that document shore bird predation on seed bivalves. Knots feed on small mussels (5–23 mm) (Zwarts and Blomert, 1992) and small cockles (5–12 mm) (Goss-Custard et al., 1977a; Zwarts et al., 1992). Reading and McGrorty (1978) found that 80% of the *Macoma balthica* eaten by the knot on an English beach were between 9 and 13 mm, and that this was correlated to the seasonally controlled depth at which these clams were found in the sediment. More than 90% of the biomass of the clams was accessible to the birds in June, but, due to growth of the prey, only about 4% was available in December.

Goss-Custard et al. (1977b) found that the redshank, *Tringa totanus* and curlew, *Numenius arquata*, fed predominately on crustaceans and worms, but in some locations, bivalves, such as *Macoma balthica*, and *Scrobicularia plana*, were consumed by the redshank and curlew, respectively. The density of these two bird species was highly correlated with the density of their preferred prey species. The two bivalves as well as the cockle, *Cerastoderma edule*, were considered to be important foods of the knot, but no data were given on this latter species (Goss-Custard et al., 1977b).

Evans et al. (1979) reported on gut contents of birds feeding on an intertidal flat in the Tees estuary. The shelduck, *Tadorna tadorna*; grey plover, *Pluvialis squatarola*; curlew, *Numenius arquata*; bar-tailed godwit, *Limosa lapponica*; redshank, *Tringa totanus*; knot, *Calidris canutus* and dunlin, *Calidris alpina*; all fed extensively on the flats. The chief bivalve on the flat was *Macoma balthica*. This bivalve was eaten by all species except the curlew. *M. balthica* was considered to be an important part of the diet of three species: bar-tailed godwit, redshank and knot. In addition, three species: redshank, knot and dunlin, were found to be consuming *Mytilus edulis* from a nearby mussel bed. Many of the species also relied on the polychaete *Nereis diversicolor* for significant portions of their diets. Whether the feeding on this worm has significant effects on the recruitment of other infauna (see Ambrose, 1984b,c), including bivalves, cannot be assessed.

Bryant (1979) also reported on the foraging of wading birds including the curlew, bar-tailed godwit, redshank, knot and dunlin. His study of 14 sites along a salinity gradient in the Forth Estuary, England examined the relationship between the invertebrate prey density, the area, configuration and exposure of the flats and the time the birds spent feeding. In general, there were statistically significant associations between feeding hours km^{-1}, numbers of birds km^{-2} and the density of at least one of the dominant prey species. Dominant prey were the polychaete *Nereis diversicolor* for curlew, redshank and dunlin, *Cerastoderma edule* for the knot and dunlin, *Macoma balthica* for the redshank and knot. The association between the bar-tailed godwit and its prey was less precise, but it was positively associated with the polychaete *Nephthys hombergi*.

Myers et al. (1980) reported on the ability of sanderlings to find crustaceans on a sandy beach. These authors utilized dead prey so they could separate effects due to prey size, prey depth, prey density and substrate characteristics. As with other authors they found that all these factors influenced the ability of the birds to find the prey. Prey that were buried more than 10 mm below the surface entered a refuge from predation. Larger prey were more vulnerable than smaller prey, but sediments that were more tightly packed (low penetrability) significantly reduced the ability of the birds to feed. A similar effect was noted by Quammen (1980) who found reduced feeding efficiency by birds when she added sand to an intertidal mudflat.

Hicklin and Smith (1979) examined the gut contents of five intertidal feeding birds (semipalmated plover, black-bellied plover, short-billed dowitcher, semipalmated sandpiper and the least sandpiper) collected on the flats of the Minas Basin, Nova Scotia. Bivalves comprised only a small portion of the diet of these species. The only bivalve found, *Macoma balthica*, was reported as a minor component of the diet of the semipalmated plover and short-billed dowitcher (Hicklin and Smith, 1979).

Quammen (1980, 1982, 1984) experimentally attempted to examine the predatory effects of shorebirds on intertidal mudflat invertebrate populations in southern California. As with the studies of Zwarts and Wanink (1989) she noted that bill length was an important variable in determining prey selection, but that the quantity of prey consumed could be controlled by the percentage of sand on the flats (Quammen, 1982). While birds were able to reduce population density of prey on muddy flats, this did not happen when sand was present (Quammen, 1982). Increasing the quantity of sand on the flat was thought to interfere with the feeding of the birds (Quammen, 1980). Quammen (1984) did not list the species consumed, but dowitchers (*Limnodromus* spp.), western sandpipers and dunlins were all reported to eat bivalves. What was striking about these data was that bivalves occupied a higher portion of the gut content of birds than they did for the species of benthic feeding fish studied in the same habitats (Quammen, 1984).

Schneider and Harrington (1981) conducted studies similar to those of Recher (1966) but in Massachusetts. Dowitchers fed primarily on polychaete worms, but also consumed some *Tellina agilis* and *Gemma*. Sanderlings focused their foraging on sand shrimp, *Crangon septemspinosa*, but *Gemma gemma* was reported from their stomachs. The semipalmated sandpiper also fed on sand shrimp and *Gemma*. Black-bellied plovers consumed sand shrimp, worms, gem clams, *Mytilus edulis* (3–10 mm), and *Mya arenaria* (3–10 mm). All of the bivalves found in the plovers stomachs were crushed. In most other shorebirds, *Gemma gemma* are found intact, suggesting that these and other clams could be important components in the black-bellied plovers diet. The data reported by Schneider and Harrington (1981) included samples inside and outside cages that excluded the birds. Reductions in prey density were noted for some invertebrate species, but for other species no differences were found inside and outside the cages. This data is similar to the lack of effect noted by Botton (1984) (see below) on intertidal flats in New Jersey, but Botton's studies were conducted on sand flats and Quammen (1980) and Myers et al. (1980) both reported that sand impedes feeding by shore birds. Food consumption by birds, relative to their body weight may be nearly 10 times greater than that for fish or invertebrate predators (Schneider and Harrington, 1981), and large flocks may be attracted temporarily to areas of food abundance.

Botton (1984) found that shorebirds: semipalmated sandpiper, knot, sanderling, *Calidris alba*; ruddy turnstone, *Arenaria interpres* and laughing gulls, *Larus atricilla*, had little effect on the benthic fauna of intertidal flats on the Delaware Bay shore near Cape May, New Jersey. These flats were an important migratory stopover for many species, and the eggs of *Limulus polyphemus* were an important food resource during the spring migration. This extremely abundant resource may be one of the reasons for the low usage of other intertidal food resources. Alternatively, the dominant macroinvertebrate species, the gem clam, *Gemma gemma*, and the mud snail, *Illyanassa obsoletus*, may be relatively difficult for the birds to utilize. While adult *Gemma gemma* occurred in extremely high densities, are similar in size to juvenile hard clams, and have been found in the gut of a number of bird species, most authors seem to believe this is often the result of accidental inclusion of the gem clams when the birds were feeding on other species.

Wilson (1991) reviewed the effects of shorebirds on prey populations and noted that in a number of sites, *Macoma balthica* was an important prey item. In three of the four sites, no statistically significant density effects were found on the infaunal prey species (Wilson, 1991). Many of the same bird species or ecological congeners occur on the East coast and they feed at the same level of the intertidal zone on both coasts (Recher, 1966). The data suggest that, at times, small wading birds such as knots, dunlin, curlews and turnstones could be important predators on seed *Mercenaria mercenaria* in the intertidal (Hibbert, 1975, Hibbert, 1977b), but for most species, the positive effects of their preying on potential clam predators, such as crustaceans, may offset any clams consumed.

Laridae

Medcof (1949) reported that herring gulls, *Larus argentatus*, 'puddled' with their feet to stir up food on intertidal flats in Nova Scotia. He did not provide a list of foods being consumed, but commented that *Mya arenaria* were abundant on the beach and shells of this species could be found in the regurgitated pellets of the gulls. Harris (1965) examined the gut contents and regurgitations of herring gull populations in Wales and reported significant differences between sites. In one area, 64% of the food came from land sources and 21% from the shore, while at another colony the land and shore proportions were 53 and 7.5%, respectively. The molluscs consumed included: *Patella* spp. *Mytilus* spp. *Cerastoderma edule*, *Tellina tenuis*, *Monodonta lineata*, *Ensis* sp. *Buccinum undatum* and *Nucella lapillis*. Dare and Edwards (1976) noted that on some occasions, one or two herring gulls took small mussels from unprotected plantings in the Menai Strait.

Harris (1965) measured the sizes the molluscan prey ingested by gulls, and all were in the size range of year 0–2 hard clam seed (*Mytilus*, 9–23 mm long; *Tellina*, 15–32 mm; and *Patella*, 5–38 mm). For this latter species the gulls ate smaller *Patella* than oystercatchers feeding on the same shore (Harris, 1965). Spaans (1971) examined feeding of herring gull populations in the Netherlands and found that they foraged in a wide variety of places including garbage dumps, fields and intertidal flats. The gulls food was as varied as the foraged habitats. In winter months, the birds fed mainly on tidal flats where cockles were the most frequently consumed items, but *Asterias rubens*, *Mytilus edulis*, *Mya arenaria*, *Crangon crangon*, and *Carcinus maenas* were locally important prey items. The dominant species consumed were related to the lunar tidal cycle and the habitats available for feeding (Spaans, 1971). A seasonal cycle of feeding on the bivalve species corresponded to the time of the year when the bivalve young reached a size normally ingested by the gulls. The diet of the herring gull chicks was 15.9% invertebrates (9.9% bivalves) and 74.2% fish. Studies showed that the adults foraged on small cockles by treading while second year cockles were picked from the substrate with the bill. Both large and small cockles were swallowed whole. The bivalves consumed (year 1 cockles = 9–12 mm, year class 2 = 21–24 mm) were relatively small and thus the birds did not have to carry the prey to areas where they could be dropped and broken. An average of 5 h of low tide cycle was available for feeding, and during that time an average gull consumed 248 second year and 1825 first year cockles (49.6 and 365 cockles h^{-1}). The daily consumption was believed to be higher than the reported rates because gulls often fed before the flats emerged and these data were collected only during the day. Conversion from the number of cockles consumed to meat indicated that these birds averaged 28.3 ml of meat h^{-1} for the 5-h exposure period (Spaans, 1971).

Verbeek (1977) examined the feeding of herring and lesser black-backed gulls, *Larus fuscus*, in a variety of habitats in England. While gulls fed in both terrestrial habitats, such as garbage dumps, fields, and in marine habitats, such as near trawlers and in harbors, both species fed on intertidal flats. Important food items were bivalves; *Mactra corallina*, *Donax vittatus*, *Macoma balthica*, *Mytilus edulis*, crabs; *Portunus depurator*, and starfish; *Asterias rubens*. On the mussel bed, herring gulls dominated the bird population and there was a strong effect of the spring and neap tide cycle on the birds movement to and from the shore. On spring tides, large numbers of herring gulls left the colony to feed on the mussel beds. The primary food was starfish (Verbeek, 1977). Only a few mussels were consumed, this suggest that the gulls may have had a positive effect on the mussels by reducing the numbers of an important predator. No data were presented to evaluate this effect. Verbeek (1977) noted that the lesser black-backed gulls were more abundant on sand flats than the herring gulls and were able to locate burrowed *Mactra* easily, but only consumed about half of the clams they found.

Mudge and Ferns (1982) also examined the importance of various feeding sites to gulls. This study also found that gulls utilized a wide variety of feeding sites including refuse dumps, sewage outfalls, fields, freshwater areas, and intertidal and oceanic areas. One portion of the study used the regurgitations of herring gull chicks, lesser black-backed and greater black-backed gulls as an estimate of the type of food and the source of the food. Only 8% by volume of the food of lesser black-backed gulls was in the form of marine or estuarine invertebrates. Similar data for herring gulls ranged from 0 at one colony to 6.1% at another, while that of greater black-backed gulls was 18.2%. These analyses do not provide specific information on the species being fed to the chicks, but studies of adult feeding sites indicated that 90% of the gulls feeding on muddy shores were greater black-backed gulls while 58% of the gulls on rocky shores were herring gulls. Observations on greater black-backed gulls indicated they were taking mainly worms and snails while herring gulls took shore crabs and *Mytilus edulis*. No data were given to indicate the number consumed or the rate of predation.

Peterson (1990) reported that overwintering herring and ringbilled gulls were important predators on intertidal populations of bay scallops, *Argopecten irradians*, near Beaufort, North Carolina. He found that in a single low tide period, the gulls could consume between 10 and 15% of the scallops exposed by the receding tide.

Hockey and Steele (1990) reported on the diets of kelp gulls, *Larus dominicanus*, in Chile and South Africa. In South Africa, *Donax serrata* was the dominant source of food. These were removed from the intertidal areas and dropped on the beach to break the shell. The modal size of those dropped was 40–49 mm. In Chile, the bivalves *Mesodesma donacium*, *Mulinia* sp. and *Gaimardia* sp. were consumed. Modal sizes for *M. donacium* varied from 70–79 to 30–39 mm dependent on the size distribution in the population. *Mulinia* sp. and *Gaimardia* sp. modal sizes were 20–29 and 50–59 mm, respectively. Smaller bivalves were swallowed whole and larger ones were dropped. Swallowing the smaller specimens allowed the gulls to consume 10–51 prey min^{-1}, but handling time increased when the birds were forced to drop the prey to open it. Younger and smaller birds were at a disadvantage when dropping prey because larger or older individuals would often steal the prey. The probability of the prey being stolen by larger birds increased with the number of drops, reaching nearly 70% if it took 6–8 drops to open a bivalve.

Ysebaert and Meire (1990) described the food selection by black-headed gulls (*Larus ridibundus*) on tidal flats in the Netherlands. The polychaete *Nereis* sp. was the preferred prey during this summer breeding season study. These authors noted that small cockles, *Cerastoderma edule* were consumed by these birds, but that the bivalves were not an important part of the diet. The selection of a presumed predator of small clams, such as *Nereis*, suggests that the predation by this gull may enhance the flats for bivalve recruitment. Little is known about such interactions.

Ward (1991) compared the size preferences of African black oystercatchers *Haematopus moquini* and kelp gulls *Larus dominicanus* feeding on the bivalve *Donax serrata.* Gulls dropped clams from about 6 m onto pebble surfaces. These birds selected clams >31 mm, but clams >51 mm were the predominant part of the diet at all of the sites studied. The selection of larger sizes of *Donax serrata* by these gulls is similar to the selection by herring gulls in England, and in Virginia, USA (see discussion below).

Chestnut (1952) provided an early report of gulls feeding on adult hard clams in the intertidal zone of the US east coast, and he noted that they were responsible for 31% of the losses in intertidal experimental plots. Most of these clams were found on a nearby dock where the gulls had dropped them to break the shell. Anecdotal reports of chowder and cherrystone size clams being dropped into parking lots, on roof tops and on other land surfaces are common throughout New England and in some areas of the mid Atlantic, but there is little scientific documentation of the extent or importance of these losses.

Mitchell (1974) reported that in England, the black-headed gull, *Larus ridibundus*, fed on hard clams over 20 mm long by dropping them from a height of 15–20 m. Hibbert (1975, 1977a,b) found that herring gulls were a very important hard clam predator on an intertidal flat in Southampton, England. On this flat, a population of about 55 gulls removed 4% of the total energy passing through the clam population (9.5% of assimilated energy). This predation was primarily on adult (>4 years old) clams, and the observed rates indicated a removal of 5–10 adult hard clams m^{-2} $year^{-1}$ from populations of approximately 3–18 adult hard clams m^{-2}. Wading birds were considered to be an important predators of smaller clams. Oystercatchers and herring gulls were expected to remove larger individuals, and diving ducks may prey on seed size clams (Hibbert, 1975, Hibbert, 1977b).

Kent (1981) examined the feeding of herring gulls on the northwestern Gulf of Mexico beaches and reported that *Argopecten irradians* was the dominant prey species, but the shells of other bivalves, such as *Mercenaria campechiensis*, *Dinocardium robustum*, *Trachycardium egmontianum*, *Lucina floridana*, and *Macrocallista nimbosa*, were also found. All of the *Mercenaria mercenaria* eaten were between 67 and 95 mm shell length (Kent, 1981).

Roberts et al. (1989) examined the vertical migration of hard clams on intertidal areas in North Carolina, and reported a displacement of about 20 mm in response to the tidal cycle. Clams were deeper in the sediment at low tide than at high tide, and laboratory experiments indicated that a change in pressure of 100 mb was sufficient to cause this vertical migration. Furthermore, the depth of burial and movement was independent of clam size within the size tested (20–40 mm thickness). These authors also placed clams on the tidal flats and restricted their depth of burrowing. Observations on these clams showed that gulls, *Larus argentatus*, consumed more of the clams that were on or near the surface (Roberts et al., 1989).

Kraeuter (unpublished) collected hard clams dropped by herring gulls on an intertidal oyster reef in Virginia. These clams were picked up from nearby sand and mudflats and dropped repeatedly until the shell shattered. Only large adult clams were represented in this assemblage (Fig. 11.22). Lastly, Micheli (1997) reported that the presence of gulls appeared to be the primary reason for seasonal reduction in blue crab predation on hard clams in specific intertidal and shallow subtidal habitats.

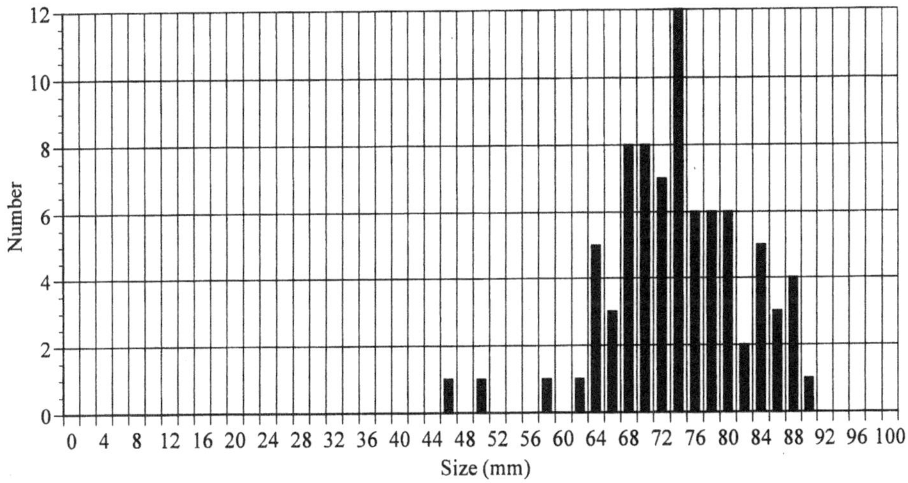

Fig. 11.22. Size–frequency distribution of the hard clam *Mercenaria mercenaria* length collected from intertidal flats in Virginia over several years. Clams had been preyed upon by gulls — chiefly herring gulls, *Larus argentatus*, but occasionally by the black backed gull, *Larus fuscus*.

Passeriformes
Corvidae

Richardson and Verbeek (1986, 1987) found that Northwestern crows *Corvus caurinus*, fed on intertidal populations of Manila (Japanese) littleneck clams *Tapes philippinarum*, in the British Columbia, Canada, and that the predators selected particular sizes of prey. These birds actively extracted clams (24 and 40 mm) from the sediment, but any of the smaller clams within this size range were rejected even after being dug up (Richardson and Verbeek, 1986, 1987). Clams were located by random search rather than visual cues, and nearly all the clams >31 mm that were found were eaten (Richardson, 1985). The random nature of the search pattern was hypothesized to be related to the crows habit of flying from the flat to a nearby area before they open their prey in much the same way they fly to rocky areas to open whelks (Zack, 1979; Richardson, 1983).

On the US east coast fish crows *Corvus ossifragus*, and boat-tailed grackles *Quiscalus major*, can be observed feeding on intertidal oyster reefs and nearby sand and mud flats. Potential effects of these species on intertidal bivalve populations have not been assessed.

11.10.1.2.1 Summary Aves

Since most migratory birds inhabit a site for a small portion of the year, predation on those organisms with short (<1 year) life cycles and fast recovery times may be insignificant, but it is difficult to discount these potential effects because birds often arrive in large numbers. Bird predation can apparently cause significant losses for species that occupy the intertidal zone, have low levels of recruitment, and cannot escape predatory activities by outgrowing the predators preferred size (Figs. 11.22 and 11.23). This suggests populations of seed to intermediate size hard clams in the intertidal zone may be the vulnerable to depletion by

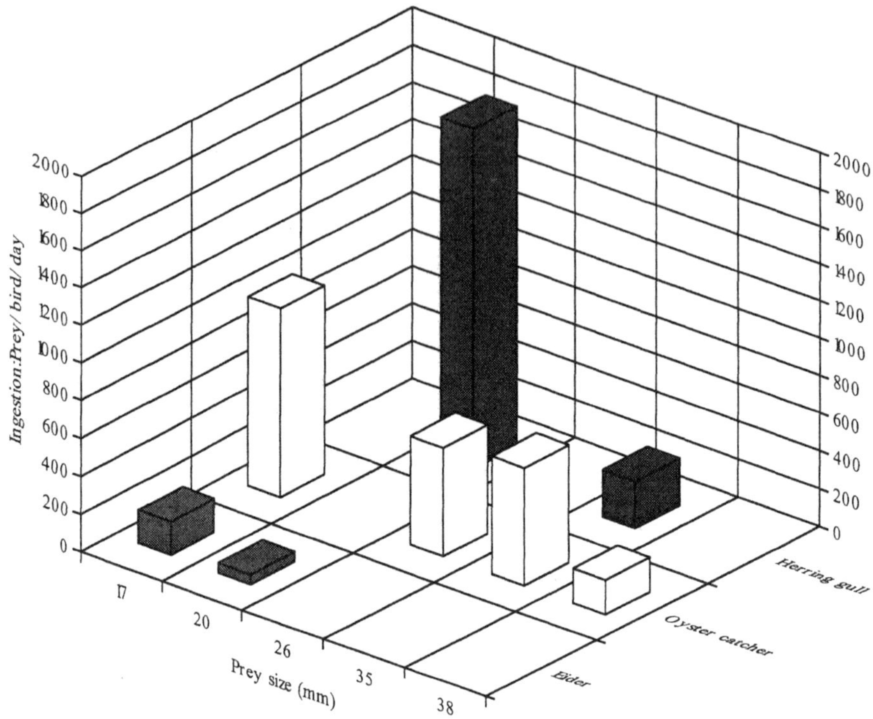

Fig. 11.23. Daily ingestion rate of bivalve prey of five sizes by three species of bird predators. Data were compiled from sources cited in the text.

avian predators. It is interesting to note that a recent review (White and Wilson, 1996) on predators of the eastern oyster, *Crassostrea virginica*, did not mention birds as predators. While some birds have been reported to prey on oysters, intertidal clam populations may be more vulnerable to avian predators than intertidal oyster populations.

It is difficult to derive daily predation rates for birds from the available data. Many of the studies were based on 7-h tidal cycles of feeding. The 827 cockles day^{-1} consumed by the oystercatcher (Brown and O'Connor, 1974) was based on the bird feeding equally during two tidal cycles. Similar feeding behavior was assumed to derive the 1000 *Macoma balthica* day^{-1} estimates for the same species, but some studies indicate that birds feeding at night are less efficient (Hulscher, 1982; Sutherland, 1982c). In addition, there are indications that most birds continue to feed to meet metabolic requirements instead of simply feeding both day and night. In spite of these differences, it has been thoroughly documented that in localized intertidal areas birds can have significant impact on bivalve populations. Whether birds are important in structuring hard clam populations has not received much attention. Diving ducks such as the eider and scoter have been reported to affect epifaunal bivalve populations, and can be serious predators on high density aquaculture plots of mussels, but there is little information to indicate whether such species affect subtidal populations of hard clams, or hard clam recruitment.

11.10.1.3 Mammalia

Other than the harvesting by humans, there is little information on mammals consuming hard clams. There are a number of reports from the Pacific Coast that document the predatory effects of sea otters, *Enhydra lutris*, feeding on a variety of clam species. This predator is capable of excavating trenches over 1.5 m long 0.5 m wide and 0.5 m deep to obtain large clams (Hines and Loughlin, 1980). In many locations, bivalves are the dominant food item in the sea otter diet and the bivalves consumed are numerous. The bivalves reported as food are: *Tivela stultorlum*, *Saxidomus giganteus*, *Saxidomus nuttalli*, *Tresus nuttallii*, *Protothaca staminea*, *Mya truncata*, *Macoma inquinata*, *Macoma incongrua*, *Mytilus edulis*, *Mytilus californianus*, *Musculus vernicosa*, *Volsella volsella*, *Pododesmus macroschisma*, *Clinocardium nuttalli*, *Hinites gigantea* (Miller et al., 1975; Stephenson, 1977; Calkins, 1978; Hines and Loughlin, 1980; Wendell et al., 1986). The latter three reports document the effect of this predator on Pismo clam, *T. stultorlum*, populations and the fishery. Kvitek et al. (1988) reported that >60% of sea otter prey was large bivalves, but they also consumed mussels, crabs and moon snails. There was a clay layer 23–40 cm below the surficial sediments that prevented some bivalves from burying to their maximum siphon depth. The sea otters utilized this sub-bottom feature and concentrated their foraging in the areas where the depth of clam burrowing was the shallowest (Kvitek et al., 1988). Kvitek et al. (1991) determined that sea otters were able to differentiate between butter clam, *Saxidomus giganteus*, that had high or low levels of the paralytic shellfish poisoning toxin, saxitoxin. At high levels of toxins, the otters discarded those parts of the clams that had the highest levels of the toxin.

Irons et al. (1986) reported on differences in foraging by gulls when sea otters were present. When sea otters depressed the populations of preferred prey species the gulls switched to more diverse prey and eventually began to feed almost exclusively on fish. Although this study concentrated on rocky shores, it seems likely such interactions could also occur in sedimentary areas as well.

There is no ecological equivalent to the sea otter on the east coast, but racoons, *Procyon lotor*, have been reported to prey on intertidal populations of soft-shell clams (Gosnor, 1979). Rheault (personal communication) has observed muskrats, *Ondatra zibethica*, breaking through 0.5 galvanized mesh (hardware cloth) covers on floating tray to consume bivalve seed.

11.11 SUMMARY

There are clearly a large number of species that are capable of consuming *Mercenaria mercenaria*, and the rates of predation reported in the literature are high enough to cause severe reductions in recruitment to hard clam populations. If the limited data on predator abundance is combined with the predation rates from most studies, it is evident that, with rare exceptions, all the hard clams that recruit each year could easily be consumed. This suggests that either our means of estimating predation rates are not directly applicable to field conditions, that there are some forms predator refuge (depth, sediment type, density), or that stochastic processes are more important than have been suggested in the literature. Throughout the review, there are a few references to such refuges, but field data that confirm their presence are lacking.

The list of predators that have been shown to feed on hard clams could easily be expanded if more taxa were examined. There is so little data on predation on larvae that it is difficult to evaluate its importance to either the predators or the prey. It is apparent that at least some cnidarians, ctenophores, polychaetes, bivalves, barnacles, and ascidians are capable of ingesting bivalve larvae including those of the hard clam. There is little evidence that fish prey on hard clam larvae, but because clam larvae are within the size range that can be removed by species such as menhaden and bay anchovy, the possibility for localized depletion of clam larvae by fish cannot be excluded. Although the rates of larval consumption are impressive, the data do not indicate that any of these larval predators have any demonstrable effect on hard clams at the population level.

Newly set hard clams (<1 mm) are certainly vulnerable to a very large number of potential predators. Protozoa, flatworms, annelids, gastropods, crustaceans (chiefly shrimp and hermit crabs, but also juveniles of crabs and perhaps amphipods), some asteroids and echinoids and juvenile fish all could consume hard clams of this size. Drills such as *Urosalpinx* and *Neverita* are the only two gastropods which have been shown to consume hard clams at rates greater than 1 day^{-1}, and these were on seed <2 mm. The highest rate of consumption of small hard clams by any molluscan predator was by adult *Mercenaria*, but these were larvae. Other predators that 'sift' through the sediment, such as horseshoe crabs, and some fish (spot and other scianids) could cause mortality even if the predators were not actively searching for the newly set seed. Unfortunately, predatory processes on this size of hard clam have received little attention, and there have been few field studies so we cannot evaluate the effect of this potential source of loss to hard clam populations. Dense assemblages of other bivalves, such as gem clams, may exert some inhibitory effects on new set, but this may be offset by processes that enhance survival of the small clams. As with many of the factors controlling predation, the enhancement effects on this size seed may depend on sediment type, density of both the adults and juveniles, interactions between various predators, etc. The data on whether there is density-dependent or density-independent mortality seems to be directly related to the investigation. Because there are a large number of variables to be considered, it is difficult to be assured that any of these individual studies provides adequate data. Modeling efforts, based on New England clam population data have shown that density-dependent post settlement mortality is important in the recruitment process, but that mortality of adult clams is density independent (Malinowski, 1985, 1993; Malinowski and Whitlatch, 1988).

In addition to direct effects, such as consumption of prey, predators can cause secondary effects. There is evidence of two forms of growth response by prey to predatory species. Irlandi and Peterson (1991) found that introducing a snail, *Busycon carica*, into a trough of hard clams significantly reduced growth rates of the clams. In contrast, Crowl and Covich (1990) reported that the freshwater snail *Physella virgata* increased its growth rate in response to predation on conspecifics by the crayfish *Orconectes virilis*. The decrease in growth may have been to simple disturbance, but the increased growth was apparently due to a chemical cue. The snails subject to this cue grew to a larger size, but had little reproduction compared to snails that had not experienced the chemical cue. Thus, there was a response that increased growth rate, age and size to maturation (Crowl and Covich, 1990), but whether either or both of these responses are significant in field populations is unknown. A number of marine bivalves react to the presence of starfish by exhibiting escape responses, and other bivalves have exhibited escape behavior from nemerteans. In at least one case, a crab was observed to elicit an escape response, but whether a continued presence of starfish, whelks, nemerteans

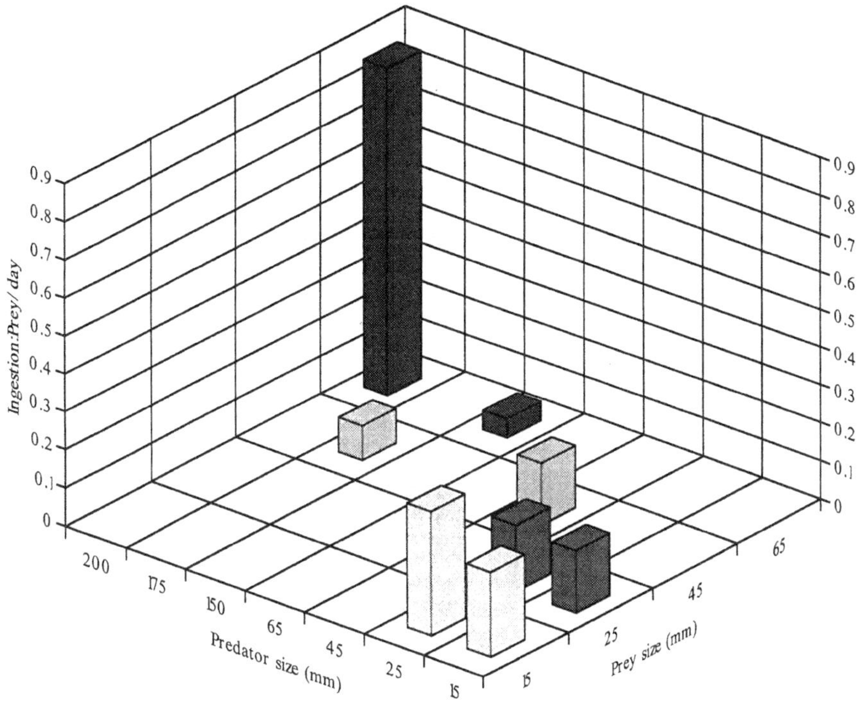

Fig. 11.24. Number of bivalve prey, by prey size, consumed daily by a variety of sizes of molluscan predators. Data were compiled from sources cited in the text.

or crabs will stimulate a response in marine bivalves similar to that described by Crowl and Covich (1990) is unknown.

Once hard clams have reached >1 to 2 mm there are more data on which to formulate hypotheses. From this size to about 20–25 mm, nemerteans, gastropods, horseshoe crabs, crabs, lobsters, echinoderms, fish and birds all become potentially important predators. In general, with the exception of larvae and newly set individuals, predation by molluscs across all sizes of predator and prey tend to be <1 prey predator^{-1} day^{-1} (Fig. 11.24). In terms of rate of consumption, arthropods appear to be major consumers of all bivalve seed up to about 20–25 mm. Rates of consumption exceeding 10 bivalves predator^{-1} day^{-1} have commonly been observed, and rates exceeding 100 bivalves predator^{-1} day^{-1} have been reported in laboratory studies (Fig. 11.25). There were enough studies on the rates of arthropod consumption on bivalves and the hard clam that they could be combined. These data, when taken as a group, indicate that predation on infaunal bivalves, such as *Mercenaria mercenaria*, is considerably higher than on epifaunal organisms (Table 11.16, Fig. 11.25). Cohen et al. (1995) provided evidence for such a response in experiments in which the green crab was allowed to select between the infaunal clams *Potamocorbula amurensis* and *Tapes philippinarum* placed in sediments in the same container as the epifaunal mussel *Mytilus* spp. The crab selected *Potamocorbula amurensis* at a greater rate than the mussel even though the clams were buried. The same was not true for the Manila clam, which even though all

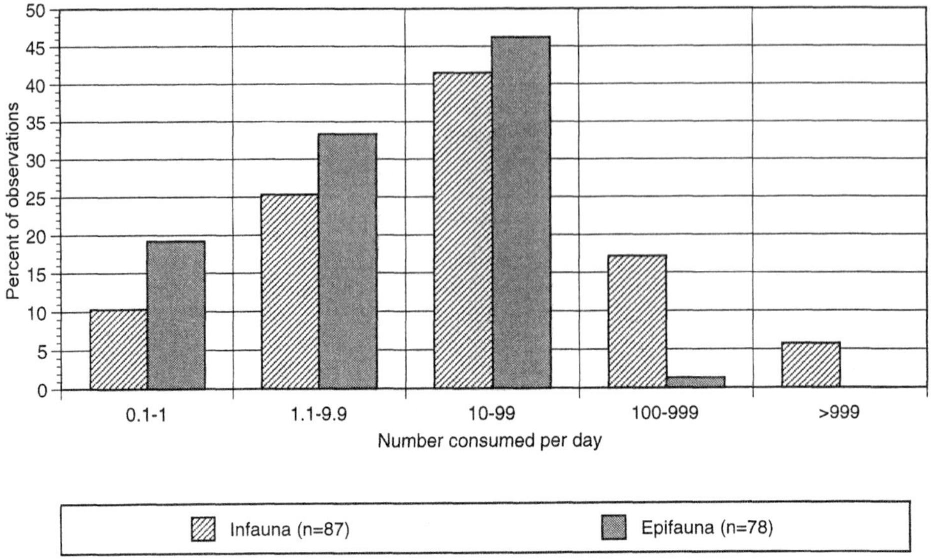

Fig. 11.25. Combined observations on arthropod predation on infaunal and epifaunal bivalves indicating the number of observations of particular classes of prey consumed per day. Infaunal observations = 87, Epifaunal observations = 78. In many cases, the predators were the same species. Data are combined from a variety of sources reviewed in the text.

species of prey were approximately the same size, was selected at a lower rate than either of the other two species. It is such interactions that make generalizations from the accumulated data difficult to interpret and very tenuous. The reader is cautioned that in order to construct the table and graphs data were compiled from sources that used a wide variety of techniques, sizes of predators, sizes of prey, and types of substrates. More importantly, predators were allowed to feed for time periods ranging from a few hours to more than a week. These were all normalized to 24 h by simple multiplication or division, and as such, the patterns depicted in the graphs should be viewed with extreme caution. In spite of these difficulties, predation on epifauna rarely exceeded 99 prey predator^{-1} day^{-1}, but data on infauna, in many cases with the same predators, contains a significant number of studies that report rates of consumption >100 prey predator^{-1} day^{-1}. In some cases, the rates on infauna exceeded 999 prey predator^{-1} day^{-1}. It might be reasonable to hypothesize that epifauna, because of their exposure to predators, have developed some mechanism to reduce size specific losses to predators. The alternative hypothesis would be that infauna may have not developed these mechanisms. Epifauna, relative to infauna of the same size, resist opening by starfish, but most high predation rates are due to simple shell crushing so it is difficult to determine that such a mechanism would be operable, and there is no direct evidence from existing experiments. A recent study (Mascaro and Seed, 2000b) showed that green crabs appeared to be selecting on the basis of prey shape and volume with emphasis on the minimum shell dimension. Interestingly, crabs selected mussels in preference to either of two species of oyster presented, and this selection was independent of prey size/frequency or abundance. When offered mussels and cockles, prey selection was based on the size/frequency of the

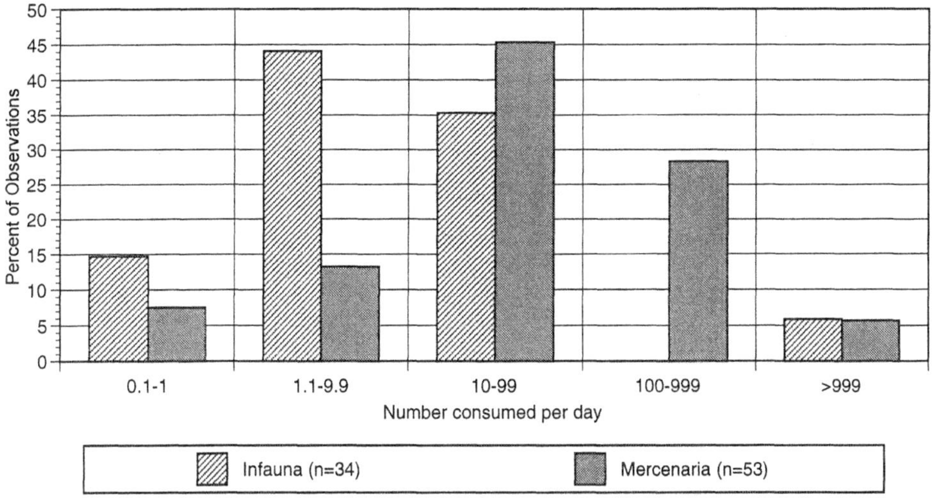

Fig. 11.26. Combined observations on arthropod predation on infaunal bivalves exclusive of hard clams and hard clams, *Mercenaria mercenaria*, indicating the number of observations of particular classes of prey consumed per day. Infaunal observations = 34, hard clam observations = 53. In many cases, the predators were the same species. Data are combined from a variety of sources reviewed in the text.

prey not by the species presented. In addition, when offered models of oysters and mussels the crabs spent more time examining the mussel shape than either of the oyster shapes. The time spent investigating cockle and mussel shapes was statistically the same, but crabs always spent slightly more time handling the cockle shape (Mascaro and Seed, 2000b). In any case, it is clear that additional work, similar to that by Mascaro and Seed (2000a,b), needs to be done to determine if the differential effects between epifauna and infauna are real or simply an artifact of methodologies.

If the compiled data for the hard clam are isolated from the remainder of the infauna (Fig. 11.26), the information seems to indicate that juveniles of this species are particularly vulnerable to arthropod predation. The number of studies that have indicated consumption rates >100 prey day^{-1} for hard clams when compared to a nearly equal number of trials for epifauna is remarkable. Although, in many cases, the predatory species were the same, there is no way to determine if this discrepancy is a methodological artifact or a substantive difference.

Clams that exceed 25–30 mm have a reduced suite of potential predators. Large predaceous snails, some crabs, asteroids, some fish and, in the intertidal zone, birds are all capable of handling these large heavy shelled prey. Studies on predation on adult hard clams are few, but certain predators such as large snails, and rays appear to specialize on mature bivalves. Birds, such as the oystercatcher and gulls have been shown to have population level effects on other bivalve species in the intertidal zone. While the oystercatcher appears to prefer individuals of smaller size, gulls are limited only by their ability to carry the clam aloft so the shell can be broken. In northern latitudes, echinoderms, such as *Asterias forbesi*, are capable of consuming prey, such as large adult hard clams (Fig. 11.17), but, in general, starfish appear to prefer the smaller individuals. Once hard clams reach adult size, man becomes an important, and in most areas, the dominant predator.

11.11.1 Latitudinal Predator Guilds

Along the US East and Gulf Coasts there are natural zoogeographic breaks such as Cape Cod and Cape Hatteras, and within the broad areas formed by these biological zones are various predator guilds. Many species, particularly estuarine species, have ranges that extend from Cape Cod through the Gulf of Mexico. North of Cape Cod the number of predator species appears to be somewhat restricted, but so are the areas occupied by extensive hard clam populations. Thus while the blue crab, most xanthids, and the busyconnine whelks are reduced in the north, predators such as the naticid gastropods, *Asterias* spp., green crabs, cancer crabs, sand shrimp, lobster, hermit crabs, and a variety of birds all persist. The mid-Atlantic is populated by all these species, at least on a seasonal basis, but the cancer crabs, lobster, green crabs and *Asterias* spp. generally decrease in importance with increasing latitude. At the same time, species numbers and populations of such predators as busyconnine whelks, horseshoe crabs, xanthid crabs, and blue crabs all dramatically increase with increasing latitude. Within this area, the effects of species that migrate between inshore and the continental shelf, such as *Cancer irroratus* and *Crangon septemspinosus*, may be seasonally important. No data are available that would allow evaluation of the effect of these predators on recruiting hard clams that did not reach a refuge size prior to growth cessation in winter. Other predators, such as blue crabs, whelks and naticids, generally move from the intertidal to the subtidal zone and reduce their activity during the colder months. Below Cape Hatteras, the number of predatory species continues to increase, but we have almost no data on which to evaluate these additional species and their potential interactions with hard clams. Perhaps, more importantly, the season over which the predators are active is extended with increasing latitude.

Generally, survivorship of hard clams increases with age when predators are considered to be the major source of losses, but Kennish (1980) examined losses of adult hard clams, and noted that there appeared to be a bimodal pattern in mortality rates with an increase after year 4. These losses of adult clams were most often associated with periods of presumed high stress such as the winter or during spawning. Whether such seasonal stress effects are more important in the north (cold, spawning), or in the south (heat, spawning), or the physiological races of the clam can compensate for these latitudinal temperature/spawning effects has not been evaluated.

11.11.2 Sediment

It is clear from the existing data that hard clam survival and growth can be affected by a variety of sediment characteristics including the presence of shell fragments and other large grain sizes, the stability of the sediment, the grain size, and the presence or height of the redox layer. Other factors, such as the water content, porosity, permeability, pH, organic content, mineralogy and the presence or absence of hard clam adults, or presence or absence of other species may have growth and survival effects on hard clams, but these have not been documented. Clearly, many of these same parameters may affect, in a positive or negative fashion, the predators ability to find and consume their prey. Some of the sedimentary effects may vary with latitude (such as the presence of carbonate sediments in Florida), while others are similar from Canada through the Gulf of Mexico. For instance, hard clam population density in muddy sediments is generally low, and the clams found there are typically of large size. There is some evidence that this distribution may reflect the importance of predators, but naticids and *Busycon carica*, two potentially important predators are themselves relatively rare in such sediments. These predators, along with *Asterias forbesi*, and the lady crab prefer high salinity sand to sandy mud sediments, and are thus usually abundant near tidal inlets. These same areas may be inhabited by large populations of hard clams. The existing data do not provide a means of evaluating the effects of such interactions on hard clam population dynamics, but we do have some data that compare relative importance of certain habitat variables to clam survival.

Peterson (1986) examined the reasons for enhancement of hard clam densities in seagrass beds in North Carolina. He concluded that at least 50% of the difference must be attributed to differential post-set survival and that the most likely cause of this differential was post-settlement predation. The study did not evaluate which predators might be responsible for these effects. Other studies (Peterson, 1982a, Peterson, 1986) found that whelks preferentially selected larger hard clams and that seagrass roots and rhizomes offered significant protection against whelk predation. Smaller clams were more likely to be lost (missing) than larger clams. This loss was attributed to either emigration or predators that consumed the clams by breaking the shell. Peterson et al. (1995) reported that the best time to plant hard clams in North Carolina was in the fall after blue crab predation had diminished. In these southern waters, clam growth continues, while predation is reduced, thus when predators become active in the spring, the clams have reached a size at which predation is reduced. What was not tested was whether overwintering the seed by some means that would increase their growth to the same size as the field planted seed would achieve the same result. Conversely, Beal (1991) and Beal et al. (1995) reported that juvenile *Mya arenaria* transplanted to the mudflats of Maine in the spring had improved survival over similar fall plantings. Apparently this manipulation allows the clams to reach a size at which they can burrow deeply enough to avoid predation by green (*Carcinus maenas*) and rock crabs (*Cancer irroratus*). Size (burrowing depth) may also be an important variable in reducing ice scour mortality (Beal, 1991; Beal et al., 1995) in northern areas.

In addition to sediment type, predation on hard clams may be affected by the presence of additional prey species, or by interactions between the predators. There is some evidence that naticid snails are preyed upon by whelks, crangonid shrimp are consumed by *Cancer* spp. crabs, and palaemonetid shrimp predation on clams was significantly reduced by the presence of mummichogs. In this latter instance, the reduction in predation was not due to reduced populations of the shrimp, but the fish caused the shrimp to shift to a different habitat. Significant portions of blue crab and cancrid crab diets are composed of juveniles of their own species, and many of these predators have well established antagonistic behavior toward other members of their own species. In other instances, the presence of more than one predator of the same species (for both *Carcinus maenas* and *Asterias forbesi*) has been show to increase predation rates. How these behaviors interact with prey species and spatial patch distribution patterns remains an area for investigation.

Some studies have shown that interactions between different taxa of predators, even where both consume many of the same prey species can be complex. These experimental investigations, such as that of Martin et al. (1989), who compared spot and blue crabs singly and in combination, have indicated that competition between two species of predators did not result in reduction in either of the predators. The combined predators effects were found to be complex, and simple summing of the two effects did not predict the effects observed when predators were introduced into the system. In this case, spot survival was enhanced by the blue crabs. The results of this and similar studies point out the danger of extrapolation from single species studies to field situations.

There appears to be a substantial number of reports (mostly anecdotal) of high recruitment of bivalves following natural (or anthropogenic) disasters. Hancock (1970) reported that there was an inverse density-dependent relationship between cockle density and recruitment. This relationship was primarily due to high recruitment after extremely cold winters had reduced the population of adult cockles (Hancock, 1970). A similar intense recruitment of surf clams (*Spisula solidissima*) took place following an extreme anoxia/hypoxia event in the Mid-Atlantic continental shelf in 1976 (Steimle and Sinderman, 1978. Recruitment of hard clams in Indian River, Florida was reported to be extremely large following a severe rain event which is believe to have substantially reduced many invertebrate populations in sections of that estuary. Unfortunately, documentation of this event and many similar events is lacking. Whether these intense recruitment periods were due to open space created by fewer adults, a reduction in disease and/or parasites in the bivalve population, a change in sediment properties, or a reduction in predation was generally unknown, but most authors speculate that reduced predation was the most likely cause of the increased recruitment.

11.12 ACKNOWLEDGMENTS

This chapter would not have been as complete without input from many individuals in both the academic and private sectors. I am particularly indebted to Dr. Sandra Shumway who cajoled me into developing the book and to Drs. Steve Fegley, Victor Kennedy, Charles Peterson, and Robert Rheault, who added substantially by reviewing and commenting on the various drafts of this chapter. I would be remiss without commenting on Melbourne Carriker and his works on *Mercenaria*. Without his field efforts in the late 1940s and early 1950s, our knowledge base on the hard clam, and its early life history would be substantially less complete. There is a great need for additional similar efforts in other locations on the hard clam, and on many other benthic species. Lastly, my co-editor Mike Castagna started me working with *Mercenaria* and our collaboration through the years has been a major factor in my interest in aquaculture. This is contribution No. 2000-(2000-21) from the Institute of Marine and Coastal Sciences, Rutgers University and New Jersey Agriculture Experiment Station Project No. K-32406. Support for some of this work has been provided by NJ State funds and the New Jersey Commission on Science and Technology.

REFERENCES

Abolmasova, G.I., 1970. The food of black sea crabs and an analysis of some elements in their energy balance. Hydrobiol. J., 6: 49–56.

Ahn, I.-Y., Lopez, G. and Malouf, R., 1993a. Effects of the gem clam *Gemma gemma* on early post-settlement emigration, growth and survival of the hard clam *Mercenaria mercenaria*. Mar. Ecol. Prog. Ser., 99: 61–70.

Ahn, I.-Y., Malouf, R. and Lopez, G., 1993b. Enhanced larval settlement of the hard clam *Mercenaria mercenaria* by the gem clam *Gemma gemma*. Mar. Ecol. Prog. Ser., 99: 51–59.

Alexander, S.K., 1986. Diet of the blue crab, *Callinectes sapidus* Rathbun, from nearshore habitats of Galveston Island, Texas. Texas J. Sci., 38: 85–89.

Allen, J.A., 1960. On the biology of *Crangon allmani* Kinahan in Northumberland waters. J. Mar. Biol. Assoc. U.K., 39: 481–508.

Ambrose Jr., W.G., 1984a. Influence of residents on the development of a marine soft-bottom community. J. Mar. Res., 42: 633–654.

Ambrose Jr., W.G., 1984b. Role of predatory infauna in structuring soft-bottom communities. Mar. Ecol. Prog. Ser., 17: 109–115.

Ambrose Jr., W.G., 1984c. Influences of predatory polychaetes and epibenthic predators on the structure of a soft-bottom community in a Maine estuary. J. Exp. Mar. Biol. Ecol., 81: 115–145.

Ambrose Jr., W.G., 1991. Are infaunal predators important in structuring marine soft-bottom communities?. Am. Zool., 31: 849–860.

Anderson, G.J., Miller, M.B. and Chew, K.K., 1982. Guide to manila clam aquaculture in Puget Sound. WA Sea Grant Tech. Rep. WSG 82–4, University of Washington, 45 pp.

Andre, C. and Rosenberg, R., 1991. Adult–larval interactions in the suspension feeding bivalves *Cerastoderma edule* and *Mya arenaria*. Mar. Ecol. Prog. Ser., 71: 227–234.

Andre, C., Jonsson, P.R. and Lindegarth, M., 1993. Predation on settling bivalve larvae by benthic suspension feeders: the role of hydrodynamics and larval behavior. Mar. Ecol. Prog. Ser., 97: 183–192.

Anger, K., Rogal, U., Schriever, G. and Valentin, C., 1977. In-situ investigations on the echinoderm *Asterias rubens* as a predator of soft-bottom communities in the western Baltic Sea. Helgolander Wiss. Meeresunters, 29: 439–459.

Ankar, S., 1980. Growth and production of *Macoma balthica* (L.) in a northern Baltic soft bottom. Ophelia, Suppl., 1: 31–48.

Ansell, A.D., 1960. Observations on predation of *Venus striatula* (Da Costa) by *Natica alderi* (Forbes). Proc. Malacol. Soc. London, 34: 157–164.

Ansell, A.D., 1982. Experimental studies of a benthic predator–prey relationship: III. Factors affecting rate of predation and growth in juveniles of the gastropod drill *Polinices catena* (Da Costa) in laboratory cultures. Malacologia, 22: 367–375.

Ansell, A.D. and Gibson, R.N., 1990. Patterns of feeding and movement of juvenile flatfishes on an open sandy beach. In: M. Barnes and R.N. Gibson (Eds.), Trophic Relationships in the Marine Environment, Proc. 24th European Mar. Biol. Symp., pp. 191–207.

Ansell, A.D. and Mace, A.M., 1978. Comparative studies of the gastropod, *Polinices alderi* from Mediterranean and North Atlantic populations. Haliotis, 9: 65–72.

ap Rheinallt, T. and Hughes, R.N., 1985. Handling methods used by the velvet swimming crab *Liocarcinus puber* when feeding on molluscs and shore crabs. Mar. Ecol. Prog. Ser., 25: 63–70.

Arnold, W.S., 1983. The effects of prey size, predator size, and sediment composition on the rate of predation of the blue crab, *Callinectes sapidus* Rathbun, on the hard clam, *Mercenaria mercenaria* (Linne). MS Thesis, University of Georgia, Athens, GA, 46 pp.

Arnold, W.S., 1984. The effects of prey size, predator size, and sediment composition on the rate of predation of the blue crab, *Callinectes sapidus* Rathbun, on the hard clam, *Mercenaria mercenaria* (Linne). J. Exp. Mar. Biol. Ecol., 80: 207–219.

Asson-Batres, M.A., 1986. The feeding behavior of the juvenile dungeness crab, *Cancer magister* Dana, on the bivalve *Transennella tantilla* (Gould), and a determination of its daily consumption rate. Calif. Fish Game, 72: 144–152.

Auster, P.J. and Crockett, L.R., 1984. Foraging tactics of several crustacean species for infaunal prey. J. Shellfish Res., 4: 139–143.

Auster, P.J. and DeGoursey, R.E., 1994. Winter predation on blue crabs, *Callinectes sapidus*, by starfish *Asterias forbesi*. J. Shellfish Res., 13: 316–366.

Baird, D. and Milne, H., 1981. Energy flow in the Ythan estuary, Aberdeenshire, Scotland. Est. Coast. Shelf Sci., 13: 455–472.

Baird, D. and Ulanowicz, R.E., 1989. The seasonal dynamics of the Chesapeake Bay ecosystem. Ecol. Monogr., 59: 329–364.

Baker, M.C. and Baker, A.E.M., 1973. Niche relationships among six species of shorebirds on their wintering and breeding ranges. Ecol. Monogr., 43: 193–212.

Baldwin, W.P., 1946. Clam catches oyster-catcher. Auk, 63: 589.

Barbeau, M.A. and Scheibling, R.E., 1994a. Behavioral mechanisms of prey size selection by sea stars (*Asterias vulgaris* Verrill) and crabs (*Cancer irroratus* Say) preying on juvenile sea scallops (*Placopecten magellanicus* (Gmelin)). J. Exp. Mar. Biol. Ecol., 180: 103–136.

Barbeau, M.A. and Scheibling, R.E., 1994b. Temperature effects on predation of juvenile sea scallops (*Placopecten magellanicus* (Gmelin)) by sea stars (*Asterias vulgaris* Verrill) and crabs (*Cancer irroratus* Say). J. Exp. Mar. Biol. Ecol., 182: 27–47.

Barbeau, M.A., Scheibling, R.E., Hatcher, B.G., Taylor, L.H. and Hennigar, A.W., 1994. Survival of tethered juvenile sea scallops *Placopecten magellanicus* in field experiments: effects of predators, scallop size and density, site and season. Mar. Ecol. Prog. Ser., 115: 243–256.

Bardach, J.E., Ryther J.H. and McLarney, W.P., 1972. Aquaculture: The Farming and Husbandry of Freshwater and Marine Organisms. John Wiley and Sons, New York, 868 pp.

Barnett, R.K., Barton, III, H. and Knutson, A.B., 1987. The use of the oyster toadfish (*Opsanus tau*) as a predator control tool in a seed clam growout project. Final Report to the NY State Urban Development Corp. By the Town of Smithtown, NY, 27 pp.

Bartoli, P., 1974. Recherches sur les Gymnophallidae Morozof, F.N., 1955 (Digenea), parasites b'oiseaux des cotes de Camargue: Systematique, Biologie et Ecologie. Ph.D. Dissertation, University Aix-Marseille, Aix-Marseille, France. (Not seen, but see Thomas et al. 1997 for information).

Bayne, B.L., 1976. The biology of mussel larvae. In: B.L. Bayne (Ed.), Marine Mussels: Their Ecology and Physiology. Cambridge Univ. Press, Cambridge, pp. 81–121.

Bayne, B.L. and Scullard, C., 1978. Rates of feeding by *Thais* (*Nucella*) *lapillis* (L.). J. Exp. Mar. Biol. Ecol., 32: 113–129.

Beal, B.F., 1983a. Effects of environment, intraspecific density, predation by snapping shrimp and other consumers on the population biology of *Mercenaria mercenaria* near Beaufort, North Carolina. MS Thesis, University of North Carolina, Chapel Hill, NC, 174 pp.

Beal, B.F., 1983b. Predation of juveniles of the hard clam *Mercenaria mercenaria* (Linne) by the snapping shrimp *Alpheus heterochaelis* Say and *Alpheus normanni* Kingsley. J. Shellfish Res., 3: 1–10.

Beal, B.F., 1991. The fate of hatchery-reared juveniles of *Mya arenaria* L. in the field: how predation and competition are affected by initial clam size and stocking density. J. Shellfish Res., 10: 292–293.

Beal, B.F., Lithgow, C.D., Shaw, D.P., Renshaw, S. and Ouellette, D., 1995. Overwintering hatchery-reared individuals of the soft-shell clam, *Mya arenaria* L.: a field test of site, clam size and intraspecific density. Aquaculture, 130: 145–158.

Belding, D.L., 1912. The quahog fishery of Massachusetts. Comm. Massachusetts Dept. Conservation. Div. Fish and Game, Mar. Fish. Ser. No. 2, 41 pp.

Bengtson, S.A. and Svensson, B., 1968. Feeding habits of *Calidris alpina* L. and *C. minuta* Leisl. (Aves) in relation to the distribution of marine shore invertebrates. Oikos, 19: 152–157.

Berg Jr., C.J., 1975. A comparison of adaptive strategies of predation among naticid gastropods. Biol. Bull., 149: 420–421.

Berg Jr., C.J. and Porter, M.E., 1974. A comparison of predatory behavior among the naticid gastropods *Lunatia heros, Lunatia triseriata* and *Polinices duplicatus*. Biol. Bull, 147: 469–470.

Bernard, F.R., 1979. The food of Hecate Strait crabs, August 1977. Manuscript Report Fisheries – Marine Serv. Canada. 1464: 1–23.

Best, B.A., 1978. The effects of suspension feeding by the bivalve, *Mercenaria mercenaria*, on community structure. M.S. Thesis, University of Florida, FL, 40 pp.

Bigelow, H.B. and Schroeder, W.C., 1953. Fishes of the Gulf of Maine. U.S. Fish and Wildlife Service Fish. Bull. 74: 1–577.

Bisker, R. and Castagna, M., 1987. Predation on single spat oysters *Crassostrea virginica* (Gmelin) by blue crabs *Callinectes sapidus* Rathbun and mud crabs *Panopeus herbstii* Milne–Edwards. J. Shellfish Res., 6: 37–40.

Black, R. and Peterson, C.H., 1988. Absence of preemption and interference competition for space between large suspension feeding bivalves and smaller infaunal macroinvertebrates. J. Exp. Mar. Biol. Ecol., 120: 183–198.

Blegvad, H., 1914. Food and conditions of nourishment among the communities of invertebrate animals found on or in the sea bottom in Danish waters. Rep. Danish Biol. Sta., 22: 41–78.

Blegvad, H., 1917. On the food of fish in the Danish waters within the Skaw. Rep. Danish Biol. Sta., 24: 17–72.

Blundon, J.A. and Kennedy, V.S., 1982a. Refuges for infaunal bivalves from blue crab, *Callinectes sapidus* (Rathbun), predation in Chesapeake Bay. J. Exp. Mar. Biol. Ecol., 65: 67–81.

Blundon, J.A. and Kennedy, V.S., 1982b. Mechanical and behavioral aspects of blue crab, *Callinectes sapidus* (Rathbun), predation on Chesapeake Bay bivalves. J. Exp. Mar. Biol. Ecol., 65: 47–65.

Boggs, C.H., Rice, J.A., Kitchell, J.A. and Kitchell, J.F., 1984. Predation at a snail's pace. What's time to a gastropod. Oecologia, 62: 13–17.

Botsford, L.W. and Wickham, D.E., 1978. Behavior of age-specific density-dependent models and the northern California Dungeness crab (*Cancer magister*) fishery. J. Fish. Res. Bd. Canada, 35: 833–843.

Botton, M.L., 1982. Predation by adult horseshoe crabs, *Limulus polyphemus*, (L.), and its effect on benthic intertidal community structure of breeding beaches in Delaware Bay, New Jersey. Ph.D. Dissertation, Rutgers University, New Brunswick, NJ, 465 pp.

Botton, M.L., 1984. Effects of laughing gull and shorebird predation on the intertidal fauna at Cape May, New Jersey. Est. Coast. Shelf Sci., 18: 209–220.

Botton, M.L. and Haskin, H.H., 1984. Distribution and feeding of the horseshoe crab, *Limulus polyphemus*, on the continental shelf off New Jersey. Fish. Bull., 82: 383–389.

Boulding, E.G., 1984. Crab-resistant features of shells of burrowing bivalves: decreasing vulnerability by increasing handling time. J. Exp. Mar. Biol. Ecol., 76: 201–223.

Boulding, E.G. and Hay, T.K., 1984. Crab response to prey density can result in density-dependent mortality of clams. Can. J. Fish. Aquatic Sci., 41: 521–525.

Boulding, E.G. and LaBarbera, M., 1986. Fatigue damage: repeated loading enables crabs to open larger bivalves. Biol. Bull., 171: 538–547.

Bourne, N., 1984. Clam predation by scoter ducks in the Strait of Georgia, British Columbia, Canada. Can. Tech. Rep. Fish. Aq. Sci. 1331, 20 pp.

Braber, L. and de Groot, S.J., 1973. The food of five flatfish species (Pleuronectiformes) in the southern North Sea. Neth. J. Sea Res., 6: 163–172.

Breese, W.P. and Phibbs, F.D., 1972. Ingestion of bivalve molluscan larvae by the polychaete annelid *Polydora ligni*. Veliger, 14: 274.

Britton, W., 1991. Clam Cultivation Manual. Aquaculture Explained, No. 8, Bord Iascaigh Mhara, Dublin, 60 pp.

Brown, K.M. and Richardson, T.D., 1987. Foraging ecology of the southern oyster drill *Thais haemastoma* (Gray): constraints on prey choice. J. Exp. Mar. Biol. Ecol., 114: 123–141.

Brown, R.A. and O'Connor, R.J., 1974. Some observations on the relationships between oystercatchers *Haematopus ostralegus* L. and cockles *Cardium edule* L. in Strangford Lough. Irish Nat. J., 18: 73–80.

Bryant, D.M., 1979. Effects of prey density and site character on estuary usage by overwintering waders (Charadrii). Est. Coast. Mar. Sci., 9: 369–384.

Buchanan, J.B. and Hedley, R.H., 1960. A contribution to the biology of *Astrorhiza limnicola* (Foraminifera). J. Mar. Biol. Assoc. U.K., 39: 549–560.

Burnett, A.L., 1960. The mechanism employed by the starfish Asterias forbesi to gain access to the interior of the bivalve *Venus mercenaria*. Ecology, 41: 583–584.

Cahn, A.R., 1951. Clam culture in Japan. U.S. Dept. Interior, Fish Wild. Serv. Fish Disease Leaflet FL-399, 103 pp.

Cake, E.W., 1970. Some predator–prey relationships involving the sunray venus clam, *Macrocallista nimbosa* (Lightfoot) (Pelecypoda: Veneridae) along the gulf coast of Florida. MS Thesis, Florida State Univ. Tallahassee, FL, 167 pp.

Calkins, D.G., 1978. Feeding behavior and major prey species of the sea otter, *Enhydra lutris*, in Montague Strait, Prince William Sound, Alaska. Fish. Bull., 76: 125–131.

Carriker, M.R., 1951. Observations on the penetration of tightly closing bivalves by *Busycon* and other predators. Ecology, 32: 73–83.

Carriker, M.R., 1955. Critical review of biology and control of oyster drills *Urosalpinx* and *Eupleura*. U.S. Dept. Interior, Fish Wild. Serv. Spec. Sci. Rep. Fisheries, No. 148, 150 pp.

Carriker, M.R., 1957. Preliminary study of behavior of newly hatched oyster drills, *Urosalpinx cinerea* (Say). J. Elisha Mitchell Sci. Soc., 73: 328–351.

Carriker, M.R., 1959. The role of physical and biological factors in the culture of *Crassostrea* and *Mercenaria* in a salt-water pond. Ecol. Monogr., 29: 219–266.

Carriker, M.R., 1961. Interrelation of functional morphology, behavior, and autecology in early stages of the bivalve *Mercenaria mercenaria*. J. Elisha Mitchell Sci. Soc., 77: 168–241.

Cantin, M., Bedard, J. and Milne, H., 1974. The food and feeding of common eiders in the St. Lawrence estuary in summer. Can. J. Zool., 52: 319–334.

Castagna, M. and Kraeuter, J.N., 1977. *Mercenaria* culture using stone aggregate for predator protection. Proc. Natl. Shellfish. Assoc., 67: 1–6.

Castagna, M. and Kraeuter, J.N., 1981. Manual for growing the hard clam, *Mercenaria*. Spec. Sci. Rep. Appl. Mar. Sci. and Ocean Engineer, No. 249. Virginia Inst. Mar. Sci. Gloucester Point, VA, 110 pp.

Castagna, M. and Kraeuter, J.N., 1994. Age, growth rate, sexual dimorphism and fecundity of knobbed whelk, *Busycon carica* (Gmelin 1791) in a western mid-Atlantic lagoon system, Virginia, USA. J. Shellfish Res., 13: 581–586.

Cayford, J.T. and Goss-Custard, J.D., 1990. Seasonal changes in the size selection of mussels, *Mytilus edulis*, by oystercatchers, *Haematopus ostralegus*: an optimality approach. Anim. Behav., 40: 609–624.

Chestnut, A.F., 1952. Growth rates and movements of hard clams, *Venus mercenaria*. Proc. Gulf Carib. Fish Inst, 4th Annual Session, pp. 49–59.

Chew, K.K., 1989. Manila clam biology and fishery development in Western North America. In: J.J. Manzi and M. Castagna (Eds.), Clam mariculture in North America. Developments in Aquaculture and Fisheries Science, No. 19, pp. 243–261.

Christensen, A.M., 1962. Some aspects of prey–predator relationships in marine level-bottom animal communities. First National Coastal and Shallow Water Research Conference, pp. 69–70.

Coen, L.D. and Heck Jr., K.L., 1991. The interacting effects of siphon nipping and habitat on bivalve (*Mercenaria mercenaria* (L.)) growth in a subtropical seagrass (*Halodule wrightii* Aschers) meadow. J. Exp. Mar. Biol. Ecol., 145: 1–13.

Cohen, A.N., Carlton, J.T. and Fountain, M.C., 1995. Introduction, dispersal and potential impacts of the green crab *Carcinus maenas* in San Francisco Bay, California. Mar. Biol., 122: 225–237.

Cole, H.A., 1942. The American whelk tingle, *Urosalpinx cinerea* (Say), on British oyster beds. J. Mar. Biol. Assoc. U.K., 25: 477–508.

Colton, H.S., 1908. How *Fulgar* and *Sycotypus* eat oysters, mussels and clams. Proc. Acad. Nat. Sci. Philadelphia, 60: 3–10.

Commito, J.A., 1982. Effects of *Lunatia heros* predation on the population dynamics of *Mya arenaria* and *Macoma balthica* in Maine, USA. Mar. Biol., 69: 187–193.

Commito, J.A., 1987. Adult–larval interactions: predictions, mussels and cocoons. Est. Coast. Shelf. Sci., 25: 599–606.

Cowden, C., Young, C.M. and Chia, F.S., 1984. Differential predation on marine invertebrate larvae by two benthic predators. Mar. Ecol. Prog. Ser., 14: 145–149.

Craig, M.A. and Bright, T.J., 1986. Abundance, age distributions and growth of the Texas hard clam, *Mercenaria mercenaria texana*, in Texas bays. Contrib. Mar. Sci., 29: 59–72.

Crowl, T.A. and Covich, A.P., 1990. Predator-induced life-history shifts in a freshwater snail. Science, 247: 949–951.

Dame, R.F., 1993. Bivalve filter feeders in estuarine and coastal ecosystem processes. NATO ASI Series, Series G: Ecological Sciences, 33: 1–597.

Dame, R.F. and Patten, B.C., 1981. Analysis of energy flows in an intertidal oyster reef model. Mar. Ecol. Prog. Ser., 5: 115–124.

Dame, R.F. and Vernberg, F.J., 1982. Energetics of a population of the mud crab *Panopeus herbstii* (Milne Edwards) in the North Inlet estuary, South Carolina. J. Exp. Mar. Biol. Ecol., 63: 183–193.

Dankers, N., 1993. Integrated estuarine management — obtaining a sustained yield of bivalve resources while maintaining environmental quality. In: R.F. Dame (Ed.), Bivalve Filter Feeders in Estuarine and Coastal Ecosystem Processes. NATO ASI Series G: Ecological Sciences 33. Springer, New York, pp. 479–511.

Dare, P.J. and Edwards, D.B., 1976. Experiments on the survival, growth and yield of relaid seed mussels (*Mytilus edulis*) in the Menai Straits, North Wales. J. Cons. Int. Explor. Mer., 37: 16–28.

Dare, P.J. and Mercer, A.J., 1973. Foods of the oystercatcher in Morecambe Bay, Lancashire. Bird Study, 20: 173–184.

Dare, P.J., Davies, G. and Edwards, D.B., 1983. Predation on juvenile Pacific oysters (*Crassostrea gigas* Thunberg) and mussels (*Mytilus edulis* L.) by shore crabs (*Carcinus maenas* (L.)). Min. Agric. Fish. Food. Direct. Fish. Res. Lowestof. Fish. Res. Tech. Rep., No. 73, 15 pp.

Daro, M.H. and Polk, P., 1973. The autecology of *Polydora ciliata* along the Belgian coast. Neth. J. Sea Res., 6: 130–140.

Davidson, P.E., 1967. A study of the oystercatcher (*Haematopus ostralegus*) in relation to the fishery for cockles

(*Cardium edule* L.) in Burry Inlet, South Wales. Min. Agric. Food Fish. Fish. Invest. London, Ser. 2, 25(7): 1–28.

Davidson, R.J., 1986. Mussel selection by the paddle crab *Ovalipes cantharus* (White): evidence of flexible foraging behavior. J. Exp. Mar. Biol. Ecol., 102: 281–299.

Davies, G., Dare, P.J. and Edwards, D.B., 1980. Fenced enclosures for the protection of seed mussels (*Mytilus edulis* L.) from predation by shore crabs (*Carcinus maenas* (L.). Min. Agric. Fish. Food, Fish. Res. Tech. Rep. 56, 14 pp.

Davis, J.P. and Sisson, R.T., 1988. Aspects of the biology relating to fisheries management of New England populations of whelks, *Busycotypus canaliculatus* and *Busycon carica*. J. Shellfish Res., 7: 453–460.

Day, E., 1987. Substrate type and predatory risk: effects on mud crab predation of juvenile hard clams. MS Thesis, Marine Sciences Research Center, State University of New York. Stony Brook, NY, 112 pp.

Day, E. and Lawton, P., 1988. Mud crab (Crustacea: Brachyura: Xanthidae) substrate preference and activity. J. Shellfish Res., 7: 421–426.

Dean, D., 1981. Predation on juvenile clams by sandworms. Proc. 8th Ann. Clam Conference, Boothbay Harbor, ME, pp. 72–78.

De Vlas, J., 1979. Annual food intake by plaice and flounder in a tidal flat area in the Dutch Wadden Sea, with special reference to consumption of regenerating parts of macrobenthic prey. Neth. J. Sea Res., 13: 117–153.

Dewar, J.M., 1922. Ability of the oystercatcher to open oysters, and its bearing upon the history of the species. Br. Birds, 16: 118–125.

Dietl, G.P. and Alexander, R.R., 1997. Predator–prey interactions between the naticids *Euspira heros* Say and *Neverita duplicata* Say and the Atlantic surfclam *Spisula solidissima* Dillwyn from Long Island to Delaware. J. Shellfish Res., 16: 413–422.

Doering, P.H., 1981a. A burrowing response of *Mercenaria mercenaria* (Linnaeus, 1758) elicited by *Asterias forbesi* (Desor, 1848). Veliger, 19: 167–176.

Doering, P.H., 1981b. Observations on the behavior of *Asterias forbesi* feeding on *Mercenaria mercenaria*. Ophelia, 20: 169–177.

Doering, P.H., 1982. Reduction of sea star predation by the burrowing response of the hard clam *Mercenaria mercenaria* (Mollusca: Bivalvia). Estuaries, 5: 310–315.

Drinnan, R.E., 1957. The winter feeding of the oystercatcher (*Haematopus ostralegus*) on the edible cockle (*Cardium edule*). J. Anim. Ecol., 26: 441–469.

Drinnan, R.E., 1958. The winter feeding of the oystercatcher (*Haematopus ostralegus*) on the edible mussel (*Mytilus edulis*) in the Conway estuary, North Wales. Min. Agric. Fish. Food. Fish. Invest. London, Ser. 2, 22: 1–15.

Dumbauld, B.R., Armstrong, D.A. and McDonald, T.L., 1993. Use of oyster shell to enhance intertidal habitat and mitigate the loss of Dungeness crab (*Cancer magister*) caused by dredging. Can. J. Fish. Aq. Sci., 50: 381–390.

Durbin, A.G. and Durbin, E.G., 1975. Grazing rates of the Atlantic menhaden, *Brevoortia tyrannus*, as a function of particle size and concentration. Estuaries, 4: 21–41.

Durbin, A.G. and Durbin, E.G., 1998. Effects of menhaden predation on plankton populations in Narragansett Bay, Rhode Island. Estuaries, 21: 449–465.

Du Preez, H.H., 1984. Molluscan predation by *Ovalipes punctatus* (De Haan) (Crustacea: Brachyura: Portunidae). J. Exp. Mar. Biol. Ecol., 84: 55–71.

Ebersole, E.L. and Kennedy, V.S., 1994. Size selection of Atlantic Rangia clams, *Rangia cuneata*, by blue crabs, *Callinectes sapidus*. Estuaries, 17: 668–673.

Ebersole, E.L. and Kennedy, V.S., 1995. Prey preference of blue crabs *Callinectes sapidus* feeding on three bivalve species. Mar. Ecol. Prog. Ser., 118: 167–177.

Edwards, D.C., 1974. Preferred prey of *Polinices duplicatus* in Cape Cod inlets. Bull. Am. Malacol. Union, pp. 17–20.

Edwards, D.C. and Huebner, J.D., 1977. Feeding and growth rates of *Polinices duplicatus* preying on *Mya arenaria* at Barnstable Harbor, Massachusetts. Ecology, 58: 1218–1236.

Eggleston, D.B., 1990a. Foraging behavior of the blue crab, *Callinectes sapidus*, on juvenile oysters, *Crassostrea virginica*: effects of prey density and size. Bull. Mar. Sci., 46: 62–82.

Eggleston, D.B., 1990b. Behavioral mechanisms underlying variable functional responses of blue crabs, *Callinectes sapidus*, feeding on juvenile oysters, *Crassostrea virginica*. J. Anim. Ecol., 59: 615–630.

Eggleston, D.B., 1990c. Functional responses of blue crabs *Callinectes sapidus* Rathbun feeding on juvenile oysters *Crassostrea virginica* (Gmelin): effects of predator sex and size, and prey size. J. Exp. Mar. Biol. Ecol., 143: 73–90.

Eggleston, D.B., Lipcius, R.N. and Hines, A.H., 1992. Density-dependent predation by blue crabs upon infaunal clam species with contrasting distribution and abundance patterns. Mar. Ecol. Prog. Ser., 85: 55–68.

Ejdung, G. and Bonsdorff, E., 1992. Predation on the bivalve *Macoma balthica* by the isopod *Saduria enotmon*: laboratory and field experiments. Mar. Ecol. Prog. Ser., 88: 207–214.

Eldridge, P.J. and Eversole, A.G., 1982. Compensatory growth and mortality of the hard clam, *Mercenaria mercenaria* (Linnaeus, 1758). Veliger, 24: 276–278.

Eldridge, P.J., Waltz, W., Gracy, R.C. and Hunt, H.H., 1976. Growth and mortality rates of hatchery seed clams, *Mercenaria mercenaria*, in protected trays in waters of South Carolina. Proc. Natl. Shellfish. Assoc., 66: 13–20.

Eldridge, P.J., Eversole, A.G. and Whetstone, J.M., 1979. Comparative survival and growth rates of hard clams *Mercenaria mercenaria*, planted in trays subtidally and intertidally at varying densities in a South Carolina estuary. Proc. Natl. Shellfish. Assoc., 69: 30–39.

Elmgren, R., Ankar, S., Marteleur, B. and Ejdung, G., 1986. Adult interference with postlarvae in soft sediments: the *Pontoporeia-Macoma* example. Ecology, 67: 827–836.

Elner, R.W., 1980. The influence of temperature, sex and chela size in the foraging strategy of the shore crab, *Carcinus maenas* (L.). Mar. Behav. Physiol., 7: 15–24.

Elner, R.W., 1981. Diet of green crab *Carcinus maenas* (L.) from Port Herbert, southwestern Nova Scotia. J. Shellfish Res., 1: 89–94.

Elner, R.W. and Hughes, R.N., 1978. Energy maximization in the diet of the shore crab, *Carcinus maenas*. J. Anim. Ecol., 47: 103–116.

Elner, R.W. and Jamieson, G.S., 1979. Predation of sea scallops, *Placopecten magellanicus*, by the rock crab, *Cancer irroratus*, and the American lobster, Homarus americanus. J. Fish. Res. Bd. Canada, 36: 537–543.

Ens, B.J. and Goss-Custard, J.D., 1984. Interference among oystercatchers, *Haematopus ostralegus*, feeding on mussels, *Mytilus edulis*, on the Exe estuary. J. Anim. Ecol., 53: 217–231.

Ertman, S.C. and Jumars, P.A., 1988. Effects of bivalve siphonal currents on the settlement of inert particles and larvae. J. Mar. Res., 46: 797–813.

Evans, P.R., Herdson, D.M., Knights, P.J. and Piekowski, M.W., 1979. Short-term effects of reclamation of part of Seal Sands, Teesmouth, on wintering waders and shelduck. I. Shorebird diets, invertebrate densities, and the impact of predation on the invertebrates. Oecologia, 41: 183–206.

Evans, S., 1983. Production, predation and food niche segregation in a marine shallow soft-bottom community. Mar. Ecol. Prog. Ser., 10: 147–157.

Feder, H.M., 1955. On the methods used by the starfish *Pisaster ochraceus* in opening three types of bivalve mollusks. Ecology, 36: 764–767.

Feder, H.M., 1956. Natural history studies on the starfish *Pisaster ochraceus* (Brandt, 1835) in the Monterey Bay area. Doctoral Dissertation, Stanford University, 294 pp.

Feder, H.M. and Paul, A.J., 1980. Food of the king crab, *Paralithodes camtschatica*, and the Dungeness crab, *Cancer magister*, in Cook Inlet, Alaska. Proc. Natl. Shellfish. Assoc., 70: 240–246.

Fernandez, M., Armstrong, D. and Iribarne, O., 1993. First cohort of young-of-the-year Dungeness crab, *Cancer magister*, reduces abundance of subsequent cohorts in intertidal shell habitat. Can. J. Fish. Aq. Sci., 50: 2100–2105.

Fitch, J.E., 1965. A relatively unexploited population of Pismo clams, *Tivela stuttorum* (Mawe, 1823) (Veneridae). Proc. Malac. Soc. London, 36: 309–312.

Fitz, H.C. and Wiegert, R.G., 1991a. Tagging juvenile blue crab, *Callinectes sapidus*, with microwire tags: retention, survival, and growth through multiple molts. J. Crust. Biol., 11: 229–245.

Fitz, H.C. and Wiegert, R.G., 1991b. Utilization of the intertidal zone of a salt marsh by the blue crab *Callinectes sapidus*: density, return frequency, and feeding habits. Mar. Ecol. Prog. Ser., 76: 249–260.

Flagg, P.J. and Malouf, R.E., 1983. Experimental plantings of juveniles of the hard clam *Mercenaria mercenaria* (Linne) in the waters of Long Island, New York. J. Shellfish Res., 3: 19–27.

Frame, D.W., 1974. Feeding habits of young winter flounder (*Pseudopleuronectes americanus*): prey availability and diversity. Trans. Am. Fish. Soc., 1974: 261–268.

Franz, D.R., 1977. Size and age-specific predation by *Lunatia heros* (Say, 1822) on the surf clam *Spisula solidissima* (Dillwyn, 1817) off western Long Island, New York. Veliger, 20: 144–150.

Frechette, M., Butman, C.A. and Geyer, W.R., 1989. The importance of boundary-layer flows in supplying phytoplankton to the benthic suspension feeder *Mytilus edulis* L. Limnol. Oceanogr., 34: 19–36.

Gallagher, E.D., Jumars, P.A. and Trueblood, D.D., 1983. Facilitation of soft-bottom benthic succession by tube builders. Ecology, 64: 1200–1216.

Galleni, L., Tongiorgi, P., Ferrero, E. and Salghetti, U., 1980. *Stylochus mediterraneus* (Turbellaria: Polycladida), predator on the mussel *Mytilus galloprovincialis*. Mar. Biol., 55: 317–326.

Galtsoff, P.S., Prytherch, H.F. and Engle, J.B., 1937. Natural history and methods of controlling the common oyster drills (*Urosalpinx cinerea* Say and *Eupleura caudata* Say). U.S. Bur. Comm. Fish. Circular, No. 25, pp. 1–24.

Garton, D. and Stickle, W.B., 1980. Effects of salinity and temperature on the predation rate of *Thais haemastoma* on *Crassostrea virginica* spat. Biol. Bull., 158: 49–57.

Gibbons, M.C., 1984. Aspects of predation by the crabs *Neopanope sayi, Ovalipes ocellatus*, and *Pagurus longicarpus* on juvenile hard clams *Mercenaria mercenaria*. Ph.D. Dissertation, Graduate School, State University of New York at Stony Brook, Stony Brook, NY, 96 pp.

Gibbons, M.C. and Blogoslawski, W.J., 1989. Predators, pests, parasites, and diseases. In: J.J. Manzi and M. Castagna (Eds.), Clam Mariculture in North America. Developments in Aquaculture and Fisheries Science, Vol. 19, Elsevier, Amsterdam, pp. 167–200.

Gibbons, M.C. and Castagna, M., 1985. Biological control of predation by crabs in bottom cultures of hard clams using a combination of crushed stone aggregate, toadfish and cages. Aquaculture, 47: 101–104.

Glude, J.B., 1955. The effects of temperature and predators on the abundance of the soft-shell clam, *Mya arenaria*, in New England. Trans. Am. Fish. Soc., 1994: 13–26.

Glude, J.B., 1964. The effect of scoter duck predation on a clam population in Dabob Bay, Washington. Proc. Natl. Shellfish. Assoc., 55: 73–86.

Goudie, R.I. and Ankney, C.D., 1986. Body size, activity budgets, and diets of sea ducks wintering in Newfoundland. Ecology, 67: 1475–1482.

Gosnor, K.L., 1979. A Field Guide to the Atlantic Seashore. Houghton Mifflin, Boston, MA, 329 pp.

Goss-Custard, J.D., Jones, R.E. and Newbery, E., 1977a. The ecology of the Wash Distribution, I. and diet of wading birds (Charadrii). J. Appl. Ecol., 14: 681–700.

Goss-Custard, J.D., Kay, D.G. and Blindell, R.M., 1977b. The density of migratory and overwintering redshank, *Tringa totanus* (L.) and curlew, *Numenius arquata* (L.),in relation to the density of their prey in south-east England. Est. Coast. Mar. Sci., 5: 497–510.

Gotshall, D.W., 1977. Stomach contents of northern California Dungeness crab, *Cancer magister*. Calif. Fish. Game, 63: 43–51.

Grant, J. and McDonald, J., 1979. Desiccation tolerance of *Eurypanopeus depressus* (Smith) (Decapoda: Xanthidae) and the exploitation of microhabitat. Estuaries, 2: 172–177.

Greene, G.T., 1978. Population structure, growth and mortality of hard clams at selected locations in Great South Bay. Masters Thesis. Marine Environmental Sciences Program, State University of New York, Stony Brook, NY, 199 pp.

Gregory, M.R., Ballance, P.F., Gibson, G.W. and Ayling, A.M., 1979. On how some rays (Elasmobranchia) excavate feeding depressions by jetting water. J. Sed. Petrol., 49: 1125–1130.

Griffiths, C.L. and Seiderer, J.L., 1980. Rock-lobsters and mussels limitations and preferences in a predator–prey interaction. J. Exp. Mar. Biol. Ecol., 44: 95–109.

Griffiths, R.J., 1981. Predation on the bivalve *Choromytilus meridionalis* (Kr.) by the gastropod *Natica* (*Tectonatica*) *tecta* Anton. J. Moll. Stud., 47: 112–120.

Grosholz, E.D. and Ruiz, G.M., 1996. Assessing the impact of introduced marine species: lessons from the multiple invasions of the European green crab *Carcinus maenas*. Biol. Conserv., 78: 59–66.

Guillou, M., 1990. Biotic interactions between predators and super-predators in the Bay of Douarnenez, Brittany. In: M. Barnes and R.N. Gibson (Eds.), Trophic Relationships in the Marine Environment, Proc. 24th European Mar. Biol. Symp., pp. 141–156.

Gulliksen, B. and Skjaeveland, H., 1973. The sea star, *Asterias rubens* L., as predator on the ascidian, *Cliona intestinalis* L., in Brogenfjorden, Nordt-Trondelag, Norway. Sarsia, 52: 15–20.

Haddon, M. and Wear, R.G., 1987. Biology and feeding in the New Zealand paddle crab *Ovalipes catharus* (Crustacea Portunidae). New Zealand J. Mar. Freshwat. Res., 21: 55–64.

Haddon, M., Wear, R.G. and Packer, H., 1987. Depth and density of burial by the bivalve *Paphies ventricosa* as refuges from predation by the crab *Ovalipes catharus*. Mar. Biol., 94: 25–30.

Haefner Jr., P.A., 1972. The biology of the sand shrimp, *Crangon septemspinosa*, at Lamoine. Maine. J. Elisha Mitchell Sci. Soc., 88: 36–42.

Hancock, D.A., 1954. The destruction of oyster spat by *Urosalpinx cinerea* (Say) on Essex oyster beds. J. Conseil Int. Explor. Mer., 1954: 186–196.

Hancock, D.A., 1955. The feeding behavior of starfish on Essex oyster beds. J. Mar. Biol. Assoc. U.K., 34: 313–331.

Hancock, D.A., 1970. The role of predators and parasites in a fishery for the mollusc *Cardium edule* L. In: Proceedings Adv. Study Inst. on dynamics of numbers in populations (Oosterbeck). Center for Agricultural Publishing and Documentation, Wageningen, Netherlands, pp. 419–439.

Hancock, D.A. and Urquhart, A.E., 1965. The determination of natural mortality and its causes in an exploited population of cockles (*Cardium edule* L.). Min. Ag. Food Fish. Fishery Investigation, London, Ser. 2, 24(2): 1–40.

Hanks, J.E., 1952. The effect of changes in water temperature and salinity on the feeding habits of the boring snails, *Polinices heros* and *Polinices duplicata*. In: Fifth report on investigations of the shellfisheries of Massachusetts. Department of Conservation, Div. Mar. Fish. Commonwealth of Massachusetts, pp. 33–37.

Hanks, J.E., 1957. The rate of feeding of the common oyster drill, *Urosalpinx cinerea* (Say), at controlled temperatures. Biol. Bull., 112: 330–335.

Hanks, J., 1961. Chemical control of the green crab, Carcinus maenas, (L.). Proc. Natl. Shellfish. Assoc., 52: 75–86.

Harris, M.P., 1965. The food of some gulls, *Larus*. Ibis, 107: 45–53.

Haskin, H.H., 1951. Progress report on hard clam studies to the Campbell Soup Company, Haskin Shellfish Res. Lab., Inst. Mar. Coastal Sci. Rutgers Univ., Bivalve, NJ, 53 pp.

Hatanaka, M. and Kosaka, M., 1959. Biological studies on the population of the starfish *Asterias amurensis*, in Sendai Bay, Tohoku. J. Agric. Res., 4: 149–178.

Haven, D.S. and Andrews, J.D., 1957. Survival and growth of *Venus mercenaria*, *Venus campechiensis*, and their hybrids suspended in trays and on natural bottoms. Proc. Natl. Shellfish. Assoc., 47: 43–49.

Heppleston, P.B., 1971. The feeding ecology of oystercatchers (*Haematopus ostralegus* L.) in winter in northern Scotland. J. Anim. Ecol., 40: 651–672.

Hibbert, C.J., 1975. Production studies of a bivalve population on an intertidal mudflat, with particular reference to the energy budget of *Mercenaria mercenaria* (Linne). Ph.D. thesis, University of Southampton, Southampton, England, 225 pp.

Hibbert, C.J., 1977a. Growth and survivorship in a tidal-flat population of the bivalve *Mercenaria mercenaria* from Southampton Water. Mar. Biol., 44: 71–76.

Hibbert, C.J., 1977b. Energy relations of the bivalve *Mercenaria mercenaria* on an intertidal mudflat. Mar. Biol., 44: 77–84.

Hicklin, P.W. and Smith, P.C., 1979. The diets of five species of migrant shorebirds in the Bay of Fundy. Proc. Nova Scotia Inst. Sci., 29: 483–488.

Higgins, K., Hastings, A., Sarvela, J.N. and Botsford, L., 1997. Stochastic dynamics and deterministic skeletons: population behavior of Dungeness crab. Science, 276: 1431–1435.

Hines, A.H. and Loughlin, T.R., 1980. Observations of sea otters digging for clams at Monterey Harbor, California. Fish. Bull., 78: 159–163.

Hines, A.H., Posey, M.H. and Haddon, P.J., 1989. Effects of adult suspension- and deposit-feeding bivalves on recruitment of estuarine infauna. Veliger, 32: 109–119.

Hines, A.H., Haddon, A.M. and Wiechert, L.A., 1990. Guild structure and foraging of blue crabs and epibenthic fish in a subestuary of Chesapeake Bay. Mar. Ecol. Prog. Ser., 67: 105–126.

Hockey, P.A.R. and Steele, W.K., 1990. Intraspecific kleptoparasitism and foraging efficiency as constraints on food selection in kelp gulls *Larus dominicanus*. In: R.N. Hughes (Ed.), Behavioral mechanisms of food selection. NATO ASI Series. G20: 679–706.

Holland, A.F., Mountford, N.K., Hiegel, M.H., Kaumeyer, K.R. and Mihursky, J.A., 1980. Influence of predation on infaunal abundance in upper Chesapeake Bay, USA. Mar. Biol., 57: 221–235.

Holmes, R.T. and Pitelka, F.A., 1968. Food overlap among coexisting sandpipers on northern Alaskan tundra. Syst. Zool., 17: 305–318.

Horwood, J.W. and Goss-Custard, J.D., 1977. Predation by the oystercatcher, *Haematopusostralegus* (L.), in relation to the cockle, *Cerastoderma edule* (L.), fishery in the Burry Inlet, Wales. J. App. Ecol., 14: 139–158.

Howard, J.D., Mayou, T.V. and Heard, R.W., 1977. Biogenic sedimentary structures formed by rays. J. Sed. Petrol., 47: 341–346.

Huebner, J.D. and Edwards, D.C., 1981. Energy budget of the predatory marine gastropod *Polinices duplicatus*. Mar. Biol., 61: 221–226.

Hughes, R.N., 1970. Population dynamics of the bivalve *Scrobicularia plana* (Da Costa) on an intertidal mud-flat in north Wales. J. Anim. Ecol., 39: 333–356.

Hughes, R.N., 1985. Predatory behavior of *Natica unifasciata* — feeding intertidally on gastropods. J. Moll. Stud., 51: 331–335.

Hughes, R.N. and Dunkin, S. de B., 1984. Effect of dietary history on selection of prey, and foraging behavior among patches of prey, by the dogwhelk, *Nucella lapillis* (L.). J. Exp. Mar. Biol. Ecol., 79: 159–172.

Hughes, R.N. and Seed, R., 1981. Size selection of mussels by the blue crab *Callinectes sapidus*: energy maximizer or time minimizer?. Mar. Ecol. Prog. Ser., 6: 83–89.

Hulscher, J.B., 1974. An experimental study of the food intake of the oystercatcher *Haematopus ostralegus* L. in captivity during the summer. Ardea, 62: 155–171.

Hulscher, J.B., 1982. The oystercatcher *Haematopus ostralegus* as a predator of the bivalve *Macoma balthica* in the Dutch Wadden Sea. Ardea, 70: 89–152.

Humphrey, E.C., 1990. Feeding studies of the muricid gastropod *Ocenebra erinacea* (Linnaeus, 1758), with ultrastructural and histochemical investigations of the ABO. Ph.D. Dissertation, Dept. Oceanogr. Univ. Southampton, England, 169 pp.

Hunt, J.H., Ambrose Jr., W.G. and Peterson, C.H., 1987. Effects of the gastropod, *Illyanassa obsoleta* (Say), and the bivalve, *Mercenaria mercenaria* (L.), on larval settlement and juvenile recruitment of infauna. J. Exp. Mar. Biol. Ecol., 108: 229–240.

Hunt, O.D., 1925. The food of the bottom fauna of the Plymouth fishing grounds. J. Mar. Biol. Assoc. U.K., 13: 560–599.

Hylleberg, J., Brock, V. and Jorgensen, F., 1978. Production of sublittoral cockles, *Cardium edule* L. with emphasis on predation by flounders and sea stars. Natura Jutlandica, 20: 183–191.

IFREMER, 1988. La Palourde-dossier d'elevage. IFREMER Doc. and Publ. Unit, Plouzane, France, 106 pp.

Irons, D.B., Anthony, R.G. and Estes, J.A., 1986. Foraging strategies of glaucous-winged gulls in a rocky intertidal community. Ecology, 67: 1460–1474.

Irlandi, E.A. and Peterson, C.H., 1991. Modification of animal habitat by large plants: mechanisms by which seagrasses influence clam growth. Oecologia, 87: 307–318.

Janssen, G.M. and Kuipers, B.R., 1980. On tidal migration in the shrimp *Crangon crangon*. Neth. J. Sea Res., 14: 339–348.

Jennings, J.B., 1957. Studies on feeding, digestion and food storage in free-living flatworms (Platyhelminthes; Turbellaria). Biol. Bull., 95: 320–332.

Jensen, K.T., 1985. The presence of the bivalve *Cerastoderma edule* affects migration, survival and reproduction of the amphipod *Corophium volutator*. Mar. Ecol. Prog. Ser., 25: 269–277.

Jensen, K.T. and Jensen, J.N., 1985. The importance of some epibenthic predators on the density of juvenile benthic macrofauna in the Danish Wadden Sea. J. Exp. Mar. Biol. Ecol., 89: 157–174.

Jewett, S.C. and Feder, H.M., 1980. Autumn food of adult starry flounders, *Platichthys stellatus*, from the northeastern Bering Sea and the southeastern Chukchi Sea. J. Cons. Int. Explor. Mer., 39: 7–14.

Johnson, J.H., Dropkin, D.S., Warkentine, B.E., Rachlin, J.W. and Andrews, W.D., 1997. Food habits of Atlantic Sturgeon off the central New Jersey coast. Trans. Am. Fish. Soc., 126: 166–170.

Johnson, L., 1994. Pattern and process in ecological systems: a step in the development of a general ecological theory. Can. J. Fish. Aquat. Sci., 51: 226–246.

Juanes, F., 1992. Why do decapod crustaceans prefer small-sized molluscan prey?. Mar. Ecol. Prog. Ser., 87: 239–249.

Juanes, F. and Hartwick, E.B., 1990. Prey size selection in Dungeness crabs: the effect of claw damage. Ecology, 71: 744–758.

Jubb, C.A., Hughes, R.N. and Rheinallt, T., 1983. Behavioral mechanisms of size-selection by crabs, *Carcinus maenas* (L.) feeding on mussels, *Mytilus edulis* L.. J. Exp. Mar. Biol. Ecol., 66: 81–87.

Kalin, R.J., 1984. Observations of a feeding method of the Atlantic ribbon worm, *Cerebratulus lacteus*. Estuaries, 7: 179–180.

Kato, K., 1944. Polyclads of Japan. Sigen kagagu kenkyusho 1: 257–318 (not seen).

Kelso, W.E., 1979. Predation on soft-shell clams, *Mya arenaria*, by the common mummichog, *Fundulus heteroclitus*. Estuaries, 2: 249–254.

Kennedy, V.S., 1995. Biology of larvae and spat. In: V.S. Kennedy, R.I.E. Newell and A.F. Eble (Eds.), The Eastern Oyster *Crassostrea virginica*. Maryland Sea Grant College, University of Maryland, College Park, MD. Publication No. UM-SG-TS-96-01, pp. 371–421.

Kennish, M.J., 1980. Shell microgrowth analysis, *Mercenaria mercenaria* as a type example for research in population dynamics. In: D.C. Rhoads and R.A. Lutz (Eds.), Skeletal Growth of Aquatic Organisms. Plenum Press, NY, pp. 255–294.

Kent, B.W., 1981. Prey dropped by herring gulls (*Larus argentatus*) on soft sediments. Auk, 98: 350–354.

Kent, B.W., 1983. Patterns of coexistence in Busyconine whelks. J. Exp. Mar. Biol. Ecol., 66: 257–283.

Kim, Y.K., 1969. An observation on the opening of bivalve molluscs by starfish, *Asterias amurensis*. Bull. Fac. Fisheries, Hokkaido Univ. 20: 60–63 + 2 plates.

Kitchell, J.A., Boggs, C.H., Kitchell, J.F. and Rice, J.A., 1981. Prey selection by naticid gastropods: experimental tests and application to the fossil record. Paleobiology, 7: 533–552.

Kneib, R.T., 1995. Behavior separates potential and realized effects of decapod crustaceans in salt marsh communities. J. Exp. Mar. Biol. Ecol., 193: 239–256.

Kneib, R.T. and Stiven, A.E., 1982. Benthic invertebrate responses to size and density manipulations of the common munnichog, *Fundulus heteroclitus*, in an intertidal salt marsh. Ecology, 63: 1518–1532.

Korringa, P., 1941. Experiments and observations on swarming, pelagic life and setting in the European flat oyster, *Ostrea edulis*. Arch. Neerl. Zool., 5: 1–249.

Kosaka, M., 1970. On the ecology of the sand shrimp, *Crangon affinis* DeHaan, as a prey of demersal fishes in Sendai Bay. J. Coll. Mar. Sci. Technol., Tokai Univ., 4: 59–80.

Kraeuter, J.N. and Castagna, M., 1977. An analysis of gravel, pens, crab traps and current baffles as protection for juvenile hard clams (*Mercenaria mercenaria*). Proc. World Mariculture Soc., 8: 581–592.

Kraeuter, J.N. and Castagna, M., 1980. Effects of large predators on the field culture of the hard clam, *Mercenaria mercenaria*. Fish. Bull., 78: 538–541.

Kraeuter, J.N. and Castagna, M., 1985a. The effects of seed size, shell bags, crab traps, and netting on the survival of the northern hard clam *Mercenaria mercenaria* (L.). J. Shellfish Res., 5: 69–72.

Kraeuter, J.N. and Castagna, M., 1985b. The effect of clam size, net size and poisoned bait treatments on survival of hard clam, *Mercenaria mercenaria*, seed in field plots. J. World Maric. Soc., 16: 337–385.

Kraeuter, J.N. and Castagna, M., 1989. Factors affecting the growth and survival of clam seed planted in the natural environment. In: J.J. Manzi and M. Castagna (Eds.), Clam Mariculture in North America. Developments in Aquaculture and Fisheries Science, Vol. 19, Elsevier, Amsterdam, pp. 149–165.

Kraeuter, J.N. and Fegley, S.R., 1994. Vertical disturbance of sediments by horseshoe crabs (*Limulus polyphemus)* during their spawning season. Estuaries, 17: 288–294.

Kraeuter, J.N., Castagna, M. and Bisker, R., 1989. Growth rate estimates for *Busycon carica* (Gmelin, 1791) in Virginia. J. Shellfish Res., 8: 219–225.

Krantz, G.E. and Chamberlin, J.V., 1978. Blue crab predation on cultchless oyster spat. Proc. Natl. Shellfish. Assoc., 68: 38–41.

Kristensen, I., 1957. Differences in density and growth in a cockle population in the Dutch Wadden Sea. Arch. Neel. Zool., 12: 351–453.

Kuipers, B.R., 1977. On the ecology of juvenile plaice on a tidal flat in the Wadden Sea. Neth. J. Sea. Res., 11: 59–91.

Kurkowski, K.P., 1981. Effects of filtration by adult *Mercenaria mercenaria* upon its larvae. MS Thesis, State University of New York, Stony Brook, NY, 70 pp.

Kvitek, R.G., Fukayama, A.K., Anderson, B.S. and Grimm, B.K., 1988. Sea otter foraging on deep-burrowing bivalves in a California coastal lagoon. Mar. Biol., 98: 157–167.

Kvitek, R.G., DeGange, A.R. and Beitler, M.K., 1991. Paralytic shellfish poisoning toxins mediate feeding behavior of sea otters. Limnol. Oceanogr., 36: 393–404.

Lake, N.C.H., Jones, M.B. and Paul, J.D., 1987. Crab predation on scallop (*Pecten maximus*) and its implication for scallop cultivation. J. Mar. Biol. Assoc. U.K., 67: 55–64.

Landers, W.S., 1954. Notes on the predation of the hard clam, *Venus mercenaria* by the mud crab, *Neopanope texana*. Ecology, 35: 422.

Landers, W.S., 1955. *Venus* predators in Rhode Island. In: Fifth Conference on Clam Research. US Fish and Wildlife Service Clam Investigations, Boothbay Harbor, ME, pp. 60–61.

Landers, W.S., 1967. Infestation of the hard clam, *Mercenaria mercenaria*, by the boring polychaete worm, *Polydora ciliata*. Proc. Natl. Shellfish. Assoc., 57: 63–66.

Landers, W.S. and Rhodes Jr., E.W., 1970. Some factors influencing predation by the flatworm, *Stylochus ellipticus* (Girard), on oysters. Chesapeake Sci., 11: 55–60.

Landry, T., Shepton, T.W. and Jones, D.A., 1993. Growth and mortality of the northern quahog, (Linnaeus, 1758) *Mercenaria mercenaria* in Prince Edward Island. J. Shellfish Res., 12: 321–327.

Laughlin, R.A., 1982. Feeding habits of the blue crab, *Callinectes sapidus* Rathbun, in the Apalachicola estuary, Florida. Bull. Mar. Sci., 32: 807–822.

Lavoie, M.S., 1956. How sea stars open bivalves. Biol. Bull., 111: 114–122.

Leber, K.M., 1985. The influence of predatory decapods, refuge, and microhabitat selection on seagrass communities. Ecology, 66: 1951–1964.

Lebour, M.V., 1922. The food of planktonic organisms. J. Mar. Biol. Assoc. U.K., 12: 644–677.

Lee, S.Y. and Kneib, R.T., 1994. Effects of biogenic structure on prey consumption by the xanthid crabs *Eurytium limosum* and *Panopeus herbstii* in a salt marsh. Mar. Ecol. Prog. Ser., 104: 39–47.

Levinton, J.S., 1977. Ecology of shallow water deposit-feeding communities Quisset Harbor, Massachusetts. In: B.C. Coull (Ed.), Ecology of Marine Benthos. Belle W. Baruch Library in Marine Science, No. 6, University of South Carolina Press, pp. 191–227.

Lin, J., 1990. Mud crab predation on ribbed mussels in salt marshes. Mar. Biol., 107: 103–109.

Lin, J., 1991. Predator–prey interactions between blue crabs and ribbed mussels living in clumps. Est. Coast. Shelf Sci., 32: 61–69.

Lipcius, R.N. and Hines, A.H., 1986. Variable functional responses of a marine predator in dissimilar homogeneous microhabitats. Ecology, 67: 1361–1371.

Littlewood, D.T.J. and Marsbe, L.A., 1990. Predation on cultivated oysters, *Crassostrea rhizophorae* (Guilding), by the polyclad turbellarian flatworm, Stylochus (Stylochus) frontalis Verrill. Aquaculture, 88: 145–150.

Lloyd, A.J. and Young, C.M., 1947. The biology of *Crangon vulgaris* L. in the Bristol channel and Severn estuary. J. Mar. Biol. Assoc. U.K., 26: 626–661.

Loosanoff, V.L., 1959. *Condylosoma* – an enemy of bivalve larvae. Science, 129: 147.

Loosanoff, V.L., 1961. Biology and methods of controlling the starfish *Asterias forbesi* (Desor). U.S. Dept. Interior. Fish Wild. Serv., Bur. of Comm. Fish., Fish. Leaflet 520, 11 pp.

Loosanoff, V.L., 1964. Variations in time and intensity of setting of the starfish, *Asterias forbesi*, in Long Island Sound during a twenty five year period. Biol Bull., 126: 423–439.

Loosanoff, V.L., 1966. Time and intensity of setting of the oyster, *Crassostrea virginica*, in Long Island Sound. Biol. Bull., 130: 211–227.

MacGinitie, G.E., 1938. Notes on the natural history of some marine animals. Am. Midl. Nat., 19: 207–219.

MacGinitie, G.E. and MacGinitie, N., 1949. Natural History of Marine Animals. McGraw Hill, New York, 473 pp.

MacKay, G.H., 1891. The scoters (*Oidemia americana, O. deglandi*, and *O. perspicillata*) in New England. Auk, 8: 279–290.

MacKay, G.H., 1892. Habits of the oldsquaw (*Clangula hyemalis*) in New England. Auk, 9: 330–337.

MacKenzie, B.R., Leggett, W.C. and Peters, R.H., 1990. Estimating larval fish ingestion rates: can laboratory derived values be reliably extrapolated to the wild?. Mar. Ecol. Prog. Ser., 67: 209–225.

MacKenzie Jr., C.L., 1969. Feeding rates of starfish, *Asterias forbesi* (Desor), at controlled water temperatures and during different seasons of the year. Fish. Bull., 68: 67–72.

MacKenzie Jr., C.L., 1970a. Oyster culture in Long Island Sound 1966–69. Comm. Fish. Rev., 32 (1): 27–40.

MacKenzie Jr., C.L., 1970b. Causes of oyster spat mortality, conditions of oyster setting beds, and recommendations for oyster bed management. Proc. Natl. Shellfish. Assoc., 60: 59–67.

MacKenzie Jr., C.L., 1977a. Sea anemone predation on larval oysters in Chesapeake Bay (Maryland). Proc. Natl. Shellfish. Assoc., 67: 113–117.

MacKenzie Jr., C.L., 1977b. Predation on hard clam (*Mercenaria mercenaria*) populations. Trans. Am. Fish. Soc., 106: 530–537.

MacKenzie Jr., C.L., 1979. Management for increasing hard clam abundance. Mar. Fish Rev., 41: 10–22.

MacKenzie Jr., C.L., 1981. Biotic potential and environmental resistance in the American oyster (*Crassostrea virginica*) in Long Island. Aquaculture, 22: 229–268.

MacKenzie Jr., C.L. and Stehlik, L.L., 1988. Aspects of predation on hard clam (*Mercenaria mercenaria*) populations in Northeastern U.S.A. estuaries. US Dept. Commerce, NMFS-NOAA Northeast Fisheries Center, Sandy Hook Laboratory, SHL Report No. 88-01, 12 pp.

Magalhaes, H., 1948. An ecological study of snails of the genus *Busycon* at Beaufort, North Carolina. Ecol. Monogr., 18: 377–409.

Malinowski, S.M., 1985. The population ecology of the hard clam, *Mercenaria mercenaria*, in eastern Long Island Sound. Ph.D. Dissertation, University of Connecticut, 101 pp.

Malinowski, S.M., 1993. Effects of density on adult growth, survivorship, fecundity, and recruitment of northern quahogs. In: M.A. Rice and D. Grossman-Garber (Eds.), Proceedings of the Second Rhode Island Shellfish Industry Conference. Rhode Island Sea Grant, Narragansett, RI, pp. 79–90.

Malinowski, S.M. and Whitlatch, R.B., 1984. Natural survivorship of young hard clams, *Mercenaria mercenaria* (Linne) in eastern Long Island Sound. J. Natl. Shellfish. Assoc., 4: 94.

Malinowski, S.M. and Whitlatch, R.B., 1988. A theoretical evaluation of shellfish resource management. J. Shellfish Res., 7: 95–100.

Mansour, R.A. and Lipcius, R.N., 1991. Density-dependent foraging and mutual interference in blue crabs preying upon infaunal clams. Mar. Ecol. Prog. Ser., 72: 239–246.

Manzi, J.J., 1970a. Combined effects of salinity and temperature on the feeding, reproductive, and survival rates of *Eupleura caudata* (Say) and *Urosalpinx cinerea* (Say) (Prosobranchia: Muricidae). Biol. Bull., 138: 35–46.

Manzi, J.J., 1970b. The effect of temperature on the feeding rate of the rough oyster drill, *Eupleura caudata* (Say). Proc. Natl. Shellfish. Assoc., 60: 54–58.

Marsh, C.P., 1986. Rocky intertidal community organization: the impact of avian predators on mussel recruitment. Ecology, 67: 771–786.

Martin, A.C. and Uhler, F.M., 1939. Food of game ducks in the United States and Canada. U.S. Dept. Agric. Tech. Bull. 634, pp. 1–147.

Martin, T.H., Wright, R.A. and Crowder, L.B., 1989. Non-additive impact of blue crabs and spot on their prey assemblages. Ecology, 70: 1935–1942.

Mascaro, M. and Seed, R., 2000a. Foraging behavior of *Carcinus maenas* (L.): Comparisons of size-selective predation on four species of bivalve prey. J. Shellfish Res., 19: 283–292.

Mascaro, M. and Seed, R., 2000b. Foraging behavior of *Carcinus maenas* (L.): Species-selective predation among four bivalve prey. J. Shellfish Res., 19: 293–300.

Mason, J., 1976. Cultivation. In: Bayne, B.L. (Ed.) Marine Mussels, their Ecology and Physiology. Cambridge Univ. Press, Cambridge, pp. 385–410.

Matthiessen, G.C., 1971. A review of oyster culture and the oyster industry in North America. Woods Hole Oceanographic Inst., Contribution No. 2528.

Mattila, J., Olafsson, E.B. and Johansson, A., 1990. Predation effects of *Crangon crangon* on benthic infauna on shallow sandy bottoms — an experimental study from Southern Sweden. In: Barnes and Gibson (Eds.), Trophic Relationships in the Marine Environment. Proc. 24th European Mar. Biol. Symp., pp. 503–516.

Maurer, D., 1983. The effect of an infaunal suspension feeding bivalve *Mercenaria mercenaria* (L.) on benthic recruitment. P. S. Z. N. Mar. Ecol., 4: 263–274.

Mauzey, K.P., Birkeland, C. and Dayton, P.K., 1968. Feeding behavior of asteroids and escape responses of their prey in the Puget Sound region. Ecology, 49: 603–619.

McClintok, J.B. and Robnett Jr., T.R., 1986. Size selective predation by the asteroid *Pisaster ochraceus* on the bivalve *Mytilus californianus*: a cost benefit analysis. P. S. Z. N. Mar. Ecol., 7: 321–332.

McCormick, 1969. Trophic relationships of the hydromedusae in Yaquina Bay, Oregon. Northwest Sci., 43: 207–217.

McDermott, J.J., 1960. The predation of oysters and barnacles by crabs of the Family Xanthidae. Proc. Pennsylvania Acad. Sci., 34: 199–211.

McDermott, J.J., 1976. Predation of the razor clam *Ensis directus* by the nemertean worm *Cerebratulus lacteus*. Chesapeake Sci., 17: 299–301.

McDermott, J.J., 1998. The western Pacific brachyuran, *Hemigrapsus sanguineus* (Grapsidae), in its new habitat along the Atlantic coast of the United States: geographic distribution and ecology. ICES J. Mar. Sci., 55: 289–298.

McDermott, J.J., Fowler, F.B., 1951. Preliminary studies of the common mud crabs on oyster beds of Delaware Bay. Natl. Shellfish. Assoc. Convention Addresses, 1952, pp. 147-150.
McDonald, J., 1982. Divergent life history patterns in the co-occurring intertidal crabs. *Panopeus herbstii* and *Eurypanopeus depressus* (Crustacea: Brachyura: Xanthidae). Mar. Ecol. Prog. Ser., 8: 173–180.
Medcof, J.C., 1949. 'Puddling' – a method of feeding by herring gulls. Auk, 66: 205.
Medcof, J.C. and Thurber, L.W., 1958. Trial control of the greater clam drill (*Lunatia heros*) by manual collection. J. Fish. Res. Bd. Canada, 15: 1355–1369.
Meire, P., 1993. The impact of bird predation on marine and estuarine bivalve populations: a selective review of patterns and underlying causes. In: R.F. Dame (Ed.), Bivalve Filter Feeders in Estuarine and Coastal Ecosystem Processes. NATO ASI Series G: Ecological Sciences 33. Springer, New York, pp. 197–243.
Meissner, J. and Brager, S., 1990. The feeding ecology of wintering eiders *Somateria mollissima* and common scoters *Melanitta nigra* on the Baltic coast of Schleswig-Holstein, FRG, Wader Study Group Bull. 58: 10–12. cited in Meire, 1993 (see above).
Menge, B.A., 1978. Predation intensity in a rocky intertidal community. Effect of an algal canopy, wave action and desiccation on predator feeding rates. Oecologia, 34: 17–35.
Menge, B.A., 1979. Coexistence between seastars *Asterias vulgaris* and *A. forbesi* in a heterogeneous environment: a non-equilibrium explanation. Oecologia, 41: 245–272.
Menzel, R.W. and Hopkins, S.W., 1955. Crabs as predators of oysters in Louisiana. Proc. Natl. Shellfish. Assoc., 46: 177–184.
Menzel, R.W. and Nichy, F.E., 1958. Studies of the distribution and feeding habits of some oyster predators in Alligator Harbor, Florida. Bull. Mar. Sci. Gulf Carib., 8: 125–145.
Menzel, R.W. and Sims, H.W., 1964. Experimental farming of hard clams, *Mercenaria mercenaria*, in Florida. Proc. Natl. Shellfish. Assoc., 53: 103–109.
Menzel, R.W., Cake, E.W., Haines, M.L., Martin, R.E. and Olsen, L.A., 1976. Clam mariculture in northwest Florida: field study on predation. Proc. Natl. Shellfish. Assoc., 65: 59–62.
Merriner, J.V. and Smith, J.W., 1979. Gear feasibility study for the cownose ray, *Rhinoptera bonasus*. Virginia Inst. Mar. Sci. Spec. Rep. Appl. Mar. Sci. Ocean Engineer. 227, 27 pp.
Meyer, D.L., 1994. Habitat partitioning between the xanthid crabs *Panopeus herbstii* and *Eurypanopeus depressus* on intertidal oyster reefs (*Crassostrea virginica*) in southeastern North Carolina. Estuaries, 17: 674–679.
Micheli, F., 1995. Behavioral plasticity in prey-size selectivity of the blue crab *Callinectes sapidus* feeding on bivalve prey. J. Anim. Ecol., 64: 63–74.
Micheli, F., 1996. Predation intensity in estuarine soft bottoms: between habitat comparisons and experimental artifacts. Marine Ecol. Prog. Ser., 141: 295–302.
Micheli, F., 1997. Effects of predator foraging behavior on patterns of prey mortality in marine soft bottoms. Ecol. Monogr., 67: 203–224.
Mileikovsky, S.A., 1959. Interrelations between pelagic larvae of *Nepthys ciliata* (O.F. Muller) and *Macoma balthica* L. and *Mya arenaria* L. of the White Sea. Zool. Zh. 38: 1889–1891 (data from English summary of Russian).
Mileikovsky, S.A., 1974. On predation of pelagic larvae and early juveniles of marine bottom invertebrates by adult benthic invertebrates and their passing through their predators. Mar. Biol., 26: 303–311.
Miller, D.J., Harwick, J.E. and Dahlstrom, W.A., 1975. Pismo clams and sea otters. Calif. Fish and Game Mar. Res. Tech. Rep. 31, 49 pp.
Mironov, G.N., 1948. Filtration work and feeding of black sea mussels. Trudy Sevastopol. Biol. Sta., 6: 338–352.
Mitchell, R., 1974. Studies on the population dynamics and some aspects of the biology of *Mercenaria mercenaria* (Linne). Ph.D. Thesis, University of Southampton, Southampton, 191 pp.
Modlin, R.F., 1980. The life cycle and recruitment of the sand shrimp, *Crangon septemspinosa*, in the Mystic River estuary, Connecticut. Estuaries, 3: 1–10.
Moller, P., 1986. Physical factors and biological interactions regulating infauna in shallow boreal areas. Mar. Ecol. Prog. Ser., 30: 33–47.
Moller, P. and Rosenberg, R., 1983. Recruitment, abundance and production of *Mya arenaria* and *Cardium edule* in marine shallow waters western Sweden. Ophelia, 22: 33–55.
Monismith, S.G., Koseff, J.R., Thompson, J.K., O'Riordan, C.A. and Nepf, H.M., 1990. A study of model bivalve siphonal currents. Limnol. Oceanogr., 35: 680–696.
Morgan, D.E., Goodsell, J., Matthiessen, G.C., Garey, J. and Jacobson, P., 1980. Release of hatchery-reared

bay scallops (*Argopecten irradians*) onto a shallow coastal bottom in Waterford, Connecticut. Proc. World. Maricult. Soc., 11: 247–261.
Mudge, G.P. and Ferns, P.N., 1982. The feeding ecology of five species of gulls (Aves: Larini) in the inner Bristol Channel. Zool. J., 197: 497–510.
Murdoch, W.W., 1971. The developmental response of predators to changes in prey density. Ecology, 52: 132–134.
Myers, J.P., Williams, S.L. and Pitelka, F.A., 1980. An experimental analysis of prey availability for sanderlings (Aves: Scolopacidae) feeding on sandy beach crustaceans. Can. J. Zool., 58: 1564–1574.
Nakaoka, M., 1996. Size-dependent survivorship of the bivalve *Yoldia notabilis* (Yokoyama, 1920): the effect of predation. J. Shellfish Res., 15: 355–362.
Nehls, G., 1989. Occurrence and consumption of the common eider, *Somateria mollissima*, in the Wadden Sea of Schleswig–Holstein. Helgolander Wiss. Meeresunters, 43: 385–393.
Nelson, J., 1903. Experimental studies in oyster propagation. Report of biologist for 1902, New Jersey Agricultural College Experiment Station, New Brunswick, NJ, 329–374.
Nelson, T.C., 1921. Report of the Department of Biology. In: Annual Report New Jersey Agriculture Experiment Station for 1920, New Brunswick, NJ, pp. 319–349.
Nelson, T.C., 1925a. On the occurrence and food habits of ctenophores in New Jersey inland coastal waters. Biol. Bull., 48: 92–111.
Nelson, T.C., 1925b. Report of the Department of Biology. In: Annual Report Agriculture Experiment Station for 1924, New Brunswick, NJ, pp. 237–246.
Newell, C., 1983. Increasing clam harvests in Maine. Maine/New Hampshire Sea Grant College Program Publication TR-MSG-83-3, University of Maine, Orono, 60 pp.
Newell, R.I.E. and Kennedy, V.S., 1992. Spatfall monitoring and partitioning the sources of oyster spat fall mortality. Final Report: Maryland Dept. Natural Resources, Annapolis, MD.
Nichy, F.E. and Menzel, R.W., 1960. Mortality of intertidal and subtidal oysters in Alligator Harbor, Florida. Proc. Natl. Shellfish. Assoc., 51: 33–41.
Nielsen, C., 1975. Observations on *Buccinum undatum* L. attacking bivalves and on prey responses, with a short review on attack methods of other prosobranchs. Ophelia, 13: 87–108.
Norton-Griffiths, M., 1967. Some ecological aspects of the feeding behavior of the oystercatcher *Haematopus ostralegus* on the edible mussel *Mytilus edulis*. Ibis, 109: 412–424.
Olafsson, E.B., 1989. Contrasting influences of suspension-feeding and deposit-feeding populations of *Macoma balthica* on infaunal recruitment. Mar. Ecol. Prog. Ser., 55: 171–179.
O'Riordan, C.A., Monismith, S.G. and Koseff, J.R., 1995. The effect of bivalve excurrent jet dynamics on mass transfer in a benthic boundary layer. Limnol. Oceanogr. 40: 330–344.
Orth, R.J., 1975. Destruction of eelgrass, *Zostera marina*, by the cownose ray, *Rhinoptera bonasus*, in the Chesapeake Bay. Chesapeake Sci., 16: 205–208.
Orth, R.J., 1977. The importance of sediment stability in seagrass communities. In: B.C. Coull (Ed.), Ecology of marine benthos, Belle W. Baruch Lib. Mar. Sci. 6. Univ. South Carolina Press, Columbia, SC, pp. 281–300.
Osman, R.W., Witlatch, R.B. and Zajac, R.N., 1989. Effects of resident species on recruitment into a community: larval settlement versus post-settlement mortality in the oyster *Crassostrea virginica*. Mar. Ecol. Prog. Ser., 54: 61–73.
Paine, R.T., 1962. Ecological diversification in sympatric gastropods of the genus *Busycon*. Evolution, 16: 515–523.
Paine, R.T., 1976. Size-limited predation: an observational and experimental approach with the *Mytilus–Pisaster* interaction. Ecology, 57: 858–873.
Palumbi, S.R. and Freed, L.A., 1998. Agonistic interactions in a keystone predatory starfish. Ecology, 69: 1624–1627.
Parker, K.M., 1975. A study of natural recruitment of *Mercenaria mercenaria*. Unpublished manuscript, North Carolina Div. Mar. Fish. 301 Causeway Drive, Wrightsville Beach, NC, 17 pp.
Pearson, W.H., Woodruff, D.L. and Sugarman, P.C., 1981. Effects of oiled sediment on predation on the littleneck clam, *Protothaca staminea*, by the Dungeness crab, *Cancer magister*. Est. Coast. Shelf Sci., 13: 445–454.
Perry, M.C. and Deller, A.S., 1996. Review of factors affecting the distribution and abundance of waterfowl in shallow-water habitats of Chesapeake Bay. Estuaries, 19: 272–278.
Perry, M.C. and Uhler, F.M., 1988. Food habits and distribution of wintering canvasbacks, *Aythya valisineria*, on Chesapeake Bay. Estuaries, 11: 57–67.

Peterson, C.H., 1975. Space utilization and community organization in two New Jersey intertidal systems. Bull. Ecol. Soc. Am., 56: 47.

Peterson, C.H., 1979. The importance of predation and competition in organizing the intertidal epifaunal communities of Barnegat Inlet, New Jersey. Oecologia, 39: 1–24.

Peterson, C.H., 1982a. Clam predation by whelks (*Busycon* spp.): experimental tests of the importance of prey size, prey density, and seagrass cover. Mar. Biol., 66: 159–170.

Peterson, C.H., 1982b. The importance of predation and intra- and interspecific competition in the population biology of two infaunal suspension-feeding bivalves, *Protothaca staminea* and *Chione undatella*. Ecol. Monogr., 52: 437–475.

Peterson, C.H., 1983. Interactions between two infaunal bivalves, *Chione undatella* (Sowerby) and *Protothaca staminea* (Conrad), and two potential enemies, *Crepidula onyx* Sowerby and *Cancer anthonyi* (Rathbun). J. Exp. Mar. Biol. Ecol., 68: 145–158.

Peterson, C.H., 1986. Enhancement of *Mercenaria mercenaria* densities in seagrass beds: is pattern fixed during settlement season or altered by subsequent differential survival?. Limnol. Oceanogr., 31: 200–205.

Peterson, C.H., 1990. On the role of ecological experimentation in resource management: managing fisheries through a mechanistic understanding of predator feeding behavior. In: R.N. Hughes (Ed.), Behavioral Mechanisms of Food Selection, NATO ASI Series, G20, pp. 821–846.

Peterson, C.H. and Beal, B.F., 1989. Bivalve growth and higher order interactions: importance of density, site and time. Ecology, 70: 1390–1404.

Peterson, C.H. and Black, R., 1987. Resource depletion by active suspension feeders on tidal flats: influence of local density and tidal elevation. Limnol. Oceanogr., 23: 143–166.

Peterson, C.H. and Black, R., 1995. Drilling by buccinid gastropods of the Genus *Cominella* in Australia. Veliger, 38: 37–42.

Peterson, C.H. and Skilleter, G.A., 1994. Control of foraging behavior of individuals within an ecosystem context: the clam *Macoma balthica*, flow environment, and siphon-cropping fishes. Oecologia, 100: 256–267.

Peterson, C.H., Summerson, H.C., Fegley, S.R. and Prescott, R.C., 1989. Timing, intensity and sources of autumn mortality of adult bay scallops *Argopecten irradians concentricus* Say. J. Exp. Mar. Biol. Ecol., 127: 121–140.

Peterson, C.H., Summerson, H.C. and Huber, J., 1995. Replenishment of hard clam stocks using hatchery seed: combined importance of bottom type, seed size, planting season and density. J. Shellfish Res., 14: 293–300.

Pihl, L., 1985. Food selection and consumption of mobile epibenthic fauna in shallow marine areas. Mar. Ecol. Prog. Ser., 22: 169–179.

Pihl, L. and Rosenberg, R., 1982. Production, abundance, and biomass of mobile epibenthic marine fauna in shallow waters, western Sweden. J. Exp. Mar. Biol. Ecol., 57: 273–301.

Pihl, L. and Rosenberg, R., 1984. Food selection and consumption of the shrimp *Crangon crangon* in some shallow marine areas in western Sweden. Mar. Ecol. Prog. Ser., 15: 159–168.

Pile, A.J., Lipcius, R.N., Van Montfrans, J. and Orth, R.J., 1996. Density-dependent settler-recruit-juvenile relationships in blue crabs. Ecol. Monogr., 66: 277–300.

Platt, N.E., 1969. A comparative survey of the gut parasites of *Callionymus lyra* L. from various regions of the British Coast, with notes on the gut parasites of *Cottus bubalis* Euphrasen in Southampton Water. Southampton University, M.Sc. Thesis (from Mitchell, 1974).

Pohle, D.G., Bricelj, V.M. and Garcia-Esquivel, Z., 1991. The eelgrass canopy: an above-bottom refuge from benthic predators for juvenile bay scallops *Argopecten irradians*. Mar. Ecol. Prog. Ser., 74: 47–59.

Posey, M.H. and Hines, A.H., 1991. Complex predator–prey interactions within an estuarine benthic community. Ecology, 72: 2155–2169.

Pratt, D.M., 1953. Abundance and growth of *Venus mercenaria* and *Callocardia morrhuana* in relation to the character of bottom sediments. J. Mar. Res., 12: 60–74.

Pratt, D.M. and Campbell, D.A., 1956. Environmental factors affecting growth in *Venus mercenaria*. Limnol. Oceanogr., 1: 2–17.

Prescott, R.C., 1990. Sources of predatory mortality in the bay scallop *Argopecten irradians* (Lamarck); interactions with seagrass and epibiotic coverage. J. Exp. Mar. Biol. Ecol., 144: 63–83.

Price Jr., K.S., 1962. Biology of the sand shrimp, *Crangon septemspinosa*, in the shore zone of the Delaware Bay region. Chesapeake Sci., 3: 244–255.

Provenzano, A.J., 1959. Effects of the flatworm *Stylochus ellipticus* (Girard) on oyster spat in two salt water ponds in Massachusetts. Proc. Natl. Shellfish. Assoc., 50: 83–88.

Purcell, J.E., Creswell, F.P., Cargo, D.G. and Kennedy, V.S., 1991. Differential ingestion and digestion of bivalve larvae by the scyphozoan *Chrysaora quinquecirrha* and the ctenophore *Mnemiopsis leidyi*. Biol. Bull., 180: 103–111.

Quaglietta, C.E., 1987. Predation by *Mnemiopsis leidyi* on hard clam larvae and other natural zooplankton in Great South Bay, New York. M.S. Thesis, State University of New York, Stony Brook, NY, 66 pp.

Quammen, M.L., 1980. The effect of sand grains on feeding by shorebirds on a southern California mudflat. Am. Zool., 20: 953.

Quammen, M.L., 1982. Influence of subtle substrate differences on feeding by shorebirds on intertidal mudflats. Mar. Biol., 71: 339–343.

Quammen, M.L., 1984. Predation by shorebirds, fish and crabs on invertebrates in intertidal mudflats: an experimental test. Ecology, 65: 529–537.

Quayle, D.B., 1964. Distribution of introduced marine mollusca in British Columbia waters. J. Fish. Res. Bd. Canada, 21: 1155–1181.

Quayle, D.B., 1988. Pacific oyster culture in British Columbia. Canadian Bull. Fish. Aquat. Sci., 218, 241 p.

Quayle, D.B. and Bourne, N., 1972. The clam fisheries in British Columbia. Fisheries Research Board of Canada, Rep. 179, 70 pp.

Raffaelli, D., Conacher, A., McLachlan, H. and Emes, C., 1989. The role of epibenthic crustacean predators in an estuarine food web. Est. Coast. Shelf Sci., 28: 149–160.

Raffaelli, D., Falcy, V. and Galbraith, C., 1990. Eider predation and the dynamics of mussel-bed communities. In: Barnes, M. and Gibson, R.N. (Eds.), Trophic Relationships in the Marine Environment, Proc. 24th European Mar. Biol. Symp. Aberdeen Univ. Press, pp. 157–169.

Reading, C.J. and McGrorty, S., 1978. Seasonal variations in the burying depth of *Macoma balthica* (L.) and its accessibility to wading birds. Est. Coast. Mar. Sci., 6: 133–144.

Recher, H.F., 1966. Some aspects of the ecology of migrant shorebirds. Ecology, 47: 393–407.

Redeke, H.C., 1906. Verslag omtrent onderzoekingen over het voedsel van eenigre visschen. Jaarboek Rijksinstituut Onderzoek der Zee, 1905, pp. 88–111. Cited in Kristensen (1957).

Reed, P.H., 1969. Culture methods and effects of temperature and salinity on survival and growth Dungeness crab (*Cancer magister*) larvae in the laboratory. J. Fish. Res. Bd. Canada, 26: 389–397.

Reise, K., 1977a. Predation pressure and community structure of an intertidal soft-bottom fauna. In: B.F, Keegan, P.O. Ceidigh and P.J.S. Boaden (Eds.), Biology of Benthic Organisms. Pergamon Press, New York, pp. 513–519.

Reise, K., 1977b. Predator exclusion experiments on an intertidal mud flat. Helgolander Wiss. Meeresunters, 30: 263–271.

Reise, K., 1979. Moderate predation on meiofauna by the macrobenthos of the Wadden Sea. Helgolander Wiss. Meeresunters, 31: 55–101.

Reise, K., 1985. Tidal Flat Ecology: An Experimental Approach to Species Interactions, Ecological Studies, Vol. 54. Springer, New York, 191 pp.

Rhoads, D.C. and Young, C.M., 1971. Animal-sediment relations in Cape Cod Bay, Massachusetts II. Reworking by *Molpadia oolitica* (Holothuroidea). Mar. Biol., 11: 255–261.

Ribi, G. and Jost, P., 1978. Feeding rate and duration of daily activity of *Astropecten aranciacus* (Echinodermata: Asteroidea) in relation to prey density. Mar. Biol., 45: 249–254.

Ribi, G., Scharer, R. and Ochsner, P., 1977. Stomach contents and size frequency distributions of two coexisting sea star species, *Astropecten aranciacus* and *A. bispinosus*, with reference to competition. Mar. Biol., 43: 181–185.

Rice, M.A., 1992. The northern quahog: the Biology of *Mercenaria mercenaria*. Rhode Island Sea Grant Publication No. RIU-B-92-001. University of Rhode Island, Narragansett, RI, 60 pp.

Richardson, H., 1983. A field study of diet selection and optimization by northwestern crows feeding on clams: tests and predictions, Ph.D. Thesis, Simon Fraser University, Burnaby, BC, Canada.

Richardson, H., 1985. Availability of buried littleneck clams (*Venerupis japonica*) to northwestern crows (*Corvus caurinus*). J. Anim. Ecol., 54: 443–457.

Richardson, H. and Verbeek, N.A., 1986. Diet selection and optimization by northwestern crows feeding on Japanese littleneck clams. Ecology, 67: 1219–1226.

Richardson, H. and Verbeek, N.A., 1987. Diet selection by yearling Northwestern Crows (*Corvus caurinus*) feeding on littleneck clams (*Venerupis japonica*). Auk, 104: 263–269.

Roberts, D., Rittschof, D., Gerhart, D.J., Schmidt, A.R. and Hill, L.G., 1989. Vertical migration of the clam

Mercenaria mercenaria (L.) (Mollusca: Bivalvia): environmental correlates and ecological significance. J. Exp. Mar. Biol. Ecol., 126: 271–280.
Roberts Jr., M.H., 1974. Larval development of *Pagurus longicarpus* Say reared in the laboratory. V. Effect of diet on survival and molting. Biol. Bull., 146: 66–77.
Robles, C., 1987. Predator foraging characteristics and prey population structure on a sheltered shore. Ecology, 68: 1502–1514.
Rodhouse, P.G., 1973. Factors affecting spatfall of the clam *Mercenaria mercenaria* L. M.Sc. Dissertation, University of Southampton, Southampton.
Rodrigues, C.L., 1986. Predation of the naticid gastropod, *Neverita didyma* (Roding), on the bivalve, *Ruditapes philippinarum* (Adams and Reeve): evidence for a preference linked functional response. Publ. Amakusa Mar. Biol. Lab., 8: 125–141.
Rodrigues, C.L., Nojima, S. and Kikuchi, T., 1987. Mechanics of prey size preference in the gastropod *Neverita didyma* preying on the bivalve *Ruditapes philippinarum*. Mar. Ecol. Prog. Ser., 40: 87–93.
Rogers, B.B. and Rogers, D.A., 1989. Can a small moonsnail, *Polinices lewisii*, eat a large clam, *Tapes japonica*?. Northwest Environ. J., 5: 177–179.
Ropes, J.W., 1968. The feeding habits of the green crab, *Carcinus maenas* (L.). Fish. Bull., 67: 183–203.
Ropes, J.W., 1988. The food habits of five crab species at Pettaquamscutt River, Rhode Island. Fish. Bull., 87: 197–204.
Rowell, T.W. and Woo, P., 1990. Predation by the nemertean worm, *Cerebratulus lacteus* Verrill, on the soft-shell clam, *Mya arenaria* Linnaeus, 1758, and its apparent role in the destruction of a clam flat. J. Shellfish Res., 9: 291–297.
Russell-Hunter, W. and Grant, D.C., 1966. Estimates of population density and dispersal in the naticid gastropod, *Polinices duplicatus*, with a discussion of computational methods. Biol. Bull., 131: 292–307.
Saila, S.B., Flowers, J.M. and Cannario, M.T., 1967. Factors affecting the relative abundance of *Mercenaria mercenaria* in the Providence River, Rhode Island. Proc. Natl. Shellfish. Assoc., 57: 83–89.
Saito, S., 1936. Chiba-Ken Tokyo naiwan (shuto shite edogawarjiri) ni okeru kairui yoshoku to tori ni yoru higai. Chiba Fishers Expt. Sta. Mimeo, 12 pp. (cited in Cahn, 1951).
Sanchez-Salizar, M.E., Griffiths, C.L. and Seed, R., 1987a. The interactive roles of predation and tidal elevation in structuring populations of the edible cockle, Cerastoderma edule. Est. Mar. Coast. Shelf Sci., 25: 245–260.
Sanchez-Salizar, M.E., Griffiths, C.L. and Seed, R., 1987b. The effect of size and temperature on the predation of cockles *Cerastoderma edule* (L.) by the shore crab *Carcinus maenas* (L.). J. Exp. Mar. Biol. Ecol., 111: 181–193.
Sandberg, E. and Bonsdorff, E., 1990. On the structuring role of *Saduria entomon* (L.) on shallow water zoobenthos. Ann. Zool. Fenn., 27: 279–284.
Sawyer, D.B., 1950. Feeding activities of the boring snail *Polinices duplicata*. In: Third report on investigations of methods of improving the shellfish resources of Massachusetts. Dept. Conserv. Div. Mar. Fish. Commonwealth of Massachusetts, pp. 16–17.
Scherer, B. and Reise, K., 1981. Significant predation on micro- and macrobenthos by the crab *Carcinus maenas* L. in the Wadden sea. Kieler Meeresforsch. Sonderh., 5: 490–500.
Schneider, D., 1982. Escape response of an infaunal clam *Ensis directus* Conrad 1843, to a predatory snail, *Polinices duplicatus* Say 1822. Veliger, 24: 371–372.
Schneider, D. and Harrington, B.A., 1981. Timing of shorebird migration in relation to prey depletion. Auk, 98: 801–811.
Schuster Jr., C.N., 1950. Observations on the natural history of the American horseshoe crab, *Limulus polyphemus*. In: H.J. Turner Jr. (Ed.), Third report on investigations of methods of improving the shellfish resources of Massachusetts. Dept. Conserv. Div. Mar. Fish. Commonwealth of Massachusetts, pp. 18–23.
Seed, R., 1980. Predator–prey relationships between the mud crab *Panopeus herbstii*, the blue crab *Callinectes sapidus* and the Atlantic ribbed mussel *Geukensia* (= *Modiolus*) *demissa*. Est. Coast. Mar. Sci., 11: 445–458.
Seed, R., 1982. Predation of the ribbed mussel *Geukensia demissa* by the blue crab *Callinectes sapidus*. Neth. J. Sea Res., 16: 163–172.
Seed, R., 1990. Behavioral and mechanical aspects of predation by the swimming crab *Thalamita danae* on the green-lipped mussel *Perna viridis*. In: Barnes and Gibson (Eds.), Trophic Relationships in the Marine Environment. Proc. 24th European Mar. Biol. Symp., pp. 528–540.
Segerstrale, S.G., 1962. Investigations on Baltic populations of the bivalve *Macoma baltica* (sic) (L.). Part II. What

are the reasons for the periodic failure of recruitment and the scarcity of *Macoma* in the deeper waters of the inner Baltic? Soc. Sci. Fenn. Comment. Biol. XXIV, 7, 26 pp.
Sellmer, G.P., 1967. Functional morphology and ecological life history of the gem clam *Gemma gemma* (Eulamellibranchia: Veneridae). Malacologia, 5: 137–223.
Skilleter, G.A. and Peterson, C.H., 1994. Control of foraging behavior of individuals within an ecosystem context: the clam *Macoma balthica* and interactions between competition and siphon cropping. Oecologia, 100: 268–278.
Sloan, N.A. and Robinson, S.M.C., 1983. Winter feeding by asteroids on a subtidal sandbed in British Columbia. Ophelia, 22: 125–140.
Smidt, E.L., 1951. Animal production in the Danish Waddensea. Medd. Komm. Danm. Fisk.-og Havunders. Ser. Fisk., 11: 1–51.
Smith, I.S., 1961. Clam-digging behavior in the starfish *Pisaster brevispinus* (Asteroidea: Echinodermata) on soft substrate. Behavior, 18: 148–153.
Smith, J.W. and Merriner, J.V., 1978. Biology and identification of rays in the Chesapeake Bay. Virginia Institute of Marine Science Education Series No. 20. Virginia Institute of Marine Science, Gloucester Point, VA, 22 pp.
Smith, L.D. and Palmer, A.R., 1994. Effects of manipulated diet on size and performance of brachyuran crab claws. Science, 264: 710–712.
Smith, O., 1953. Notes on the ability of the horseshoe crab, *Limulus polyphemus* to locate soft-shell clams, *Mya arenaria*. Ecology, 34: 636–637.
Smith, O. and Chin, E., 1951. The effects of predation on soft clams, *Mya arenaria*. National Shellfisheries Assoc., Convention Addresses, 1951, pp. 37–44.
Smith, O.R., Baptist, J.P. and Chin, E., 1955. Experimental farming of the soft-shell clam, *Mya arenaria*, in Massachusetts, 1949–1953. Comm. Fish. Rev., 17 (6): 1–16.
Spaans, A.L., 1971. On the feeding ecology of the herring gull *Larus argentatus* Pont. in the northern part of the Netherlands. Ardea, 59: 73–188.
Spear, H.S., 1955. Notes on laboratory experiments on feeding habits of green crabs. In: Fifth conference on clam research. US. Fish and Wildlife Service Clam Investigations. Boothbay Harbor, ME, 45.
Spencer, B.E., 1991. Predators and methods of control in molluscan shellfish cultivation in north European waters. In: N. DePauw and J. Joyce (Eds.), Aquaculture and the Environment. European Aquaculture Soc. Spec. Publ. No. 16, pp. 309–337.
Spencer, B.E., Edwards, D.B. and Millican, P.F., 1992. Protecting Manila clam (*Tapes philippinarum*) beds with plastic netting. Aquaculture, 105: 251–268.
Sponaugle, S. and Lawton, P., 1990. Portunid crab predation on juvenile hard clams: effects of substrate type and prey density. Mar. Ecol. Prog. Ser., 67: 43–53.
Steimle, F.W. and Sinderman, C.J., 1978. Review of oxygen depletion and associated mass mortalities of shellfish in the Middle Atlantic Bight in 1976. Mar. Fish. Rev., 40 (12): 17–26.
Steinberg, P.D. and Kennedy, V.S., 1979. Predation upon *Crassostrea virginica* (Gmelin) larvae by two invertebrate species common to Chesapeake Bay oyster bars. Veliger, 22: 78–84.
Stephenson, M.D., 1977. Sea otter predation on Pismo clams in Monterey Bay. Calif. Fish Game, 63: 117–120.
Stevens, B.G., Armstrong, D.A. and Cusimano, R., 1982. Feeding habits of the Dungeness crab *Cancer magister* as determined by the index of relative importance. Mar. Biol., 72: 135–145.
Stickney, A.P. and Stringer, L.D., 1957. A study of the invertebrate bottom fauna of Greenwich Bay, Rhode Island. Ecology, 38: 111–122.
Strieb, M.D., Bricelj, V.M. and Bauer, S.I., 1995. Population biology of the mud crab, *Dyspanopeus sayi*, an important predator of juvenile bay scallops in Long Island (USA) eelgrass beds. J. Shellfish Res., 14: 347–357.
Sutherland, W.J., 1982a. Spatial variation in the predation of cockles by oystercatchers at Traeth Melynog, Anglesey. I. The cockle population. J. Anim. Ecol., 51: 481–489.
Sutherland, W.J., 1982b. Spatial variation in the predation of cockles by oystercatchers at Traeth Melynog, Anglesey. II. The pattern of mortality. J. Anim. Ecol., 51: 491–500.
Sutherland, W.J., 1982c. Do oystercatchers select the most profitable cockles?. Anim. Behav., 30: 857–861.
Swennen, C., 1976. Populatie-struktuur en voedsel van de eidereend *Somateria mollissima* Moll. in de Nederlandse Waddenzee. Ardea, 64: 171–311.
Talent, L.G., 1982. Food habits of the gray smoothhound, *Mustelus californicus*, the brown smoothhound, *Mustelus*

henlei, the shovelnose guitarfish, *Rhinobatos productus*, and the bat ray, *Mylobatis californica*, in Elkhorn Slough, California. . Calif. Fish Game, 68: 224–234.

Tamburri, M.N. and Zimmer-Faust, R.K., 1996. Suspension feeding: Basic mechanisms controlling recognition and ingestion of larvae. Limnol. Oceanogr., 41: 1188–1197.

Taylor, J.L. and Saloman, C.H., 1972. Nereid shell blisters in the southern quahog clam. Quart. J. Florida Acad. Sci., 35: 21–26.

Thorson, G., 1946. Reproduction and larval development of Danish marine bottom invertebrates. Medd. Komm. Danm. Fiskeri-og Havunders Ser. Plankton, 4: 1–523.

Thorson, G., 1958. Parallel level bottom communities, their temperature adaptation, and their 'balance' between predators and food animals. In: A.A. Buzzati-Traverso (Ed.), Perspectives in Marine Biology. University of California Press, pp. 67–86.

Thorson, G., 1966. Some factors influencing the recruitment and establishment of marine benthic communities. Neth. J. Sea Res., 3: 267–293.

Thomas, F., Cezilly, F., DeMeeus, T., Crivelli, A. and Renaud, F., 1997. Parasitism and ecology of wetlands: a review. Estuaries, 20: 646–654.

Thompson, D.S., 1995. Substrate additive studies for the development of hardshell clam habitat in waters of Puget Sound in Washington State: An analysis of effects on recruitment, growth and survival of the Manila clam, *Tapes philippinarum*, and on existing species diversity and abundance of existing benthic organisms. Estuaries, 18: 91–107.

Thrush, S.F., Hewitt, J.E. and Pridmore, R.D., 1989. Patterns in the spatial arrangements of polychaetes and bivalves in intertidal flats. Mar. Biol., 102: 529–535.

Tiller, R.E., Glude, J.B. and Stringer, L.D., 1952. Hard clam fishery of the Atlantic coast. Comm. Fish. Rev., 14 (10): 1–25.

Toba, D.R., Tompson, D.S., Chew, K.K., Anderson, G.J. and Miller, M.B., 1992. Guide to Manila clam culture in Washington. Univ. Washington, Sea Grant Program, 80 pp.

Tomkins, I.R., 1947. The oyster-catcher of the Atlantic coast of North America and its relation to oysters. Wilson Bull., 59: 204–208.

Turner Jr., H.J., 1948. Report on investigations of the propagation of the soft-shell clam, *Mya arenaria*. Woods Hole Oceanographic Inst. Cont. 462, Woods Hole, MA, pp. 3–10.

Turner Jr., H.J., 1949. Report on investigations of methods of improving the shellfish resources of Massachusetts. Dept. Conserv. Div. Mar. Fish., Commonwealth of Massachusetts, 22 pp.

Turner Jr., H.J., 1950. Investigations on the soft-shell clam, *Mya arenaria*. In: H.J. Turner Jr. (Ed.). Third report on investigations of methods of improving the shellfish resources of Massachusetts. Dept. Conserv. Commonwealth of Massachusetts, pp. 5–15.

Turner Jr., H.J. and George, C.J., 1955. Some aspect of the behavior of the quahaug, *Venus mercenaria*, during the early stages. 8th Report Investigations Shellfish of Massachusetts. Dept. Nat. Res. Div. Mar. Fish. Commonwealth of Massachusetts, pp. 5–14.

Turner Jr., H.J., Ayers, J.C. and Wheeler, C.L., 1948a. The horseshoe crab and boring snail as factors limiting the abundance of the soft-shell clam. Report on investigations of the propagation of the soft-shell clam, *Mya arenaria*. Woods Hole Oceanographic Inst. Contrib. 462, Woods Hole, MA, pp. 43–45.

Turner Jr., H.J., Ayers, J.C. and Wheeler, C.L., 1948b. Further observations on predators of the soft-shell clam. Report on investigations of the propagation of the soft-shell clam, *Mya arenaria*. Woods Hole Oceanographic Inst. Contrib. 462, Woods Hole, MA, pp. 46–49.

Tyler, A.V., 1972. Food resource division among northern, marine, demersal fishes. J. Fish. Res. Bd. Canada, 29: 997–1003.

Uguccioni, D.M. and Posey, M.H., 1992. Predation by grass shrimp, *Palaemonetes pugio*, on hard clams, *Mercenaria mercenaria*: size and substrate interactions. J. Elisha Mitchell Sci. Soc., 108: 29–34.

Vadas, R.L., Burrows, M.T. and Hughes, R.N., 1994. Foraging strategies of dogwhelks, *Nucella lapillis* (L.); interacting effects of age, diet and chemical cues to the threat of predation. Oecologia, 100: 439–450.

VanBlaricom, G.R., 1982. Experimental analyses of structural regulation in a marine sand community exposed to oceanic swell. Ecol. Monogr., 52: 283–305.

Van Breemen, P.J. and Redeke, H.C., 1907. Bijdrage tot de kennis van de flora en fauna der Zuiderzee. Duiderzeerapport, Bijlage v: 1–21. Cited in Kristensen (1957).

van Montfrans, J., Ryer, C.H. and Orth, R.J., 1991. Population dynamics of blue crab *Callinectes sapidus* Rathbun in a lower Chesapeake Bay tidal marsh creek. J. Exp. Mar. Biol. Ecol., 153: 1–14.

Verbeek, N.A.M., 1977. Comparative feeding ecology of herring gulls *Larus argentatus* and lesser black-backed gulls *Larus fuscus*. Ardea, 65: 25–42.

Virnstein, R.W., 1977. The importance of predation by crabs and fishes on benthic infauna in Chesapeake Bay. Ecology, 58: 1199–1217.

Virnstein, R.W., 1979. Predation on estuarine infauna: response patterns of component species. Estuaries, 2: 69–86.

Walker, R.L., 1983. Population dynamics of the hard clam, *Mercenaria mercenaria* (Linne), and its relation to the Georgia hard clam fishery. MS Thesis, Georgia Inst. Tech., 121 pp.

Walker, R.L., 1988. Observations on intertidal whelk (*Busycon* and *Busycotypus*) populations in Wassaw Sound, Georgia. J. Shellfish Res., 7: 473–478.

Walker, R.L. and Tenore, K.R., 1984. The distribution and production of the hard clam *Mercenaria mercenaria* in Wassaw Sound, Georgia. Estuaries, 7: 19–27.

Walker., R.L., Fleetwood, M.A. and Tenore, K.R., 1980. The distribution of the hard clam, *Mercenaria mercenaria* (Linne'), and clam predators in Wassaw Sound, Georgia. Georgia Marine Science Center, Tech. Rept., 80-8, 59 pp.

Walne, P.R. and Dean, G.J., 1972. Experiments on predation by the shore crab, *Carcinus maenas* L., on *Mytilus* and *Mercenaria*. J. Cons. Int. Explor. Mer, 34: 190–199.

Wanink, J. and Zwarts, L., 1985. Does an optimally foraging oystercatcher obey the functional response. Oceologia, 67: 98–106.

WAPORA, 1982. Impact assessment on shellfish resources of Great South Bay, South Oyster Bay and Hempstead Bay, New York. Draft Report, US EPA Region II, New York, 115 pp.

Ward, D., 1991. The size selection of clams by African black oystercatcher and kelp gulls. Ecology, 72: 513–522.

Warren, S., 1916. The feeding habits of *Busycon*. Nautilus, 30: 66–68.

Watzin, M.C., 1983. The effects of meiofauna on settling macrofauna: meiofauna may structure macrofaunal communities. Oecologia, 59: 163–166.

Webster, J.R. and Medford, R.Z., 1959. Flatworm distribution and associated oyster mortality in Chesapeake Bay. Proc. Natl. Shellfish. Assoc., 50: 89–95.

Wells, H.W., 1957. Abundance of the hard clam, *Mercenaria mercenaria* (L.) in relation to environmental factors. Ecology, 38: 123–128.

Wells, H.W., 1958a. Predation of pelecypods and gastropods by *Fasciolaria hunteria* (Perry). Bull. Mar. Sci. Gulf and Caribbean., 8: 152–166.

Wells, H.W., 1958b. Feeding habits of *Murex fulvescens*. Ecology, 39: 556–558.

Wells, H.W., Wells, M.J. and Gray, I.E., 1961. Food of the sea-star *Astropecten articulatus*. Biol. Bull., 120: 265–271.

Wendell, F.E., Hardy, R.A., Ames, J.A. and Burge, R.T., 1986. Temporal and spatial patterns in sea otter (*Enhydra lutris*) range expansion and in the loss of the Pismo clam fisheries. Calif. Fish Game, 72: 197–212.

Whetstone, J.M., 1978. Predation on hard clams, *Mercenaria mercenaria*, by mud crabs, *Panopeus herbstii*. M.S. Thesis, Clemson University, 58 pp.

Whetstone, J.M. and Eversole, A.G., 1978. Predation on hard clams, *Mercenaria mercenaria*, by mud crabs, *Panopeus herbstii*. Proc. Natl. Shellfish. Assoc., 68: 42–48.

Whetstone, J.M. and Eversole, A.G., 1981. Effects of size and temperature on mud crab, *Panopeus herbstii*, predation on hard clams, *Mercenaria mercenaria*. Estuaries, 4: 153–156.

White, M.E. and Wilson, E.A., 1996. Predators, pests, and competitors. In: V.S. Kennedy, R.I.E. Newell and A.F. Eble. (Eds.), The Eastern Oyster *Crassostrea virginica*. University of Maryland Sea Grant, College Park, MD, pp. 559–579.

Wilcox, J.R. and Jeffries, H.P., 1974. Feeding habits of the sand shrimp, *Crangon septempsinosa*. Biol. Bull., 146: 424–434.

Williams, A.B., 1983. The mud crab, *Panopeus herbstii*, S.L. partition into six species (Decapoda: Xanthidae). Fish. Bull., 81: 863–882.

Williams, J.G., 1980. The influence of adults on the settlement of the spat of the clam, *Tapes japonica*. J. Mar. Res., 38: 729–741.

Wilson Jr., H.H., 1991. The foraging ecology of migratory shorebirds in marine soft-sediment communities: the effects of episodic predation on prey populations. Syst. Zool., 31: 840–848.

Wiltse, W., 1978. Effects of predation by *Polinices duplicatus* on community structure. Ph.D. Dissertation, University of Massachusetts, Amherst, 130 pp.

Wiltse, W., 1980a. Effects of *Polinices duplicatus* (Gastropoda: Naticidae) on infaunal community structure at Barnstable Harbor, Massachusetts, USA. Mar. Biol., 56: 301–310.

Wiltse, W., 1980b. Predation by juvenile *Polinices duplicatus* (Say) on *Gemma gemma* (Totten). J. Exp. Mar. Biol. Ecol., 42: 187–199.

Wood, L., 1968. Physiological and ecological aspects of prey selection by the marine gastropod *Urosalpinx cinerea* (Prosobranchia: Muricidae). Malacologia, 6: 267–320.

Yocom, C.F. and Keller, M., 1961. Correlation of food habits and abundance of waterfowl, Humboldt Bay, California. Calif. Fish Game, 47: 41–54.

Young, C.M., 1989. Larval depletion by ascidians has little effect on settlement of epifauna. Mar. Biol., 102: 481–489.

Young, C.M. and Gotelli, N.J., 1988. Larval predation by barnacles: effects on patch colonization in a shallow subtidal community. Ecology, 69: 624–634.

Young, D.K. and Young, M.W., 1978. Regulation of species density of seagrass-associated macrobenthos: Evidence from field experiments in the Indian River estuary, Florida. J. Mar. Res., 36: 569–593.

Ysebaert, T.J. and Meire, P.M., 1990. Factors affecting food selection and foraging behavior on mudflats by breeding black-headed gulls, *Larus ridibundus*. In: Barnes, M. and Gibson, R.N. (Eds.), Trophic Relationships in the Marine Environment. Proc. 24th European Mar. Biol. Symp, Aberdeen Univ. Press, pp. 250–265.

Zachary, A. and Haven, D.S., 1973. Survival and activity of the oyster drill *Urosalpinx cinerea* under conditions of fluctuating salinity. Mar. Biol., 22: 45–52.

Zack, R., 1979. Shell dropping: decision-making and optimal foraging in northwestern crows. Behavior, 58: 106–117.

Zelickman, E.A., Gelfand, V.I. and Shifrin, M.A., 1969. Growth, reproduction and nutrition of some Barents Sea hydromedusae in natural aggregations. Mar. Biol., 4: 167–173.

Zwarts, L. and Blomert, A.M., 1992. Why Knot (*Claidris canutus*) take medium-sized *Macoma balthica* when six prey species are available. Mar. Ecol. Prog. Ser., 83: 113–128.

Zwarts, L. and Drent, R.H., 1981. Prey depletion and the regulation of predator density: Oystercatchers (*Haematopus ostralegus*) feeding on mussels (*Mytilus edulis*). In: N.V. Jones and W.J. Wolff (Eds.), Feeding and Survival Strategies of Estuarine Organisms. Plenum Press, New York, pp. 193–216.

Zwarts, L. and Wanink, J., 1989. Siphon size and burying depth in deposit- and suspension feeding benthic bivalves. Mar. Biol., 100: 227–240.

Zwarts, L., Blomert, A.M. and Wanink, J.H., 1992. Annual and seasonal variation in the food supply harvestable for Knot *Claidris canutus* staging in the Wadden Sea in late summer. Mar. Ecol. Prog. Ser., 83: 128–139.

Biology of the Hard Clam
J.N. Kraeuter and M. Castagna (Eds.),

Chapter 12

Pests, Parasites, Diseases, and Defense Mechanisms of the Hard Clam, Mercenaria mercenaria

Susan E. Ford

12.1 INTRODUCTION

Clams, mussels, scallops, and oysters have been harvested commercially for many centuries, and have undoubtedly suffered from disease-caused mortalities throughout that time. The study of bivalve parasites, diseases, and defense mechanisms is relatively recent, however, driven largely by epizootic mortalities of oysters in the United States and Europe in the last half century (Ford and Tripp, 1996). Investigations have shown that each molluscan group becomes infected by a similar array of organisms from viruses to copepods, although relatively few cause disease (Bower et al., 1994). The distinction between infection and disease is important. *Infection* refers to the establishment of a foreign organism (e.g., a parasite) in host tissues. *Disease* indicates damage to a body part, organ, or system, which may or may not be caused by an infectious agent, such that the affected organism no longer functions normally. In fact, an infection does not necessarily lead to disease. Many infectious agents are parasites that may cause localized tissue damage, but relatively little overall harm to their hosts. Infectious agents capable of causing disease are termed pathogens, of which there are several types. An *obligate* pathogen *must* live in another organism. It is not capable of sustained life or proliferation outside of a host. A *facultative* pathogen, on the other hand, is able to live freely in the natural environment, but it can also live off the tissues of another organism. An *opportunistic* pathogen is one that lives in the tissues of another organism, but is usually found at numbers low enough that it does not harm that host. Host organisms that are otherwise "healthy" can prevent infection by, or control proliferation of, facultative or opportunistic pathogens through structural (e.g., shell or epithelial barriers) or biological (physiological activity or the internal defense system) mechanisms. Both facultative and opportunistic pathogens, however, may proliferate and cause disease if the host is compromised in some manner so that it can no longer effectively defend itself or if the number of these pathogens in the environment is large enough to overwhelm host defenses.

Disease is not necessarily caused by an infectious agent. Environmental, genetic, and nutritional problems can also result in disease. The factors that cause disease, alone or in combination, are referred to as the *etiological agents* of that disease. Disease, in turn, does not necessarily result in death although it can cause significant sublethal effects. In their review of shellfish diseases, Sindermann and Rosenfield (1967) relate numerous instances of mass mortalities of bivalve species for which causes were never found. Although some appeared to be contagious, many others probably resulted from transient, unfavorable environmental conditions. Harmful environmental conditions could also make the animals more susceptible to facultative or opportunistic pathogens. A number of recent reviews describe and discuss parasites, pathogens, and diseases of commercial molluscs (Lauckner, 1983; Sparks, 1985; Fisher, 1988; Gibbons and Blogoslawski, 1989; Sindermann, 1990a; Getchell, 1991; Bower, 1992a; Elston, 1993; Perkins, 1993; Bower et al., 1994; Ford and Tripp, 1996).

That infectious diseases can result in devastating losses to wild populations of adult molluscs is amply illustrated by recent oyster epizootics. The pathogenic protozoans *Perkinsus marinus* and *Haplosporidium nelsoni* in the United States, and *Bonamia ostreae* and *Marteilia refringens* in Europe, can cause death rates in excess of 50% per year (Grizel, 1983; Ford and Tripp, 1996). Similar "mass mortalities" of mussels, clams, and scallops of various species have been reported from time to time, sometimes associated with potential disease agents (Sindermann and Rosenfield, 1967). With few exceptions, however, the consistent, predictable relationship found between oysters and certain pathogens has not been reported in other bivalves (Bower et al., 1994). Among commercially harvested species, the hard clam, *Mercenaria mercenaria*, appears to be exceptionally "healthy". A number of studies of captive populations in which attempts have been made to exclude predators, or at least to measure their contribution to clam deaths, shows that non-predation losses are typically only about 5–10% per year (Table 12.1). Many of the occasional large-scale mortalities of adult hard clams, which have been reported from the northeastern states and Canada, appear to be related to the effects of extreme cold weather on clams exposed at low tide or in shallow water (Dow and Wallace, 1951; Haven and Andrews, 1957), although parasites and other (unknown) stressors are also associated (Drinnan and Henderson, 1963; Greene and Becker, 1977; Smolowitz et al., 1998).

Diseases of larval and juvenile hard clams held under intensive culture conditions are usually caused by viruses, bacteria, or fungi that are also common in other cultured bivalves (Tubiash, 1975; Elston, 1984). For the most part, these disease agents are opportunistic or facultative invaders of animals stressed by high-density culture conditions.

The remainder of this chapter is organized into four sections: (1) parasites, symbionts, and pests found occasionally in hard clams, but not known to be associated with disease or mortality; (2) pathogens and diseases of wild and cultured hard clams; (3) the internal defense system of hard clams, including the use of clams in experimental studies of molluscan defense mechanisms; and (4) hypotheses and discussion of why the hard clam has so few known diseases. The emphasis, where possible, will be on comparative aspects, especially with the eastern oyster, *Crassostrea virginica*, which is present in many of the same habitats as hard clams, but suffers from many more infectious diseases.

12.2 OCCASIONAL PARASITES, SYMBIONTS, AND PESTS

Like all other marine bivalves that have been collected from the field, occasional parasites are reported in hard clams in numbers too low, or without associated pathology, to suggest potential for mortality. In a 10-year (1974–1984) pathological survey involving 1200 hard clams from 30 field locations and hatcheries from New Jersey to Maine, Leibovitz (1985) found a variety of symbiotic organisms (those living with the clams while not causing obvious harm). He reported no pathogens, and all of the organisms described as being abundant were from specimens obtained from hatchery or nursery culture.

TABLE 12.1

Mortality rates of hard clams in studies where predators have been excluded or where the proportion of predator mortality has been identified

Location	Starting size (mm)	Species	Duration	Mortality (%)	Comments	Reference
Florida	33–44	*M. mercenaria*	7 months	5–18	predators responsible for 90%	Menzel and Sims (1964)
Georgia	10	*M. mercenaria*	1 year	70	mostly crab predation	Walker and Humphrey (1984)
Georgia	6	*M. mercenaria*	6 months	35	clams from MA, 3 mm vexar-screen cages	Walker (1984)
	22–24	*M. mercenaria*	12 months	1	clams from MA, 3 mm vexar-screen cages	
Georgia	20	*M. mercenaria*	15 months	12–25	clams from MA and DE	Walker and Heffernan (1990)
S. Carolina	12–14	*M. mercenaria*	21 months	49–55	mostly crab predation	Eldridge et al. (1977)
S. Carolina	13	*M. mercenaria*	36 months	50–60	mostly crab predation	Eldridge et al. (1979)
	>25	*M. mercenaria*	24 months	6–9		
S. Carolina	3 years old?	*M. mercenaria*	2 years	~3	survivors of earlier experiment?	Eldridge and Eversole (1982)
Virginia	11	*M. mercenaria* ×	2 years	0–8	in sand-filled, suspended trays	Haven and Andrews (1957)
		M. campechiensis	2 years	70–96	mostly overwinter	
		Hybrids	2 years	5–25	mostly overwinter	
Virginia	10–14	*M. mercenaria*	18 months	15	protected by net, shell bag, crab trap	Kraeuter and Castagna (1985)
New York	29	*M. mercenaria*	2 months	14–23	gravel protection	Flagg and Malouf (1983)
England	~4 year classes	*M. mercenaria*	1 year	0–17	introduced population	Hibbert (1977)

MA = Massachusetts; DE = Delaware.

12.2.1 Viruses and Bacteria

Viruses have been isolated from numerous molluscan species, including hard clams (Hill, 1976), but with no evidence of disease. They have also been reported in Chlamydia, which themselves were found as obligate intracellular parasites in several bivalve species, including hard clams, in Chesapeake Bay (Harshbarger et al., 1977). Bacteria are known associates of hard clam disease only in hatchery cultures of larvae and juveniles (Tubiash et al., 1965; Brown, 1974; Elston et al., 1982; Brown and Tettelbach, 1988).

12.2.2 Protozoans

Similarly, parasitic protozoans are conspicuously absent from wild populations of hard clams. Organisms similar to *Perkinsus marinus*, the protozoan that causes Dermo disease in eastern oysters, have been isolated from hard clams, as well as from a variety of other bivalves, but appeared to be only an incidental inhabitant. Hard clams were one of thirteen bivalve species from the lower York River, Virginia, found to harbor small numbers of a *P. marinus*-like organism (Andrews, 1955). Because of this finding, both Ray (1954) and Andrews and Hewatt (1957) attempted to experimentally infect hard clams with this parasite. Exposure of clams in enzootic waters in Louisiana failed to cause detectable infections in either live or dead clams collected over the following 4 months, whereas adjacent oysters became infected within 1 to 3 weeks (Ray, 1954). Feeding of tissue minces from infected oysters also failed to initiate infections (Ray, 1954; Andrews and Hewatt, 1957). Injections of *P. marinus* from heavily infected oysters into clams did produce localized infections near the site of the injection, but little evidence that they spread beyond this location (Ray, 1954). Andrews and Hewatt (1957) also inoculated large doses of parasites from infected oysters and examined tissues distant from the site of injections to avoid the localization artifact. They found no evidence of infection in either live or gaping clams after 1 month. In a recent publication, Cheng et al. (1995) reported the transmission of *P. marinus* from infected oysters to hard clams after a 10-day proximity experiment in which the clams and oysters were held together in beakers of seawater. Infections were diagnosed histologically and the parasites in the clams, which were morphologically indistinguishable from those in the oysters, were found primarily in the connective tissue surrounding the digestive tract. Interestingly, at least one other species of *Perkinsus* is present in areas where clams and oyster grow. It infects the clam *Macoma balthica* (Kleinschuster et al., 1994). Other species of *Perkinsus* may soon be identified with gene probes. Taken together, these results indicate that a *Perkinsus* sp., perhaps *P. marinus*, can infect hard clams in the wild, but that it does not cause noticeable ill effects on the clams. In addition, *P. marinus* from infected oysters can also infect hard clams under experimental conditions, but there is no evidence that parasites spread from initial lesions or even remain viable.

12.2.3 Cestodes, Trematodes, and Nemerteans

Larval cestodes infect hard clams in the Gulf of Mexico and heavy infestations reduce meat condition (Cake, 1977). Also, hard clams can be experimentally infected with trematodes and nematodes (Cheng and Burton, 1965; Cheng et al., 1966a), although these do not appear to be common parasites in natural populations.

The nemertean worm *Malacobdella grossa* inhabits the mantle cavity of numerous clam species, including *M. mercenaria* and *M. campechiensis* throughout most of their ranges (Ropes, 1963; Porter, 1964). In *M. mercenaria* from Nantucket Sound, Massachusetts, Ropes (1963) found summer prevalences ranging from 5 to 35% according to location within the Sound. Only one worm was found in each infested clam. Porter (1964) obtained more extensive data in a year-long survey of 2100 *M. campechiensis* off the North Carolina coast. The prevalence of *M. grossa* ranged from 61 to 100% according to season and sampling site, but the worms caused no obvious damage to mantle or gills, and no loss of meat condition. Ropes (1963) hypothesized that site-specific variation in prevalence of *M. grossa* was associated with the direction of transport of larval worms by water currents. His argument was supported by Porter's (1964) evidence that recruitment of young *M. grossa* occurred at some sites and not others. Porter (1964) also noted that the highest infestation rates were offshore and the lowest, inshore in estuarine environments. Prevalence was lowest in the autumn, increased over the winter, reached a peak in spring, and remained fairly high over the summer (temperature from 9° to 26°C). Ninety-three percent of the clams contained only one worm. Multiple infestations occurred primarily during the recruitment period, from midwinter into autumn, and typically consisted of small and medium-sized worms. Experiments in which large and small *M. grossa* were placed together in vitro failed to substantiate the hypothesis that the predominance of single worm infestations resulted from attacks by large on small individuals, although worms living together appeared less active than those kept singly (Porter, 1964).

In a study of growth and survival of hard clams in Prince Edward Island, Canada, Landry et al. (1993) ascribed high mortality at one site to the presence of another nemertean, *Cerabratulus lacteus*, although no direct evidence was provided that the worm was actually involved.

12.2.4 Copepods

Parasitic copepods infest many commercial bivalves, including hard clams. Members of the family Myicolidae were first reported on the gills and mantle of hard clams from the vicinity of Beaufort, North Carolina by Pearse (1947). This survey included a number of potential host species and found that copepods were more abundant on fish than in molluscs. Two subsequent surveys of bivalves for copepods, although also limited, were quantitative and involved several bivalve species. Humes (1954) found a new species, which he named *Mytilicola porrecta*, in the intestines of mussels *Geukensia demissa* and *Mytilus (= Brachidontes) recurvus*, and in *M. mercenaria* in the salt marshes of Barataria Bay, Louisiana. Prevalence of copepods was quite variable among species, although all were living in approximately the same area and were collected during the month of June. One copepod was found in the single hard clam examined, a sample size that precludes valid comparisons. In a more comprehensive seasonal survey conducted in New England, Humes and Cressey (1960) found that the copepod *Myocheres major* (considered to be the same species as found in North Carolina by Pearse) (1) was found frequently in the mantle cavities of stout tagelus clams, *Tagelus gibbus* (= *plebius*) and soft-shell clams, *Mya arenaria* in the warm months, (2) was present occasionally during the summer in *M. mercenaria* and Atlantic jackknife clams, *Ensis directus*, and (3) was never found in mussels *Mytilus edulis* and *G. demissa*, scallops, *Pecten irradians*, and false angel wings, *Petricola pholadiformis*. In Europe, copepod infestations are much more prevalent in mussels (80–100% prevalence with numerous copepods per mussel), in which they cause some localized tissue damage; however, there is no evidence that they cause loss of meat condition or mortality (Davey, 1989; Figueras et al., 1991). Thus, it is improbable that copepods cause problems for hard clams at either the individual or population level.

12.2.5 Polychaetes

Polychaetes in the genus *Polydora* are frequent inhabitants in and on the shells of epifaunal bivalves like oysters and scallops, but are rare in subsurface dwelling organisms like hard clams. Landers (1967) exposed juvenile hard clams (5 to 32 mm) to *P. ciliata* in a series of experimental studies. The worms readily attacked and bored into non-buried clams, penetrating completely through the shell within 20 days for 5- to 10-mm clams and 40 days for 32-mm clams. Buried clams were not attacked and exposed individuals that subsequently buried themselves lost their worms. There was no consistent evidence that *Polydora* infestation elevated clam mortality, but the fact that the worms could penetrate completely through the shell indicates that if clams were for some reason unable to bury themselves, they would be left more susceptible to predators even if the borings alone did not cause death. Davis (1969) reported that a number of large hard clams caught by scallop dredges in Nantucket Harbor, Massachusetts, were found to be infested by *Polydora* excavations. The coloration of the clams' valves indicated that they had been at least partly exposed above the sediment surface, perhaps unable to rebury after recent storms, and "in each case the more exposed valve exhibited excavations characteristic of *Polydora* infestation". Similarly, Jeffries (1972) associated the presence of *Polydora* spp. blisters in 5–10% of hard clams at a polluted site in Narragansett Bay, Rhode Island, with emergence of the clams from the bottom sediments in response to irritants and low dissolved oxygen.

12.3 PATHOGENS AND DISEASES

12.3.1 In Culture

12.3.1.1 Bacterial and fungal diseases

Bacteria are the principal agents implicated in diseases of hard clams in hatcheries and nurseries, where larvae and juveniles are grown at high density. Similar problems affect all molluscs cultured under similar conditions. The most common bacterial agents have been identified as members of the genera *Vibrio* (including *V.* [now *Listonella*] *anguillarum* and *V. alginolyticus*) and *Pseudomonas*. The vibrios typically cause a disease in larval cultures that is characterized by rapid onset (a few hours) and evident swarming of bacteria in and around moribund larvae (Tubiash et al., 1970). Concentrations of 10^6 CFU (colony forming units) mL^{-1} in larval cultures is sufficient to cause mortality. Brown and Tettelbach (1988) described another bacterial disease of hard clam larvae with different characteristics, although the etiological agent resembled *L. anguillarum*. This isolate caused mortalities in concentrations as low as 10–100 CFU mL^{-1}, but only after about 10 days in commercial hatcheries and without evidence of the typical bacterial swarming syndrome. Brown (1974) identified a probable *Pseudomonas* sp. as the cause of a pink coloration at the bottom of hard clam larval containers. This species caused no larval mortalities at 10^3 CFU mL^{-1}, but abnormal development and mortality at 10^4 CFU mL^{-1}. Fungi can also cause problems in hatchery cultures, although they do not appear to be as severe as those involving bacteria (Davis et al., 1954).

Bacteria causing larval diseases can be controlled with antibiotics (Elston, 1984; Kraeuter and Castagna, 1984), but hatchery operators are cautioned against over reliance on this type of treatment because of the potential for development of antibiotic resistance. Bacteria that affect larval cultures may originate in algal cultures, from incoming water, or from pipes, tanks, and other hatchery equipment that are not properly and regularly cleaned (Elston, 1984). For most purposes, improved animal husbandry and hatchery cleanliness are the most appropriate means to combat bacterial diseases of larvae.

Juvenile clams grown in dense, nursery cultures harbor a variety of epiphytes on their shell surfaces. *Vibrio* spp. are probably the most common, sometimes progressing inward through cracks and fissures in the shell to reach the mantle and gills (Elston et al., 1982; Leibovitz, 1985). Cytophaga-like bacteria attack and destroy the hinge ligament of juvenile bivalves, including hard clams, in nursery cultures (Dungan and Elston, 1988). Loss of the hinge attachment prevents normal feeding and respiration, eventually killing affected animals. Filamentous bacteria, blue–green algae, entoprocts, vorticellids, and folliculinids are also found on shell surfaces, and diatoms may cause impaction of the digestive tract (Leibovitz, 1985). A preliminary report of shell deformities in "newly planted stunted hatchery reared hard clams", which was associated with mortality and with microscopic lesions (mantle atrophy and epithelial necrosis, intranuclear inclusions, bacterial masses, and adductor muscle inflammation and degeneration) (Leibovitz et al., 1976) was not pursued, but may well have been a problem associated with poor culture conditions. In fact, hard clams seem to be less sensitive to shell diseases than are oysters. Elston et al. (1982) reported anomalous calcification and protein deposition in juvenile oysters with shell infestations by *Vibrio* spp. and other epiphytes. In contrast, juvenile hard clams had "dense growth of epiphytes and filamentous bacteria on their shells, but no grossly detectable shell damage". A recently described disease, of unknown etiology, that results in an anomalous conchiolin deposit on the inner shell and subsequent mortalities of juvenile oysters grown in high-density nursery floats does not affect juvenile hard clam held in identical containers in the same system (Bricelj et al., 1992).

12.3.1.2 Winter mortality

Juvenile clams, in the size range of 2 to 10 mm, often suffer high mortalities in the spring (water temperature $\leq$10°C) after being held overwinter in the hatchery or nursery (Kraeuter and Castagna, 1984). *Vibrio* spp. were isolated from dying clams in one such incident and treatment of affected clams with antibiotics (Combiotic and Chloromycetin) reduced mortality. Combiotic worked equally well in fresh and in salt water, which was interpreted as an indication that bacteria on the outside of the shells were involved. (Internal bacteria should not have been affected in fresh water because the clams would have remained closed during the treatment.) Commercial bleach (5.25% sodium hypochlorite) was also very effective in treating the problem and a daily, 2-h treatment in an 18 ppm bleach solution for

3 to 4 weeks was recommended as a readily available and economical substitute for antibiotics (Kraeuter and Castagna, 1984). A commercially available concentrate of nitrifying bacteria added to flowing seawater in a clam nursery was also reported to improve survival, although the mechanism of action is not known (Castagna et al., 1989).

Although these observations suggested a bacterial pathogen, introducing diseased whole clams or their homogenates into nursery tanks failed to cause mortalities of exposed clams (J.N. Kraeuter, Haskin Shellfish Research Laboratory, personal communication, 1995). Consequently, an intensive study of other potential factors involved in winter mortality was initiated concurrently in New Jersey and New York (J.N. Kraeuter, Haskin Shellfish Research Laboratory, personal communication, 1999). Several factors were manipulated prior to, or at, autumn field deployment: nutrition (starved, ambient, or ambient supplemented with cultured algae), temperature at deployment (14° or 12°C), sediment type (sand or mud), and size (6, 8, 10, and 12 mm average shell width). The only consistent correlate with overwinter mortality was size: larger clams survived better than smaller ones. Neither temperature at deployment nor sediment type had any effect. Further, mortality was not associated with any measure of stored energy reserves (lipid or carbohydrate) or with any pathological condition. The data were inconsistent with the hypothesis that winter mortality was due to low energy reserves of small clams entering the winter dormancy period. Thus, supplemental feeding of small clams just before field deployment would be an unneeded expense. Nevertheless, the consistently better survival of the larger clams indicates that scheduling nursery operations so that clams are of adequate size ($\geq$10 mm) by the end of the growing season would minimize the problem.

12.3.1.3 QPX

Pathological examination of adult and juvenile hard clams experiencing mortalities of up to 100% in a Prince Edward Island, Canada hatchery led to the finding and description of a protistan named QPX (Quahog Parasite X) in the tissues of dead and dying clams (Whyte et al., 1994). What was probably the same parasite had been recognized in the late 1950s, associated with mortalities of wild-stock clams in New Brunswick, Canada (Drinnan and Henderson, 1963). The parasite was first found in tissues of moribund clams held in cold dry storage and subsequently in field populations from which these clams had come. Clams transplanted to affected beds became infected and experienced high mortality. A similar organism was found in dead and dying clams collected in Barnegat Bay, New Jersey, in early December, 1976, after a particularly severe early freeze (unpublished report by T. Keller, Haskin Shellfish Research Laboratory, 1977). In 1995, QPX or QPX-like organisms were associated with mortality in two locations in Massachusetts, although clam growers at one site had been noticing mortalities for several years (Smolowitz et al., 1998). The organism was detected in Massachusetts as early as 1991, although it could not be tied to mortalities at the time (Smolowitz et al., 1998). Subsequent surveys in 1996 and 1997 found the organism in New Jersey (Kraeuter et al., 1998a,b) and Virginia (Ragone Calvo et al., 1998) where high prevalences were associated with mortality in some locations.

Based on 18s rDNA and small-subunit rRNA (SSURNA) sequence comparisons, Mass et al. (1999) have classified QPX as a primitive member of the phylum Labyrinthulomycota (Kingdom Protista) (see Pokorny, 1985). Using morphological characteristics, Whyte et al. (1994) had earlier noted its similarity to the Thraustochytriales and Labyrinthulales, and had also placed it in the phylum Labyrinthomorpha. These are the "slime-net" protists, which are common saprophytic organisms in marine and estuarine environments (Porter, 1990). Members of the group have been associated with high mortalities of cultured juvenile abalones on the west coast of Canada (Bower, 1987). In wild stocks, these organisms have been occasionally found in nudibranchs (shell-less snails), octopus, and squid (Polglase, 1980; McLean and Porter, 1982; Jones and O'Dor, 1983). Although apparently rare in natural populations, they are associated with mortality of the infected molluscs and can be transmitted to members of the same species held in captivity with the diseased molluscs.

In histological section, three forms of the QPX organism have been identified and named using classical mycological terminology (Ragone Calvo et al., 1998; Smolowitz et al., 1998). Thalli are usually the most common forms and range from about 2 to 20 μm in diameter (Fig. 12.1A). Larger (10–48 μm) cells (sporangia), each contain 20 to 40 endospores with diameters of 2–5 μm (Fig. 12.1B). Inside the clam, thalli develop into sporangia, which eventually rupture, releasing the endospores (Fig. 12.1B). The endospores are immature thalli that develop and repeat the cycle. A characteristic feature of QPX in tissue sections, in fresh preparations, and in cell culture is a clear, "halo"-like area around parasites (Fig. 12.1A,B). Net-like projections extending from the cell surface into the halo (Fig. 12.1B) are common in preparations of clams from U.S. locations, but were not reported in Canada (Whyte et al., 1994). Their high affinity for Alcian blue/PAS stains indicates that the extensions are composed in part of mucopolysaccharides and are more appropriately termed a "mucofilamentous halo" (R. Smolowitz, Marine Biological Laboratory, personal communication, 1999). Viewed by transmission electron microscopy, QPX cells have certain morphological differences from other common Labyrinthomorphids, including fibrogranular, rather than plate-like, cell walls, and the lack of sagenogenetosomes or the ectoplasmic net they produce.

The parasite has been propagated in vitro in MEM (Minimal Essential Medium) (Whyte et al., 1994; Kleinschuster et al., 1998). In culture, QPX shows the same developmental stages as in tissue sections of clams, including the same halo-like areas made of the same mucofilamentous material surrounding the parasites. In addition, a zoospore stage developed when endospores were placed in seawater. Interestingly, zoospores developed into thalli when returned to MEM, suggesting that the zoospores may be a transmission form (Kleinschuster et al., 1998). Presumptive ectoplasmic nets were also reported surrounding cells placed in seawater. The use of cell culture as a diagnostic method has been investigated and has been found to be much more sensitive than histology (G. Bacon, Gulf Fisheries Centre, Moncton, NB, Canada, personal communication, 1999). For instance, in one sample of wild clams, only 6.7% were detected as having QPX by standard histological sections whereas 96% of the same clams were diagnosed positive using the culture method. One possible explanation for the disparity between methods may be the use of much larger tissue quantities for culture, including the mantle fluid, in which Labyrinthomorphids are known to reside (Perkins, 1973; Porter, 1990). Although the culture method may give positive results based on the detection of QPX (or other Labyrinthomorphids) that are not actually infecting clams, it is valuable in that it shows the nearly ubiquitous association of these organisms with the clams.

The demonstrated transmissibility of Labyrinthulids (Bower, 1987) raised early concerns that the parasite might be transmitted via seed clams, which are often shipped from hatcheries in one geographic location to grow-out sites in another. To explore this possibility, tissue

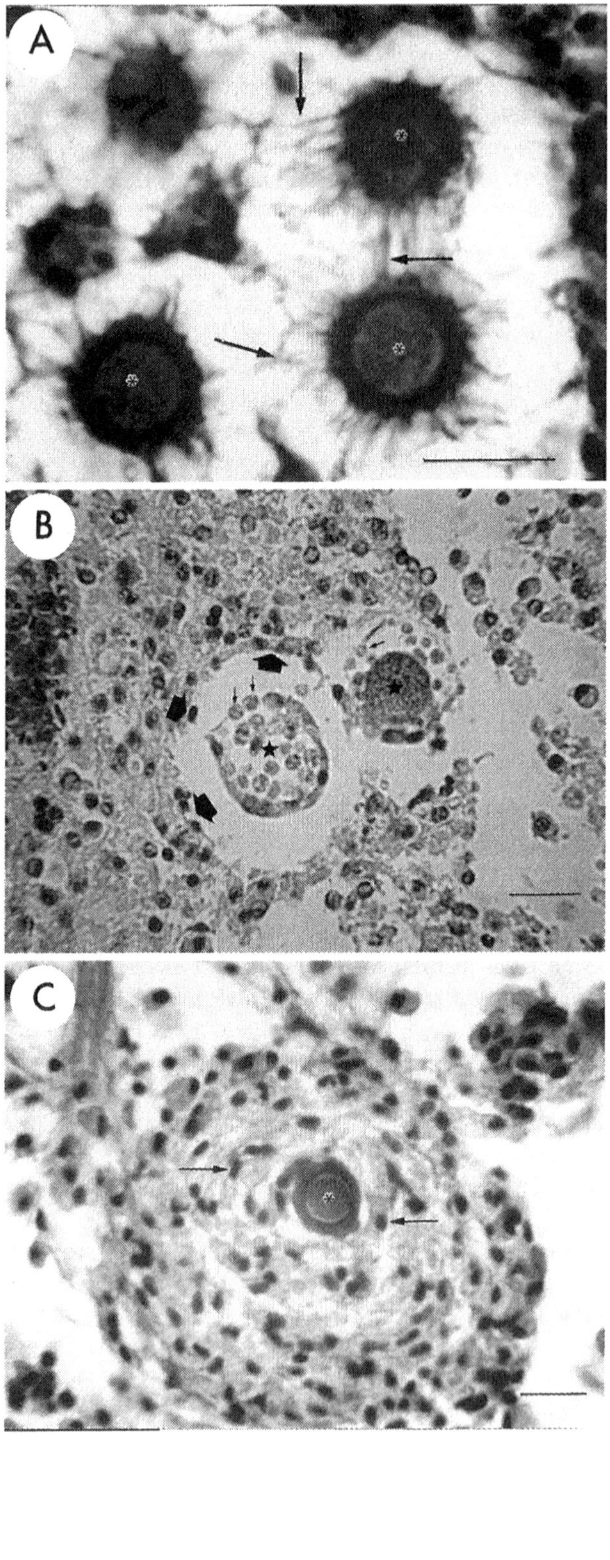
A
B
C

sections of 2203 seed clams (<1 to 20 mm) from 13 different hatcheries in six states from Massachusetts to Florida, collected from 1995 to 1997, were examined by pathologists in three laboratories (Ford et al., 1997). No QPX or QPX-like organisms were found. Further, QPX was not detected in a total of 756 hatchery-produced clams examined during their first year of field grow-out. Continued examination of seed since that study confirms the initial findings and indicates that seed is an unlikely source of the parasite. It is much more probable that clams become infected in planting areas because the earliest histologically detectable infections are not found until a year after seed has been planted (R. Smolowitz, Marine Biological Laboratory, personal communication, 1998). Infections become increasingly severe with time. In Massachusetts, macroscopic signs of the disease (see below) appeared after about two years and were followed shortly by mortalities.

Clams with moderate to severe infections had lower growth and meat condition compared to those with light or no detectable infections (Smolowitz et al., 1998). In sandy sediments, heavily infected and dead clams were often found at the surface and typically showed marked chipping of the shell margin. The shell chips are thought to be caused by sand grains, trapped in mucous, that lodge between the valves of weakened clams and crack the shell as the clams attempt to close their valves. Shell chipping and the presence of moribund clams at the surface were not characteristic of a site where QPX-infected clams were growing in a finer-grained sediment.

Other macroscopic symptoms of QPX are a retracted, swollen mantle edge, occasionally displaying nodules (Smolowitz et al., 1998). These are macroscopic manifestations of the host inflammatory response. Dense concentrations of hemocytes surround the parasites in formations that range from loose aggregations of cells (swollen mantle) to tight, well-defined encapsulations (nodules). Phagocytosis has also been observed (Ragone Calvo et al., 1998; Smolowitz et al., 1998). Smolowitz et al. (1998) found a positive association between the relative lack of delimiting halos and the effectiveness of inflammation in ridding the clams of QPX parasites, and have argued that the mucofilamentous halo forms a physical barrier that prevents hemocytes from attacking the parasites.

Certain differences between the QPX and its effects on clams along a latitudinal cline are noteworthy. In Canadian clams, QPX cells were found most often in the connective tissue of the digestive gland and in the foot muscle tissue, where they were reported to cause necrosis (Whyte et al., 1994). In contrast, the most frequently infected organs of clams in U.S. sites are the mantle and gills (Ragone Calvo et al., 1998; Smolowitz et al., 1998). The mucofilamentous halo described in U.S. clams by Smolowitz et al. (1998) and Ragone Calvo et al. (1998) was not reported in the Canadian study (Whyte et al., 1994). Within the U.S., there are also geographical differences. Parasite sizes were somewhat smaller in the Massachusetts than in the Virginia studies (Ragone Calvo et al., 1998; Smolowitz et al., 1998). Ragone Calvo et al.

Fig. 12.1. Photomicrographs of Quahog Parasite X (QPX) in hard clam tissues. (A) Thalli (*) showing filamentous, mucopolysaccharide structures (arrows) present in the clear areas around the parasites. (B) Sporangium (⋆) showing endospores (small arrows) being released from ruptured parent cells. Hemocytes (large arrows) surround parasites at periphery of clear, halo-like areas. (C) Single thallus (*) encapsulated by hemocytes (arrows). Scale bars = 20 μm. Histological preparations courtesy of S. McGladdery and R. Smolowitz.

(1998) reported that parasites frequently appeared moribund in Virginia clams and a similar observation has been made for clams in New Jersey (R. Barber, Haskin Shellfish Research Laboratory, personal communication, 1998). Moribund parasites were rare in the Canadian and Massachusetts studies. Nodules are rarely associated with infections in New Jersey or Virginia. The differences in morphological features among the sites, including the degree of mucofilamentous halo formation, could be explained by environmental differences, genetic differences in the clams, variations in tissue fixation procedures, or the presence of two or more parasite species.

The epizootiological aspects of QPX disease are being clarified, although many uncertainties remain. Temperature is probably important in the seasonal pattern and in the only report to date with frequent sampling over a 1-year period, a bimodal infection pattern was described on the seaside of Virginia (Ragone Calvo et al., 1998). The most numerous and severe infections occurred in November and again in May. An unusual mortality coincided with the May peak. Sampling in other areas has not been sufficiently frequent to confirm a similar pattern; however, Smolowitz et al. (1998) reported mortalities associated with QPX in Massachusetts to occur in summer and fall. Ragone Calvo et al. (1998) sampled in both high- and low-salinity regions of Virginia and found QPX only in the high-salinity bays along the Atlantic coast (30–34 ppt) and not in Chesapeake Bay (15–25 ppt). All other sites where the parasite has been found are in high salinity (generally 30 ppt; R. Smolowitz, Marine Biological Laboratory and J.N. Kraeuter, Haskin Shellfish Research Laboratory, personal communications, 1998), although hard clams are found at lower salinities (see Chapter 8). Although most cases of QPX disease have been reported in cultured clams, natural populations are also infected and can experience mortality (Drinnan and Henderson, 1963; Ragone Calvo et al., 1998; MacCallum et al., 2000; R. Smolowitz, Marine Biological Laboratory, personal communication, 1999).

Labyrinthomorphids are common in the ambient waters inhabited by hard clams and often present in their mantle fluid (Perkins, 1973; Porter, 1990). Consequently, the possibility arises that QPX is a facultative or opportunistic pathogen that invades already compromised clams. It is not yet known whether the QPX organism falls into this category. Much current evidence suggests that infections by QPX and QPX-like organisms, and subsequent mortalities, have occurred in molluscs growing under conditions that could foster the proliferation of such pathogens in stressed hosts: very high-density in culture or on natural beds; unusually low temperature; maladapted stocks; and generally suboptimal growing or holding situations. It is noteworthy that clams from southern broodstock grown in more northern locations have higher QPX prevalence and intensity, and suffer higher mortality, than same-age clams from local broodstock grown in the same location (Kraeuter et al., 1998a,b). The southern stocks may be less well adapted to ambient northern conditions, which leaves them more susceptible than the local groups to infection by QPX.

12.3.1.4 Gas bubble disease

Bivalves held in the hatchery or laboratory are affected by a condition known as "gas bubble disease", which occurs when cold ambient water is heated and becomes supersaturated with atmospheric gases. When the gases come out of solution, they cause blood vessel emboli and may result in death. Although hard clams are affected by Gas Bubble Disease, the species appears to be less susceptible than oysters or other clam species. Experimental exposure of eastern oysters and hard clams to supersaturated water caused bubbles to appear in mantle and gill tissue, and blisters to form on the shell (Malouf et al., 1972). Ten percent of the oysters died and the remainder were in poor condition. In contrast, hard clam deaths were rare and the only gross symptom was "an obvious lightening of the color of the gills". Juvenile (8 to 12 mm) coot clams, *Mulinia lateralis*, and soft-shelled clams, *M. arenaria*, had mortalities or were induced to float because of trapped bubbles, at gas concentrations of $\geq$108% and $\geq$114% saturation, respectively (Bisker and Castagna, 1985). Juvenile hard clams survived and did not float, although their growth rate was reduced at $\geq$115% saturation. The best preventive measure for gas bubble disease is to equilibrate incoming water with air before it is introduced into rearing tanks.

12.3.2 In Nature

12.3.2.1 Chlamydiales and Rickettsiales

Chlamydiales and Rickettsiales, which are bacteria-like obligate intracellular parasites of most species, were first reported in bivalves, including hard clams, from wild stocks in Chesapeake Bay (Harshbarger et al., 1977). The proliferation of these organisms within cells results in the formation of *inclusion bodies* easily visible under the light microscope (Fig. 12.2A). Organisms that reacted to an anti-Chlamydia antibody were found in hard clams from Great South Bay, New York (Meyers, 1979). The microbes were identified in epithelial cells of the digestive diverticulum in about 8% of wild clams and 25–50% of hatchery-reared individuals planted on commercial beds. This difference suggested that the parasites might have been acquired in the hatchery, although transmission in aquaria was not demonstrated. There were no seasonal or sex-related differences in prevalence, but infection intensity was statistically ($p < 0.001$) higher in females than in males. In some individuals, the inclusions were large enough to fill an entire digestive tubule lumen (Fig. 12.2B). Infection resulted in lysis of subcellular organelles, followed by rupture of the host cell itself, and release of the microbes into the digestive tract lumen of the clam. Despite the obvious tissue damage, no host response was evident. Similar organisms have been found in mantle and siphon epithelia of up to half of the juvenile clams (8–12 mm) sampled from New Jersey, New York, and Massachusetts hatcheries in 1995 (R.D. Barber, Haskin Shellfish Research Laboratory, personal communication, 1995; R. Smolowitz, Marine Biological Laboratory, Wood Hole, personal communication, 1996). Other than the infected cells themselves, which were swollen and ruptured by the inclusion body, clam tissues retained their normal architecture (Fig. 12.2C). Organisms resembling rickettsia were also identified in gill epithelial cells and hemocytes of hard clams from Rehoboth and Indian River Bays, Delaware (Fries and Grant, 1991; Fries and Grant, 1992). Holding clams in the laboratory for up to 3 months caused infections to intensify. Limited water turnover was also implicated in the finding of high Rickettsiales infections in hatchery-held giant clams, *Hippopus hippopus* (Norton et al., 1993). The association of infection prevalence with holding conditions was further illustrated in a recent study at the Haskin Shellfish Research Laboratory. Juvenile clams (6–8 mm) were examined histologically after being held in the hatchery and nursery for approximately 6 months from spawning. Twenty-three percent of the clams were diagnosed

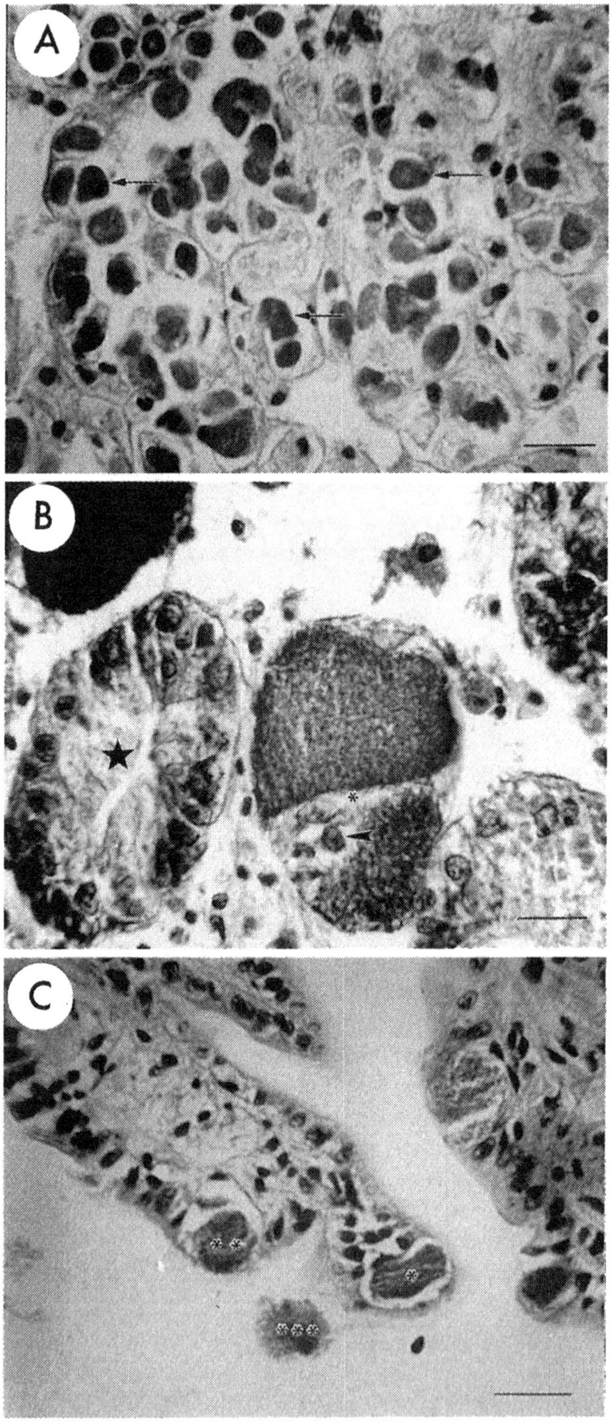
A
B
C

with rickettsiales/chlamydiales-like bodies; 92% of the inclusions were in the siphon or mantle epithelia and the remainder were in the digestive diverticular epithelium. The clams were transferred to the field in late October and sampled again in early April at which time only 4% of the clams had detectable infections. Of the remaining inclusion bodies, >95% were in the digestive diverticular epithelium. At no time were bodies found in the gill.

Rickettsiales and chlamydiales-like bodies appear to be common in epithelia of marine bivalves. Their heightened presence in cultured molluscs may reflect density-related ease of transmission or stress on hosts, or both. Apparently they are readily shed from external epithelia (see Fig. 12.2C). Infected cells are typically distorted and probably eventually rupture, and parasite loads can be locally heavy. The pathological evidence suggests potential abnormal function of heavily infected organs, but rickettsiales and chlamydiales-like organisms are generally considered benign in bivalves. In several instances, however, gill infections have been associated with mortalities (Gulka et al., 1983; Elston, 1986; Le Gall et al., 1988; Norton et al., 1993; Villalba et al., 1999). Subsequent investigation, including transmission experiments, of organisms in sea scallops, *Placopecten magellanicus* (Gulka et al., 1983), concluded that the protist had probably not been the causative agent (Gulka and Chang, 1984). On the other hand, Pacific razor clams, *Siliqua patula*, are infected with a rickettsia-like organism (NIX) that increases in prevalence just before mortalities (Elston, 1986; Bower et al., 1994). Despite the localized damage they cause and their association with mortalities, the role of rickettsiales- and chlamydiales-like organisms as disease-causing agents in marine bivalves has not been conclusively demonstrated.

12.3.2.2 Neoplasms

Two basic types of neoplasms (tumors) have been described in molluscs: (1) disseminated forms, particularly those thought to be of hemic (blood stem cell) origin (Elston et al., 1992); and (2) gonadal forms (Peters et al., 1994). The latter were first described in hard clams collected in 1969 and 1970 from Narragansett Bay, Rhode Island (Yevich and Barry, 1969; Barry and Yevich, 1972). Of 316 female clams examined, 12 (3.8%) had tumors whereas only 2 of 223 (0.1%) males were affected. The neoplasms were of germ-cell origin and all but one was confined to the gonadal area. In one female, the neoplastic cells had metastasized (spread) into the kidney, heart, and pericardial cavity. A far more prevalent neoplasm arising from the germinal epithelium has been found in hard clams from the southeastern United States. In advanced lesions, neoplastic cells completely fill the gonadal follicles and normal gametogenesis ceases (Fig. 12.3A–C). In a 2-year survey of clams in the Indian River, Florida, 147 of 1263 individuals (11.6%) were found with the condition (Hesselman et al., 1988). All age/size (28–102 mm) classes were affected, and females were twice as likely as males to

Fig. 12.2. Photomicrographs of Chlamydiales/Rickettsiales-like organisms in hard clam tissues. (A) Inclusion bodies (arrows) in epithelial cells of heavily infected digestive diverticula. (B) Epithelial cells distended by inclusion bodies almost completely occluding lumen (*) of digestive tubule. Note nucleus of uninfected epithelial cell (arrow) and adjacent unaffected tubule (★). (C) Inclusion bodies (*) in mantle epithelial cells, showing rupture (**) and discharge (***) of bodies. Scale bars = 20 μm.

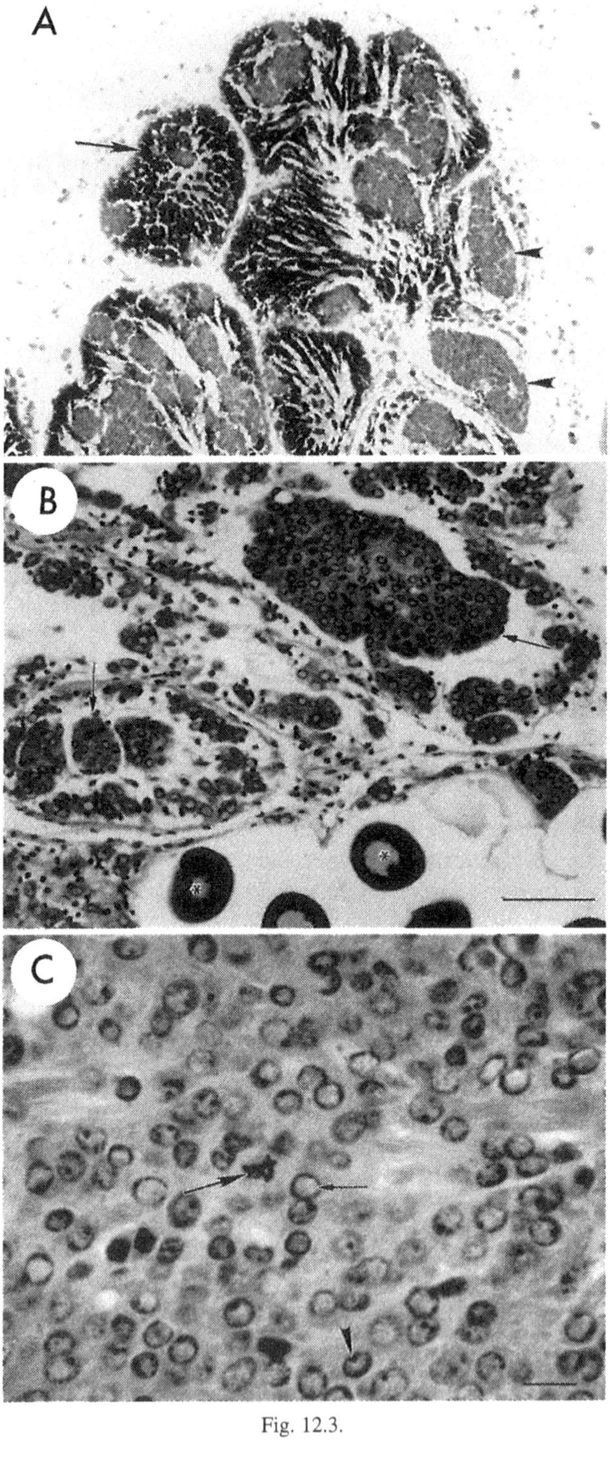

Fig. 12.3.

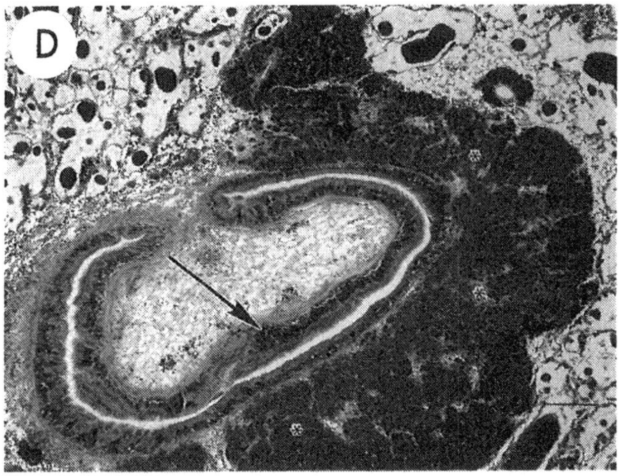

Fig. 12.3 (continued). Photomicrographs of gonadal neoplasia in hard clam tissues. (A) Male gonad showing neoplastic cells (arrowheads) and sperm (arrow). (B) Female gonad showing masses of neoplastic cells (arrows) and eggs (*). (C) Neoplastic cells showing nuclei with emarginated chromatin (small arrow), nucleolus (arrowhead), and mitotic figure (large arrow). (D) Neoplastic cells (*) invading connective tissue surrounding the intestine (arrow). Scale bars = 50 μm in A and B; 10 μm in C; 100 μm in D. Courtesy of D. Hesselman and with permission of the Journal of Invertebrate Pathology.

have the condition. The only case in which neoplastic cells invaded non-gonadal tissues was in a single female. A distinct seasonal pattern suggested that the onset of neoplasia occurred during the spring and summer (May to September) when the highest prevalences (~30%) occurred and when 80–90% of lesions were in the early stage. Lowest prevalences (<5%) occurred in winter. Neoplasms were most frequent in clams that were in the late-spawning or spent stages of reproduction. Despite their high prevalence, the gonadal tumors are not known to depress growth or cause mortality (Hesselman et al., 1988; Eversole and Heffernan, 1995).

Many sites where neoplastic bivalves occur also contain toxic compounds (Elston et al., 1992; Peters et al., 1994) and digestive gland enzymes in hard clams have the potential to transform environmental contaminants into carcinogens (Anderson and Döös, 1983; Anderson and Angel, 1986). In preliminary studies, Van Beneden et al. (1993) and Van Beneden (1994) found that DNA isolated from neoplastic hard clam cells was capable of inducing tumor formation in mammalian cells. They hypothesized that oncogenes in the clams might be activated by the presence of environmental contaminants. Neoplasia, however, also affects bivalves in unpolluted sites and Hesselman et al. (1988) discounted pollutants as a cause for the hard clam gonadal tumors, suggesting that they might be associated with high temperature (30°C) stress.

More recent studies indicate that a genetic mechanism is involved in gonadal neoplasia affecting hard clams in the southeastern United States. Hybrids between *M. mercenaria* and *M. campechiensis* had 2–3 times higher prevalence, and more severe intensities, than did pure-species (Bert et al., 1993; Eversole and Heffernan, 1995). In a South Carolina study, Eversole and Heffernan (1995) found prevalences up to 95–100% in 1-year-old hybrid clams, which also had significantly more severe neoplasia than did their pure-species counterparts

TABLE 12.2

Mean monthly prevalence (%) and severity values (1–5, no neoplasia to heavy neoplasia) for *Mercenaria mercenaria* (MM) and *M. campechiensis* (CC) parental species, hybrids from *M. mercenaria* female × *M. campechiensis* male cross (MC), and the reciprocal cross (CM)

Cross	Prevalence			Severity		
	N [a]	X [b]	SD	N [a]	X [b]	SD
MM	11	32.5 *,**	24.6	78	1.41 *	0.65
CC	12	15.6 *	20.3	113	1.33 *	0.84
MC	13	47.0 **	25.3	120	1.87 **	1.08
CM	13	82.0 ***	15.7	129	2.68 ***	1.12

Clams were spawned in late 1985 and early 1986, planted in the field in October 1986, and collected between September 1987 and October 1988. From Eversole and Heffernan (1995).

[a] Number of monthly samples for prevalence and number of individuals for severity.

[b] Values not sharing the same superscript are significantly different at $\alpha = 0.05$.

(Table 12.2). Disease intensity showed a bi-phasic seasonal pattern in the hybrids, with peaks during the spawning period in May–July and September–October. As in earlier studies, these investigators found that pure-species females had higher prevalence and intensities than did males, but that the highest levels were in undifferentiated clams. In hybrid clams, however, which overall had much higher disease levels, there was no association between sex or sexual differentiation and neoplasia. Interestingly, shell size was positively correlated with disease intensity in hybrids, but not in the pure-species. Gametogenic inhibition by neoplasia may have allowed energy normally used for reproduction to be used for somatic growth (Eversole and Heffernan, 1995) and the relative scarcity of advanced cases could explain the lack of observed size–disease correlation among pure-species clams.

Although no mortality has been associated with gonadal neoplasia of hard clams, occasional metastases into non-gonadal tissues of a small percentage of clams suggest that the tumors can be lethal to these individuals (Fig. 12.3D). Further, advanced neoplasia disrupts gametogenesis (Hesselman et al., 1988; Eversole and Heffernan, 1995). The preponderance of neoplasia in hybrids appears to represent an example of reduced hybrid fitness and a mechanism that may help to maintain species purity (Bert et al., 1993). Whether the association between gonadal tumors and hybridization is unique to the *M. mercenaria–M. campechiensis* combination or may occur in other molluscs remains to be seen. Practices that deliberately or accidentally result in hybridization, however, may decrease the fitness of natural or cultured stocks.

12.4 NONSPECIFIC DISEASE SYMPTOMS

Molluscs often display anomalies not associated with just a single disease, but with generalized distress. A biochemical shift, considered by many investigators as an index of stress in bivalves, was first described in hard clams by Jeffries (1972). He compared various measures of condition in hard clams from a site polluted by industrial and domestic wastes with those in clams from an uncontaminated location. The gill and mantle tissues of clams at the contaminated site had reduced carbohydrate and total free amino acid concentrations.

The molar ratio of taurine to glycine increased, however. The ratio in clams from the uncontaminated site was 2.5 to 2.9, that in clams from the contaminated site was 5.0 to 6.6. Further, Jeffries (1972) re-analyzed hemolymph free amino acid data from eastern oysters parasitized by the protozoan *H. nelsoni* and the trematode *Bucephalus* sp. (Feng et al., 1970). The taurine-to-glycine ratio increased with infection, leading Jeffries to propose this ratio as a general indicator of stress: a ratio of <3 would indicate a "normal" population; a ratio of 3 to 5 would indicate chronic stress; and a ratio >5 would signify acute stress. Subsequent studies with other molluscs have described a similar relationship (Soniat and Keonig, 1982; Bayne et al., 1985), although it is not a universal finding (Paynter et al., 1995).

Other nonspecific symptoms described for hard clams include (1) mantle retraction, which results in a calcified ridge-like deposit inside of, but parallel to the pallial line, and indicates response to mantle irritation (Jeffries, 1972; Greene and Becker, 1977), (2) dark-colored meats that makes clams unmarketable (Jeffries, 1972; Kraeuter et al., 1997), and (3) abundant brown cells, which contain cellular degradation products (Jeffries, 1972; Zaroogian et al., 1992; Kraeuter et al., 1997). These symptoms are present in other bivalve species where they are associated with both infectious and noninfectious diseases, and adverse environmental conditions (Mackin, 1951; Farley, 1968; Bricelj et al., 1992; Zaroogian and Yevich, 1994).

12.5 DEFENSE MECHANISMS

In order to establish themselves in a host and then to propagate, infectious agents must penetrate certain barriers and then survive various internal defenses. The shell and epithelium are the principal, and highly effective, external barriers of molluscs. Breaches in one or both allow easy access of potential pathogens into host tissues. Mucus on the external epithelium may further discourage entry of parasites. In addition, it is possible that the sorting mechanism on the gill and palps may allow the host to reject parasites before they enter the digestive tract (Chintala et al., 1995).

The internal defense system of molluscs, used to protect themselves against foreign organisms that have breached the epithelium, has been reviewed in a number of recent publications (Fisher, 1988; Adema et al., 1991; Ford and Tripp, 1996). With relatively few exceptions (Loker, 1994), investigations have involved challenging this system, either in vivo or in vitro, with biotic or abiotic foreign particles that are not pathogenic to the particular host involved. Recent studies using oyster pathogens are adding new dimensions to the understanding of molluscan internal defenses (Mourton et al., 1992; Anderson, 1996; Ford and Tripp, 1996), but similar studies have not been conducted with hard clams, largely because so few pathogens have been described from this species. Nevertheless, hard clams have been the subject of a number of important studies of molluscan defense mechanisms, the results of which can be compared to those from other bivalve species.

12.5.1 Components of the Internal Defense System

12.5.1.1 Hemocytes

The internal defense system is usually described as residing in the hemolymph and consisting of hemocytes and noncellular, soluble components. Like other molluscs, hard clams have two principal types of hemocytes, granular and agranular, which vary in proportion and within-class morphology (Cheng, 1981). Granular hemocytesgranular hemocyte (granulocytes) are typically the most numerous cell type in clams and are also the most phagocytic (Foley and Cheng, 1974; Cheng and Foley, 1975; Moore and Eble, 1977; Tripp, 1992a). The granules, which are contained in the endoplasm, consist of several types of organelles: mitochondria, lysosomes, and lipid storage centers (Moore and Eble, 1977). A variety of enzymes are found in clam hemocytes, including β glucuronidase, acid and alkaline phosphatase, amylase, lipase, lysozyme and aminotransferases (Cheng and Rodrick, 1975). At least some are contained in the membrane-bound lysosomes of the granulocytes (Yoshino and Cheng, 1976) and are used for the intracellular killing and digestion of ingested particles. Agranular cells, often called hyalinocytes, are small cells with scant cytoplasm, few or no granules, a relatively large nucleus, and a poorly understood function.

Granulocytes and hyalinocytes are readily identified by all investigators. Differences in description, as well as terminology, are associated with a type of cell that is often agranular or with few granules, but otherwise has the general appearance and nuclear size of granulocytes. For instance, Cheng and Foley (1975) described a cell, which they termed "fibrocyte", in addition to granulocytes and hyalinocytes, in hard clams. "Fibrocytes" have no granules, but contain large aggregates of rosette-pattern glycogen granules and bodies consisting of concentric arrays of lamellae resembling residual bodies left after the degradation of cellular contents. Based on similarities to the ultrastructure of oyster granulocytes known to have recently digested bacteria (Cheng and Cali, 1974), Cheng and Foley (1975) suggested that the "fibrocytes" were degranulated granulocytes. Moore and Eble (1977) described both small and large granulocytes, in addition to agranulocytes (= hyalinocytes). Their small granulocytes had abundant granules and were highly motile in vitro, and were essentially the "typical" granulocytes. The large granulocytes had fewer granules and were less motile. Based on similarities in nuclear and granule morphology, Moore and Eble (1977) suggested that the small and large granulocytes were part of the same lineage and that the larger cells may have been senescent. Following the arguments of Cheng and Foley, it is likely that the large granulocytes of Moore and Eble were in an early stage of degranulation and hence similar to the "fibrocyte".

Substantial variability in the proportions of hemocyte subpopulations identified by various researchers is apparent. For instance, Tripp (1992a) reported that granulocytes from clams collected from Rehoboth Bay, Delaware, comprised 75% of the hemocyte population. Foley and Cheng (1974) found a similar percentage, 60 to 68%, in clams from Buzzard's Bay, Massachusetts and Great Bay, New York, and stated that the hemocytes were "quantitatively and qualitatively" the same in clams from both sites. In contrast, Moore and Eble (1977), using clams from Buzzard's Bay and Great Bay, New Jersey, reported that granular hemocytes, which they divided into large and small subsets, were 98% of all hemocytes. Friedl and Alvarez (1990) described a unimodal distribution when they examined hemocytes of both *M. mercenaria* and *M. campechiensis* from Florida for size, DNA content, and protein content. Similar variability has been described in other marine bivalves and attributed to individual, seasonal, and geographic factors, including disease, that affect the physiological state of the animal at the time of sampling (Huffman and Tripp, 1982; Ford et al., 1994; Oliver and Fisher, 1995). Consequently, great care must be exercised in assessing the results of any single examination. In fact, Hawkins et al. (1993) reported very short-term changes in numbers and behavior of hard clam hemocytes. They demonstrated a correlation of experimentally induced tidal rhythms with both measures: cell numbers were elevated and locomotory activity was reduced in tidally exposed clams compared to submerged animals.

An interesting addition to the knowledge of molluscan blood cells was made by Anderson (1987), who described the formation of multinucleated giant cells (MGC) during in vitro observations of hard clam hemocytes. These bodies were formed by the fusion of macrophage-like cells (= granular hemocytes), which over the course of 48 h in vitro, produced large cells containing 50 to >100 nuclei. The lectin Concanavalin A increased the rate of MGC formation, but phagocytic stimulation, by exposing hemocytes to various particles, did not. The phenomenon varied considerably from clam to clam, but with no relationship to season, temperature, or animal size, and MGCs were always a very small proportion (<0.01%) of the total number of cells. Because of this, Anderson concluded that MGC formation was not solely the result of in vitro treatment, but was a trait inherent to some individuals and not others. In vivo, multinucleated giant cells have been reported in tissue sections of oysters during postmortem examinations (Sparks and Pauley, 1964), associated with allographs and xenografts in freshwater snails (Cheng and Galloway, 1970), and phagocytosing QPX cells (Ragone Calvo et al., 1998; Smolowitz et al., 1998). They are very similar in appearance to forms seen in vertebrates, which are associated with neoplasms, infections, or implanted material. Anderson (1987) suggested that the ability of clam hemocytes to form vertebrate-like MGCs substantiates the theory of an early phylogenetic origin for macrophage-like cells.

12.5.1.2 Noncellular elements

Studies of soluble hemolymph components in hard clams are fewer than those involving hemocytes; however, cell-free enzymes are the same as those found in hemocytes and include β glucuronidase, acid and alkaline phosphatase, amylase, lipase, lysozyme and aminotransferases (Cheng and Rodrick, 1975). Lysozyme and β glucuronidase are biochemically similar to enzymes in oysters and soft-shell clams (Cheng, 1976a; Cheng, 1983).

In addition to lysosomal enzymes, two other categories of soluble elements thought to play a role in defense have been identified in the hemolymph of hard clams and many other marine bivalves: hemolysins and agglutinins (see Chu, 1988). Hemolysins, substances that lyse vertebrate erythrocytes, are present in the hemocytes, the cell-free hemolymph, and the shell fluid of hard clams (Anderson, 1981). Hemolytic activity required the presence of divalent cations, was equally high at 1°C and 22°C and very low at 37°C, and could be enhanced by pre-injection of erythrocytes. Obviously, vertebrate erythrocytes are not among the foreign elements typically encountered by marine molluscs. Consequently, the significance of hemolysins to molluscs is unclear, but they may represent a lytic activity that could be activated against other non-self cells. Experimentally, hemolysins are useful in studying the specificity of soluble factors because their activity is often restricted to the species providing the hemolysin and to the erythrocyte donor (Anderson, 1981).

Substances that bind foreign particles are present in hemolymph of all molluscs and are known as agglutinins. A subset of agglutinins, called lectins, are multivalent carbohydrate-binding proteins thought to function in molecular recognition by invertebrates, which do not possess antibodies. Arimoto and Tripp (1977) identified a substance in the hemolymph of hard clams that agglutinated 4 of 30 species of bacteria tested (many of which were isolated from clam hemolymph or containers in which clams had been held) and one of two species of microalgae. To determine if the agglutinin enhanced recognition by hemocytes, three bacterial species, one that was agglutinated and two that were not, were further tested to see if incubation in clam hemolymph caused them to be more readily phagocytosed. Incubation of one of the agglutination-positive species enhanced phagocytosis by clam hemocytes; incubation of the other two did not. The agglutinin was at least partly proteinaceous, with a subunit size of 21,000 Da. Calcium was required for activity and several sugars, which typically compete for binding sites with lectins, reduced agglutination. Based on these results, Arimoto and Tripp (1977) concluded that the agglutinin, which had the characteristics of a lectin, might have evolved along a dual pathway involving both nutrition and defense.

Tripp (1992b) conducted further studies on hard clam agglutinins using a greater variety of target particles, including red blood cells and yeast, as well as bacteria. He documented a wide range of agglutinating activity, depending on target cell type and treatment (live, fixed, or heated), which suggested some degree of specificity. Pre-incubation of target erythrocytes with each of ten different sugars failed to substantially inhibit agglutination. Further, phagocytosis of erythrocytes and the two species of bacteria tested was not enhanced after incubation with agglutinin-containing clam serum. From these results, Tripp concluded that "the serum lectins of *M. mercenaria* do not serve as recognition molecules that promote phagocytosis of foreign particles". He noted that hard clams are at one extreme of a range of lectin–phagocyte–target interactions in molluscs, and that most reports indicate at least some enhancement of phagocytosis by serum in other species (Olafsen, 1988).

Recently, prophenoloxidase activity has been found in cell-free hard clam hemolymph, as well as in lysed hemocytes (Deaton and Dankert, 1999). The prophenoloxidase activation cascade produces melanin and is believed to be part of the internal defense system of arthropods (Smith and Söderhäll, 1991). Its role in molluscs is unclear, although it has been found in the extrapallial fluid of edible mussels, *Mytilus edulis*, and Manila clams, *Tapes philippinarum*. It may play a role in producing melanin to sequester and inactivate external pathogens such as bacteria, as well as for normal shell formation (Smith and Söderhäll, 1991; Paillard et al., 1994).

As with the cellular components of the internal defense system, there is much variability in the soluble constituents. For instance, two analyses of hemolysin activity in hard clams gave different results for erythrocyte specificity, temperature optima for activity, and absorption/inactivation of activity by homologous erythrocytes (Graham, 1968; Anderson, 1981). Tripp (1992b) also documented considerable variability among individual clams as well as between studies. He found no evidence of hemolytic activity by hard clam serum against any of the four types of red blood cells tested, which contrasted markedly with the findings of Anderson (1981). Both Anderson (1981) and Tripp (1992b) acknowledged that differences in clam physiological status, erythrocyte source, or laboratory procedures could result in these discrepancies.

12.5.2 Activities of the Internal Defense System

Hard clams have been the subject of a number of studies into the functional aspects of the molluscan internal defense system. Phagocytosis is often described as the principal defense mechanism and there is ample evidence that molluscan hemocytes avidly ingest foreign materials, including food particles (Cheng, 1981; Feng, 1988). Foley and Cheng (1975) analyzed phagocytosis of heat-killed bacteria by hard clam and eastern oyster hemocytes in vitro. In both species, bacteria were observed to be associated (i.e., contact or ingestion) most frequently with granular hemocytes, which were judged the most phagocytic. In vitro ingestion rates were temperature dependent. Clam hemocytes were equally phagocytic at 22° and 37°C, but there was almost no adherence or phagocytosis at 4°C. Similarly, Tripp (1992a) described a strong temperature dependency for phagocytosis, with rates severely depressed, but not completely eliminated, at 3°C. Alvarez et al. (1989) reported a comparable result for eastern oysters, with ingestion of polystyrene beads high and constant between 30° and 8°C and falling sharply by 4°C.

Recognition that a particle is non-self must precede phagocytosis. Lectins, which are present on hemocyte surfaces and in cell-free hemolymph, are thought to play a role in this process (Parish, 1977; Olafsen, 1988). Incubation of target particles in serum often increases phagocytic rates presumably by adding these "recognition factors", termed opsonins, to cell surfaces (reviewed by Chorney and Cheng, 1980). Specific differences in surface molecules may permit further discrimination among various non-self particles. Tripp (1992a) studied recognition aspects of phagocytosis by hard clam hemocytes in vitro using yeast and erythrocytes from several vertebrate species as target particles. The addition of cell-free hemolymph did enhanced phagocytosis of yeast, but only when the temperature and the yeast-to-hemocyte ratio were low. Phagocytosis of erythrocytes was unaffected by addition of hemolymph. Eight different sugars (competitive inhibitors of lectins) failed to impede phagocytosis, suggesting that lectins were not involved in the process. Tripp concluded that phagocytosis by hard clams appeared to be relatively nonspecific. The absence of sugar inhibition notwithstanding, this conclusion is surprising considering that the proportion of erythrocytes phagocytosed varied from 10 to 100% according to donor species. It is likely that surface molecules would vary at least somewhat according to species and that phagocytic rates specific to target-cell species would reflect an ability to discriminate among specific surface receptors. A similar conclusion can be drawn from the differences in phagocytic response between yeast and erythrocytes.

Infiltration and accumulation of blood cells are typical responses to infection, tissue damage, or both, in invertebrates as well as vertebrates. Analysis of elements that stimulate directed movement (chemotaxis) of hemocytes helps to elucidate the factors that draw them to these sites. In vitro measurements of chemotaxis by Fawcett and Tripp (1994) demonstrated that hemocytes of the hard clam migrate towards live bacteria (*Escherichia coli* and *Staphylococcus aureus*). They were also attracted to media in which the bacteria had grown, but not to dead bacteria. The presence or absence of cell-free hemolymph did not affect chemotaxis and only granular cells migrated. The process was receptor mediated, as in vertebrates, and most probably involved small proteins or peptides. Thus, molecules secreted by foreign organisms are one element that attracts clam blood cells to sites of infection. A similar phenomenon occurs in oysters (Cheng and Howland, 1979), although secretions from damaged tissues alone can also stimulate chemotaxis of molluscan cells (Des Voigne and Sparks, 1968; Stefano et al., 1989).

Killing of foreign organisms is a primary function of defense cells. Among the mechanisms that contribute to killing and subsequent digestion are intracellular lysosomal enzymes (Cheng, 1983). Experiments in which bacteria were incubated in vitro in whole hemolymph from hard clams demonstrated an increase, in the serum, of the lysosomal enzyme, lysozyme, concomitant with a decrease in the same enzyme in hemocytes. From these results, Cheng et al. (1975) deduced that the serum enzymes were released from phagocytosing cells. Similar results were subsequently reported for other marine bivalves (eastern oysters and soft-shelled clams) and freshwater gastropods (Cheng, 1983). Also using hard clams as their experimental animal, Foley and Cheng (1977) and Mohandas et al. (1985) later showed quantitatively that lysosomes are released from granulocytes by degranulation. Subsequent experiments demonstrated that injections of sterile seawater could also elicit the same response, indicating that the mechanism was not solely a reaction to biotic stimuli (Cheng et al., 1977). Also, it is not clear that hemocytes are the only source of soluble enzymes because numerous other cell types contain hydrolytic enzymes (Yoshino and Cheng, 1977). Cheng (1983) hypothesized that lysosomal enzymes released into the serum could aid in the destruction of invading organisms, alter surface molecules that might make them more recognizable as foreign, or, by causing autolysis of host cells, enhance an inflammatory-type protective response at sites of injury. Hydrolytic enzyme are also important, of course, in digestion of food particles (Cheng, 1977).

Using hard clam hemocytes in an in vitro system, Cheng (1976b) demonstrated that the cells did not utilize oxygen during phagocytosis of bacteria. Rather, the depletion of glucose and glycogen, the production of lactate, and the failure of potassium cyanide to inhibit the system, all indicated that glycolysis was the energy-producing pathway. During this study, Cheng found no evidence of a reactive oxygen species (ROS) antimicrobial process that was subsequently demonstrated in many other bivalve species, including eastern oysters (Adema et al., 1991; Anderson, 1994). The absence of such a system in hard clam hemocytes is supported by studies of Hawkins et al. (1993) and Anderson et al. (1994). Neither investigation could detect generation of hydrogen peroxide when blood cells were challenged with foreign particles. Hawkins et al. (1993) did find endogenous hydrogen peroxide that they concluded was a by-product of general metabolism. Anderson (1994) also found oxygen radical production lacking in blood cells of soft-shell clams, *Mya arenaria*, but present in ribbed mussels, *Geukensia demissa*, indicating clear intraspecies variation in the presence of this intracellular killing mechanism.

The exact role of ROS in marine bivalves is unclear. The initially measured response (to nonliving stimuli) prompted the argument that these molecules killed microorganisms, as they do in vertebrates. A recent series of experiments comparing the ROS response of eastern oyster and striped bass (*Morone saxatilis*) blood cells to bacteria calls this into question (Bramble and Anderson, 1997; Bramble and Anderson, 1998). The striped bass cells produced a significant ROS when challenged by live bacteria, whereas the oyster cells produced no measurable response, possibly because the bacterial enzyme catalase suppressed it. Nevertheless, the oyster cells readily killed the bacteria, even when exposed to an inhibitor of the ROS response (Bramble and Anderson, 1999). The bass cells also killed the bacteria, but not after exposure to the same inhibitor.

Cheng et al. (1966a,b) investigated, both in vivo and in vitro, the responses of eight marine bivalves, including hard clams, to cercaria of the trematode *Himasthla quissetensis*. Sea gulls are the primary host for this parasite, but several species of marine gastropods and pelecypods can be second intermediate hosts. For the in vivo assay, cercaria were introduced into the shell cavity of live bivalves (Cheng et al., 1966a). After 32–34 h, the animals were shucked, fixed, and examined by tissue-section histology to determine the numbers of cercaria that had penetrated the tissues. None was found in the tissues of Pacific or eastern oysters (although a few were located in the circulatory system of the latter). Cercaria were found encysted in hard clam tissues, but in the lowest numbers of all six bivalves in which the parasites were found. In vitro experiments were designed to explain differences in compatibility between parasite and the various host species (Cheng et al., 1966b). Cercaria were incubated in hemolymph and tissue extracts from the different molluscs; survival time and time required for the trematodes to encyst were recorded. In vitro, cercaria exposed to oyster plasma and tissue extracts had the shortest encystment time of all hosts tested. Relating this to the finding that cercaria rarely penetrated soft tissues of oysters, the authors argued that substances leaking into the shell cavity (of oysters) stimulated encystment of the trematodes before they could penetrate the soft tissues. This argument, however, could not be made for hard clams. Even though very few trematodes penetrated clam tissues, cercaria incubated in clam fluids had the longest in vitro survival time and the longest time to encystment of all species. Thus, soluble factors did not appear to be a common mechanism responsible for compatibility differences between parasite and hosts in this study.

The existence of antibacterial factors in molluscan hemolymph has been demonstrated in hard clams. Hartland and Timoney (1979) showed that bacteria with human health significance (*Salmonella typhimurium*, *Shigella flexineri*, and *Escherichia coli*) injected intracardially into hard clams and eastern oysters were cleared within 24 h at 20°C and only slightly slower at 6°C. Overall, uptake of bacteria from the incubation water appeared to be somewhat lower, and subsequent clearance somewhat faster, in hard clams, but the differences were not large and statistical comparisons were not made. Anderson et al. (1981) documented that a *Flavobacterium* sp. isolated from oyster holding tanks was also rapidly eliminated by hard clams. More than 90% of the injected dose present at 30 min was cleared from hemolymph after 4 h. Pre-exposure of clams to sublethal concentrations of environmental contaminants (benzo[*a*]pyrene, hexachlorobenzene, and pentachlorophenol), significantly decreased clearance rates and, under these conditions, some individual clams failed to rid themselves of bacteria at all.

As might be expected from the variability associated with the components of the internal defense system, its activity also fluctuates. Hawkins et al. (1993) linked their findings of variation in putative defense-related activities (hemocyte behavior and counts, and lysozyme activity) with metabolic activity (indicated by hydrogen peroxide concentration) during tidal exposure and re-immersion. They suggested that defense activities are elevated when clams are submerged, feeding, and mostly likely to be exposed to pathogen, and that they diminish at low tide when the animals are closed and less likely to encounter pathogens.

12.6 WHY DO HARD CLAMS HAVE SO FEW RECOGNIZED DISEASES?

In an early review, Sindermann and Rosenfield (1967) commented on the scarcity of mass mortalities in clams, compared to mussels and oysters. A more recent and very comprehensive synopsis by Bower et al. (1994) indicates that clams do have fewer diseases than oysters, although scallops and mussels also have relatively few (Table 12.3). Even when clam parasites are associated with mortality, however, it is not always clear whether they are the only causative agent or simply microbes that proliferate in hosts already compromised by some other factor.

TABLE 12.3

Parasites, pathogens, and diseases of commercial molluscs (adapted from Bower et al., 1994)

Pathogenicity	Oyster	Clam	Scallop	Mussel	Abalone
Causing/associated with mortalities in wild and cultured molluscs	12 (33%)	4 (19%)	3 (23%)	1 (9%)	1 (25%)
Causing/associated with mortalities only in cultured molluscs	6 (16%)	4 (19%)	3 (23%)	1 (9%)	2 (50%)
Causing localized pathology or prevalence very low — mortality not observed	15 (42%)	11 (52%)	7 (54%)	8 (73%)	1 (25%)
Parasite reported, but effect on host not described	3 (8%)	2 (10%)	0	1 (9%)	
TOTAL:	36	21	13	11	4

Each molluscan category includes all species for which data are available.

As a group, clams have the same general classes of parasites as do oysters, scallops, and mussels (Bower et al., 1994). All have copepods, trematodes, ciliates and chlamydiales/ rickettsiales, which are not typically detrimental. In hatcheries all bivalve larvae seem equally susceptible to bacteria and viruses that become pathogenic in high-density cultures. In fact, species of most commercial bivalves studied to date are parasitized by protozoans capable of causing epizootic mortalities in some hosts: *Perkinsus* spp., *Haplosporidium* spp., *Minchinia* spp., *Marteilia* spp., *Bonamia* spp. and *Mikrocytos* spp. With few exceptions (e.g., *P. atlanticus* in Manila clams); however, epizootic mortalities known to be directly *caused* by these pathogens occur only in oysters.

Even among commercial clam species, the hard clam, *M. mercenaria*, has notably few diseases. It harbors only one reported parasite associated with mortality, the QPX organism (Whyte et al., 1994), but assessment of its exact role in mortality is complicated by the observation that certain environmental or culture situations appear to promote infection and disease. Thus, compared to the eastern oyster, *C. virginica*, which inhabits roughly the same geographic region, the hard clam seems remarkably resistant to pathogen-caused mortality. The obvious question is why? Three hypotheses are examined below.

12.6.1 Mortalities are not Seen or Documented because Clams are Infaunal

hieroIn noting the scarcity of reported clam mortalities, Sindermann and Rosenfield (1967) suggested that epizootics might be occurring but not documented because the clams are infaunal and more difficult to observe than are epifaunal bivalves like oysters. Although this proposition seems logical, there are a number of arguments against it. Numerous experimental studies conducted after the Sindermann and Rosenfield review have investigated survival of captive hard clam populations and have provided reasonable estimates showing that annual non-predation mortality rates are typically $<10\%$ (Table 12.1).

Further, mortalities of clams can be observed in both wild and cultured populations because diseased or otherwise stressed individuals typically rise to the sediment surface. For instance, widespread anoxia off the mid-Atlantic coast in 1976 killed large numbers of surf clams,

Spisula solidissima (Ropes et al., 1979). The mortalities were almost immediately discovered by sports divers and were subsequently quantified by researchers. In intertidal locations, dead and dying clams accumulate at the sediment surface (Bower, 1992b; Paillard et al., 1994). Thus, while mortalities of infaunal populations may be somewhat obscured by their location, it seems unlikely that any major mortality episode in a fished or cultured clam would go unnoticed or unreported.

12.6.2 Clams are Less Suitable Hosts or have Better Protective Mechanisms than Oysters

Hard clams are typically found in waters with salinity of 20 to 32 ppt, whereas eastern oysters grow in waters from about 5 to 32 ppt. Both tolerate a wide range of temperatures, although eastern oysters are found farther south than hard clams. This suggests that eastern oysters' habitat would overlap that of more potential pathogens than would the habitat of hard clams. However, the major disease organisms that affect eastern oysters are most abundant and destructive at high salinities and in intermediate temperature regimes (Ford and Tripp, 1996).

There is no inherent reason to conclude that hard clams have better defense mechanisms or are less suitable hosts than oysters, but there are very few data with which to test this assumption. Their infaunal habitat does protect clams from certain pests like the mud worm *Polydora* spp., which infest the shells of epifaunal species; however, clams' exposure to water-borne microbes should be similar to other bivalves, because size-specific filtration rates, which help to determine encounter rates between host and parasite, are equivalent for all bivalves (Powell et al., 1992). The fact that clam species harbor the range of parasite types found in other bivalves supports this contention.

Studies of potential defense mechanisms have sometimes compared hard clams with other bivalves, specifically eastern oysters (Cheng et al., 1966a,b; Cheng and Rodrick, 1975; Foley and Cheng, 1975; Hartland and Timoney, 1979). The values measured show no consistent differences between clams and oysters (Table 12.4). In fact, it would be risky to draw conclusions about the relative effectiveness of clam and oyster internal defenses from these studies because of high variance about means and difficulties in distinguishing species-related variation from variation in protocol or due to previous history of experimental organisms (e.g.: Anderson, 1981; Tripp, 1992b). Further, the exact role of these potential defense activities remains unclear, especially in connection with protozoan diseases (Ford et al., 1993; La Peyre et al., 1995; Volety and Chu, 1995).

12.6.3 Clams have not been Transported to the Extent that Oysters have, thus Limiting Potential Spread of Pathogens

All species of mollusc harbor some parasites. The extent of parasitism ranges from undetectable to epizootic. A parasite that causes epizootic disease may do so over only part of the host's range if the host tolerates greater environmental extremes than the parasite or if some geographic barrier prevents the parasite's spread. A change in environment or a circumvention of the barrier may allow the parasite to spread. A parasite that is rare, and without apparent effect, in one particular host (carrier species) may be pathogenic in another (naive) species that has never encountered the parasite. If the carrier population and the

TABLE 12.4

Comparison of potential defense activities and molecules in hard clams, *Mercenaria mercenaria*, and eastern oysters, *Crassostrea virginica*

Parameter	*M. mercenaria*		*C. virginica*		Reference
Bacterial clearance time:					
6°C (intracardial)	23–28 h		72 h		Hartland and Timoney (1979)
20°C (intracardial)	6–18 h		13–19 h		Hartland and Timoney (1979)
20°C (feeding)	6–14 h		12–21 h		Hartland and Timoney (1979)
In vitro phagocytosis (22°C):					
Hemocytes associated with bacteria	50–90% [a]		15–60% [b]		Foley and Cheng (1975)
Bacteria $granulocyte^{-1}$	10.9 (1.5)[a]		6.4 (5.6)[c]–22.5 (19.2)[d]		Foley and Cheng (1975)
Response to *Himasthla quissetensis*:					
In vivo (number in tissue)	2.1 (1.9)		0.4 (0.9)		Cheng et al. (1966a)
In vitro (L_{50} [min] in tissue extract)	55 (6.2)		39 (4.1)		Cheng et al. (1966b)
In vitro (min to encystment in plasma)	53 (4.2)		21 (3.4)		Cheng et al. (1966b)
	Supernatant	Pellet	Supernatant	Pellet	
Enzymes:					
β-glucuronidase (Sigma U mg $protein^{-1}$)	305 (68)	224 (40)	12 (6.6)	309 (183)	Cheng and Rodrick (1975)
Acid phosphatase (mU mg $protein^{-1}$)	1.5 (1.5)	10.3 (5.4)	0.7 (0.4)	46.8 (5.2)	Cheng and Rodrick (1975)
Alkaline phosphatase (mU mg $protein^{-1}$)	4.3 (1.3)	14.1 (12.0)	1.5 (1.3)	16.3 (8.6)	Cheng and Rodrick (1975)
Amylase (Somogyi U mg $protein^{-1}$)	0	0	0 (0.4)	0	Cheng and Rodrick (1975)
Lipase (Sigma-Tietz U mg $protein^{-1}$)	0.08 (0.05)	0.02 (0.01)	0.02 (0.02)	0.06 (0.04)	Cheng and Rodrick (1975)
Lysozyme (ΔOD_{540} min^{-1} mg $protein^{-1}$)	0	0.01 (0.01)	2.17 (0.35)	0.14 (0.02)	Cheng and Rodrick (1975)

Values are given as ranges or means (±S.D.).
[a] *Bacillus megaterium.*
[b] *Escherichia coli* or *Staphylococcus aureus.*
[c] *Escherichia coli.*
[d] *Staphylococcus aureus.*

naive population (of the same or different species) come in contact through intentional or accidental introduction of one or the other into a new area, the parasite can be transmitted to the naive population, where it may cause disease and mortality (Sindermann, 1990b). Thus, introductions and transfers of molluscs risk the spread of disease (Rosenfield and Kern, 1979; Farley, 1992; Ford, 1992).

Two cases illustrate the spread of a pathogen within a single host species, but into new geographic regions. Ford (1996) hypothesized that the recent northward extension of *Perkinsus marinus*, cause of Dermo disease in the eastern oyster, was linked to historical transplantation of infected southern oysters combined with a warming period that provided a favorable environment for the parasite. Another case in which long-distance transportation of hosts is suspected of spreading a disease organism involves the European (flat) oyster, *Ostrea edulis*. A possible route of introduction for the pathogenic protozoan, *Bonamia ostreae*, was traced through shipments of flat oysters from Connecticut to California in the 1960s, and later from California to the state of Washington and then to France (Elston et al., 1986; Farley et al., 1988). The pathogen has been especially devastating to the flat oyster industry in western Europe, causing epizootic mortalities beginning in the 1970s (Grizel, 1983).

The Pacific oyster, *Crassostrea gigas*, is often cited as a suspected source of between-species disease introduction, although the evidence is largely circumstantial. Pacific oysters have been introduced to a number of locations around the world because of their fast growth and good survival (Andrews, 1980; Mann et al., 1991). During the period when imports of this species were being made into France (1966 to 1975) three epizootic diseases appeared in native oysters (*C. angulata* and *O. edulis*) that have severely depleted them as commercial species (Andrews, 1980; Grizel and Héral, 1991). The Pacific oyster, which is resistant to all three diseases, now dominates the oyster industry in western Europe. Despite the suspicion of many, there is no direct proof that any of the three diseases (one viral and two protozoan [*Marteilia refringens* and *Bonamia ostreae*]) were carried by the Pacific oysters. Only the close timing between the import of the exotic species and the onset of diseases in the indigenous ones supports the importation hypothesis. In fact, as previously noted, there is some evidence that *B. ostreae* was brought to Europe from the United States in shipments of flat oysters (Elston et al., 1986). It is not known, however, whether the parasite was indigenous to the United States (where it has not been reported in native molluscs) or whether it accompanied the flat oyster to the United States in introductions to New England from 1949 through the 1950s (Welch, 1966; Friedman and Perkins, 1994). *Haplosporidium nelsoni*, a protozoan that has caused epizootic mortalities of eastern oysters in the mid-Atlantic since 1957 has also been linked to the Pacific oyster, *Crassostrea gigas*, which harbors a morphologically similar parasite, at very low levels in the population (Kern, 1976). Recent evidence from small subunit ribosomal RNA gene sequences supports the hypothesis that *H. nelsoni* may have been brought to the east coast in infected *C. gigas* (Burreson et al., 2000).

Parasites are often host specific, that is, a certain species of parasite infects only a single host species, or a group of closely related hosts. Thus, it can be argued that the more a particular species is moved from one geographic location to another, the greater its chances of spreading a pathogen to a similar host species in the new location. Alternatively, the introduced species has a greater chance of becoming infected by a pathogen that infects a similar host resident to the new location. That is, if oysters have been transported more frequently than clams, oysters should be transmitting or acquiring disease agents more often.

Historically, it does appear that oysters have been moved more often and in larger quantities than clams. Eastern oyster seed was being transplanted from areas of abundance on the mid-Atlantic coast of the United States to depleted areas in New England and around New York City at least as early as the 1830s (Ingersoll, 1887; Kochiss, 1974). This species was also shipped to the west coast of the United States and to Europe for consumption and planting in 1870s (Ingersoll, 1887; Andrews, 1980). Flat oysters were introduced to New England over a 12-year period beginning in 1949 (Welch, 1966), and from there to the west coast of the United States (Elston et al., 1986). Pacific oysters have been distributed for culture around the world (Chew, 1990). In addition, adult oysters can be transported affixed to the hulls of ships, which clams cannot. So even in the absence of deliberate or accidental introductions associated with shellfish commerce, oysters or their parasites could be unintentionally introduced via ships. It is interesting that several major disease outbreaks in oysters (e.g., those caused by *H. nelsoni*, *B. ostreae*, and *M. refringens*) have occurred since the end of World War II, a period during which international shipping has expanded greatly, with larger vessels and shorter transit times.

Clams, too, have a history of long-distance shipment, although it is not as extensive as that for oysters. *Tapes* (= *Raditapes* = *Venerupis*) *philippinarum* (= *semidecussatus* = *japonica*) is native to the western Pacific, but was accidentally introduced to the west coast of the U.S. in the 1930s and 1940s (Carlton, 1992). From there it was deliberately brought to western Europe between 1972 and 1985 for mariculture (Paillard et al., 1994). There is no evidence that these introductions spread disease. *T. philippinarum* is sensitive to low temperature in some of its new range (Bower, 1992b) and is more susceptible than the native *R. decussatus* to the etiological agent of "brown ring disease", a marine bacterium (*Vibrio tapetis*) that causes disease at relatively low temperatures of 5° to 20°C (Paillard et al., 1994).

Hard clams were introduced to California in the 1870s, either deliberately or with eastern oysters, and planted in San Francisco Bay where they "thrive..., increase and grow rapidly and are figuring largely in local markets" (Ingersoll, 1887). They became established in Europe after introductions in the mid-1800s (Heppell, 1961). Populations on the north and east coast of Puerto Rico have a high genetic similarity to those from Long Island Sound, New York, and are thought to have been introduced from these waters over 100 years ago (Juste, 1992). More recently, it has become routine to transplant hard clams up and down the east coast of the United States for experimental and commercial purposes (Haven and Andrews, 1957; Walker and Humphrey, 1984; Walker and Heffernan, 1990; personal observation). Also, clams from multiple locations are frequently planted temporarily in a "storage" site until they can be marketed (J.N. Kraeuter, personal communication, 1996).

In many cases, large-scale transplantations and introductions of both oysters and clams occurred before the establishment of guidelines by the International Council for the Exploration of the Seas (ICES), which are intended to minimize the spread of disease in aquatic species. The ICES guidelines stipulate that the original introduction be held and spawned in quarantine, and that only its progeny be placed in the new environment (see Sindermann, 1990a). Thus, large-scale introductions of natural stocks of both clams and oysters had the potential of introducing their pathogens, too.

In summary, none of the three hypotheses examined here offers a convincing explanation for the observed differences between disease incidence in clams (or scallops and mussels) and oysters. Perhaps the strongest evidence is for dispersion of pathogens through transfers and introductions, either accidental or for commerce; however, the evidence is mostly circumstantial. At present, the intriguing question of why clams have so few diseases compared to oysters is still unanswered and a plausible hypothesis remains elusive.

12.7 ACKNOWLEDGMENTS

I am indebted to J.N. Kraeuter, C.A. Farley, and F.G. Kern for discussion and suggestions on early drafts of the manuscript. This is contribution No. 00-20 from the Institute of Marine and Coastal Sciences at Rutgers University and New Jersey Agricultural Experiment Station Publication No. D 32504-2-00, supported by state funds.

REFERENCES

Adema, C.M., Van der Knaap, W.P.W. and Sminia, T., 1991. Molluscan hemocyte-mediated cytotoxicity: the role of reactive oxygen intermediates. Rev. Aquat. Sci., 2: 201–223.

Alvarez, M.R., Friedl, F.E., Johnson, J.S. and Hinsch, G.W., 1989. Factors affecting *in vitro* phagocytosis in oyster hemocytes. J. Invertebr. Pathol., 54: 233–241.

Anderson, R.S., 1981. Inducible hemolytic activity in *Mercenaria mercenaria* hemolymph. Dev. Comp. Immunol., 5: 575–586.

Anderson, R.S., 1987. Polykaryon formation by *Mercenaria mercenaria* hemocytes. Biol. Bull., 172: 236–245.

Anderson, R.S., 1994. Hemocyte-derived reactive oxygen intermediate production in four bivalve mollusks. Dev. Comp. Immunol., 18: 89–96.

Anderson, R.S., 1996. Interactions of *Perkinsus marinus* with humoral factors and hemocytes of *Crassostrea virginica*. J. Shellfish Res., 15: 127–134.

Anderson, R.S. and Angel, M.A., 1986. Biotransformation of Benzy(*a*)pyrene by *Mercenaria mercenaria* and *Crassostrea virginica*. In: T.M. Poston and R. Purdy (Eds.), Aquatic Toxicology and Environmental Fate, 9th Volume, ASTM STP 921. American Society for Testing and Materials, Philadelphia, PA, pp. 241–251.

Anderson, R.S. and Döös, J.E., 1983. Activation of mammalian carcinogens to bacterial mutagens by microsomal enzymes from a pelecypod mollusk, *Mercenaria mercenaria*. Mut. Res., 116: 246–247.

Anderson, R.S., Giam, C.S., Ray, L.E. and Tripp, M.R., 1981. Effects of environmental pollutants on immunological competency of the clam *Mercenaria mercenaria*: impaired bacterial clearance. Aquat. Toxicol., 1: 187–195.

Andrews, J.D., 1955. Notes on fungus parasites of bivalve mollusks in Chesapeake Bay. Proc. Natl. Shellfish. Assoc., 45: 157–163.

Andrews, J.D., 1980. A review of introductions of exotic oysters and biological planning for new importations. U.S. Natl. Mar. Fish. Serv. Mar. Fish. Rev., 42: 1–11.

Andrews, J.D. and Hewatt, W.G., 1957. Oyster mortality studies in Virginia II. The fungus disease caused by *Dermocystidium marinum* in oysters of Chesapeake Bay. Ecol. Monogr., 27: 1–26.

Arimoto, R. and Tripp, M.R., 1977. Characterization of a bacterial agglutinin in the hemolymph of the hard clam, *Mercenaria mercenaria*. J. Invertebr. Pathol., 30: 406–413.

Barry, M.M. and Yevich, P.P., 1972. Incidence of gonadal cancer in the quahog, *Mercenaria mercenaria*. Oncogenesis, 26: 96–97.

Bayne, B.L., Brown, D.A., Burns, K., Dixon, D.R., Ivanovici, A., Livingston, D.R., Lowe, D.M., Moore, M.N., Stebbing, A.R.D., and Widdows, J., 1985. The Effects of Stress and Pollution on Marine Animals. Praeger, New York, 384 pp.

Bert, M.M., Hesselman, D.M., Arnold, W.S., Moore, W.S., Cruz-Lopez, H. and Marelli, D.C., 1993. High frequency of gonadal neoplasia in a hard clam (*Mercenaria* spp.) hybrid zone. Mar. Biol., 117: 97–104.

Bisker, R. and Castagna, M., 1985. The effect of various levels of air-supersaturated seawater on *Mercenaria mercenaria* Linné, *Mulinia lateralis* Say, and *Mya arenaria* Linné with reference to gas-bubble disease. J. Shellfish Res., 5: 97–102.

Bower, S.M., 1987. *Labyrinthuloides haliotidis* n. sp. (Protozoa: Labyrinthomorpha), a pathogenic parasite of small juvenile abalone in a British Columbia mariculture facility. Can. J. Zool., 65: 1996–2007.

Bower, S.M., 1992a. Diseases and parasites of mussels. In: E. Gosling (Ed.), The Mussel Mytilus: Ecology, Physiology, Genetics and Culture. Developments in Aquaculture and Fisheries Science, Vol. 25, Elsevier, Amsterdam, pp. 543–563.

Bower, S.M., 1992b. Winter mortalities and histopathology in Japanese littlenecks *Tapes philippinarum* (A. Adams and Reeve, 1850) in British Columbia due to freezing temperatures. J. Shellfish Res., 11: 255–264.

Bower, S.M., McGladdery, S.E. and Price, I.M., 1994. Synopsis of infectious disease and parasites of commercially exploited shellfish. Annu. Rev. Fish Dis., 4: 1–200.

Bramble, L.H. and Anderson, R.S., 1997. Modulation of *Crassostrea virginica* hemocyte reactive oxygen species production by *Listonella anguillarum*. Dev. Comp. Immunol., 21: 337–348.

Bramble, L.H. and Anderson, R.S., 1998. A comparison of the chemiluminescent response of *Crassostrea virginica* and *Morone saxatilis* phagocytes to zymosan and viable *Listonella anguillarum*. Dev. Comp. Immunol., 22: 55–61.

Bramble, L.H. and Anderson, R.S., 1999. Lack of involvement of reactive oxygen species in the bactericidal activity of *Crassostrea virginica* haemocytes in contrast to *Morone saxatilis* phagocytes. Fish Shellfish Immunol., 9: 109–123.

Bricelj, V.M., Ford, S.E., Borrero, F.J., Perkins, F.O., Rivara, G., Hillman, R.E., Elston, R.A. and Chang, J., 1992. Unexplained mortalities of hatchery-reared, juvenile oysters, *Crassostrea virginica* (Gmelin). J. Shellfish Res., 11: 331–347.

Brown, C., 1974. A pigment-producing pseudomonad which discolors culture containers of embryos of a bivalve mollusk. Chesapeake Sci., 15: 17–21.

Brown, C. and Tettelbach, L.P., 1988. Characterization of a nonmotile *Vibrio* sp. pathogenic to larvae of *Mercenaria mercenaria* and *Crassostrea virginica*. Aquaculture, 74: 195–204.

Burreson, E.M., Stokes, N.A. and Friedman, C.S., 2000. Increased virulence in an introduced pathogen: *Haplosporidium nelsoni* (MSX) in the eastern oyster *Crassostrea virginica*. J. Aquat. Animal Health, 12: 1–8.

Cake, E.W., 1977. Larval cestode parasites of edible mollusks of the northeastern Gulf of Mexico. Gulf Res. Rep., 6: 1–8.

Carlton, J.T., 1992. Introduced marine and estuarine mollusks of North America: An end-of-the-20th century perspective. J. Shellfish Res., 11: 489–505.

Castagna, M., Luckenbach, M. and Kelley, P., 1989. Use of concentrated bacteria to enhance survival of early post set *Mercenaria mercenaria* in a flowing seawater nursery (abstr.). J. Shellfish Res., 8: 444.

Cheng, T.C., 1976a. Beta-glucuronidase in the serum and hemolymph cells of *Mercenaria mercenaria* and *Crassostrea virginica* (Mollusca: Pelecypoda). J. Invertebr. Pathol., 27: 125–128.

Cheng, T.C., 1976b. Aspects of substrate utilization and energy requirement during molluscan phagocytosis. J. Invertebr. Pathol., 27: 263–268.

Cheng, T.C., 1977. Biochemical and ultrastructural evidence for the double role of phagocytosis in molluscs: defense and nutrition. In: L.A. Bulla, Jr. and T.C. Cheng (Eds.), Comparative Pathobiology, Vol. 3. Plenum, New York, pp. 22–30.

Cheng, T.C., 1981. Bivalves. In: N.A. Ratcliffe and A.F. Rowley (Eds.), Invertebrate Blood Cells, Vol. 1. Academic Press, London, pp. 233–299.

Cheng, T.C., 1983. The role of lysosomes in molluscan inflammation. Am. Zool., 23: 129–144.

Cheng, T.C. and Burton, R.W., 1965. The American oyster and clam as experimental intermediate hosts of *Angiostrongylus cantonensis*. J. Parasitol., 51: 296.

Cheng, T.C. and Cali, A., 1974. An electron microscope study of the fate of bacteria phagocytosed by granulocytes of *Crassostrea virginica*. Contemp. Top. Immunobiol., 4: 25–35.

Cheng, T.C. and Foley, D.A., 1975. Hemolymph cells of the bivalve mollusc *Mercenaria mercenaria*: an electron microscopical study. J. Invertebr. Pathol., 26: 341–351.

Cheng, T.C. and Galloway, P.C., 1970. Transplantation immunity in mollusks: the histoincompatibility of *Helisoma duryi normale* with allographs and xenografts. J. Invertebr. Pathol., 15: 177–192.

Cheng, T.C. and Howland, K.H., 1979. Chemotactic attraction between hemocytes of the oyster, *Crassostrea virginica*, and bacteria. J. Invertebr. Pathol., 33: 304–310.

Cheng, T.C. and Rodrick, G.E., 1975. Lysosomal and other enzymes in the hemolymph of *Crassostrea virginica* and *Mercenaria mercenaria*. Comp. Biochem. Physiol., 52B: 443–447.

Cheng, T.C., Shuster Jr., C.N. and Anderson, A.H., 1966a. A comparative study of the susceptibility and response of eight species of marine pelecypods to the trematode *Himasthla quissetensis*. Trans. Am. Microsc. Soc., 85: 284–295.

Cheng, T.C., Shuster Jr., C.N. and Anderson, A.H., 1966b. Effects of plasma and tissue extracts of marine pelecypods on the cercaria of *Himasthla quissetensis*. Exp. Parasitol., 19: 9–14.

Cheng, T.C., Rodrick, G.E., Foley, D.A. and Loehler, S.A., 1975. Release of lysozyme from hemolymph cells of *Mercenaria mercenaria* during phagocytosis. J. Invertebr. Pathol., 25: 261–265.

Cheng, T.C., Chorney, M.J. and Yoshino, T.P., 1977. Lysozyme-like activity in the hemolymph of *Biomphalaria glabrata* challenged with bacteria. J. Invertebr. Pathol., 29: 170–174.

Cheng, T.C., Burrell, V.G., Manzi, J.J. and Perkins, F.O., 1995. Natural and experimental infection of *Mercenaria mercenaria* by *Perkinsus marinus* from *Crassostrea virginica*. Res. Rev. Parasitol., 55: 31–34.

Chew, K.K., 1990. Global bivalve shellfish introductions. World Aquacult., 21: 9–22.

Chintala, M.M., Alcox, K.A., Ford, S.E. and Bushek, D., 1995. Cultured and natural *Perkinsus marinus* cells: a possible mechanisms for virulence differences (abstr.). J. Shellfish Res., 14: 240.

Chorney, M.J. and Cheng, T.C., 1980. Discrimination of self and non-self in invertebrates. In: J.J. Marchalonis and N. Cohen (Eds.), Contemporary Topics in Immunobiology, Vol. 9. Plenum, New York, pp. 37–54.

Chu, F.-L.E., 1988. Humoral defense factors in marine bivalves. In: W.S. Fisher (Ed.), Disease Processes in Marine Bivalve Molluscs, Vol. 18. American Fisheries Society, Bethesda, MD, pp. 178–188.

Davey, J.T., 1989. *Mytilicola intestinalis* (Copepoda: Cyclopoida): a ten year survey of infested mussels in a Cornish estuary, 1978–1988. J. Mar. Biol. Assoc. U.K., 69: 823–836.

Davis, J.D., 1969. Polydora infestation of *Mercenaria mercenaria*. Nautilus, 83: 74.

Davis, H.C., Loosanoff, V.L., Weston, W.H. and Martin, C., 1954. A fungus disease in clam and oyster larvae. Science, 120: 36–38.

Deaton, L.E. and Dankert, J.R., 1999. Phenoloxidase activity in the hemolymph of bivalve mollusks. J. Shellfish Res., 18: 223–226.

Des Voigne, D.M. and Sparks, A.K., 1968. The process of wound healing in the Pacific oyster, *Crassostrea gigas*. J. Invertebr. Pathol., 2: 53–65.

Dow, R.L. and Wallace, D.E., 1951. A method of reducing winter mortalities of *Venus mercenaria* in Maine waters. Proc. Natl. Shellfish. Assoc., 1951: 15–21.

Drinnan, R.E. and Henderson, E.B., 1963. 1962 mortalities and a possible disease organism in Neguac quahaugs. Biological Station, St. Andrews, NB, Annual Report, pp. B18–20.

Dungan, C.F. and Elston, R.A., 1988. Histopathological and ultrastructural characteristics of bacterial destruction of hinge ligaments of cultured juvenile Pacific oysters, *Crassostrea gigas*. Aquaculture, 72: 1–14.

Eldridge, P.J. and Eversole, A.G., 1982. Compensatory growth and mortality of the hard clam *Mercenaria mercenaria*. Veliger, 24: 276–278.

Eldridge, P.J., Waltz, W., Gracy, R.C. and Hunt, H.H., 1977. Growth and mortality rates of hatchery seed clams *Mercenaria mercenaria* in protected trays in waters of South Carolina, USA. Proc. Natl. Shellfish. Assoc., 66: 13–20.

Eldridge, P.J., Eversole, A.G. and Whetstone, J.M., 1979. Comparative survival and growth rates of hard clams, *Mercenaria mercenaria*, planted in trays subtidally and intertidally at varying densities in a South Carolina estuary. Proc. Natl. Shellfish. Assoc., 69: 30–39.

Elston, R., 1984. Prevention and management of infectious disease in intensive mollusc husbandry. J. World Maricult. Soc., 5: 284–300.

Elston, R., Elliot, E.L. and Colwell, R.R., 1982. Conchiolin infection and surface coating *Vibrio*: shell fragility, growth depression and mortalities in cultured oysters and clams, *Crassostrea virginica*, *Ostrea edulis* and *Mercenaria mercenaria*. J. Fish Dis., 5: 265–284.

Elston, R.A., 1986. An intranuclear pathogen [Nuclear Inclusion X (NIX)] associated with massive mortalities of the pacific razor clam, *Siliqua patula*. J. Invertebr. Pathol., 47: 93–104.

Elston, R.A., 1993. Infectious diseases of the Pacific oyster, *Crassostrea gigas*. Ann. Rev. Fish Dis., 2: 259–276.

Elston, R.A., Farley, C.A. and Kent, M.L., 1986. Occurrence and significance of bonamiasis in European flat oysters *Ostrea edulis* in North America. Dis. Aquat. Org., 2: 49–54.

Elston, R.A., Moore, J.D. and Brooks, K., 1992. Disseminated neoplasia of bivalve molluscs. Rev. Aquat. Sci., 6: 405–466.

Eversole, A.G. and Heffernan, P.B., 1995. Gonadal neoplasia in northern *Mercenaria mercenaria* (Linnaeus, 1758),

southern *Mercenaria campechiensis* (Gmelin, 1791) and their hybrids cultured in South Carolina. J. Shellfish Res., 14: 33–39.
Farley, C.A., 1968. *Minchinia nelsoni* (Haplosporida) disease syndrome in the American oyster *Crassostrea virginica*. J. Protozool., 15: 585–599.
Farley, C.A., 1992. Dispersal of pathogens, parasites, pests, predators and competitors. In: A. Rosenfield and R. Mann (Eds.), Dispersal of Living Organisms into Aquatic Ecosystems. Maryland Sea Grant Publications, College Park, MD, pp. 139–154.
Farley, C.A., Wolf, P.H. and Elston, R.A., 1988. A long term study of "Microcell" Disease in oysters with a description of a new genus — *Mikrocytos* (g. n.) and two new species — *Mikrocytos mackini* (sp. n.) and *Mikrocytos roughleyi* (sp. n.). Fish. Bull., 86: 581–593.
Fawcett, L.B. and Tripp, M.R., 1994. Chemotaxis of *Mercenaria mercenaria* hemocytes to bacteria in vitro. J. Invertebr. Pathol., 63: 275–284.
Feng, S.Y., 1988. Cellular defense mechanisms of oysters and mussels. In: W.S. Fisher (Ed.), Disease Processes in Marine Bivalve Molluscs, Vol. 18. American Fisheries Society, Bethesda, MD, pp. 153–168.
Feng, S.Y., Khairallah, E.A. and Canzonier, W.J., 1970. Hemolymph-free amino acids and related nitrogenous compounds of *Crassostrea virginica* infected with *Buchephalus* sp. and *Minchinia nelsoni*. Comp. Biochem. Physiol., 34: 547–556.
Figueras, A.J., Jardon, C.F. and Caldas, J.R., 1991. Diseases and parasites of rafted mussels (*Mytilus galloprovincialis* Lmk): preliminary result. Aquaculture, 99: 17–33.
Fisher, W.S. (Ed.), 1988. Disease Processes in Marine Bivalve Molluscs, Vol. 18, American Fisheries Society, Bethesda, MD, 315 pp.
Flagg, R.J. and Malouf, R.E., 1983. Experimental plantings of juveniles of the hard clam *Mercenaria mercenaria* in the waters of Long Island, New York, USA. J. Shellfish Res., 3: 19–28.
Foley, D.A. and Cheng, T.C., 1974. Morphology, hematologic parameters, and behavior of hemolymph cells of the quahog clam *Mercenaria mercenaria*. Biol. Bull., 146: 393–456.
Foley, D.A. and Cheng, T.C., 1975. A quantitative study of phagocytosis by hemolymph cells of the pelecypods *Crassostrea virginica* and *Mercenaria mercenaria*. J. Invertebr. Pathol., 25: 189–197.
Foley, D.A. and Cheng, T.C., 1977. Degranulation and other changes of molluscan granulocytes associated with phagocytosis. J. Invertebr. Pathol., 29: 321–325.
Ford, S.E., 1992. Avoiding the spread of disease in commercial culture of molluscs, with special reference to *Perkinsus marinus* (Dermo) and *Haplosporidium nelsoni* (MSX). J. Shellfish Res., 11: 539–546.
Ford, S.E., 1996. Range extension by the oyster parasite *Perkinsus marinus* into the northeastern US: response to climate change?. J. Shellfish Res., 15: 45–56.
Ford, S.E. and Tripp, M.R., 1996. Diseases and defense mechanisms. In: R.I.E. Newell, V.S. Kennedy and A.F. Eble (Eds.), The Eastern Oyster *Crassostrea virginica*. Maryland Sea Grant College, College Park, MD, pp. 383–450.
Ford, S.E., Kanaley, S.A. and Ashton-Alcox, K.A., 1993. *In vitro* interactions between bivalve hemocytes and the oyster pathogen *Haplosporidium nelsoni* (MSX). J. Parasitol., 79: 255–265.
Ford, S.E., Alcox, K.A. and Kanaley, S.A., 1994. Comparative cytometric and microscopic analyses of oyster hemocytes. J. Invertebr. Pathol., 64: 114–122.
Ford, S.E., Smolowitz, R., Ragone-Calvo, L., Barber, R.D. and Kraeuter, J.N., 1997. Evidence that QPX (Quahog Parasite Unknown) is not present in hatchery-produced hard clam seed. J. Shellfish Res., 16: 519–521.
Friedl, F.E. and Alvarez, M.R., 1990. Cytometric studies on *Mercenaria* hemocytes. In: F.O. Perkins and T.C. Cheng (Eds.), Pathology in Marine Science. Academic Press, San Diego, CA, pp. 489–500.
Friedman, C.S. and Perkins, F.O., 1994. Range extension of *Bonamia ostreae* to Maine, U.S.A. J. Invertebr. Pathol., 64: 179–181.
Fries, C.R. and Grant, D.M., 1991. Rickettsiae in gill epithelial cells of the hard clam *Mercenaria mercenaria*. J. Invertebr. Pathol., 57: 166–171.
Fries, C.R. and Grant, D.M., 1992. *Ehrlichia*-like microorganisms in hemocytes in the gills of the marine bivalve *Mercenaria mercenaria*. J. Invertebr. Pathol., 59: 210–211.
Getchell, R.G., 1991. Diseases and parasites of scallops. In: S.E. Shumway (Ed.), Scallops: Biology, Ecology and Aquaculture, Vol. 21. Elsevier, Amsterdam, pp. 471–494.
Gibbons, M.C. and Blogoslawski, W.J., 1989. Predators, Pests, Parasites, and Diseases. In: J.J. Manzi and M. Castagna (Eds.), Clam Mariculture in North America, Vol. 19. Elsevier, Amsterdam, pp. 167–200.

Graham, M.A., 1968. A hemolytic enzyme in the hemolymph of the clam *Mercenaria mercenaria*. Master's Thesis, University of Delaware.

Greene, G.T. and Becker, D.S., 1977. Winterkill of hard clams in Great South Bay, N.Y. 1976–77. Marine Science Research Center, State University of New York, Special Report, 23 pp.

Grizel, H., 1983. Impact of *Marteilia refringens* and *Bonamia ostreae* on oyster culture of Brittany. ICES C.M. 1983; Gen:9/Mini Symp., Goteborg, October, 29 pp.

Grizel, H. and Héral, M., 1991. Introduction into France of the Japanese oysters (*Crassostrea gigas*). J. Cons. Int. Explor. Mer, 47: 399–403.

Gulka, G. and Chang, P.W., 1984. Pathogenicity and infectivity of a rickettsia-like organism in the sea scallop, *Placopecten magellanicus*. J. Fish Dis., 8: 309–318.

Gulka, G., Chang, P.W. and Marti, K.A., 1983. Prokaryotic infection associated with a mass mortality of the sea scallop, *Placopecten magellanicus*. J. Fish Dis., 6: 355–364.

Harshbarger, J.C., Chang, S.C. and Otto, S.V., 1977. Chlamydiae (with phages), mycoplasms, and rickettsiae in Chesapeake Bay bivalves. Science, 196: 666–668.

Hartland, B.J. and Timoney, J.F., 1979. *In vivo* clearance of enteric bacteria from the hemolymph of the hard clam and the American oyster. Appl. Environ. Microbiol., 37: 517–520.

Haven, D. and Andrews, J.D., 1957. Survival and growth of *Venus mercenaria*, *Venus campechiensis*, and their hybrids in suspended trays and on natural bottoms. Proc. Natl. Shellfish. Assoc., 47: 43–49.

Hawkins, L.E., Brooks, J.D., Brooks, S. and Hutchinson, S., 1993. The effect of tidal exposure on aspects of metabolic and immunological activity in the hard clam *Mercenaria mercenaria* Linnaeus. Comp. Biochem. Physiol. A, 104: 225–228.

Heppell, D., 1961. The naturalization in Europe of the quahog, *Mercenaria mercenaria* (L.). J. Conchol., 25: 21–34.

Hesselman, D.M., Blake, N.J. and Peters, E.C., 1988. Gonadal neoplasms in hard shell clams *Mercenaria* spp., from the Indian River, Florida: occurrence, prevalence, and histopathology. J. Invertebr. Pathol., 52: 436–446.

Hibbert, C.J., 1977. Growth and survivorship in a tidal flat population of the bivalve *Mercenaria mercenaria* from Southampton England water. Mar. Biol., 44: 71–76.

Hill, B.J., 1976. Molluscan viruses: their occurrence, culture and relationships. In: Proc. 1st Int. Colloq. Invertebrate Pathology, Kingston, ON, pp. 25–29.

Huffman, J.E. and Tripp, M.R., 1982. Cell types and hydrolytic enzymes of soft shell clam (*Mya arenaria*) hemocytes. J. Invertebr. Pathol., 40: 68–74.

Humes, A.G., 1954. *Mytilicola porrecta* n. sp. (Copepoda: Cyclopoida) from the intestine of marine pelecypods. J. Parasitol., 40: 186–194.

Humes, A.G. and Cressey, R.F., 1960. Seasonal population changes and host relationships of *Myocheres major* (Williams), a cyclopoid copepod from pelecypods. Crustaceana, 1: 307–325.

Ingersoll, E., 1887. The oyster, scallop, clam, mussel, and abalone industries. In: G.B. Goode (Ed.), The Fisheries and Fishery Industries of the United States, II, Section V. United States Commission of Fish and Fisheries, Washington, DC, pp. 507–626.

Jeffries, H.P., 1972. A stress syndrome in the hard clam, *Mercenaria mercenaria*. J. Invertebr. Pathol., 20: 242–251.

Jones, G. and O'Dor, R.K., 1983. Ultrastructural observations on a Thraustochytrid fungus parasitic in the gills of squid (*Illex illecebrosus* Lesueur). J. Parasitol., 69: 903–911.

Juste, V., 1992. Study on the population genetics, using electrophoretic analysis, of seven populations of the hard clam *Mercenaria mercenaria* (Linn.) from the east coast of the United States and Puerto Rico. PhD Dissertation, University of Puerto Rico, Mayaguez, Puerto Rico, 127 pp.

Kern, F.G., 1976. Sporulation of *Minchinia* sp. (Haplosporida, Haplosporidiidae) in the Pacific oyster *Crassostrea gigas* (Thunberg) from the republic of Korea. J. Protozool., 23: 498–500.

Kleinschuster, S.J., Perkins, F.O., Dykstra, M.J. and Swink, S.L., 1994. The *in vitro* life cycle of a *Perkinsus* species (Apicomplexa, Perkinsidae) isolated from *Macoma balthica* (Linneaus, 1958). J. Shellfish Res., 13: 461–465.

Kleinschuster, S.J., Smolowitz, R. and Parent, J., 1998. In vitro life cycle and propagation of Quahog parasite unknown. J. Shellfish Res., 17: 75–78.

Kochiss, J.M., 1974. Oystering from New York to Boston. Wesleyan University Press, Middletown, CT, 251 pp.

Kraeuter, J.N. and Castagna, M., 1984. Disease treatment in hard clams, *Mercenaria mercenaria*. J. World Maricult. Soc., 15: 310–317.

Kraeuter, J.N. and Castagna, M., 1985. The effects of seed size, shell bags, crab traps, and netting on the survival of the northern hard clam *Mercenaria mercenaria* Linné. J. Shellfish Res., 5: 69–72.

Kraeuter, J.N., Flimlin, G.E., Canzonier, W., Ford, S., Zodl, J. and Parsons, D., 1997. Declining quality of hard clams in Little Egg Harbor Bay: effects of a long-term salinity reduction? In: G.E. Flimlin and M.J. Kennish (Eds.), Proceedings of The Barnegate Bay Workshop, Barnegat Bay Estuary Program and Rutgers University, pp. 123–138.

Kraeuter, J., Ford, S., Barber, A.R. and Mathis, G., 1998a. Effects of planting density and depth on QPX infections in hard clams (abstr.). J. Shellfish Res., 17: 357.

Kraeuter, J.N., Ford, S.E., Smolowitz, R., Leavitt, D. and Ragone Calvo, L.M., 1998b. QPX, a protistan parasite of hard clam (*Mercenaria mercenaria*) and its importance to rehabilitation efforts. In: Proceedings of the International Conference on Shellfish Restoration, Hilton Head, SC, Abstr., p. 62.

La Peyre, J.F., Chu, F.-L.E. and Movern, J.A., 1995. Hemocytic and humoral activities of eastern and Pacific oysters following challenge by the protozoan *Perkinsus marinus*. Fish Shellfish Immunol., 5: 179–190.

Landers, W.S., 1967. Infestation of the hard clam, *Mercenaria mercenaria*, by the boring polychaete worm, *Polydora ciliata*. Proc. Natl. Shellfish. Assoc., 57: 63–66.

Landry, T., Sephton, T.W. and Jones, D.A., 1993. Growth and mortality of northern quahog (Linneaeus, 1758) *Mercanaria mercanaria* in Prince Edward Island. J. Shellfish Res., 2: 321–327.

Lauckner, G., 1983. Diseases of Mollusca: Bivalvia. In: O. Kinne (Ed.), Diseases of Marine Animals, Vol. 2. Biologische Anstalt Helgoland, Hamburg, pp. 477–963.

Le Gall, G., Chagot, D., Mialhe, E. and Grizel, H., 1988. Branchial Rickettsiales-like infection associated with a mass mortality of sea scallop *Pecten maximus*. Dis. Aquat. Org., 4: 229–232.

Leibovitz, L., 1985. Retrograde diseases of fouled cultured juvenile hard clams (*Mercenaria mercenaria*). J. World Maricult. Soc., 16: 387–395.

Leibovitz, L., Meyers, T.R. and Frey, M., 1976. A shell deforming disease of hard clams (*Mercenaria mercenaria*). In: Proc. 1st Int. Colloq. Invertebrate Pathology, Kingston, ON, pp. 386–387.

Loker, E.S., 1994. On being a parasite in an invertebrate host: a short survival course. J. Parasitol., 80: 728–747.

MacCallum, G.S. and McGladdery, S.E., 2000. Quahog Parasite Unknown (QPX) in the northern quahog *Mercenaria mercenaria* (Linneaus, 1758) and *M. mercenaria* var. *notata* from Atlantic Canada, survey results from three maritime provinces. J. Shellfish Res., 19: 43–50.

Mackin, J.G., 1951. Histopathology of infection of *Crassostrea virginica* (Gmelin) by *Dermocystidium marinum* Mackin, Owen, and Collier. Bull. Mar. Sci. Gulf Caribb., 1: 72–87.

Malouf, R., Keck, R., Maurer, D. and Epifanio, C., 1972. Occurrence of gas bubble disease in 3 species of bivalve mollusks. J. Fish. Res. Board Can., 29: 588–589.

Mann, R., Burreson, E.M. and Baker, P.K., 1991. The decline of the Virginia oyster fishery in Chesapeake Bay: considerations for introduction of a non-endemic species, *Crassostrea gigas* (Thunberg, 1793). J. Shellfish Res., 10: 379–388.

Mass, P.A.Y., Kleinschuster, S.J., Dykstra, M.J., Smolowitz, R. and Parent, J., 1999. Molecular characterization of QPX (Quahog parasite Unknown), a pathogen of *Mercenaria mercenaria*. J. Shellfish Res., submitted.

McLean, N. and Porter, D., 1982. The yellow spot disease of *Tritonia diomedea* Bergh, 1894 (Mollusca: Gastropoda; Nudibranchia): Encapsulation of the Traustochytreaceous parasite by host amoebocytes. J. Parasitol., 68: 243–252.

Menzel, R.W. and Sims, H.W., 1964. Experimental farming of hard clams, *Mercenaria mercenaria*, in Florida. Proc. Natl. Shellfish. Assoc., 53: 103–109.

Meyers, T.R., 1979. Preliminary studies on a chlamydial agent in the digestive diverticular epithelium of hard clams, *Mercenaria mercenaria* (L.) from Great South Bay, New York. J. Fish Dis., 2: 179–189.

Mohandas, A., Cheng, T.C. and Cheng, J.B., 1985. Mechanism of lysosomal enzyme release from *Mercenaria mercenaria* granulocytes: a scanning electron microscope study. J. Invertebr. Pathol., 46: 189–197.

Moore, C.A. and Eble, E.F., 1977. Cytochemical aspects of *Mercenaria mercenaria* hemocytes. Biol. Bull., 152: 105–119.

Mourton, C., Boulo, V., Chagot, D., Hervio, D., Bachère, E., Mialhe, E. and Grizel, H., 1992. Interactions between *Bonamia ostreae* (Protozoa: Ascetospora) and hemocytes of *Ostrea edulis* and *Crassostrea gigas* (Mollusca: Bivalvia): in vitro system establishment. J. Invertebr. Pathol., 59: 235–241.

Norton, J.H., Shepherd, M.A., Abdon-Naguit, M.R. and Lindsay, S., 1993. Mortalities in the giant clam *Hippopus hippopus* associated with Rickettsiales-like organisms. J. Invertebr. Pathol., 62: 207–209.

Olafsen, J.A., 1988. Role of lectins in invertebrate humoral defense. In: W.S. Fisher (Ed.), Disease Processes in Marine Bivalve Molluscs, Vol. 18. American Fisheries Society, Bethesda, MD, pp. 189–205.

Oliver, L.M. and Fisher, W.S., 1995. Comparative form and function of oyster *Crassostrea virginica* hemocytes from Chesapeake Bay (Virginia) and Apalachicola Bay (Florida). Dis. Aquat. Org., 22: 217–225.

Paillard, C., Maes, P. and Oubella, R., 1994. Brown ring disease in clams. Annu. Rev. Fish Dis., 4: 219–240.

Parish, C.R., 1977. Simple model for self–non-self-discrimination in invertebrates. Nature, 267: 711–713.

Paynter, K.T., Pierce, S.K. and Burreson, E.M., 1995. Levels of intracellular free amino acids used for salinity tolerance by oyster (*Crassostrea virginica*) are altered by protozoan (*Perkinsus marinus*) parasitism. Mar. Biol., 122: 67–72.

Pearse, A.S., 1947. Parasitic copepods from Beaufort, North Carolina. J. Elisha Mitchell Sci. Soc., 63: 1–16.

Perkins, F.O., 1973. A new species of marine Labyrinthulid. Arch. Mikrobiol., 90: 1–17.

Perkins, F.O., 1993. Infectious diseases of molluscs. In: J.A. Couch and J.W. Fournie (Eds.), Pathobiology of Marine and Estuarine Organisms. CRC Press, Boca Raton, FL, pp. 255–287.

Peters, E.C., Yevich, P.P., Harshbarger, J.C. and Zaroogian, G.E., 1994. Comparative histopathology of gonadal neoplasms in marine bivalve molluscs. Dis. Aquat. Org., 20: 59–76.

Pokorny, K.S., 1985. Phylum labyrinthomorpha. In: J.J. Lee, S.H. Hutner and E.C. Bovee (Eds.), Illustrated Guide to the Protozoa. Society of Protozoologists, Lawrence, KS, pp. 318–321.

Polglase, J.L., 1980. A preliminary report on the Thraustochytrid(s) and Labyrinthulid(s) associated with a pathological condition in the lesser octopus *Eledone cirrhosa*. Bot. Mar., 23: 699–706.

Porter, D., 1990. Phylum labyrinthomycota. In: L. Margulis, J.O. Corliss, M. Melkonian and D.J. Chapman (Eds.), Handbook of Protoctista. Jones and Bartlett, Boston, MA, pp. 388–398.

Porter, H.J., 1964. Incidence of *Malacobdella* in *Mercenaria campechiensis* off Beaufort Inlet, North Carolina. Proc. Natl. Shellfish. Assoc., 53: 133–145.

Powell, E.N., Hofmann, E.E., Klinck, J.M. and Ray, S.M., 1992. Modeling oyster populations I. A commentary on filtration rate. Is faster always better?. J. Shellfish Res., 11: 387–398.

Ragone Calvo, L.M., Walker, J.G. and Burreson, E.M., 1998. Prevalence and distribution of QPX, Quahog Parasite Unknown, in hard clams, *Mercenaria mercenaria*, in Virginia, USA. Dis. Aquat. Org., 33: 209–219.

Ray, S.M., 1954. Biological studies of *Dermocystidium marinum*, a fungus parasite of oysters. Rice Institute. Houston, TX, 114 pp.

Ropes, J.W., 1963. The incidence of *Malacobdella grossa* in hard clams from Nantucket Sound, Massachusetts. Limnol. Oceanogr., 8: 353–355.

Ropes, J.W., Merrill, A.S., Murawski, S.A., Chang, S. and MacKenzie Jr., C.L., 1979. Impact on clams and scallops. Field survey assessments. Chapter 11, Part 1. In: R.L. Swanson and C.J. Sindermann (Eds.), Oxygen Depletion and Associated Benthic Mortalities in New York Bight, 1976. US Department of Commerce, National Oceanic and Atmospheric Administration, Prof. Pap. 11, pp. 263–275.

Rosenfield, A. and Kern, F.G., 1979. Molluscan imports and the potential for introduction of disease organisms. In: R. Mann (Ed.), Exotic Species in Mariculture. MIT Press, Cambridge, MA, pp. 165–191.

Sindermann, C.J., 1990a. Diseases of Marine Fish. Academic Press, San Diego, CA, 521 pp.

Sindermann, C.J., 1990b. Diseases of Marine Shellfish. Academic Press, San Diego, CA, 516 pp.

Sindermann, C.J. and Rosenfield, A., 1967. Principal diseases of commercially important marine bivalve mollusca and crustacea. Fish. Bull., 66: 335–385.

Smith, V.J. and Söderhäll, K., 1991. A comparison of phenoloxidase activity in the blood of marine invertebrates. Dev. Comp. Immunol., 15: 251–261.

Smolowitz, R., Leavitt, D. and Perkins, F., 1998. Observations of a protistan disease similar to QPX in *Mercenaria mercenaria* (hard clams) from the coast of Massachusetts. J. Invertebr. Pathol., 71: 9–25.

Soniat, T.M. and Keonig, M.L., 1982. The effects of parasitism by *Perkinsus marinus* on the free amino acid composition of *Crassostrea virginica* mantle tissue. J. Shellfish Res., 2: 25–28.

Sparks, A.K., 1985. Synopsis of Invertebrate Pathology Exclusive of Insects. Elsevier, Amsterdam, 423 pp.

Sparks, A.K. and Pauley, G.B., 1964. Studies of the normal postmortem changes in the oyster *Crassostrea gigas* (Thunberg). J. Invertebr. Pathol., 6: 78–101.

Stefano, G.B., Leung, M.K., Zhoa, X. and Scharrer, B., 1989. Evidence for the involvement of opioid neuropeptides in adherence and migration of immunocompetent invertebrate hemocytes. Proc. Natl. Acad. Sci. USA, 86: 626–630.

Tripp, M.R., 1992a. Phagocytosis by hemocytes of the hard clam *Mercenaria mercenaria*. J. Invertebr. Pathol., 59: 222–227.

Tripp, M.R., 1992b. Agglutinins in the hemolymph of the hard clam *Mercenaria mercenaria*. J. Invertebr. Pathol., 59: 228–234.

Tubiash, H.S., 1975. Bacterial pathogens associated with cultured bivalve mollusk larvae. In: W.L. Smith and M.H. Chanley (Eds.), Culture of Marine Invertebrate Animals. Plenum, New York, pp. 61–71.

Tubiash, H.S., Chanley, P. and Leifson, E., 1965. Bacillary necrosis disease of larval and juvenile bivalve molluscs. I. Etiology and epizootiology. J. Bacteriol., 90: 1036–1044.

Tubiash, H.S., Colwell, R.R. and Sakazaki, R., 1970. Marine vibrios associated with bacillary necrosis, a disease of larval and juvenile bivalve mollusks. J. Bacteriol., 103: 272–273.

Van Beneden, R.J., 1994. Molecular analysis of bivalve tumors: models for environmental/genetic interactions. Environ. Health Perspect., 102: 81–83.

Van Beneden, R.J., Gardner, G.R., Blake, N.J. and Blair, D.G., 1993. Implications for the presence of transforming genes in gonadal tumors in two bivalve mollusk species. J. Cancer Res., 53: 2976–2979.

Villalba, A., Carballa, M.J., López, C., Cabada, A., Corral do, L. and Azevedo, C., 1999. Branchial rickettsia-like infection associated with clam *Venerupis rhomboides* mortality. Dis. Aquat. Org., 36: 53–60.

Volety, A.K. and Chu, F.-L.E., 1995. Suppression of chemiluminescence of eastern oyster (*Crassostrea virginica*) hemocytes, by the protozoan parasite *Perkinsus marinus*. Dev. Comp. Immunol., 19: 135–142.

Walker, R.L., 1984. Effects of density and sampling time on the growth of the hard clam, *Mercenaria mercenaria*, planted in predator-free cages in coastal Georgia. Nautilus, 98: 114–119.

Walker, R.L. and Heffernan, R.B., 1990. Intertidal growth and survival of northern quahogs *Mercenaria mercenaria* Linnaeus 1758 and Atlantic surf clams *Spisula solidissima* Dillwyn 1817 in Georgia, USA. J. World Aquacult. Soc., 21: 307–313.

Walker, R.L. and Humphrey, C.M., 1984. Growth and survival of the northern hard clam *Mercenaria mercenaria* from Georgia, Virginia, and Massachusetts in coastal waters of Georgia, USA. J. Shellfish Res., 4: 125–130.

Welch, W.R., 1966. The European oyster, *Ostrea edulis*, in Maine. Proc. Natl. Shellfish. Assoc., 54: 7–23.

Whyte, S.K., Cawthorn, R.J. and McGladdery, S.E., 1994. QPX (Quahaug Parasite X) a pathogen of northern quahaug *Mercenaria mercenaria* from the Gulf of St. Lawrence, Canada. Dis. Aquat. Org., 19: 129–136.

Yevich, P.P. and Barry, M.M., 1969. Ovarian tumors in the quahog *Mercenaria mercenaria*. J. Invertebr. Pathol., 14: 266–267.

Yoshino, T.P. and Cheng, T.C., 1976. Fine structural localization of acid phosphatase EC-3.1.3.2 in granulocytes of the pelecypod *Mercenaria mercenaria*. Trans. Am. Microsc. Soc., 95: 215–220.

Yoshino, T.P. and Cheng, T.C., 1977. Aminopeptidase activity in the hemolymph and body tissues of the pulmonate gastropod *Biomphalaria glabrata*. J. Invertebr. Pathol., 30: 76–79.

Zaroogian, G. and Yevich, P., 1994. The nature and function of the brown cells in *Crassostrea virginica*. Mar. Environ. Res., 37: 355–373.

Zaroogian, G., Anderson, S. and Voyer, R.A., 1992. Individual and combined cytotoxic effects of cadmium copper and nickel on brown cells of *Mercenaria mercenaria*. Ecotoxicol. Environ. Safety, 24: 328–337.

Section 3

Fisheries, Aquaculture and Human Interactions

Biology of the Hard Clam
J.N. Kraeuter and M. Castagna (Eds.),

Chapter 13

Management of Hard Clam Stocks, Mercenaria mercenaria

J.L. McHugh [1]

13.1 INTRODUCTION

Records of total landings of hard clams (*Mercenaria mercenaria*) in the United States, from Maine to Florida (Fig. 13.1) as far as records are available, began in 1888 at about 8.4 million pounds of meats, rose and fell irregularly to a maximum of about 20.8 million in 1950. Landings fell and rose again after that, but in 1992 they had dropped to 9.1 million pounds. This was a decline from 1950 to 1992 of about 56%, and in general the northern states dropped the most. What were the reasons for this decline in the last 42 years? According to Buckner (1984), the decline in New York, which was by far the greatest producer during this time, was caused primarily by over-fishing. More clams were harvested than were replaced by new spawnings.

The second factor was deterioration of water quality, which caused a considerable number of clam grounds to be closed for production. Closure of these areas was not always respected by clammers, however, because many did not believe the standards set by the Surgeon General of the United States. They argued that public health had not been affected by clams harvested from many grounds, and thus there was little respect for the law in some quarters.

The harvest of seed clams, less than minimum size, is also a problem. This is equally bad in certified or uncertified waters, because small clams are worth more than larger ones, and the risk is worth getting caught and paying a fine.

Predation is a problem also (see Chapter 11). A number of marine animals are important predators on hard clams. If management is successful, the population of predators may increase also, so that predation may increase, and actually take a much heavier toll than before.

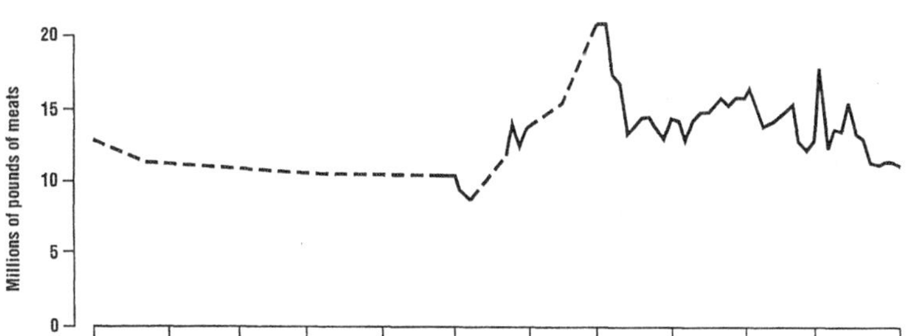

Fig. 1. Historic landings of hard clam meats along the Atlantic coast of North America from Maine to the west coast of Florida. Where data were missing or incomplete the points were joined by broken lines.

[1] Deceased.

Reduced hard clam reproductive success will affect the size of the harvest. Reproductive success varies greatly from time to time, and some of the causes are not well known. Changes in water salinity can be important. Hard clams require somewhat higher salinity than oysters, and storms that open or close access to the ocean can alter salinity and thus affect abundance. Variations in temperature, in water quality, and many other factors can affect reproduction, and cannot easily be identified or controlled. Management must try to be aware, and be prepared as far as possible to alter adverse effects.

13.2 INCREASING THE HARVEST

Spawner sanctuaries are one way of assuring a regular supply of larvae. Some areas may be better than others, depending on currents. This may be settled by dye studies. Some bays or certain areas of bays may be better for sanctuaries than others. This must be determined by research. Once suitable locations are determined the grounds must be covered with large spawners, and covered with low-cost materials such as iron gratings, rocks, or other materials that do not interfere with hard clam survival, but make it difficult or impossible to remove clams from the area. New large clams must be replaced periodically as the old clams die off. There may be merit in dividing areas of bottom into three or four parts, based on number of clams rather than size of bottom, and opening only one area at a time. A regular schedule of surveillance should be planned for each area to determine the success of spawning, recruitment, and growth, so that the time to open each is known. The number of areas of bottom needed to rotate will vary with location. For the sake of simplicity we will assume that rotation time is three years, each area to be opened for one year. Details will be worked out as information accumulates.

Grounds may at first need to be closed before the end of the year, if the stock of harvestable clams is reduced too much by clammers. If so, individual areas may have been chosen too small, and areas of bottom may need to be increased in size, or decreased in size if areas are too large. These matters will be worked out in time, as information accumulates.

Another question to be determined is the matter of transplants. Is there merit in transplanting clams from uncertified areas to approved areas for cleansing, to be harvested at a later date when they are known to be safe to take? Should this be done by the state, or would it be better to have reliable baymen, under careful supervision, do the transplanting to clean bottom, to be harvested by them from the bottom to which they had transplanted the clams?

The establishment of hatcheries is important. These are in operation already in some places, some by private individuals, and some by states. Hatcheries must be large enough to produce the millions of clams needed if the harvest is to be increased by much. Hatcheries are expensive to maintain and operate. There is some question as to whether they actually increase the value of the harvest, or merely are in fact a subsidy to clammers. It is expensive to maintain clams in the hatchery until they are large enough to plant, which means that hard clams may have to be planted in closed areas until they are large enough to be harvested. They will have to be protected during that period. The costs and benefits of hatchery operation will have to be worked out very carefully.

Stock assessments will have to be done frequently. Quotas are one way of controlling the harvest, but quotas, especially if they mean closing an area before the year is over, or if they could continue to produce significant harvests safely after the year is up, will not be ideal.

Another way of controlling harvests is to control entry into the commercial fishery. This will require accurate assessment of hard clam stocks, so that the numbers of vessels or clammers are not too high or two low. This will take time. Control of entry into the recreational fishery will also be important. What level of harvest will be determined for sport clammers? Will it be increased as the human population increases? That will affect the commercial harvest adversely. Or will it be based on present harvests? These are very difficult questions, with no single answer, and answers will never satisfy everybody.

What about catch-per-unit-of-effort (CPUE)? If management succeeds in holding CPUE at a variable level, what balance should prevail between commercial and recreational catches? If human population and demand for shellfish increases, as it certainly will, how does industry cope? If numbers of clammers are allowed to increase proportionally, then CPUE will have to decrease, and if profit is to remain in clamming, profit will have to increase also, and the consumers will have to pay more for their clams.

Alternate opening and closing of harvest grounds has been discussed already, and probably needs no further discussion.

Enforcement of laws, and penalties for violations, will have to be made much more severe. Enforcement will require more boats and men, and possibly more court appointments also. Penalties for violations will need to be greater, to pay the increased costs alone, but further increases must be written into the law to make it a serious question whether violations are worth the penalty. Public and private hard clam aquaculture activities need to be examined closely to determine whether they increase the possibility of violation of laws, and whether it is worth the risk.

Marine water quality and sewage treatment plant monitoring need to be increased, to improve the condition of clam bottoms, and generally improve the quality of the product. And improved monitoring of bottoms, not only to catch violators, but also to get better understanding of the population dynamics of the hard clam. Feasibility studies are needed to detect trends in quality of marine waters, and the levels and sources of pollutants.

And, finally, the possibility of establishing more decontamination facilities, for taking clams from closed areas, treating them in clean waters for a sufficient time to make them fit for human consumption, and then harvesting them. The costs of this method also need to be closely watched to be sure that there is profit in the method, and not merely a subsidy from public funds to clammers.

13.3 TREATMENT BY STATES

There follows a discussion state by state, to see how these subjects have been treated in the past, how successful they have been, and whether much improvement has resulted.

13.3.1 Maine–New Hampshire

In Maine the center of clam digging is Brunswick and Casco Bay, near the southeast corner of the state. Hard clam production has never been great. It fell from a peak of about 590 thousand pounds of meats in 1949 irregularly to zero from 1988 on, and was very irregular from 1931 to 1987 with 1000 pounds or less before 1931. It is near the lower limit of its temperature range, varies widely in abundance, and good sets are obtained only once in seven to ten years. North of Cape Cod it is rare and local, and is found in small, sheltered bays where the water is shallow and warm in summer. Commercial sets are used locally, and transplanting and other devices to improve production are of local interest.

13.3.2 Massachusetts

Although the hard clam occurs in commercial abundance in certain areas north of Cape Cod, its chief region of abundance extends from Cape Cod southward. Since the turn of the century annual production of hard clam meats has ranged from one and two million pounds of meats. But production figures for 1951 to 1992 suggest an unmistakable downward trend. Hard clams support an active recreational and commercial fishery. In 1990 the recreational harvest exceeded 36,000 bushels, with an estimated market value of about $2 million. In some years recreational landings may be more than one-half the commercial catch. The number of family permits issued each year corresponds with the growing resident year-round population, which has doubled, as has the number of recreational permits.

Most hard clams are fished in shallow waters, with hand rakes, bull rakes, or tongs. In deeper water offshore they are harvested by hydraulic dredges from large fishing vessels.

The contribution of "cultured" hard clams has increased in recent years, and now amounts to over 20% of the total value. Most farming activity takes place in the Towns of Wellfleet and Provincetown, which have extensive tidal flats and waters favorable for growth. Seed clams, usually less than one-half inch in size, are purchased by growers from private hatcheries. These are spread on prepared beds, protected from predators by a cover of plastic mesh screening, and harvested when of marketable size.

Hard clams in Massachusetts have been declining for about 40 years. Closure of areas during the 1970s was relatively minor. By 1980, approximately 20,000 acres were closed, and by 1990, 39,000 acres, nearly doubled the 1980 figure. The total area for the entire State of Massachusetts designated as uncertified on the basis of bacterial analysis reached a record high in 1992 of about 115,000 acres. Only 54% of estuarine waters were classified as approved in 1990, as compared with 92% in 1985. It is therefore not surprising that hard clam landings have declined.

One reason for the sharp increase in closures was because under the National Shellfish Sanitation Program (NSSP) of 1988, all shellfish-growing areas had to be examined for possible sources of pollution. Known producing areas were monitored first. Known areas not surveyed by the 1989 deadline were placed under a management closure system to comply with NSSP.

Development of the Massachusetts coastline has been proceeding at a rapid rate. The impact of pollution on the shell fisheries is exemplified in areas such as Cape Cod, which historically has been Massachusetts leading hard clam producing region. Since 1970, summer and year-round populations of the area have nearly doubled. In 1970 only 421 acres of hard clam grounds were uncertified. By 1990 this figure had increased to 6000 acres.

Massachusetts has conducted a hard clam relaying in the past. This has involved transfer of hard clams from polluted waters of Mount Hope Bay, Taunton River, and New Bedford Harbor to certified areas. But this program has nearly terminated because it has been difficult to supervise the program under existing budget limitations.

Massachusetts' hard clam industry has been increasingly dependent on privately owned shellfish farms. Culture of hard clams on private grounds is regulated at the local level under Chapter 130 (Section 57-68b) of the Massachusetts General Laws, and private leases are issued only by the Town Board of Selectmen for that particular town.

As wild stocks continue to decline, the importance of shellfish culture can only increase, particularly as the product has a stronger claim to having been grown in clean water and should command a higher price. Some towns, however, are strongly opposed to the concept of private control. Most leases are small, ranging in size from one to five acres, but they are intensely managed and highly productive. Only 1 to 2% of Massachusetts' open shellfishing grounds are currently under lease, but about 20% of hard clams in Massachusetts are cultured on private grounds.

Prime responsibility for supervision rests with the Division of Marine Fisheries in Massachusetts. The Division issues permits, oversees leasing of beds, and exercises control over uncertified areas. It is responsible for monitoring water quality and for classifying areas in accordance with NSSP, and in cooperation with the Massachusetts Department of Public Health and the Federal Food and Drug Administration. One of its major responsibilities is to assist coastal communities to develop and implement shellfish conservation programs.

Local control is traditional in Massachusetts, and the towns have jurisdiction over their hard clam resources in certified waters. Most towns have shellfish wardens or conservation officers who implement conservation measures and enforce town or state fishing regulations. They are assisted by members of the Massachusetts Division of Environmental Law Enforcement.

13.3.3 Rhode Island

Rhode Island's coastline lacks the expansive tidal flats and abundance of shallow coves and harbors that are so favorable for hard clams. Collapse of the oyster industry by the 1940s coincided with the growth of Rhode Island's hard clam industry, which is one of the largest in the nation. Fishermen quickly found beds in Narragansett Bay and Rhode Island's salt ponds to be productive. The hard clam industry has never depended on out-of-state seed stock, and this species can thrive in areas inhospitable to oysters. The fishery, which extends into the deeper, cleaner water of Narragansett Bay, is still a major fishery, although threatened by deterioration of water quality and destruction of natural habitat. In 1990 over 99% of the value of commercial shellfish landings was hard clam.

A distinguishing feature of Rhode Island's shellfisheries dates to the charter given in 1683 by Charles II. This gave the public right to fish and shellfish resources, considered public property "free and common to all". The public must benefit from conservation of the resources. Since the disappearance of the oyster industry, private control of shellfisheries has been strongly, and largely successfully, resisted.

Hard clam production rose steadily during the 1930s and 1940s, peaked at over 5 million pounds during the 1950s, declined to less than 1 million pounds in the 1960s and 1970s, rose again to about 4.5 million pounds, but by the 1990s it had declined to about 2.5 million pounds. The drop in the early 1970s was caused primarily by an increase in the level of pollution in upper Narragansett Bay, which closed important hard clam grounds. Over exploitation and poor recruitment also were contributing factors. The sharp decline also coincided with prohibition of the use of dredge boats in areas worked by hand tongs and rakes. The second rise in production was attributed to introduction of the bull rake into the fishery. This enabled fishermen to harvest hard clams down to 26 ft, and nearly doubled the amount of workable bottom available. But production has fallen since the mid-1980s.

The hard clam fishery in Rhode Island supports about 800 full-time fishermen, and about three times that number take hard clams for a supplementary income. An estimated 50,000 recreational fishermen take hard clams as well. Hard clam harvesting by mechanical dredge is prohibited in Rhode Island inshore waters, although mechanical dredges are allowed by special license for other molluscs.

The upper portions of Narragansett Bay are the most heavily fished. Providence River and the Rhode Island sector of Mount Hope Bay are uncertified because they are polluted. The upper bay immediately south of the river is conditionally opened to shell-fishing depending on rainfall. This region is open to fishing on an average of 40 to 50% of the time, and in some years considerably less, but the harvest averages 70% of the average catch.

Greenwich Bay is also important for hard clams, but many cannot be harvested because the areas are uncertified. When over-fishing of hard clams resulted in serious depletion in the early 1980s, clams from closed areas were relayed in a joint effort by clammers and the State Division of Fish and Wildlife to restore the fishery. The bay now functions as a hard clam management area, stocked annually with transplants. Harvesting is done in winter with reduced daily catch limits. Clams are taken from Greenwich Cove to Greenwich Bay in May by clammers, under state supervision. They are not harvested until six months later. Roughly 2500 bushels have been moved each year since the program started. Clammers are paid 10¢ per pound. This provides a winter ground which can be fished when winter in other places is too bad.

The waters of lower Narragansett Bay are less productive, but are clean and open to harvesting all year round.

13.3.4 Connecticut

Today's Connecticut shellfish industry is based for the most part on hard clams and oysters. Since the early 1970s both these fisheries have undergone a remarkable upswing, largely as a result of effective management of the resource, and coinciding with improved water quality in Long Island Sound. In 1992 the harvest of hard clams was about 1.4 million pounds of meats, by far the highest since 1951.

The production of hard clams in Connecticut was of relatively little importance until the late 1950s. As oyster populations declined, commercial clammers began to harvest hard clams more intensively. Prior to 1951 almost all hard clams were harvested from public bottom with hand tongs or rakes. Introduction of the dredge allowed harvesters to fish more efficiently, and to work in deeper, cleaner water. By 1958 hard clam production in Connecticut had surpassed oyster production in pounds of meats. Hard clam production began to increase rapidly in the 1980s with development of the hydraulic dredge. Most landings are harvested by this method in Long Island Sound. The species relays well. Private lease holders transplant clams from contaminated beds inshore to private offshore grounds for purification, under state control. But a major drawback to increased production of marketable hard clams is the limited amount of productive bottom located in unpolluted water.

The remarkable improvement in the status of commercial hard clams can be attributed to cooperative efforts between the industry, the State Department of Agriculture's Aquaculture

Division, and the State General Assembly. The combined efforts on the part of the state and private growers to restore the industry were assisted by passage and implementation of the Clean Water Act of 1972. There seems little question that the Act has resulted in a marked reduction in the amount of untreated sewage and industrial wastes entering Connecticut's estuaries and Long Island Sound.

The Division of Aquaculture, established in 1972 under the State Department of Agriculture, maintains jurisdiction over the state's shellfisheries in waters outside municipal jurisdiction. This includes coordination of the state leasing program, assisting municipalities in developing shellfish management plans, conducting shellfish assessment studies for the industry, administering seed oyster restoration programs, and administering the NSSP. Shellfisheries in waters of West Haven, New Haven, Milford, and Westport, are managed by the Division of Agriculture. But other towns have jurisdiction over their own shellfish resources under CGS 26-257a, and may regulate their shellfisheries as they see fit. Management by the towns is through appointed shellfish commissions empowered to regulate catch limits, seasons, minimum sizes and methods of harvest. Some towns have developed management plans through which, with certain stipulations, commercial fishing in closed areas is allowed as part of a relaying program.

13.3.5 New York

In the early days, New York shellfisheries were also dominated by the oyster. As the oyster industry declined hard clams became of major importance. Most hard clams harvested in New York were taken in Great South Bay, on the south shore of Long Island. The other major harvesting areas are Huntington Bay, Moriches and Shinnecock bays, Peconic and Gardiners bays, the north shore bays of Brookhaven and Smithtown, and the far western bays of Little Neck, Raritan, and Westchester. Landings remained relatively insignificant until the 1930s, following the decline of the oyster and an increase in the salinity of Great South Bay. Landings increased sharply during World War II when meat was rationed but fishery products were not.

After the war, as meat products became more available, hard clam landings decreased, but increased again from the early 1950s to the mid-1970s.

In 1976 a record harvest of about 750,000 bushels accounted for over 58% of total landings of hard clams in the United States. Landings again declined in the late 1970s, caused primarily by over-fishing. Between 1976 and 1986, the acreage closed to shellfishing in Great South Bay nearly doubled, from 3870 to 7626 acres. From then on, production has remained fairly stable, but only at about 25% of production of the late 1970s.

The hard clam remains New York's most valuable fishery. It provides the greatest landed value and number of jobs of any marine fishery in the state. The dockside value of hard clam landings was over $13 million in 1992. It is estimated by the New York State Department of Environmental Conservation that the economic return to local communities falls between $45 and $72 million, and nearly 3000 permits to harvest were issued in 1990. More than 10% of the hard clam harvest was derived from a transplanting program run by the department, taking over one-half million bushels from uncertified areas in western Long Island Sound.

From 1966 to 1986 an average of 87% of hard clams harvested in New York came from Great South Bay. By 1986, however, Great South Bay accounted for only 57% of total New

York landings. Between 1970 and 1986 the number of acres closed or restricted to fishing because bacterial counts were high, tripled from 2495 to 7626. Thus, some of the decline in production was caused by degradation in water quality, but over-fishing also contributed importantly to the decline in production.

Most hard clams fished commercially inshore are taken by hand rake or bull rake. In shallow water some recreational fishermen may use their bare feet.

To qualify as an "approved" source of hard clams, New York, like the New England state, must comply with NSSP, which makes each state responsible for classification of hard clam growing areas, and issuance of permits for interstate shipment. In New York, the agency with legal authority to do this is the Department of Environmental Conservation. The Federal Food and Drug Administration periodically reviews each state's program for conformity with guidelines specified by NSSP. Recent changes in New York State law have increased the penalty for illegal harvesting. There still is a need for greater numbers of enforcement officers at all levels.

Towns in New York State have important responsibility for hard clam management. These towns have hard clam programs of various magnitudes, especially Islip and Brookhaven on Great South Bay. Islip, for example, operates a fairly large hatchery, which at present raises about 40 million young clams per year. After rearing hard clams to a reasonable size in the hatchery, it then transplants them to a protected area until they are large enough to harvest, when they are transplanted again to certain bottoms. The town also has divided its hard clam bottoms into a reasonable number, only one of which is opened at a time. It also monitors these areas frequently to watch for setting, growth, recruitment, and so on, so that it knows when to open and close each.

13.3.6 New Jersey

New Jersey's landings of hard clams reached a peak of over 5 million pounds of meats in 1950. After that they fell irregularly to about 1.2 million pounds in 1992. Any development that would result in destruction of hard clam grounds is prohibited, unless the development is of national interest and no prudent and feasible alternative sites exist. Any development, which would result in contaminating or condemnation of hard clam beds, is prohibited. Development, which would significantly alter water quality, salinity, substrate characteristics, natural water circulation pattern, or natural functioning of hard clam beds during construction, or operation of the development, is prohibited. New dredging within hard clam beds is prohibited. Maintenance dredging of existing navigation channels is conditionally acceptable if state-managed hard clam recovery programs are encouraged prior to dredging.

Hard clams are harvested by commercial and recreational fishermen. The commercial harvest is estimated to support employment of 1500 persons and recreational clammers purchased 13,179 licenses in 1988. It is estimated also that there are about 10,000 senior citizen recreational clammers.

All hard clam growing waters of Monmouth County are closed to harvest of hard clams for direct marketing. The waters of Sandy Hook and Raritan bays, the Navesink and Shrewsbury rivers, are extremely productive hard clam growing areas: consequently illegal harvest of hard clams is a chronic problem. Ocean County is also an area for relay problems. Other counties have illegal harvest problems too, mostly because the numbers of patrol officers are too few.

A hard clam inventory was carried out from 1985 to 1989. Hard clam densities ranged from zero to 2.82 clams per square foot. Estimates of the hard clam resource in the Shark, Manasquan, and Metedeconk rivers were 6.9, 11.6, and 1.2 million clams, respectively. Initial analysis of the size–frequency distributions for hard clams showed that approximately seven to ten year-classes were represented in the Manasquan River, and approximately nine in the Shark River.

Size–frequency distributions for hard clams in the Metedeconk River were not plotted because the numbers per station were too low. Recruitment rates at the moderate to high density stations in the Manasquan River ranged from 3.0 to 31.1%, with a mean of 21.7%. In the Shark River recruitment rates ranged from zero to 11.6% with an average of 4.8%. The recruitment rate of 11.9% in the Manasquan River is the highest observed since the inventory began in 1983. All the areas sampled, except the Metedeconk River, have been utilized as harvest sites for relay and/or depuration programs.

Since the relay program began in 1970, nearly 66 million clams have been utilized from condemned waters. The relay program was expanded in 1983 to include areas of the Navesink and Shrewsbury rivers, and Sandy Hook Bay. This expansion has provided the first harvest of hard clams from these waters since they were closed to clamming in 1961. The total state hard clam landings attributable to relays amounted to about 13 million pounds of meats from 1980 to 1984. This amounted to about 0.5 million pounds to 4.3 million pounds of meats depending on the source of the relays. This came to 1.9% to 20.3% of the total state harvest.

Dr. Bonnie J. McCay, an anthropologist and Associate Professor in the Department of Human Ecology and Anthropology at Rutgers University in New Jersey, and Bill Jenks, a clammer and bayman, began a New Jersey "spawner sanctuary" program in 1985. McCay stated, however, that it is doubtful whether any more than 200 or so people steadily clam for a living any more. Commercial landings of hard clam have decreased steadily during the past 35 years. Hard clams are now difficult to find in the state's bays and tidal rivers, and so are clammers.

Clam spawner transplants originated in the bays of Long Island, New York, in the early 1960s, to increase the length of time that clam larvae were present in the bay. The "spawner sanctuary" is a refinement of this strategy, developed in the early 1980s by the State University of New York at Stony Brook, and implemented by two Long Island townships in Great South Bay. An area is located, set aside, and stocked with fecund, low market value, large adult clams. This maximizes the probability of spat in these previously selected certified areas, identified as good for clam development.

So what has New Jersey done in the past two years? It has completed two "hard clam spawner sanctuaries" and has started a new trend towards biological enhancement. The term "hard clam spawner sanctuary" is now being used daily along the entire New Jersey coast. One major feature of the project so far was the way in which people of vastly different ideas and education, scientists, state employees, baymen and others worked together, for the most part, cooperatively. Who knows? It might work! (ed note: the results of this effort were summarized in McCay, 1988).

The state's management style was to develop its own plans first, without official involvement of baymen, then present the plans to the advisory shellfish council, or implement them on the water. This generated almost ritualistic hostility and allegations of favoritism and hidden political agendas on one side or another. The state agencies now formally involve a few respected baymen in the program at an early stage, preventing this right at the beginning.

13.3.7 Delaware

Hard clam landings in Delaware were less than one tenth of a million pounds of meats until 1948, and reached a peak in 1956 at about one-sixth of a million pounds. They then fell irregularly and by 1992 had dropped to about 0.04 million pounds, a decline of more than 90%. Currently management practices include a 1.5 inch lower size limit (umbo to margin). Residents may harvest 100 clams per day, and non-residents 50 clams per day, using non-commercial harvesting gear. Commercial harvesters may use bull rakes or tongs only, and are limited to 2500 clams per day. The Division of Fish and Wildlife enforcement staff enforce these laws, and also restricted areas closed for public health reasons.

All harvesting presently occurs in two coastal bays (Indian River and Rehoboth Bays). Delaware Bay supported a significant dredge hard clam fishery during the 1960s. Oyster boats put out of work by oyster disease problems survived by clamming. Abundant surf clam stocks in the early 1970s reduced the value of large "chowder" clams, and there has been no significant harvest of hard clams from Delaware Bay in the last 20 years.

There have been three baywide clam surveys in coastal bays. In 1968, 363 stations were arranged in a random grid in both bays. The same stations were visited in 1977–1978, and sampled with a SCUBA-operated venturi sampler. In 1987 the Indian River survey was repeated using diver-operated gear. Surveys have shown modest density to be 2 to 3 clams per m^2 and a declining standing stock. A creel survey of recreational clammers has been conducted in 1968, 1979, 1985, and 1987. Effort was variable but catch-per-unit-of-effort remained relatively consistent. Recreational catch varies from 40 to 60% of total harvest.

During the 1980s the department attempted to examine the reasons for limited reproductive success, because successful year-classes were occurring only rarely. Gametogenesis and spawning were occurring normally. But a larval trap study showed much lower numbers than were observed earlier. A recruitment survey documented only one significant year-class, in 1985.

13.3.8 Maryland

The fishery in Maryland is concentrated in two areas: (1) the lower portion of Chesapeake Bay near the Town of Crisfield; and (2) the seaside bays. The fishery for hard clams was never very large, reaching a peak in 1969 of about 800,000 pounds of meats, and dropping to zero beginning in 1982.

Commercial harvest is prohibited in summer, but recreational clamming, with a one-bushel limit, is permitted in the closed commercial season. Hydraulic escalator dredges may be used in Chesapeake Bay, but most bay harvest is with patent tongs. Hydraulic escalator dredges are used in the seaside bays.

During the last harvest year slightly more than 3000 bags of hard clams were harvested. Each bag contains 105 clams, with legal size of one-inch transverse dimension. The State of Maryland imposes a 25¢ tax per bag. A harvest license is required to harvest for sale. A clammer may purchase one of three licenses: (1) a $50 license to use patent tongs; (2) a $100 license to use a hydraulic dredge; or (3) a $300 license that covers all commercial fisheries. Most clammers have the $300 all-inclusive license.

The Department of Natural Resources manages the fishery with input from industry. There is no town management system. The primary advisory body is the Tidal Fish Advisory Commission.

13.3.9 Virginia

In most clam waters of the state there is no season, and no minimum or maximum size of hard clams. The fishery began rather late, and reached a peak in 1981 of slightly over 5 million pounds of meats. There is some question whether this is the correct figure, because landings the year after and also the year before were less than 1 million pounds. The second highest landings were in 1965 at about 2.5 million pounds, and the landings rose irregularly to this level and then fell irregularly to about 1 million pounds in 1992. There has been an increase in clam licenses issued and good evidence of a decline of catch-per-unit-of-effort. A survey of clam grounds was made in the mid-1970s, and the state hopes to complete a survey in 1993–1994 to compare the results with the earlier survey.

Laws are set by the State Legislature and regulations promulgated by the Marine Resources Commission. Beginning in 1993, commercial fishing in Virginia waters will be controlled by a Delayed Entry System that controls entry into all commercial fisheries in the state. Before getting a clamming license, a fisherman must have a $150 commercial harvester's license. If that license was not purchased in January or February of 1993, the individual must wait two years to get a license. If a fisherman has the commercial harvester's license, he can get any clam license: hand or hand tongs ($10), patent tongs ($35), or dredge ($50).

Baylor oyster grounds (set aside in the late 19th century as natural grounds that were public property) were mapped by Lt. Baylor of the Coast and Geodetic Survey. Some of these grounds are not open to clamming with patent tongs or dredges. A summer relay season for polluted clams in the Hampton Roads area runs from May 1 to August 15, and in that area there is a maximum size of 2 7/8 inches above which clams must be thrown back. There are also two small management areas for hard clams in the York and Poquoson rivers in which there is a relay season from January 1 to March 31. These two areas have a 2 p.m. daily time limit.

13.3.10 North Carolina

Hard clam landings from North Carolina to Florida have been relatively small and for the most part peaked relatively late. In North Carolina the peak was in 1982 at about 1.7 million pounds of meats, and then fell off to just over 7/10 of a million pounds in 1992. Broad legal authority to manage hard clam stocks is given to the North Carolina Marine Fisheries Commission by the North Carolina General Assembly. The general purpose of the plan is to (1) summarize existing information on the biology and habitat of hard clam stocks, (2) describe past and present commercial and recreational fisheries and conflicts within them, (3) review economic and social considerations and existing rules affecting the fishery, and (4) to propose management measures with their consequences, for implementation by the division or for consideration of the commission. In North Carolina, hard clams have been harvested in the immediate vicinity of Oregon Inlet, and trawlers occasionally take a few clams when trawling for shrimp or oysters in western Pamlico Sound. They are far more abundant, however, in higher-salinity waters just inside the barrier islands; there they are harvested in commercial quantities.

Spawning occurs from spring to fall, commencing when water temperatures reach 20°C. Fecundity estimates vary widely and depend on clam size and condition. Growth rates of hard clams are highly variable and depend on water temperature, available food supply, and to some extent genetics. Annually deposited growth lines may be used with a fairly high degree of accuracy for aging hard clams. Sexual maturity is usually reached during the second year of life at a shell length of 33 mm. Legally harvestable size of 44.6 mm is reached at one and one-half years of age.

Present trends in total landings suggest that, despite increasing fishing effort, a threshold in availability of the resource has been reached. An increase in price of meats is associated with the apparent decline in landings, and encourages heavier, rather than reduced fishing effort by clammers. From 1986 to the present, a steady decline in CPUE is indicated as the mechanical harvest season progresses.

Currently there are about 327,775 acres of suitable hard clam bottoms in North Carolina, of which about 39,701 acres are closed to harvest because the waters are polluted. Carteret County's closed acreage has increased by 49% from 1980 to 1988. During the same period the human population has increased by nearly 23%. Some of the richest hard clam areas are being affected by continued deterioration, and this trend probably will continue as human population and development increase.

Hurricanes that open inlets appear to have a beneficial effect on hard clam populations in North Carolina. Hurricanes have not been shown to have negative effects on hard clams, although the closing of an inlet by a storm will probably cause hard clam stocks to decline by lowering the salinity.

Public opposition to the mechanical hard clam fishery seems to be growing. Depressions and tracks made by mechanical harvesters constitute habitat alteration, and the North Carolina Coastal Resources Commission is of the opinion that the practice should be regulated.

13.3.11 South Carolina

Landings in South Carolina have been relatively small. The best year on record was 1982, when just over half a million pounds of meats were produced. By 1992 this had been reduced to about 0.15 million.

It is unlawful to take, harvest, possess, sell, purchase or import a hard clam of less than one inch in thickness, measured as the maximum depth of the intact clam from the exterior surface of one valve of the shell to the exterior surface of the opposite valve. Any person may gather for personal use not more than one-half U.S. bushel of hard clams in any one day from state bottoms designated for public shellfishing. When bottoms are under permit by the state for shellfish cultivation, written permission for gathering shellfish must be obtained from the permit holder. Owners of riparian authority may gather hard clams for personal use in the amounts specified from bottoms adjoining their property; if written permission is obtained from the department. No person can take or harvest hard clams for commercial purposes from state-owned bottoms without an individual harvesting permit issued by the department. The department may limit the number of harvesting permits in accordance with sound fishery management practices. No person may harvest hard clams between one-half hour after official sunset and one-half before official sunrise.

No person may remove, take, or harvest any hard clams from the coastal waters and bottoms of the state from May 15 to September 15, inclusive. The commission, acting upon the advice of the department, has the authority to open or close any area of state waters or bottoms for the removal, taking, or harvesting of hard clams, for specified periods at any time during the year when biological and other conditions warrant the action. Nothing in this article may be construed to alter the authority of the Department of Health and Environmental Control to open and close hard clam grounds for public health reasons.

No person 16 years of age or older may engage in fishing for recreation in South Carolina's tidal waters without a marine recreational fishing stamp issued pursuant to this chapter. The stamp and permit must be available for inspection at all times.

13.3.12 Georgia

This state had the smallest hard clam landings of all the Atlantic coast states. The largest landings were 0.064 million pounds of meats in 1991. No person may take or possess hard clams in commercial quantities or for commercial purposes without first having obtained a master collecting permit, or without proof of purchase that such hard clams were purchased from a certified shellfish dealer. Master collecting permits shall specify that the permittee is authorized to take hard clams and shall only be issued to persons certified by the Department of Agriculture to handle hard clams, unless permission to take and possess hard clams for aquaculture purposes has been granted by the department. No person shall take or possess hard clams from unauthorized locations and during unauthorized periods of taking. The state prohibits taking hard clams except between the hours of one-half hour before sunrise and one-half hour after sunset. No person shall take any quantity of hard clams for commercial purposes from public recreational areas. Recreational quantities of hard clams in the shell shall be one bushel or less per person with no more than one bushel per boat per day. The department may permit any person to harvest hard clams recreationally with two exceptions: (1) areas designated by the commissioner; (2) private property owners may harvest recreational quantities of hard clams if they have in their possession a letter of permission from the property owner stating the dates allowed to take hard clams, and a description of the area.

Unless authorized by the department, no person shall take or possess for commercial purposes any hard clams taken from the salt waters of this state except by hand or hand-held implement. No person shall take or possess hard clams for recreational purposes using any instrument other than hand or hand-held implement.

13.3.13 Florida, East Coast

The east coast of Florida produced very few hard clams until 1984, and the peak was reached in 1985 with about 3.7 million pounds of meats. Subsequently, it declined to between about 0.3 and 0.85 million pounds by 1992.

Responsibility in Florida is now shared by the Department of Natural Resources and the Marine Fisheries Commission. The latter, appointed by the governor, has rule-making authority on gear specifications, prohibited gear, bag limits, closed areas, quality control codes, seasons, and some other details. The department enforces commission rules, regulates the hard clam fishery, and maintains quality. The department's marine patrol is the enforcement authority. Growing areas are classified as approved, conditionally approved, and prohibited, on the basis of bacteriological and sanitary surveys. Harvesting is not permitted in unclassified areas pending surveys.

Rules have been adopted to protect hard clam resources from over-harvesting and depletion. These rules established a minimum harvesting size of 7/8 inches across the hinge, protect the environment by restricting harvesting hard clams in grassbeds, protect environmentally sensitive areas by regulating mechanical harvesting gear, and ensure public health. Bottoms are leased at $5 per acre.

Recently, clam production has increased. Access to leases is not restricted except for harvesting shellfish. One method of increasing production is through relaying hard clams to depuration facilities. Individuals or companies which hold shellfish leases can apply for permits to relay hard clams. This is closely monitored, and supervised by law enforcement officers. There is no closed season for harvesting hard clams, but harvesting is allowed only during daylight.

13.3.14 Florida, West Coast

The most productive and extensive hard clam bed in the United States for many years was from Cape Romano southward to the Ten Thousand Islands along Florida's Gulf of Mexico's coast. It supported a sizable fishery from 1913 to 1947. Harvesting with mechanical conveyor type dredges began in 1905 and continued until 1947. Landings plummeted by 1950 and the resource has not recovered. The reasons for the decline are obscure, but an outbreak of red tide, and completion of the Tamiami Trail were coincident with the decline. Over-fishing also may have been a cause.

Southern hard clams are less suitable for the half-shell trade because they readily gape when held in cold storage. The peak in landings on Florida's Gulf coast was reached in 1932, when a little over 1.1 million pounds of meats were landed. There was a smaller increase from 1966 to 1973, but landings have fallen off almost to nothing since then.

13.4 CONCLUSIONS

Although there are many similarities in management by the individual states, there are also many differences. There is no doubt that landings of hard clams along the Atlantic coast have been going down for some time, and the two most probable causes have been over-fishing and a decline in water quality. The latter has forced hard clam grounds to be closed in increasing amounts. The states have taken steps of various kinds to try to deal with these problems, but in my opinion these steps have not been sufficient to halt the decline, at least as yet.

Most, if not all, of the states do not have sufficient enforcement officers to deal effectively with violators of the laws that exist, and the courts do not have harsh enough authority to penalize those who break the law. Clammers themselves, for the most part, do not believe that there is a problem, and are responsible for many violations that are never detected.

It is fairly clear that harvesting levels cannot, and should not, be restored to the levels of maximum landings reached in the past. For one thing, abundance of hard clams, like all other marine creatures, varies naturally from time to time, and *optimum* rather than *maximum* sustainable yield, must somehow be determined. For another thing, when plans are being laid, reliable and respected hard clammers must be brought in to the planning, right at the start, for their viewpoints to be listened to, and acted upon. Many programs have failed because respected clammers were not involved.

As a minimum, information needed for management of a wild fishery resource must include measures of setting, growth, recruitment, standing stock, and natural and fishing mortality. These parameters, which vary with time, must be measured periodically. For successful management, fluctuations must be noted, and due allowance must be made. Fluctuations in recruitment are especially important, for they could be caused by natural forces or by harvesting. It is possible that fishery-related variations in recruitment can be controlled; this is a principal objective of management. Variations in recruitment caused by environmental fluctuations possibly are amenable to control by man; variations in setting probably are not. Thus, the successful manager must be able to cope with a variable hard clam supply if he is to achieve optimal yields.

Management according to textbooks and model-makers appears to be relatively simple. Management against the "background noise" of natural fluctuations in abundance, complicated by social–political interference, is much more difficult. The relatively slow downward trend in production of hard clams along the Atlantic coast suggests that hard clam resources, some of them at least, are remarkably resilient, although hard clam production in the New England area since the 1950s, and in the New York Bight region since then have declined substantially. At present levels of scientific knowledge, and under present management regimes, the future of the hard clam stocks is not promising in the long run.

13.5 BASIC REQUIREMENTS FOR MANAGEMENT

In New England, as in the Middle Atlantic States, hard clam management is complicated by subdivision of authority among local communities. In some respects it is an advantage to have authority localized, but this is not conducive to uniformity of policy and objectives. Some coordination is advisable to develop compatible plans, and this need not come from higher authority. It is better that the initiatives arise at the local level.

A first requirement is that objectives be defined clearly. Conservation of the resource must have high priority, but economic goals also must be considered. For example, management action will be different depending upon whether maximum employment or maximum individual earnings is desired. A particular goal, however desirable it may seem for industry, cannot succeed without public and official support, including the baymen themselves.

Management will be most successful if it is based on demonstrated facts. Transplants of spawner clams are popular on Long Island, New York. The practice has been followed for years by managers of public and private grounds. The philosophy is that hard clams brought in from colder waters after local clams have spawned, will be stimulated by warmer waters to spawn, thereby improving chances for a successful set. Other than the demonstrated fact that hard clams can be stimulated to spawn by an appropriate increase in water temperature, I am not aware that there is any direct evidence that the cost of transplanting spawners is recovered by increased production.

Full-time hard clam diggers in Great South Bay, New York, once the major producer of hard clams along the Atlantic coast, are now convinced that the resource is being over-harvested. Studies of size and age distribution of hard clams on the public grounds tend to support this view. In fact, for the most part, the industry is subsisting on annual recruitment to the beds. It is also significant that total annual recruitment has been falling recently, although the number of hard clam permits has continued to increase. Under such conditions, the future health of the resource would be enhanced, and incomes of diggers would be improved in the long run, if somehow the numbers of clammers could be reduced. Most full-time clammers now recognize the potential benefits of limited entry, and some are asking why steps are not being taken by the towns.

Unfortunately, a sound basis for planning a limited entry program is not yet available. It is not yet possible to make a convincing case for optimum level of harvest, or for optimum numbers of clammers. The Town of Islip is closer to this stage than any other, but until a convincing plan can be presented, many clammers and the general public will oppose it. Denial of a premature proposal could delay the chances of success later. Essential ingredients of a successful limited entry program, in addition to the scientific facts required to make a convincing case, will be adequate surveillance, enforcement, and information gathering capabilities. These are not available at present. Another way of limiting effort is to place a quota on landings, and clamming must stop altogether when the quota is reached. But how is a quota arranged? Thought must be given to spreading the supply of clams over the full year so that buyers have an equal opportunity to satisfy their customers. But in some years, because recruitment has been especially good, there may be an excess, which can be taken. Is the quota then extended, or is some other means of taking the excess supply of clams found? This must be worked out.

Another essential ingredient of a successful management program is to gain public understanding of the issues and public support of management plans. It is rare to find a hard clam producing region where such understanding and support exist. The need to bring baymen into the planning from the beginning has already been mentioned. Lack of understanding by baymen can be illustrated by two flagrant present examples of disregard for the law. It is common knowledge that many baymen disregard prohibitions on harvesting hard clams in uncertified waters, closed to shellfishing for reasons of public health.

It is equally well known that many hard clams less than minimum size are harvested and sold. No enforcement program, however efficient, can entirely prevent evasion of the law. When the program lacks adequate equipment and manpower, and when the courts hand out totally inadequate penalties, as often happens in Great South Bay, infractions are far too numerous. It is common gossip around the waterfront, and there is no reason to reject it, that some people have grown wealthy from illegal hard clamming.

Particularly disturbing is the fact that some of this illegal traffic totally disregards public health standards. In fact there is some justification for this lack of regard. Baymen are aware that no epidemic of infectious hepatitis or other viral disease has ever been traced to hard clams originating in Great South Bay. Thus, they conclude that regulations are too strict and that the risk of disease has been exaggerated. They are only dimly aware of the variability of coliform counts in a particular locality, and of the variability from day to day in response to weather conditions, and the consequent need to allow a margin of safety in the interest of public health. Many of them are not that sophisticated. Where returns are great enough to justify the risk, most of them simply do not care.

Traffic in undersized clams is almost equally disturbing because it reflects disrespect for the law, although it has no public health significance per se. Better public understanding of the need for management, and of the benefits that management can bring, are needed. Fishermen in general are resistant to management, and hard clammers are no exception. Industry–government cooperation in lawmaking and enforcement is desirable. It is most likely to come about if the public, and especially baymen, are well informed.

Transplanting hard clams from closed areas to clean waters, to be held for a reasonable period and then harvested, must be understood. Or transplanting to depuration facilities to be cleansed and then harvested after a reasonable time, demonstrated by careful research. This must be done under close state supervision, to be sure that all regulations are strictly observed.

13.6 RECOMMENDATIONS

Probably the most important way to manage hard clam stocks for optimum yield is to limit the catch in some way. Limited entry is considered to be the best alternative, that is, limiting the number of boats and clammers so that the average fisherman can make a reasonable income. This can be done by quota, but unless there is also a limit of some sort in number of boats and licenses, there is likely to be an increase in the number of clammers, so that returns to individual clammers are not adequate. It is important also to conduct adequate stock assessments, so that information on spawning, growth, mortality, and recruitment can be determined at reasonable intervals. If done adequately this will provide information on future harvests, and allow managers to adjust numbers of vessels and licenses up or down.

Selective closures are also an important matter. The clamming area in each locality should be divided up according to the number of clams in each division. Not by area of bottom, but by the number of hard clams in each, so that if possible, each area should provide a fraction of a year's harvest for every vessel and number of licenses. Each year that area should be closed and another opened, so that clammers could continue to operate. The number of areas will depend on information obtained from stock assessments and the number of licenses issued would depend upon stock assessments also.

The suitability of each area should be determined for spawner sanctuaries. This may be done by dye studies, and areas that consistently receive good sets and places that consistently provide larvae for such sets, must be identified. Then the place of origin of consistently good sets should be closed, large spawners should be placed there, and the bottom should be covered with iron gratings, rocks, or other cheap materials that do not interfere with the clams but make it difficult or impossible to harvest them. This should be a good technique for improving abundance of hard clams, but if bottoms do not consistently receive good sets, nor the origin of larvae is consistent, spawner sanctuaries will not be worth establishing.

Transplanting hard clams from uncertified waters to clean waters under close cooperation with the state, held for a suitable time to cleanse themselves, also under close supervision, will increase the supply of commercial clams. This also can be used to improve recreational clamming.

Is aquaculture possible? Clams can be raised in hatcheries, but may have to be removed to closed areas under adequate supervision before they reach commercial size. If this can be done at reasonable cost, aquaculture is feasible, and is currently being conducted in many east coast states. A location that is feasible for all these conditions must be found.

Predator control may be feasible in some areas. The various methods of control, and costs versus benefits must be reviewed in great detail. In some areas it may be worth trying.

It may be worthwhile also to establish a maximum legal size as well as a minimum legal size. Hard clams are less valuable if they are big, but their spawning potential is greatest. There may be a real benefit to clammers if the larger sizes are saved. It is also worthwhile to determine if the hard clam resource is significantly limited by natural physical factors. Does salinity, temperature, and other factors, which vary considerably from time to time, have a large effect, and if so, can they be controlled?

Evaluate coastal construction practices, with an idea of mitigating their effects on water quality and hard clams. Enhance monitoring to detect trends in water quality and characteristics, and levels and sources of pollutants. Evaluate impact of improvements in sewage treatment and disposal on hard clam growing areas. Take steps to ensure that there is no further alteration in water quality and, in fact, try very hard to improve water quality and pollution.

And, last but by no means least, enhance enforcement of hard clam laws, increasing patrols, intensifying prosecution of violators, and make prosecution severe enough that it is no longer worthwhile to take a chance on breaking the law.

13.7 COSTS

The costs of these various management measures have never been fully examined. All will cost money if they are to be done adequately, and it is most important that these costs be determined. Some management measures, no matter how important they are, may cost more than the return to clammers, in which case they may be merely a subsidy to clam diggers from public funds, and therefore of questionable value. Perhaps this is the most important first step in beginning to draw up a reasonable management plan. Once costs have been determined, then it is a matter of deciding whether management is really practical, or if a partial plan, considering costs, can help clammers.

13.8 THE HUMAN POPULATION PROBLEM

Perhaps the most difficult problem of all, and one which comes back time and time again in all fishery studies, is the alarming growth of the human population. This not only brings more people into the world to compete for a limited supply of living resources, but also creates more pollution and industrial wastes. These wastes tend to reduce the supply of living resources, not only by killing them directly, but also by denying these resources to be used by man because they affect public health. This "population explosion" as it has been termed by Ehrlich and Ehrlich (1990), is being addressed in some parts of the world, although resisted in many places. But control of so-called "progress", the economic growth side of the problem, is a side that nobody even wants to hear about. These are certainly basic elements of hard clam management, but is there any possibility that they can be coped with?

13.9 ACKNOWLEDGMENTS

Most of this chapter was based on replies received to letters written to the heads of fishery agencies in each of the states from Maine to Florida. Many thanks are due to the recipients of these letters, and to the staff members who were assigned the task of putting the information together. The studies on which this chapter is based were supported by grants from the New York Sea Grant Institute. This paper is contribution number 965 of the Marine Sciences Research Center, State University of New York at Stony Brook, New York 11794-5000.

REFERENCES

Buckner, S.A., 1984. Aspects of the population dynamics of the hard clam, *Mercenaria mercenaria* L., in Great South Bay, New York. A dissertation presented to the Graduate School in partial fulfillment of the requirements for the Degree of Doctor of Philosophy in Coastal Oceanography, State University of New York at Stony Brook: xiii + 217 pp.

Ehrlich, P.R., and Ehrlich, A.H., 1990. The Population Explosion. Simon and Schuster, New York, 320 pp.

Koppelman, L.E., and Davies, D.S., 1987. Strategies and Recommendations for Revitalizing the hard clam fisheries in Suffolk County. Suffolk County Planning Department, Hauppauge, New York, vii + 58 pp., appendices I-4, II-19.

Malouf, R.E., 1989. Clam culture as a resource management tool. In: J.J. Manzi and M. Castagna (Eds.), Clam Mariculture in North America. Elsevier, Amsterdam, Ch. 18, 427–447 pp.

McCay, B.J., 1988. Muddling through the clam beds: cooperative management of New Jersey's hard clam spawner sanctuaries. J. Shellfish Res., 7: 327–340.

Schubel, J.R. (Ed.), 1985. Suffolk County's Hard Clam Industry: An Overview and an Analysis of Management Alternatives (COSMA) Program of the Marine Sciences Research Center, State University of New York, Spec. Rep. 63, Reference 85-19: xvii + numerous pages.

Biology of the Hard Clam
J.N. Kraeuter and M. Castagna (Eds.),

Chapter 14

A History of Hard Clamming

Clyde L. MacKenzie Jr., David L. Taylor and William S. Arnold

14.1 INTRODUCTION

Hard clams of the genus *Mercenaria* have been harvested in the Canadian Maritimes and in the United States in all states from Maine to Florida. Native American Indians harvested clams, known in the industry as quahogs from Rhode Island northward and hard clams from Connecticut southward, by treading (searching with the feet or hands) in wading depths (Ingersoll, 1887; Quitmyer and Jones, 1992). The Indians ate the meats and used the shells as tools and utensils. The European colonists exploited shallow-water clam populations by treading and probably using metal rakes with wooden handles. Partly due to legal constraints, the technologies used to harvest clams have since remained as simple rakes, tongs, or dredges, depending on location. As a result, the fishery for clams remains easy to enter because the rakes and boats are inexpensive or can be borrowed from other fisheries and little skill is required for harvesting. Newcomers can simply follow regular fishermen to the beds and harvest nearly as many clams as they do. The main requirements for making good catches are physical strength and determination. The daily activity of a fisherman typically consists of leaving the shore in his boat for a clam bed in the early morning when winds are light, raking clams for 5–7 h, and then returning to the shore and delivering the clams to a dealer. The fisherman is paid then or once every 1–2 weeks. The dealer sells some clams locally to restaurants and ships the bulk to large population centers, such as New York City and vicinity, for sale. The clams are eaten raw on the half-shell, in stews with milk, potatoes, and onions or with tomatoes and other vegetables, or as stuffed clams.

14.2 BACKGROUND

According to Belding (1912), hard clams have been eaten for at least as long as European colonists have occupied North America. However, commercial production did not become substantial until the late 1880s, when the demand for littlenecks arose and inland markets in the midwest developed (The Fishing Gazette, 1987). During that period, oysters on the half-shell were popular, and littlenecks filled the summer gap when oysters (eaten mainly in the "R" months) were not available. The increasing popular demand for littlenecks spurred rapid development of the clam industry, thereby providing employment for hundreds of fishermen and giving the hard clam an important value as a seafood (Belding, 1912; Tressler and Lemon, 1951).

Since the late 1880s, clams have been sorted by shell size in response to market demand. Originally, clams measuring 2–2.25 inches in shell length (SL = maximum distance across

Fig. 14.1. This fisherman's clam harvest in Rhode Island is a mixture of littlenecks, topnecks and cherrystones, 1998 (photograph by C.L. MacKenzie Jr.).

the shell) were classified as "littlenecks", those measuring 2.25–3 inches SL were classified as "cherrystones", and those measuring greater than 3 inches SL were classified as "mediums" or "chowders". In recent years, the demand for clams has increased and this has been accompanied by changes in the size categories used in the trade. Larger clams now carry the size names of the smaller ones and are sold at their prices. The size range for littlenecks now is roughly 1.5–2.5 inches SL, clams in the size range 2.6–2.9 inches SL are now classified as "topnecks", and the 3.0–3.25 inch SL clams are now sold as cherrystones. Most chowders range in size from 3.25 to 3.5 inches SL, although now many are sold as cherrystones (Fig. 14.1). Due to their low prices, chowders are commonly returned to the beds. The size ranges listed above are subject to considerable variation and interpretation dependent upon local conditions and market demand.

State and local regulatory agencies have limited clam harvesting effort by placing restrictions on the gear type and time of harvest. Some states have additional limits on the daily harvest per fisherman. Most regulations have been implemented since the early 1900s (Belding, 1912). In many areas, only hand rakes or tongs are allowed in the public fishery, which comprises the vast majority of harvest grounds in most states (excepting New York and Connecticut), although in some southern states mechanical harvesting is allowed in some public areas.

14.3 HISTORY OF HARVESTING METHODS AND GEAR

At least fourteen methods and types of gear have been used to harvest hard clams (Table 14.1). The first four methods we describe are most effective on exposed flats during low tide or at wading depths. The remaining methods are most commonly used in waters at least 1 m deep and sometimes dredges are used in waters exceeding 5 m deep.

14.3.1 Treading

Treading is the most ancient method used to harvest clams, but it was first documented in the 1870s (Ingersoll, 1887). By the late 1800s, treading had become a common clam harvesting method around Highlands, New Jersey, where women and men treaded in the warm summer months (Kobbe, 1982). The harvesters placed their feet close together and parallel, twisting them as they moved sideways to feel for the clams and gathering them with their hands. The best treading was in water less than 30 cm deep, and each harvester could collect as many as four bushels of clams in two hours (MacKenzie, 1992a). Treading continues to be effectively practiced in areas such as: Barnegat Bay, New Jersey; Chincoteague Bay, Virginia; Cove and Bogue Sounds, North Carolina; and the Indian River lagoon, Florida.

14.3.2 Hand Picking

Harvesting clams by using the hands for detection and collection is commonly practiced on Prince Edward Island, Canada. Using rubber gloves for protection, clammers get on their hands and knees in knee-deep water and sweep their hands through the firm muddy bottom, retrieving the discovered clams and placing them in a bushel box floating in a tire tube. Successful fishermen can collect up to 2.5 bushels of clams on a single low tide.

14.3.3 Short-Raking

The short rake (i.e., "scratch rake") probably was the earliest gear used to harvest clams (Fig. 14.2). Fishermen have used the short rake by wading in water up to 1.2 m deep and pulling the rake through the sediments. Short rakers can harvest two to four bushels of clams during a single low-tide fishing session.

14.3.4 "Signing"

Signing is most commonly practiced on intertidal flats within the bays and estuaries of coastal Virginia and North Carolina. Fishermen look for siphon holes and fecal pellets indicative of clams, then use a pick or rake for collection. Each harvester can collect two to four bushels of clams during a low tide.

14.3.5 Hand Tonging

Hand tongs were originally devised for harvesting oysters, and their application to oyster harvesting was first recorded from Maryland in 1701 (Witty and Johnson, 1988) and

TABLE 14.1

Types of boats and gear used in harvesting hard clams

Boat type	Length (m)	Location	Gear	Time period	Source
Rowboat [a]	3.7–4.25	Massachusetts, Raritan Bay	Short rake, tongs, bull rakes	1800s to mid-1900s	MacKenzie, 1992a,b
		P.E.I. [b]	picking	1970s to present	Jenkins et al., 1997
Sailing sharpie	6–8	Cape Cod and islands [c]	Short rake	1800s to early 1900s	MacKenzie, 1992b
Catboat	5.5–8.5	Cape Cod and islands	Basket and bull rakes	1880s to 1950s	Belding, 1912; MacKenzie, 1992b
		Raritan Bay	Sail dredge	1800s–1961	MacKenzie, 1992a
Sloop	11	Great South Bay	Tongs	1800s–1970s	[d]
	10.7–12.2	Raritan Bay	Sail dredge	1875–1961	Ingersoll, 1887; MacKenzie, 1992a
Garvey	6	Great South Bay	Tongs	1900s	[d]
		Raritan–Barnegat bays	Bull rake	1900s	Moonsammy et al., 1987
Fiberglass boat	5.5	Narragansett Bay	Bull rake	1970s–present	Boyd, 1991
Full-sized oyster boat	15–20	Connecticut	Rocking chair dredge	1946–1957	
		Ches. Bay	Hydraulic dredge	1958–present	MacKenzie, 1997a
Working boat	12	Ches. Bay, VA	Patent tongs	Early 1900s–present	MacKenzie, 1997b

[a] After mid-1940s, rowboats were propelled by outboard motors.
[b] Prince Edward Island.
[c] Islands: Martha's Vineyard and Nantucket.
[d] Fishermen interviews and personal observations.

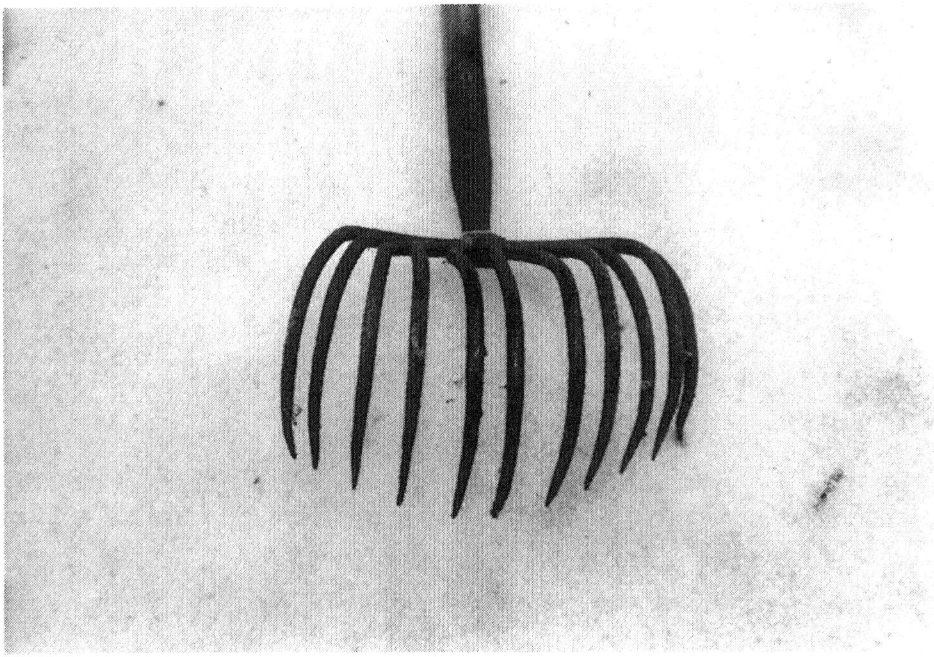

Fig. 14.2. Type of short rake ("eagle claw") used in the early 1900s (photograph by C.L. MacKenzie Jr.).

from Nova Scotia in 1721 (De Charlevoix, 1744). Clam tongs are made with longer and more steeply angled teeth than oyster tongs to penetrate more deeply into sediments. This modification requires greater effort from the fisherman. Tongs continue to be used in many localities from the Canadian Maritimes to Florida; their use is limited to depths less than 3.7 m (Fig. 14.3).

14.3.6 Patent Tongs

Patent tongs were first used for harvesting clams in Virginia, probably in the early 1900s. By the 1920s, patent tongs were operational on at least 25 vessels in Virginia waters. Early versions of patent tongs were large (approximately 1.3 m by 1.3 m) and heavy (55 kg or more), and were powered by a hand-winch requiring two men for effective operation. Tonging was most often done during slack water; one man worked the winch while the other handled the tongs and culled. An effective crew required three to five minutes to complete a "lick" (a complete drop and retrieval of the tongs) and usually collected about 100 clams per lick. During the late 1920s and early 1930s, hand-operated patent tongs were replaced by engine-powered tongs, increasing operational speed to about three licks per minute in sandy bottoms and two licks per minute in muddy bottoms where the clams have to be washed by raising and lowering the tongs in the water (Fig. 14.4). Although powered patent tongs remain large and heavy, most boats now operate with a crew of one due to the mechanical efficiency of the gear (MacKenzie, 1997b).

Fig. 14.3. Tonging clams in Great South Bay, New York, 1885 (from Frank Leslie's Illustrated Newspaper, 1885).

Fig. 14.4. Patent tongs on vessel at Gwynn Island, Virginia, 1996 (photograph by C.L. MacKenzie Jr.).

14.3.7 Bull Raking

Long-handled bull ("shinnecock") rakes originated in the Raritan Bay area of New York and New Jersey in approximately 1863 (Leonard, 1923). Bull rakes were constructed of a

Fig. 14.5. Sideview of modern bull rake, 1998 (photograph by C.L. MacKenzie Jr.).

rake head having 25–30 curved teeth spaced approximately 20 mm apart. Early rakes had a wooden handle ranging in length from 7.5 to 12 m and with a 30 cm cross-head handle at the fisherman end. Bull rakes are an effective harvesting gear in 1.8–7.5 m water depth (Belding, 1912). About 20 years ago, the bull rake was redesigned. It has been patterned after the basket rake described by Belding (1912), and an extendable aluminum handle has been used instead of a wooden handle (Figs. 14.5 and 14.6). Though they are easier to use and far more effective than hand tongs, they did not become the harvesting tool for most clammers in Great South Bay, New York, until the early 1960s (Anonymous, 1985), in Narragansett Bay, Rhode Island, until 1971 (Pratt, 1988), and in North Carolina until the mid-1970s. In the recently developed Indian River lagoon hard clam fishery, the bull rake has been and remains the gear used by most clammers.

Fig. 14.6. Bull rakes in use in Raritan Bay, New York, 1998. In foreground, "hauler" is retrieving rake containing clams, while raker guides handle; in background, raker is pulling rake, while "hauler" culls the previous catch (photograph by C.L. MacKenzie Jr.).

14.3.8 Sail Dredging

The sail dredge was developed by 1870 as a modification of the bull rake, and remained in common usage until the early 1970s (Fig. 14.7). The sail dredge was restricted in usage to Raritan Bay. It was similar in design to the bull rake but was about four teeth wider and was constructed with a stout wooden handle about 1.5 m long that allowed for towing under sail from a catboat, sloop, or small schooner. As they sailed almost sideward in the water, catboats towed two or three dredges, sloops towed four dredges, and schooners towed up to six dredges at a time. Most catboats and sloops had a crew of two, with each man hauling two dredges (one at a time), whereas schooners often had a crew of three when six dredges were used. Typical catches were 9–10 bushels of clams per boat per day (MacKenzie, 1992a).

14.3.9 Basket Rake

The basket rake (Fig. 14.8) was used on Cape Cod and the offshore island of Martha's Vineyard from at least the early 1900s (Belding, 1912) into the 1950s (MacKenzie, 1992b), mostly in 1–2 m of water. To harvest clams with a basket rake, the fisherman anchored his boat, put the rake out to the full length of its wooden handle, put the handle against his shoulder, and worked the rake through the sand toward him. This forced its teeth through the bottom dislodging the clams into the basket.

Fig. 14.7. Harvesting clams with a sail dredge, Raritan Bay, 1940s (photograph courtesy of W. Thompsen).

14.3.10 Rocking-Chair Dredge

The rocking-chair ("Fall River") dredge (Fig. 14.9) was developed for use in Narragansett Bay in 1945–1946, but its usage quickly spread to Connecticut and Raritan Bay. In Narragansett Bay, about 40 boats, 9–10.5 m long, with crews of two used the dredges during 1945–1958. The boats dredged throughout the year, and each vessel was permitted up to 30 bushels of clams per day. In Connecticut, rocking-chair dredges were installed on four boats owned by oyster companies and were in use through 1957. The Connecticut vessels harvested clams during the late summer and winter when oystering was slack. In the New Jersey portion of Raritan Bay, about 20 boats harvested clams in 6–8 m of water using the dredges. Each vessel, manned by a captain and two deckhands, harvested about 40 bushels of clams per day. Dredging operations were confined to November through February because at other times the dredging forced sediment into the actively pumping clams and reduced their marketability. Use of the dredges in Raritan Bay continued through 1961, when New

Fig. 14.8. Basket rake used in Massachusetts in first half of 1900s (from Belding, 1912).

Jersey temporarily banned all clamming in the bay because of pollution (MacKenzie, 1992a). Finally, in 1960 about twelve shrimp trawlers used rocking-chair dredges to harvest clams (*M. campechiensis*) in 9–12 m of water in the Atlantic Ocean between Cape Lookout and Beaufort Inlet, North Carolina. The fishery continued through 1962 when the clam resource became scarce (Porter and Chestnut, 1962).

14.3.11 Hydraulic Dredging

The hydraulic dredge was first used for clam harvesting in 1958 in Connecticut, where it replaced the rocking-chair dredge. The hydraulic dredge is powered by a water pump mounted on the deck of the fishing vessel. The pump delivers high-pressure water through jets mounted on a manifold in front of the dredge bar (Fig. 14.10). The water jetted into the sand bottom loosens the sediments and embedded clams; the clams then are collected as the dredge is towed along the bottom. About 30 boats currently use hydraulic dredges for hard clam harvesting in Connecticut, as clams became abundant during the 1980s and 1990s.

14.3.12 Escalator Harvester Dredge

The escalator harvester dredge was first used for harvesting hard clams in Great South Bay, New York, in the early 1950s. The clams are washed out of the bottom and onto a mesh

Fig. 14.9. Rocking chair dredge used from Massachusetts through New Jersey, 1945–1960s (photograph by C.L. MacKenzie Jr.).

belt that carries them to the boat. Workers pick the clams off the belt and sort them into baskets. Escalator harvestersescalator harvester are also used for clam harvesting in Maryland, North Carolina, and South Carolina. Now illegal in Florida, hydraulic escalator dredges were once used in the Marco Island clam fishery, which harvested *M. campechiensis*. The fishery, extant from 1905 to 1947 (Schroeder, 1924; Godcharles and Jaap, 1973), was one of the most productive hard clam fisheries ever in the United States, with peak daily landings of 1800 bushels of mostly chowder size hard clams (Godcharles and Jaap, 1973).

14.3.13 Kicking

In North Carolina, clam "kicking" is effected by washing clams out of the bottom by propeller-driven inboard engines mounted on boats measuring 6–9 m long. A heavily chained trawl net, 3.7–6 m wide, collects the clams as the boat that tows it moves slowly forward. This fishery, first developed in about 1940, is popular with the fishermen because it does not require elaborate equipment and the boats are relatively easily equipped for the operation (Guthrie and Lewis, 1982).

Kicking is restricted to water depths less than 3 m. The fishermen position their boat propellers about 30–35 cm above the bottom and use the downwardly deflected prop wash to eject the clams from the sediments. Boats with drafts up to 2 m can harvest clams in water depths up to 3 m, while boats with shallow draft and a tunnel for the propeller shaft can

Fig. 14.10. Hydraulic dredge for clams being lifted out of water with water streaming from its jets, 1998 (photograph by C.L. MacKenzie Jr.).

harvest at depths of 30–60 cm. Additional weight, in the form of barrels of water, bags of clams, or flooded stern compartments, can be added to the stern or shifted around the boat to achieve the optimum propeller angle and depth above the bottom.

The kicking gear has undergone modifications. Deflector plates have been attached to the rudder to deflect the propwash downward, and a steel cage has replaced the tailbag. The cage is a 60 × 60 cm rectangular box made of 1 cm steel rod spaced 2 cm apart and with sled runners attached to the bottom. A latched rear door can be released to dump the catch contents onto the culling tray of the vessel. Because the kicking method harvests clams faster than other methods used in North Carolina, the state limits daily harvests for kicking operations to 20 bushels per vessel as a conservation method (Guthrie and Lewis, 1982).

14.3.14 SCUBA Picking

Since the early 1970s some SCUBA divers have been commercially harvesting clams in Narragansett Bay, Rhode Island. Some use a garden rake with a handle about 20 cm long, and a bag tied around their waste to hold the catch. While lying on the bottom, they reach ahead and loosen the sediments with the rake, then set the rake down and feel for any uncovered clams which are then placed in the catch bag. Other divers fan the bottom with their hands to sweep away some sand, and then pick up the clams that have been exposed. Each dive lasts about 30 min and the divers remain in the water for a total of 3–4 h per day. Individual SCUBA clammers can often harvest more clams than individual bull rakers (Fleet, 1992).

SCUBA picking has also become a popular harvesting technique in the Indian River lagoon, Florida. The warm, shallow water coupled with soft sediments make this an ideal approach to clam harvesting there. The pickers rarely need to surface if they use a "surface-supply" air system rather than SCUBA. SCUBA picking is controversial in Florida. Surface-based clammers feel that SCUBA pickers infringe on their territory, because they cover large areas and they cannot be easily observed from the surface. Additionally, SCUBA pickers are difficult to regulate; illegal nighttime fishing is virtually impossible to observe and prevent, and even during the day it is difficult to monitor the catch because the diver may simply leave the catch on the lagoon bottom for later retrieval. Despite those concerns, SCUBA picking remains a legal and effective means of clam harvesting in Florida.

14.4 RELATION OF HARD CLAMMING TO OTHER FISHERIES

The lengthy life span of the hard clam coupled with its relatively slow growth at least in the northern states allows good sets of clams to remain harvestable for many years. Hard clam stocks therefore tend to be more stable over the medium term than do stocks of bay scallops (*Argopecten irradians*) and soft-shell clams (*Mya arenaria*). Men caught without a job, or those desiring additional income, have long been able to rely on clamming as a viable income alternative or supplement. Over the distributional range of clams in eastern North America, about 5000 fishermen currently harvest clams in the summer (MacKenzie and Burrell, 1997). In many areas, most fishermen are involved in clamming on a part-time basis, either working ashore during part of the year or shifting to other fisheries.

In the Canadian Maritime provinces, many fishermen harvest clams during the spring and summer and harvest oysters from September through December. Although some remain unemployed during winter, many trap smelts (*Osmerus mordax*) to generate additional income. In southern New England and eastern Long Island, fishermen once dug clams in the spring and summer (Figs. 14.11 and 14.12) but shifted to bay scalloping in the fall and winter. In recent years, bay scallop populations have declined in many areas, and the fishermen have shifted to whelk harvesting or leave fishing entirely during the fall and winter months.

In New York and New Jersey, many clam fishermen shifted effort to a variety of alternative pursuits. For example, in Great South Bay during the early 1900s, fishermen harvested seed oysters during spring and then sailed their sloops to the western part of the bay to harvest clams (Taylor, 1983). Even during the best oystering years, clams remained important to the economy of the fishermen and the local communities. During the winter of 1924–1925, a typhoid epidemic caused the near closure of the oyster fishery. Men laid off by the oyster companies supported themselves by harvesting clams from the bay (New York Times, 1925). In recent years, most Great South Bay fishermen harvest clams throughout the year although a few harvest the smooth whelk (*Busycotypus canaliculatus*) in the spring.

During 1946 through 1961 in Raritan Bay, New Jersey, fishermen who harvested clams with rocking-chair dredges during the cold months shifted to scup (*Stenotomus chrysops*) fishing in the summer, and fishermen who operated pound nets in the spring and early summer would shift to sail dredging for clams during late summer and then dredge for blue crabs (*Callinectes sapidus*) in the winter (MacKenzie, 1992a). McCay (1984) noted

Fig. 14.11. Clam fleet at anchor, Wellfleet, Massachusetts (from Belding, 1912).

Fig. 14.12. House of clam dealer and typical clam boat, Wellfleet, Massachusetts; the fisherman passes his clams from his boat through door at center (from Belding, 1912).

that the relatively predictable abundance of clams in Raritan Bay coupled with the minimal capital and technology required to harvest that resource provided an effective buffer against downturns in other fisheries. Clamming, like crabbing and, from about 1825 to 1925, oystering, provided a suitable year-round fishing opportunity. The ability to turn to clamming and other low-investment activities also meant that fishermen who worked for owners of pound nets and menhaden seiners had some control over the conditions of their labor because they had alternative fisheries to which they could turn if necessary.

In mid-Atlantic coastal areas from Delaware Bay to North Carolina, clamming is a part- or full-time endeavor primarily during the summer and fall. During the colder winter and spring months, many fishermen shift to oystering while others fish for blue crabs or beach seine for mullet. In North Carolina, clam "kicking" is a primarily wintertime occupation. In that state, there have been many conflicts between hand and mechanical harvesters due to disagreements over resource access. Hand harvesters want to keep mechanical harvesters out of shallow-water areas accessible to their hand rakes; those areas also tend to support healthy seagrass beds that can be destroyed by the kicking methodology. Of course, kickers want to maximize the area available to them for harvest. Such conflicts between fishermen over resource access are becoming more common as pollution closes once-productive shellfish beds and as competition for limited resources continues to increase (Taylor, 1995).

In Florida, especially in the Indian River lagoon, clams are harvested year-round and the most successful fishermen are active year-round (Fig. 14.13). Because the clam fishermen in northern states have switched to harvesting bay scallops and some clam beds there are inaccessible during the winter due to ice covers, the value of Florida clams peaks during winter. At that time, year-round clammers must share the resource with seasonal harvesters. In recent years, seasonal participation in the Indian River clam fishery has become restricted because recently enacted licensing requirements prevent a large influx of out-of-state clammers. During hard times (e.g., low clam abundance or a low price), even the most dedicated clammer must find alternative work. Before 1995, net fishing for mullet and trout was a popular alternative, but the constitutional net ban in Florida has severely restricted that activity. Instead, many clammers now shift to blue crabbing or land-based work, while others culture clams as a supplement to or a substitute for the natural clam fishery.

14.5 CHARACTERISTICS OF CLAM FISHERMEN

Hard clam fishermen would likely be tradesman or laborers if they worked on land, but they prefer to work independently and out-of-doors (Moonsammy et al., 1987; Gates, 1991). Most clammers entered the fishery when they were in their late teens or twenties because the pay was relatively good, the training minimal, and the initial investment for equipment is low. In every locality, the fishermen consist of a mixture of full-timers and part-timers. The part-timers harvest between alternative jobs or may work on their day off, and during summers students often harvest clams as their summer job. Many fishermen remain in the fishery even when clam abundances are low and their income is low because they have limited training in other vocations. The fishermen feel that the "bad times" are offset by the advantages gained during the "good times", those advantages being self-supervision, healthy outdoor work, and an excellent ratio of income to training.

Fishermen have disliked regulations imposed by government. They usually have opposed

Fig. 14.13. Some fishermen in the Indian River lagoon, Florida, stand on stilts while raking clams in waters about 2 m deep (photograph by W.S. Arnold).

regulations that would limit their fishing effort and income or alter their mode of operation. Although they remain outspoken and generally defensive of their harvest techniques, in recent years clam fishermen have become more cognizant of the need for regulation. Fishermen have different priorities than do resource managers, asserting that the objective of management should be to ensure fishing success rather than the maintenance of resource that is temporally unpredictable under any circumstances. Nevertheless, clammers realize that their vocation can provide a steady income only if the resource is conserved and maintained.

In Great South Bay, fishermen have long been perceived as an important political force and often have been extremely vocal in their opinions. Local governments in that area see the fishermen as a major constituency and many politicians feel that the fishermen do know what is best for the resource (Anonymous, 1985). Elected officials rarely further any proposals by shellfish managers and researchers that lack the popular support of the fishing community (Kassner, 1988). Fishermen remain critical of the government's failure to enhance depleted clam stocks. Unfortunately, fishermen's organizations rarely present a unified voice, making agreement on any issue nearly impossible. When proposals are made that further their interests, clammers want immediate action and are skeptical of the need for additional scientific research. Because they perceive that practical results rarely emanate from research, they are suspicious or hostile to any action that would place research before practical action (McCay, 1988).

Poaching remains a problem in the clam fishery. A tiny minority of fishermen knowingly violate laws designed to protect public health and the viability of the fishery by harvesting clams from uncertified or closed beds where the income potential is high. Poachers threaten the continued existence of the entire shellfishery.

14.6 COMMUNITY VIEW OF CLAMMERS

Local residents are aware that clamming involves much more physical labor than most other professions and it can also be an unreliable money-earner over a long term. Nonetheless, in localities where clam fishing supports a large number of fishermen, residents view the fishery positively as a major supplier of jobs and income both directly to the fishermen and indirectly to those who supply the support structure for those fishermen. They want the fishery to support as many people as possible in a stable, prosperous condition. The clam fishery has a large "multiplier effect". For example, in Rhode Island the shellfishing industry has the highest multiplier, 4.5, of any industry (Kadri, 1991). When a clammer earns money, he spends most of it in the local economy on fishing equipment, food, clothing, and real estate. The shellfish dealers sell clams to local restaurants where many tourists eat, and the profits of the dealers and restaurant owners pump more money into the regional economy.

Clam harvesting effort tends to respond inversely to changes in community unemployment rates. For example, from World War II to 1970, the correlation between the number of clam licenses issued and the unemployment rate in Rhode Island was about 0.8 (Gates, 1991). The fishery acted as a "sponge" to soak up unemployed workers and provide them with an employment opportunity. Of course, this effect is ultimately limited by the availability of clams and the size of the clam market.

14.7 EFFECTS OF HEAVY SETS OF CLAMS IN FOUR REGIONS

14.7.1 Edgartown, Massachusetts, in the 1930s

In 1930 or 1931, Katama Bay in Edgartown received a dense and widespread set of clams. By the mid-1930s, littlenecks were so abundant that the fishermen had trouble getting their short rakes into the bottom between them, and they could collect as many as 50 clams in their rakes after pulling them only about 60 cm through the bottom. Fishermen using bull rakes from catboats were also successful. About 30 fishermen harvested the clams; half were regular fishermen and half were part-timers who went out early in the morning and got their two-bushel limit of littlenecks before 8 a.m. and then proceeded to their regular jobs. Catches remained good until 1938, when a hurricane washed sand into the bay that smothered many of the clam beds (MacKenzie, 1992b).

14.7.2 Raritan Bay, New Jersey, in the 1930s

During the 1920s, clamming was depressed in Raritan Bay due to pollution. Clamming was banned in New York waters which comprise the northern half of the bay, and only about twelve men were bull raking for clams on the New Jersey side of the bay. In the 1930s that situation changed substantially due to a heavy clam set over vast areas of the bay. This set was detected by local fishermen, and markets quickly developed for this seed resource. To collect the seed, clammers inserted fine mesh screens over the teeth of their bull rakes, and each harvested several bushels of seed per day. The seed was sold for $1.00 to $1.50 per bushel to leaseholders in Barnegat and Chincoteague Bays for relay and later harvest. Between 500 and 600 clammers, mostly in row boats, were involved in the fishery on a daily basis and each earned as much as $10 per day. As the clams grew, other markets developed and many of the larger clams were sold to coal truckers who resold them in markets in their home towns in Pennsylvania (MacKenzie, 1992a).

In the mid-1930s, when the clams attained littleneck and cherrystone size, authorities in New York briefly banned the importation of Raritan Bay clams due to public health concerns. However, in 1935 the U.S. Health Service certified the beds as safe for clam harvest, and on October 15, 1935, New York City lifted the ban. A substantial decrease in a high New Jersey unemployment rate was associated with the lifting of that ban as new markets opened and fishing opportunities increased. In 1939, New York opened some beds in its waters for clam harvesting. Landings of marketable clams rose from 11,560 bushels (worth $13,029) in 1933 to 141,167 bushels (worth $164,930) in 1938 (MacKenzie, 1992a).

By the late 1930s, New Jersey bull rakers were harvesting eight to ten bushels of cherrystone and chowder clams per day. During the 1940s and 1950s, many of the New Jersey clammers migrated into New York waters for harvesting at night to avoid detection, because New York residency laws restricted harvest by out-of-state clammers. Individuals' catches were as high as 20 bushels per night, thus explaining the willingness of those clammers to risk incarceration to gain access to the New York resource.

As mentioned, a fleet of about 20 fishing boats used rocking-chair dredges to harvest deep-water clams in Raritan Bay that were not being exploited by the rakers. The operations yielded 40 bushels of clams per boat per trip from 1946 through 1961. All harvesting in the bay temporarily ended in 1961 because of public health concerns related to an outbreak of hepatitis attributed to the Raritan Bay clam industry (MacKenzie, 1992a).

14.7.3 Great South Bay, New York, in the 1960s

In the late 1950s and early 1960s, two dense sets of clams occurred in Great South Bay. Because clams grow slowly in the bay, they remained as seed and littlenecks for several years. In the late 1960s the number of clam fishermen increased to several thousand and each was harvesting up to eight bushels per day. At that time, the bay produced approximately 45% of the clams being landed in the United States. Stocks declined after the late 1970s.

14.7.4 Indian River Lagoon, Florida, in the 1980s and 1990s

Before the 1980s, clam landings from the Indian River lagoon, in Brevard County, Florida, averaged less than 12,500 bushels per year (Adams, 1988). However, beginning in 1981 dense clam sets occurred in the lagoon; survival was good and growth rate rapid (Jones et al., 1990) and a fishery rapidly developed. Although Ryther (1988) suggested that the sets resulted from above-normal rainfall during 1982 through 1984, it is more likely that several years of below normal rainfall before 1982 provided the ideal salinity for successful hard clam recruitment and growth. By 1984 landings were over 175,000 bushels valued at $4.4 million. They represented over 80% of the Florida hard clam landings (Barile, 1988) and about 10% of the U.S. total (Pratt, 1988). Though initially productive, the fishery was short-lived. The peak landings were 475,000 bushels valued at $8.1 million in 1985, but production declined thereafter because most clams died due to freshwater flowing into the clam beds from flood control canals (Barile and Rathjen, 1986).

The economic success of clamming in the lagoon attracted fishermen from many other states, and much of that immigration was solicited by local processors who were anxious to exploit the clams. They advertised in northern newspapers to attract experienced clammers. A daily earnings potential of $300 per man attracted clammers from Massachusetts, Rhode Island, New York, and North Carolina. At that time, Florida did not have a residency law (as do many northern states), and so non-resident clammers had free access to the resource. Although some clammers remained in Florida and continued clamming in the lagoon after the fishery collapsed in the late 1980s, most returned to their home states when beds there reopened (Busby, 1988). The northern fishermen left behind the knowledge and technology (mainly modern bull rakes) for continued clam harvesting in the lagoon and throughout Florida (Barile, 1988).

The influx of northern clammers placed additional burdens on the Florida Marine Patrol, the enforcement branch of the Florida Department of Environmental Protection. With a limited number of officers to patrol the extensive Indian River lagoon clam beds, and with over 1000 clammers on the water on any given day, the Marine Patrol had difficulty monitoring open-water clam operations and overseeing clam relay operations.

The advent and practical application of depuration technology contributed substantially to the expansion and continuation of the Indian River clam fishery. Depuration technology allowed for clam harvests from many unapproved shellfish waters throughout the lagoon, based upon subsequent transfer to approved shellfish waters or upland facilities for subsequent depuration. The effect of depuration was twofold, in that it allowed for continued harvests after the loss of the large clam beds in the southern lagoon due to freshwater inputs and it increased the clammers awareness of other areas of the lagoon that contained a large clam resource. The availability of depuration and the opening of additional shellfish harvesting grounds in the northern lagoon provided an opportunity for several hundred clammers to remain employed in the lagoon through the late 1980s and early 1990s.

A second major clam set occurred, this time in the northern lagoon, during 1990–1991. The set was exploited beginning in 1992 and lasted through 1996. Low salinity on the beds again caused high clam mortalities and substantially decreased clam landings. During the peak of this fishery, as many as 1200 licensed fishermen accounted for up to $8 million in dockside landings (Fig. 14.14). Experience gained during the early 1980s in the southern

Fig. 14.14. Harvesting clams with a bull rake in the Indian River lagoon, Florida, 1990s (photograph by W.S. Arnold).

lagoon minimized both clammer–clammer and clammer–regulator conflicts. In addition, a pre-existing license reduced the numbers of immigrant clammers. A new conflict arose between fishers of the natural resource and aquaculturists. This conflict was centered on space, as aquaculturists leased and occupied space previously available for open-water clamming. That conflict has been minimized by the development and application of regulations that prohibit aquaculture leases in naturally productive clam areas.

14.8 EFFECT OF SURFCLAM FISHERY

Surf clams (*Spisula solidissima*) began to be landed in large quantities from oceanic beds in the eastern United States beginning in the mid-1940s. Soup companies soon replaced chowder hard clam meats with surfclam meats in canned New England and Manhattan chowders, bringing a loss of most of the chowder clam market. Many fishermen did not retain chowder clams when they were collected with littlenecks and cherrystones due to a poor demand. In recent years, some chowder clams have been retained to be sold as cherrystones, or, as practiced in the past, their meats have been diced and made into stuffed clams.

14.9 EFFECTS OF AQUACULTURE DEVELOPMENT

In the 1970s, considerably reduced wild stocks of littlenecks and cherrystones in beds along the eastern seaboard of the United States, coupled with increased demand and prices for them and the development of clam culture technology, stimulated the development of hard clam culture. Individuals and firms from Massachusetts to Florida now are producing large quantities of littlenecks by growing hatchery-reared seed on leased grounds.

Clammers harvesting in public beds in some areas have been uneasy about the development of private shellfish culture, because their lifestyle is threatened by the private control of some beds and competition for clam markets will increase. For example, in the early 1990s the price of littlenecks fell by a few cents apiece and open-water clammers blamed the influx of cultured clams for that price depression. However, Hsiao et al. (1986) found that the disposable income of consumers was the most important factor in determining the dockside value of clams and not clam availability. Demand usually exceeds supply for hard clams during most times of the year. The allocation of bottom-land may be a more substantial long-term source of conflict between open-water clammers and aquaculturists.

14.10 CLAM MANAGEMENT ACTIONS

State agencies have tried to manage clam fisheries in a way that will protect the resource from depletion while providing optimum economic benefit to the fishery. The somewhat contradictory goals can be difficult to reconcile. In Massachusetts and Rhode Island, state agencies have conducted clam transplants from polluted areas to certified public areas to enhance stock abundances. In North Carolina, the primary management goals are to provide equal access to all user groups, protect critical habitats from destructive harvesting practices, and still maintain a viable and beneficial fishery. Management regulations include a minimum harvestable size of 2 inches SL, daily bag limits per person or vessel, and potentially a limited entry system. In Florida, a limited entry system has been established for the Indian River lagoon clam fishery. New clam licenses will not be granted until the current number of 1200 licensed clammers falls below 500. The clammers currently holding a license may keep it, but the license must be renewed every three years and renewal requires a certificate of completion of a clam education course. The requirement restricts license availability to itinerant clammers.

14.11 ACKNOWLEDGMENTS

This document benefitted greatly from information provided by Indian River clammers Peter Barile, Artie Feldman, Nick Hill, Bill Leeming, and Doug Telgen.

REFERENCES

Adams, C., 1988. No title. In: D. Busby (Ed.), An Overview of the Indian River Clamming Industry and the Indian River Lagoon. Fla. Sea Grant Extension Program Tech. Pap. 44, pp. 7–8.

Anonymous, 1985. Suffolk County's hard clam industry: an overview and an analysis of management objectives. Marine Sciences Research Center, SUNY, Stony Brook, Spec. Rep. 63, 374 pp.

Barile, D.D., 1988. History of the Indian River lagoon. In: D. Busby (Ed.), An overview of the Indian River Clamming Industry and the Indian River Lagoon. Fla. Sea Grant Extension Program Tech. Pap. 44, pp. 7–8.

Barile, D.D. and Rathjen, W., 1986. Report on the rainfall event of September and October 1985 and the impact of storm discharge on salinity and the clam population (*Mercenaria mercenaria*) of the Indian River lagoon. Marine Resources Council, Florida Institute of Technology, Melbourne, Fla., 171 pp.

Belding, D.L., 1912. The quahog fishery of Massachusetts, including the natural history of the quahog and a discussion of quahog farming. Commonw. Mass., Mar. Fish. Ser. 2, 41 pp.

Boyd, J.R., 1991. The Narragansett Bay shellfish industrie: A historical perspective and an overview of problems in the 1990s. In: M.A. Rice, M. Grady and M.L. Schwartz (Eds.), Proceedings of the first Rhode Island Shellfisheries conference, Rhode Island Sea Grant.

Busby, D.S., 1988. Overview of the industry. In: D. Busby (Ed.), An Overview of the Indian River Clamming Industry and the Indian River Lagoon. Fla. Sea Grant Extension Program. Tech. Pap. 44, pp. 1–6.

De Charlevoix, P., 1744. Journal of a Voyage to North America, 1. March of America Facsimile Series, 36, 383 pp.

Fleet, S., 1992. R.I. quahog diggers demand ban on commercial diving. Commer. Fish. News, Mar. 27A.

Gates, J.M., 1991. Clam market trends. In: M.A. Rice, M. Grady and M.L. Swartz (Eds.), Proceedings of the First Rhode Island Shellfisheries Conference. Rhode Island Sea Grant, Narragansett, pp. 93–103.

Godcharles, M.F. and Jaap, W.C., 1973. Exploratory clam survey of Florida nearshore and estuarine waters with commercial hydraulic dredging gear. Florida Department of Natural Resources, Marine Research Laboratory, Prof. Pap. Ser. 21, 77 pp.

Guthrie, J.F., Lewis, C.W., 1982. The clam-kicking fishery of North Carolina, Mar. Fish. Rev., 44: 16–21.

Hsiao, Y., Johnson, T. and Easley, J.E., 1986. An economic analysis of a potential overfishing problem: the North Carolina hard clam fishery. UNC Sea Grant Publ. UNC-SG-86-11.

Ingersoll, E., 1887. The oyster, scallop, calm, mussel, and abalone industries. In: G.B. Goode (Ed.), The Fisheries and Fishery Industries of the United States, II. U.S. Government Printing Office, Washington, DC.

Jenkins, J.A., Morrison, A. and MacKenzie, Jr., C.L., 1997. The molluscan fisherien of the Canadian Maritimes. In: C.L. MacKenzie, Jr., V.G. Burrell, A. Rosenfield and W.L. Hobart (Eds.), The History, Present Condition, and Future of the Molluscan fisheries of North and Central America and Europe. Volume 1, Atlantic and Guld coasts. U.S. Dept. Commer., NOAA Tech. Rep., 127.

Jones, D.S., Quitmyer, I.R., Arnold, W.S., Marelli, D.C., 1990. Annual shell banding, age, and growth rate of hard clams (*Mercenaria* spp.) from Florida, J. Shellfish Res., 9: 215–225.

Kadri, J., 1991. A raw deal: combined sewer outflow pollution in Narragansett Bay. In: M.A. Rice, M. Gradt and M.L. Swartz (Eds.), Proceedings of the First Rhode Island Shellfisheries Conference. Rhode Island Sea Grant, Narragansett.

Kassner, J., 1988. The consequence of baymen: the hard clam (*Mercenaria nercenaria*) management situation in Great South Bay, New York, J. Shellfish Res., 7: 289–293.

Kobbe, G., 1982. New Jersey Coast and Pines. Reprint of 1889 edition, Walker News, P.O. Box 352, New York.

Leonard, T.H., 1923. From Indian Trail to Electric Rail. Atl. Highlands (N.J.) Historical Society, 665 pp.

MacKenzie Jr., C.L., 1992a. The Fisheries of Raritan Bay. Rutgers University Press, New Brunswick, NJ, 304 pp.

MacKenzie Jr., C.L., 1992b. Shellfisheries on Martha's Vineyard. The Dukes County Intelligencer [Dukes County Hist. Soc., Mass.] 34(1): 1–34.

MacKenzie Jr., C.L., 1997a. The U.S. molluscan fisheries from Massachusetts Bay through Raritan Bay. In: C.L. MacKenzie Jr., V.G. Burrell Jr., A. Rosenfield and W.L. Hobart (Eds.), The History, Present Condition, and Future of the Mollusk Fisheries of North and Central America and Europe. NOAA Tech. Rep. NMFS.

MacKenzie Jr., C.L., 1997b. The molluscan fisheries of Chesapeake Bay. In: C.L. MacKenzie Jr., V.G. Burrell Jr., A. Rosenfield and W.L. Hobart (Eds.), The History, Present Condition, and Future of the Mollusk Fisheries of North and Central America and Europe. NOAA Tech. Rept. NMFS.

MacKenzie Jr., C.L. and Burrell, Jr., V.G., 1997. Trends and status of molluscan fisheries in North and Central America and Europe — a synopsis. In: C.L. MacKenzie Jr., V.G. Burrell Jr., A. Rosenfield and W.L. Hobart (Eds.), The History, Present Condition, and Future of the Mollusk Fisheries of North and Central America and Europe. NOAA Tech. Rep. NMFS.

McCay, B.J., 1984. The pirates of Piscary: ethnohistory of illegal fishing in New Jersey, Ethnohistory, 31: 17–37.

McCay, B.J., 1988. Muddling through the clam beds: cooperative management of New Jersey's hard clam spawner sanctuaries, J. Shellfish Res., 7: 327–340.

Moonsammy, R.S., Cohen, D.S. and Hufford, M.T., 1987. Living with the landscape: folklife in the environmental subregions of the Pinelands. In: R.Z. Moonsammy, D.S. Cohen and L.E. Williams (Eds.), Pinelands Folklife. Rutgers University Press.

Porter, H.J. and Chestnut, A.F., 1962. The offshore clam fishery of North Carolina, Proc. Natl. Shellfish. Assoc., 51: 67–76.

Pratt, S.D., 1988. Status of the hard quahog fishery in Narragansett Bay. Final Rep. NBP 88-07 for Narragansett Bay Project. Grad. School Oceanogr. Univ. Rhode Island, Narragansett, 89 pp.

Quitmyer, I.R. and Jones, D.S., 1992. Calendars of the coast: seasonal growth increment patterns in shells of modern and archaeological southern quahogs, *Mercenaria campechiensis*, from Charlotte Harbor, Florida. In: W.H. Marquardt (Ed.), Culture and Environment in the Domain of the Calusa. Monograph 1, Institute of Archaeology and Paleoenvironmental Studies, Univ. Florida, Gainsville.

Ryther, J.H., 1988. Untitled. In: D. Busby (Ed.), An Overview of the Indian River Clamming Industry and the Indian River lagoon. Fla. Sea Grant Extension Program Tech. Pap. 44, p. 14.

Schroeder, W.C., 1924. Fisheries of Key West and the clam industry of southern Florida. U.S. Dep. of Commerce, Bureau of Fisheries Document 962, 74 pp.

Taylor, L.J., 1983. Dutchmen on the Bay. University of Pennsylvania Press, Philadelphia, 206 pp.

Taylor, D.L., 1995. North Carolina fishery management plan hard clam. Unpubl. Rep., N.C. Dep. of Environment, Health, and Natural Resources, Div. of Marine Fisheries, Morehead City, NC, 38 pp.

Tressler, D.K. and Lemon, J.M., 1951. The clam industry of the United States. In: Marine Products of Commerce. Reinhold, New York.

Witty, A. and Johnson, P.J., 1988. An introduction to the catalog of artifacts. In: P.J. Johnson (Ed.), Working the Water, the Commercial Fisheries of Maryland's Patuxent River. The University Press of Virginia, Charlottesville.

Biology of the Hard Clam
J.N. Kraeuter and M. Castagna (Eds.),

Chapter 15

Aquaculture of the Hard Clam, Mercenaria mercenaria

Michael Castagna

15.1 HISTORY

Aquaculture of mollusks has been practiced for thousands of years in various parts of the world (Iverson, 1968; Milne, 1972) but the culture of clams is relatively new, with *Mercenaria mercenaria* being a prime candidate species. In precolonial times wild clams were harvested by native Americans and moved to more accessible areas for use in winter months. Midden heaps give ample testimony to clam utilization and feasts at social gatherings. Oyster culture methods practiced in Europe, some of which were applicable to clams, were probably brought to the new world with the early settlers. Crude forms of clam culture, such as small stockades of trimmed saplings or tree branches placed around relayed clams to protect them from larger predators, were practiced.

The start of modern day molluscan culture in the United States dates to 1879 when William K. Brooks, from Johns Hopkins University in Maryland, was successful in obtaining reproductive products from oyster gonads, fertilizing the ova and rearing the larvae to metamorphosis (Brooks, 1879, 1880, 1891). There are no published records that indicate any further success by other investigators of that period in rearing mollusk larvae.

The first successful *Mercenaria mercenaria* rearing experiments were carried out by William Firth Wells (1924, 1925, 1926), reported in the Annual Reports of the Commission, New York State Conservation Commission, and were republished by the State of New York Conservation Department, Division of Marine and Coastal Resources (Wells, 1969). Wells was not only successful in growing *M. mercenaria*, but between 1920 and 1926 he also cultured the eastern oyster *Crassostrea virginica*, the mussel *Mytilus edulis*, soft clams *Mya arenaria*, surf clams *Spisula solidissima* and bay scallops *Argopecten irradians*. This serendipitous success came about while Wells was experimenting with a De Laval milk clarifier, a recent invention of that time used to separate cream from milk. The clarifier is essentially a continuous feed centrifuge. Seawater was clarified and a sample of the denser separated portion was examined. Wells found microscopic clam-like animals, undoubtedly veligers, in the resuspended sample and decided these were probably oyster larvae. Attempts to grow these larvae in containers of natural seawater failed. Wells then changed his tactics and used the milk clarifier to remove particulates from raw seawater, obtained gametes by the method described by Brooks and placed them in clarified seawater. Each day the clarifier was used to clarify a fresh volume of natural bay water and then to centrifuge the previous day's cultured water to concentrate the larvae. The larvae were then resuspended in the new volume of clarified water. Using this technique, Wells was able to grow all five species to metamorphosis. Most of the species were later grown to market size. Wells carried out his experiments in one of the Bluepoint Oyster Company's shucking houses in West Sayville, New York, then later he moved this operation to the southern bridge tender's house on the bridge between Oyster Bay and Bayville, New York (Manzi and Castagna, 1989a). Wells patented this method (Wells, 1933). Despite Wells' successful experiments there are no published records to indicate that anyone continued this line of study. Adequate stocks of wild populations of clams were still available in most areas so there was no great impetus for hatchery culture.

Victor Loosanoff and his associates at the Bureau of Commercial Fisheries Laboratory of the U.S. Fish and Wildlife Service in Milford Connecticut, USA (now the National Marine Fisheries Service) renewed interest in bivalve culture in the early 1950s. Some of the techniques developed at the Milford lab were: (1) methods for conditioning mollusks for spawning; (2) holding broodstock for delayed spawning and to prevent spontaneous spawning; (3) use of thermal shock to induce spawning; (4) the use and culture of unicellular algae as food for larvae; and (5) the identification of some common diseases and methods for controlling them. These as well as a great number of other contributions are now commonly used wherever mollusk culture is carried out (Loosanoff and Davis, 1963). Perhaps the most significant contribution was the cadre of personnel that were trained by this program (Manzi and Castagna, 1989a).

A number of small commercial hatcheries were started in the late 1950s based on the technology developed by the Milford Laboratory. Previously salt water hatcheries used lead pipes and lead-lined wooden tanks or ceramic containers. The recent availability of plastic or plastic resin (fiberglass) revolutionized seawater handling and containment and greatly reduced the startup costs. The first commercial hatchery was started by Richard L. Kelly in Atlantic, Virginia, USA, in 1956. Mr. Kelly was an oyster planter and packer, a farmer, poultry processor, shell button manufacturer and entrepreneur. Having heard about the Milford Laboratory's success in rearing quahogs, Kelly established a clam hatchery in his oyster shucking building. He also constructed a new building that contained a hatchery and nursery. He attempted to follow the Milford Laboratory's technology including the culture of unicellular algae *Pavlova (Monochrysis) lutheri*, *Isochrysis galbana* and *Phaeodactylum tricornutum* (Manzi and Castagna, 1989a). The algal culture was later modified by replacing the unicellular cultures with mixed wild algae culture. Natural seawater was filtered and stored in a tank. Inorganic fertilizer was added and the fertilized algae were allowed to bloom, similar to the method described by Loosanoff and Engle (1942). Once the clams were set (metamorphosed) they were held in a series of painted wooden trays containing beach sand. Seawater was continuously pumped to the trays with no supplemental food. The production from this hatchery–nursery was sporadic but reasonably successful. When the clams were about 3–4 mm in height, they were planted in the sand–mud bottom of Watts Bay, Accomack County, Virginia. Early field plantings were complete failures, probably due to predation. The nursery was expanded to include two 3 $\times$ 5 m shallow ponds with dirt bottoms and concrete sides, under a shade roof. Due to the low pumping capacity to these ponds, they were only marginally successful. He finally reverted to field planting using both old and new technology. A stockade, constructed of small sapling trees shoved into the bay bottom, was erected and seed clams three to five mm in height were placed in this enclosure. Fish carcasses injected with a concentrated farm pesticide (Lindane) were periodically thrown into the enclosure to attract and poison predacious crabs. This system, although effective, would be unacceptable under present regulations. Within a few years and before a significant amount of clams were produced, Richard Kelly died: The hatchery was sold and converted to a shucking and packing plant (Manzi and Castagna, 1989a).

Another pioneer in clam culture was Mr. Joseph Glancy, who started a hatchery in West Sayville, New York on the shore of Great South Bay in 1959. Glancy, an industrial biology student at Massachusetts Institute of Technology, had spent his summers working with William Firth Wells. Glancy's hatchery was an old carriage shed equipped with crockery spawning jars, wooden troughs and tables acquired from the Cold Springs Fish Hatchery, Cold Springs, New York. He also acquired a secondhand monocular compound light microscope to complete a low cost mollusk hatchery (Manzi and Castagna, 1989a). After completing preliminary attempts of growing *Mercenaria* larvae using the methods developed at the Milford Laboratory, he decided those methods were too complicated and costly (personal communication with Joseph Glancy, 1961). He then purchased a used milk clarifier and reverted to the methods patented by Wells, with which he was very familiar. This relieved him of the cost of unicellular algal production. It should be remembered that algal culture in those days required heat sterilization, either boiling, autoclaving or pasteurization. This restricted the size of the culture container to about 20 L so it would fit in an autoclave and greatly increased the cost of algal production. Glancy made many important discoveries that he shared with other culturists. The most important part of his method was to clarify seawater removing most of the unwanted zooplankters, silt, large diatoms, etc., and allow the remaining phytoplankton to bloom in sunlight-exposed tanks (Castagna and Kraeuter, 1981; Manzi and Castagna, 1989a). Glancy serendipitously discovered this innovation when he added a temporary expansion to his carriage house to make additional room for clam production. The addition was constructed of a wooden frame covered with clear polyethylene plastic (personal communication with Joseph Glancy, 1959). Although successful in producing seed clams, the seed was usually broadcast over leased planting bottoms with no protection from predators. No attempt was made to differentiate between natural and hatchery-reared seed planted in this manner and it is doubtful that the hatchery-reared seed made any serious contribution to the clam harvest. Glancy became terminally ill shortly after he started his hatchery, but before he died he patented his methods (Glancy, 1965) and trained Mr. Charles Hart (a son-in-law) how to use his methods. Glancy's work had a great impact on shellfish culture, most of the east coast hatcheries used and are still using his innovative ideas. Bluepoint Oyster Company used this method and Hart's talent to grow clams. Subsequently Charles Hart and his brother William established another hatchery called Shellfish Inc. in West Sayville, New York (Manzi and Castagna, 1989a). Bluepoint, after using Glancy's methods for a number of years for culturing both clams and oysters, converted to growing unicellular algae. To provide abiotic water for this culture, they used bore hole water with a high enough salinity for algal and larval production. This had the advantage of being free from bacteria and high in nitrogen, but had the disadvantage of being free from food or dissolved organic material. Therefore, an adequate production of phytoplankton was necessary for larviculture. At first seed clams were broadcast over their subaqueous planting bottoms at relatively small sizes. Today information indicates that this is an ineffective method. Bluepoint now grows clams to a larger size and plants in trays or other exclusion devices to protect the seed from predators (Manzi and Castagna, 1989a).

John Plock started a hatchery in Greenport, New York, and hired Paul and Matoria Chanley to operate the facility using the technology developed at the Milford Laboratory. The most significant development at this facility was the development of the protocol for the chemical sterilization of seawater by Matoria Chanley. Chanley's method used commercially available laundry bleach (5.25% hypochlorite) followed by neutralization with sodium thiosulfate. This method was much more economical than pasteurization or heat sterilization and could be used with any sized container (i.e. 5000 L tanks). It was immediately accepted by most culturists and is still the most common method used in algal production. Unfortunately the Plock clam growing operation was never successful because of the field grow out problems. The post-set clams were planted in a series of tidal ponds where the clams failed to grow to a plantable size (personal communication P.E. Chanley, 1974). The Chanleys moved to other endeavors and the operation closed.

The Frank M. Flowers and Sons Oyster Company of Oyster Bay, New York, was another early pioneer in molluscan culture. They primarily produced oysters, but also grew clam seed for sale and for broadcasting on their leases. This hatchery, one of the most successful in the northeast, started as a Glancy operation but later changed to the use of bore hole nutrient rich, abiotic, low salinity water. This allowed a more complete control of the larval environment (Manzi and Castagna, 1989a).

Another pioneer was Long Island Oyster Farms Incorporated, a subsidiary of Inmont Chemical Company. This company constructed a fiberglass solarium over an acre in size at Northport, New York. This facility adjoined a Long Island electric company electrical generating station. Although the company started as an oyster hatchery, it grew a number of species and ended up growing clams. Juveniles were grown in the warm seawater of the power plant cooling lagoon to accelerate growth. The hatchery used cultured algae and wild algae (similar to the Glancy method) for growing larval food and conditioning broodstock. The hatchery, which operated for about ten years, left a history of unsolved problems (Manzi and Castagna, 1989a). The hatchery was never financially successful, and has since been consumed by fire.

Two state-owned aquaculture facilities that should be noted are the University of Delaware at Lewes, Delaware and The College of William and Mary, Wachapreague, Virginia. The former was unique because it attempted to develop a closed system or recirculating system hatchery and nursery. The major problem of low and sporadic production was never solved. Bolton (1982) reported that this type of hatchery was not economically feasible. The College of William and Mary's hatchery was perhaps the most successful hatchery on the East Coast. The hatchery used either the Glancy method (Glancy, 1965) or the Brown Water method (Castagna and Kraeuter, 1981; Ogle, 1982). The staff of this facility developed user friendly hatchery and nursery techniques and successful field grow out methods. A short course was offered in clam aquaculture methods for almost twenty years. Technical assistance and direction were freely given to commercial growers, students, teachers and extension agents. The roster of students reads like a who's who of clam farming (Appendix 15.A). In addition to growing *Mercenaria*, over 80 bivalve species were grown (Manzi and Castagna, 1989a) (Appendix 15.B). A number of bivalves were grown for multiple generations and *Mercenaria mercenaria* were selected for fast growth over eight generations.

Hatchery techniques are now well established, dependable and produce fairly consistent yields. Technicians deviating from these procedures may lose some cohorts of larvae, but these problems can usually be solved by using the tried and true methods. Over the last two decades attention has turned to the problem of nursery and grow out (Castagna, 1984; Kraeuter and Castagna, 1985a,b). The nursery stage grows seed from metamorphosis to a size large enough so that a high percentage will survive when planted under predator control systems in the field. Nurseries are primarily based on pumping bay water to trays, troughs or impoundments holding the clams. The water furnishes food and dissolved oxygen while removing the waste products (Manzi and Castagna, 1989b). The technology for removing the waste products and adjusting the pH, salinity, temperature and replacing the dissolved oxygen is well known and could be applied to a closed or recycled seawater system. However, the technology for formulated diets made from agricultural products is still lacking. Mass culture of suitable algae has been greatly improved over the last two decades, but the costs are still too high for commercial success in growing seed beyond small size. Pumping natural water or planting in natural waters has been the only way of controlling cost. Finfish aquaculture in both fresh and seawater has successfully used recirculated water, adding inexpensive food. It is probable that shellfish culture will use this method in the near future. There are some interesting plans for closed system nurseries that may replace flow through nurseries and perhaps allow some early grow out (Castagna et al., 1996).

Early attempts to plant clam seed on natural bottom without predator control proved fruitless. It was not until R. Winston Menzel and his students at Florida State University, Tallahassee, Florida addressed this problem that some predator protection was achieved (Menzel and Sims, 1964). Menzel and his co-workers fenced in plots with fish net to exclude the larger predators and planted seed that had reached sanctuary size from the smaller predators. Richard Crema, Oceanville, NJ, one of the pioneer clam culturists, was the first to introduce the method of growing clams under light density (buoyant) horizontal plastic nets. This was a significant new farming method. Researchers at the College of William and Mary Laboratory at Wachapreague, Virginia developed grow out systems and produced data to predict the expected survival of different size seed in various types of protected field planting (Kraeuter and Castagna, 1977, 1980; Castagna and Kraeuter, 1977; Castagna, 1978, 1983a,b, 1984; Gibbons and Castagna, 1984, 1985). Seed smaller than 8 to 10 mm in height experienced high predation losses even when planted using predator exclusion nets. In recent years intermediate culture nurseries have been developed to grow seed from metamorphosis, about 200 μm (0.200 mm), to the 10 or 15 mm height required for successful field planting. Manzi (1985) reviewed the type of nurseries commercial growers were using. Upweller nurseries (Manzi et al., 1986) are commonly used, but part of the industry is returning to shallow seawater tables or shallow raceways where an adequate dissolved oxygen level and good food distribution can be more easily maintained. Technological advances in the last twenty years have transformed the industry. Contributions of a number of scientists and inventive culturists have brought clam culture to a commercially viable industry.

15.2 CURRENT PRACTICES

The following sections will discuss clam culture as it is now practiced. The technology is common to most clam hatcheries, nurseries and grow out facilities. When appropriate, more than one method will be discussed even when they are used somewhat exclusively. Clam culture as practiced today is essentially a three-phase process, commonly designated as hatchery, nursery and grow out. Most entrepreneurs design or modify these steps to better utilize the physical factors in their immediate areas. Needless to say that the site of an operation could dictate the success or limitations of an operation. Most culture operations are sited where waterfront, subaqueous leases, permitting, etc., are most easily available, often on land already owned or leased by the individual or the company. The requirements, especially for the hatchery, are more stringent than would be necessary for a nursery.

15.3 CULTURE

As mentioned above, *Mercenaria* culture is carried out in three phases — hatchery, nursery, and grow out. It is recommended that entrepreneurs entering the field purchase seed from an existing hatchery and start with the grow out phase. This requires the least capital and has the potential for the highest return on investment. This chapter will follow a more logical sequence of hatchery, nursery and finally grow out.

15.4 PHYSICAL PLANT

This is a description of the usual equipment and methods that were successfully used by a number of culture operations. Virtually every operator will modify either the method or equipment to fit the specific location or to solve specific problems. Many of the early hatcheries were started in existing buildings, often storage sheds or oyster shucking houses. Fiberglass or plastic solariums were installed to take advantage of natural sunlight to produce the algal food needed for the larval clams. Hatchery buildings are now well-insulated frame or cinder block buildings with limited window and door openings so that temperature can be more easily controlled. Algae are usually grown using artificial light. Onshore nurseries can be out of doors, under a shade cover or in a storage building. Nurseries with flowing water are usually unheated or have minimal heat and few amenities. The flowing water in a nursery acts as a heat sump making it very expensive to heat or cool most nurseries. A place to dock a workboat and room to make and store the protective devices is necessary for grow out but need not be adjacent to the rest of the facility.

15.5 SEAWATER SYSTEM

A dual seawater system is the most common system. The advantage of this system is its simplicity in controlling fouling within the pipes (Castagna and Kraeuter, 1981; Castagna, 1983a,b, 1987). Duplicate seawater systems from the intakes to the delivery valves allow one intake pump and piping to be used for about a week to ten days and then shut off. The seawater remains in the system where it becomes anaerobic and stagnant while the duplicate system is used for the next week. The following week, the first intake, pump, and pipe are flushed and reactivated. If we designate the dual system as lines A and B, line A is active for one week, then shut off to become stagnant while line B is used the following week. Then line A is reactivated in the third week while line B becomes stagnant, etc. Anaerobic conditions kill fouling organisms that may have set in the inactive line during the preceding week. Such fouling organisms are usually microscopic in size after seven to ten days of growth and are easily killed and flushed out as soon as seawater is introduced through the system at the next active interval. The intake is lifted out of the water for the week the line is inactive so fouling organisms on its surfaces will be killed by desiccation or anoxia.

Another common system uses an open sluice box fed from one station. In this system the sluice box can be drained and scrubbed to control fouling.

15.6 INTAKES, PUMPS AND PIPES

The seawater intake screen is constructed from a 10 to 15 cm polyvinyl chloride (PVC) pipe into which multiple saw cuts or drill holes are made. One end of the pipe is capped or plugged and the other end reduced to receive a smaller 7.5 cm PVC fitting. The intakes are connected to flexible non-collapsible intake hoses that are connected in turn by unions fitted on PVC pipes that are plumbed onto the pumps. The intakes are suspended from a frame constructed on a catamaran or other floating platform. A rope attached to the plug or cap end is designed to hold the intakes suspended a preset distance below the surface of the water. The intake can then be lifted out of the water and up to the frame to air dry. The float structure is usually moored to a pier or to pilings installed for this purpose or secured with a multiple anchor arrangement. A floating intake has several advantages over a fixed intake: (1) it rises and falls with the tides maintaining the opening a set distance below the surface, where the most suitable water temperature, food or dissolved oxygen are found; (2) there are fewer problems with entrainment of resuspended silt and associated toxins; (3) less problems with clogging by free floating macroalgae and debris; and (4) free swimming organisms such as shrimp and minnows are less likely to be entrained and trapped against the intake screen. Some hatcheries use a fixed intake with hard PVC plumbing attached to a screen box set either on the bottom or a set distance above the bottom.

Virtually any type pump can be used in a flow through (non-recirculating) system provided it has an adequate flow rate at the required dynamic head pressure. Economical 5-cm cast iron or thermoplastic centrifugal pumps are often used, but fiberglass, plastic, or resin-lined pumps are valid alternatives. Brass, bronze or aluminum pumps can be used in a flow-through system but are not suitable in a recirculating system. Submersible pumps suspended from a float or davit are also used. Submersible pumps should be rated for use in seawater or corrosive environment; otherwise the pumps may be plagued with electrolysis problems.

Dual seawater lines and drainpipes are constructed of PVC pipe. The drainpipe is usually constructed of 15 cm PVC pipe. The drain pipe not in use is closed with a commercially available plastic and rubber soil plug, which allows the water to become anoxic preventing fouling just as it does in the dual delivery pipes.

15.7 FILTERS AND WATER PURIFICATION

Inshore seawater can contain large amounts of silt or be contaminated due to runoff or biotoxins from blooms. Water provided to eggs and embryos may be treated to alleviate these problems. Seawater is passed through standard pressurized fiberglass or plastic swimming pool filters connected in series. The first filter is filled with sand and operates like an ordinary sand filter to mechanically remove particulates. The second filter has only a base of sand with the rest of the column filled with activated carbon that acts as a semifluidized charcoal bed that removes chemical contaminants by adsorption. Many hatcheries build their own charcoal filters using plastic barrels fitted with a coil of drip irrigation hose buried in charcoal granules so that water enters the barrel through the irrigation hose. This distributes the water through the carbon before leaving the barrel. The activated carbon is replaced at intervals of two to three weeks based on water use. One kilogram of activated carbon will treat approximately 100,000 L of seawater. Activated coal charcoal or coke has proven to be the most efficient and economical for this use. The average pore size of the carbon is an important factor; coke has the proper pore size for efficiently treating seawater (TIGG Corp., 1988). This water is also used for a base for algal production via the Milford method.

After embryogenesis, the larvae of clams are usually hearty enough to withstand some dilute contaminants, thereby allowing the carbon bed to be bypassed. Polypropylene bag filters are also used. These are commercially available in various sizes. The bag filters, usually 10 to 25 μm, are used either as the only filter or as a final filter when some type of prefiltering (i.e. sand filter) is used. They are normally installed at the final point of delivery to the tank. These are replaced daily and the used filters are washed, dried and reused. Care must be taken to be sure any detergent or soap is completely rinsed off so there are no toxic effects and the filters are completely dry to reduce bacterial growth. Bacterial contamination is perhaps the major problem in hatcheries. All containers and equipment must be carefully washed with a mild biodegradable detergent and fresh water (Castagna and Kraeuter, 1981). A few culturists use 5.25% sodium hypochlorite, but this practice can sometimes cause more problems. Larvae are sensitive to chlorine gas and chlorine residuals, and will often die when exposed to them.

15.8 BROODSTOCK SELECTION

Broodstock can be selected for desired characteristics such as fast growth, shell markings or environmental adaptability (Chanley, 1961). The advantages of faster growth are obvious. Selection of notata shell markings develops a biological marker that separates harvested cultured clams from harvested wild clams. This has even been used as evidence in court to prosecute clam poachers.

Sexual dimorphism is lacking in *M. mercenaria*. The shell size, shape, color or siphon pigmentation has no distinct differences between males and females. Sexes are usually determined by microscopic examination (Eversole, 1989). It is not possible to select clams by sex unless they have been previously spawned and their sex identified. Alternatively, a gonad sample can be removed and checked under a microscope to identify ova or sperm. A small tissue biopsy can be removed without sacrificing the spawner. A small hole, drilled through the shell over the gonad, will allow insertion of a 21 or 22 gauge needle. This can be pushed through the mantle into the gonad. A syringe fastened to the needle will cause enough negative pressure to pull a small piece of tissue into the needle bore. This tissue is placed on a slide and a squash of the tissue can be examined under a microscope. Ova or sperm can be easily identified at 100× magnification. In general practice, sexing is not worth the effort since populations of adult clams are about equally divided between sexes and most hatcheries utilize mass spawnings.

15.9 BROODSTOCK CONDITIONING AND DELAYED SPAWNING

There are obvious advantages of manipulating the gonadal cycle and spawning period of the broodstock so that some can be spawned either earlier or later than the normal spawning period. This gives the producer an early start on seed production and allows better utilization of larval and nursery systems by extending the period when larvae are available. Clam gonads ripen naturally as ambient water temperature and food levels increase. Early gonad development can be achieved by collecting spawners and subjecting them to higher water temperatures and adequate amounts of suitable food.

A typical protocol for broodstock handling would be as follows. Adult clams are collected either from natural stocks or from planted beds in January or February (in the mid-latitudes of the Northern Hemisphere) and placed in a tank of seawater held between 20° and 23°C. About 250 adult clams can be held in a 1200 L tank with aeration. The water can be heated with a heat exchanger made of inert or non-toxic material or by using electric immersion heaters made with glass, quartz, Teflon, carbon or any other inert or stabile exterior material. Provisions should be made for complete water exchange three times per week. It is important that temperatures do not fluctuate during the water changes since that might cause a spontaneous spawning. A simple solution is to use two tanks, so that water could be brought up to the same temperature in both the new and the used tank. The clams, held in a container like a plastic milk crate, can be transferred from one tank to the next as soon as the water temperature has equalized between tanks. The used tank is then drained, washed and rinsed and filled with fresh seawater of approximately the same salinity (or adjusted to the required salinity) and immersion heaters or a heat exchanger are used to bring the tank to the proper temperature before the next scheduled change.

Gonads will usually ripen in four to six weeks depending on temperature, water quality, food quality and food quantity. Conditioning can take place at 19–28°C. Lower temperatures require a longer period. Higher temperatures are quicker, but may induce a spontaneous spawn in the conditioning tank. Temperatures of 22–23°C are usually a good trade off. After about two weeks it is wise to move to a lower temperature so that spontaneous spawning does not take place in the conditioning tank. Temperatures between 16° and 18°C will usually prevent an accidental spawn but will not encourage reabsorption. Cultured algae should be added to the conditioning tank at a rate of 0.5 to 1 L per 50 clams/day of 10×10^5 cells/mL of the unicellular algae available. Tahitian *Isochrysis* is often used. Algae are best utilized in multiple feedings, two to four times per day.

As the normal spring/summer spawning season nears, ripe individuals are available for broodstock directly from planted beds or from natural stock. The conditioning tanks can then be converted to use for delaying spawning to extend the season. Ripe broodstock held in seawater at temperatures between 15° and 19°C and fed daily will not spawn or absorb their gametes. Seawater can be cooled by a chiller and heat exchanger or simply held in a cold room. Cold rooms are often built into a hatchery as a small well-insulated room with two room air conditioners cooling the air. One conditioning unit is set slightly lower than the other so the higher temperature can act as a back up should the primary unit fail. If standing water is used, it must be changed three times a week with pre-cooled water. The same sort of protocol described above for static heated water can be used. Months after ripening, clams held in this manner can be warmed to 23°C for 24 to 72 h and then induced to spawn. These procedures extend the period for obtaining ripe individuals and frees the culturist of the restrictions of the natural spawning period.

15.10 OBTAINING GAMETES

Gametes are usually obtained by inducing ripe adult clams to spawn. Ripe clams are placed in a spawning trough or tank containing filtered seawater at 22–24°C. The clams are left undisturbed until most of them appear to be pumping vigorously. The seawater temperature is then increased to about 28°C by passing the water through a heat exchanger or heating the water with aquarium heaters. The clams are held at 28°C for about 45 min. If no spawning activity is observed, the water temperature is lowered to about 24°C for 30 min, then raised again to 28°C. Temperature cycling continues until an adequate number of clams spawn and the needed number of eggs are collected. Quite often other stimulation is necessary to trigger spawning. The most commonly used stimulus employs gametes stripped from a clam sacrificed to provide gonadal products. The clam is opened and its gametes are collected by lacerating the gonad with a scalpel or razor blade and rinsing the gonadal material into a beaker of filtered seawater. The suspension of gametes is poured into the water over the spawners or a pipet is used to release some suspension near individual clams that are pumping vigorously.

An alternate spawning protocol requires only small quantities of heated filtered seawater. Ripe clams are stored overnight in a refrigerator (ca. 6°C), dry in a container with a wet cloth or wet paper cover to prevent desiccation. The following morning they are placed in the spawning trough and 30°C filtered seawater is added to a depth of 10 to 15 cm. The clams are left undisturbed for up to 3 h as the water cools to room temperature (ca. 22–24°C). The water is drained and refilled with fresh 30°C seawater. Sperm obtained from either a previous spawn or from stripping as described above is added. Temperature cycling process can be repeated hourly until spawning occurs (Castagna et al., 1996). Sperm from a previous spawn, stored overnight in the refrigerator, will act as a stimulant but will usually perform poorly in fertilizing ova.

A third alternative is the use of serotonin (5HT) as a spawning stimulus, especially for male clams. Serotonin, mixed with 1 μm filtered seawater and about 0.4 mL of 2.0 mM solution, is injected into the clam's adductor muscle by inserting a hypodermic needle through a notch filed into the edge of the shell. Within about 15 min after the injection, a ripe male clam will spawn (Gibbons and Castagna, 1984, 1985). Females are usually slower in responding to serotonin. Recent experiments on scallops indicate that prostaglandin injections will enhance female spawning when used in conjunction with serotonin (Martinez et al., 1996). Fong et al. (1996) working with *Mactra chinensis* investigated how serotonergic ligands function and reviewed the literature on this subject. Serotonin needs to be handled carefully because it can be absorbed through human skin and is a suspected teratogen.

Normally the culturist will simply allow fertilization to take place in the spawning container, but if special crosses are required or a defined stock improvement program is being practiced, individual males and females can be placed in separate containers and allowed to spawn. After sufficient spawning is complete, the adult is removed and fertilization can start by the addition of gonadal products from the proper contributor to the cross. Fertilization success decreases with time so crosses should be made within the first hour; delaying for 3 h yields relatively poor levels of fertilization. Within 15 min of fertilization a sample of eggs can be checked under a microscope to determine polar body formation. If fertilization rate is low, additional sperm may be added.

When individuals or a select group of spawners are needed for a specific genetic cross, inactivated sperm or eggs of an unselected wild clam can be used as stimulus to prevent unwanted fertilization. Thus an individual from the select group does not have to be sacrificed to obtain gametes for a stimulus. The sperm or egg suspension from the unselected clams is rendered inactive by freezing, pasteurizing, or microwaving. For example, a 500 mL suspension of gametes in filtered seawater requires about 90 s in a 700 W microwave at full power. If pasteurization is preferred, temperatures in excess of 60°C for 15 min are usually sufficient. Freezing requires about 2 h in a household freezer or until ice crystals form on the surface. Gamete viability should be checked under a microscope before use. Inactivated sperm after it is warmed to room temperature (a small sample on a slide will reach room temperature in minutes) will show none of the vigorous movement of living sperm. Inactive ova will become more consistently opaque without the less opaque central area normally observed around the cell nucleus.

Clams with well developed or ripe gonads will usually spawn when subjected to the usual temperature shock or temperature cycling and stimuli as discussed above. However, if spawning is not achieved, gametes can be stripped from the gonad as described earlier (Castagna and Kraeuter, 1981). Some culturists remove the meat from the shell of five to ten clams, trim excess tissue from the gonadal area and place the trimmed portions in a blender with filtered seawater. The tissue is blended for 3–5 s and the liquefied tissue poured through a 73 μm sieve into a container with 10 L of filtered seawater. The eggs can then be collected on a 44 μm sieve, washed with filtered seawater and inspected to be sure fertilization has occurred. The eggs can then be treated in the usual method. Eggs obtained from stripping will usually produce fewer larvae than a natural spawn because immature eggs are included. Despite this drawback it is sometimes necessary to revert to this heroic method.

15.11 REARING LARVAE

Fertilization takes place during mass spawning and the fertilized eggs are collected. These are separated by size by rinsing through appropriate descending sized sieves. Larger eggs have higher lipid content and produce more viable larvae (Kraeuter et al., 1982; Gallager and Mann, 1986; Gallager et al., 1986). Eggs retained on a 53 μm or larger mesh sieve will have better survival than smaller sizes. Embryogenesis is more successful in seawater that has been purified as described earlier. The eggs are started in culture at densities as high as 60 mL. Water temperatures should be 20–28°C. Culture containers, constructed of fiberglass, range in size from 100 to 50,000 L and often have bottoms sloped to promote draining.

Embryogenesis occurs in a relatively short time, and eggs develop into embryos in 4–8 h at 24°C. Trochophore stages usually develop within 12 h after fertilization and straight-hinge veliger larvae within 24 h. An initial water change is usually performed 24 to 48 h after fertilization. Virtually no feeding takes place until the straight-hinge stage is reached so there is no need to change water or add algal food until that time. Straight-hinge larvae are much hardier than embryonic and trochophore stages. They are held at densities of about 15 per mL in filtered seawater. At this stage chemical sterilization is seldom used, but water must be filtered to 10–15 μm range. Some culturists routinely use 4500 Å UV light to reduce bacteria, but most do not. Rearing tanks are drained to change the water three times each week and larvae are collected and sorted through a descending size series of sieves (200–64 μm). When water temperatures are 27°C or above, it is sometimes necessary to revert to daily water changes to reduce bacterial problems and increase food levels. Gentle aeration is some times added. The larvae should be sampled, counted and inspected under the microscope for evidence of disease. Different sizes are segregated when they are returned to the culture containers, and slow growers are often culled. *Mercenaria* larvae grown at 24° to 28°C develop to the pediveliger stage in 8 to 14 days depending on water quality, temperature and food.

The larvae usually metamorphose at about 200 to 210 μm in length, but this may occur as small as 175 μm (Loosanoff and Davis, 1963). Most culturists move all larvae harvested on a 130 or larger mesh size sieve to the nursery. Upon reaching the proper size, clam larvae will set or metamorphose with no special treatment. The use of various chemicals such as L-dopa or epinephrine can trigger metamorphosis (Coon et al., 1985), but most culturists will simply allow setting to occur. Increasing water temperature or increased feeding is sometimes used to encourage a more concentrated setting period. Clams do not require any special substrate or stimulant to set and are usually simply found attached with byssal threads to the side and bottom of the container. At this point the pediveligers are moved to a nursery system.

15.12 NURSERY SYSTEM

Once the clams metamorphose or set they need to be grown to a larger size before they can be planted into beds in the bay or lagoon. This is the function of a nursery. There are a wide variety of nursery methods.

15.13 POST-SET MAINTENANCE

Perhaps the most critical period in the production of clam seed occurs at the time the larvae metamorphose. Most commercial hatchery operators indicate that this period has consistently presented a challenge to their operation. This was verified by Manzi and Castagna (1989b). There are a variety of problems that must be addressed. Metamorphosis is a major physical and physiological change for any bivalve. Suddenly its biological systems and requirements have changed, and these changes dictate new culture conditions. Pediveliger (post-set) clams or setting size larvae are light tan in color and even though they are visible without optical enhancement, they are still concentrated or transferred using sieves. Post-set clams or pediveligers can be moved to ambient seawater if the water temperature is high enough to promote growth. This opens up a wide variety of treatments. The following are some of the methods commonly practiced. Care and attention to detail are still very important.

15.14 SEAWATER REQUIREMENTS

Good-quality seawater is still a requisite. Post-set clams are much more tolerant than larvae or embryos but they can still be lost to toxins, inadequate dissolved oxygen (DO), rapid salinity changes, poor food or overcrowding. Short-term exposures to low DO, salinity or other conditions are usually tolerated.

15.15 POST-SET REQUIREMENTS

Post-set clams can obtain adequate food from natural phytoplankton and the phytoplankton volume is usually available if the temperature has been high enough for the past 48 to 72 h. They also utilize dissolved organic material (DOM) found in most natural seawater (Manahan, 1983).

Clams that were from winter or early spring spawns require further care if the ambient seawater temperature is below 12° to 15°C. These early post-sets will not feed or grow very well at low temperatures. In the interest of supplying higher temperatures and adequate food at an affordable price, a static system, holding them in tanks with triweekly water changes and algal food additions or a closed or semi-closed recirculating system, can be used. Since tanks are usually in short supply in hatcheries, most culturists move to a more efficient recirculating system. This amounts to a tank, wet table or trough filled with filtered seawater. Post-set clams are placed in sieves at relatively high densities and placed in the tank. The sieves are placed so that the tops are above the water. This can be achieved by floating, suspending or placing them on a PVC pipe rack or scaffold placed on the bottom of the tank. Water from the tank is then circulated into the sieve (downweller) or from inside the sieve back to the tank (upweller) using air-lifts or a small recirculating pump. This system is usually drained, cleaned and refilled daily or triweekly with algal food added once or twice a day at a rate of about 10^5–10^6 cells mL^{-1} day^{-1}. The water is sometimes heated with a heat exchanger or allowed to reach room temperature. When ambient seawater temperatures are above 12° to 15°C, this step can be omitted, although some culturists always use this system for newly set clams.

When water temperatures have reached >15°C, clams can be placed on a wet table or raceway at densities of approximately 150 per cm^2. About 3 L of seawater per minute should be pumped to each 2 million clams. Supplemental algae may be added for a few days to a week. A 130 μm mesh sieve should be positioned below the overflow to capture any pediveligers not byssally attached. The sieve should be rinsed back into the raceway at least every morning and evening or any other time it is convenient. After the first 48 h the flow rate should be increased daily until a rate of 12–20 L/min is reached. Many culturists use a 50–200 μm bag filter or a sand filter to prevent competitors from setting on the table but this practice is not as common as it once was.

Presently, many culturists spray the water onto the tables to increase the dissolved oxygen. Some form of filtering may be necessary to prevent large particles (>3 mm) from entering the spray system. A manifold constructed from three-quarter inch PVC pipe with 4 mm holes drilled in a line about every 45 cm is suspended a few inches above the water level in the tank. The water entering the tank increases the dissolved oxygen and reduces the supersaturation problems that occur at cooler temperatures (Bisker and Castagna, 1985). Ambient seawater is subject to a diurnal dissolved oxygen cycle (McConnaughey and Zottoli, 1989). In estuarine systems, typical for most hatcheries, this can mean that the incoming seawater has a relatively low DO level during the night. Pratt and Campbell (1956) demonstrated that hard clams' growth was unaffected by DO concentrations above 5 mg/L. Even clam larvae can stand low DO, but growth is curtailed below 2.4 mg/L (Morrison, 1971). The introduction of manifold sprayers has reduced the growing time for producing 12 mm seed by 3 to 6 weeks, suggesting that these low DO conditions are common. Similar systems are being constructed for upwellers which will be discussed later.

Competitors are controlled by using routine fresh water soaks (Castagna and Kraeuter, 1981; Castagna and Manzi, 1989; Castagna et al., 1996). After the clams are about 10 days post-set, they are subjected to a 30-min fresh water soak every 7 to 10 days. This is a simple operation. Raceways are usually divided into five groups so all can be treated during the normal workweek. The raceways to be treated are drained by lifting the standpipe and the water is passed through an appropriate sized mesh sieve. No attempt is made to wash clams or detritus into the sieve, but a few clams will be carried out with draining water. The standpipe is then replaced, the raceway is flooded to the standpipe height with fresh water (chlorinated water is acceptable) and the few clams caught on the sieve are rinsed into the fresh water. This is allowed to stand for 30 to 60 min (2 mm or larger clams can withstand 60 min but smaller sizes are given 30 or 45 min) and then the seawater flow is restarted. Seawater being denser will sink to the bottom and float the fresh water to the top and out the standpipe. Raceways or troughs can promote excellent growth and high survival when ambient conditions are suitable and as stated above if the DO is maintained at a high level with sprayers. Once clams reach 1–2 mm they are typically placed in systems supplied with pumped seawater. No attempt is made to add cultured algae.

15.16 ONSHORE NURSERY SYSTEMS

Hatcheries have used onshore nursery systems to grow young seed to commercially salable or usable sizes. These nurseries have traditionally used raceways or shallow trays, continuously supplied with seawater pumped from an adjacent source on a pass-through system. A number of hatcheries use upflow culture systems instead of raceways or trays; however, the trend is returning toward raceways and trays which often improves the DO levels. In general, onshore nursery systems provide a semi-controlled environment that is easy to monitor for rearing young bivalve seed. They require adequate siting and sufficient technical staff to insure that uninterrupted and appropriately high quality of estuarine waters are available. To this end a dependable plumbing system (dual seawater system or an easy access for clearing and cleaning the lines) and an auxiliary pumping system that can be used during periods of electrical stoppage are absolute necessities. Depending on their size and density, clams can stand short periods without flowing water (about 1–3 h) and longer periods (5–10 h) in air (if protected from direct sunlight). However, it is still a good idea to have a gasoline-powered pump to fill in when electricity is off. Some operations have standby generators, which automatically cut in when electricity is interrupted.

Raceways are the least expensive culture units and are found in many forms and configurations. They are normally long shallow tanks or troughs, constructed of wood sealed with epoxy coating, fiberglass or lined with a plastic pond liner material. A popular raceway design is a concrete pad with sides constructed of cinder block. Some of these systems are coated, but others are lined with plastic pond liner material. Raceways of wood or plastic are often placed on trestles or racks, sometimes stacked in layers. This latter design uses less room but access for maintenance and cleaning is sometimes limited.

Seawater, pumped from an adjacent source, enters the raceway either at one end or through multiple ports (in longer raceways) or sprayed into the tank utilizing a manifold, and flows horizontally along the tank to drains at the opposite end. In shallow raceways, a thin layer of clam seed is spread over the bottom and the water level is adjusted by standpipes so that the horizontal flow of water is just deep enough to cover the seed. This insures good water mixing, food distribution and efficient water use per unit biomass of seed. Many on the ground raceways have sand bottoms to give the seed a substrate to bury into after the siphons develop. This allows clams growing in a substrate to have a less inflated shell and sometimes grow faster. Other growers simply allow sediment from the source water to accumulate in their tanks. This is usually fine grained and can be easily washed from the clams when they are graded. In deeper raceways, tiers of plastic mesh trays are used to make use of a three-dimensional design to take advantage of the increased tank capacity. Sometimes baffles are used to mix the horizontal water flow as it passes through the structure. The increased and continued use of raceways attests to their functional utility. Manzi and Castagna (1989b) stated that the general use of raceway culture had decreased but this trend has reversed and raceways have regained their popularity. This is due in some measure to the increased use of spray manifolds.

Upflow nursery systems for seed production have been in use since the mid-1960s (Bayes, 1981), but it is only in the past decade that commercial interest in these systems has been revived (Manzi and Whetstone, 1981; Manzi et al., 1985, 1986; Manzi and Castagna, 1989b). Upwellers require less space than raceways because water flow is directed up through a seed mass rather than horizontally across the seed clams. Two types of upweller systems are used in flow-through bivalve nurseries: active, forces water up through the clam mass; passive, pulls water up through the seed mass. Active systems are constructed of closed bottom cylinders plumbed to accept water under pressure at or near the bottom of the cylinder. The seed clams are suspended above the water intake by a plastic screen with a small enough mesh size to retain the seed. The water is forced up through the screen, partially fluidizing the clam seed mass, and exits through the drains positioned near the top of the cylinder. Active or force flow cylinders are normally used primarily for smaller-sized (<3 mm) seed clams (Manzi and Castagna, 1989b).

Passive upweller systems use open-ended containers suspended in a tank or reservoir. Usually the cylinders are made from PVC pipe or plastic buckets, although box shapes constructed of coated wood or plastic boxes have been used. A screen with an appropriate mesh size for retention of clam seed forms the bottom of the container and supports the clam seed mass which covers the screen up to a depth of several centimeters. When plastic buckets are used, the bucket lid, with the entire center removed, locks the screen to the top. The bottom is removed to form a truncated cone that holds the seed on the larger end with the screen. Seawater entering the reservoir moves up through the seed mass and drains out through a pipe near the top of the upweller. Incoming water enters the reservoir and is drawn up into each upweller, accelerating as it passes through the seed mass and decelerating upon reaching the overlying water between the seed and the surface. If flow rates are appropriately adjusted, wastes and silt are swept through the seed mass and settle as a loose layer on the surface of the seed mass. The active upwellers are more difficult to monitor than the passive upwellers in that they must be disassembled in order to see the clam seed and there is some danger of eroding the valves from the seed if the water is allowed to flow through the seed mass at too high a velocity. In the passive upwellers the clam seed can be observed and handled anytime since the top is open, and too high a velocity would float the seed clams out the drain which could be observed during the initial flow velocity adjustment. The passive system is much more commonly used in commercial operations. Dissolved oxygen can be increased in upwellers by using spray manifolds over the reservoir tank or even into the tank and upwellers. Care must be taken to make sure air bubbles do not accumulate under the screen bottom of the upwellers and prevent water to flow upward. This has improved growth rates in the systems where it has been introduced.

Today some large commercial nurseries use floating upweller systems. These are large rafts or floating units that hold trays or upwellers. The rafts are moored or anchored near the dock so that electricity can be furnished. The upwellers are piped to a seawater trunk that is evacuated by a suitable pump, causing the seawater to flow up through the upweller, into the trunk and pumped overboard. These are modifications of the system described by John C. Bayes, in a special publication of the European Mariculture Society (Bayes, 1981). These systems go by the acronym of FLUPSY for floating upweller system.

15.17 FIELD NURSERY SYSTEMS

Field nurseries are systems utilizing natural areas for the intermediate growth of clam seed. Shallow subtidal or low intertidal areas are suitable for a number of types of bottom nursery culture. Castagna and Kraeuter (1981) describe low intertidal bottom nurseries using aggregate cover and plastic mesh tenting. The aggregate cover and mesh combine to provide protection from predators and deployments of this system proved successful. However, most culturists forgo the use of aggregate because it is expensive and requires more labor to carry and spread it on the beds. The use of light density plastic covers properly installed and tended has shown excellent results and is the most commonly used protection. Another popular method used by one of the larger clam farms is to place seed in plastic trays 9 × 1.2 × 0.15 m filled with sand and covered with plastic mesh. The seed is between 2 and 3 mm in height when placed in the trays. The trays are placed in an easily retrievable pattern in shallow subtidal areas. Maintenance is minimal (occasional brushing of the top mesh) and in a relatively short time the trays are retrieved and the seed and sand are mechanically sieved. The larger seed (>12 mm) are sorted for planting and the smaller size seed and sand are recovered for reuse. Another popular field nursery method is a commercially available 1 × 0.5 m plastic 5 mm square mesh lay flat tube that can be made into trays by folding and fastening the ends. These are filled with 3 to 4 mm seed and placed on an intertidal or subtidal flat so they can be recovered when the seed reaches the appropriate size for planting (<9 mm). Other home-built trays have been used with equal success. These are usually constructed with wood frames and plastic tops and either plastic or plywood bottom. Some are quite large, 1.2 × 2.4 × 0.1 m, some with 0.2 m legs to hold them off the bottom. Some are smaller and can be used either on the bottom or suspended in multiple layers beneath a float in deeper water. William Mook, Mook Sea Farm, Damariscotta, Maine, developed a tidal powered upweller. This is a raft-like device that directs the tidal water up through a series of upwellers. This proved to be economically and commercially successful (Mook, 1988; Mook and Johnson, 1988). This has been redesigned and improved by a South Carolina Sea Grant project (Baldwin et al., 1995).

Vaughan et al. (1988) developed a bag system for growing clams. The system was constructed so that the mesh bags, approximately 1.0 by 0.5 m in size, were fastened to a pair of ropes laid onto each end of the bags so they could be handled like a continuous conveyor belt. These have proven to be successful in many areas of Florida. This system was tested in Georgia and was found to be a biologically feasible means of culturing clams in soft bottom areas. The system worked well for growing seed clams to approximately 30 to 35 mm in length, but growth to market size appeared to be retarded (Walker and Hurley, 1995). There are a great number of innovations and variations to these methods. The only thing all these methods have in common is that they all retain and protect the small seed until it reaches a sanctuary size where it can be placed in a grow out bed and it can be easily recovered or harvested.

15.18 FIELD GROW OUT

There have been attempts to grow clams to market size in tanks, raceways and ponds. These have been unsuccessful either because of the high costs of pumping seawater, the costs of growing supplemental phytoplankton food or the slow growth or stunting that occurs when the physical and biological needs of the clams are not met. For instance, clams need 65% oxygen or 5.7 mg/L per clam at 17°C. This amounts to between 22 and 28 L/h of seawater for each bushel of clams (Hamwi and Haskin, 1969; Canzonier, 1971) and does not address the food and flow rate requirements. It is difficult maintaining food and oxygen in ponds, and there have been problems with predators or competitors invading the ponds. There have been several attempts to grow clams in trays both with and without substrate, but these were not successful. Clams planted in substrate at low enough densities to grow to market size required too many trays to be economically viable. Growing clams to market size without substrate can result in shells covered with fouling. This leaves them vulnerable to weakened shells from boring sponge infestations. Both fouling and the sponge infestation reduce value.

It is universally accepted that the best method of growing clams is in a suitable natural bottom, planting seed large enough that the smaller predators cannot prey on them and utilizing a predator exclusion device to foil the larger predators (Castagna, 1984). Clams will grow under a wide variety of conditions (i.e. salinity, temperature, food, etc.) which can be measured, but the surest way to find suitable planting ground is to sample the natural clam population in the proposed site. If no clams or very few clams are found for whatever reason, this will be at best a very marginal planting area. It is best to find a variety of sizes of healthy looking clams displaying good growth margins and relatively sharp, not blunt, shells. The condition of the natural population will indicate that the physical and biological factors for growing clams are available. The area should be shallow enough so that it can be inspected and maintained (Kraeuter and Castagna, 1985a,b, 1989). Clams grow well in deep water, but it is difficult to farm areas that are not intertidal or shallow enough for comfortable wading for inspections, cleaning nets or inspecting the crop.

15.19 PLANTING PROCEDURE

After a site has been selected and all the necessary permits, licenses, and applications have been processed and seed is acquired, the following procedure is often used. The bed is established using poles or PVC pipe to mark the corners and subareas. It is usually worthwhile to rake or harvest the bed to remove clumps of shell or organisms that can compete with the clam seed. A decision should be made on the density of the planting, usually between 550 and 1650 per m^2 (50 to 150 per square foot) and the seed divided for distribution. Generally the seed are over 9 mm in height, and most culturists prefer 10 to 15 mm seed with 12 mm the most common size. If the seed has not reached that size, some nursery procedure would be followed until the proper size is reached as described in the previous section. The seed are spread as evenly as possible at the proper density, and a buoyant plastic net is spread over the beds. The netting is light density polyethylene (Durethene, Internet, Tenax are some of the brand names, preferably black containing UV inhibitors) and is sold as predator and barrier netting. This netting was originally developed for grape growers to protect the ripening grapes from marauding birds. Some clam farmers use racks constructed of PVC pipe that fit over the bed so the netting can be spread exactly right. Three edges are secured and the rack removed, then the fourth side is secured. A number of methods are used to secure the netting. Perhaps the most popular method is to construct tubes of plastic filter barrier cloth commonly used by road construction crews to control soil erosion near construction sites. The tubes are about 2 m long and 10 to 15 cm in diameter. The tube is filled with sand or gravel, the ends secured, and placed along the edge of the netting to hold the net on the bottom so no crabs or other predator can get under the net. This tubing is flexible and it follows the contours of the bottom so it allows the barrier to work efficiently. Other materials used are reinforcement steel rods (rebar), chain, thin-walled PVC pipe filled with sand, lead core rope, or virtually any other material that can hold the net in place. Some farmers will dig a trench along the edge of the bed and bury the edge of the net by caving in the trench. Whatever is used, it is important to follow the contour of the bottom so that the net will not bridge dips in the bottom allowing openings for predators.

15.20 PREDATORS

This will be covered more thoroughly in Chapter 11, but generally the most important predators are crabs, rays, muricid and naticid snails and birds (Gibbons and Blogoslawski, 1989). Two decades of experience has established that nets properly placed and maintained will deter most predators.

15.21 MONITORING AND MAINTENANCE

Nets should be monitored once or twice a week if possible. If the net is moved, lifted, torn or shifted, remedial action should be taken. Perhaps the greatest menace to culturing clams under nets is burial of the nets. Although clams can survive considerable depth of burial (Kranz, 1974), they will smother if prevented from returning to siphonal depth. In water temperatures above 15°C it is wise to check the nets twice weekly and if the net is partially buried, it should be lifted above the sand or silt so it can float during flood tides. At lower temperatures, weekly checks should be adequate. Fouling by macroalgae and soft organisms such as bryozoans, hydroids, tunicates and sponges will often occur. This will reduce water exchange to the clams and encourage silting and burial. In some areas tube-producing amphipods (*Ampelisca*) can foul the mesh and cause suffocation of the clams. This process can happen within a few weeks so continued vigilance is necessary. *Crassostrea*, *Mytilus*, *Crepidula*, and *Illyanassa* will sometimes foul nets and must be controlled. Fouling can usually be controlled by brushing or sweeping the top of the nets to dislodge the fouling community. In severe cases the net should be replaced and the old net taken to shore to dry before reuse. It is well to occasionally look at the clams, measure their growth, and estimate their survival. Discrete samples should be taken using a core sample or a measuring frame so that accurate estimates of growth and survival can be made. Clams that display slow growth rate are probably too dense and should be harvested and replanted or at least drastically thinned to a more desirable density. It is usually more efficient to simply harvest the entire bed and establish new beds with the proper density. Routine monitoring of clam beds on a daily or weekly schedule not only reduces losses by illuminating problems so they can be immediately addressed, but it apparently also discourages poaching. It is commonly agreed among clam farmers that the best deterrent to poaching is quick arrests and a high conviction rate, especially if severe sentences are imposed on the perpetrators.

15.22 HARVEST

Clams should be harvested as soon as a high enough percent reach market size, about 50 mm in length. Most mechanical graders grade clams by thickness. The percent marketable (about 22 mm thick) should be about 70% unless there is reason to harvest sooner such as an outstanding demand that inflates the price.

Harvesting is carried out by a variety of methods. Perhaps the most commonly used method is hand raking using a bull rake, shinnicock rake or oversized clam rake. At the present time more clams are harvested by hand than any other method. Some growers use a modified clam hack or potato rake with a shortened (sawed off) handle. At least one clam farm rents clam rakes for a harvest your own operation (Faris, 1996). This fee fishing can be very profitable if enough tourists are in the area. Mechanical harvesters are almost always based on a hydraulic manifold jetting water just ahead of a blade similar to a surf clam dredge (Medcof and MacPhail, 1955; MacPhail, 1961; Haven et al., 1979). The clams can then be washed back into a box or onto an escalator or a venturi device that vacuums the clams to a belt or chain escalator and up to the deck of the boat. Unfortunately, many states have restrictions on the type of harvesters that are allowed.

In intertidal areas such as those found in Virginia's seaside lagoons, the clam beds are dug using the propeller wash of an outboard motor. This is done by anchoring the stern of a boat next to the clam bed and washing away the substrate to expose the clams. The exposed clams are then harvested by hand or with an oyster fork. This usually works best on a falling tide so that the clams are visible for harvest.

15.23 PACKING AND SHIPPING

Because of Hazard Analysis Critical Control Point (HACCP) regulations, clams are supposed to be held at 10°C until shipped in a refrigerated transport. Before this recent ruling, clams were normally held at room temperature and shipped on a bed of ice or in a refrigerated truck. Clams have an excellent shelf life provided they are not exposed to direct sunlight or allowed to dry out. They will withstand a few days of freezing provided they are not shaken or dropped. Shelf life of clams is considered to be 10 to 20 days. At the present time almost all cultured clams are sold live in the shell. There appear to be opportunities for developing value added products, such as shucked, canned, frozen, or packaged precooked clams.

15.24 SUMMARY

Mercenaria culture has progressed to where it is a viable and rapidly growing industry, especially in the western mid Atlantic. There are about 300 clam farms operating along the east coast of the United States. Clam culture is carried out in a three step operation — hatchery, nursery, and field grow out. More useful technology is being developed annually which will make the industry more efficient and profitable.

APPENDIX A

Avrenica, Galicana T. PHILIPPINES
Aesch, Harold W. FL
Amor, Analia ARGENTINA
Anders, Merri NY
Anderson, Phillip NJ
Bagwell, Susan E. VA
Bagwell, Tom VA
Bagwell, Yvonne VA
Ballard, Ronald L. MD
Beckley, Judith NJ
Beckley, Richard NJ
Birk, Robert CT
Bondi, Ken CT
Booth, Roy VA
Borrero, Francisco J. SC
Bradford, Brenda CANADA
Buckner, Stuart NY
Burke, Terence M. NY
Burton, Wilbur E. MD
Bush, James FL
Butler, Charles E. NJ
Byrnes, Martin NY
Campos, Bernardita CHILE
Capehart, Tony NC
Carrick, Robert VA
Ceely, Martin MA
Chadwick, David L. MA
Chamberlin, John MD
Chatry, Mark LA
Cherrix, Carroll B. Jr. VA
Common, Tod C. CO
Connell, Bob NJ
Cormier, Paul CANADA
Crema, Barbara NJ
Crema, Richard NJ
Daiger, Richard H. VA
Day, Godfrey G. MA
DeAlteris, Joseph T. RI
Dean, George FL
Denues, John VA
DiCosimo, Jane VA
Donaire, Tiburcio C. PHILIPPINES
Dougherty, Jim CT
Duer, Andy VA
Duncan, Pat VA
Escario, Dale D. PHILIPPINES
Farnam, Thomas M. MD
Faucett, Linford P. III DE
Feldeisen, Donald NJ
Ferrel, Lance SC
Feustel, Kenneth E. NY
Freeman, Don Jr. NC
Frosch, Alfred VA
Fulcher, Joe NC
Garcia, Miguel A. FL
Garlo, Elizabeth V. NC
Garrison, J. Shermer DE
Giberson, Daniel NJ
Gieg, Chuck MA
Gilgo, Julian NC
Ginger, Sydney Dee NC
Gleszer, Kenneth M. CT
Gleszer, K.M. Mrs. CT
Goff, Greg NJ
Goff, Renee NJ
Gonzalez, Aljadys VA
Goodman, Len PR
Gresham, Roy NJ
Grimes, Ellen VA
Hall, Billy VA
Halpin, G.T. VA
Hambley, Richard NC
Handy, Gilbert VA
Healey, Terence D. NY
Hennessee, G. Dillon NC
Hoet, Edgard VENEZUELA
Hoffmann, Charlotte VA
Hoffmann, Fred VA
Holzapfel, John NY
Hopkins, Sewell VA
Horne, Peter J. ME
Howard, Paul E. VA
Howard, Paul E. Jr. VA
Howard, Raymond NC
Howard, Rena NC
Hsu, Ya-Ke KOREA
Hu, Ya-ping P.R. CHINA
Hughes, Pat MA
Hurst, Frank P. VA
Hurwitz, Samuel PA
Huskey, Earl NJ
Jarvis, George J. RI
Jennings, George B. VA
Johnson, JoAnne NY
Jones, David NC
Jordan, Sam NJ
Kamens, Todd C. DE
Keefer, Verne M. VA
Kurkowski, Kenneth VA
Kvaternik, Andre VA
Lamb, Peggy Sue FL
Landry, Thomas CANADA
Laron, Elena L. PHILIPPINES
Lawson, Russ ME

Lemarie, David P. MD
Lewin, Lucas VA
Lewis, Beverly P. VA
Lilliston, William MD
Lilliston, William Jr. MD
Luhrssen, William NJ
Luckenbach, Mark SC
Mac Glaflin, James E. CT
MacLane Kathryn A. VA
Madison, Jeffrey MA
Mahler, Charlie VA
Martin, John O. VA
Martin, Michael NJ
Mattson, James VA
Maurer, Susan NY
McCauley, Jack NC
McCloy, Thomas W. NJ
McGowan, Jay D. NJ
McIntyre, Thomas M. VA
Mears, Benjamin W. III VA
Merritt, Frank VA
Midgett, Billy VA
Midgett, Harry L. VA
Midgett, Kevin W. VA
Midgett, M.M. VA
Midgett, Murray VA
Miller, Anne B. SC
Monahan, Gerald F. NC
Monte, Dave DE
Moore, Mug DE
Mote, Karen NJ
Muller, George R. VA
Murphy, Jack NY
Naeder, Stephen NY
Naeder, Ms S. NY
Neikirk, Chip VA
Nesheim, Martin VA
Nickel, Bill VA
Noonan, Patrick F. DC
Nordstrom, Chip VA
Oesterling, Mike VA
O'Keefe, Jim NJ
Oliver, Stuart L. VA
Palazzolo, Richard NY
Parks, Roy G. Jr. VA
Peirson, Lee VA
Peirson, Michael W. VA
Petre, Preston O. VA
Petro, Janet NJ
Pfeuffer, Ray NJ
Phoel, W.C. NJ
Rae, Robert NJ
Realy, William B. FL
Resler, Steven C. NY
Robinson, Tom CT
Rogers, Bruce A. RI
Rollison, Stephen VA
Rose, Amaranth VA
Rose, R.E. VA
Rucker, William VA
Saetta, Bobby NY
Safrit, Glenn Jr. NC
Sage, Roland GA
Sanderlin, Steve VA
Schneider, Randall R. MD
Scott, Roy F. MD
Scull, Bob NJ
Shoemaker, David S. NJ
Simmons, George M. Jr. VA
Simpson, John P. MD
Spence, Stephanie A. NC
Sterling, Newton NJ
Stevens, Robert P. VA
Stocker, Lee NY
Stroup, C. Foster NJ
Swan, William H. NY
Sweeney, Brian VA
Sweiden, Phil NJ
Sykes, Robert O. VA
Sziklas, Robert W. MA
Takacs, Michael MA
Tamblyn, Richard F. Jr. FL
Terry, Kenneth S. VA
Terry, Wec VA
Thomas, Barbara VA
Tinsman, Jeff DE
Trahan, Robert CT
Tucker, Beverly R. VA
Turner, David ENGLAND
Tyler, Glen A. VA
Valentine, Edward D. VA
Vanderhoop, David E. MA
Vanderhoop, Lenny MA
Van Housen, Garret NC
Vause, H.L. FL
Vause, Joyce FL
Velez, Anibal VENEZUELA
Vouglitois, James J. NJ
Walker, Randal L. GA
Walker, Thomas VA
Walker, Wade VA
Walsh, Bill VA
Wells, Morgan MD
Whetstone, Jack SC
White, Edward MD
Whorton, Everett L. Jr. MA
Wigger, John NC
Wigger, Terri NC

Williams, Armistead VA
Williams, Emanuel Eion FL
Wilson, Robert NJ
Wilson, Sharon NJ
Wolleyhan, Elizibeth M. DE
Womack, Harry MD
Wustner, Michael DC
Wynne, David GA
Zodl, Jerry NJ

APPENDIX B

GASTROPODS
Busycon carica (Gmelin, 1791).
Busycotypus canaliculatum (Linné, 1758).
Crepidula convexa Say, 1822.
Crepidula fornicata (Linné, 1758).
Euplera caudata (Say, 1822).
Eupleura caudata etteri B. Baker, 1951.
Ilyanassa obsoleta (Say, 1822).
Littorina littorea (Linné, 1758)
Nassarius vibex (Say, 1822).
Polinices duplicata (Say, 1822).
Urosalpinx cinerea (Say, 1822).
Urosalpinx cinera folyensis B. Baker, 1951.

NUDIBRANCHS
Acanthodoris pilosa (Abilgaard, 1789).
Aldeeria modesta (Loven, 1844).
Cratena kaoruae Marcus, 1957.
Cratena pilata (Gould, 1870).
Doris verrucosa (Linné, 1858).
Doridella obscura Verrill, 1870.
Doriopsilla pharpa Marcus, 1961.
Elysia chlorotica Gould, 1970.
Ercolania vanellus (Marcus, 1957).
Hermaea cruciata Gould, 1870.
Okenia cupella (Vogel and Schultz, 1970).
Placida dendritica (Alder and Hancock, 1855).
Polycerella emertoni Verrill, 1881).
Stiliger fuscatus (Gould, 1870).
Tenellia aspersa (Nordmann, 1845).
Tenellia fuscata (Gould, 1870).
Tenellia sp.

BIVALVES
Amygdalum papyrium (Conrad, 1856).
Anomia simplex Orbigny, 1842.
Anadara ovalis (Bruguiere, 1789).
Anadara transversa (Say, 1822).
Arctica islandica (Linné, 1767).
Argopecten gibbus (Linné, 1758).
Argopecten irradians amplicostata Dall, 1898.
Argopecten irradians concentricus (Say, 1822).
Argopecten irradians irradians (LMexk, 1819).
Bankia gouldi Bartsch, 1908.
Barnea truncata (Say, 1822).
Chione cancellata (Linné, 1767).
Congeria leucophaeta (Conrad, 1831).
Crassostrea virginica (Gmelin, 1791).
Crassostrea gigas (Thunberg, 1793).
Corbicula manilensis Phillippi.
Cyrtopleura costata (Linné, 1758).
Diplothyra smithii Tryon, 1862.
Donax fosser Say, 1822.
Donax variabilis Say, 1822.
Donax variabilis roemeri Philippi, 1849.
Dosinia discus (Reeve, 1850).
Ensis directus Conrad, 1843.
Gemma gemma (Totten, 1834).
Geukensia demissa (Dillwyn, 1817).
Ischadium recurvum (Rafinesque, 1820).
Laevicardium mortoni (Conrad, 1830).
Lithophaga bisulcata (Orbigny, 1842).
Lyonsia hyalina Conrad, 1831.
Macoma balthica (Linné, 1758).
Macrocallista nimbosa (Lightfoot, 1786).
Mercenaria campechiensis (Gmelin, 1791).
Mercenaria mercenaria (Linné, 1758).
Mercenaria mercenaria notata (Say, 1822).
Mercenaria mercenaria texana (Dall, 1902).
Mulinia lateralis (Say, 1822).
Mya arenaria Linné, 1758.
Mytilus edulis Linné, 1758.
Noetia ponderosa (Say, 1822).
Ostrea edulis Linné, 1758.
Ostrea equestris Say, 1834.
Pandora gouldiana Dall, 1886.
Petricola pholadiformis (Lamarck, 1818).
Pitar marrhuanus (Linsley, 1848).
Rangea cuniata (Sowerby, 1831).
Solemya velum Say, 1822.
Solen viridis Say, 1821.
Spisula solidissima (Dillwyn, 1817).
Tagelus plebeius (Lightfoot, 1786).
Tellina agilis Stimpson, 1857.
Teredo navalis Linné, 1758.

CEPHALOPOD
Loligo pealeii Lesueur, 1821.

MISCELLANEOUS
Arbacia punctulata (Lamarck).
Fundulus heteroclitus.

REFERENCES

Baldwin, R.B., Mook, W., Hadley, N.H., Rhodes, R.J., and Devoe R.M., 1995. Construction and operational manual for a tidal-powered upwelling nursery system. South Carolina Sea Grant Consortium, Marine Extension Program Publication, April 1995, Charleston, SC, 44 pp.

Bayes, J.C., 1981. Forced upwelling nurseries for oysters and clams using impounded water systems. In: C. Claus, N. DePauw and E. Jaspers (Eds.), Spec. Publ. 7, European Mariculture Society, Bredene, pp. 73–83.

Bisker, R. and Castagna, M., 1985. The effect of various levels of air-supersaturated seawater on *Mercenaria mercenaria* (Linné), *Mulinia lateralis* (Say) and *Mya arenaria* Linné, with reference to gas bubble disease. J. Shellfish Res., 5: 97–102.

Bolton, E.T., 1982. Introduction. In: E.T. Bolton (Ed.), Marine Bivalve Cultivation in a Controlled Recirculating Seawater Prototype System. DEL-SG-07-82. University of Delaware Sea Grant College Program, Newark, pp. 7–14.

Brooks, W.K., 1879. Abstracts of observations upon artificial fertilization of oyster eggs and embryology of American oysters. Am. J. Sci., 18: 425–427.

Brooks, W.K., 1880. The development of the American oyster. Stud. Biol. Lab. Johns Hopkins Univ. 4: 1–104.

Brooks, W.K., 1891. The Oyster, A Popular Summary of a Scientific Study. The Johns Hopkins Press, 230 pp.

Canzonier, W.J., 1971. Measurement of the oxygen requirements of a large population of *Mercenaria*. Proc. Natl. Shellfish. Assoc. 61: 2 (abstr.).

Castagna, M., 1978. Need and use of low technology aquaculture. Rep. Conf. Marine Resources of the Coastal Plains States, Williamsburg, VA, Dec. 8–9, 1977, pp. 59–60.

Castagna, M., 1983a. Review of recent bivalve culture methods. J. World Maricul. Soc., 14: 567–575.

Castagna, M., 1983b. Culture methods for growing the clam *Mercenaria mercenaria*. Mem. Asoc. Latinoam. Acuicult., 5: 283–288.

Castagna, M., 1984. Methods of growing *Mercenaria mercenaria* from postlarval to preferred-size seed for field planting. Aquaculture, 39: 355–359.

Castagna, M., 1987. Mollusk culture for the Chesapeake Bay. In: S.K. Majumdar, L.W. Hall, Jr. and H.M. Austin (Eds.), Contaminant Problems and Management of Living Chesapeake Bay Resources. Academy of Sciences, Easton, PA, 573 pp.

Castagna, M. and Kraeuter, J.N., 1977. *Mercenaria* culture using stone aggregate for predator protection. Proc. Natl. Shellfish. Assoc., 67: 1–6.

Castagna, M., and Kraeuter, J.N., 1981. Manual for growing the hard clam *Mercenaria*. VIMS Special Report in Applied Marine Science and Ocean Engineering 249, 110 pp.

Castagna, M., and Manzi, J.J., 1989. Clam culture in North America: Hatchery production of nursery stock clams. In: J.J. Manzi and M. Castagna (Eds.), Clam Mariculture in North America. Elsevier, New York, Ch. 4, pp. 111–125.

Castagna, M., Gibbons, M.C., and Kurkowski, K., 1996. Culture: Application. In: V.S. Kennedy, R.E.I. Newell and A.E. Eble (Eds.), The Eastern Oyster *Crassostrea virginica*. Maryland Sea Grant College Publication UM-SG-TS-96-01. Ch. 19, pp. 675–690.

Chanley, P.E., 1961. Inheritance of shell markings and growth in the hard clam, *Venus mercenaria*. Proc. Natl. Shellfish. Assoc., 50: 163–169.

Coon, S.L., Bonar, D.B., and Weiner, R.M., 1985. Control of oyster settlement and metamorphosis by L-DOPA, epinephrine, and other related compounds. In: The Fate and Effects of Pollutants, a Symposium. Maryland Sea Grant Publ., UM-SG-TS-85-02, Univ. of Maryland, College Park, MD P. 58 (Abstr.).

Eversole, A.G., 1989. Gametogenesis and spawning in North American clam populations: Implications for culture. In: J.J. Manzi and M. Castagna (Eds.), Clam Mariculture in North America. Elsevier, New York, Ch. 3, pp. 75–109.

Faris, J., 1996. Rakin' (sic) in the clams. Coastwatch, North Carolina Sea Grant, May/June, pp. 2–9.

Fong, P.P., Deguchi, R. and Kyozuka, K., 1996. Serotonergic ligands induce spawning but not oocyte maturation in the bivalve *Mactra chinensis* from central Japan. Biol. Bull., 191: 27–32.

Gallager, S.M. and Mann, R., 1986. Growth and survival of larvae of *Mercenaria mercenaria* (L.) and *Crassostrea virginica* (Gmelin) relative to broodstock conditioning and lipid content of eggs. Aquaculture, 56: 105–121.

Gallager, S.M., Mann, R. and Sasaki, G.C., 1986. Lipid as an index of growth and viability in three species of bivalve larvae. Aquaculture, 56: 81–103.

Gibbons, M.C., and Blogoslawski, W.J., 1989. Predators, pests, parasites and diseases. In: J.J. Manzi and M. Castagna (Eds.), Clam Mariculture in North America. Elsevier, New York. Ch. 7, pp. 167–200.
Gibbons, M.C. and Castagna, M., 1984. Serotonin as an inducer of spawning in six bivalve species. Aquaculture, 40: 189–191.
Gibbons, M.C. and Castagna, M., 1985. Responses of the hard clam *Mercenaria mercenaria* (Linné) to induction of spawning by serotonin. J. Shellfish Res., 5: 65–67.
Glancy, J.B., 1965. Method of raising shellfish seed in a simulated habitat. Patent No. 3,196,833, July 27.
Hamwi, A. and Haskin, H.H., 1969. Oxygen consumption and pumping rates in the hard clam *Mercenaria mercenaria*: a direct method. Science, 163 (3869): 823–824.
Haven, D.S., Whitcomb, J.P. and Davis, Q.C., 1979. A mechanical escalator harvester for live oysters and shell. Mar. Fish. Rev., 41 (12): 17–20.
Iverson, E.S., 1968. Farming the Edge of the Sea. Fishing News (Books), London, 301 pp.
Kraeuter, J.N. and Castagna, M., 1977. An analysis of gravel, pens, crab traps and current baffles as protection for juvenile hard clams *Mercenaria mercenaria*. Proc. World Maricul. Soc., 8: 581–592.
Kraeuter, J.N. and Castagna, M., 1980. Effects of large predators on the field culture of the hard clam, *Mercenaria mercenaria*. Fish. Bull., 78: 538–541.
Kraeuter, J.N. and Castagna, M., 1985a. The effect of clam size, net size, and poisoned bait treatments on survival of hard clam, *Mercenaria mercenaria*, seed in field plots. J. World Maricul. Soc., 16: 377–385.
Kraeuter, J.N. and Castagna, M., 1985b. The effects of seed size, shell bags, crab traps and netting on the survival of the northern hard clam *Mercenaria mercenaria* (Linné). J. Shellfish Res., 5: 69–72.
Kraeuter, J.N., and Castagna, M., 1989. Factors affecting the growth and survival of clam seed planted in the natural environment. In: J.J. Manzi and M. Castagna (Eds.), Clam Mariculture in North America. Elsevier, New York, Ch. 6, pp. 149–165.
Kraeuter, J.N., Castagna, M. and Van Dessel, R., 1982. Egg size and larval survival of *Mercenaria mercenaria* and *Argopecten irradians*. J. Exp. Mar. Biol. Ecol., 56: 3–8.
Kranz, P.M., 1974. The anastrophic burial of bivalves and its paleoecological significance. J. Geol., 82: 237–265.
Loosanoff, V.L. and Davis, H.C., 1963. Rearing of bivalve mollusks. Adv. Mar. Biol., 1: 1–136.
Loosanoff, V.L. and Engle, J.B., 1942. Use of complete fertilizers in cultivation of microorganisms. Science, 95: 487–488.
MacPhail, J.S., 1961. A hydraulic escalator shellfish harvester. Fish. Res. Board Can., Bull. 128, 24 pp.
Manahan, D.T., 1983. The uptake and metabolism of dissolved amino acids by bivalve larvae. Biol. Bull., 164: 236–250.
Manzi, J.J., 1985. Clam aquaculture. In: J. Huner and E. Brown (Eds.), Crustacean and Mollusk Aquaculture in the United States. AVI Publishing, Westport, CT, Ch. 7, pp. 275–310.
Manzi, J.J., and Castagna, M., 1989a. History of clam culture in North America. In: J.J. Manzi and M. Castagna (Eds.), Clam Mariculture in North America. Elsevier, New York, Ch. 1, pp. 11–16.
Manzi, J.J., and Castagna, M., 1989b. Nursery culture of clams in North America. In: J.J. Manzi and M. Castagna (Eds.), Clam Mariculture in North America. Elsevier, New York, Ch. 5, pp. 127–147.
Manzi, J.J., and Whetstone, J.M., 1981. Intensive hard clam mariculture: a primer for South Carolina watermen. South Carolina Sea Grant Consortium, MAS Publ. No. 81-01, Charleston, SC, 20 pp.
Manzi, J.J., Hadley, N.H., Batley, C., Haggerty, R., Hamilton, R. and Carter, M., 1985. Hard clam, *Mercenaria mercenaria*, culture in a commercial-scale upflow nursery system. J. Shellfish Res., 4 (2): 119–124.
Manzi, J.J., Hadley, N.H. and Maddox, M.B., 1986. Seed clam, *Mercenaria mercenaria*, culture in an experimental-scale upflow nursery system. Aquaculture, 54: 301–311.
Martinez, G., Garrote, C., Metligogo, L., Perez, H. and Vribe, E., 1996. Monoamines and prostaglandin E_2 as inducers of the spawning of the scallop *Argopecten purpuratus* Lamarck. J. Shellfish Res., 15 (2): 245–249.
McConnaughey, B.H., and Zottoli, R., 1989. Introduction to Marine Biology. Waveland Press, Prospect Heights, IL. Issued 1983, reissued 1989.
Medcof, J.C., and MacPhail, J.S., 1955. Survey of bar clam resources of the Maritime Provinces. Fish. Res. Board Can., Bull. 102, 6 pp.
Menzel, R.W. and Sims, H.W., 1964. Experimental farming of hard clams, *Mercenaria mercenaria*, in Florida. Proc. Natl. Shellfish. Assoc., 53: 103–109.
Milne, P.H., 1972. Fish and Shellfish Farming in Coastal Waters. Whitefriars Press, London, 208 pp.
Mook, W., 1988. Guide to construction of a tidal upweller. Mook Sea Farm, Damariscotta, Maine, 23 pp.

Mook, W., and Johnson, N.C., 1988. Utilization of low-cost, tidal-powered floating nurseries to rear bivalve seed. Mook Sea Farm, Damariscotta, Maine, 28 pp.
Morrison, G., 1971. Dissolved oxygen requirements for embryonic and larval development of the hard shell clam *Mercenaria mercenaria*. J. Fish. Res. Board, Can., 28: 379–381.
Ogle, J.T., 1982. Operation of an oyster hatchery utilizing a brown water culture technique. J. Shellfish Res., 2: 153–156.
Pratt, D.M. and Campbell, D.A., 1956. Environmental factors affecting growth of *Mercenaria*. Limnol. Oceanogr., 1: 2–17.
TIGG Corporation, 1988. Selection criteria for granular activated carbon (GAC). TIGG Corp. Box 11661, Pittsburgh, PA 15228. TIGG Corp. Copyright, 1988, 5 pp.
Vaughan, D.L., Creswell, L., and Pardee, M., 1988. A manual for farming the hard shell clam in Florida. Aquaculture Report Series, Florida Dep. Agriculture and Consumer Services, Tallahassee, FL, 42 pp.
Walker, R.L., and Hurley, D.H., 1995. Biological feasibility of mesh bag culture of the northern quahog *Mercenaria mercenaria* (L.) in soft-bottom sediments in coastal waters of Georgia. Georgia Sea Grant College Program, The University of Georgia, Athens, GA. Mar. Extension Bull. 16, August 1995, 35 pp.
Wells, W.F., 1924. First Report of Station for Shellfish Culture. Annual Report to the Legislature, New York State Conserv. Comm., pp. 98–117.
Wells, W.F., 1925. Opening of New Chapter in Shellfish Culture. Annual Report to the Legislature, New York State Conserv. Comm., pp. 93–126.
Wells, W.F., 1926. Report of Experimental Shellfish Station. Annual Report to the Legislature, New York State Conserv. Comm., pp. 112–130.
Wells, W.F., 1933. Methods of Shellfish Culture. Patent No. 1,933,950, Nov. 14, 1933.
Wells, W.F., 1969. Early Oyster Culture Investigations by the New York State Conserv. (1920–1926). Reprinted by State of New York Conserv. Dep., Division of Marine and Coastal Resource, 119 pp.

Chapter 16

Introduction of the Hard Clam (Mercenaria mercenaria) *to the Pacific Coast of North America with Notes on its Introduction to Puerto Rico, England, and France*

Kenneth K. Chew

16.1 INTRODUCTION

The native range of *Mercenaria mercenaria* is from the Gulf of Maine to Florida and the Gulf Coast of the United States (Loosanoff, 1946; Abbott, 1974). This bivalve is a suspension feeder and its depth distribution ranges from the intertidal to about 15 m. It is most common on flats near the low water line (Loosanoff, 1944), lives in most substrates, but prefers sandy and soft bottoms with some shell, but is often abundant in seagrass beds (Pratt, 1953; Pratt and Campbell, 1956; Wells, 1957; Rhoads and Panella, 1970). Large populations are most commonly found in bays or protected areas where temperature is between 2 and 28°C and salinity ranges from 17 to 32 ppt (Castagna and Chanley, 1973; Kraeuter and Castagna, 1977).

This paper covers the status of introduced populations of *M. mercenaria* to the Pacific coast of North America and other countries as Puerto Rico, England, and France. No doubt other countries have received introductions but will not be discussed here.

16.2 PACIFIC COAST OF NORTH AMERICA

From a historical point of view, it was difficult to find information concerning the transport and introduction of the quahog or hard clam *M. mercenaria* to the Pacific coast. In checking with key researchers along the Pacific coast, there was no indication of transplants into British Columbia or the state of Oregon. However, the state of California and Washington did receive clams for testing at various times, and this will be discussed. Hanna (1966) reported that there was an introduction of the hard clam into Newport Bay, California in 1940, but there was no evidence that this was successful.

16.2.1 State of California

The information of hard clam transplants into the state of California was gathered by Mr. Emil J. Smith Jr., Associate Marine Biologist for the Marine Resources Division of the California Department of Fish and Game. He noted that documented introduction of *M. mercenaria* into the San Francisco Bay was reported in an investigation conducted by American Shellfish Corporation on behalf of Ideal Basic Industries and Westbay Community Associates in 1980–1992. Portions of a draft report on this introduction were obtained from the California Department of Fish and Game headquarters in Sacramento, California.

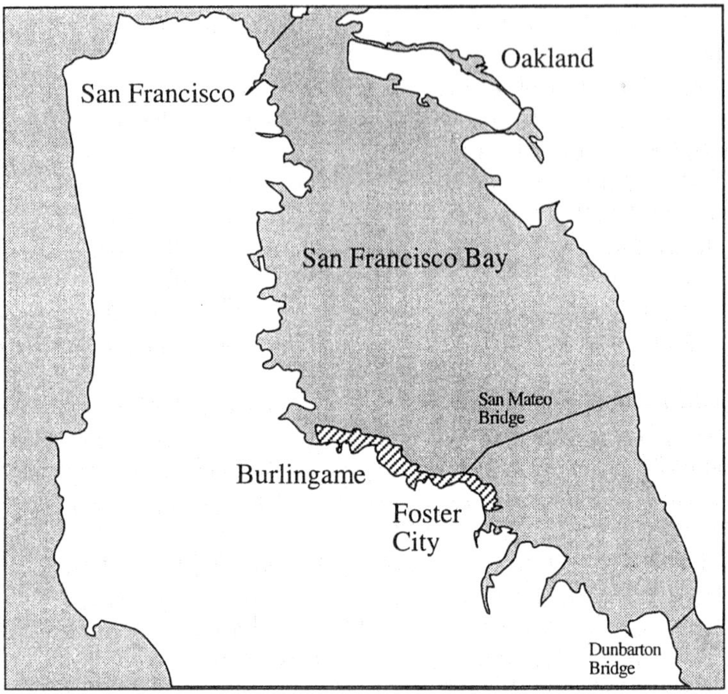

Fig. 16.1. Southern San Francisco Bay Study Areas (hatch mark area) where hard clams were introduced between March 27, 1981 through May 20, 1981.

Mr. Smith indicated that perhaps the late Mr. Walter Dahlstrom, former biologist for the California Department of Fish and Game in charge of shellfish, may have planted some hard clams in the late 1950s and early 1960s, but this cannot be verified. During the 1950s and early 1960s the state of Washington had introduced the hard clam at various times and it is known that Mr. Dahlstrom had regular contact with biologists from the Washington Department of Fisheries (now the Washington Department of Fish and Wildlife) and could have had an opportunity to introduce the hard clam in California during the same period.

The introduction of hard clams between 1980 and 1982 in a location at the southwest corner of San Francisco Bay, between Burlingame and Foster City (Fig. 16.1), was experimental. Plots were established on the north and south sides of the San Mateo Bridge. Most of the experimental plots were located towards the northern part of the San Mateo Bridge. The following data on quantities of clams received and planted was made available by Mr. Smith.

March 27, 1981. 2700 seed clams (*M. mercenaria*) were planted in furrows with no protective netting on top. Planting took place on several plots near Foster City.

April 21, 1981. Introduction of 19,000 seed clams which were planted on 12 × 12 ft plots with protective netting on top. These seed clams were planted just north of the San Mateo Bridge.

March 22, 1981. 54,000 small seed clams were received with 24,000 placed in long rows under netting and 30,000 planted without netting.

March 23, 1981. 22,500 seed clams were planted into various habitats located north of the San Mateo Bridge.

May 20, 1981. 57,500 seed clams were introduced in experimental plots between Burlingame and the San Mateo Bridge. Of these numbers, 20,000 were planted under netting and 37,500 were planted in open areas without netting.

More specific information concerning these plantings is not available. One would assume the seed clams came from Maine or Massachusetts. This planting study was later followed up by the Department of Fish and Game for the San Francisco Bay Regional Water Quality Board to determine if any of the clams survived or reproduced. None were found, but even if success in spawning had occurred, the settlement of larval clams would not necessarily be at the study area. The currents could have carried the pelagic larvae to a distant location before setting.

Mr. Smith also recalls that a small reproducing population of *M. mercenaria* was found in the late 1950s or early 1960s in a tidal basin off Long Beach near the Marine Stadium. This population was apparently well established by 1967 (Salchak and Haas, 1971), and in 1969 Cerritos College conducted a survey of the intertidal area in the lagoon. This survey of 31 m^2 quadrats found 267 specimens ranging in size from 8 mm to 101 mm (Salchak and Haas, 1971). This population was later utilized by Murphy (1985) in his study of the effects of bioturbation on *Mercenaria*. Quantitative surveys of Colorado Lagoon and Marine Stadium at the Long Beach area indicated that the bivalve communities in the Lagoon and the Stadium were dramatically different, even though water exchanged freely between the two systems for most of the year (Murphy, 1985). The Lagoon's total bivalve density was $143/m^2$, inclusive of a hard clam density of $78/m^2$. In the Stadium the mean total clam density was $57/m^2$ and the *M. mercenaria* was absent. It was further noted that in the Lagoon, bivalve populations were dominated by suspension-feeders (which includes *M. mercenaria)*, while in the Stadium deposit-feeders were most abundant. The absence in the Stadium of hard clams was apparently related to the more abundant burrowing ghost shrimp, *Callianassa californiensis*. This shrimp burrows deep tunnels into the substrate and as water is drawn through, it can create levels of turbidity and sediment destabilization to sufficiently reduce survival and growth of hard clams as noted by Murphy (1985). Although not confirmed, there could still be populations of *M. mercenaria* in Colorado Lagoon from natural spawning.

16.2.2 State of Washington

Archived information from the Washington Department of Fish and Wildlife was made available through the efforts of Mr. Doug Thompson, former Fisheries Biologist for the Marine Fish and Shellfish Program of the Department. Mr. Thompson is Hatchery Manager for the Coast Seafood Company shellfish operations near Quilcene, Washington. As shown in Fig. 16.2, the hard clams that were brought over from the east coast were planted as juveniles or adults in seven different areas over the time span between 1954 and 1962. A chronological review of how the hard clams were brought in and their distribution are as follows.

September 1954. Arrangements were made to receive hatchery-produced juveniles through the efforts of the late Dr. Victor Loosanoff of the National Marine Fisheries Service Milford Laboratory in Connecticut. During this time approximately 7500 small hard clams produced from the hatchery were sent over to Washington State. Of the 7500, 5000 were

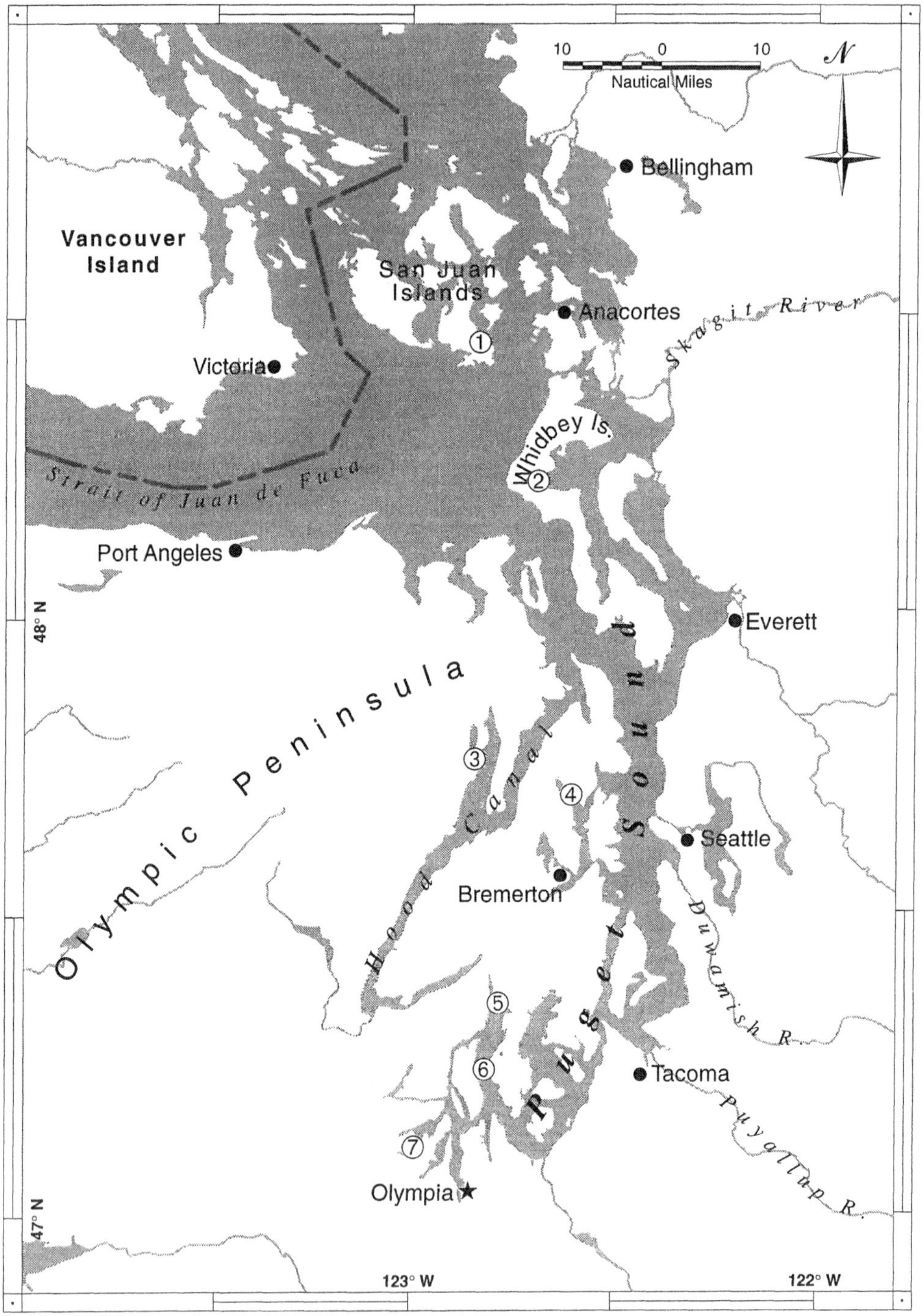

10 0 10
Nautical Miles
N
Bellingham
Vancouver Island
San Juan Islands
Anacortes
Skagit River
Victoria
Whidbey Is.
Strait of Juan de Fuca
Port Angeles
Everett
48° N
Olympic Peninsula
Hood Canal
Puget Sound
Seattle
Bremerton
Duwamish R.
Tacoma
Puyallup R.
Olympia
47° N
123° W
122° W
①
②
③
④
⑤
⑥
⑦

small *M. mercenaria*, 500 small *M. campechiensis*, 1000 hybrids (*M. mercenaria* × *M. campechiensis*), and 1000 hybrids (*M. campechiensis* × *M. mercenaria*). These were held in Point Whitney lagoon and most were moved and introduced to the North Bay reserve in southern Puget Sound (Fig. 16.2).

May 1958. The Maine Department of Sea and Shore Fisheries sent over 7.5 kegs of small *M. mercenaria*. Each keg is estimated to be 1.5 bushels in size. Three of these kegs of small clams were planted at Penn Cove on May 7, 1958. Also, on May 20 two kegs were introduced to Oyster Bay in southern Puget Sound. One and a half kegs of the small clams were kept at Point Whitney.

April 1959. Communication from Mr. Dana Wallace indicated that more small clams were shipped over from the Maine Department of Sea and Shores Fisheries. The quantity is unknown, but apparently they were planted in Mud Bay, Lopez Island.

June 1959. Another keg of less than 2-inch *M. mercenaria* was brought in from the Maine Department of Sea and Shores Fisheries. This batch was apparently introduced into Case Inlet in southern Puget Sound.

January 1960. Up to 50 lb. of live adult clams were shipped in from the Massachusetts Department of Natural Resources, Division of Marine Fisheries. These clams were kept at Point Whitney for a short time and the bulk of them were planted in Case Inlet in southern Puget Sound. A small portion apparently was planted at Keyport Lagoon in trays on March 9, 1960.

After this last shipment in January 1960, the Director of the Washington Department of Fisheries at the time, Mr. Milo Moore, noted in his correspondence that there should be no further introduction of the hard clam due to concerns with disease. He also noted that it was time that work should be started to produce the hard clam seed in the established Washington State shellfish hatchery. Once the seed is acquired from the hatchery, it could be distributed for further testing.

Summer 1962. Although the quantity is unknown, seed was produced at the Point Whitney Shellfish Laboratory of the Washington State Department of Fisheries. Some of these seed clams were planted at Keyport Lagoon, Mud Bay on Lopez Island, and Case Inlet in southern Puget Sound. Subsequent memos related to this planting have shown that the best survival and growth occurred at Case Inlet.

Later correspondence (September 7, 1970) by Mr. Cedric Lindsay, former Director of the Washington State Department of Fisheries Shellfish Laboratory at Point Whitney in Hood Canal, indicated that there could be another possible transfer from the Long Island Oyster Farm in West Sayville, New York to the Pacific Northwest. A sample of the clams was to be shipped to the then National Marine Fisheries Service Oxford Laboratory in Maryland to determine if they were disease-free. The disease check was apparently not made, and no clams were shipped to the state of Washington.

Fig. 16.2. Locations where clams were introduced from 1954–1962 in Washington State. (1) Mud Bay, Lopez Island: April, 1959; summer, 1962. (2) Penn Cove, Whidbey Island: May 7, 1958; May 20, 1958. (3) Point Whitney Lagoon, Hood Canal: September, 1954; May, 1958; January, 1960. (4) Keyport Lagoon, Liberty Bay: March 9, 1960; summer, 1962. (5) North Bay Reserve: September, 1954. (6) Case Inlet, southern Puget Sound: June, 1959; January, 1960; summer, 1962. (7) Oyster Bay, southern Puget Sound: May 29, 1958.

Recent discussions with Mr. Lindsay indicated that the only time there was any recovery of clams occurred within a year after planting. He recollected recovering some clams after one year in Mud Bay at Lopez Island. He does not recollect hearing of anyone recovering any hard clams after several years. He noted that perhaps the water may be too cold for reproduction and spawning in all areas where clams were planted. Mr. Lindsay also remembered that there was a planting on the east side of Grays Harbor near Bay City back in the 1960s. However, he was not clear about what may have happened to them.

Finally, it can be assumed that all the plantings that have taken place in the state of Washington in the late 1950s and early 1960s have led to failures.

16.3 PUERTO RICO

Between July 1985 and October 1989, populations of *M. mercenaria* were discovered in Puerto Rico (Juste and Cortes, 1990). During this study period, the authors found live clams at maximum and minimum temperature and salinity of 34–28.9°C and 34–15 ppt, respectively. These temperatures approach the maximum where the species has been reported to survive (Ansell, 1968, and see Chapter 8). Maximum hard clam density in Puerto Rico ranged from 0.2 to $0.8/m^2$, but previous sampling had revealed population maxima of $0.4/m^2$. Only *M. mercenaria* were reported. In Laguna Torrecillas, on the east side of the island, the authors found both *M. mercenaria* and *M. campechiensis*. Approximately 10% of the clams had notable markings. All clams found in Playa Santa on the southwestern portion of the island were *M. mercenaria* (Juste and Cortes, 1990). These populations were heavily fished, and at the time of the report, the authors noted that the population at this site appeared to be declining.

The genus *Mercenaria* has not previously been reported in Puerto Rico, and does not appear in the island's fossil records (Dahl and Simpson, 1901; McLean, 1951; Warmke and Abbott, 1961; Abbott, 1974; Morris, 1975; Rehder, 1981). The presence of the species in Puerto Rico represented a new distribution record for the Caribbean (Juste and Cortes, 1990), and was undoubtedly due to some form of introduction.

16.4 ENGLAND

Heppell (1961) provides a historical account of the introduction of the species into England, and additional data on European populations. According to his report, the shell of a *Mercenaria* from Bootle was exhibited at the Literary and Philosophical Society of Liverpool in 1850. Live specimens were found in the Humber in 1864, 1868, and by 1889 there were reports of the species being harvested commercially.

Moore (1884, 1886) reported on attempts to introduce the hard clam to England in 1869 and another attempt in 1883. In the first attempt they were placed in enclosures in Kent, others were placed into the ocean at Southport, some near the mouth of the Mersey and the remainder at Crosby. These introductions were not successful. The second attempt in 1883 introduced the species into the Dee estuary, and again the individuals doing the introduction were not able to demonstrate success.

Heppell (1961) further records that in the 1920s shells were found in the Menai Straits where American oysters had been imported and stored. Colonies of hard clams were found in the Solent area in 1956, in Southampton Water at Marchwood in 1958, in Weston and Hamble

Spit in 1959. Utting and Spencer (1992) reported on introductions of commercial mollusks to England, and indicated that there had been more species of bivalves introduced into England than any other group of marine animals. Most of these introductions were deliberate. The early history presented by Heppell (1961) clearly suggests multiple attempts at introduction of the hard clam. Other possible attempts have been attributed to American servicemen during the first world war, ballast in sailing ships from New York, or discards from transatlantic liners. Utting and Spencer (1992) report a persistent rumor that some leftover clams were discarded from the galley larder of a cruise ship, and lastly, there is always the possibility of the transport via bilge water. There is no doubt that, whatever the form of the introduction, the populations in Southampton Water reproduced and became established. Heppell (1961) recorded visiting Weston in May 1960 and found many hundreds of single and double valves and live specimens. Ansell (1964) reported that *Mercenaria* had been found in the Solent and that it "... probably occurs along the whole eastern shore of the Southampton Water and the Solent...". Ansell (1964) provides length–frequency histograms for the population in three locations. These histograms clearly show clams of a variety of ages. Subsequent to this effort, Rodhouse (1973), Mitchell (1974) and Hibbert (1975, 1977a,b) documented recruitment, production and predation on the Southampton Water hard clam population. These studies provided evidence that this population was well established and recruitment was occurring.

The first major survey of the Southampton Water hard clam population in 1979 suggested a population of about 1500 t. Landings increased until the late 1980s when the stock declined (Utting and Spencer, 1992). In the early 1970s, the Marchwood power station at the head of Southampton Water closed. After that, irregular recruitment occurred, and some attribute this to the lack of thermal addition that was supplied by the power plant (Utting and Spencer, 1992). The status of this population is uncertain.

16.5 FRANCE

According to literature cited in Heppell (1961), attempts were made to introduce hard clams to the Arcachon basin in 1861. While these clams grew, there was no evidence that they established a reproducing population. The first recorded successful introduction of the hard clam into France was in the Seudre basin by Prunier in 1910. This population became established, and within a few years the population became sufficiently large to support a fishery (Ruckenbusche, 1949; Heppell, 1961). Introduction of these species into other locations such as Concarneau, St-Armel, Sarzeau by Prunier were not successful (Lambert, 1949; Heppell, 1961). Heppell (1961) noted that clam fisheries on this species were being conducted in Mornac and Marennes.

Hard clams were imported into Brittany from 1936 until WWII stopped this practice. These clams were placed in oyster fattening ponds, primarily near the river Belon (Heppell, 1961). In general, these clams were only held for a short time before marketing, but some were missed, and specimens have been found in areas that were not used for holding hard clams for market. Marteil (1956) provided an account of clams from the oyster *claires* near the rivers Etel and Penerf, and in the Gulf of Morbihan. Other large clams were found in areas where oysters had been imported such as the rivers Bono and Auray. Small individuals were also found in the river Bono and were interpreted to indicate that hard clams had recruited during the warm years of 1945, 1947, 1949, and 1952 to form the local population (Marteil, 1956).

16.6 SUMMARY

It is apparent that *M. mercenaria* has been intentionally introduced into the Pacific coast waters of North America in the states of California and Washington, and in England and France, and has probably been transported to England, France, and the West Coast of North America with imports of the American oyster (*Crassostrea virginica*). The source of the Puerto Rico populations is uncertain. It is very likely that more hard clams have been imported into these and other countries than have been recorded. It is also clear that most of these introductions do not lead to established, viable populations, but the exact reasons for success or failure of these introductions are not clear. Examples were noted where reproducing populations were established under appropriate hydrographic and bottom substrate conditions.

16.7 ACKNOWLEDGMENTS

Special thanks are extended to Dr. John Kraeuter of Rutgers University and Mr. Michael Castagna of Virginia Institute of Marine Sciences for information on Puerto Rico, England and France, and to Mr. Emil J. Smith Jr., Associate Marine Biologist, Marine Resources Division of the California Department of Fish and Game, for information on the California introductions. Mr. Doug Thompson of Coast Seafoods was instrumental in pulling together letters and internal reports relative to introductions in the state of Washington. Special thanks are also extended to Mr. Cedric Lindsay, former Director of the Washington State Department of Fisheries (now the Washington Department of Fish and Wildlife), Shellfish Laboratory at Point Whitney for his recollection of introductions, and to Mr. Blake Feist and Ms. Cathy Schwartz for the production of the Washington and San Francisco Bay maps, respectively.

REFERENCES

Abbott, R.T., 1974. American Seashells, 2nd ed. Van Nostrand and Reinhold, New York, 663 pp.

Ansell, A.D., 1964. *Venus mercenaria* (L.) in Southampton Water. Ecology, 44: 396–397.

Ansell, A.D., 1968. The rate of growth of the hard clam *Mercenaria mercenaria* (L.) throughout the geographic range. J. Cons. Perm. Intl. Explor. Mer, 31: 364–409.

Ansell, A.D., Lander, K.F., Coughlan, J. and Loosmore, F.A., 1964. Studies on the hard shell clam, *Venus mercenaria*, in British waters, I. Growth and reproduction in natural and experimental colonies. J. Appl. Ecol., 1: 63–82.

Castagna, M. and Chanley, P., 1973. Salinity tolerance of some marine bivalves from inshore and estuarine environments in Virginia waters on the western mid-Atlantic coast. Malacologia, 12: 47–96.

Dahl, W.H. and Simpson, N., 1901. The Mollusca of Puerto Rico. Bull. U.S. Fish. Comm. 1900, 20: 351–524.

Hanna, G.D., 1966. Introduced mollusks of western North America. Occas. Pap. Calif. Acad. Sci., 7 (48): 1–108.

Heppell, D., 1961. The naturalization in Europe of the quahog *Mercenaria mercenaria*. J. Conchol. London, 25: 21–33.

Hibbert, C.J., 1975. Production studies of a bivalve population on an intertidal mudfat, with particular reference to the energy budget of *Mercenaria mercenaria* (Linné). Ph.D. thesis, University of Southampton, Southampton, 225 pp.

Hibbert, C.J., 1977a. Growth and survivorship in a tidal-flat population of the bivalve *Mercenaria mercenaria* from Southampton Water. Mar. Biol., 44: 71–76.

Hibbert, C.J., 1977b. Energy relations of the bivalve *Mercenaria mercenaria* on an intertidal mudflat. Mar. Biol., 44: 77–84.

Juste, V. and Cortes, R., 1990. Distribution and biological aspects of the hard clam *Mercenaria mercenaria*

(Linnaeus), *M. Mercenaria notata* (Say), and *M. campechiensis* (Gmelin) in Puerto Rico. Caribb. J. Sci., 26: 136–140.

Kraeuter, J.N. and Castagna, M., 1977. An analysis of gravel, pens, crab traps and current baffles as protection for juvenile hard clams (*Mercenaria mercenaria*). Proc. World Maricult. Soc., 8: 167–173.

Lambert, L., 1949. L'ostréiculture en Amérique du Nord. Rev. Trav. Off. Peches Marit., 15: 123–152.

Loosanoff, V.L., 1944. Soft and hard clams of the Atlantic Coast of the United States. U.S. Dept Interior Fish Wildlife Serv., Fish. Leafl., 13: 1–11.

Loosanoff, V.L., 1946. Commercial clams of the Atlantic coast of the United States. U.S. Dep. Interior, Fish Wildlife Serv., Fish. Leafl., 13: 1–12.

Marteil, L., 1956. Acclimation du clam (*Venus mercenaria* L.) en Bretagne. Rev. Trav. Inst. Peches Marit., 20: 157–160.

McLean, R.A., 1951. The pelecypods or bivalve mollusks of Puerto Rico and the Virgin Islands. N.Y. Acad. Sci., Sci. Surv. Puerto Rico Virgin Islands, 17: 1–183.

Moore, T.J., 1884. Note on a further local attempt to naturalize the American clam (*Venus mercenaria*). Proc. Lit. Philos. Soc. Liverpool 38: xc.

Moore, T.J., 1886. Note on the possible naturalization of the American clam *Venus mercenaria* on the coasts of Lancashire and Cheshire. First report on the fauna of Liverpool Bay. Proc. Lit. Philos. Soc. Liverpool, 40: 368–370.

Morris, P.A., 1975. A Field Guide to the Shells of the Atlantic and Gulf Coast, and the West Indies, 3rd ed. Houghton Mifflin, 330 pp.

Mitchell, R., 1974. Studies on the population dynamics and some aspects of the biology of *Mercenaria mercenaria* (Linné). Ph.D. thesis, University of Southampton, Southampton, 191 pp.

Murphy, R.C., 1985. Factors affecting the distribution of the introduced bivalve, *Mercenaria mercenaria*, in a California lagoon — the importance of bioturbation. J. Mar. Res., 43: 673–692.

Pratt, D.M., 1953. Abundance and growth of *Venus mercenaria* and *Callocardia morrhuana* in relation to the character of bottom sediments. J. Mar. Res., 12: 60–74.

Pratt, D.M. and Campbell, D., 1956. Environmental factors affecting growth in *Venua mercenaria*. Limnol. Oceangr., 1(1): 2–17.

Rehder, H.A., 1981. The Audubon Society Field Guide to North American Seashells. Alfred A. Knopf, 894 pp.

Rhoads, D.C. and Panella, G., 1970. The use of molluscan shell for growth patterns in ecology and paleoecology. Lethaia, 3: 143–161.

Rodhouse, P.G., 1973. Factors affecting spatfall of the clam *Mercenaria mercenaria* L. M.Sc. dissertation, University of Southampton, Southampton.

Ruckenbusche, H., 1949. Le clam. Note sur *Vénus mercenaria* L. Son introduction et son élévage dans le bassin de la Sudre. Rev. Trav. Off. Peches Marit., 15: 99–117.

Salchak, A. and Haas, J., 1971. Occurrence of the northern quahog, *Mercenaria mercenaria*, in Colorado Lagoon, Long Beach California. Calif. Fish Game, 72: 126–128.

Utting, S.D. and Spencer, B.E., 1992. Introductions of marine bivalve molluscs into the United Kingdom for commercial culture — case histories. ICES Mar. Sci. Symp., 194: 84–91.

Warmke, G.L., and Abbott, R.T., 1961. Caribbean Seashells. Dover Publications, New York, 348 pp.

Wells, H.W., 1957. Abundance of the hard clam, *Mercenaria mercenaria* (L.) in relation to environmental factors. Ecology, 38: 123–128.

References Index

Species Index

General Index

Kolibri 360 GmbH
Ferdinand-Jühlke-Straße 7
99095 Erfurt, Deutschland
produktsicherheit@kolibri360.de